The Tree Shrew (*Tupaia belangeri chinensis*) Brain in Stereotaxic Coordinates

Jiang-Ning Zhou • Rong-Jun Ni

The Tree Shrew (*Tupaia belangeri chinensis*) Brain in Stereotaxic Coordinates

Science Press
Beijing

Springer

Jiang-Ning Zhou
School of Life Sciences
Universities of Science and Technology of China
Hefei
Anhui
China

Rong-Jun Ni
School of Life Sciences
Universities of Science and Technology of China
Hefei
Anhui
China

ISBN 978-981-10-0610-4 ISBN 978-981-10-0611-1 (eBook)
DOI 10.1007/978-981-10-0611-1

The print edition is not for sale in China Mainland. Customers from China Mainland please order the print book from: Science Press.
ISBN of the China Mainland edition: 978-7-03-049297-5

Library of Congress Control Number: 2016949467

Printed on acid-free paper

This Springer imprint is published by Springer Nature
The registered company is Springer Science+Business Media Singapore Pte Ltd.

Foreword

In 1991 I received a letter from a Chinese student. It was long before we were all using email. This student, Jiang-Ning Zhou, told me, much to my amazement, that he had been following my publications and asked me whether it was possible to do a PhD project on the human hypothalamus in my group at the Netherlands Institute for Brain Research, as it was called then an Institute of the Royal Netherlands Academy of Arts and Sciences, in Amsterdam. I wrote back, as was my custom in those cases, that this would only be possible if he obtained a fellowship. Normally, that would be the end of it, but this time it was different. Six months later I received another letter from Zhou, in which he informed me that he had done as I asked and obtained a Nuffic fellowship, but that he was going to postpone his stay in the Netherlands until he had a certain proficiency in Dutch! I told him that this would be a waste of time. Even then, the Institute could already boast 20 nationalities and the working language in the lab was bad English spoken with a great variety of accents. I have to add that 3 years later, when a press release in Dutch appeared about his work, Zhou had to take my word for it that it did indeed pertain to his research, and he told me that he still regretted not learning Dutch before he came to Amsterdam.

In 1992, Zhou travelled to Amsterdam from China, where in those days strict rules applied about sexual behavior. Yet, Zhou had come to the conclusion that despite his fascination with Alzheimer's disease, his main interest had shifted to a highly controversial topic, namely sexual differentiation of the brain in relation to homosexuality and transsexuality. His finding that the sex difference in the bed nucleus of the stria terminalis was reversed in transsexuals and thus matched how they themselves experienced their gender identity (Zhou et al., *Nature* 378: 68–70, 1995) established his reputation worldwide. This paper was selected as one of the top 100 science stories in the world in 1995. In addition, this work had societal relevance: it was used to change the birth certificate and passport of transsexuals in the European court in Strasbourg and to get a law about this topic accepted in the UK. After his 4 years as a PhD student, culminating in his thesis, and two more years at our Institute as a postdoc, Zhou was appointed Professor in the School of Life Sciences of the University of Science and Technology of China, Hefei, and became one of the leading scientists in the field of neurobiology in China. In 1998, he invited me over and introduced me to Chinese traditions, culture, nature, and scientific research. Since then I have been working in China for a couple of months per year and have come to think of it as my second home country.

Jiang-Ning Zhou's research at UCST illustrates his broad interest and involves a great number of topics in basic research and translational neuroscience. The tree shrew has turned out to be a suitable animal for both these lines of research.

Tree shrews are small monogamous diurnal primate-like animals, native to the tropical forests of Southeast Asia that are easy to keep in laboratory conditions. They are omnivorous, feeding on insects, small vertebrates, fruit, and seeds. They have a well-developed visual system and relatively large brains. Tree shrews were first considered to belong to the order of the Insectivora, until they were classified as primates, among other things because of the similarities in brain, skull, and muscle anatomy, as was highlighted by Sir Wilfrid Le Gros Clark (1924), which made them very popular as an alternative for expensive primate research. Data from the whole genome of this species supported the idea that the tree shrew is genetically closely related to primates (Xu et al. 2013). However, since 1972 it has been customary to allocate tree shrews to their own small order, the Scandentia.

Brain research starts with detailed knowledge of the microscopic anatomy of the species under investigation. There is so far only one neuroanatomical atlas of this species, showing 64 Nissl-stained frontal sections of the brain of the common tree shrew (*Tupaia glis*), see: http://www.brainmuseum.org/Specimens/scandentia/treeshrew/index.html.

The present atlas of the Chinese tree shrew (*Tupaia belangeri chinensis*) brain, created with cutting edge technologies, offers a much more detailed picture. It shows both Nissl-stained and acetyl-cholinesterase-stained sections in the frontal plane (192 figures), horizontal plane (49 figures), and sagittal plane (36 figures) and has diagrams of each section and a coordinate system for stereotaxic research. What is unique is that the light microscopic images were captured at a single-cell resolution using a series scanning microscope camera. Image resolution was about 1–3 GB (pixel size) for each section. Each section consisted of hundreds of small segments that were captured using a 10× objective. In addition, high-resolution MR images were made. Zhou and his colleagues have, moreover, already published immunocytochemical maps of the distribution of neuropeptide Y, vasopressin, oxytocin, vasoactive intestinal polypeptide, and corticotropin-releasing factor in the tree shrew brain (Ni et al. (2014, 2015), Shu et al. (2015), and Xu (2013)) that could be a useful start for further research.

The great diversity of tree shrew studies published in the last few years shows how popular this animal is becoming in basic and translational neuroscience. There are publications on topics ranging from brain development to the ultrastructure of synaptic connections. Recent papers, using molecular, genetic, pharmacological, electrophysiological, and noninvasive scanning techniques, range from the function of the visual system, including eye diseases, to sleep architecture, memory, energy metabolism, and temperature regulation. The tree shrew is now also emerging as a model for human brain diseases in the field of neurodegeneration, psycho-social stress, addiction, pain, and anesthesia.

I want to congratulate Drs Jiang-Ning Zhou, Rong-Jun Ni, and the other contributors with this excellent, timely, and very useful atlas that will, without any doubt, be the basis for many neurobiological investigations in the upcoming decades.

Dick F. Swaab, MD, PhD
Neurobiology
University of Amsterdam
Amsterdam, The Netherlands

Group Leader Neuropsychiatric Disorders
Netherlands Institute for Neuroscience, KNAW
Meibergdreef 47, Amsterdam, 1105, BA, The Netherlands

References

Le Gros Clark WE. On the brain of the tree-shrew (*Tupaia minor*). Proc Zool Soc Lond. 1924;94:1053–74.

Ni RJ, Shu YM, Luo PH, Fang H, Wang Y, Yao L, Zhou JN. Immunohistochemical mapping of neuropeptide Y in the tree shrew brain. J Comp Neurol. 2015;523:495–529.

Ni RJ, Shu YM, Wang J, Yin JC, Xu L, Zhou JN. Distribution of vasopressin, oxytocin and vasoactive intestinal polypeptide in the hypothalamus and extrahypothalamic regions of tree shrews. Neuroscience. 2014;265:124–36.

Shu YM, Ni RJ, Sun YJ, Fang H, Zhou JN. Distribution of corticotropin-releasing factor in the tree shrew brain. Brain Res. 2015;1618:270–85.

Xu L, Fan Y, Jiang XL, Yao YG. Molecular evidence on the phylogenetic position of tree shrews. Zoological Res. 2013;34:70–6.

Acknowledgements

We thank Dr. Lin Xu and Jing Wang for their generous donation of tree shrew samples and their support during these experiments.

We thank Dr. Li Cai, Fan-Tao Meng, Jun Zhao, and Pu Hu for their expert assistance during the histological and surgical work in this study.

We thank Dr. Ji Liu, who applied his enormous knowledge to assist us in describing the hypothalamus of the tree shrew.

We also thank Ms. Qu Huang for feeding the tree shrews each day.

We express deep appreciation for our colleagues in the Chinese Academy of Science Key Laboratory of Brain Function and Diseases for their support.

We thank the Olympus Company for providing us with a series of scanning microscopes for a trial period.

We are very grateful to Dr. Dick F. Swaab (Netherlands Institute for Brain Research, Amsterdam, Netherlands) for his support and assistance.

We acknowledge support by the Strategic Priority Research Program of the Chinese Academy of Science (XDB02030001), the National Natural Science Foundation of China (31500859, 91432305, and 91132304) and the Fundamental Research Funds for the Central Universities (WK2070000058).

Finally, we thank our families for their support.

About the Authors

Jiang-Ning Zhou earned his doctoral degrees at the University of Amsterdam in 1996. From 1998 to 2003, he was Professor of Geriatrics and Vice Dean of The First Hospital of Anhui Medical University in China. Since 2003 he is Professor of Neurobiology at Department of Neurobiology, School of Life Sciences, University of Science and Technology of China. His major research interests focus on the pathogenesis of Alzheimer's disease and depression. He has published over 100 papers on SCI journals, including *Nature, Archives General Psychiatry, Molecular Psychiatry, Brain and Biological Psychiatry*, etc. Zhou mentored 40 Ph.D. students. In 2011, he obtained Excellent Scholar Awards from Chinese Neuroscience Society. In 2014, he received Natural Science Awards from the Ministry of Education of China.

Rong-Jun Ni obtained his undergraduate degree in 2009 from Anhui University. He earned his Doctor of Philosophy degree from the University of Science and Technology of China in 2014 and began the postdoctoral training in the Department of Neurobiology, University of Science and Technology of China from 2014. Dr. Ni's research focuses on the brain anatomy, neural circuit, neuropeptides, and behaviors of tree shrews. He has published several papers in SCI journals and authored one chapter in the *Basic Biology and Disease Models of Tree Shrews*. These studies are funded by the Natural Science Foundation of China and the Fundamental Research Funds for the Central Universities.

Contents

Contributors

Zhao-Huan Huang University of Science and Technology of China, Hefei, China

Peng-Hao Luo University of Science and Technology of China, Hefei, China

Jun-Chao Qian Hefei Institutes of Physical Science Chinese Academy of Sciences, Hefei, China

Xin-Ya Qin University of Science and Technology of China, Hefei, China

Yu-Mian Shu University of Science and Technology of China, Hefei, China

Xu Wu University of Science and Technology of China, Hefei, China

Lei Yao University of Science and Technology of China, Hefei, China

Kai Zhong Hefei Institutes of Physical Science Chinese Academy of Sciences, Hefei, China

Ke-Qing Zhou University of Science and Technology of China, Hefei, China

Part I

Tree Shrew Whole-Brain Anatomy: Drawings and Nomenclature

Background and Methods

1

1.1 Introduction

Tree shrews belong to the order Scandentia; they are day-active animals that live in arboreal habitats in South and Southeast Asia (Peng et al. 1991). Recently, whole-genome sequencing of the Chinese tree shrew (*Tupaia belangeri chinensis*) was completed (Fan et al. 2013). According to recent studies, tree shrews are the closest living relatives of primates (Janecka et al. 2007; Kriegs et al. 2007; Fan et al. 2013; McCollum and Roberts 2014). The tree shrew is a good model species because they are day-active animals, they are small in size (12–18 cm and 110–180 g), they are easy to handle and feed, and they have a relatively short gestation period (41–55 days) and length of puberty (2–6 months). They are also a good choice for reasons related to ethical considerations. There have been many recent attempts to use tree shrews as an animal model. They have been used, for example, in studies of viral infections (Yan et al. 1996; Zhao et al. 2002; Tsukiyama-Kohara and Kohara 2014), myopia (Norton et al. 2006), stress-related disorders (Czeh et al. 2005; Fuchs 2005), and pharmacological tests (Zhao et al. 2014).

Because there is a lack of knowledge concerning the brain anatomy and the precise locations of brain structures in tree shrews, it has been a challenge for scientists to establish this species as an animal model system in preclinical and mechanistic studies of central nervous system-related disorders. In 1970, Marrocco et al. (1970) published camera lucida drawings of the tree shrew brain at 10 levels using the Horsley-Clarke system of coordinates. In this stereotaxic coordinate system, the horizontal zero is defined as the plane that passes through the ear bars and the infraorbital ridge, and the vertical zero is defined as the plane that is perpendicular to the horizontal zero and that passes through the interaural line. In the stereotaxic atlas of the brain of the tree shrew (*Tupaia belangeri chinensis*) published by Yang et al. (1990), the horizontal zero was positioned parallel to the plane that passes through the ear bars and the incisor bar at 4 mm above the interaural line. The vertical zero was defined as the plane that lies perpendicular to the horizontal zero and that passes through the interaural line. Their work contains 37 figures that show tissues separated by approximately 500-μm intervals. However, no further versions of these atlases have been published.

In our version of "*The Tree Shrew (Tupaia belangeri chinensis) Brain in Stereotaxic Coordinates*," we provide complete series of coronal sections, sagittal sections, and horizontal sections that were labeled using Nissl staining or acetylcholinesterase (AChE) histochemistry. The coronal sections are presented at 0.13-mm intervals, which is sufficient to include all major structures in the tree shrew brain that are relevant to modern neuroscience research purposes. Furthermore, we have provided a series of diagrams besides microphotographs of the Nissl- and AChE-stained sections. In this atlas, we have used coordinates that reference the interaural line and a position we have newly defined as bregma (brain origin). In addition, we also referenced magnetic resonance (MR) images that were acquired at 0.25-mm intervals using a 9.4 T MR scanner.

In this atlas, we have attempted to improve the accuracy of our coordinates and descriptions as far as possible, and we have also incorporated recent findings regarding tree shrew brain anatomy. We welcome any valuable comments and corrections that might help us to improve our diagrams in the future.

J.-N. Zhou, R.-J. Ni, *The Tree Shrew (Tupaia belangeri chinensis) Brain in Stereotaxic Coordinates*,
DOI 10.1007/978-981-10-0611-1_1

1.2 Methods

1.2.1 Surgery and Stereotaxic Approach

Sixty-two adult male tree shrews (*Tupaia belangeri chinensis*) (from the breeding colony in the Animal House Center of the Kunming Institute of Zoology, Kunming, P. R. China) were used. The use and care of the animals in the present study were performed accordance with international guidelines and protocols that were approved by the Animal Care and Use Committee of the University of Science and Technology of China. All efforts were made to minimize animal suffering and the number of animals that were used. The tree shrews were deeply anesthetized using pentobarbital sodium (80 mg/kg, i.p.) and then positioned in a digital rat stereotaxic instrument (Stoelting Company, USA). The following operations were performed under a stereoscope. We adjusted the incisor bar until the heights corresponding to lambda and bregma were equal to achieve a flat skull position (Fig. 1.1). At this position, the incisor bar was decreased to 5.01 ± 0.01 mm below zero (Table 1.1). The anteroposterior position of bregma was considered to be a point that was 7.65 ± 0.11 mm and 9.00 ± 0.12 mm anterior to the coronal plane that passed through the interaural line and lambda, respectively (Table 1.1). The surface of the skull was 12.77 ± 0.09 mm dorsal to the interaural zero plane at both bregma and lambda (Table 1.1). For the brain specimens that were sectioned in the coronal, horizontal, and sagittal planes, reference needle tracks were made to confirm the stereotaxic orientation and the position of the sections. For tissue sections in the coronal plane, a vertical needle was inserted at the following position relative to bregma: anteroposterior, −1.00 mm; mediolateral, −6.00 mm; and dorsoventral, −3.00 mm. For tissue sectioned in the horizontal plane, a horizontal needle was inserted perpendicular to the sagittal plane at the following distances relative to bregma: anteroposterior, 0.00 mm; mediolateral, −14.00 mm; and dorsoventral, −3.00 mm. For tissue sectioned in the sagittal plane, two vertical needles were inserted at the following positions relative to bregma: anteroposterior, 0.00 mm; mediolateral, ±3.00 mm; and dorsoventral, −6.00 mm.

After stereotaxic surgery, the tree shrews were perfused with 0.9 % saline followed by 4 % paraformaldehyde (PFA) in phosphate buffer (PB, 0.1 M, pH 7.4). All brains sectioned in the coronal, horizontal, and sagittal planes were separated into two parts while the skull was positioned in the stereotaxic instrument, and they were then removed from the skull. Some brains were sectioned (at 60 μm) using a vibratome (Leica VT1200S, Germany). Other brains were dehydrated and sectioned (at 40 μm) using a Leica microtome (Leica CM1950, Germany). All sections were mounted on gelatin-coated slides and dried overnight.

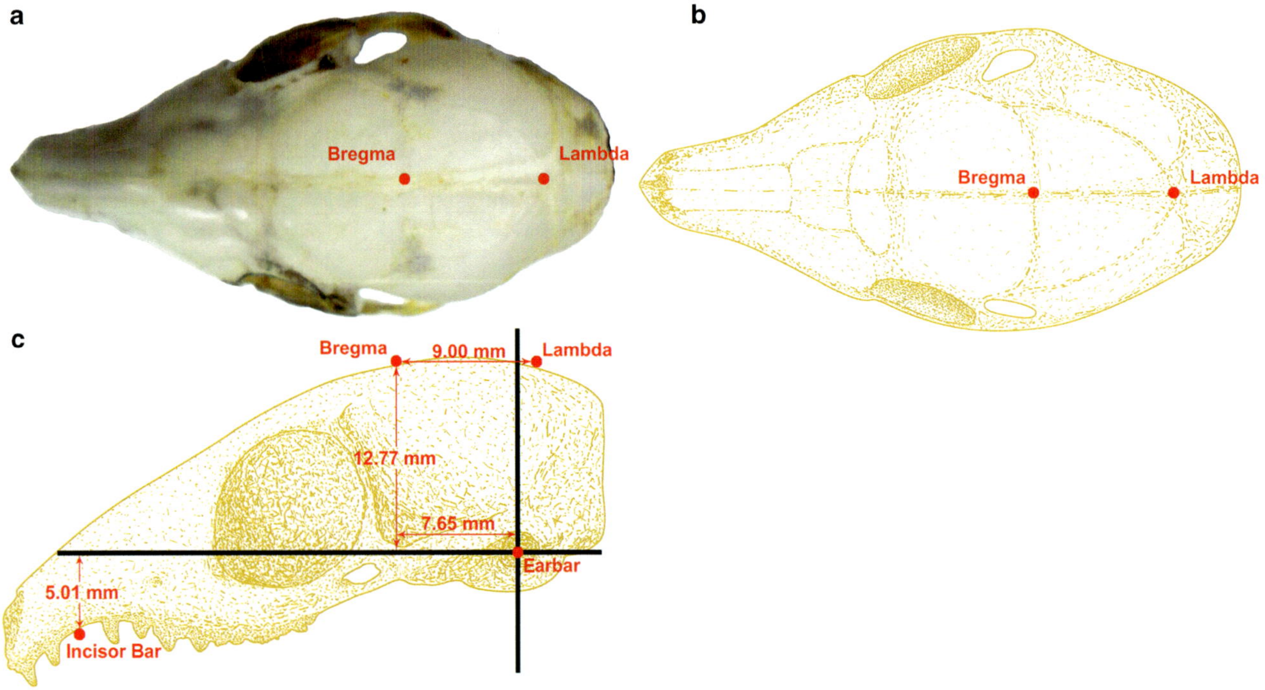

Fig. 1.1 Dorsal (**a**, **b**) and lateral (**c**) views of the skull of a male adult tree shrew. (**b**) Is a drawing of (**a**). The positions of bregma, lambda, the incisor bar, and the plane of the interaural line (defined by the ear bar) are shown. When the heights of lambda and bregma were equal, the incisor bar was decreased to 5.01 mm below the zero plane (**c**). The anteroposterior position of bregma was 7.65 mm and 9.00 mm anterior to the coronal plane that passed through the interaural line and lambda, respectively (**c**). The surface of the skull was 12.77 mm dorsal to the interaural zero plane at both bregma and lambda (**c**)

Table 1.1 Stereotaxic data (means ± SEM) for tree shrews including age and weight

NO.	Age (months)	Weight (g)	L – B (mm)	I – B (mm)	Incisor bar – I (mm)	I – B (mm)
			AP	AP	DV	DV
TS-13SEP29	9	142.5	9.00	8.40	5.00	
TS-13SEP15	9	103.5	9.02	7.43	5.00	
TS-13DEC09	7	136.3	9.17	8.10	5.10	
TS-13DEC03	7	115.3	9.38	7.25	5.00	
TS-13DEC12	8	126.7	8.52	7.59	5.02	
TS-13DEC11	8	139.5	10.62	8.91	5.10	13.19
TS-13SEP26	13	137.8	9.06	7.69	5.00	12.46
TS-13SEP17	13	108.4	9.89	7.98	5.02	12.40
TS-13DEC07	9	145.8	8.03	7.31	5.00	12.61
TS-13DEC08	9	147	8.94	8.11	5.00	12.85
TS-13DEC15	14	145.3	8.21	6.22	5.00	13.62
TS-15APR12	9	122.6	8.58	6.86	5.00	13.40
TS-15 NO.1	7	140.8	8.31	7.04	4.88	13.02
TS-15APR11	8	120.9	9.39	7.62	4.98	12.73
TS-15APR08	8	135.7	8.31	7.24	4.99	13.36
TS-15APR29	13	116.3	9.22	7.95	5.07	12.34
TS-15APR24	15	131	9.19	7.42	5.00	12.36
TS-15 NO.2	6	160	8.97	7.80	5.01	12.58
TS-14APR03	10	139.9	9.48	7.76	5.00	12.82
TS-15 NO.4	8	143.7	9.54	8.53	4.96	11.97
TS-15 NO.3	7	139.5	9.25	7.49	5.03	13.04
TS-15APR14	10	126.5	8.29	7.67	4.98	12.94
TS-15APR07	8	116.2	7.85	7.71	5.00	12.10
TS-15 NO.6	7	170	9.87	7.58	5.02	12.51
TS-15APR03	10	120.4	9.06	8.09	4.98	13.06
TS-15 NO.5	8	140.6	8.96	7.22	5.04	12.88
means ± SEM		133.55 ± 3.02	9.00 ± 0.12	7.65 ± 0.11	5.01 ± 0.01	12.77 ± 0.09

AP anterior-posterior, *B* bregma, *DV* dorsal-ventral, *I* interaural line, *L* lambda

1.3 Histology

1.3.1 Thionin Staining (Nissl Staining)

One series of sections was rinsed twice with phosphate-buffered saline (PBS, 0.1 M; pH 7.4) and then stained in 0.02 % thionin acetate salt (Sigma-Aldrich) solution for 20 min. The sections were then quickly rinsed in PBS, dehydrated in ethanol-xylene, and cover-slipped. They were then visualized using an Olympus VS120-S1-E microscope (Olympus U-TV0.5XC-3, Japan).

1.3.2 Acetylcholinesterase (AChE) Staining

AChE histochemistry was performed as previously described by Paxinos and Watson (2006). Briefly, the other set of slides was incubated for 15 h in 100 mL of a stock solution (50 mM sodium acetate buffer, 4.0 mM copper sulfate, 16 mM glycine, pH 5.0) to which 116 mg of acetylthiocholine iodide and 3.0 mg of ethopropazine hydrochloride (Sigma-Aldrich) were added immediately prior to incubation. The slides were rinsed with distilled water and developed for 10 min in a solution of 1 % sodium sulfide (pH 7.5). The slides were then rinsed with water, immersed in 4 % PFA in PB for 8 h, and rinsed with water again. The slides were then dehydrated in ethanol-xylene and cover-slipped. The finished slides were visualized using an Olympus VS120-S1-E microscope (Olympus U-TV0.5XC-3, Japan).

1.3.3 Digital Photomicrographs and Drawings

Light microscopic images were captured at a single-cell resolution using an Olympus VS120-S1-E series scanning microscope camera (Olympus U-TV0.5×C-3, Japan). All digital photomicrographs of sections were adjusted for brightness/contrast, cropping, and image size using Adobe Photoshop CS5 (Adobe Systems, USA). All files were processed in the .tif format. To avoid confusion, missing or severely distorted portions of sections were deleted in our atlas (e.g., Figs. 2.3, 2.61, 2.88, and 3.15). Nissl-stained and AChE-stained sections are alternately presented.

All camera lucida drawings were edited using Adobe Illustrator (Adobe Systems, USA) by tracing the digital photomicrographs. For the coronal, sagittal, and horizontal sections, we drew the outlines of the structures on only the left side of each section. We then derived the right side using mirror image construction. As far as possible, we have attempted to accurately draw the outlines of structures in each section. Small adjustments have been made to some sections in which the brain slices were distorted. Otherwise, the drawings demonstrate the asymmetries that were present in the sections. When portions of brain slices were missing or severely distorted, the missing parts were drawn in after a consideration of adjacent sections in the same brain or slices in the same position in other brains. Each abbreviation was placed in the center of the neuroanatomical structure it describes or alongside the structure with a leader line. We added labels to both sides of the diagram whenever possible.

1.3.4 Magnetic Resonance Imaging (MRI)

Adult male tree shrews were anesthetized using pentobarbital sodium (80 mg/kg, i.p.) and positioned in a digital rat stereotaxic instrument (Stoelting Company, USA). We adjusted the incisor bar until the heights of lambda and bregma were equal to achieve a flat skull position. For tree shrew brains that were scanned in the coronal, sagittal, and horizontal planes, four vertical needles were inserted into both hemispheres at 0.00 mm and −5.00 mm from bregma (±3.0 mm lateral to the midline and −10 mm from the skull). The tree shrews were then decapitated, and the skulls were immersed in 4 % PFA in PB. All MRI experiments were performed using a 9.4 Tesla 40-cm bore MR scanner (Agilent Technologies, Inc., Santa Clara, CA, USA) with a quadrature birdcage coil (inner diameter: 40 mm). A two-dimensional (2D) spoiled gradient echo (GE) sequence was used to acquire high-resolution MR images of tree shrew brains that were immersed in a 4 % PFA solution. The following acquisition parameters were chosen.

The parameters for the coronal scans were as follows: repetition time (TR) = 3100 ms, echo time (TE) = 12.5 ms, acquisition bandwidth (BW) = 12 kHz, nominal excitation flip angle (FA) = 80°, field of view (FOV) = 23 mm × 23 mm, matrix = 256 × 256, spatial resolution = 90 μm × 90 μm (pixel size), slice thickness = 0.25 mm (120 slices, gap = 0), and number of averages = 30. The scan time was 397 min.

The parameters for the horizontal scans were as follows: TR = 2600 ms, TE = 18 ms, acquisition BW = 12 kHz, nominal excitation FA = 80°, FOV = 34.5 mm × 23 mm; matrix = 384 × 256; spatial resolution = 90 μm × 90 μm (pixel size), slice thickness = 0.25 mm (70 slices, gap = 0), and number of averages = 30. The scan time was 333 min.

The parameters for the sagittal scans were as follows: TR = 2900 ms, TE = 18 ms, acquisition BW = 12 kHz; nominal excitation FA = 80°, FOV = 39 mm × 26 mm, matrix = 384 × 256, spatial resolution = 102 μm × 102 μm (pixel size), slice thickness = 0.25 mm (80 slices, gap = 0), and number of averages = 30. The scan time was 371 min.

All MRI images were adjusted by cropping and for image size using Adobe Photoshop CS5 (Adobe Systems, USA). In our atlas, all of the presented MRI images were obtained from one tree shrew brain, no. TS-13DEC15.

1.3.5 Stereotaxic Reference System and Stereotaxic Accuracy

Stereotaxic coordinates were measured only when the heights of lambda and bregma were equal. Occasionally, the bregma was difficult to visualize, even when the skull was observed using a stereoscope. According to the methods described in Paxinos and Watson (2006), two coronal, two horizontal, and one sagittal zero planes were used in our stereotaxic reference system. The two coronal zero planes cut through bregma and the interaural line, respectively. One of the horizontal zero planes cut through bregma and lambda at the top of the skull. The other horizontal zero plane cut through the level of the interaural line. The sagittal zero plane cut through the midline of the two brain hemispheres. The anteroposterior position of bregma was assigned a position 7.65 ± 0.11 mm and 9.00 ± 0.12 mm anterior to the coronal plane that passed through the interaural line and lambda, respectively. The surface of the skull at bregma was 12.77 ± 0.09 mm dorsal to the horizontal zero plane that passed through the interaural line. The interaural line can be used as an alternative reference point during stereotaxic surgery. The levels of the sections in the stereotaxic reference are shown on the skull diagram (top left corner). The stereotaxic reference grid is shown using 0.2-mm intervals.

1.3.5.1 Drawings of Coronal Brain Sections

In our atlas, all coronal sections were selected from two tree shrew brains, no. TS-13Sep27 (Figs. 2.1–2.88 and 2.97–2.192) and no. TS-15Apr12 (Figs. 2.89–2.96). In each of the coronal drawings, the large number at the bottom right indicates the anteroposterior distance of the plate from the vertical coronal plane that passed through bregma. The large number at the bottom left indicates the anteroposterior distance of the plate from the vertical coronal plane that passed through the interaural line. The levels of the sections in the stereotaxic reference are shown in the top left corner. The figure numbers are shown in the top right corner. The small numbers in the left and right margins indicate the dorsoventral distance from the horizontal plane that passed through bregma at the surface of the skull. The small numbers in the top and bottom margins indicate the mediolateral distance of the section from the midline sagittal plane between the two brain hemispheres.

1.3.5.2 Drawings of Horizontal Brain Sections

In our atlas, all horizontal sections were selected from one tree shrew brain, no. TS-14APR11 (Figs. 3.1–3.49). In each of the horizontal drawings, the large number at the bottom right indicates the distance of the plate below the horizontal plane that passed through bregma. The large number at the bottom left indicates the distance of the plate below the horizontal plane that passed through the interaural line. The levels of the sections in the stereotaxic reference are shown in the top left corner. The figure numbers are shown in the top right corner. The small numbers in the left and right margins indicate the distance of the section from the midline sagittal plane. The small numbers in the top margins indicate the anteroposterior distance from the vertical plane that passed through bregma. The small numbers in the bottom margins indicate the anteroposterior distance from the vertical plane that passed through the interaural line.

1.3.5.3 Drawings of Sagittal Brain Sections

In our atlas, all sagittal sections were selected from one tree shrew brain, no. TS-13DEC11 (Figs. 4.1–4.36). In each of the sagittal drawings, the large number at the bottom left indicates the distance of the plate from the midline sagittal plane. The levels of the sections in the stereotaxic reference are shown at the top left corner. The figure numbers are shown in the top right corner. The small numbers in the left margins indicate the dorsoventral distance from the horizontal plane that passed through the interaural line. The small numbers in the right margins indicate the dorsoventral distance from the horizontal plane that passed through bregma and lambda at the surface of the skull. The small numbers in the top margins indicate the distance of the section from the vertical coronal plane that passed through bregma. The small numbers on the bottom margins indicate the distance of the section from the vertical coronal planes that passed through the interaural line.

1.3.5.4 Stereotaxic Accuracy

In each of the section drawings, the distance of structures from the reference points were measured in series of thionin- and AChE-stained photomicrographs that were obtained at a single-cell resolution. These measurements were repeated in a large number of tree shrew brains. To verify our measurements, we also referred to whole tree shrew brain MR images that were scanned rostrocaudally, mediolaterally, and dorsoventrally using a 9.4 T MR scanner. After measurements and calculations were performed, the anteroposterior distance of the sections from bregma was moved posterior by approximately 0.38 mm. The variability in the other stereotaxic coordinates of the structures was very small.

1.3.6 Nomenclature

In previous reports, Latin and English nomenclatures used to describe multiple brain structures in other animals have been used to describe single structures in tree shrews. To allow scientists to accurately and efficiently communicate information, a stable terminology and system of abbreviations is necessary. After considering some of the stable neuroanatomical nomenclatures used in various other species, the nomenclature and abbreviations we chose made use, as often as possible, of the excellent atlases by Paxinos for the Rhesus monkey and rodents (Paxinos et al. 2000; Paxinos and Franklin 2001; Paxinos and Watson 2006).

1.3.7 Methods Used to Determine the Positions of Structures

Few studies have been published to describe the systematic and comprehensive organization of tree shrew brain structures. In this atlas, we have mainly relied upon our previous papers (Ni et al. 2014, 2015; Shu et al. 2015) and comparisons between series of our own stained tree shrew sections and those shown in publications describing brain structures in other mammals (Paxinos et al. 2000; Paxinos and Franklin 2001; Paxinos and Watson 2006; Xavier and Nicole 2008). We also referred to several publications and papers that describe the precise positions of tree shrew brain structures (Marrocco et al. 1970; Airaksinen et al. 1989; Yang et al. 1990; Agarwala et al. 1992; Conley and Friederich-Ecsy 1993; Flugge et al. 1994; Wolter et al. 1999; Keuker et al. 2003; Lyon et al. 2003a; Kosaka and Kosaka 2004; Chomsung et al. 2010; Day-Brown et al. 2010; Rice et al. 2011; Baldwin et al. 2013).

1.3.7.1 Rhinencephalon

Determinations of the locations of structures in the main olfactory bulb were guided by the works of Malz et al. (2000) and Kosaka K and Kosaka T (2004). Our previous studies included a general description of the anterior olfactory nucleus (Ni et al. 2015; Shu et al. 2015). The positions of the accessory olfactory bulb and the olfactory tubercle were determined according to our Nissl and AChE staining data and in reference to Paxinos et al. (2000) and Paxinos and Watson (2006).

1.3.7.2 Telencephalon

Neocortex. Wong and Kaas (2009) described the architectonic subdivisions of the neocortex in the tree shrew. The organization of the frontoparietal cortex has also been described in previous papers (Remple et al. 2006, 2007). These subdivisions are identified in our tree shrew atlas; however, we did not mark the boundaries of these subdivisions of the neocortex. The hippocampal formation of the tree shrew has been described in great detail in previous studies (Wolter et al. 1999; Keuker et al. 2003; Ni et al. 2015).

Basal ganglia. Recently, the basal ganglia of tree shrews was characterized from rostral to caudal (Day-Brown et al. 2010; Rice et al. 2011; McCollum and Roberts 2014; Ni et al. 2015). We identified distinct subdivisions of the globus pallidus and the substantia nigra in Nissl- and AChE-stained sections. The striatum was located using high-density AChE staining.

Amygdala. We delineated the distinct subregions of the amygdala according to the works of Flugge et al. (1994) and Airaksinen et al. (1989). Their findings are in agreement with our previous studies of tree shrew anatomy (Ni et al. 2015; Shu et al. 2015).

Septal and preoptic regions. We identified the tree shrew homologue of the monkey septal and preoptic regions (Paxinos et al. 2000). In recent papers, we identified and described distinct subregions within these brain structures (Ni et al. 2014, 2015; Shu et al. 2015). In this atlas, the septal nucleus and the bed nucleus of the stria terminalis were described in more detail in coronal, sagittal, and horizontal sections.

1.3.7.3 Diencephalon

Hypothalamus. By taking advantage of immunoreactivity for vasopressin, oxytocin, neuropeptide Y, corticotropin-releasing factor, and vasoactive intestinal polypeptide, we were able to identify subregions within the hypothalamus in the tree shrew brain (Ni et al. 2014, 2015; Shu et al. 2015).

Thalamus. Several subdivisions were delineated in the thalamus of the tree shrew brain using previous studies (Agarwala et al. 1992; Conley and Friederich-Ecsy 1993; Lyon et al. 2003a, b; Chomsung et al. 2010; Baldwin et al. 2011; Ni et al. 2015; Shu et al. 2015). In addition, we also identified other subdivisions by the presence of Nissl and AChE reactivity. Most subregions in the thalamus were identified using high-density AChE staining.

1.3.7.4 Mesencephalon

The mesencephalon consists of the interpeduncular nucleus, rhabdoid nucleus, ventral tegmental area, superior colliculus, inferior colliculus, Edinger-Westphal nucleus, mesencephalic reticular formation, oculomotor nucleus, periaqueductal gray, median raphe nucleus, and dorsal raphe nucleus. We identified these brain regions based upon the results of neuropeptide Y and corticotropin-releasing factor immunohistochemical analyses (Ni et al. 2015; Shu et al. 2015) and retrograde tracing studies (Hashikawa et al. 1986; Chomsung et al. 2008; Baldwin et al. 2013).

1.3.7.5 Metencephalon

Refer to Ni et al. (2015) and Shu et al. (2015) for a general delineation of the parabrachial nucleus, pontine nuclei, tegmental nucleus, and locus coeruleus. The locations of other brain regions in the metencephalon were determined based upon descriptions of these nuclei in monkey and rat brains (Paxinos et al. 2000; Paxinos and Watson 2006; Xavier and Nicole 2008).

1.3.7.6 Myelencephalon

The brain regions in the myelencephalon were delineated according to recent works aimed at mapping the tree shrew brain (Schroeder and Jane 1971; Ni et al. 2015; Shu et al. 2015). The myelencephalon consists of the spinal trigeminal nucleus, vestibular nucleus, abducens nucleus, facial nucleus, motor trigeminal nucleus, cochlear nucleus, gigantocellular reticular nucleus, parvicellular reticular nucleus, nucleus of the solitary tract, hypoglossal nucleus, dorsal motor nucleus of vagus, cuneate nucleus, gracile nucleus, medial lemniscus, medullary reticular nucleus, reticular nucleus, and inferior olivary nucleus.

Conclusion

In this tree shrew brain atlas, we have provided a series of 192 coronal sections, 36 sagittal sections, and 49 horizontal sections that were stained using Nissl or acetylcholinesterase histochemistry in addition to a series of diagrams. We have also presented MR images that were acquired using a 9.4 T MR scanner. This atlas will enable scientists to perform experiments using accurate coordinates for brain regions of interest, which will assist in establishing the tree shrew as an animal model that can be used in preclinical and mechanistic studies aimed at exploring central nervous system-related disorders.

References

Agarwala S, Gunluk AE, May 3rd JG, Petry HM. Immunohistochemical organization of the ventral lateral geniculate nucleus in the tree shrew. J Comp Neurol. 1992;318:267–76.

Airaksinen MS, Flugge G, Fuchs E, Panula P. Histaminergic system in the tree shrew brain. J Comp Neurol. 1989;286:289–310.

Baldwin MK, Wei H, Reed JL, Bickford ME, Petry HM, Kaas JH. Cortical projections to the superior colliculus in tree shrews (*Tupaia belangeri*). J Comp Neurol. 2013;521:1614–32.

Baldwin MK, Wong P, Reed JL, Kaas JH. Superior colliculus connections with visual thalamus in gray squirrels (*Sciurus carolinensis*): evidence for four subdivisions within the pulvinar complex. J Comp Neurol. 2011;519:1071–94.

Chomsung RD, Petry HM, Bickford ME. Ultrastructural examination of diffuse and specific tectopulvinar projections in the tree shrew. J Comp Neurol. 2008;510:24–46.

Chomsung RD, Wei H, Day-Brown JD, Petry HM, Bickford ME. Synaptic organization of connections between the temporal cortex and pulvinar nucleus of the tree shrew. Cereb Cortex. 2010;20:997–1011.

Conley M, Friederich-Ecsy B. Functional organization of the ventral lateral geniculate complex of the tree shrew (*Tupaia belangeri*): I Nuclear subdivisions and retinal projections. J Comp Neurol. 1993;328:1–20.

Czeh B, Pudovkina O, van der Hart MG, Simon M, Heilbronner U, Michaelis T, Watanabe T, Frahm J, Fuchs E. Examining SLV-323, a novel NK1 receptor antagonist, in a chronic psychosocial stress model for depression. Psychopharmacology (Berl). 2005;180:548–57.

Day-Brown JD, Wei H, Chomsung RD, Petry HM, Bickford ME. Pulvinar projections to the striatum and amygdala in the tree shrew. Front Neuroanat. 2010;4:143.

Fan Y, Huang ZY, Cao CC, Chen CS, Chen YX, Fan DD, He J, Hou HL, Hu L, Hu XT, Jiang XT, Lai R, Lang YS, Liang B, Liao SG, Mu D, Ma YY, Niu YY, Sun XQ, Xia JQ, Xiao J, Xiong ZQ, Xu L, Yang L, Zhang Y, Zhao W, Zhao XD, Zheng YT, Zhou JM, Zhu YB, Zhang GJ, Wang J, Yao YG. Genome of the Chinese tree shrew. Nat Commun. 2013;4:1426.

Flugge G, Ahrens O, Fuchs E. Monoamine receptors in the amygdaloid complex of the tree shrew (*Tupaia belangeri*). J Comp Neurol. 1994;343:597–608.

Fuchs E. Social stress in tree shrews as an animal model of depression: an example of a behavioral model of a CNS disorder. CNS Spectr. 2005;10:182–90.

Hashikawa T, Van Lieshout D, Harting JK. Projections from the parabigeminal nucleus to the dorsal lateral geniculate nucleus in the tree shrew Tupaia glis. J Comp Neurol. 1986;246:382–94.

Janecka JE, Miller W, Pringle TH, Wiens F, Zitzmann A, Helgen KM, Springer MS, Murphy WJ. Molecular and genomic data identify the closest living relative of primates. Science. 2007;318:792–4.

Keuker JI, Rochford CD, Witter MP, Fuchs E. A cytoarchitectonic study of the hippocampal formation of the tree shrew (*Tupaia belangeri*). J Chem Neuroanat. 2003;26:1–15.

Kosaka K, Kosaka T. Organization of the main olfactory bulbs of some mammals: musk shrews, moles, hedgehogs, tree shrews, bats, mice, and rats. J Comp Neurol. 2004;472:1–12.

Kriegs JO, Churakov G, Jurka J, Brosius J, Schmitz J. Evolutionary history of 7SL RNA-derived SINEs in Supraprimates. Trends Genet. 2007;23:158–61.

Lyon DC, Jain N, Kaas JH. The visual pulvinar in tree shrews I. Multiple subdivisions revealed through acetylcholinesterase and Cat-301 chemoarchitecture. J Comp Neurol. 2003a;467:593–606.

Lyon DC, Jain N, Kaas JH. The visual pulvinar in tree shrews II. Projections of four nuclei to areas of visual cortex. J Comp Neurol. 2003b;467:607–27.

Malz CR, Knabe W, Kuhn HJ. Pattern of calretinin immunoreactivity in the main olfactory system and the vomeronasal system of the tree shrew, *Tupaia belangeri*. J Comp Neurol. 2000;420:428–36.

Marrocco RT, De Valois RL, Boles JI. A stereotaxic atlas of the brain of the tree shrew (*Tupaia glis*). J Hirnforsch. 1970;12:307–12.

McCollum LA, Roberts RC. Ultrastructural localization of tyrosine hydroxylase in tree shrew nucleus accumbens core and shell. Neuroscience. 2014;271:23–34.

Ni RJ, Shu YM, Luo PH, Fang H, Wang Y, Yao L, Zhou JN. Immunohistochemical mapping of neuropeptide Y in the tree shrew brain. J Comp Neurol. 2015;523:495–529.

Ni RJ, Shu YM, Wang J, Yin JC, Xu L, Zhou JN. Distribution of vasopressin, oxytocin and vasoactive intestinal polypeptide in the hypothalamus and extrahypothalamic regions of tree shrews. Neuroscience. 2014;265:124–36.

Palazzi X, Bordier N. The marmoset brain in stereotaxic coordinates. New York: Springer; 2008.

Norton TT, Amedo AO, Siegwart Jr JT. Darkness causes myopia in visually experienced tree shrews. Invest Ophthalmol Vis Sci. 2006;47:4700–7.

Paxinos G, Franklin KBJ. The mouse brain in stereotaxic coordinates. New York: Academic; 2001.

Paxinos G, Huang XF, Toga AW. The rhesus monkey brain in stereotaxic coordinates. San Diego: Academic; 2000.

Paxinos G, Watson C. The rat brain in stereotaxic coordinates. New York: Academic; 2006.

Peng YZ, Ye ZZ, Zou RJ, Wang YX, Tian BP, Ma YY, Shi LM. Biology of Chinese Tree Shrews (*Tupaia belangeri chinensis*). Yunnan Science and Technology Press; 1991.

Remple MS, Reed JL, Stepniewska I, Kaas JH. Organization of frontoparietal cortex in the tree shrew (*Tupaia belangeri*). I. Architecture, microelectrode maps, and corticospinal connections. J Comp Neurol. 2006;497:133–54.

Remple MS, Reed JL, Stepniewska I, Lyon DC, Kaas JH. The organization of frontoparietal cortex in the tree shrew (*Tupaia belangeri*): II. Connectional evidence for a frontal-posterior parietal network. J Comp Neurol. 2007;501:121–49.

Rice MW, Roberts RC, Melendez-Ferro M, Perez-Costas E. Neurochemical characterization of the tree shrew dorsal striatum. Front Neuroanat. 2011;5:53.

Schroeder DM, Jane JA. Projection of dorsal column nuclei and spinal cord to brainstem and thalamus in the tree shrew, *Tupaia glis*. J Comp Neurol. 1971;142:309–50.

Shu YM, Ni RJ, Sun YJ, Fang H, Zhou JN. Distribution of corticotropin-releasing factor in the tree shrew brain. Brain Res. 2015;1618:270–85.

Tsukiyama-Kohara K, Kohara M. Tupaia belangeri as an experimental animal model for viral infection. Exp Anim. 2014;63(4):367–74.

Wolter R, Tauer U, Fuchs E, Volk B. Mapping of cytoskeletal components in the hippocampal formation of the tree shrew (*Tupaia belangeri*). J Chem Neuroanat. 1999;17:65–74.

Wong P, Kaas JH. Architectonic subdivisions of neocortex in the tree shrew (*Tupaia belangeri*). Anat Rec (Hoboken). 2009;292: 994–1027.

Yan RQ, Su JJ, Huang DR, Gan YC, Yang C, Huang GH. Human hepatitis B virus and hepatocellular carcinoma. I Experimental infection of tree shrews with hepatitis B virus. J Cancer Res Clin Oncol. 1996;122:283–8.

Yang WG, Liu J-L, Luo Y-Q, Chen P-L, Zhang C-R, Wan X-C, Tao Z-L. A seterotaxic atlas of the brain of tupaia belangeri and macaque monkey living in Guangxi. Guangxi: Guangxi Science & Technology Publishing House; 1990.

Zhao F, Guo X, Wang Y, Liu J, Lee WH, Zhang Y. Drug target mining and analysis of the Chinese tree shrew for pharmacological testing. PLoS One. 2014;9:e104191.

Zhao X, Tang ZY, Klumpp B, Wolff-Vorbeck G, Barth H, Levy S, von Weizsacker F, Blum HE, Baumert TF. Primary hepatocytes of *Tupaia belangeri* as a potential model for hepatitis C virus infection. J Clin Invest. 2002;109:221–32.

Part II

Figures of the Tree Shrew Whole Brain

Coronal Sections of the Tree Shrew Brain

2

J.-N. Zhou, R.-J. Ni, *The Tree Shrew (Tupaia belangeri chinensis) Brain in Stereotaxic Coordinates*,
DOI 10.1007/978-981-10-0611-1_2

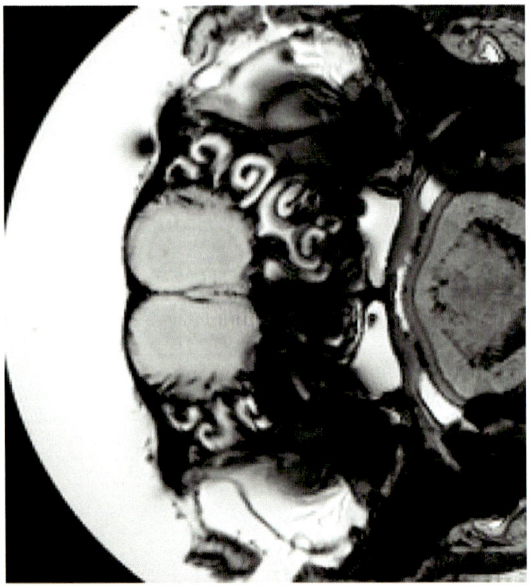

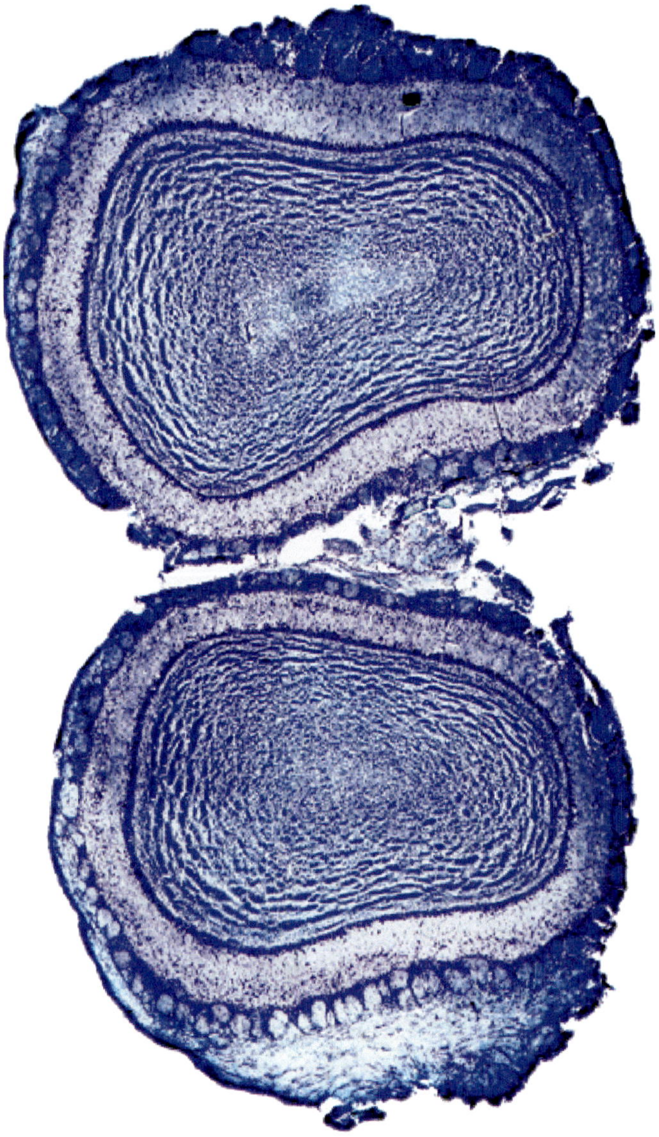

EPl external plexiform layer of the olfactory bulb
GCL granule cell layer of the olfactory bulb
GL glomerular layer of the olfactory bulb
IPl internal plexiform layer of the olfactory bulb
MCL mitral cell layer of the olfactory bulb
ONL olfactory nerve layer

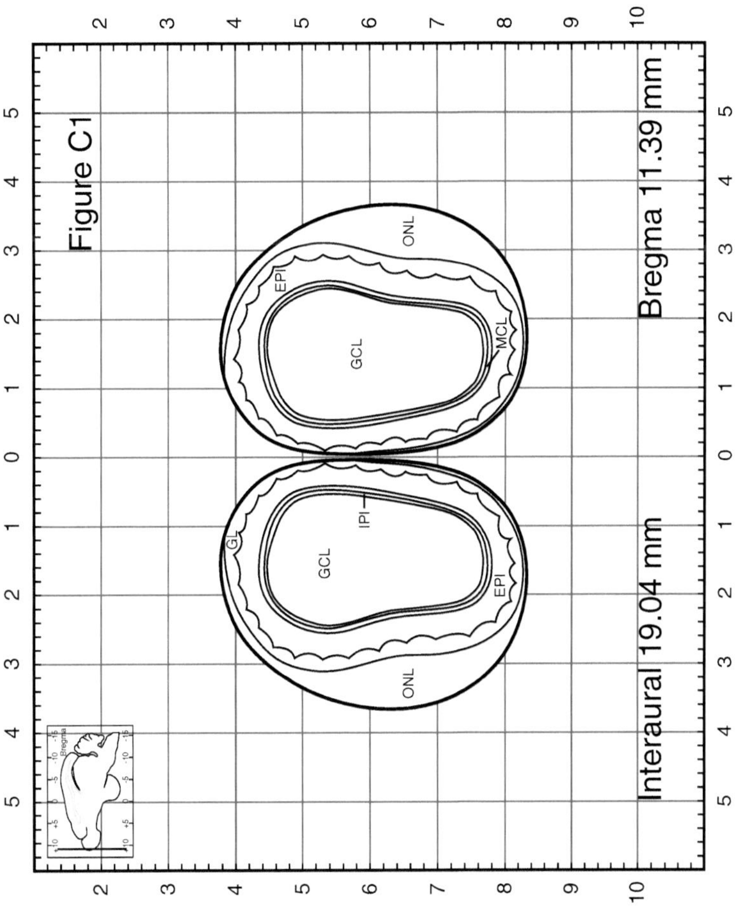
Figure C1
Bregma 11.39 mm
Interaural 19.04 mm
ONL
EPl
GCL
MCL
IPl
GL
GCL
EPl
ONL

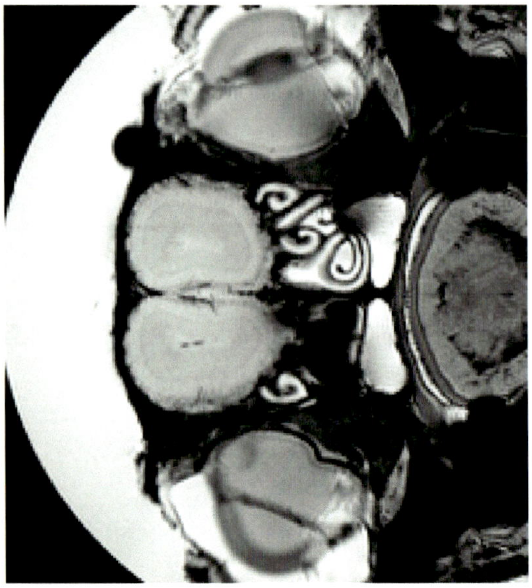

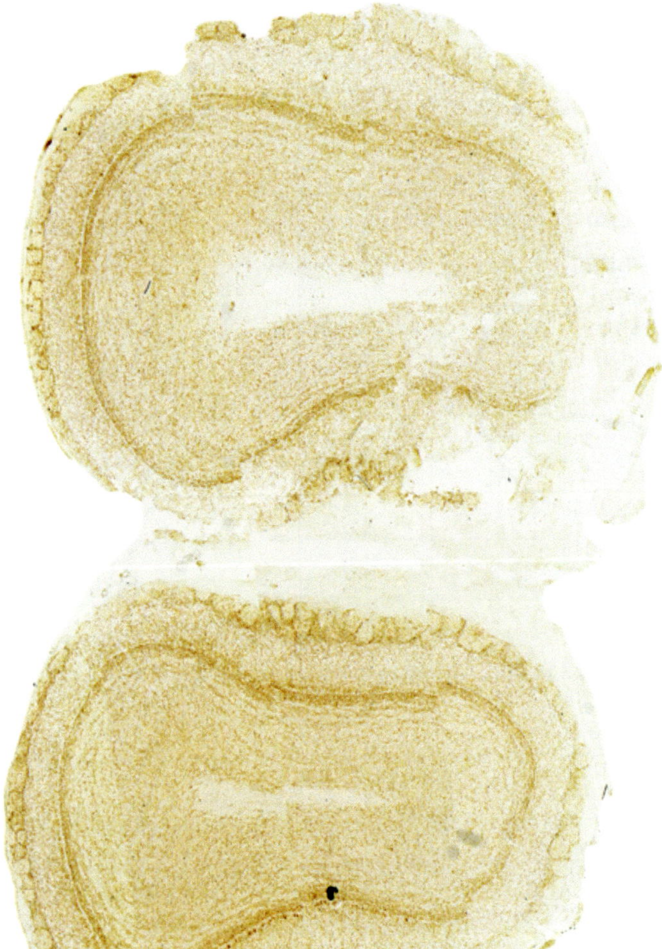

E ependyma and subependymal layer
EPl external plexiform layer of the olfactory bulb
GCL granule cell layer of the olfactory bulb
GL glomerular layer of the olfactory bulb
IPl internal plexiform layer of the olfactory bulb
MCL mitral cell layer of the olfactory bulb
ONL olfactory nerve layer
OV olfactory ventricle (olfactory part of lateral ventricle)

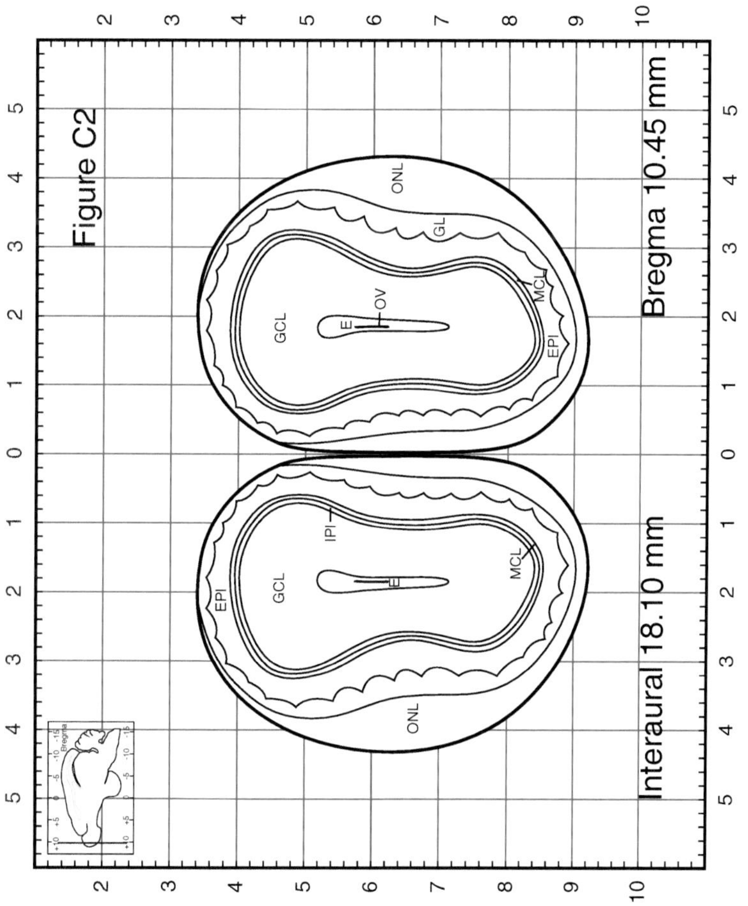
Figure C2
Bregma 10.45 mm
Interaural 18.10 mm
ONL
GL
MCL
EPI
GCL
E
OV
IPI

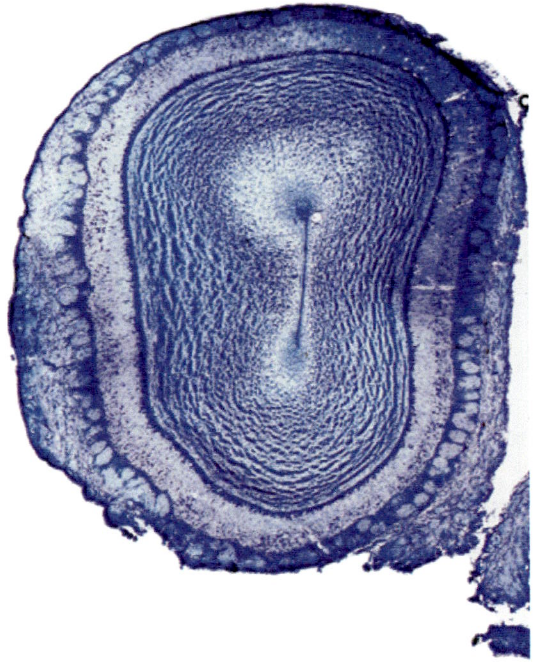

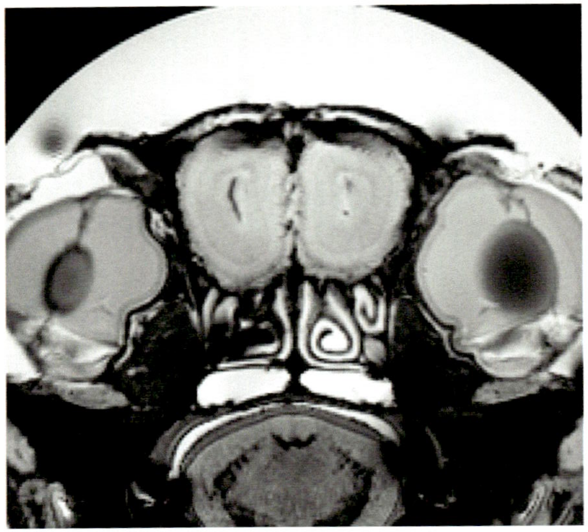

E	ependyma and subependymal layer
EPI	external plexiform layer of the olfactory bulb
GCL	granule cell layer of the olfactory bulb
GL	glomerular layer of the olfactory bulb
IPI	internal plexiform layer of the olfactory bulb
MCL	mitral cell layer of the olfactory bulb
ONL	olfactory nerve layer
OV	olfactory ventricle (olfactory part of lateral ventricle)

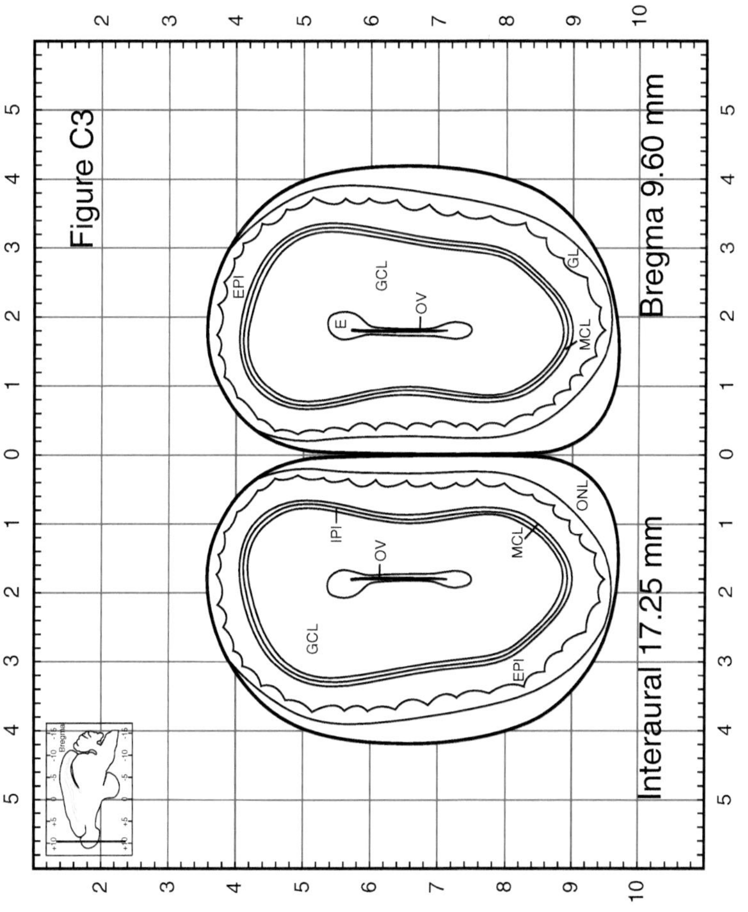
Figure C3
Bregma 9.60 mm
Interaural 17.25 mm
EPI
GCL
OV
E
GL
MCL
ONL
IPI

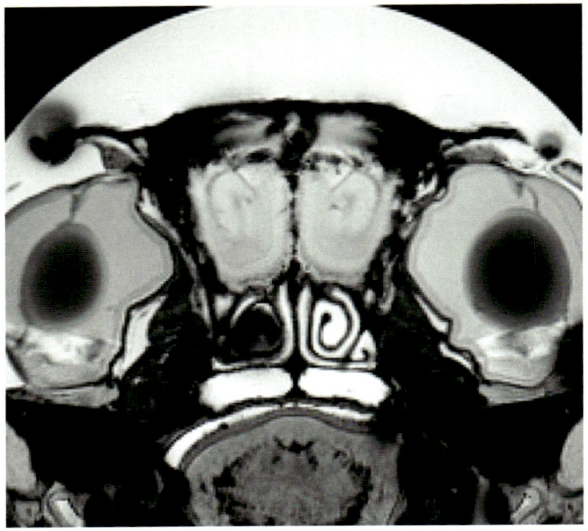

aci	anterior commissure, intrabulbar part
AOL	anterior olfactory nucleus, lateral part
E	ependyma and subependymal layer
EPl	external plexiform layer of the olfactory bulb
EPlA	external plexiform layer of the accessory olfactory bulb
GCL	granule cell layer of the olfactory bulb
GL	glomerular layer of the olfactory bulb
GlA	glomerular layer of the accessory olfactory bulb
GrA	granule cell layer of the accessory olfactory bulb
IPl	internal plexiform layer of the olfactory bulb
MCL	mitral cell layer of the olfactory bulb
MiA	mitral cell layer of the accessory olfactory bulb
OFC	orbital frontal cortex
ONL	olfactory nerve layer
OV	olfactory ventricle (olfactory part of lateral ventricle)

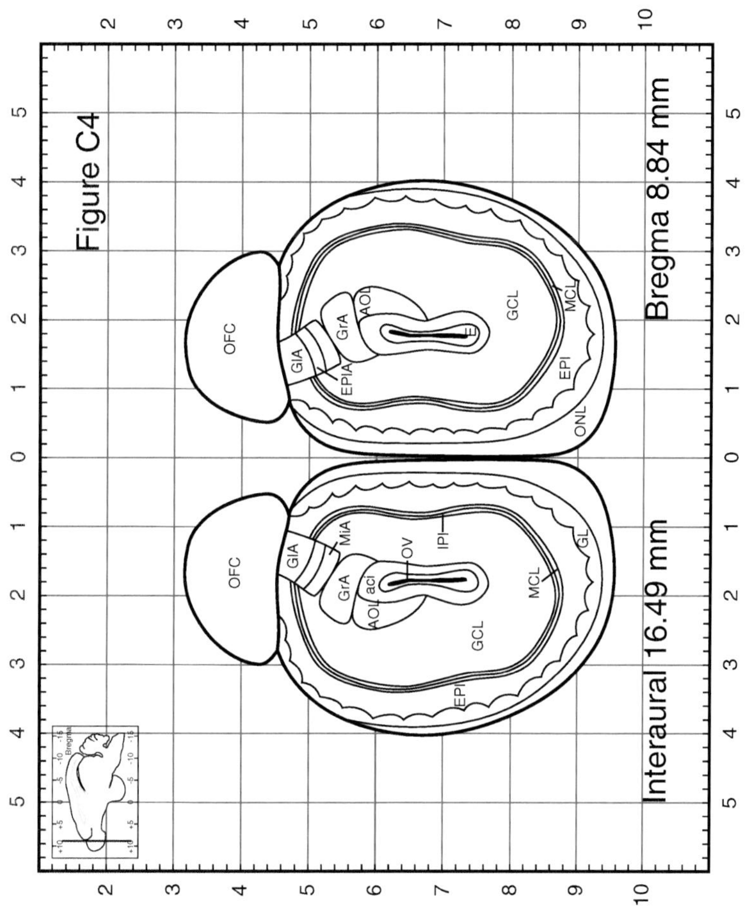
Figure C4
OFC
GIA
EPIA
GrA
AOL
E
GCL
MCL
EPI
ONL
MiA
aci
OV
IPl
GL
Bregma 8.84 mm
Interaural 16.49 mm

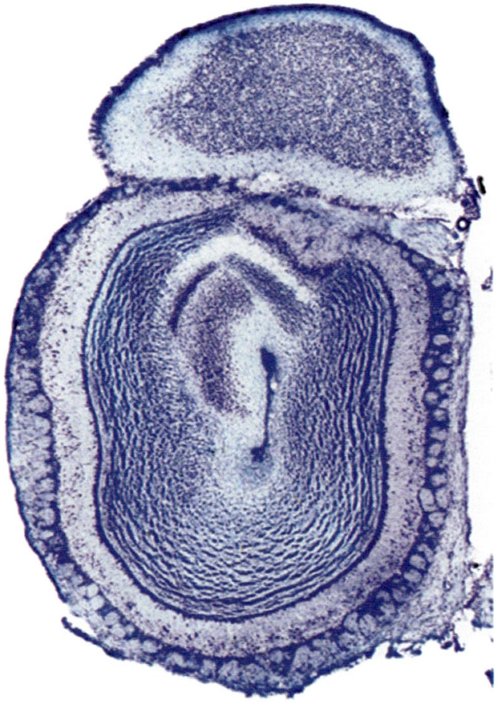

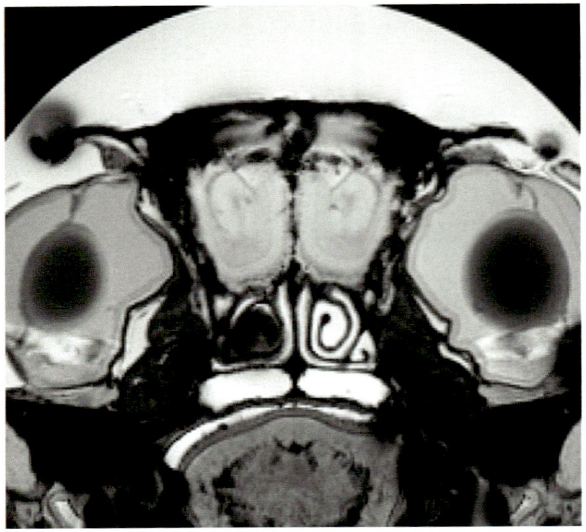

aci	anterior commissure, intrabulbar part
AOE	anterior olfactory nucleus, external part
AOL	anterior olfactory nucleus, lateral part
E	ependyma and subependymal layer
EPI	external plexiform layer of the olfactory bulb
EPIA	external plexiform layer of the accessory olfactory bulb
GCL	granule cell layer of the olfactory bulb
GL	glomerular layer of the olfactory bulb
GlA	glomerular layer of the accessory olfactory bulb
GrA	granule cell layer of the accessory olfactory bulb
IPI	internal plexiform layer of the olfactory bulb
lo	lateral olfactory tract
MCL	mitral cell layer of the olfactory bulb
MiA	mitral cell layer of the accessory olfactory bulb
OFC	orbital frontal cortex
ONL	olfactory nerve layer
OV	olfactory ventricle (olfactory part of lateral ventricle)

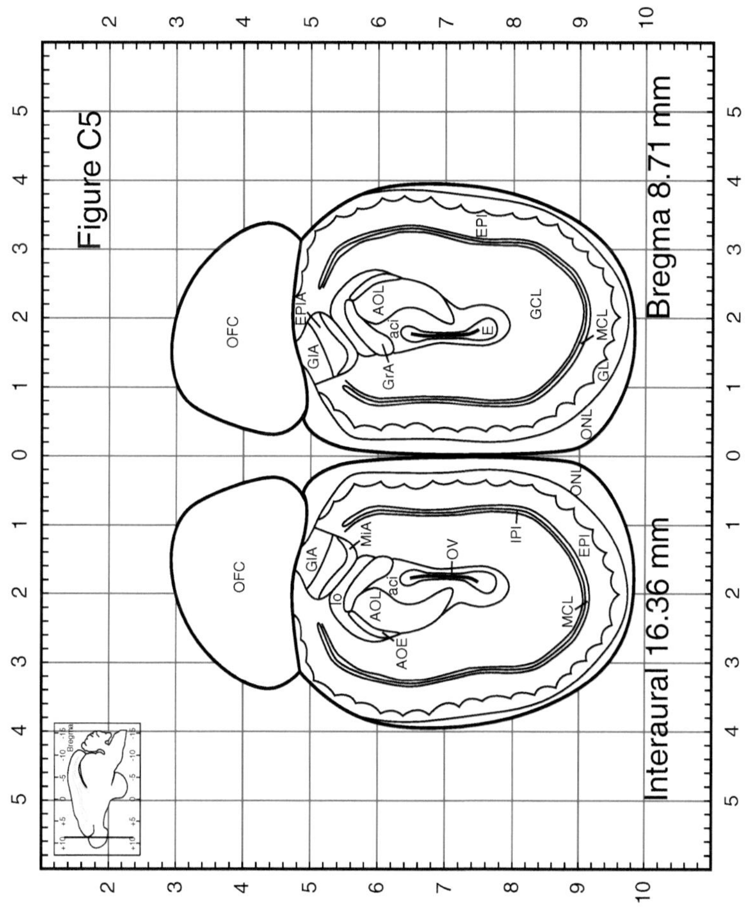
Figure C5
Bregma 8.71 mm
Interaural 16.36 mm
OFC
GIA
EPIA
AOL
aci
GrA
E
GCL
EPI
MCL
GL
ONL
MIA
lo
AOE
OV
IPI

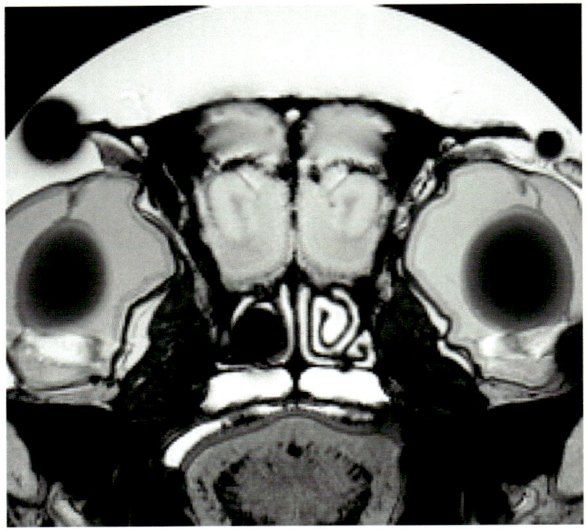

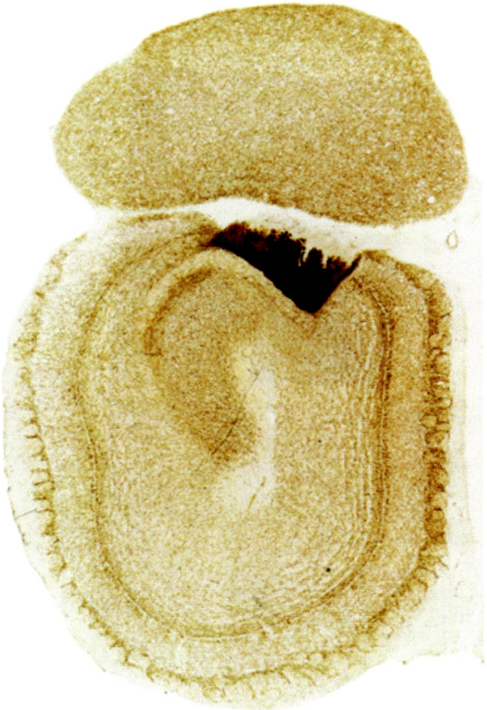

aci anterior commissure, intrabulbar part
AOE anterior olfactory nucleus, external part
AOL anterior olfactory nucleus, lateral part
DFC dorsal frontal cortex
E ependyma and subependymal layer
EPl external plexiform layer of the olfactory bulb
EPlA external plexiform layer of the accessory olfactory bulb
GCL granule cell layer of the olfactory bulb
GL glomerular layer of the olfactory bulb
GlA glomerular layer of the accessory olfactory bulb
GrA granule cell layer of the accessory olfactory bulb
IPl internal plexiform layer of the olfactory bulb
lo lateral olfactory tract
MCL mitral cell layer of the olfactory bulb
MiA mitral cell layer of the accessory olfactory bulb
OFC orbital frontal cortex
ONL olfactory nerve layer
OV olfactory ventricle (olfactory part of lateral ventricle)

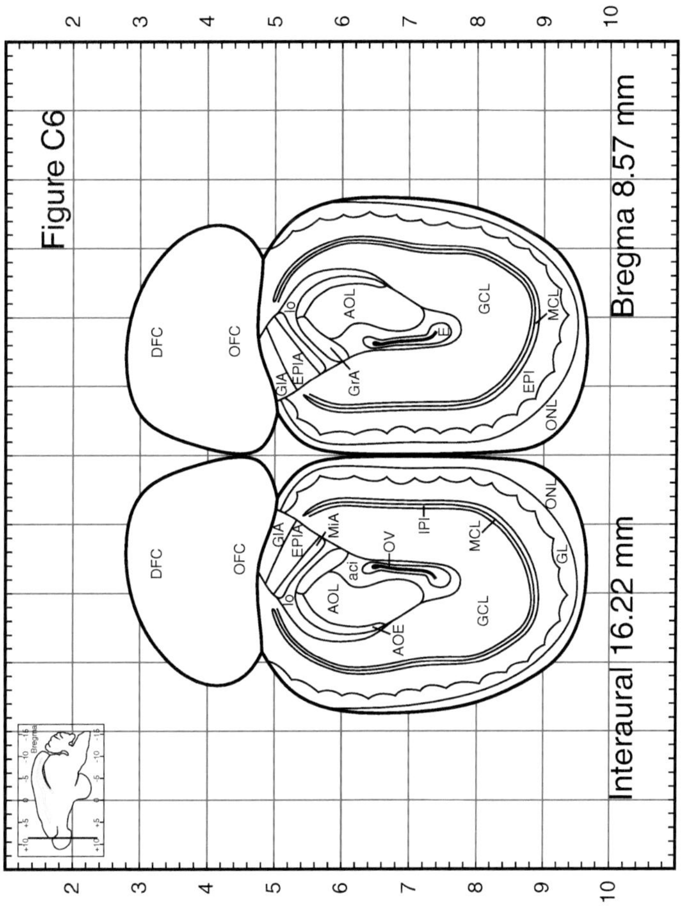
Figure C6
Bregma 8.57 mm
Interaural 16.22 mm
DFC
OFC
GIA
EPIA
lo
AOL
GrA
E
GCL
MCL
EPI
ONL
MiA
aci
OV
IPl
AOE
GL

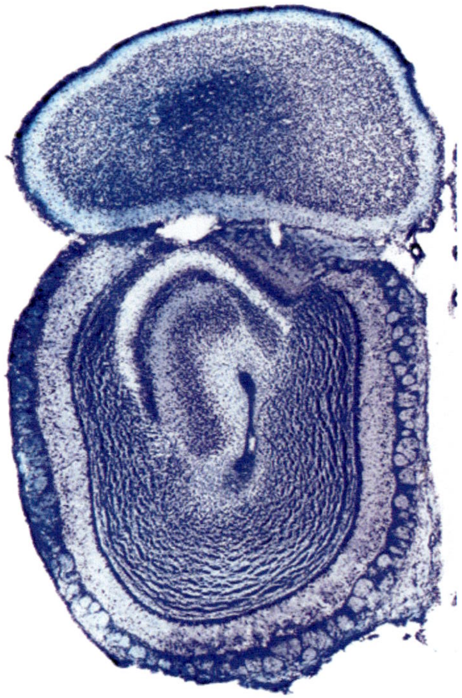

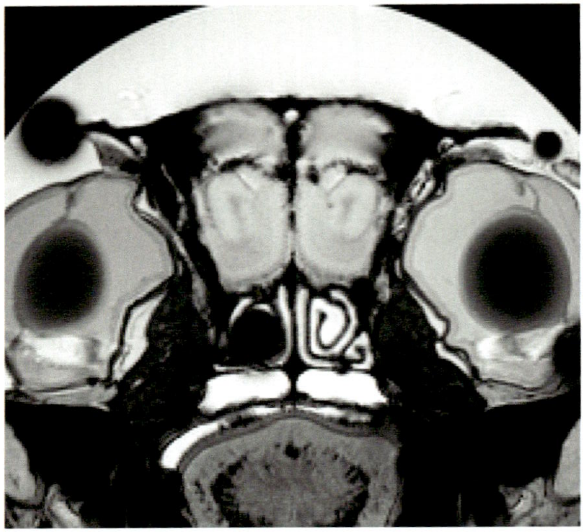

aci	anterior commissure, intrabulbar part
AOD	anterior olfactory nucleus, dorsal part
AOE	anterior olfactory nucleus, external part
AOL	anterior olfactory nucleus, lateral part
AOV	anterior olfactory nucleus, ventral part
DFC	dorsal frontal cortex
E	ependyma and subependymal layer
EPl	external plexiform layer of the olfactory bulb
EPlA	external plexiform layer of the accessory olfactory bulb
GCL	granule cell layer of the olfactory bulb
GL	glomerular layer of the olfactory bulb
GlA	glomerular layer of the accessory olfactory bulb
GrA	granule cell layer of the accessory olfactory bulb
IPl	internal plexiform layer of the olfactory bulb
lo	lateral olfactory tract
MCL	mitral cell layer of the olfactory bulb
MiA	mitral cell layer of the accessory olfactory bulb
OFC	orbital frontal cortex
ONL	olfactory nerve layer
OV	olfactory ventricle (olfactory part of lateral ventricle)

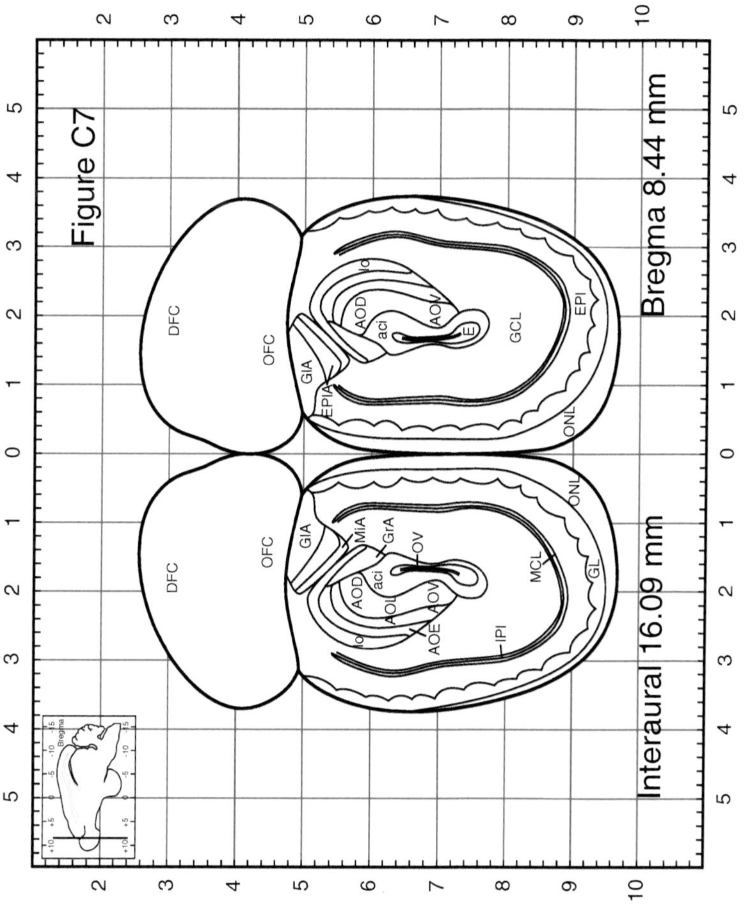
Figure C7
Bregma 8.44 mm
Interaural 16.09 mm
DFC
OFC
GIA
EPIA
AOD
aci
AOV
lo
E
GCL
EPI
ONL
MiA
GrA
OV
AOL
AOE
IPI
MCL
GL

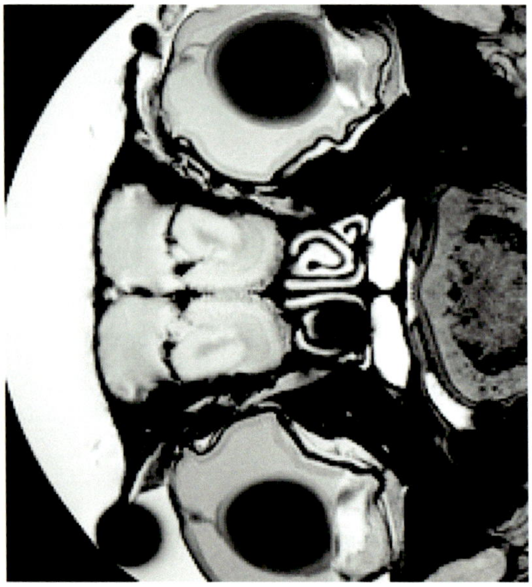

aci anterior commissure, intrabulbar part
AOD anterior olfactory nucleus, dorsal part
AOE anterior olfactory nucleus, external part
AOL anterior olfactory nucleus, lateral part
AOV anterior olfactory nucleus, ventral part
DFC dorsal frontal cortex
E ependyma and subependymal layer

EPl external plexiform layer of the olfactory bulb
EPlA external plexiform layer of the accessory olfactory bulb
GCL granule cell layer of the olfactory bulb
GL glomerular layer of the olfactory bulb
GlA glomerular layer of the accessory olfactory bulb
GrA granule cell layer of the accessory olfactory bulb
IPl internal plexiform layer of the olfactory bulb

lo lateral olfactory tract
MCL mitral cell layer of the olfactory bulb
MFC medial frontal cortex
MiA mitral cell layer of the accessory olfactory bulb
OFC orbital frontal cortex
ONL olfactory nerve layer
OV olfactory ventricle (olfactory part of lateral ventricle)

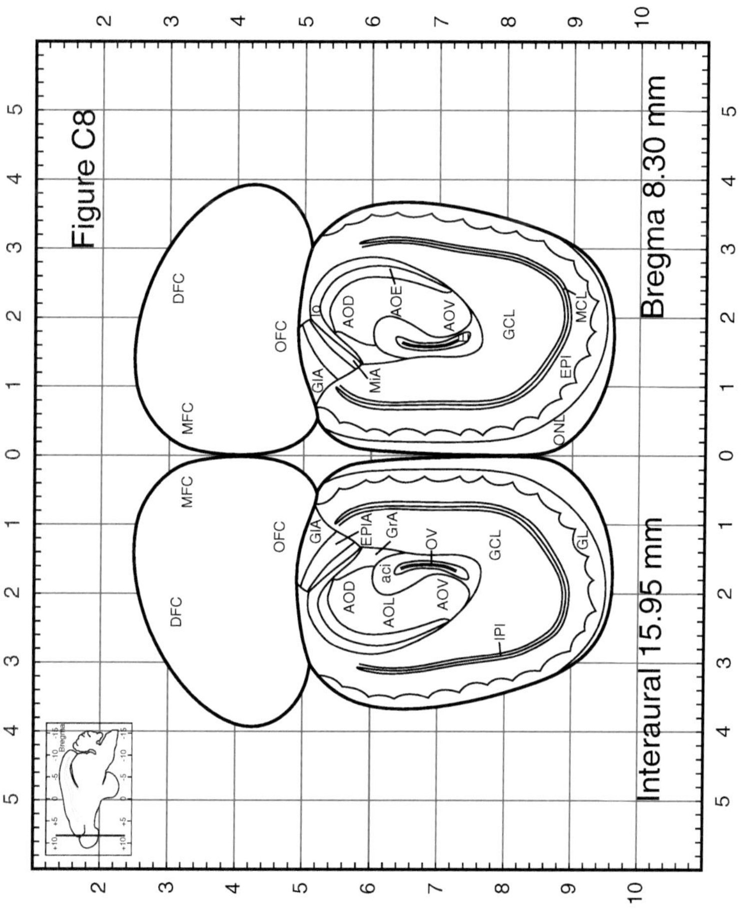
Figure C8
Bregma 8.30 mm
Interaural 15.95 mm
DFC
MFC
OFC
GIA
IO
AOD
AOE
AOV
MiA
GCL
MCL
EPI
ONL
EPIA
GrA
OV
aci
AOL
IPI
GL

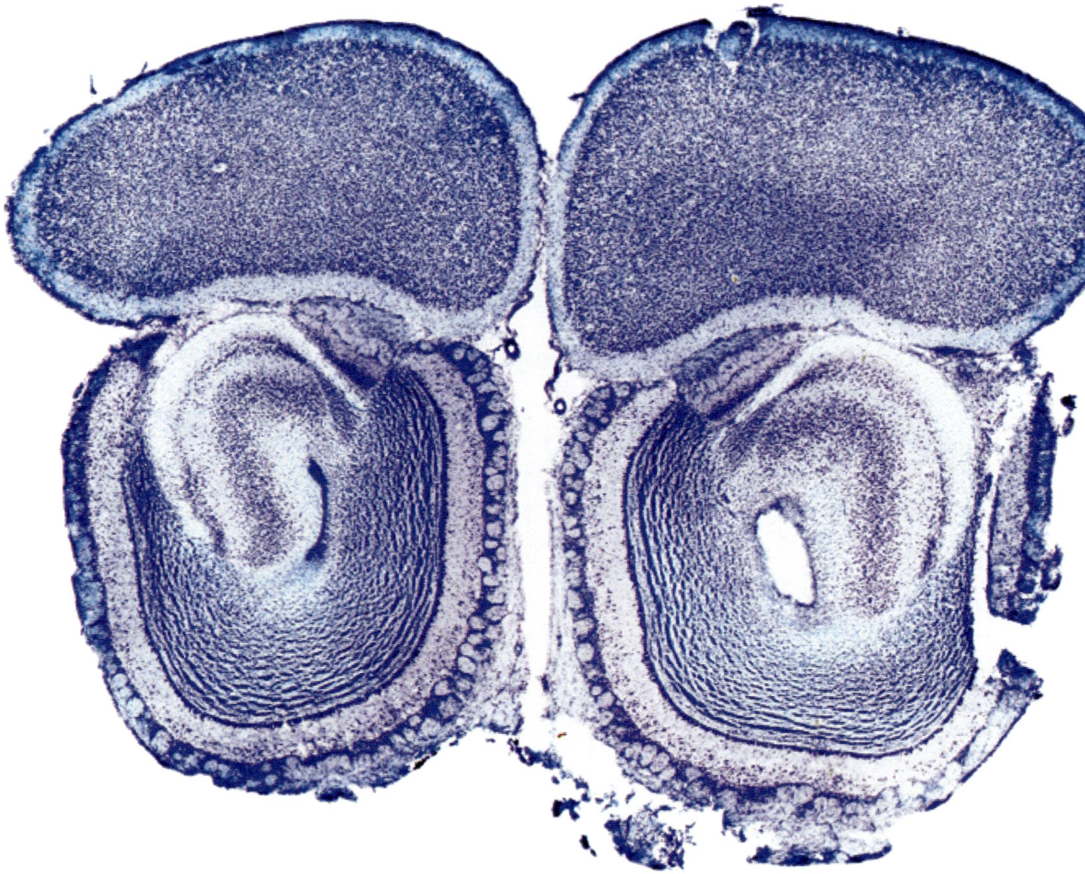

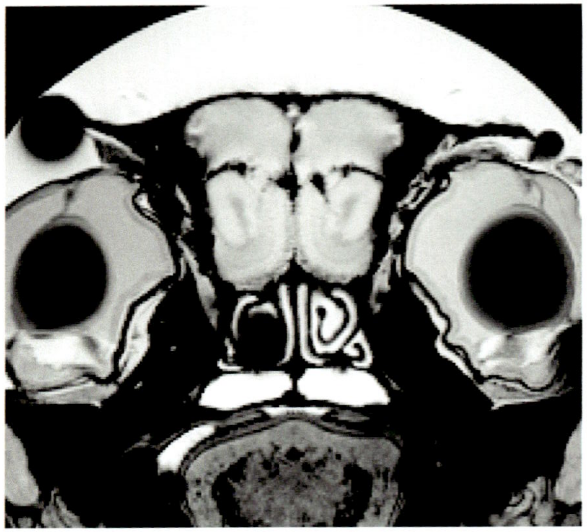

aci anterior commissure, intrabulbar part
AOD anterior olfactory nucleus, dorsal part
AOE anterior olfactory nucleus, external part
AOL anterior olfactory nucleus, lateral part
AOV anterior olfactory nucleus, ventral part
DFC dorsal frontal cortex
E ependyma and subependymal layer

EPl external plexiform layer of the olfactory bulb
EPlA external plexiform layer of the accessory olfactory bulb
GCL granule cell layer of the olfactory bulb
GL glomerular layer of the olfactory bulb
GlA glomerular layer of the accessory olfactory bulb
GrA granule cell layer of the accessory olfactory bulb
IPl internal plexiform layer of the olfactory bulb

lo lateral olfactory tract
MCL mitral cell layer of the olfactory bulb
MFC medial frontal cortex
MiA mitral cell layer of the accessory olfactory bulb
OFC orbital frontal cortex
ONL olfactory nerve layer
OV olfactory ventricle (olfactory part of lateral ventricle)

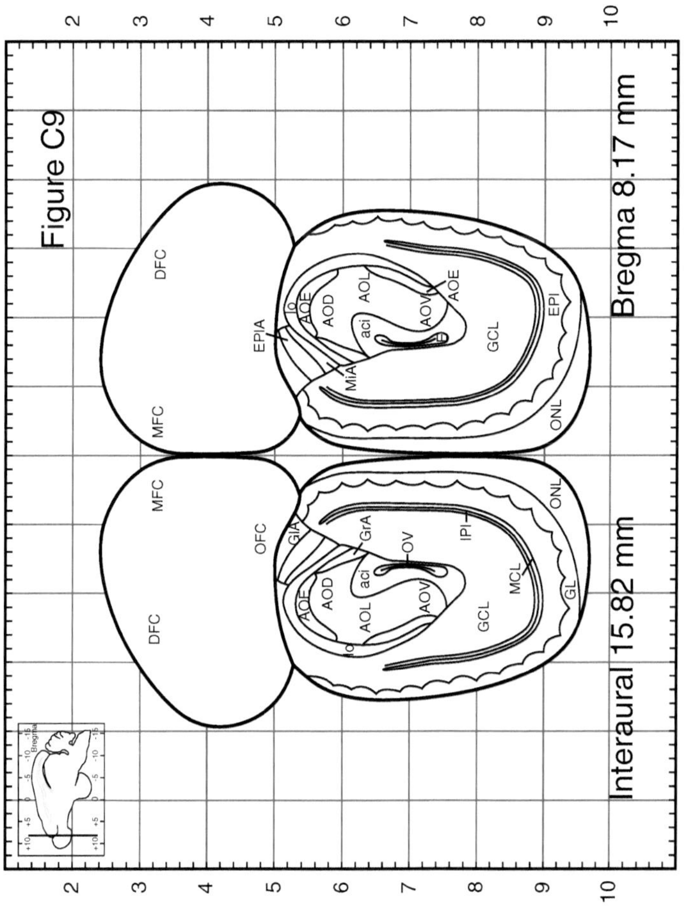
Figure C9
Bregma 8.17 mm
Interaural 15.82 mm
DFC
MFC
OFC
EPlA
MiA
lo
AOE
AOD
AOL
AOV
aci
GCL
EPl
ONL
GlA
GrA
OV
IPl
MCL
GL

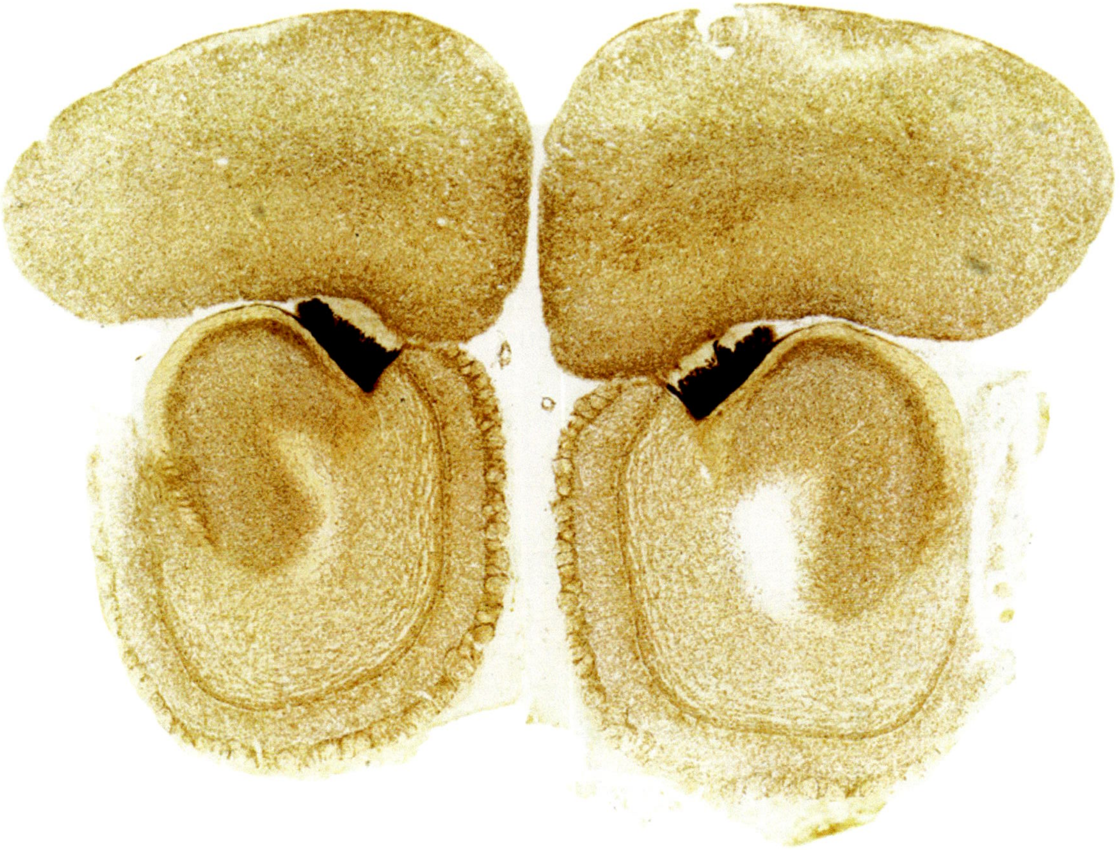

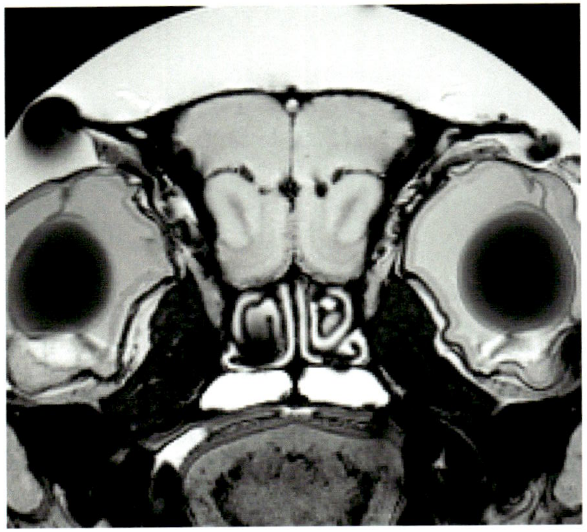

aci	anterior commissure, intrabulbar part
AOD	anterior olfactory nucleus, dorsal part
AOE	anterior olfactory nucleus, external part
AOL	anterior olfactory nucleus, lateral part
AOV	anterior olfactory nucleus, ventral part
DFC	dorsal frontal cortex
E	ependyma and subependymal layer
EPl	external plexiform layer of the olfactory bulb
EPlA	external plexiform layer of the accessory olfactory bulb
GCL	granule cell layer of the olfactory bulb
GL	glomerular layer of the olfactory bulb
GlA	glomerular layer of the accessory olfactory bulb
GrA	granule cell layer of the accessory olfactory bulb
IPl	internal plexiform layer of the olfactory bulb
lo	lateral olfactory tract
MCL	mitral cell layer of the olfactory bulb
MFC	medial frontal cortex
MiA	mitral cell layer of the accessory olfactory bulb
OFC	orbital frontal cortex
ONL	olfactory nerve layer
OV	olfactory ventricle (olfactory part of lateral ventricle)

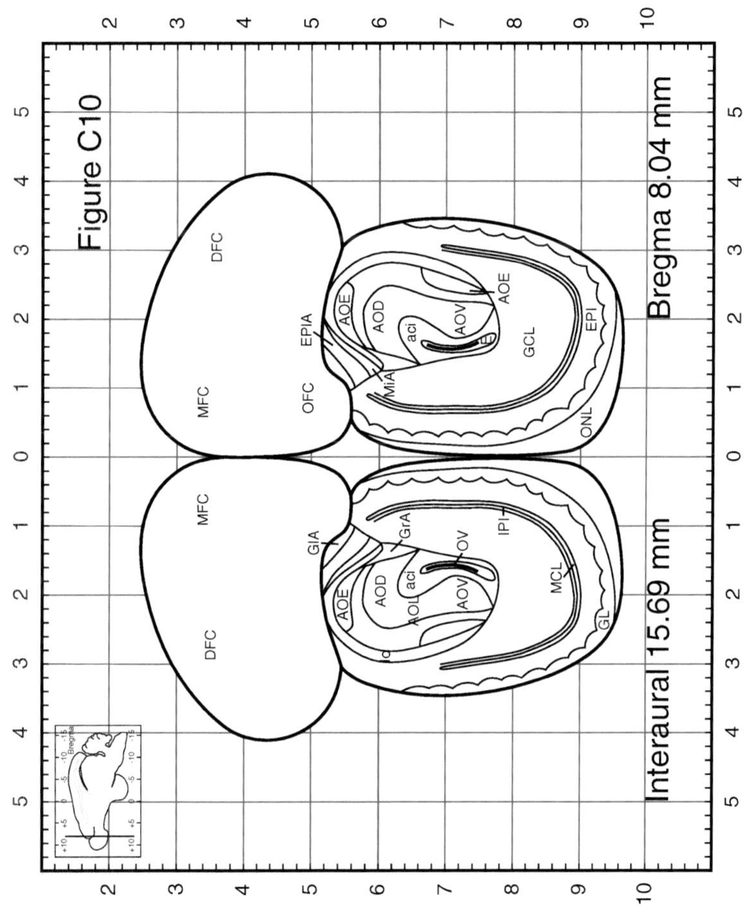
Figure C10
Bregma 8.04 mm
Interaural 15.69 mm
DFC
MFC
OFC
EPIA
AOE
AOD
aci
AOV
GCL
EPl
MiA
ONL
GIA
GrA
OV
IPl
AOL
MCL
GL
lo

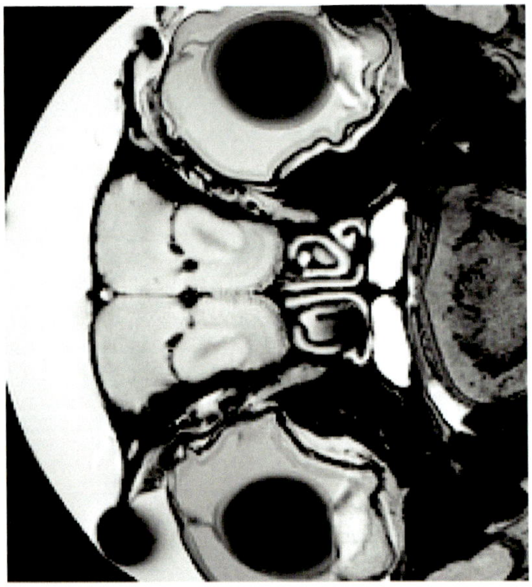

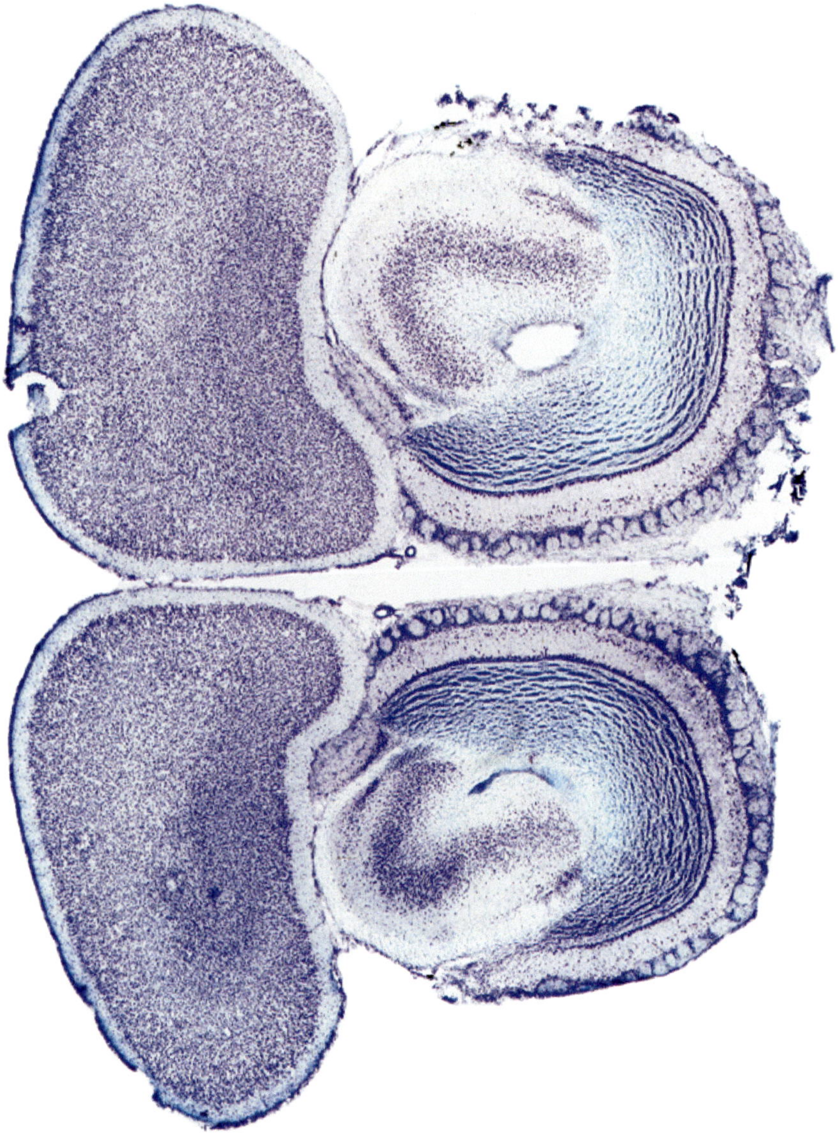

aci anterior commissure, intrabulbar part
AOD anterior olfactory nucleus, dorsal part
AOE anterior olfactory nucleus, external part
AOL anterior olfactory nucleus, lateral part
AOM anterior olfactory nucleus, medial part
AOV anterior olfactory nucleus, ventral part
Cg cingulate cortex
DFC dorsal frontal cortex

E ependyma and subependymal layer
EPl external plexiform layer of the olfactory bulb
EPlA external plexiform layer of the accessory olfactory bulb
GCL granule cell layer of the olfactory bulb
GL glomerular layer of the olfactory bulb
GlA glomerular layer of the accessory olfactory bulb
GrA granule cell layer of the accessory olfactory bulb
IPl internal plexiform layer of the olfactory bulb

lo lateral olfactory tract
MCL mitral cell layer of the olfactory bulb
MFC medial frontal cortex
MiA mitral cell layer of the accessory olfactory bulb
OFC orbital frontal cortex
ONL olfactory nerve layer
OV olfactory ventricle (olfactory part of lateral ventricle)

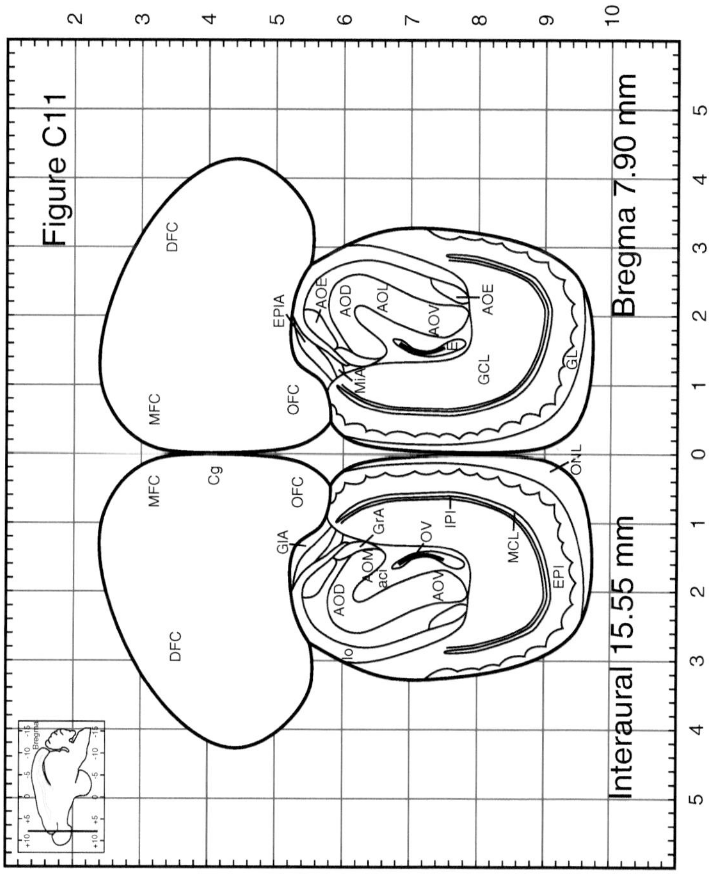
Figure C11
Bregma 7.90 mm
Interaural 15.55 mm
DFC
MFC
Cg
OFC
EPlA
AOE
AOD
AOL
AOV
E
GCL
GL
MiA
ONL
GlA
GrA
AOM
aci
OV
IPl
MCL
EPl
lo

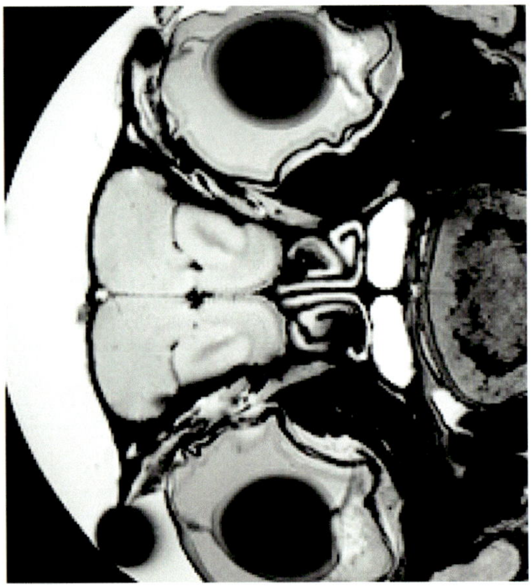

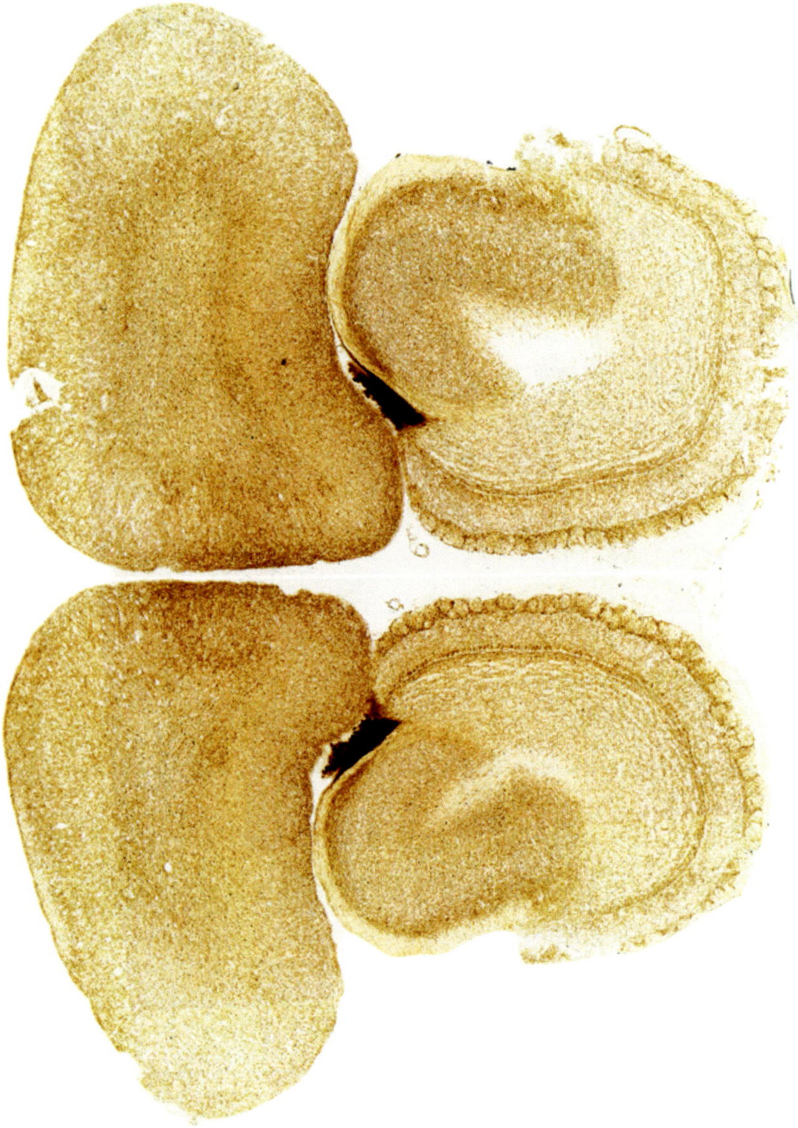

aci anterior commissure, intrabulbar part
AOD anterior olfactory nucleus, dorsal part
AOE anterior olfactory nucleus, external part
AOL anterior olfactory nucleus, lateral part
AOM anterior olfactory nucleus, medial part
AOV anterior olfactory nucleus, ventral part
Cg cingulate cortex
DFC dorsal frontal cortex

E ependyma and subependymal layer
EPI external plexiform layer of the olfactory bulb
EPIA external plexiform layer of the accessory olfactory bulb
GCL granule cell layer of the olfactory bulb
GL glomerular layer of the olfactory bulb
GrA granule cell layer of the accessory olfactory bulb
IPI internal plexiform layer of the olfactory bulb
lo lateral olfactory tract

MCL mitral cell layer of the olfactory bulb
MFC medial frontal cortex
MiA mitral cell layer of the accessory olfactory bulb
OFC orbital frontal cortex
ONL olfactory nerve layer
OV olfactory ventricle (olfactory part of lateral ventricle)

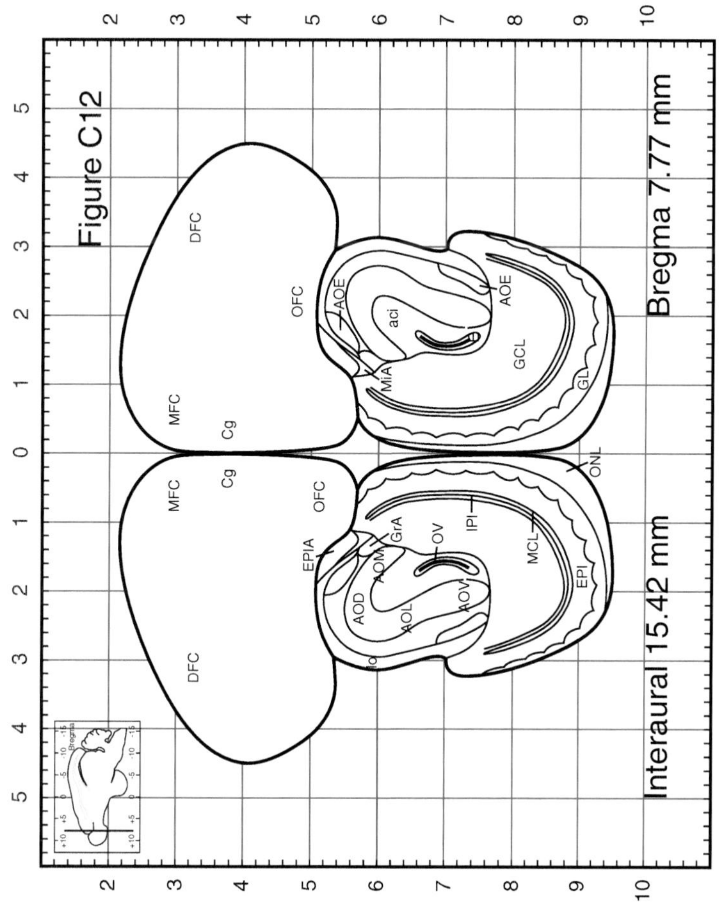
Figure C12
Bregma 7.77 mm
Interaural 15.42 mm
DFC
MFC
Cg
OFC
AOE
aci
MiA
GCL
GL
ONL
IPI
MCL
EPI
EPIA
AOM
GrA
OV
AOD
AOL
AOV

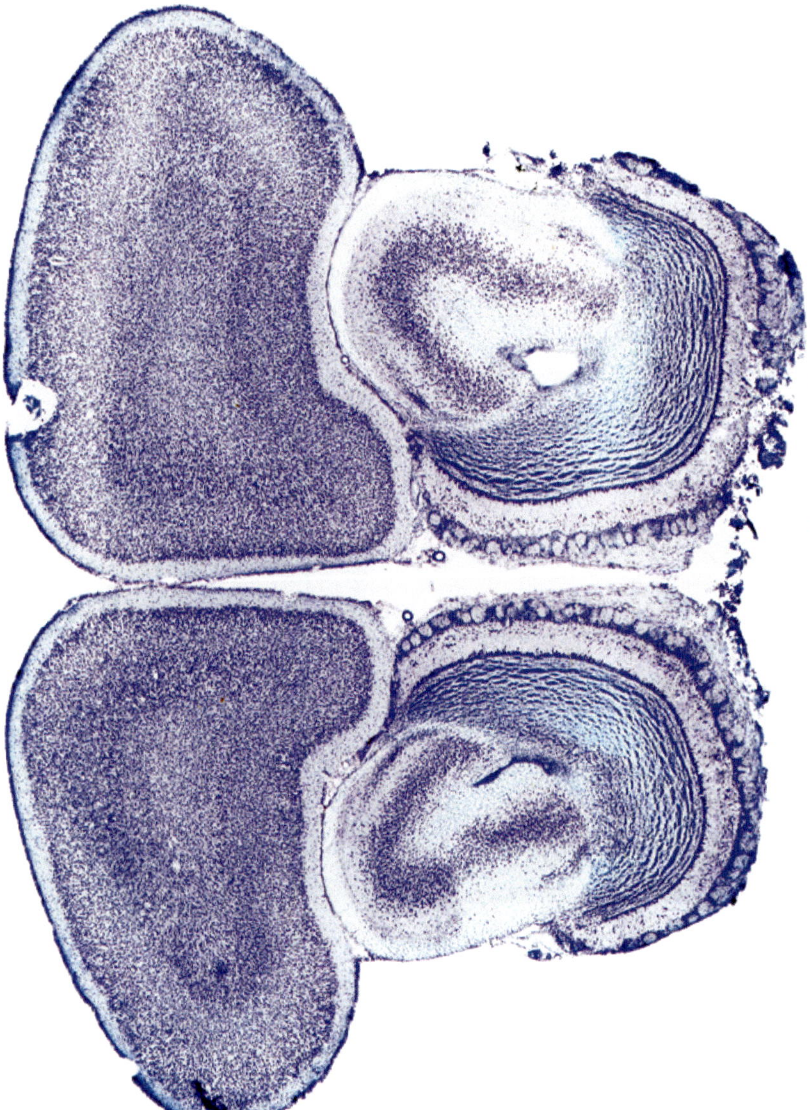

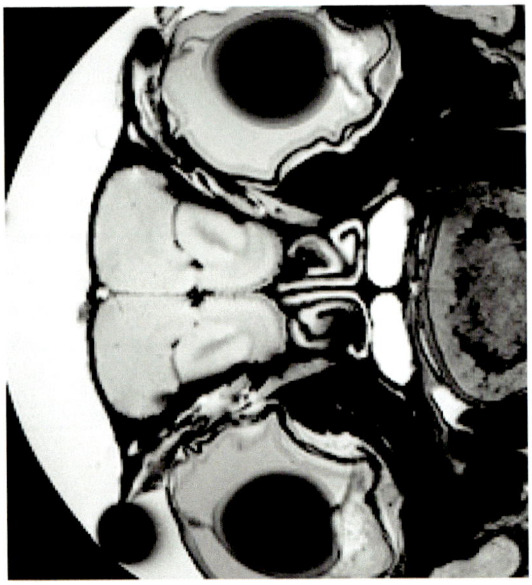

aci anterior commissure, intrabulbar part
AOD anterior olfactory nucleus, dorsal part
AOE anterior olfactory nucleus, external part
AOL anterior olfactory nucleus, lateral part
AOM anterior olfactory nucleus, medial part
AOV anterior olfactory nucleus, ventral part
Cg cingulate cortex

DFC dorsal frontal cortex
E ependyma and subependymal layer
EPl external plexiform layer of the olfactory bulb
GCL granule cell layer of the olfactory bulb
GL glomerular layer of the olfactory bulb
GrA granule cell layer of the accessory olfactory bulb
Ins insular cortex

IPl internal plexiform layer of the olfactory bulb
lo lateral olfactory tract
MCL mitral cell layer of the olfactory bulb
MFC medial frontal cortex
OFC orbital frontal cortex
ONL olfactory nerve layer
OV olfactory ventricle (olfactory part of lateral ventricle)

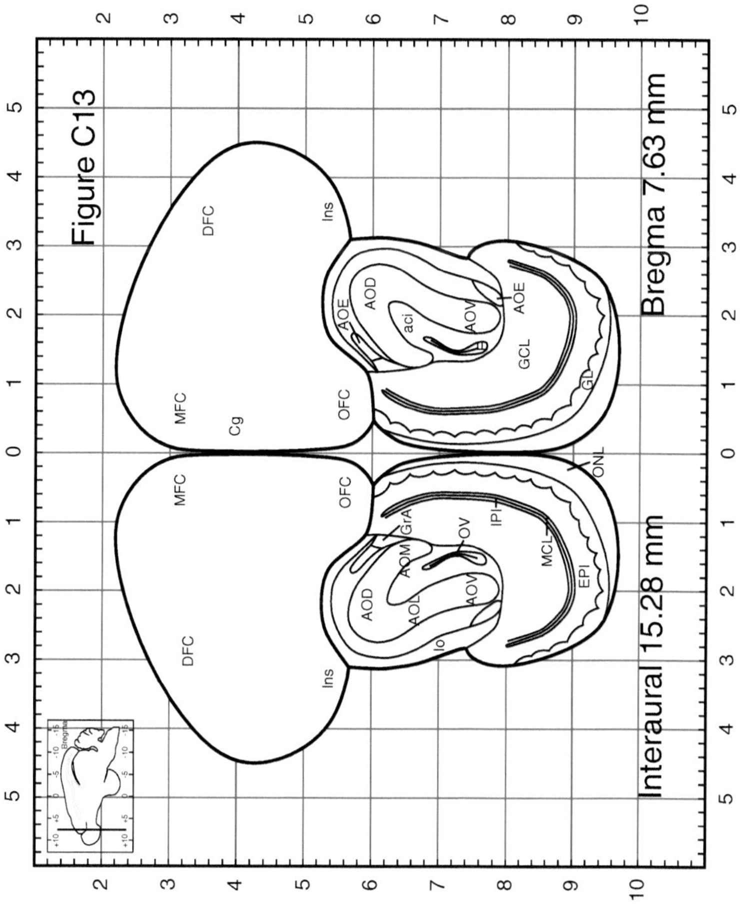
Figure C13
Bregma 7.63 mm
Interaural 15.28 mm
DFC
Ins
MFC
Cg
OFC
AOE
AOD
aci
AOV
GCL
GL
ONL
GrA
AOM
AOL
lo
OV
IPl
MCL
EPl

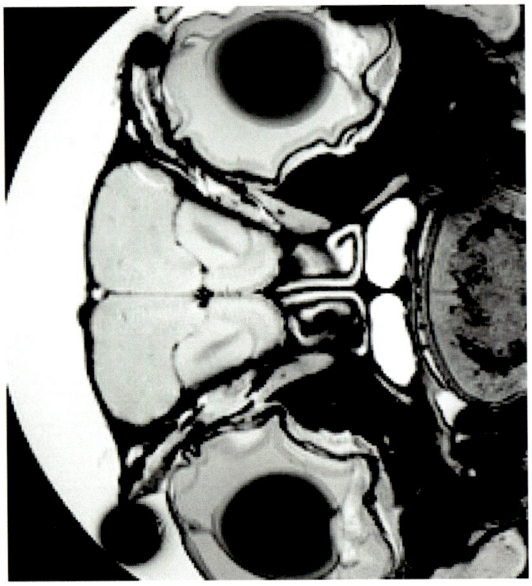

aci anterior commissure, intrabulbar part
AOD anterior olfactory nucleus, dorsal part
AOE anterior olfactory nucleus, external part
AOL anterior olfactory nucleus, lateral part
AOM anterior olfactory nucleus, medial part
AOV anterior olfactory nucleus, ventral part
Cg cingulate cortex

DFC dorsal frontal cortex
E ependyma and subependymal layer
EPl external plexiform layer of the olfactory bulb
GCL granule cell layer of the olfactory bulb
GL glomerular layer of the olfactory bulb
GrA granule cell layer of the accessory olfactory bulb
Ins insular cortexa

IPl internal plexiform layer of the olfactory bulb
lo lateral olfactory tract
MCL mitral cell layer of the olfactory bulb
MFC medial frontal cortex
OFC orbital frontal cortex
ONL olfactory nerve layer
OV olfactory ventricle (olfactory part of lateral ventricle)

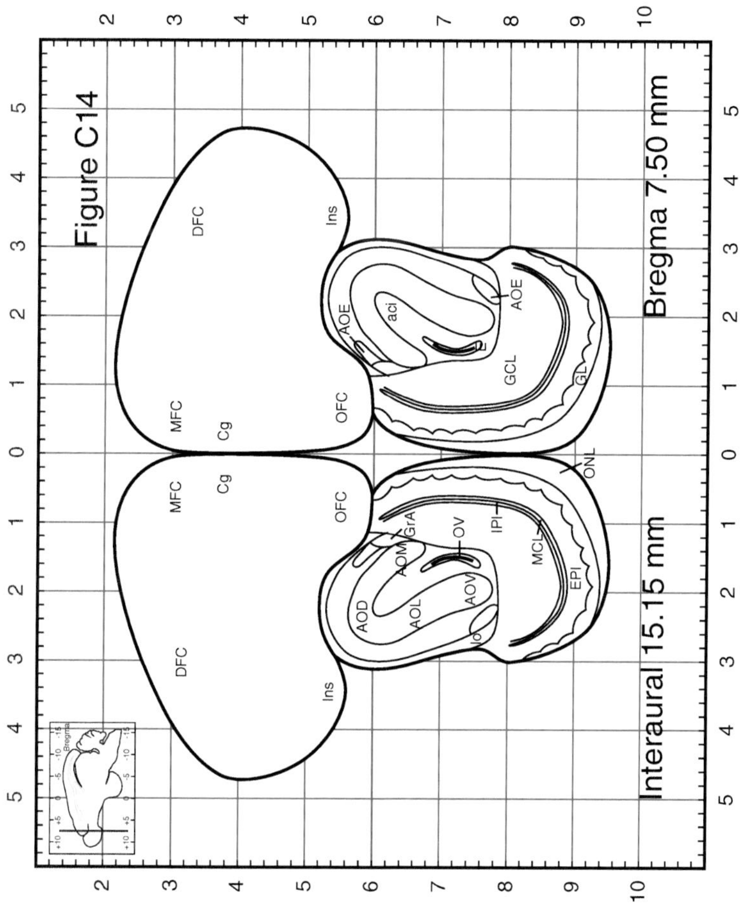
Figure C14
Bregma 7.50 mm
Interaural 15.15 mm
DFC
Ins
MFC
Cg
OFC
AOE
aci
AOE
GCL
GL
ONL
IPl
MCL
EPl
OV
GrA
AOM
AOD
AOL
AOV

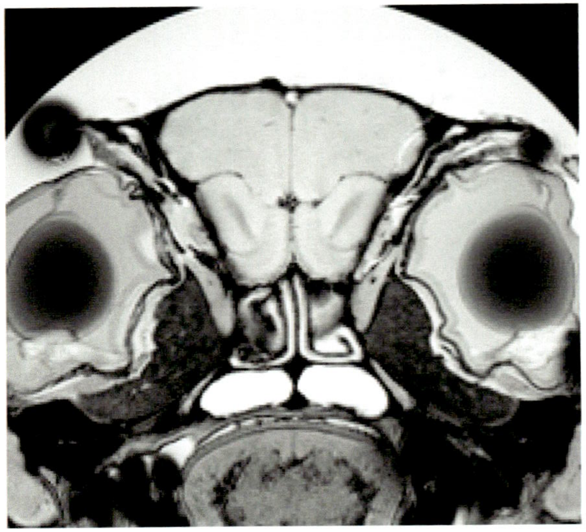

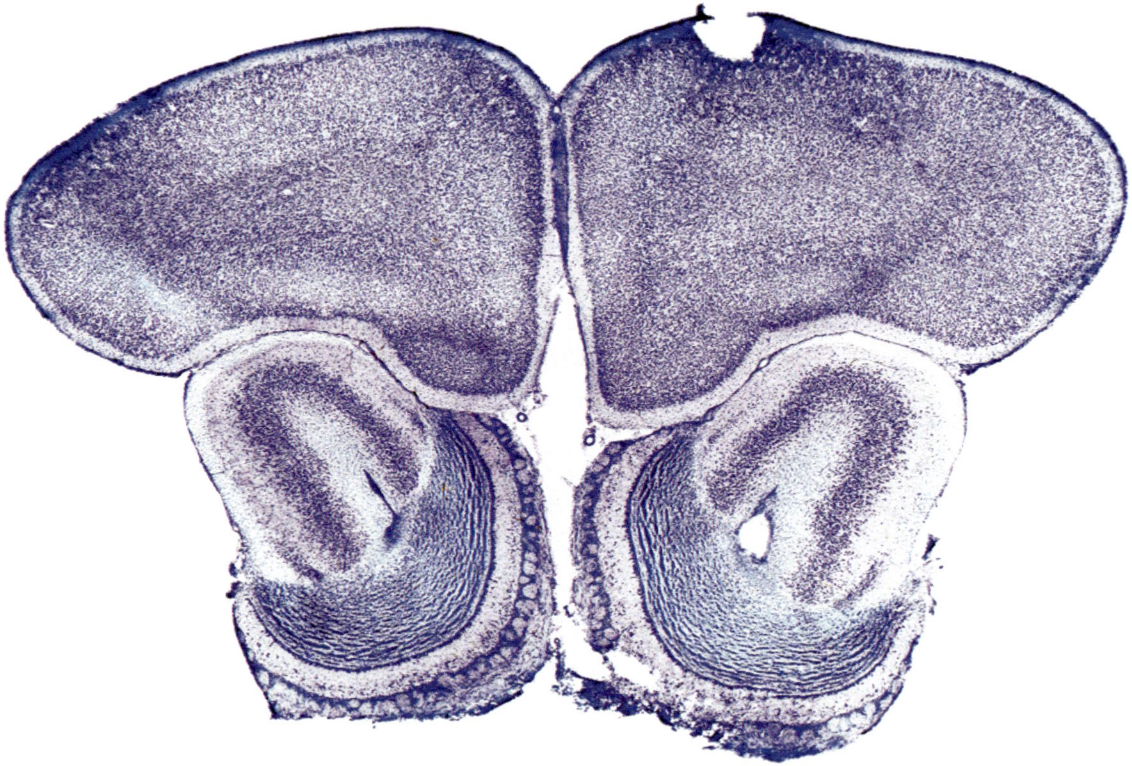

aci anterior commissure, intrabulbar part
AOD anterior olfactory nucleus, dorsal part
AOE anterior olfactory nucleus, exteral part
AOL anterior olfactory nucleus, lateral part
AOM anterior olfactory nucleus, medial part
AOV anterior olfactory nucleus, ventral part
Cg cingulate cortex
DFC dorsal frontal cortex
E ependyma and subependymal layer
EPl external plexiform layer of the olfactory bulb
GCL granule cell layer of the olfactory bulb
GL glomerular layer of the olfactory bulb
Ins insular cortex
IPl internal plexiform layer of the olfactory bulb
lo lateral olfactory tract
MCL mitral cell layer of the olfactory bulb
MFC medial frontal cortex
OFC orbital frontal cortex
ONL olfactory nerve layer
OV olfactory ventricle (olfactory part of lateral ventricle)

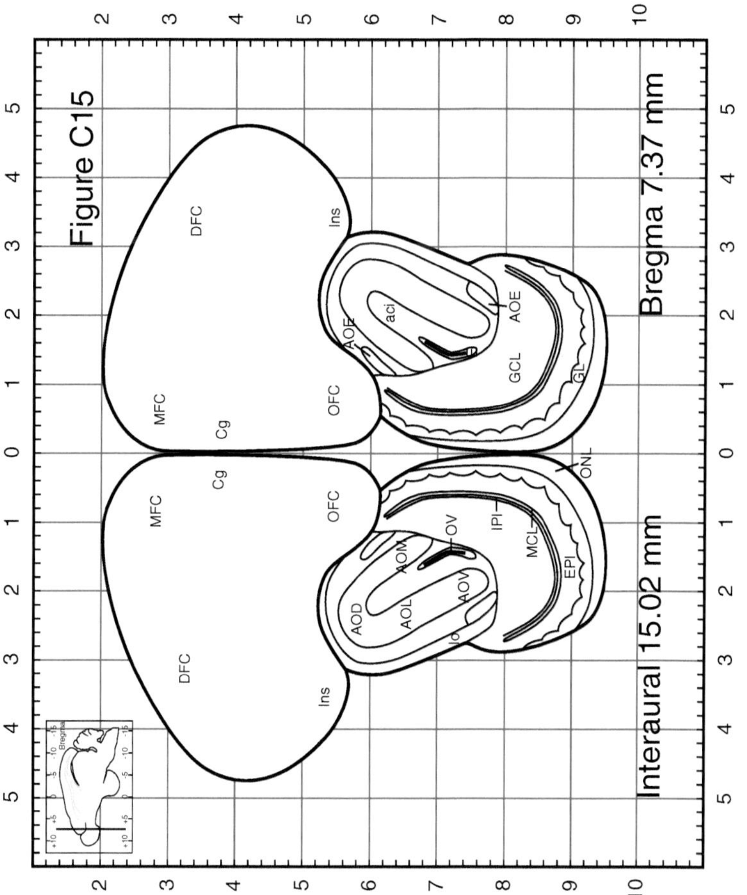
Figure C15
Bregma 7.37 mm
Interaural 15.02 mm
DFC
MFC
Cg
OFC
Ins
AOE
aci
GCL
GL
ONL
IPl
MCL
EPl
OV
AOM
AOV
AOL
AOD
lo

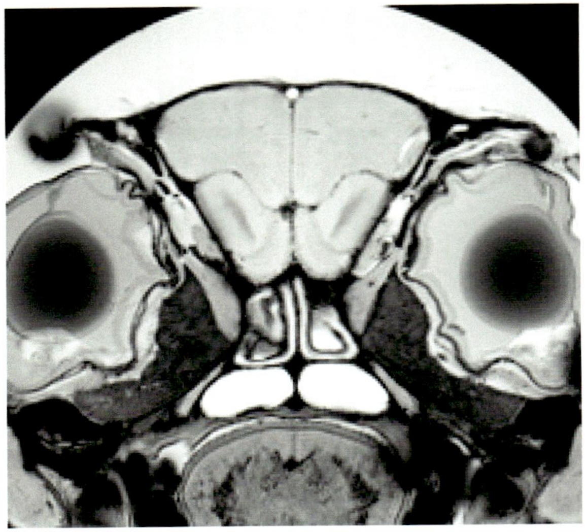

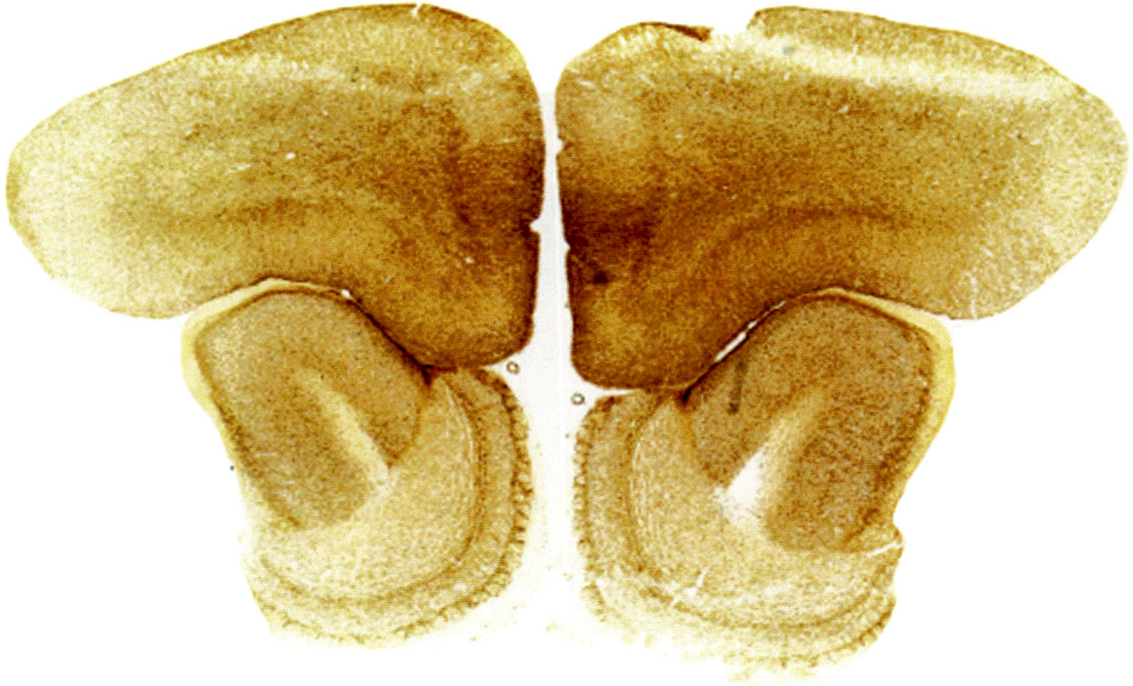

aci	anterior commissure, intrabulbar part
AOD	anterior olfactory nucleus, dorsal part
AOE	anterior olfactory nucleus, exteral part
AOL	anterior olfactory nucleus, lateral part
AOM	anterior olfactory nucleus, medial part
AOV	anterior olfactory nucleus, ventral part
Cg	cingulate cortex
DFC	dorsal frontal cortex
E	ependyma and subependymal layer
EPI	external plexiform layer of the olfactory bulb
GCL	granule cell layer of the olfactory bulb
GL	glomerular layer of the olfactory bulb
Ins	insular cortex
IPI	internal plexiform layer of the olfactory bulb
lo	lateral olfactory tract
MCL	mitral cell layer of the olfactory bulb
MFC	medial frontal cortex
OFC	orbital frontal cortex
ONL	olfactory nerve layer
OV	olfactory ventricle (olfactory part of lateral ventricle)

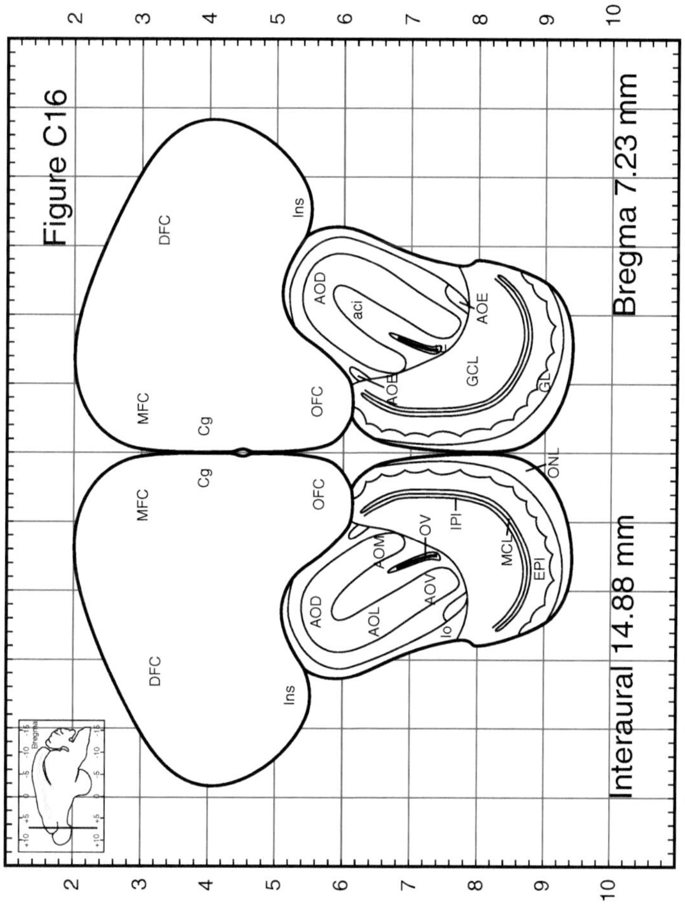
Figure C16
Bregma 7.23 mm
Interaural 14.88 mm
DFC
Ins
MFC
Cg
OFC
AOD
aci
AOE
GCL
AOB
GL
ONL
IPl
OV
AOM
MCL
EPl
AOV
AOL
lo

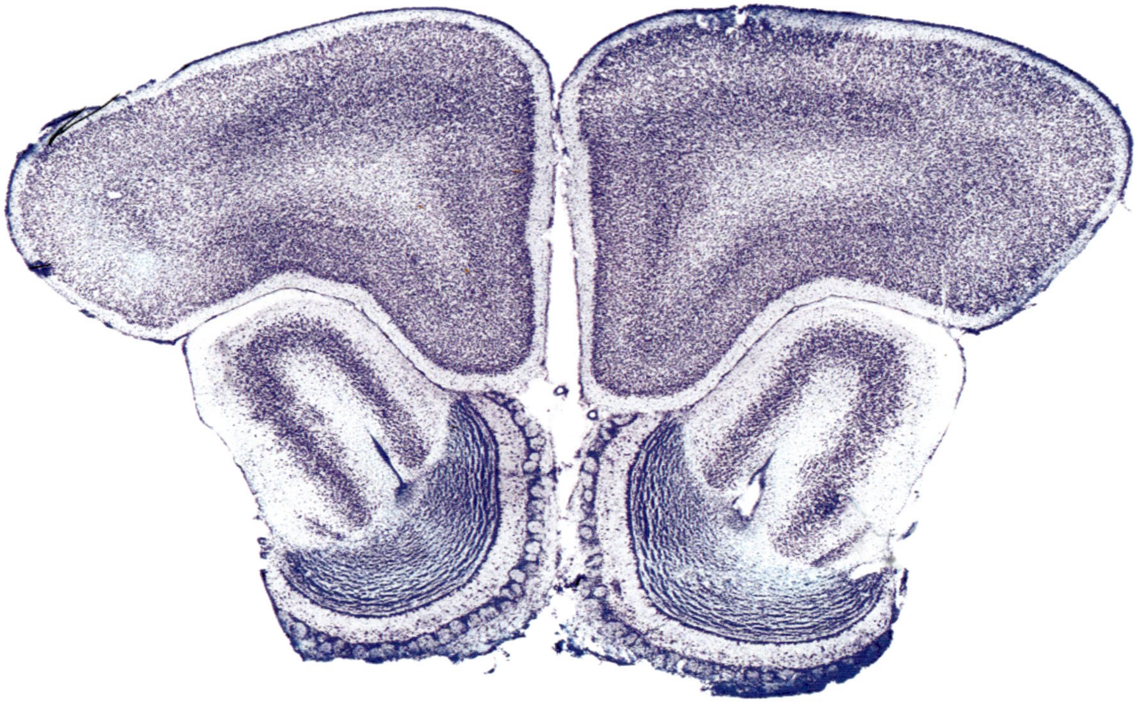

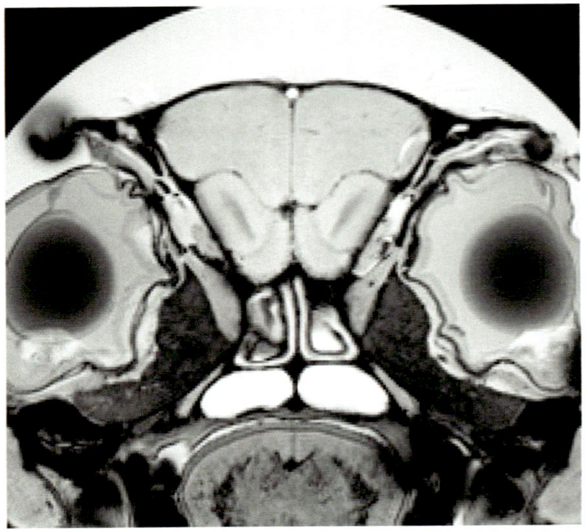

aci anterior commissure, intrabulbar part
AOD anterior olfactory nucleus, dorsal part
AOE anterior olfactory nucleus, exteral part
AOL anterior olfactory nucleus, lateral part
AOM anterior olfactory nucleus, medial part
AOV anterior olfactory nucleus, ventral part
Cg cingulate cortex
DFC dorsal frontal cortex
E ependyma and subependymal layer
EPl external plexiform layer of the olfactory bulb
fmi forceps minor of the corpus callosum
GCL granule cell layer of the olfactory bulb
GL glomerular layer of the olfactory bulb
Ins insular cortex
IPl internal plexiform layer of the olfactory bulb
lo lateral olfactory tract
MCL mitral cell layer of the olfactory bulb
MFC medial frontal cortex
OFC orbital frontal cortex
ONL olfactory nerve layer
OV olfactory ventricle (olfactory part of lateral ventricle)

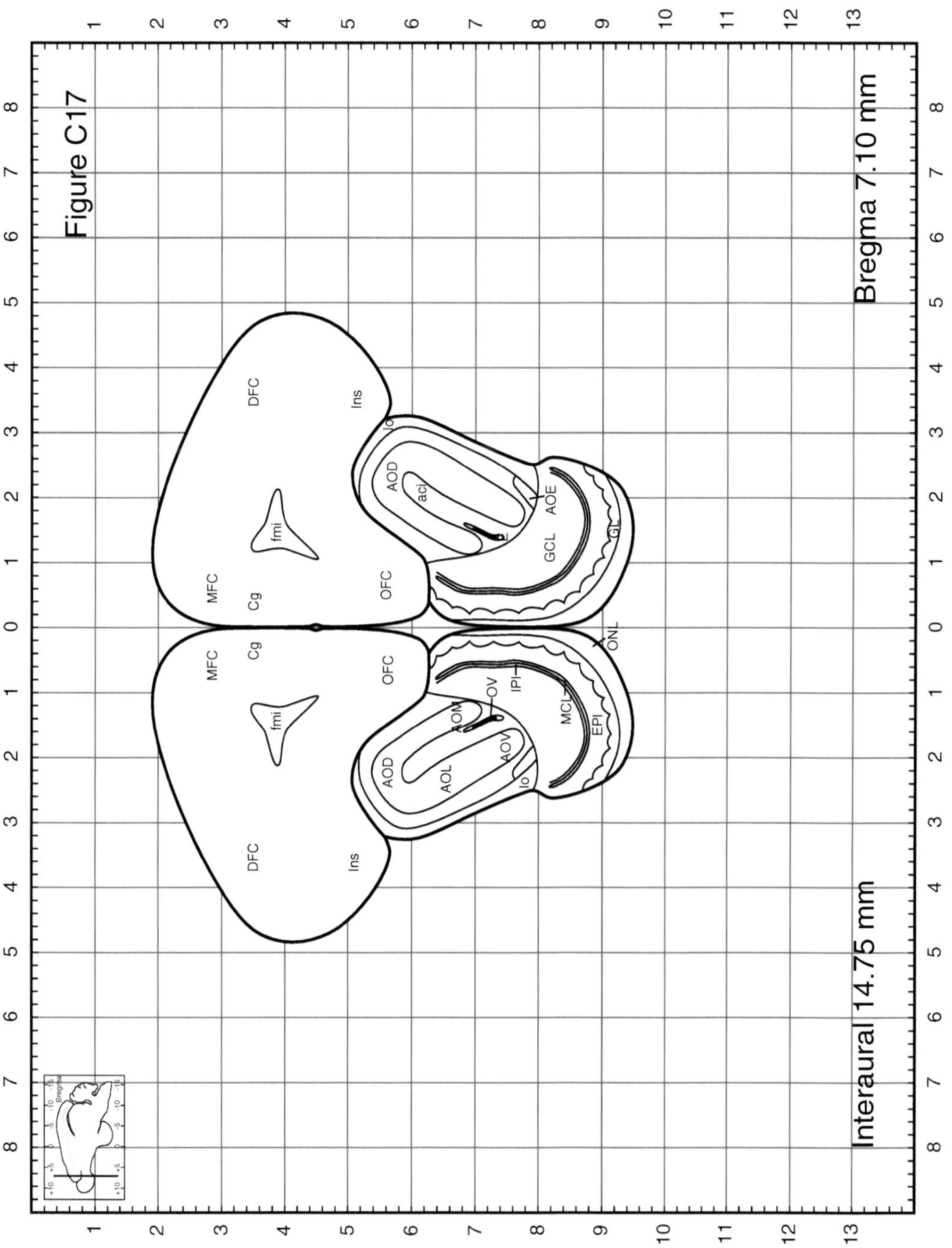
Figure C17
Bregma 7.10 mm
Interaural 14.75 mm
DFC
Ins
MFC
Cg
fmi
OFC
lo
AOD
aci
AOE
GCL
GL
ONL
IPl
OV
AOM
MCL
EPl
AOV
AOL

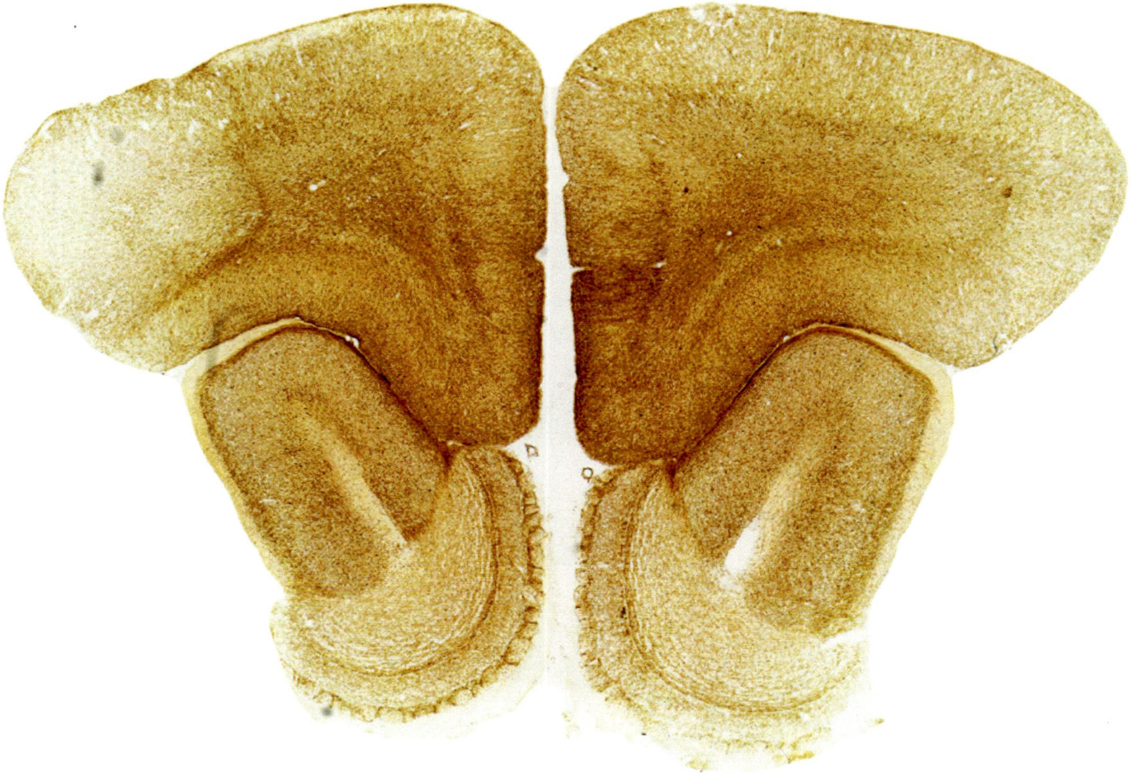

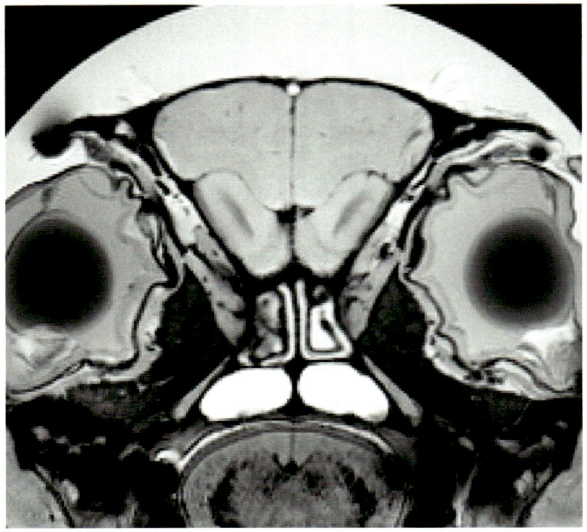

aci	anterior commissure, intrabulbar part
AOD	anterior olfactory nucleus, dorsal part
AOE	anterior olfactory nucleus, exteral part
AOL	anterior olfactory nucleus, lateral part
AOM	anterior olfactory nucleus, medial part
AOV	anterior olfactory nucleus, ventral part
Cg	cingulate cortex
DFC	dorsal frontal cortex
E	ependyma and subependymal layer
EPI	external plexiform layer of the olfactory bulb
fmi	forceps minor of the corpus callosum
GCL	granule cell layer of the olfactory bulb
GL	glomerular layer of the olfactory bulb
Ins	insular cortex
IPI	internal plexiform layer of the olfactory bulb
lo	lateral olfactory tract
M2	secondary motor cortex
MCL	mitral cell layer of the olfactory bulb
MFC	medial frontal cortex
OFC	orbital frontal cortex
ONL	olfactory nerve layer
OV	olfactory ventricle (olfactory part of lateral ventricle)

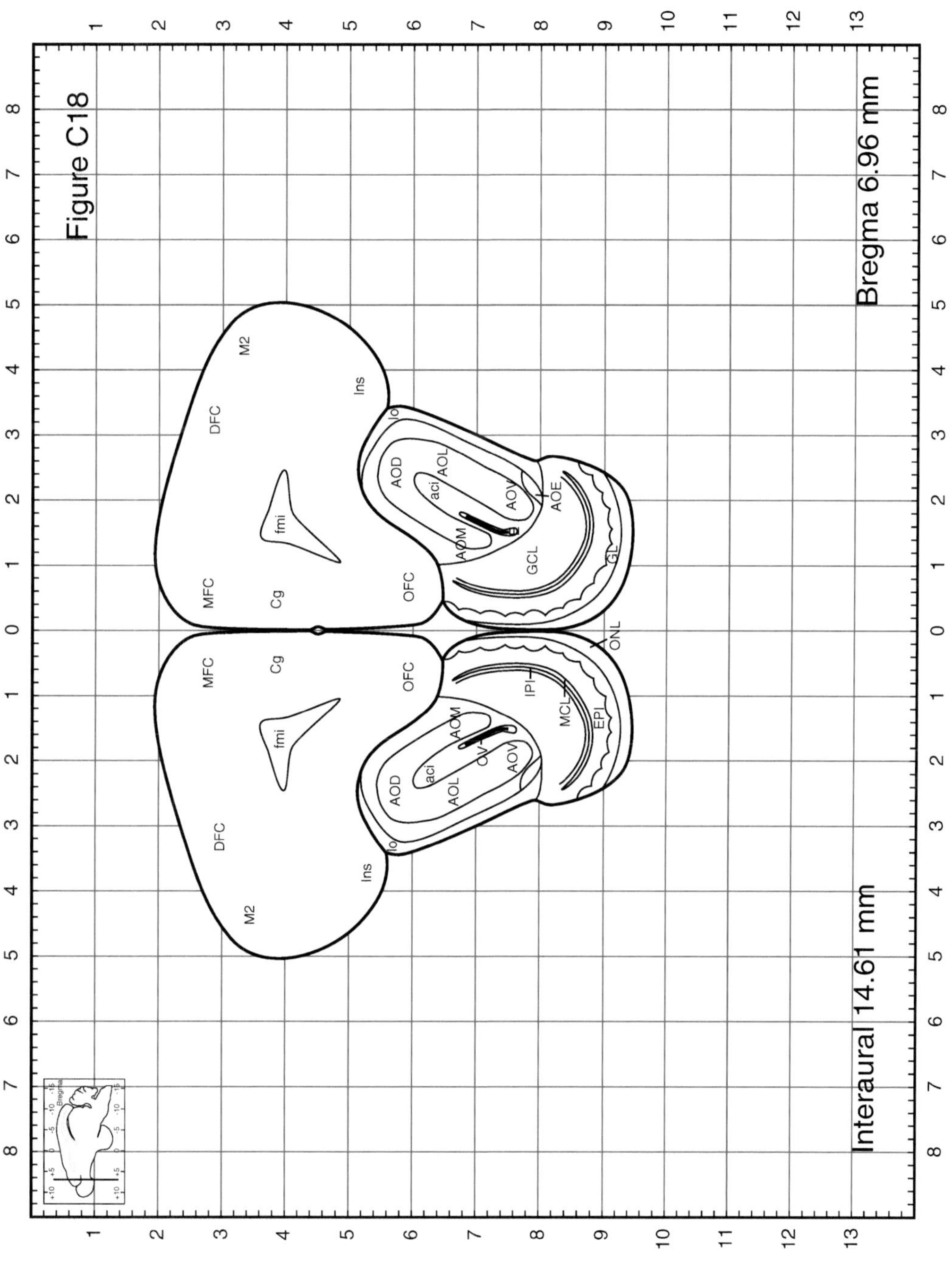
Figure C18
Bregma 6.96 mm
Interaural 14.61 mm
M2
DFC
MFC
Cg
fmi
Ins
OFC
lo
AOD
AOL
aci
AOV
AOM
AOE
GCL
GL
ONL
IPl
MCL
EPl
CV

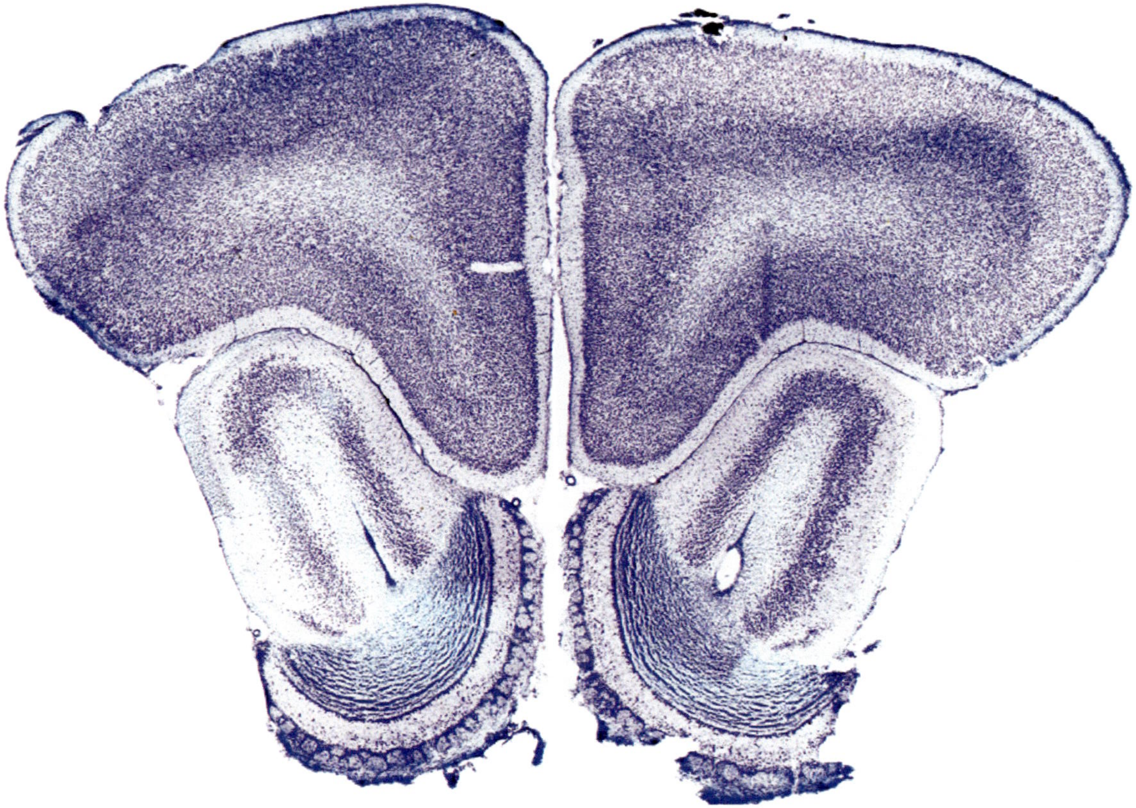

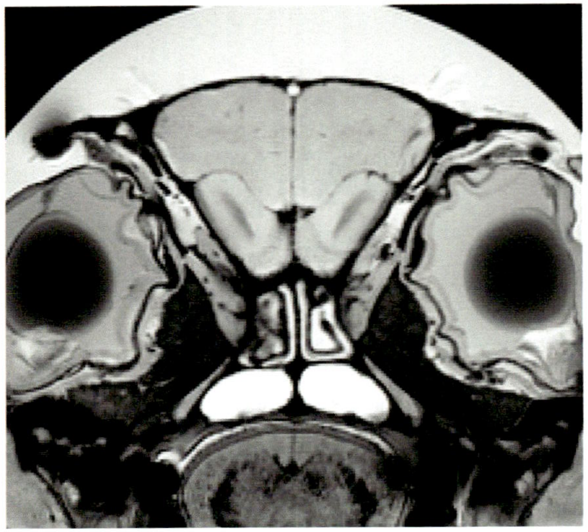

aci	anterior commissure, intrabulbar part
AOD	anterior olfactory nucleus, dorsal part
AOE	anterior olfactory nucleus, exteral part
AOL	anterior olfactory nucleus, lateral part
AOM	anterior olfactory nucleus, medial part
AOV	anterior olfactory nucleus, ventral part
Cg	cingulate cortex
DFC	dorsal frontal cortex
EPl	external plexiform layer of the olfactory bulb
fmi	forceps minor of the corpus callosum
GCL	granule cell layer of the olfactory bulb
GL	glomerular layer of the olfactory bulb
IL	infralimbic cortex
Ins	insular cortex
IPl	internal plexiform layer of the olfactory bulb
lo	lateral olfactory tract
M2	secondary motor cortex
MCL	mitral cell layer of the olfactory bulb
MFC	medial frontal cortex
OFC	orbital frontal cortex
ONL	olfactory nerve layer
OV	olfactory ventricle (olfactory part of lateral ventricle)
PrL	prelimbic cortex

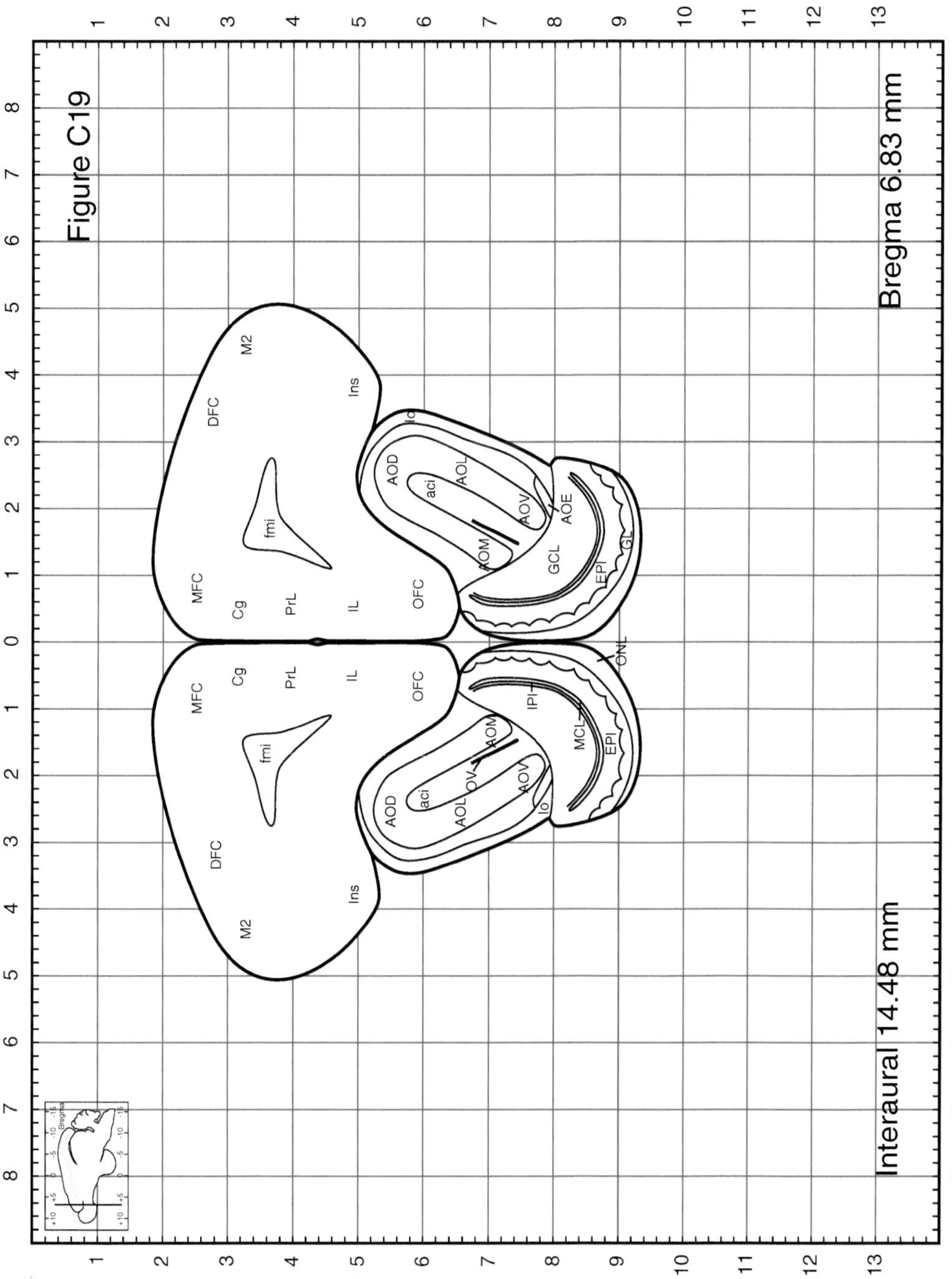
Figure C19
Bregma 6.83 mm
Interaural 14.48 mm
M2
DFC
Ins
MFC
Cg
PrL
IL
OFC
fmi
AOD
aci
AOL
AOV
AOM
AOE
lo
GCL
EPl
GL
ONL
IPl
MCL

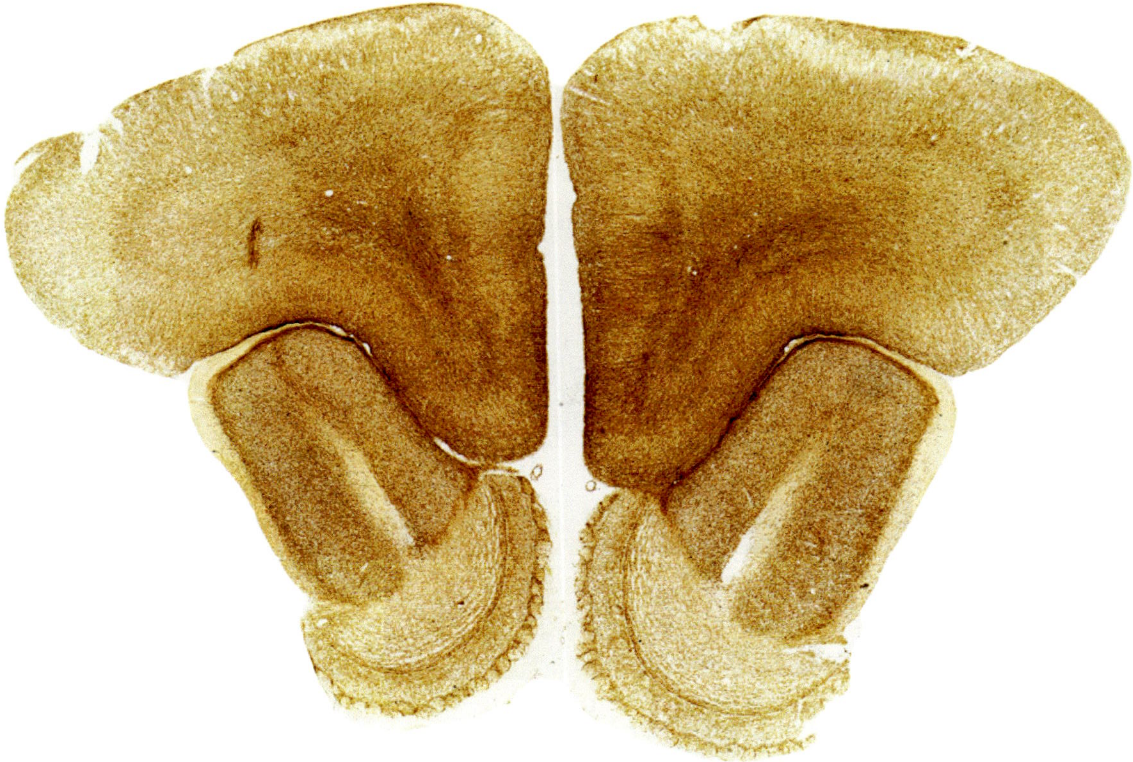

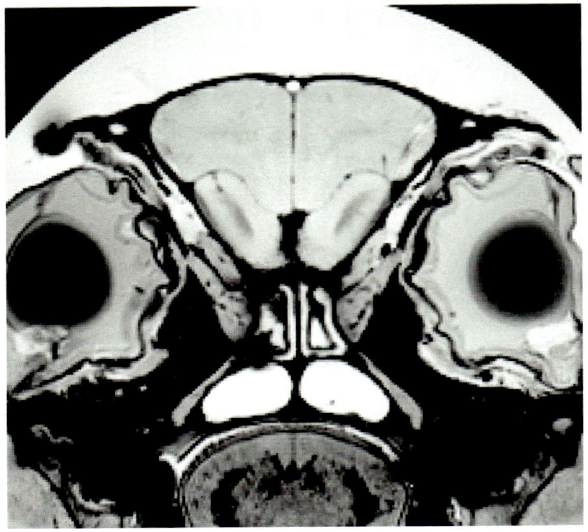

aci	anterior commissure, intrabulbar part
AOD	anterior olfactory nucleus, dorsal part
AOE	anterior olfactory nucleus, exteral part
AOL	anterior olfactory nucleus, lateral part
AOM	anterior olfactory nucleus, medial part
AOV	anterior olfactory nucleus, ventral part
Cg	cingulate cortex
DFC	dorsal frontal cortex
EPl	external plexiform layer of the olfactory bulb
fmi	forceps minor of the corpus callosum
GCL	granule cell layer of the olfactory bulb
GL	glomerular layer of the olfactory bulb
IL	infralimbic cortex
Ins	insular cortex
IPl	internal plexiform layer of the olfactory bulb
lo	lateral olfactory tract
M2	secondary motor cortex
MCL	mitral cell layer of the olfactory bulb
MFC	medial frontal cortex
OFC	orbital frontal cortex
ONL	olfactory nerve layer
OV	olfactory ventricle (olfactory part of lateral ventricle)
PrL	prelimbic cortex

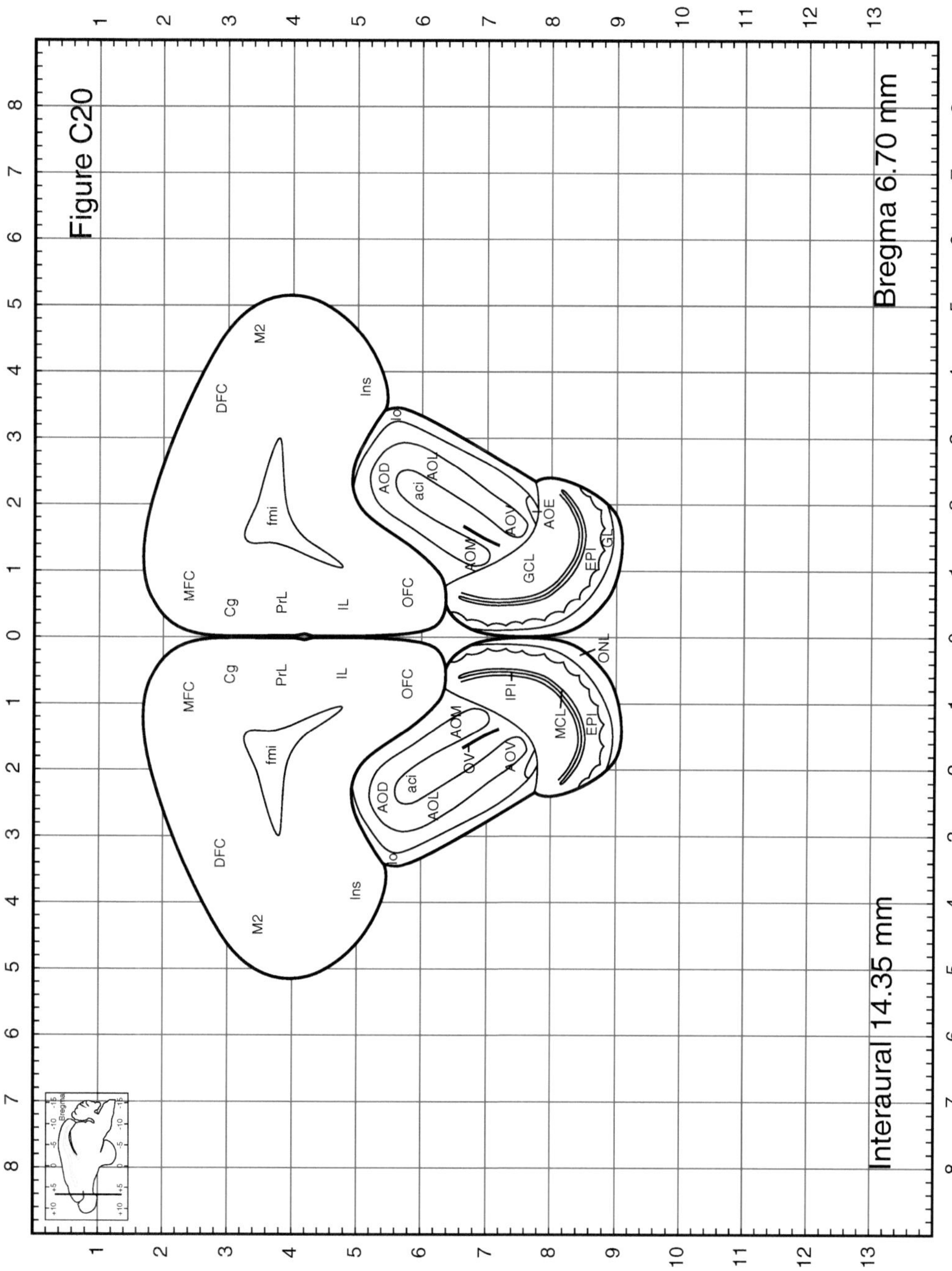
Figure C20
Bregma 6.70 mm
Interaural 14.35 mm
M2
DFC
Ins
fmi
MFC
Cg
PrL
IL
OFC
lo
AOD
aci
AOL
AOV
AOM
AOE
GCL
EPl
GL
IPl
MCL
ONL
OV

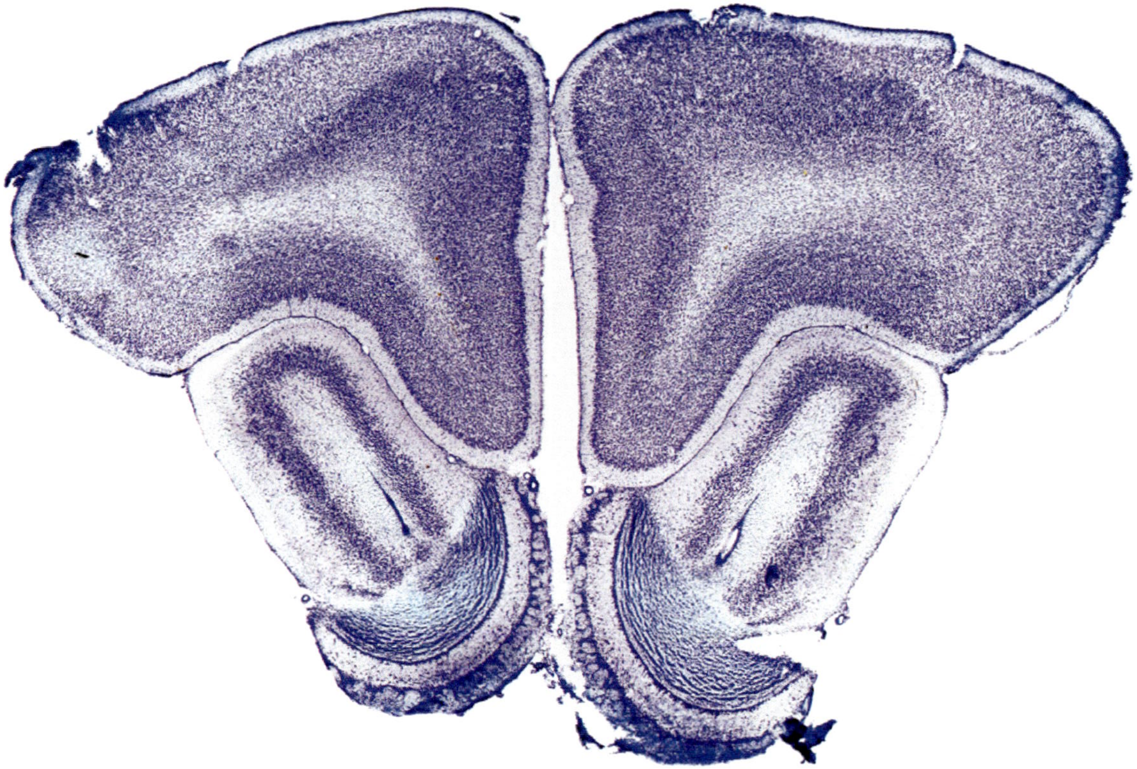

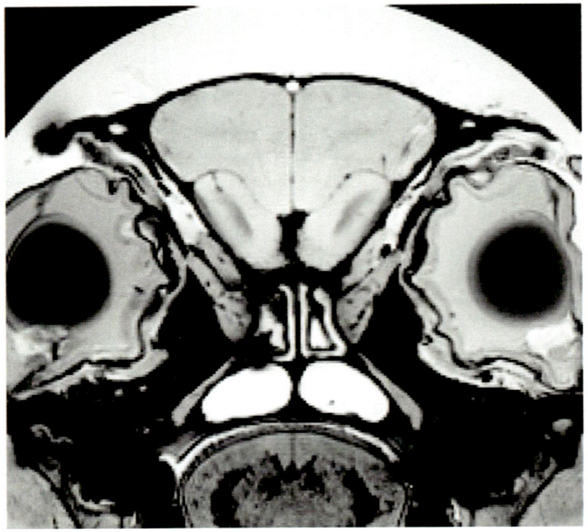

aci	anterior commissure, intrabulbar part
AOD	anterior olfactory nucleus, dorsal part
AOE	anterior olfactory nucleus, exteral part
AOL	anterior olfactory nucleus, lateral part
AOM	anterior olfactory nucleus, medial part
AOV	anterior olfactory nucleus, ventral part
Cg	cingulate cortex
DFC	dorsal frontal cortex
EPl	external plexiform layer of the olfactory bulb
fmi	forceps minor of the corpus callosum
GCL	granule cell layer of the olfactory bulb
GL	glomerular layer of the olfactory bulba
IL	infralimbic cortex
Ins	insular cortex
IPl	internal plexiform layer of the olfactory bulb
lo	lateral olfactory tract
M2	secondary motor cortex
MCL	mitral cell layer of the olfactory bulb
MFC	medial frontal cortex
OFC	orbital frontal cortex
ONL	olfactory nerve layer
OV	olfactory ventricle (olfactory part of lateral ventricle)
PrL	prelimbic cortex

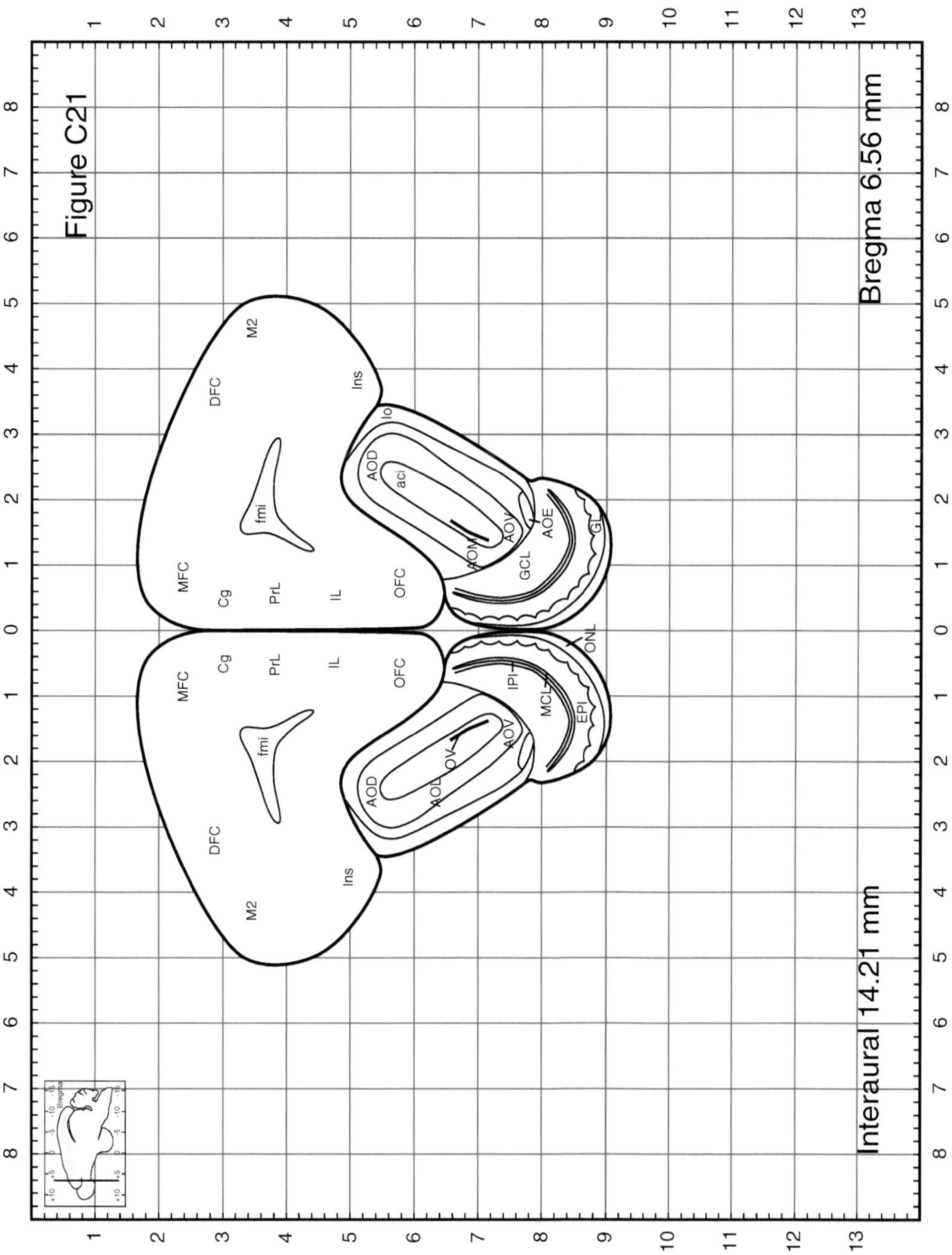

Figure C21
Bregma 6.56 mm
Interaural 14.21 mm
MFC
Cg
PrL
IL
OFC
DFC
M2
Ins
fmi
lo
AOD
aci
AOV
AOM
AOE
GCL
GL
ONL
IPl
MCL
EPl
AOL
OV

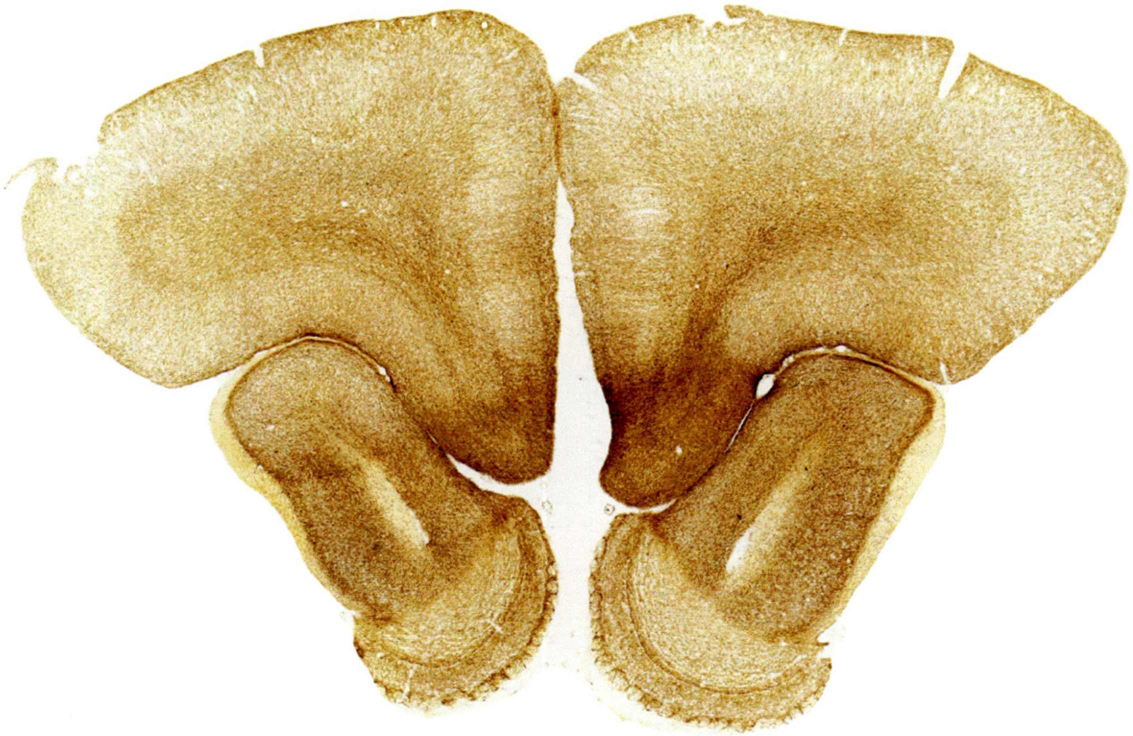

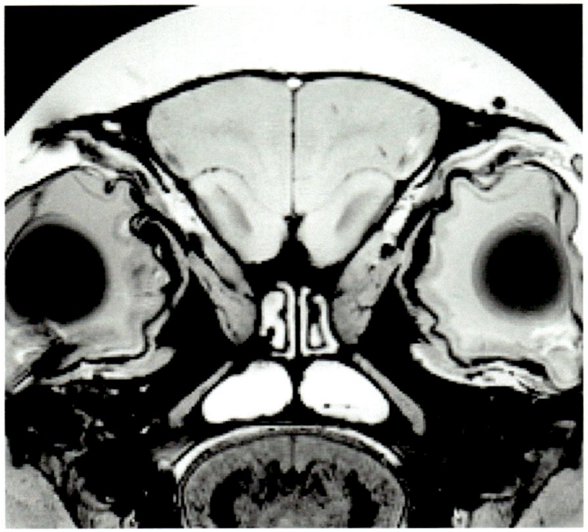

aci	anterior commissure, intrabulbar part
AOD	anterior olfactory nucleus, dorsal part
AOE	anterior olfactory nucleus, exteral part
AOL	anterior olfactory nucleus, lateral part
AOM	anterior olfactory nucleus, medial part
AOV	anterior olfactory nucleus, ventral part
Cg	cingulate cortex
DFC	dorsal frontal cortex
EPl	external plexiform layer of the olfactory bulb
fmi	forceps minor of the corpus callosum
GCL	granule cell layer of the olfactory bulb
GL	glomerular layer of the olfactory bulb
IL	infralimbic cortex
Ins	insular cortex
IPl	internal plexiform layer of the olfactory bulb
lo	lateral olfactory tract
M2	secondary motor cortex
MCL	mitral cell layer of the olfactory bulb
MFC	medial frontal cortex
OFC	orbital frontal cortex
ONL	olfactory nerve layer
OV	olfactory ventricle (olfactory part of lateral ventricle)
PrL	prelimbic cortex

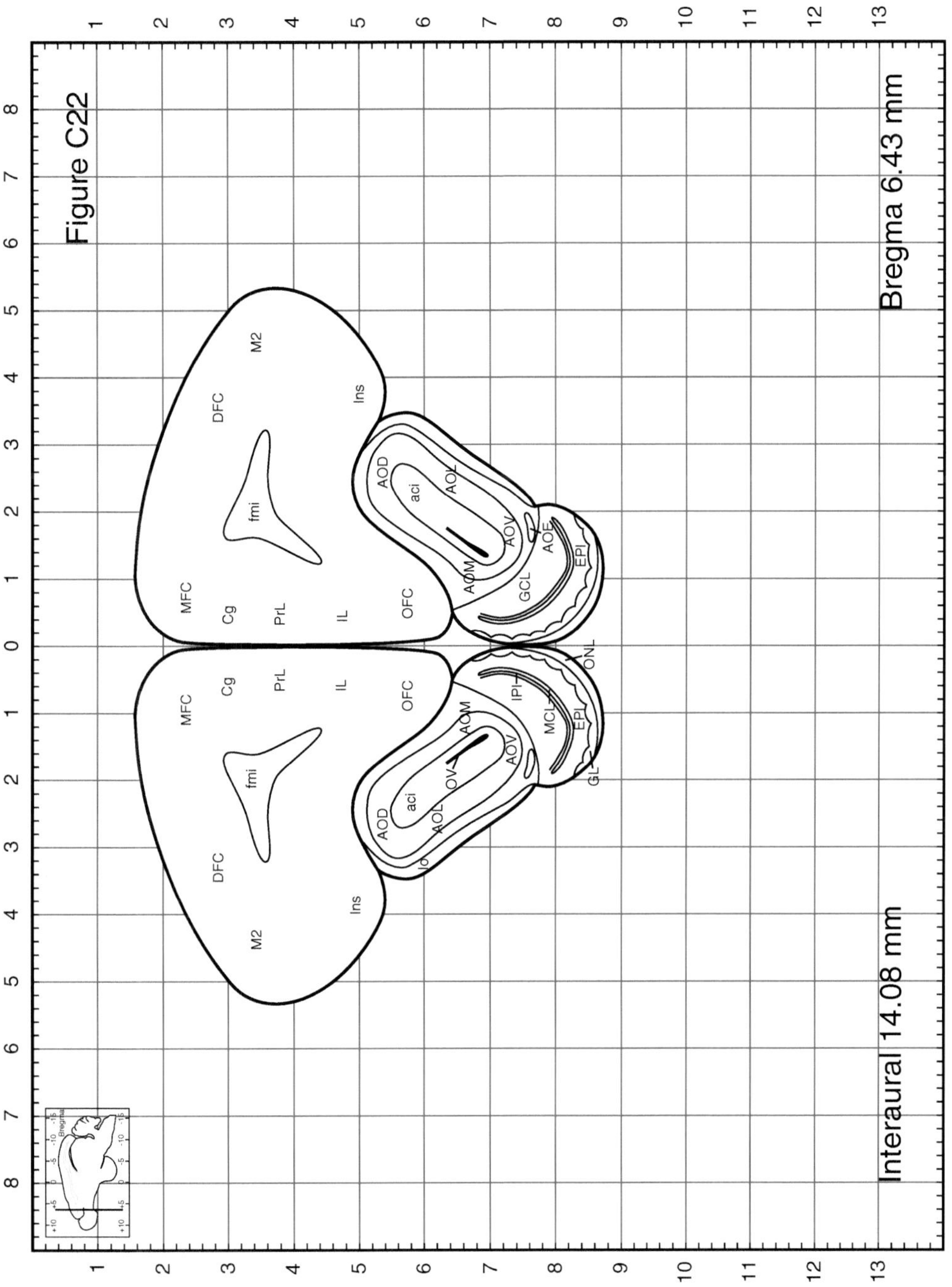
Figure C22
Bregma 6.43 mm
Interaural 14.08 mm
MFC
Cg
PrL
IL
OFC
DFC
M2
Ins
fmi
AOD
aci
AOL
AOV
AOM
AOE
GCL
EPl
IPl
MCL
GL
ONL
OV

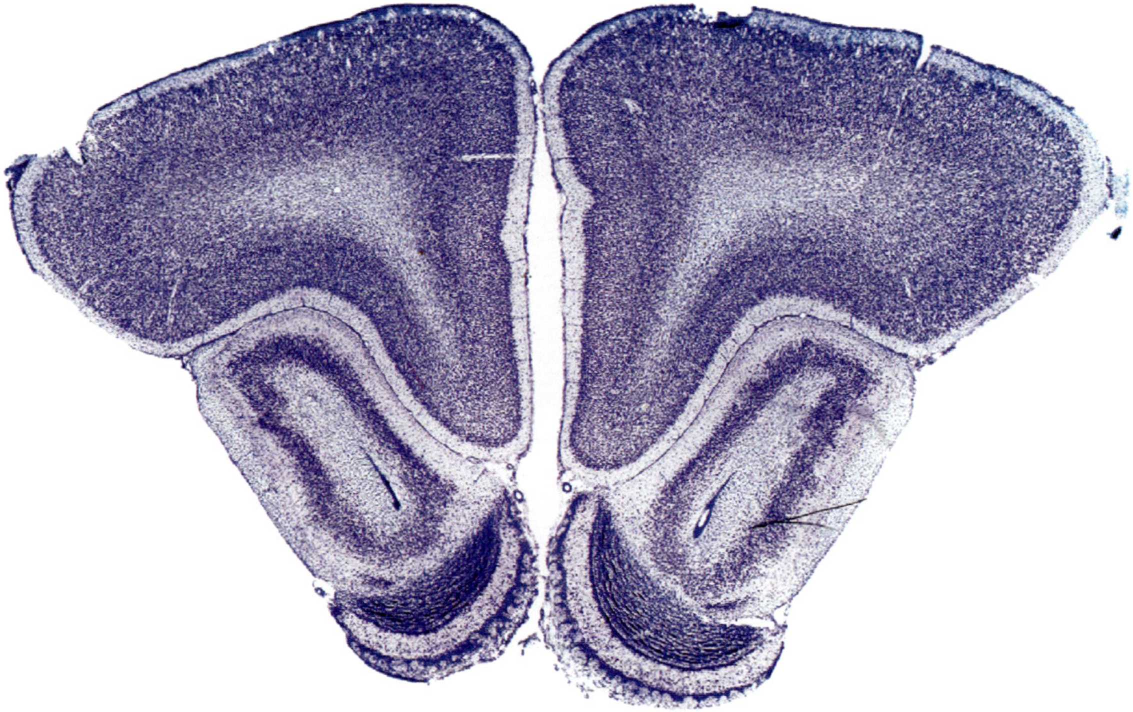

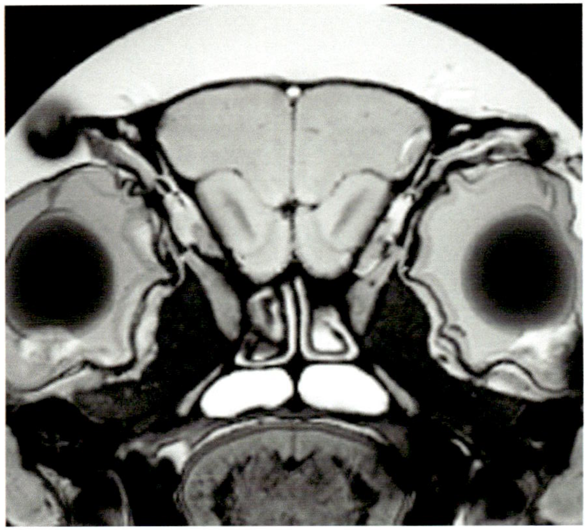

aci	anterior commissure, intrabulbar part
AOD	anterior olfactory nucleus, dorsal part
AOE	anterior olfactory nucleus, exteral part
AOL	anterior olfactory nucleus, lateral part
AOM	anterior olfactory nucleus, medial part
AOV	anterior olfactory nucleus, ventral part
Cg	cingulate cortex
DFC	dorsal frontal cortex
EPl	external plexiform layer of the olfactory bulb
fmi	forceps minor of the corpus callosum
GCL	granule cell layer of the olfactory bulb
GL	glomerular layer of the olfactory bulb
IL	infralimbic cortex
Ins	insular cortex
IPl	internal plexiform layer of the olfactory bulb
lo	lateral olfactory tract
M2	secondary motor cortex
MCL	mitral cell layer of the olfactory bulb
MFC	medial frontal cortex
OFC	orbital frontal cortex
ONL	olfactory nerve layer
OV	olfactory ventricle (olfactory part of lateral ventricle)
PrL	prelimbic cortex

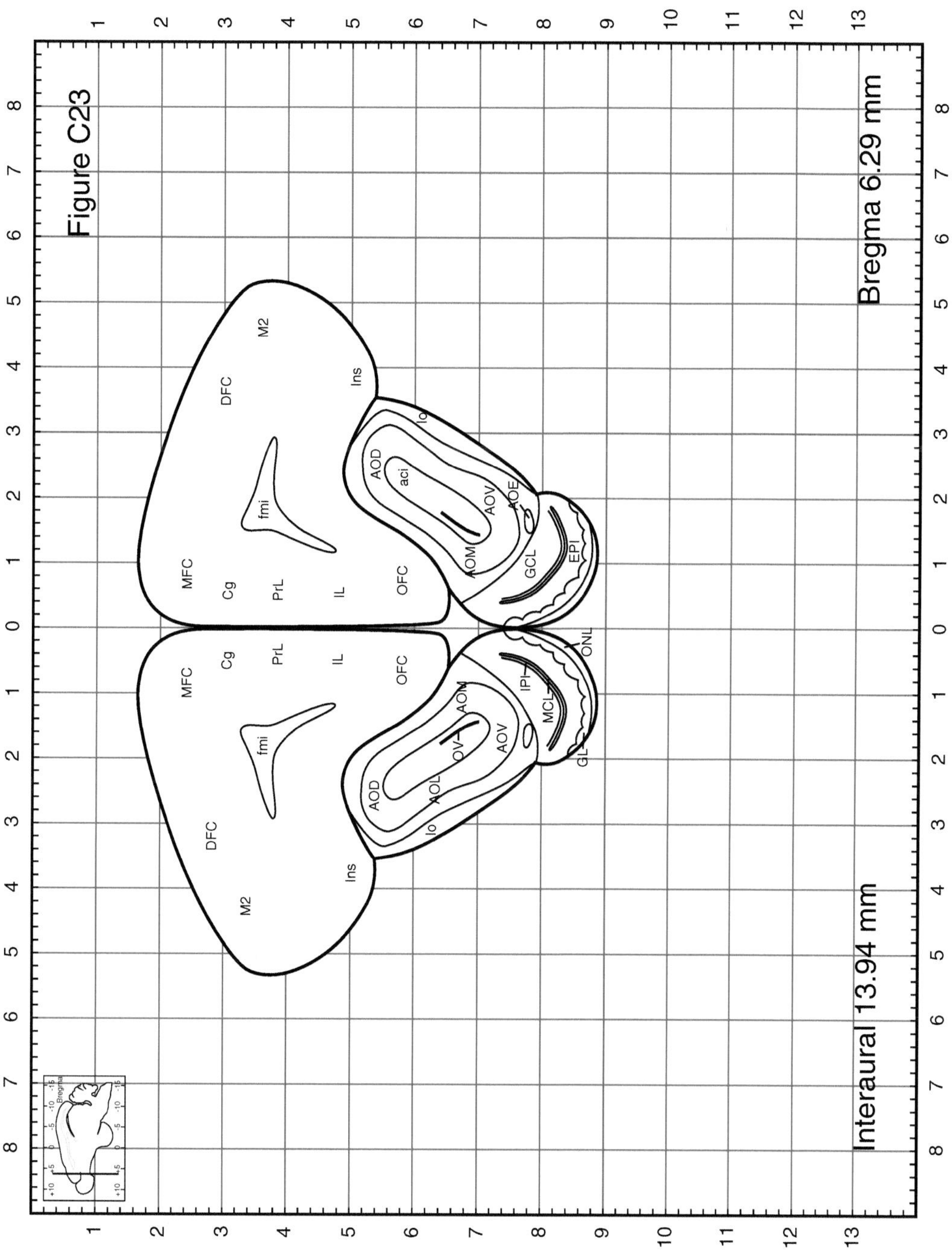
Figure C23
Bregma 6.29 mm
Interaural 13.94 mm
MFC
Cg
PrL
IL
OFC
DFC
M2
Ins
fmi
lo
AOD
aci
AOV
AOE
AOM
AOL
GCL
EPl
IPl
MCL
ONL
GL

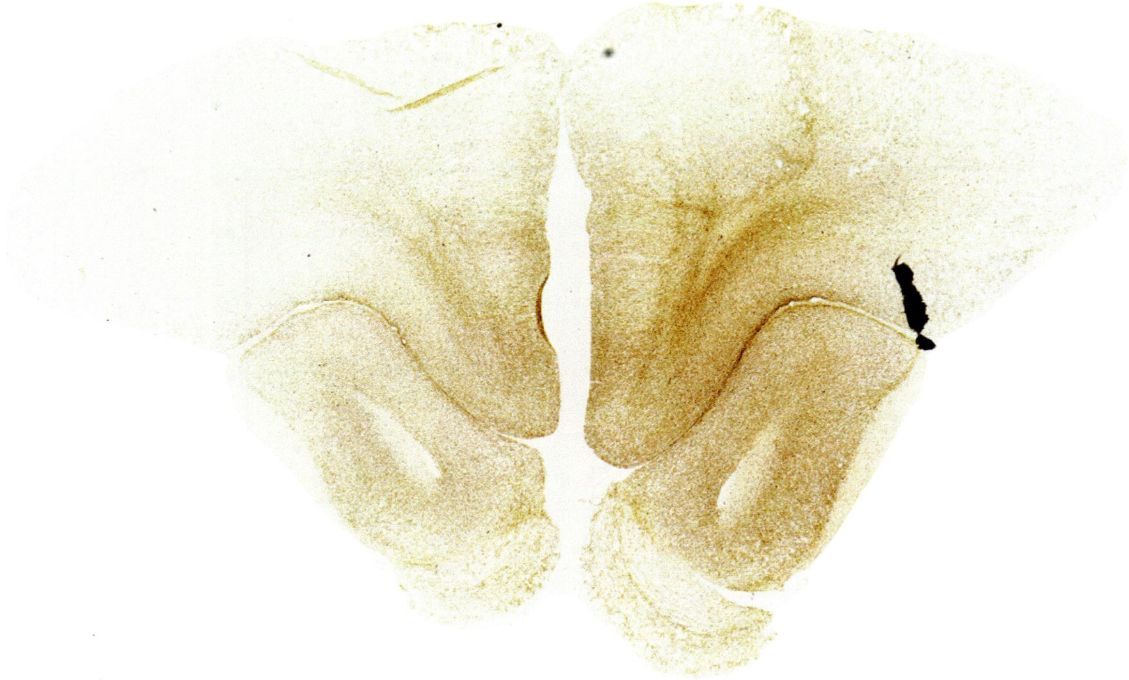

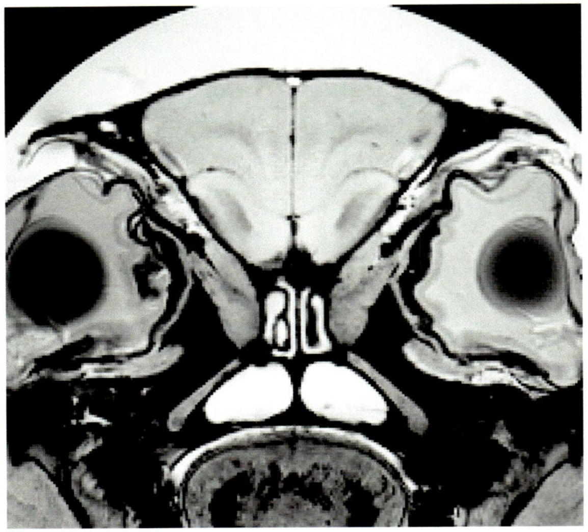

aci anterior commissure, intrabulbar part
AOD anterior olfactory nucleus, dorsal part
AOL anterior olfactory nucleus, lateral part
AOM anterior olfactory nucleus, medial part
AOV anterior olfactory nucleus, ventral part
Cg cingulate cortex
DFC dorsal frontal cortex
EPI external plexiform layer of the olfactory bulb
fmi forceps minor of the corpus callosum
GCL granule cell layer of the olfactory bulb
GL glomerular layer of the olfactory bulb
IL infralimbic cortex
Ins insular cortex
IPI internal plexiform layer of the olfactory bulb
lo lateral olfactory tract
M2 secondary motor cortex
MCL mitral cell layer of the olfactory bulb
MFC medial frontal cortex
OFC orbital frontal cortex
ONL olfactory nerve layer
OV olfactory ventricle (olfactory part of lateral ventricle)
PrL prelimbic cortex

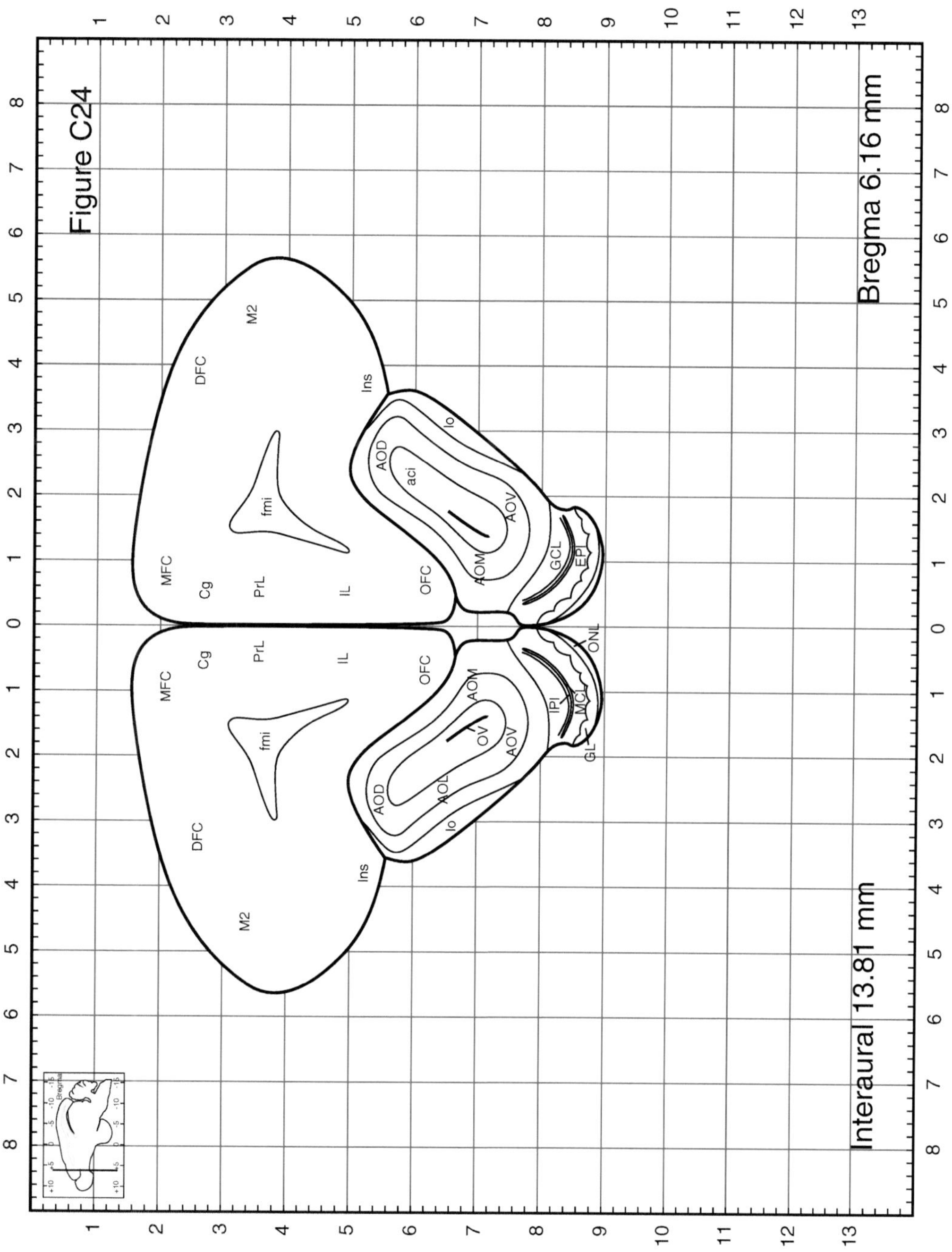
Figure C24
Bregma 6.16 mm
Interaural 13.81 mm
M2
DFC
Ins
lo
AOD
aci
fmi
AOV
MFC
Cg
PrL
IL
OFC
AOM
GCL
EPl
ONL
IPl
MCL
GL
OV
AOL

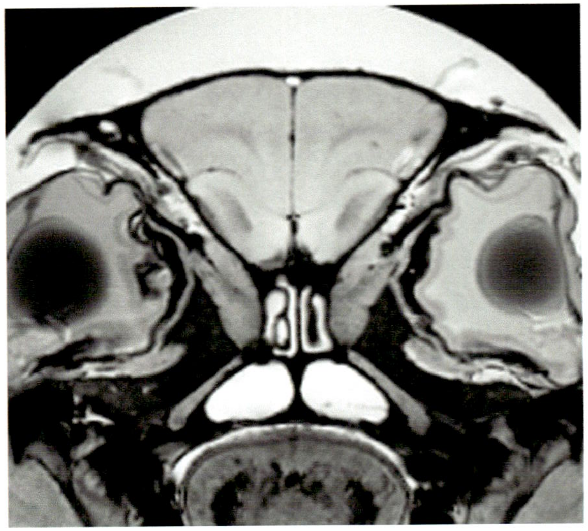

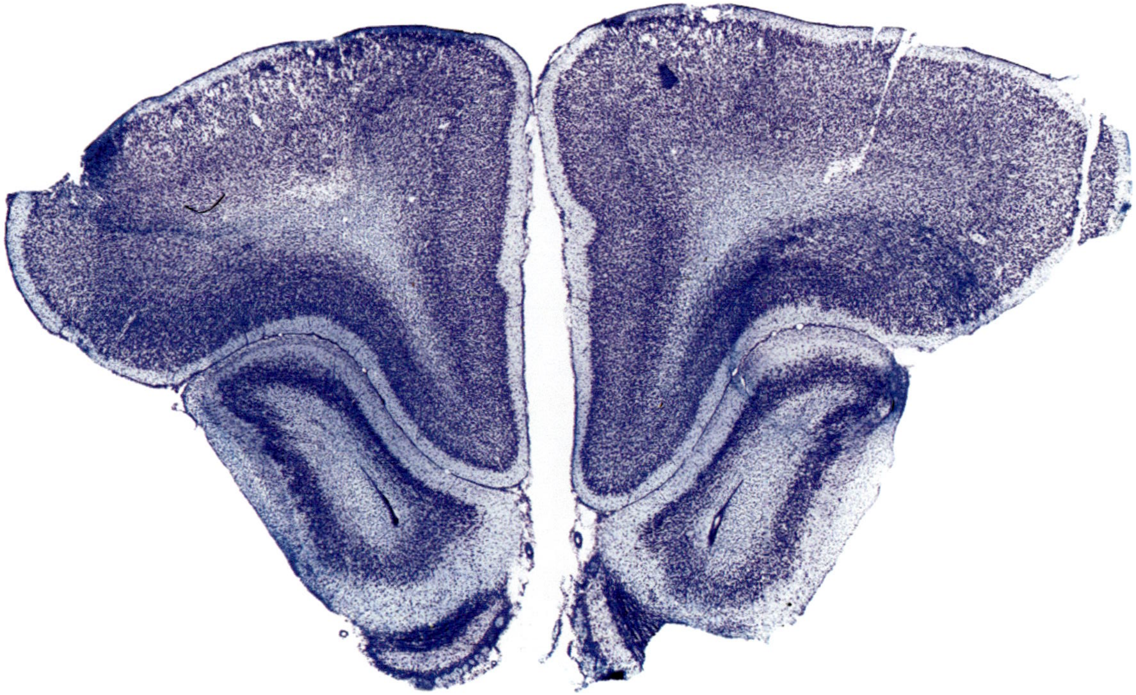

aci	anterior commissure, intrabulbar part
AOD	anterior olfactory nucleus, dorsal part
AOL	anterior olfactory nucleus, lateral part
AOM	anterior olfactory nucleus, medial part
AOV	anterior olfactory nucleus, ventral part
Cg	cingulate cortex
DFC	dorsal frontal cortex
EPl	external plexiform layer of the olfactory bulb
fmi	forceps minor of the corpus callosum
GCL	granule cell layer of the olfactory bulb
GL	glomerular layer of the olfactory bulb
IL	infralimbic cortex
Ins	insular cortex
IPl	internal plexiform layer of the olfactory bulb
lo	lateral olfactory tract
M2	secondary motor cortex
MCL	mitral cell layer of the olfactory bulb
MFC	medial frontal cortex
OFC	orbital frontal cortex
OV	olfactory ventricle (olfactory part of lateral ventricle)
PrL	prelimbic cortex

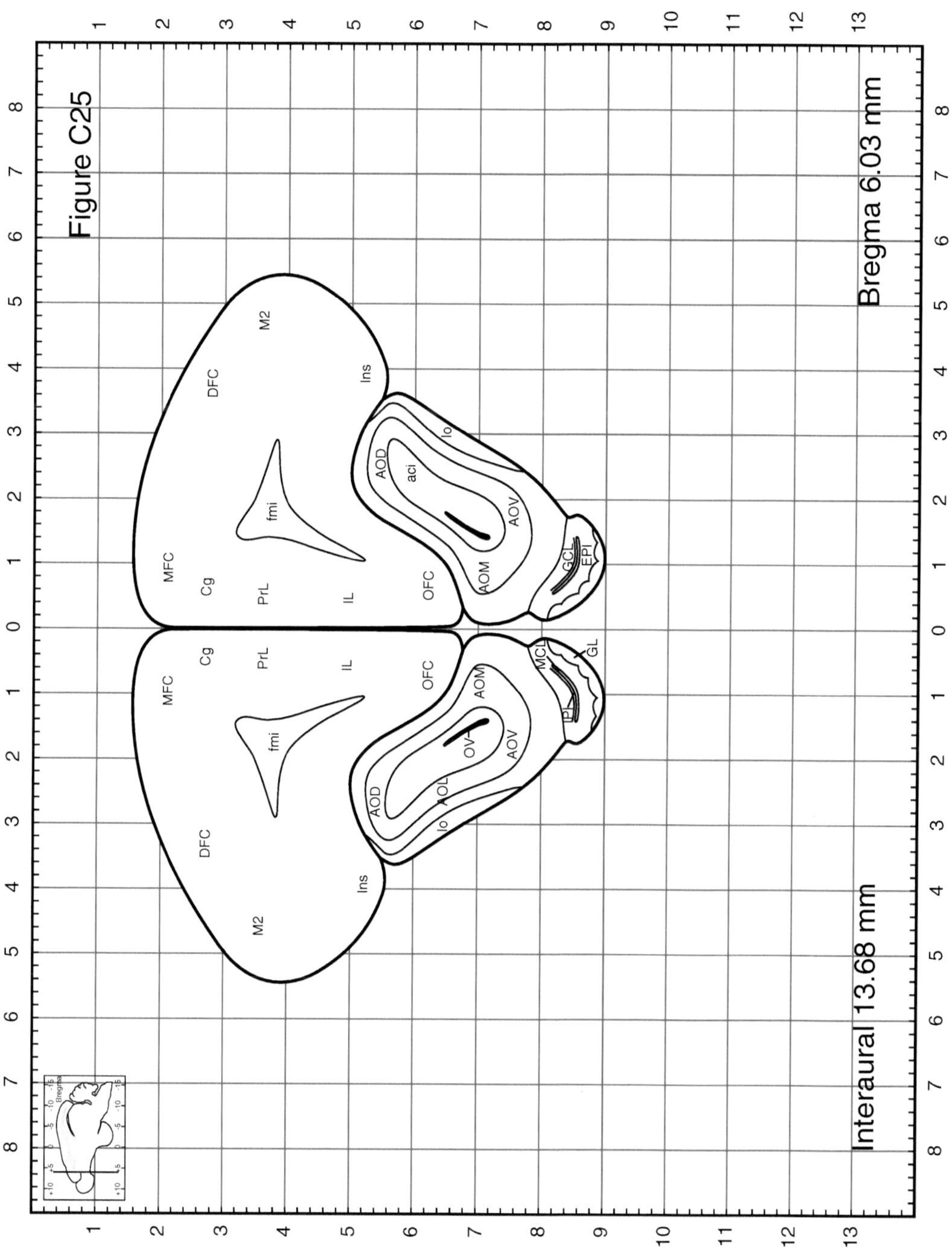
Figure C25
Bregma 6.03 mm
Interaural 13.68 mm
M2
DFC
Ins
MFC
Cg
PrL
IL
OFC
fmi
AOD
aci
lo
AOV
AOM
GCL
EPl
MCL
GL
IPl
OV
AOL

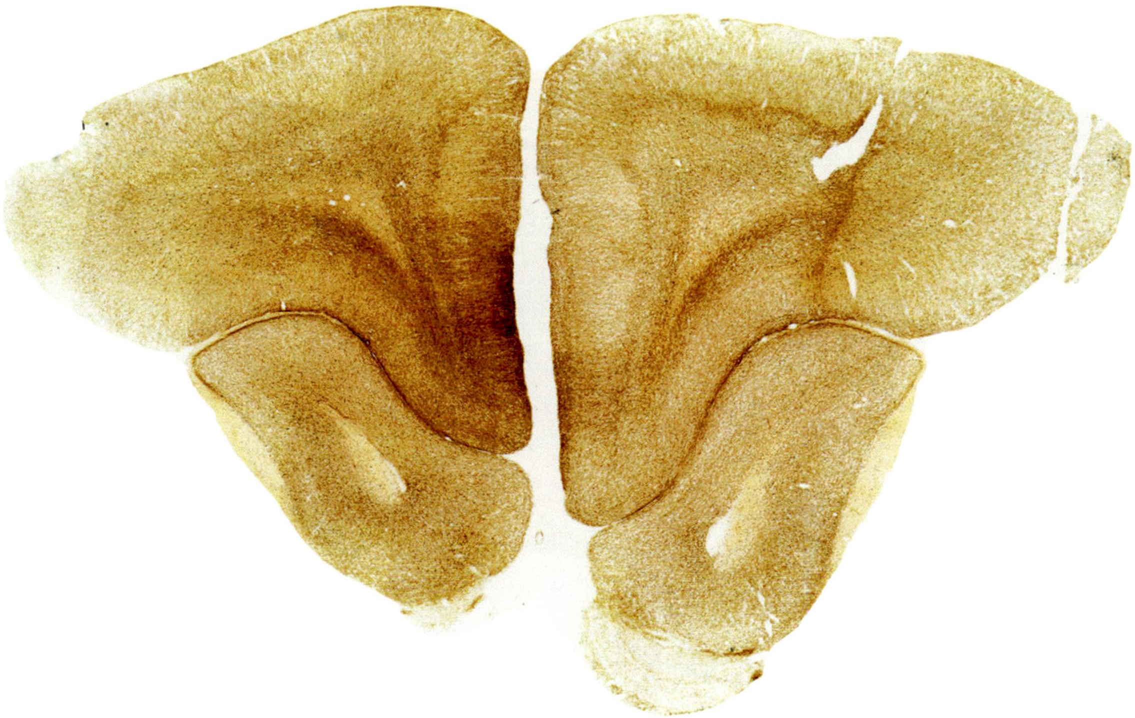

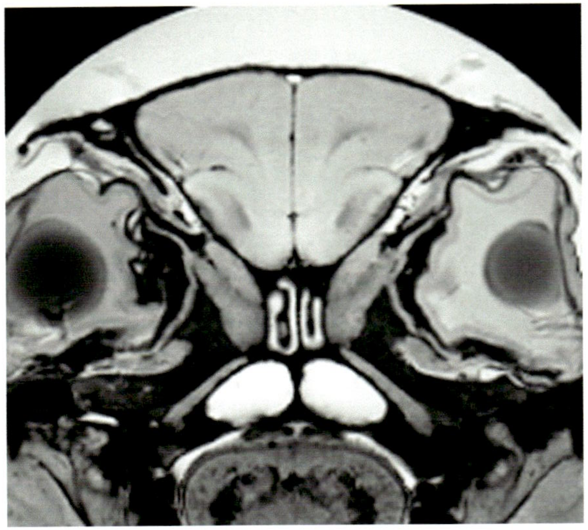

aci	anterior commissure, intrabulbar part
AOD	anterior olfactory nucleus, dorsal part
AOL	anterior olfactory nucleus, lateral part
AOM	anterior olfactory nucleus, medial part
AOV	anterior olfactory nucleus, ventral part
Cg	cingulate cortex
DFC	dorsal frontal cortex
fmi	forceps minor of the corpus callosum
GCL	granule cell layer of the olfactory bulb
IL	infralimbic cortex
Ins	insular cortex
lo	lateralolfactory tract
M2	secondary motor cortex
MFC	medial frontal cortex
OFC	orbital frontal cortex
OV	olfactory ventricle (olfactory part of lateral ventricle)
PrL	prelimbic cortex

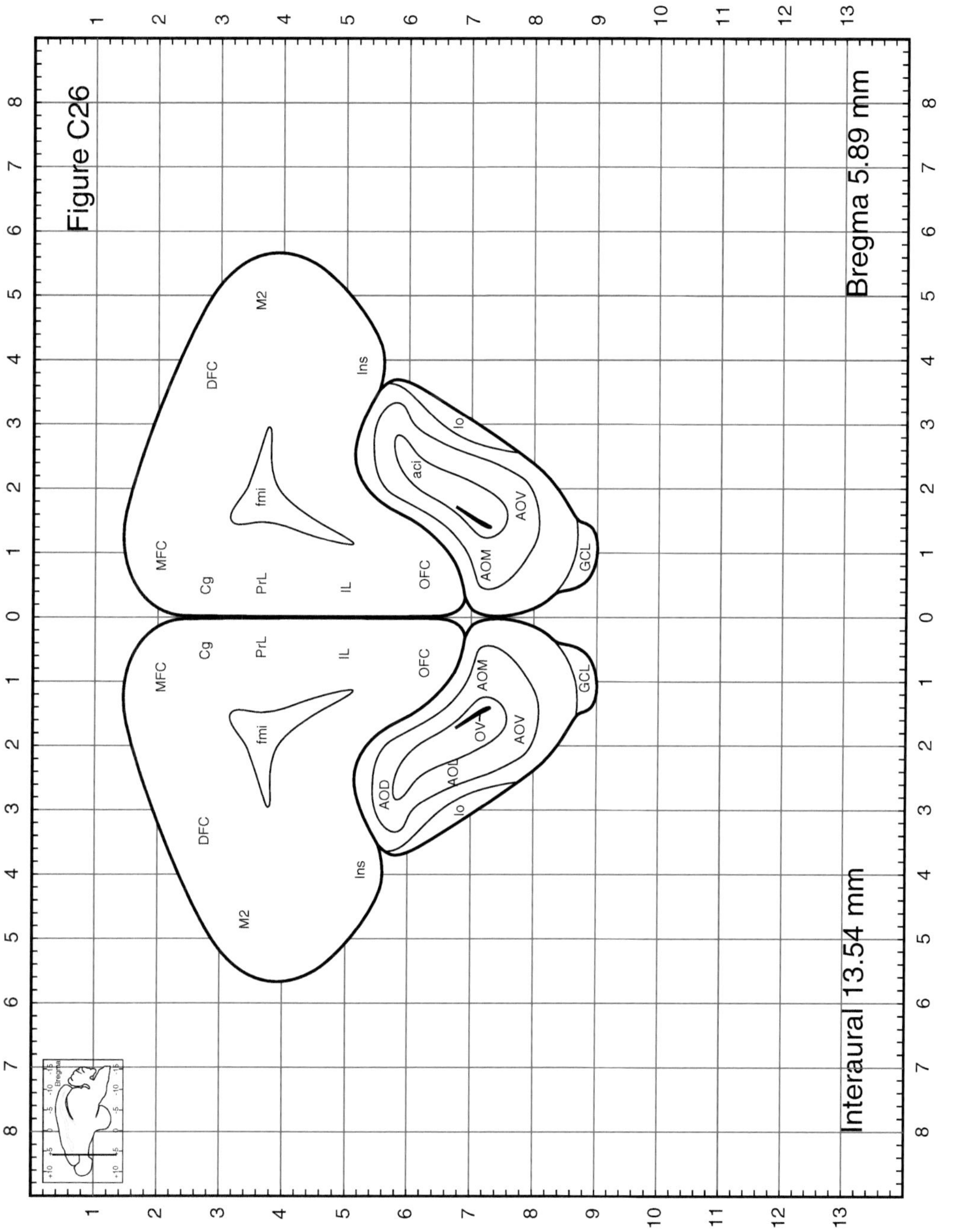
Figure C26
Bregma 5.89 mm
Interaural 13.54 mm
M2
DFC
Ins
MFC
Cg
PrL
IL
OFC
fmi
aci
lo
AOV
AOM
AOD
AOL
OV
GCL

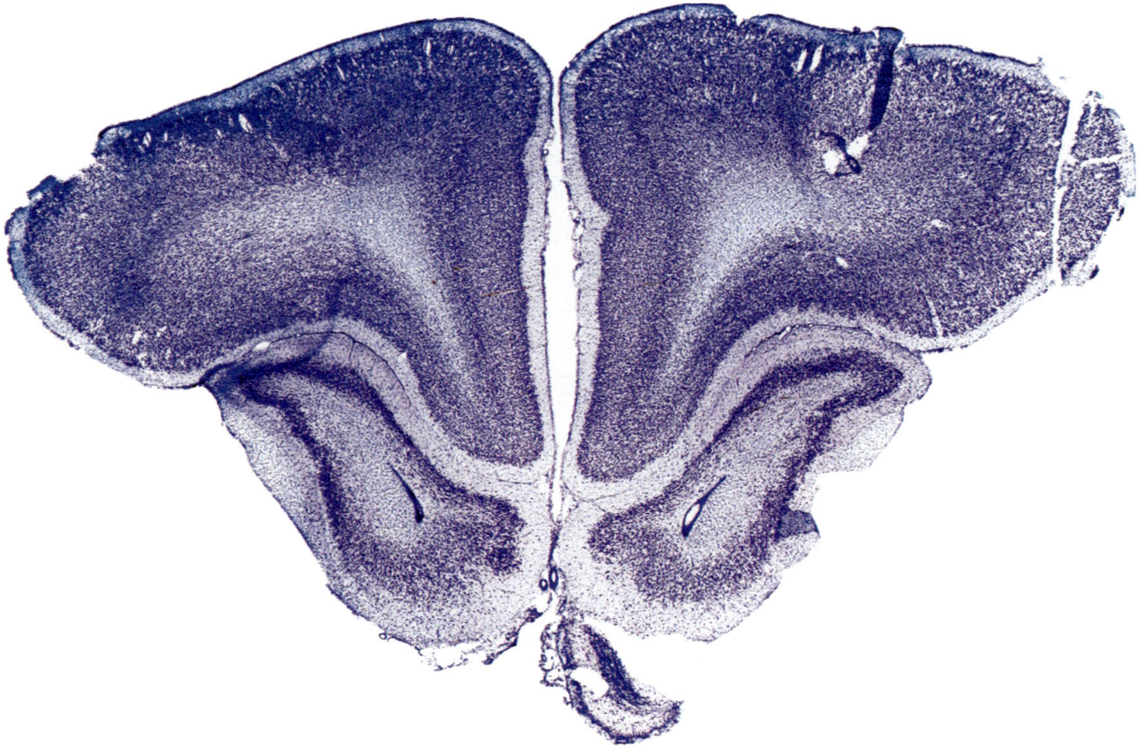

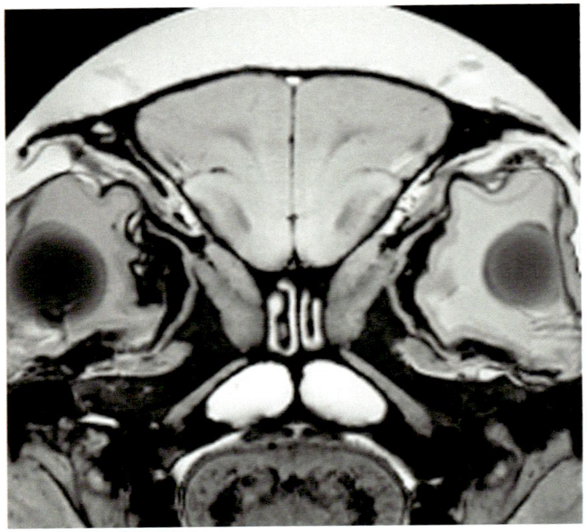

aci	anterior commissure, intrabulbar part
AOD	anterior olfactory nucleus, dorsal part
AOL	anterior olfactory nucleus, lateral part
AOM	anterior olfactory nucleus, medial part
AOV	anterior olfactory nucleus, ventral part
Cg	cingulate cortex
Cl	claustrum
DFC	dorsal frontal cortex
fmi	forceps minor of the corpus callosum
IL	infralimbic cortex
Ins	insular cortex
lo	lateral olfactory tract
M2	secondary motor cortex
MFC	medial frontal cortex
OFC	orbital frontal cortex
OV	olfactory ventricle (olfactory part of lateral ventricle)
PrL	prelimbic cortex

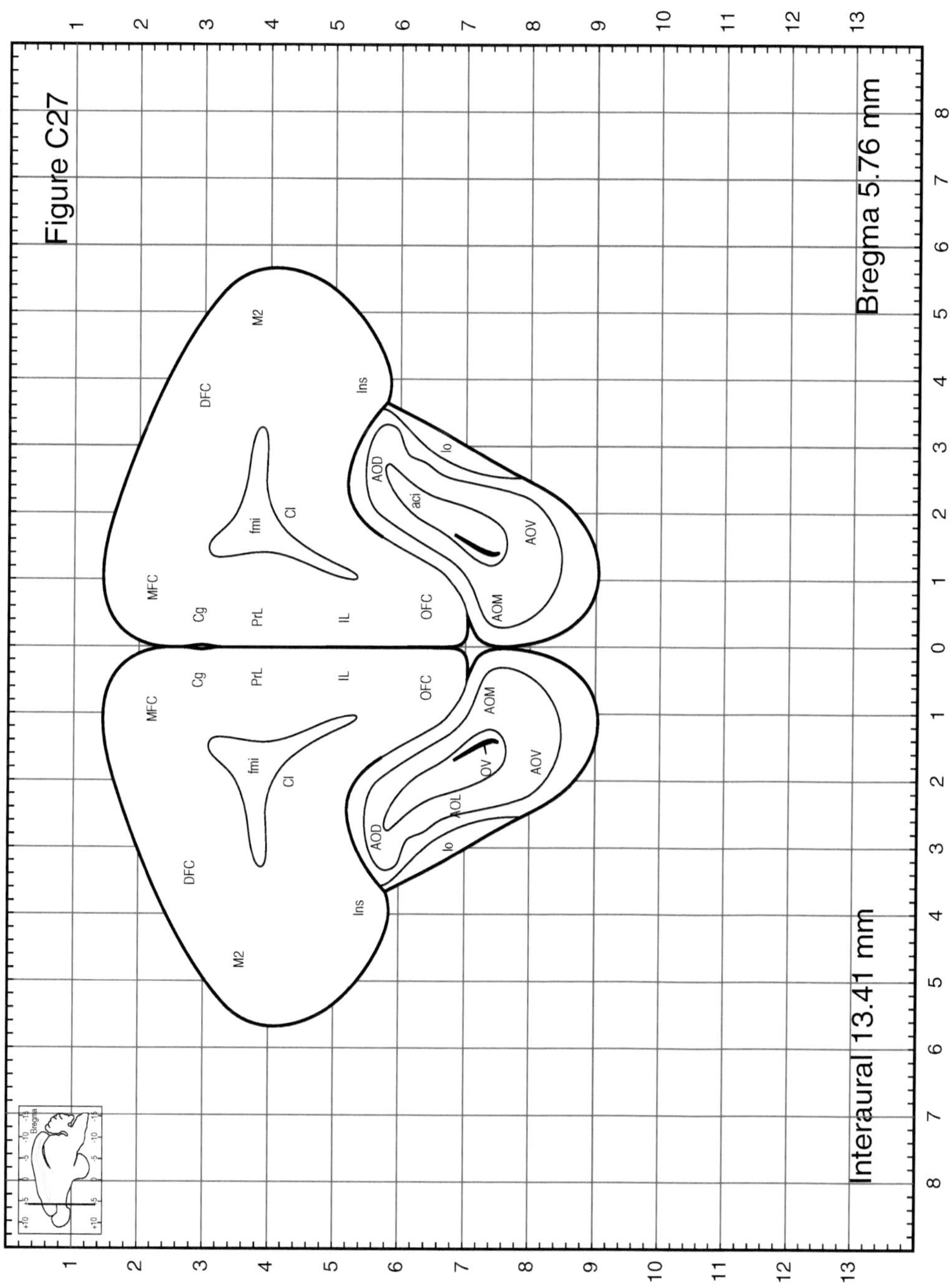
Figure C27
Bregma 5.76 mm
Interaural 13.41 mm
M2
DFC
Ins
MFC
Cg
PrL
IL
OFC
fmi
Cl
AOD
AOM
AOV
AOL
lo
aci
OV

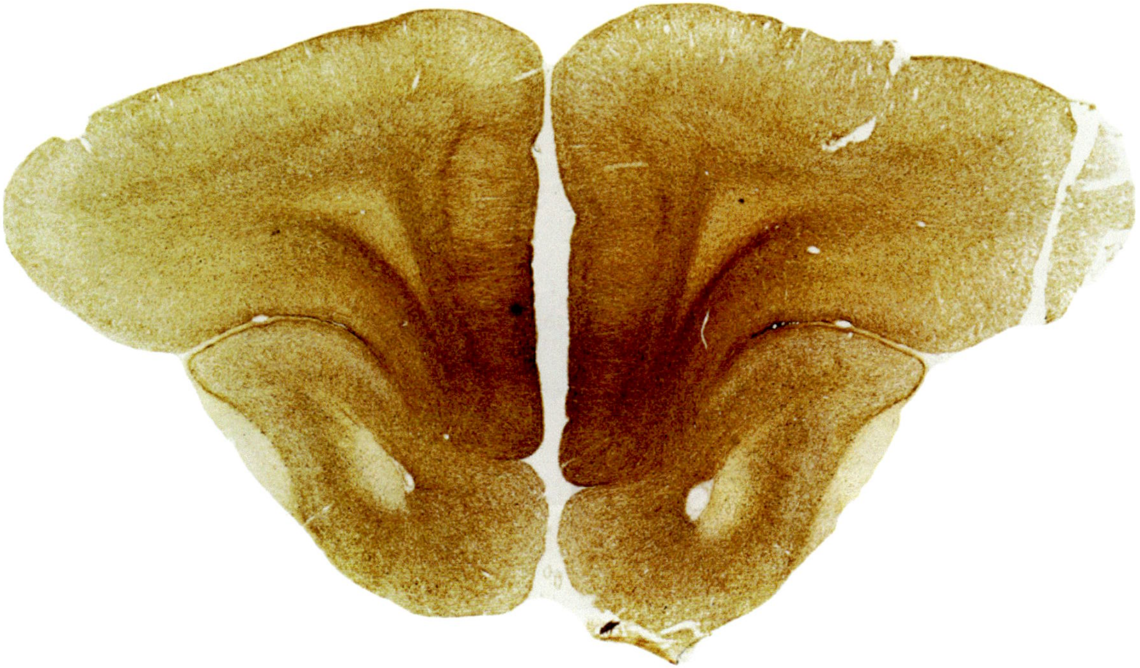

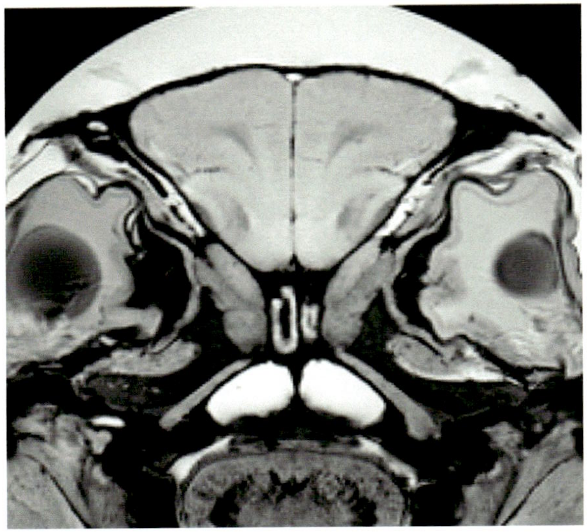

2 layer 2 of cortex
3 layer 3 of cortex
aci anterior commissure, intrabulbar part
Cg cingulate cortex
Cl claustrum
DFC dorsal frontal cortex
fmi forceps minor of the corpus callosum
IL infralimbic cortex
Ins insular cortex
lo lateral olfactory tract
M2 secondary motor cortex
MFC medial frontal cortex
OFC orbital frontal cortex
OV olfactory ventricle (olfactory part of lateral ventricle)
Pir1 piriform cortex, layer 1
PrL prelimbic cortex

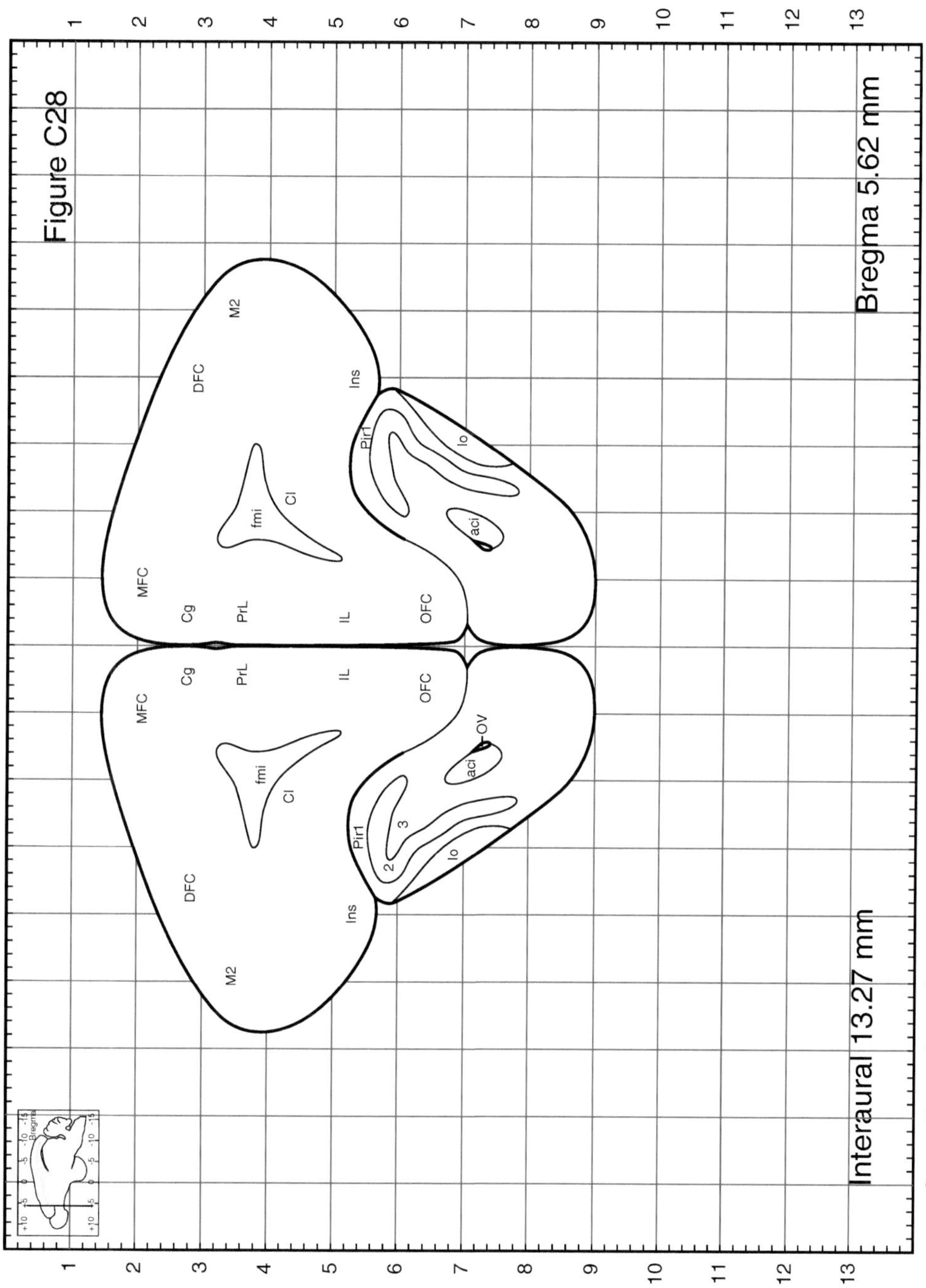
Figure C28
Bregma 5.62 mm
Interaural 13.27 mm
M2
DFC
Ins
MFC
Cg
PrL
IL
OFC
fmi
Cl
Pir1
lo
aci
OV
2
3

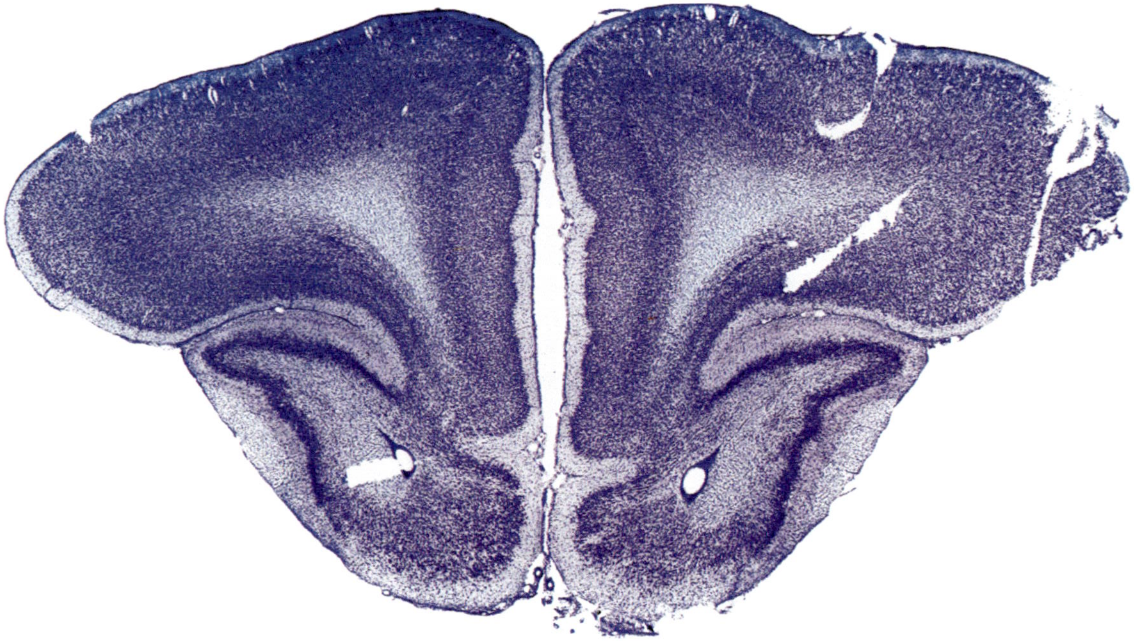

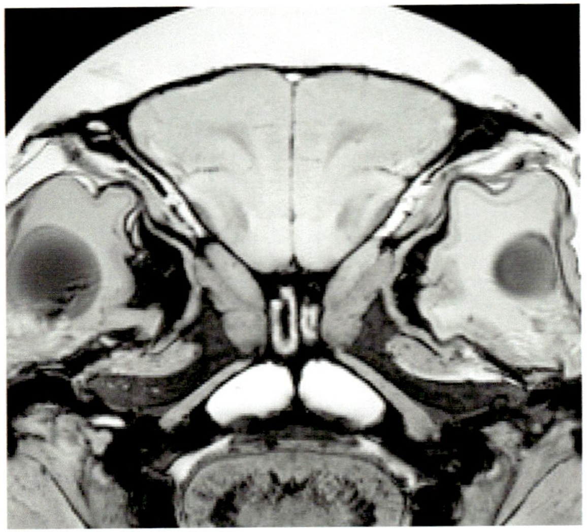

2	layer 2 of cortex
3	layer 3 of cortex
aci	anterior commissure, intrabulbar part
Cg	cingulate cortex
Cl	claustrum
DFC	dorsal frontal cortex
fmi	forceps minor of the corpus callosum
IL	infralimbic cortex
Ins	insular cortex
lo	lateral olfactory tract
M2	secondary motor cortex
MFC	medial frontal cortex
OV	olfactory ventricle (olfactory part of lateral ventricle)
Pir1	piriform cortex, layer 1
PrL	prelimbic cortex

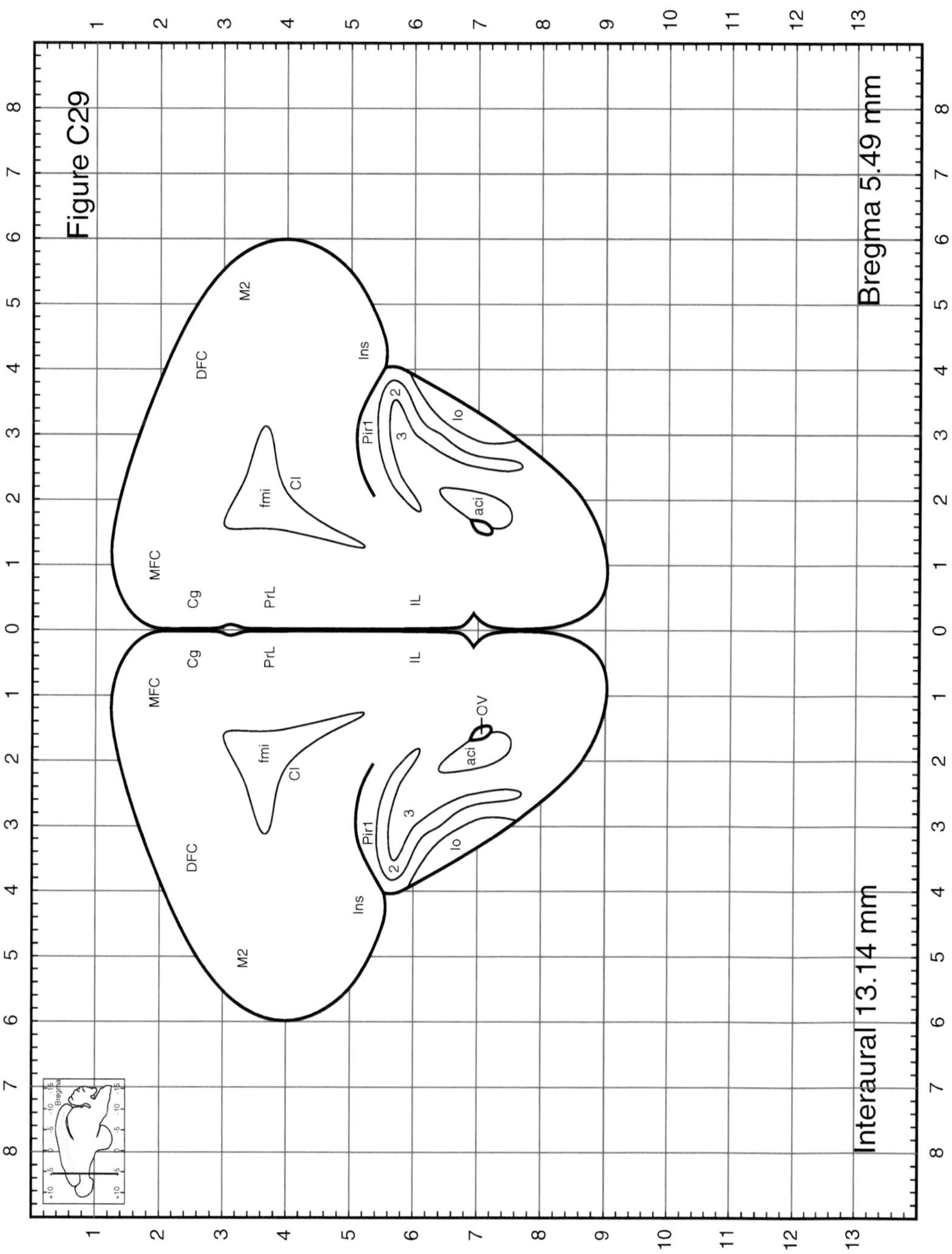
Figure C29
Bregma 5.49 mm
Interaural 13.14 mm
M2
DFC
Ins
MFC
Cg
PrL
IL
fmi
Cl
Pir1
2
3
lo
aci
OV

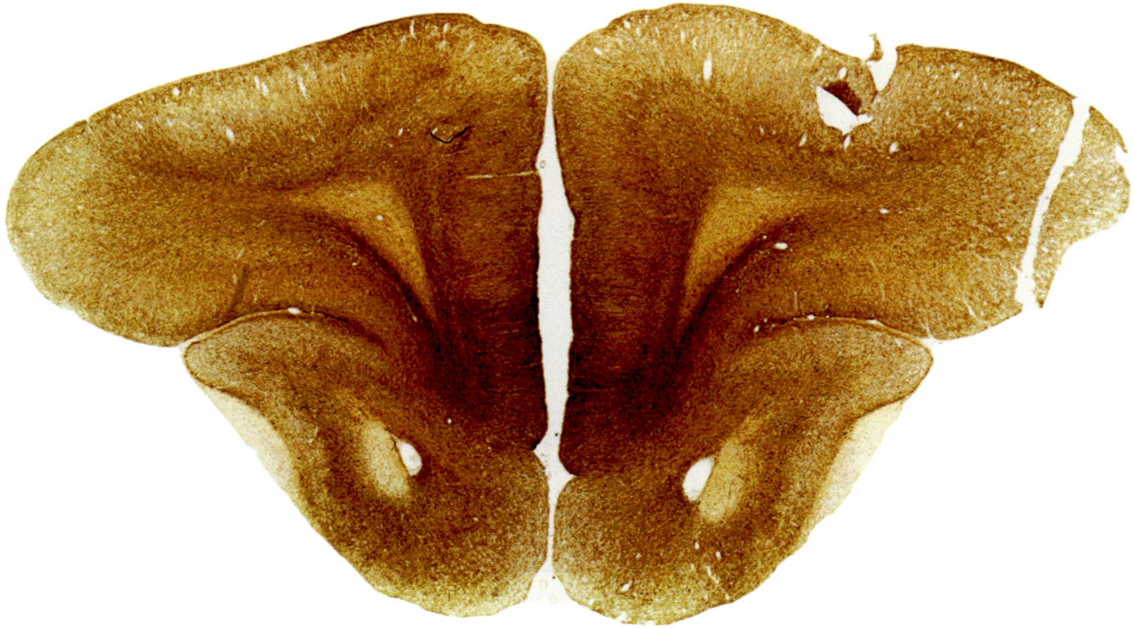

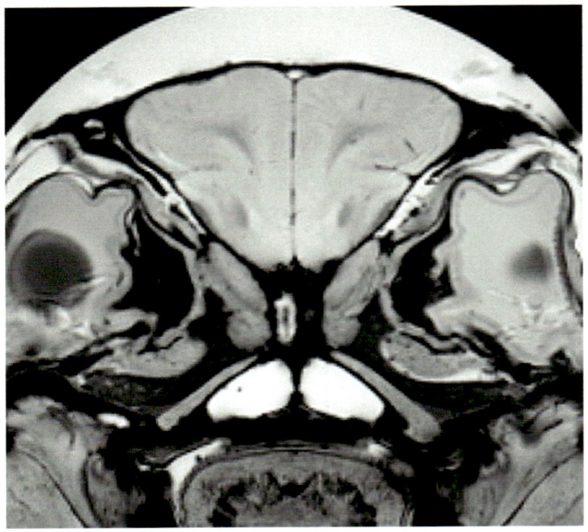

2	layer 2 of cortex
3	layer 3 of cortex
aci	anterior commissure, intrabulbar part
Cg	cingulate cortex
Cl	claustrum
DFC	dorsal frontal cortex
fmi	forceps minor of the corpus callosum
IL	infralimbic cortex
Ins	insular cortex
lo	lateral olfactory tract
M2	secondary motor cortex
MFC	medial frontal cortex
OV	olfactory ventricle (olfactory part of lateral ventricle)
Pir1	piriform cortex, layer 1
PrL	prelimbic cortex

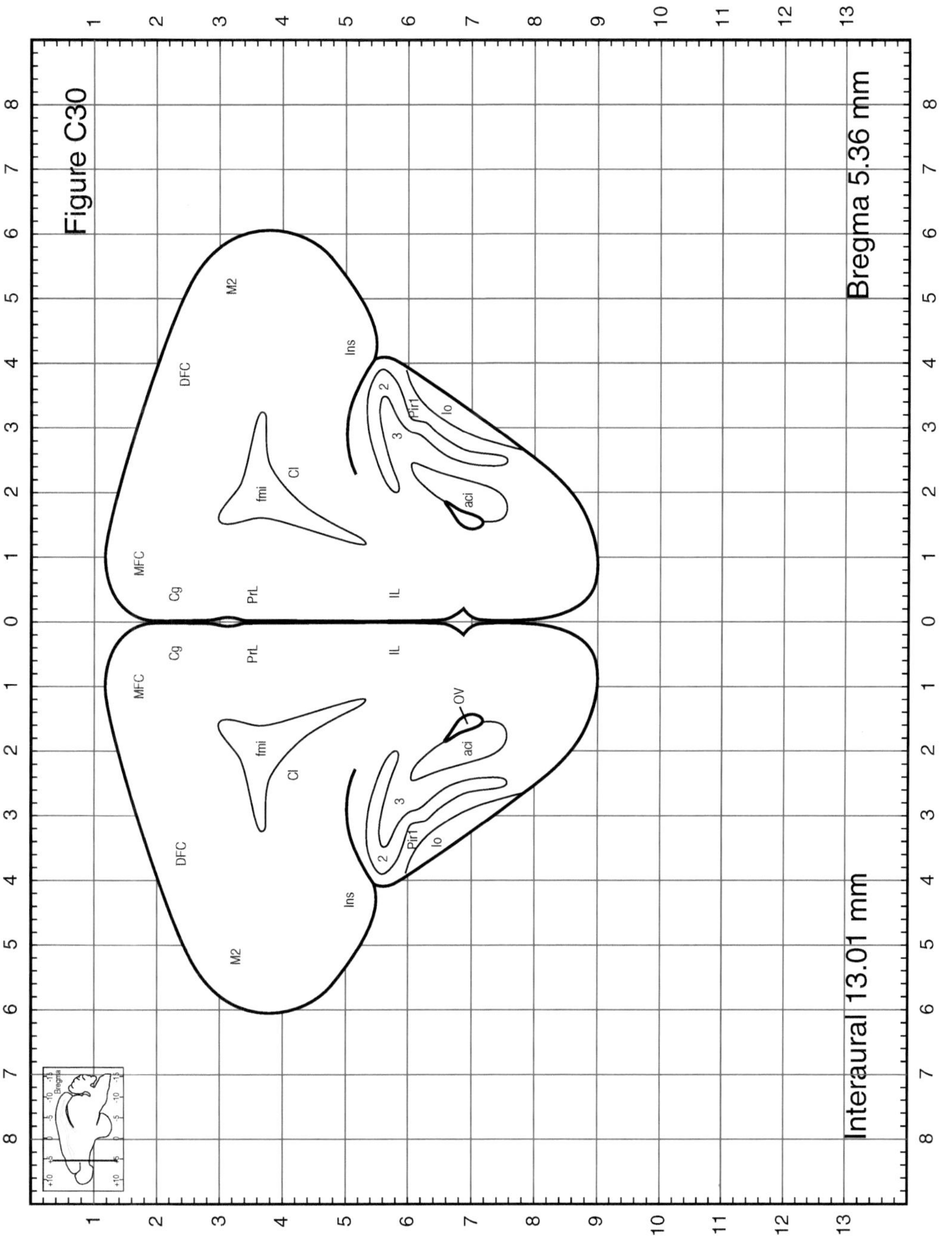
Figure C30
Bregma 5.36 mm
Interaural 13.01 mm
M2
Ins
DFC
fmi
Cl
MFC
Cg
PrL
IL
Pir1
lo
aci
OV
2
3

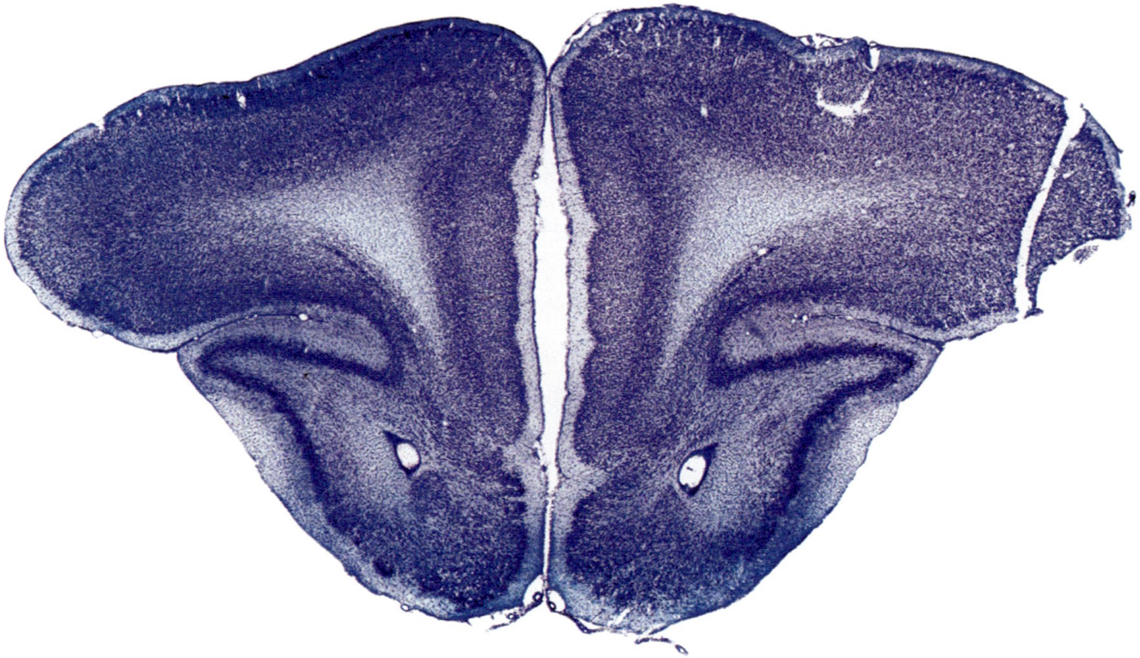

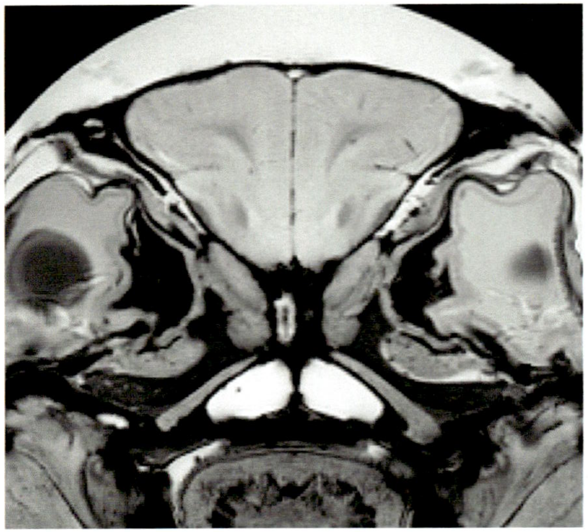

2	layer 2 of cortex
3	layer 3 of cortex
aci	anterior commissure, intrabulbar part
AOP	anterior olfactory nucleus, posterior part
Cg	cingulate cortex
Cl	claustrum
DFC	dorsal frontal cortex
fmi	forceps minor of the corpus callosum
IL	infralimbic cortex
Ins	insular cortex
lo	lateral olfactory tract
M2	secondary motor cortex
MFC	medial frontal cortex
OV	olfactory ventricle (olfactory part of lateral ventricle)
Pir1	piriform cortex, layer 1
PrL	prelimbic cortex
Tu1	layer 1 of olfactory tubercle

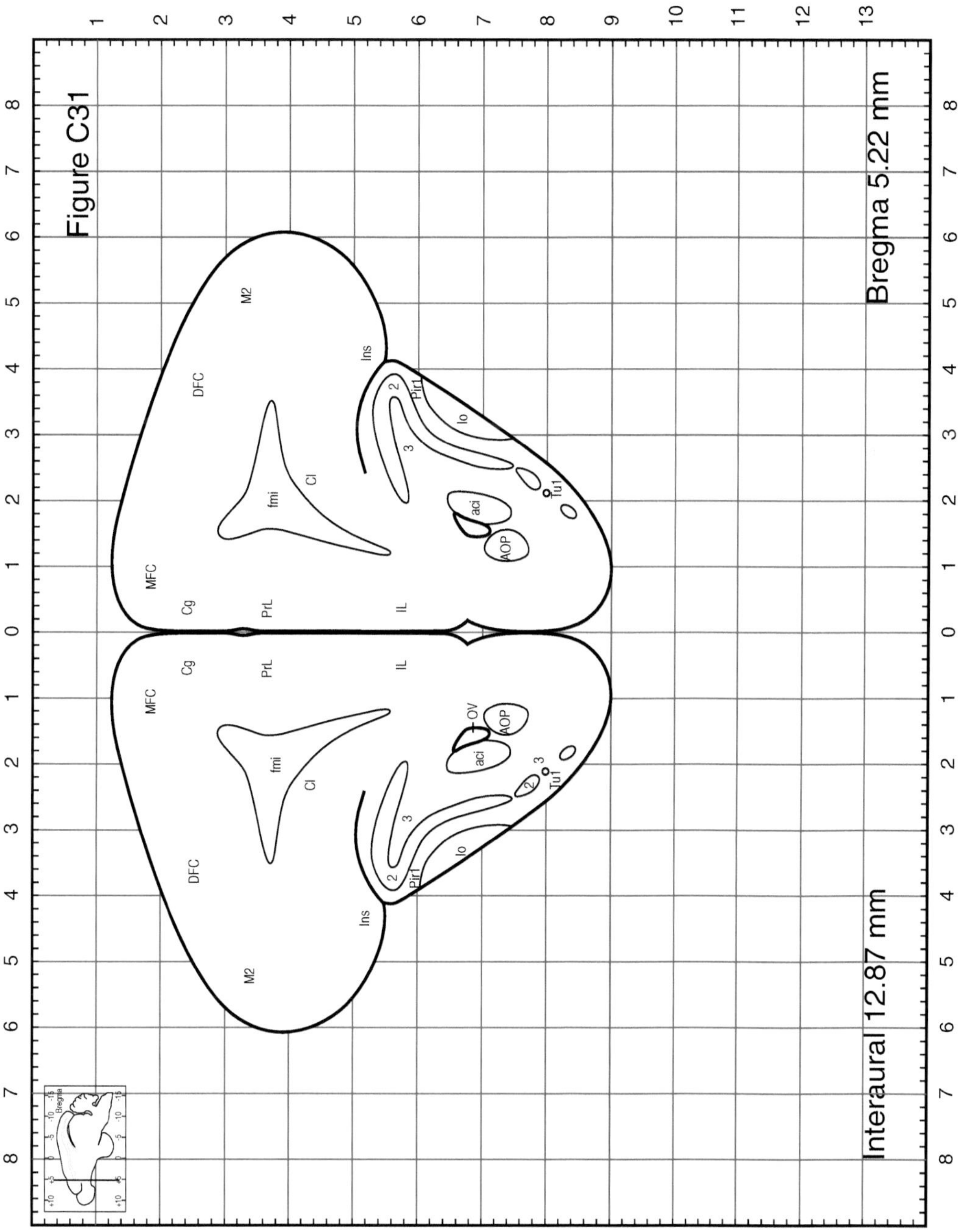

Figure C31
Bregma 5.22 mm
Interaural 12.87 mm
M2
DFC
MFC
Cg
PrL
IL
Ins
fmi
Cl
Pir1
lo
aci
AOP
OV
Tu1

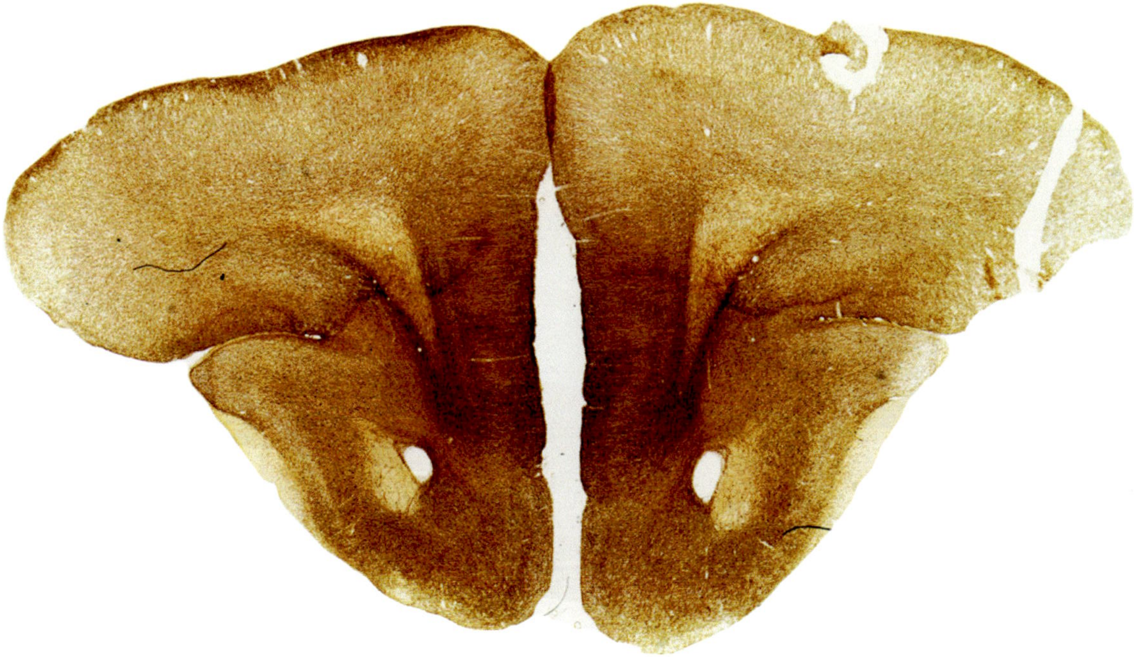

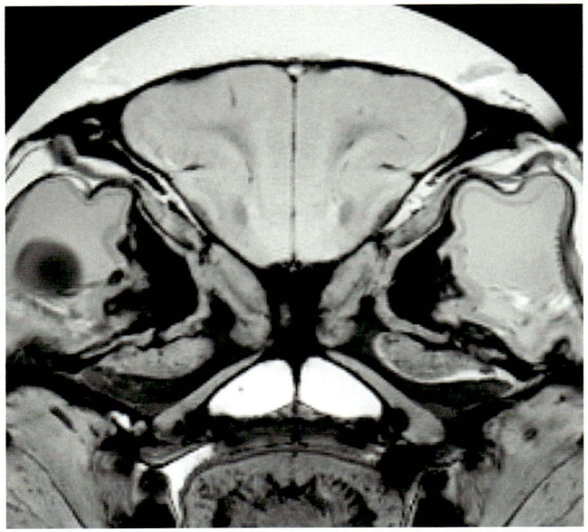

2	layer 2 of cortex		DFC	dorsal frontal cortex		M2	secondary motor cortex
3	layer 3 of cortex		fmi	forceps minor of the corpus callosum		MFC	medial frontal cortex
aci	anterior commissure, intrabulbar part		IL	infralimbic cortex		OV	olfactory ventricle (olfactory part of lateral ventricle)
AOP	anterior olfactory nucleus, posterior part		Ins	insular cortex		Pir1	piriform cortex, layer 1
Cg	cingulate cortex		lo	lateral olfactory tract		PrL	prelimbic cortex
Cl	claustrum		M1	primary motor cortex		Tu1	layer 1 of olfactory tubercle

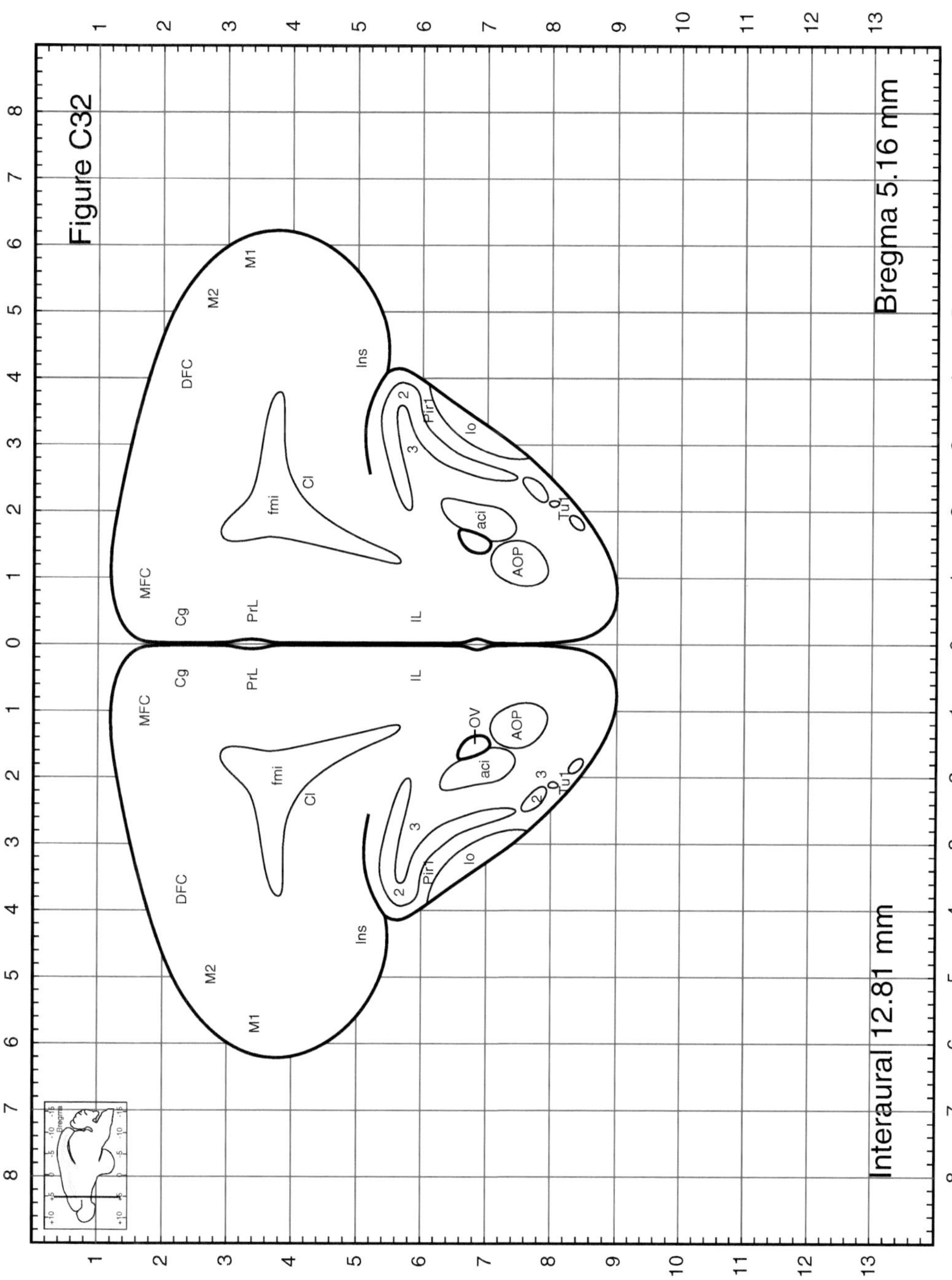
Figure C32
Bregma 5.16 mm
Interaural 12.81 mm
MFC
Cg
PrL
IL
DFC
M2
M1
Ins
fmi
Cl
Pir1
lo
aci
AOP
Tu1
OV
2
3

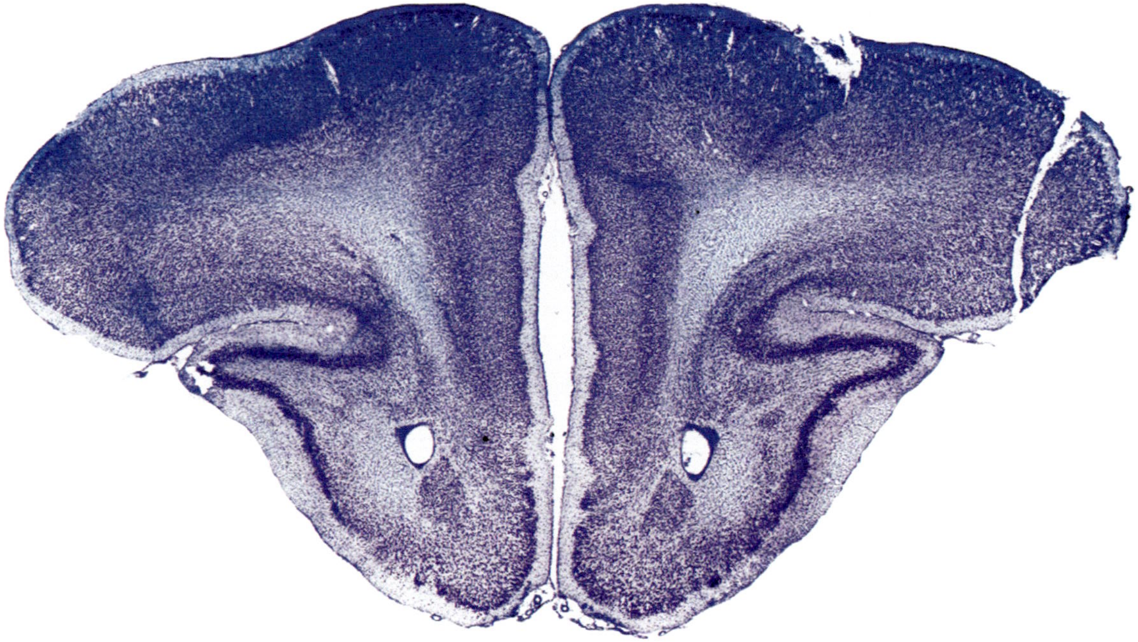

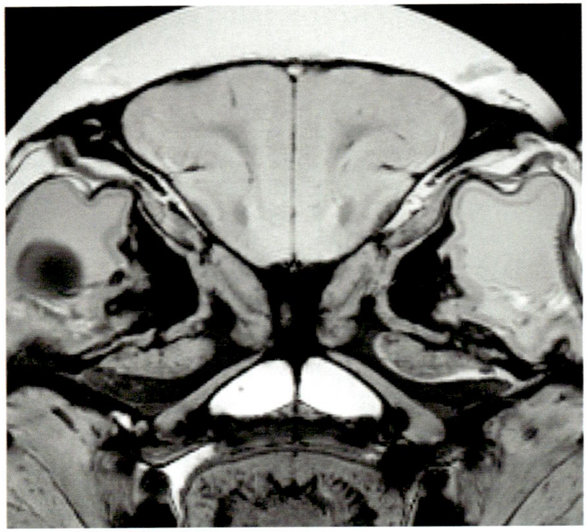

2	layer 2 of cortex	DFC	dorsal frontal cortex	M2	secondary motor cortex
3	layer 3 of cortex	fmi	forceps minor of the corpus callosum	MFC	medial frontal cortex
aci	anterior commissure, intrabulbar part	IL	infralimbic cortex	OV	olfactory ventricle (olfactory part of lateral ventricle)
AOP	anterior olfactory nucleus, posterior part	Ins	insular cortex	Pir1	piriform cortex, layer 1
Cg	cingulate cortex	lo	lateral olfactory tract	PrL	prelimbic cortex
Cl	claustrum	M1	primary motor cortex	Tu1	layer 1 of olfactory tubercle

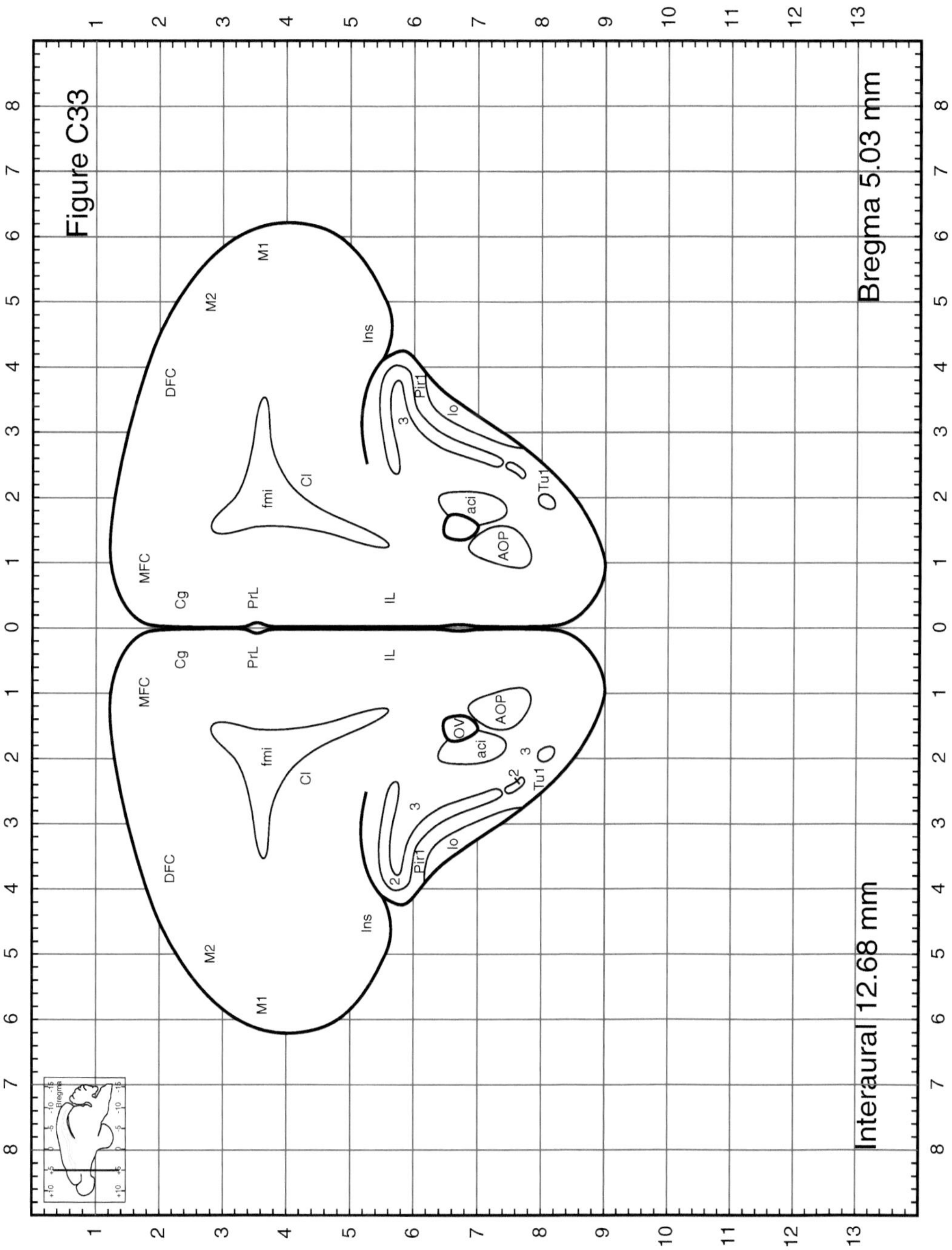
Figure C33
Bregma 5.03 mm
Interaural 12.68 mm
MFC
Cg
PrL
IL
DFC
M2
M1
Ins
fmi
Cl
Pir1
lo
3
2
aci
OV
AOP
Tu1

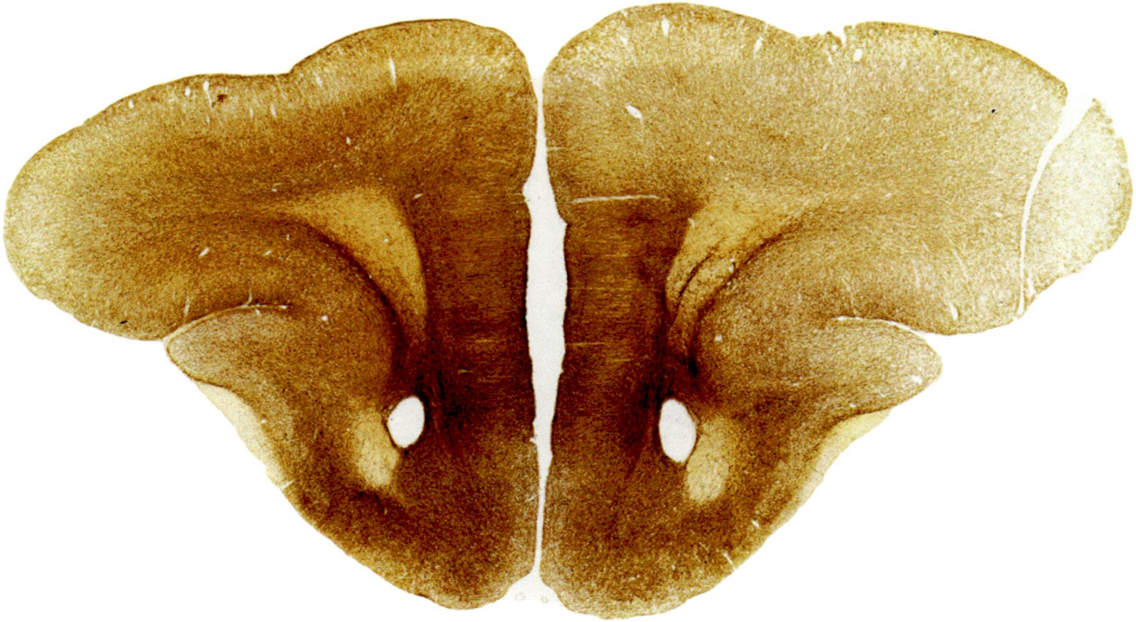

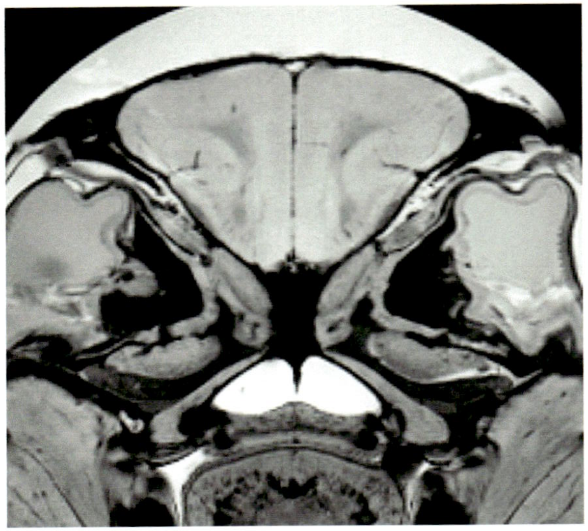

2	layer 2 of cortex
3	layer 3 of cortex
aci	anterior commissure, intrabulbar part
AOP	anterior olfactory nucleus, posterior part
Cg	cingulate cortex
Cl	claustrum
DFC	dorsal frontal cortex
fmi	forceps minor of the corpus callosum
IL	infralimbic cortex
Ins	insular cortex
lo	lateral olfactory tract
M1	primary motor cortex
M2	secondary motor cortex
MFC	medial frontal cortex
OV	olfactory ventricle (olfactory part of lateral ventricle)
Pir1	piriform cortex, layer 1
PrL	prelimbic cortex
Tu1	layer 1 of olfactory tubercle

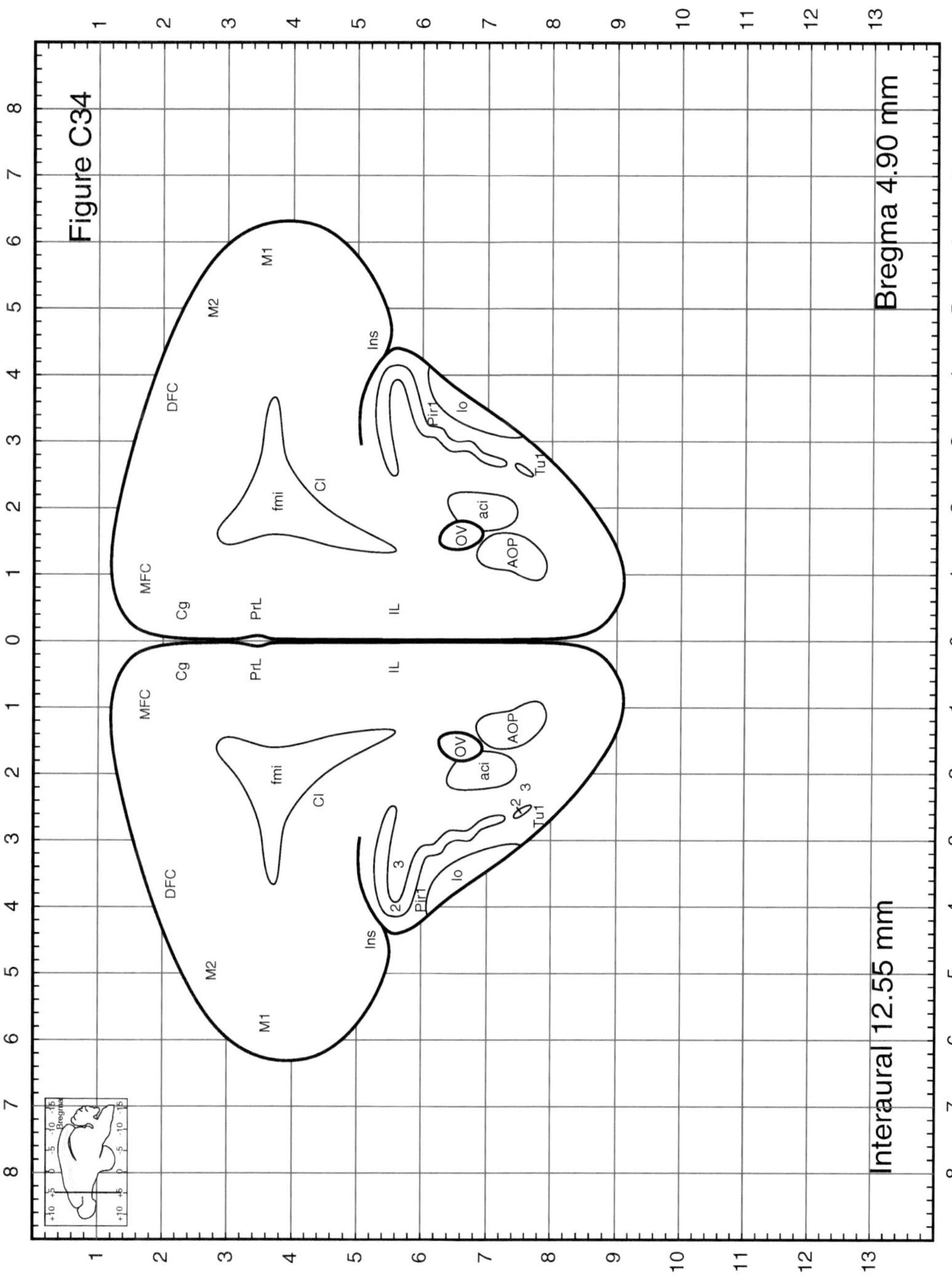
Figure C34
Bregma 4.90 mm
Interaural 12.55 mm
MFC
Cg
PrL
IL
DFC
M2
M1
Ins
fmi
Cl
Pir1
lo
Tu1
aci
OV
AOP
2
3

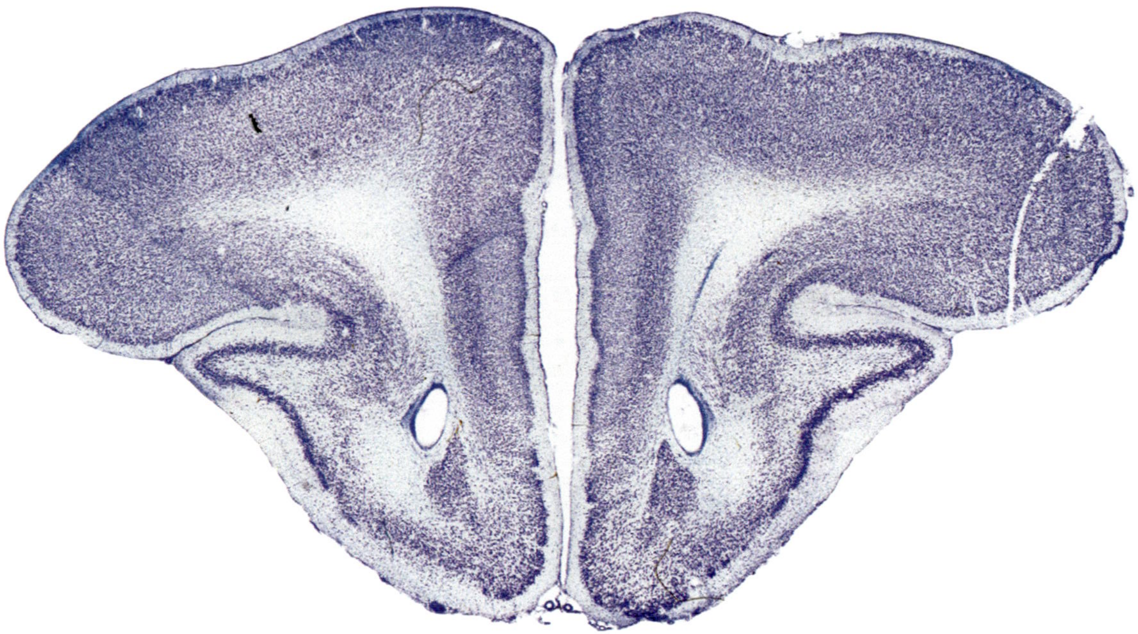

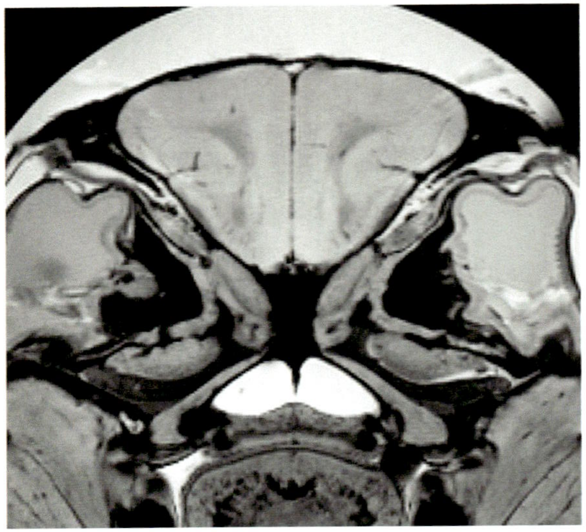

2	layer 2 of cortex
3	layer 3 of cortex
ac	anterior commissure
AOP	anterior olfactory nucleus, posterior part
Cg	cingulate cortex
Cl	claustrum
fmi	forceps minor of the corpus callosum
IL	infralimbic cortex
Ins	insular cortex
lo	lateral olfactory tract
LV	lateral ventricle
M1	primary motor cortex
M2	secondary motor cortex
Pir1	piriform cortex, layer 1
PrL	prelimbic cortex
Pu	putamen
Tu1	layer 1 of olfactory tubercle

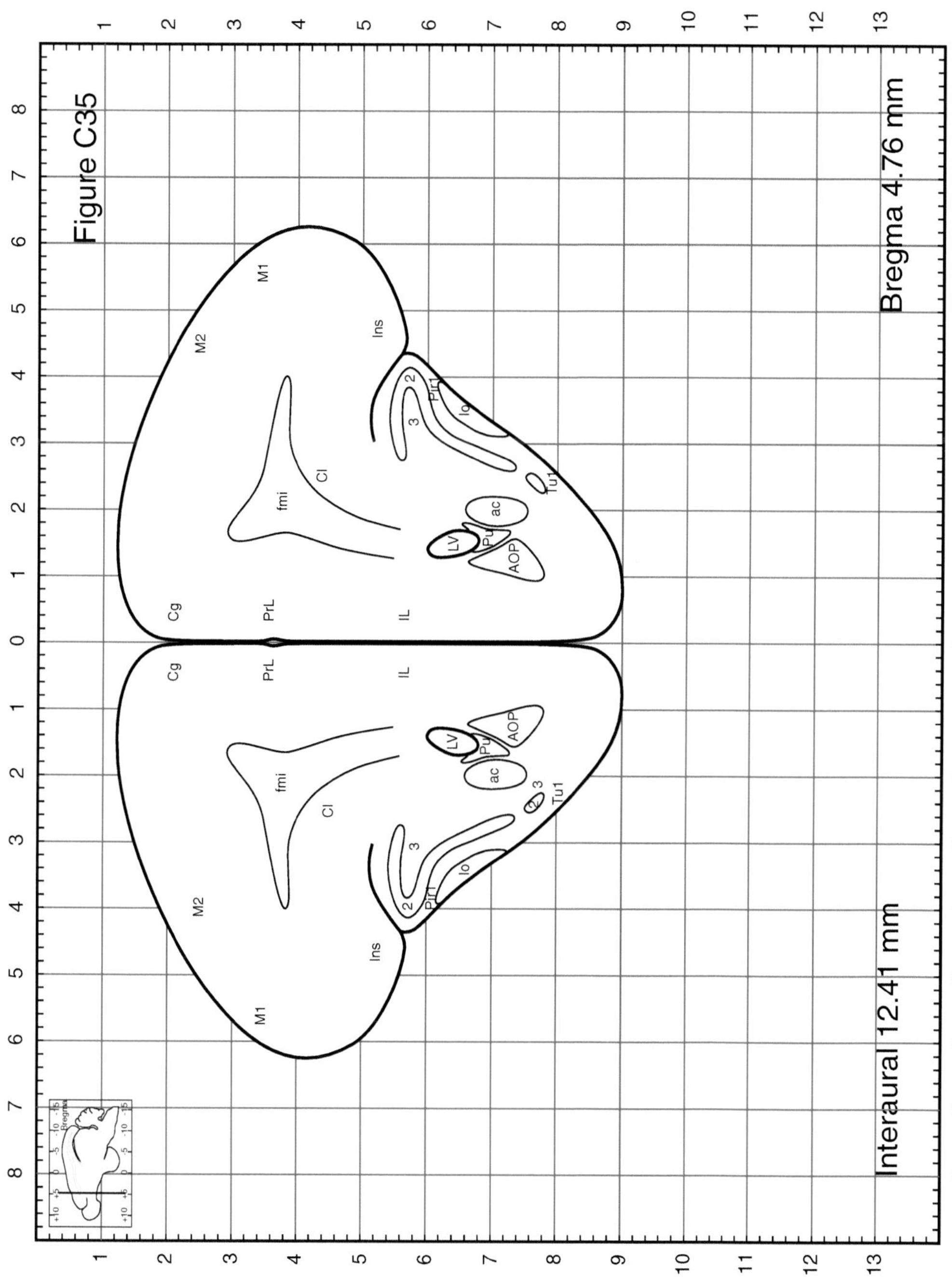
Figure C35
Bregma 4.76 mm
Interaural 12.41 mm
Cg
PrL
IL
M1
M2
Ins
fmi
Cl
LV
ac
Pu
AOP
Pir1
lo
Tu1

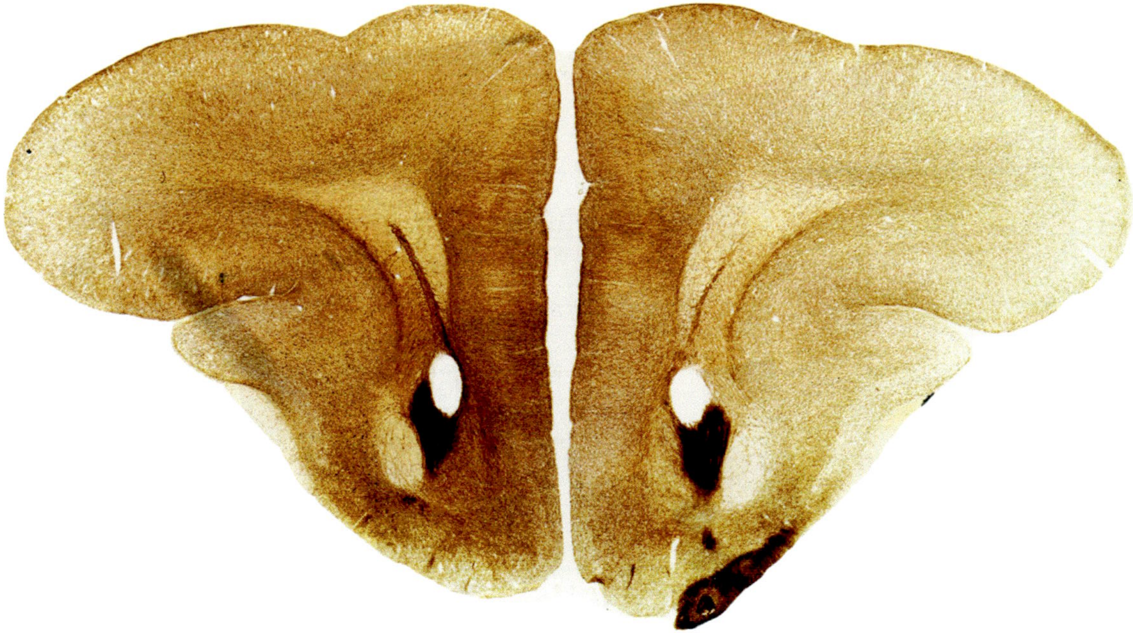

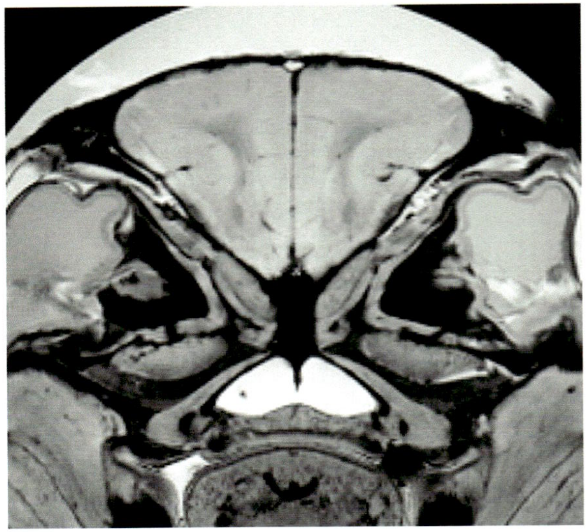

2	layer 2 of cortex	fmi	forceps minor of the corpus callosum	M2	secondary motor cortex
3	layer 3 of cortex	IL	infralimbic cortex	Pir1	piriform cortex, layer 1
ac	anterior commissure	Ins	insular cortex	PrL	prelimbic cortex
AOP	anterior olfactory nucleus, posterior part	lo	lateral olfactory tract	Pu	putamen
Cg	cingulate cortex	LV	lateral ventricle	Tu1	layer 1 of olfactory tubercle
Cl	claustrum	M1	primary motor cortex		

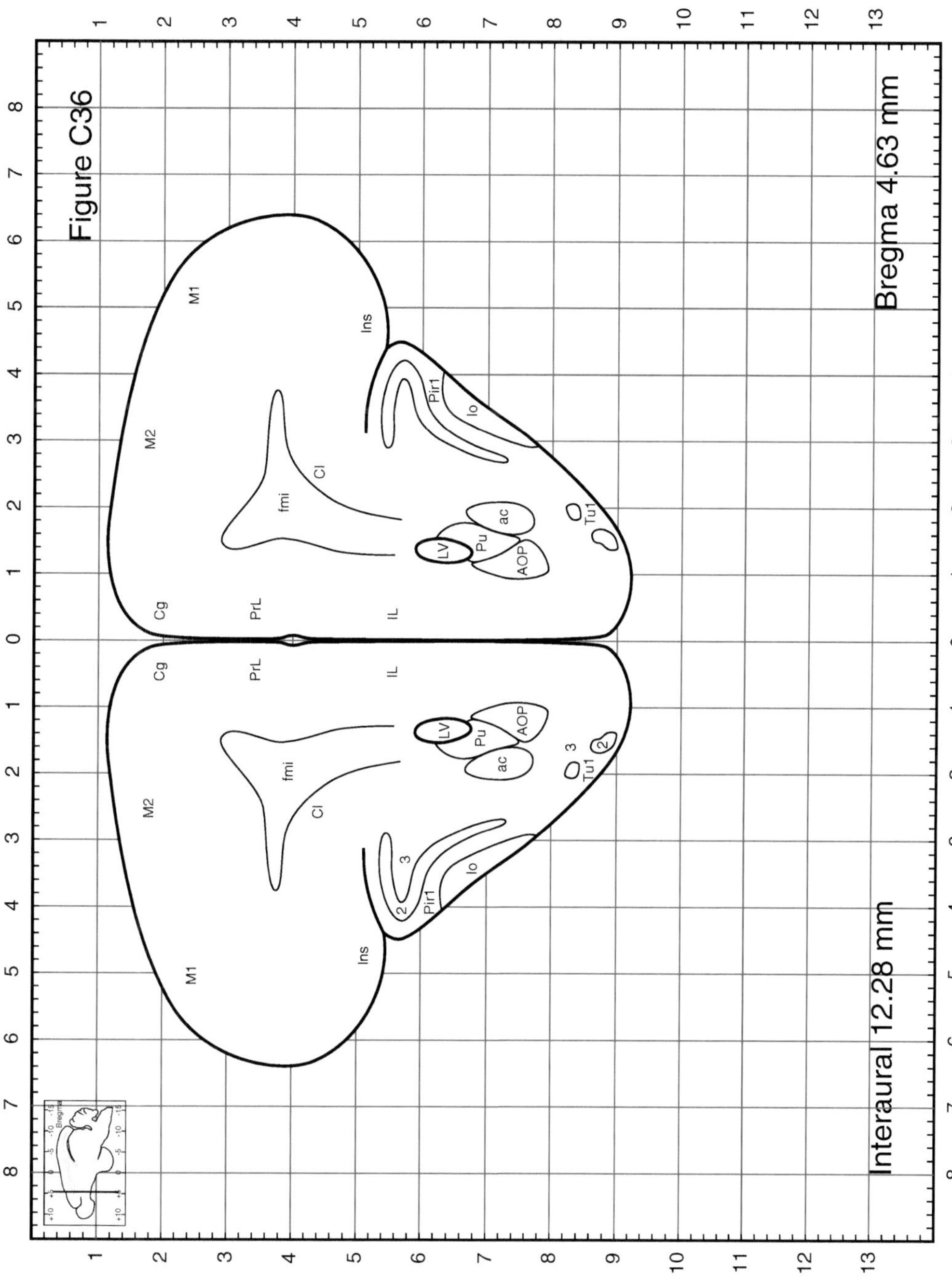
Figure C36
Bregma 4.63 mm
Interaural 12.28 mm
M1
M2
Cg
PrL
IL
Ins
fmi
Cl
Pir1
lo
LV
Pu
ac
AOP
Tu1
2
3

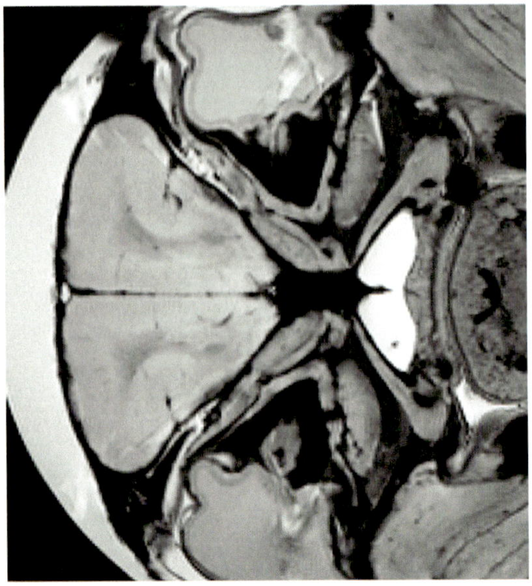

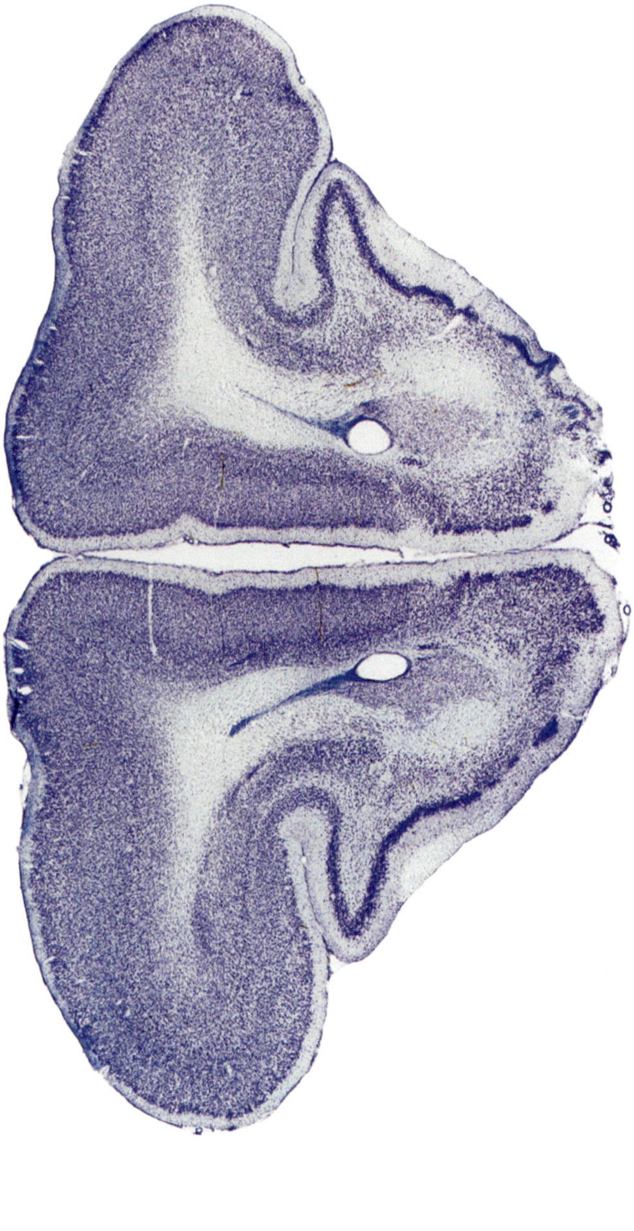

2	layer 2 of cortex
3	layer 3 of cortex
ac	anterior commissure
Acb	accumbens nucleus
Cg	cingulate cortex
Cl	claustrum
fmi	forceps minor of the corpus callosum
IL	infralimbic cortex
Ins	insular cortex
lo	lateral olfactory tract
LV	lateral ventricle
M1	primary motor cortex
M2	secondary motor cortex
Pir1	piriform cortex, layer 1
PrL	prelimbic cortex
Pu	putamen
Tu1	layer 1 of olfactory tubercle

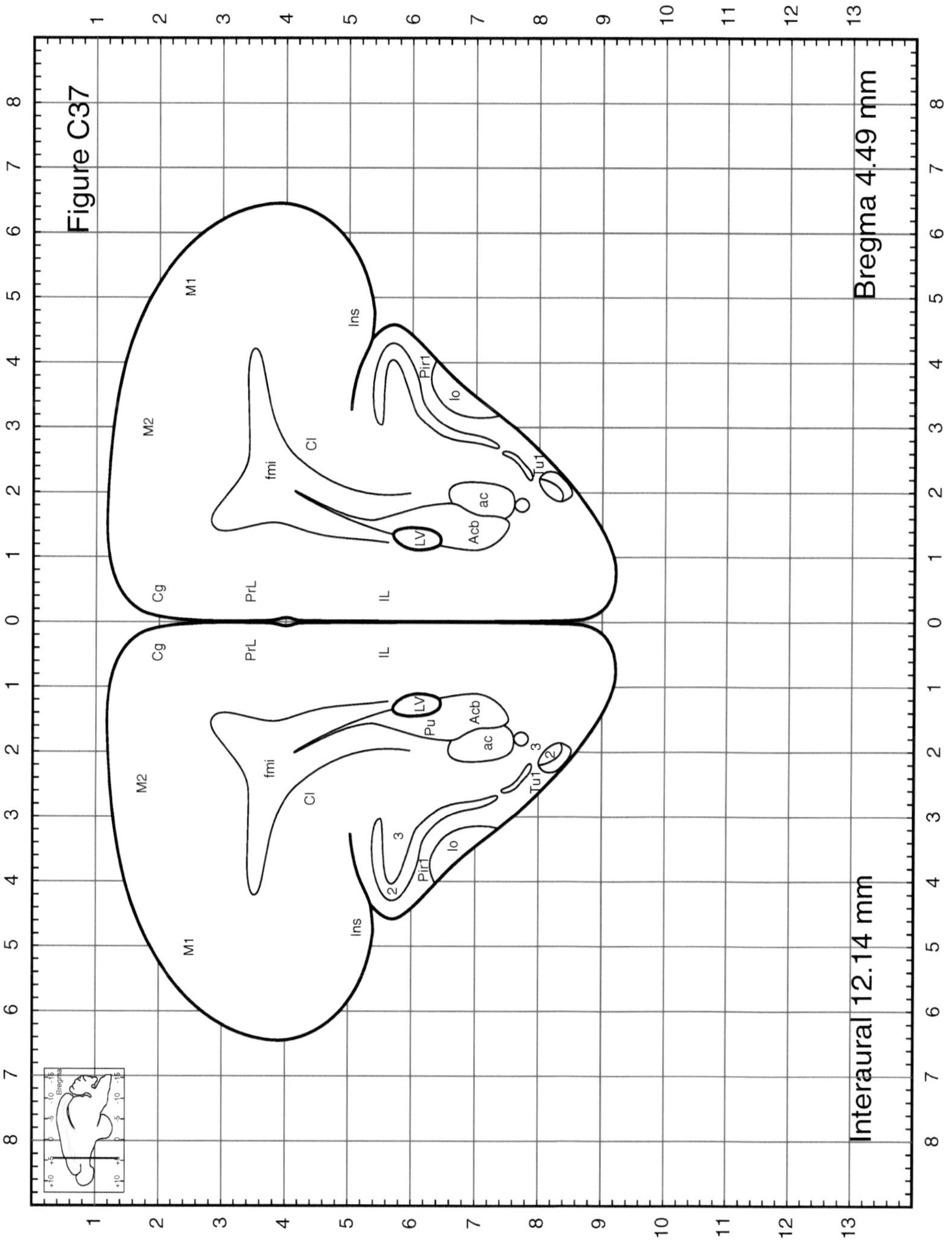
Figure C37
Bregma 4.49 mm
Interaural 12.14 mm
M1
M2
Cg
PrL
IL
Ins
Cl
fmi
LV
Pu
Acb
ac
Pir1
lo
Tu1
2
3

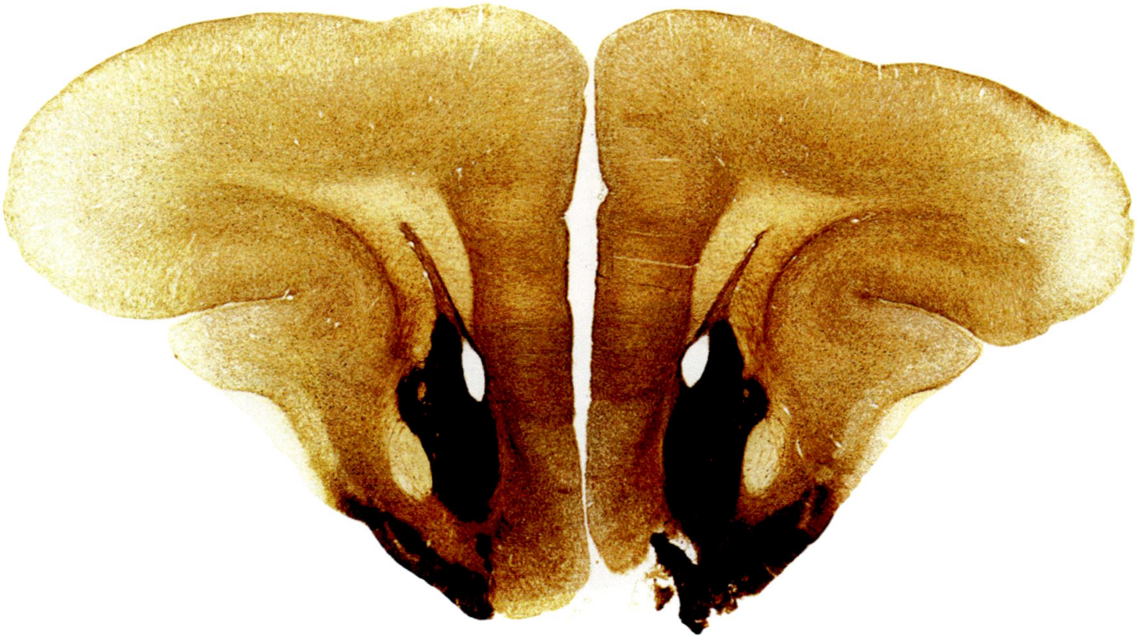

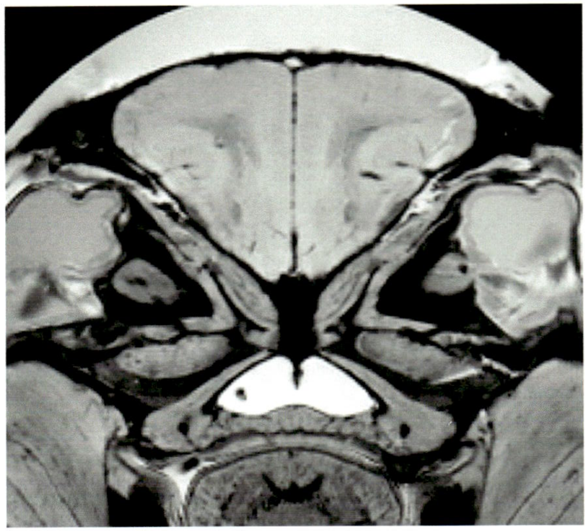

2	layer 2 of cortex
3	layer 3 of cortex
ac	anterior commissure
Acb	accumbens nucleus
Cg	cingulate cortex
Cl	claustrum
fmi	forceps minor of the corpus callosum
IL	infralimbic cortex
Ins	insular cortex
lo	lateral olfactory tract
LV	lateral ventricle
M1	primary motor cortex
M2	secondary motor cortex
Pir1	piriform cortex, layer 1
PrL	prelimbic cortex
Pu	putamen
S1	primary somatosensory cortex
Tu1	layer 1 of olfactory tubercle

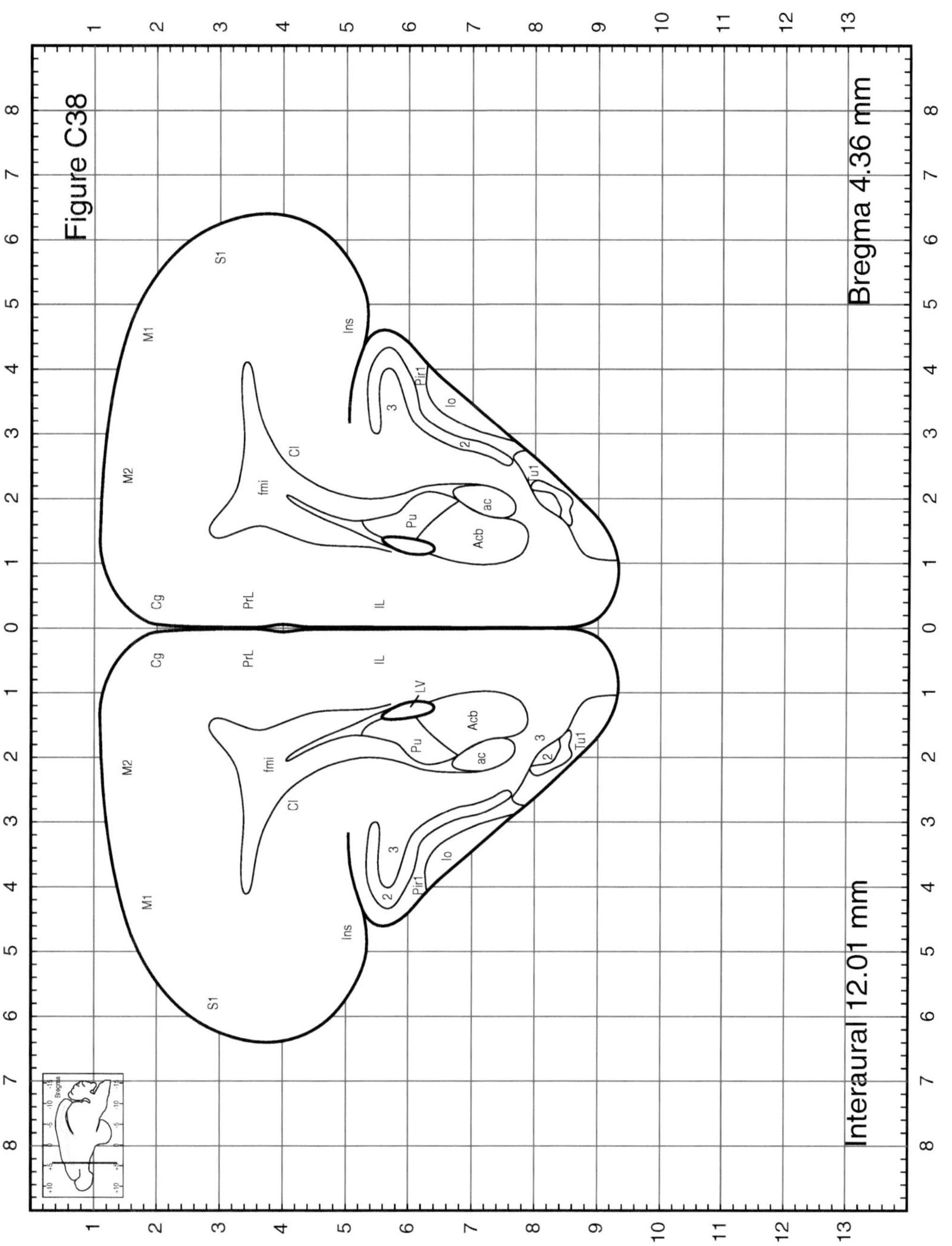
Figure C38
Bregma 4.36 mm
Interaural 12.01 mm
S1
M1
M2
Cg
PrL
IL
Ins
Cl
fmi
Pu
Acb
ac
LV
Pir1
lo
Tu1
2
3

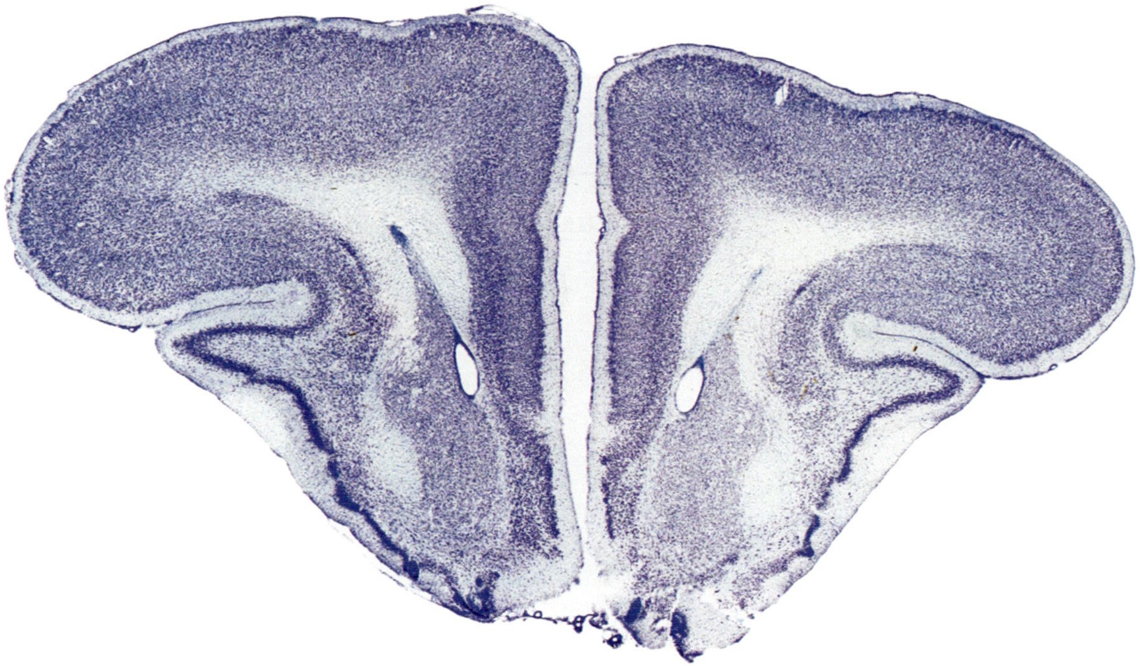

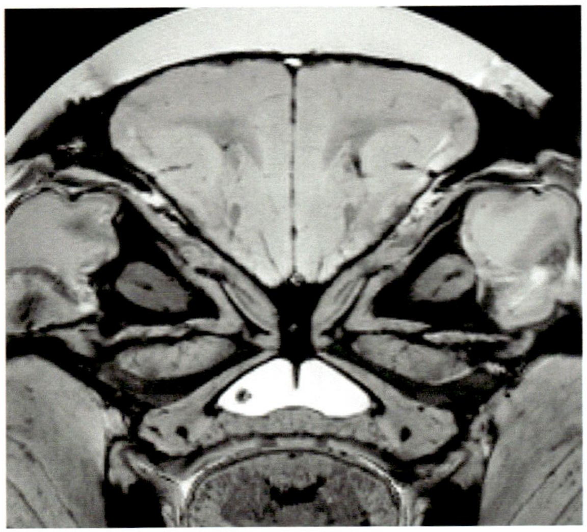

2	layer 2 of cortex
3	layer 3 of cortex
ac	anterior commissure
Acb	accumbens nucleus
Cd	caudate nucleus
Cg	cingulate cortex
Cl	claustrum
fmi	forceps minor of the corpus callosum
IL	infralimbic cortex
Ins	insular cortex
lo	lateral olfactory tract
LV	lateral ventricle
M1	primary motor cortex
M2	secondary motor cortex
Pir1	piriform cortex, layer 1
PrL	prelimbic cortex
Pu	putamen
S1	primary somatosensory cortex
Tu1	layer 1 of olfactory tubercle

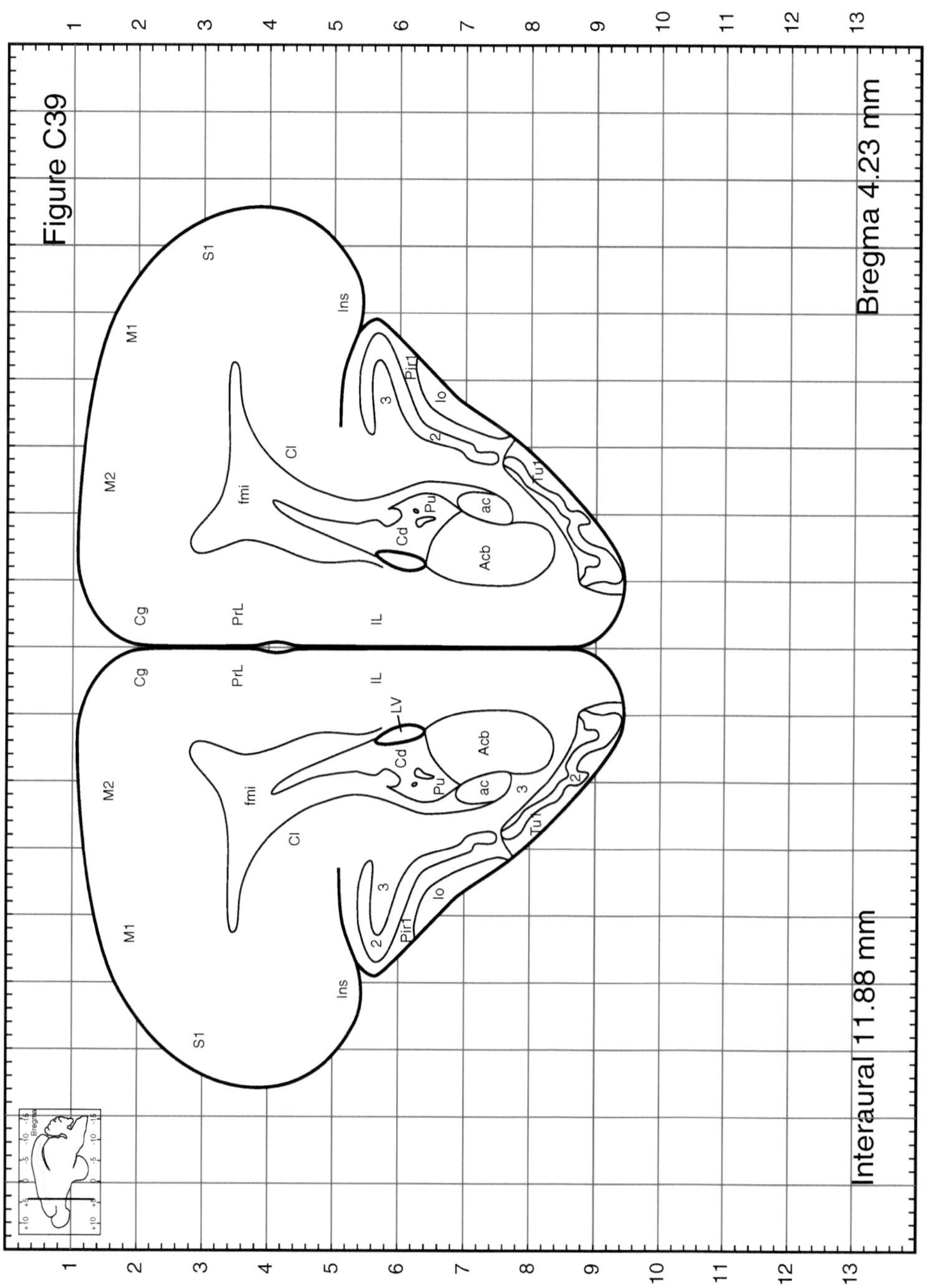
Figure C39
Bregma 4.23 mm
Interaural 11.88 mm
S1
M1
M2
Cg
PrL
IL
Ins
Cl
fmi
Cd
Pu
ac
Acb
LV
Pir1
lo
Tu1
2
3

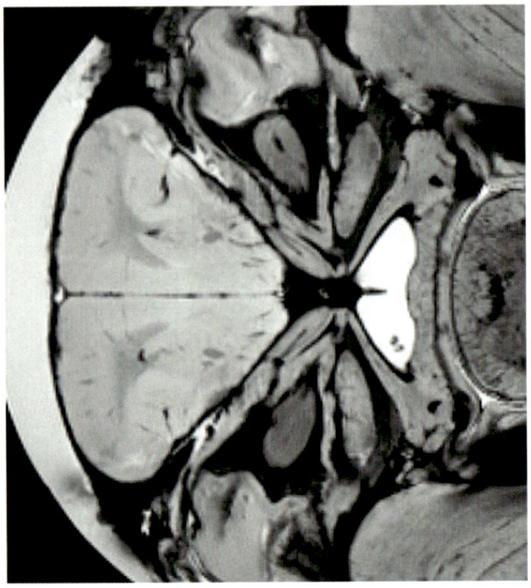

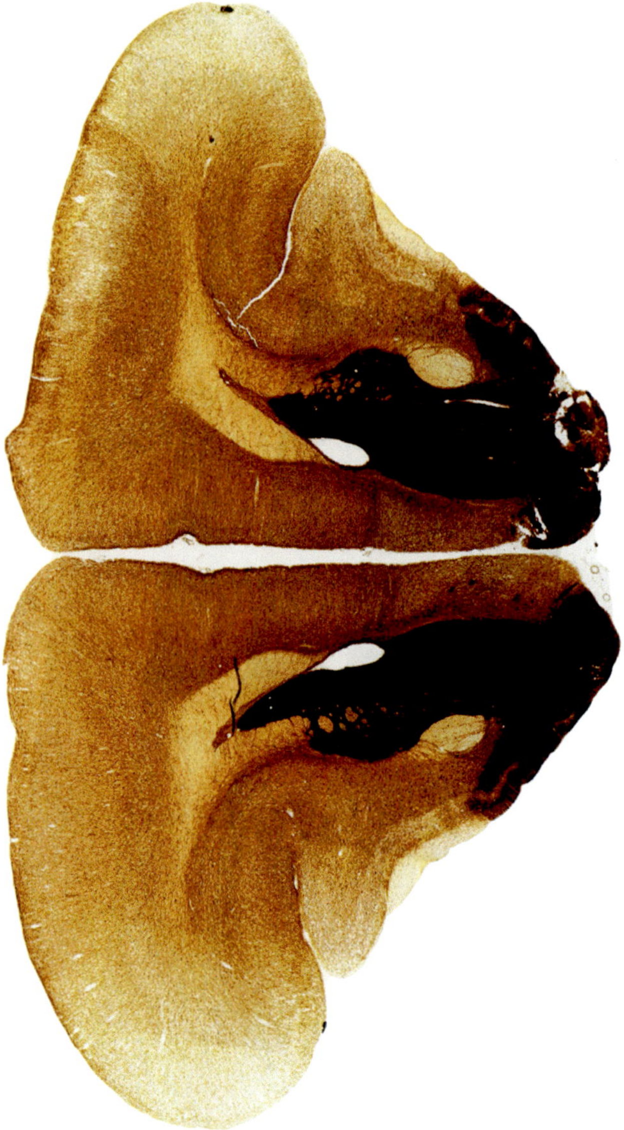

2	layer 2 of cortex
3	layer 3 of cortex
ac	anterior commissure
Acb	accumbens nucleus
Cd	caudate nucleus
Cg	cingulate cortex
Cl	claustrum
fmi	forceps minor of the corpus callosum
ic	internal capsule
IL	infralimbic cortex
Ins	insular cortex
lo	lateral olfactory tract
LV	lateral ventricle
M1	primary motor cortex
M2	secondary motor cortex
Pir1	piriform cortex, layer 1
PrL	prelimbic cortex
Pu	putamen
S1	primary somatosensory cortex
Tu1	layer 1 of olfactory tubercle

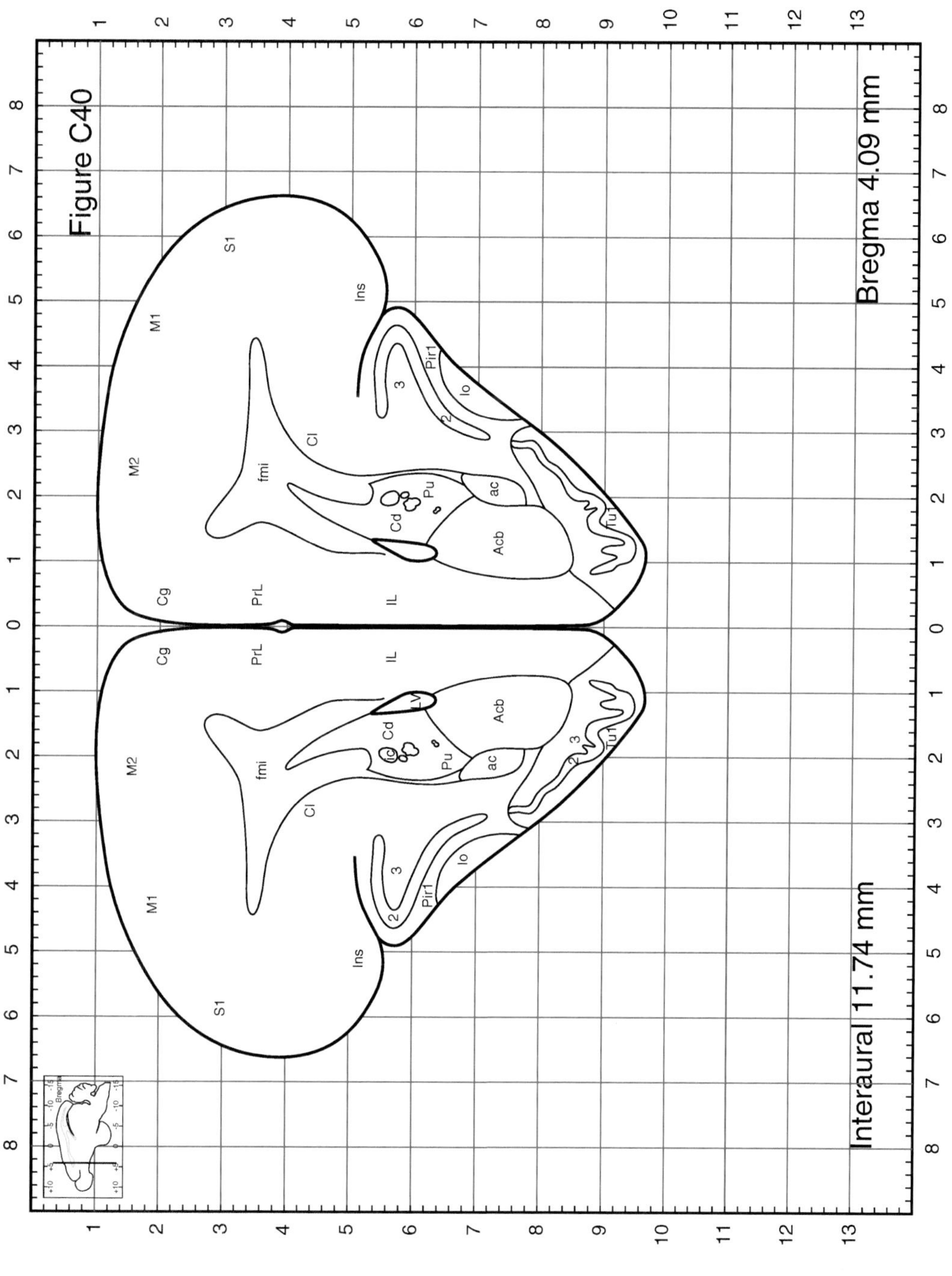
Figure C40
Bregma 4.09 mm
Interaural 11.74 mm
S1
M1
M2
Cg
PrL
IL
Ins
Cl
fmi
Cd
Pu
ac
Acb
LV
Pir1
lo
Tu1
2
3

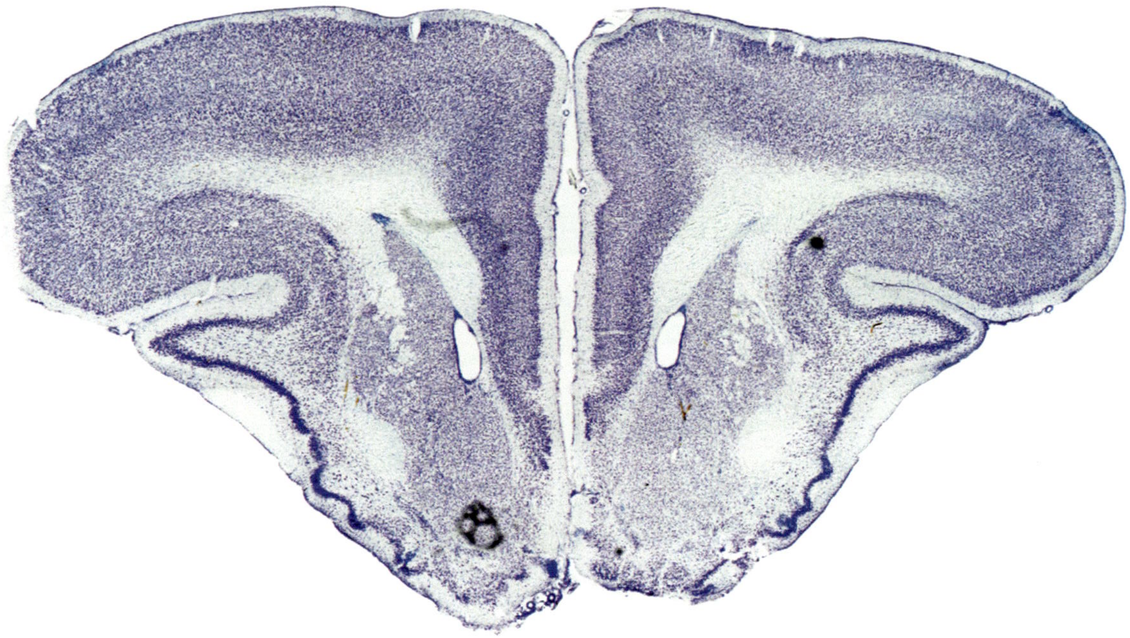

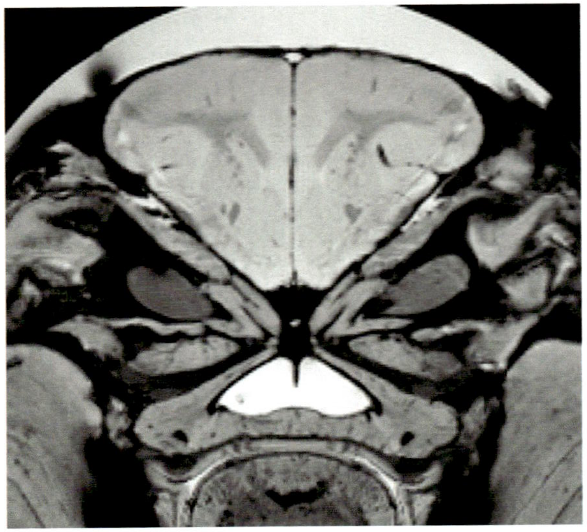

2	layer 2 of cortex		fmi	forceps minor of the corpus callosum		M1	primary motor cortex
3	layer 3 of cortex		ic	internal capsule		M2	secondary motor cortex
ac	anterior commissure		ICj	islands of Calleja		Pir1	piriform cortex, layer 1
Acb	accumbens nucleus		IL	infralimbic cortex		PrL	prelimbic cortex
Cd	caudate nucleus		Ins	insular cortex		Pu	putamen
Cg	cingulate cortex		lo	lateral olfactory tract		S1	primary somatosensory cortex
Cl	claustrum		LV	lateral ventricle		Tu1	layer 1 of olfactory tubercle

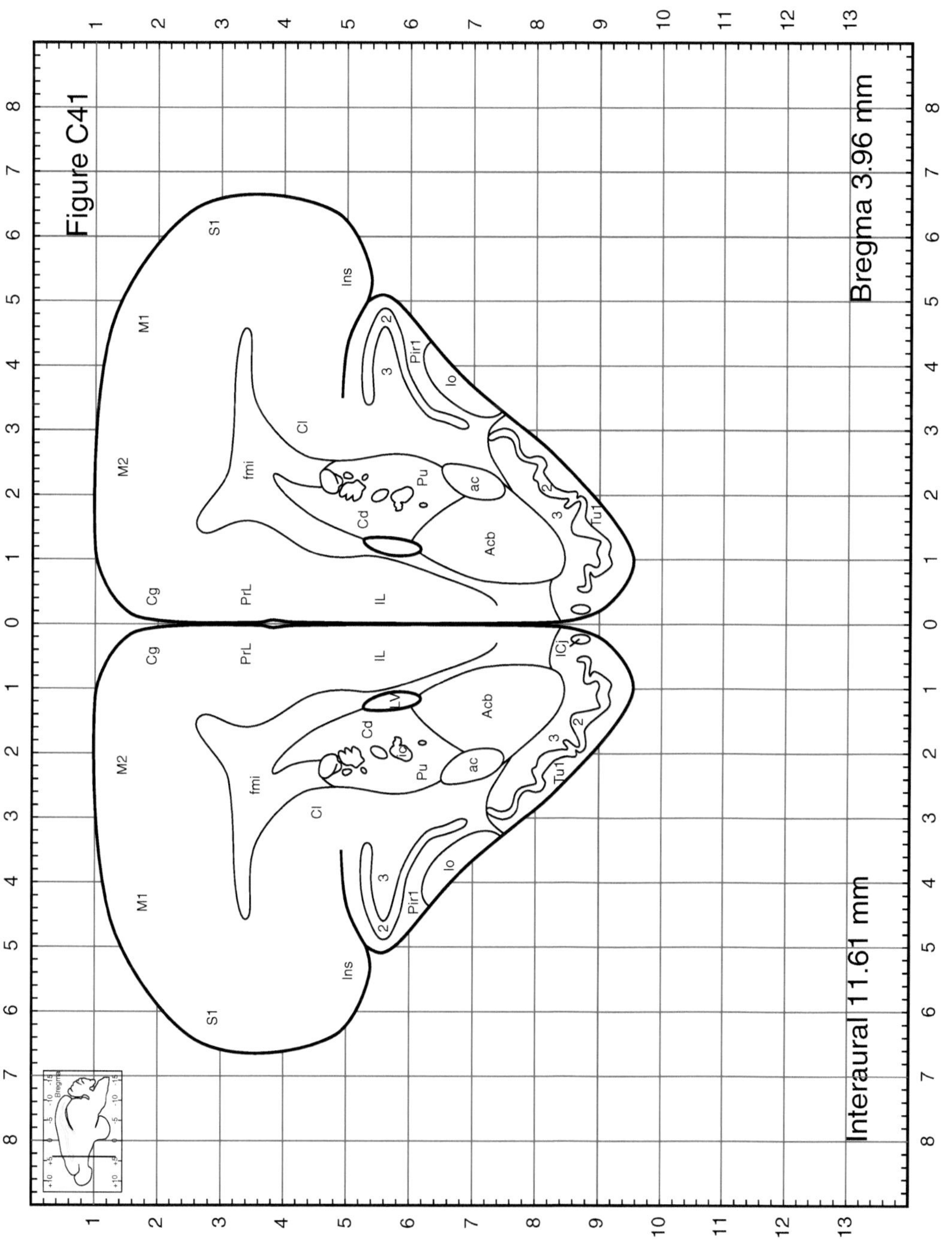
Figure C41
Bregma 3.96 mm
Interaural 11.61 mm
S1
Ins
M1
M2
Cg
PrL
IL
fmi
Cl
Cd
Pu
ac
Acb
LV
Pir1
lo
Tu1
ICj
2
3

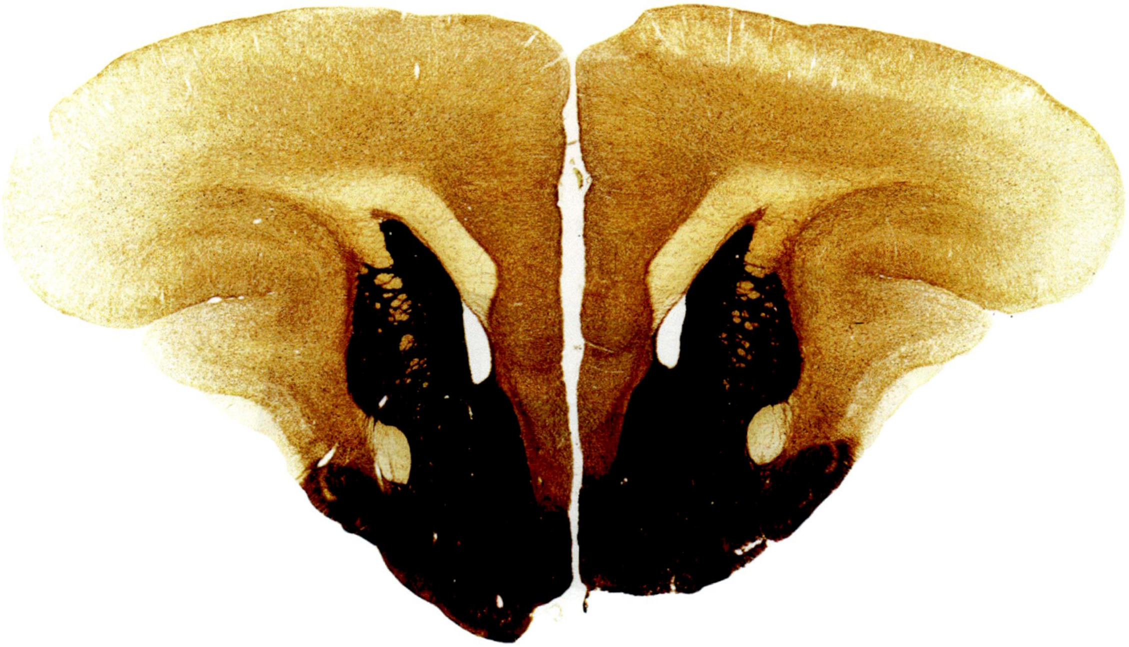

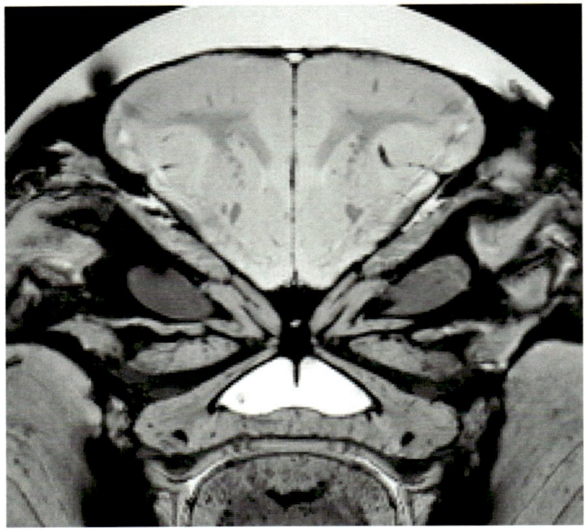

2	layer 2 of cortex
3	layer 3 of cortex
ac	anterior commissure
Acb	accumbens nucleus
Cd	caudate nucleus
Cg	cingulate cortex
Cl	claustrum
fmi	forceps minor of the corpus callosum
ic	internal capsule
ICj	islands of Calleja
IL	infralimbic cortex
Ins	insular cortex
lo	lateral olfactory tract
LV	lateral ventricle
M1	primary motor cortex
M2	secondary motor cortex
Pir1	piriform cortex, layer 1
PrL	prelimbic cortex
Pu	putamen
S1	primary somatosensory cortex
Tu1	layer 1 of olfactory tubercle

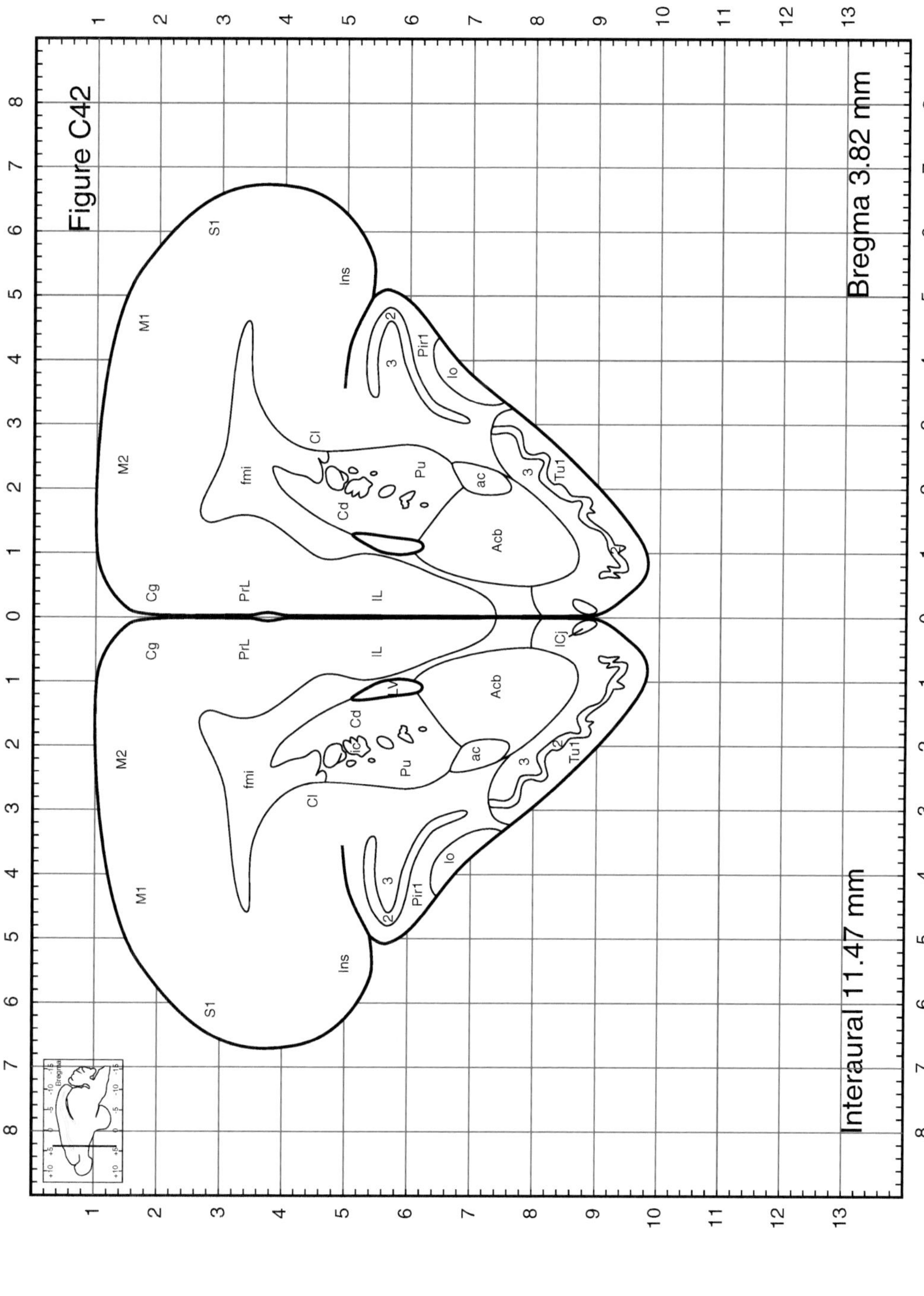

Figure C42
Bregma 3.82 mm
Interaural 11.47 mm
M1
M2
S1
Ins
Cg
PrL
IL
fmi
Cl
Cd
Pu
ac
Acb
Pir1
lo
Tu1
ICj
LV

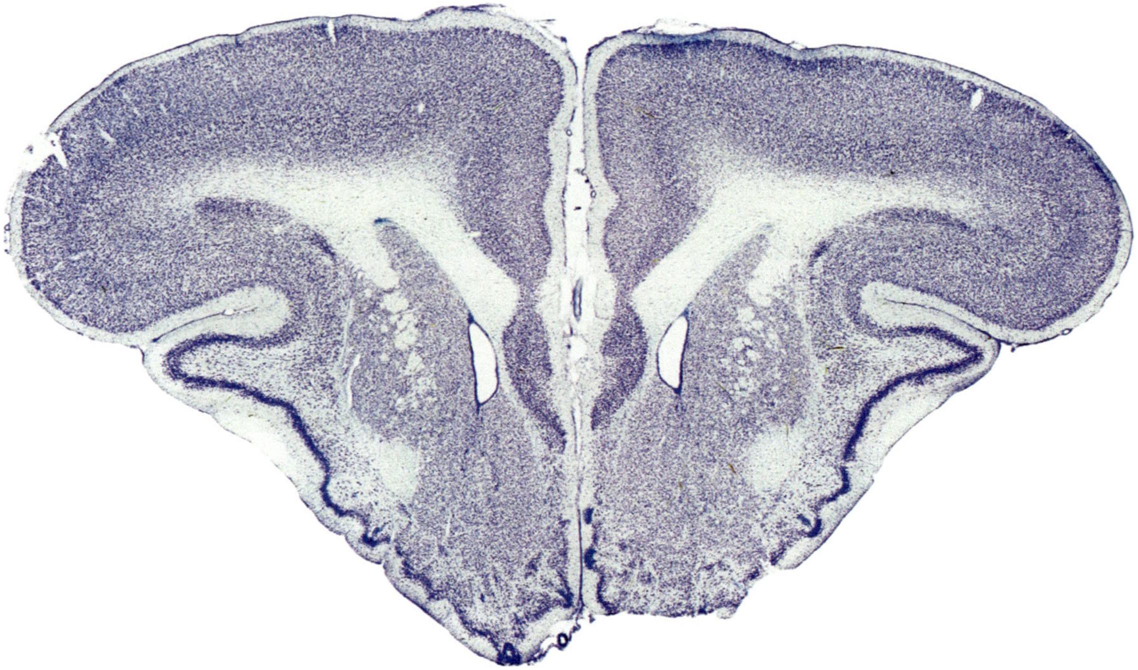

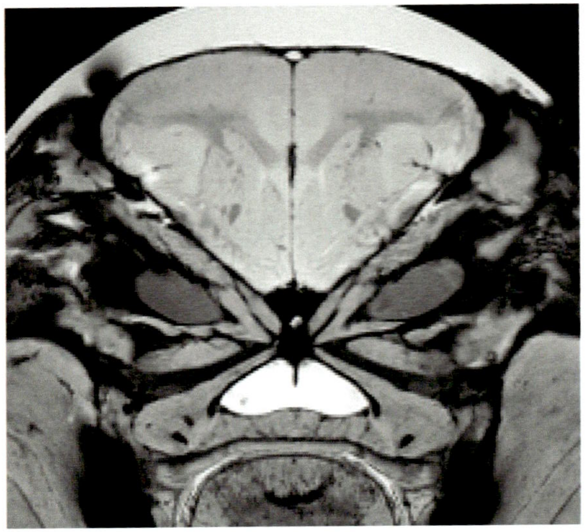

2	layer 2 of cortex
3	layer 3 of cortex
ac	anterior commissure
Acb	accumbens nucleus
Cd	caudate nucleus
Cg	cingulate cortex
Cl	claustrum
fmi	forceps minor of the corpus callosum
ic	internal capsule
ICj	islands of Calleja
IL	infralimbic cortex
Ins	insular cortex
lo	lateral olfactory tract
LV	lateral ventricle
M1	primary motor cortex
M2	secondary motor cortex
Pir1	piriform cortex, layer 1
PrL	prelimbic cortex
Pu	putamen
S1	primary somatosensory cortex
Tu1	layer 1 of olfactory tubercle

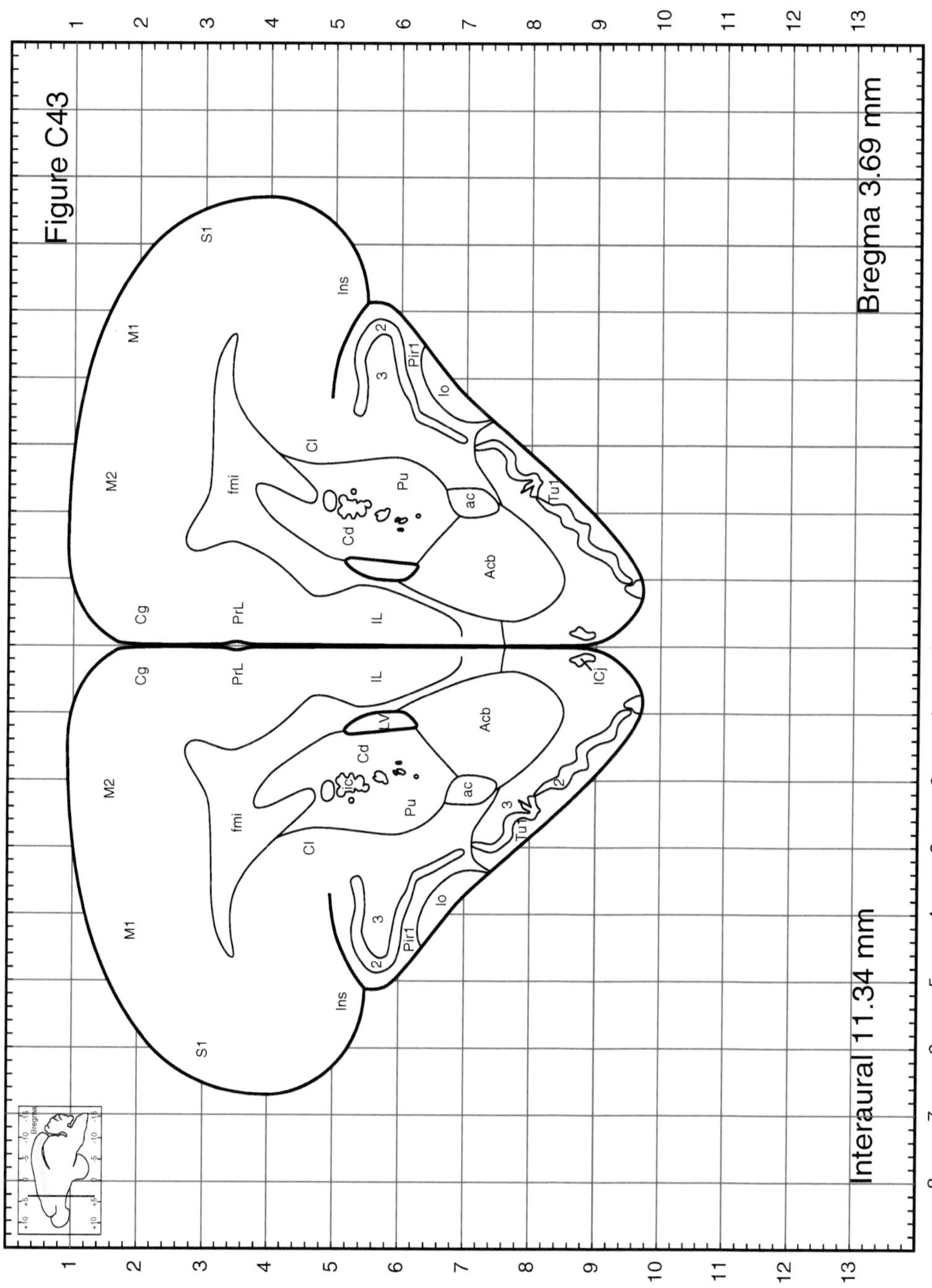
Figure C43
Bregma 3.69 mm
Interaural 11.34 mm
S1
Ins
M1
M2
Cg
PrL
IL
fmi
Cl
Cd
Pu
ic
LV
ac
Acb
Pir1
lo
Tu1
ICj
2
3

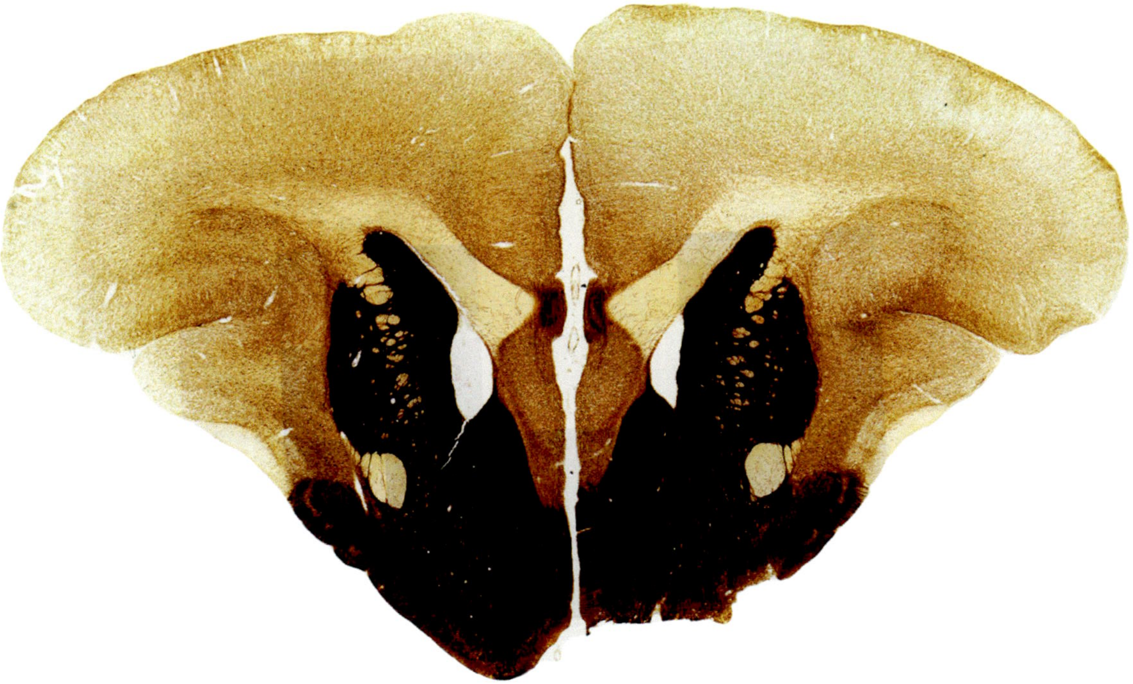

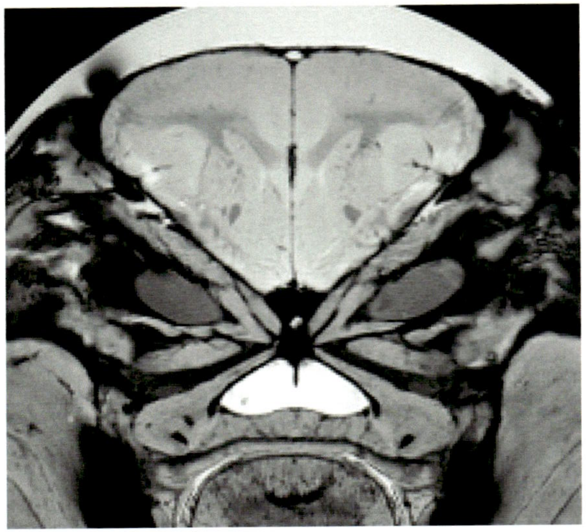

2	layer 2 of cortex
3	layer 3 of cortex
ac	anterior commissure
Acb	accumbens nucleus
Cd	caudate nucleus
Cg	cingulate cortex
Cl	claustrum
fmi	forceps minor of the corpus callosum
ic	internal capsule
ICj	islands of Calleja
IL	infralimbic cortex
Ins	insular cortex
lo	lateral olfactory tract
LV	lateral ventricle
M1	primary motor cortex
M2	secondary motor cortex
Pir1	piriform cortex, layer 1
PrL	prelimbic cortex
Pu	putamen
S1	primary somatosensory cortex
Tu1	layer 1 of olfactory tubercle

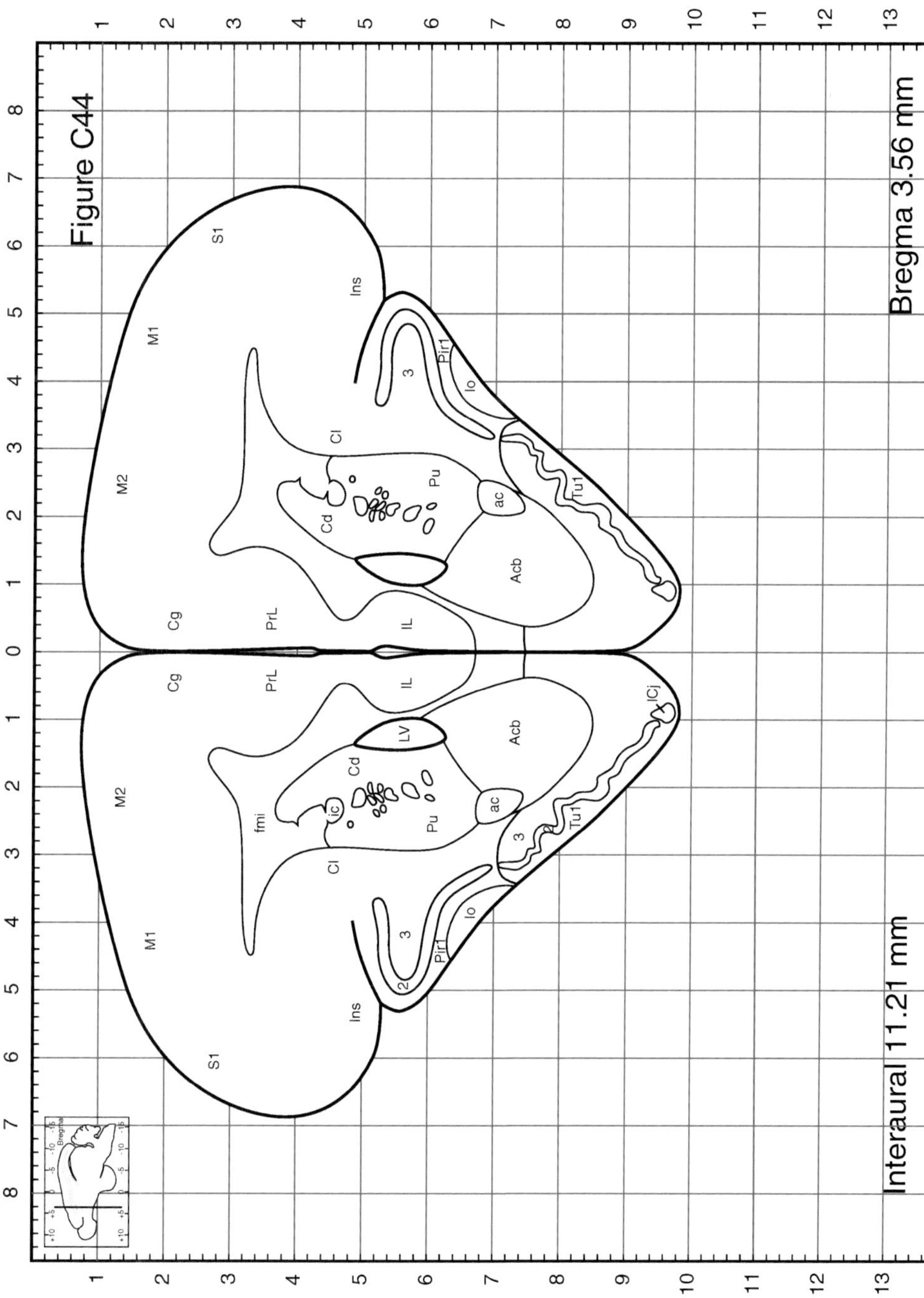

Figure C44
Bregma 3.56 mm
Interaural 11.21 mm
S1
M1
M2
Cg
PrL
IL
Ins
Cl
Cd
Pu
ac
Acb
Tu1
3
Pir1
lo
fmi
ic
LV
ICj
2

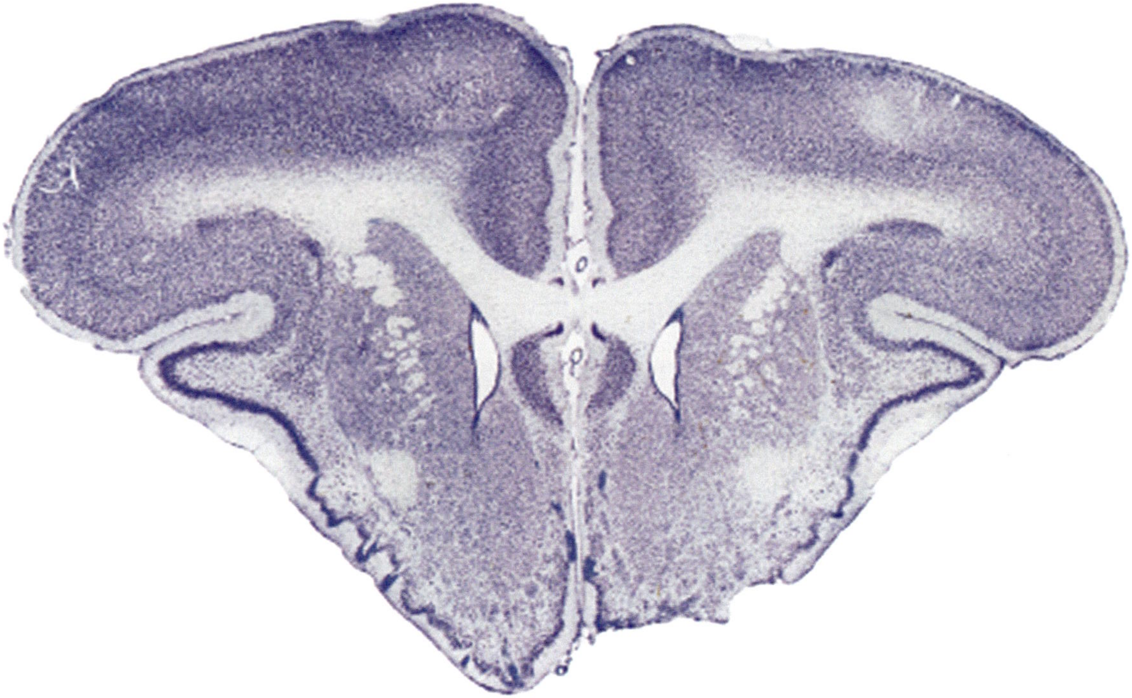

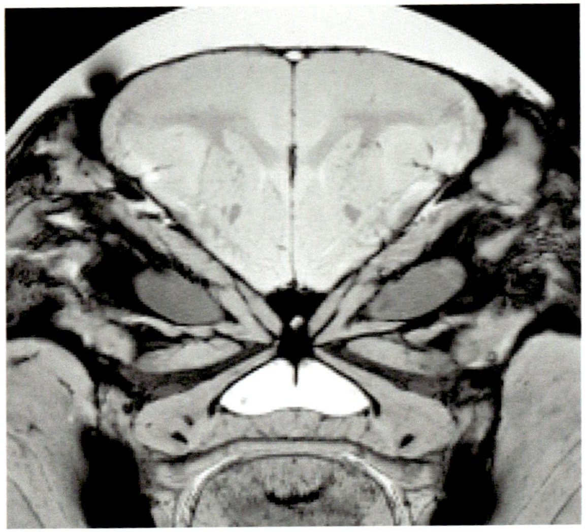

2	layer 2 of cortex
3	layer 3 of cortex
ac	anterior commissure
AcbC	accumbens nucleus, core
AcbSh	accumbens nucleus, shell
cc	corpus callosum
Cd	caudate nucleus
Cg	cingulate cortex
Cl	claustrum
DP	dorsal peduncular cortex
ic	internal capsule
ICj	islands of Calleja
IG	indusium griseum
Ins	insular cortex
lo	lateral olfactory tract
LV	lateral ventricle
M1	primary motor cortex
M2	secondary motor cortex
Pir1	piriform cortex, layer 1
Pu	putamen
S1	primary somatosensory cortex
Tu1	layer 1 of olfactory tubercle

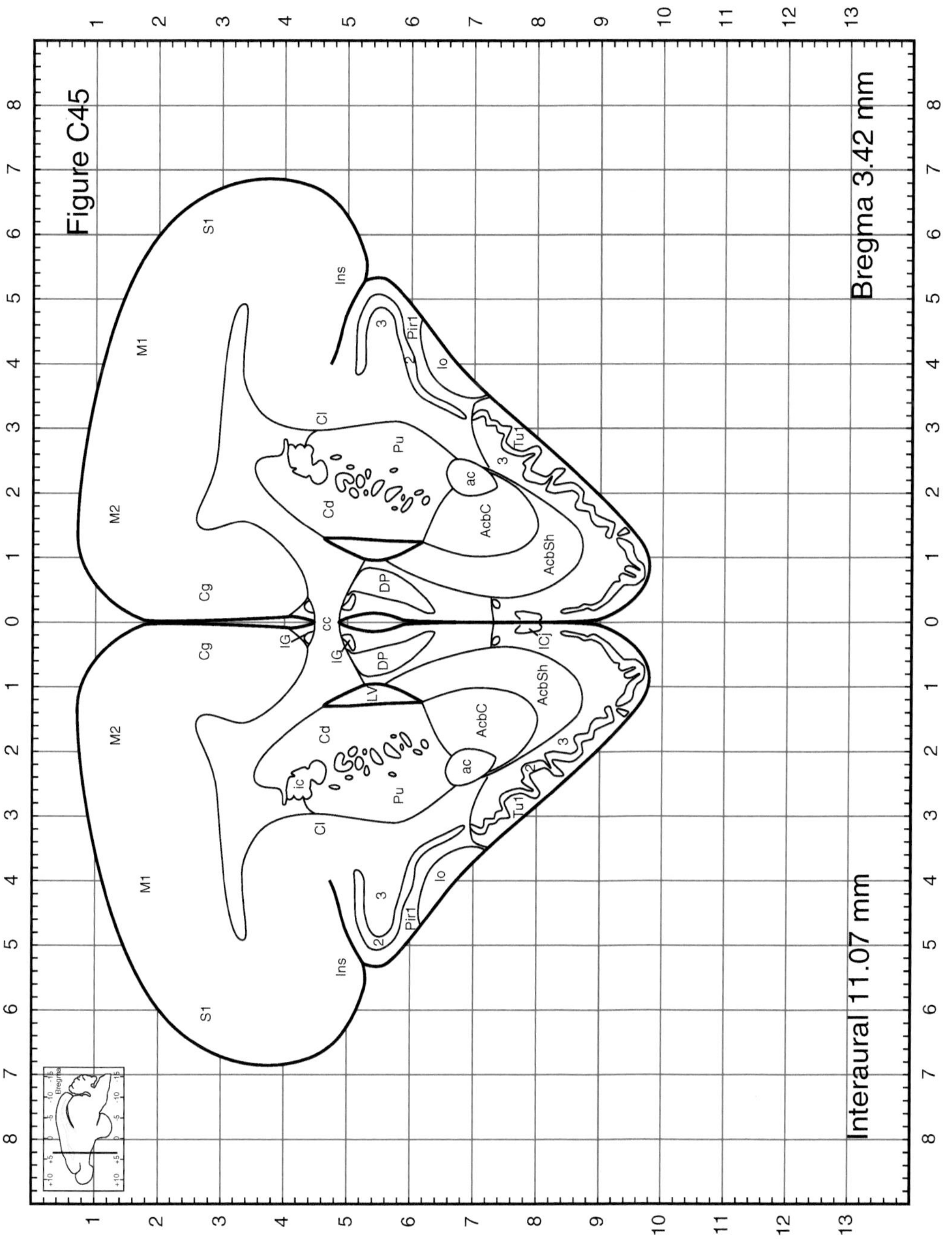
Figure C45
Bregma 3.42 mm
Interaural 11.07 mm
S1
Ins
M1
M2
Cg
Cl
Pu
Cd
ic
ac
AcbC
AcbSh
DP
IG
cc
LV
ICj
Tu1
Pir1
lo
3
2

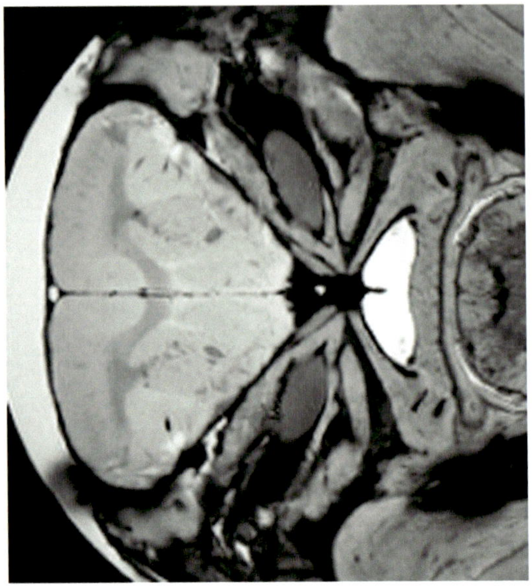

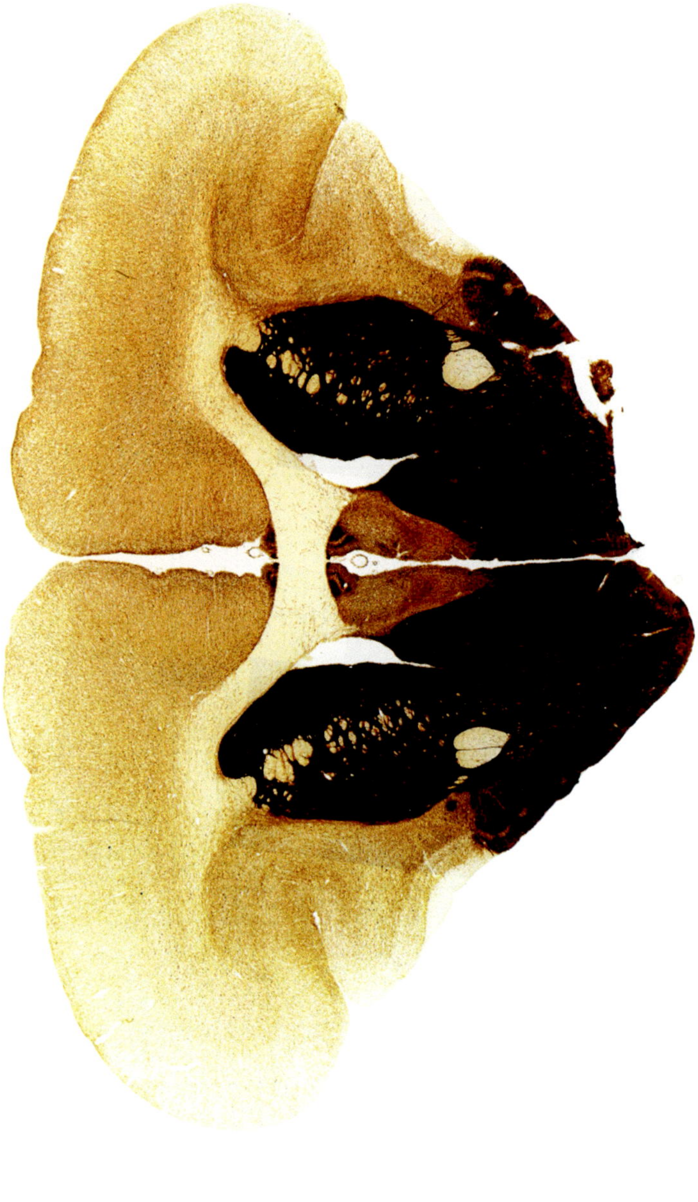

2	layer 2 of cortex
3	layer 3 of cortex
ac	anterior commissure
AcbC	accumbens nucleus, core
AcbSh	accumbens nucleus, shell
cc	corpus callosum
Cd	caudate nucleus
Cg	cingulate cortex
Cl	claustrum
DP	dorsal peduncular cortex
ic	internal capsule
ICj	islands of Calleja
IG	indusium griseum
Ins	insular cortex
lo	lateral olfactory tract
LV	lateral ventricle
M1	primary motor cortex
M2	secondary motor cortex
Pir1	piriform cortex, layer 1
Pu	putamen
S1	primary somatosensory cortex
Tu1	layer 1 of olfactory tubercle

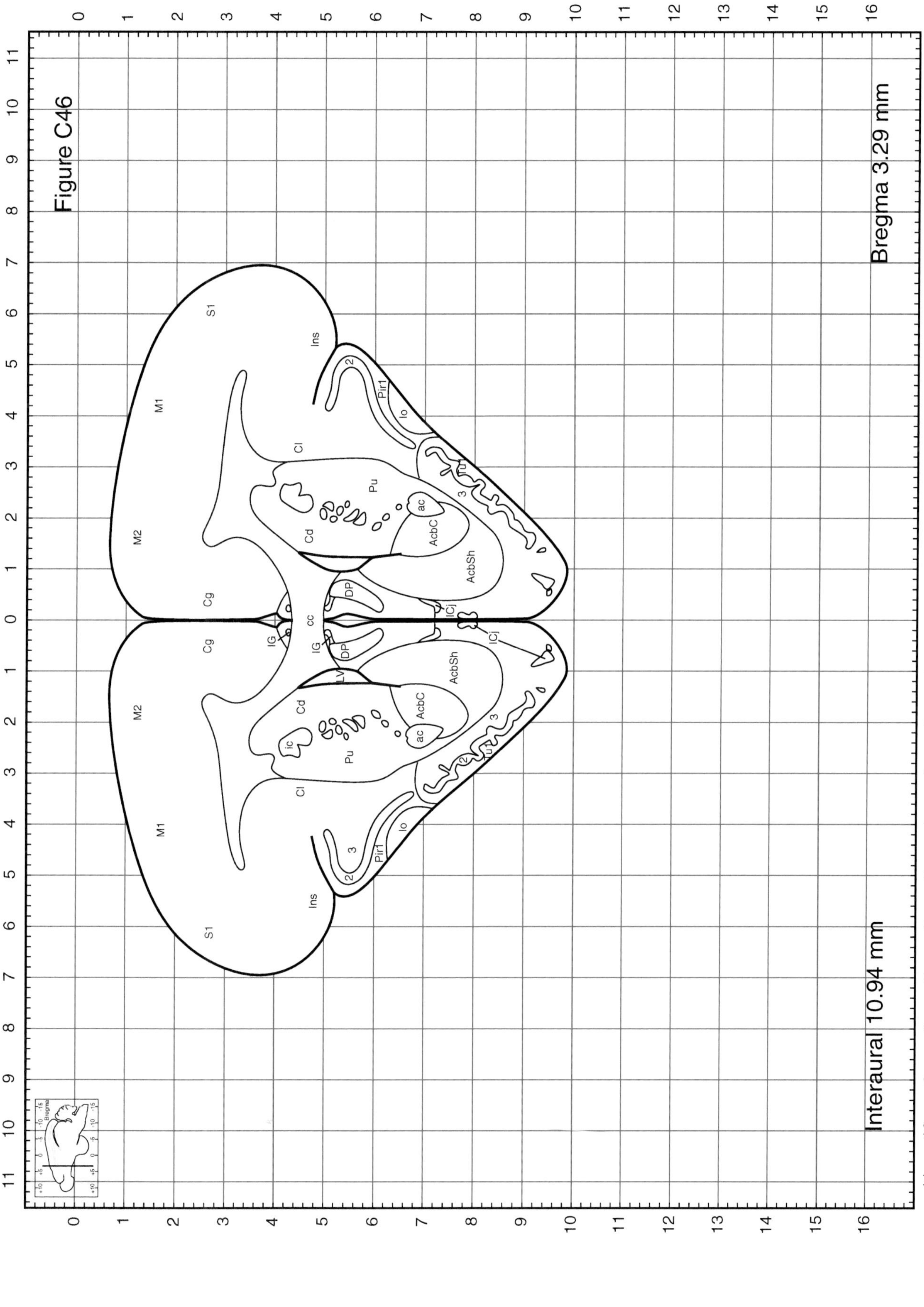
Figure C46
Bregma 3.29 mm
Interaural 10.94 mm
S1
Ins
M1
M2
Cg
IG
cc
Cl
Cd
Pu
ic
ac
AcbC
AcbSh
DP
LV
Pir1
lo
Tu
ICj
2
3

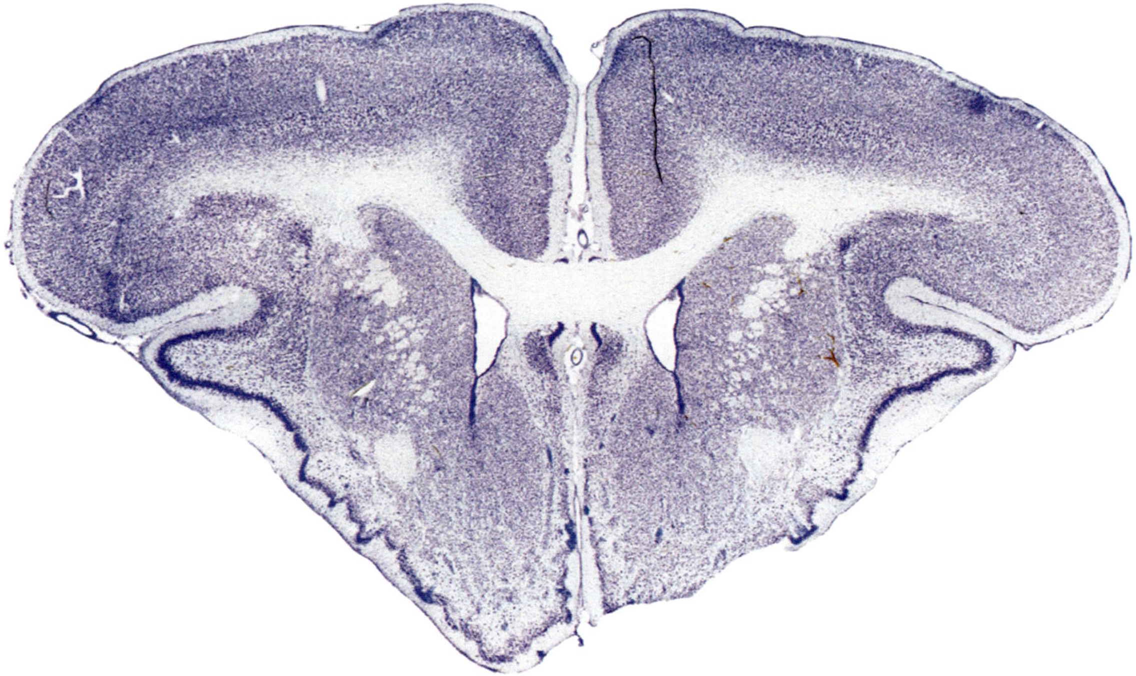

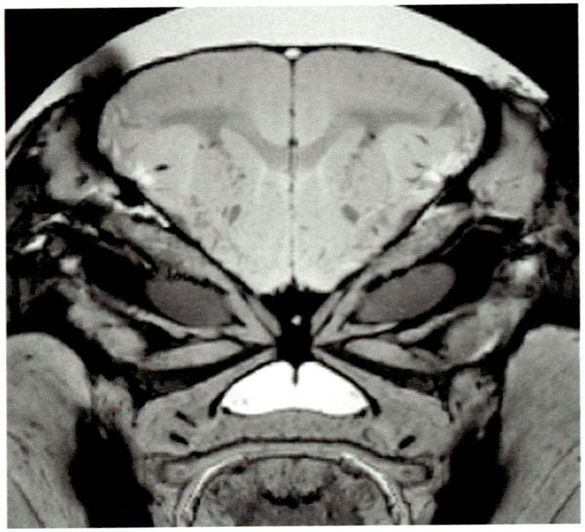

2	layer 2 of cortex
3	layer 3 of cortex
ac	anterior commissure
AcbC	accumbens nucleus, core
AcbSh	accumbens nucleus, shell
cc	corpus callosum
Cd	caudate nucleus
Cg	cingulate cortex
Cl	claustrum
DP	dorsal peduncular cortex
ic	internal capsule
ICj	islands of Calleja
IG	indusium griseum
Ins	insular cortex
lo	lateral olfactory tract
LV	lateral ventricle
M1	primary motor cortex
M2	secondary motor cortex
Pir1	piriform cortex, layer 1
Pu	putamen
S1	primary somatosensory cortex
Tu1	layer 1 of olfactory tubercle

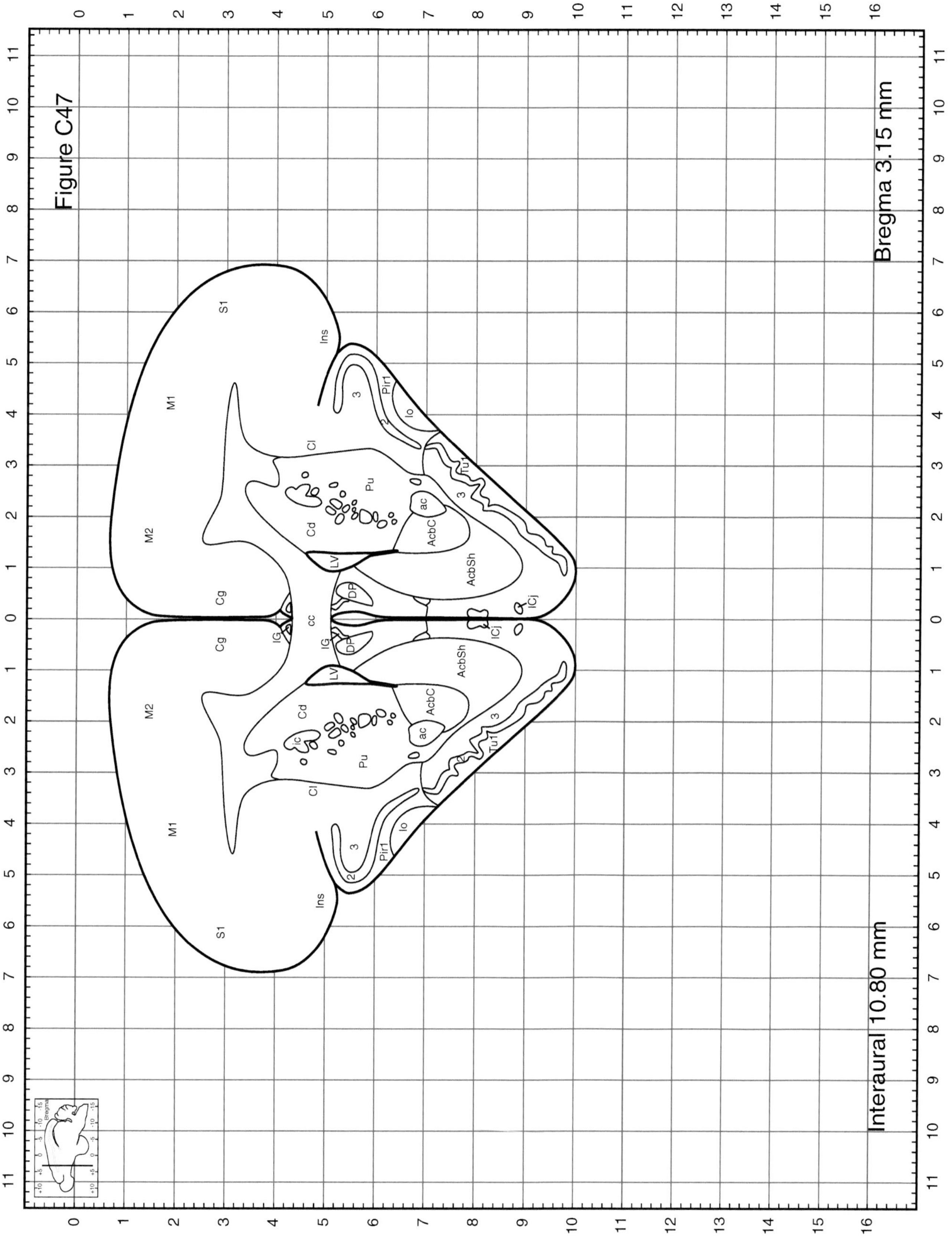
Figure C47
Bregma 3.15 mm
Interaural 10.80 mm
S1
Ins
M1
M2
Cg
cc
IG
Cl
Pu
Cd
ic
LV
DP
ac
AcbC
AcbSh
Pir1
lo
Tu1
ICj
3
2

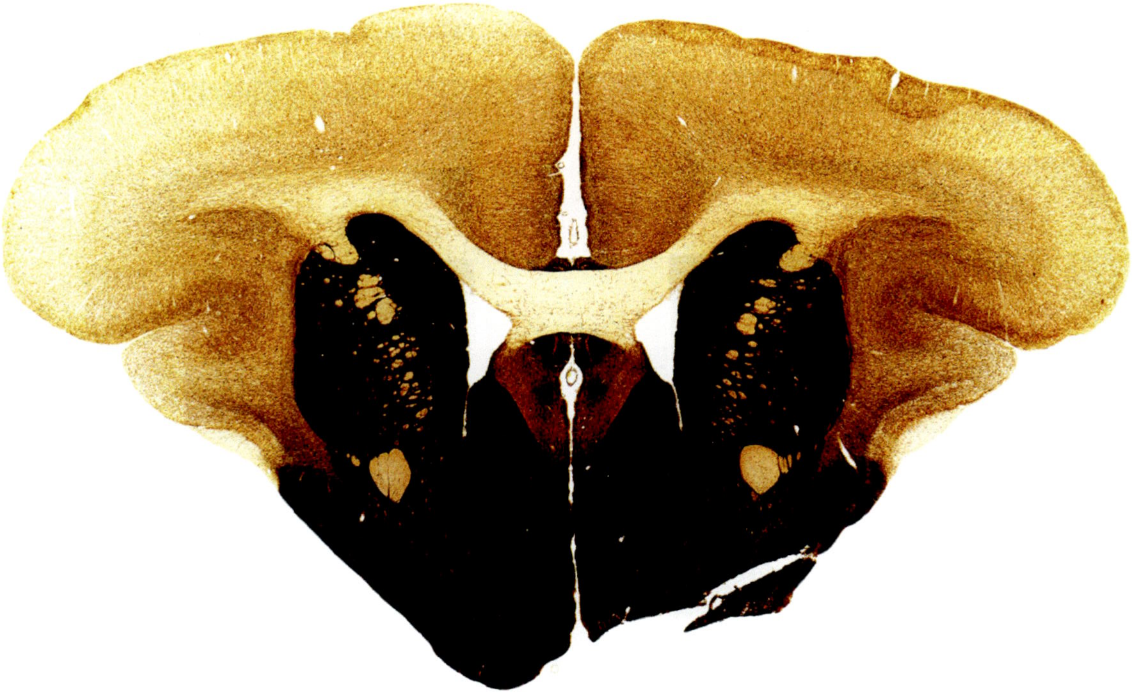

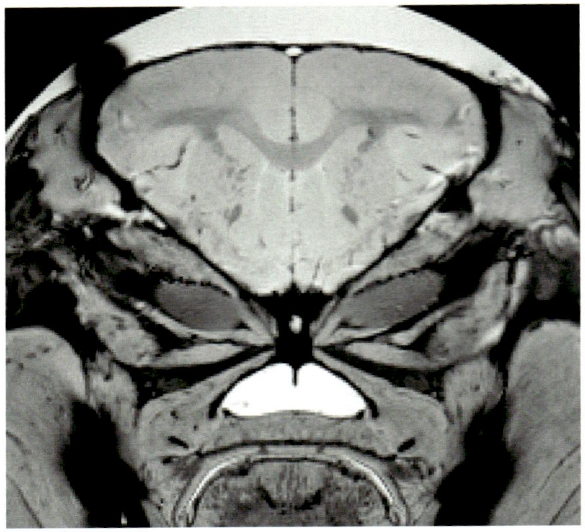

2	layer 2 of cortex
3	layer 3 of cortex
ac	anterior commissure
AcbC	accumbens nucleus, core
AcbSh	accumbens nucleus, shell
cc	corpus callosum
Cd	caudate nucleus
Cg	cingulate cortex
Cl	claustrum
DP	dorsal peduncular cortex
ic	internal capsule
ICj	islands of Calleja
IG	indusium griseum
Ins	insular cortex
lo	lateral olfactory tract
LV	lateral ventricle
M1	primary motor cortex
M2	secondary motor cortex
Pir1	piriform cortex, layer 1
Pu	putamen
S1	primary somatosensory cortex
Tu1	layer 1 of olfactory tubercle

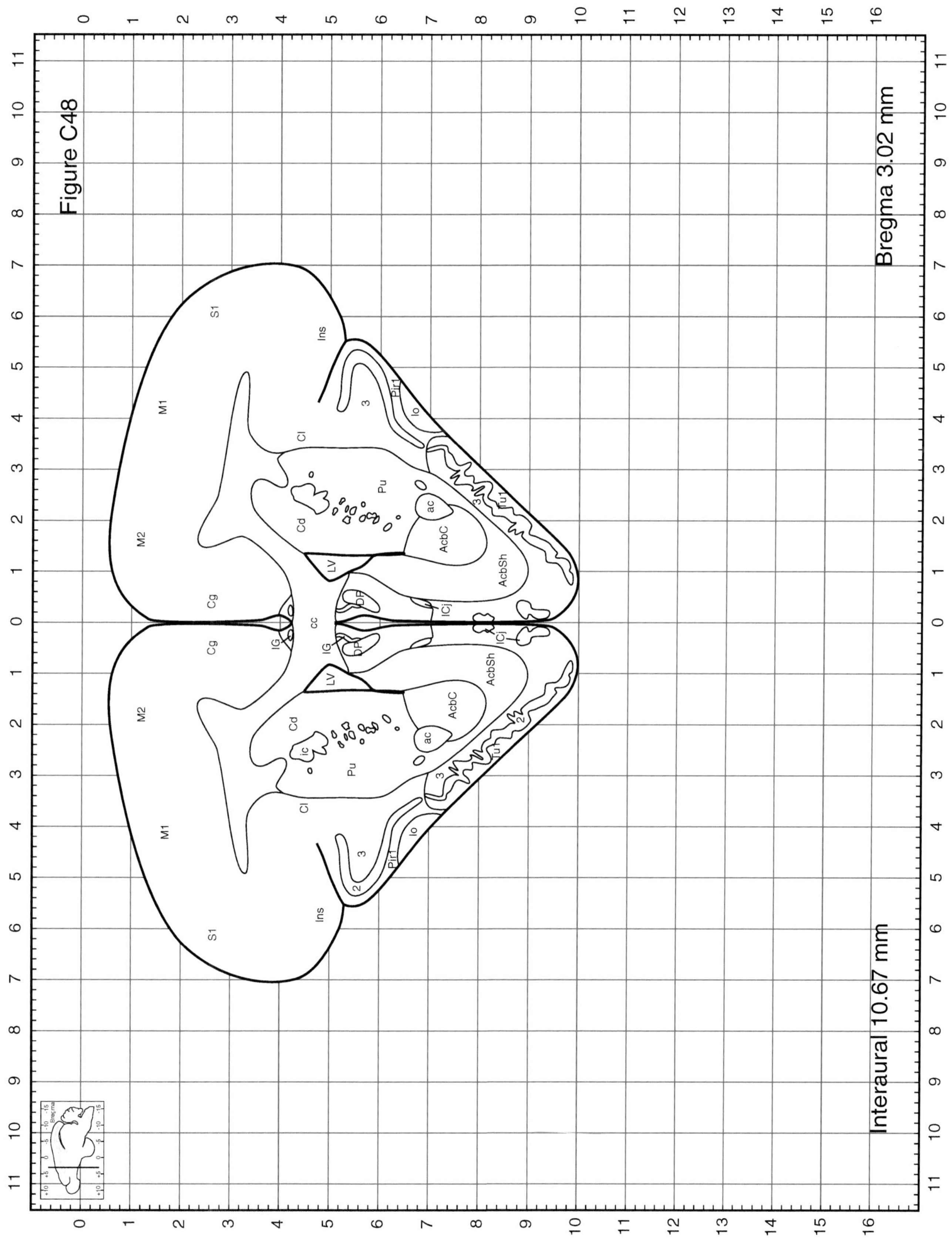
Figure C48
Bregma 3.02 mm
Interaural 10.67 mm
S1
Ins
M1
M2
Cg
Cl
Cd
Pu
ic
ac
AcbC
AcbSh
LV
cc
IG
DP
ICj
Pir1
lo
Tu1
3
2

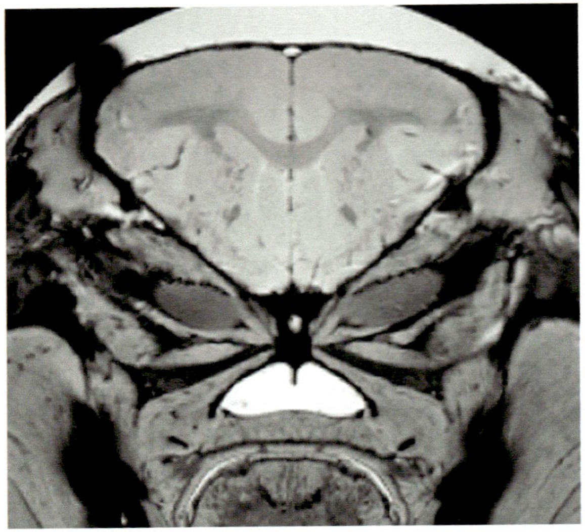

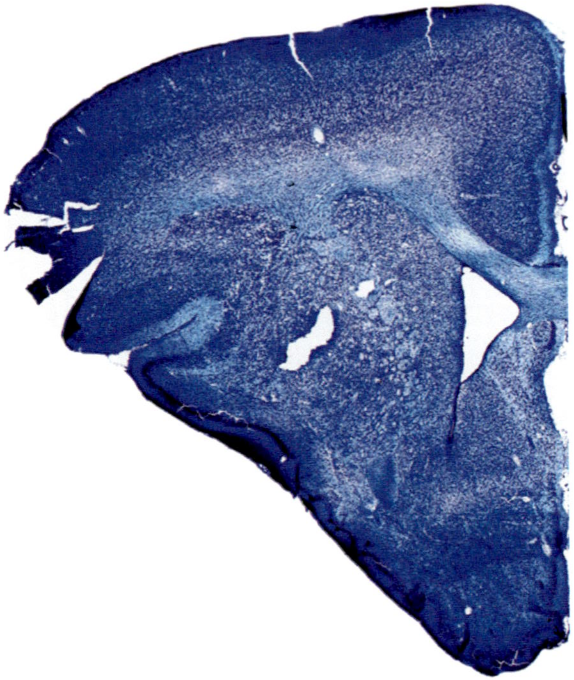

2	layer 2 of cortex
3	layer 3 of cortex
ac	anterior commissure
AcbC	accumbens nucleus, core
AcbSh	accumbens nucleus, shell
cc	corpus callosum
Cd	caudate nucleus
Cg	cingulate cortex
Cl	claustrum
DP	dorsal peduncular cortex
ic	internal capsule
ICj	islands of Calleja
IG	indusium griseum
Ins	insular cortex
lo	lateral olfactory tract
LSi	lateral septal nucleus, intermediate part
LV	lateral ventricle
M1	primary motor cortex
M2	secondary motor cortex
Pir1	piriform cortex, layer 1
Pu	putamen
S1	primary somatosensory cortex
Tu1	layer 1 of olfactory tubercle

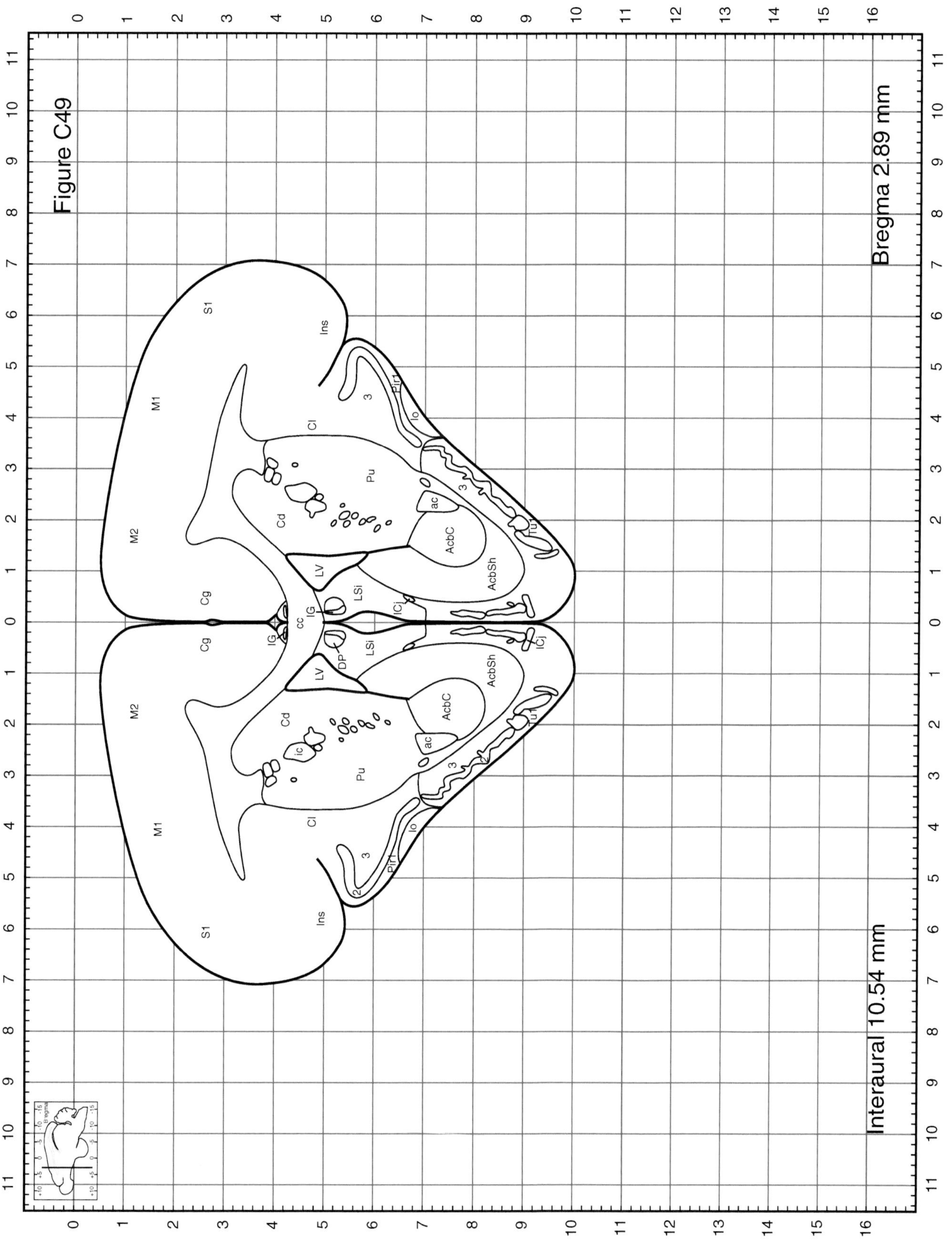
Figure C49
Bregma 2.89 mm
Interaural 10.54 mm
S1
Ins
M1
M2
Cg
Cl
Cd
Pu
ac
ic
AcbC
AcbSh
LV
LSi
ICj
IG
cc
DP
Pir1
lo
Tu
3
2

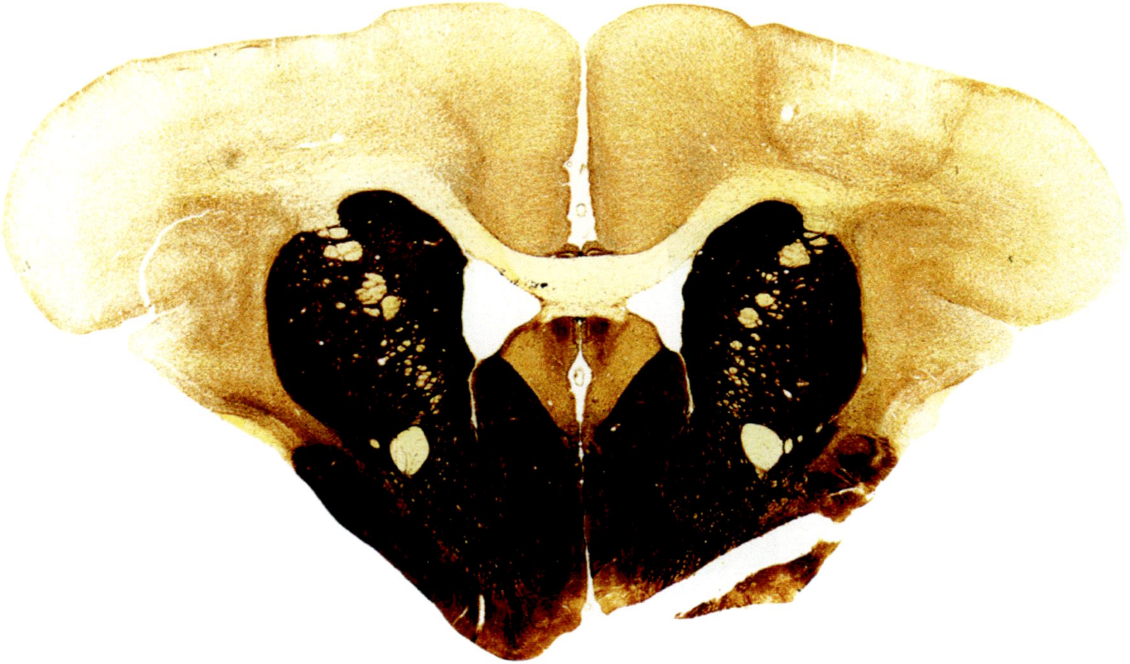

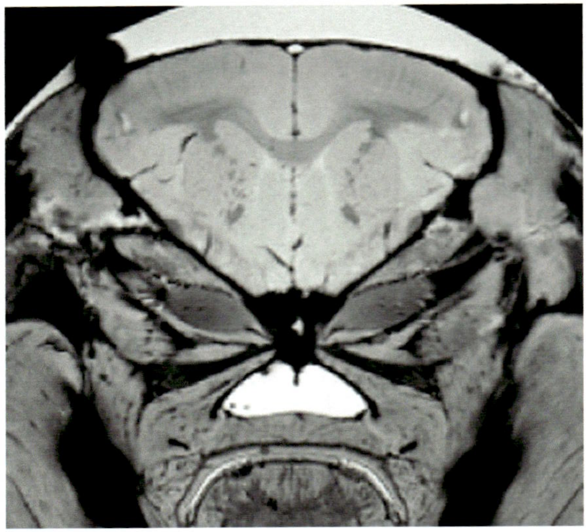

2	layer 2 of cortex
3	layer 3 of cortex
ac	anterior commissure
AcbC	accumbens nucleus, core
AcbSh	accumbens nucleus, shell
cc	corpus callosum
Cd	caudate nucleus
Cg	cingulate cortex
Cl	claustrum
DP	dorsal peduncular cortex
ic	internal capsule
ICj	islands of Calleja
IG	indusium griseum
Ins	insular cortex
lo	lateral olfactory tract
LSi	lateral septal nucleus, intermediate part
LV	lateral ventricle
M1	primary motor cortex
M2	secondary motor cortex
Pir1	piriform cortex, layer 1
Pu	putamen
S1	primary somatosensory cortex
Tu1	layer 1 of olfactory tubercle

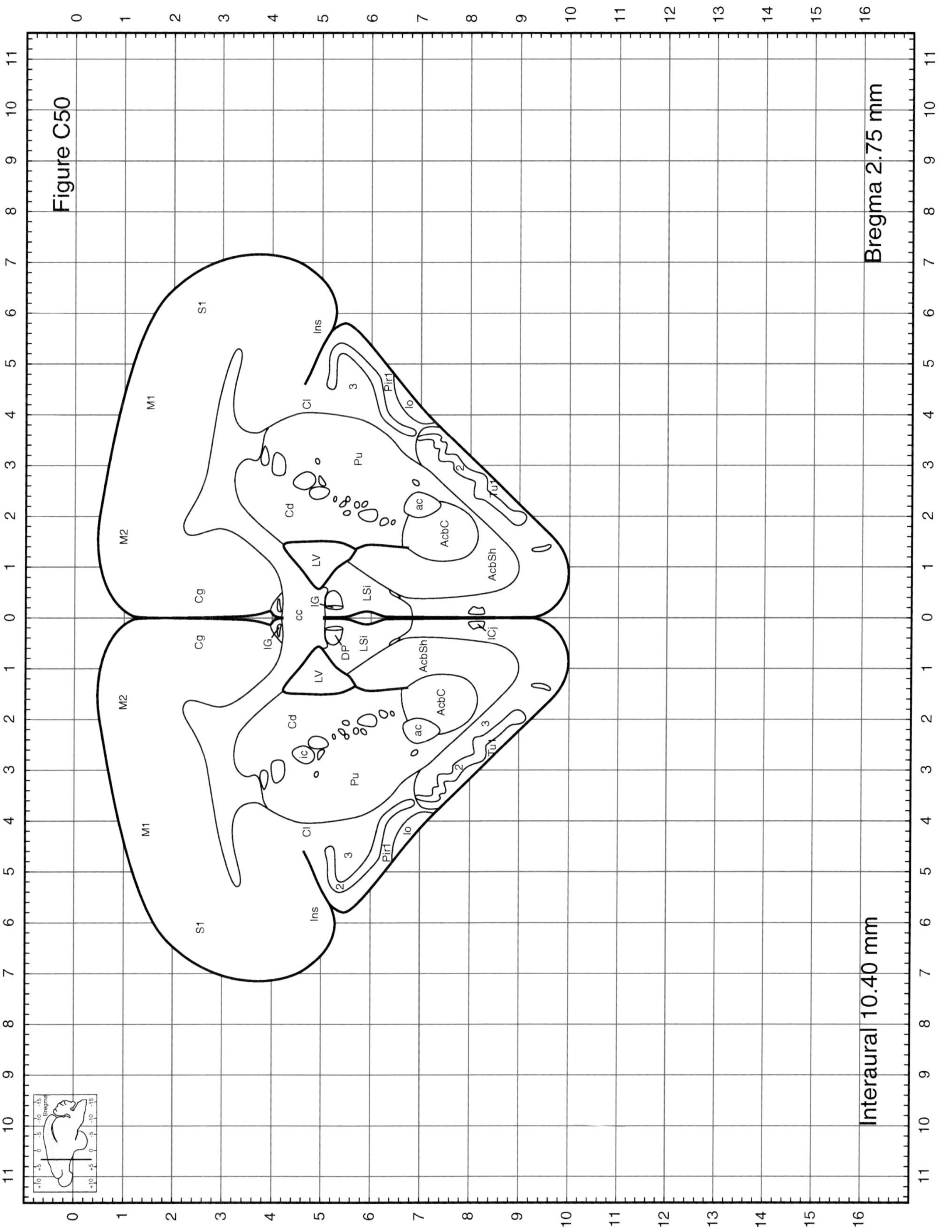
Figure C50
Bregma 2.75 mm
Interaural 10.40 mm
S1
Ins
M1
M2
Cg
IG
cc
LV
DP
LSi
Cd
Pu
Cl
3
2
Pir1
lo
ac
ic
AcbC
AcbSh
Tu1
ICj

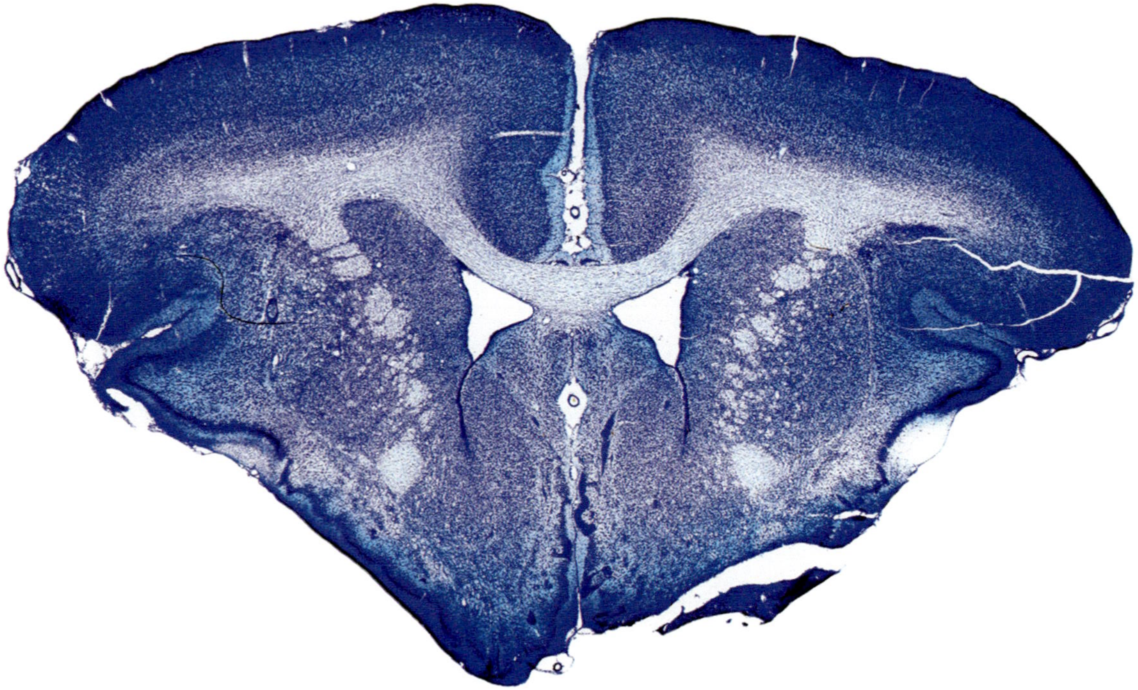

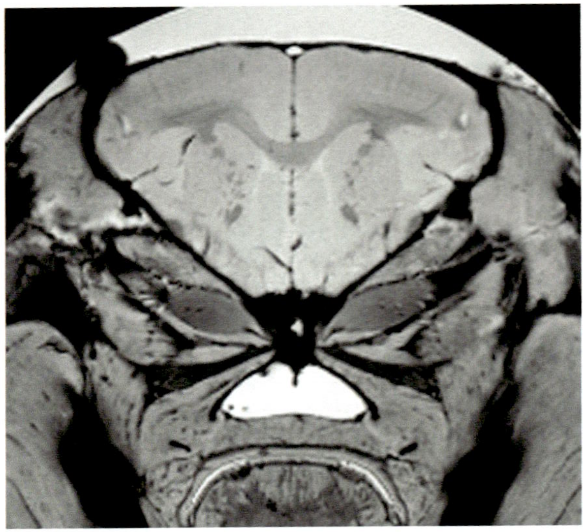

2	layer 2 of cortex
3	layer 3 of cortex
ac	anterior commissure
AcbC	accumbens nucleus, core
AcbSh	accumbens nucleus, shell
cc	corpus callosum
Cd	caudate nucleus
Cg	cingulate cortex
Cl	claustrum
ic	internal capsule
ICj	islands of Calleja
Ins	insular cortex
lo	lateral olfactory tract
LSd	lateral septal nucleus, dorsal part
LSi	lateral septal nucleus, intermediate part
LV	lateral ventricle
M1	primary motor cortex
M2	secondary motor cortex
Pir1	piriform cortex, layer 1
Pu	putamen
S1	primary somatosensory cortex
SHi	septohippocampal nucleus
Tu1	layer 1 of olfactory tubercle

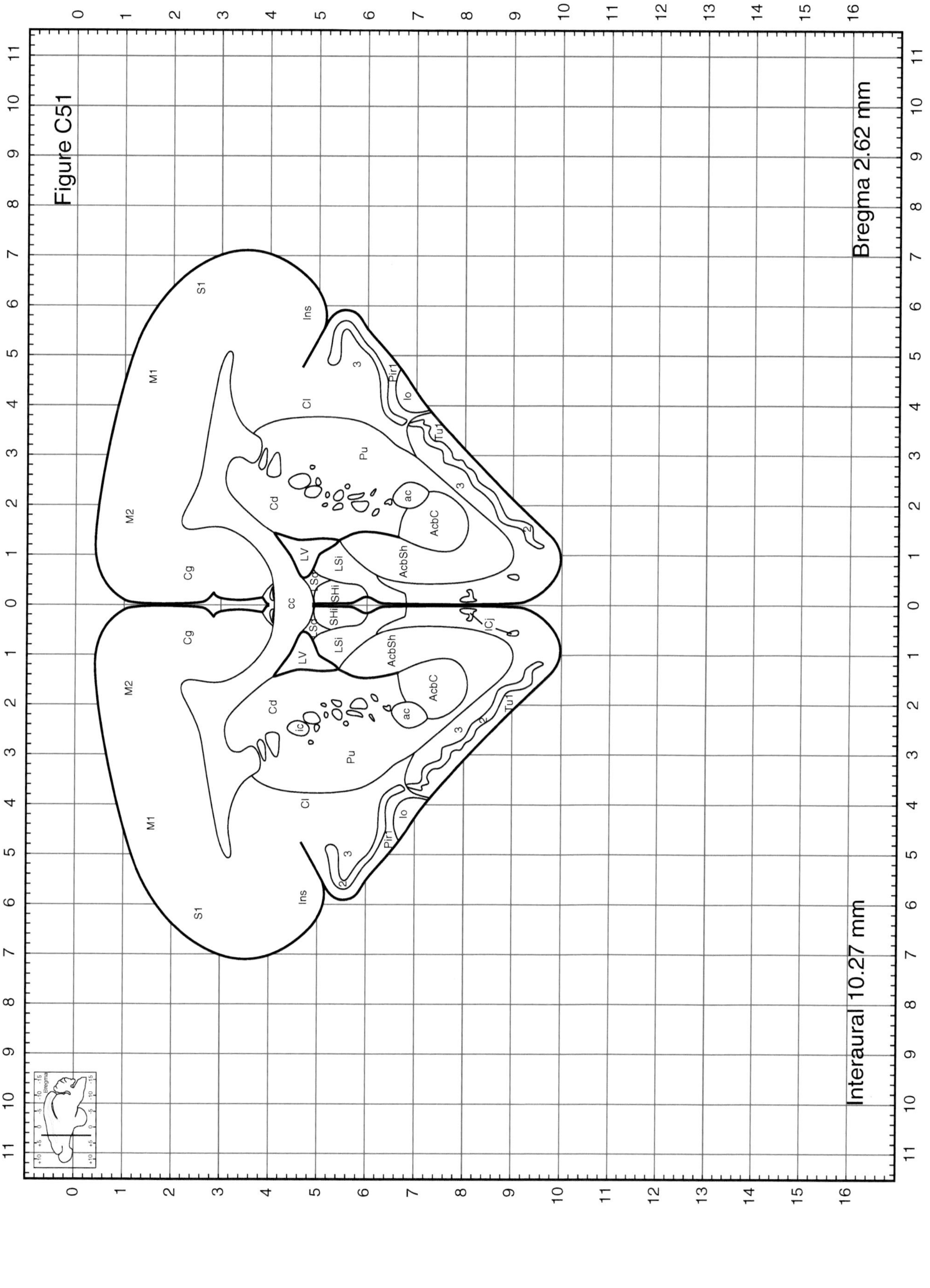
Figure C51
Bregma 2.62 mm
Interaural 10.27 mm
S1
Ins
M1
M2
Cg
Cl
Cd
Pu
LV
cc
LSi
SHi
LSc
AcbSh
AcbC
ac
ic
ICj
Pir1
lo
Tu1
3
2

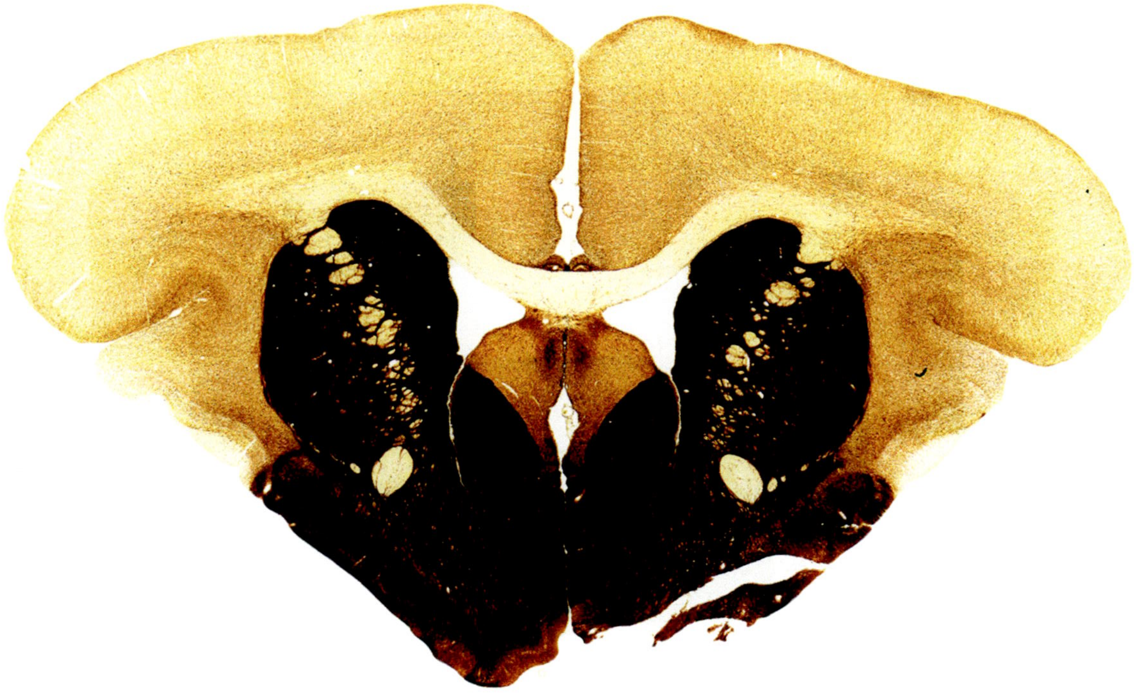

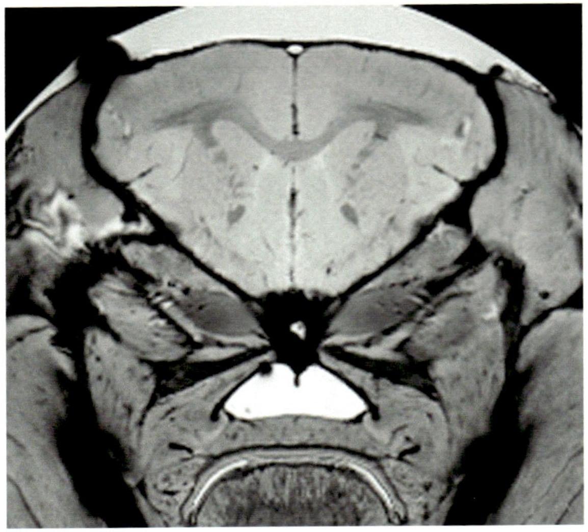

2	layer 2 of cortex
3	layer 3 of cortex
ac	anterior commissure
AcbC	accumbens nucleus, core
AcbSh	accumbens nucleus, shell
cc	corpus callosum
Cd	caudate nucleus
Cg	cingulate cortex
Cl	claustrum
ic	internal capsule
ICj	islands of Calleja
IG	indusium griseum
Ins	insular cortex
lo	lateral olfactory tract
LSd	lateral septal nucleus, dorsal part
LSi	lateral septal nucleus, intermediate part
LSv	lateral septal nucleus, ventral part
LV	lateral ventricle
M1	primary motor cortex
M2	secondary motor cortex
Pir1	piriform cortex, layer 1
Pu	putamen
S1	primary somatosensory cortex
SHi	septohippocampal nucleus
Tu1	layer 1 of olfactory tubercle

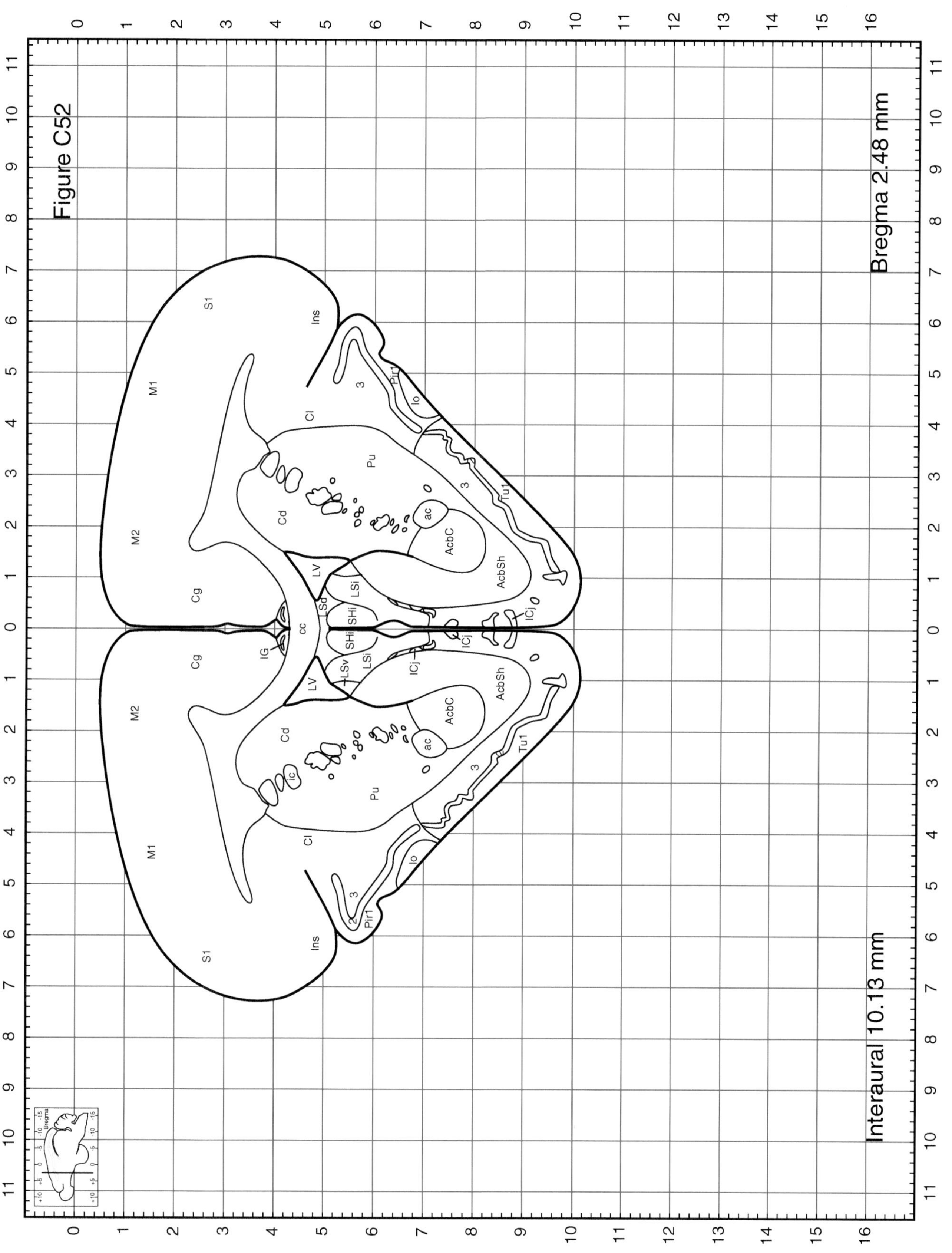
Figure C52
Bregma 2.48 mm
Interaural 10.13 mm
S1
Ins
M1
M2
Cg
Cl
Cd
Pu
ic
3
Pir1
lo
Tu1
ac
AcbC
AcbSh
LV
LSd
LSi
LSv
SHi
cc
IG
ICj

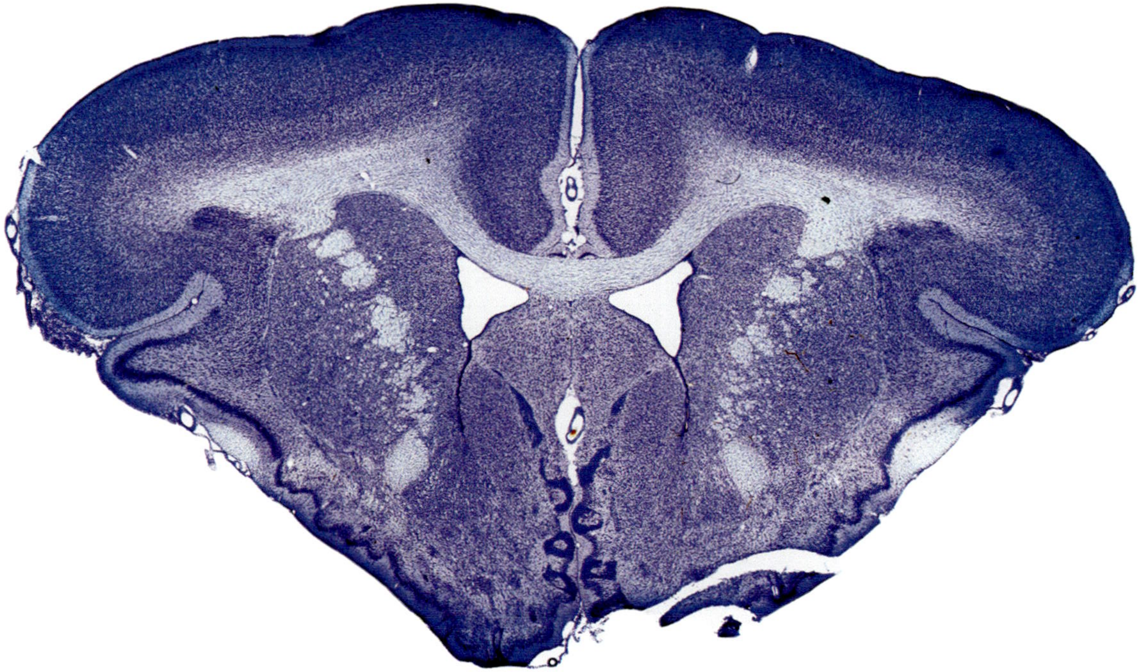

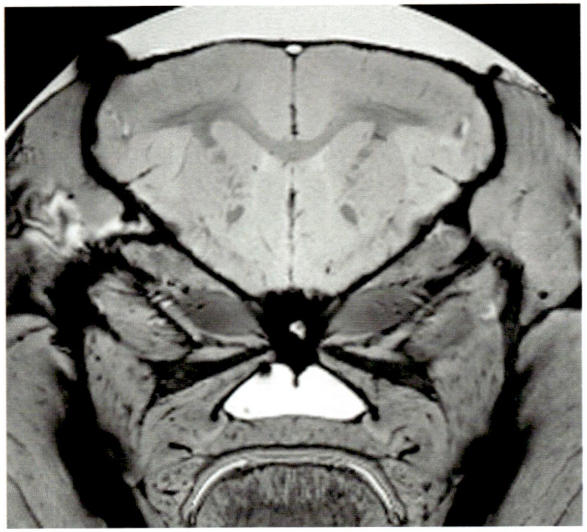

2	layer 2 of cortex	Cl	claustrum	LV	lateral ventricle
3	layer 3 of cortex	ic	internal capsule	M1	primary motor cortex
ac	anterior commissure	ICj	islands of Calleja	M2	secondary motor cortex
AcbC	accumbens nucleus, core	IG	indusium griseum	Pir1	piriform cortex, layer 1
AcbSh	accumbens nucleus, shell	Ins	insular cortex	Pu	putamen
cc	corpus callosum	lo	lateral olfactory tract	S1	primary somatosensory cortex
Cd	caudate nucleus	LSd	lateral septal nucleus, dorsal part	SHi	septohippocampal nucleus
Cg	cingulate cortex	LSi	lateral septal nucleus, intermediate part	Tu1	layer 1 of olfactory tubercle

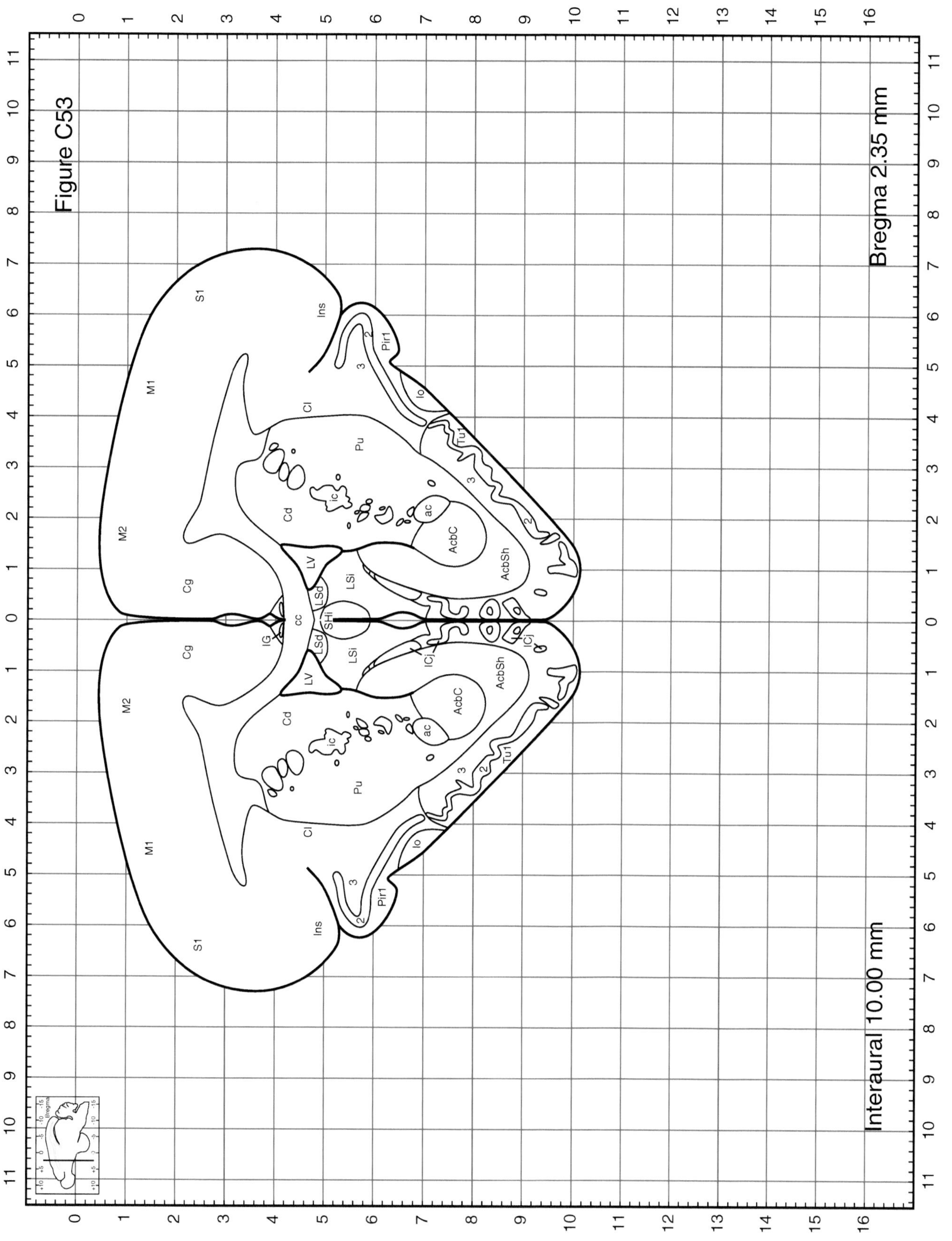
Figure C53
Bregma 2.35 mm
Interaural 10.00 mm
S1
Ins
M1
M2
Cg
Cl
Pu
Cd
ic
ac
AcbC
AcbSh
Tu1
lo
Pir1
LV
LSd
LSi
SHi
IG
cc
ICj
Bregma

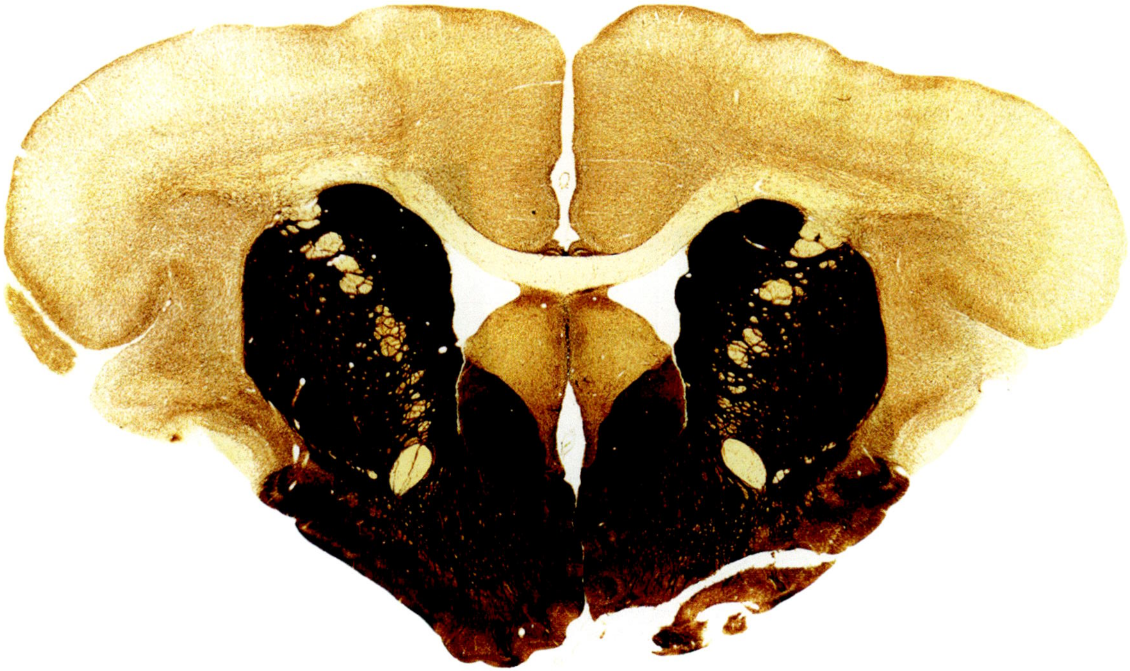

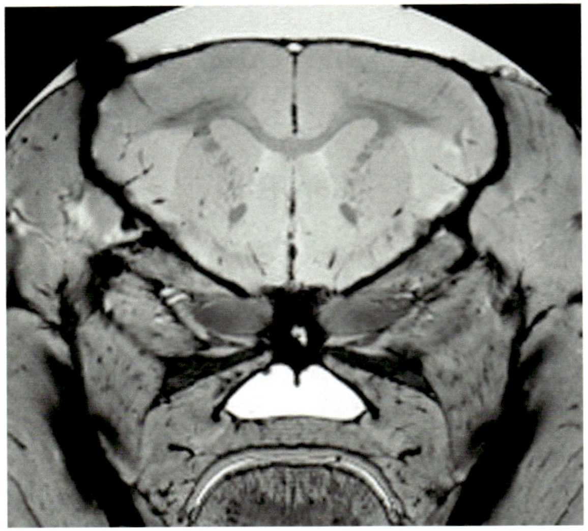

2	layer 2 of cortex
3	layer 3 of cortex
ac	anterior commissure
AcbC	accumbens nucleus, core
AcbSh	accumbens nucleus, shell
cc	corpus callosum
Cd	caudate nucleus
Cg	cingulate cortex
Cl	claustrum
ic	internal capsule
ICj	islands of Calleja
IG	indusium griseum
Ins	insular cortex
lo	lateral olfactory tract
LSd	lateral septal nucleus, dorsal part
LSi	lateral septal nucleus, intermediate part
LV	lateral ventricle
M1	primary motor cortex
M2	secondary motor cortex
Pir1	piriform cortex, layer 1
Pu	putamen
S1	primary somatosensory cortex
SHi	septohippocampal nucleus
Tu1	layer 1 of olfactory tubercle

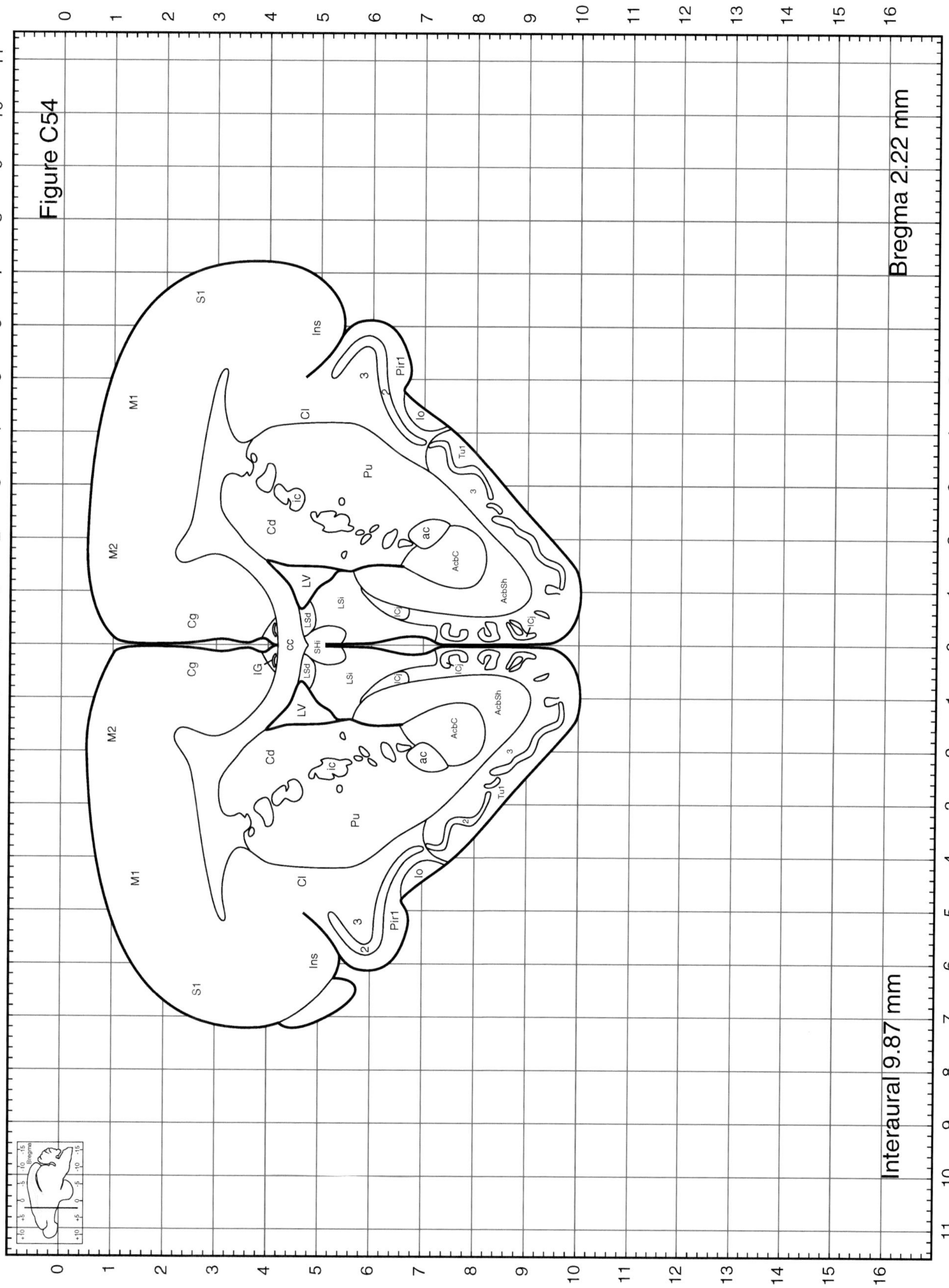
Figure C54
Bregma 2.22 mm
Interaural 9.87 mm
S1
Ins
M1
M2
Cg
IG
cc
LV
LSd
LSi
SHi
Cl
Cd
Pu
ic
ac
AcbC
AcbSh
ICj
Tu1
Pir1
lo
3
2

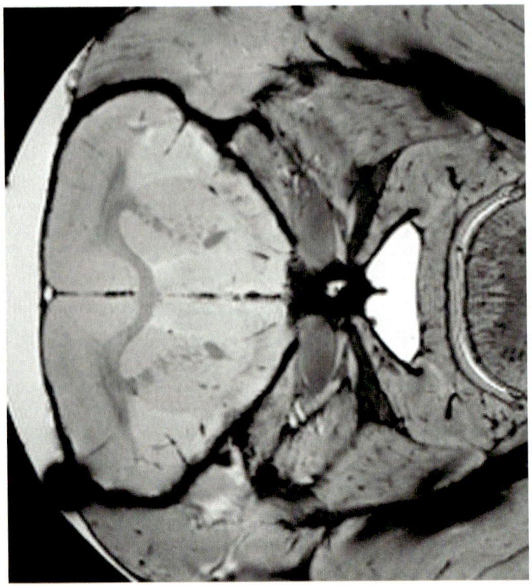

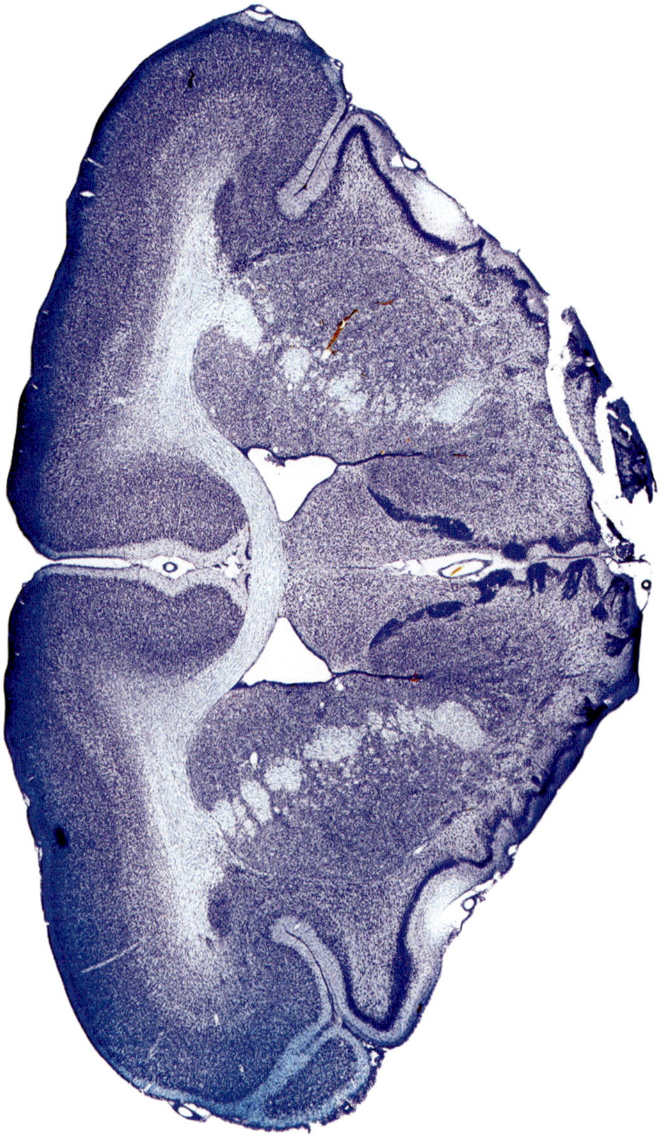

2 layer 2 of cortex
3 layer 3 of cortex
ac anterior commissure
AcbC accumbens nucleus, core
AcbSh accumbens nucleus, shell
cc corpus callosum
Cd caudate nucleus
Cg cingulate cortex
Cl claustrum
ic internal capsule
ICj islands of Calleja
IG indusium griseum
Ins insular cortex
lo lateral olfactory tract
LSd lateral septal nucleus, dorsal part
LSi lateral septal nucleus, intermediate part
LV lateral ventricle
M1 primary motor cortex
M2 secondary motor cortex
Pir1 piriform cortex, layer 1
Pu putamen
S1 primary somatosensory cortex
S2 secondary somatosensory cortex
SHi septohippocampal nucleus
Tu1 layer 1 of olfactory tubercle

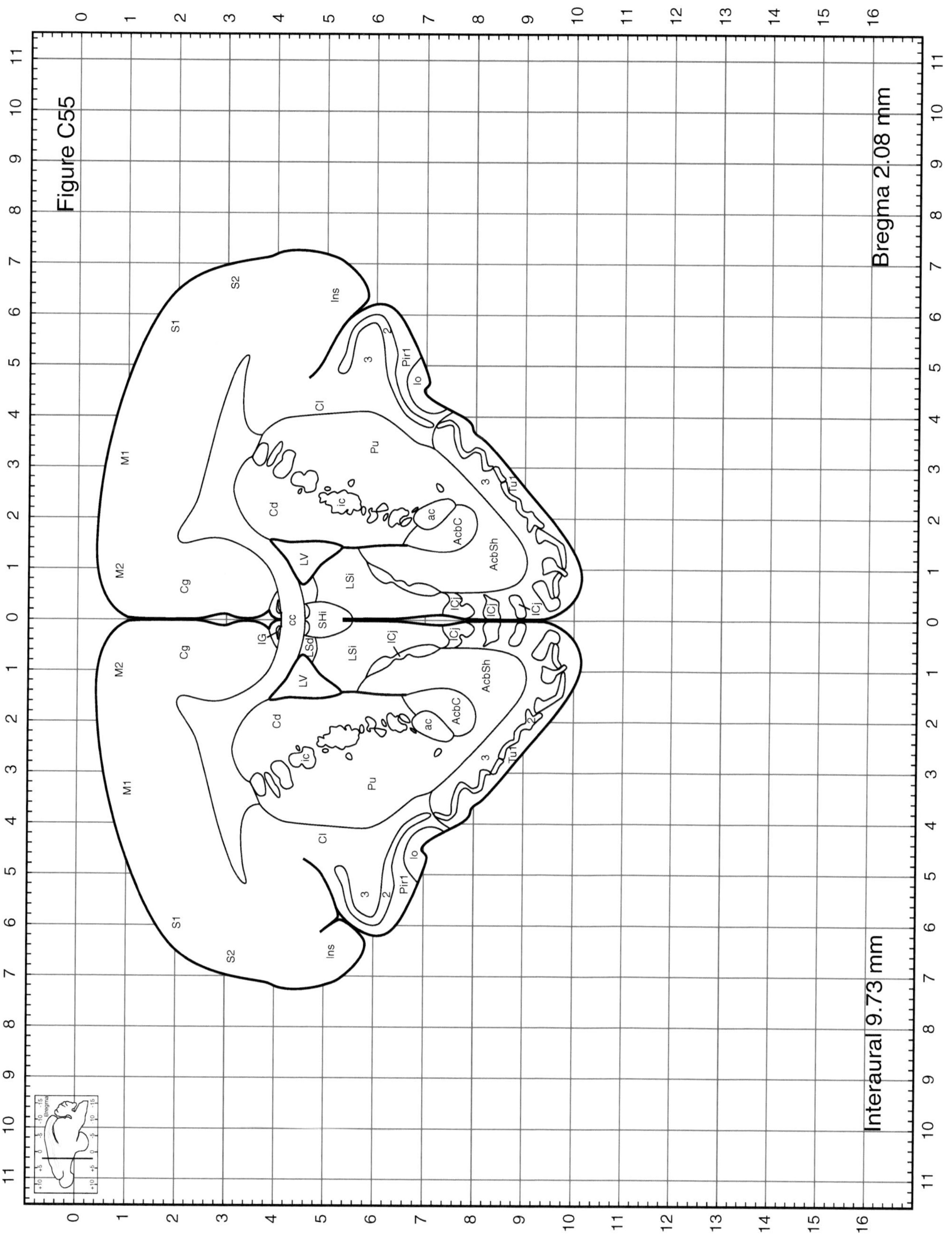
Figure C55
Bregma 2.08 mm
Interaural 9.73 mm
M1
M2
S1
S2
Ins
Cg
Cl
Pu
Cd
ic
ac
AcbC
AcbSh
LV
LSi
LSd
SHi
cc
IG
ICj
Tu1
Pir1
lo
2
3

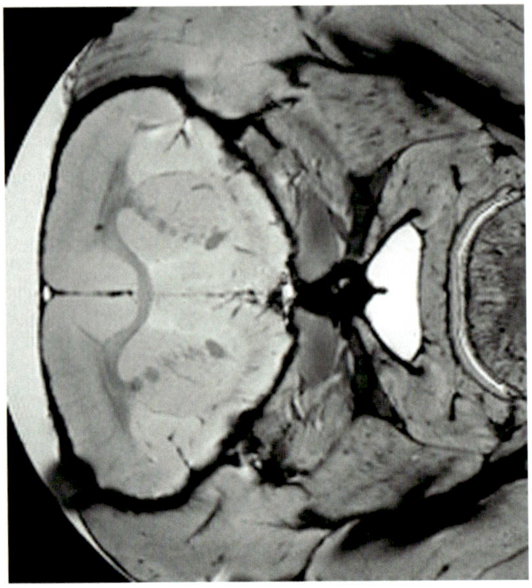

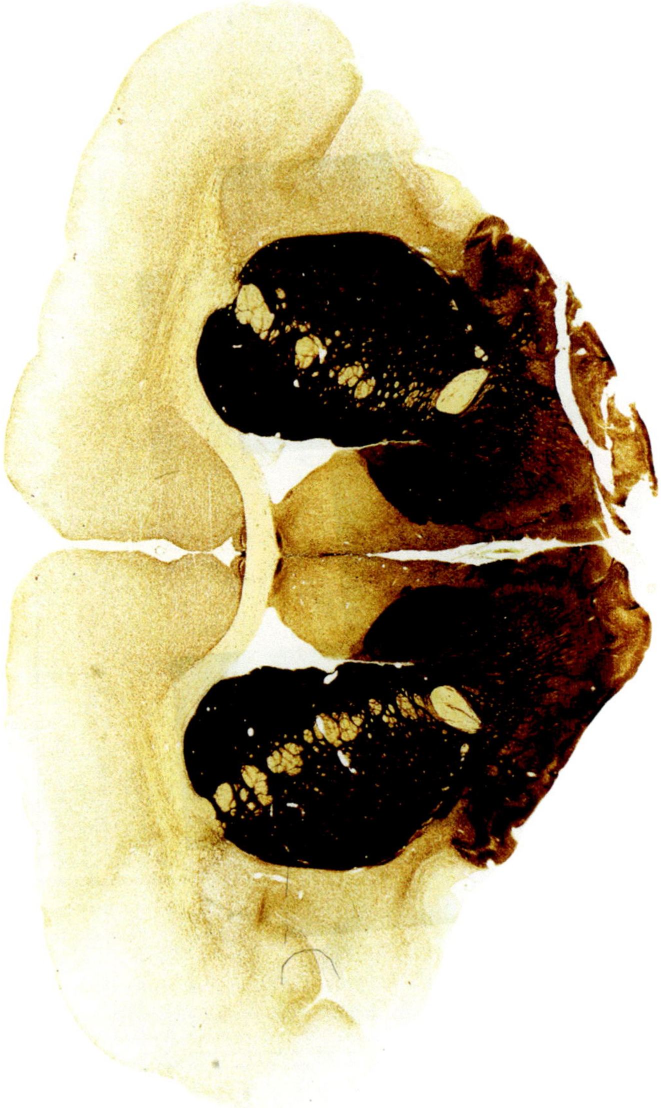

2	layer 2 of cortex
3	layer 3 of cortex
ac	anterior commissure
AcbC	accumbens nucleus, core
AcbSh	accumbens nucleus, shell
cc	corpus callosum
Cd	caudate nucleus
Cg	cingulate cortex
Cl	claustrum
ic	internal capsule
ICj	islands of Calleja
IG	indusium griseum
Ins	insular cortex
lo	lateral olfactory tract
LSd	lateral septal nucleus, dorsal part
LSi	lateral septal nucleus, intermediate part
LSv	lateral septal nucleus, ventral part
LV	lateral ventricle
M1	primary motor cortex
M2	secondary motor cortex
MS	medial septal nucleus
Pir1	piriform cortex, layer 1
Pu	putamen
S1	primary somatosensory cortex
S2	secondary somatosensory cortex
SHi	septohippocampal nucleus
Tu1	layer 1 of olfactory tubercle

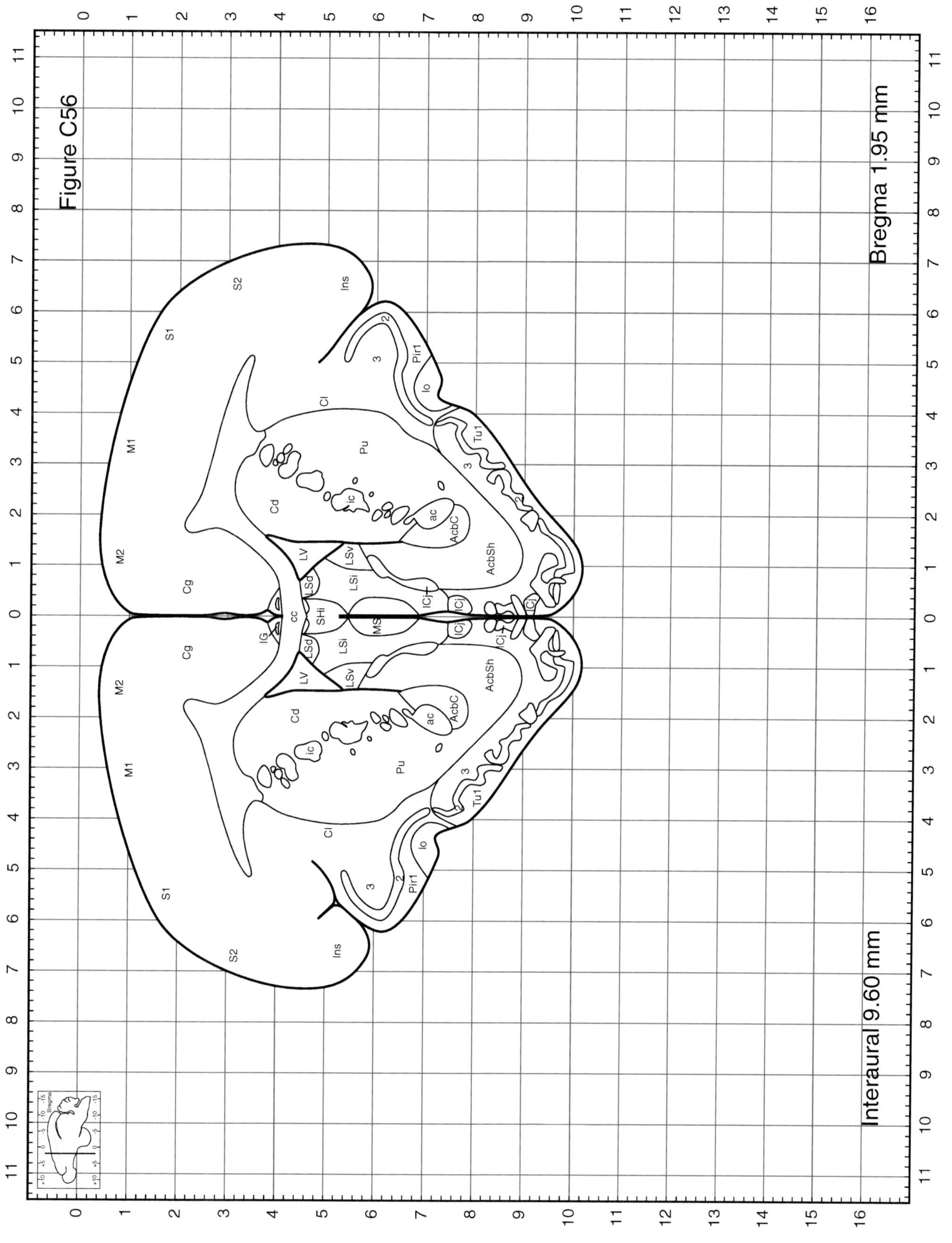
Figure C56
Bregma 1.95 mm
Interaural 9.60 mm
M1
M2
S1
S2
Cg
Ins
Cl
Cd
Pu
ic
ac
AcbC
AcbSh
Tu1
Pir1
lo
3
2
LV
LSd
LSi
LSv
SHi
MS
IG
cc
ICj

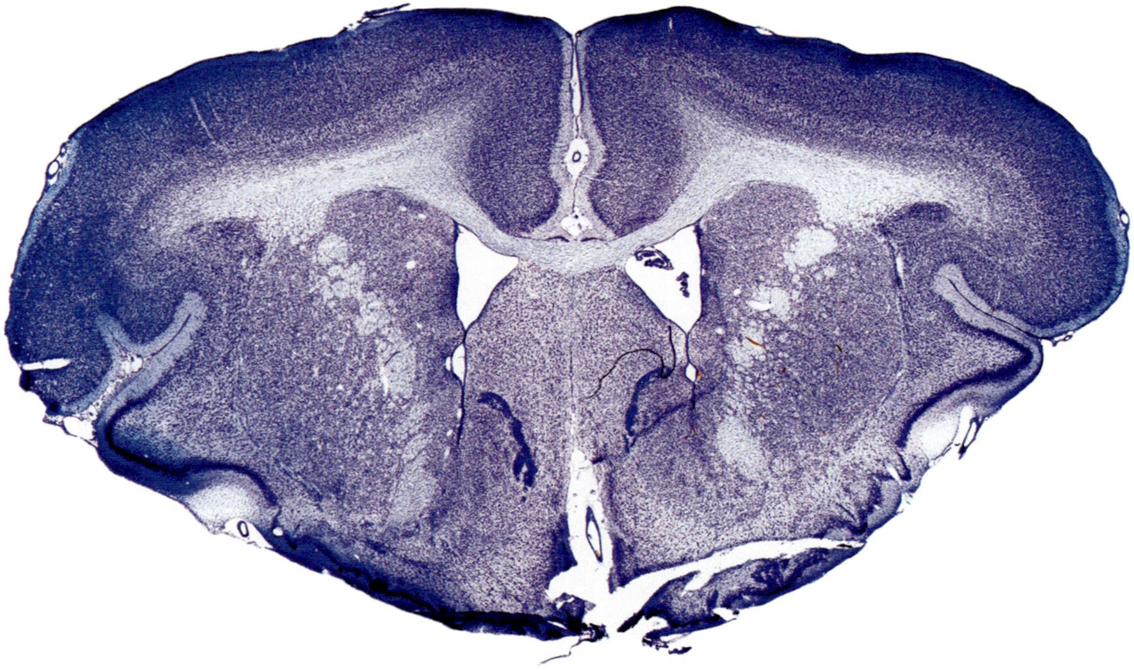

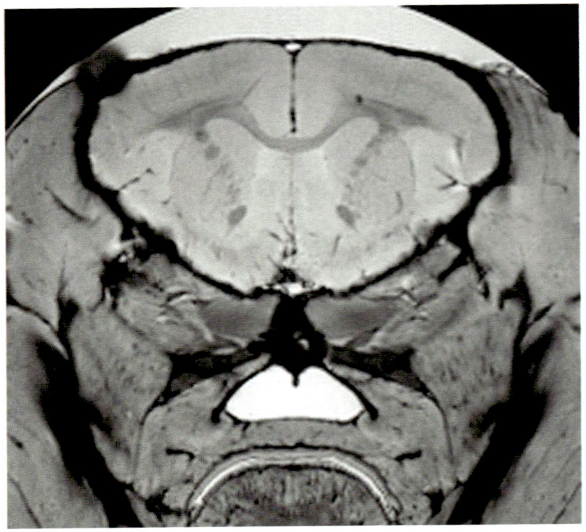

2	layer 2 of cortex
3	layer 3 of cortex
ac	anterior commissure
AcbC	accumbens nucleus, core
AcbSh	accumbens nucleus, shell
cc	corpus callosum
Cd	caudate nucleus
Cg	cingulate cortex
Cl	claustrum
ic	internal capsule
ICj	islands of Calleja
IG	indusium griseum
Ins	insular cortex
lo	lateral olfactory tract
LSd	lateral septal nucleus, dorsal part
LSi	lateral septal nucleus, intermediate part
LSv	lateral septal nucleus, ventral part
LV	lateral ventricle
M1	primary motor cortex
MS	medial septal nucleus
Pir1	piriform cortex, layer 1
Pu	putamen
S1	primary somatosensory cortex
S2	secondary somatosensory cortex
SHi	septohippocampal nucleus
Tu1	layer 1 of olfactory tubercle
VDB	nucleus of the vertical limb of the diagonal band

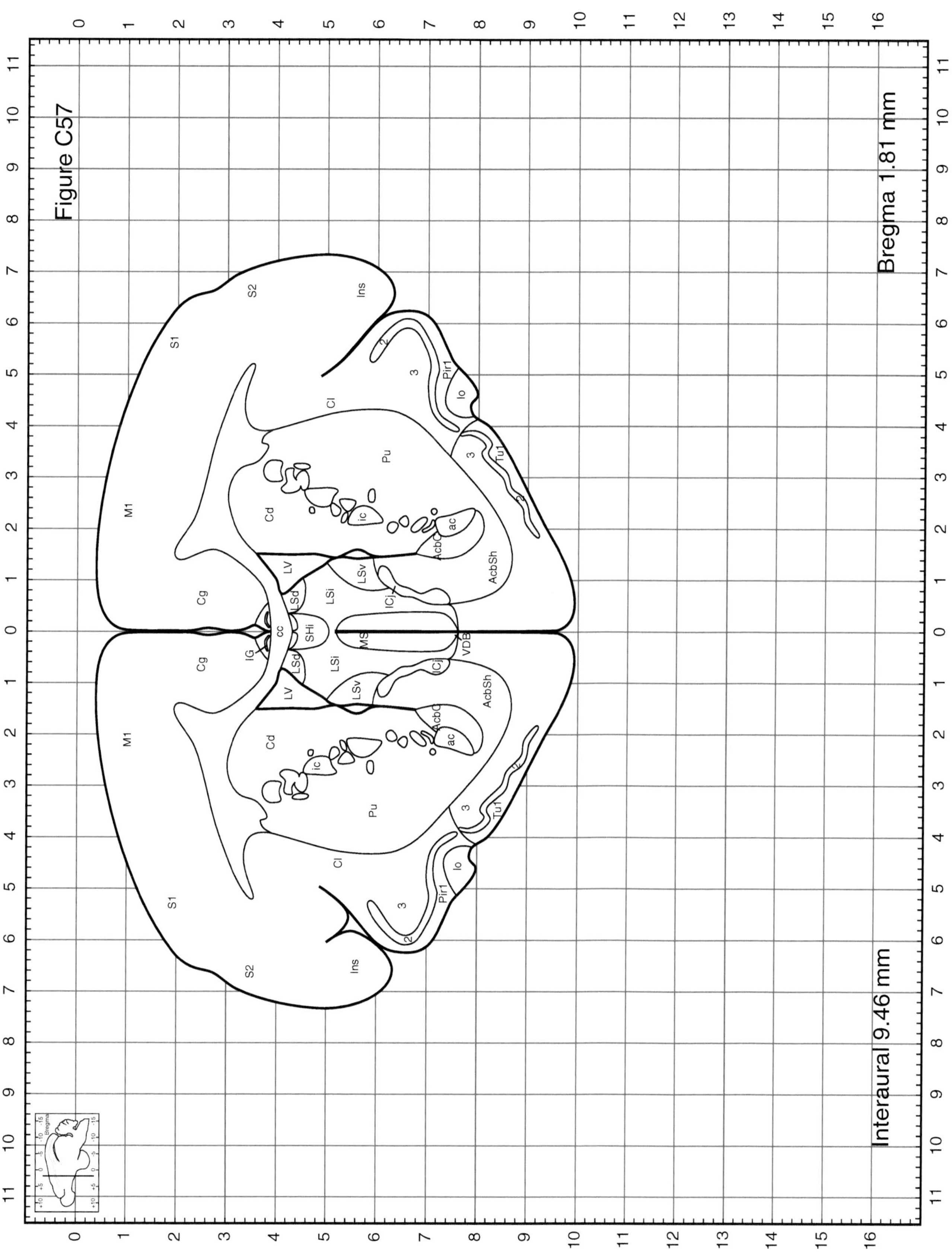
Figure C57
Bregma 1.81 mm
Interaural 9.46 mm
M1
S1
S2
Ins
Cl
Cg
IG
cc
LV
LSd
LSi
LSv
SHi
MS
ICj
VDB
Cd
Pu
ic
ac
AcbC
AcbSh
Tu1
Pir1
lo
2
3

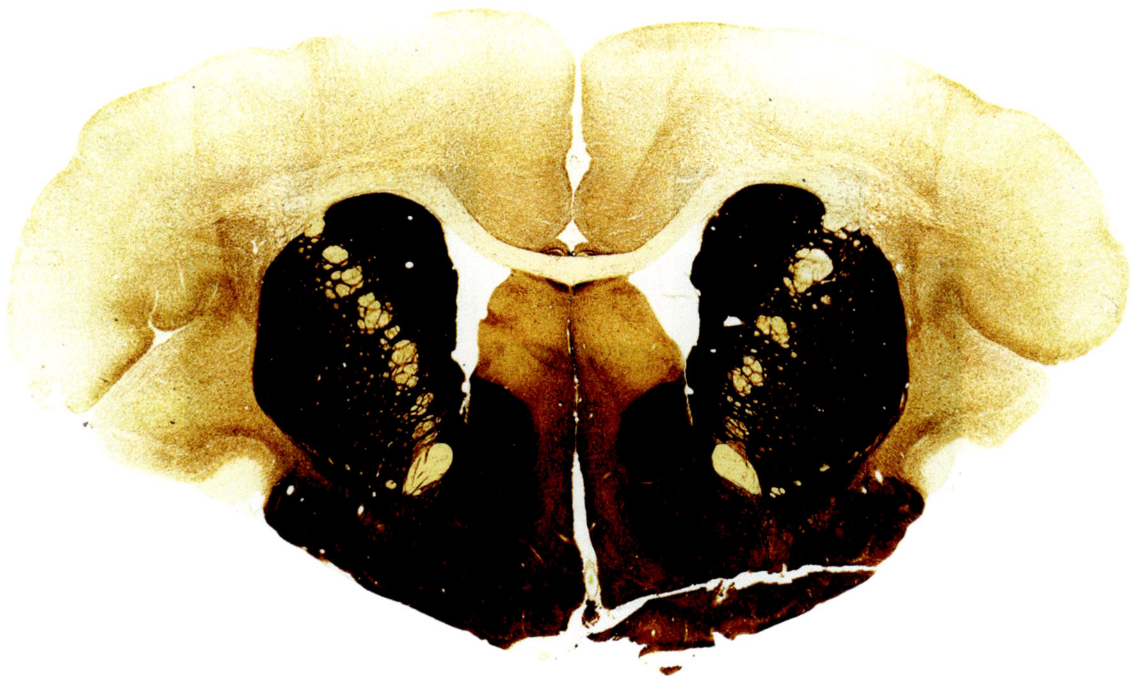

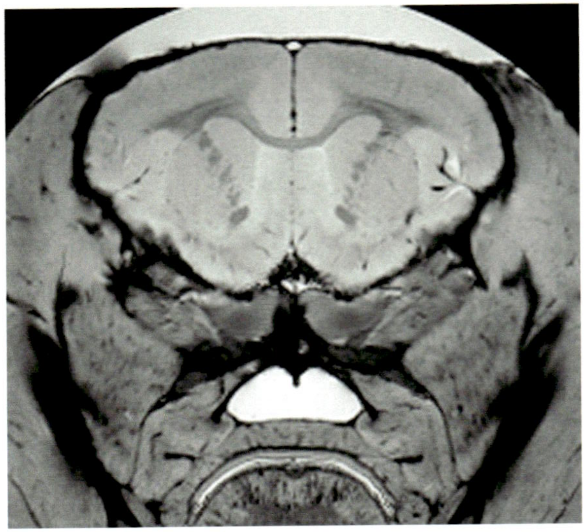

2	layer 2 of cortex	ic	internal capsule	M1	primary motor cortex
3	layer 3 of cortex	ICj	islands of Calleja	MS	medial septal nucleus
ac	anterior commissure	IG	indusium griseum	Pir1	piriform cortex, layer 1
AcbC	accumbens nucleus, core	Ins	insular cortex	Pu	putamen
AcbSh	accumbens nucleus, shell	lo	lateral olfactory tract	S1	primary somatosensory cortex
cc	corpus callosum	LSd	lateral septal nucleus, dorsal part	S2	secondary somatosensory cortex
Cd	caudate nucleus	LSi	lateral septal nucleus, intermediate part	SHi	septohippocampal nucleus
Cg	cingulate cortex	LSv	lateral septal nucleus, ventral part	Tu1	layer 1 of olfactory tubercle
Cl	claustrum	LV	lateral ventricle	VDB	nucleus of the vertical limb of the diagonal band

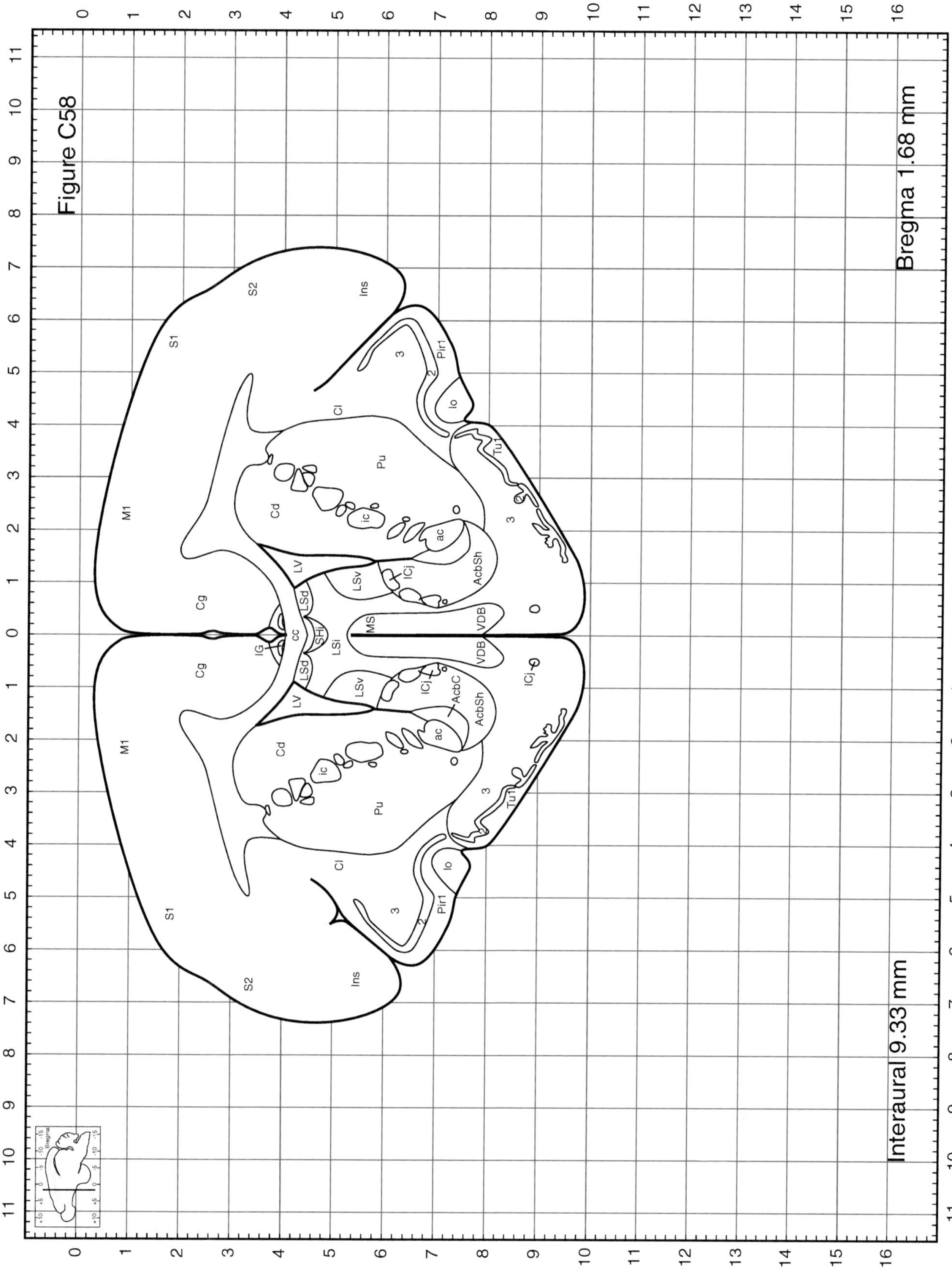
Figure C58
Bregma 1.68 mm
Interaural 9.33 mm
M1
S1
S2
Ins
Cg
Cl
Cd
Pu
ic
ac
LV
LSd
LSv
LSi
SHi
cc
IG
MS
VDB
ICj
AcbC
AcbSh
Tu1
Pir1
lo
3
2

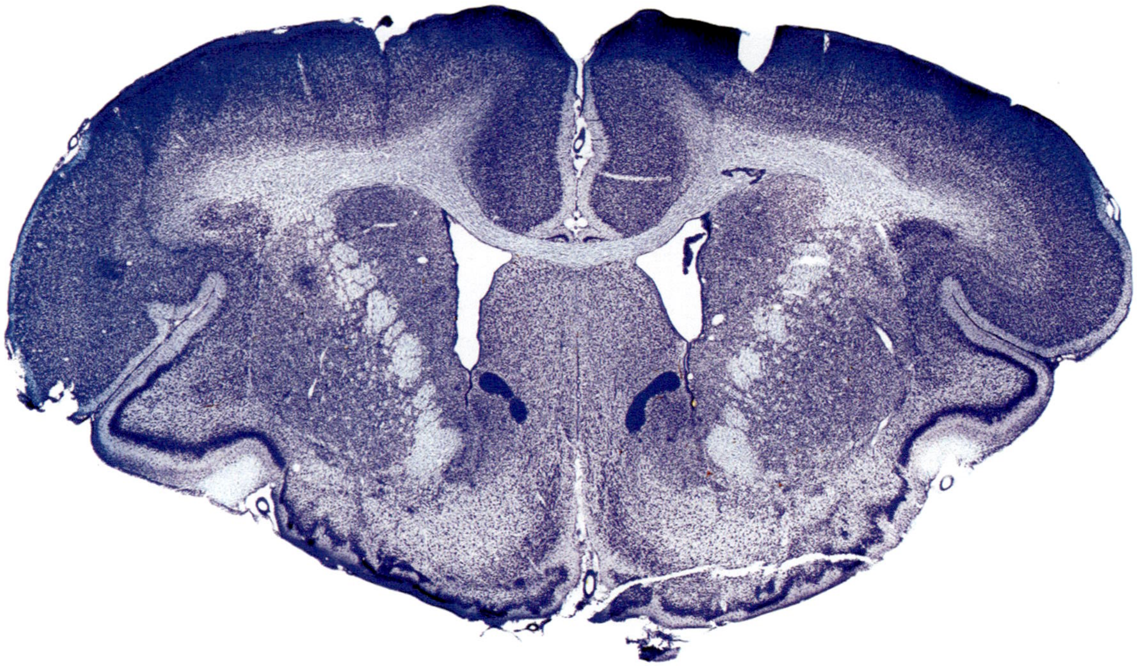

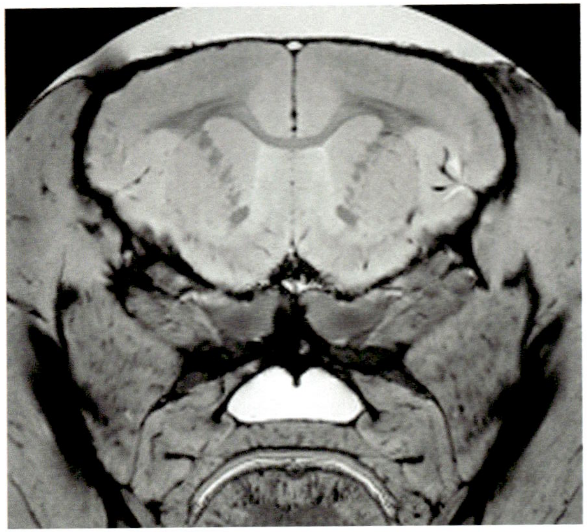

2	layer 2 of cortex
3	layer 3 of cortex
ac	anterior commissure
AcbC	accumbens nucleus, core
AcbSh	accumbens nucleus, shell
cc	corpus callosum
Cd	caudate nucleus
Cg	cingulate cortex
Cl	claustrum
HDB	nucleus of the horizontal limb of the diagonal band
ic	internal capsule
ICj	islands of Calleja
IG	indusium griseum
Ins	insular cortex
lo	lateral olfactory tract
LSd	lateral septal nucleus, dorsal part
LSi	lateral septal nucleus, intermediate part
LSv	lateral septal nucleus, ventral part
LV	lateral ventricle
M1	primary motor cortex
MS	medial septal nucleus
Pir1	piriform cortex, layer 1
Pu	putamen
S1	primary somatosensory cortex
S2	secondary somatosensory cortex
SHi	septohippocampal nucleus
STMA	bed nucleus of the stria terminalis, medial division, anterior part
Tu1	layer 1 of olfactory tubercle
VDB	nucleus of the vertical limb of the diagonal band

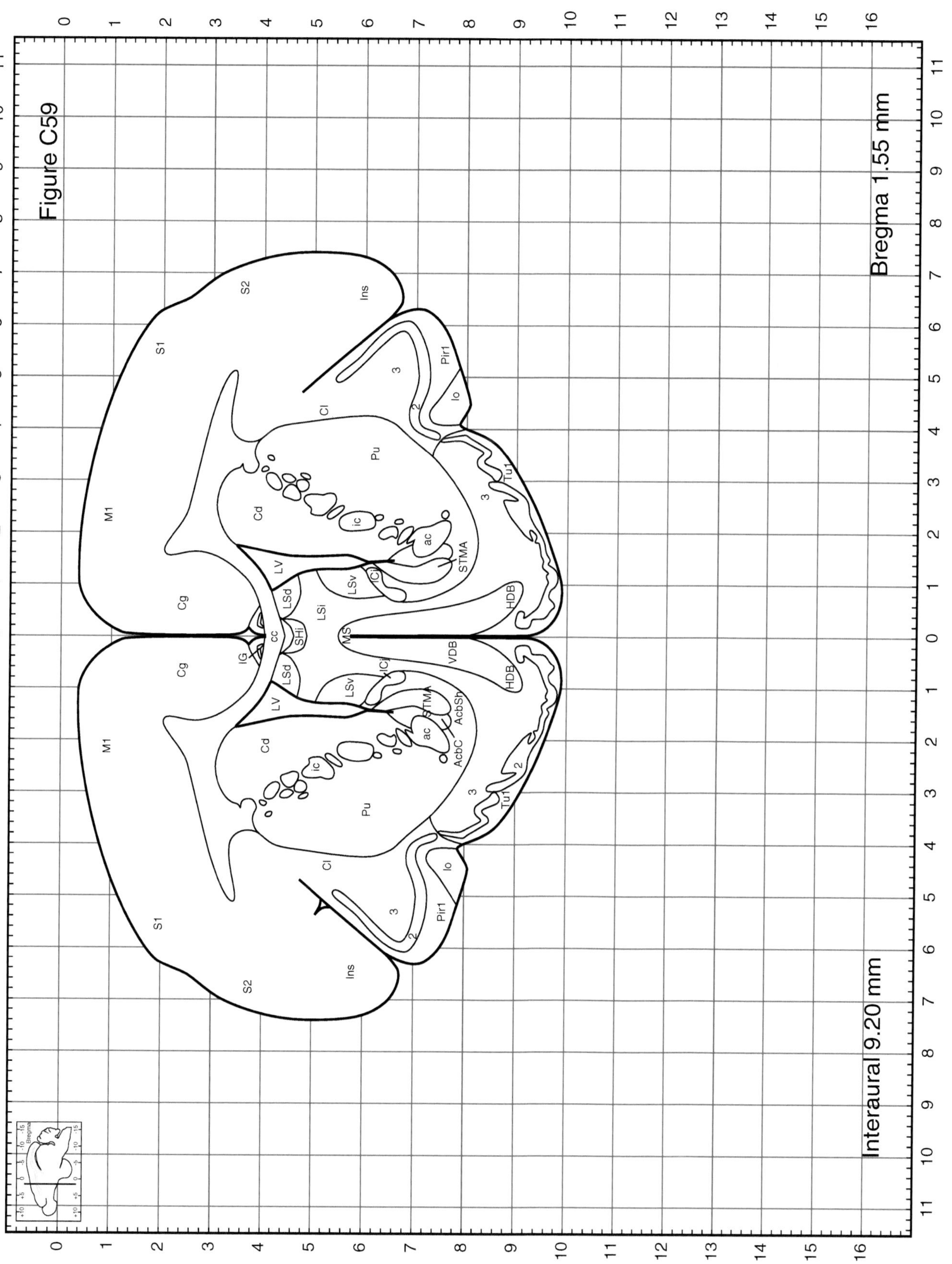
Figure C59
Bregma 1.55 mm
Interaural 9.20 mm
M1
S1
S2
Ins
Cg
Cl
Pir1
3
2
lo
Pu
Cd
ic
ac
STMA
LV
LSv
LSd
LSi
ICj
IG
cc
SHi
MS
VDB
HDB
Tu1
AcbSh
AcbC

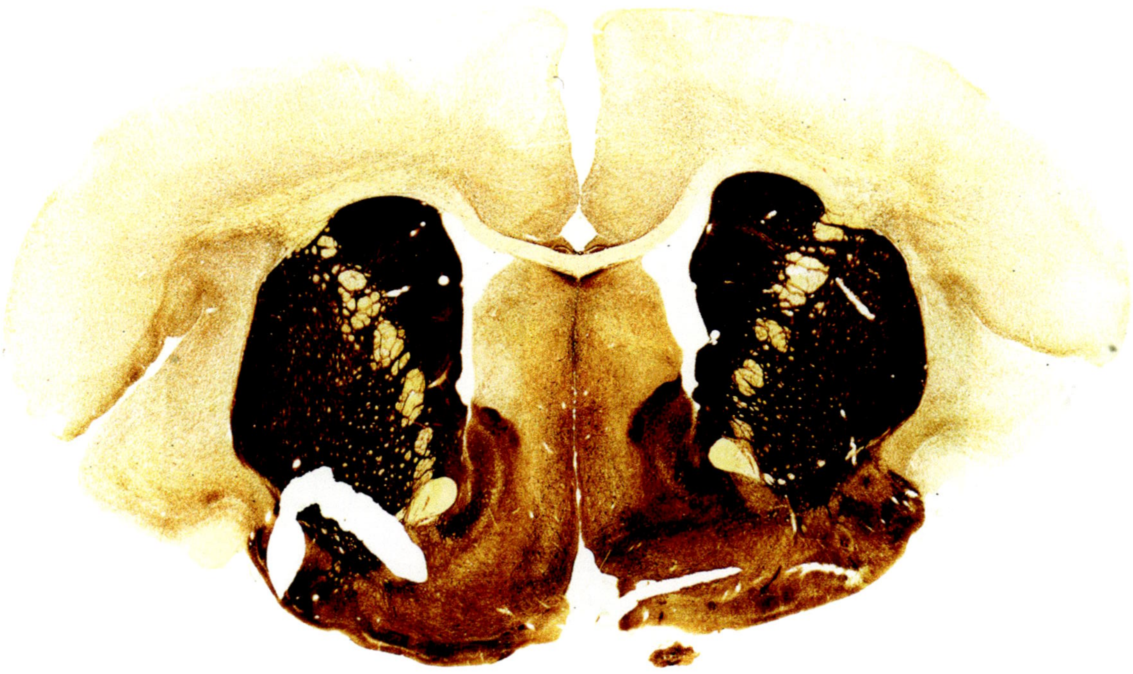

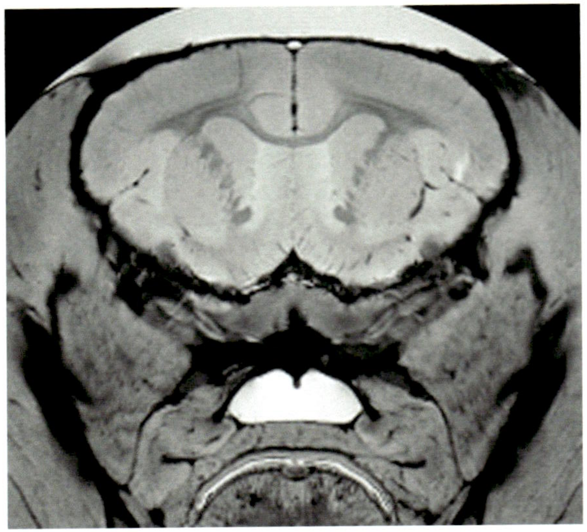

2	layer 2 of cortex
3	layer 3 of cortex
ac	anterior commissure
AcbC	accumbens nucleus, core
AcbSh	accumbens nucleus, shell
cc	corpus callosum
Cd	caudate nucleus
Cg	cingulate cortex
Cl	claustrum
HDB	nucleus of the horizontal limb of the diagonal band
ic	internal capsule
ICj	islands of Calleja
IG	indusium griseum
Ins	insular cortex
lo	lateral olfactory tract
LSd	lateral septal nucleus, dorsal part
LSi	lateral septal nucleus, intermediate part
LSv	lateral septal nucleus, ventral part
LV	lateral ventricle
M1	primary motor cortex
MS	medial septal nucleus
Pir1	piriform cortex, layer 1
Pu	putamen
S1	primary somatosensory cortex
S2	secondary somatosensory cortex
SHi	septohippocampal nucleus
STMA	bed nucleus of the stria terminalis, medial division, anterior part
Tu1	layer 1 of olfactory tubercle
VDB	nucleus of the vertical limb of the diagonal band

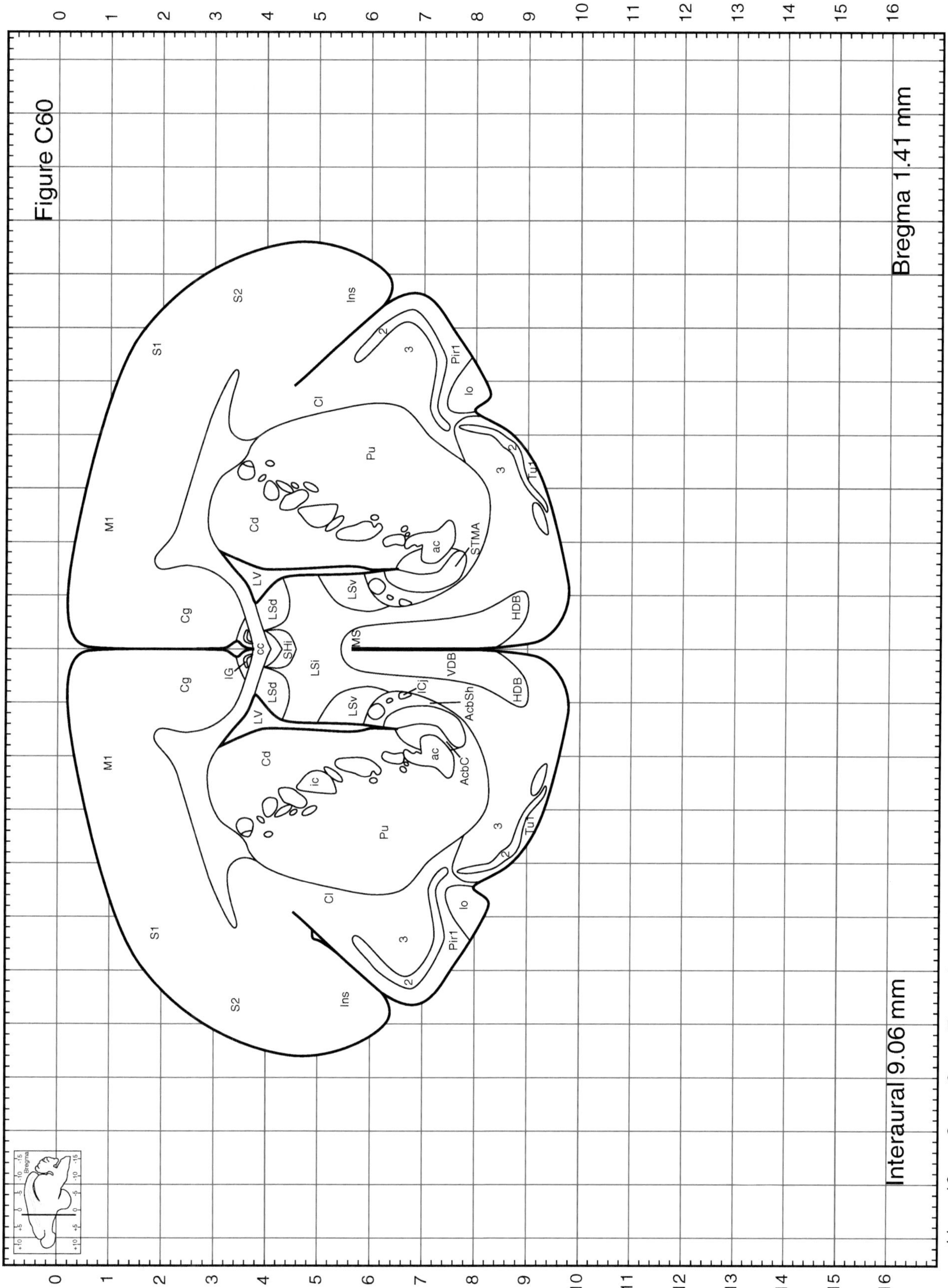
Figure C60
Bregma 1.41 mm
Interaural 9.06 mm
M1
S1
S2
Ins
Cg
Cl
Pu
Cd
LV
LSd
LSv
LSi
SHi
cc
IG
MS
VDB
HDB
ac
ic
STMA
AcbSh
AcbC
ICj
Pir1
lo
Tu1
2
3

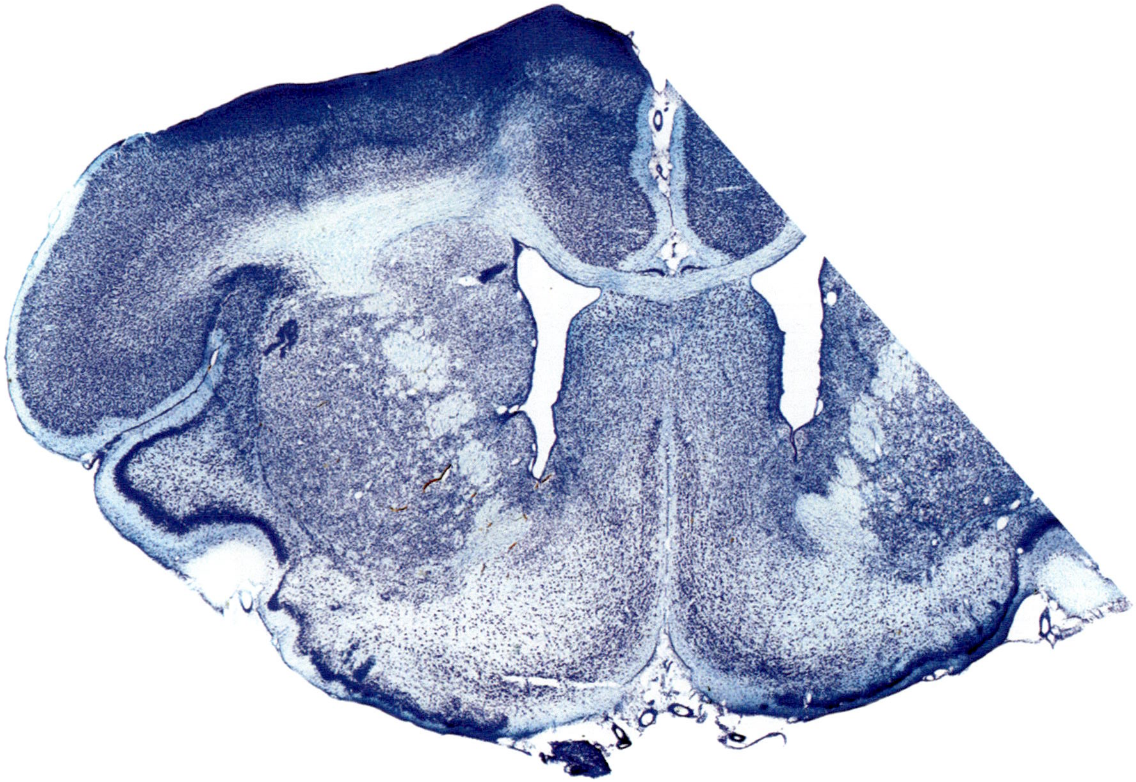

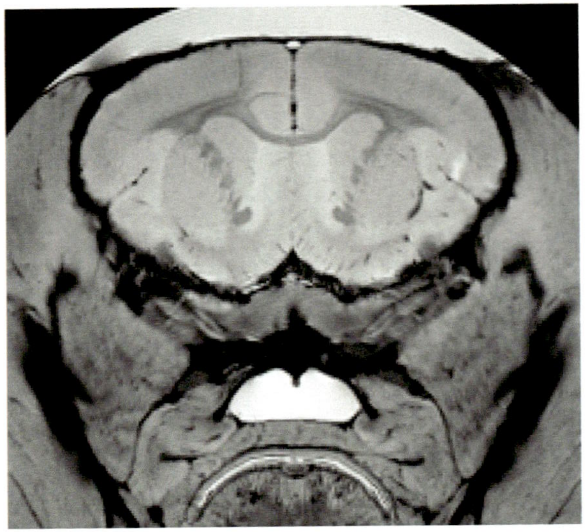

2	layer 2 of cortex
3	layer 3 of cortex
ac	anterior commissure
Acb	accumbens nucleus
cc	corpus callosum
Cd	caudate nucleus
Cg	cingulate cortex
Cl	claustrum
HDB	nucleus of the horizontal limb of the diagonal band
ic	internal capsule
IG	indusium griseum
Ins	insular cortex
lo	lateral olfactory tract
LSd	lateral septal nucleus, dorsal part
LSi	lateral septal nucleus, intermediate part
LSv	lateral septal nucleus, ventral part
LV	lateral ventricle
M1	primary motor cortex
MS	medial septal nucleus
Pir1	piriform cortex, layer 1
Pu	putamen
S1	primary somatosensory cortex
S2	secondary somatosensory cortex
SHi	septohippocampal nucleus
STMA	bed nucleus of the stria terminalis, medial division, anterior part
Tu1	layer 1 of olfactory tubercle
VDB	nucleus of the vertical limb of the diagonal band

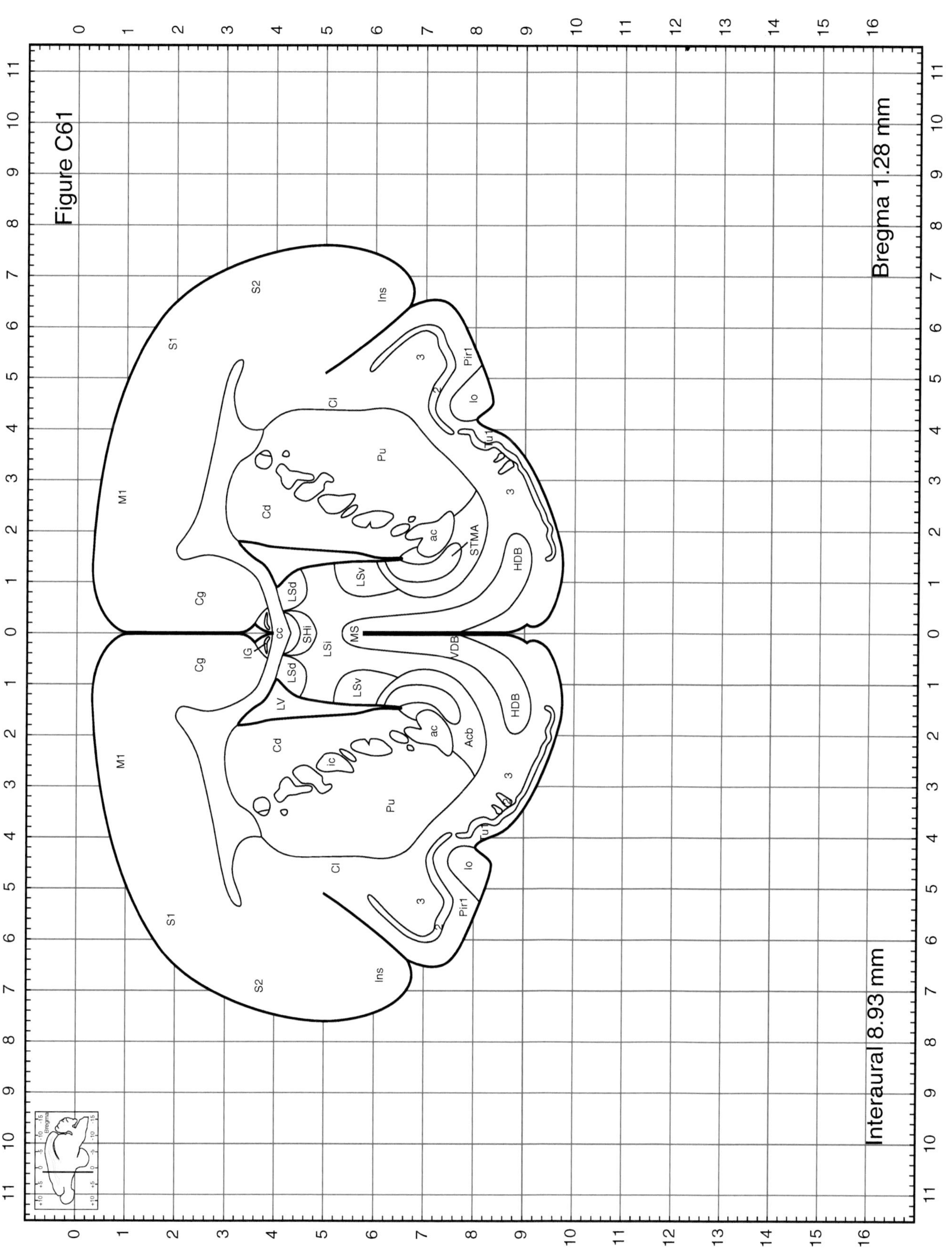
Figure C61
Bregma 1.28 mm
Interaural 8.93 mm
M1
S1
S2
Ins
Cg
Cl
Cd
Pu
ac
ic
LV
LSd
LSv
LSi
SHi
IG
cc
MS
VDB
HDB
STMA
Acb
Tu1
Pir1
lo
3
2

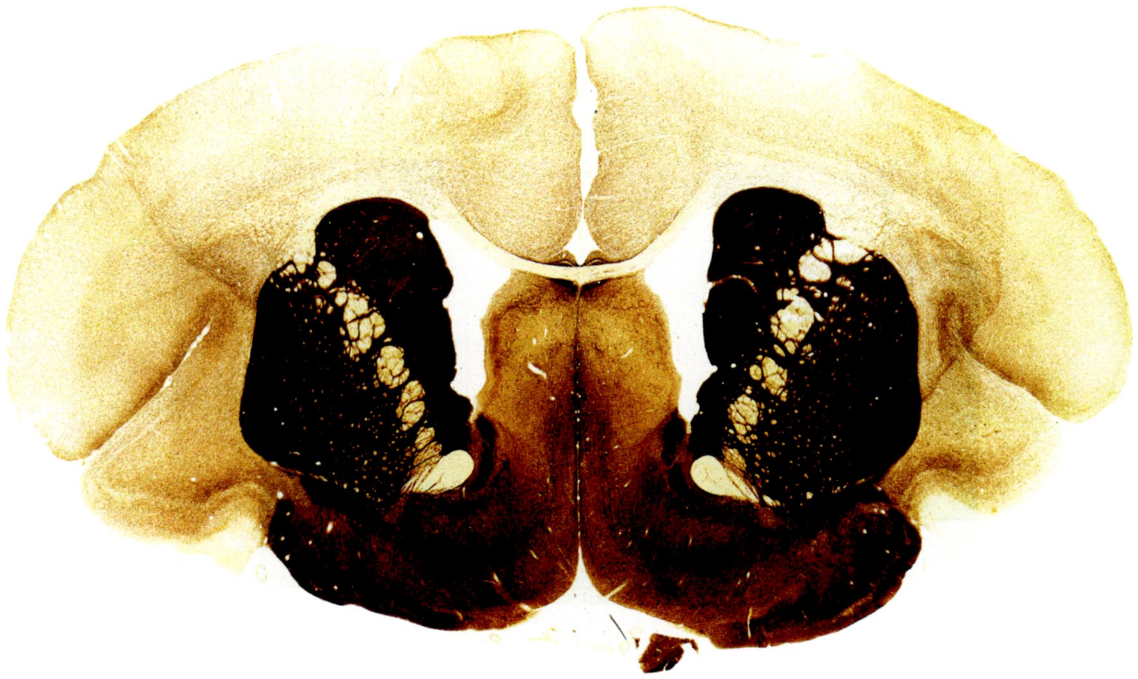

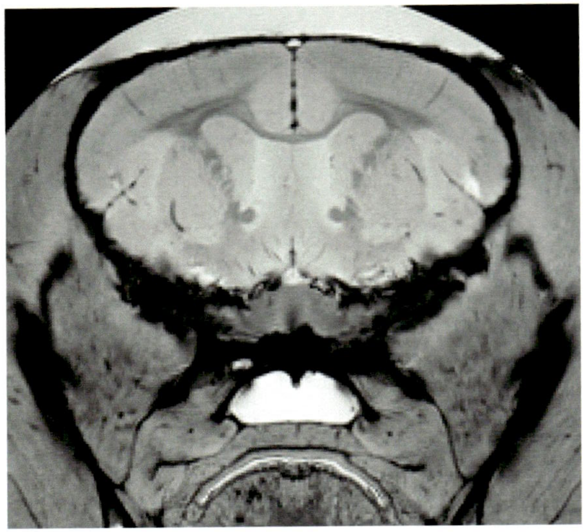

2	layer 2 of cortex
3	layer 3 of cortex
ac	anterior commissure
Acb	accumbens nucleus
cc	corpus callosum
Cd	caudate nucleus
Cg	cingulate cortex
Cl	claustrum
HDB	nucleus of the horizontal limb of the diagonal band
ic	internal capsule
IG	indusium griseum
Ins	insular cortex
IPA	interpeduncular nucleus, apical subnucleus
lo	lateral olfactory tract
LSd	lateral septal nucleus, dorsal part
LSi	lateral septal nucleus, intermediate part
LSv	lateral septal nucleus, ventral part
LV	lateral ventricle
M1	primary motor cortex
MS	medial septal nucleus
Pir1	piriform cortex, layer 1
Pu	putamen
S1	primary somatosensory cortex
S2	secondary somatosensory cortex
SHi	septohippocampal nucleus
STMA	bed nucleus of the stria terminalis, medial division, anterior part
Tu1	layer 1 of olfactory tubercle
VDB	nucleus of the vertical limb of the diagonal band

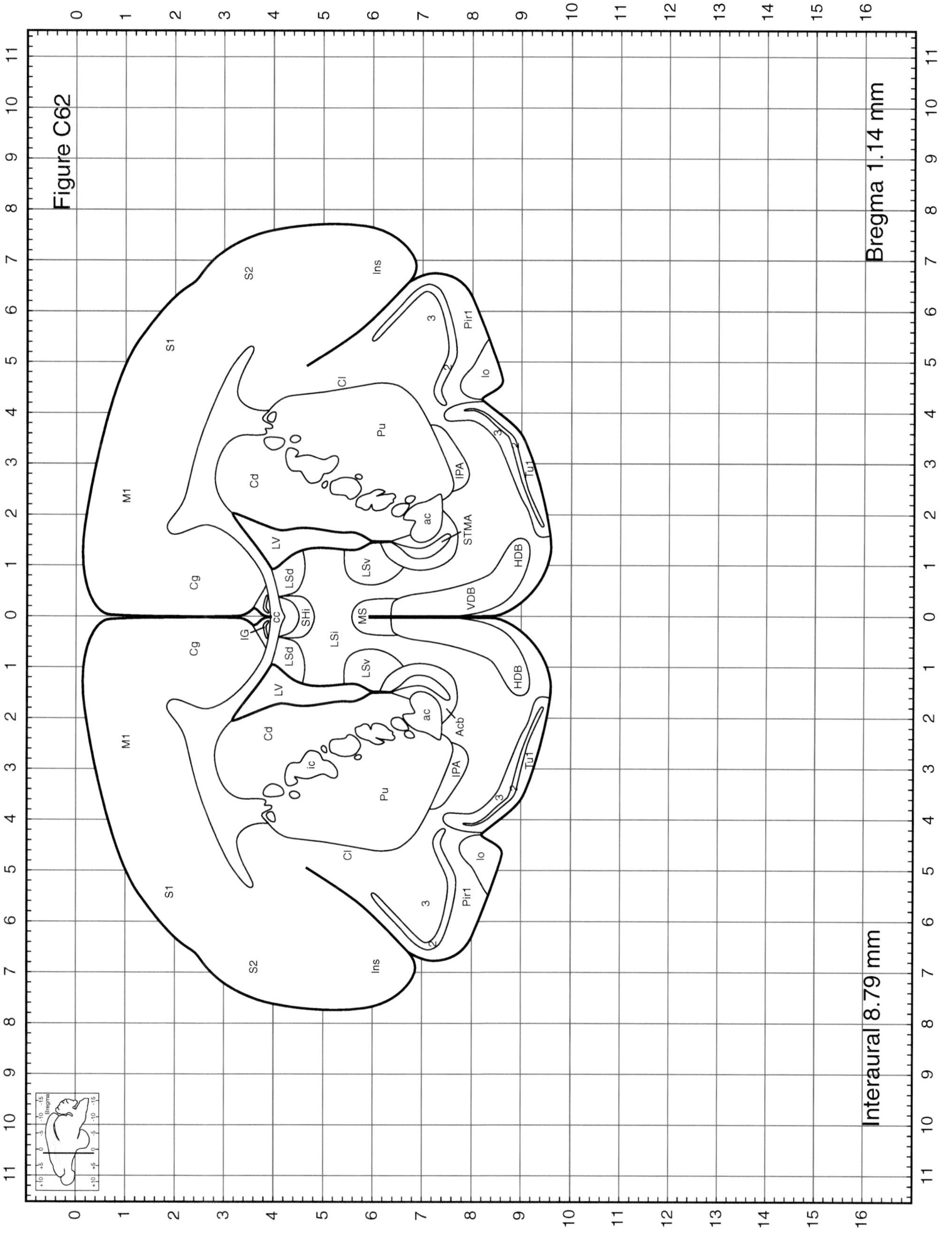
Figure C62
Bregma 1.14 mm
Interaural 8.79 mm
M1
S1
S2
Ins
Cg
Cl
Cd
Pu
LV
LSd
LSv
LSi
SHi
MS
IG
cc
ac
ic
IPA
STMA
Acb
HDB
VDB
Tu1
Pir1
lo
3
2

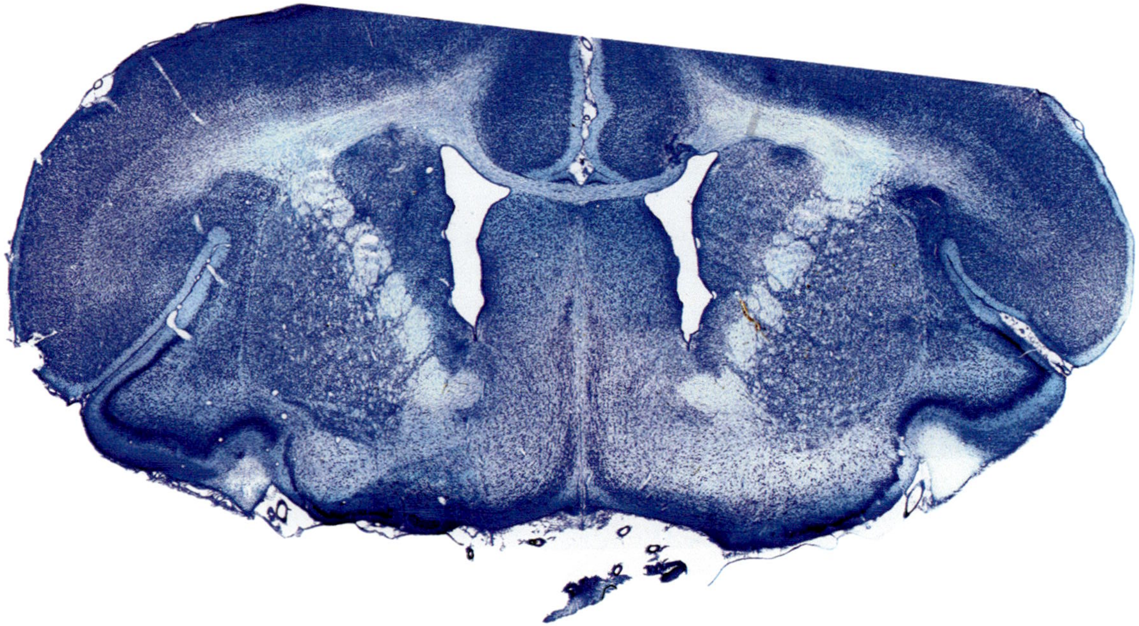

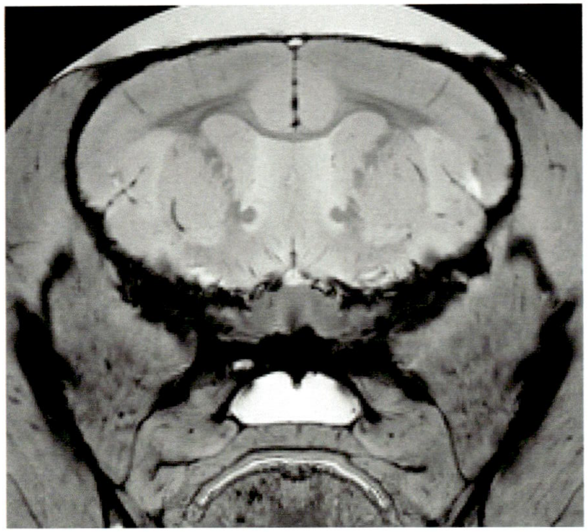

2	layer 2 of cortex
3	layer 3 of cortex
ac	anterior commissure
Acb	accumbens nucleus
cc	corpus callosum
Cd	caudate nucleus
Cg	cingulate cortex
Cl	claustrum
HDB	nucleus of the horizontal limb of the diagonal band
ic	internal capsule
IG	indusium griseum
Ins	insular cortex
IPA	interpeduncular nucleus, apical subnucleus
lo	lateral olfactory tract
LSd	lateral septal nucleus, dorsal part
LSi	lateral septal nucleus, intermediate part
LSv	lateral septal nucleus, ventral part
LV	lateral ventricle
M1	primary motor cortex
MS	medial septal nucleus
Pir1	piriform cortex, layer 1
Pu	putamen
S1	primary somatosensory cortex
S2	secondary somatosensory cortex
SHi	septohippocampal nucleus
STMA	bed nucleus of the stria terminalis, medial division, anterior part
Tu1	layer 1 of olfactory tubercle
VDB	nucleus of the vertical limb of the diagonal band

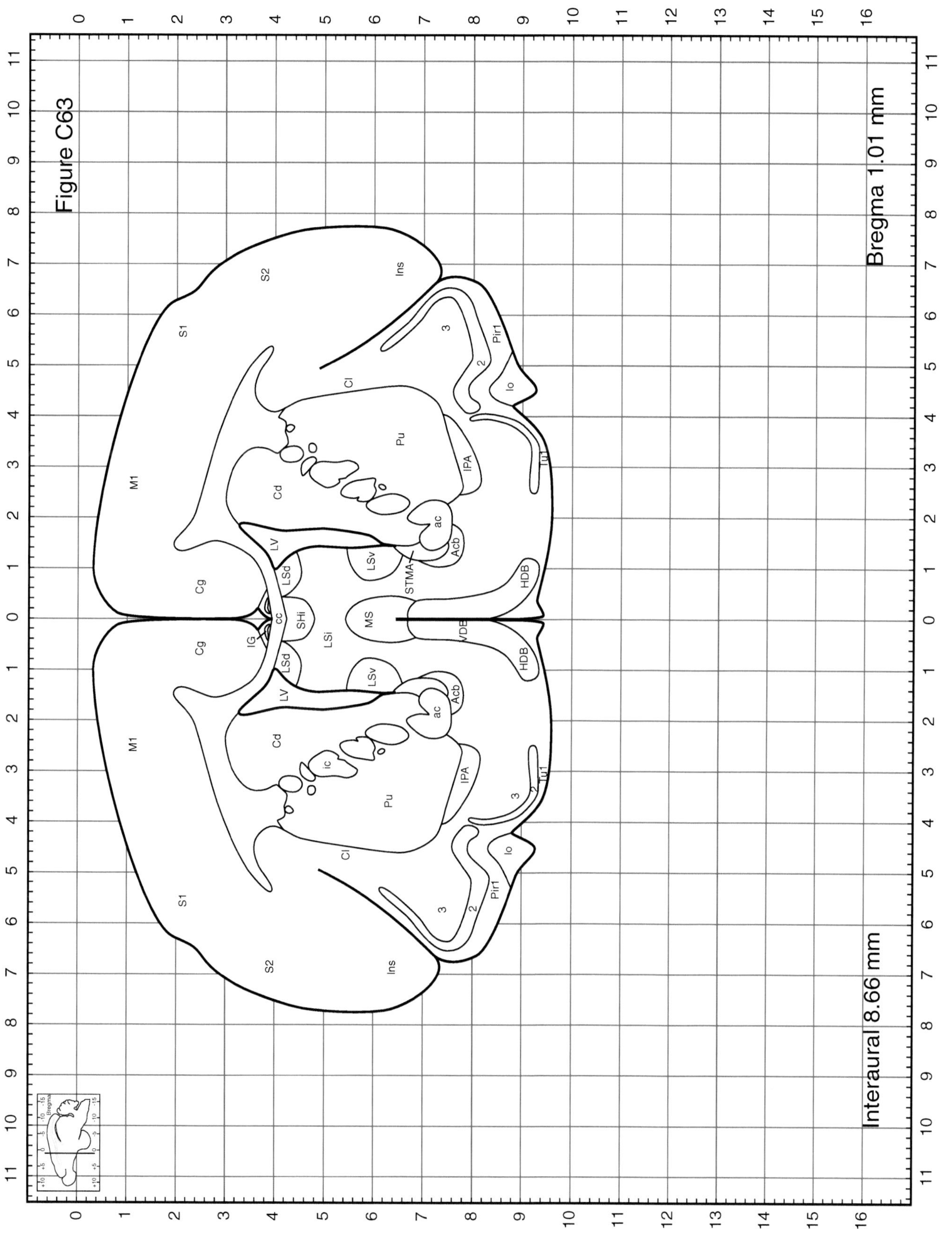
Figure C63
Bregma 1.01 mm
Interaural 8.66 mm
M1
S1
S2
Ins
Cg
Cd
Pu
Cl
LV
LSd
LSv
LSi
SHi
cc
IG
MS
VDB
HDB
STMA
ac
Acb
IPA
Tu1
Pir1
lo
ic
2
3

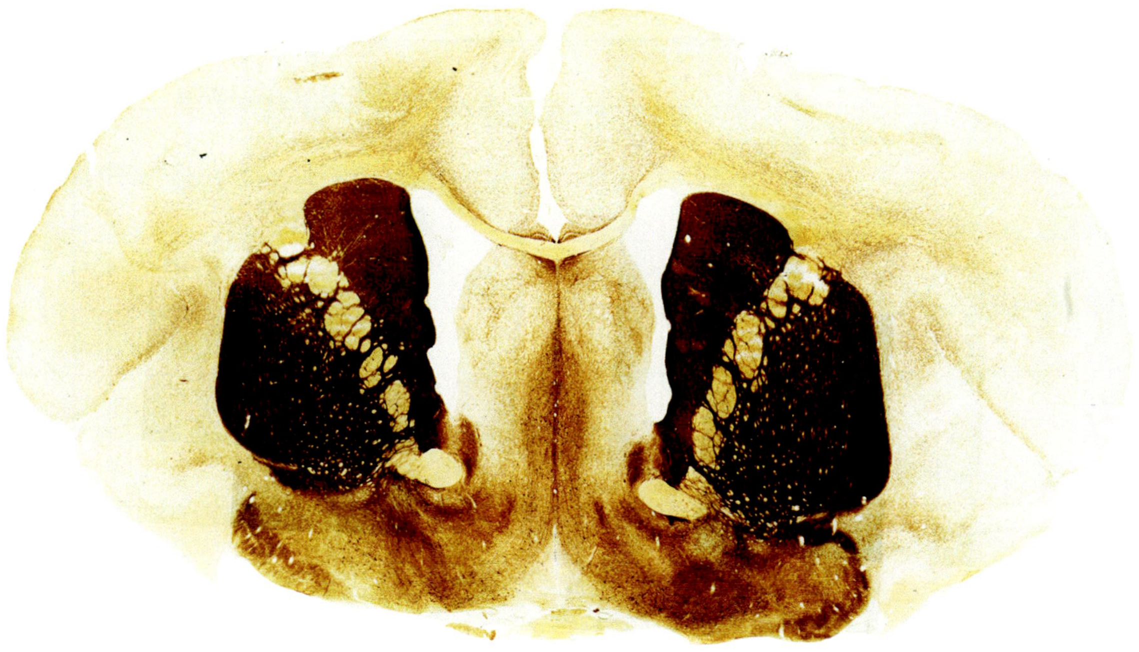

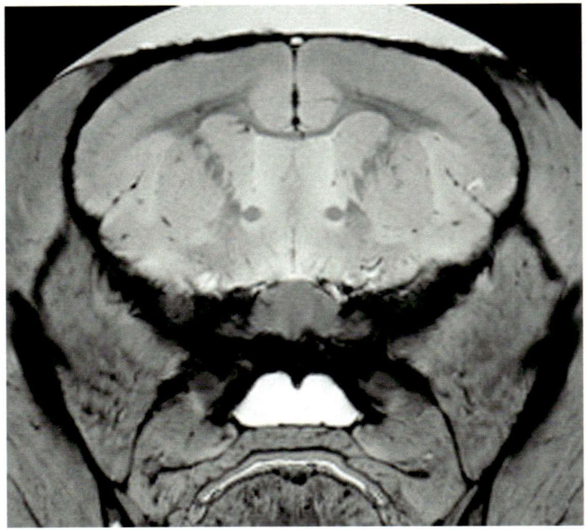

2	layer 2 of cortex
3	layer 3 of cortex
ac	anterior commissure
cc	corpus callosum
Cd	caudate nucleus
Cg	cingulate cortex
Cl	claustrum
HDB	nucleus of the horizontal limb of the diagonal band
ic	internal capsule
IG	indusium griseum
Ins	insular cortex
IPA	interpeduncular nucleus, apical subnucleus
lo	lateral olfactory tract
LSd	lateral septal nucleus, dorsal part
LSi	lateral septal nucleus, intermediate part
LSv	lateral septal nucleus, ventral part
LV	lateral ventricle
M1	primary motor cortex
MS	medial septal nucleus
och	optic chiasm
Pir1	piriform cortex, layer 1
Pu	putamen
S1	primary somatosensory cortex
S2	secondary somatosensory cortex
SHi	septohippocampal nucleus
STMA	bed nucleus of the stria terminalis, medial division, anterior part
Tu1	layer 1 of olfactory tubercle
VDB	nucleus of the vertical limb of the diagonal band

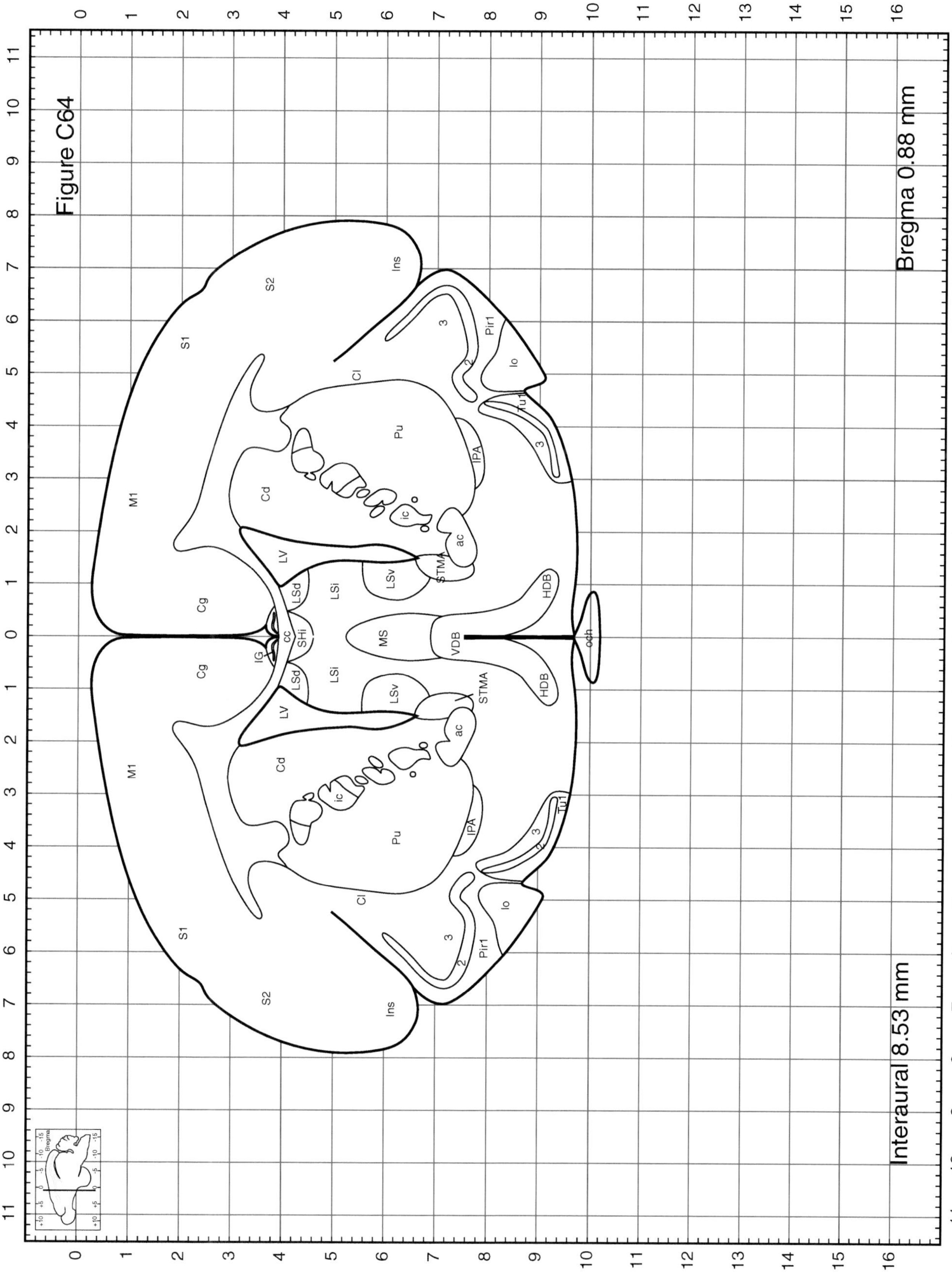
Figure C64
Bregma 0.88 mm
Interaural 8.53 mm
M1
S1
S2
Ins
Cg
Cl
Cd
Pu
ic
LV
LSd
LSi
LSv
cc
SHi
IG
MS
VDB
HDB
STMA
ac
IPA
Tu1
Pir1
lo
och
2
3

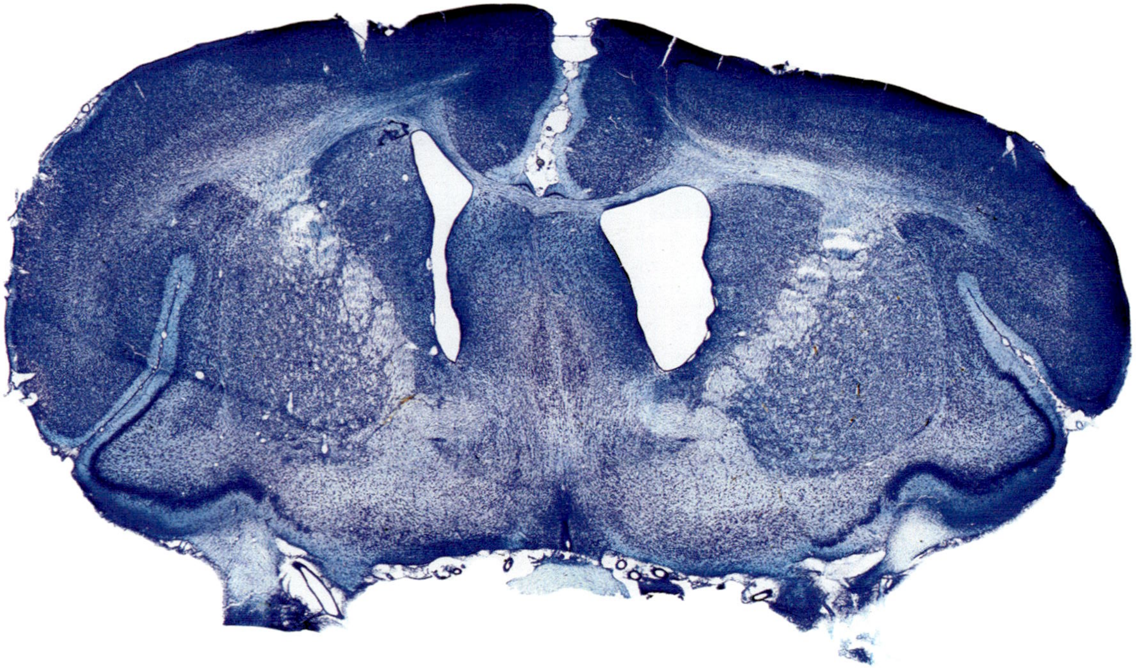

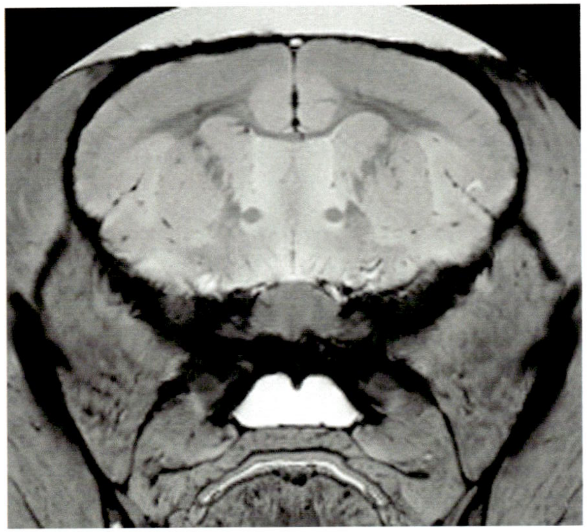

2	layer 2 of cortex
3	layer 3 of cortex
ac	anterior commissure
cc	corpus callosum
Cd	caudate nucleus
Cg	cingulate cortex
Cl	claustrum
HDB	nucleus of the horizontal limb of the diagonal band
ic	internal capsule
IG	indusium griseum
Ins	insular cortex
IPAC	interstitial nucleus of the posterior limb of the anterior commissure
lo	lateral olfactory tract
LSd	lateral septal nucleus, dorsal part
LSi	lateral septal nucleus, intermediate part
LSv	lateral septal nucleus, ventral part
LV	lateral ventricle
M1	primary motor cortex
MPA	medial preoptic area
MS	medial septal nucleus
och	optic chiasm
Pir1	piriform cortex, layer 1
Pu	putamen
PvA	parietal ventral area
S1	primary somatosensory cortex
S2	secondary somatosensory cortex
SHi	septohippocampal nucleus
STMA	bed nucleus of the stria terminalis, medial division, anterior part
STMV	bed nucleus of the stria terminalis, medial division, ventral part
Tu1	layer 1 of olfactory tubercle
VDB	nucleus of the vertical limb of the diagonal band
VP	ventral pallidum

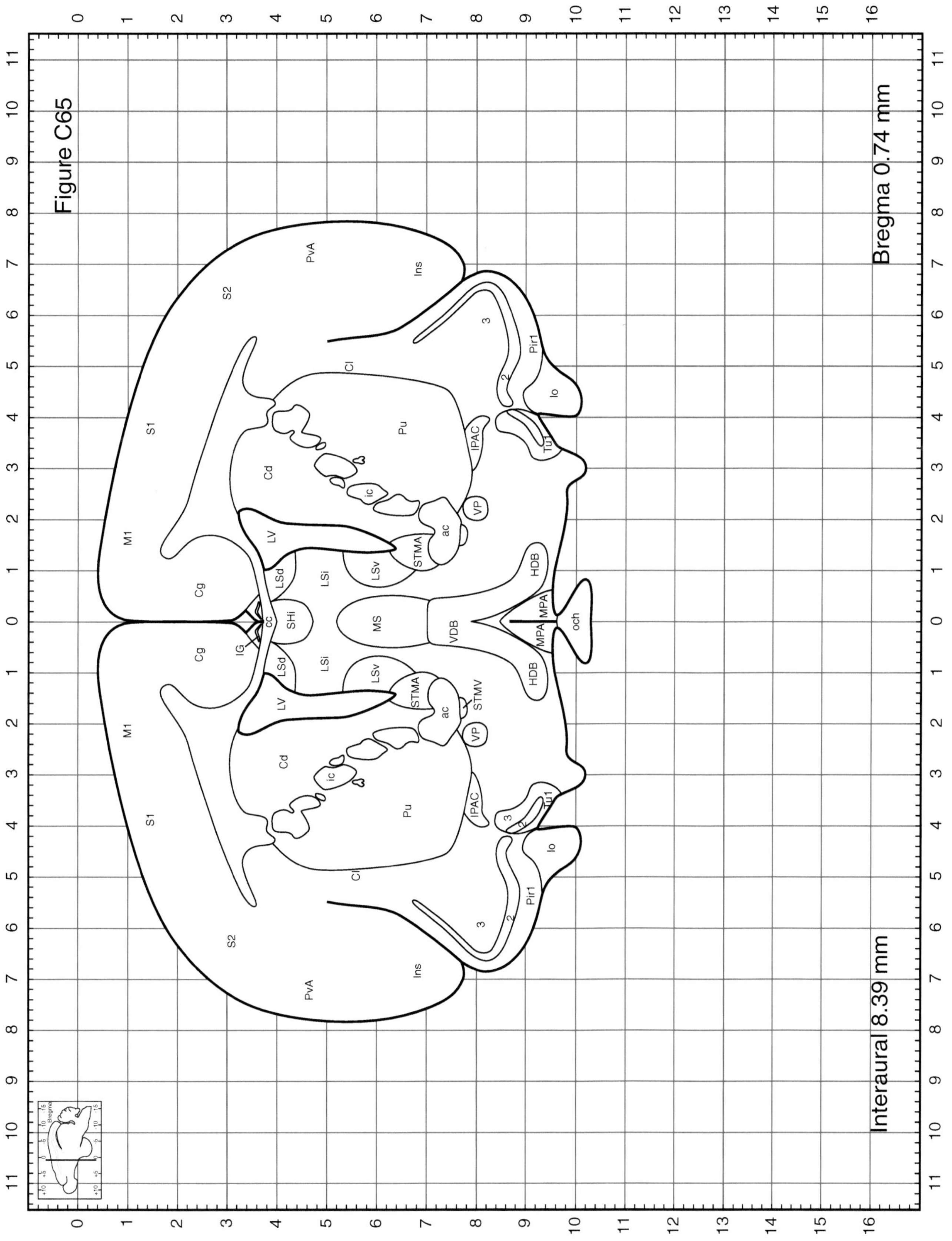
Figure C65
Bregma 0.74 mm
Interaural 8.39 mm
M1
S1
S2
PvA
Ins
Cg
IG
cc
LSd
LSi
LSv
SHi
MS
LV
Cd
Pu
Cl
ic
ac
STMA
STMV
VP
IPAC
VDB
HDB
MPA
och
Tu1
lo
Pir1
3
2

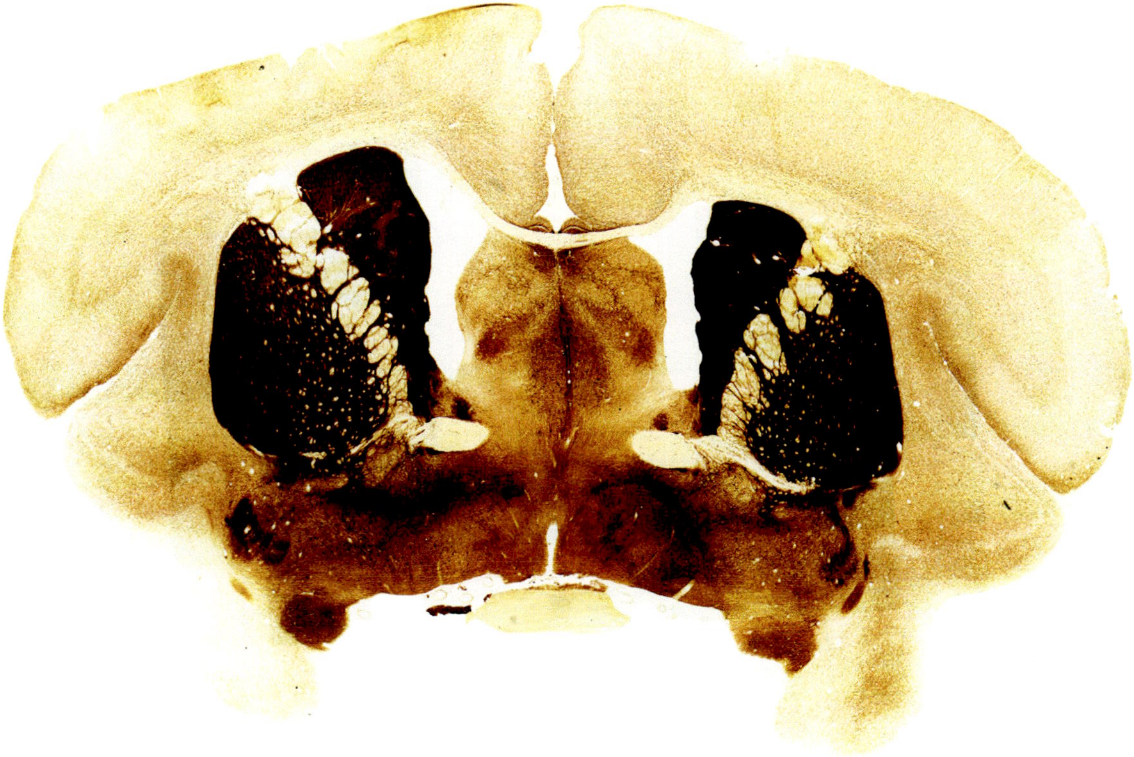

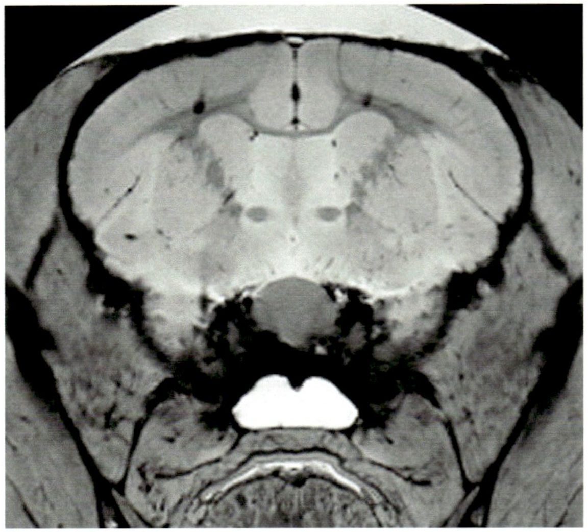

2	layer 2 of cortex
3	layer 3 of cortex
3V	3rd ventricle
ac	anterior commissure
cc	corpus callosum
Cd	caudate nucleus
Cg	cingulate cortex
Cl	claustrum
HDB	nucleus of the horizontal limb of the diagonal band
ic	internal capsule
IG	indusium griseum
Ins	insular cortex
IPAC	interstitial nucleus of the posterior limb of the anterior commissure
lo	lateral olfactory tract
LOT	nucleus of the lateral olfactory tract
LSd	lateral septal nucleus, dorsal part
LSi	lateral septal nucleus, intermediate part
LSv	lateral septal nucleus, ventral part
LV	lateral ventricle
MPA	medial preoptic area
MS	medial septal nucleus
och	optic chiasm
Pir1	piriform cortex, layer 1
Pu	putamen
PvA	parietal ventral area
S1	primary somatosensory cortex
S2	secondary somatosensory cortex
SHi	septohippocampal nucleus
STMA	bed nucleus of the stria terminalis, medial division, anterior part
STMV	bed nucleus of the stria terminalis, medial division, ventral part
Tu1	layer 1 of olfactory tubercle
VDB	nucleus of the vertical limb of the diagonal band
VP	ventral pallidum

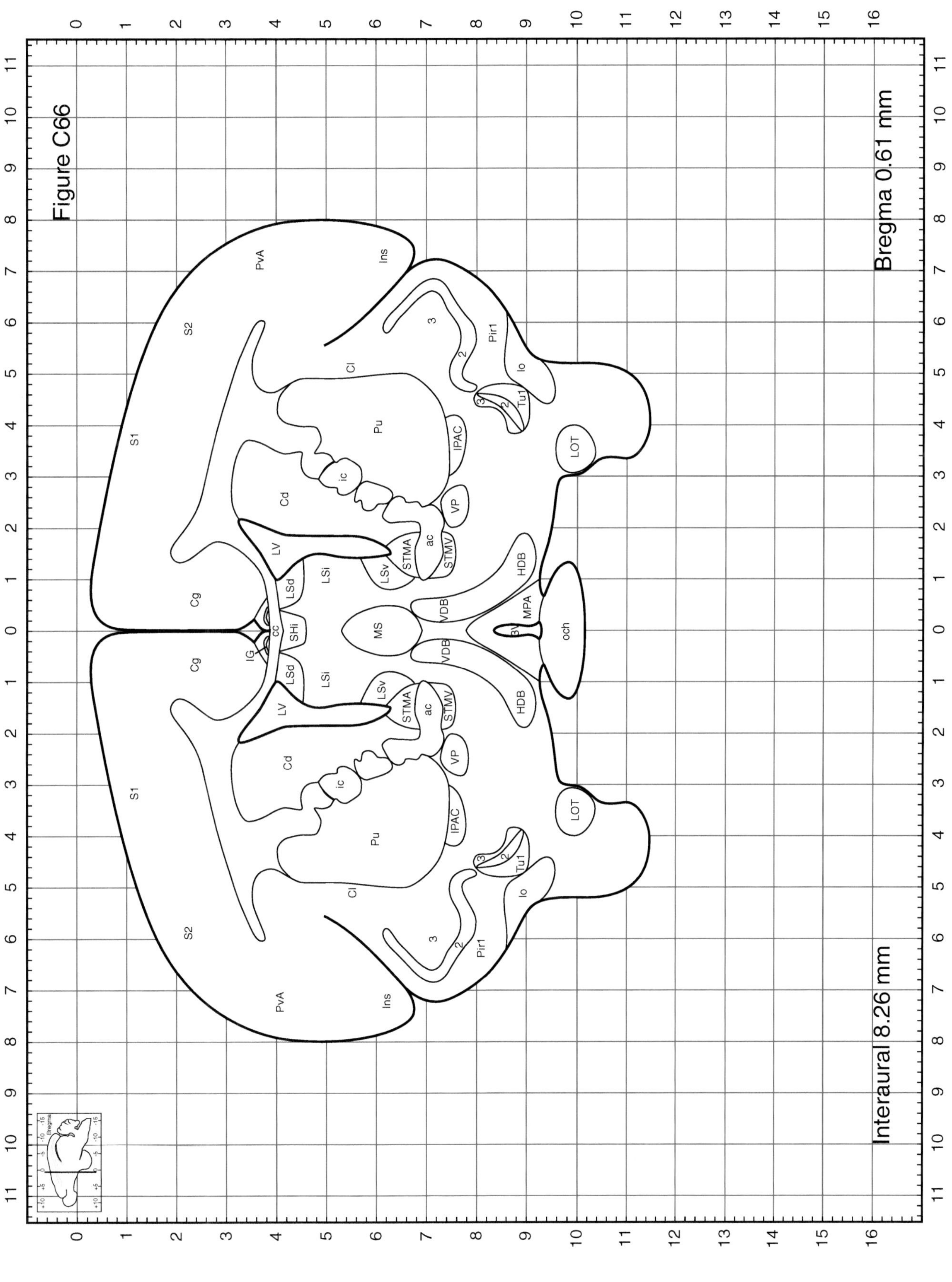
Figure C66
Bregma 0.61 mm
Interaural 8.26 mm
PvA
Ins
S2
S1
Cl
Pir1
3
2
lo
Tu1
Pu
IPAC
LOT
ic
Cd
VP
ac
STMA
STMV
LV
LSd
LSi
LSv
HDB
VDB
MPA
3V
och
MS
SHi
cc
IG
Cg

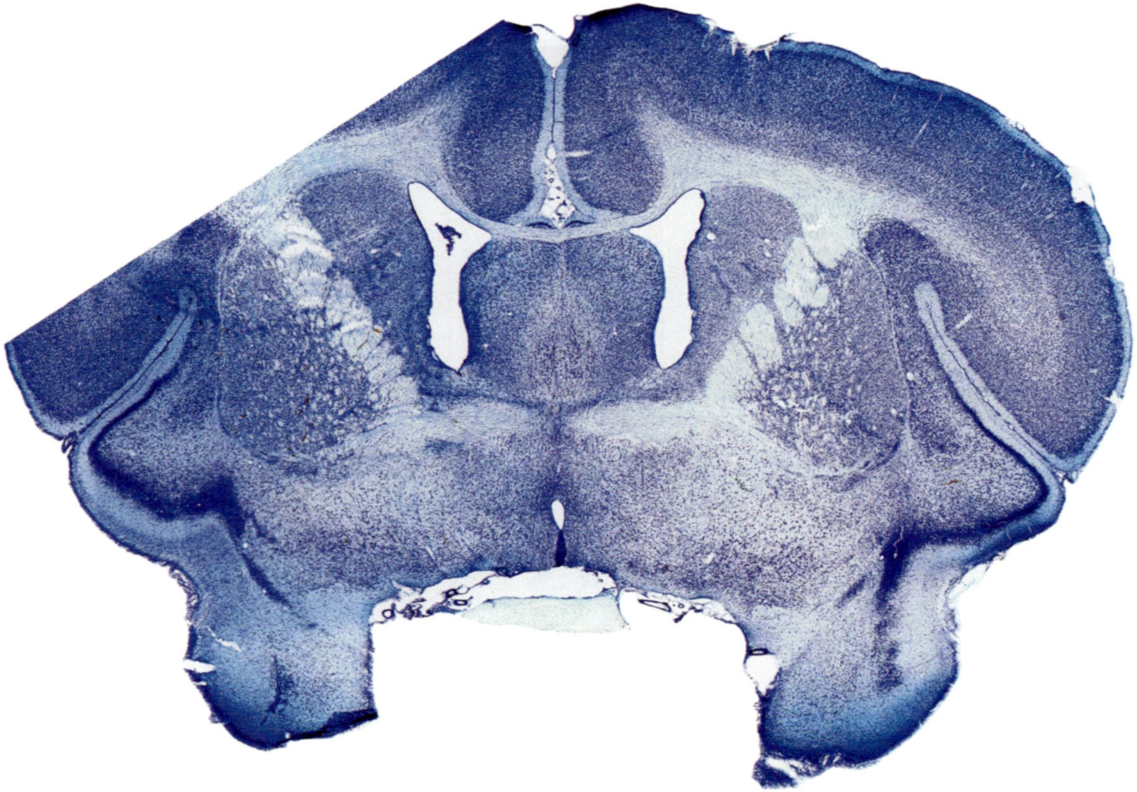

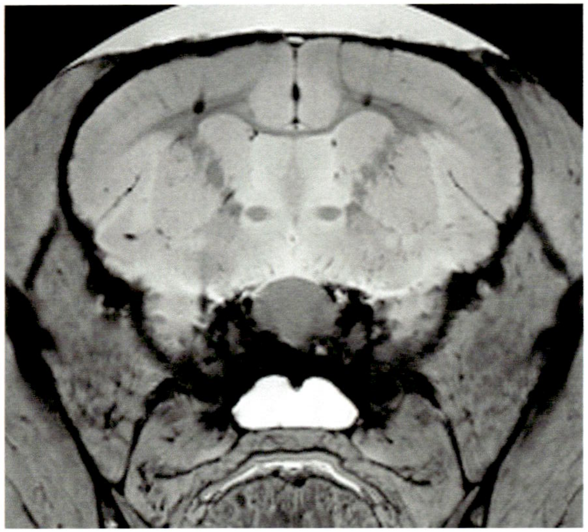

2	layer 2 of cortex
3	layer 3 of cortex
3V	3rd ventricle
ac	anterior commissure
AVPe	anteroventral periventricular nucleus
cc	corpus callosum
Cd	caudate nucleus
Cg	cingulate cortex
Cl	claustrum
ic	internal capsule
IG	indusium griseum
Ins	insular cortex
IPAC	interstitial nucleus of the posterior limb of the anterior commissure
lo	lateral olfactory tract
LOT	nucleus of the lateral olfactory tract
LPO	lateral preoptic area
LSd	lateral septal nucleus, dorsal part
LSi	lateral septal nucleus, intermediate part
LSv	lateral septal nucleus, ventral part
LV	lateral ventricle
MPA	medial preoptic area
MS	medial septal nucleus
och	optic chiasm
Pir1	piriform cortex, layer 1
Pu	putamen
PvA	parietal ventral area
S2	secondary somatosensory cortex
SCC	somatosensory caudal cortex
SHi	septohippocampal nucleus
SON	supraoptic nucleus
STMA	bed nucleus of the stria terminalis, medial division, anterior part
STMV	bed nucleus of the stria terminalis, medial division, ventral part
Tu	olfactory tubercle
VP	ventral pallidum

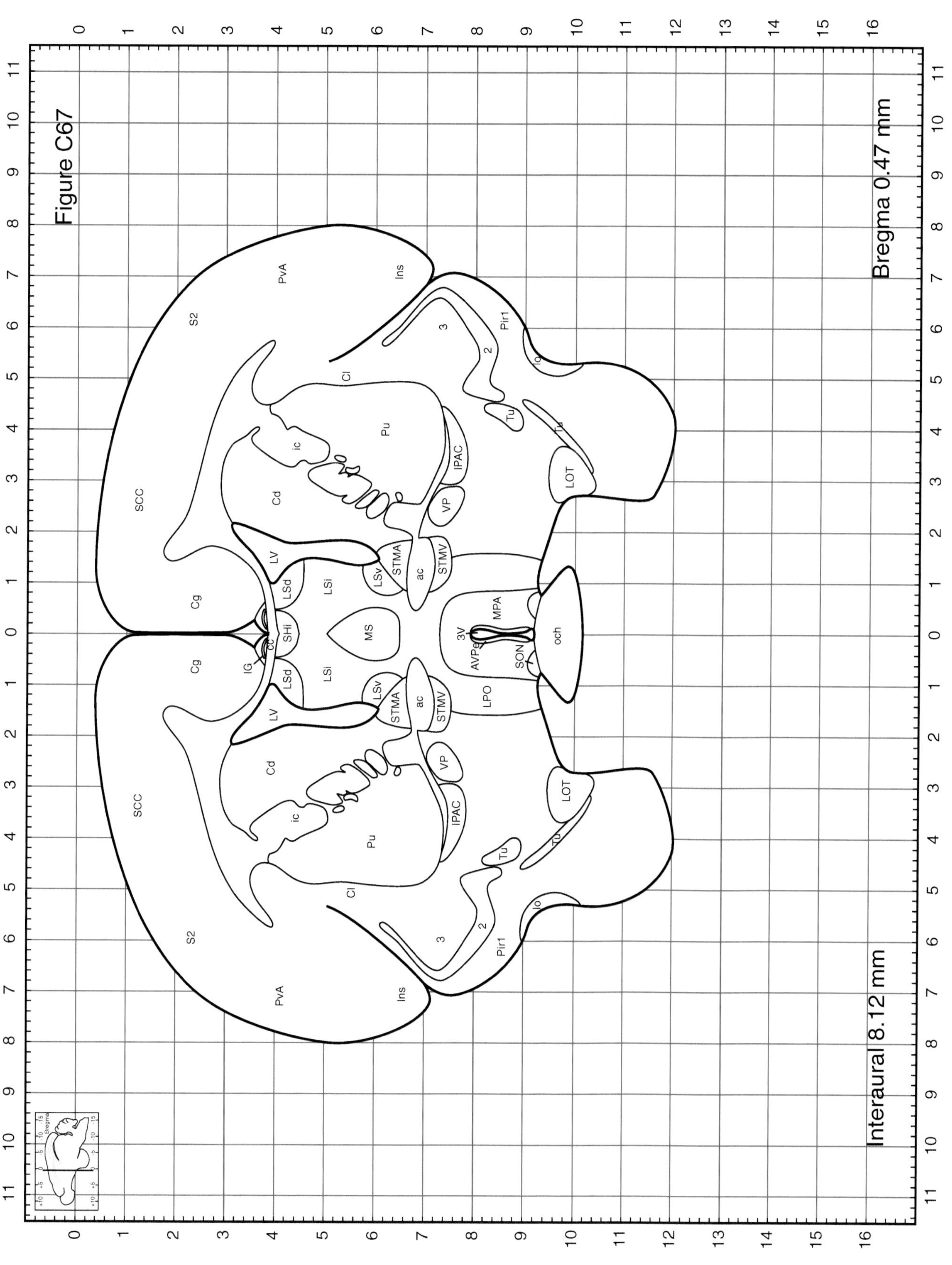

Figure C67
Bregma 0.47 mm
Interaural 8.12 mm
PvA
Ins
S2
Pir1
3
2
lo
Cl
Tu
Pu
ic
IPAC
LOT
SCC
Cd
VP
LV
STMA
STMV
ac
LSv
LSd
LSi
Cg
MPA
MS
SHi
3V
och
cc
IG
AVPe
SON
LPO
Bregma

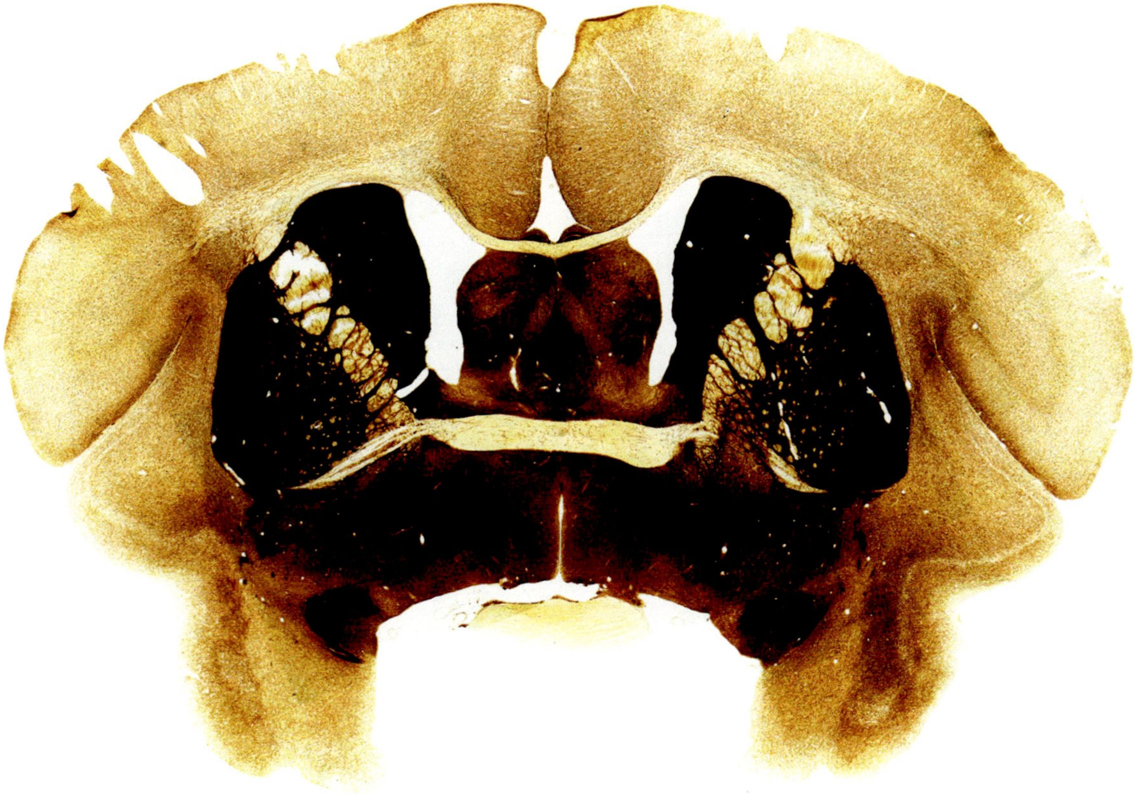

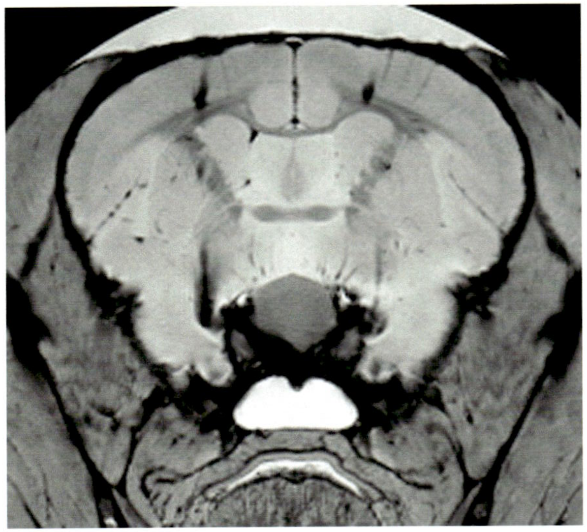

2	layer 2 of cortex
3	layer 3 of cortex
3V	3rd ventricle
AA	anterior amygdaloid area
ac	anterior commissure
Aco	anterior cortical amygdaloid nucleus
AVPe	anteroventral periventricular nucleus
cc	corpus callosum
Cd	caudate nucleus
Cl	claustrum
ic	internal capsule
IG	indusium griseum
Ins	insular cortex
IPAC	interstitial nucleus of the posterior limb of the anterior commissure
IRd	infraradiata dorsalis
IRv	infraradiata ventral area
lo	lateral olfactory tract
LOT	nucleus of the lateral olfactory tract
LPO	lateral preoptic area
LSd	lateral septal nucleus, dorsal part
LSi	lateral septal nucleus, intermediate part
LSv	lateral septal nucleus, ventral part
LV	lateral ventricle
MPA	medial preoptic area
MS	medial septal nucleus
och	optic chiasm
Pir1	piriform cortex, layer 1
Pu	putamen
PvA	parietal ventral area
S2	secondary somatosensory cortex
SCC	somatosensory caudal cortex
SHi	septohippocampal nucleus
SON	supraoptic nucleus
STMA	bed nucleus of the stria terminalis, medial division, anterior part
STMV	bed nucleus of the stria terminalis, medial division, ventral part
Tu	olfactory tubercle
VP	ventral pallidum

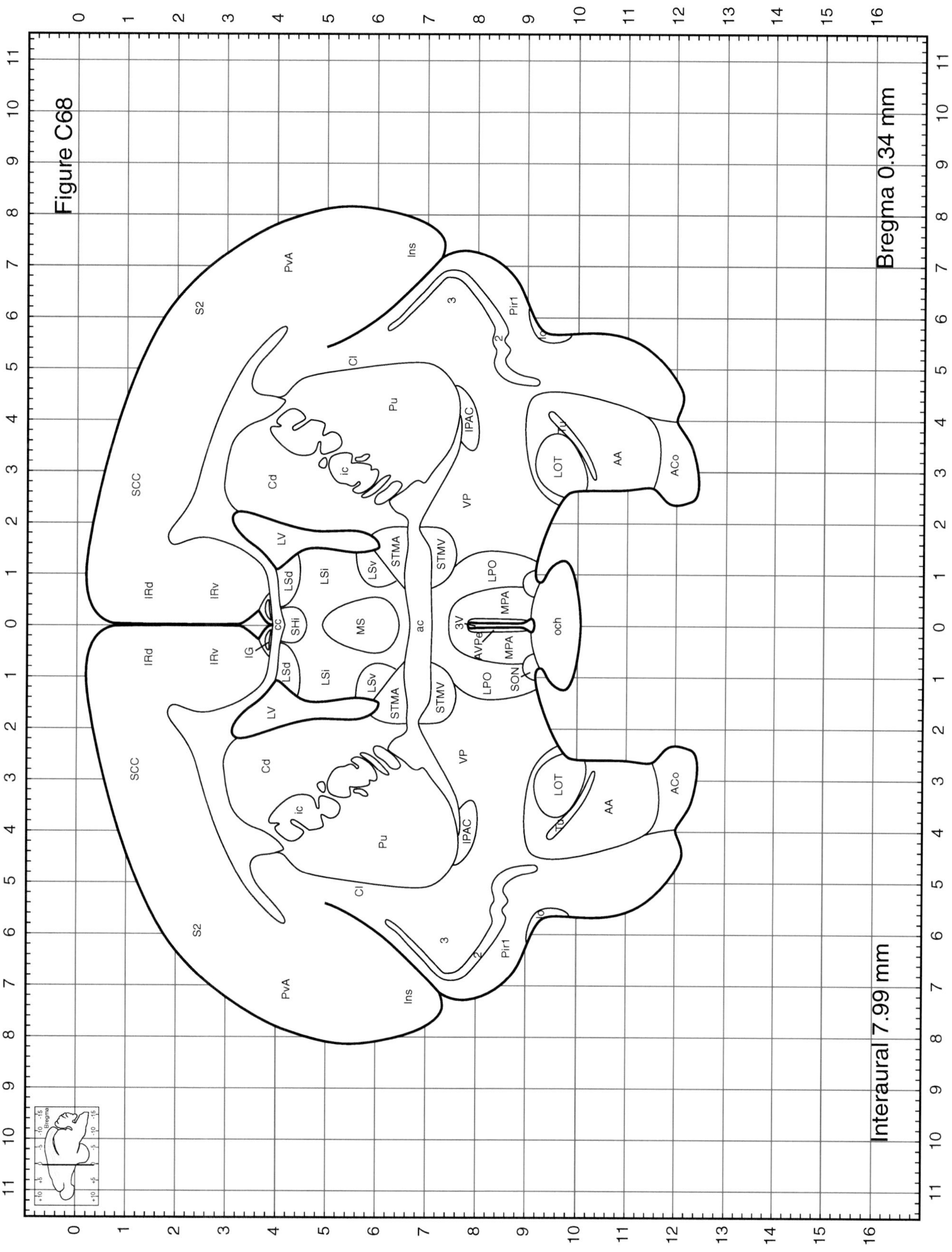
Figure C68
Bregma 0.34 mm
Interaural 7.99 mm
PvA
Ins
S2
3
Pir1
2
lo
Cl
Pu
IPAC
SCC
Cd
ic
LOT
Tu
AA
ACo
VP
LV
STMA
STMV
LSv
LSd
LSi
LPO
IRd
IRv
MPA
cc
SHi
MS
ac
3V
AVPe
och
IG
SON

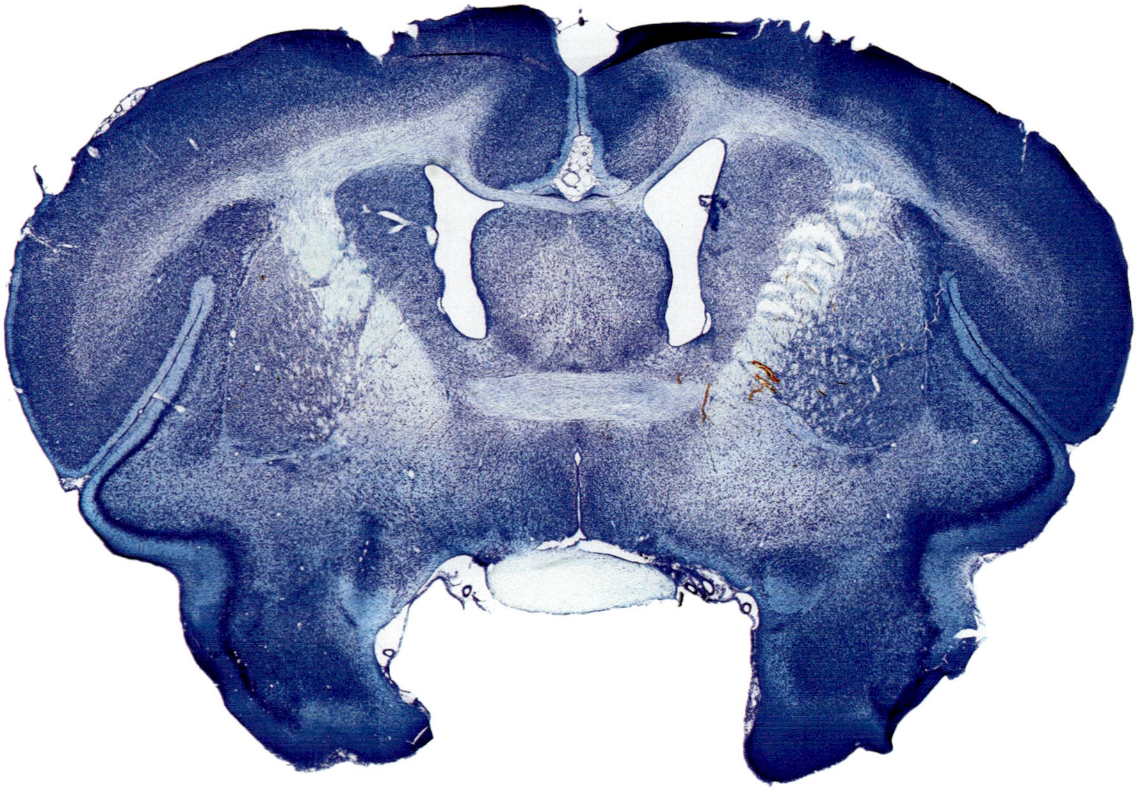

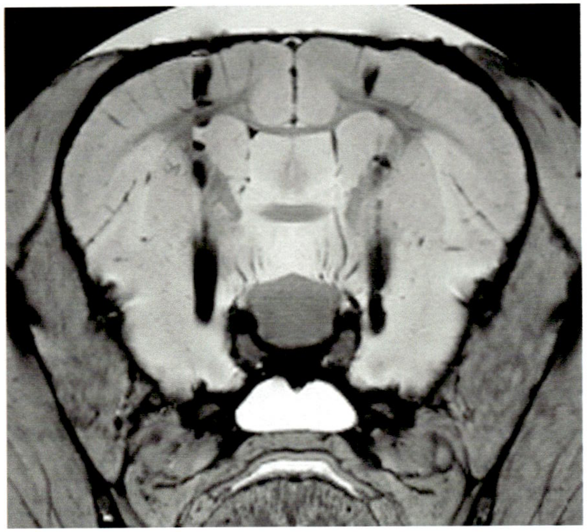

2	layer 2 of cortex
3	layer 3 of cortex
3V	3rd ventricle
AA	anterior amygdaloid area
ac	anterior commissure
ACo	anterior cortical amygdaloid nucleus
acp	anterior commissure, posterior part
AVPe	anteroventral periventricular nucleus
cc	corpus callosum
Cd	caudate nucleus
Cl	claustrum
EGP	external globus pallidus
ic	internal capsule
IG	indusium griseum
Ins	insular cortex
IPAC	interstitial nucleus of the posterior limb of the anterior commissure
IRd	infraradiata dorsalis
IRv	infraradiata ventral area
lo	lateral olfactory tract
LOT	nucleus of the lateral olfactory tract
LPO	lateral preoptic area
LSd	lateral septal nucleus, dorsal part
LSi	lateral septal nucleus, intermediate part
LSS	lateral stripe of the striatum
LSv	lateral septal nucleus, ventral part
LV	lateral ventricle
MnPO	median preoptic nucleus
MPO	medial preoptic nucleus
MS	medial septal nucleus
och	optic chiasm
Pir1	piriform cortex, layer 1
Pu	putamen
PvA	parietal ventral area
S2	secondary somatosensory cortex
SCC	somatosensory caudal cortex
SFi	septofimbrial nucleus
SHi	septohippocampal nucleus
SHy	septohypothalamic nucleus
SON	supraoptic nucleus
STMAL	bed nucleus of the stria terminalis, medial division, posterolateral part
STMAM	bed nucleus of the stria terminalis, medial division, anteromedial part
STMV	bed nucleus of the stria terminalis, medial division, ventral part
Tu	olfactory tubercle
VP	ventral pallidum

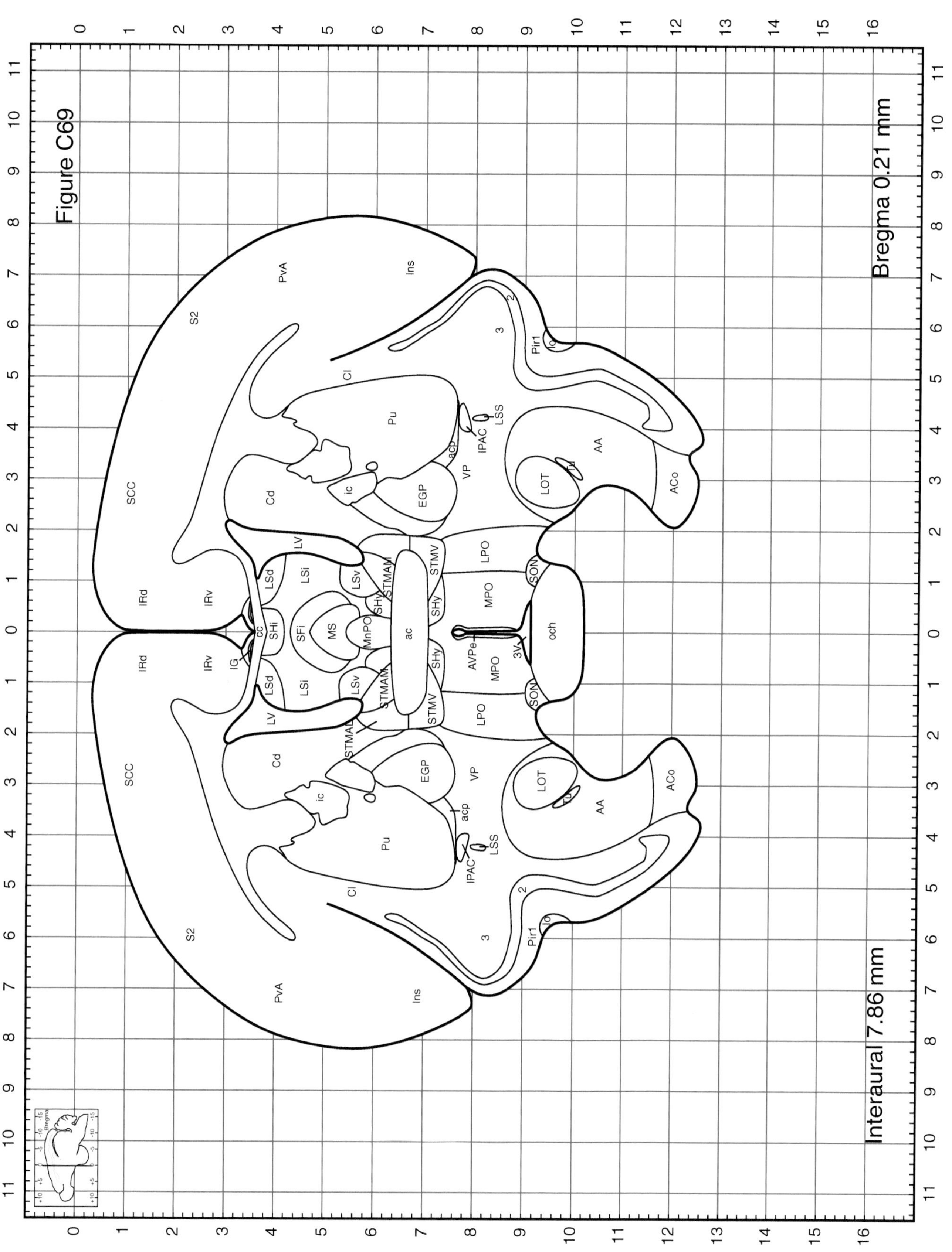
Figure C69
Bregma 0.21 mm
Interaural 7.86 mm
PvA
Ins
S2
SCC
IRd
IRv
IG
cc
Cl
Pu
Cd
ic
EGP
acp
IPAC
LSS
VP
AA
LOT
Tu
ACo
Pir1
lo
2
3
LV
LSd
LSi
LSv
SHi
SFi
MS
MnPO
SHy
STMAM
STMAL
STMV
ac
LPO
MPO
AVPe
3V
SON
och

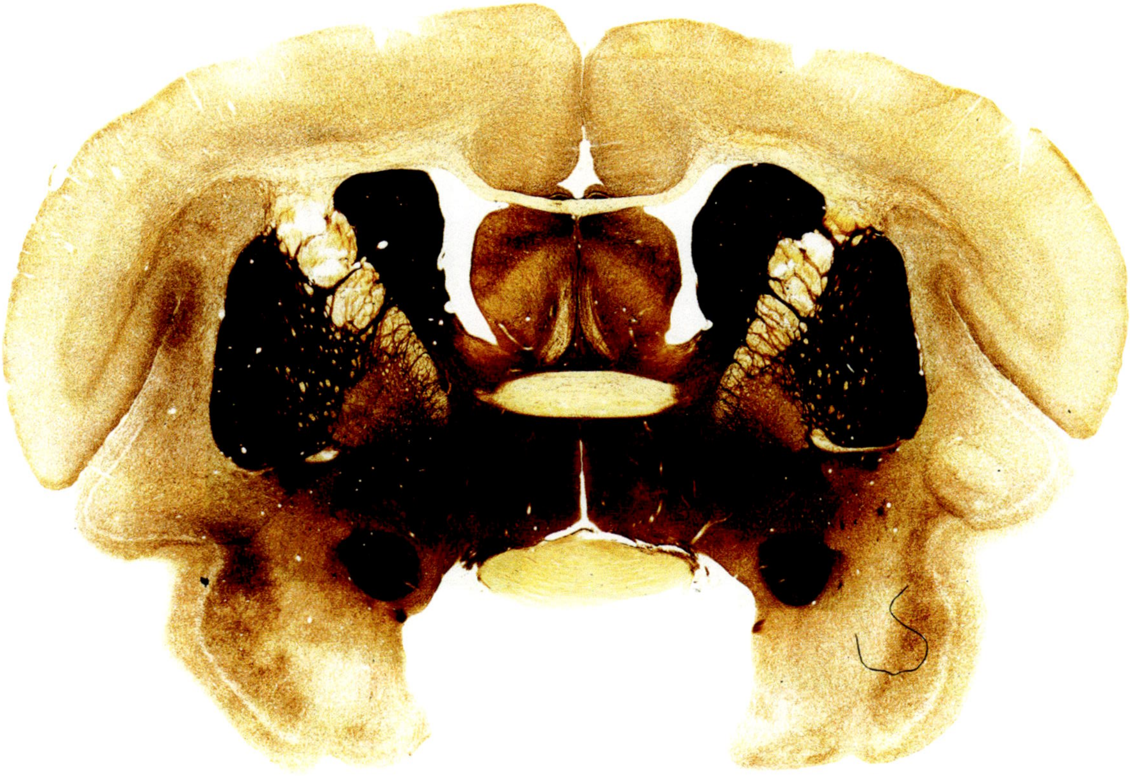

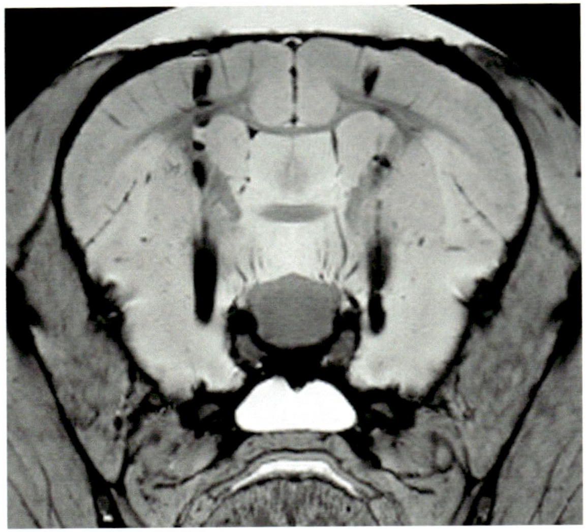

2	layer 2 of cortex
3	layer 3 of cortex
3V	3rd ventricle
AA	anterior amygdaloid area
ac	anterior commissure
Aco	anterior cortical amygdaloid nucleus
acp	anterior commissure, posterior part
AVPe	anteroventral periventricular nucleus
cc	corpus callosum
Cd	caudate nucleus
Cl	claustrum
EGP	external globus pallidus
f	fornix
ic	internal capsule
IG	indusium griseum
Ins	insular cortex
IPAC	interstitial nucleus of the posterior limb of the anterior commissure
IRd	infraradiata dorsalis
IRv	infraradiata ventral area
lo	lateral olfactory tract
LOT	nucleus of the lateral olfactory tract
LPO	lateral preoptic area
LSd	lateral septal nucleus, dorsal part
LSi	lateral septal nucleus, intermediate part
LSS	lateral stripe of the striatum
LSv	lateral septal nucleus, ventral part
LV	lateral ventricle
MnPO	median preoptic nucleus
MPO	medial preoptic nucleus
och	optic chiasm
Pir1	piriform cortex, layer 1
Pu	putamen
PvA	parietal ventral area
S2	secondary somatosensory cortex
SCC	somatosensory caudal cortex
SFi	septofimbrial nucleus
SHi	septohippocampal nucleus
SHy	septohypothalamic nucleus
SON	supraoptic nucleus
STMAL	bed nucleus of the stria terminalis, medial division, posterolateral part
STMAM	bed nucleus of the stria terminalis, medial division, anteromedial part
STMV	bed nucleus of the stria terminalis, medial division, ventral part
TS	triangular septal nucleus
Tu	olfactory tubercle
VP	ventral pallidum

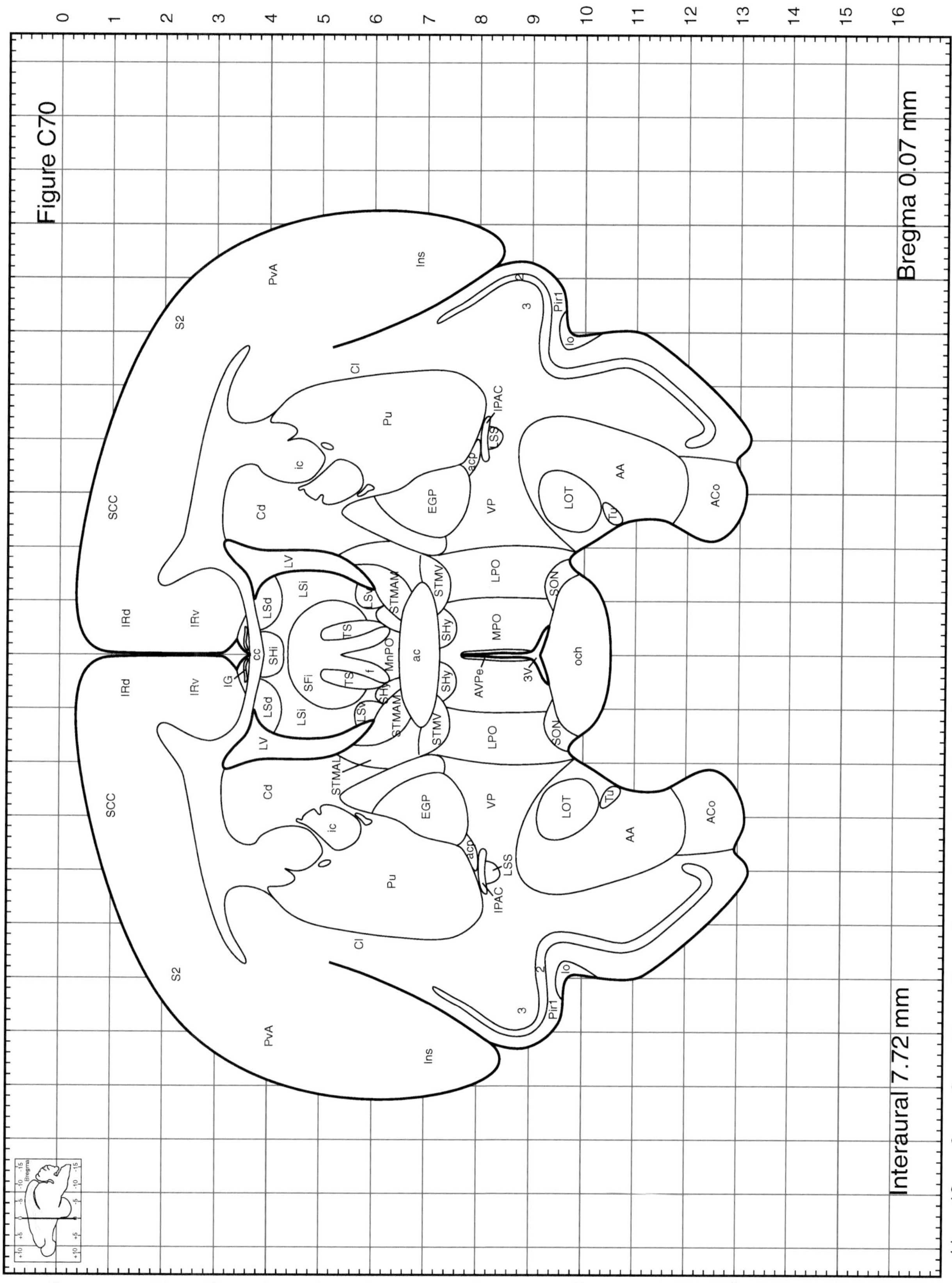
Figure C70
Bregma 0.07 mm
Interaural 7.72 mm
SCC
S2
PvA
Ins
Cl
Pu
ic
Cd
EGP
VP
LOT
AA
ACo
Tu
IPAC
LSS
acp
Pir1
lo
2
3
IRd
IRv
LV
LSd
LSi
LSv
SHi
SFi
cc
IG
TS
f
MnPO
SHy
STMAM
STMAL
STMV
ac
LPO
MPO
AVPe
3V
SON
och

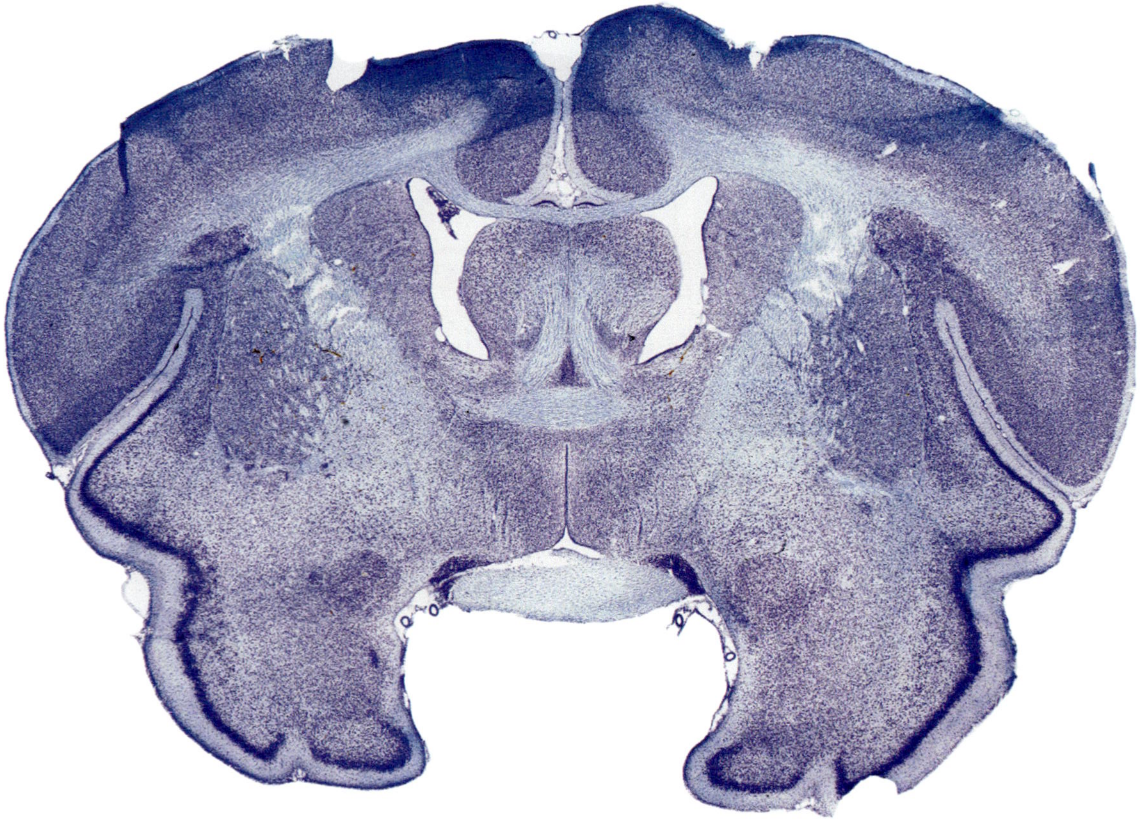

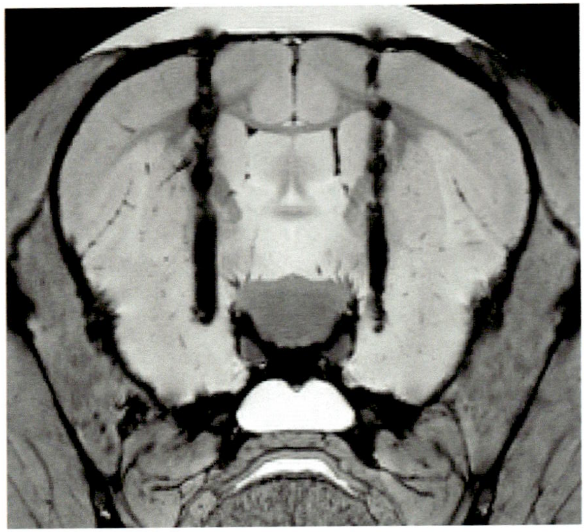

2	layer 2 of cortex
3	layer 3 of cortex
3V	3rd ventricle
AA	anterior amygdaloid area
Ab	auditory belt area
ac	anterior commissure
Aco	anterior cortical amygdaloid nucleus
acp	anterior commissure, posterior part
AVPe	anteroventral periventricular nucleus
cc	corpus callosum
Cd	caudate nucleus
Cl	claustrum
EGP	external globus pallidus
f	fornix
ic	internal capsule
IG	indusium griseum
IPAC	interstitial nucleus of the posterior limb of the anterior commissure
IRd	infraradiata dorsalis
IRv	infraradiata ventral area
lo	lateral olfactory tract
LOT	nucleus of the lateral olfactory tract
LPO	lateral preoptic area
LSd	lateral septal nucleus, dorsal part
LSi	lateral septal nucleus, intermediate part
LSS	lateral stripe of the striatum
LSv	lateral septal nucleus, ventral part
LV	lateral ventricle
MnPO	median preoptic nucleus
MPO	medial preoptic nucleus
och	optic chiasm
Pir1	piriform cortex, layer 1
PPd	posterior parietal dorsal area
PPr	posterior parietal rostral area
Pu	putamen
PvA	parietal ventral area
S2	secondary somatosensory cortex
SFi	septofimbrial nucleus
SHi	septohippocampal nucleus
SHy	septohypothalamic nucleus
SON	supraoptic nucleus
STMAL	bed nucleus of the stria terminalis, medial division, posterolateral part
STMAM	bed nucleus of the stria terminalis, medial division, anteromedial part
STMV	bed nucleus of the stria terminalis, medial division, ventral part
TS	triangular septal nucleus
VP	ventral pallidum

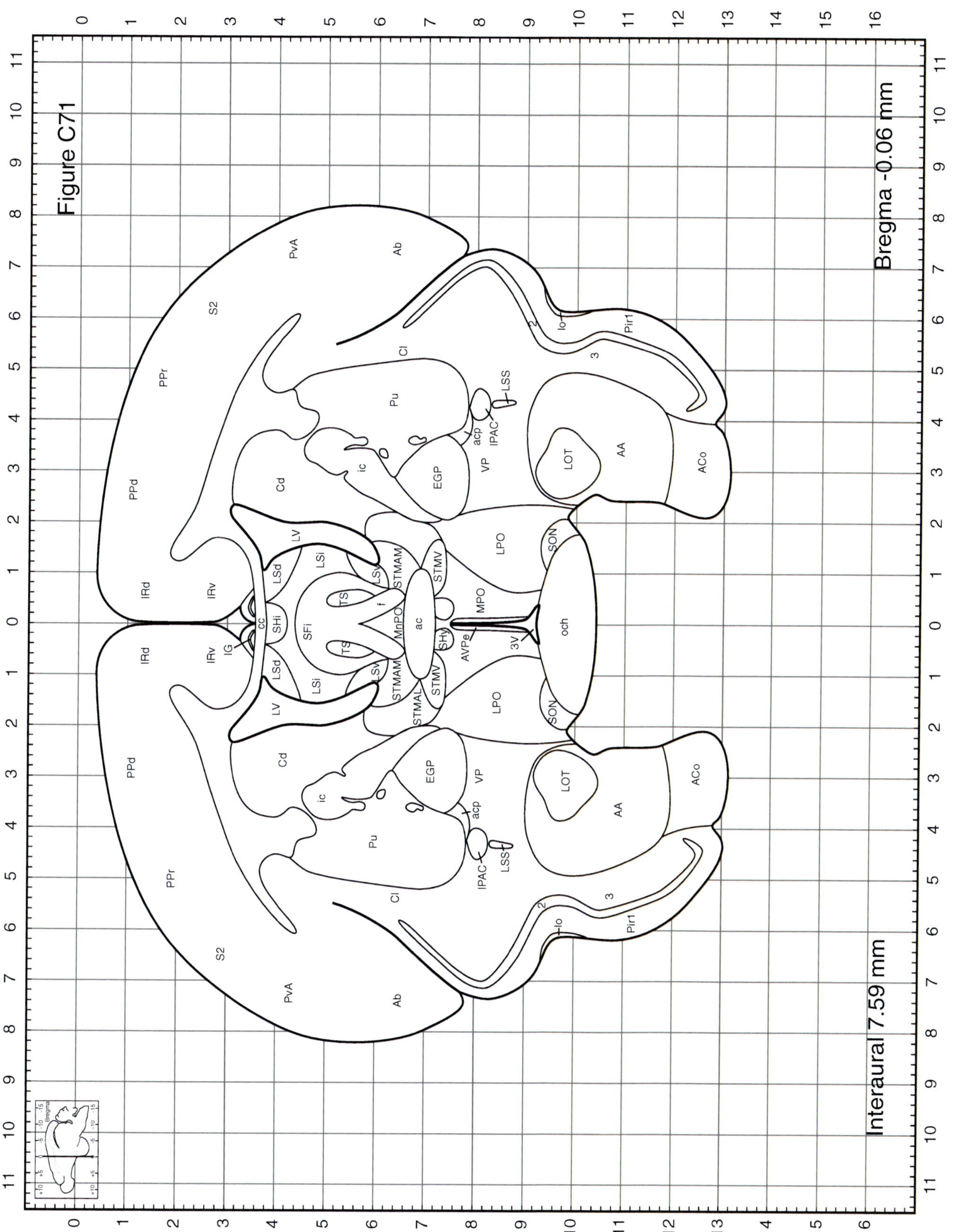
Figure C71
Bregma -0.06 mm
Interaural 7.59 mm
PvA
Ab
S2
PPr
PPd
IRd
IRv
Cl
Pu
Cd
ic
EGP
VP
acp
IPAC
LSS
lo
Pir1
AA
ACo
LOT
LPO
SON
MPO
och
3V
AVPe
STMV
STMAM
STMAL
LSv
LSd
LSi
LV
SFi
SHi
SHy
TS
f
MnPO
ac
cc
IG
2
3

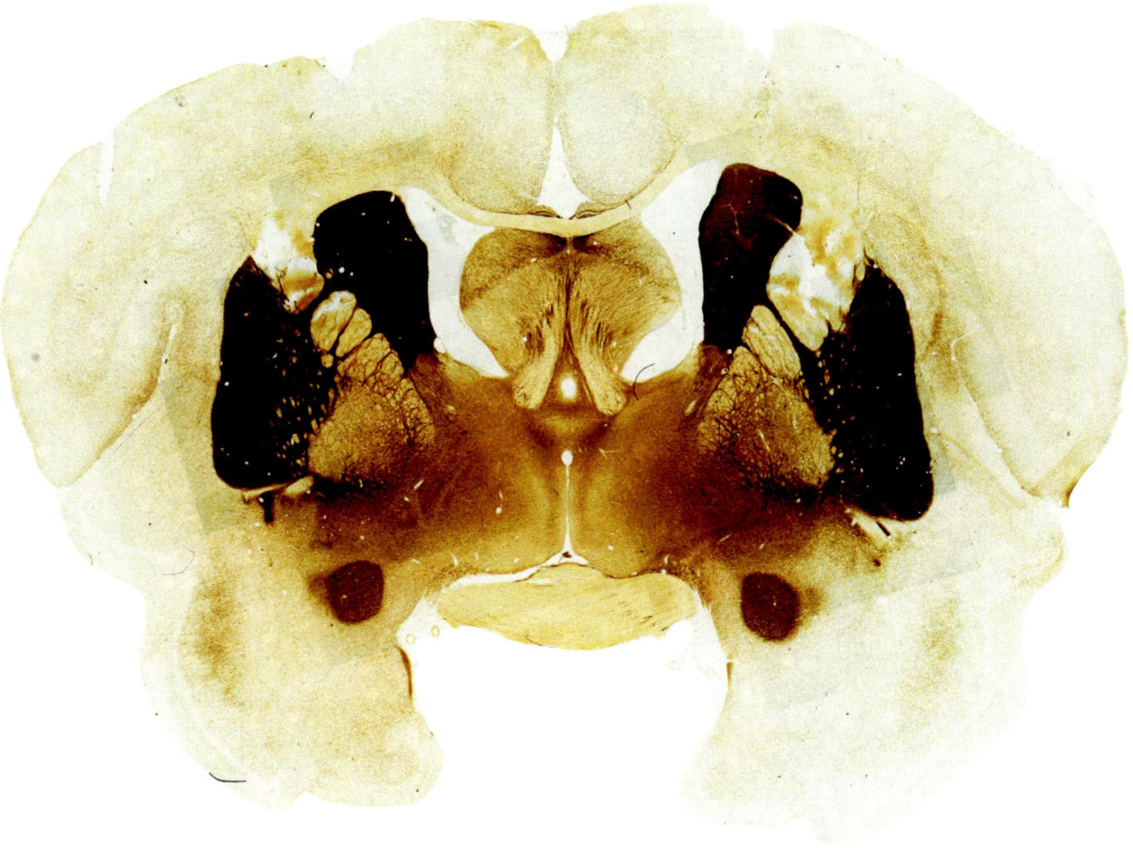

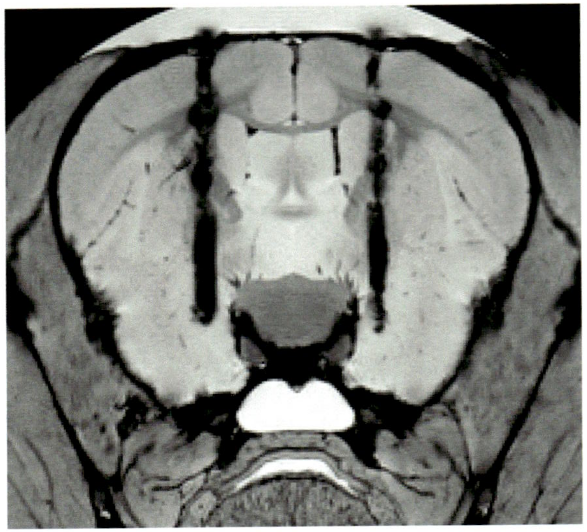

2	layer 2 of cortex
3	layer 3 of cortex
3V	3rd ventricle
AA	anterior amygdaloid area
Ab	auditory belt area
ac	anterior commissure
Aco	anterior cortical amygdaloid nucleus
acp	anterior commissure, posterior part
cc	corpus callosum
Cd	caudate nucleus
Cl	claustrum
EAC	sublenticular extended amygdala, central part
EGP	external globus pallidus
f	fornix
ic	internal capsule
IG	indusium griseum
IPAC	interstitial nucleus of the posterior limb of the anterior commissure
IRd	infraradiata dorsalis
IRv	infraradiata ventral area
LOT	nucleus of the lateral olfactory tract
LPO	lateral preoptic area
LSd	lateral septal nucleus, dorsal part
LSi	lateral septal nucleus, intermediate part
LSS	lateral stripe of the striatum
LSv	lateral septal nucleus, ventral part
LV	lateral ventricle
MnPO	median preoptic nucleus
MPO	medial preoptic nucleus
och	optic chiasm
Pe	periventricular hypothalamic nucleus
Pir1	piriform cortex, layer 1
PPd	posterior parietal dorsal area
PPr	posterior parietal rostral area
Pu	putamen
PvA	parietal ventral area
S2	secondary somatosensory cortex
SFi	septofimbrial nucleus
SHi	septohippocampal nucleus
SHy	septohypothalamic nucleus
SON	supraoptic nucleus
STMAL	bed nucleus of the stria terminalis, medial division, posterolateral part
STMAM	bed nucleus of the stria terminalis, medial division, anteromedial part
STMV	bed nucleus of the stria terminalis, medial division, ventral part
TS	triangular septal nucleus
VP	ventral pallidum

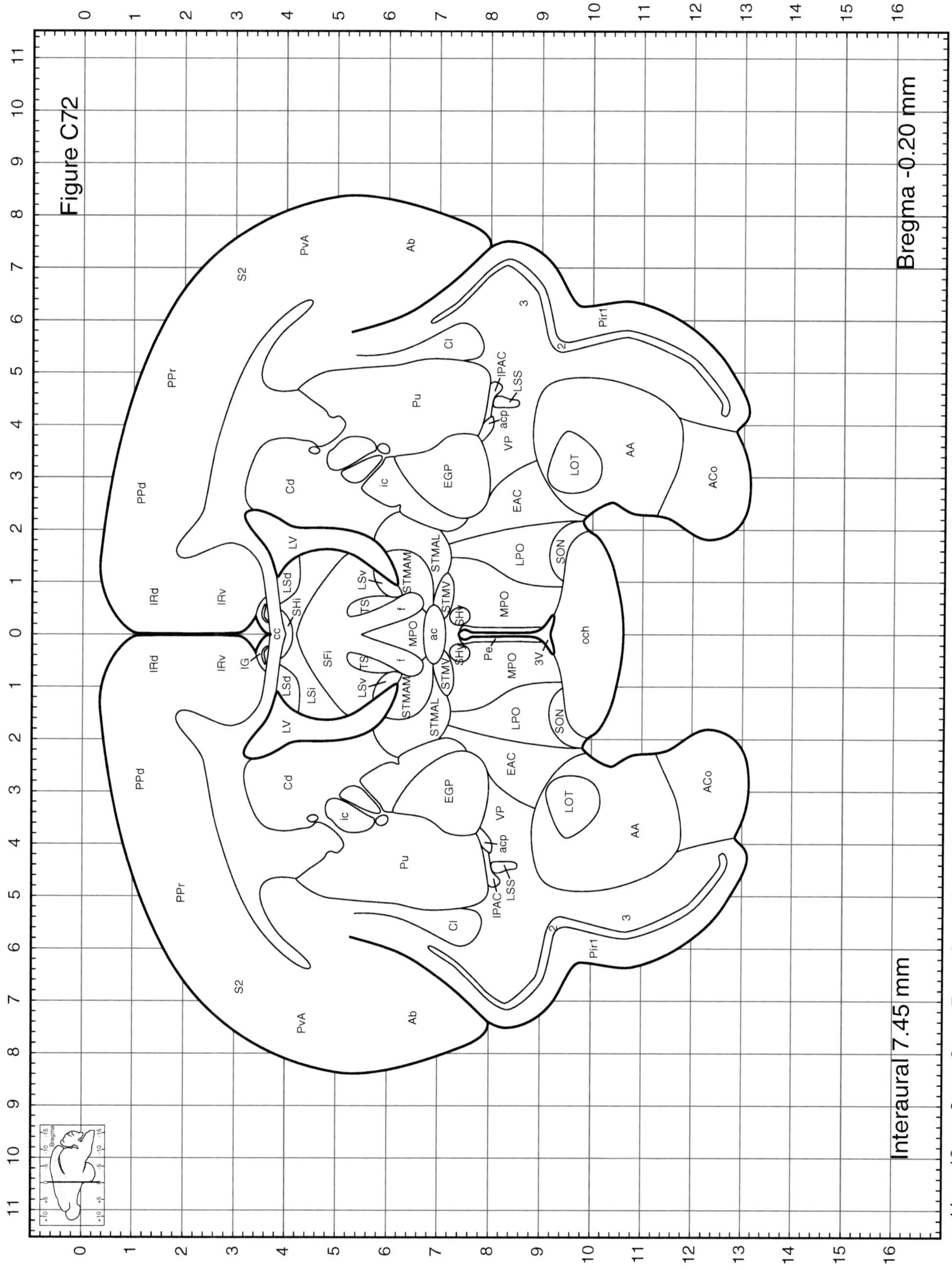

Figure C72
Bregma -0.20 mm
Interaural 7.45 mm
PvA
Ab
S2
PPr
PPd
IRd
IRv
Cd
LV
LSd
LSi
LSv
SHi
cc
IG
SFi
TS
f
STMAM
STMAL
STMV
MPO
ac
SHy
Pe
3V
LPO
SON
och
EGP
ic
Pu
Cl
VP
acp
IPAC
LSS
EAC
LOT
AA
ACo
Pir1
2
3

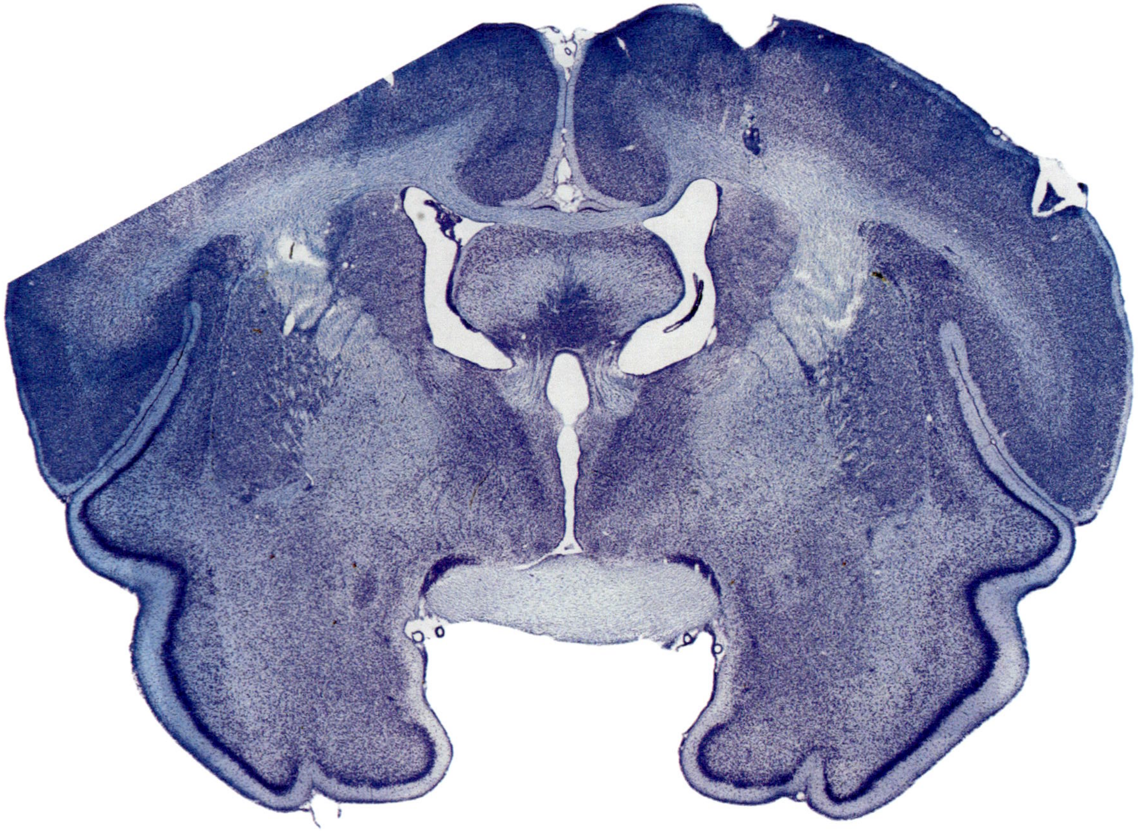

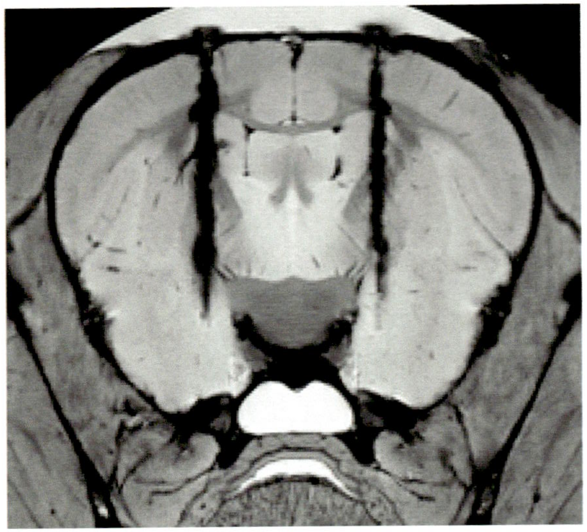

2	layer 2 of cortex
3	layer 3 of cortex
3V	3rd ventricle
AA	anterior amygdaloid area
Ab	auditory belt area
Aco	anterior cortical amygdaloid nucleus
acp	anterior commissure, posterior part
cc	corpus callosum
Cd	caudate nucleus
Cl	claustrum
EAC	sublenticular extended amygdala, central part
EGP	external globus pallidus
f	fornix
ic	internal capsule
IG	indusium griseum
IPAC	interstitial nucleus of the posterior limb of the anterior commissure
IRd	infraradiata dorsalis
IRv	infraradiata ventral area
LOT	nucleus of the lateral olfactory tract
LPO	lateral preoptic area
LSd	lateral septal nucleus, dorsal part
LSS	lateral stripe of the striatum
LV	lateral ventricle
MPO	medial preoptic nucleus
och	optic chiasm
PaA	paraventricular hypothalamic nucleus, anterior part
Pe	periventricular hypothalamic nucleus
Pir1	piriform cortex, layer 1
PPd	posterior parietal dorsal area
PPr	posterior parietal rostral area
Pu	putamen
PvA	parietal ventral area
S2	secondary somatosensory cortex
SCN	suprachiasmatic nucleus
SFi	septofimbrial nucleus
SHi	septohippocampal nucleus
SON	supraoptic nucleus
STMPL	bed nucleus of the stria terminalis, medial division, posterolateral part
STMPM	bed nucleus of the stria terminalis, medial division, posteromedial part
TS	triangular septal nucleus
VP	ventral pallidum

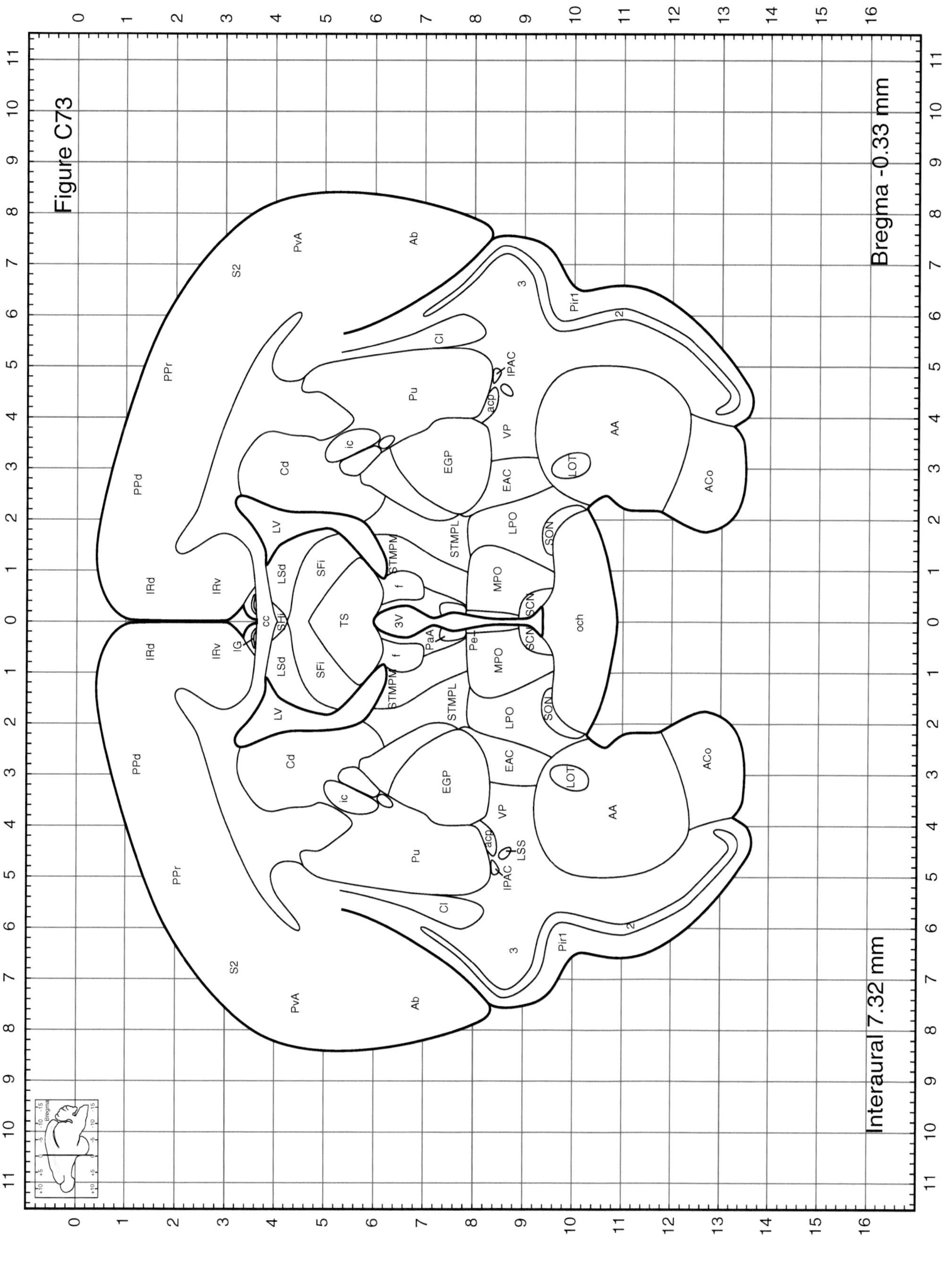
Figure C73
Bregma -0.33 mm
Interaural 7.32 mm
PvA
Ab
S2
3
Pir1
2
Cl
PPr
IPAC
Pu
acp
VP
AA
ic
Cd
EGP
PPd
LOT
EAC
ACo
LV
STMPM
STMPL
LPO
SON
LSd
SFi
MPO
IRd
IRv
f
SCN
cc
SHi
TS
3V
och
IG
PaA
Pe
LSS

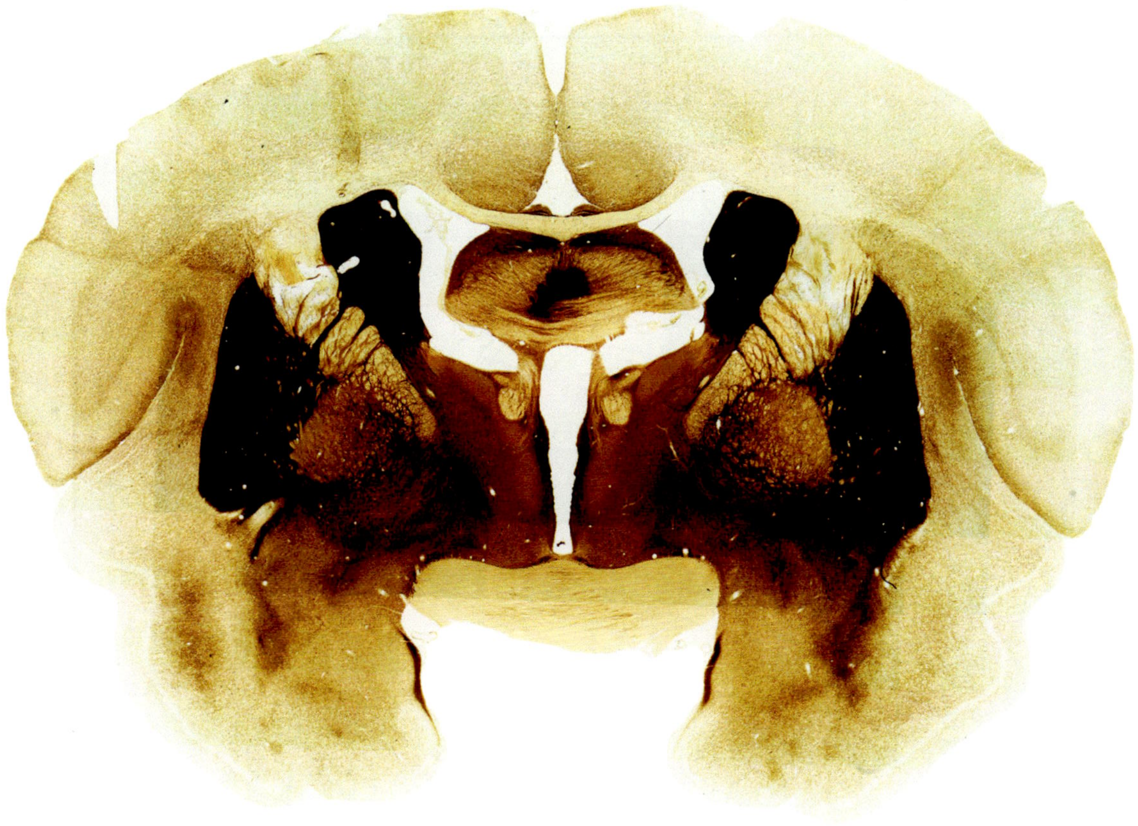

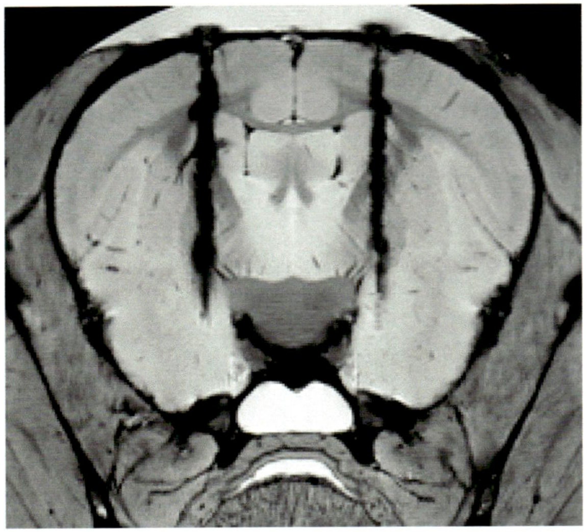

2	layer 2 of cortex
3	layer 3 of cortex
3V	3rd ventricle
AA	anterior amygdaloid area
Ab	auditory belt area
Aco	anterior cortical amygdaloid nucleus
acp	anterior commissure, posterior part
BLA	basolateral amygdaloid nucleus
cc	corpus callosum
Cd	caudate nucleus
Cl	claustrum
EA	sublenticular extended amygdala
EGP	external globus pallidus
f	fornix
ic	internal capsule
IG	indusium griseum
IPAC	interstitial nucleus of the posterior limb of the anterior commissure
IRd	infraradiata dorsalis
IRv	infraradiata ventral area
LOT	nucleus of the lateral olfactory tract
LPO	lateral preoptic area
LSd	lateral septal nucleus, dorsal part
LSS	lateral stripe of the striatum
LV	lateral ventricle
MPO	medial preoptic nucleus
och	optic chiasm
PaA	paraventricular hypothalamic nucleus, anterior part
Pe	periventricular hypothalamic nucleus
Pir1	piriform cortex, layer 1
PPd	posterior parietal dorsal area
PPr	posterior parietal rostral area
Pu	putamen
PvA	parietal ventral area
S2	secondary somatosensory cortex
SCN	suprachiasmatic nucleus
SFi	septofimbrial nucleus
SHi	septohippocampal nucleus
sm	stria medullaris of the thalamus
SON	supraoptic nucleus
STMPL	bed nucleus of the stria terminalis, medial division, posterolateral part
STMPM	bed nucleus of the stria terminalis, medial division, posteromedial part
TS	triangular septal nucleus
VP	ventral pallidum

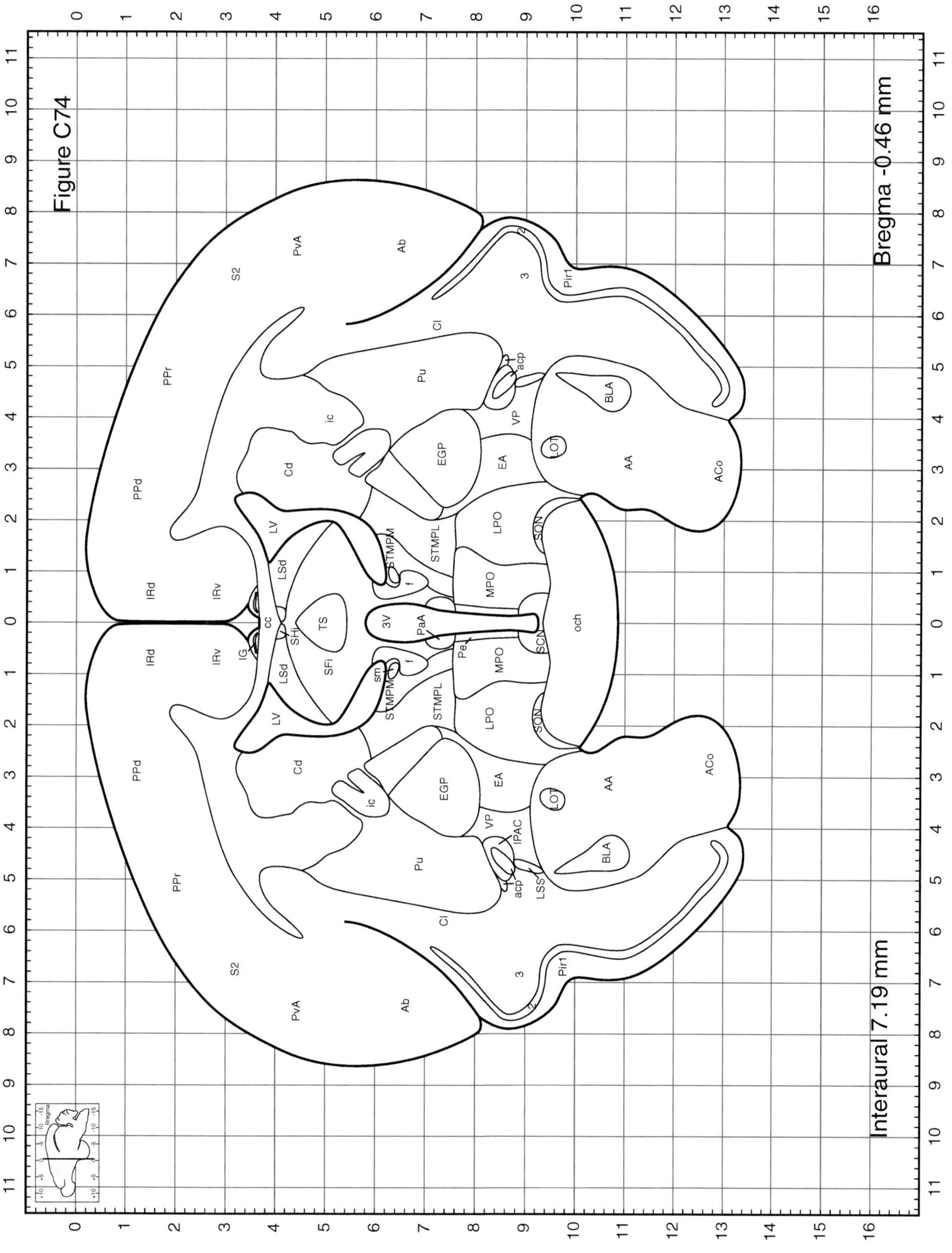
Figure C74
Bregma -0.46 mm
Interaural 7.19 mm
S2
PvA
Ab
PPr
PPd
IRd
IRv
Cl
Pu
ic
Cd
EGP
LV
LSd
cc
IG
SHi
TS
SFi
sm
STMPM
STMPL
f
3V
PaA
Pe
LPO
MPO
SCN
SON
och
EA
VP
IPAC
acp
LSS
LOT
AA
ACo
BLA
Pir1
3
2

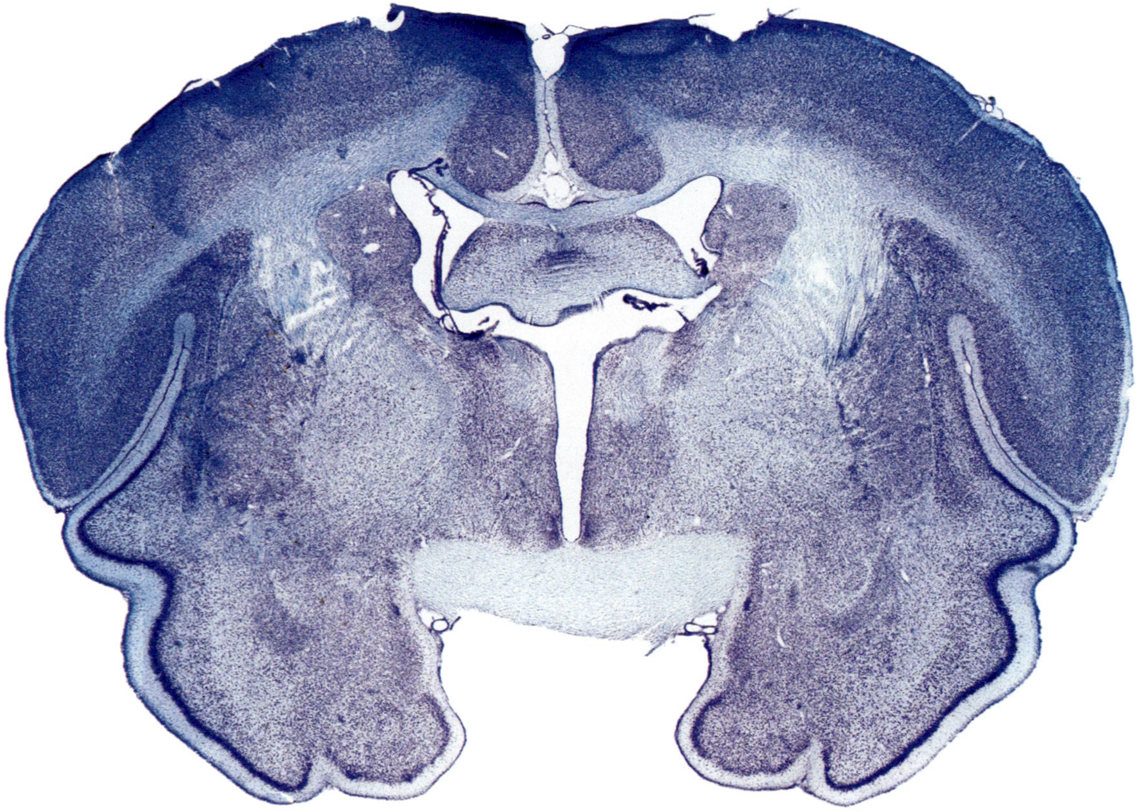

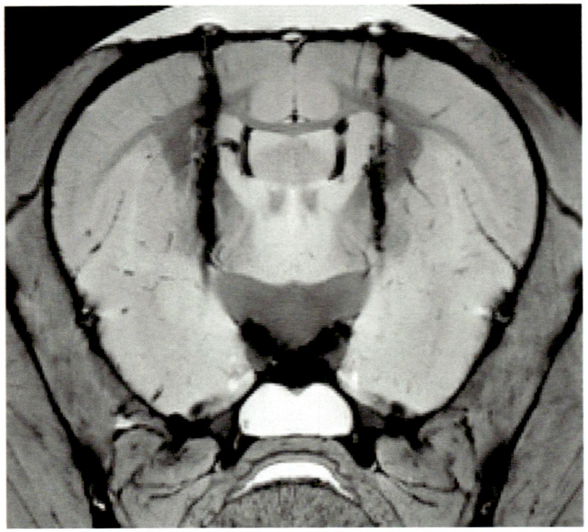

2	layer 2 of cortex
3	layer 3 of cortex
3V	3rd ventricle
AA	anterior amygdaloid area
Ab	auditory belt area
Ac	auditory core area
Aco	anterior cortical amygdaloid nucleus
acp	anterior commissure, posterior part
BLA	basolateral amygdaloid nucleus
cc	corpus callosum
Cd	caudate nucleus
Cl	claustrum
EAC	sublenticular extended amygdala, central part
EGP	external globus pallidus
f	fornix
ic	internal capsule
IG	indusium griseum
IPAC	interstitial nucleus of the posterior limb of the anterior commissure
IRd	infraradiata dorsalis
IRv	infraradiata ventral area
LOT	nucleus of the lateral olfactory tract
LPO	lateral preoptic area
LSd	lateral septal nucleus, dorsal part
LV	lateral ventricle
MPO	medial preoptic nucleus
och	optic chiasm
PaA	paraventricular hypothalamic nucleus, anterior part
Pe	periventricular hypothalamic nucleus
Pir1	piriform cortex, layer 1
PPd	posterior parietal dorsal area
PPr	posterior parietal rostral area
Pu	putamen
PvA	parietal ventral area
PVA	paraventricular thalamic nucleus, anterior part
Rt	reticular thalamic nucleus
SCN	suprachiasmatic nucleus
SFi	septofimbrial nucleus
sm	stria medullaris of the thalamus
SON	supraoptic nucleus
st	stria terminalis
StHy	striohypothalamic nucleus
STMP	bed nucleus of the stria terminalis, medial division, posterior part
TS	triangular septal nucleus
VP	ventral pallidum

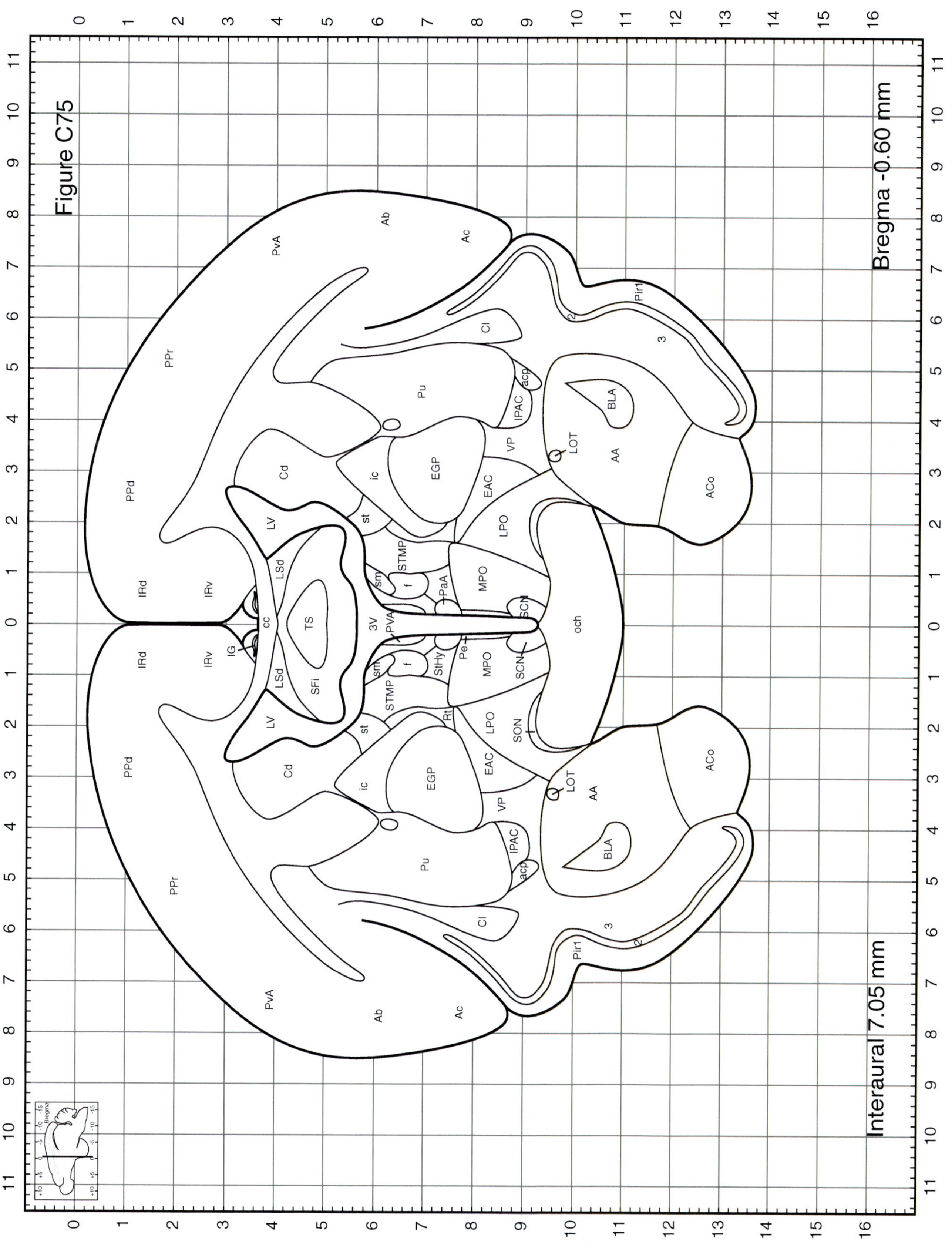
Figure C75
Bregma -0.60 mm
Interaural 7.05 mm
PvA
Ab
Ac
PPr
PPd
IRd
IRv
Cd
LV
LSd
cc
TS
SFi
IG
3V
PVA
sm
f
STMP
st
ic
EGP
Pu
Cl
acp
IPAC
VP
EAC
LPO
MPO
PaA
Pe
StHy
Rt
SCN
SON
och
LOT
AA
BLA
ACo
Pir1
2
3

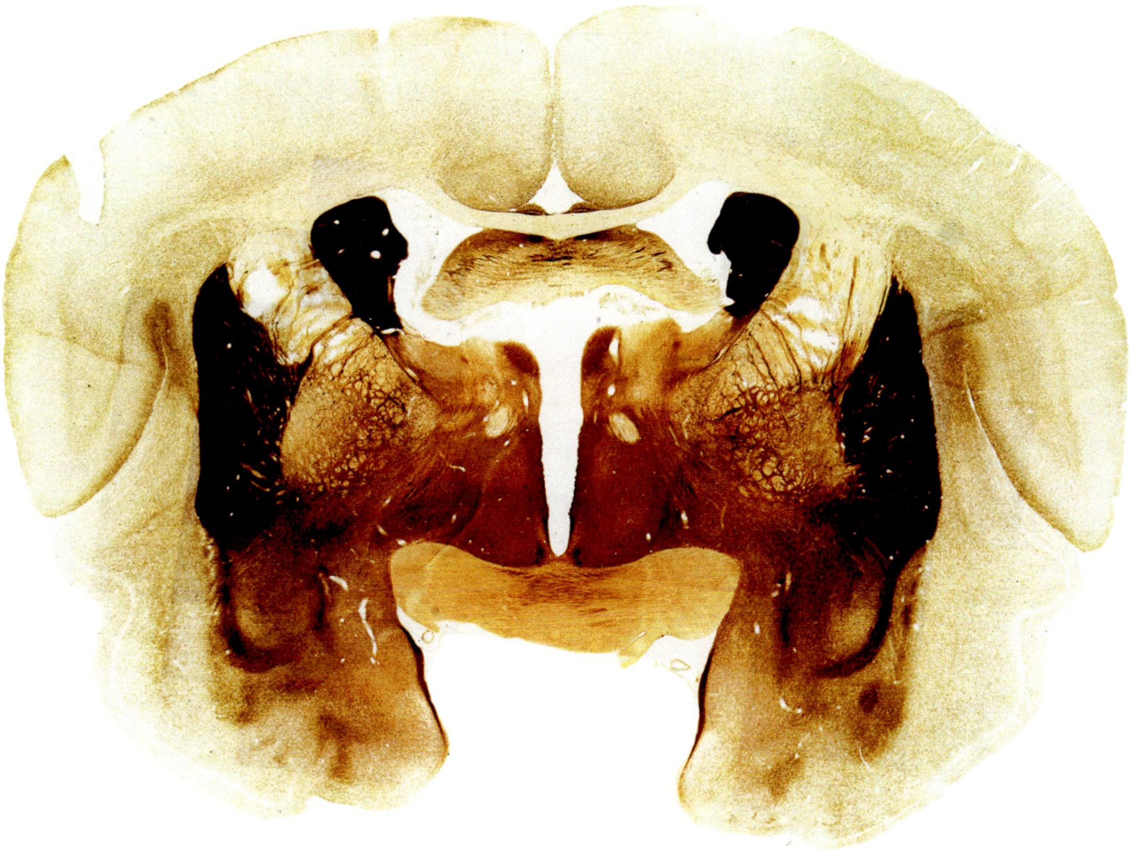

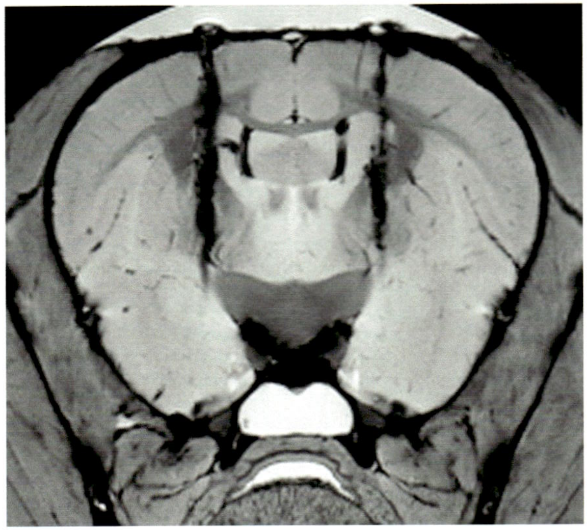

2 layer 2 of cortex
3 layer 3 of cortex
3V 3rd ventricle
Ab auditory belt area
Ac auditory core area
Aco anterior cortical amygdaloid nucleus
BLA basolateral amygdaloid nucleus
BMA basomedial amygdaloid nucleus
cc corpus callosum
Cd caudate nucleus
CeA central amygdaloid nucleus
Cl claustrum
EAC sublenticular extended amygdala, central part
EGP external globus pallidus
f fornix
ic internal capsule
IG indusium griseum
IPAC interstitial nucleus of the posterior limb of the anterior commissure
IRd infraradiata dorsalis
IRv infraradiata ventral area
LPO lateral preoptic area
LSd lateral septal nucleus, dorsal part
LV lateral ventricle
MeA medial amygdaloid nucleus
MPO medial preoptic nucleus
och optic chiasm
PaA paraventricular hypothalamic nucleus, anterior part
Pe periventricular hypothalamic nucleus
Pir1 piriform cortex, layer 1
PPd posterior parietal dorsal area
PPr posterior parietal rostral area
Pu putamen
PvA parietal ventral area
PVA paraventricular thalamic nucleus, anterior part
Rt reticular thalamic nucleus
SCN suprachiasmatic nucleus
SFi septofimbrial nucleus
sm stria medullaris of the thalamus
SON supraoptic nucleus
st stria terminalis
StHy striohypothalamic nucleus
STMP bed nucleus of the stria terminalis, medial division, posterior part
TS triangular septal nucleus
VP ventral pallidum

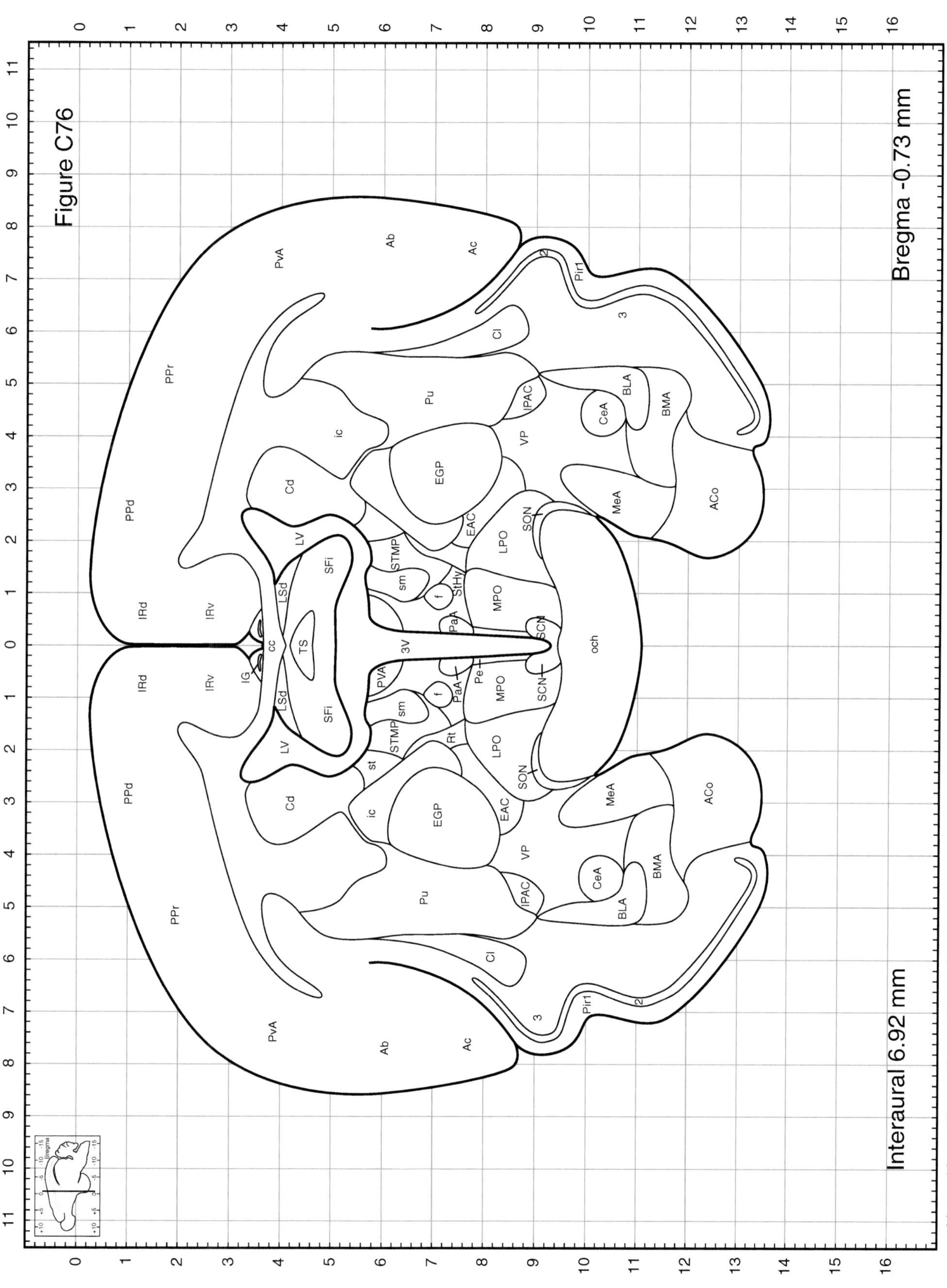
Figure C76
Bregma -0.73 mm
Interaural 6.92 mm
PvA
Ab
Ac
PPr
PPd
IRd
IRv
Pir1
2
3
Cl
Pu
ic
Cd
EGP
IPAC
VP
CeA
BLA
BMA
MeA
ACo
EAC
SON
LPO
MPO
StHy
STMP
sm
f
PaA
SCN
och
3V
PVA
Pe
Rt
st
LV
LSd
SFi
TS
cc
IG

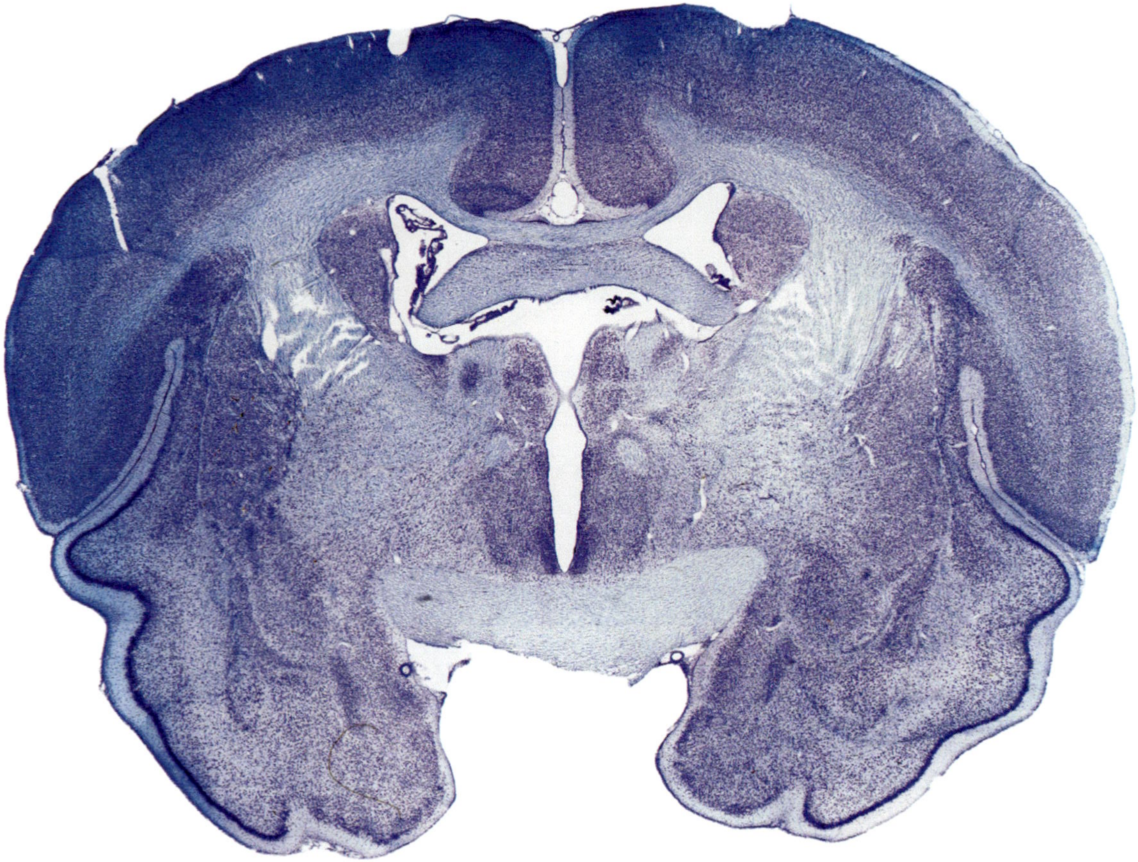

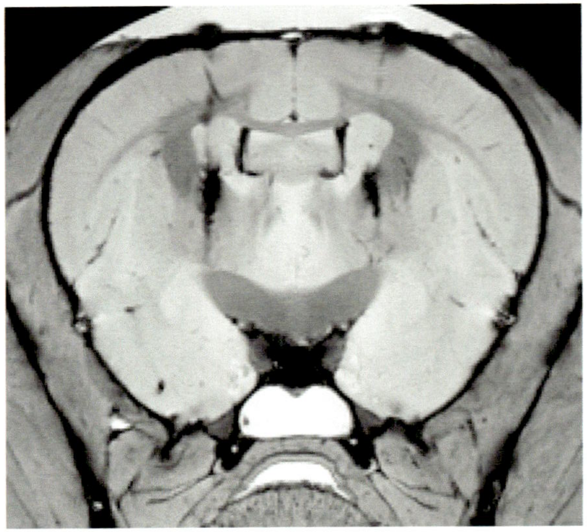

2	layer 2 of cortex
3	layer 3 of cortex
3V	3rd ventricle
Ab	auditory belt area
Ac	auditory core area
Aco	anterior cortical amygdaloid nucleus
AHy	anterior hypothalamic area
AM	anteromedial thalamic nucleus
BLA	basolateral amygdaloid nucleus
BMA	basomedial amygdaloid nucleus
cc	corpus callosum
Cd	caudate nucleus
CeA	central amygdaloid nucleus
Cl	claustrum
EAC	sublenticular extended amygdala, central part
EGP	external globus pallidus
f	fornix
ic	internal capsule
IG	indusium griseum
IPAC	interstitial nucleus of the posterior limb of the anterior commissure
IRd	infraradiata dorsalis
IRv	infraradiata ventral area
LHA	lateral hypothalamic area
LSd	lateral septal nucleus, dorsal part
LV	lateral ventricle
MeA	medial amygdaloid nucleus
MPO	medial preoptic nucleus
och	optic chiasm
PaA	paraventricular hypothalamic nucleus, anterior part
Pe	periventricular hypothalamic nucleus
Pir1	piriform cortex, layer 1
PPd	posterior parietal dorsal area
PPr	posterior parietal rostral area
PT	paratenial thalamic nucleus
Pu	putamen
PvA	parietal ventral area
PVA	paraventricular thalamic nucleus, anterior part
Re	reuniens thalamic nucleus
Rt	reticular thalamic nucleus
SCN	suprachiasmatic nucleus
SFi	septofimbrial nucleus
sm	stria medullaris of the thalamus
SON	supraoptic nucleus
st	stria terminalis
STMP	bed nucleus of the stria terminalis, medial division, posterior part
TS	triangular septal nucleus
VP	ventral pallidum

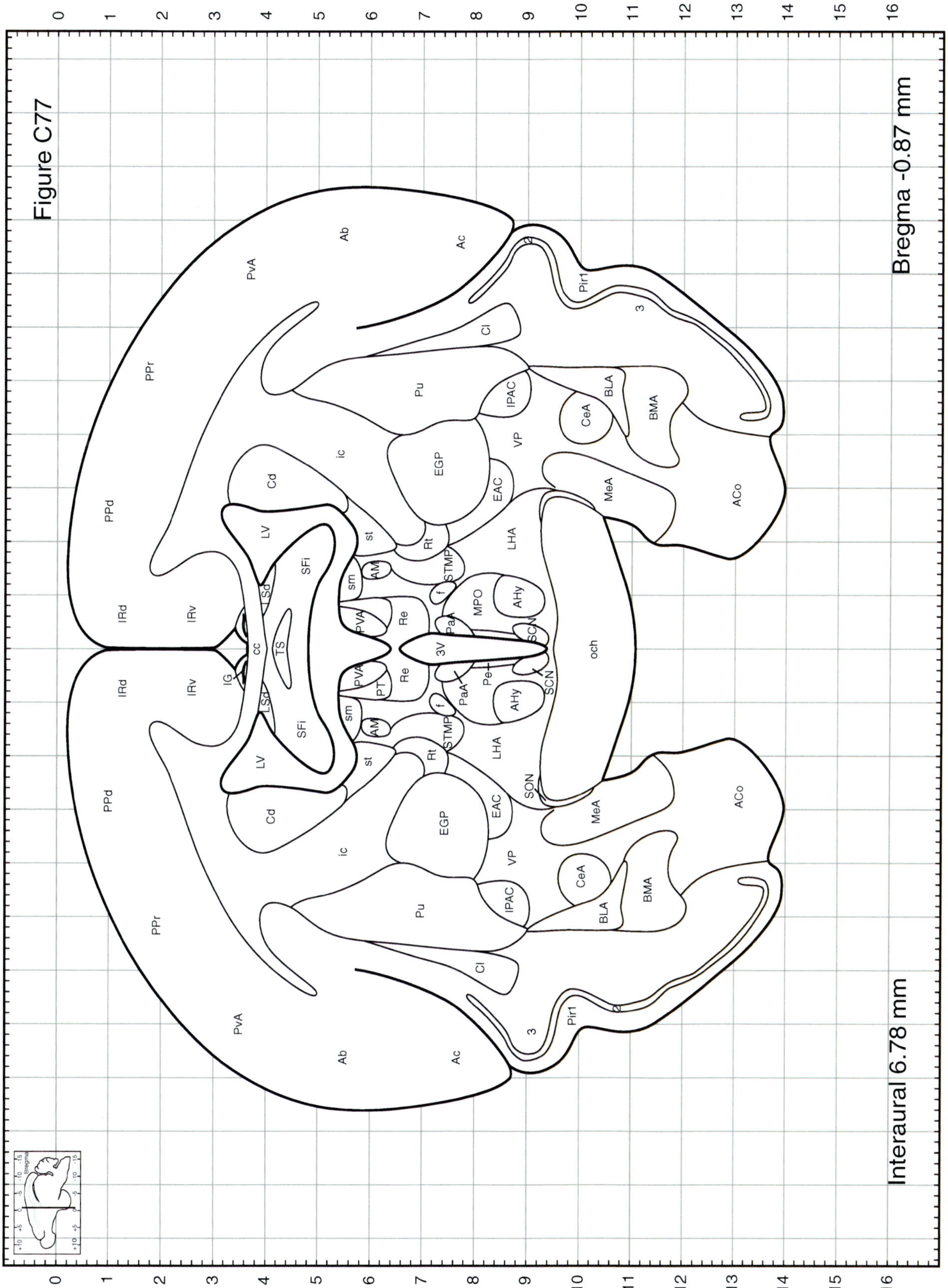
Figure C77
Bregma -0.87 mm
Interaural 6.78 mm

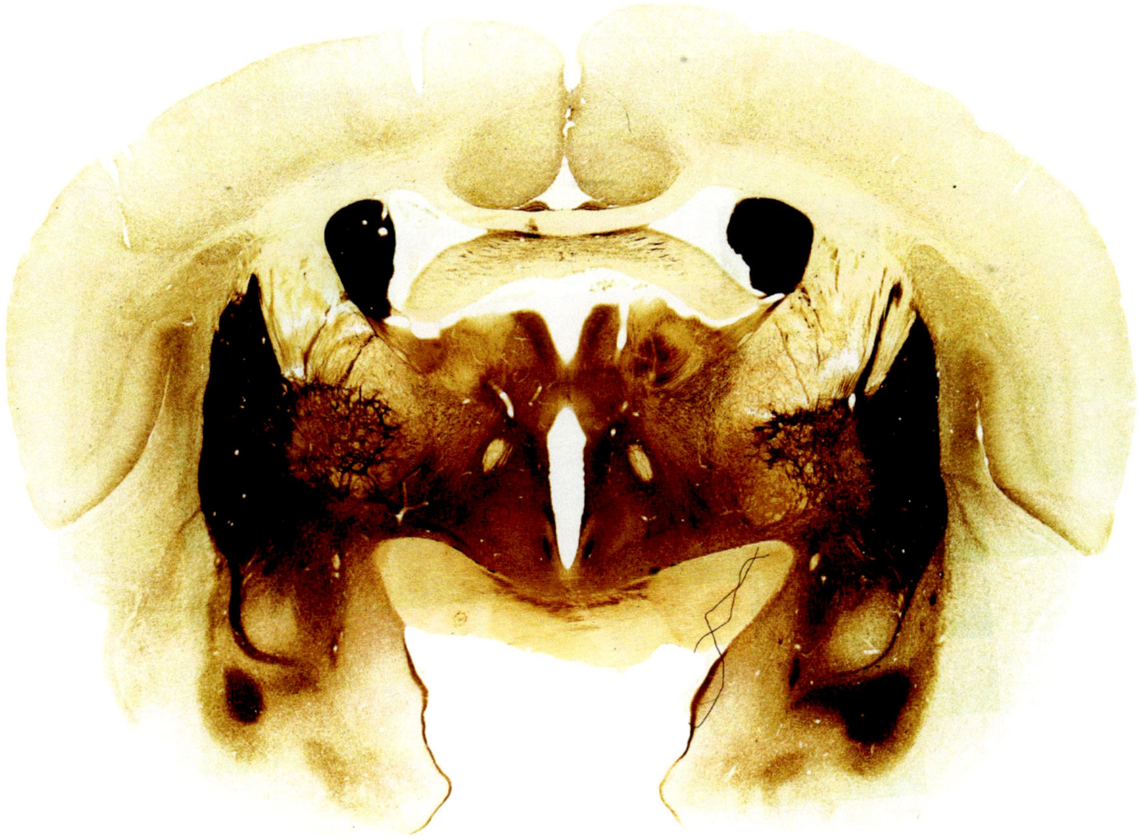

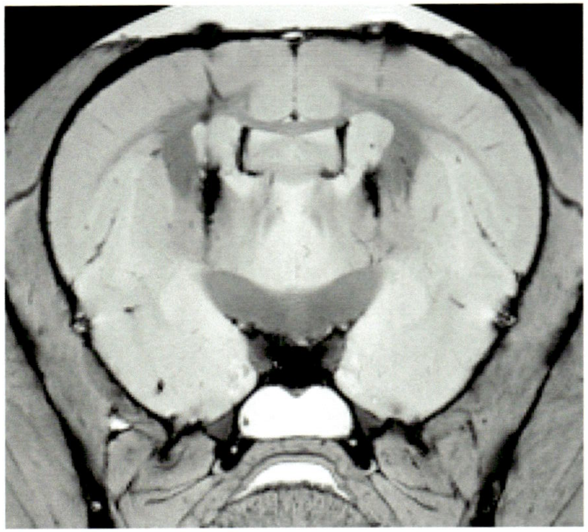

2	layer 2 of cortex
3	layer 3 of cortex
3V	3rd ventricle
Ab	auditory belt area
AB	accessory basal nucleus
Ac	auditory core area
Aco	anterior cortical amygdaloid nucleus
AHy	anterior hypothalamic area
AM	anteromedial thalamic nucleus
BLA	basolateral amygdaloid nucleus
BMA	basomedial amygdaloid nucleus
cc	corpus callosum
Cd	caudate nucleus
CeA	central amygdaloid nucleus
Cl	claustrum
EAC	sublenticular extended amygdala, central part
EGP	external globus pallidus
f	fornix
ic	internal capsule
IG	indusium griseum
IM	intercalated amygdaloid nucleus, main part
IRd	infraradiata dorsalis
IRv	infraradiata ventral area
LHA	lateral hypothalamic area
LV	lateral ventricle
MeA	medial amygdaloid nucleus
och	optic chiasm
PaA	paraventricular hypothalamic nucleus, anterior part
Pe	periventricular hypothalamic nucleus
Pir1	piriform cortex, layer 1
PPd	posterior parietal dorsal area
PPr	posterior parietal rostral area
PT	paratenial thalamic nucleus
Pu	putamen
PvA	parietal ventral area
PVA	paraventricular thalamic nucleus, anterior part
Re	reuniens thalamic nucleus
Rt	reticular thalamic nucleus
SCN	suprachiasmatic nucleus
SFi	septofimbrial nucleus
sm	stria medullaris of the thalamus
SPa	subparaventricular zone of the hypothalamus
st	stria terminalis
STMP	bed nucleus of the stria terminalis, medial division, posterior part
TI	temporal inferior area
VP	ventral pallidum

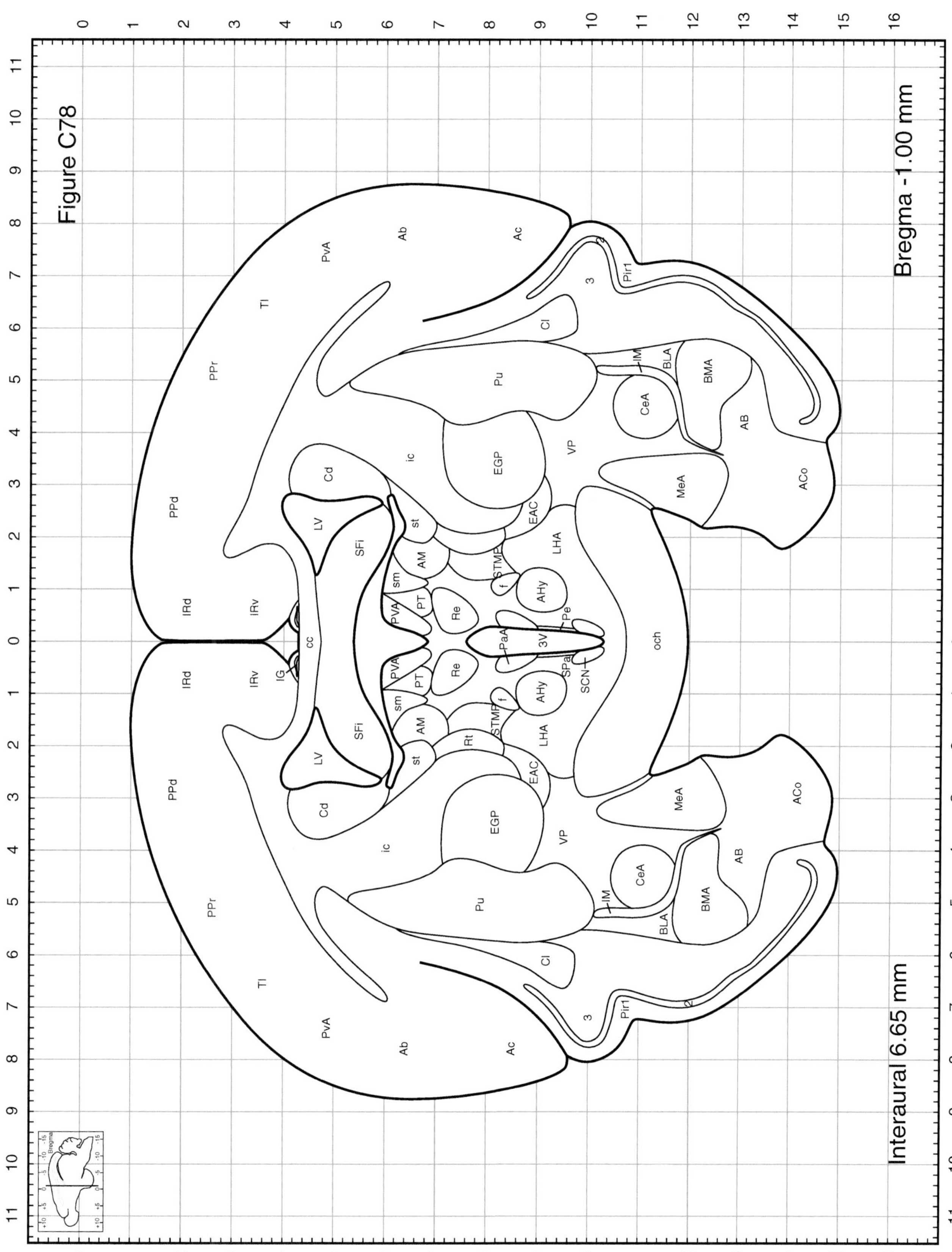

Figure C78
Bregma -1.00 mm
Interaural 6.65 mm
PvA
Ab
Ac
TI
PPr
PPd
IRd
IRv
IG
cc
LV
Cd
SFi
ic
st
sm
AM
PVA
PT
Re
Rt
STMP
f
AHy
PaA
3V
Pe
SPa
SCN
LHA
EAC
EGP
Pu
Cl
VP
och
MeA
CeA
IM
BLA
BMA
AB
ACo
Pir1
3
2

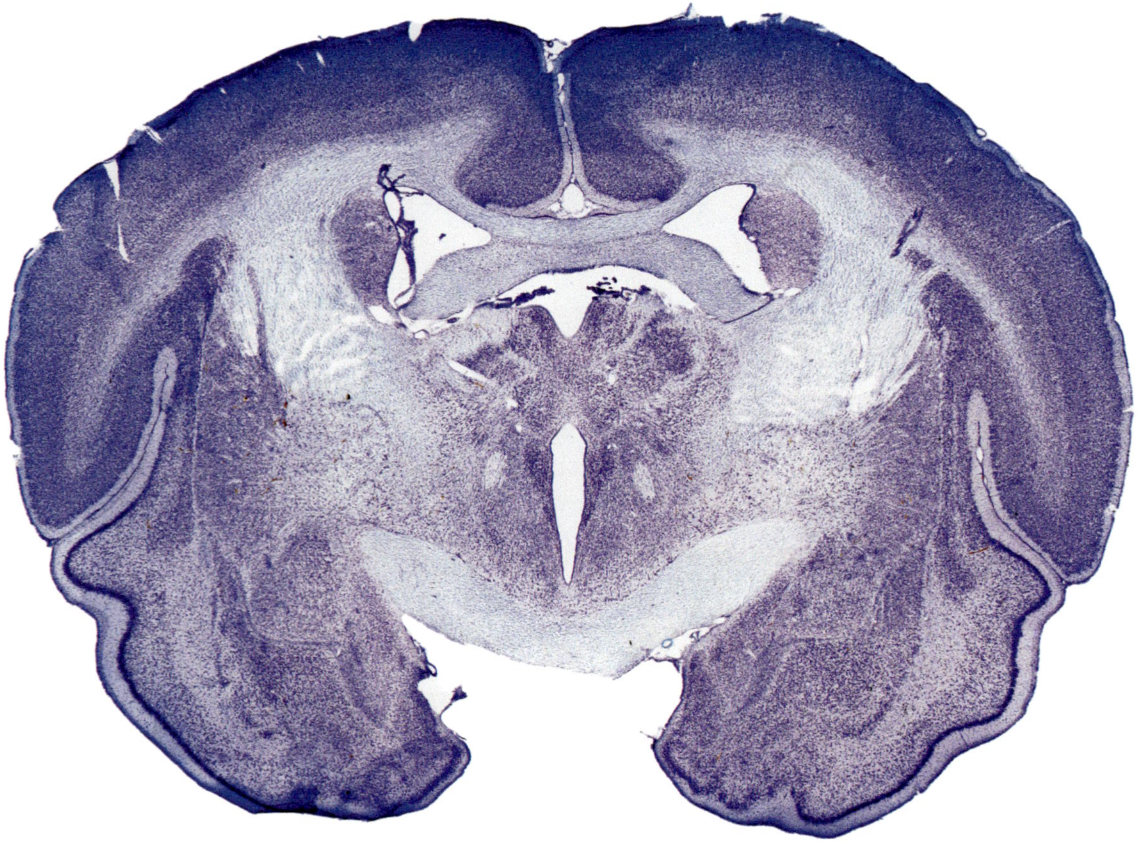

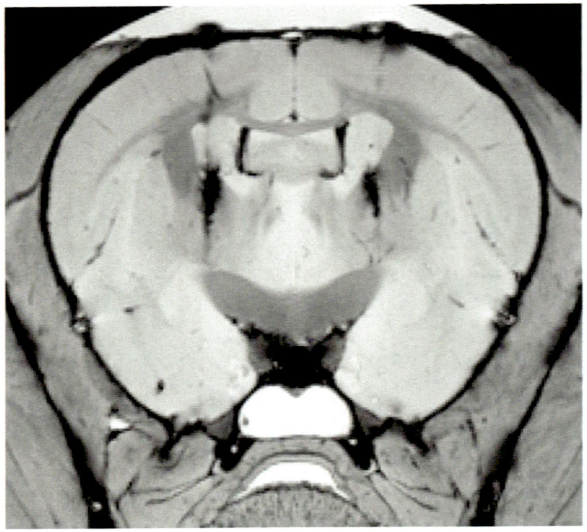

2	layer 2 of cortex
3	layer 3 of cortex
3V	3rd ventricle
Ab	auditory belt area
AB	accessory basal nucleus
Ac	auditory core area
Aco	anterior cortical amygdaloid nucleus
AD	anterodorsal thalamic nucleus
AHy	anterior hypothalamic area
AM	anteromedial thalamic nucleus
AV	anteroventral thalamic nucleus
BLA	basolateral amygdaloid nucleus
BMA	basomedial amygdaloid nucleus
cc	corpus callosum
Cd	caudate nucleus
CeA	central amygdaloid nucleus
Cl	claustrum
CM	central medial thalamic nucleus
D3V	dorsal 3rd ventricle
EA	sublenticular extended amygdala
EGP	external globus pallidus
f	fornix
ic	internal capsule
IG	indusium griseum
IGP	internal globus pallidus
IM	intercalated amygdaloid nucleus, main part
IRd	infraradiata dorsalis
IRv	infraradiata ventral area
LHA	lateral hypothalamic area
LV	lateral ventricle
MeA	medial amygdaloid nucleus
opt	optic tract
PaV	paraventricular hypothalamic nucleus, ventral part
Pe	periventricular hypothalamic nucleus
Pir1	piriform cortex, layer 1
PPd	posterior parietal dorsal area
PPr	posterior parietal rostral area
PT	paratenial thalamic nucleus
Pu	putamen
PvA	parietal ventral area
PVA	paraventricular thalamic nucleus, anterior part
PVN	paraventricular hypothalamic nucleus
Re	reuniens thalamic nucleus
Rt	reticular thalamic nucleus
SCN	suprachiasmatic nucleus
SFi	septofimbrial nucleus
sm	stria medullaris of the thalamus
st	stria terminalis
TI	temporal inferior area
VL	ventrolateral thalamic nucleus
VMH	ventromedial hypothalamic nucleus

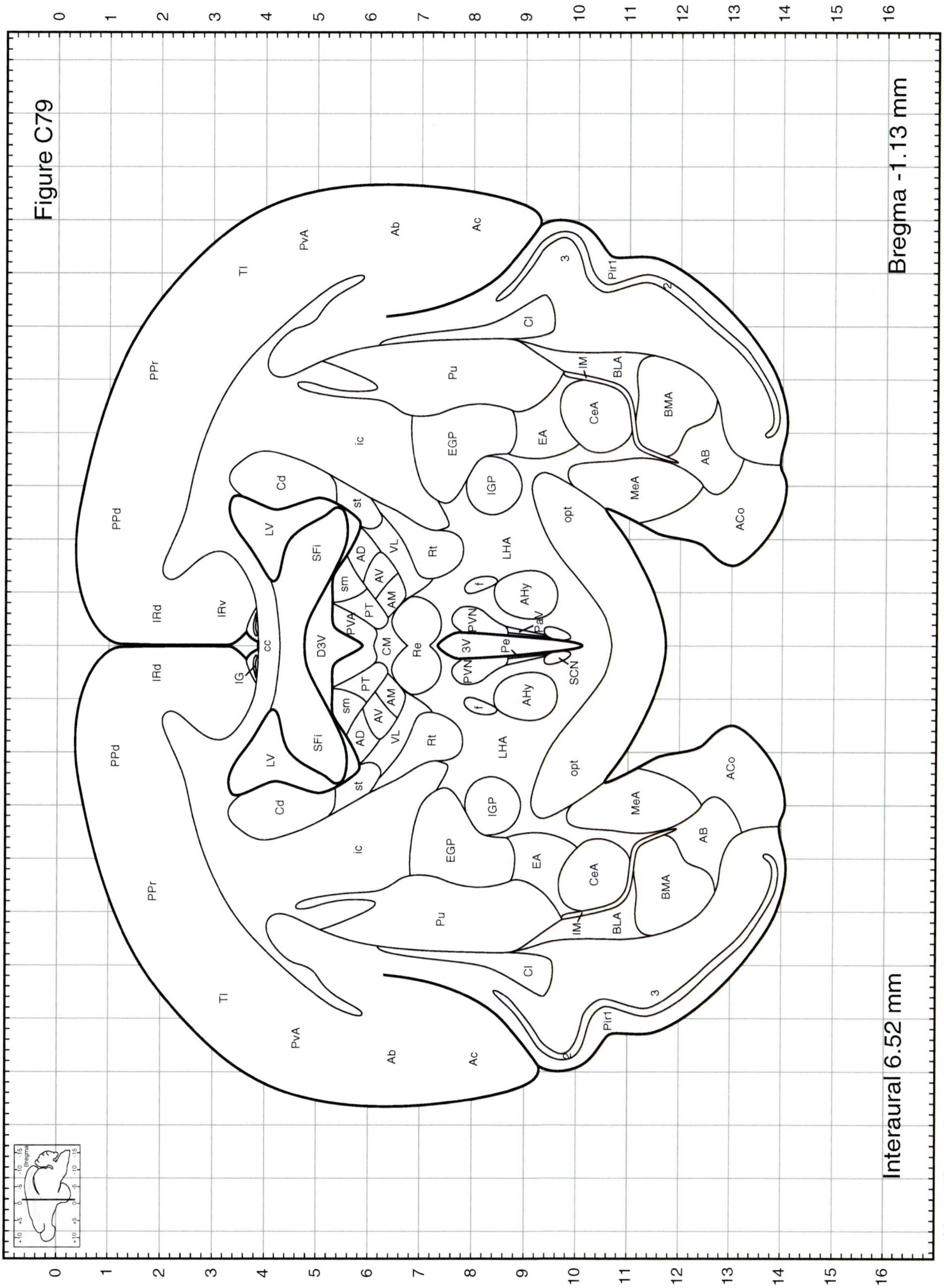
Figure C79
Bregma -1.13 mm
Interaural 6.52 mm
PvA
Ab
Ac
TI
PPr
PPd
IRd
IRv
Cd
LV
SFi
st
AD
AV
VL
AM
PT
PVA
sm
D3V
cc
IG
CM
Re
Rt
ic
Pu
EGP
IGP
LHA
f
AHy
PVN
3V
Pe
PaV
SCN
opt
EA
CeA
IM
BLA
BMA
AB
MeA
ACo
Cl
Pir1
3
2

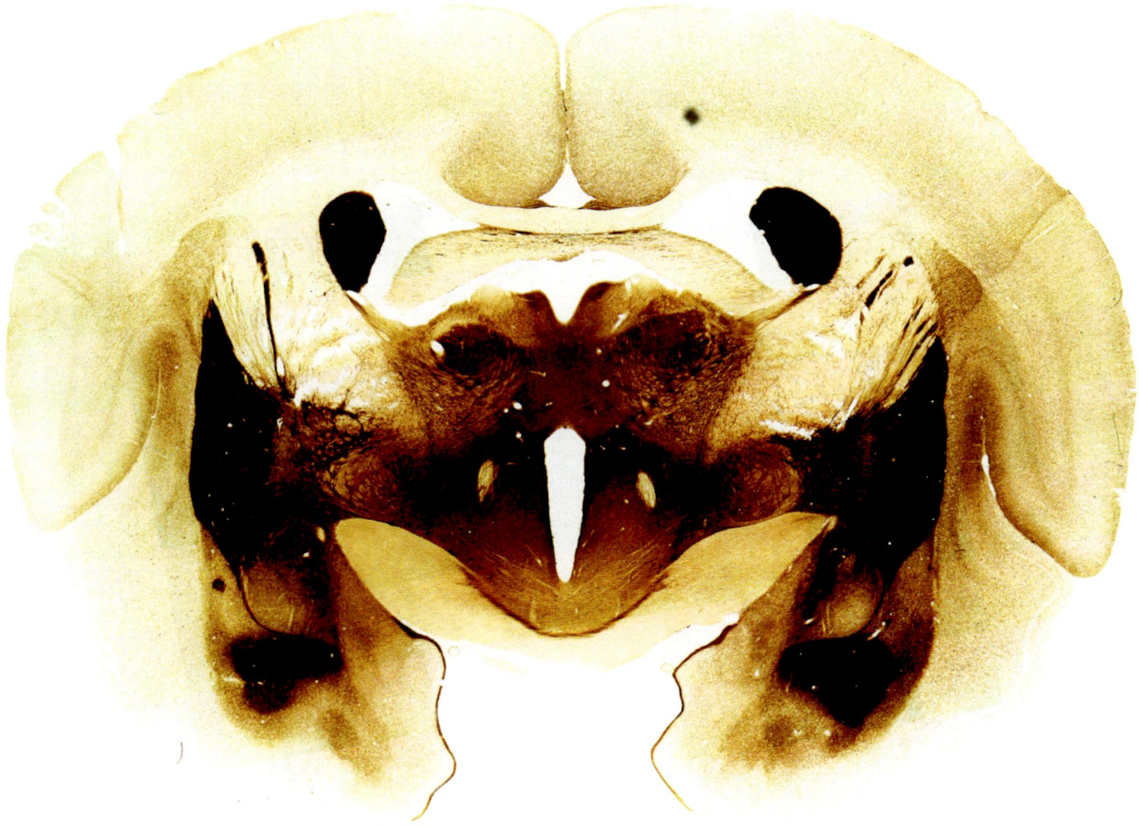

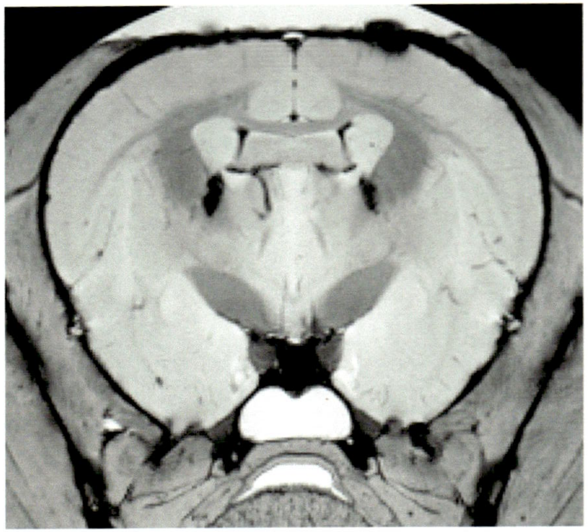

2 layer 2 of cortex
3 layer 3 of cortex
3V 3rd ventricle
Ab auditory belt area
AB accessory basal nucleus
Ac auditory core area
Aco anterior cortical amygdaloid nucleus
AD anterodorsal thalamic nucleus
AHy anterior hypothalamic area
AM anteromedial thalamic nucleus
AV anteroventral thalamic nucleus
BLA basolateral amygdaloid nucleus
BMA basomedial amygdaloid nucleus
cc corpus callosum
Cd caudate nucleus
CeA central amygdaloid nucleus
Cl claustrum
CM central medial thalamic nucleus
D3V dorsal 3rd ventricle
EA sublenticular extended amygdala
EGP external globus pallidus
f fornix
Hip hippocampus
ic internal capsule
IG indusium griseum
IGP internal globus pallidus
IM intercalated amygdaloid nucleus, main part
IRd infraradiata dorsalis
IRv infraradiata ventral area
IT inferior temporal cortex
LHA lateral hypothalamic area
LV lateral ventricle
MeA medial amygdaloid nucleus
opt optic tract
Pir1 piriform cortex, layer 1
PPd posterior parietal dorsal area
PPr posterior parietal rostral area
PT paratenial thalamic nucleus
Pu putamen
PvA parietal ventral area
PVA paraventricular thalamic nucleus, anterior part
PVN paraventricular hypothalamic nucleus
Re reuniens thalamic nucleus
Rt reticular thalamic nucleus
SFi septofimbrial nucleus
SFO subfornical organ
sm stria medullaris of the thalamus
st stria terminalis
TI temporal inferior area
VL ventrolateral thalamic nucleus
VMH ventromedial hypothalamic nucleus

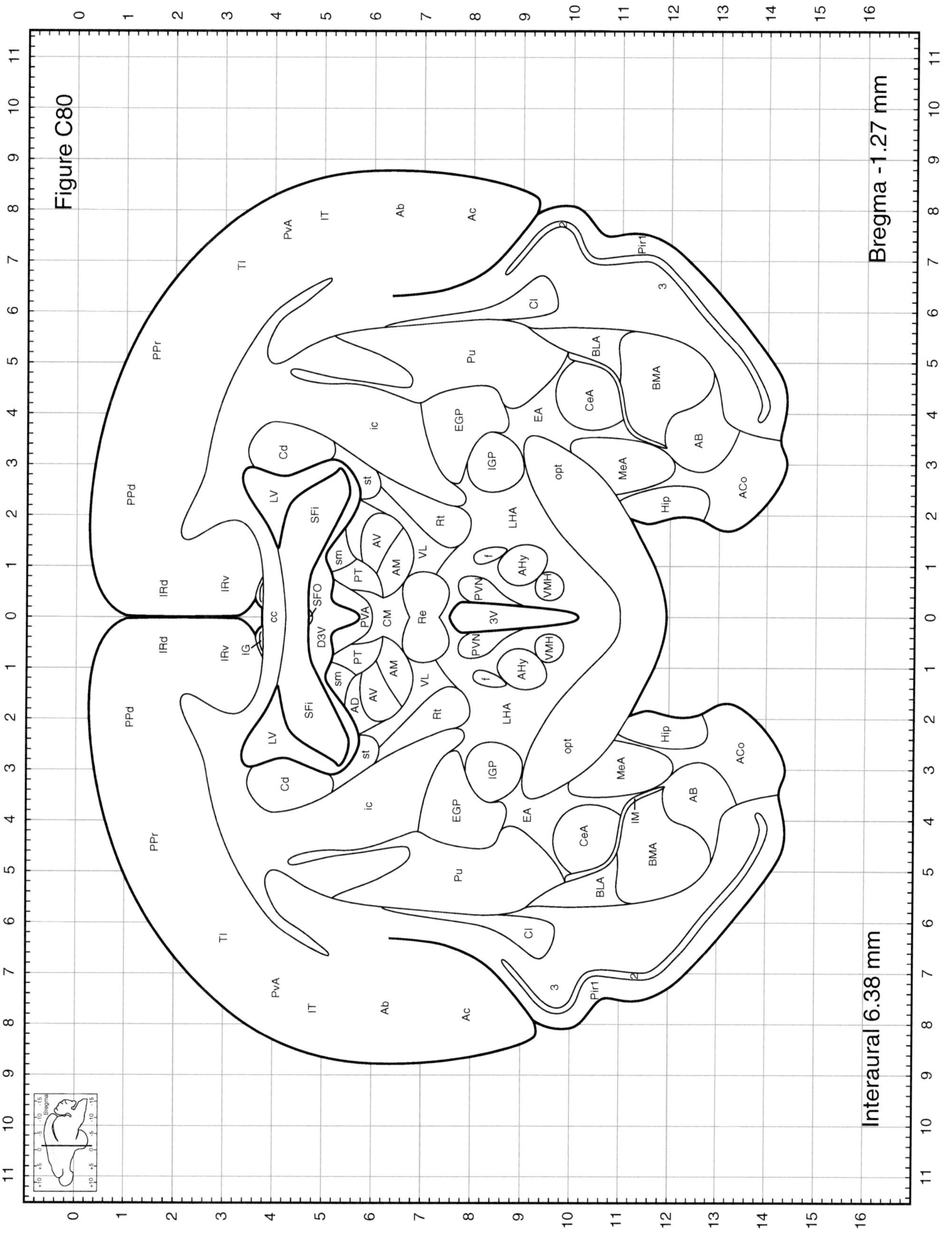
Figure C80
Bregma -1.27 mm
Interaural 6.38 mm
PPd
PPr
IRd
IRv
TI
PvA
IT
Ab
Ac
Cd
LV
SFi
cc
IG
SFO
D3V
sm
PT
PVA
CM
AV
AM
AD
VL
Re
st
Rt
ic
EGP
IGP
Pu
Cl
EA
LHA
opt
f
AHy
PVN
VMH
3V
CeA
BLA
BMA
MeA
AB
ACo
Hip
IM
Pir1
2
3

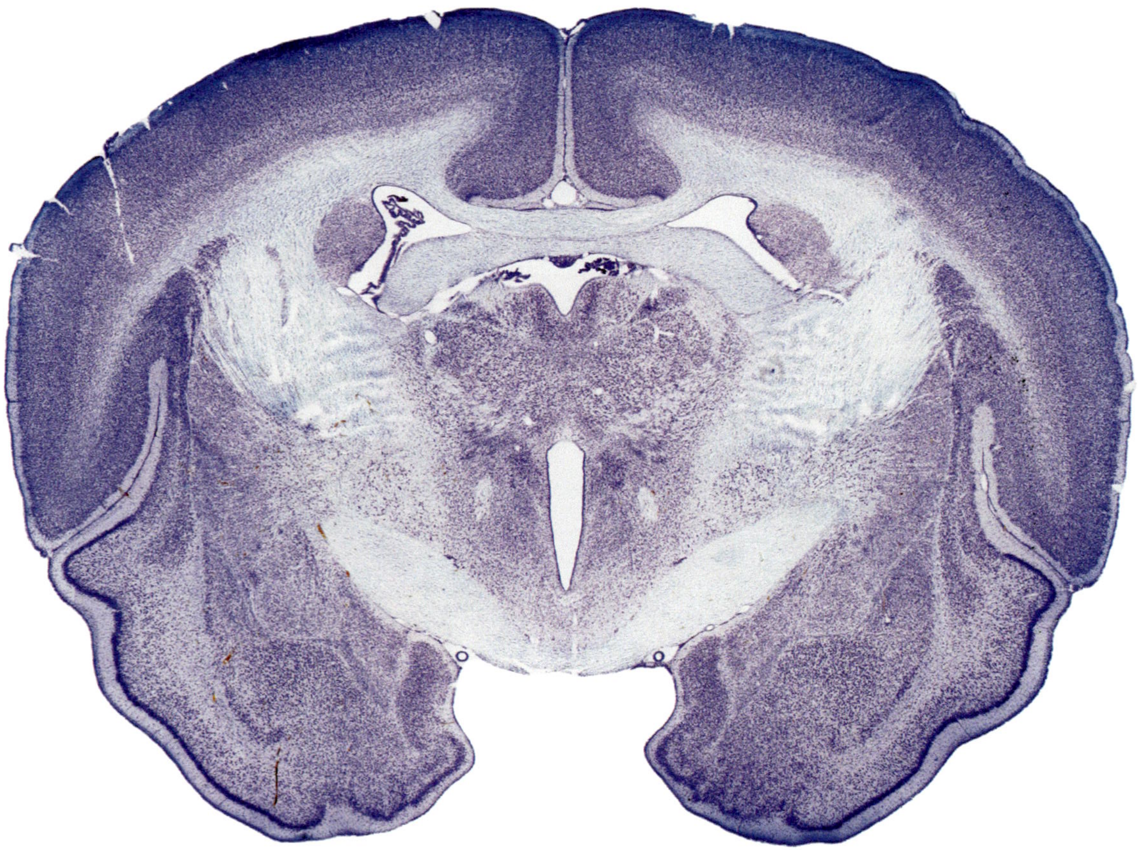

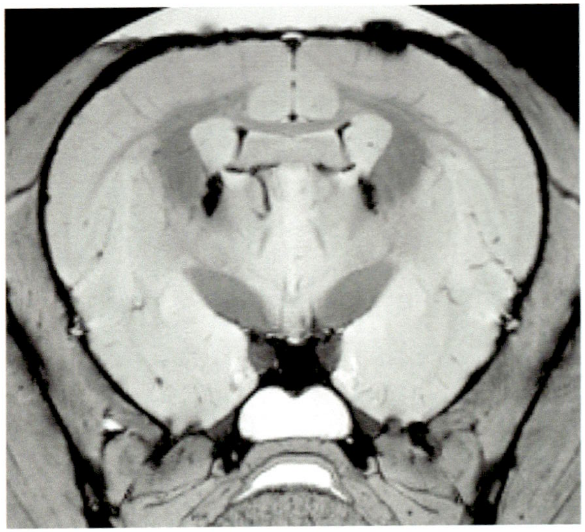

2	layer 2 of cortex
3	layer 3 of cortex
3V	3rd ventricle
Ab	auditory belt area
AB	accessory basal nucleus
Ac	auditory core area
Aco	anterior cortical amygdaloid nucleus
AD	anterodorsal thalamic nucleus
AHy	anterior hypothalamic area
AM	anteromedial thalamic nucleus
AV	anteroventral thalamic nucleus
BLA	basolateral amygdaloid nucleus
BMA	basomedial amygdaloid nucleus
cc	corpus callosum
Cd	caudate nucleus
CeA	central amygdaloid nucleus
Cl	claustrum
CM	central medial thalamic nucleus
D3V	dorsal 3rd ventricle
EA	sublenticular extended amygdala
EGP	external globus pallidus
f	fornix
Hip	hippocampus
IAM	interanteromedial thalamic nucleus
ic	internal capsule
IG	indusium griseum
IGP	internal globus pallidus
IM	intercalated amygdaloid nucleus, main part
IRd	infraradiata dorsalis
IRv	infraradiata ventral area
IT	inferior temporal cortex
LHA	lateral hypothalamic area
LV	lateral ventricle
MeA	medial amygdaloid nucleus
opt	optic tract
Pir1	piriform cortex, layer 1
PPd	posterior parietal dorsal area
PPr	posterior parietal rostral area
PT	paratenial thalamic nucleus
Pu	putamen
PvA	parietal ventral area
PVA	paraventricular thalamic nucleus, anterior part
PVN	paraventricular hypothalamic nucleus
Re	reuniens thalamic nucleus
Rt	reticular thalamic nucleus
SFi	septofimbrial nucleus
SFO	subfornical organ
sm	stria medullaris of the thalamus
TC	temporal cortex
TI	temporal inferior area
VL	ventrolateral thalamic nucleus
VMH	ventromedial hypothalamic nucleus

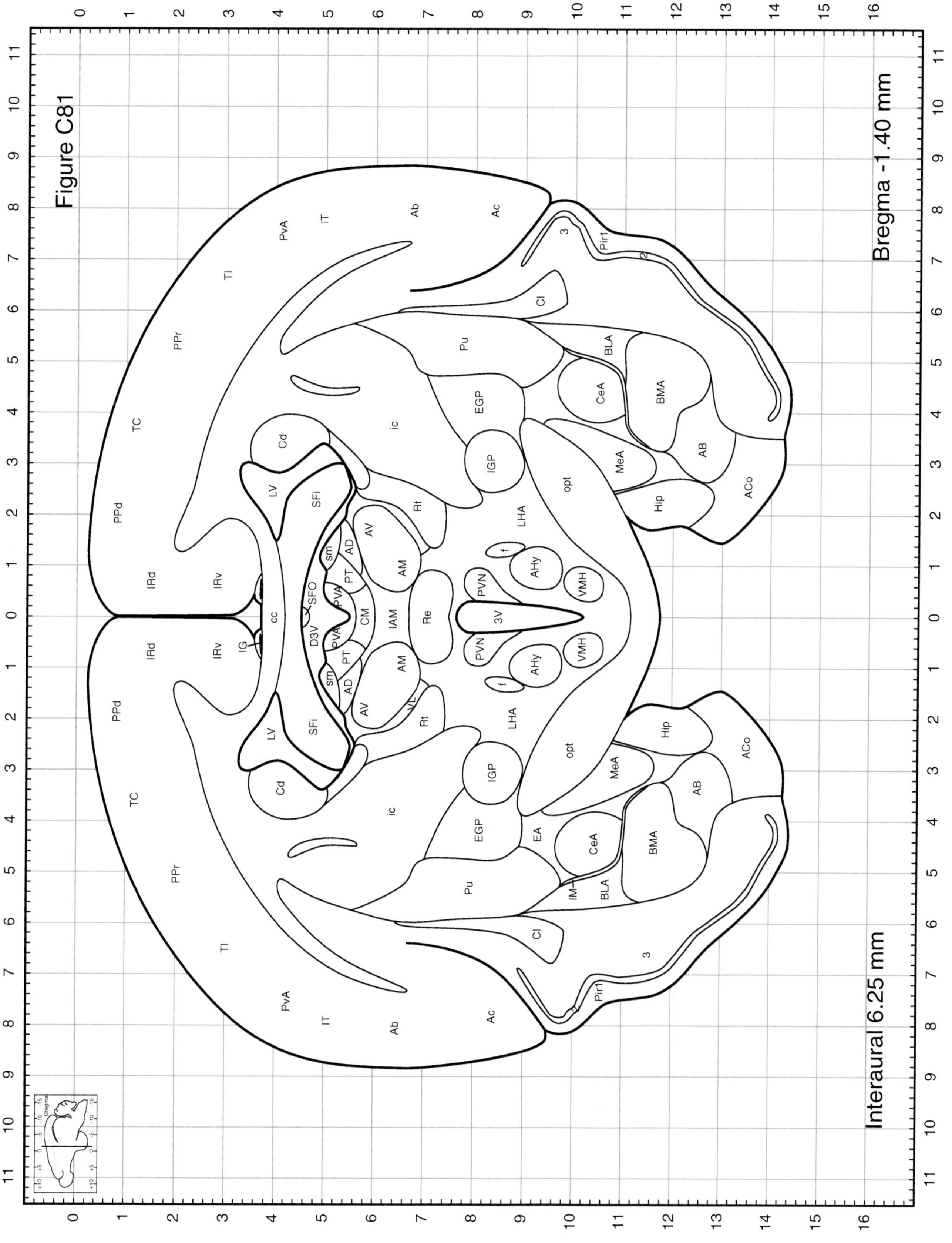
Figure C81
Bregma -1.40 mm
Interaural 6.25 mm
PPd
TC
PPr
TI
PvA
IT
Ab
Ac
IRd
IRv
IG
cc
LV
Cd
SFi
SFO
D3V
sm
AD
AV
PT
PVA
CM
IAM
AM
Re
Rt
VL
ic
EGP
IGP
Pu
Cl
LHA
opt
f
PVN
3V
AHy
VMH
MeA
CeA
BLA
BMA
AB
ACo
Hip
Pir1
EA
IM

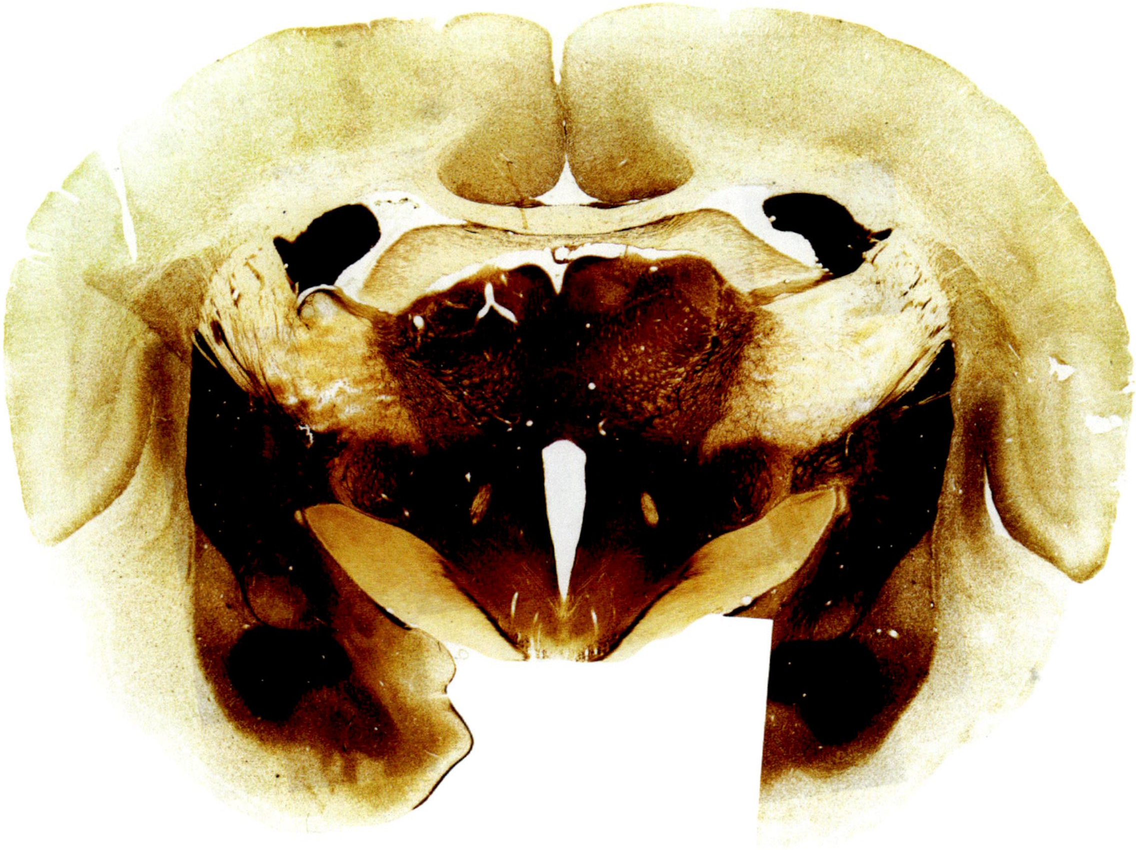

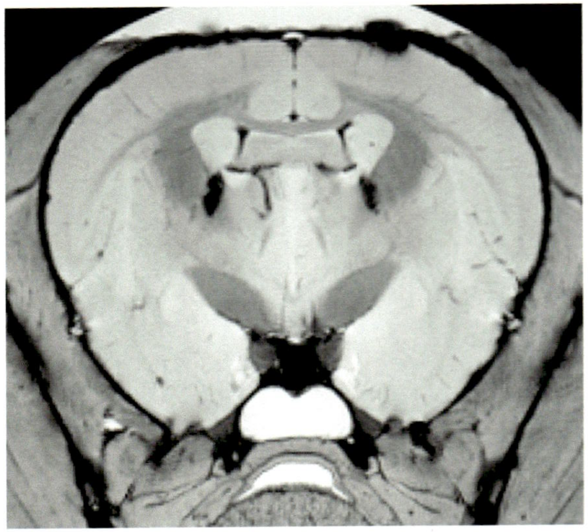

2	layer 2 of cortex
3	layer 3 of cortex
3V	3rd ventricle
Ab	auditory belt area
AB	accessory basal nucleus
Ac	auditory core area
Aco	anterior cortical amygdaloid nucleus
AD	anterodorsal thalamic nucleus
AHy	anterior hypothalamic area
AM	anteromedial thalamic nucleus
AV	anteroventral thalamic nucleus
BLA	basolateral amygdaloid nucleus
BMA	basomedial amygdaloid nucleus
cc	corpus callosum
Cd	caudate nucleus
CeA	central amygdaloid nucleus
Cl	claustrum
CM	central medial thalamic nucleus
D3V	dorsal 3rd ventricle
EA	sublenticular extended amygdala
EGP	external globus pallidus
f	fornix
Hip	hippocampus
IAM	interanteromedial thalamic nucleus
ic	internal capsule
IG	indusium griseum
IGP	internal globus pallidus
IM	intercalated amygdaloid nucleus, main part
IRd	infraradiata dorsalis
IRv	infraradiata ventral area
IT	inferior temporal cortex
JPLH	juxtaparaventricular part of lateral hypothalamus
LHA	lateral hypothalamic area
LV	lateral ventricle
MeA	medial amygdaloid nucleus
opt	optic tract
Pir1	piriform cortex, layer 1
PPc	posterior parietal caudal area
PPd	posterior parietal dorsal area
PT	paratenial thalamic nucleus
Pu	putamen
PVA	paraventricular thalamic nucleus, anterior part
PVN	paraventricular hypothalamic nucleus
Re	reuniens thalamic nucleus
Rt	reticular thalamic nucleus
SFi	septofimbrial nucleus
sm	stria medullaris of the thalamus
SON	supraoptic nucleus
st	stria terminalis
TC	temporal cortex
TI	temporal inferior area
VL	ventrolateral thalamic nucleus
VMH	ventromedial hypothalamic nucleus

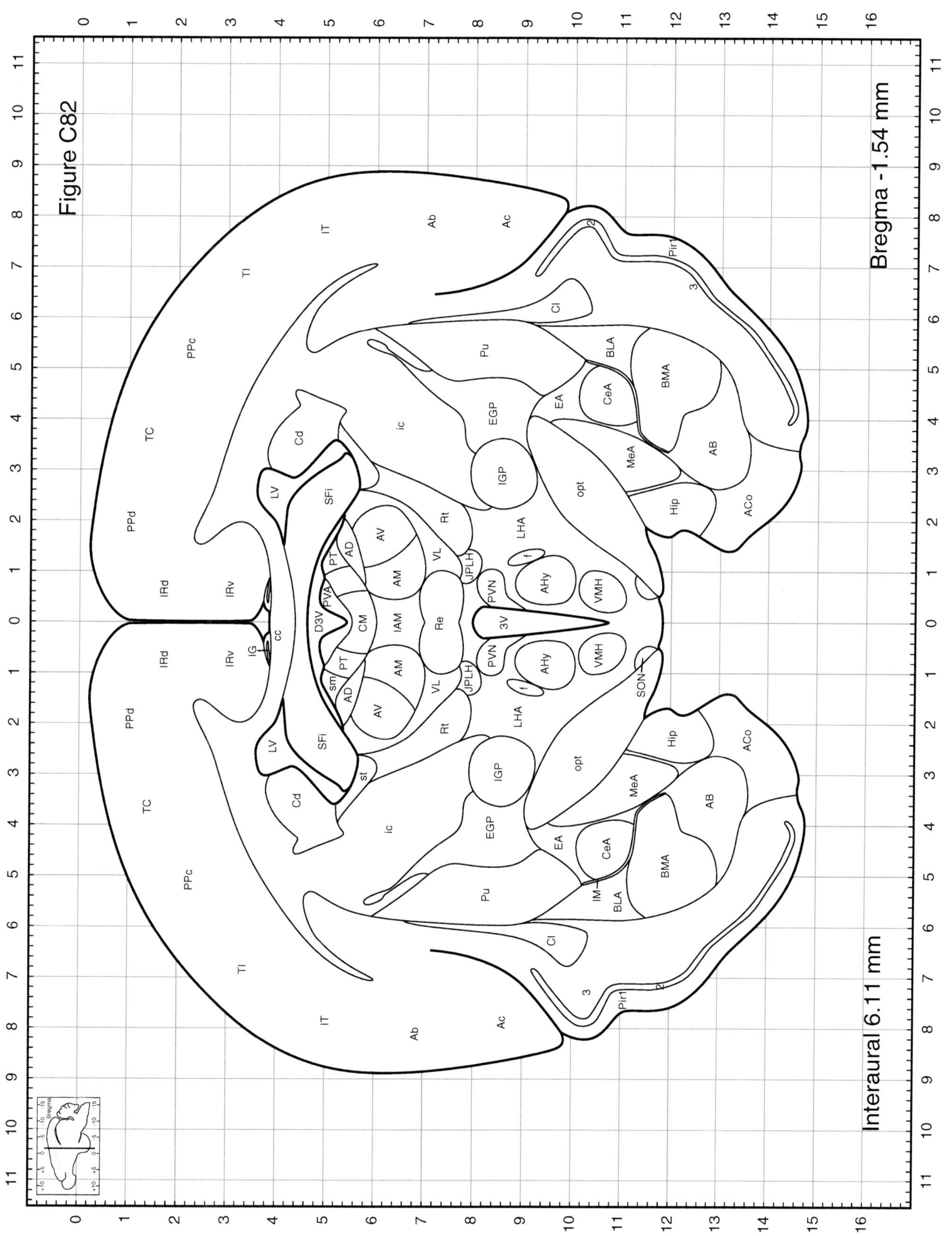
Figure C82
Bregma -1.54 mm
Interaural 6.11 mm
IT
Ab
Ac
TI
PPc
TC
PPd
IRd
IRv
Cd
LV
SFi
cc
IG
D3V
PVA
PT
AD
AV
AM
CM
IAM
Re
VL
Rt
JPLH
PVN
3V
f
AHy
VMH
LHA
IGP
EGP
ic
Pu
Cl
opt
EA
CeA
BLA
BMA
MeA
AB
Hip
ACo
Pir1
SON
IM
sm
st
2
3

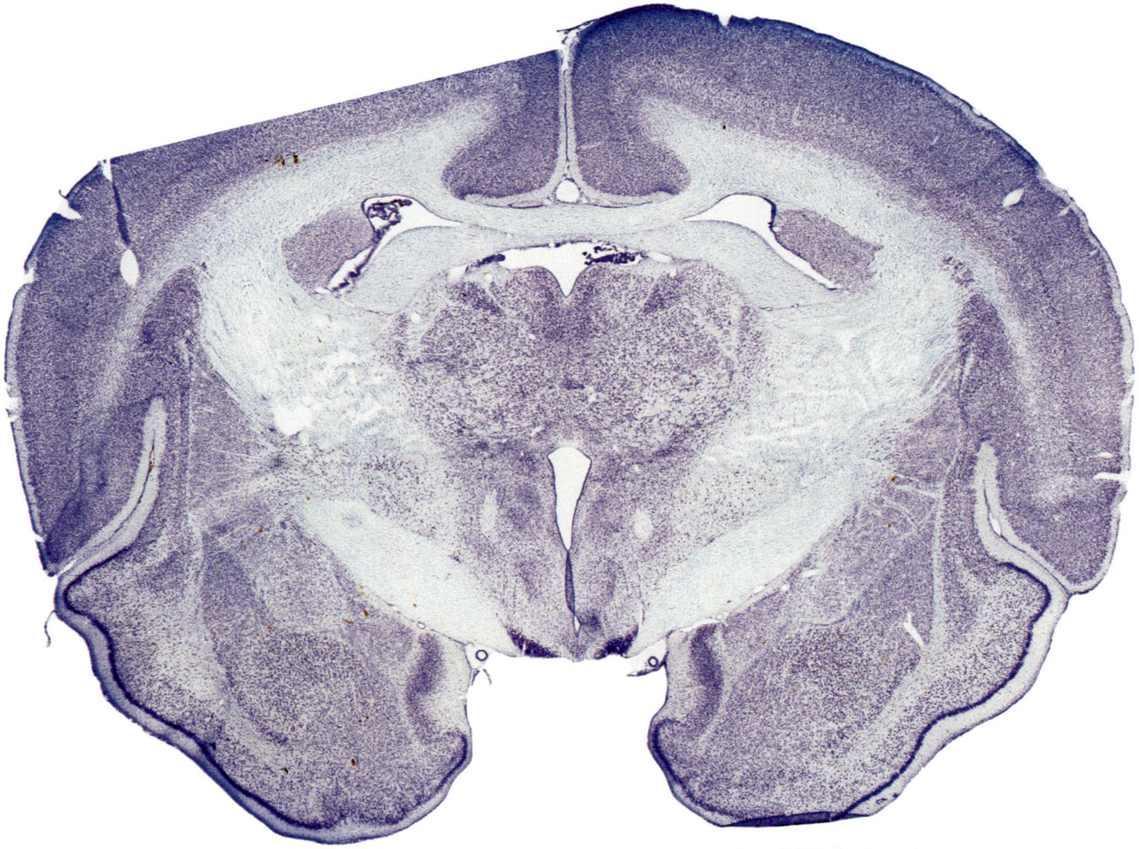

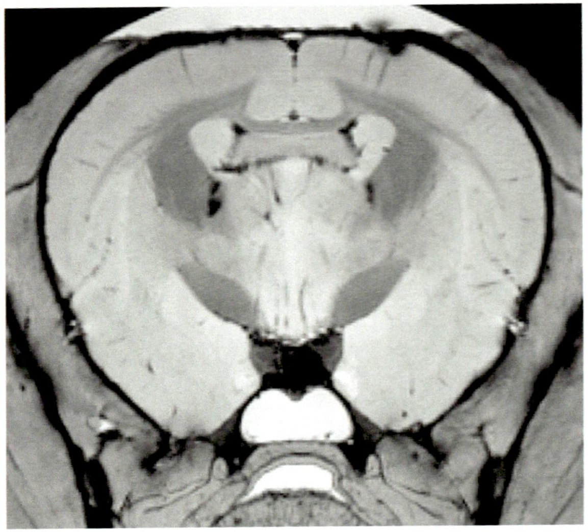

2	layer 2 of cortex
3	layer 3 of cortex
3V	3rd ventricle
Ab	auditory belt area
AB	accessory basal nucleus
Ac	auditory core area
Aco	anterior cortical amygdaloid nucleus
AD	anterodorsal thalamic nucleus
AHy	anterior hypothalamic area
AM	anteromedial thalamic nucleus
Arc	arcuate hypothalamic nucleus
AV	anteroventral thalamic nucleus
BLA	basolateral amygdaloid nucleus
BMA	basomedial amygdaloid nucleus
cc	corpus callosum
Cd	caudate nucleus
CeA	central amygdaloid nucleus
Cl	claustrum
CM	central medial thalamic nucleus
D3V	dorsal 3rd ventricle
EA	sublenticular extended amygdala
EGP	external globus pallidus
f	fornix
Hip	hippocampus
IAM	interanteromedial thalamic nucleus
ic	internal capsule
IG	indusium griseum
IGP	internal globus pallidus
IM	intercalated amygdaloid nucleus, main part
IRd	infraradiata dorsalis
IRv	infraradiata ventral area
IT	inferior temporal cortex
JPLH	juxtaparaventricular part of lateral hypothalamus
LHA	lateral hypothalamic area
LV	lateral ventricle
MeA	medial amygdaloid nucleus
opt	optic tract
Pir1	piriform cortex, layer 1
PPc	posterior parietal caudal area
PPd	posterior parietal dorsal area
PT	paratenial thalamic nucleus
Pu	putamen
PVA	paraventricular thalamic nucleus, anterior part
PVN	paraventricular hypothalamic nucleus
Re	reuniens thalamic nucleus
Rt	reticular thalamic nucleus
SFi	septofimbrial nucleus
sm	stria medullaris of the thalamus
SON	supraoptic nucleus
st	stria terminalis
TC	temporal cortex
TI	temporal inferior area
VL	ventrolateral thalamic nucleus
VMH	ventromedial hypothalamic nucleus

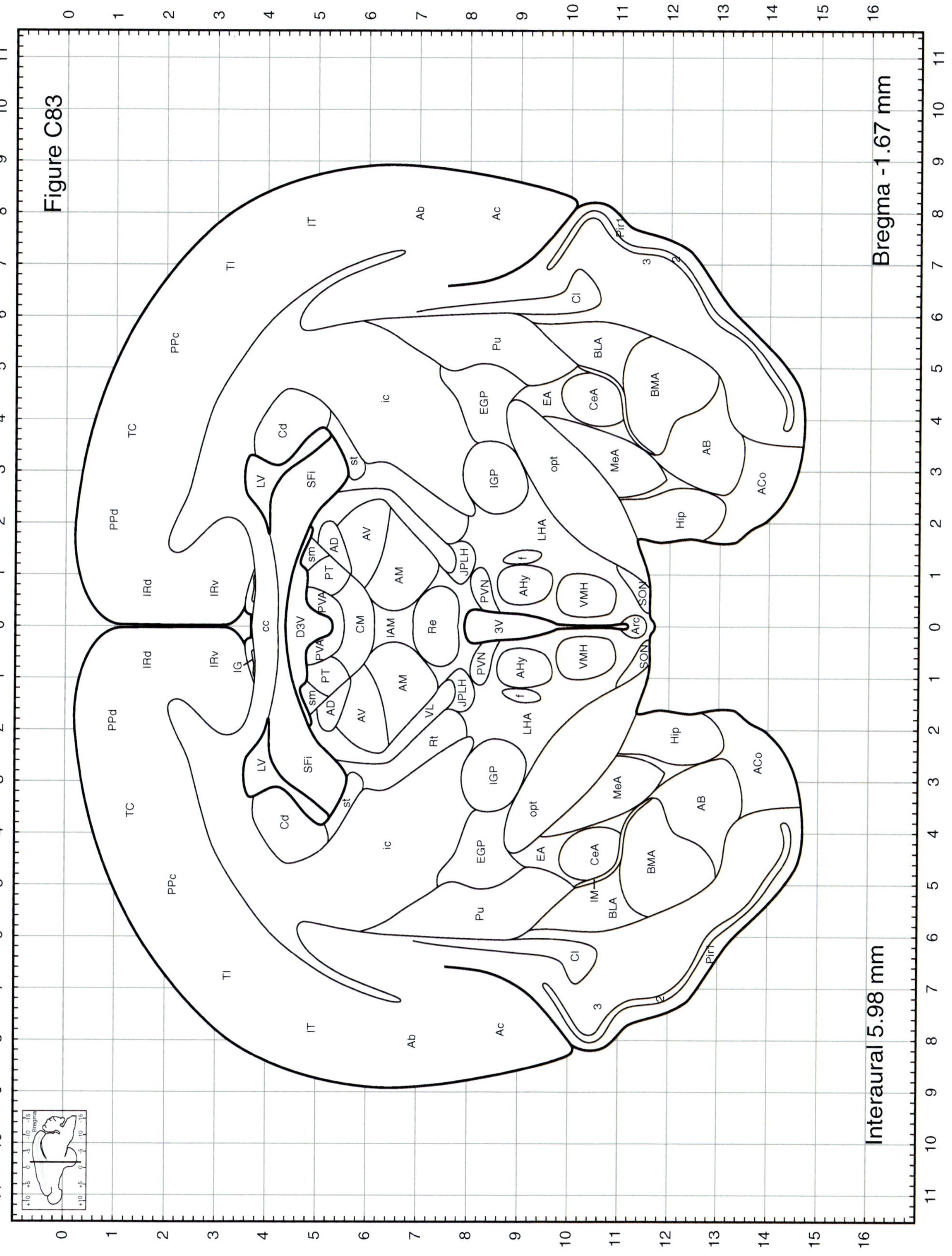
Figure C83
Bregma -1.67 mm
Interaural 5.98 mm
IT
Ab
Ac
TI
PPc
TC
PPd
IRd
IRv
Pir1
3
2
Cl
Pu
BLA
BMA
EGP
EA
CeA
ic
Cd
st
LV
SFi
IGP
opt
MeA
AB
ACo
Hip
AV
AD
sm
PT
AM
JPLH
LHA
f
PVN
AHy
VMH
SON
cc
D3V
PVA
CM
IAM
Re
3V
Arc
IG
VL
Rt
IM

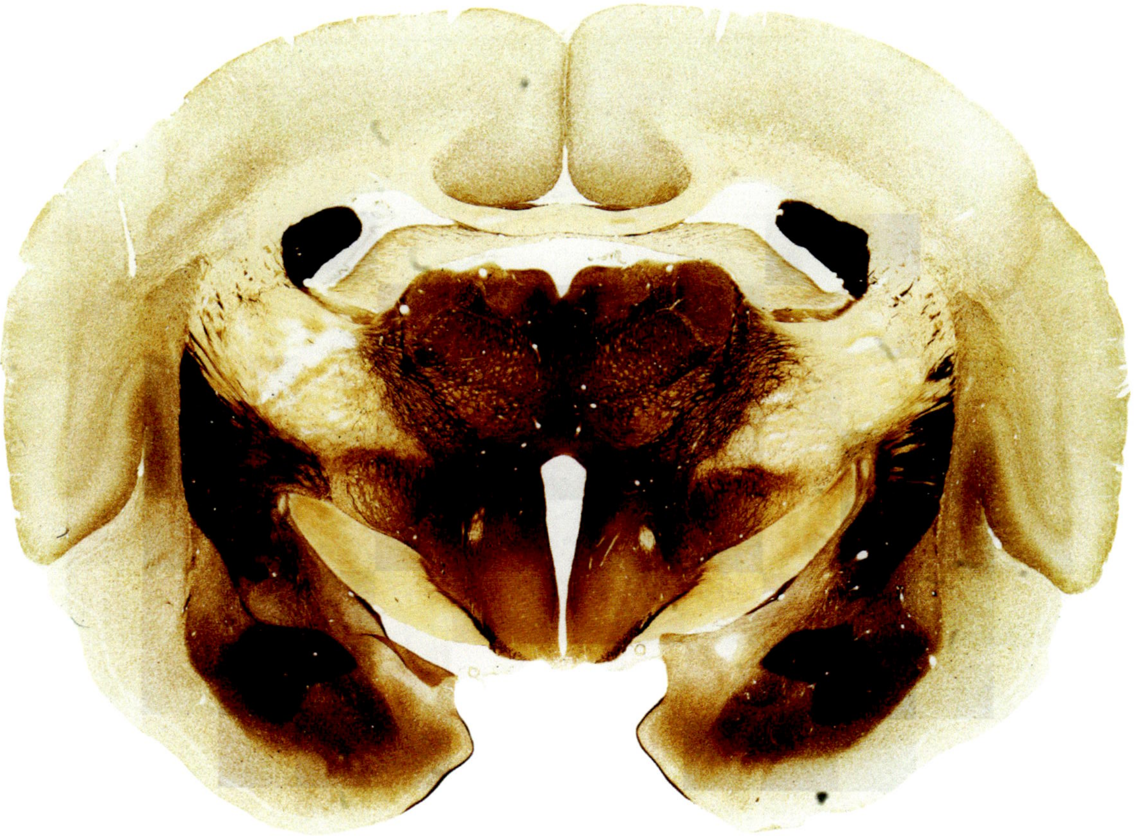

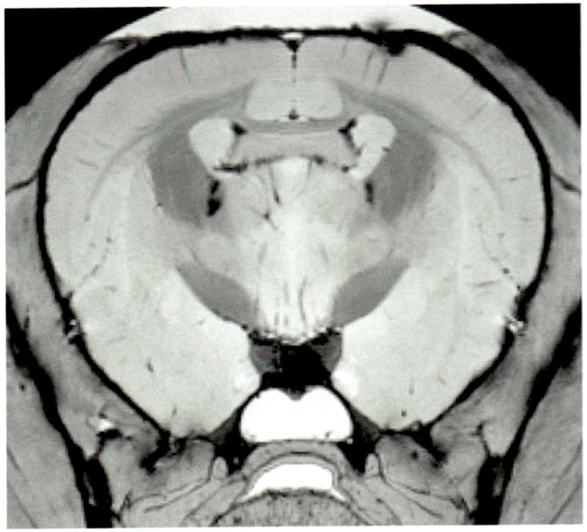

2	layer 2 of cortex
3	layer 3 of cortex
3V	3rd ventricle
Ab	auditory belt area
AB	accessory basal nucleus
Ac	auditory core area
Aco	anterior cortical amygdaloid nucleus
AD	anterodorsal thalamic nucleus
AHy	anterior hypothalamic area
AM	anteromedial thalamic nucleus
Arc	arcuate hypothalamic nucleus
AV	anteroventral thalamic nucleus
BLA	basolateral amygdaloid nucleus
BMA	basomedial amygdaloid nucleus
BN	basal nucleus (Meynert)
cc	corpus callosum
Cd	caudate nucleus
CeA	central amygdaloid nucleus
Cl	claustrum
CM	central medial thalamic nucleus
D3V	dorsal 3rd ventricle
EA	sublenticular extended amygdala
EGP	external globus pallidus
f	fornix
Hip	hippocampus
IAM	interanteromedial thalamic nucleus
ic	internal capsule
IG	indusium griseum
IGP	internal globus pallidus
IM	intercalated amygdaloid nucleus, main part
IRd	infraradiata dorsalis
IRv	infraradiata ventral area
IT	inferior temporal cortex
LD	laterodorsal thalamic nucleus
LHA	lateral hypothalamic area
LV	lateral ventricle
ME	median eminence
MeA	medial amygdaloid nucleus
opt	optic tract
Pir1	piriform cortex, layer 1
PPc	posterior parietal caudal area
PPd	posterior parietal dorsal area
PT	paratenial thalamic nucleus
Pu	putamen
PVA	paraventricular thalamic nucleus, anterior part
PVN	paraventricular hypothalamic nucleus
Re	reuniens thalamic nucleus
Rt	reticular thalamic nucleus
SFi	septofimbrial nucleus
sm	stria medullaris of the thalamus
SON	supraoptic nucleus
st	stria terminalis
TC	temporal cortex
TI	temporal inferior area
VL	ventrolateral thalamic nucleus
VMH	ventromedial hypothalamic nucleus

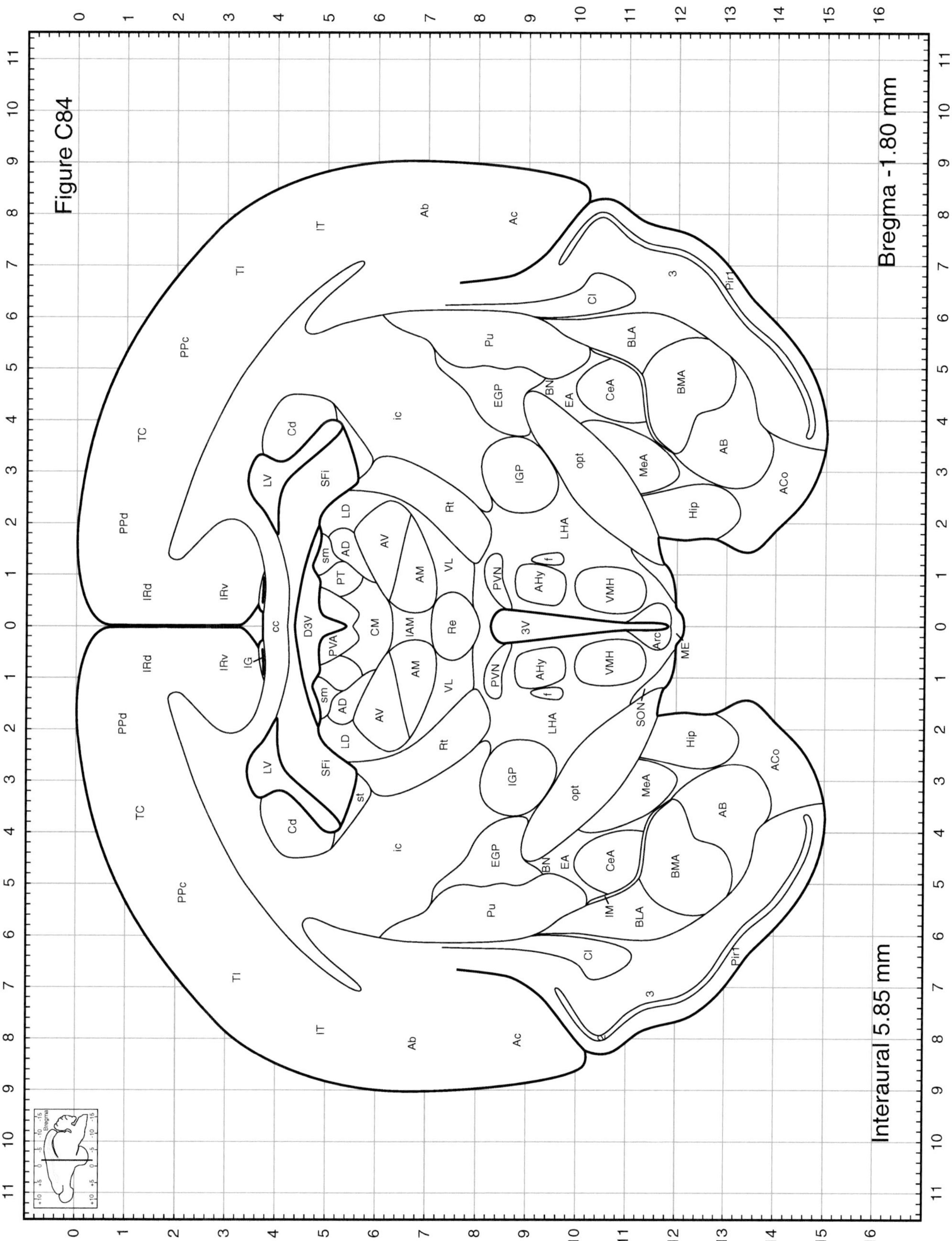
Figure C84
Bregma -1.80 mm
Interaural 5.85 mm
PPd
IRd
IRv
TC
PPc
TI
IT
Ab
Ac
Cd
LV
SFi
cc
IG
D3V
PVA
CM
IAM
Re
sm
AD
PT
LD
AV
AM
VL
Rt
ic
st
EGP
IGP
Pu
BN
EA
CeA
BLA
BMA
IM
Cl
3
Pir1
AB
ACo
MeA
Hip
opt
LHA
PVN
AHy
f
VMH
3V
Arc
ME
SON
Bregma

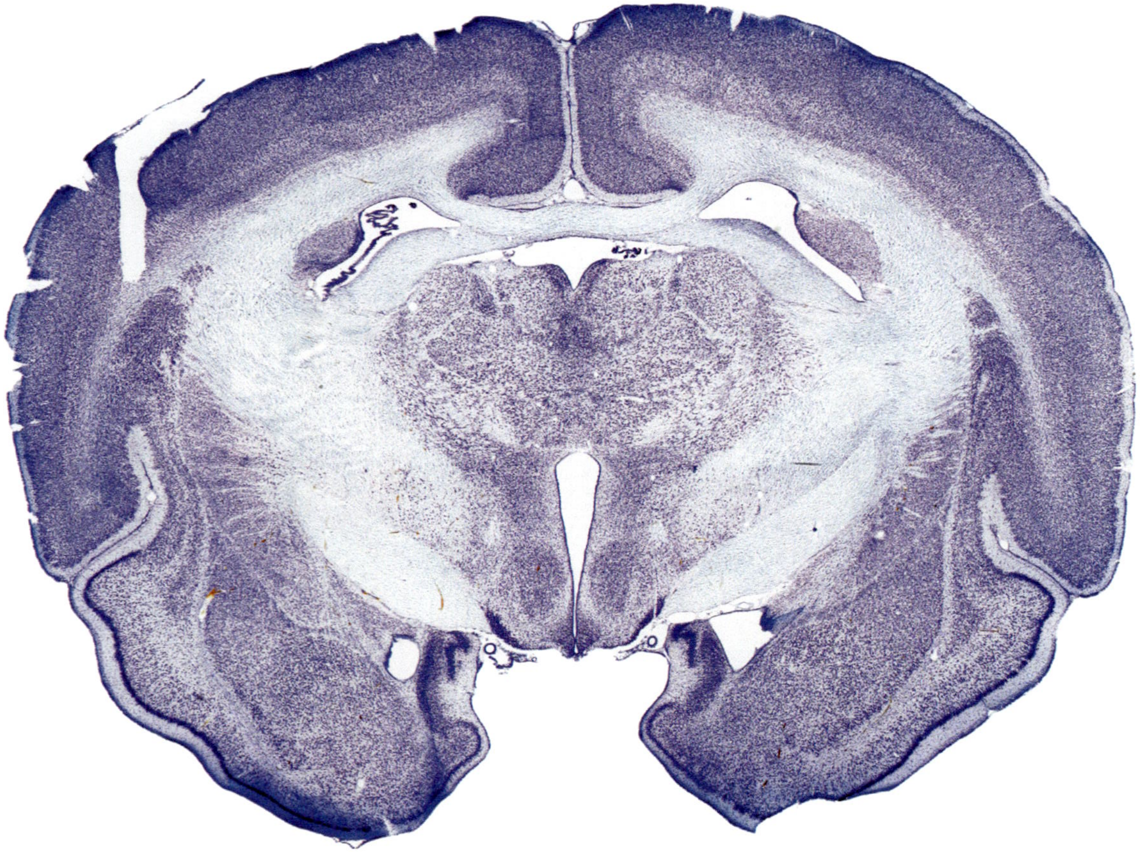

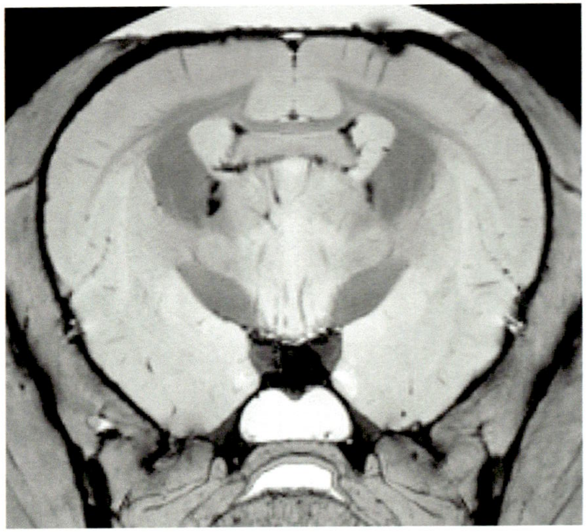

2	layer 2 of cortex
3	layer 3 of cortex
3V	3rd ventricle
Ab	auditory belt area
AB	accessory basal nucleus
Ac	auditory core area
Aco	anterior cortical amygdaloid nucleus
AD	anterodorsal thalamic nucleus
AHy	anterior hypothalamic area
AM	anteromedial thalamic nucleus
Arc	arcuate hypothalamic nucleus
AV	anteroventral thalamic nucleus
BLA	basolateral amygdaloid nucleus
BMA	basomedial amygdaloid nucleus
BN	basal nucleus (Meynert)
CA3	field CA3 of the hippocampus
cc	corpus callosum
Cd	caudate nucleus
CeA	central amygdaloid nucleus
Cl	claustrum
CM	central medial thalamic nucleus
D3V	dorsal 3rd ventricle
DG	dentate gyrus of the hippocampus
EA	sublenticular extended amygdala
EGP	external globus pallidus
f	fornix
IAM	interanteromedial thalamic nucleus
ic	internal capsule
IG	indusium griseum
IGP	internal globus pallidus
IM	intercalated amygdaloid nucleus, main part
IMD	intermediodorsal thalamic nucleus
IRd	infraradiata dorsalis
IRv	infraradiata ventral area
IT	inferior temporal cortex
LD	laterodorsal thalamic nucleus
LHA	lateral hypothalamic area
LV	lateral ventricle
MD	mediodorsal thalamic nucleus
ME	median eminence
MeA	medial amygdaloid nucleus
opt	optic tract
PaP	paraventricular hypothalamic nucleus, posterior part
Pir1	piriform cortex, layer 1
PPc	posterior parietal caudal area
PPd	posterior parietal dorsal area
PT	paratenial thalamic nucleus
Pu	putamen
PVA	paraventricular thalamic nucleus, anterior part
Re	reuniens thalamic nucleus
Rh	rhomboid thalamic nucleus
Rt	reticular thalamic nucleus
SFi	septofimbrial nucleus
sm	stria medullaris of the thalamus
SON	supraoptic nucleus
st	stria terminalis
Sub	submedius thalamic nucleus
TC	temporal cortex
TI	temporal inferior area
VL	ventrolateral thalamic nucleus
VMH	ventromedial hypothalamic nucleus

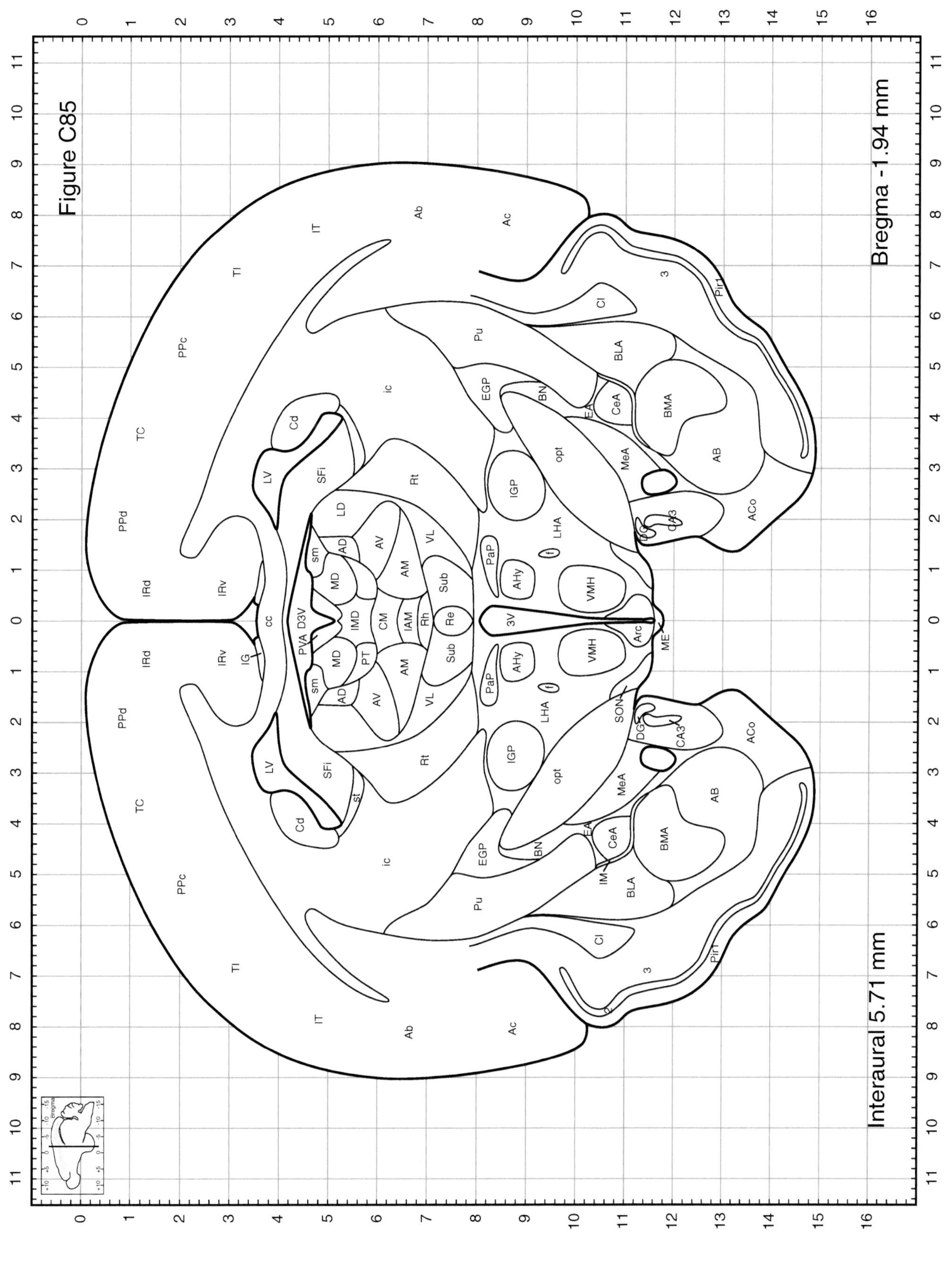
Figure C85
Bregma -1.94 mm
Interaural 5.71 mm
PPd
IRd
IRv
PPc
TC
TI
IT
Ab
Ac
Cd
LV
SFi
cc
IG
st
sm
PVA
D3V
MD
IMD
PT
AD
AV
LD
AM
CM
IAM
Rh
Re
Sub
VL
Rt
ic
Pu
EGP
BN
IGP
opt
PaP
AHy
f
LHA
3V
VMH
Arc
ME
SON
EA
CeA
MeA
BLA
BMA
AB
ACo
DG
CA3
Cl
IM
3
Pir1

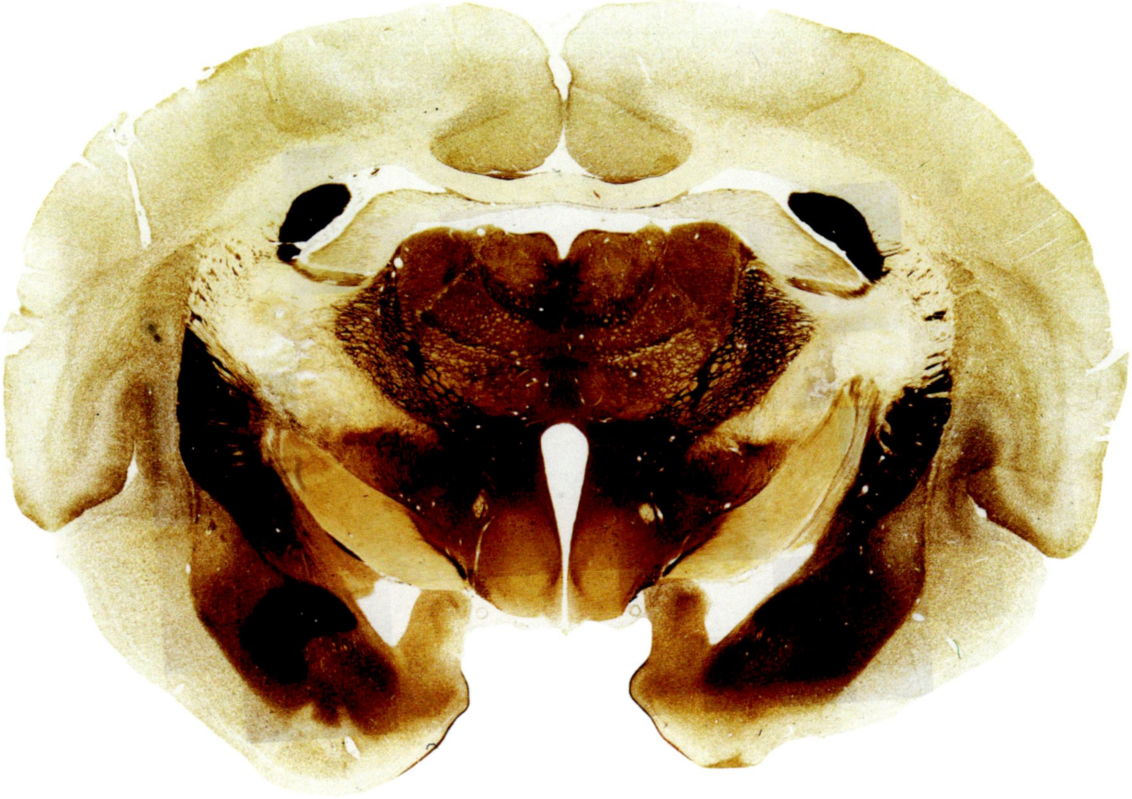

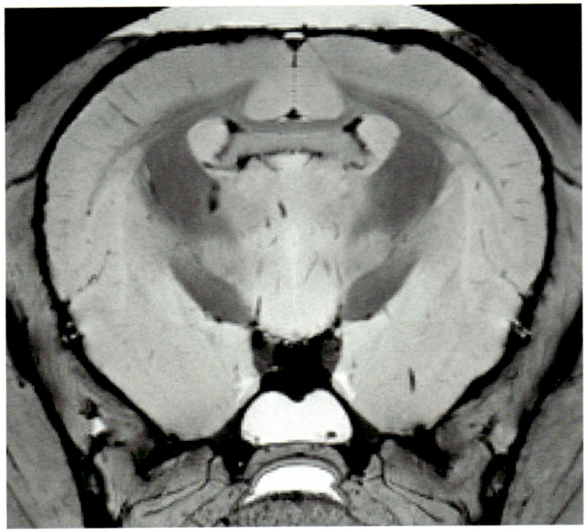

2	layer 2 of cortex	D3V	dorsal 3rd ventricle	opt	optic tract
3	layer 3 of cortex	DG	dentate gyrus of the hippocampus	PaP	paraventricular hypothalamic nucleus, posterior part
3V	3rd ventricle	DMH	dorsomedial hypothalamic nucleus	Pir1	piriform cortex, layer 1
Ab	auditory belt area	EGP	external globus pallidus	PPc	posterior parietal caudal area
AB	accessory basal nucleus	f	fornix	PPd	posterior parietal dorsal area
Ac	auditory core area	IAM	interanteromedial thalamic nucleus	PT	paratenial thalamic nucleus
Aco	anterior cortical amygdaloid nucleus	ic	internal capsule	Pu	putamen
AD	anterodorsal thalamic nucleus	IG	indusium griseum	PVA	paraventricular thalamic nucleus, anterior part
AM	anteromedial thalamic nucleus	IGP	internal globus pallidus	Re	reuniens thalamic nucleus
Arc	arcuate hypothalamic nucleus	IM	intercalated amygdaloid nucleus, main part	Rh	rhomboid thalamic nucleus
AV	anteroventral thalamic nucleus	IMD	intermediodorsal thalamic nucleus	Rt	reticular thalamic nucleus
BLA	basolateral amygdaloid nucleus	IRd	infraradiata dorsalis	SFi	septofimbrial nucleus
BMA	basomedial amygdaloid nucleus	IRv	infraradiata ventral area	sm	stria medullaris of the thalamus
BN	basal nucleus (Meynert)	IT	inferior temporal cortex	SON	supraoptic nucleus
CA3	field CA3 of the hippocampus	LD	laterodorsal thalamic nucleus	st	stria terminalis
cc	corpus callosum	LHA	lateral hypothalamic area	Sub	submedius thalamic nucleus
Cd	caudate nucleus	LV	lateral ventricle	TC	temporal cortex
CeA	central amygdaloid nucleus	MD	mediodorsal thalamic nucleus	TI	temporal inferior area
Cl	claustrum	ME	median eminence	VL	ventrolateral thalamic nucleus
CM	central medial thalamic nucleus	MeA	medial amygdaloid nucleus	VMH	ventromedial hypothalamic nucleus

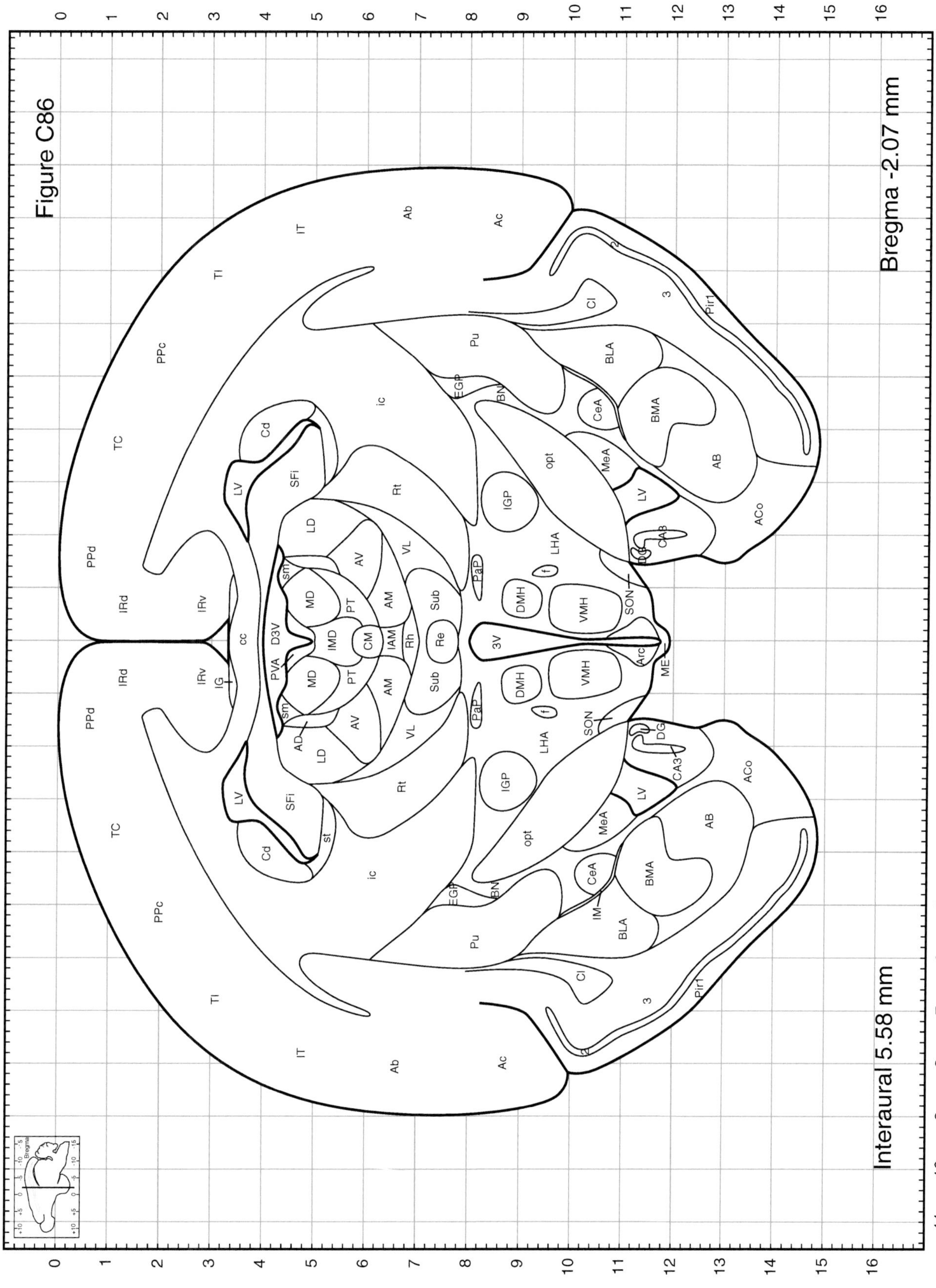
Figure C86
Bregma -2.07 mm
Interaural 5.58 mm
PPd
PPc
TC
TI
IT
Ab
Ac
IRd
IRv
IG
cc
LV
Cd
SFi
st
ic
Pu
EGP
BN
Cl
Pir1
BLA
BMA
CeA
IM
MeA
AB
ACo
CA3
DG
SON
Arc
ME
LHA
lGP
opt
f
DMH
VMH
3V
PaP
Rt
LD
AD
AV
VL
sm
MD
IMD
PT
AM
CM
IAM
Rh
Re
Sub
D3V
PVA

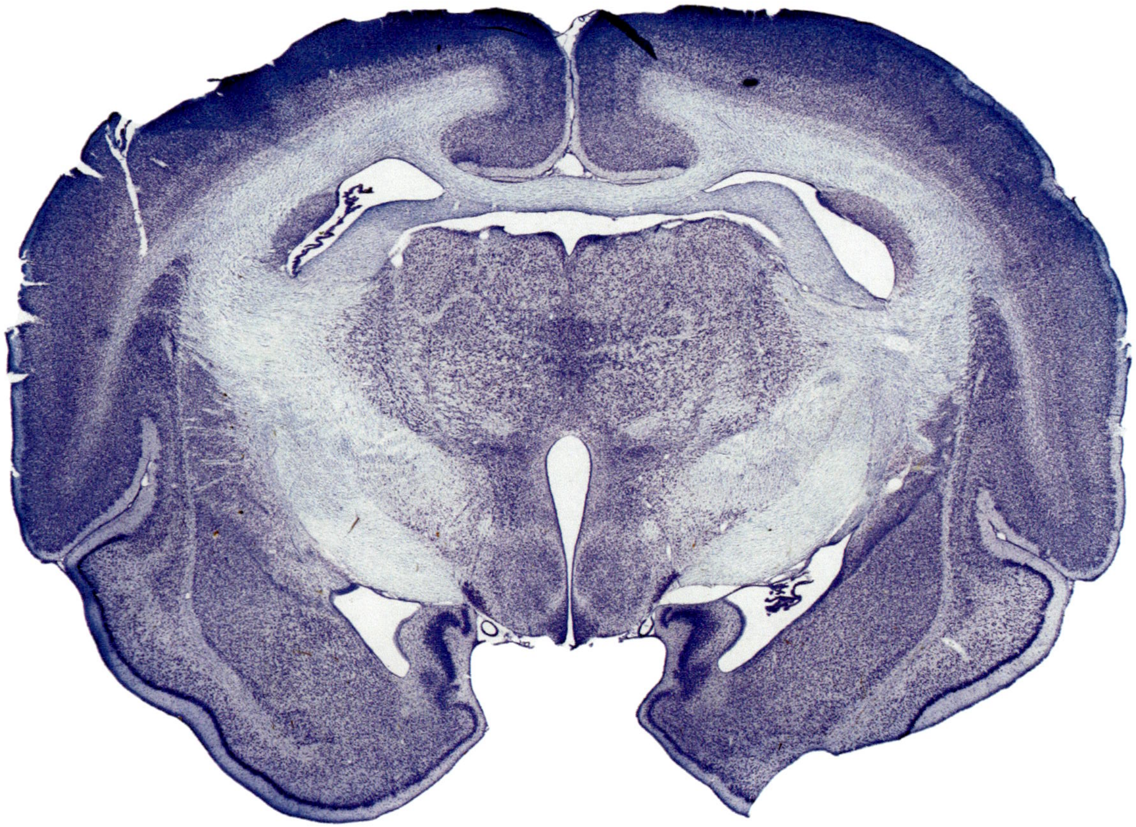

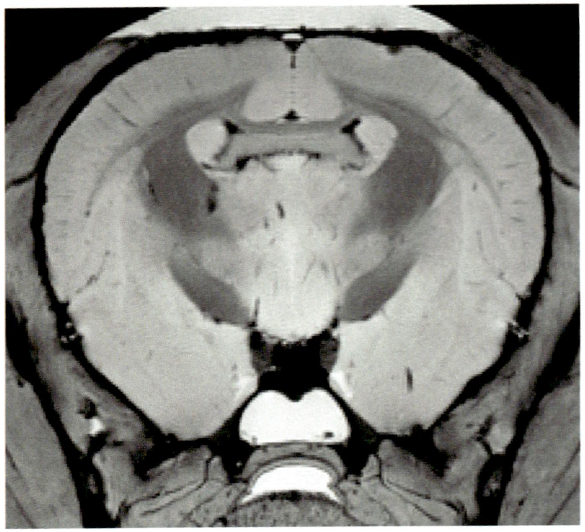

2	layer 2 of cortex
3	layer 3 of cortex
3V	3rd ventricle
Ab	auditory belt area
AB	accessory basal nucleus
Ac	auditory core area
Aco	anterior cortical amygdaloid nucleus
AD	anterodorsal thalamic nucleus
AM	anteromedial thalamic nucleus
Arc	arcuate hypothalamic nucleus
AV	anteroventral thalamic nucleus
BLA	basolateral amygdaloid nucleus
BMA	basomedial amygdaloid nucleus
CA3	field CA3 of the hippocampus
cc	corpus callosum
Cd	caudate nucleus
Cl	claustrum
CM	central medial thalamic nucleus
D3V	dorsal 3rd ventricle
DG	dentate gyrus of the hippocampus
DMH	dorsomedial hypothalamic nucleus
f	fornix
IAM	interanteromedial thalamic nucleus
ic	internal capsule
IMD	intermediodorsal thalamic nucleus
IRd	infraradiata dorsalis
IRv	infraradiata ventral area
IT	inferior temporal cortex
LD	laterodorsal thalamic nucleus
LHA	lateral hypothalamic area
LV	lateral ventricle
MD	mediodorsal thalamic nucleus
ME	median eminence
opt	optic tract
Pir1	piriform cortex, layer 1
PPc	posterior parietal caudal area
PPd	posterior parietal dorsal area
PT	paratenial thalamic nucleus
Pu	putamen
PVA	paraventricular thalamic nucleus, anterior part
Re	reuniens thalamic nucleus
Rh	rhomboid thalamic nucleus
Rt	reticular thalamic nucleus
SFi	septofimbrial nucleus
sm	stria medullaris of the thalamus
SON	supraoptic nucleus
st	stria terminalis
Sub	submedius thalamic nucleus
TC	temporal cortex
TI	temporal inferior area
VL	ventrolateral thalamic nucleus
VMH	ventromedial hypothalamic nucleus

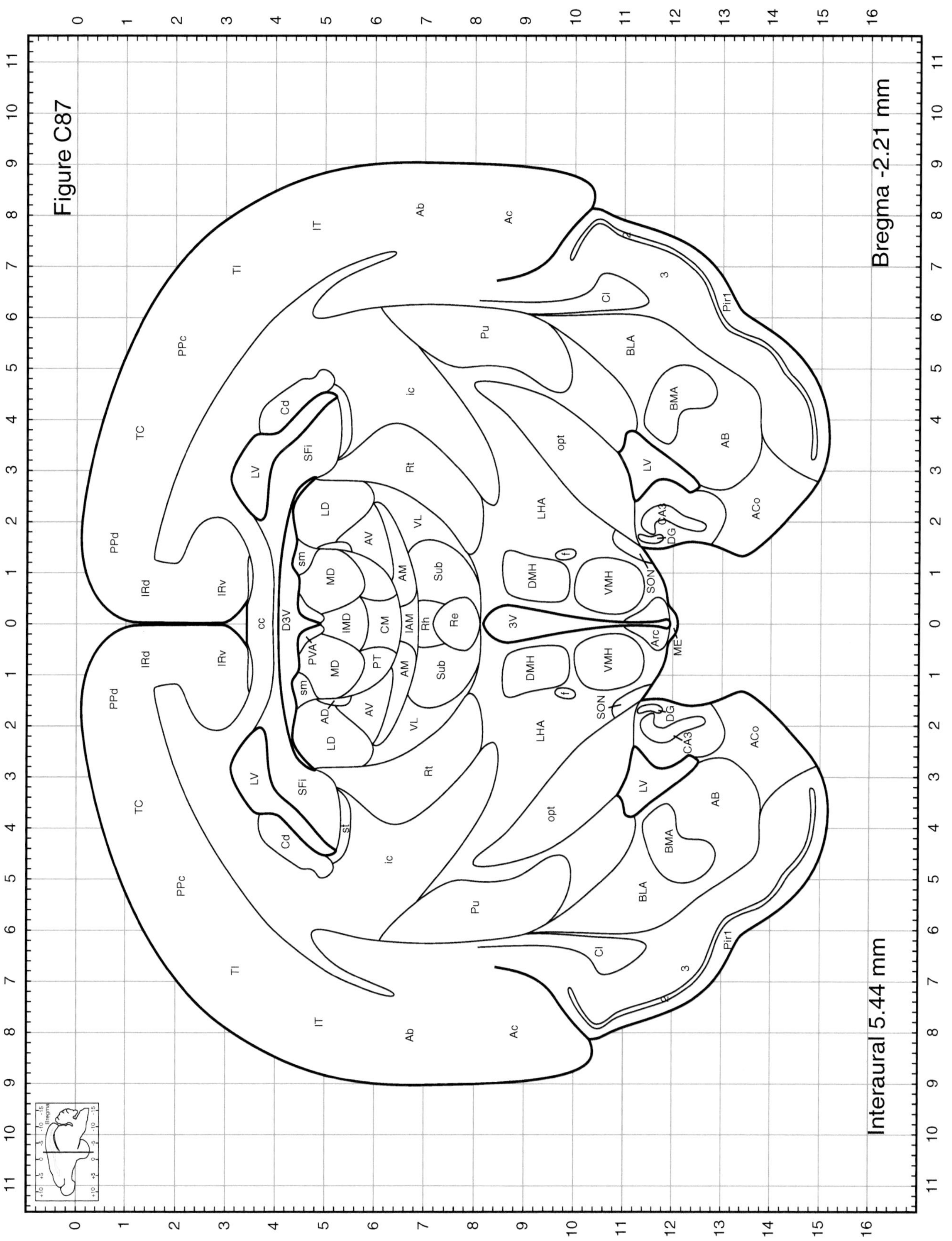
Figure C87
Bregma -2.21 mm
Interaural 5.44 mm
IT
Ab
Ac
TI
PPc
TC
PPd
IRd
IRv
cc
D3V
Cd
LV
SFi
st
ic
Pu
Cl
BLA
BMA
AB
ACo
Pir1
2
3
opt
LHA
LD
AD
AV
VL
MD
sm
PVA
IMD
CM
PT
AM
IAM
Rh
Re
Sub
Rt
DMH
VMH
f
3V
SON
Arc
ME
CA3
DG

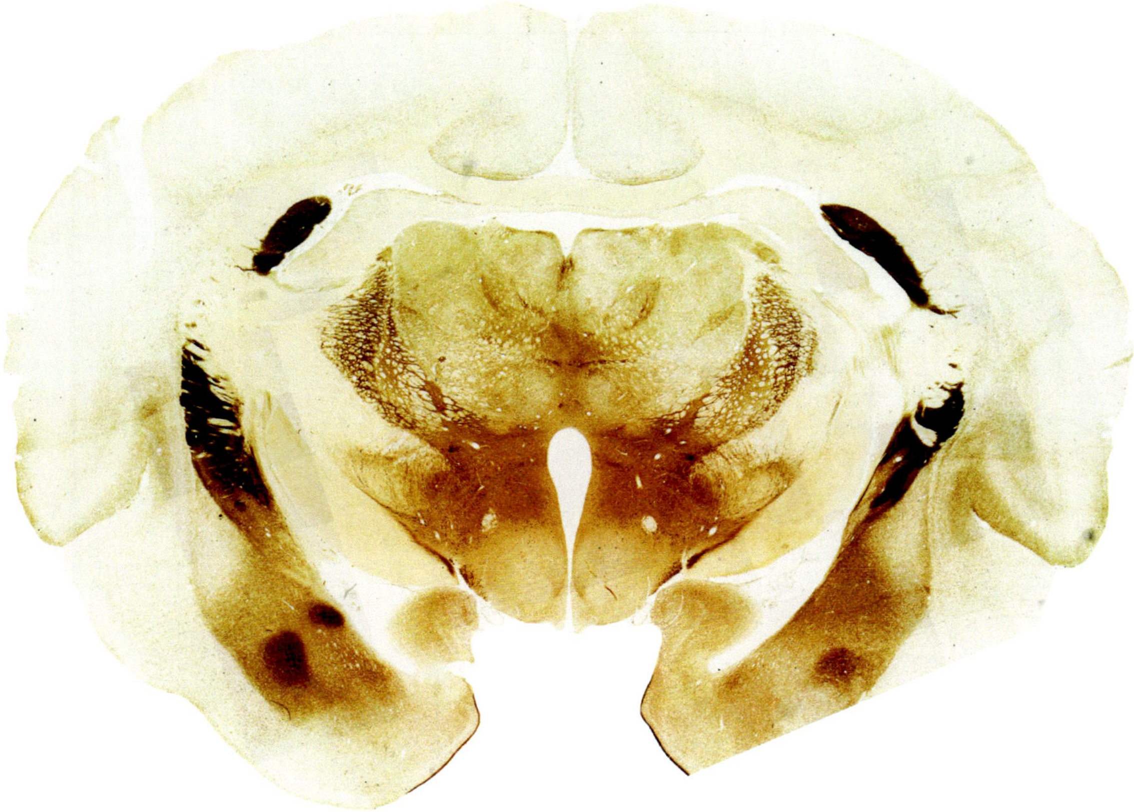

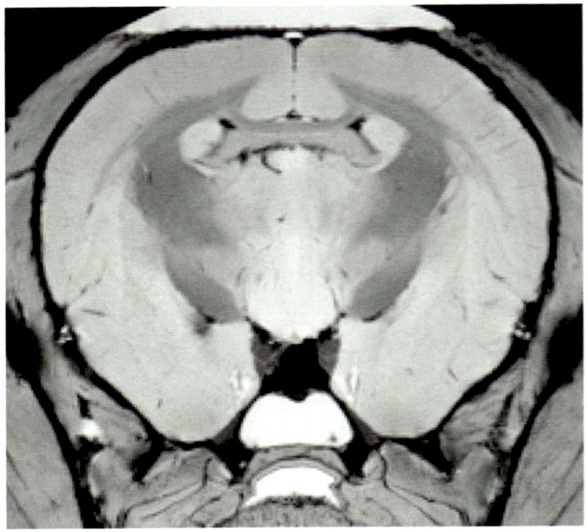

2	layer 2 of cortex
3	layer 3 of cortex
3V	3rd ventricle
Ab	auditory belt area
AB	accessory basal nucleus
Ac	auditory core area
Aco	anterior cortical amygdaloid nucleus
AM	anteromedial thalamic nucleus
Arc	arcuate hypothalamic nucleus
ASt	amygdalostriatal transition area
BLA	basolateral amygdaloid nucleus
BMA	basomedial amygdaloid nucleus
CA3	field CA3 of the hippocampus
cc	corpus callosum
Cd	caudate nucleus
Cl	claustrum
CM	central medial thalamic nucleus
D3V	dorsal 3rd ventricle
DG	dentate gyrus of the hippocampus
DMH	dorsomedial hypothalamic nucleus
f	fornix
ic	internal capsule
IMD	intermediodorsal thalamic nucleus
IRd	infraradiata dorsalis
IRv	infraradiata ventral area
IT	inferior temporal cortex
LD	laterodorsal thalamic nucleus
LHA	lateral hypothalamic area
LV	lateral ventricle
MD	mediodorsal thalamic nucleus
ME	median eminence
mt	mammillothalamic tract
opt	optic tract
Pir1	piriform cortex, layer 1
PPc	posterior parietal caudal area
PPd	posterior parietal dorsal area
PT	paratenial thalamic nucleus
Pu	putamen
PVA	paraventricular thalamic nucleus, anterior part
Re	reuniens thalamic nucleus
Rh	rhomboid thalamic nucleus
Rt	reticular thalamic nucleus
SFi	septofimbrial nucleus
sm	stria medullaris of the thalamus
SON	supraoptic nucleus
st	stria terminalis
Sub	submedius thalamic nucleus
TC	temporal cortex
TI	temporal inferior area
VL	ventrolateral thalamic nucleus
VMH	ventromedial hypothalamic nucleus
ZI	zona incerta

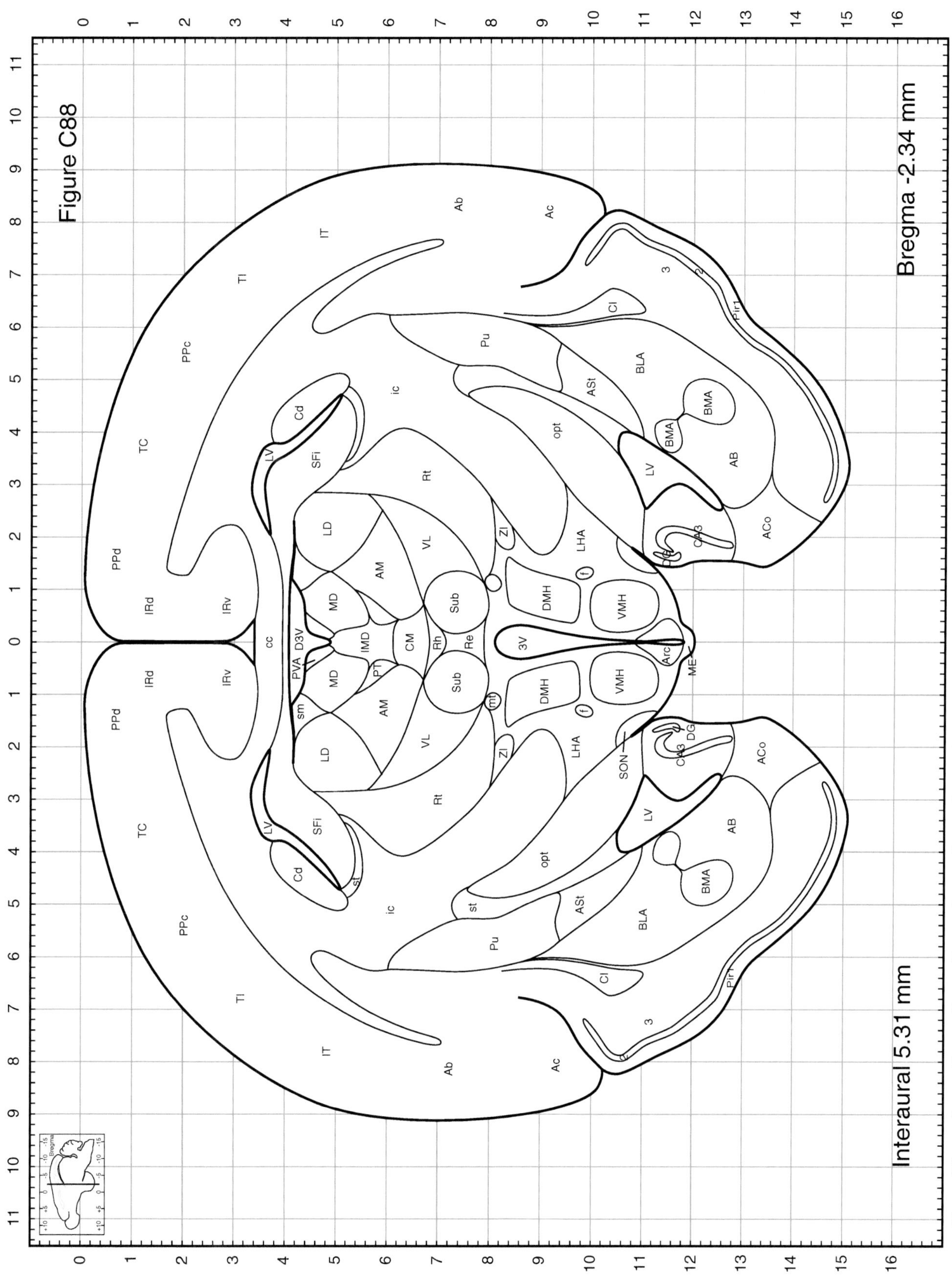
Figure C88
Bregma -2.34 mm
Interaural 5.31 mm

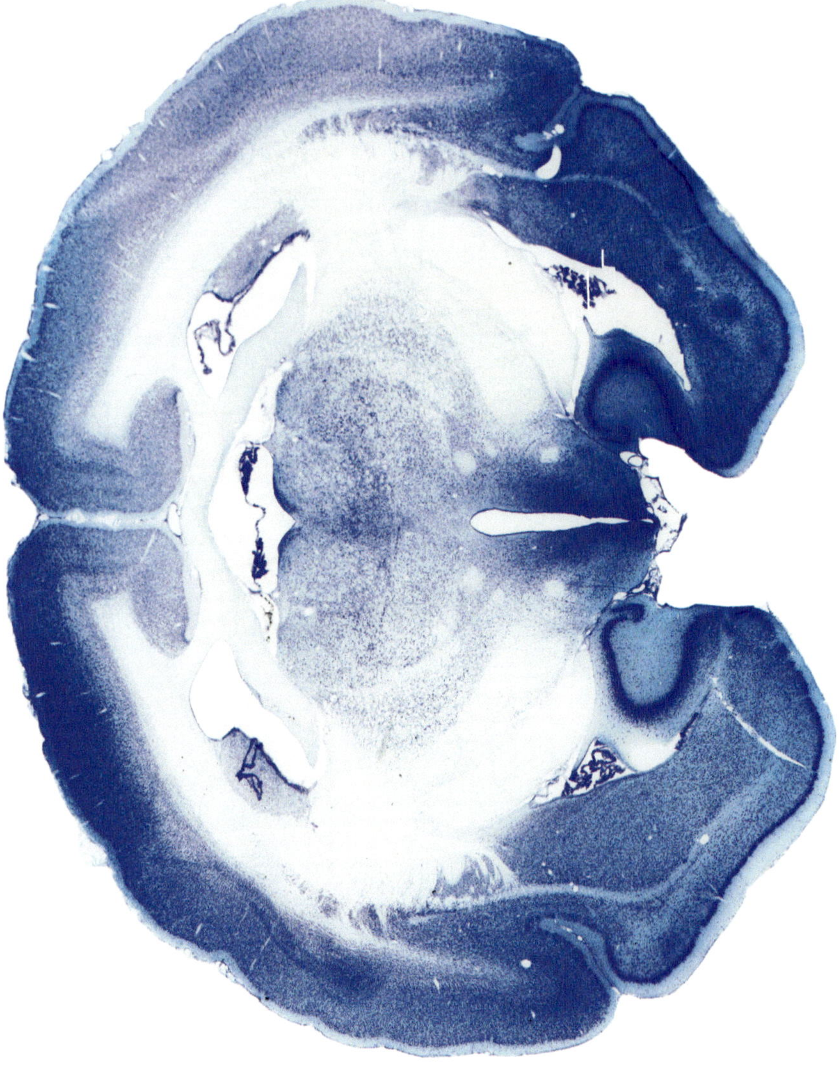

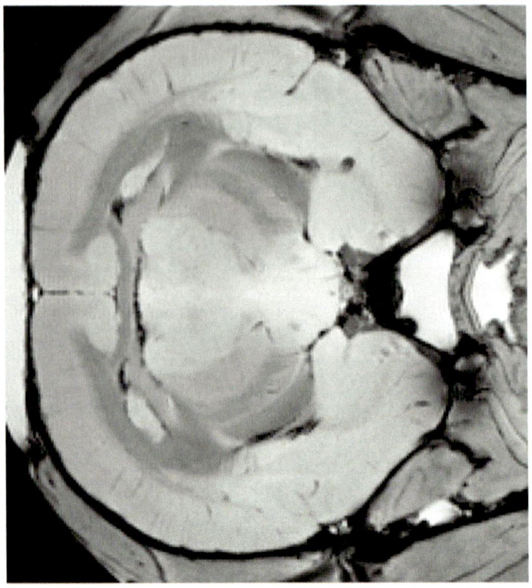

2	layer 2 of cortex
3	layer 3 of cortex
3V	3rd ventricle
Ab	auditory belt area
Ac	auditory core area
ACo	anterior cortical amygdaloid nucleus
Arc	arcuate hypothalamic nucleus
ASt	amygdalostriatal transition area
AV	anteroventral thalamic nucleus
BLA	basolateral amygdaloid nucleus
BMA	basomedial amygdaloid nucleus
CA1	field CA1 of the hippocampus
CA2	field CA2 of the hippocampus
CA3	field CA3 of the hippocampus
cc	corpus callosum
Cd	caudate nucleus
Cl	claustrum
CM	central medial thalamic nucleus
cp	cerebral peduncle
D3V	dorsal 3rd ventricle
DG	dentate gyrus of the hippocampus
DMH	dorsomedial hypothalamic nucleus
f	fornix
ic	internal capsule
IMD	intermediodorsal thalamic nucleus
IRd	infraradiata dorsalis
IRv	infraradiata ventral area
IT	inferior temporal cortex
LD	laterodorsal thalamic nucleus
LHA	lateral hypothalamic area
LHb	lateral habenular nucleus
LV	lateral ventricle
MD	mediodorsal thalamic nucleus
ME	median eminence
mt	mammillothalamic tract
opt	optic tract
PC	paracentral thalamic nucleus
PeF	perifornical nucleus
Pir1	piriform cortex, layer 1
PPc	posterior parietal caudal area
PPd	posterior parietal dorsal area
Pu	putamen
PVA	paraventricular thalamic nucleus, anterior part
Re	reuniens thalamic nucleus
Rh	rhomboid thalamic nucleus
Rt	reticular thalamic nucleus
SFi	septofimbrial nucleus
sm	stria medullaris of the thalamus
SON	supraoptic nucleus
st	stria terminalis
Sub	submedius thalamic nucleus
TC	temporal cortex
TI	temporal inferior area
VL	ventrolateral thalamic nucleus
VMH	ventromedial hypothalamic nucleus
ZI	zona incerta

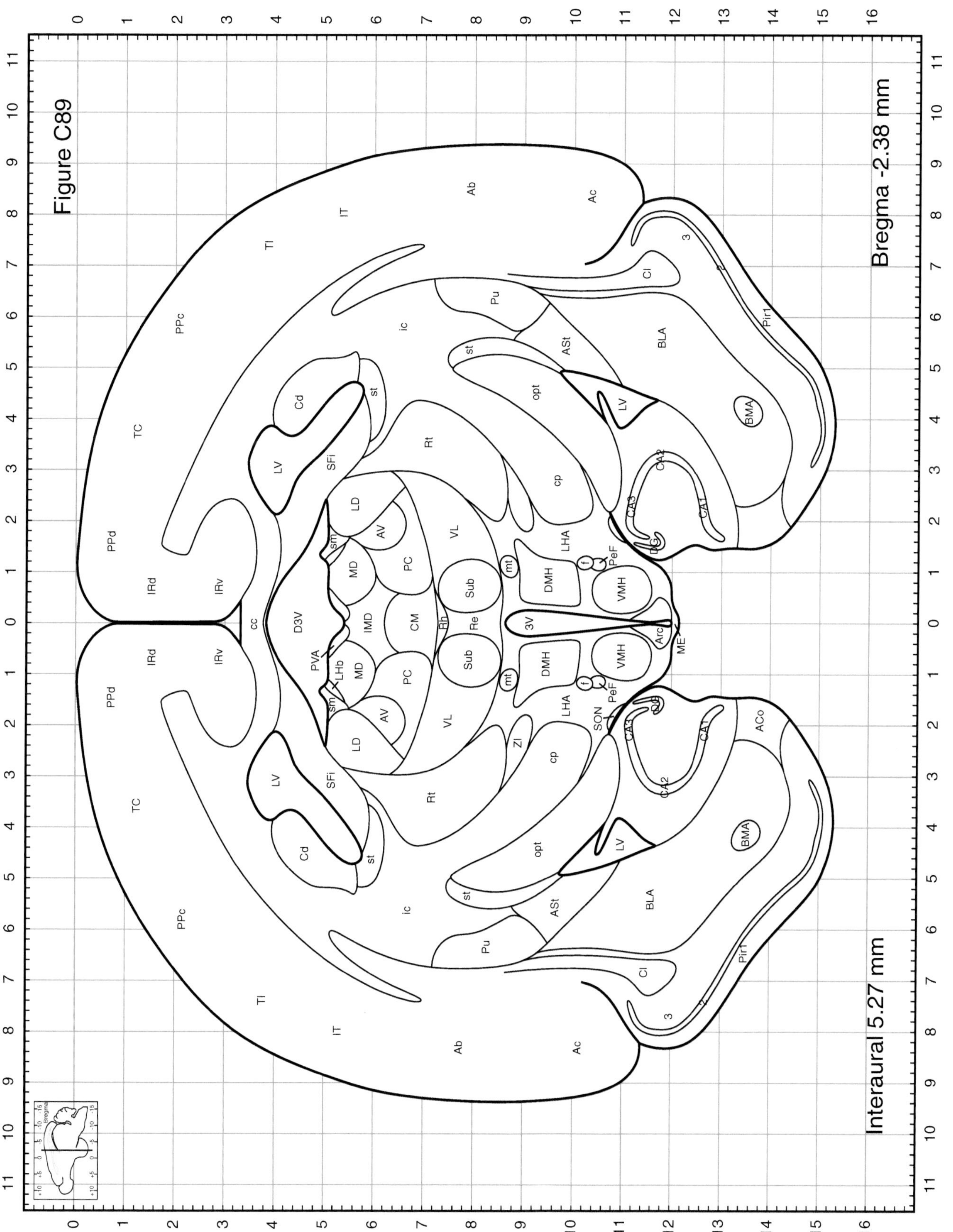
Figure C89
Bregma -2.38 mm
Interaural 5.27 mm
Ab
Ac
IT
TI
PPc
TC
PPd
IRd
IRv
cc
D3V
Cd
LV
SFi
st
ic
Pu
ASt
opt
Rt
cp
LD
AV
sm
MD
PC
IMD
CM
PVA
LHb
VL
Sub
Re
Rh
mt
DMH
VMH
3V
LHA
f
PeF
Arc
ME
SON
ZI
CA1
CA2
CA3
DG
Cl
BLA
BMA
ACo
Pir1
3
2

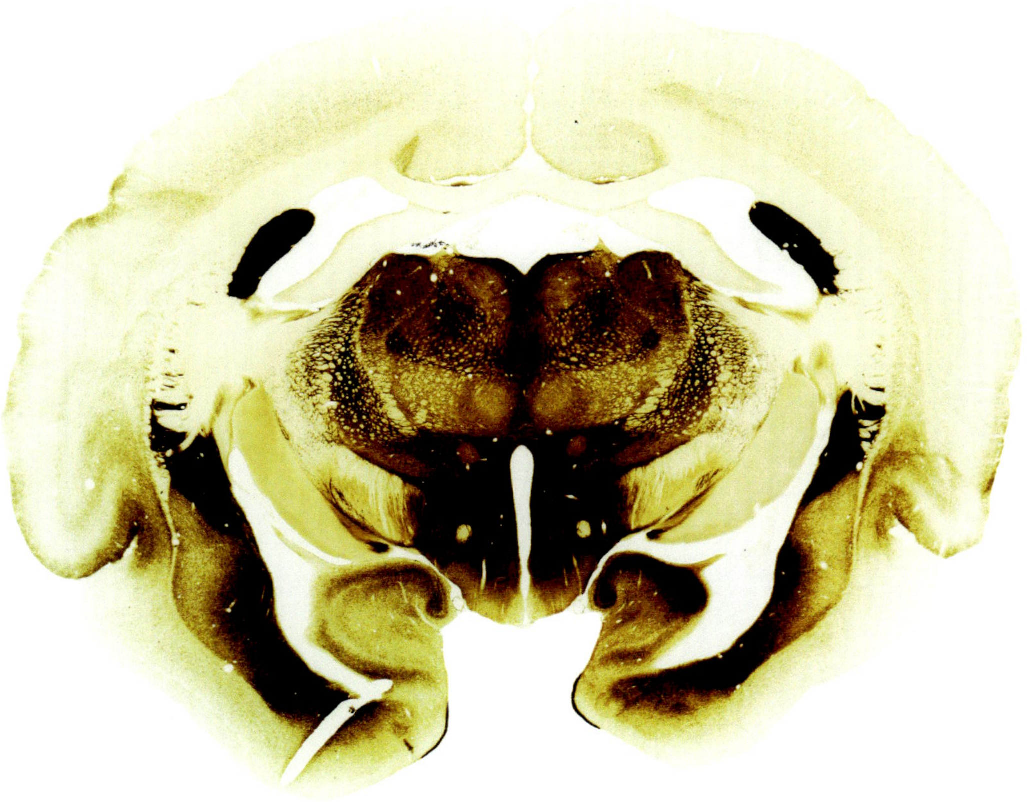

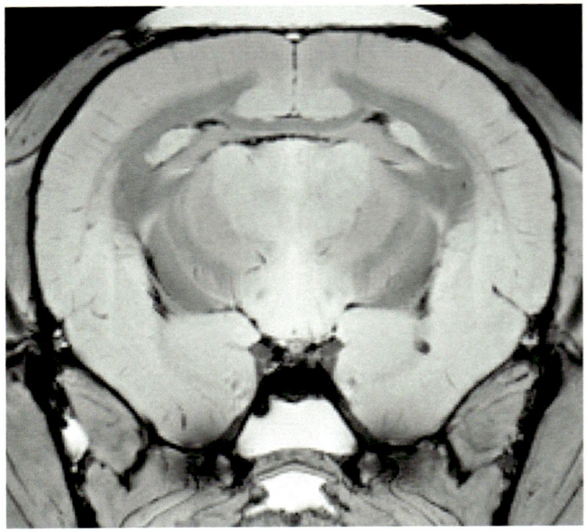

2	layer 2 of cortex
3	layer 3 of cortex
3V	3rd ventricle
Ab	auditory belt area
Ac	auditory core area
ACo	anterior cortical amygdaloid nucleus
Arc	arcuate hypothalamic nucleus
ASt	amygdalostriatal transition area
AV	anteroventral thalamic nucleus
BLA	basolateral amygdaloid nucleus
BMA	basomedial amygdaloid nucleus
CA1	field CA1 of the hippocampus
CA2	field CA2 of the hippocampus
CA3	field CA3 of the hippocampus
cc	corpus callosum
Cd	caudate nucleus
Cl	claustrum
CM	central medial thalamic nucleus
cp	cerebral peduncle
D3V	dorsal 3rd ventricle
DG	dentate gyrus of the hippocampus
DMH	dorsomedial hypothalamic nucleus
f	fornix
fi	fimbria of the hippocampus
ic	internal capsule
IMD	intermediodorsal thalamic nucleus
IRd	infraradiata dorsalis
IRv	infraradiata ventral area
IT	inferior temporal cortex
LD	laterodorsal thalamic nucleus
LHA	lateral hypothalamic area
LHb	lateral habenular nucleus
LV	lateral ventricle
MD	mediodorsal thalamic nucleus
ME	median eminence
mt	mammillothalamic tract
opt	optic tract
Or	oriens layer of the hippocampus
PC	paracentral thalamic nucleus
PeF	perifornical nucleus
Pir1	piriform cortex, layer 1
PPc	posterior parietal caudal area
PPd	posterior parietal dorsal area
Pu	putamen
PVA	paraventricular thalamic nucleus, anterior part
Py	pyramidal cell layer of the hippocampus
Rad	radiatum layer of the hippocampus
Re	reuniens thalamic nucleus
Rh	rhomboid thalamic nucleus
Rt	reticular thalamic nucleus
SFi	septofimbrial nucleus
sm	stria medullaris of the thalamus
st	stria terminalis
Sub	submedius thalamic nucleus
TC	temporal cortex
TI	temporal inferior area
VL	ventrolateral thalamic nucleus
VM	ventromedial thalamic nucleus
VMH	ventromedial hypothalamic nucleus
ZI	zona incerta

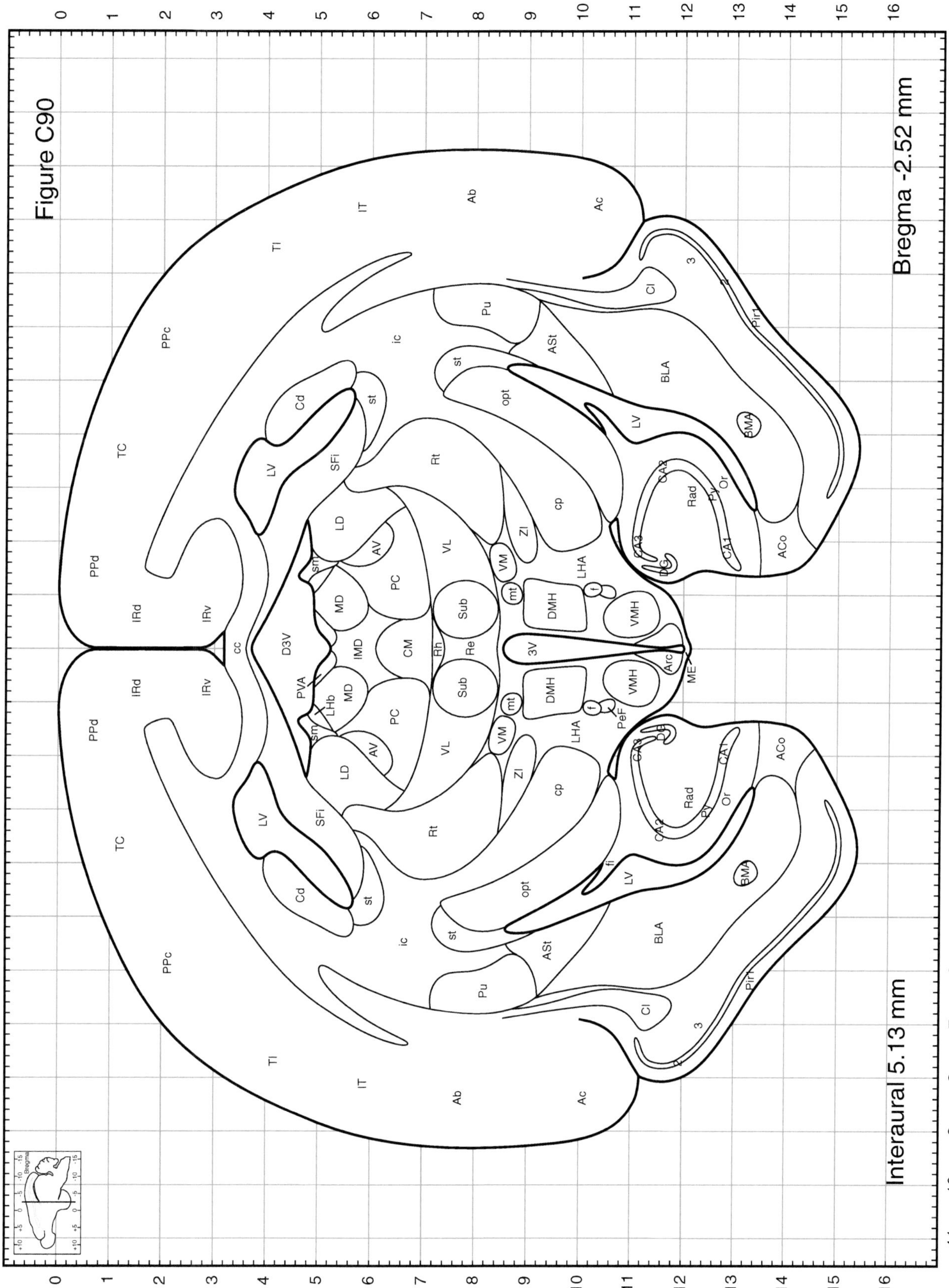
Figure C90
Bregma -2.52 mm
Interaural 5.13 mm
Ab
Ac
IT
TI
PPc
TC
PPd
IRd
IRv
cc
D3V
PVA
sm
LHb
MD
IMD
CM
Rh
Re
Sub
PC
AV
LD
VL
Rt
SFi
LV
Cd
st
ic
Pu
ASt
opt
cp
ZI
VM
mt
DMH
LHA
f
PeF
VMH
3V
Arc
ME
fi
CA1
CA2
CA3
DG
Rad
Py
Or
ACo
BMA
BLA
Cl
Pir1
2
3

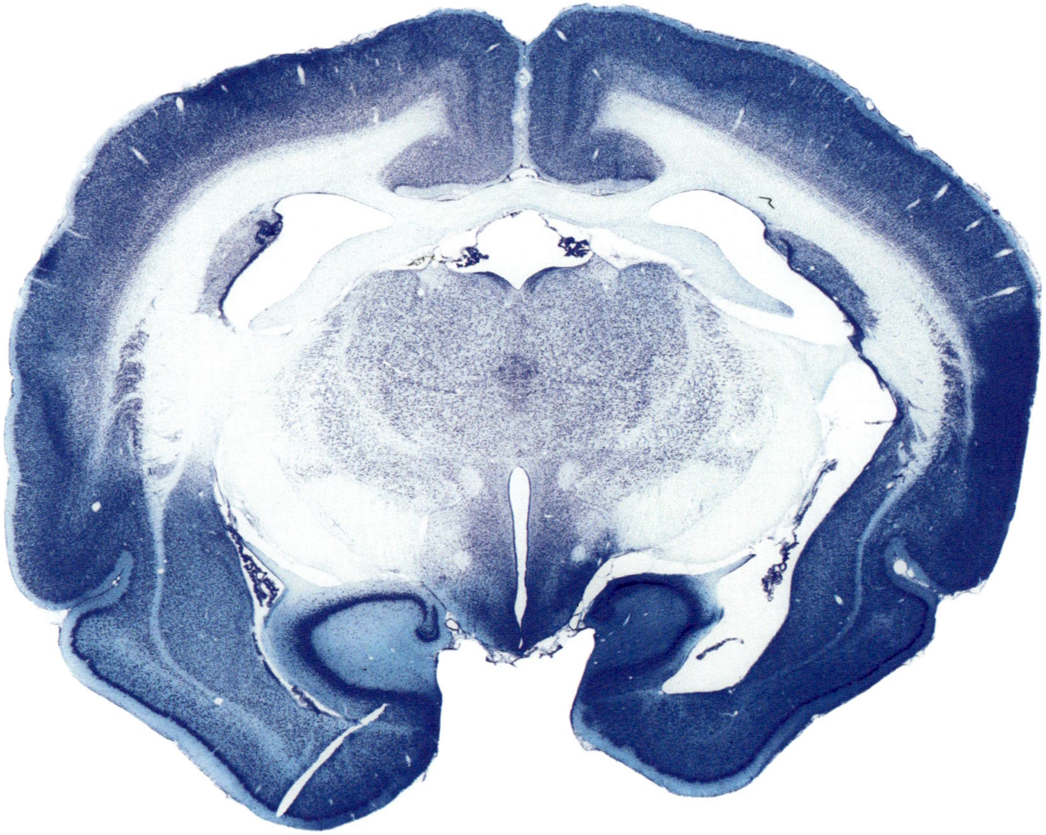

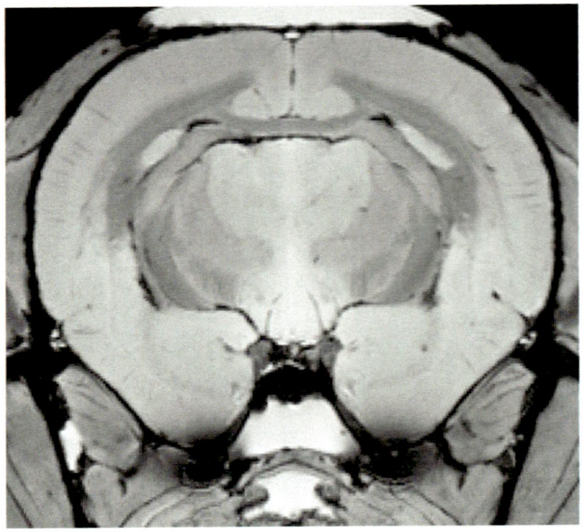

2	layer 2 of cortex
3	layer 3 of cortex
3V	3rd ventricle
Ab	auditory belt area
Ac	auditory core area
ACo	anterior cortical amygdaloid nucleus
alv	alveus of the hippocampus
Arc	arcuate hypothalamic nucleus
ASt	amygdalostriatal transition area
BLA	basolateral amygdaloid nucleus
CA1	field CA1 of the hippocampus
CA2	field CA2 of the hippocampus
CA3	field CA3 of the hippocampus
cc	corpus callosum
Cd	caudate nucleus
Cl	claustrum
CM	central medial thalamic nucleus
cp	cerebral peduncle
D3V	dorsal 3rd ventricle
DG	dentate gyrus of the hippocampus
DMH	dorsomedial hypothalamic nucleus
f	fornix
fi	fimbria of the hippocampus
ic	internal capsule
IMD	intermediodorsal thalamic nucleus
IRd	infraradiata dorsalis
IRv	infraradiata ventral area
IT	inferior temporal cortex
LD	laterodorsal thalamic nucleus
LHA	lateral hypothalamic area
LHb	lateral habenular nucleus
LV	lateral ventricle
MD	mediodorsal thalamic nucleus
ME	median eminence
mt	mammillothalamic tract
opt	optic tract
Or	oriens layer of the hippocampus
PC	paracentral thalamic nucleus
Pir1	piriform cortex, layer 1
PPc	posterior parietal caudal area
PPd	posterior parietal dorsal area
Pu	putamen
PVA	paraventricular thalamic nucleus, anterior part
Py	pyramidal cell layer of the hippocampus
Rad	radiatum layer of the hippocampus
Re	reuniens thalamic nucleus
Rh	rhomboid thalamic nucleus
Rt	reticular thalamic nucleus
SFi	septofimbrial nucleus
sm	stria medullaris of the thalamus
st	stria terminalis
Sub	submedius thalamic nucleus
TC	temporal cortex
TI	temporal inferior area
VL	ventrolateral thalamic nucleus
VLG	ventral lateral geniculate nucleus
VM	ventromedial thalamic nucleus
VMH	ventromedial hypothalamic nucleus
ZI	zona incerta

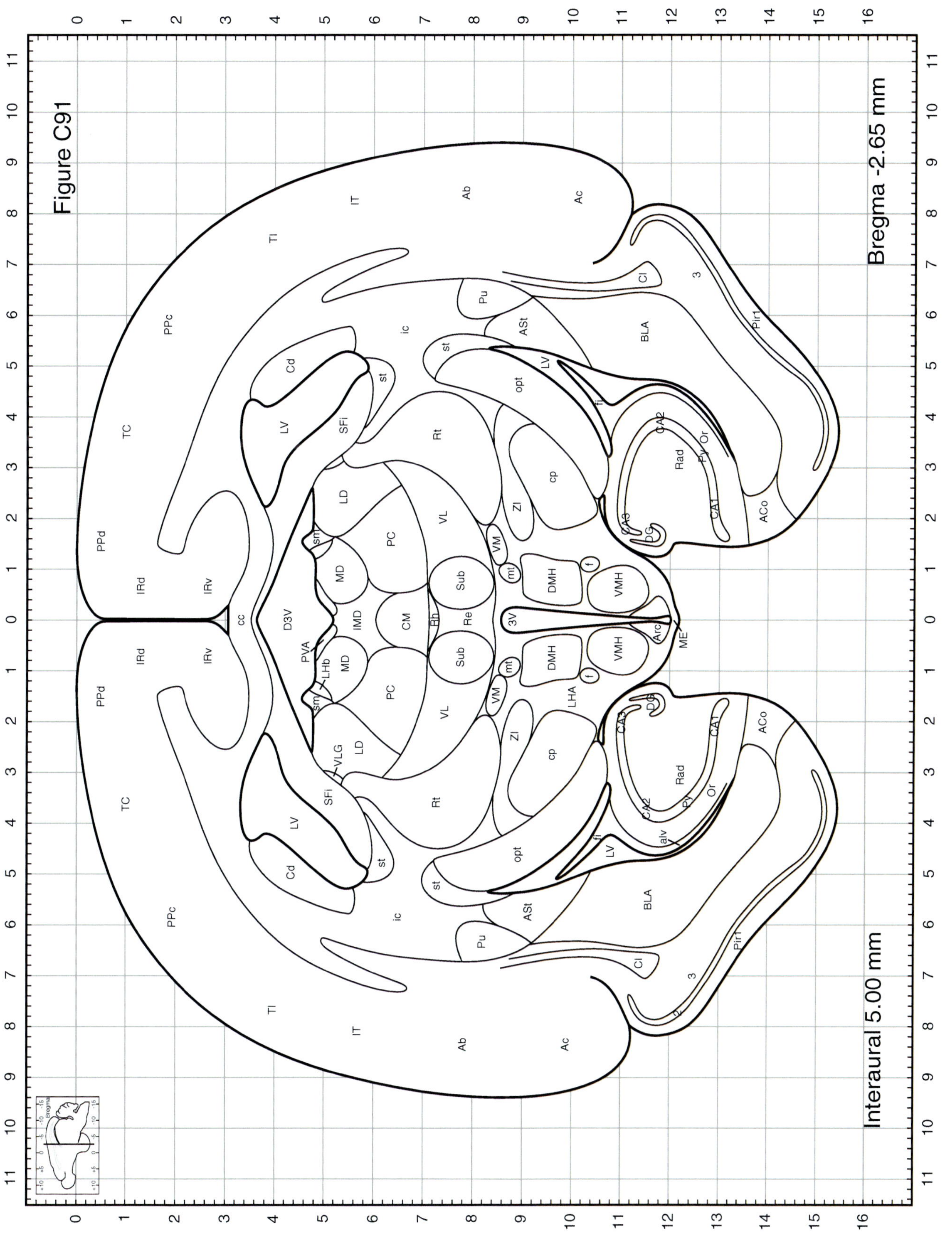
Figure C91
Bregma -2.65 mm
Interaural 5.00 mm
PPd
IRd
IRv
TC
PPc
TI
IT
Ab
Ac
cc
Cd
LV
st
SFi
ic
Pu
ASt
opt
Rt
LD
PC
VL
ZI
cp
sm
MD
IMD
CM
D3V
PVA
LHb
VLG
Sub
Rh
Re
VM
mt
3V
DMH
VMH
f
LHA
Arc
ME
fi
CA1
CA2
CA3
DG
Rad
Py
Or
alv
BLA
CI
3
Pir1
ACo

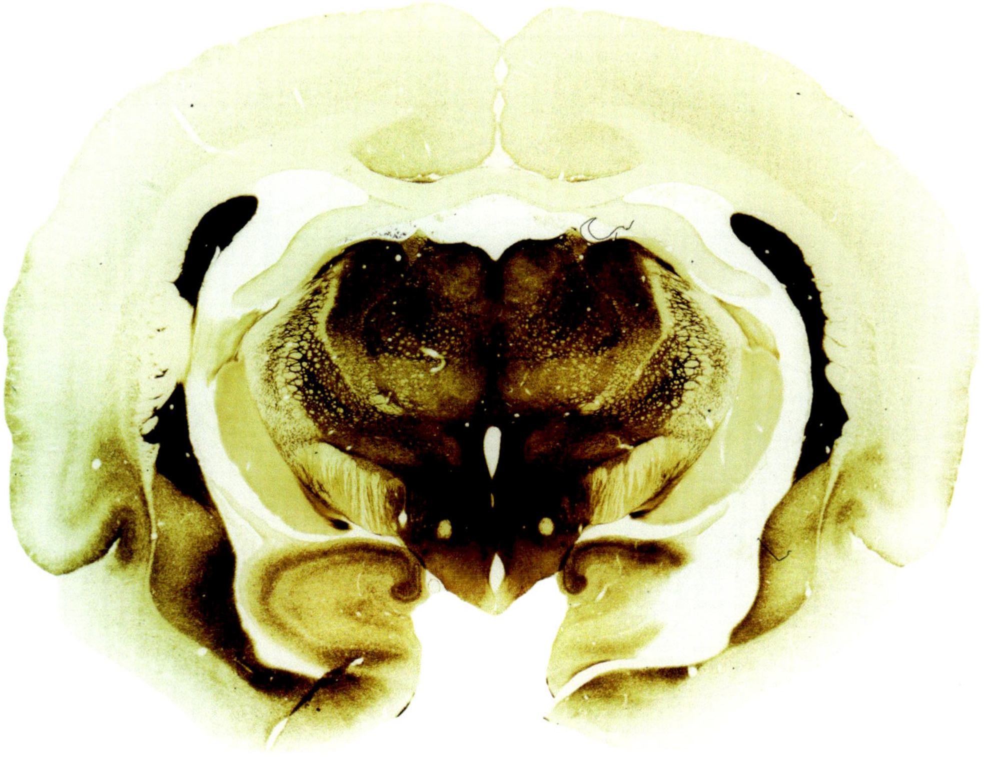

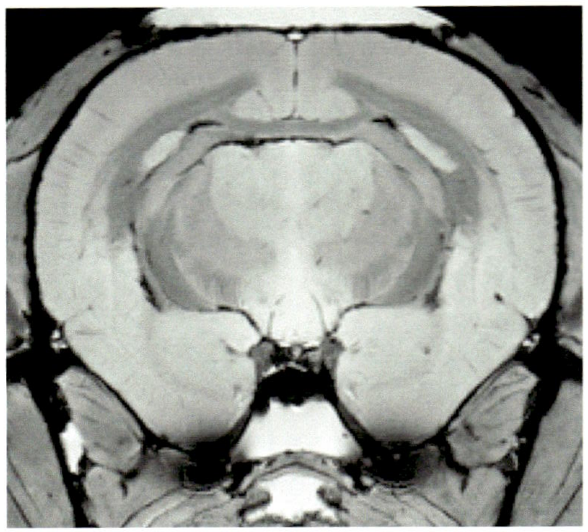

2	layer 2 of cortex
3	layer 3 of cortex
3V	3rd ventricle
Ab	auditory belt area
Ac	auditory core area
ACo	anterior cortical amygdaloid nucleus
alv	alveus of the hippocampus
Arc	arcuate hypothalamic nucleus
ASt	amygdalostriatal transition area
BLA	basolateral amygdaloid nucleus
CA1	field CA1 of the hippocampus
CA2	field CA2 of the hippocampus
CA3	field CA3 of the hippocampus
cc	corpus callosum
Cd	caudate nucleus
Cl	claustrum
CM	central medial thalamic nucleus
cp	cerebral peduncle
D3V	dorsal 3rd ventricle
DG	dentate gyrus of the hippocampus
DMH	dorsomedial hypothalamic nucleus
f	fornix
fi	fimbria of the hippocampus
ic	internal capsule
IMD	intermediodorsal thalamic nucleus
IRd	infraradiata dorsalis
IRv	infraradiata ventral area
IT	inferior temporal cortex
LD	laterodorsal thalamic nucleus
LHA	lateral hypothalamic area
LHb	lateral habenular nucleus
LV	lateral ventricle
MD	mediodorsal thalamic nucleus
ME	median eminence
mt	mammillothalamic tract
opt	optic tract
Or	oriens layer of the hippocampus
PC	paracentral thalamic nucleus
Pir1	piriform cortex, layer 1
PPc	posterior parietal caudal area
Pu	putamen
Pul	pulvinar nuclei
PVA	paraventricular thalamic nucleus, anterior part
Py	pyramidal cell layer of the hippocampus
Rad	radiatum layer of the hippocampus
Re	reuniens thalamic nucleus
Rh	rhomboid thalamic nucleus
Rt	reticular thalamic nucleus
SFi	septofimbrial nucleus
sm	stria medullaris of the thalamus
st	stria terminalis
STh	subthalamic nucleus
str	superior thalamic radiation
Sub	submedius thalamic nucleus
TC	temporal cortex
TI	temporal inferior area
V2	secondary visual cortex
VL	ventrolateral thalamic nucleus
VLG	ventral lateral geniculate nucleus
VM	ventromedial thalamic nucleus
VMH	ventromedial hypothalamic nucleus
ZI	zona incerta

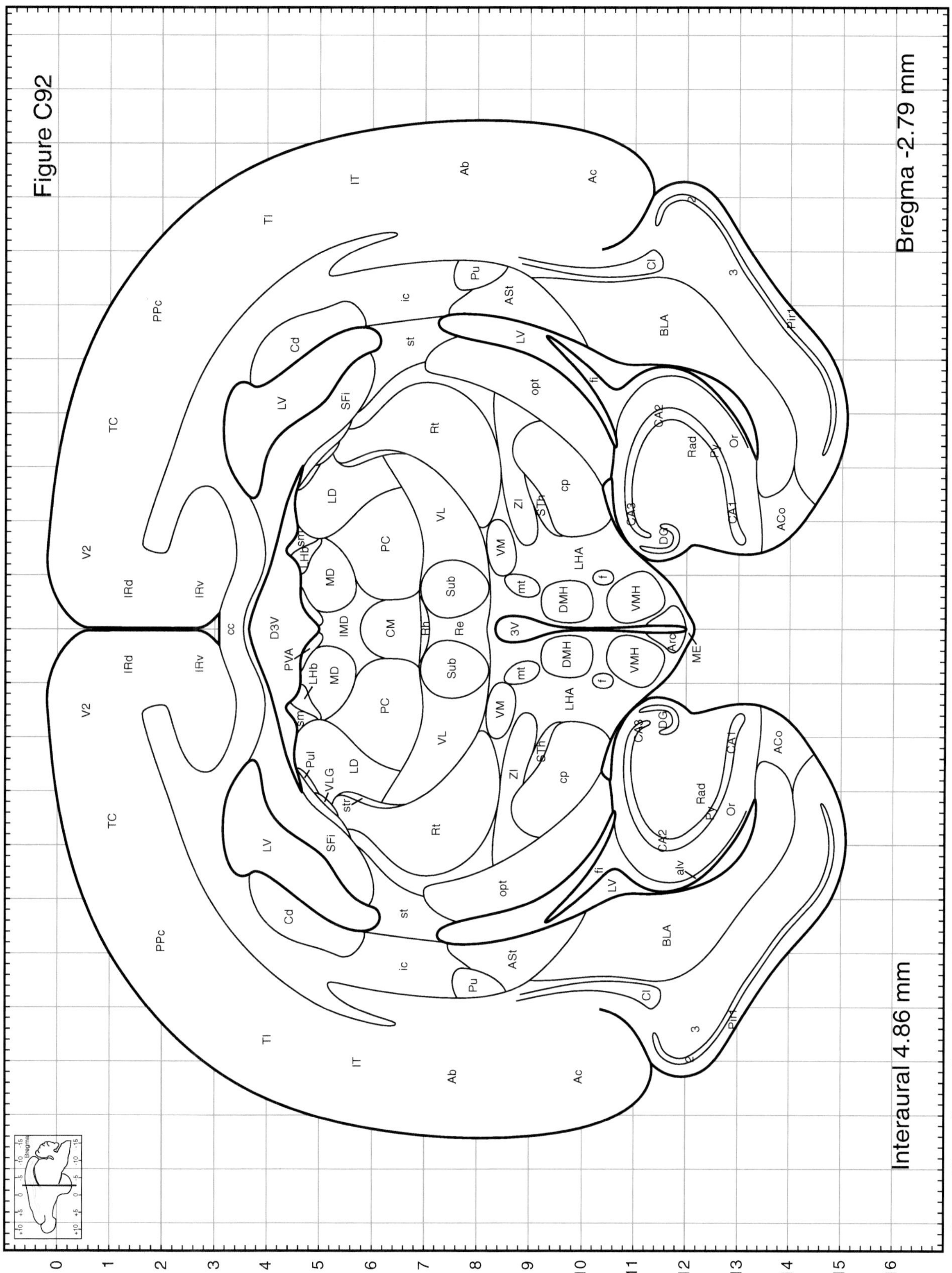
Figure C92
Bregma -2.79 mm
Interaural 4.86 mm
TI
IT
Ab
Ac
PPc
TC
V2
IRd
IRv
cc
D3V
PVA
LHb
sm
MD
IMD
CM
PC
LD
VL
Rt
Sub
Re
Rh
VM
ZI
STh
cp
opt
LHA
mt
DMH
VMH
f
3V
Arc
ME
LV
Cd
SFi
st
ic
Pu
ASt
fi
BLA
Cl
Pir1
ACo
CA1
CA2
CA3
DG
Rad
Py
Or
alv
Pul
VLG
str

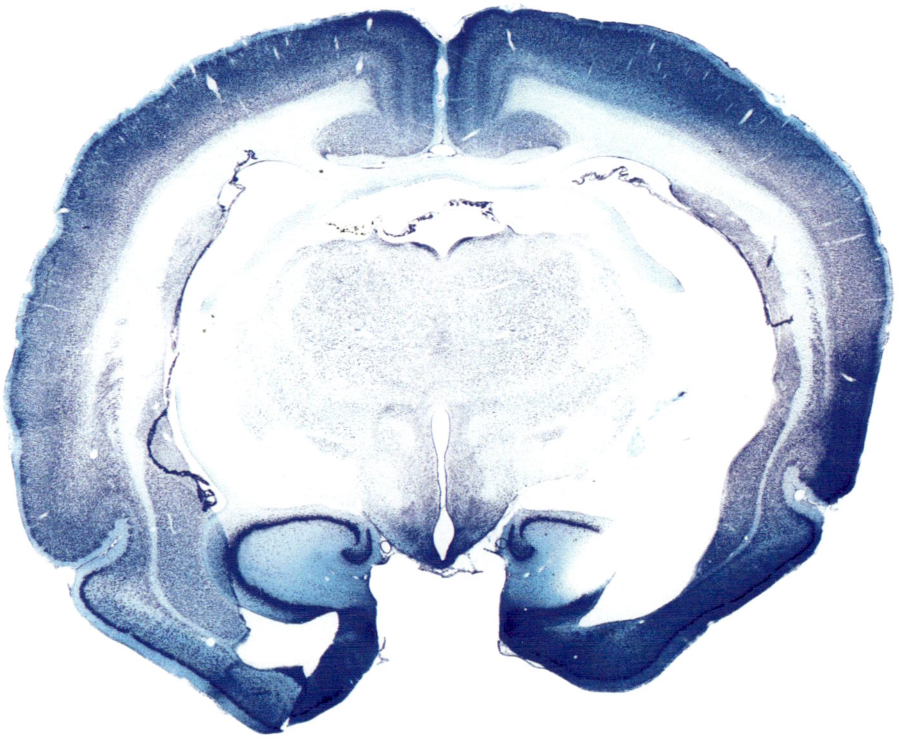

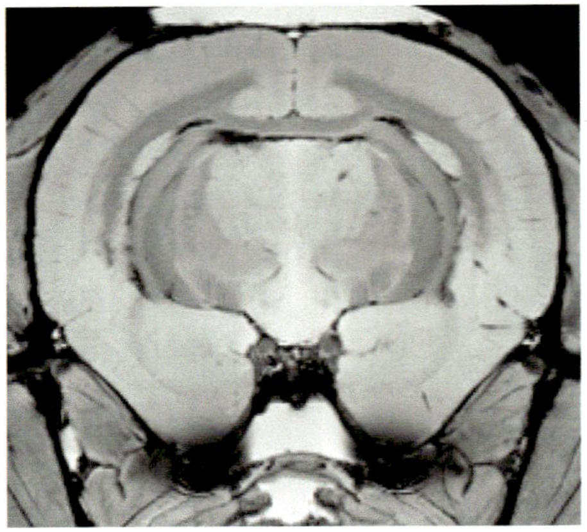

2	layer 2 of cortex	fi	fimbria of the hippocampus	PPc	posterior parietal caudal area
3	layer 3 of cortex	GrDG	granular layer of the dentate gyrus	Pul	pulvinar nuclei
3V	3rd ventricle	ic	internal capsule	PVP	paraventricular thalamic nucleus, posterior part
Ab	auditory belt area	IMD	intermediodorsal thalamic nucleus	Py	pyramidal cell layer of the hippocampus
Ac	auditory core area	IRd	infraradiata dorsalis	Rad	radiatum layer of the hippocampus
ACo	anterior cortical amygdaloid nucleus	IRv	infraradiata ventral area	Re	reuniens thalamic nucleus
alv	alveus of the hippocampus	IT	inferior temporal cortex	Rt	reticular thalamic nucleus
Arc	arcuate hypothalamic nucleus	LD	laterodorsal thalamic nucleus	SFi	septofimbrial nucleus
ASt	amygdalostriatal transition area	LHb	lateral habenular nucleus	st	stria terminalis
BLA	basolateral amygdaloid nucleus	LV	lateral ventricle	STh	subthalamic nucleus
CA1	field CA1 of the hippocampus	MD	mediodorsal thalamic nucleust	str	superior thalamic radiation
CA2	field CA2 of the hippocampus	ME	median eminence	Sub	submedius thalamic nucleus
CA3	field CA3 of the hippocampus	MoDG	molecular layer of the dentate gyrus	TC	temporal cortex
cc	corpus callosum	mt	mammillothalamic tract	TI	temporal inferior area
Cd	caudate nucleus	opt	optic tract	V2	secondary visual cortex
Cl	claustrum	Or	oriens layer of the hippocampus	VL	ventrolateral thalamic nucleus
CM	central medial thalamic nucleus	PC	paracentral thalamic nucleus	VLG	ventral lateral geniculate nucleus
cp	cerebral peduncle	PH	posterior hypothalamic nucleus	VM	ventromedial thalamic nucleus
D3V	dorsal 3rd ventricle	Pir1	piriform cortex, layer 1	VMH	ventromedial hypothalamic nucleus
DMH	dorsomedial hypothalamic nucleus	PLH	peduncular part of lateral hypothalamus	ZI	zona incerta
f	fornix	PoDG	polymorph layer of the dentate gyrus		

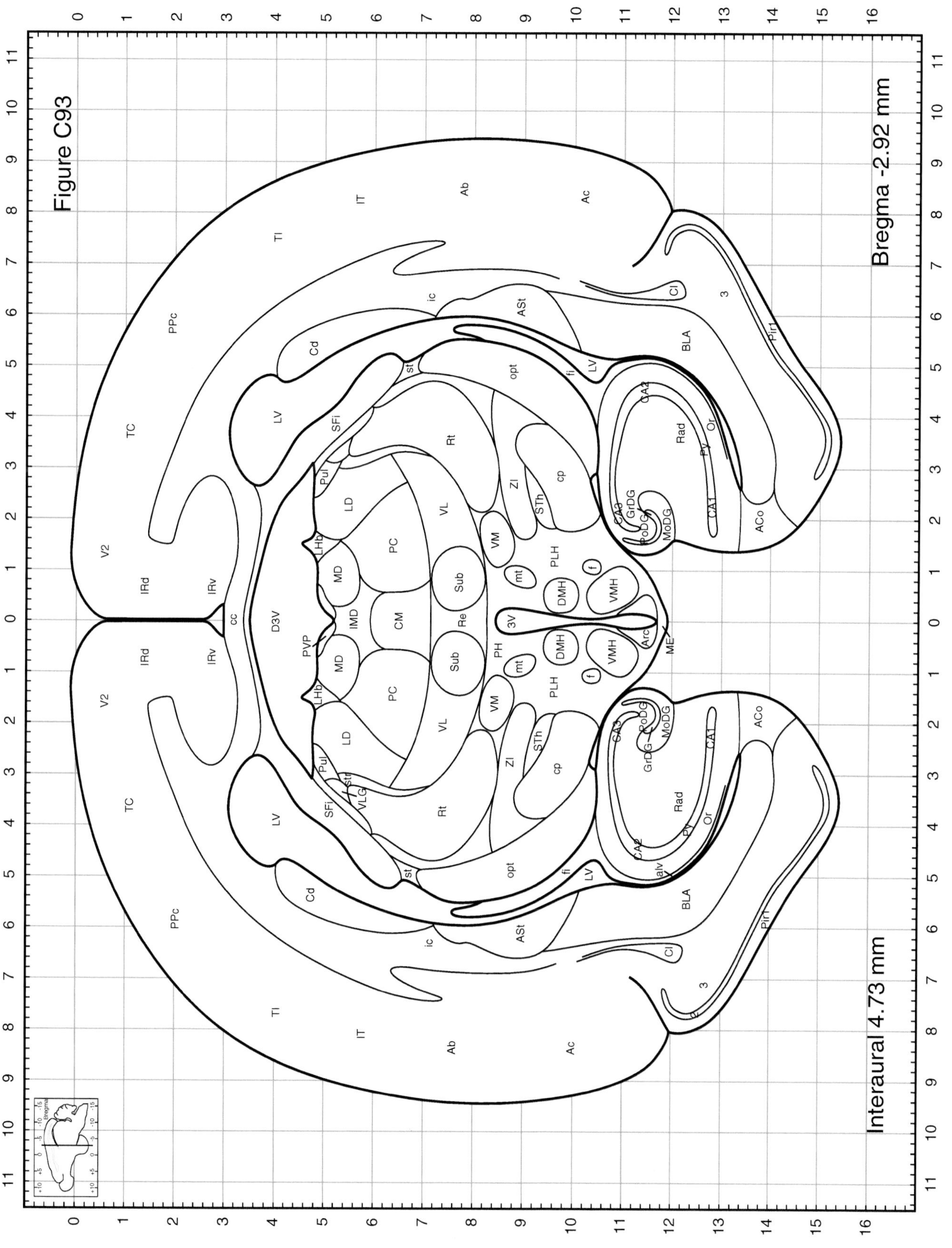
Figure C93
Bregma -2.92 mm
Interaural 4.73 mm
V2
IRd
IRv
TC
PPc
TI
IT
Ab
Ac
ic
ASt
Cd
LV
SFi
st
Pul
LD
LHb
MD
IMD
PVP
D3V
cc
PC
CM
Sub
Re
VL
Rt
opt
ZI
STh
cp
VM
PH
mt
PLH
DMH
VMH
f
3V
Arc
ME
fi
CA1
CA2
CA3
GrDG
PoDG
MoDG
Rad
Py
Or
alv
BLA
ACo
Cl
Pir1
3
VLG
Str

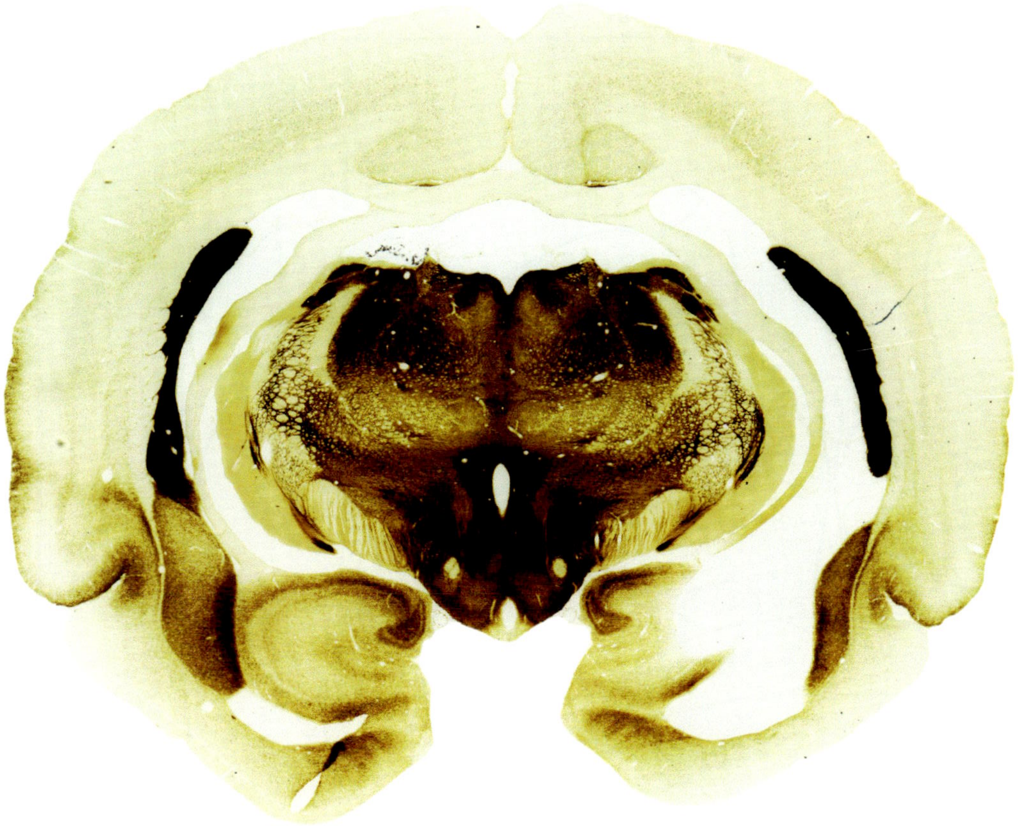

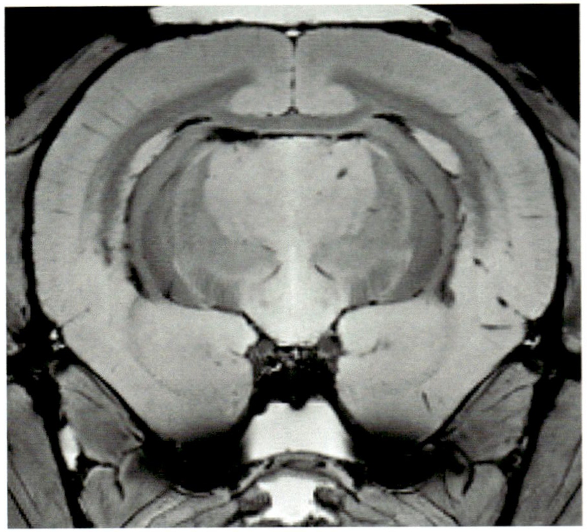

2	layer 2 of cortex
3	layer 3 of cortex
3V	3rd ventricle
Ab	auditory belt area
Ac	auditory core area
ACo	anterior cortical amygdaloid nucleus
alv	alveus of the hippocampus
Arc	arcuate hypothalamic nucleus
ASt	amygdalostriatal transition area
BLA	basolateral amygdaloid nucleus
CA1	field CA1 of the hippocampus
CA2	field CA2 of the hippocampus
CA3	field CA3 of the hippocampus
cc	corpus callosum
Cd	caudate nucleus
CM	central medial thalamic nucleus
cp	cerebral peduncle
D3V	dorsal 3rd ventricle
DLG	dorsal lateral geniculate nucleus
Ect	ectorhinal cortex
f	fornix
fi	fimbria of the hippocampus
GrDG	granular layer of the dentate gyrus
ic	internal capsule
IMD	intermediodorsal thalamic nucleus
IRd	infraradiata dorsalis
IRv	infraradiata ventral area
IT	inferior temporal cortex
LD	laterodorsal thalamic nucleus
LHb	lateral habenular nucleus
LMoL	lacunosum moleculare layer of the hippocampus
LV	lateral ventricle
MD	mediodorsal thalamic nucleus
MoDG	molecular layer of the dentate gyrus
mt	mammillothalamic tract
opt	optic tract
Or	oriens layer of the hippocampus
PC	paracentral thalamic nucleus
Pc	central nucleus of the pulvinar
Pd	dorsal nucleus of the pulvinar
PH	posterior hypothalamic nucleus
Pir1	piriform cortex, layer 1
PLH	peduncular part of lateral hypothalamus
PMD	premammillary nucleus, dorsal part
PMV	premammillary nucleus, ventral part
PoDG	polymorph layer of the dentate gyrus
PPc	posterior parietal caudal area
PVP	paraventricular thalamic nucleus, posterior part
Py	pyramidal cell layer of the hippocampus
Rad	radiatum layer of the hippocampus
Re	reuniens thalamic nucleus
Rt	reticular thalamic nucleus
SFi	septofimbrial nucleus
STh	subthalamic nucleus
str	superior thalamic radiation
Sub	submedius thalamic nucleus
TC	temporal cortex
TI	temporal inferior area
V2	secondary visual cortex
VL	ventrolateral thalamic nucleus
VLG	ventral lateral geniculate nucleus
VM	ventromedial thalamic nucleus
VPPC	ventral posterior nucleus of the thalamus, parvicellular part
VS	ventral subiculum
ZI	zona incerta

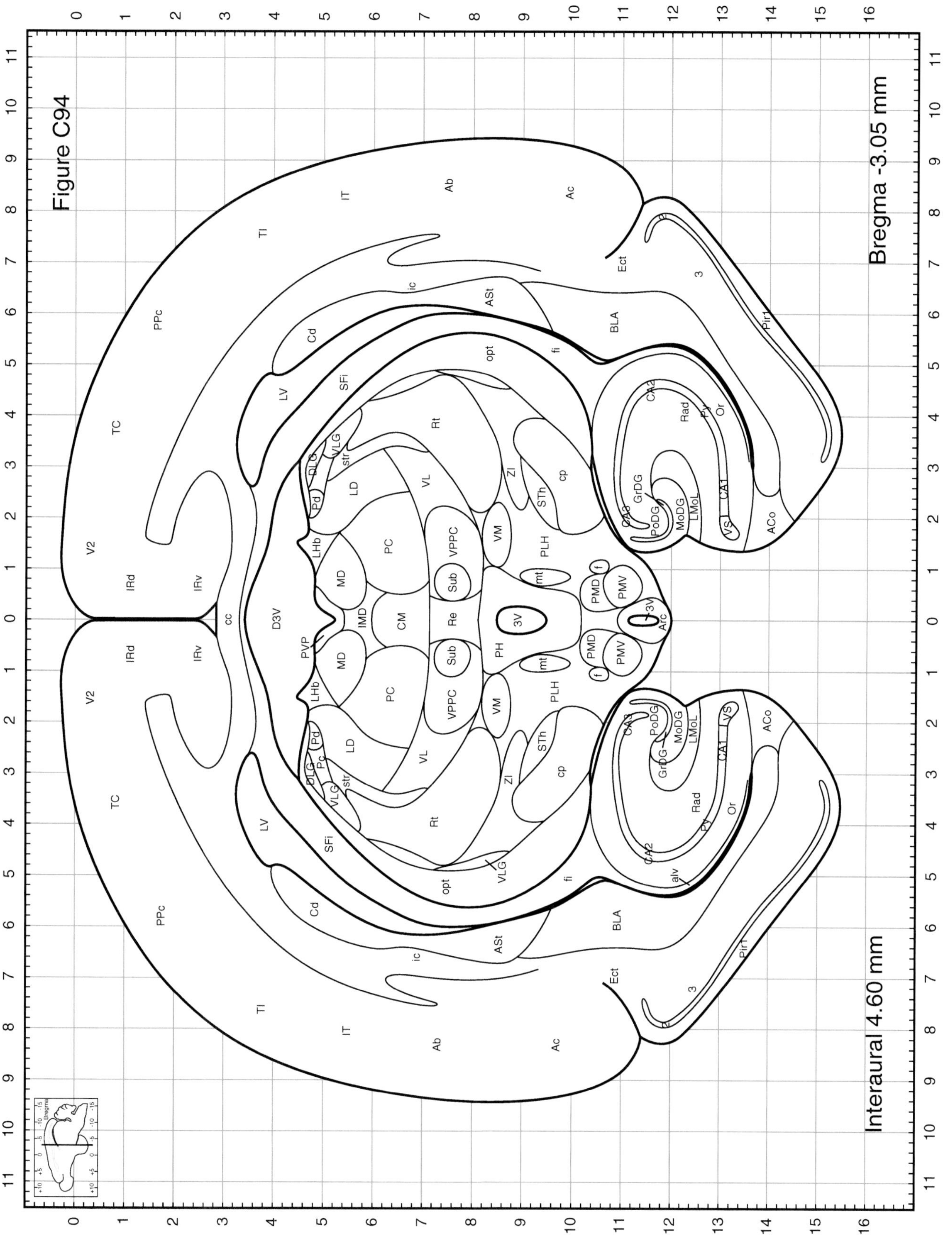
Figure C94
Bregma -3.05 mm
Interaural 4.60 mm
V2
IRd
IRv
TC
PPc
TI
IT
Ab
Ac
Ect
3
Pir1
BLA
ASt
ic
Cd
LV
SFi
opt
fi
cc
D3V
PVP
LHb
MD
IMD
CM
Re
PC
VPPC
Sub
LD
Pd
DLG
Pc
VLG
str
Rt
VL
ZI
VM
PH
3V
mt
PLH
STh
cp
PMD
PMV
f
Arc
CA1
CA2
CA3
Rad
Py
Or
GrDG
PoDG
MoDG
LMoL
VS
alv
ACo
Bregma

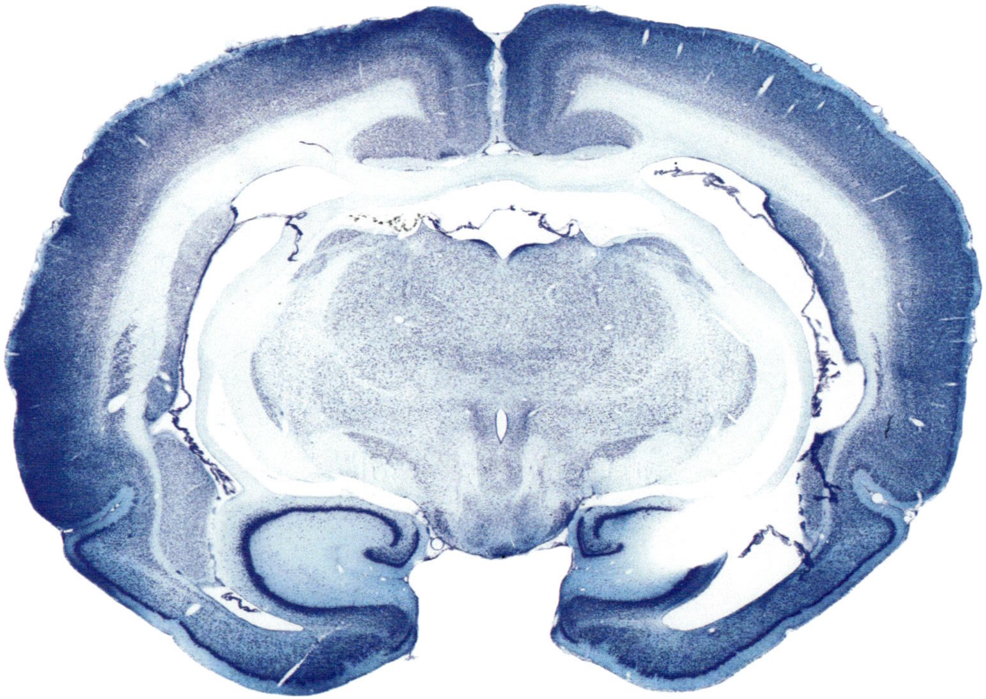

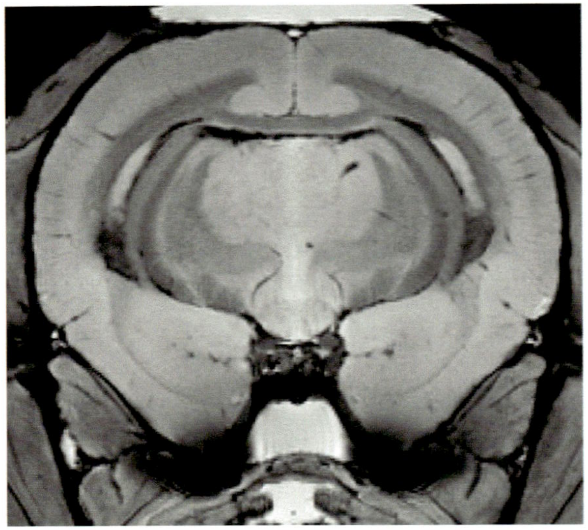

2	layer 2 of cortex
3	layer 3 of cortex
3V	3rd ventricle
Ab	auditory belt area
Ac	auditory core area
AHi	amygdalohippocampal area
alv	alveus of the hippocampus
ArcP	arcuate hypothalamic nucleus, posterior part
BLA	basolateral amygdaloid nucleus
CA1	field CA1 of the hippocampus
CA2	field CA2 of the hippocampus
CA3	field CA3 of the hippocampus
cc	corpus callosum
Cd	caudate nucleus
CM	central medial thalamic nucleus
cp	cerebral peduncle
D3V	dorsal 3rd ventricle
DLG	dorsal lateral geniculate nucleus
Ect	ectorhinal cortex
f	fornix
fi	fimbria of the hippocampus
GrDG	granular layer of the dentate gyrus
IGL	intergeniculate leaf
IMD	intermediodorsal thalamic nucleus
IRd	infraradiata dorsalis
IRv	infraradiata ventral area
IT	inferior temporal cortex
LD	laterodorsal thalamic nucleus
LHb	lateral habenular nucleus
LM	lateral mammillary nucleus
LMoL	lacunosum moleculare layer of the hippocampus
LV	lateral ventricle
MD	mediodorsal thalamic nucleus
MoDG	molecular layer of the dentate gyrus
MRe	mammillary recess of the 3rd ventricle
mt	mammillothalamic tract
opt	optic tract
Or	oriens layer of the hippocampus
PC	paracentral thalamic nucleus
Pc	central nucleus of the pulvinar
Pd	dorsal nucleus of the pulvinar
PH	posterior hypothalamic nucleus
Pir1	piriform cortex, layer 1
PLH	peduncular part of lateral hypothalamus
PMD	premammillary nucleus, dorsal part
PMV	premammillary nucleus, ventral part
PoDG	polymorph layer of the dentate gyrus
PPc	posterior parietal caudal area
Pv	ventral nucleus of the pulvinar
PVP	paraventricular thalamic nucleus, posterior part
Py	pyramidal cell layer of the hippocampus
Rad	radiatum layer of the hippocampus
Re	reuniens thalamic nucleus
Rt	reticular thalamic nucleus
scp	superior cerebellar peduncle
SFi	septofimbrial nucleus
SNR	substantia nigra, reticular part
STh	subthalamic nucleus
str	superior thalamic radiation
Sub	submedius thalamic nucleus
TC	temporal cortex
TI	temporal inferior area
V2	secondary visual cortex
VL	ventrolateral thalamic nucleus
VLG	ventral lateral geniculate nucleus
VM	ventromedial thalamic nucleus
VPPC	ventral posterior nucleus of the thalamus, parvicellular part
VS	ventral subiculum
VTM	ventral tuberomammillary nucleus
ZI	zona incerta

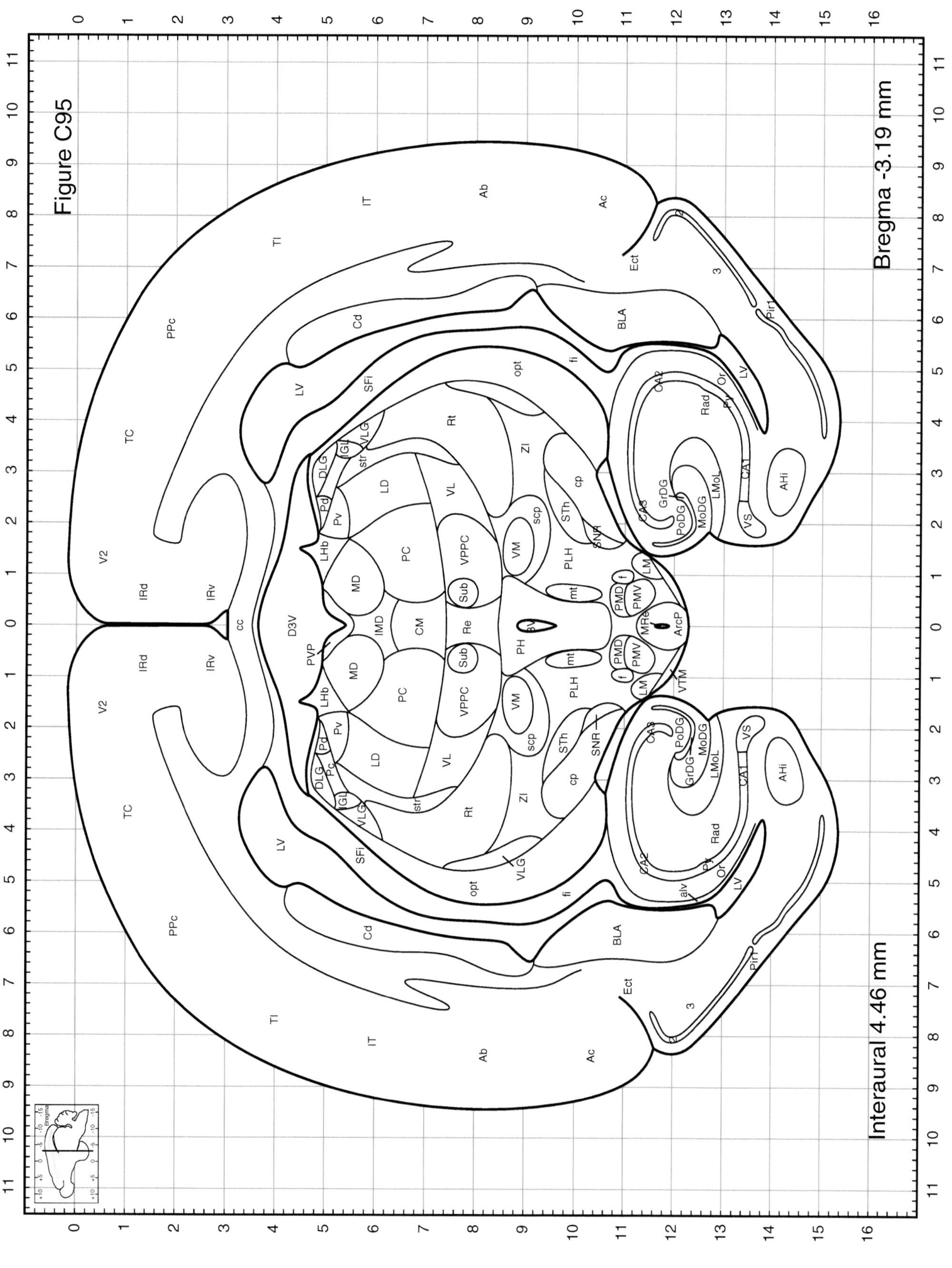
Figure C95
Bregma -3.19 mm
Interaural 4.46 mm
V2
IRd
IRv
TC
PPc
TI
IT
Ab
Ac
Ect
3
Pir1
BLA
Cd
LV
SFi
opt
fi
cc
D3V
PVP
LHb
Pd
Pv
Pc
DLG
IGL
VLG
str
LD
VL
Rt
ZI
MD
PC
VPPC
IMD
CM
Sub
Re
PH
3V
VM
scp
STh
cp
SNR
PLH
mt
f
PMD
PMV
MRe
ArcP
LM
VTM
CA1
CA2
CA3
GrDG
PoDG
MoDG
LMoL
Rad
Or
Py
alv
VS
AHi

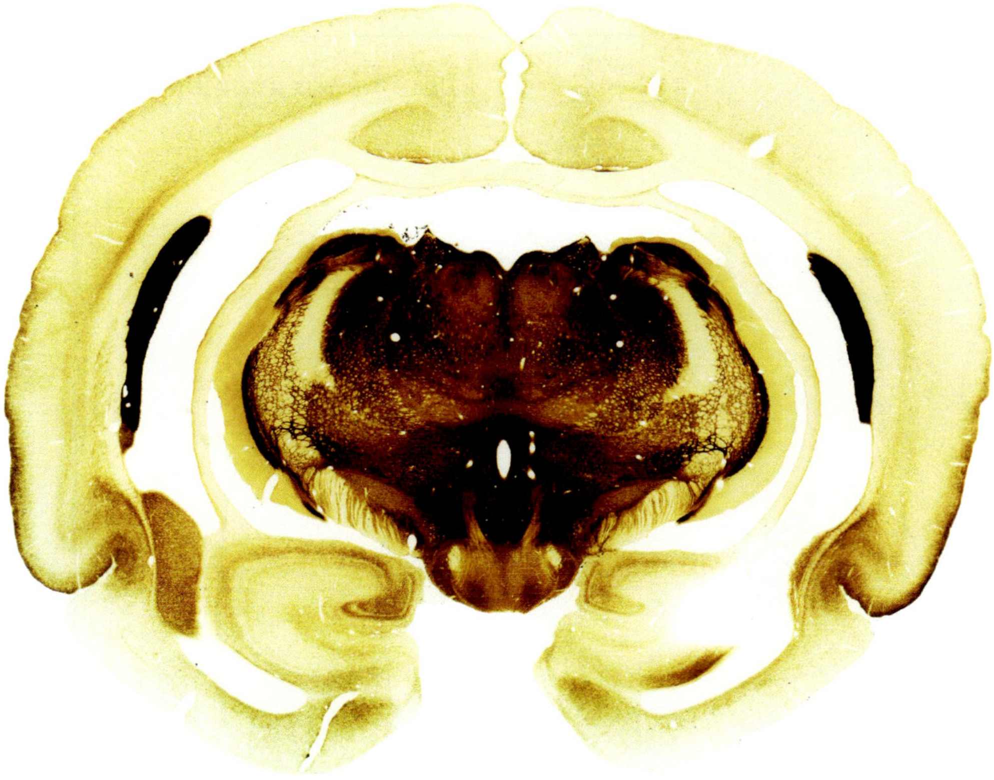

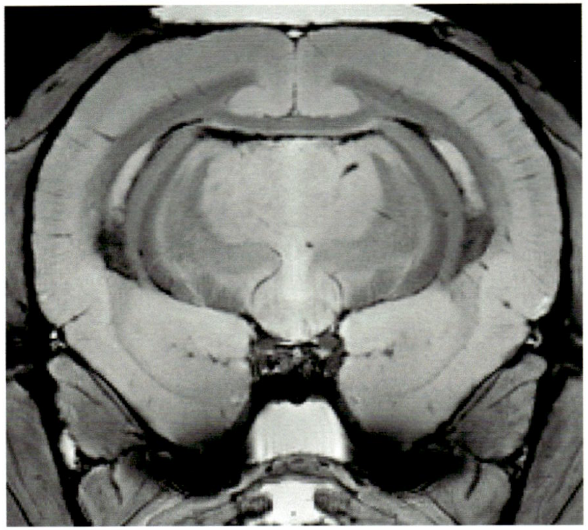

2	layer 2 of cortex
3	layer 3 of cortex
3V	3rd ventricle
Ab	auditory belt area
Ac	auditory core area
AHi	amygdalohippocampal area
alv	alveus of the hippocampus
ArcP	arcuate hypothalamic nucleus, posterior part
BLA	basolateral amygdaloid nucleus
CA1	field CA1 of the hippocampus
CA2	field CA2 of the hippocampus
CA3	field CA3 of the hippocampus
cc	corpus callosum
Cd	caudate nucleus
CM	central medial thalamic nucleus
cp	cerebral peduncle
D3V	dorsal 3rd ventricle
DLG	dorsal lateral geniculate nucleus
Ect	ectorhinal cortex
f	fornix
fi	fimbria of the hippocampus
GrDG	granular layer of the dentate gyrus
IGL	intergeniculate leaf
IMD	intermediodorsal thalamic nucleus
IRd	infraradiata dorsalis
IRv	infraradiata ventral area
IT	inferior temporal cortex
LD	laterodorsal thalamic nucleus
LHbL	lateral habenular nucleus, lateral part
LHbM	lateral habenular nucleus, medial part
LM	lateral mammillary nucleus
LMoL	lacunosum moleculare layer of the hippocampus
LV	lateral ventricle
MD	mediodorsal thalamic nucleus
MHb	medial habenular nucleus
ML	medial mammillary nucleus, lateral part
MoDG	molecular layer of the dentate gyrus
MRe	mammillary recess of the 3rd ventricle
mt	mammillothalamic tract
opt	optic tract
Or	oriens layer of the hippocampus
Pc	central nucleus of the pulvinar
PC	paracentral thalamic nucleus
Pd	dorsal nucleus of the pulvinar
PH	posterior hypothalamic nucleus
Pir1	piriform cortex, layer 1
PLH	peduncular part of lateral hypothalamus
PoDG	polymorph layer of the dentate gyrus
PPc	posterior parietal caudal area
Pv	ventral nucleus of the pulvinar
PVP	paraventricular thalamic nucleus, posterior part
Py	pyramidal cell layer of the hippocampus
Rad	radiatum layer of the hippocampus
Re	reuniens thalamic nucleus
Rt	reticular thalamic nucleus
scp	superior cerebellar peduncle
SFi	septofimbrial nucleus
SNL	substantia nigra, lateral part
SNR	substantia nigra, reticular part
STh	subthalamic nucleus
str	superior thalamic radiation
SuM	supramammillary nucleus
TC	temporal cortex
TI	temporal inferior area
V2	secondary visual cortex
VL	ventrolateral thalamic nucleus
VLG	ventral lateral geniculate nucleus
VM	ventromedial thalamic nucleus
VPPC	ventral posterior nucleus of the thalamus, parvicellular part
VS	ventral subiculum
VTM	ventral tuberomammillary nucleus
ZI	zona incerta

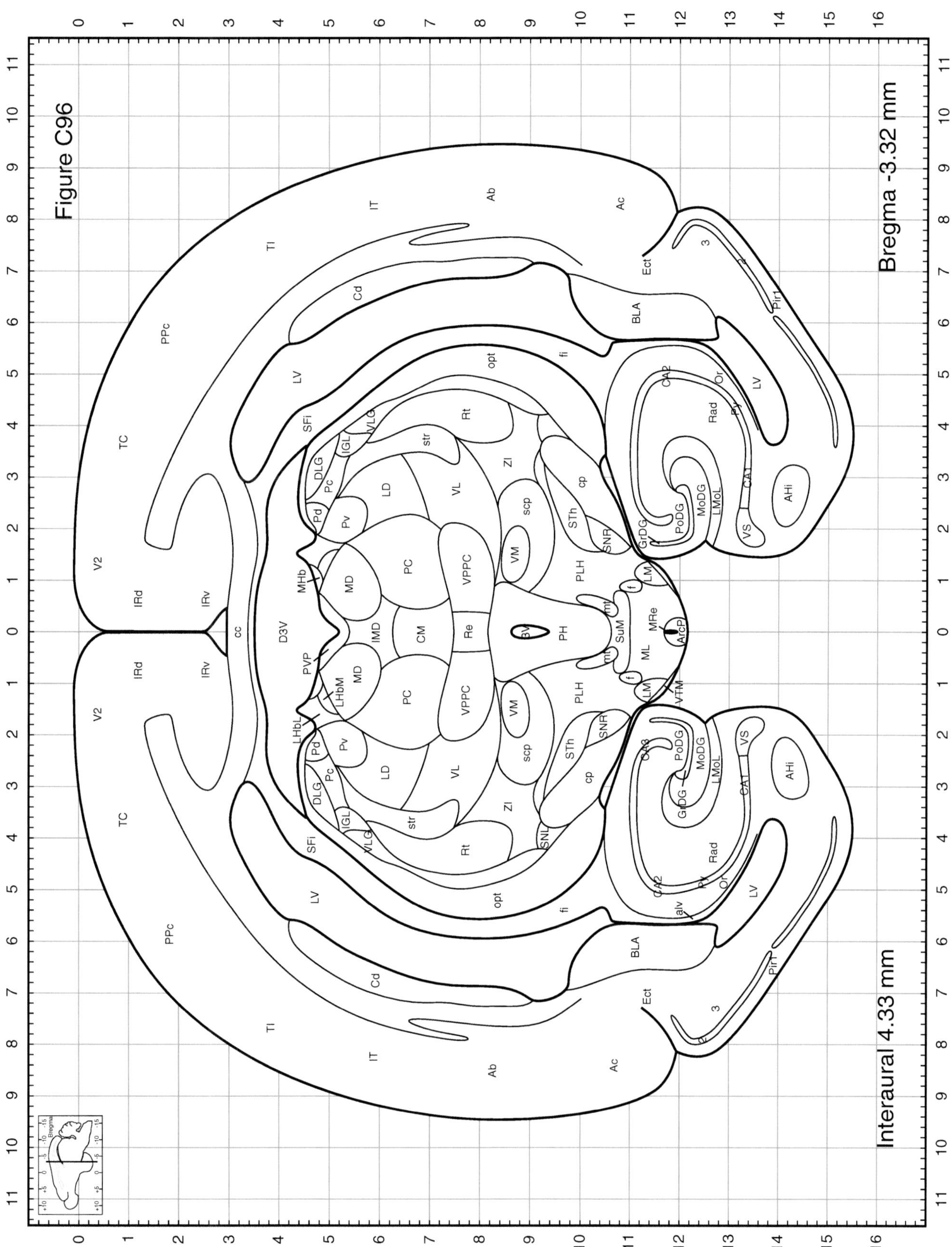
Figure C96
Bregma -3.32 mm
Interaural 4.33 mm
V2
IRd
IRv
TC
PPc
TI
IT
Ab
Ac
Ect
BLA
Cd
LV
SFi
cc
D3V
MHb
PVP
LHbM
LHbL
DLG
Pc
Pd
Pv
IGL
VLG
str
LD
VL
Rt
opt
fi
ZI
scp
STh
cp
SNR
SNL
VM
VPPC
PC
MD
IMD
CM
Re
3V
PH
PLH
mt
SuM
ML
MRe
ArcP
LM
f
VTM
GrDG
PoDG
MoDG
LMoL
CA1
CA2
CA3
VS
AHi
Rad
Py
Or
alv
Pir1
2
3
Bregma

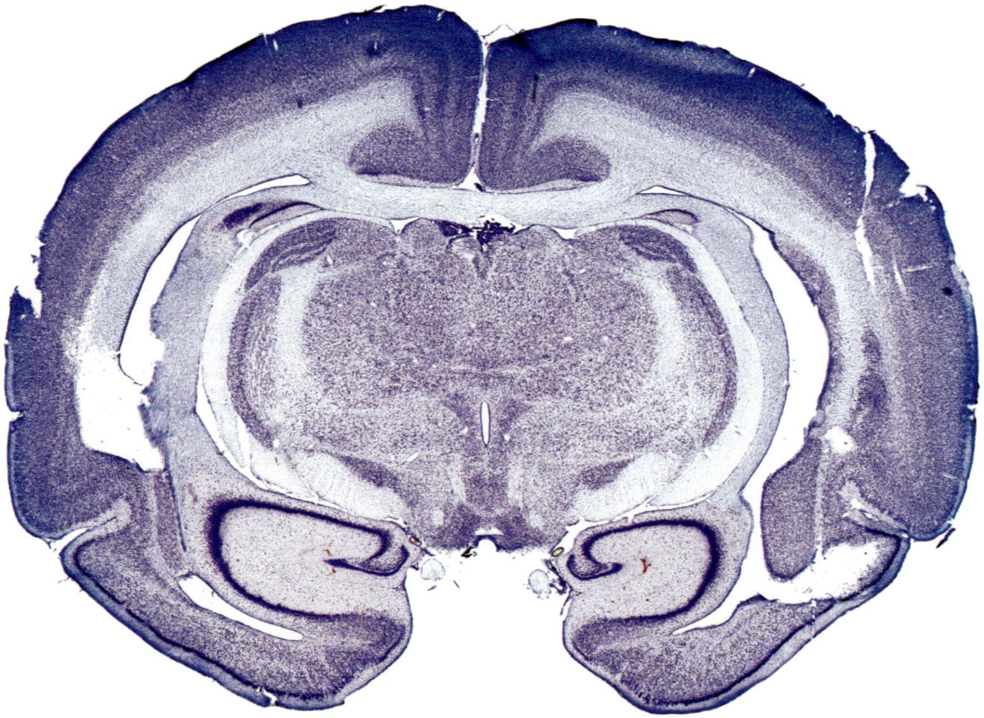

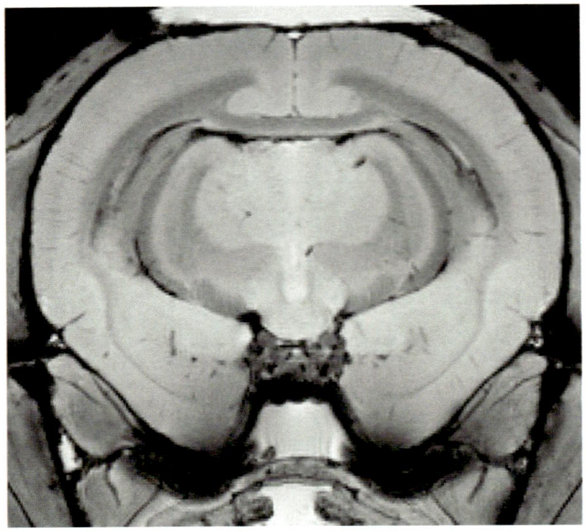

2	layer 2 of cortex
3	layer 3 of cortex
3V	3rd ventricle
Ab	auditory belt area
Ac	auditory core area
AHi	amygdalohippocampal area
alv	alveus of the hippocampus
ArcP	arcuate hypothalamic nucleus, posterior part
BLA	basolateral amygdaloid nucleus
CA1	field CA1 of the hippocampus
CA2	field CA2 of the hippocampus
CA3	field CA3 of the hippocampus
cc	corpus callosum
CM	central medial thalamic nucleus
cp	cerebral peduncle
D3V	dorsal 3rd ventricle
DG	dentate gyrus of the hippocampus
DLG	dorsal lateral geniculate nucleus
Ect	ectorhinal cortex
f	fornix
fi	fimbria of the hippocampus
fr	fasciculus retroflexus
GrDG	granular layer of the dentate gyrus
IGL	intergeniculate leaf
IMD	intermediodorsal thalamic nucleus
IT	inferior temporal cortex
LD	laterodorsal thalamic nucleus
LHbL	lateral habenular nucleus, lateral part
LHbM	lateral habenular nucleus, medial part
LM	lateral mammillary nucleus
LMoL	lacunosum moleculare layer of the hippocampus
LV	lateral ventricle
MD	mediodorsal thalamic nucleus
MHb	medial habenular nucleus
ML	medial mammillary nucleus, lateral part
MM	medial mammillary nucleus, medial part
MoDG	molecular layer of the dentate gyrus
MRe	mammillary recess of the 3rd ventricle
mt	mammillothalamic tract
opt	optic tract
Or	oriens layer of the hippocampus
PC	paracentral thalamic nucleus
Pc	central nucleus of the pulvinar
Pd	dorsal nucleus of the pulvinar
PF	parafascicular thalamic nucleus
PH	posterior hypothalamic nucleus
Pir1	piriform cortex, layer 1
PLH	peduncular part of lateral hypothalamus
PoDG	polymorph layer of the dentate gyrus
PoMn	posteromedian thalamic nucleus
PPc	posterior parietal caudal area
Pv	ventral nucleus of the pulvinar
PVP	paraventricular thalamic nucleus, posterior part
Py	pyramidal cell layer of the hippocampus
Rad	radiatum layer of the hippocampus
Re	reuniens thalamic nucleus
RSg	retrosplenial granular cortex
scp	superior cerebellar peduncle
SNL	substantia nigra, lateral part
SNR	substantia nigra, reticular part
STh	subthalamic nucleus
str	superior thalamic radiation
SuML	supramammillary nucleus, lateral par
SuMM	supramammillary nucleus, medial part
TC	temporal cortex
TI	temporal inferior area
V2	secondary visual cortex
VL	ventrolateral thalamic nucleus
VLG	ventral lateral geniculate nucleus
VM	ventromedial thalamic nucleus
VPPC	ventral posterior nucleus of the thalamus, parvicellular part
VS	ventral subiculum
VTM	ventral tuberomammillary nucleus
ZI	zona incerta

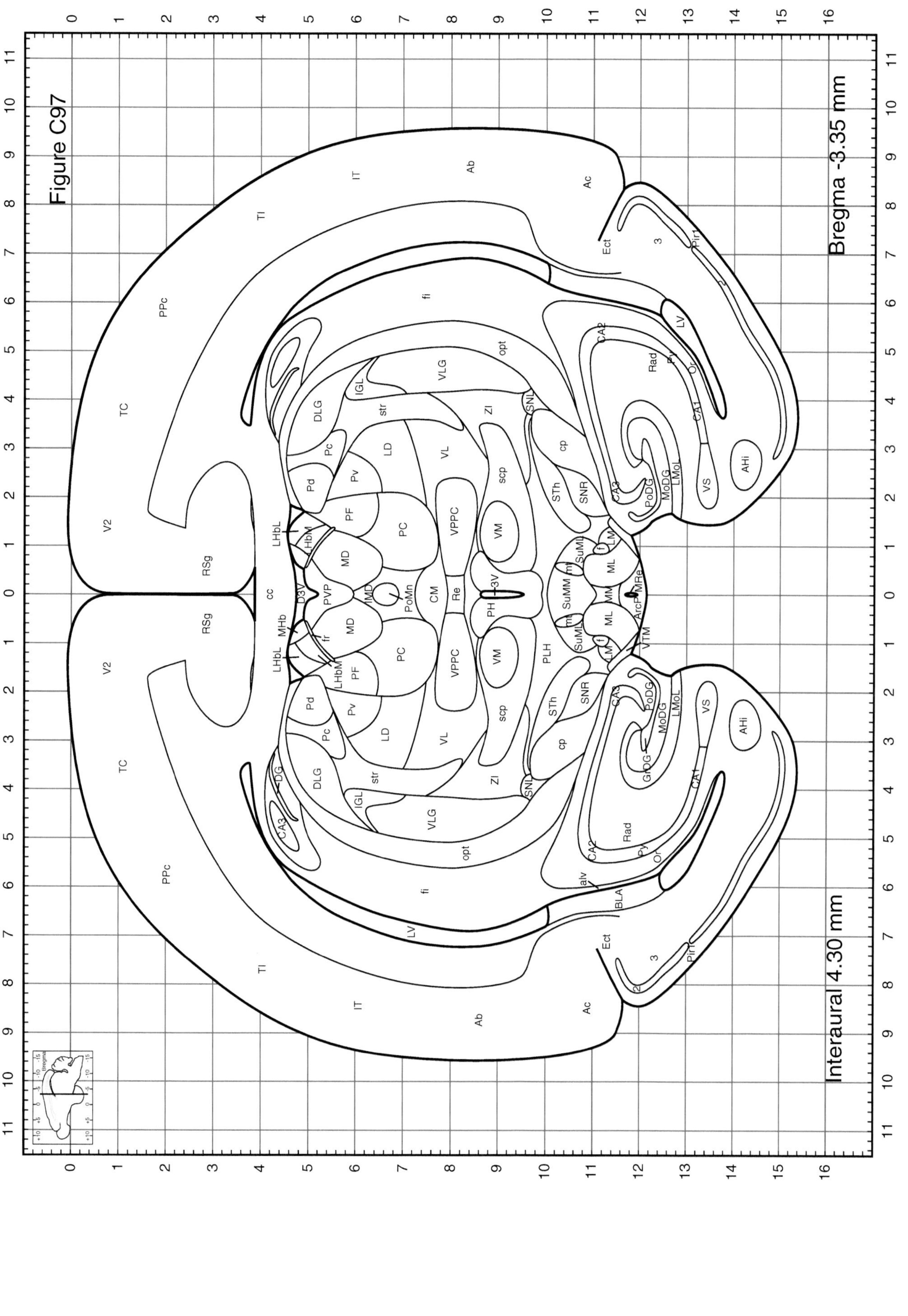
Figure C97
Bregma -3.35 mm
Interaural 4.30 mm
IT
Ab
Ac
TI
Ect
3
Pir1
PPc
fi
LV
CA2
Rad
Py
Or
opt
VLG
IGL
DLG
str
ZI
SNL
TC
CA1
cp
LD
VL
Pc
Pv
Pd
scp
STh
SNR
CA3
PoDG
MoDG
LMol
VS
AHi
V2
LHbL
MHb
PF
PC
VPPC
VM
RSg
MD
SuML
LM
ML
cc
D3V
PVP
IMD
PoMn
CM
Re
PH
3V
SuMM
MM
MRe
ArcP
f
fr
ml
LHbM
VTM
PLH
GrDG
DG
alv
BLA
2

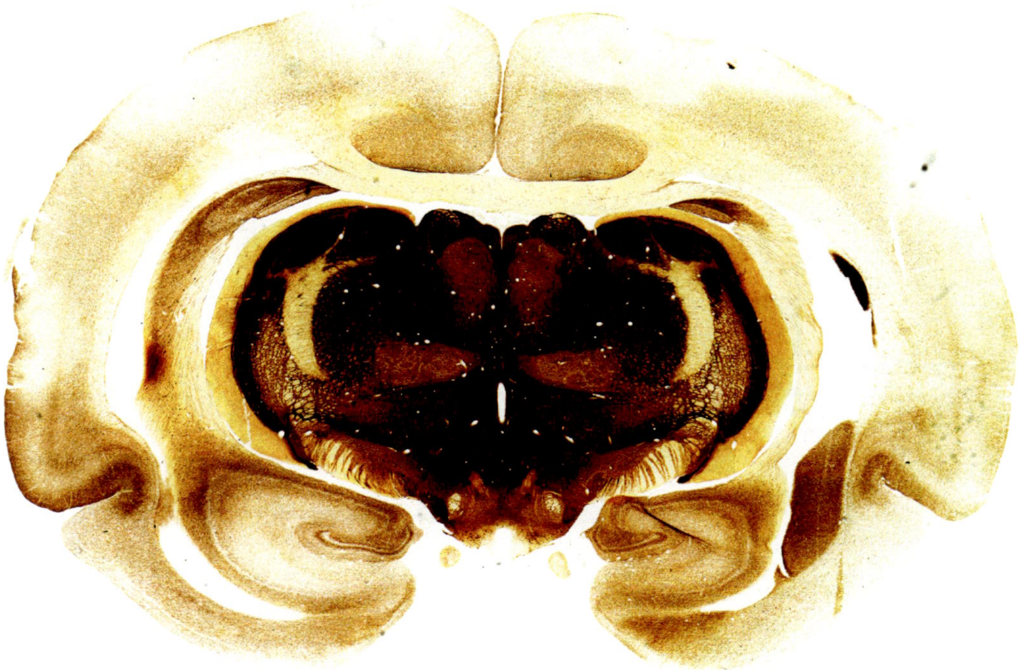

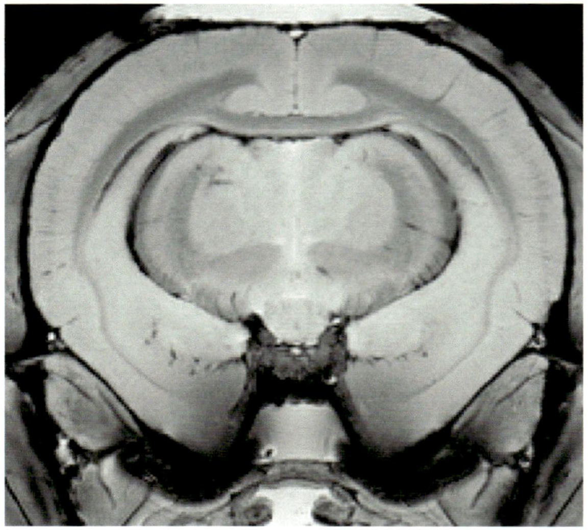

2	layer 2 of cortex
3	layer 3 of cortex
3V	3rd ventricle
Ab	auditory belt area
Ac	auditory core area
AHi	amygdalohippocampal area
alv	alveus of the hippocampus
ArcP	arcuate hypothalamic nucleus, posterior part
CA1	field CA1 of the hippocampus
CA2	field CA2 of the hippocampus
CA3	field CA3 of the hippocampus
cc	corpus callosum
CM	central medial thalamic nucleus
cp	cerebral peduncle
D3V	dorsal 3rd ventricle
DG	dentate gyrus of the hippocampus
DLG	dorsal lateral geniculate nucleus
Ect	ectorhinal cortex
f	fornix
fi	fimbria of the hippocampus
fr	fasciculus retroflexus
GrDG	granular layer of the dentate gyrus
IGL	intergeniculate leaf
IMD	intermediodorsal thalamic nucleus
IT	inferior temporal cortex
LHbL	lateral habenular nucleus, lateral part
LHbM	lateral habenular nucleus, medial part
LM	lateral mammillary nucleus
LMoL	lacunosum moleculare layer of the hippocampus
LV	lateral ventricle
MD	mediodorsal thalamic nucleus
MHb	medial habenular nucleus
ML	medial mammillary nucleus, lateral part
MM	medial mammillary nucleus, medial part
MoDG	molecular layer of the dentate gyrus
MRe	mammillary recess of the 3rd ventricle
mt	mammillothalamic tract
opt	optic tract
Or	oriens layer of the hippocampus
PC	paracentral thalamic nucleus
Pc	central nucleus of the pulvinar
Pd	dorsal nucleus of the pulvinar
PF	parafascicular thalamic nucleus
PH	posterior hypothalamic nucleus
Pir1	piriform cortex, layer 1
PLH	peduncular part of lateral hypothalamus
PoDG	polymorph layer of the dentate gyrus
PoMn	posteromedian thalamic nucleus
PPc	posterior parietal caudal area
Pv	ventral nucleus of the pulvinar
PVP	paraventricular thalamic nucleus, posterior part
Py	pyramidal cell layer of the hippocampus
Rad	radiatum layer of the hippocampus
Re	reuniens thalamic nucleus
RSg	retrosplenial granular cortex
scp	superior cerebellar peduncle
SNL	substantia nigra, lateral part
SNR	substantia nigra, reticular part
st	stria terminalis
STh	subthalamic nucleus
str	superior thalamic radiation
SuML	supramammillary nucleus, lateral part
SuMM	supramammillary nucleus, medial part
sumx	supramammillary decussation
TC	temporal cortex
TI	temporal inferior area
V2	secondary visual cortex
VL	ventrolateral thalamic nucleus
VLG	ventral lateral geniculate nucleus
VPPC	ventral posterior nucleus of the thalamus, parvicellular part
VS	ventral subiculum
VTM	ventral tuberomammillary nucleus
ZI	zona incerta

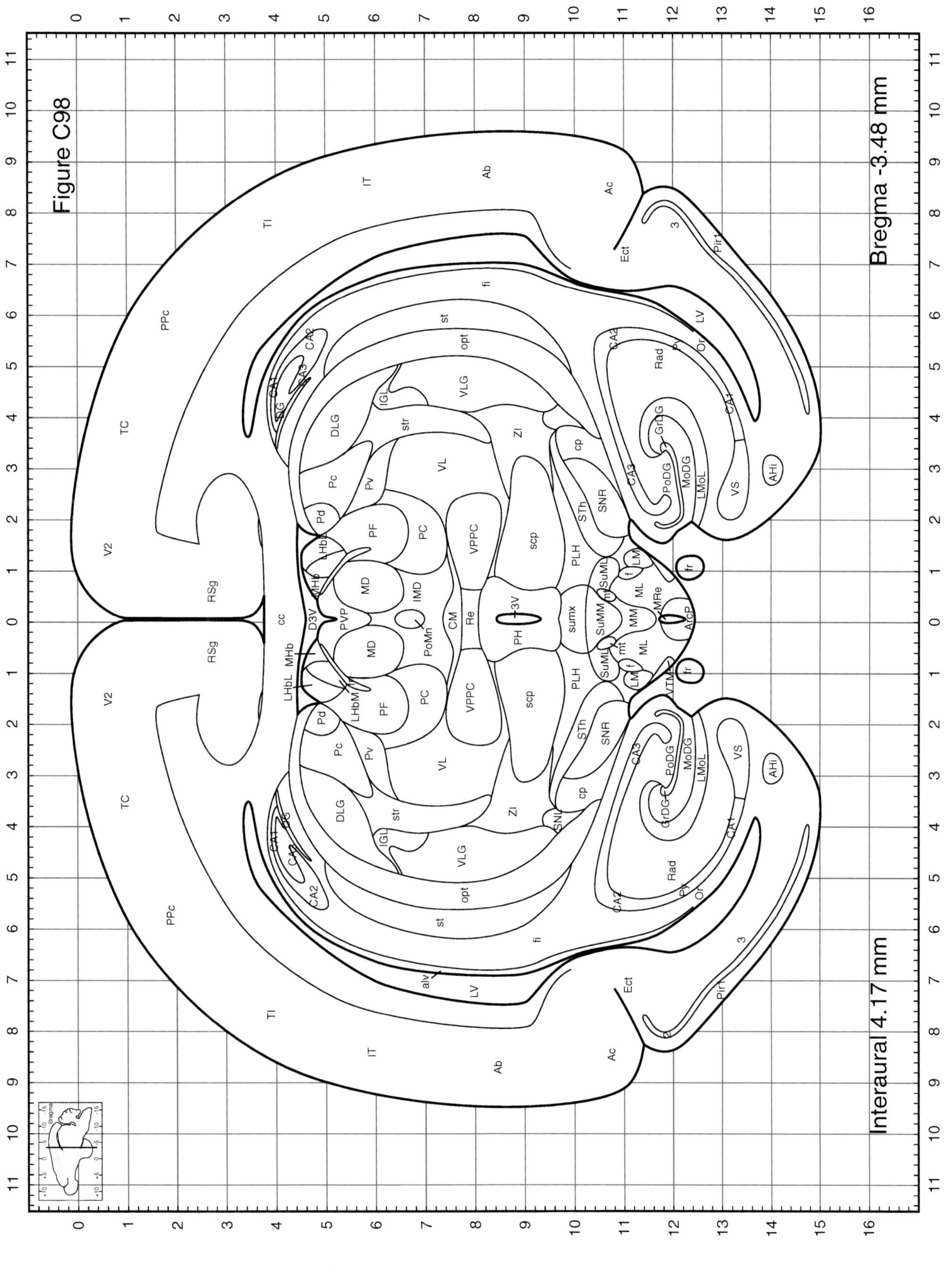
Figure C98
Bregma -3.48 mm
Interaural 4.17 mm
PPc
TC
V2
RSg
cc
TI
IT
Ab
Ac
Ect
Pir1
LV
fi
st
opt
VLG
IGL
DLG
str
Pc
Pv
Pd
PF
PC
VL
ZI
VPPC
scp
LHb
MHb
LHbL
LHbM
MD
IMD
D3V
PVP
PoMn
CM
Re
3V
PH
sumx
PLH
STh
SNR
SNL
cp
SuML
SuMM
mt
ML
MM
LM
MRe
ArcP
VTM
fr
f
CA1
CA2
CA3
DG
Rad
Py
Or
GrDG
PoDG
MoDG
LMoL
VS
AHi
alv
Bregma

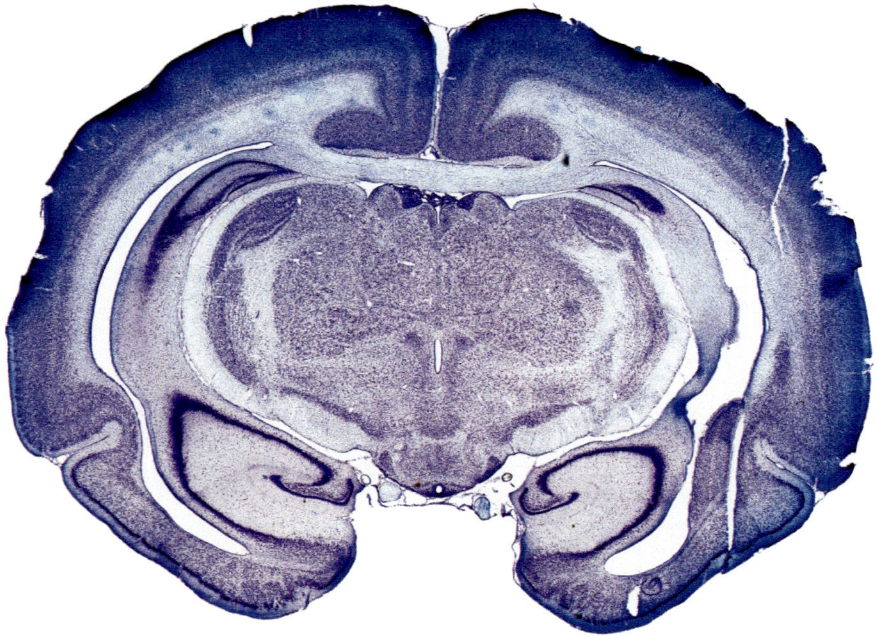

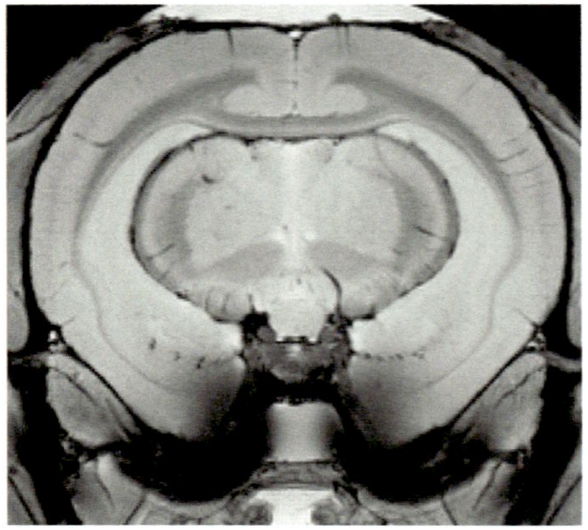

2 layer 2 of cortex
3 layer 3 of cortex
3V 3rd ventricle
Ab auditory belt area
Ac auditory core area
AHi amygdalohippocampal area
alv alveus of the hippocampus
ArcP arcuate hypothalamic nucleus, posterior part
CA1 field CA1 of the hippocampus
CA2 field CA2 of the hippocampus
CA3 field CA3 of the hippocampus
cc corpus callosum
CM central medial thalamic nucleus
cp cerebral peduncle
D3V dorsal 3rd ventricle
DG dentate gyrus of the hippocampus
DLG dorsal lateral geniculate nucleus
Ect ectorhinal cortex
f fornix
fi fimbria of the hippocampus
fr fasciculus retroflexus
Gem gemini hypothalamic nucleus
GrDG granular layer of the dentate gyrus
IGL intergeniculate leaf
IMD intermediodorsal thalamic nucleus
IT inferior temporal cortex
LHbL lateral habenular nucleus, lateral part
LHbM lateral habenular nucleus, medial part
LM lateral mammillary nucleus
LMoL lacunosum moleculare layer of the hippocampus
LV lateral ventricle
MD mediodorsal thalamic nucleus
MHb medial habenular nucleus
ml medial lemniscus
ML medial mammillary nucleus, lateral part
MM medial mammillary nucleus, medial part
MoDG molecular layer of the dentate gyrus
MRe mammillary recess of the 3rd ventricle
mt mammillothalamic tract
opt optic tract
Or oriens layer of the hippocampus
PC paracentral thalamic nucleus
Pc central nucleus of the pulvinar
Pd dorsal nucleus of the pulvinar
PF parafascicular thalamic nucleus
PH posterior hypothalamic nucleus
Pir1 piriform cortex, layer 1
PLH peduncular part of lateral hypothalamus
PoDG polymorph layer of the dentate gyrus
PoMn posteromedian thalamic nucleus
PPc posterior parietal caudal area
Pv ventral nucleus of the pulvinar
PVP paraventricular thalamic nucleus, posterior part
Py pyramidal cell layer of the hippocampus
Rad radiatum layer of the hippocampus
Re reuniens thalamic nucleus
RSg retrosplenial granular cortex
scp superior cerebellar peduncle
SNL substantia nigra, lateral part
SNR substantia nigra, reticular part
st stria terminalis
STh subthalamic nucleus
str superior thalamic radiation
SuML supramammillary nucleus, lateral part
SuMM supramammillary nucleus, medial part
sumx supramammillary decussation
TC temporal cortex
TI temporal inferior area
V2 secondary visual cortex
VL ventrolateral thalamic nucleus
VLG ventral lateral geniculate nucleus
VPPC ventral posterior nucleus of the thalamus, parvicellular part
VS ventral subiculum
VTM ventral tuberomammillary nucleus
ZI zona incerta

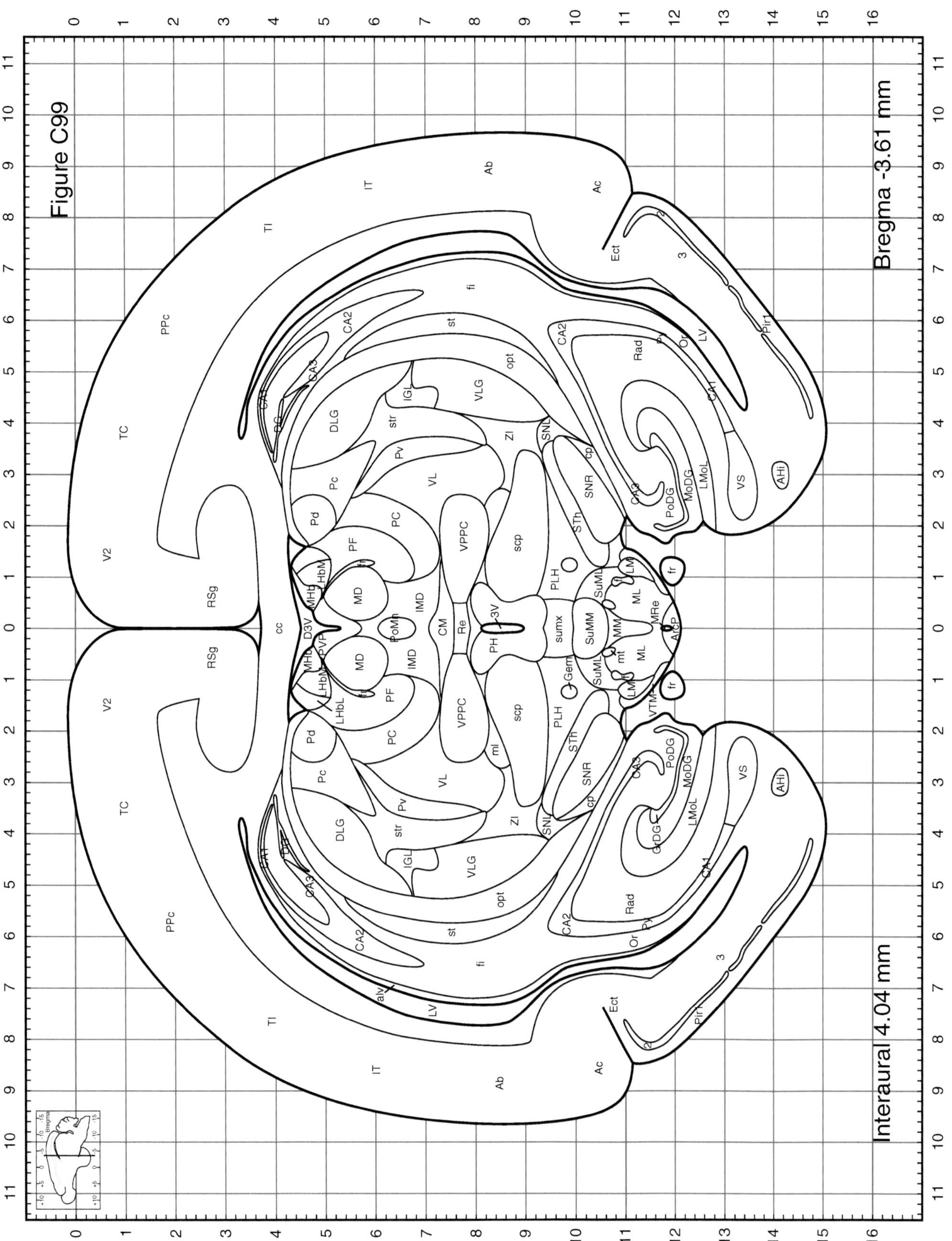
Figure C99
Bregma -3.61 mm
Interaural 4.04 mm
V2
RSg
TC
PPc
TI
IT
Ab
Ac
Ect
Pir1
cc
CA1
CA2
CA3
DG
fi
st
opt
DLG
IGL
VLG
str
Pv
Pc
Pd
PC
PF
VL
VPPC
ZI
SNL
SNR
cp
STh
scp
ml
PLH
Gem
sumx
SuMM
SuML
MM
ML
LM
mt
MRe
ArcP
VTM
fr
MD
IMD
PoMn
CM
Re
PH
3V
D3V
PVP
MHb
LHbM
LHbL
alv
LV
Rad
Py
Or
PoDG
GrDG
MoDG
LMoL
VS
AHi

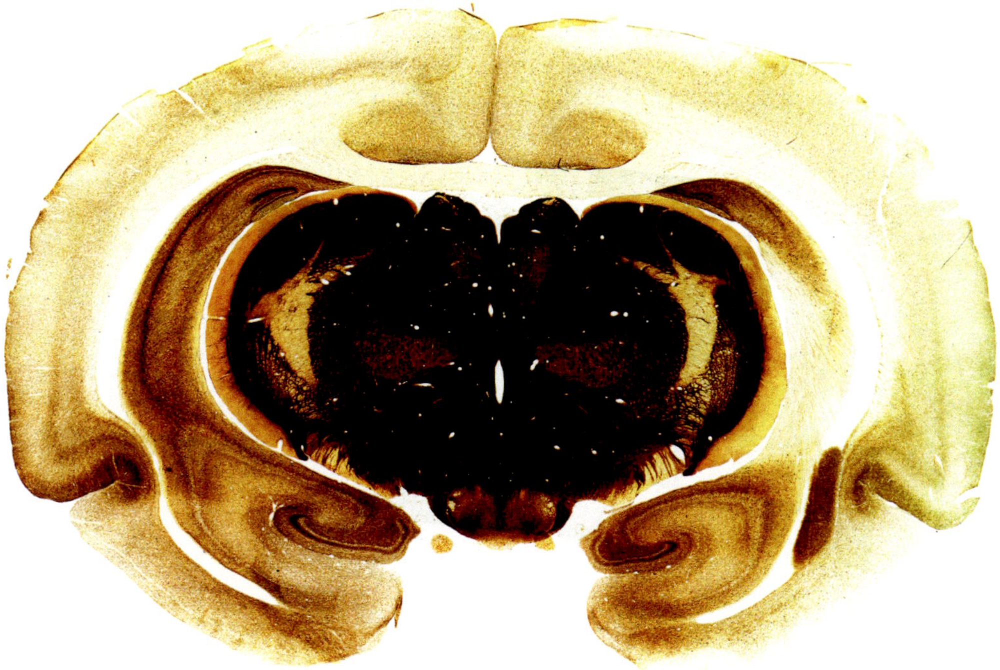

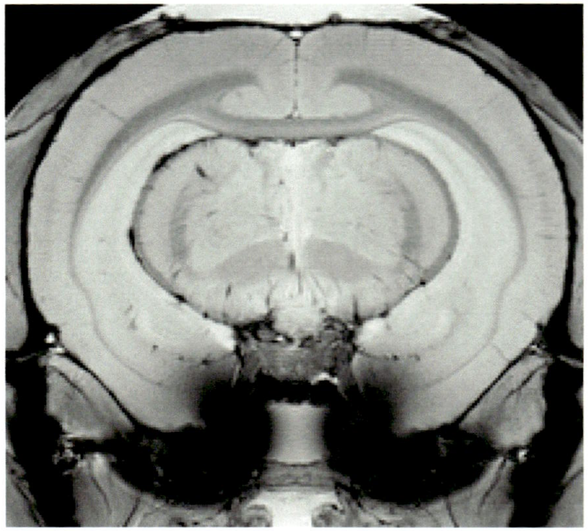

2	layer 2 of cortex
3	layer 3 of cortex
3V	3rd ventricle
Ab	auditory belt area
Ac	auditory core area
alv	alveus of the hippocampus
ArcP	arcuate hypothalamic nucleus, posterior part
CA1	field CA1 of the hippocampus
CA2	field CA2 of the hippocampus
CA3	field CA3 of the hippocampus
cc	corpus callosum
CM	central medial thalamic nucleus
cp	cerebral peduncle
D3V	dorsal 3rd ventricle
DG	dentate gyrus of the hippocampus
DLG	dorsal lateral geniculate nucleus
Ect	ectorhinal cortex
f	fornix
fr	fasciculus retroflexus
GrDG	granular layer of the dentate gyrus
IGL	intergeniculate leaf
IT	inferior temporal cortex
LHbL	lateral habenular nucleus, lateral part
LHbM	lateral habenular nucleus, medial part
LM	lateral mammillary nucleus
LMoL	lacunosum moleculare layer of the hippocampus
LV	lateral ventricle
MD	mediodorsal thalamic nucleus
MHb	medial habenular nucleus
ml	medial lemniscus
ML	medial mammillary nucleus, lateral part
MM	medial mammillary nucleus, medial part
MoDG	molecular layer of the dentate gyrus
mt	mammillothalamic tract
opt	optic tract
Or	oriens layer of the hippocampus
PC	paracentral thalamic nucleus
Pc	central nucleus of the pulvinar
Pd	dorsal nucleus of the pulvinar
PF	parafascicular thalamic nucleus
PH	posterior hypothalamic nucleus
Pir1	piriform cortex, layer 1
PLH	peduncular part of lateral hypothalamus
PoDG	polymorph layer of the dentate gyrus
PoMn	posteromedian thalamic nucleus
PPc	posterior parietal caudal area
Pv	ventral nucleus of the pulvinar
PVP	paraventricular thalamic nucleus, posterior part
Py	pyramidal cell layer of the hippocampus
Rad	radiatum layer of the hippocampus
Re	reuniens thalamic nucleus
RSg	retrosplenial granular cortex
scp	superior cerebellar peduncle
SNC	substantia nigra, compact part
SNL	substantia nigra, lateral part
SNR	substantia nigra, reticular part
STh	subthalamic nucleus
str	superior thalamic radiation
SuML	supramammillary nucleus, lateral part
SuMM	supramammillary nucleus, medial part
sumx	supramammillary decussation
TC	temporal cortex
TI	temporal inferior area
V1	primary visual cortex
V2	secondary visual cortex
VL	ventrolateral thalamic nucleus
VLG	ventral lateral geniculate nucleus
VPPC	ventral posterior nucleus of the thalamus, parvicellular part
VS	ventral subiculum
VTM	ventral tuberomammillary nucleus
ZI	zona incerta

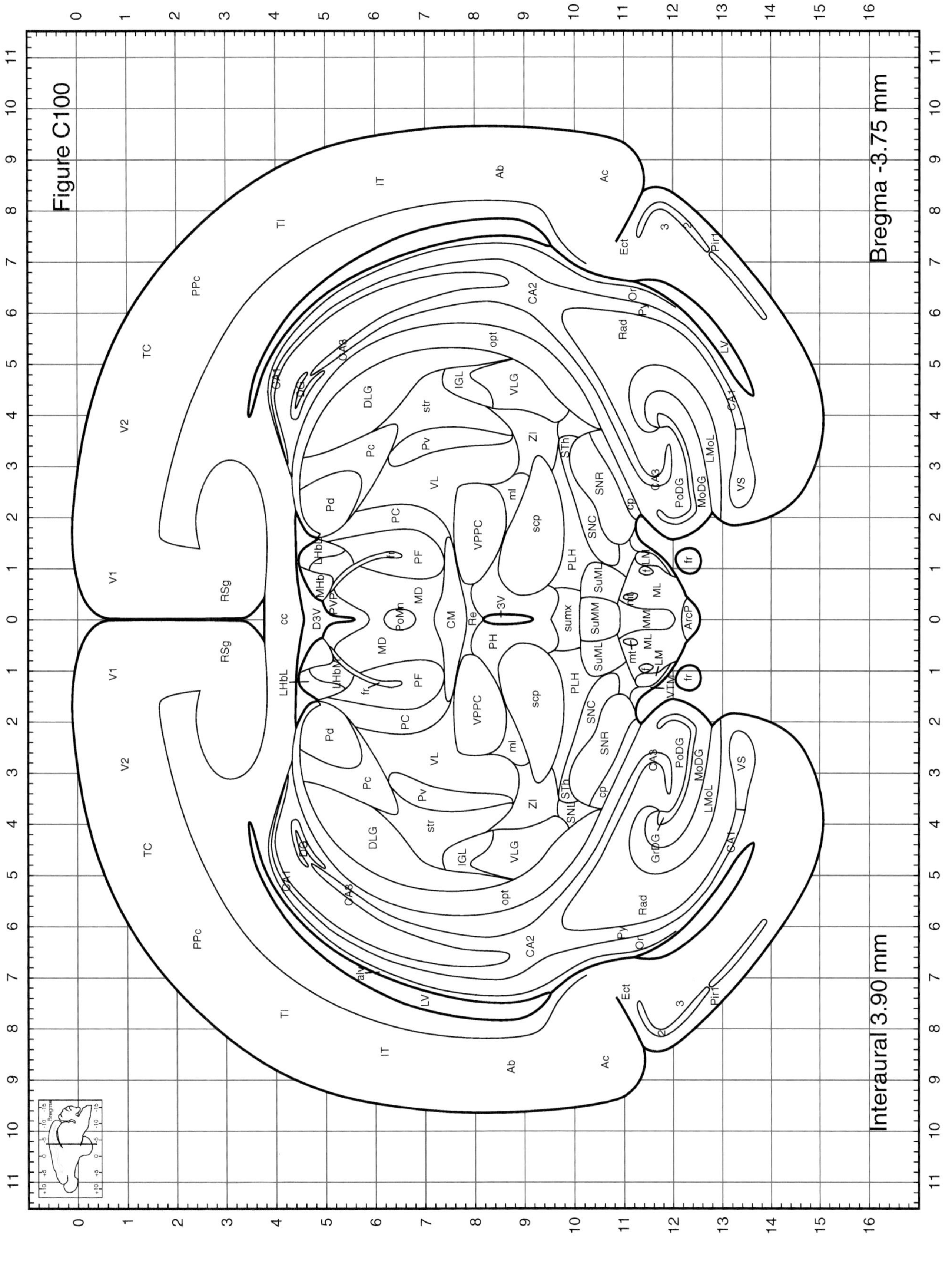

Figure C100
Bregma -3.75 mm
Interaural 3.90 mm

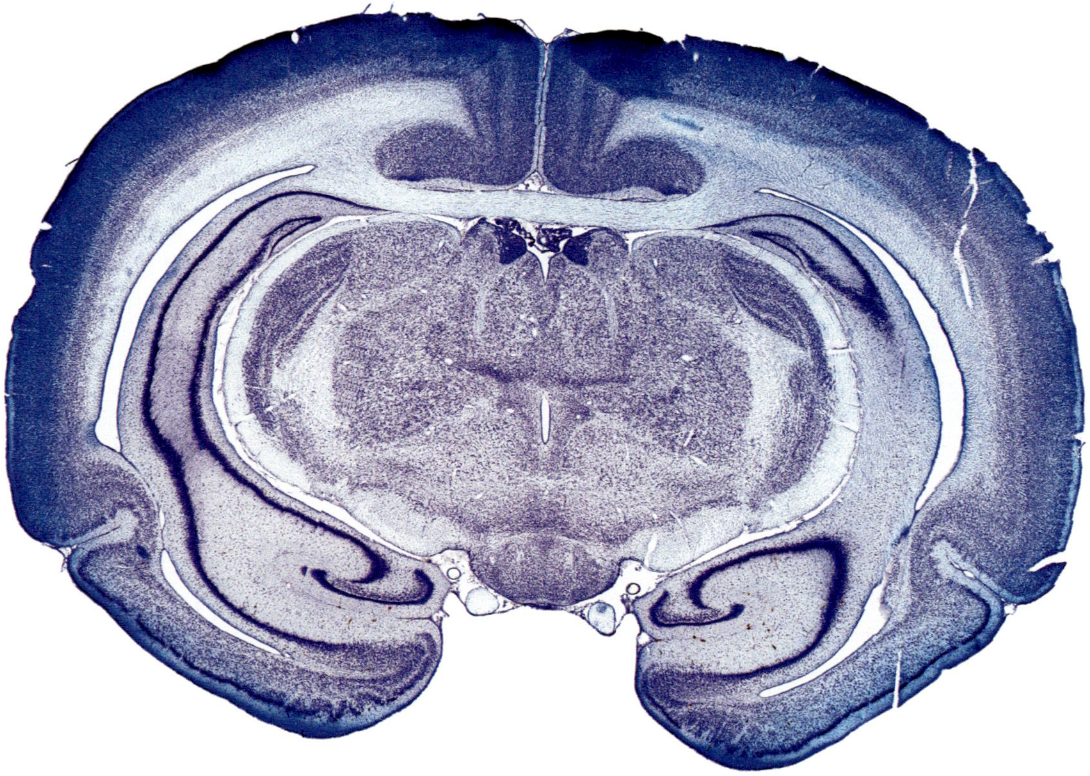

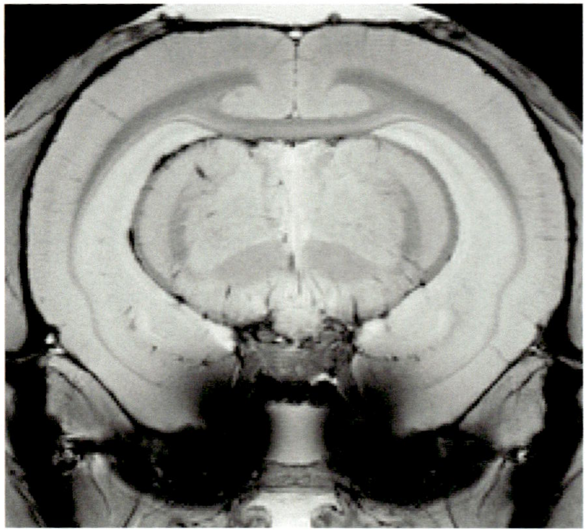

2 layer 2 of cortex
3 layer 3 of cortex
3V 3rd ventricle
alv alveus of the hippocampus
APT anterior pretectal nucleus
ArcP arcuate hypothalamic nucleus, posterior part
CA1 field CA1 of the hippocampus
CA2 field CA2 of the hippocampus
CA3 field CA3 of the hippocampus
cc corpus callosum
CM central medial thalamic nucleus
cp cerebral peduncle
D3V dorsal 3rd ventricle
DG dentate gyrus of the hippocampus
DLG dorsal lateral geniculate nucleus
Ect ectorhinal cortex
f fornix
fr fasciculus retroflexus
GrDG granular layer of the dentate gyrus
IGL intergeniculate leaf
IT inferior temporal cortex
LHb lateral habenular nucleus
LM lateral mammillary nucleus
LMoL lacunosum moleculare layer of the hippocampus
LV lateral ventricle
MD mediodorsal thalamic nucleus
MHb medial habenular nucleus
ml medial lemniscus
ML medial mammillary nucleus, lateral part
MM medial mammillary nucleus, medial part
MoDG molecular layer of the dentate gyrus
opt optic tract
Or oriens layer of the hippocampus
PC paracentral thalamic nucleus
Pc central nucleus of the pulvinar
Pd dorsal nucleus of the pulvinar
PF parafascicular thalamic nucleus
PH posterior hypothalamic nucleus
Pir1 piriform cortex, layer 1
PLH peduncular part of lateral hypothalamus
PoDG polymorph layer of the dentate gyrus
PRh perirhinal cortex
Pv ventral nucleus of the pulvinar
PVP paraventricular thalamic nucleus, posterior part
Py pyramidal cell layer of the hippocampus
Rad radiatum layer of the hippocampus
RSg retrosplenial granular cortex
scp superior cerebellar peduncle
SNC substantia nigra, compact part
SNL substantia nigra, lateral part
SNR substantia nigra, reticular part
str superior thalamic radiation
SuML supramammillary nucleus, lateral par
SuMM supramammillary nucleus, medial part
sumx supramammillary decussation
TC temporal cortex
TI temporal inferior area
V1 primary visual cortex
V2 secondary visual cortex
VLG ventral lateral geniculate nucleus
VPPC ventral posterior nucleus of the thalamus, parvicellular part
VPT ventral posterior thalamic nucleus
VS ventral subiculum
VTM ventral tuberomammillary nucleus
ZI zona incerta

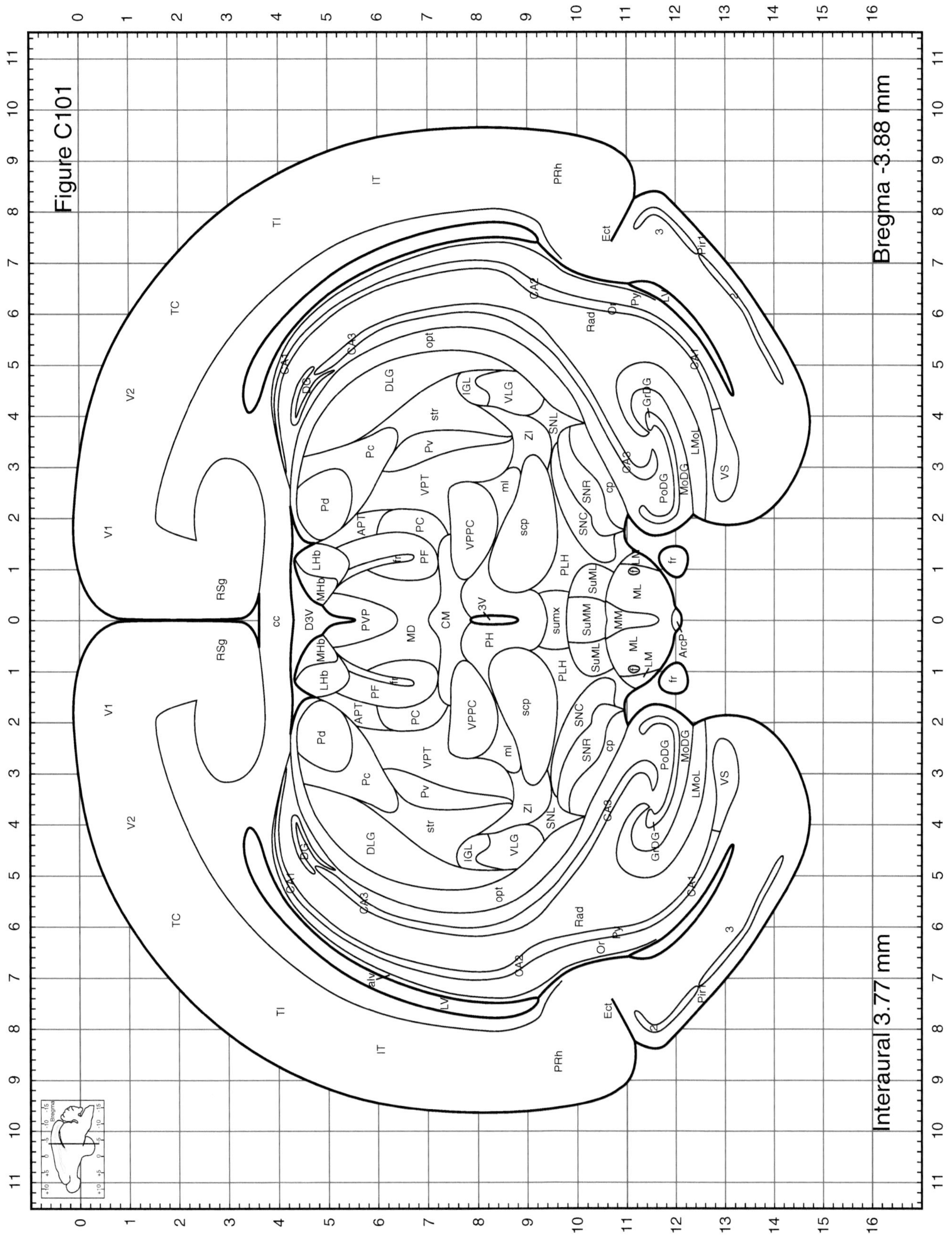
Figure C101
Bregma -3.88 mm
Interaural 3.77 mm
V1
V2
TC
TI
IT
PRh
Ect
RSg
cc
D3V
MHb
LHb
PVP
MD
CM
3V
PH
sumx
SuMM
SuML
MM
ML
LM
PLH
ArcP
fr
Pd
Pc
Pv
APT
PC
PF
VPPC
VPT
scp
ml
ZI
SNL
SNC
SNR
cp
str
IGL
VLG
DLG
opt
DG
CA1
CA2
CA3
Rad
Or
Py
LV
alv
GrDG
PoDG
MoDG
LMol
VS
Pir1
3
2

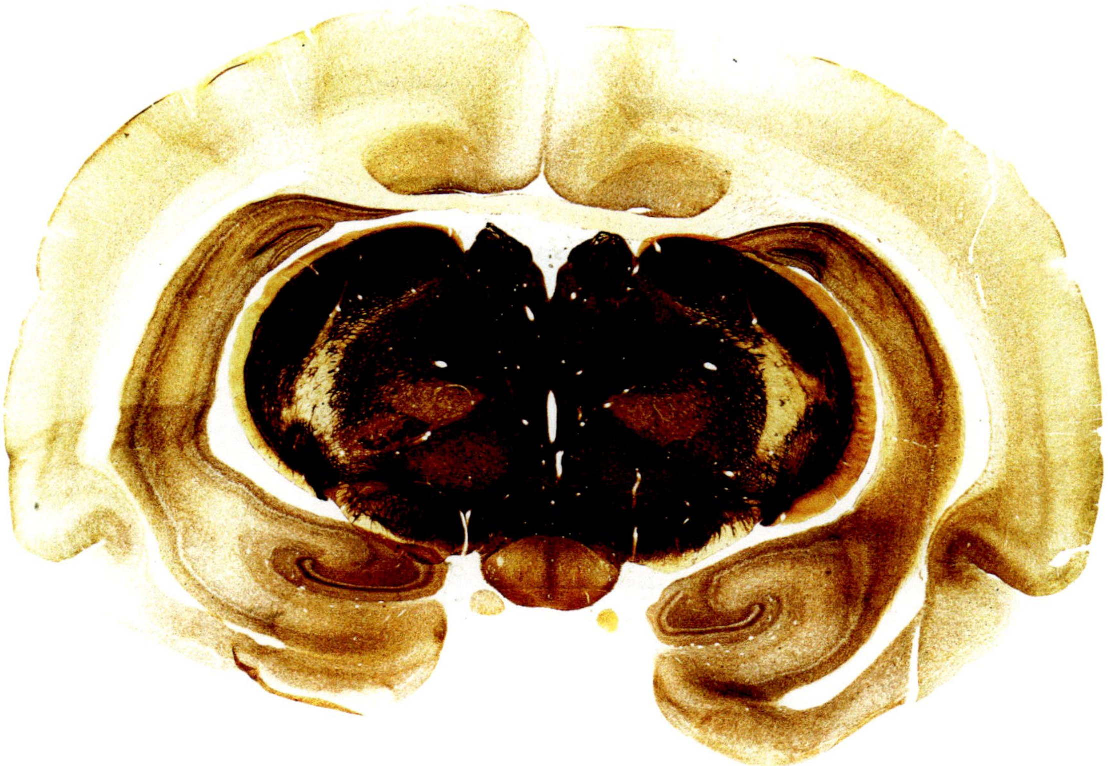

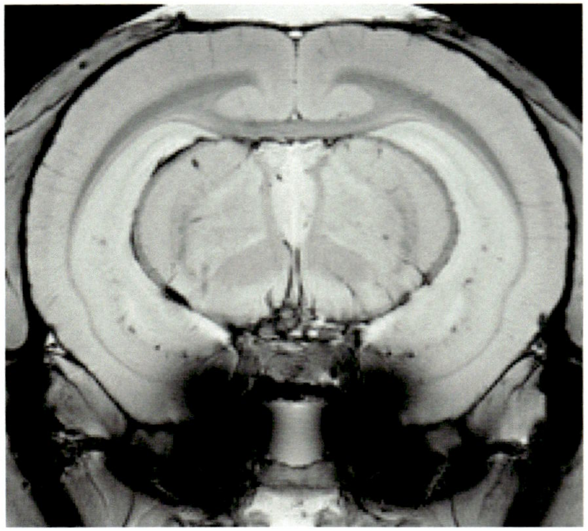

2 layer 2 of cortex
3 layer 3 of cortex
3V 3rd ventricle
alv alveus of the hippocampus
APT anterior pretectal nucleus
CA1 field CA1 of the hippocampus
CA2 field CA2 of the hippocampus
CA3 field CA3 of the hippocampus
cc corpus callosum
cp cerebral peduncle
D3V dorsal 3rd ventricle
DG dentate gyrus of the hippocampus
DLG dorsal lateral geniculate nucleus
Ect ectorhinal cortex
Ent entorhinal cortex
f fornix
fr fasciculus retroflexus
GrDG granular layer of the dentate gyrus
IGL intergeniculate leaf
IT inferior temporal cortex
LHb lateral habenular nucleus
LMoL lacunosum moleculare layer of the hippocampus
LV lateral ventricle
MD mediodorsal thalamic nucleus
MHb medial habenular nucleus
ml medial lemniscus
ML medial mammillary nucleus, lateral part
MM medial mammillary nucleus, medial part
MoDG molecular layer of the dentate gyrus
opt optic tract
Or oriens layer of the hippocampus
PAG periaqueductal gray
PC paracentral thalamic nucleus
Pc central nucleus of the pulvinar
Pd dorsal nucleus of the pulvinar
PF parafascicular thalamic nucleus
PH posterior hypothalamic nucleus
PHA posterior hypothalamic area
Pir1 piriform cortex, layer 1
PoDG polymorph layer of the dentate gyrus
PRh perirhinal cortex
Pv ventral nucleus of the pulvinar
PVP paraventricular thalamic nucleus, posterior part
Py pyramidal cell layer of the hippocampus
Rad radiatum layer of the hippocampus
RSg retrosplenial granular cortex
scp superior cerebellar peduncle
SNC substantia nigra, compact part
SNL substantia nigra, lateral part
SNR substantia nigra, reticular part
str superior thalamic radiation
SuML supramammillary nucleus, lateral par
SuMM supramammillary nucleus, medial part
sumx supramammillary decussation
TC temporal cortex
TI temporal inferior area
V1 primary visual cortex
V2 secondary visual cortex
VLG ventral lateral geniculate nucleus
VPPC ventral posterior nucleus of the thalamus, parvicellular part
VPT ventral posterior thalamic nucleus
VS ventral subiculum
VTAR ventral tegmental area, rostral part
ZI zona incerta

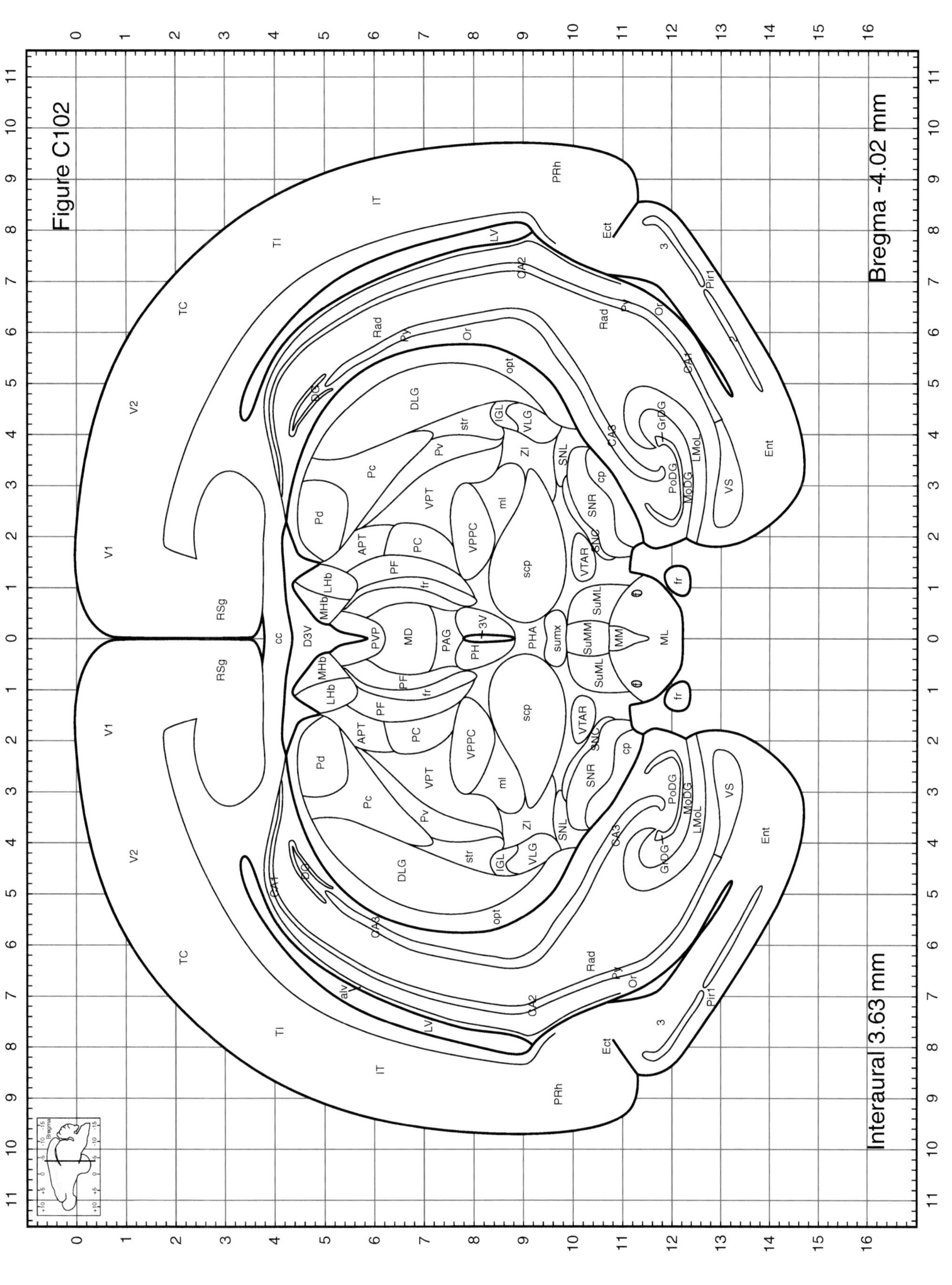
Figure C102
Bregma -4.02 mm
Interaural 3.63 mm
V1
V2
TC
TI
IT
PRh
Ect
Ent
VS
RSg
cc
D3V
MHb
LHb
PVP
MD
PAG
PH
3V
PHA
sumx
SuMM
SuML
MM
ML
fr
f
scp
VTAR
SNC
SNR
SNL
cp
ZI
ml
VPPC
VPT
PC
PF
APT
Pd
Pc
Pv
str
IGL
VLG
DLG
opt
Or
Py
Rad
CA1
CA2
CA3
DG
GrDG
PoDG
MoDG
LMoL
LV
alv
Pir1
3

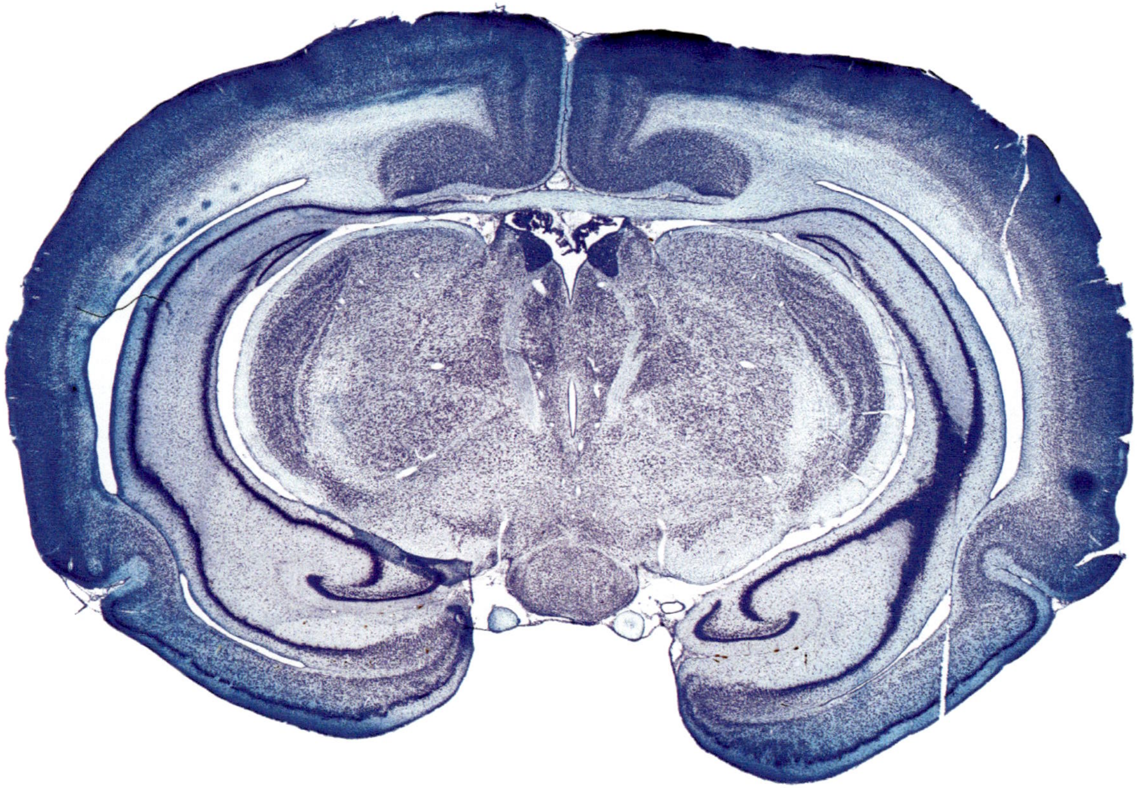

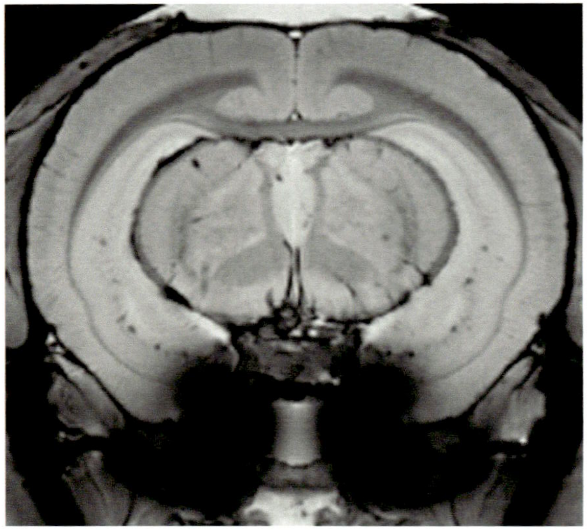

3V	3rd ventricle
alv	alveus of the hippocampus
APT	anterior pretectal nucleus
CA1	field CA1 of the hippocampus
CA2	field CA2 of the hippocampus
CA3	field CA3 of the hippocampus
cc	corpus callosum
cp	cerebral peduncle
D3V	dorsal 3rd ventricle
DG	dentate gyrus of the hippocampus
DLG	dorsal lateral geniculate nucleus
Ect	ectorhinal cortex
Ent	entorhinal cortex
fr	fasciculus retroflexus
GrDG	granular layer of the dentate gyrus
IGL	intergeniculate leaf
IT	inferior temporal cortex
LHb	lateral habenular nucleus
LMoL	lacunosum moleculare layer of the hippocampus
LV	lateral ventricle
MD	mediodorsal thalamic nucleus
MHb	medial habenular nucleus
ml	medial lemniscus
ML	medial mammillary nucleus, lateral part
MoDG	molecular layer of the dentate gyrus
MT	medial terminal nucleus of the accessory optic tract
opt	optic tract
Or	oriens layer of the hippocampus
PAG	periaqueductal gray
PC	paracentral thalamic nucleus
Pc	central nucleus of the pulvinar
Pd	dorsal nucleus of the pulvinar
PF	parafascicular thalamic nucleus
PH	posterior hypothalamic nucleus
PHA	posterior hypothalamic area
PoDG	polymorph layer of the dentate gyrus
PRh	perirhinal cortex
Pv	ventral nucleus of the pulvinar
PVP	paraventricular thalamic nucleus, posterior part
Py	pyramidal cell layer of the hippocampus
Rad	radiatum layer of the hippocampus
RSg	retrosplenial granular cortex
scp	superior cerebellar peduncle
SNC	substantia nigra, compact part
SNL	substantia nigra, lateral part
SNR	substantia nigra, reticular part
str	superior thalamic radiation
SuML	supramammillary nucleus, lateral par
SuMM	supramammillary nucleus, medial part
sumx	supramammillary decussation
TC	temporal cortex
TI	temporal inferior area
V1	primary visual cortex
V2	secondary visual cortex
VLG	ventral lateral geniculate nucleus
VPPC	ventral posterior nucleus of the thalamus,parvicellular part
VPT	ventral posterior thalamic nucleus
VS	ventral subiculum
VTAR	ventral tegmental area, rostral part
VTM	ventral tuberomammillary nucleus
ZI	zona incerta

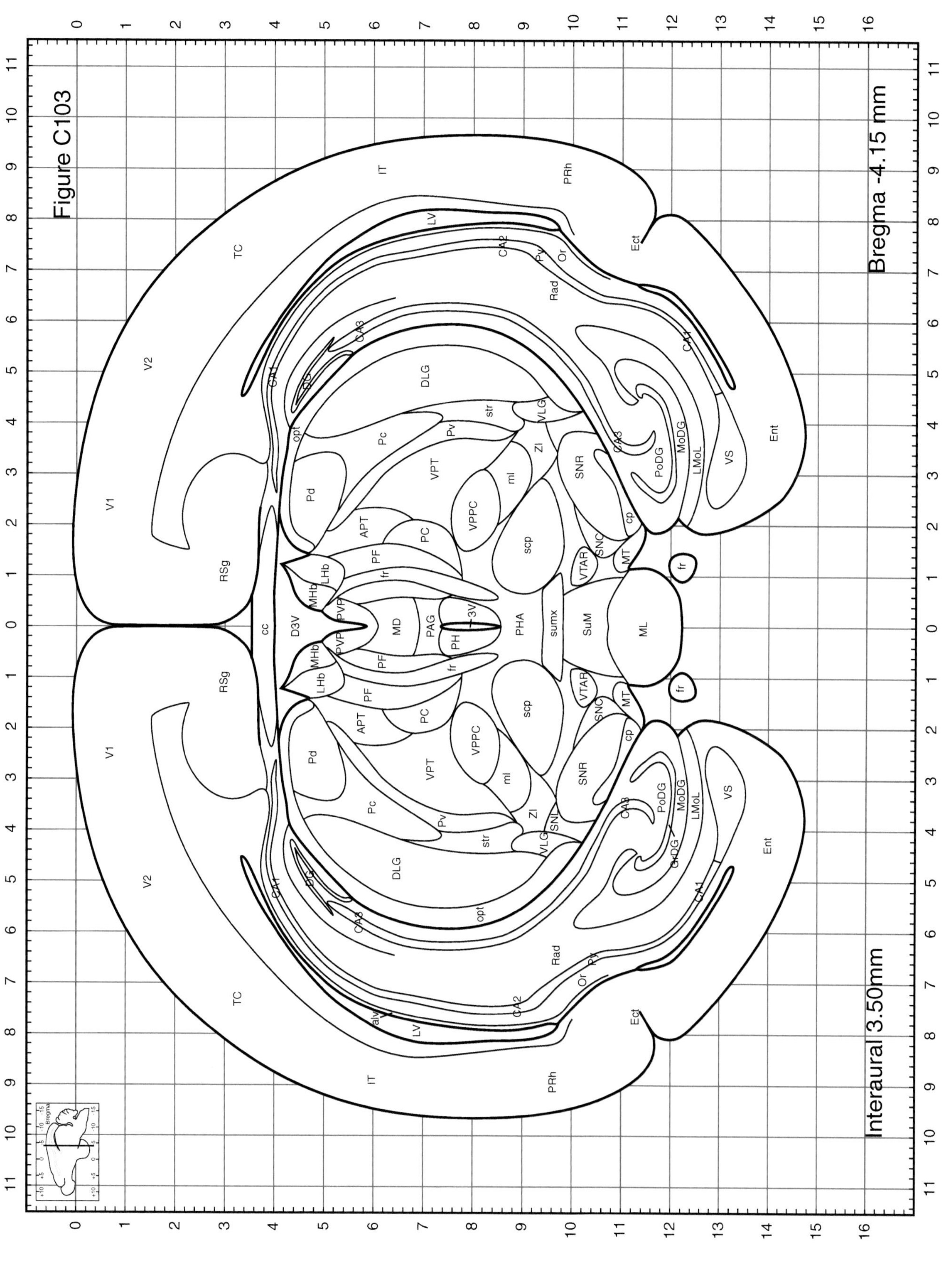
Figure C103
Bregma -4.15 mm
Interaural 3.50mm
TC
V2
V1
RSg
IT
PRh
Ect
Ent
LV
CA1
CA2
CA3
DG
Rad
Or
Py
VS
LMoL
MoDG
PoDG
cc
D3V
MHb
LHb
PVP
APT
PF
PC
Pd
Pc
DLG
VPT
VPPC
Pv
str
opt
VLG
ZI
ml
scp
SNR
SNC
cp
MT
VTAR
fr
MD
PAG
PH
3V
PHA
sumx
SuM
ML

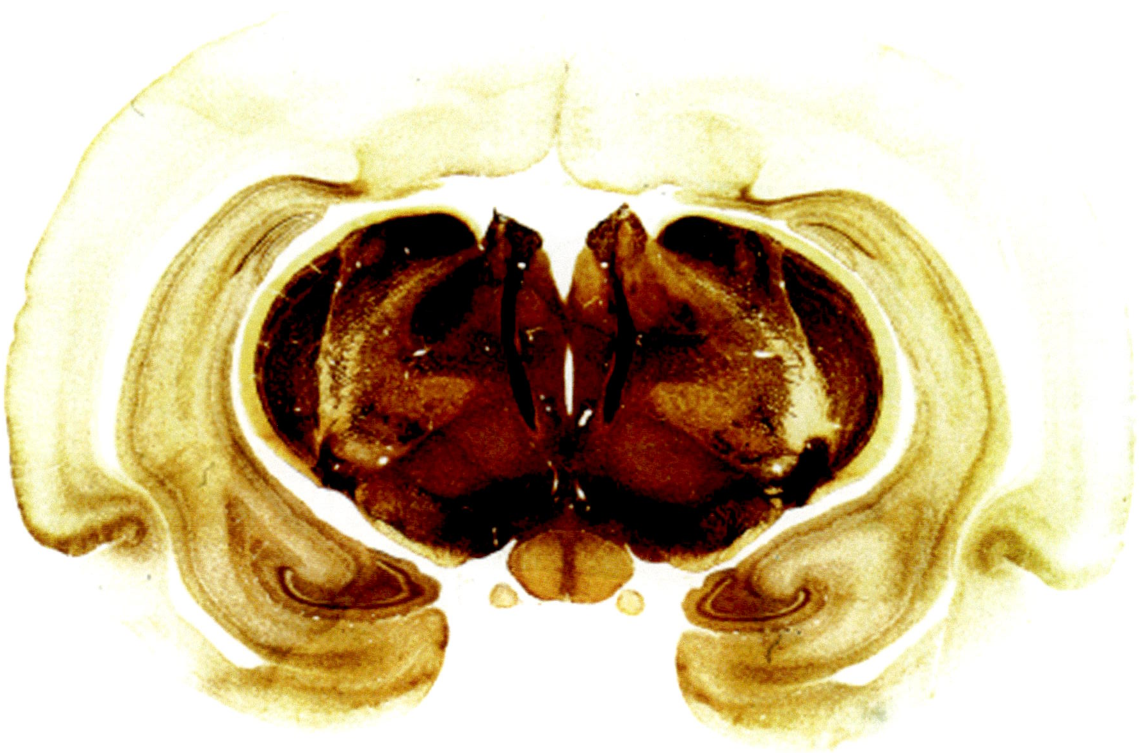

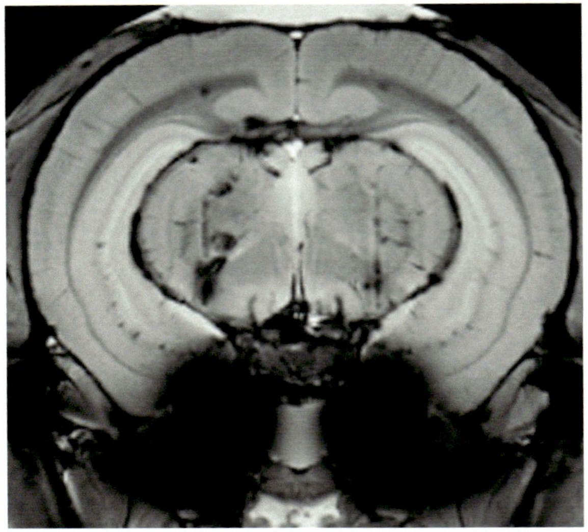

3V 3rd ventricle
alv alveus of the hippocampus
APT anterior pretectal nucleus
CA1 field CA1 of the hippocampus
CA2 field CA2 of the hippocampus
CA3 field CA3 of the hippocampus
cp cerebral peduncle
D3V dorsal 3rd ventricle
DG dentate gyrus of the hippocampus
DLG dorsal lateral geniculate nucleus
Ect ectorhinal cortex
Ent entorhinal cortex
fr fasciculus retroflexus
GrDG granular layer of the dentate gyrus
IT inferior temporal cortex
LHb lateral habenular nucleus
LMoL lacunosum moleculare layer of the hippocampus
LV lateral ventricle
MD mediodorsal thalamic nucleus
MG medial geniculate nucleus
MHb medial habenular nucleus
ml medial lemniscus
ML medial mammillary nucleus, lateral part
MoDG molecular layer of the dentate gyrus
MT medial terminal nucleus of the accessory optic tract
opt optic tract
Or oriens layer of the hippocampus
PC paracentral thalamic nucleus
Pc central nucleus of the pulvinar
Pd dorsal nucleus of the pulvinar
PF parafascicular thalamic nucleus
PH posterior hypothalamic nucleus
PHA posterior hypothalamic area
PoDG polymorph layer of the dentate gyrus
pPAG p1 periaqueductal gray
PRh perirhinal cortex
pRt p1 reticular formation
Pv ventral nucleus of the pulvinar
PVP paraventricular thalamic nucleus, posterior part
Py pyramidal cell layer of the hippocampus
Rad radiatum layer of the hippocampus
RSg retrosplenial granular cortex
scp superior cerebellar peduncle
SNC substantia nigra, compact part
SNL substantia nigra, lateral part
SNR substantia nigra, reticular part
str superior thalamic radiation
SuM supramammillary nucleus
sumx supramammillary decussation
TC temporal cortex
V1 primary visual cortex
V2 secondary visual cortex
VLG ventral lateral geniculate nucleus
VPPC ventral posterior nucleus of the thalamus, parvicellular part
VPT ventral posterior thalamic nucleus
VS ventral subiculum
VTAR ventral tegmental area, rostral part
ZI zona incerta

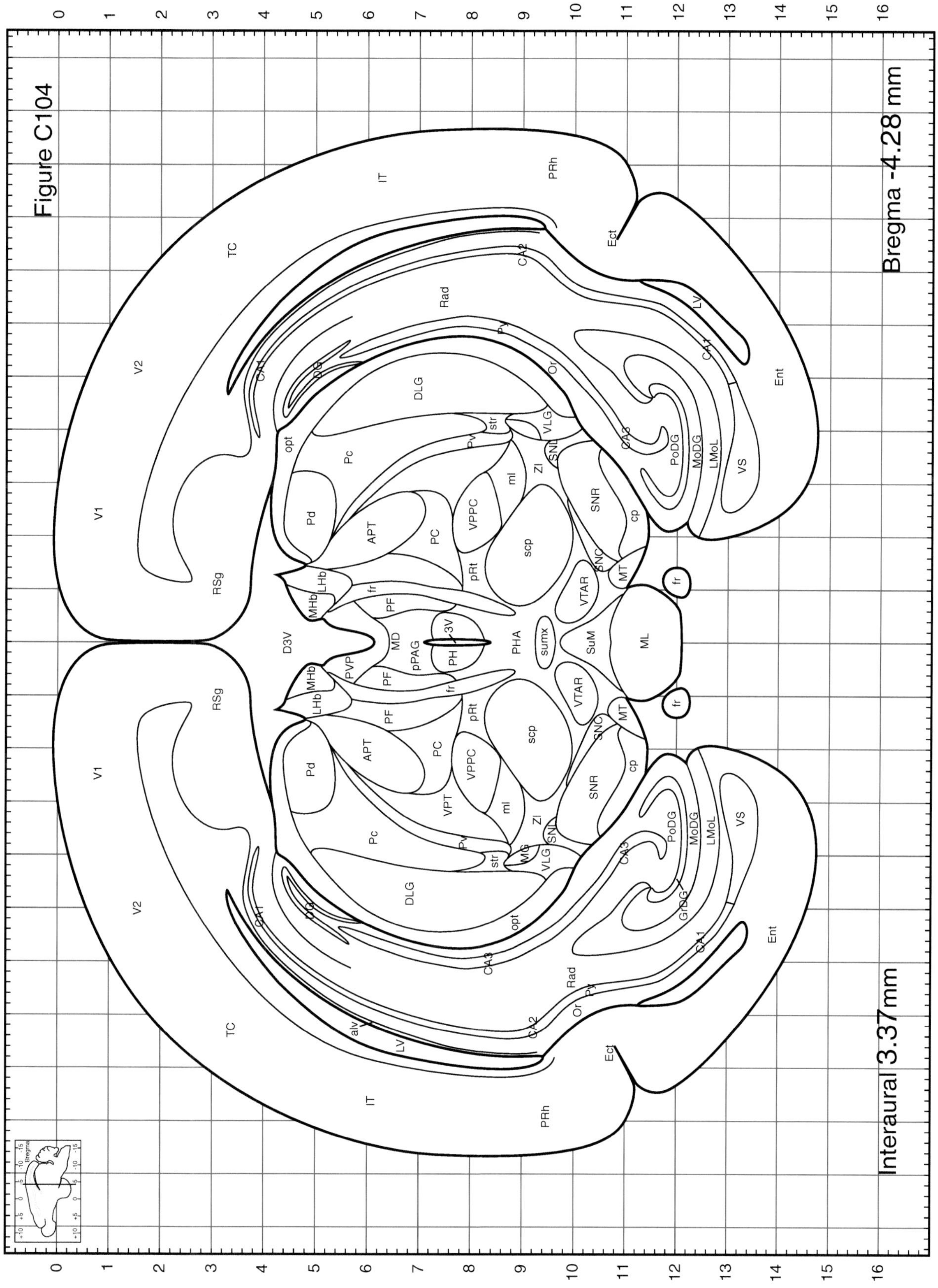
Figure C104
Bregma -4.28 mm
Interaural 3.37 mm
TC
IT
PRh
V2
V1
RSg
Ect
CA2
Rad
Py
CA1
DG
Or
LV
Ent
DLG
opt
str
VLG
CA3
PoDG
MoDG
LMoL
VS
Pc
PV
ml
ZI
SNL
SNR
cp
Pd
APT
PC
VPPC
scp
SNC
MT
pRt
VTAR
fr
LHb
MHb
PF
MD
3V
D3V
PVP
pPAG
PH
PHA
sumx
SuM
ML
VPT
MG
GrDG
alv

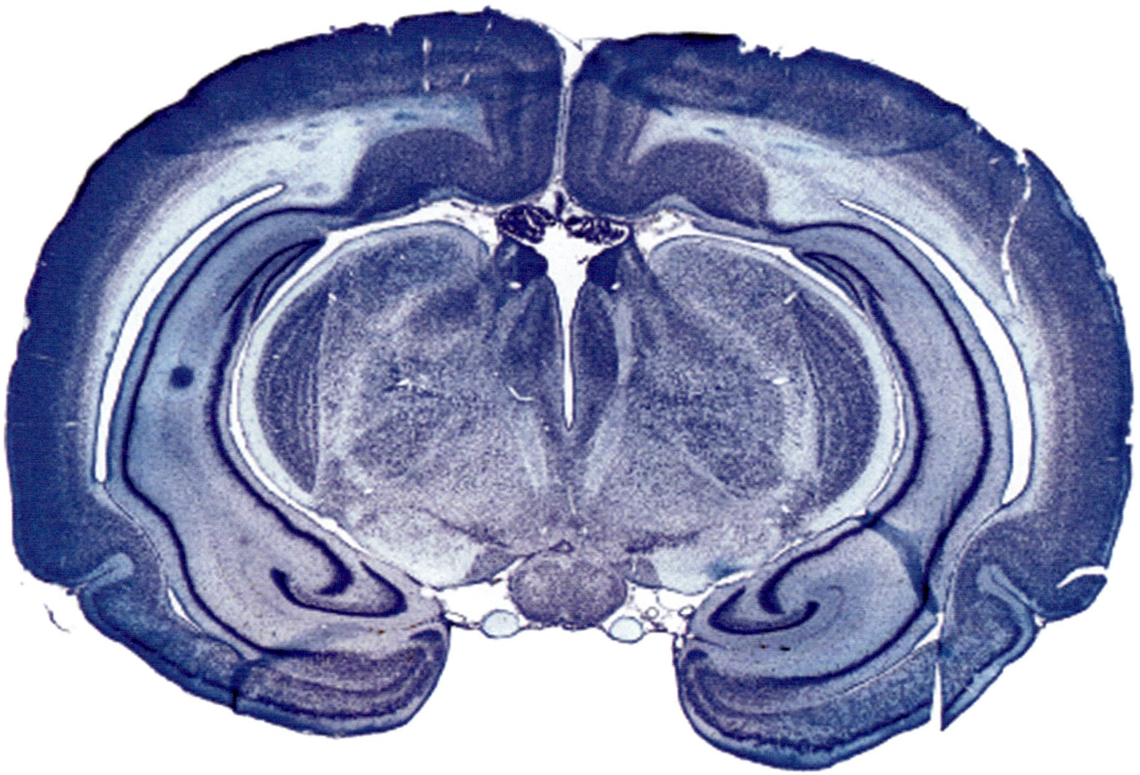

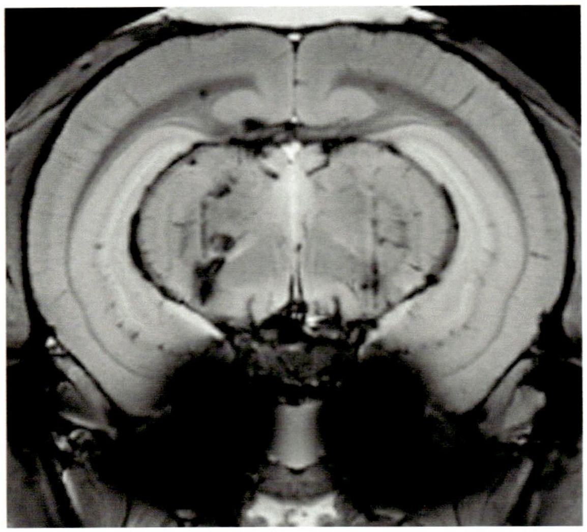

alv	alveus of the hippocampus
APTD	anterior pretectal nucleus, dorsal part
APTV	anterior pretectal nucleus, ventral part
CA1	field CA1 of the hippocampus
CA2	field CA2 of the hippocampus
CA3	field CA3 of the hippocampus
cp	cerebral peduncle
D3V	dorsal 3rd ventricle
DG	dentate gyrus of the hippocampus
DLG	dorsal lateral geniculate nucleus
DS	dorsal subiculum
Ect	ectorhinal cortex
Ent	entorhinal cortex
fr	fasciculus retroflexus
GrDG	granular layer of the dentate gyrus
IF	interfascicular nucleus
IT	inferior temporal cortex
LHb	lateral habenular nucleus
LMoL	lacunosum moleculare layer of the hippocampus
LV	lateral ventricle
MG	medial geniculate nucleus
MHb	medial habenular nucleus
ml	medial lemniscus
ML	medial mammillary nucleus, lateral part
MoDG	molecular layer of the dentate gyrus
MT	medial terminal nucleus of the accessory optic tract
opt	optic tract
Or	oriens layer of the hippocampus
PC	paracentral thalamic nucleus
Pc	central nucleus of the pulvinar
Pd	dorsal nucleus of the pulvinar
PF	parafascicular thalamic nucleus
PHA	posterior hypothalamic area
PoDG	polymorph layer of the dentate gyrus
pPAG	p1 periaqueductal gray
PrC	precommissural nucleus
PRh	perirhinal cortex
pRt	p1 reticular formation
Pv	ventral nucleus of the pulvinar
PVP	paraventricular thalamic nucleus, posterior part
Py	pyramidal cell layer of the hippocampus
Rad	radiatum layer of the hippocampus
RLi	rostral linear nucleus of the raphe
RSg	retrosplenial granular cortex
scp	superior cerebellar peduncle
SNC	substantia nigra, compact part
SNR	substantia nigra, reticular part
SuM	supramammillary nucleus
TC	temporal cortex
V1	primary visual cortex
V2	secondary visual cortex
VLG	ventral lateral geniculate nucleus
VPPC	ventral posterior nucleus of the thalamus, parvicellular part
VPT	ventral posterior thalamic nucleus
VS	ventral subiculum
VTAR	ventral tegmental area, rostral part
ZI	zona incerta

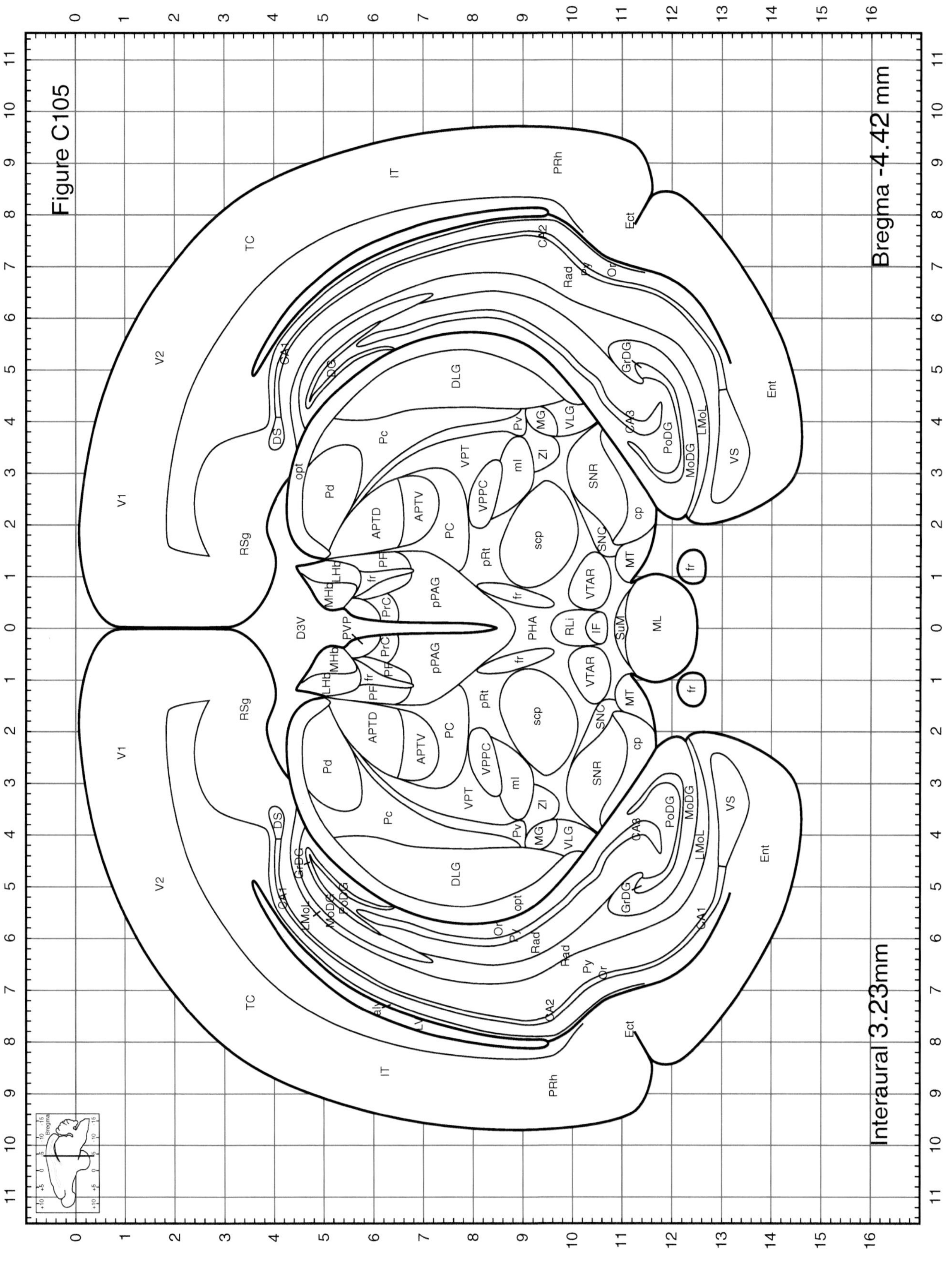
Figure C105
Bregma -4.42 mm
Interaural 3.23mm
IT
PRh
TC
Ect
CA2
Rad
Py
Or
V2
CA1
DG
DLG
GrDG
Ent
DS
Pc
Pv
MG
VLG
CA3
PoDG
LMoL
MoDG
VS
opt
VPT
ml
ZI
SNR
V1
Pd
VPPC
APTV
APTD
PC
scp
cp
SNC
RSg
pRt
MT
fr
PF
LHb
MHb
PrC
pPAG
VTAR
D3V
PVP
PHA
RLi
IF
SuM
ML
LV
alv

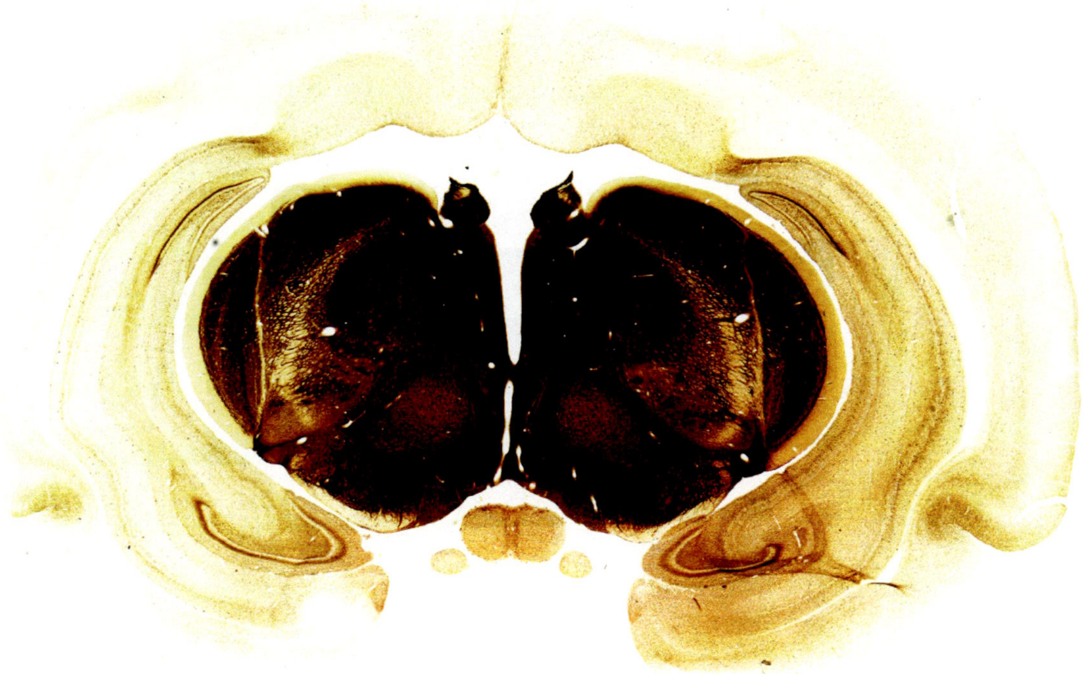

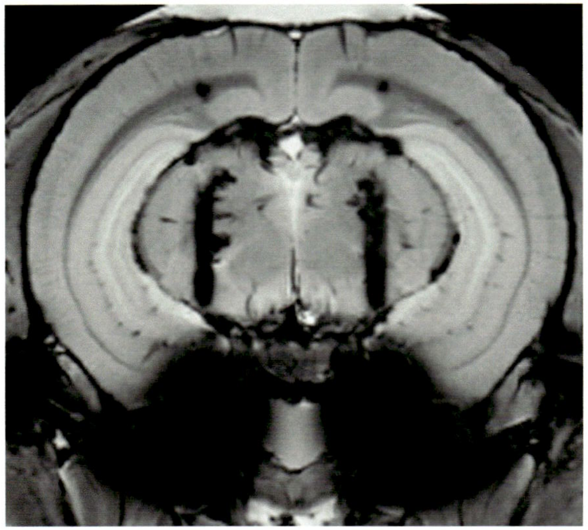

3V	3rd ventricle
alv	alveus of the hippocampus
APTD	anterior pretectal nucleus, dorsal part
APTV	anterior pretectal nucleus, ventral part
CA1	field CA1 of the hippocampus
CA2	field CA2 of the hippocampus
CA3	field CA3 of the hippocampus
cp	cerebral peduncle
DG	dentate gyrus of the hippocampus
DLG	dorsal lateral geniculate nucleus
DS	dorsal subiculum
Ect	ectorhinal cortex
Ent	entorhinal cortex
fr	fasciculus retroflexus
GrDG	granular layer of the dentate gyrus
IF	interfascicular nucleus
IT	inferior temporal cortex
LHb	lateral habenular nucleus
LMoL	lacunosum moleculare layer of the hippocampus
LV	lateral ventricle
MG	medial geniculate nucleus
MHb	medial habenular nucleus
ml	medial lemniscus
ML	medial mammillary nucleus, lateral part
MoDG	molecular layer of the dentate gyrus
MPT	medial pretectal nucleus
MT	medial terminal nucleus of the accessory optic tract
OPT	olivary pretectal nucleus
opt	optic tract
Or	oriens layer of the hippocampus
PAG	periaqueductal gray
PC	paracentral thalamic nucleus
Pc	central nucleus of the pulvinar
Pd	dorsal nucleus of the pulvinar
PH	posterior hypothalamic nucleus
Po	posterior complex of the thalamus
PoDG	polymorph layer of the dentate gyrus
PrC	precommissural nucleus
PRh	perirhinal cortex
pRt	p1 reticular formation
Pv	ventral nucleus of the pulvinar
Py	pyramidal cell layer of the hippocampus
Rad	radiatum layer of the hippocampus
RLi	rostral linear nucleus of the raphe
RSg	retrosplenial granular cortex
scp	superior cerebellar peduncle
sm	stria medullaris of the thalamus
SNC	substantia nigra, compact part
SNR	substantia nigra, reticular part
TC	temporal cortex
V1	primary visual cortex
V2	secondary visual cortex
VLG	ventral lateral geniculate nucleus
VPPC	ventral posterior nucleus of the thalamus, parvicellular part
VPT	ventral posterior thalamic nucleus
VS	ventral subiculum
VTAR	ventral tegmental area, rostral part

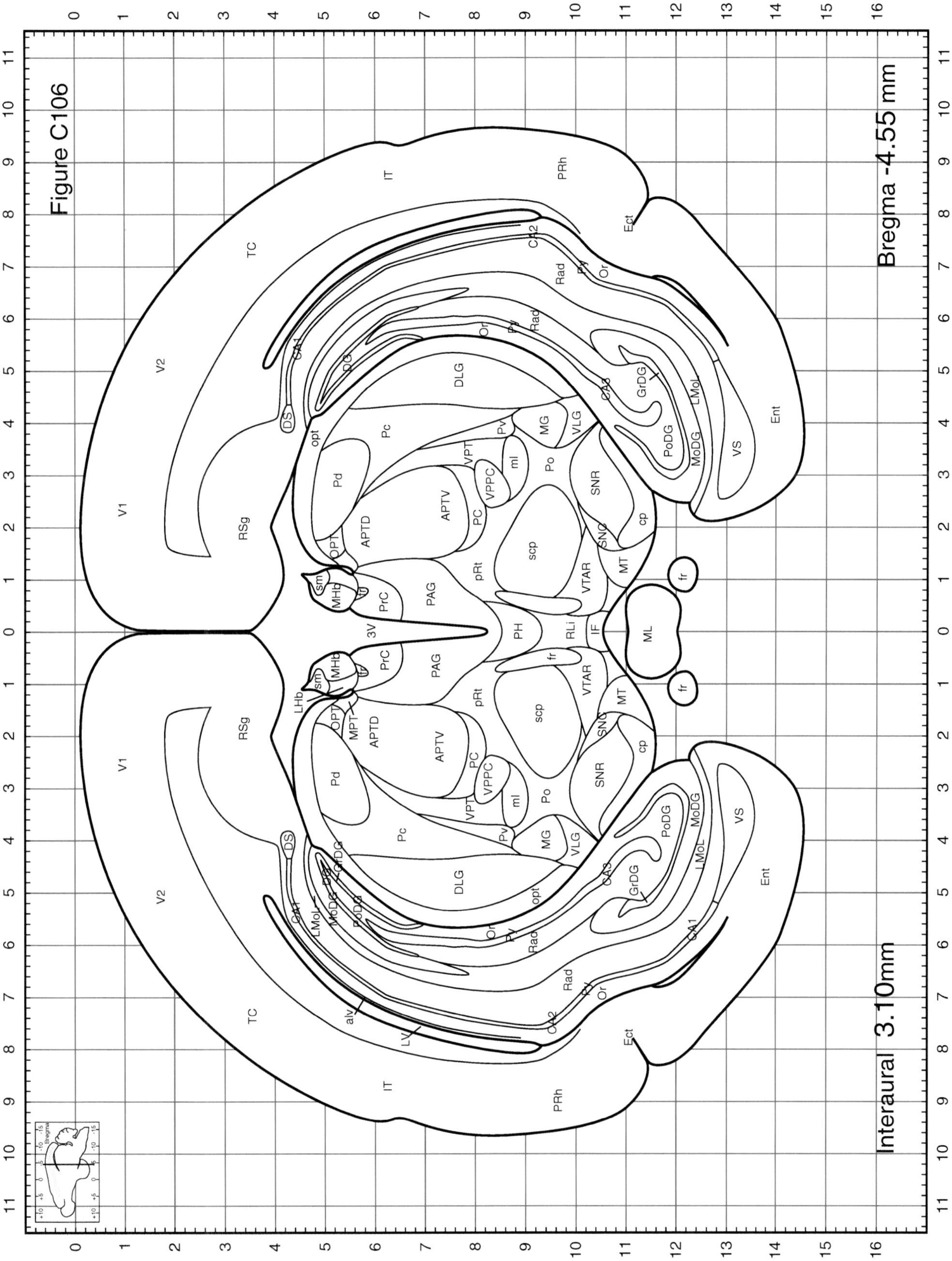

Figure C106
Bregma -4.55 mm
Interaural 3.10mm
V1
V2
RSg
TC
IT
PRh
Ect
Ent
VS
CA1
CA2
CA3
DG
GrDG
PoDG
MoDG
LMol
Rad
Py
Or
DS
opt
alv
LV
DLG
Pc
Pd
APTD
APTV
OPT
MPT
PC
VPPC
VPT
Pv
MG
VLG
ml
Po
SNR
SNC
cp
scp
pRt
PAG
PrC
MHb
LHb
sm
fr
3V
PH
RLi
IF
VTAR
MT
ML

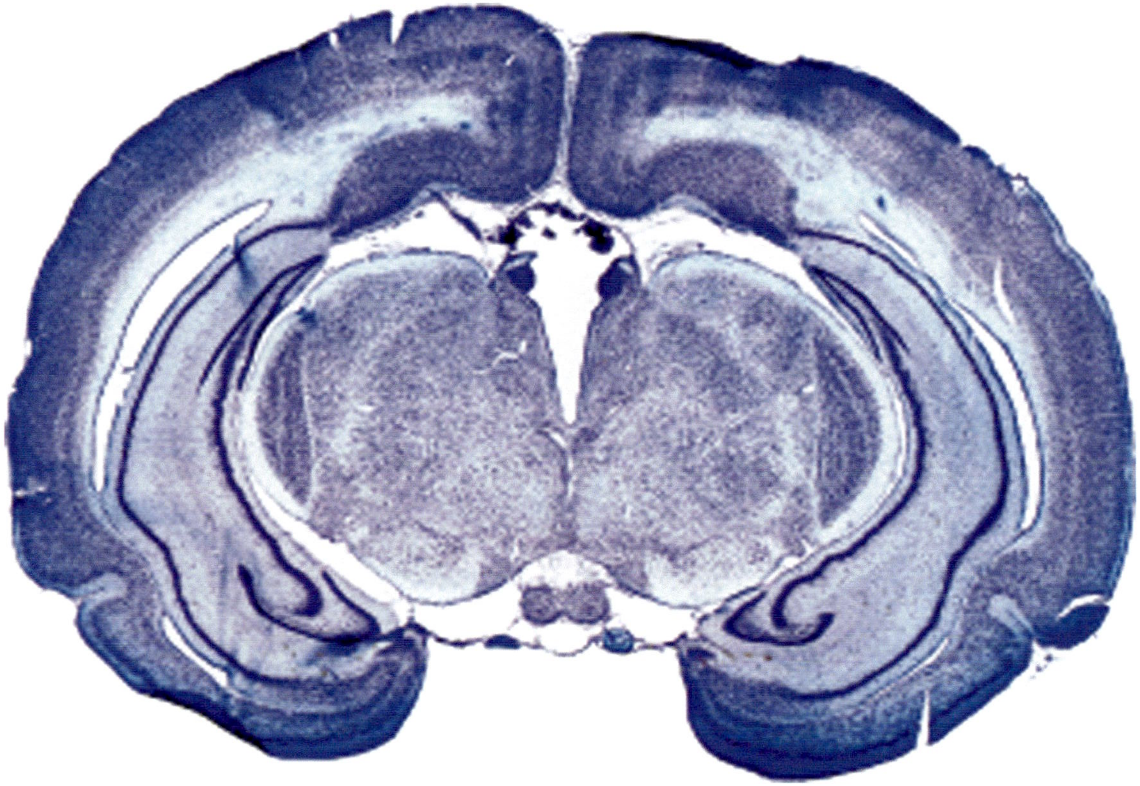

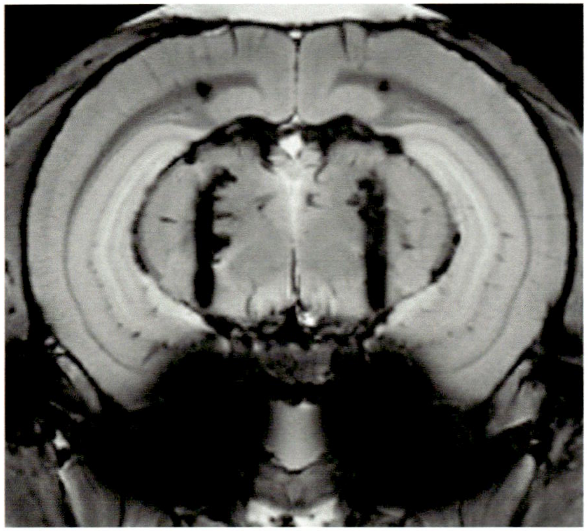

3V	3rd ventricle
alv	alveus of the hippocampus
APTD	anterior pretectal nucleus, dorsal part
APTV	anterior pretectal nucleus, ventral part
B	brachium of the superior colliculus
CA1	field CA1 of the hippocampus
CA2	field CA2 of the hippocampus
CA3	field CA3 of the hippocampus
cp	cerebral peduncle
DG	dentate gyrus of the hippocampus
DLG	dorsal lateral geniculate nucleus
DS	dorsal subiculum
Ect	ectorhinal cortex
Ent	entorhinal cortex
fr	fasciculus retroflexus
GrDG	granular layer of the dentate gyrus
IF	interfascicular nucleus
IT	inferior temporal cortex
LHb	lateral habenular nucleus
LMoL	lacunosum moleculare layer of the hippocampus
LV	lateral ventricle
MG	medial geniculate nucleus
MHb	medial habenular nucleus
ml	medial lemniscus
ML	medial mammillary nucleus, lateral part
MoDG	molecular layer of the dentate gyrus
MPT	medial pretectal nucleus
MT	medial terminal nucleus of the accessory optic tract
OPT	olivary pretectal nucleus
opt	optic tract
Or	oriens layer of the hippocampus
Pc	central nucleus of the pulvinar
Pd	dorsal nucleus of the pulvinar
PH	posterior hypothalamic nucleus
Po	posterior complex of the thalamus
PoDG	polymorph layer of the dentate gyrus
pPAG	p1 periaqueductal gray
PrC	precommissural nucleus
PRh	perirhinal cortex
pRt	p1 reticular formation
Py	pyramidal cell layer of the hippocampus
Rad	radiatum layer of the hippocampus
RLi	rostral linear nucleus of the raphe
RSg	retrosplenial granular cortex
scp	superior cerebellar peduncle
sm	stria medullaris of the thalamus
SNC	substantia nigra, compact part
SNR	substantia nigra, reticular part
SUG	superficial gray layer of the superior colliculus
TC	temporal cortex
V1	primary visual cortex
V2	secondary visual cortex
VLG	ventral lateral geniculate nucleus
VS	ventral subiculum
VTAR	ventral tegmental area, rostral part

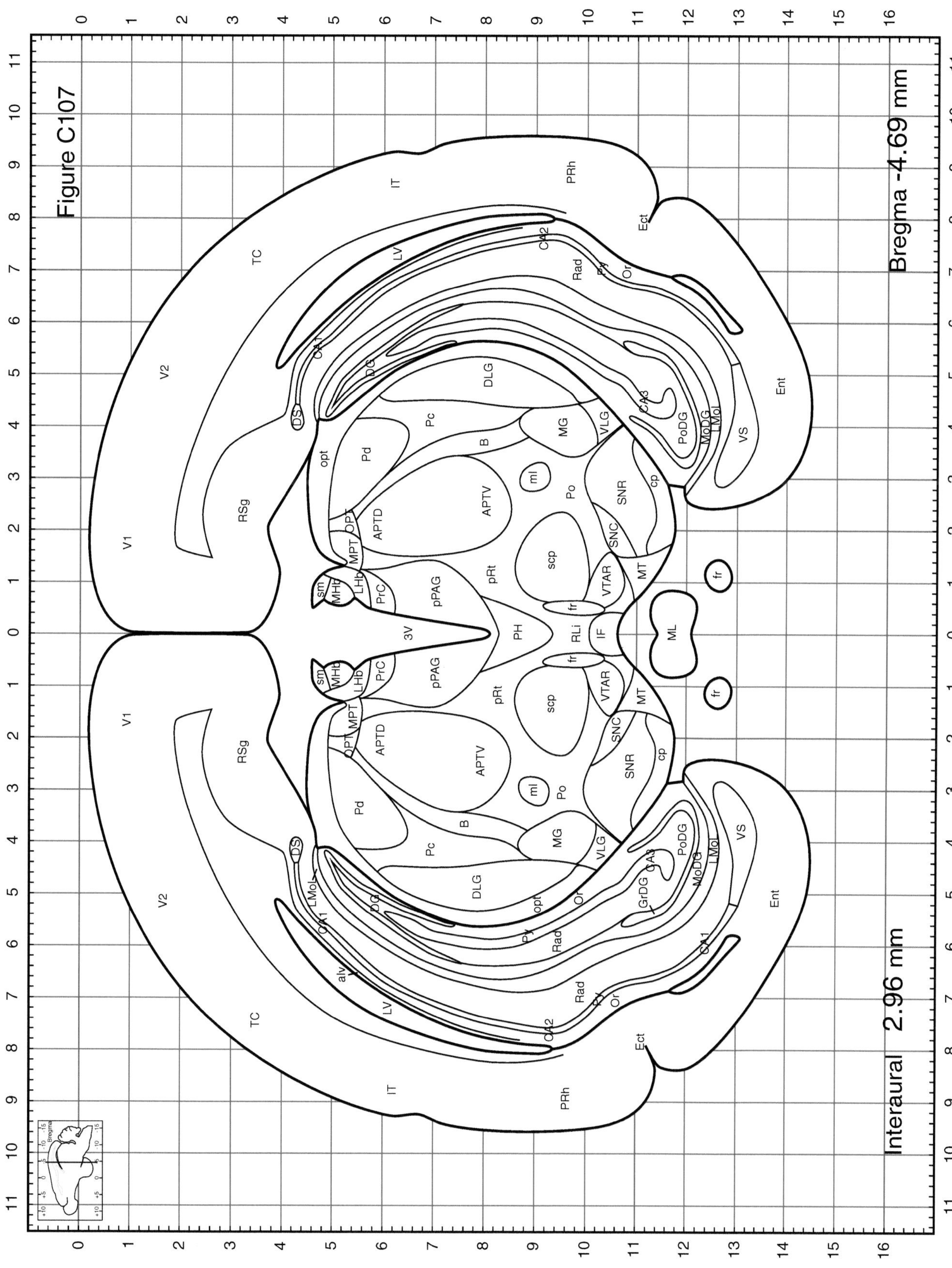
Figure C107
Bregma -4.69 mm
Interaural 2.96 mm
V1
V2
TC
IT
PRh
Ect
Ent
VS
RSg
LV
CA1
CA2
CA3
DG
DS
Rad
Py
Or
alv
LMol
MoDG
PoDG
GrDG
opt
Pd
Pc
DLG
MG
VLG
B
OPT
APTD
APTV
MPT
LHb
MHb
sm
PrC
pPAG
3V
PH
pRt
ml
Po
scp
SNR
SNC
cp
VTAR
MT
fr
RLi
IF
ML

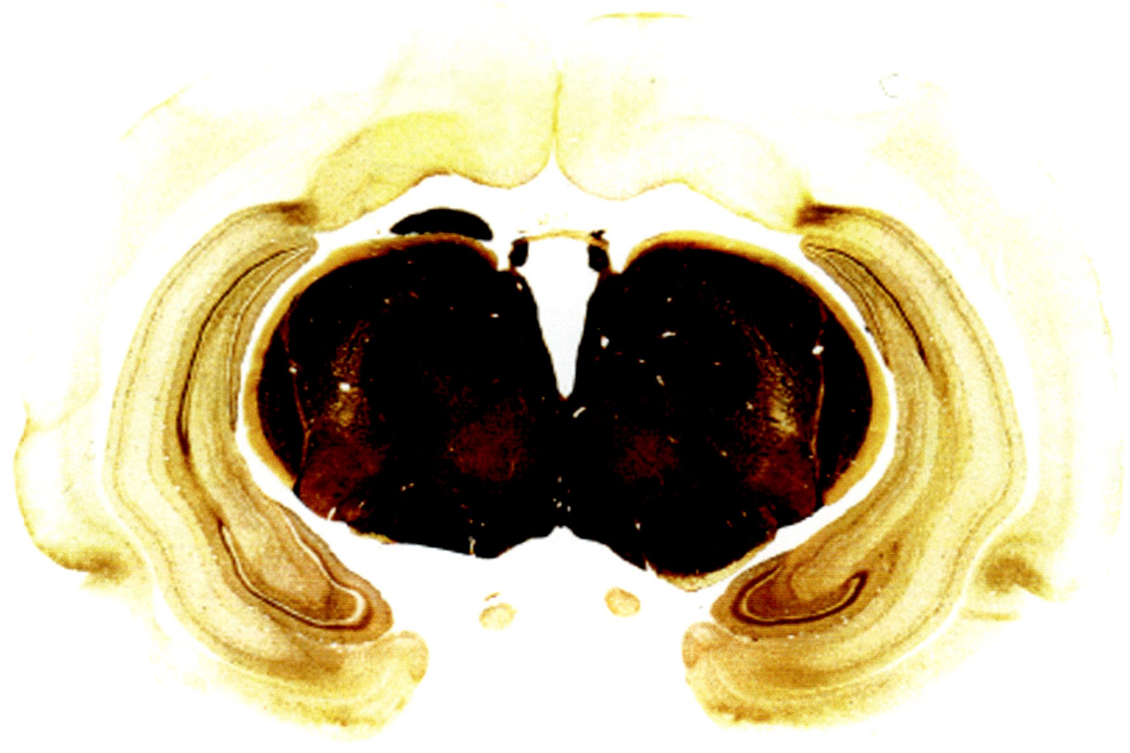

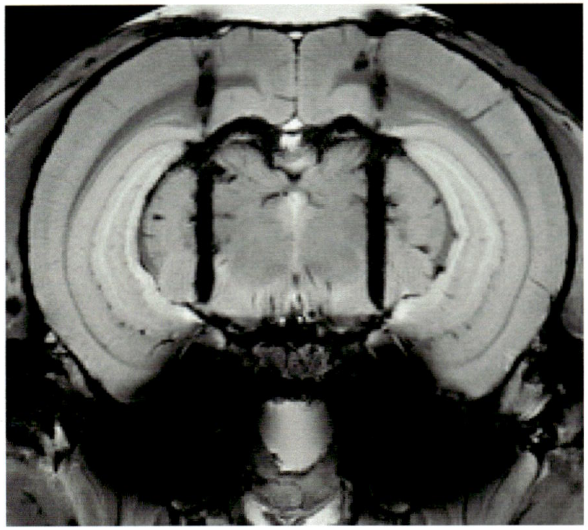

3V	3rd ventricle
alv	alveus of the hippocampus
APTD	anterior pretectal nucleus, dorsal part
APTV	anterior pretectal nucleus, ventral part
B	brachium of the superior colliculus
CA1	field CA1 of the hippocampus
CA2	field CA2 of the hippocampus
CA3	field CA3 of the hippocampus
cp	cerebral peduncle
DG	dentate gyrus of the hippocampus
DLG	dorsal lateral geniculate nucleus
DS	dorsal subiculum
Ect	ectorhinal cortex
Ent	entorhinal cortex
fr	fasciculus retroflexus
GrDG	granular layer of the dentate gyrus
IF	interfascicular nucleus
IT	inferior temporal cortex
LMoL	lacunosum moleculare layer of the hippocampus
LV	lateral ventricle
MG	medial geniculate nucleus
MHb	medial habenular nucleus
MoDG	molecular layer of the dentate gyrus
MPT	medial pretectal nucleus
MT	medial terminal nucleus of the accessory optic tract
OPT	olivary pretectal nucleus
opt	optic tract
Or	oriens layer of the hippocampus
PAG	periaqueductal gray
Pc	central nucleus of the pulvinar
Pd	dorsal nucleus of the pulvinar
Po	posterior complex of the thalamus
PoDG	polymorph layer of the dentate gyrus
PrC	precommissural nucleus
PrEW	pre-Edinger-Westphal nucleus
PRh	perirhinal cortex
pRt	p1 reticular formation
Py	pyramidal cell layer of the hippocampus
Rad	radiatum layer of the hippocampus
RLi	rostral linear nucleus of the raphe
RSg	retrosplenial granular cortex
SC	superior colliculus
scp	superior cerebellar peduncle
sm	stria medullaris of the thalamus
SNC	substantia nigra, compact part
SNR	substantia nigra, reticular part
TC	temporal cortex
V1	primary visual cortex
V2	secondary visual cortex
VLG	ventral lateral geniculate nucleus
VPM	ventral posteromedial thalamic nucleus
VS	ventral subiculum
VTAR	ventral tegmental area, rostral part

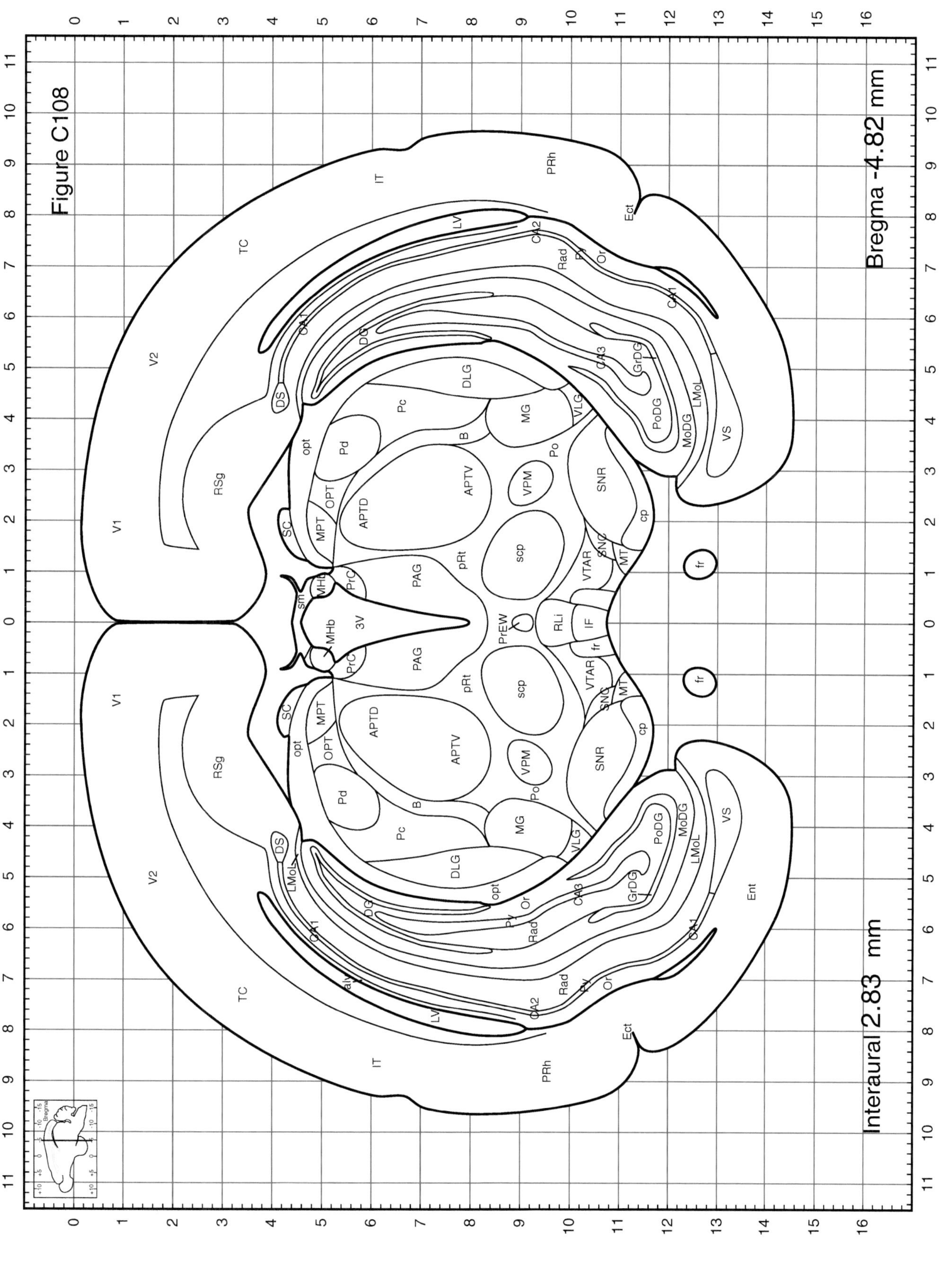

Figure C108
Bregma -4.82 mm
Interaural 2.83 mm
V1
V2
TC
IT
PRh
Ect
RSg
DS
SC
opt
Pd
Pc
OPT
MPT
APTD
APTV
B
DLG
MG
VLG
Po
VPM
SNR
cp
SNC
MT
VTAR
scp
pRt
PAG
PrC
MHb
sm
3V
PrEW
RLi
IF
fr
CA1
CA2
CA3
DG
LV
Rad
Py
Or
GrDG
PoDG
MoDG
LMol
VS
Ent
alv

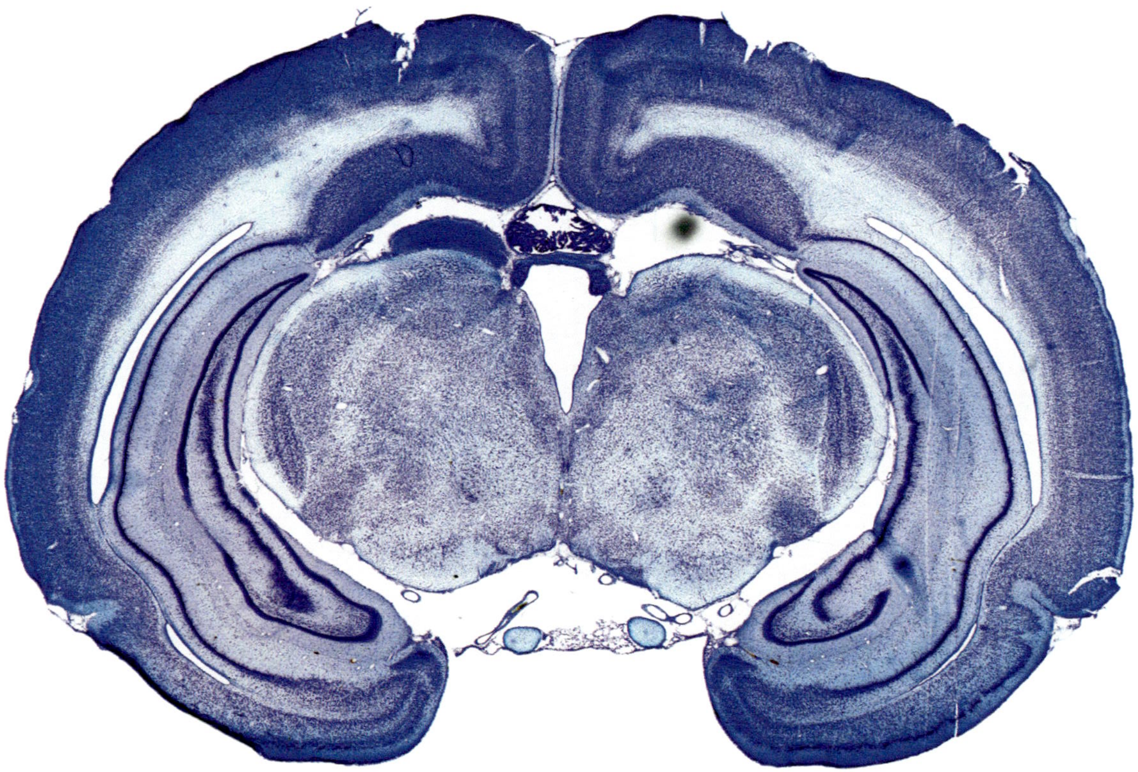

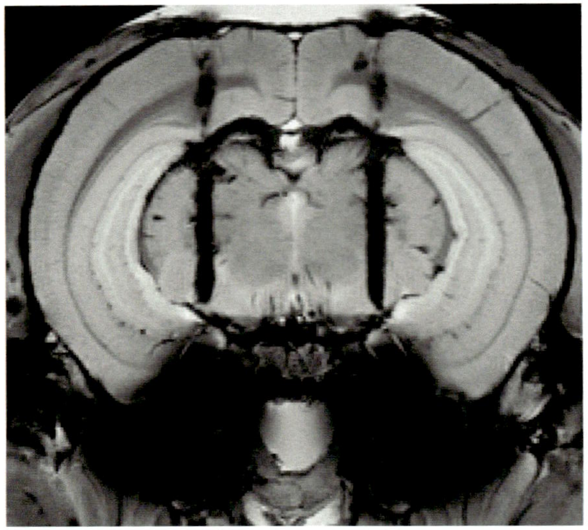

3V 3rd ventricle
alv alveus of the hippocampus
APTD anterior pretectal nucleus, dorsal part
APTV anterior pretectal nucleus, ventral part
B brachium of the superior colliculus
CA1 field CA1 of the hippocampus
CA2 field CA2 of the hippocampus
CA3 field CA3 of the hippocampus
cp cerebral peduncle
DG dentate gyrus of the hippocampus
DLG dorsal lateral geniculate nucleus
DS dorsal subiculum
Ect ectorhinal cortex
Ent entorhinal cortex
fr fasciculus retroflexus
GrDG granular layer of the dentate gyrus
IF interfascicular nucleus
InC interstitial nucleus of Cajal
IT inferior temporal cortex
LMoL lacunosum moleculare layer of the hippocampus
LV lateral ventricle
MG medial geniculate nucleus
MHb medial habenular nucleus
MoDG molecular layer of the dentate gyrus
MPT medial pretectal nucleus
MT medial terminal nucleus of the accessory optic tract
Op optic nerve layer of the superior colliculus
OPT olivary pretectal nucleus
opt optic tract
Or oriens layer of the hippocampus
Pc central nucleus of the pulvinar
Pd dorsal nucleus of the pulvinar
Po posterior complex of the thalamus
PoDG polymorph layer of the dentate gyrus
pPAG p1 periaqueductal gray
PrC precommissural nucleus
PrEW pre-Edinger-Westphal nucleus
PRh perirhinal cortex
pRt p1 reticular formation
Py pyramidal cell layer of the hippocampus
Rad radiatum layer of the hippocampus
RLi rostral linear nucleus of the raphe
RMC red nucleus, magnocellular part
RSg retrosplenial granular cortex
scp superior cerebellar peduncle
sm stria medullaris of the thalamus
SNC substantia nigra, compact part
SNR substantia nigra, reticular part
SUG superficial gray layer of the superior colliculus
TC temporal cortex
V1 primary visual cortex
V2 secondary visual cortex
VPM ventral posteromedial thalamic nucleus
VS ventral subiculum
VTAR ventral tegmental area, rostral part
Zo zonal layer of the superior colliculus

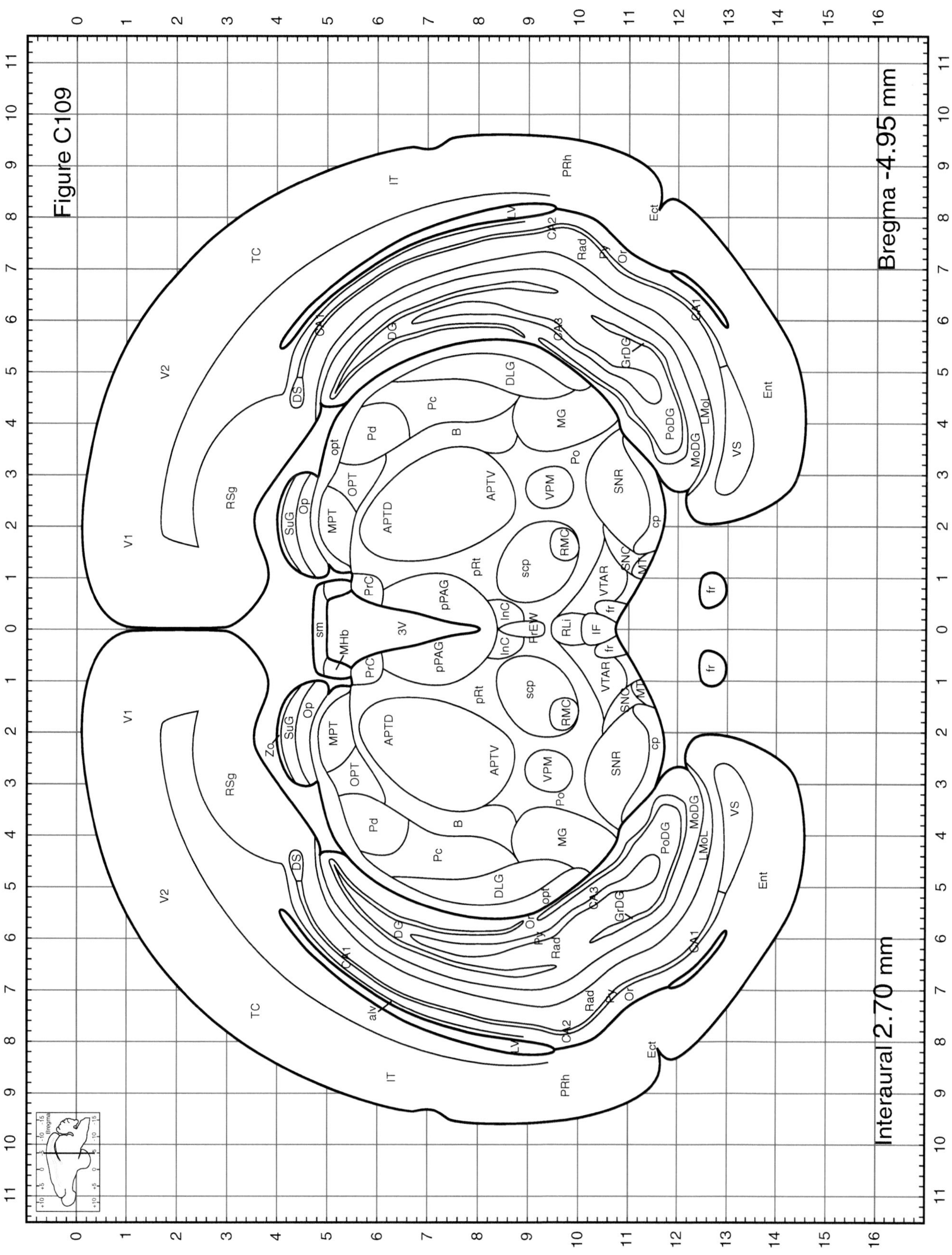
Figure C109
Bregma -4.95 mm
Interaural 2.70 mm
V1
V2
RSg
TC
IT
PRh
Ect
Ent
VS
LMoL
MoDG
PoDG
GrDG
CA1
CA2
CA3
DG
Or
Py
Rad
LV
alv
DS
SuG
Op
Zo
MPT
OPT
opt
Pd
Pc
B
DLG
MG
Po
APTD
APTV
VPM
SNR
cp
SNC
MT
RMC
scp
pRt
pPAG
PrC
sm
MHb
3V
InC
PrEW
RLi
IF
fr
VTAR

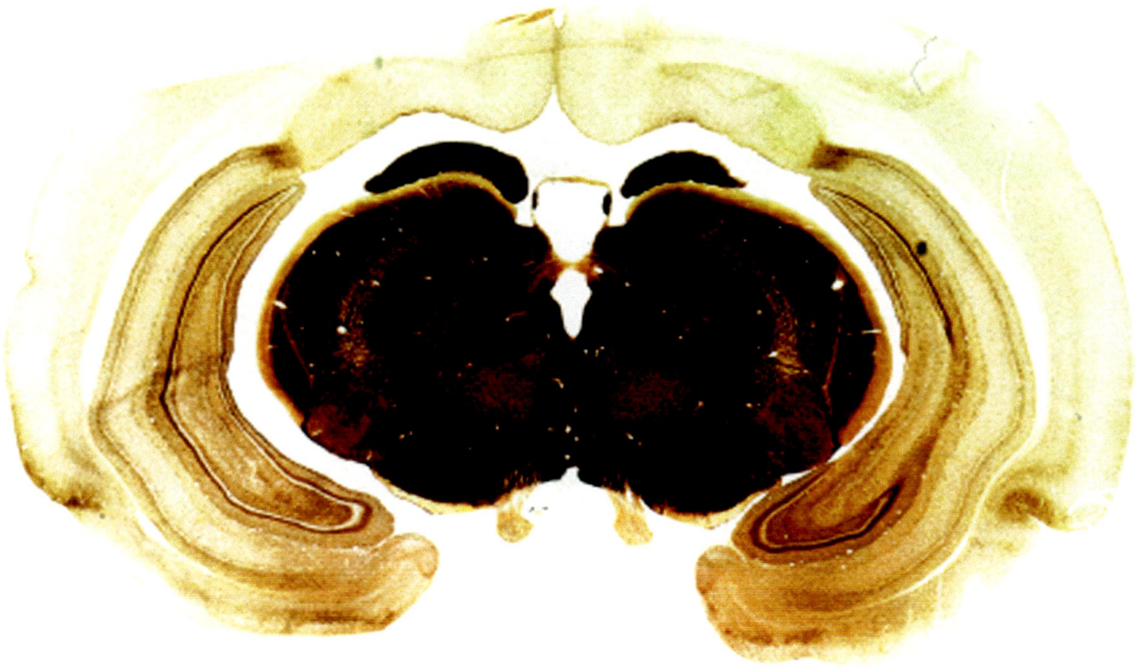

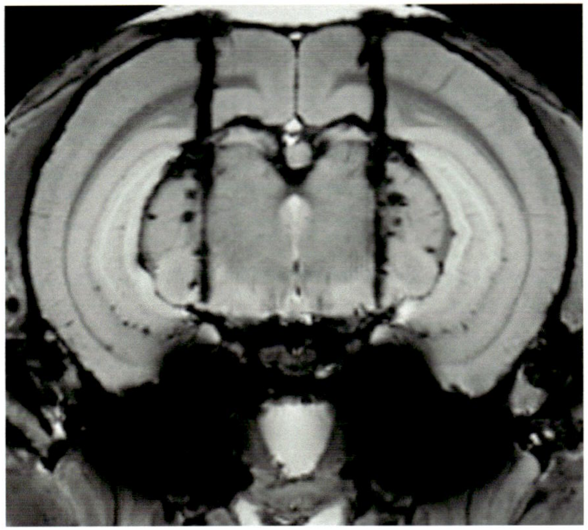

alv alveus of the hippocampus
APTD anterior pretectal nucleus, dorsal part
APTV anterior pretectal nucleus, ventral part
Aq aqueduct
B brachium of the superior colliculus
CA1 field CA1 of the hippocampus
CA2 field CA2 of the hippocampus
CA3 field CA3 of the hippocampus
cp cerebral peduncle
D3V dorsal 3rd ventricle
DG dentate gyrus of the hippocampus
DLG dorsal lateral geniculate nucleus
DpG deep gray layer of the superior colliculus
DS dorsal subiculum
dtgx dorsal tegmental decussation
Ent entorhinal cortex
fr fasciculus retroflexus
GrDG granular layer of the dentate gyrus
IF interfascicular nucleus
InC interstitial nucleus of Cajal
InG intermediate gray layer of the superior colliculus
InWh intermediate white layer of the superior colliculus
IP interpeduncular nucleus
IT inferior temporal cortex
LMoL lacunosum moleculare layer of the hippocampus
LV lateral ventricle
MG medial geniculate nucleus
MoDG molecular layer of the dentate gyrus
MPT medial pretectal nucleus
Op optic nerve layer of the superior colliculus
OPT olivary pretectal nucleus
opt optic tract
Or oriens layer of the hippocampus
PAG periaqueductal gray
PBP parabrachial pigmented nucleus of the VTA
pc posterior commissure
Pc central nucleus of the pulvinar
Pd dorsal nucleus of the pulvinar
Po posterior complex of the thalamus
PoDG polymorph layer of the dentate gyrus
PrEW pre-Edinger-Westphal nucleus
PRh perirhinal cortex
pRt p1 reticular formation
Py pyramidal cell layer of the hippocampus
Rad radiatum layer of the hippocampus
RLi rostral linear nucleus of the raphe
RMC red nucleus, magnocellular part
RSg retrosplenial granular cortex
scp superior cerebellar peduncle
sm stria medullaris of the thalamus
SNC substantia nigra, compact part
SNR substantia nigra, reticular part
SuG superficial gray layer of the superior colliculus
TC temporal cortex
V1 primary visual cortex
V2 secondary visual cortex
VPM ventral posteromedial thalamic nucleus
VS ventral subiculum
VTAR ventral tegmental area, rostral part
Zo zonal layer of the superior colliculus

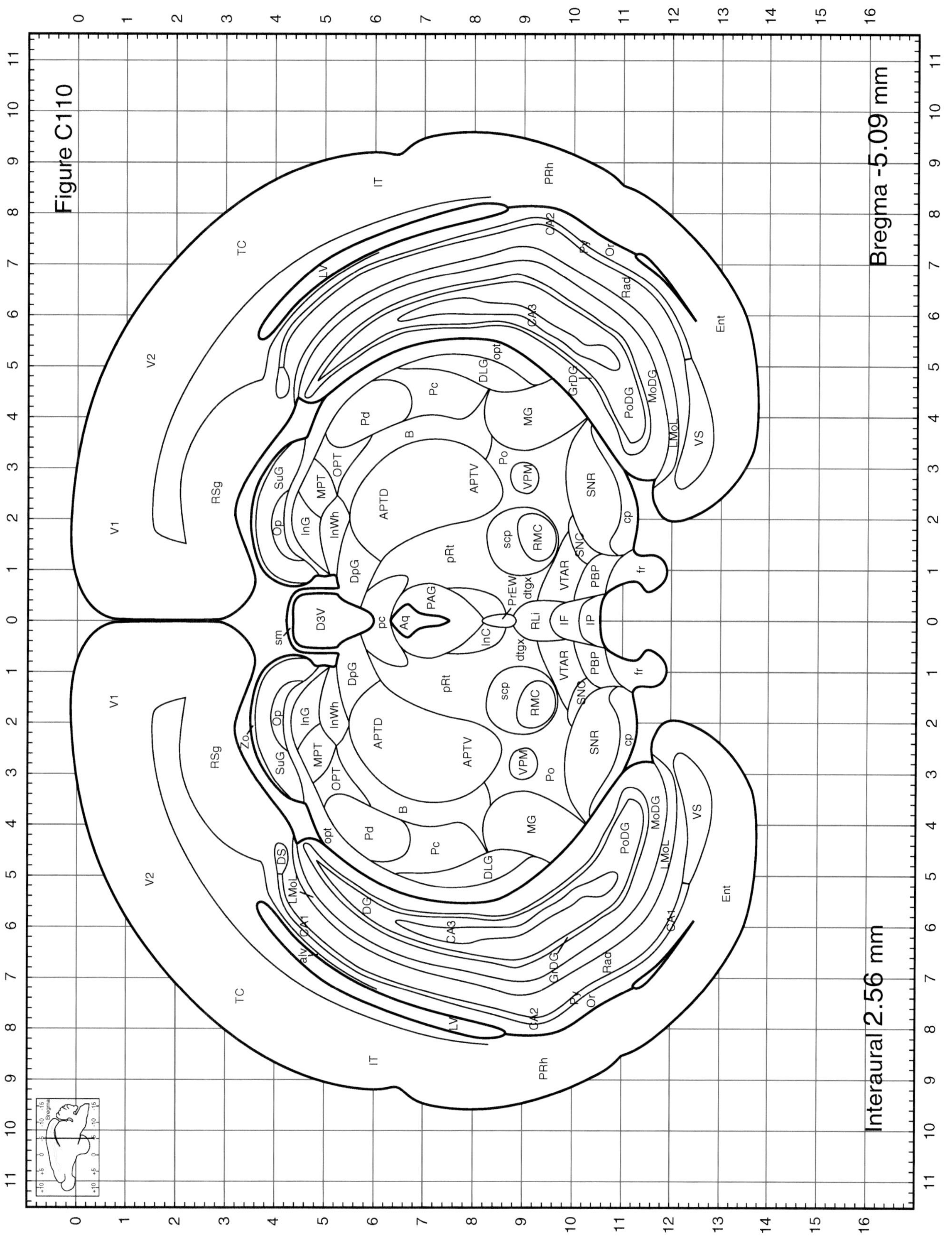
Figure C110
Bregma -5.09 mm
Interaural 2.56 mm
V1
V2
RSg
TC
IT
PRh
Ent
VS
LMoL
MoDG
PoDG
GrDG
DG
CA1
CA2
CA3
Py
Or
Rad
LV
alv
DS
Zo
SuG
Op
InG
InWh
MPT
OPT
opt
DpG
APTD
APTV
Pd
Pc
B
DLG
MG
Po
VPM
pRt
scp
RMC
SNR
SNC
cp
VTAR
PBP
fr
IF
IP
RLi
dtgx
PrEW
InC
PAG
Aq
pc
D3V
sm

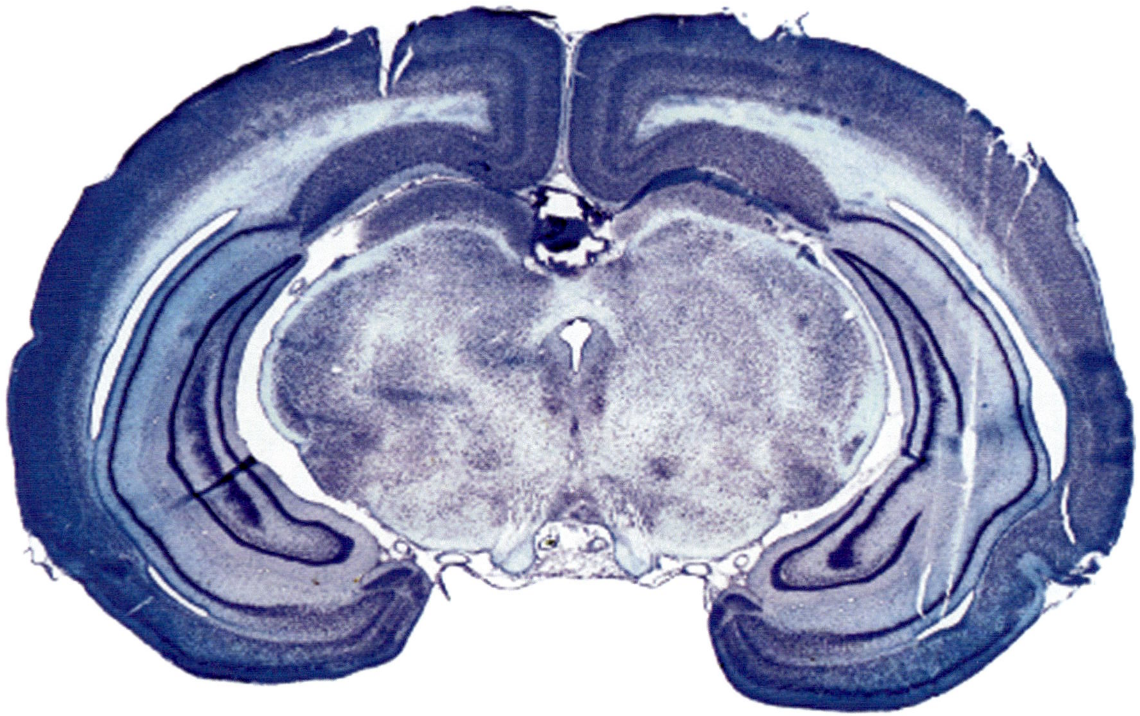

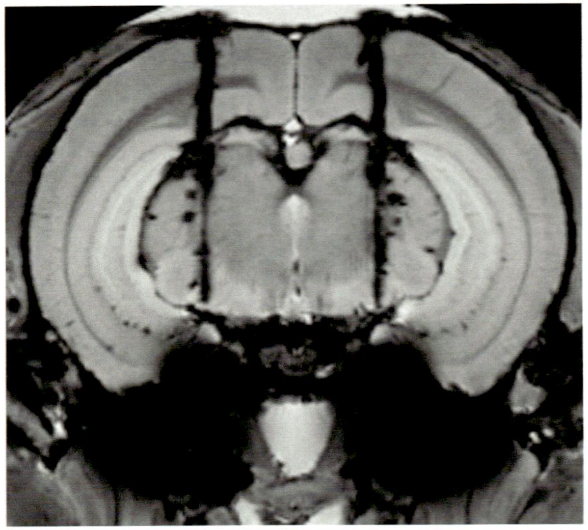

alv	alveus of the hippocampus
APTD	anterior pretectal nucleus, dorsal part
APTV	anterior pretectal nucleus, ventral part
Aq	aqueduct
B	brachium of the superior colliculus
CA1	field CA1 of the hippocampus
CA2	field CA2 of the hippocampus
CA3	field CA3 of the hippocampus
cp	cerebral peduncle
D3V	dorsal 3rd ventricle
DG	dentate gyrus of the hippocampus
DLG	dorsal lateral geniculate nucleus
DpG	deep gray layer of the superior colliculus
DS	dorsal subiculum
dtgx	dorsal tegmental decussation
Ent	entorhinal cortex
fr	fasciculus retroflexus
GrDG	granular layer of the dentate gyrus
IF	interfascicular nucleus
InC	interstitial nucleus of Cajal
InG	intermediate gray layer of the superior colliculus
InWh	intermediate white layer of the superior colliculus
IP	interpeduncular nucleus
IT	inferior temporal cortex
LMoL	lacunosum moleculare layer of the hippocampus
LV	lateral ventricle
MG	medial geniculate nucleus
MoDG	molecular layer of the dentate gyrus
MPT	medial pretectal nucleus
Op	optic nerve layer of the superior colliculus
OPT	olivary pretectal nucleus
opt	optic tract
Or	oriens layer of the hippocampus
PAG	periaqueductal gray
PBP	parabrachial pigmented nucleus of the VTA
pc	posterior commissure
Pc	central nucleus of the pulvinar
Pd	dorsal nucleus of the pulvinar
PG	pineal gland
Po	posterior complex of the thalamus
PoDG	polymorph layer of the dentate gyrus
Post	postsubiculum
PrEW	pre-Edinger-Westphal nucleus
PRh	perirhinal cortex
pRt	p1 reticular formation
Py	pyramidal cell layer of the hippocampus
Rad	radiatum layer of the hippocampus
RLi	rostral linear nucleus of the raphe
RMC	red nucleus, magnocellular part
RSg	retrosplenial granular cortex
scp	superior cerebellar peduncle
SNC	substantia nigra, compact part
SNR	substantia nigra, reticular part
SuG	superficial gray layer of the superior colliculus
TC	temporal cortex
V1	primary visual cortex
V2	secondary visual cortex
VPM	ventral posteromedial thalamic nucleus
VS	ventral subiculum
VTAR	ventral tegmental area, rostral part
Zo	zonal layer of the superior colliculus

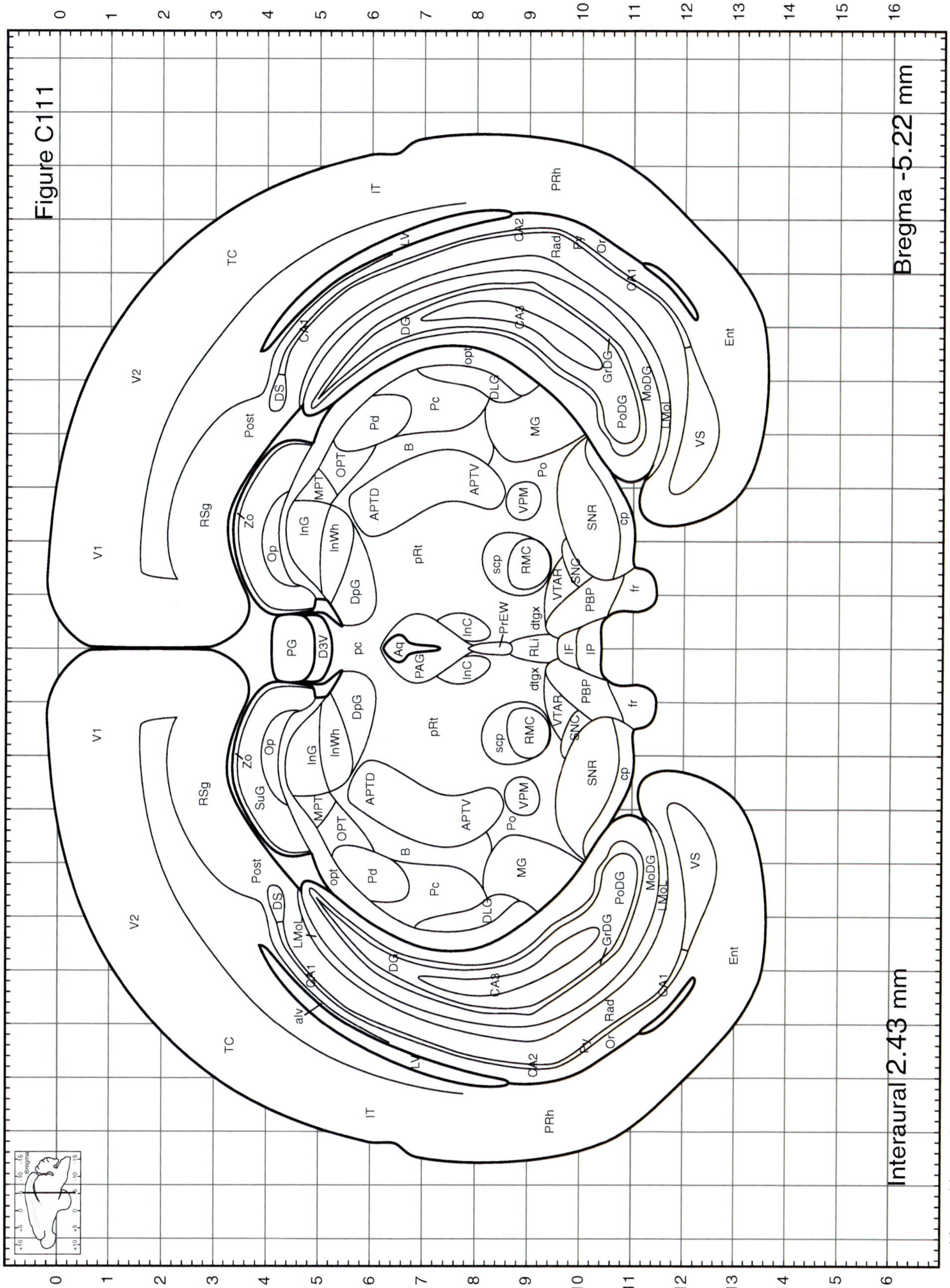
Figure C111
Bregma -5.22 mm
Interaural 2.43 mm
V1
V2
TC
IT
PRh
Ent
RSg
Post
DS
CA1
CA2
CA3
DG
LV
Rad
Py
Or
GrDG
MoDG
PoDG
LMol
VS
alv
opt
DLG
Pc
Pd
MG
B
OPT
MPT
APTD
APTV
Po
VPM
SNR
cp
Zo
SuG
Op
InG
InWh
DpG
pRt
scp
RMC
SNC
VTAR
PBP
fr
PrEW
dtgx
InC
Aq
PAG
PG
D3V
pc
RLi
IF
IP

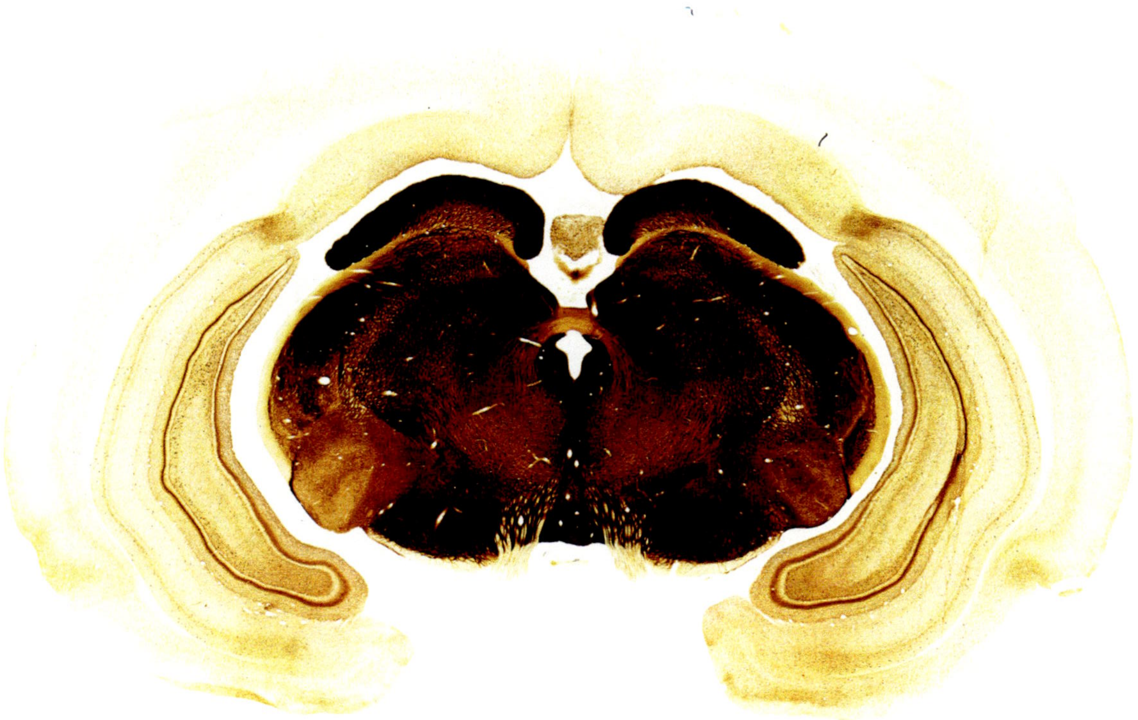

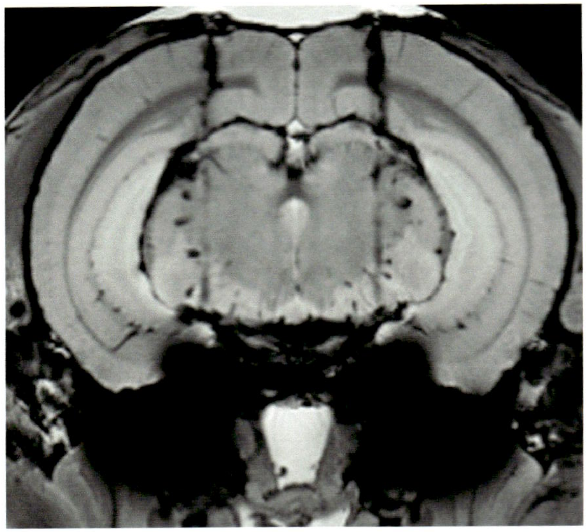

alv	alveus of the hippocampus
APTD	anterior pretectal nucleus, dorsal part
APTV	anterior pretectal nucleus, ventral part
Aq	aqueduct
CA1	field CA1 of the hippocampus
CA2	field CA2 of the hippocampus
CA3	field CA3 of the hippocampus
cp	cerebral peduncle
csc	commissure of the superior colliculus
DG	dentate gyrus of the hippocampus
DLG	dorsal lateral geniculate nucleus
DpG	deep gray layer of the superior colliculus
DpWh	deep white layer of the superior colliculus
DS	dorsal subiculum
dtgx	dorsal tegmental decussation
Ent	entorhinal cortex
fr	fasciculus retroflexus
GrDG	granular layer of the dentate gyrus
IF	interfascicular nucleus
InC	interstitial nucleus of Cajal
InG	intermediate gray layer of the superior colliculus
InWh	intermediate white layer of the superior colliculus
IPC	interpeduncular nucleus, caudal subnucleus
IPL	interpeduncular nucleus, lateral subnucleus
IPR	interpeduncular nucleus, rostral subnucleus
IT	inferior temporal cortex
LMoL	lacunosum moleculare layer of the hippocampus
LV	lateral ventricle
MGD	medial geniculate nucleus, dorsal part
MGV	medial geniculate nucleus, ventral part
MoDG	molecular layer of the dentate gyrus
MPT	medial pretectal nucleus
Op	optic nerve layer of the superior colliculus
OPT	olivary pretectal nucleus
opt	optic tract
Or	oriens layer of the hippocampus
PAG	periaqueductal gray
PBP	parabrachial pigmented nucleus of the VTA
pc	posterior commissure
Pc	central nucleus of the pulvinar
Pd	dorsal nucleus of the pulvinar
PG	pineal gland
Po	posterior complex of the thalamus
PoDG	polymorph layer of the dentate gyrus
Post	postsubiculum
PrEW	pre-Edinger-Westphal nucleus
PRh	perirhinal cortex
pRt	p1 reticular formation
Py	pyramidal cell layer of the hippocampus
Rad	radiatum layer of the hippocampus
RLi	rostral linear nucleus of the raphe
RMC	red nucleus, magnocellular part
RPC	red nucleus, parvicellular part
RSg	retrosplenial granular cortex
SNC	substantia nigra, compact part
SNR	substantia nigra, reticular part
SuG	superficial gray layer of the superior colliculus
TC	temporal cortex
V1	primary visual cortex
V2	secondary visual cortex
VS	ventral subiculum
VTAR	ventral tegmental area, rostral part
vtgx	ventral tegmental decussation
Zo	zonal layer of the superior colliculus

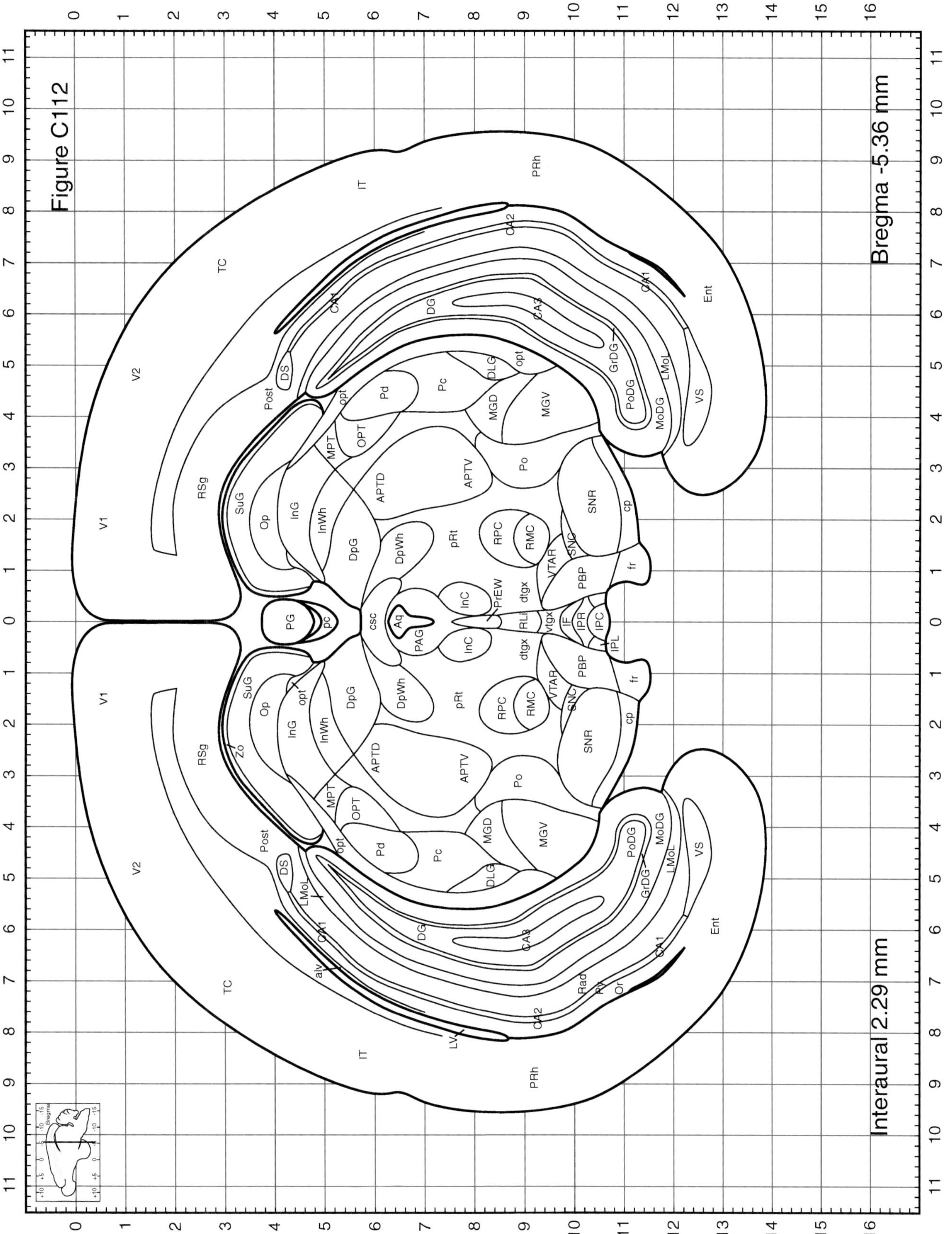
Figure C112
Bregma -5.36 mm
Interaural 2.29 mm
V1
V2
RSg
TC
IT
PRh
Ent
VS
LMol
MoDG
PoDG
GrDG
CA1
CA2
CA3
DG
Rad
Py
Or
alv
LV
DS
Post
SuG
Zo
Op
InG
InWh
DpG
DpWh
PG
pc
csc
Aq
PAG
InC
PrEW
dtgx
RLi
vtgx
IF
IPR
IPC
IPL
PBP
VTAR
SNC
SNR
RPC
RMC
pRt
APTD
APTV
MPT
OPT
opt
Pd
Pc
DLG
MGD
MGV
Po
cp
fr

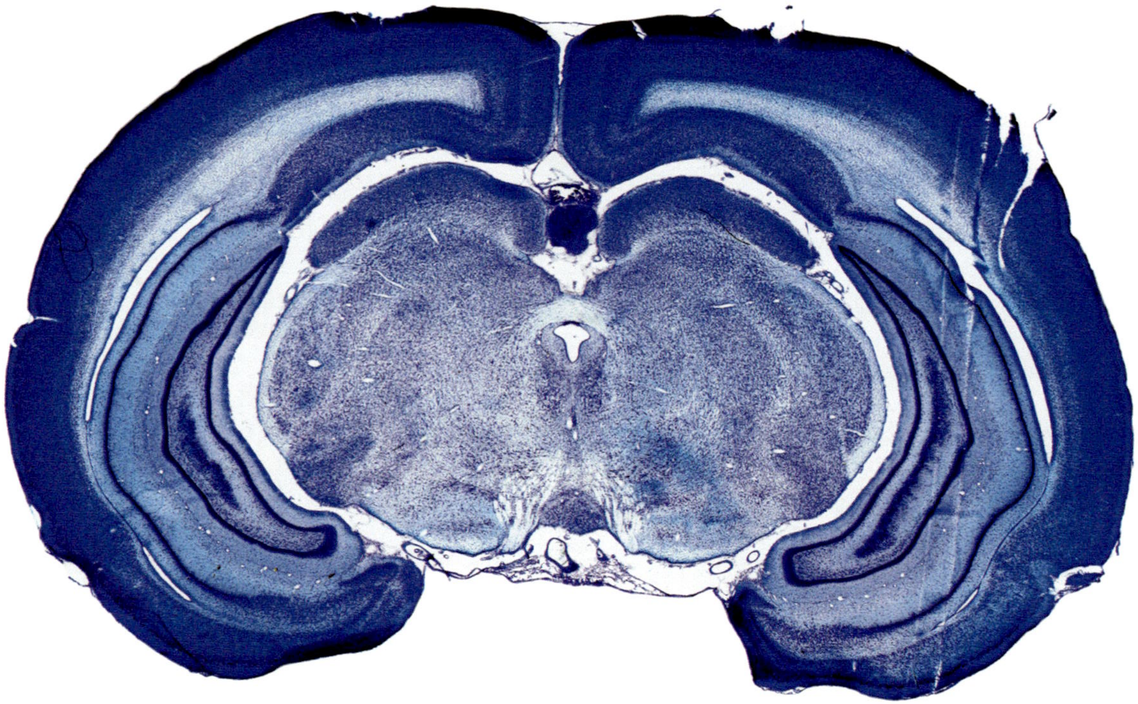

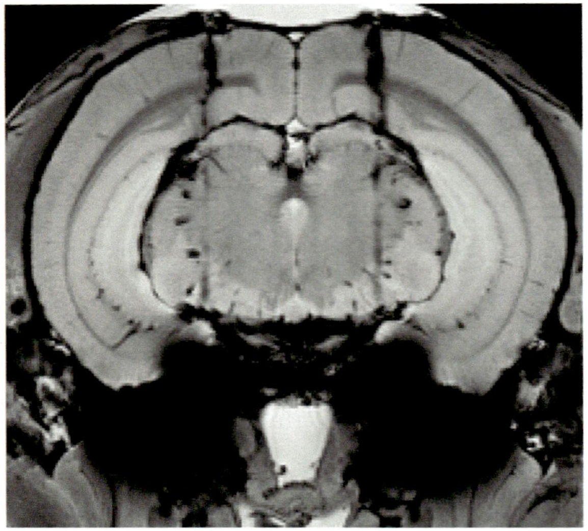

alv alveus of the hippocampus
APTD anterior pretectal nucleus, dorsal part
APTV anterior pretectal nucleus, ventral part
Aq aqueduct
CA1 field CA1 of the hippocampus
CA2 field CA2 of the hippocampus
CA3 field CA3 of the hippocampus
cp cerebral peduncle
csc commissure of the superior colliculus
DG dentate gyrus of the hippocampus
DLG dorsal lateral geniculate nucleus
DpG deep gray layer of the superior colliculus
DpWh deep white layer of the superior colliculus
DS dorsal subiculum
dtgx dorsal tegmental decussation
Ent entorhinal cortex
fr fasciculus retroflexus
GrDG granular layer of the dentate gyrus
IF interfascicular nucleus
InC interstitial nucleus of Cajal
InG intermediate gray layer of the superior colliculus
InWh intermediate white layer of the superior colliculus
IPC interpeduncular nucleus, caudal subnucleus
IPL interpeduncular nucleus, lateral subnucleus
IPR interpeduncular nucleus, rostral subnucleus
IT inferior temporal cortex
LMoL lacunosum moleculare layer of the hippocampus
LV lateral ventricle
MGD medial geniculate nucleus, dorsal part
MGV medial geniculate nucleus, ventral part
MoDG molecular layer of the dentate gyrus
Op optic nerve layer of the superior colliculus
OPT olivary pretectal nucleus
opt optic tract
Or oriens layer of the hippocampus
PAG periaqueductal gray
PBP parabrachial pigmented nucleus of the VTA
Pc central nucleus of the pulvinar
Pd dorsal nucleus of the pulvinar
PG pineal gland
Po posterior complex of the thalamus
PoDG polymorph layer of the dentate gyrus
Post postsubiculum
PrEW pre-Edinger-Westphal nucleus
PRh perirhinal cortex
PrS presubiculum
pRt p1 reticular formation
Py pyramidal cell layer of the hippocampus
Rad radiatum layer of the hippocampus
RLi rostral linear nucleus of the raphe
RMC red nucleus, magnocellular part
RPC red nucleus, parvicellular part
RSg retrosplenial granular cortex
SNC substantia nigra, compact part
SNR substantia nigra, reticular part
SuG superficial gray layer of the superior colliculus
TC temporal cortex
V1 primary visual cortex
V2 secondary visual cortex
VS ventral subiculum
VTAR ventral tegmental area, rostral part
vtgx ventral tegmental decussation
Zo zonal layer of the superior colliculus

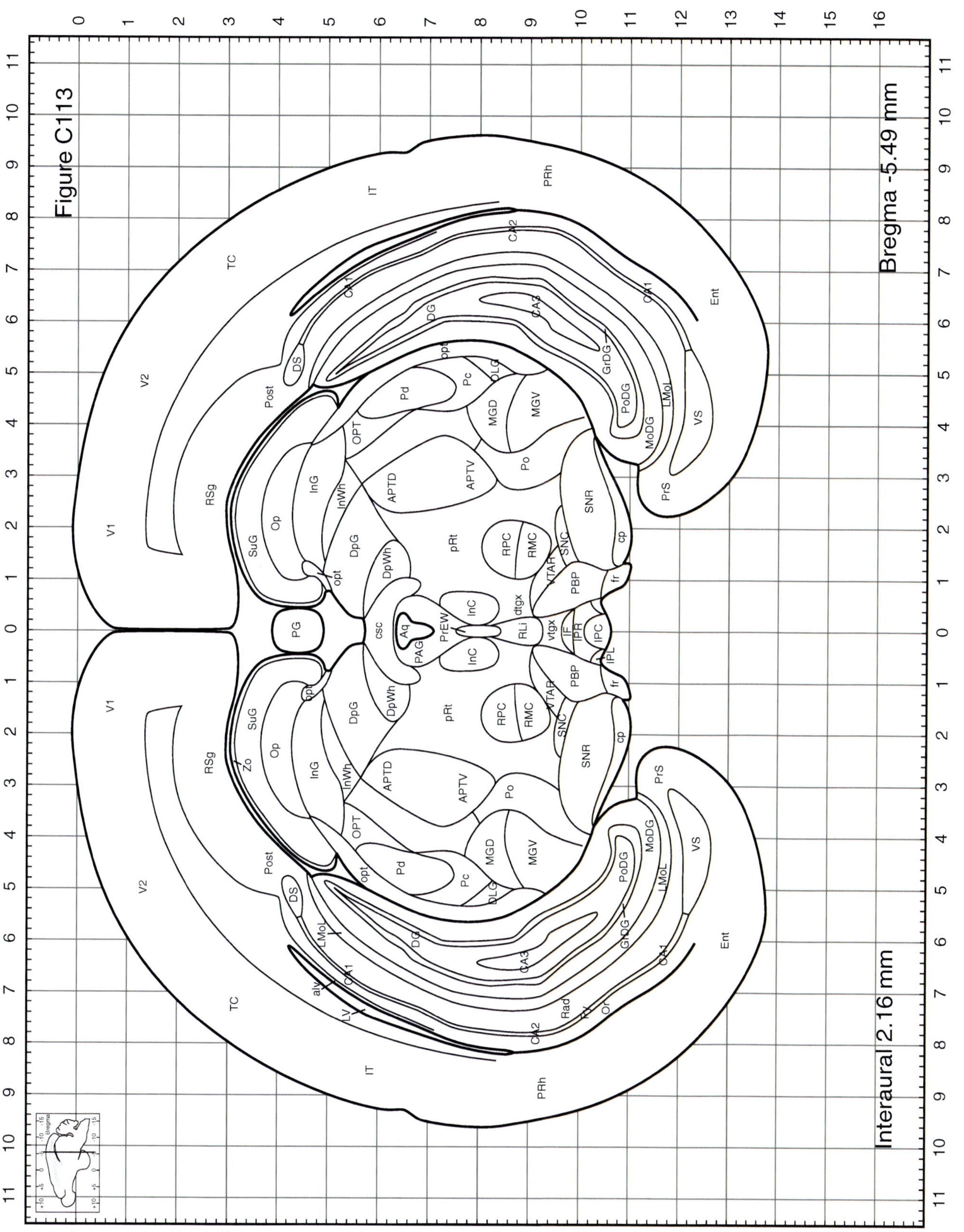
Figure C113
Bregma -5.49 mm
Interaural 2.16 mm
V1
V2
TC
IT
PRh
Ent
RSg
Post
DS
CA1
CA2
CA3
DG
GrDG
PoDG
MoDG
LMol
VS
PrS
SuG
Op
InG
InWh
DpG
DpWh
opt
OPT
APTD
APTV
Pd
Pc
DLG
MGD
MGV
Po
pRt
RPC
RMC
SNC
SNR
VTAR
PBP
cp
fr
PG
csc
Aq
PAG
PrEW
InC
dtgx
RLi
vtgx
IF
IPR
IPC
IPL
Zo
alv
LV
Rad
Py
Or

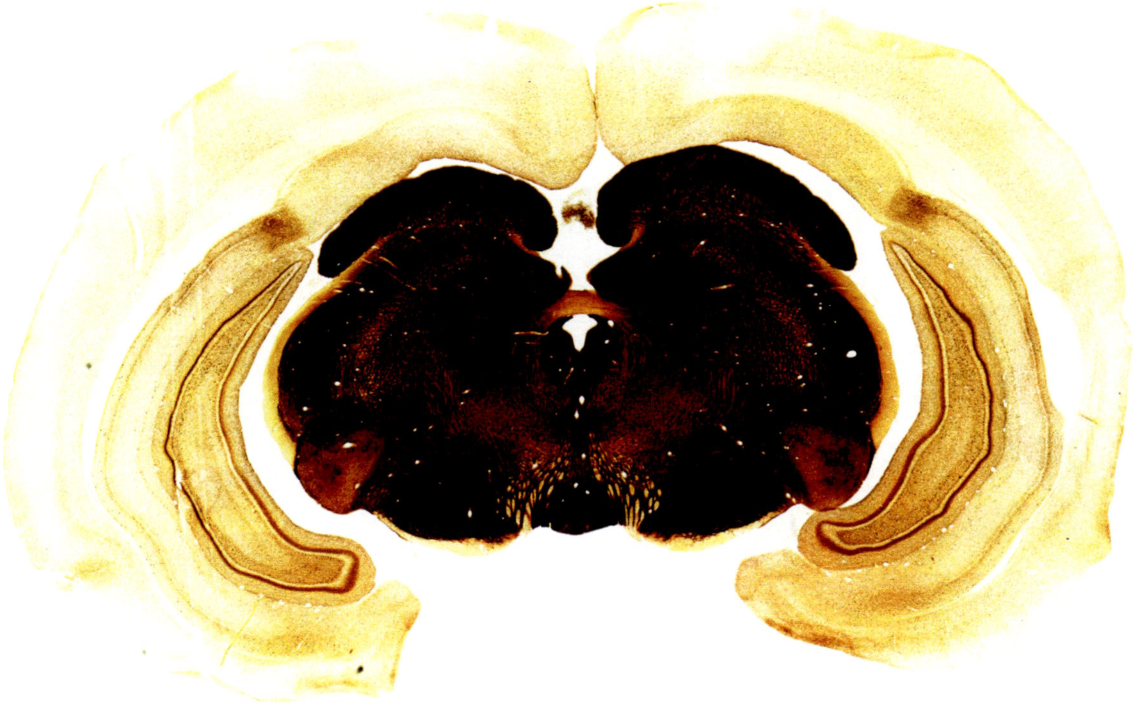

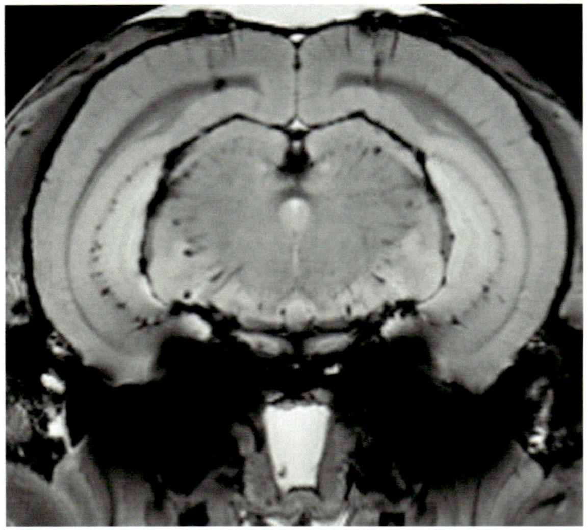

APTD	anterior pretectal nucleus, dorsal part	IPL	interpeduncular nucleus, lateral subnucleus	PrEW	pre-Edinger-Westphal nucleus
APTV	anterior pretectal nucleus, ventral part	IPR	interpeduncular nucleus, rostral subnucleus	PRh	perirhinal cortex
Aq	aqueduct	IT	inferior temporal cortex	PrS	presubiculum
CA1	field CA1 of the hippocampus	LMoL	lacunosum moleculare layer of the hippocampus	pRt	p1 reticular formation
CA2	field CA2 of the hippocampus	MGD	medial geniculate nucleus, dorsal part	Py	pyramidal cell layer of the hippocampus
CA3	field CA3 of the hippocampus	MGV	medial geniculate nucleus, ventral part	Rad	radiatum layer of the hippocampus
cp	cerebral peduncle	ml	medial lemniscus	RLi	rostral linear nucleus of the raphe
csc	commissure of the superior colliculus	MoDG	molecular layer of the dentate gyrus	RMC	red nucleus, magnocellular part
DG	dentate gyrus of the hippocampus	Op	optic nerve layer of the superior colliculus	RPC	red nucleus, parvicellular part
DpG	deep gray layer of the superior colliculus	OPT	olivary pretectal nucleus	RSg	retrosplenial granular cortex
DpWh	deep white layer of the superior colliculus	opt	optic tract	SNC	substantia nigra, compact part
DS	dorsal subiculum	Or	oriens layer of the hippocampus	SNR	substantia nigra, reticular part
dtgx	dorsal tegmental decussation	PAG	periaqueductal gray	SuG	superficial gray layer of the superior colliculus
Ent	entorhinal cortex	PBP	parabrachial pigmented nucleus of the VTA	TC	temporal cortex
fr	fasciculus retroflexus	Pc	central nucleus of the pulvinar	V1	primary visual cortex
GrDG	granular layer of the dentate gyrus	Pd	dorsal nucleus of the pulvinar	V2	secondary visual cortex
IF	interfascicular nucleus	PG	pineal gland	VS	ventral subiculum
InC	interstitial nucleus of Cajal	PN	paranigral nucleus of the VTA	vtgx	ventral tegmental decussation
InG	intermediate gray layer of the superior colliculus	Po	posterior complex of the thalamus	Zo	zonal layer of the superior colliculus
InWh	intermediate white layer of the superior colliculus	PoDG	polymorph layer of the dentate gyrus		
IPC	interpeduncular nucleus, caudal subnucleus	Post	postsubiculum		

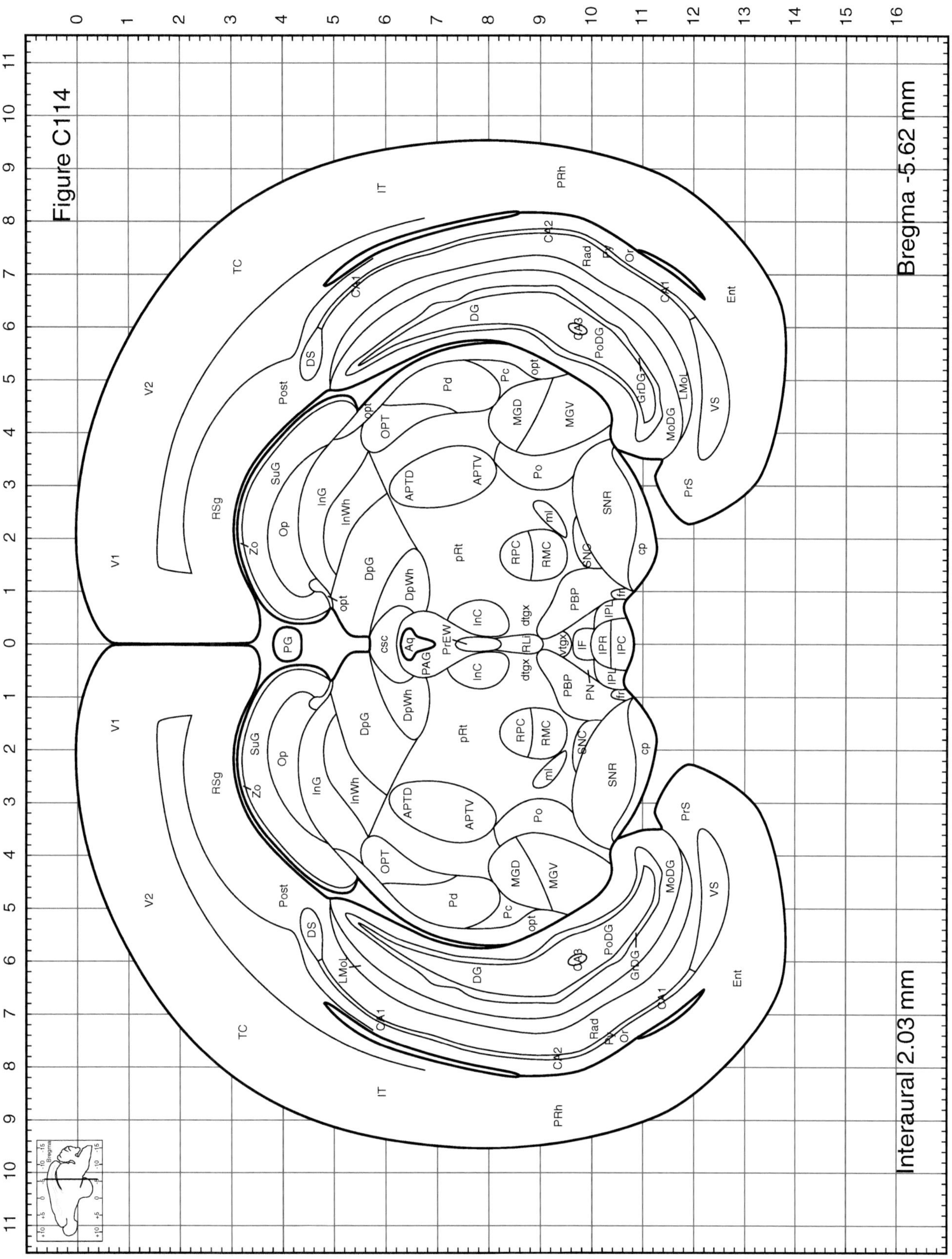
Figure C114
Bregma -5.62 mm
Interaural 2.03 mm
V1
V2
TC
IT
PRh
Ent
VS
PrS
RSg
Post
DS
SuG
Op
InG
InWh
Zo
opt
OPT
DpG
DpWh
pRt
APTD
APTV
Pd
Pc
MGD
MGV
Po
SNR
SNC
cp
ml
RPC
RMC
PBP
PN
IPL
IPR
IPC
IF
fr
vtgx
dtgx
RLi
InC
PrEW
PAG
Aq
csc
PG
CA1
CA2
CA3
Rad
Py
Or
DG
PoDG
GrDG
MoDG
LMol

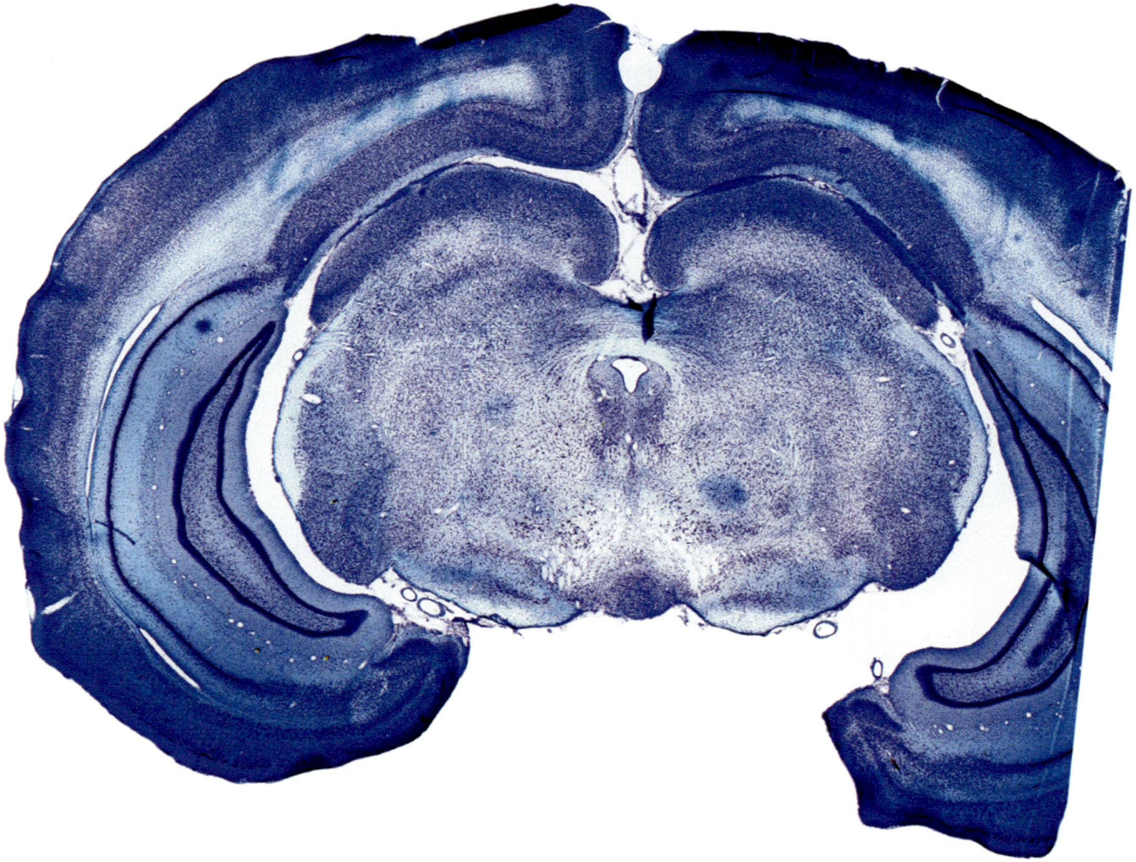

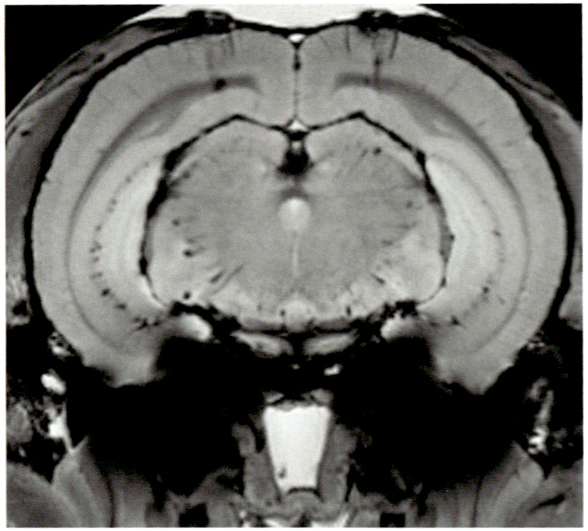

APT	anterior pretectal nucleus
Aq	aqueduct
CA1	field CA1 of the hippocampus
cp	cerebral peduncle
csc	commissure of the superior colliculus
DG	dentate gyrus of the hippocampus
DpG	deep gray layer of the superior colliculus
DpWh	deep white layer of the superior colliculus
DS	dorsal subiculum
dtgx	dorsal tegmental decussation
Ent	entorhinal cortex
GrDG	granular layer of the dentate gyrus
IF	interfascicular nucleus
InC	interstitial nucleus of Cajal
InG	intermediate gray layer of the superior colliculus
InWh	intermediate white layer of the superior colliculus
IPC	interpeduncular nucleus, caudal subnucleus
IPL	interpeduncular nucleus, lateral subnucleus
IPR	interpeduncular nucleus, rostral subnucleus
LMoL	lacunosum moleculare layer of the hippocampus
MGD	medial geniculate nucleus, dorsal part
MGV	medial geniculate nucleus, ventral part
ml	medial lemniscus
MoDG	molecular layer of the dentate gyrus
mRt	mesencephalic reticular formation
Op	optic nerve layer of the superior colliculus
OPT	olivary pretectal nucleus
opt	optic tract
Or	oriens layer of the hippocampus
PAG	periaqueductal gray
PBP	parabrachial pigmented nucleus of the VTA
Pd	dorsal nucleus of the pulvinar
PN	paranigral nucleus of the VTA
Po	posterior complex of the thalamus
PoDG	polymorph layer of the dentate gyrus
Post	postsubiculum
PrEW	pre-Edinger-Westphal nucleus
PRh	perirhinal cortex
PrS	presubiculum
Py	pyramidal cell layer of the hippocampus
Rad	radiatum layer of the hippocampus
RLi	rostral linear nucleus of the raphe
RMC	red nucleus, magnocellular part
RPC	red nucleus, parvicellular part
RSg	retrosplenial granular cortex
SNC	substantia nigra, compact part
SNR	substantia nigra, reticular part
SuG	superficial gray layer of the superior colliculus
TC	temporal cortex
V1	primary visual cortex
V2	secondary visual cortex
VS	ventral subiculum
vtgx	ventral tegmental decussation
Zo	zonal layer of the superior colliculus

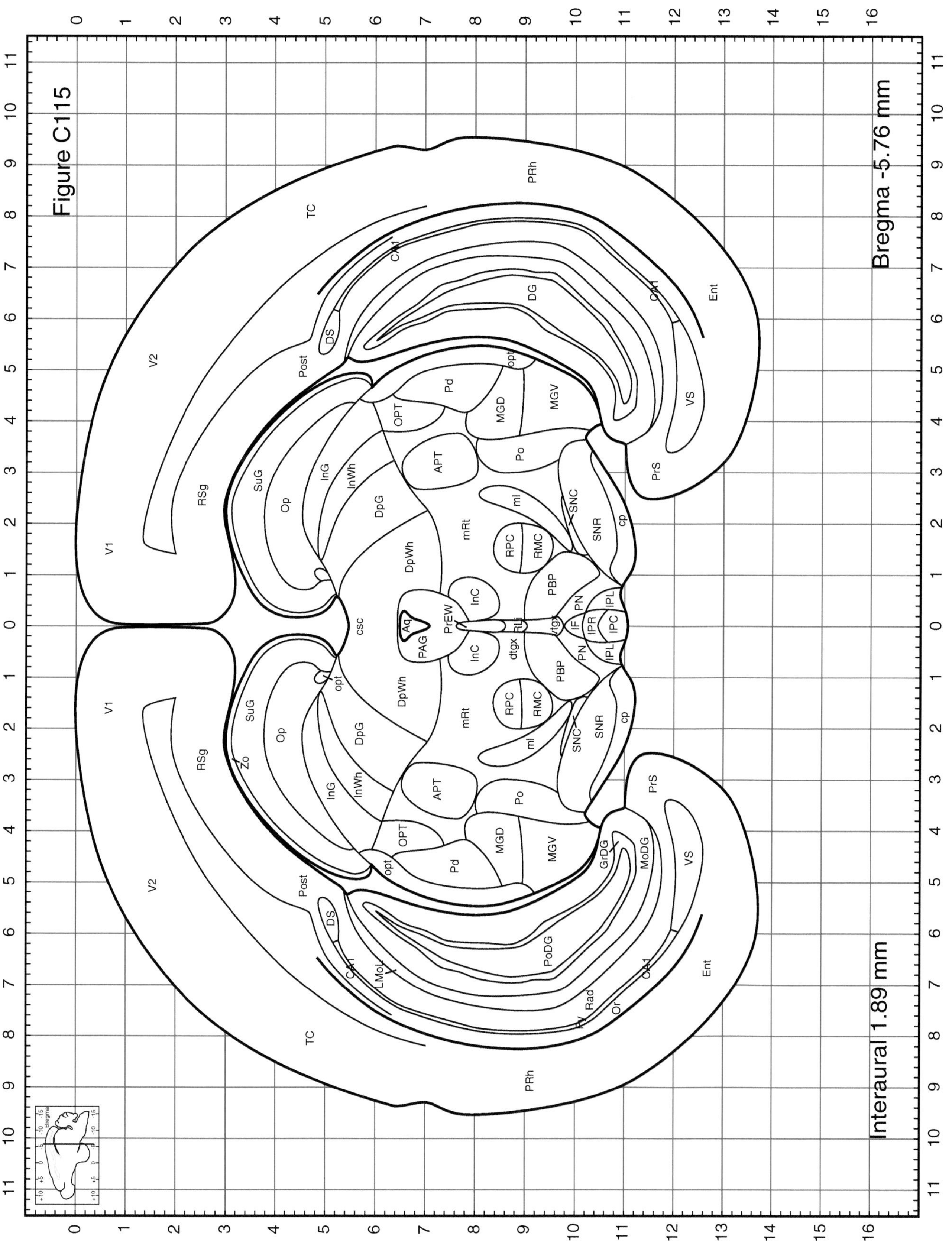
Figure C115
Bregma -5.76 mm
Interaural 1.89 mm
TC
PRh
CA1
DG
Ent
DS
Post
V2
opt
Pd
MGD
MGV
VS
OPT
APT
Po
PrS
SuG
InG
InWh
RSg
Op
DpG
mRt
ml
SNC
SNR
cp
V1
DpWh
RPC
RMC
PBP
InC
IPL
PN
IPR
IPC
IF
csc
Aq
PrEW
PAG
RLi
dtgx
vtgx
Zo
GrDG
MoDG
PoDG
LMol
Rad
Or
Py
Bregma

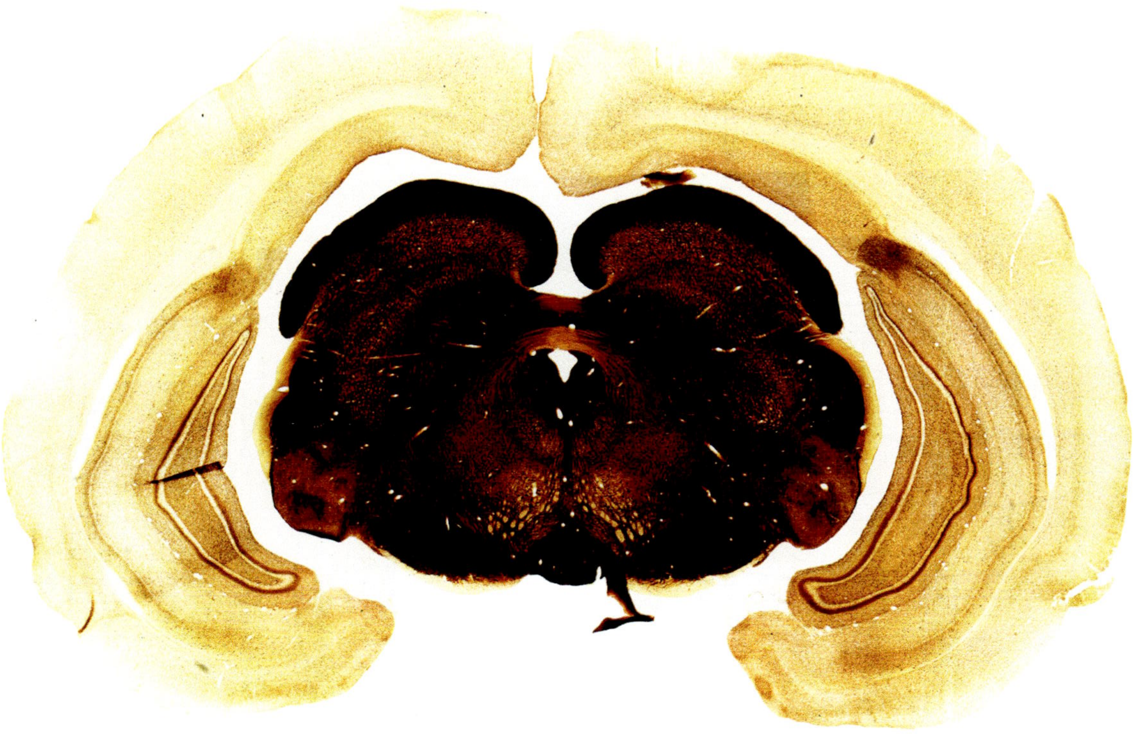

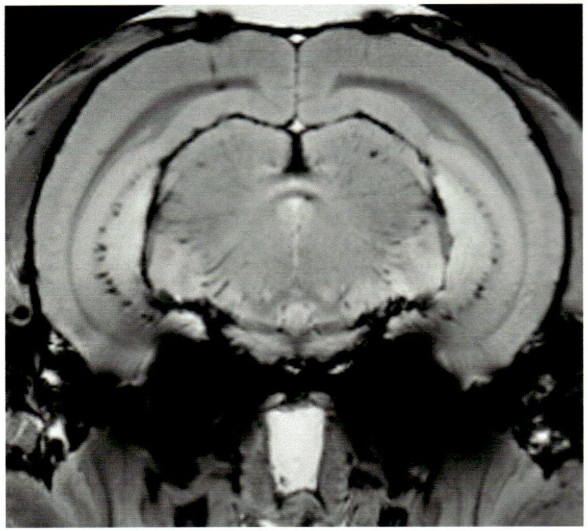

APT	anterior pretectal nucleus
Aq	aqueduct
CA1	field CA1 of the hippocampus
cp	cerebral peduncle
csc	commissure of the superior colliculus
DG	dentate gyrus of the hippocampus
DpG	deep gray layer of the superior colliculus
DpWh	deep white layer of the superior colliculus
DS	dorsal subiculum
dtgx	dorsal tegmental decussation
Ent	entorhinal cortex
GrDG	granular layer of the dentate gyrus
IF	interfascicular nucleus
InC	interstitial nucleus of Cajal
InG	intermediate gray layer of the superior colliculus
InWh	intermediate white layer of the superior colliculus
IPC	interpeduncular nucleus, caudal subnucleus
IPL	interpeduncular nucleus, lateral subnucleus
IPR	interpeduncular nucleus, rostral subnucleus
LMoL	lacunosum moleculare layer of the hippocampus
MGD	medial geniculate nucleus, dorsal part
MGV	medial geniculate nucleus, ventral part
ml	medial lemniscus
mlf	medial longitudinal fasciculus
MoDG	molecular layer of the dentate gyrus
mRt	mesencephalic reticular formation
Op	optic nerve layer of the superior colliculus
OPT	olivary pretectal nucleus
opt	optic tract
Or	oriens layer of the hippocampus
PAG	periaqueductal gray
PBP	parabrachial pigmented nucleus of the VTA
Pd	dorsal nucleus of the pulvinar
PN	paranigral nucleus of the VTA
Po	posterior complex of the thalamus
PoDG	polymorph layer of the dentate gyrus
Post	postsubiculum
PrEW	pre-Edinger-Westphal nucleus
PRh	perirhinal cortex
PrS	presubiculum
Py	pyramidal cell layer of the hippocampus
Rad	radiatum layer of the hippocampus
RLi	rostral linear nucleus of the raphe
RMC	red nucleus, magnocellular part
RPC	red nucleus, parvicellular part
RSg	retrosplenial granular cortex
SNC	substantia nigra, compact part
SNR	substantia nigra, reticular part
SuG	superficial gray layer of the superior colliculus
TC	temporal cortex
V1	primary visual cortex
V2	secondary visual cortex
VS	ventral subiculum
vtgx	ventral tegmental decussation
Zo	zonal layer of the superior colliculus

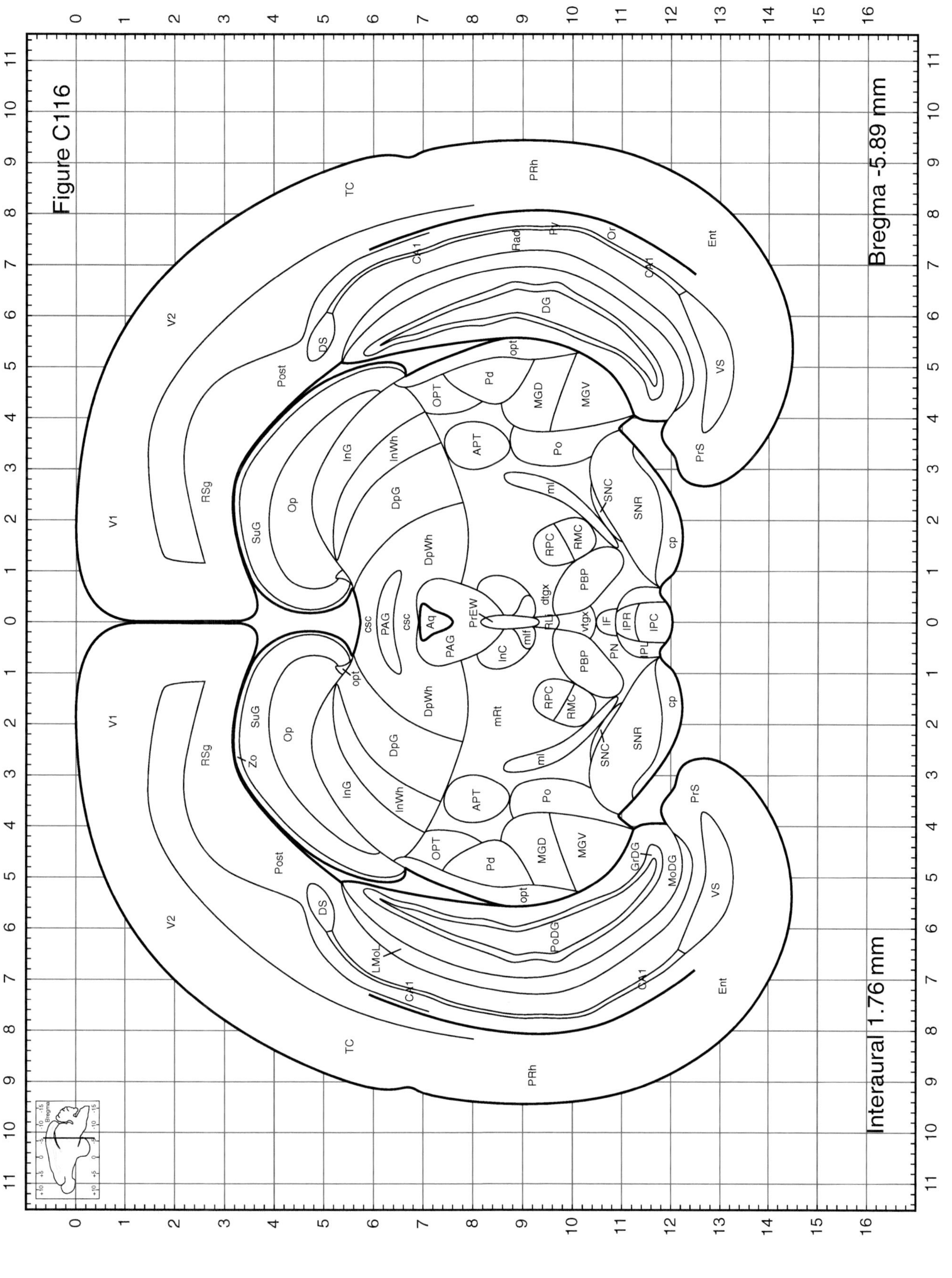
Figure C116
Bregma -5.89 mm
Interaural 1.76 mm
V1
V2
RSg
Post
TC
PRh
Ent
VS
PrS
CA1
Rad
Py
Or
DG
DS
opt
Pd
MGD
MGV
OPT
APT
Po
SNC
SNR
cp
ml
InG
InWh
DpG
DpWh
Op
SuG
Zo
csc
PAG
Aq
PrEW
InC
mlf
RLi
dtgx
vtgx
PBP
IF
IPR
IPC
IPL
PN
RPC
RMC
mRt
GrDG
MoDG
PoDG
LMol

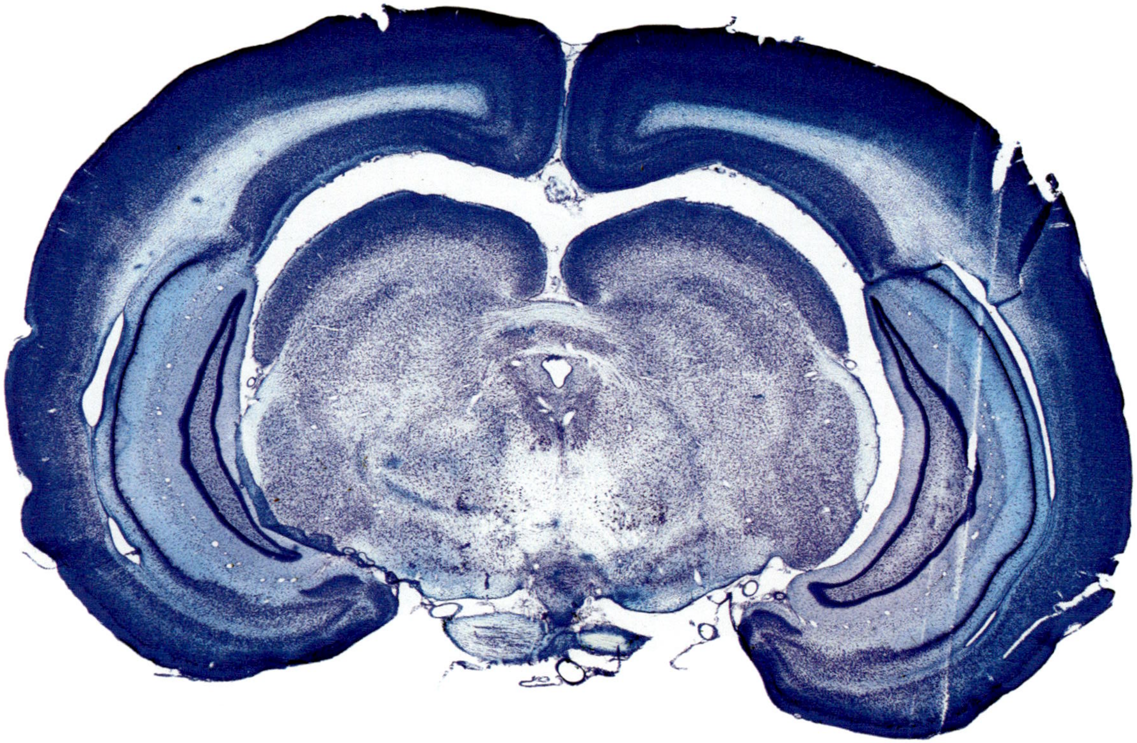

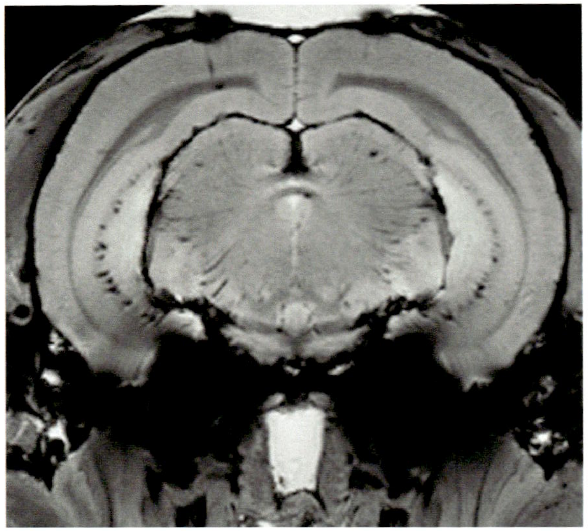

3N	oculomotor nucleus
APT	anterior pretectal nucleus
Aq	aqueduct
CA1	field CA1 of the hippocampus
cp	cerebral peduncle
csc	commissure of the superior colliculus
DG	dentate gyrus of the hippocampus
DpG	deep gray layer of the superior colliculus
DpWh	deep white layer of the superior colliculus
DS	dorsal subiculum
Ent	entorhinal cortex
EW	Edinger-Westphal nucleus
GrDG	granular layer of the dentate gyrus
IF	interfascicular nucleus
InC	interstitial nucleus of Cajal
InG	intermediate gray layer of the superior colliculus
InWh	intermediate white layer of the superior colliculus
IPC	interpeduncular nucleus, caudal subnucleus
IPL	interpeduncular nucleus, lateral subnucleus
IPR	interpeduncular nucleus, rostral subnucleus
LMoL	lacunosum moleculare layer of the hippocampus
MGD	medial geniculate nucleus, dorsal part
MGV	medial geniculate nucleus, ventral part
ml	medial lemniscus
mlf	medial longitudinal fasciculus
Mo	DGmolecular layer of the dentate gyrus
mRt	mesencephalic reticular formation
Op	optic nerve layer of the superior colliculus
OPT	olivary pretectal nucleus
opt	optic tract
Or	oriens layer of the hippocampus
PAG	periaqueductal gray
PBP	parabrachial pigmented nucleus of the VTA
Pd	dorsal nucleus of the pulvinar
PN	paranigral nucleus of the VTA
Pn	pontine nuclei
Po	posterior complex of the thalamus
PoDG	polymorph layer of the dentate gyrus
Post	postsubiculum
PRh	perirhinal cortex
PrS	presubiculum
Py	pyramidal cell layer of the hippocampus
Rad	radiatum layer of the hippocampus
RLi	rostral linear nucleus of the raphe
RMC	red nucleus, magnocellular part
RPC	red nucleus, parvicellular part
RSg	retrosplenial granular cortex
SNR	substantia nigra, reticular part
SuG	superficial gray layer of the superior colliculus
TC	temporal cortex
tfp	transverse fibers of the pons
V1	primary visual cortex
V2	secondary visual cortex
VS	ventral subiculum
vtgx	ventral tegmental decussation
Zo	zonal layer of the superior colliculus

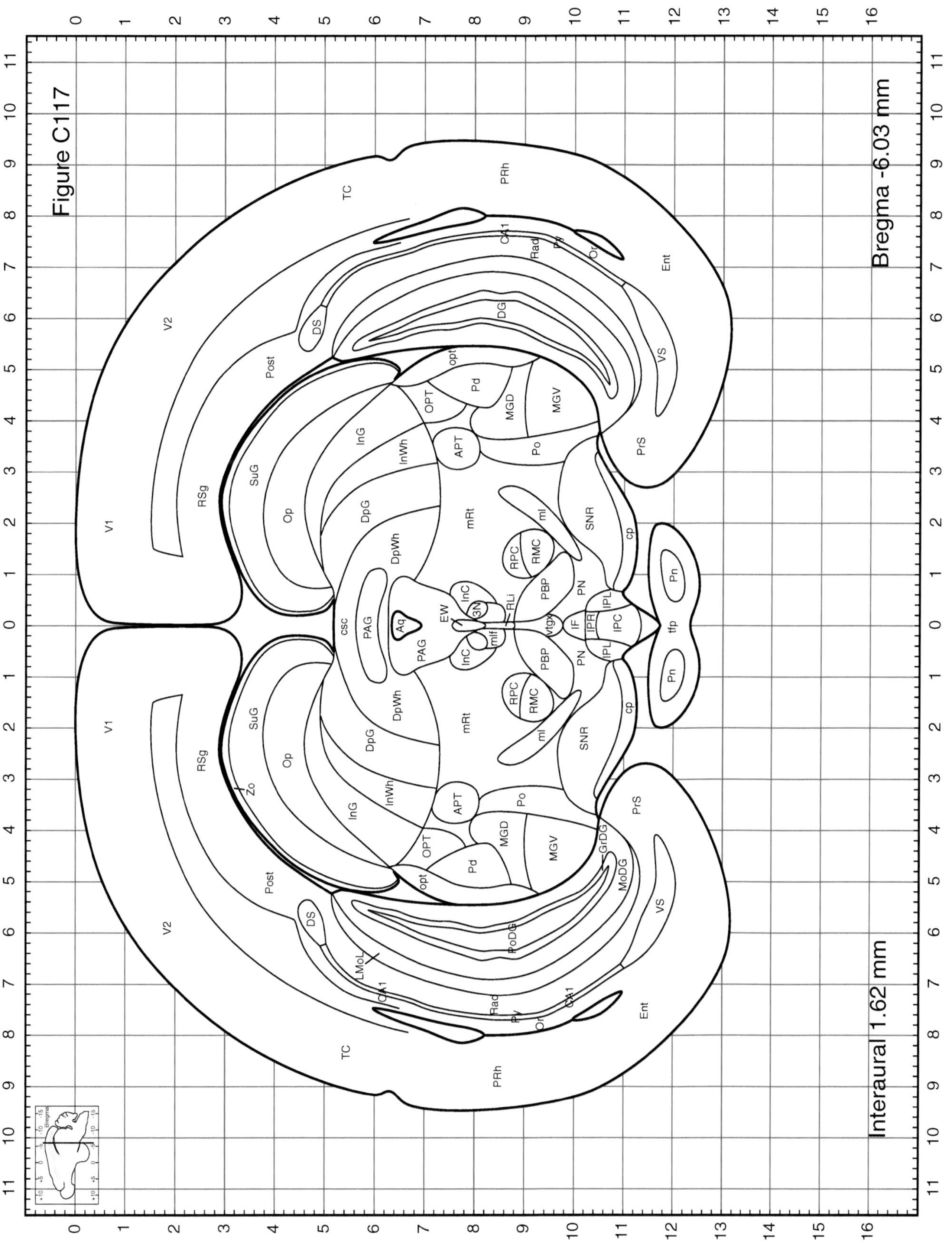
Figure C117
Bregma -6.03 mm
Interaural 1.62 mm
V1
V2
RSg
Post
TC
PRh
DS
CA1
Rad
Py
Or
Ent
DG
VS
PrS
SuG
Op
InG
InWh
DpG
DpWh
Zo
opt
OPT
Pd
MGD
MGV
APT
Po
mRt
ml
SNR
cp
Pn
tfp
RPC
RMC
PBP
PN
IPL
IPR
IPC
IF
vtgx
RLi
mlf
3N
InC
EW
Aq
PAG
csc
GrDG
MoDG
PoDG
LMoL
Bregma

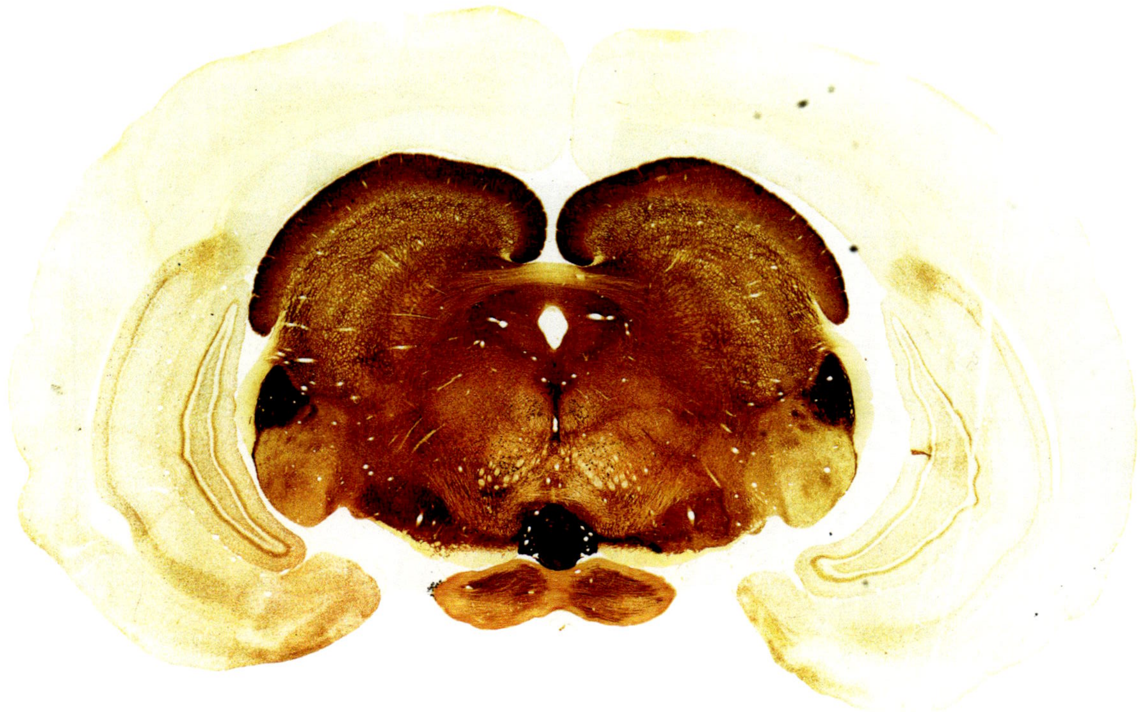

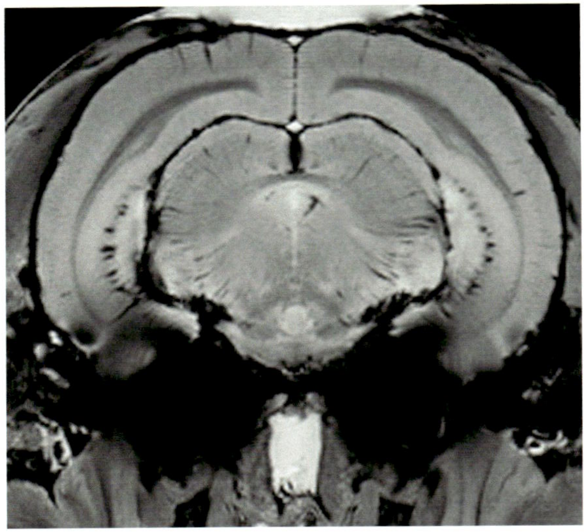

3N	oculomotor nucleus
3PC	oculomotor nucleus, parvicellular part
Aq	aqueduct
CA1	field CA1 of the hippocampus
CLi	caudal linear nucleus of the raphe
cp	cerebral peduncle
csc	commissure of the superior colliculus
DG	dentate gyrus of the hippocampus
DpG	deep gray layer of the superior colliculus
DpWh	deep white layer of the superior colliculus
DS	dorsal subiculum
Ent	entorhinal cortex
EW	Edinger-Westphal nucleus
GrDG	granular layer of the dentate gyrus
IF	interfascicular nucleus
InC	interstitial nucleus of Cajal
InG	intermediate gray layer of the superior colliculus
InWh	intermediate white layer of the superior colliculus
IPC	interpeduncular nucleus, caudal subnucleus
IPL	interpeduncular nucleus, lateral subnucleus
IPR	interpeduncular nucleus, rostral subnucleus
LMoL	lacunosum moleculare layer of the hippocampus
Me5	mesencephalic trigeminal nucleus
MGD	medial geniculate nucleus, dorsal part
MGV	medial geniculate nucleus, ventral part
ml	medial lemniscus
mlf	medial longitudinal fasciculus
MoDG	molecular layer of the dentate gyrus
mRt	mesencephalic reticular formation
Op	optic nerve layer of the superior colliculus
OPT	olivary pretectal nucleus
opt	optic tract
Or	oriens layer of the hippocampus
PAG	periaqueductal gray
PBP	parabrachial pigmented nucleus of the VTA
Pd	dorsal nucleus of the pulvinar
PN	paranigral nucleus of the VTA
Pn	pontine nuclei
Po	posterior complex of the thalamus
PoDG	polymorph layer of the dentate gyrus
Post	postsubiculum
PRh	perirhinal cortex
PrS	presubiculum
Py	pyramidal cell layer of the hippocampus
Rad	radiatum layer of the hippocampus
RMC	red nucleus, magnocellular part
RSg	retrosplenial granular cortex
SNR	substantia nigra, reticular part
SuG	superficial gray layer of the superior colliculus
TC	temporal cortex
tfp	transverse fibers of the pons
tth	trigeminothalamic tract
V1	primary visual cortex
V2	secondary visual cortex
VS	ventral subiculum
xscp	decussation of the superior cerebellar peduncle
Zo	zonal layer of the superior colliculus

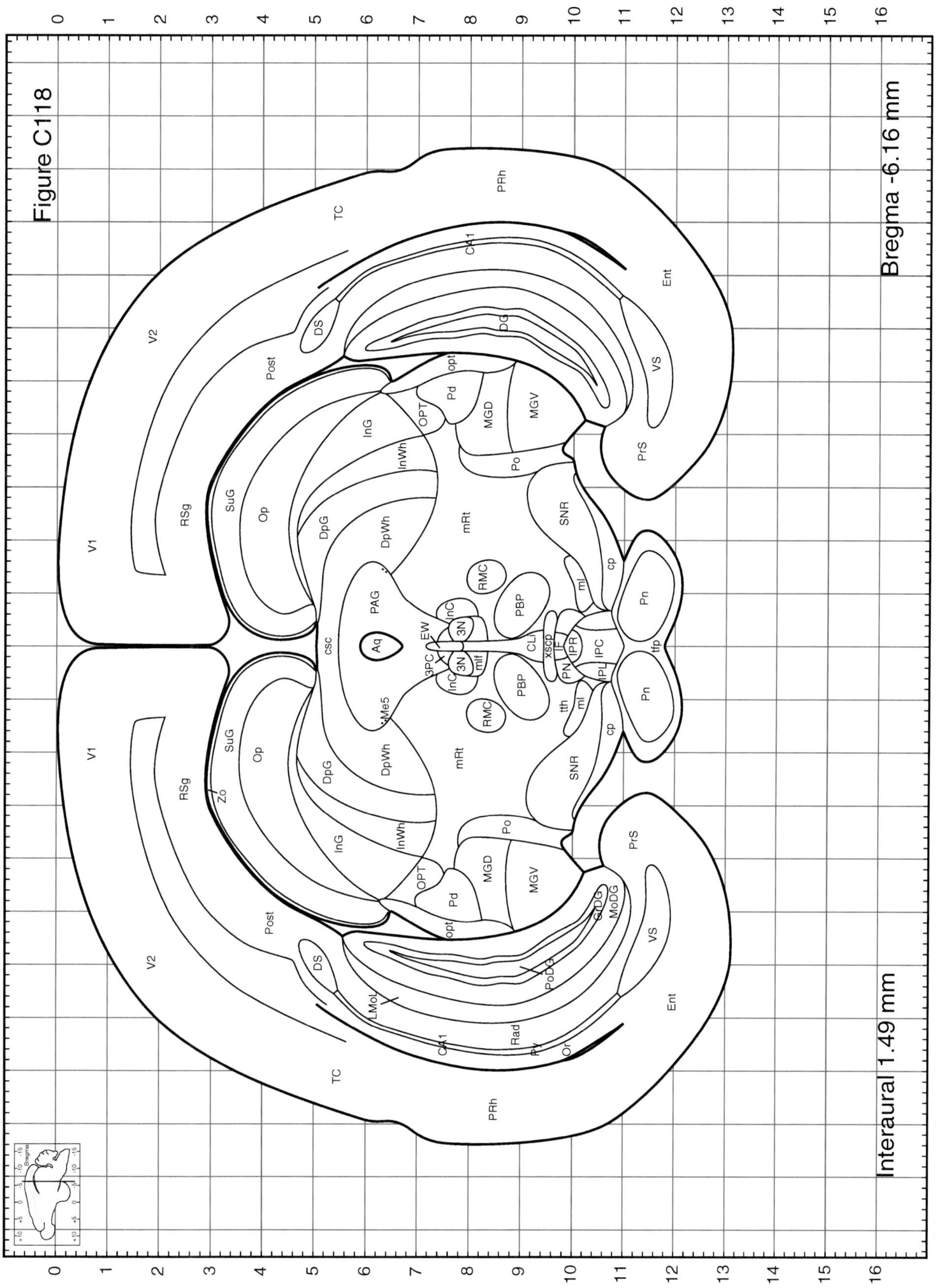
Figure C118
Bregma -6.16 mm
Interaural 1.49 mm
V1
V2
RSg
SuG
Op
Zo
Post
DS
TC
PRh
CA1
DG
Ent
VS
PrS
LMol
Rad
Py
Or
PoDG
GrDG
MoDG
opt
OPT
Pd
MGD
MGV
Po
InG
InWh
DpG
DpWh
mRt
SNR
PAG
Aq
csc
Me5
EW
InC
3N
3PC
mlf
RMC
PBP
CLi
xscp
IF
IPR
IPC
IPL
PN
tth
ml
cp
Pn
tfp

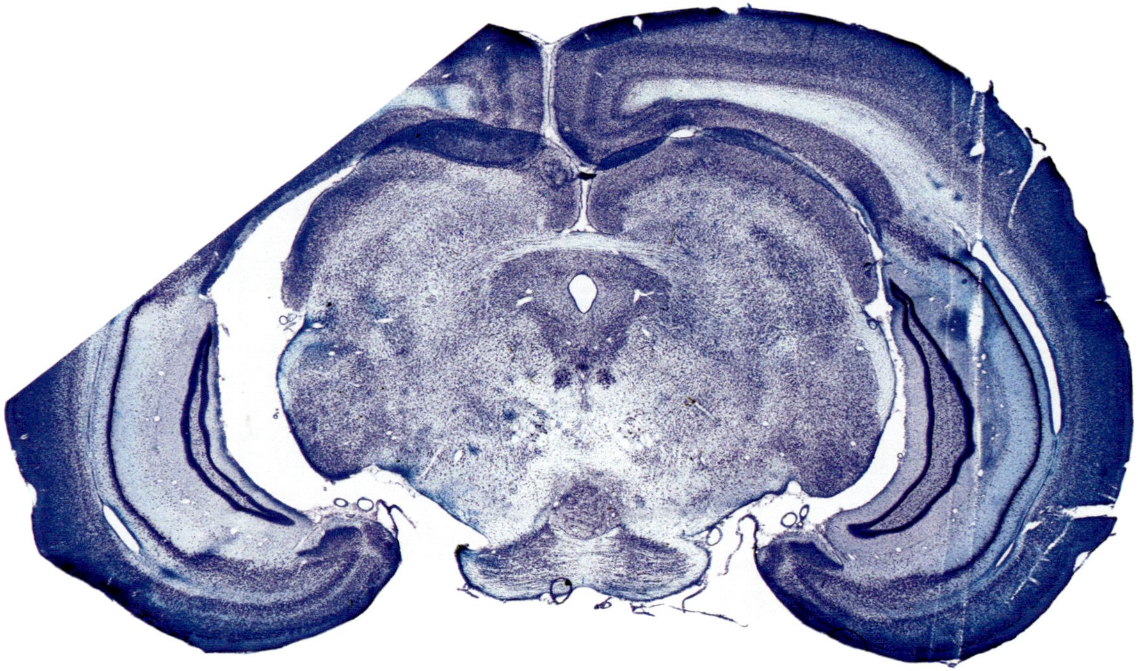

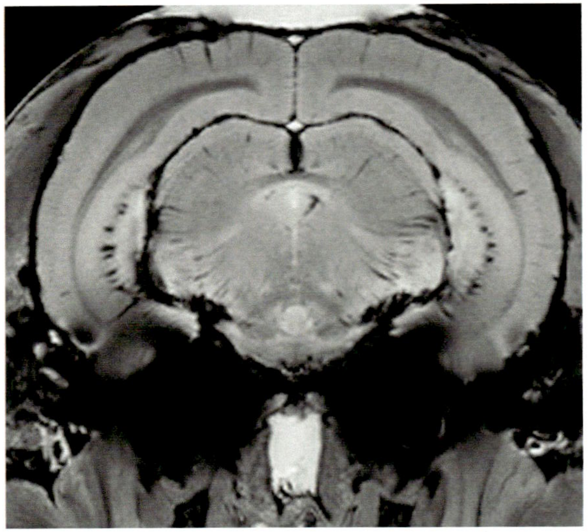

3N	oculomotor nucleus
3PC	oculomotor nucleus, parvicellular part
Aq	aqueduct
CA1	field CA1 of the hippocampus
CLi	caudal linear nucleus of the raphe
cp	cerebral peduncle
csc	commissure of the superior colliculus
DG	dentate gyrus of the hippocampus
DpG	deep gray layer of the superior colliculus
DpWh	deep white layer of the superior colliculus
DS	dorsal subiculum
Ent	entorhinal cortex
EW	Edinger-Westphal nucleus
GrDG	granular layer of the dentate gyrus
IF	interfascicular nucleus
InC	interstitial nucleus of Cajal
InG	intermediate gray layer of the superior colliculus
InWh	intermediate white layer of the superior colliculus
IPC	interpeduncular nucleus, caudal subnucleus
IPL	interpeduncular nucleus, lateral subnucleus
IPR	interpeduncular nucleus, rostral subnucleus
LMoL	lacunosum moleculare layer of the hippocampus
LV	lateral ventricle
Me5	mesencephalic trigeminal nucleus
MGD	medial geniculate nucleus, dorsal part
MGV	medial geniculate nucleus, ventral part
ml	medial lemniscus
mlf	medial longitudinal fasciculus
MoDG	molecular layer of the dentate gyrus
mRt	mesencephalic reticular formation
Op	optic nerve layer of the superior colliculus
OPT	olivary pretectal nucleus
opt	optic tract
Or	oriens layer of the hippocampus
PAG	periaqueductal gray
PBP	parabrachial pigmented nucleus of the VTA
Pd	dorsal nucleus of the pulvinar
PN	paranigral nucleus of the VTA
Pn	pontine nuclei
Po	posterior complex of the thalamus
PoDG	polymorph layer of the dentate gyrus
Post	postsubiculum
PRh	perirhinal cortex
PrS	presubiculum
Py	pyramidal cell layer of the hippocampus
Rad	radiatum layer of the hippocampus
RMC	red nucleus, magnocellular part
RRF	retrorubral field
RSg	retrosplenial granular cortex
SNR	substantia nigra, reticular part
SuG	superficial gray layer of the superior colliculus
TC	temporal cortex
tfp	transverse fibers of the pons
tth	trigeminothalamic tract
V1	primary visual cortex
V2	secondary visual cortex
VS	ventral subiculum
xscp	decussation of the superior cerebellar peduncle
Zo	zonal layer of the superior colliculus

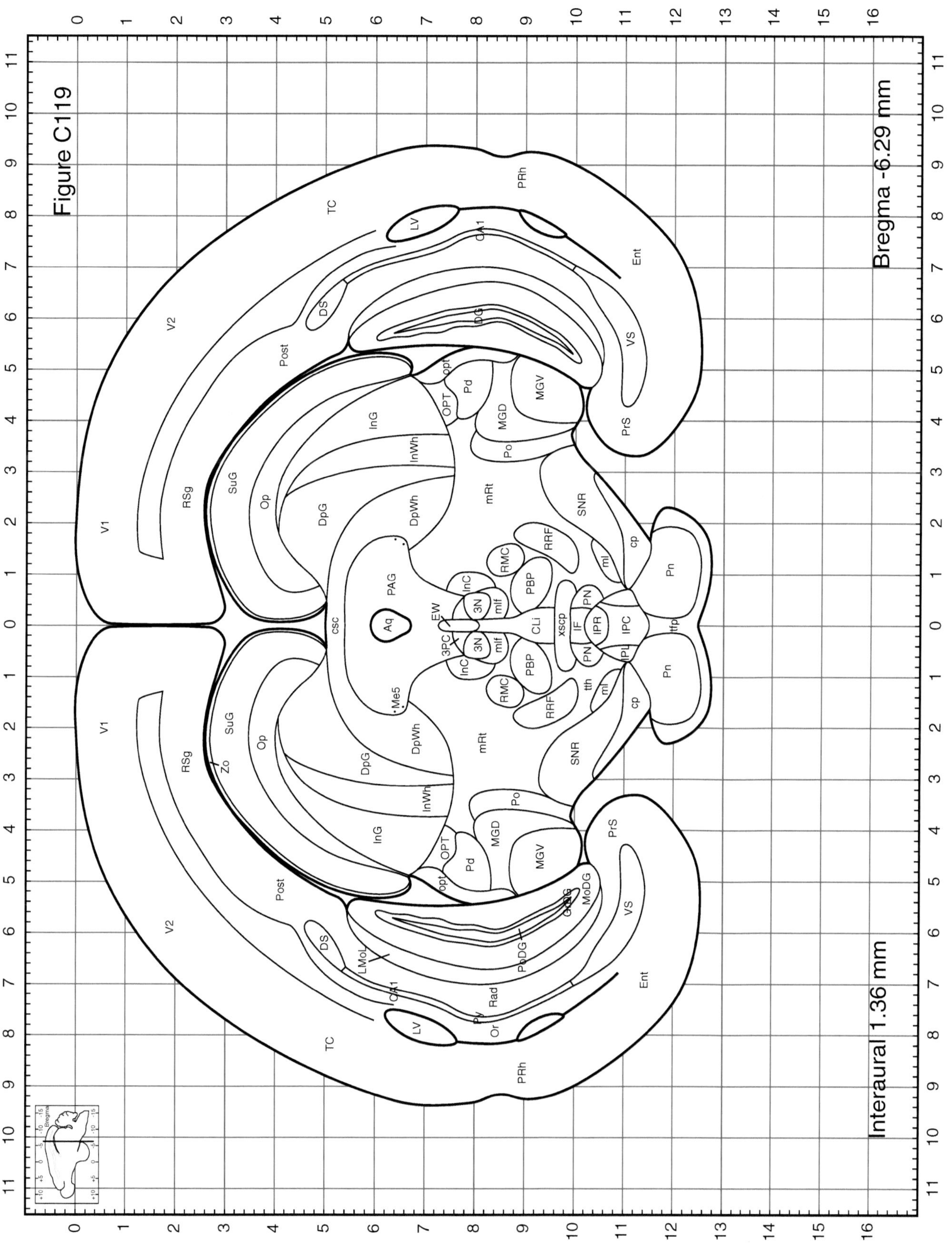
Figure C119
Bregma -6.29 mm
Interaural 1.36 mm
V1
V2
TC
PRh
Ent
LV
CA1
DS
DG
VS
PrS
Post
RSg
SuG
Zo
Op
InG
InWh
DpG
DpWh
PAG
Aq
csc
Me5
EW
3PC
3N
InC
mlf
OPT
opt
Pd
MGV
MGD
Po
mRt
SNR
RRF
RMC
PBP
CLi
xscp
IF
IPR
IPC
IPL
PN
ml
tth
cp
Pn
tfp
LMoL
PoDG
GrDG
MoDG
Rad
Py
Or

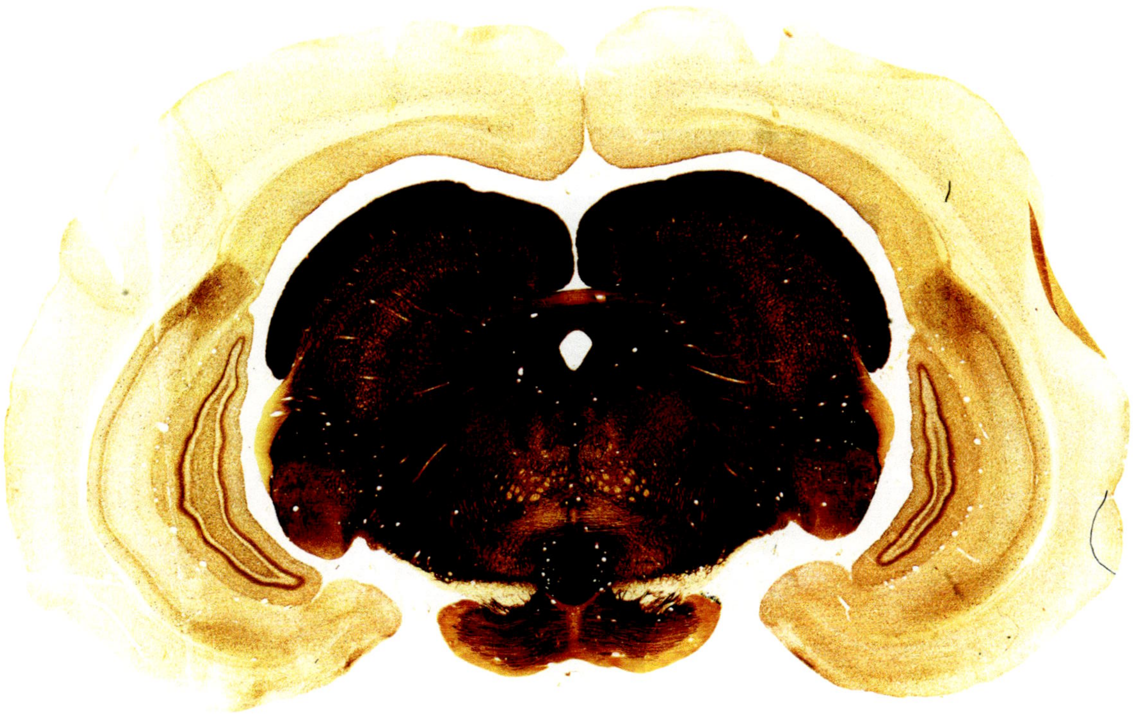

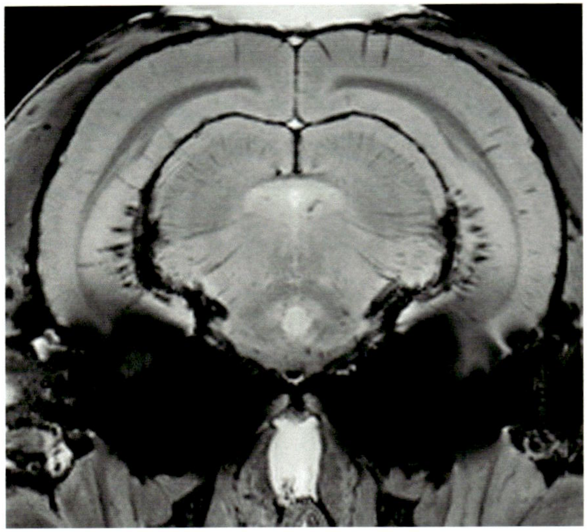

3N	oculomotor nucleus
3PC	oculomotor nucleus, parvicellular part
Aq	aqueduct
CA1	field CA1 of the hippocampus
CLi	caudal linear nucleus of the raphe
cp	cerebral peduncle
csc	commissure of the superior colliculus
DG	dentate gyrus of the hippocampus
DpG	deep gray layer of the superior colliculus
DpWh	deep white layer of the superior colliculus
DS	dorsal subiculum
Ent	entorhinal cortex
EW	Edinger-Westphal nucleus
GrDG	granular layer of the dentate gyrus
InC	interstitial nucleus of Cajal
InG	intermediate gray layer of the superior colliculus
InWh	intermediate white layer of the superior colliculus
IPA	interpeduncular nucleus, apical subnucleus
IPC	interpeduncular nucleus, caudal subnucleus
IPL	interpeduncular nucleus, lateral subnucleus
LMoL	lacunosum moleculare layer of the hippocampus
me5	mesencephalic trigeminal tract
Me5	mesencephalic trigeminal nucleus
MGD	medial geniculate nucleus, dorsal part
MGV	medial geniculate nucleus, ventral part
ml	medial lemniscus
mlf	medial longitudinal fasciculus
MoDG	molecular layer of the dentate gyrus
mRt	mesencephalic reticular formation
Op	optic nerve layer of the superior colliculus
OPT	olivary pretectal nucleus
opt	optic tract
Or	oriens layer of the hippocampus
PAG	periaqueductal gray
PBP	parabrachial pigmented nucleus of the VTA
Pd	dorsal nucleus of the pulvinar
PN	paranigral nucleus of the VTA
Pn	pontine nuclei
Po	posterior complex of the thalamus
Post	postsubiculum
PrS	presubiculum
Py	pyramidal cell layer of the hippocampus
Rad	radiatum layer of the hippocampus
RRF	retrorubral field
RSg	retrosplenial granular cortex
SNR	substantia nigra, reticular part
SuG	superficial gray layer of the superior colliculus
TC	temporal cortex
tfp	transverse fibers of the pons
tth	trigeminothalamic tract
V1	primary visual cortex
V2	secondary visual cortex
VS	ventral subiculum
xscp	decussation of the superior cerebellar peduncle
Zo	zonal layer of the superior colliculus

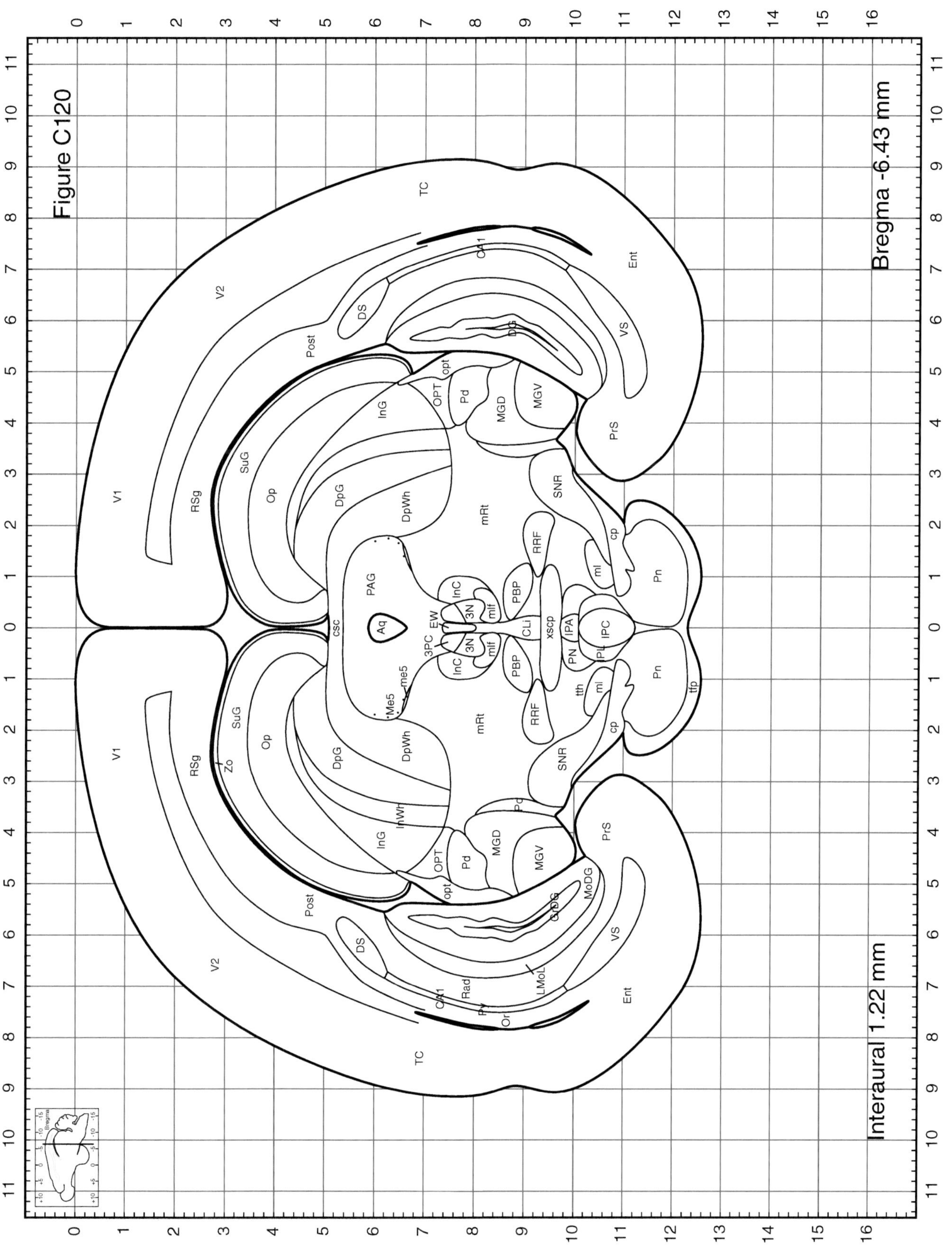
Figure C120
Bregma -6.43 mm
Interaural 1.22 mm
TC
CA1
Ent
V2
DS
DG
VS
Post
opt
OPT
Pd
MGD
MGV
InG
PrS
SuG
V1
RSg
Op
DpG
DpWh
mRt
SNR
RRF
cp
PAG
InC
PBP
ml
Pn
3N
mlf
csc
Aq
EW
CLi
xscp
IPA
IPC
3PC
PN
PL
me5
Me5
tth
tfp
Zo
InWh
MoDG
GrDG
LMol
Rad
Py
Or

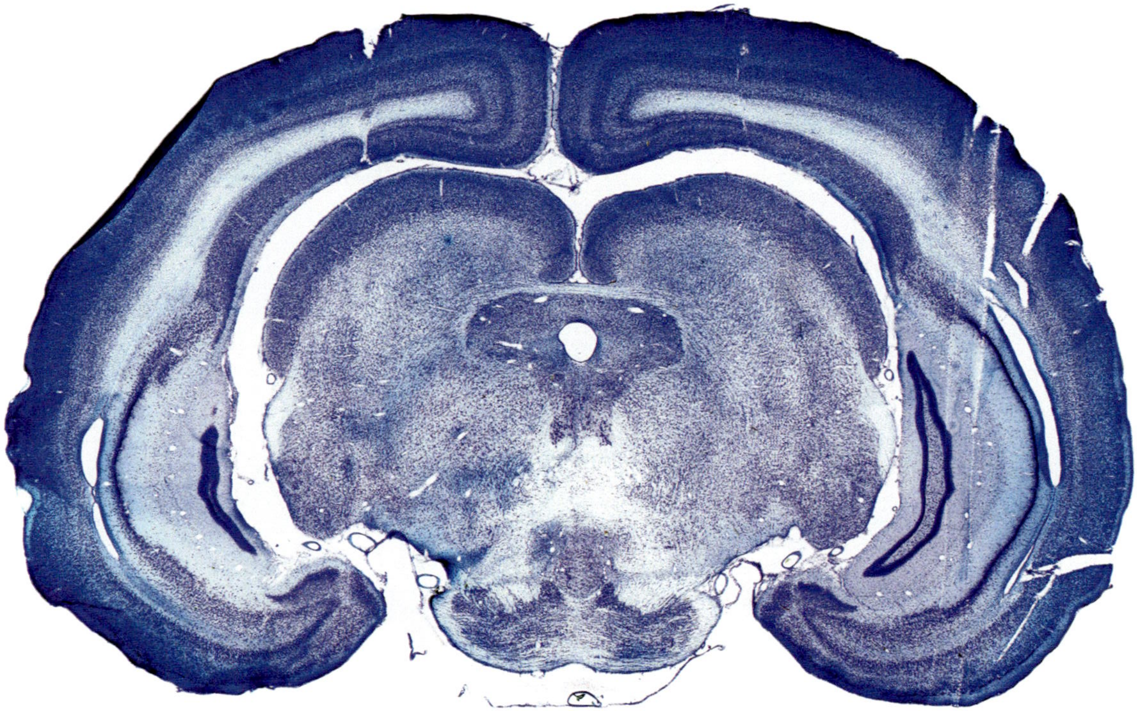

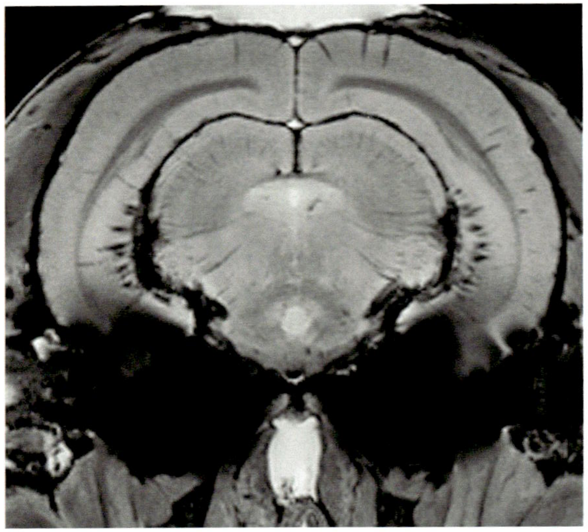

3N	oculomotor nucleus
3PC	oculomotor nucleus, parvicellular part
Aq	aqueduct
CA1	field CA1 of the hippocampus
CLi	caudal linear nucleus of the raphe
cp	cerebral peduncle
csc	commissure of the superior colliculus
DG	dentate gyrus of the hippocampus
DpG	deep gray layer of the superior colliculus
DpWh	deep white layer of the superior colliculus
DS	dorsal subiculum
Ent	entorhinal cortex
EW	Edinger-Westphal nucleus
GrDG	granular layer of the dentate gyrus
InC	interstitial nucleus of Cajal
InG	intermediate gray layer of the superior colliculus
InWh	intermediate white layer of the superior colliculus
IPA	interpeduncular nucleus, apical subnucleus
IPC	interpeduncular nucleus, caudal subnucleus
IPL	interpeduncular nucleus, lateral subnucleus
LMoL	lacunosum moleculare layer of the hippocampus
me5	mesencephalic trigeminal tract
Me5	mesencephalic trigeminal nucleus
MGD	medial geniculate nucleus, dorsal part
MGV	medial geniculate nucleus, ventral part
ml	medial lemniscus
mlf	medial longitudinal fasciculus
MoDG	molecular layer of the dentate gyrus
mRt	mesencephalic reticular formation
Op	optic nerve layer of the superior colliculus
OPT	olivary pretectal nucleus
opt	optic tract
Or	oriens layer of the hippocampus
PAG	periaqueductal gray
PBP	parabrachial pigmented nucleus of the VTA
Pd	dorsal nucleus of the pulvinar
PN	paranigral nucleus of the VTA
Pn	pontine nuclei
Po	posterior complex of the thalamus
Post	postsubiculum
PrS	presubiculum
Py	pyramidal cell layer of the hippocampus
Rad	radiatum layer of the hippocampus
RRF	retrorubral field
RSg	retrosplenial granular cortex
SNR	substantia nigra, reticular part
SuG	superficial gray layer of the superior colliculus
TC	temporal cortex
tfp	transverse fibers of the pons
tth	trigeminothalamic tract
V1	primary visual cortex
V2	secondary visual cortex
VS	ventral subiculum
xscp	decussation of the superior cerebellar peduncle
Zo	zonal layer of the superior colliculus

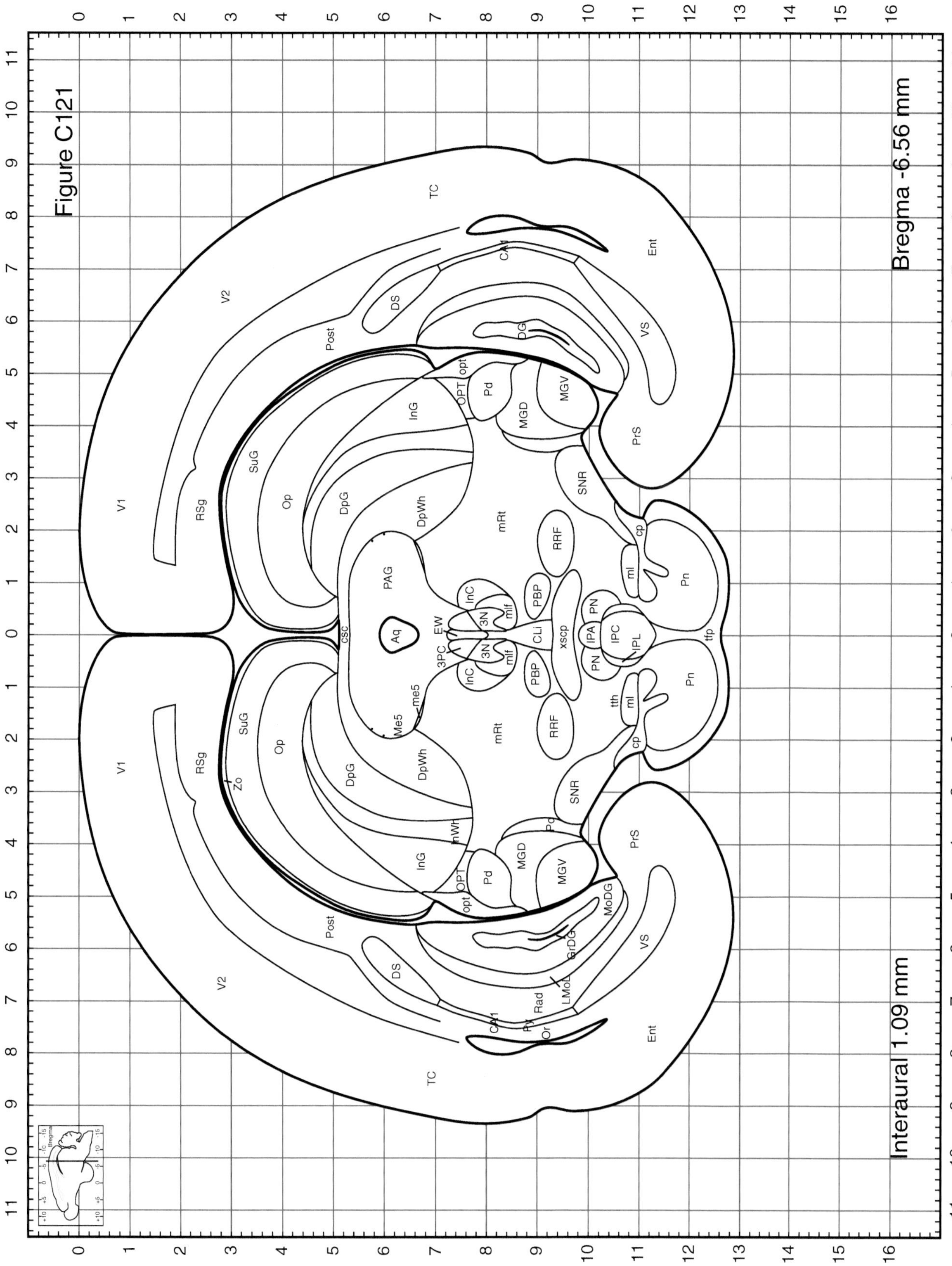
Figure C121
Bregma -6.56 mm
Interaural 1.09 mm
TC
V2
V1
RSg
Zo
SuG
Op
DpG
DpWh
InG
InWh
Post
DS
CA1
Rad
Py
Or
LMol
GrDG
MoDG
DG
Ent
VS
PrS
opt
OPT
Pd
MGD
MGV
Po
SNR
mRt
RRF
PAG
Aq
csc
EW
3PC
3N
InC
mlf
me5
Me5
PBP
CLi
xscp
PN
IPA
IPC
IPL
ml
cp
Pn
tfp
tth

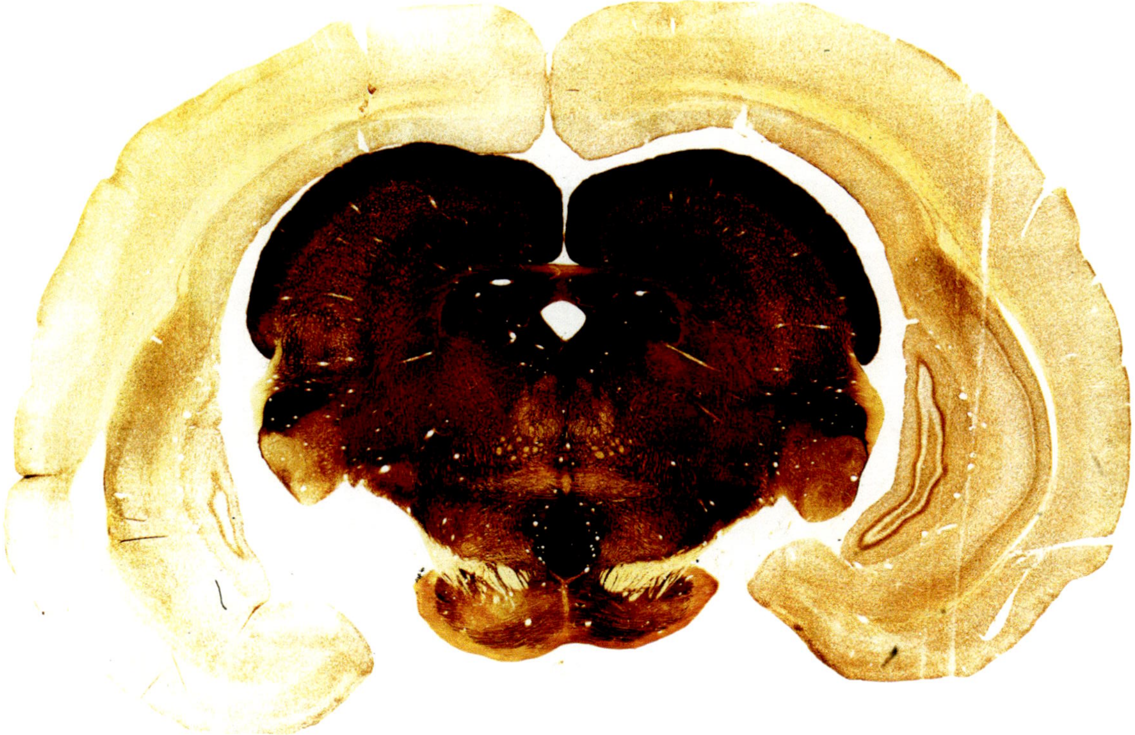

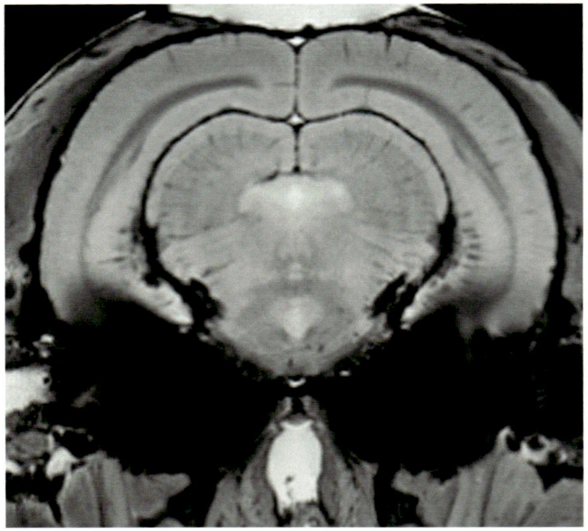

3N	oculomotor nucleus
3PC	oculomotor nucleus, parvicellular part
Aq	aqueduct
CA1	field CA1 of the hippocampus
CLi	caudal linear nucleus of the raphe
cp	cerebral peduncle
csc	commissure of the superior colliculus
DG	dentate gyrus of the hippocampus
DpG	deep gray layer of the superior colliculus
DpWh	deep white layer of the superior colliculus
DS	dorsal subiculum
Ent	entorhinal cortex
EW	Edinger-Westphal nucleus
GrDG	granular layer of the dentate gyrus
InC	interstitial nucleus of Cajal
InG	intermediate gray layer of the superior colliculus
InWh	intermediate white layer of the superior colliculus
IPA	interpeduncular nucleus, apical subnucleus
IPC	interpeduncular nucleus, caudal subnucleus
IPL	interpeduncular nucleus, lateral subnucleus
LMoL	lacunosum moleculare layer of the hippocampus
LV	lateral ventricle
me5	mesencephalic trigeminal tract
Me5	mesencephalic trigeminal nucleus
MGD	medial geniculate nucleus, dorsal part
MGV	medial geniculate nucleus, ventral part
ml	medial lemniscus
mlf	medial longitudinal fasciculus
MoDG	molecular layer of the dentate gyrus
mRt	mesencephalic reticular formation
Op	optic nerve layer of the superior colliculus
OPT	olivary pretectal nucleus
opt	optic tract
Or	oriens layer of the hippocampus
PAG	periaqueductal gray
PBP	parabrachial pigmented nucleus of the VTA
Pd	dorsal nucleus of the pulvinar
PN	paranigral nucleus of the VTA
Pn	pontine nuclei
Post	postsubiculum
PrS	presubiculum
PTg	pedunculopontine tegmental nucleus
Py	pyramidal cell layer of the hippocampus
Rad	radiatum layer of the hippocampus
RRF	retrorubral field
RSg	retrosplenial granular cortex
SuG	superficial gray layer of the superior colliculus
TC	temporal cortex
tfp	transverse fibers of the pons
tth	trigeminothalamic tract
V1	primary visual cortex
V2	secondary visual cortex
VS	ventral subiculum
xscp	decussation of the superior cerebellar peduncle
Zo	zonal layer of the superior colliculus

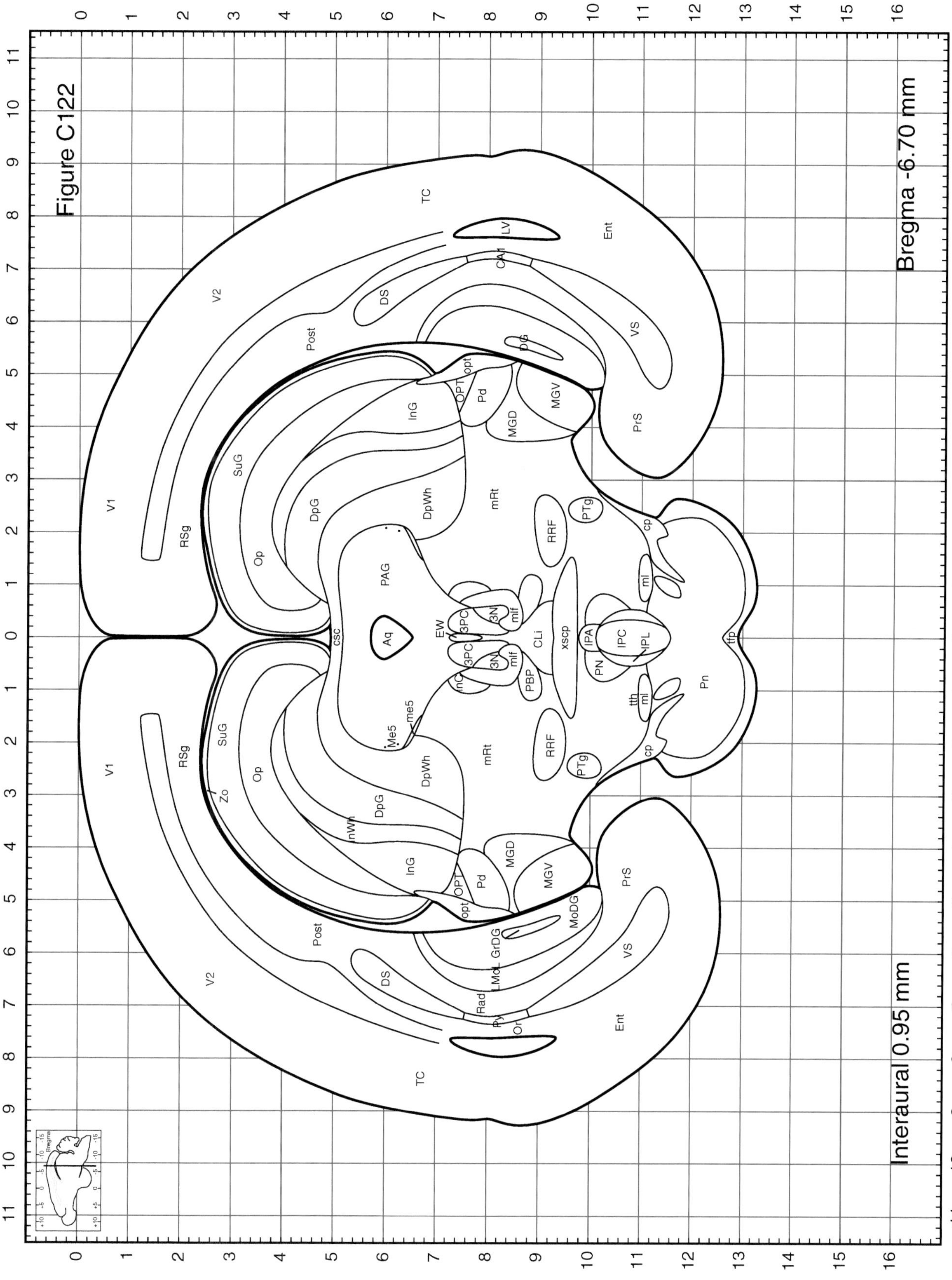
Figure C122
Bregma -6.70 mm
Interaural 0.95 mm
TC
LV
CA1
Ent
DS
V2
Post
VS
DG
opt
OPT
Pd
MGV
MGD
InG
PrS
SuG
V1
DpWh
mRt
DpG
RSg
Op
PTg
RRF
cp
PAG
ml
3N
mlf
3PC
EW
Aq
csc
CLi
xscp
IPA
IPC
IPL
PN
PBP
InC
tfp
Pn
tth
Me5
me5
Zo
InWh
MoDG
GrDG
LMol
Rad
Py
Or

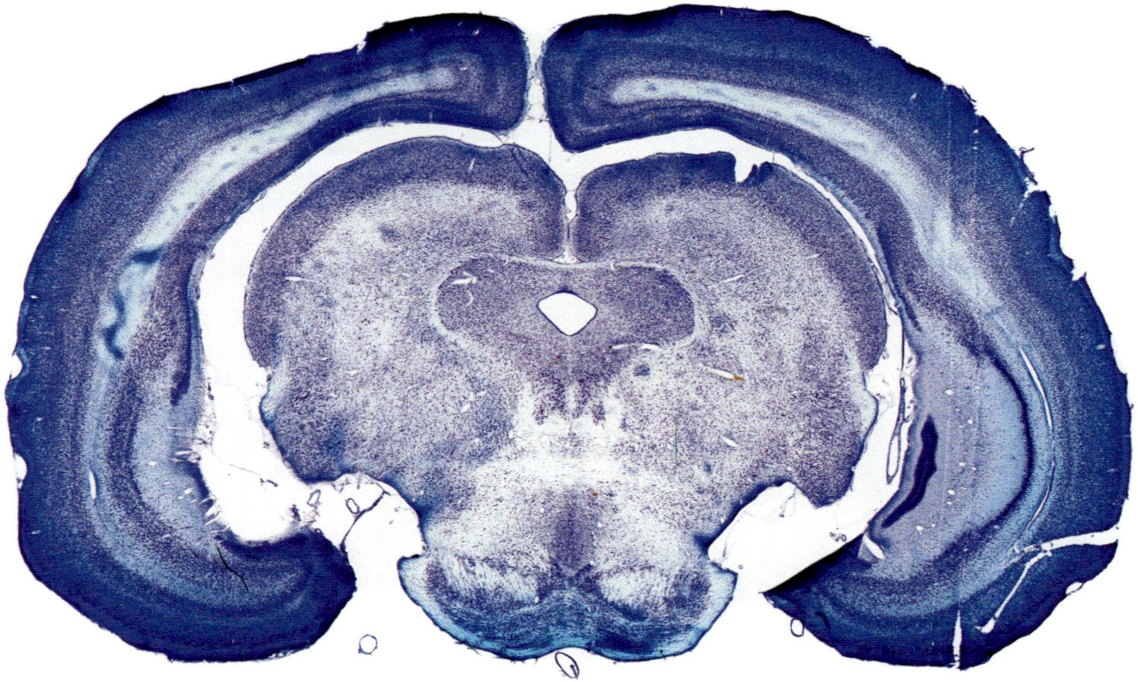

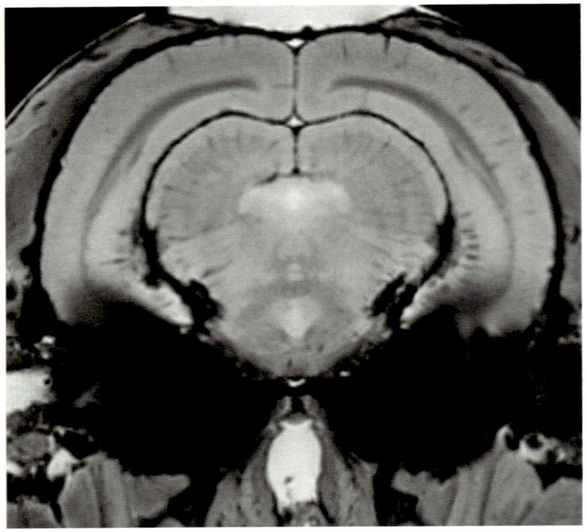

3N	oculomotor nucleus
3PC	oculomotor nucleus, parvicellular part
Aq	aqueduct
CLi	caudal linear nucleus of the raphe
cp	cerebral peduncle
csc	commissure of the superior colliculus
DpG	deep gray layer of the superior colliculus
DpWh	deep white layer of the superior colliculus
Ent	entorhinal cortex
EW	Edinger-Westphal nucleus
InC	interstitial nucleus of Cajal
InG	intermediate gray layer of the superior colliculus
InWh	intermediate white layer of the superior colliculus
IPA	interpeduncular nucleus, apical subnucleus
IPC	interpeduncular nucleus, caudal subnucleus
IPL	interpeduncular nucleus, lateral subnucleus
LMoL	lacunosum moleculare layer of the hippocampus
mcp	middle cerebellar peduncle
me5	mesencephalic trigeminal tract
Me5	mesencephalic trigeminal nucleus
MGD	medial geniculate nucleus, dorsal part
MGV	medial geniculate nucleus, ventral part
ml	medial lemniscus
mlf	medial longitudinal fasciculus
MoDG	molecular layer of the dentate gyrus
mRt	mesencephalic reticular formation
Op	optic nerve layer of the superior colliculus
OPT	olivary pretectal nucleus
opt	optic tract
PAG	periaqueductal gray
PBP	parabrachial pigmented nucleus of the VTA
Pd	dorsal nucleus of the pulvinar
PN	paranigral nucleus of the VTA
Pn	pontine nuclei
Post	postsubiculum
PrS	presubiculum
PTg	pedunculopontine tegmental nucleus
RRF	retrorubral field
RSg	retrosplenial granular cortex
STr	subiculum, transition area
SuG	superficial gray layer of the superior colliculus
TC	temporal cortex
tfp	transverse fibers of the pons
tth	trigeminothalamic tract
V1	primary visual cortex
V2	secondary visual cortex
xscp	decussation of the superior cerebellar peduncle
Zo	zonal layer of the superior colliculus

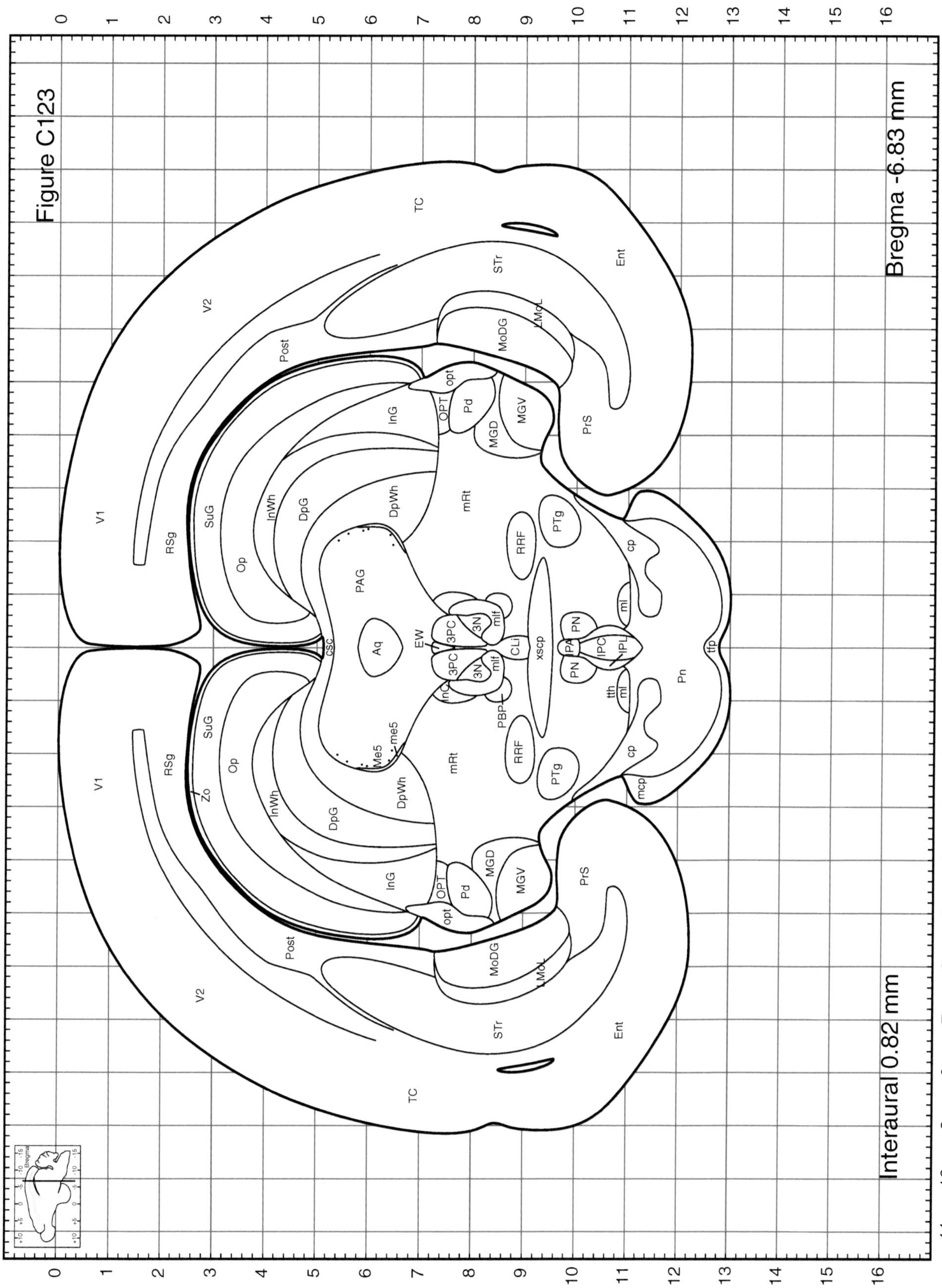
Figure C123
Bregma -6.83 mm
Interaural 0.82 mm
TC
V2
V1
STr
Ent
Post
MoDG
LMoL
opt
OPT
Pd
MGV
MGD
PrS
InG
DpWh
mRt
DpG
InWh
SuG
RSg
Op
Zo
PTg
RRF
cp
mcp
PAG
ml
Aq
EW
3PC
3N
mlf
CLi
xscp
PN
IPA
IPC
IPL
tfp
Pn
csc
InC
PBP
tth
me5
Me5

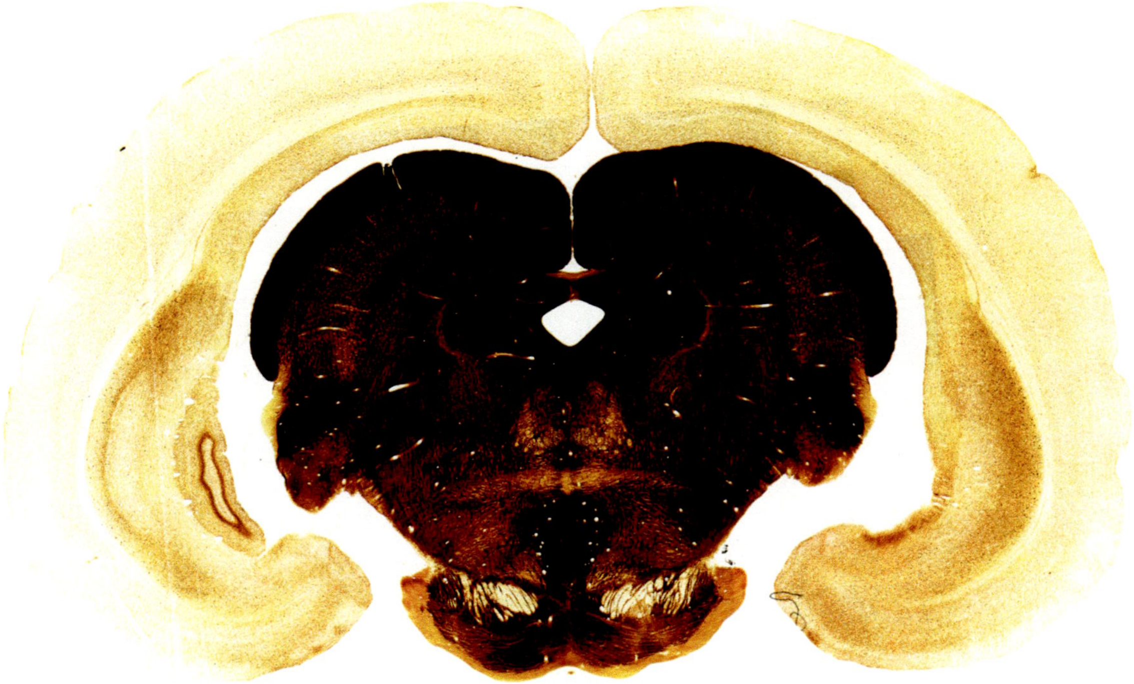

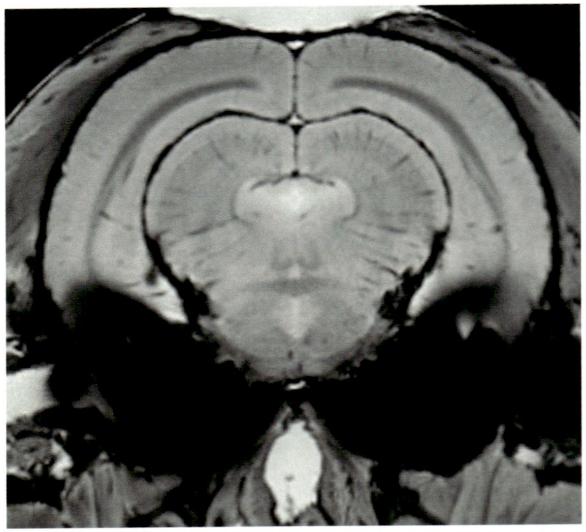

3N	oculomotor nucleus
3PC	oculomotor nucleus, parvicellular part
Aq	aqueduct
CLi	caudal linear nucleus of the raphe
csc	commissure of the superior colliculus
DpG	deep gray layer of the superior colliculus
DpWh	deep white layer of the superior colliculus
Ent	entorhinal cortex
EW	Edinger-Westphal nucleus
Hip	hippocampus
InC	interstitial nucleus of Cajal
InG	intermediate gray layer of the superior colliculus
InWh	intermediate white layer of the superior colliculus
IPA	interpeduncular nucleus, apical subnucleus
IPC	interpeduncular nucleus, caudal subnucleus
IPL	interpeduncular nucleus, lateral subnucleus
lfp	longitudinal fasciculus of the pons
mcp	middle cerebellar peduncle
me5	mesencephalic trigeminal tract
Me5	mesencephalic trigeminal nucleus
MGV	medial geniculate nucleus, ventral part
ml	medial lemniscus
mlf	medial longitudinal fasciculus
mRt	mesencephalic reticular formation
Op	optic nerve layer of the superior colliculus
OPT	olivary pretectal nucleus
opt	optic tract
PAG	periaqueductal gray
Pd	dorsal nucleus of the pulvinar
Pn	pontine nuclei
Post	postsubiculum
PrS	presubiculum
PTg	pedunculopontine tegmental nucleus
RRF	retrorubral field
RSg	retrosplenial granular cortex
SuG	superficial gray layer of the superior colliculus
TC	temporal cortex
tfp	transverse fibers of the pons
tth	trigeminothalamic tract
V1	primary visual cortex
V2	secondary visual cortex
xscp	decussation of the superior cerebellar peduncle

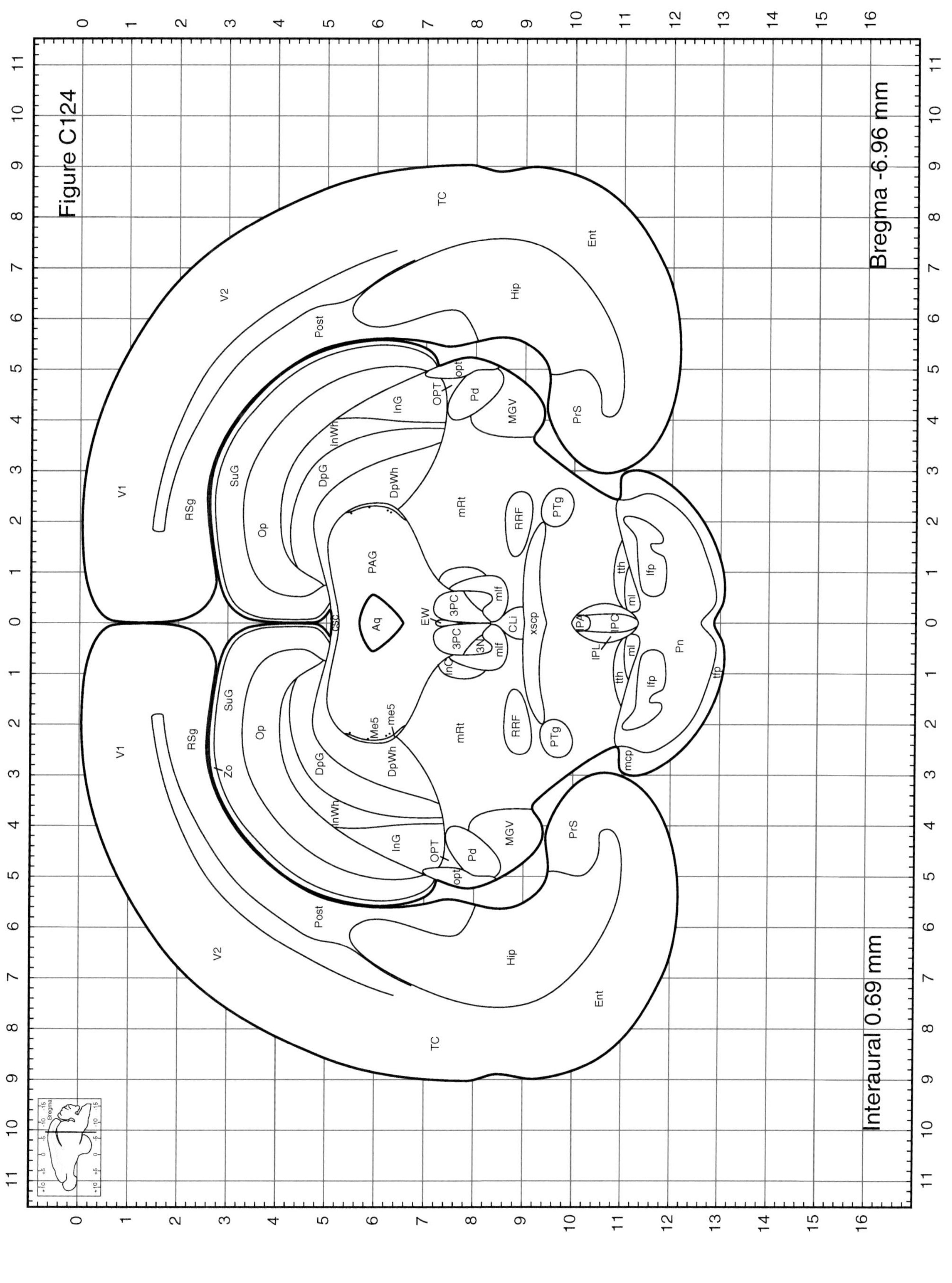
Figure C124
Bregma -6.96 mm
Interaural 0.69 mm
V1
V2
RSg
Post
TC
Ent
Hip
PrS
SuG
Op
Zo
InWh
InG
DpG
DpWh
OPT
opt
Pd
MGV
PAG
Aq
csc
Me5
me5
mRt
RRF
PTg
EW
3PC
3N
InC
mlf
CLi
xscp
PA
IPC
IPL
ml
tth
lfp
Pn
tfp
mcp

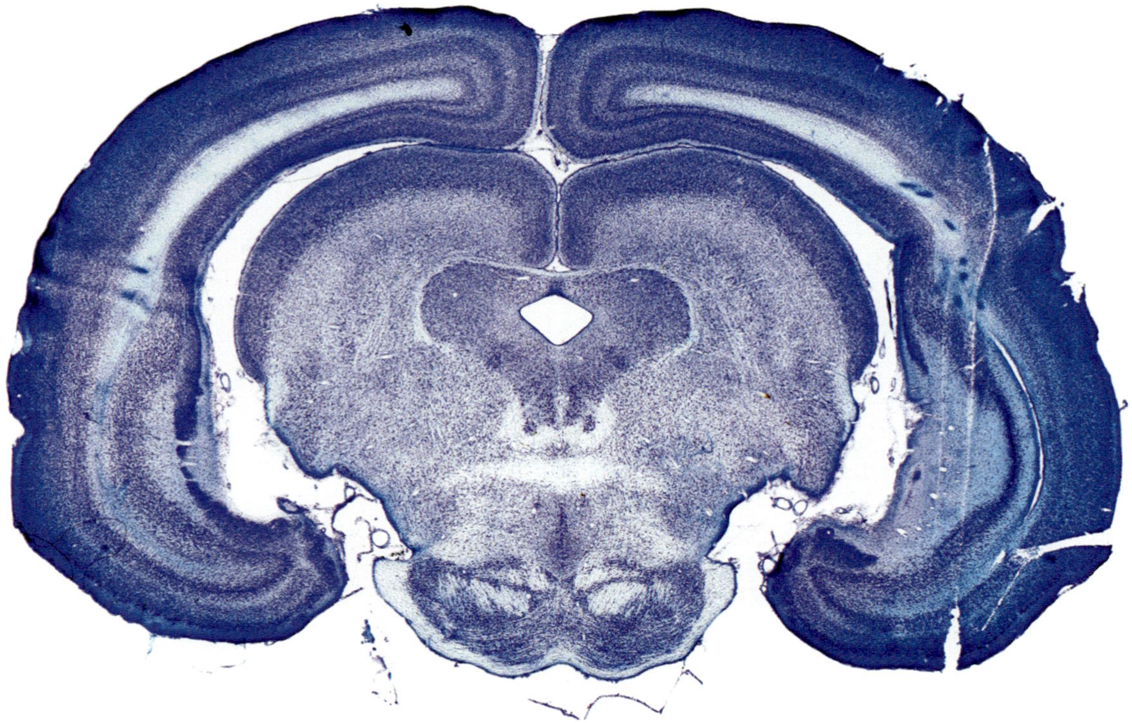

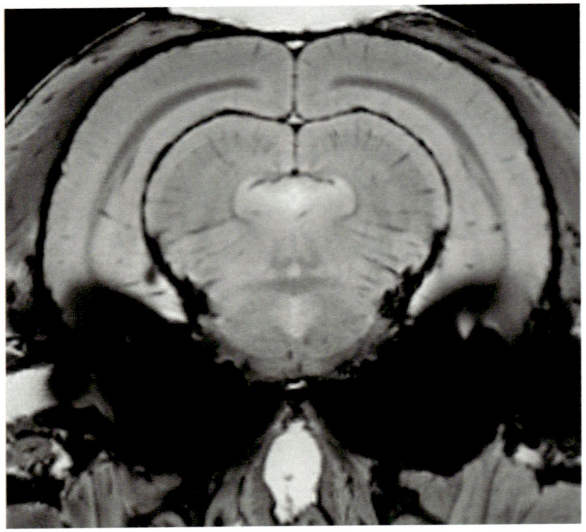

3N	oculomotor nucleus
3PC	oculomotor nucleus, parvicellular part
Aq	aqueduct
CLi	caudal linear nucleus of the raphe
csc	commissure of the superior colliculus
DpG	deep gray layer of the superior colliculus
DpWh	deep white layer of the superior colliculus
DR	dorsal raphe nucleus
Ent	entorhinal cortex
Hip	hippocampus
InC	interstitial nucleus of Cajal
InG	intermediate gray layer of the superior colliculus
InWh	intermediate white layer of the superior colliculus
IPA	interpeduncular nucleus, apical subnucleus
IPC	interpeduncular nucleus, caudal subnucleus
lfp	longitudinal fasciculus of the pons
mcp	middle cerebellar peduncle
me5	mesencephalic trigeminal tract
Me5	mesencephalic trigeminal nucleus
MGV	medial geniculate nucleus, ventral part
ml	medial lemniscus
mlf	medial longitudinal fasciculus
MnR	median raphe nucleus
mRt	mesencephalic reticular formation
Op	optic nerve layer of the superior colliculus
OPT	olivary pretectal nucleus
opt	optic tract
PAG	periaqueductal gray
Pd	dorsal nucleus of the pulvinar
Pn	pontine nuclei
Post	postsubiculum
PrS	presubiculum
PTg	pedunculopontine tegmental nucleus
RRF	retrorubral field
RSg	retrosplenial granular cortex
SuG	superficial gray layer of the superior colliculus
TC	temporal cortex
tfp	transverse fibers of the pons
tth	trigeminothalamic tract
V1	primary visual cortex
V2	secondary visual cortex
xscp	decussation of the superior cerebellar peduncle
Zo	zonal layer of the superior colliculus

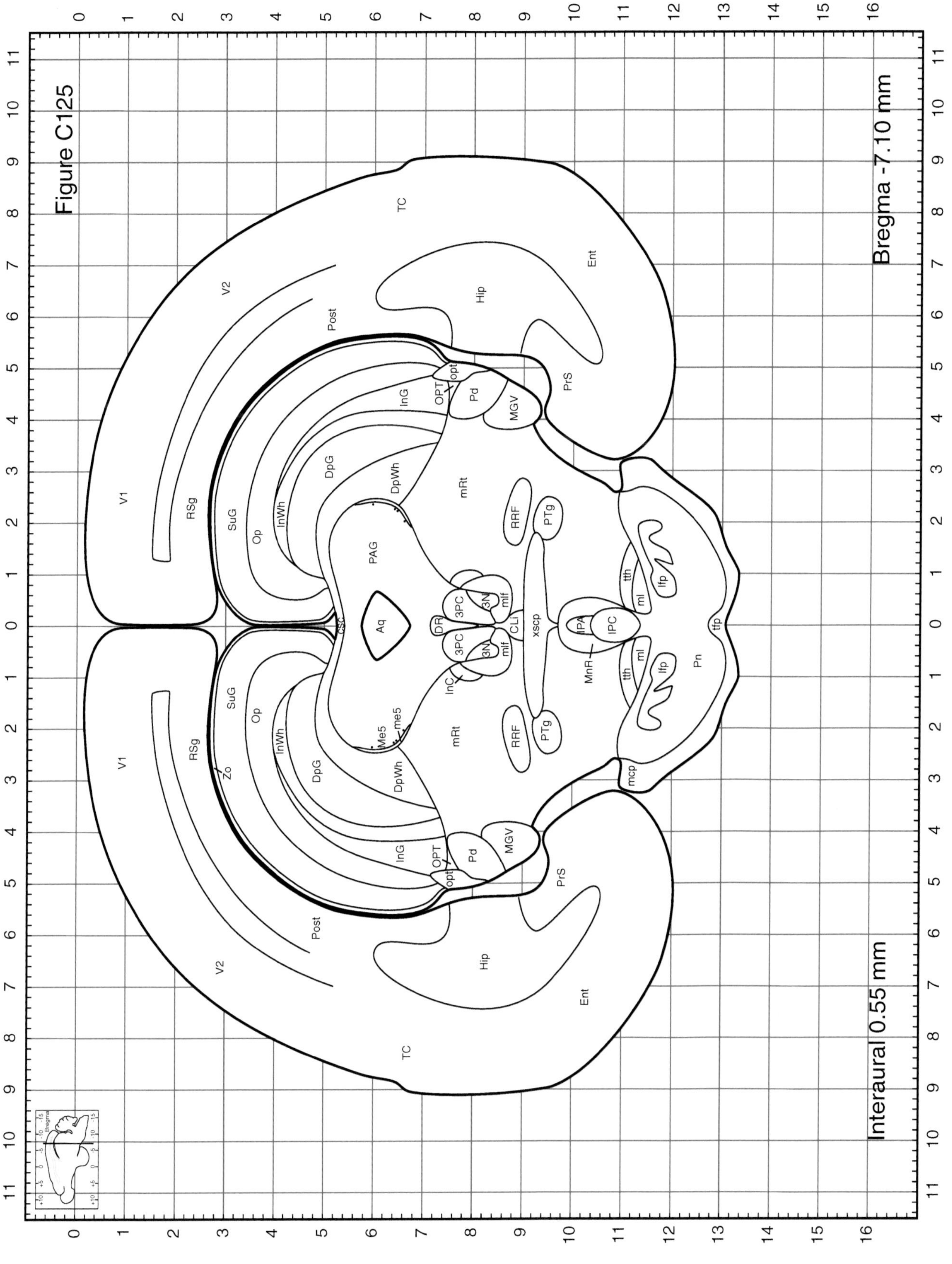
Figure C125
Bregma -7.10 mm
Interaural 0.55 mm
V1
V2
RSg
Post
TC
Ent
Hip
PrS
SuG
Op
InWh
InG
DpG
DpWh
OPT
opt
Pd
MGV
mRt
PAG
Aq
csc
DR
3PC
3N
mlf
CLi
xscp
RRF
PTg
InC
Me5
me5
IPA
IPC
MnR
ml
tth
lfp
tfp
Pn
mcp
Zo

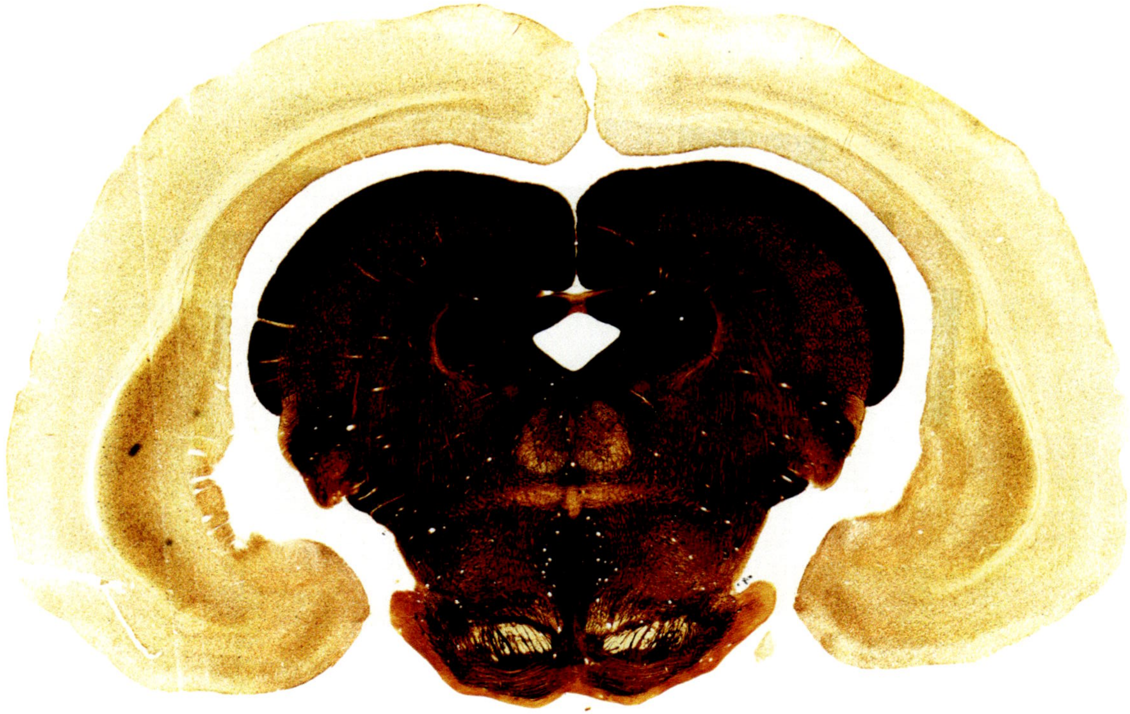

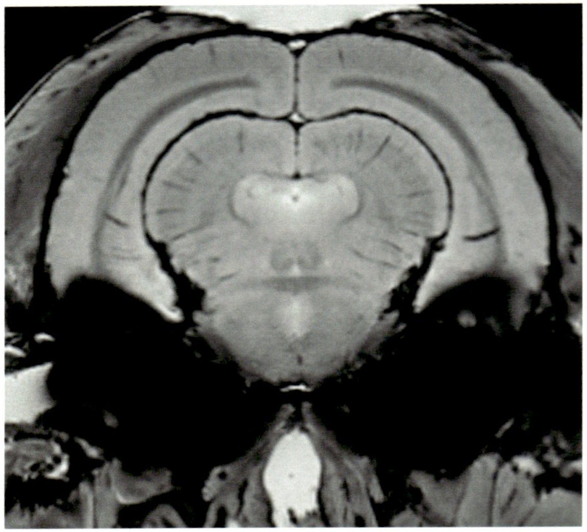

3PC	oculomotor nucleus, parvicellular part
Aq	aqueduct
CLi	caudal linear nucleus of the raphe
csc	commissure of the superior colliculus
DpG	deep gray layer of the superior colliculus
DpWh	deep white layer of the superior colliculus
DR	dorsal raphe nucleus
Ent	entorhinal cortex
Hip	hippocampus
InG	intermediate gray layer of the superior colliculus
InWh	intermediate white layer of the superior colliculus
lfp	longitudinal fasciculus of the pons
mcp	middle cerebellar peduncle
me5	mesencephalic trigeminal tract
Me5	mesencephalic trigeminal nucleus
MGV	medial geniculate nucleus, ventral part
MiTg	microcellular tegmental nucleus
ml	medial lemniscus
mlf	medial longitudinal fasciculus
MnR	median raphe nucleus
mRt	mesencephalic reticular formation
Op	optic nerve layer of the superior colliculus
opt	optic tract
PAG	periaqueductal gray
PBG	parabigeminal nucleus
Pd	dorsal nucleus of the pulvinar
Pn	pontine nuclei
PnO	pontine reticular nucleus, oral part
Post	postsubiculum
PrS	presubiculum
PTg	pedunculopontine tegmental nucleus
Rbd	rhabdoid nucleus
RRF	retrorubral field
RSg	retrosplenial granular cortex
SuG	superficial gray layer of the superior colliculus
TC	temporal cortex
tfp	transverse fibers of the pons
tth	trigeminothalamic tract
V1	primary visual cortex
V2	secondary visual cortex
xscp	decussation of the superior cerebellar peduncle
Zo	zonal layer of the superior colliculus

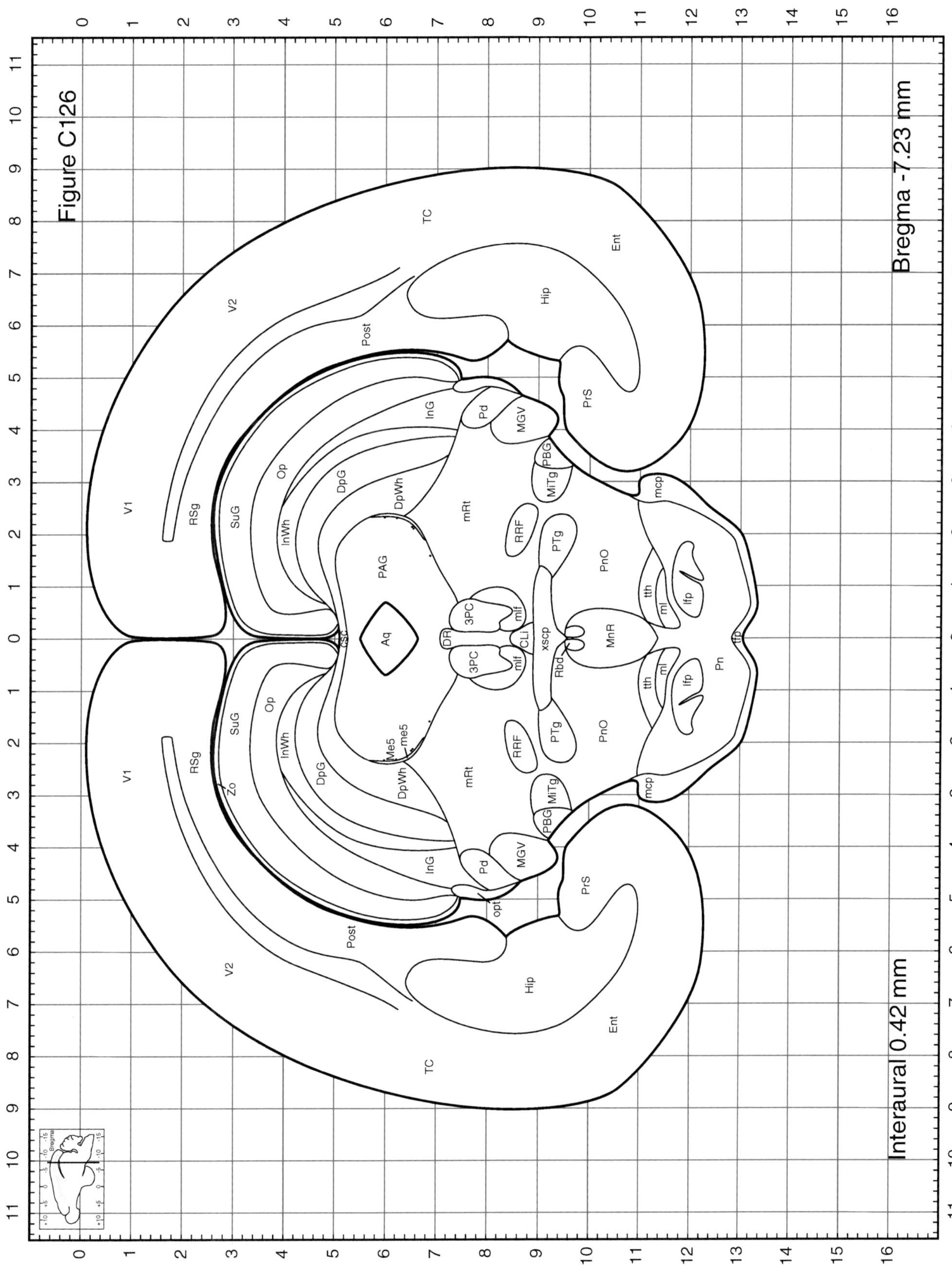
Figure C126
Bregma -7.23 mm
Interaural 0.42 mm
TC
V2
V1
Ent
Hip
Post
PrS
RSg
SuG
Op
InWh
DpG
DpWh
InG
Pd
MGV
PBG
MiTg
mRt
PAG
Aq
CSC
DR
3PC
mlf
CLi
xscp
Rbd
RRF
PTg
PnO
MnR
mcp
tth
ml
lfp
Pn
Me5
me5
Zo
opt

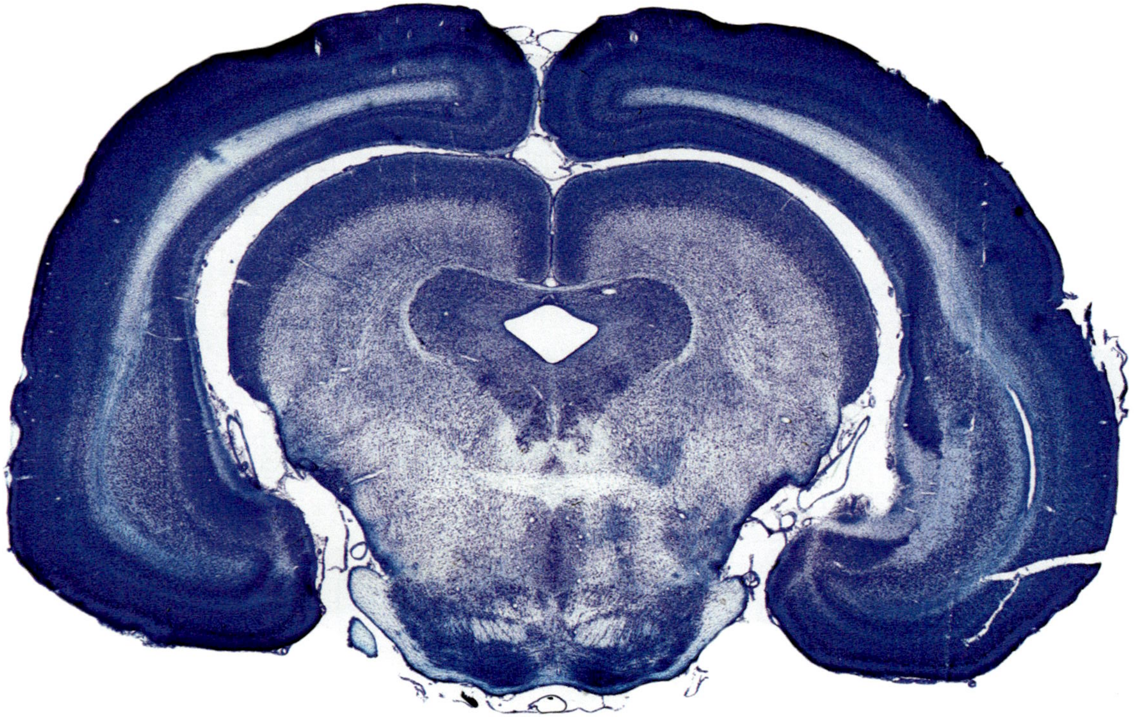

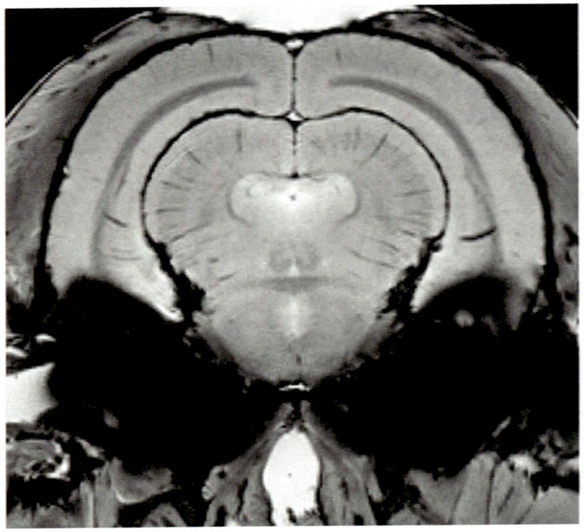

3PC	oculomotor nucleus, parvicellular part
6n	root of abducens nerve
Aq	aqueduct
CLi	caudal linear nucleus of the raphe
csc	commissure of the superior colliculus
DpG	deep gray layer of the superior colliculus
DpWh	deep white layer of the superior colliculus
DR	dorsal raphe nucleus
Ent	entorhinal cortex
Hip	hippocampus
InG	intermediate gray layer of the superior colliculus
InWh	intermediate white layer of the superior colliculus
lfp	longitudinal fasciculus of the pons
mcp	middle cerebellar peduncle
me5	mesencephalic trigeminal tract
Me5	mesencephalic trigeminal nucleus
MGV	medial geniculate nucleus, ventral part
MiTg	microcellular tegmental nucleus
ml	medial lemniscus
mlf	medial longitudinal fasciculus
MnR	median raphe nucleus
mRt	mesencephalic reticular formation
Op	optic nerve layer of the superior colliculus
opt	optic tract
PAG	periaqueductal gray
PBG	parabigeminal nucleus
Pd	dorsal nucleus of the pulvinar
Pn	pontine nuclei
PnO	pontine reticular nucleus, oral part
Post	postsubiculum
PrS	presubiculum
PTg	pedunculopontine tegmental nucleus
Rbd	rhabdoid nucleus
RRF	retrorubral field
RSg	retrosplenial granular cortex
s5	sensory root of the trigeminal nerve
SuG	superficial gray layer of the superior colliculus
TC	temporal cortex
tfp	transverse fibers of the pons
tth	trigeminothalamic tract
V1	primary visual cortex
V2	secondary visual cortex
xscp	decussation of the superior cerebellar peduncle
Zo	zonal layer of the superior colliculus

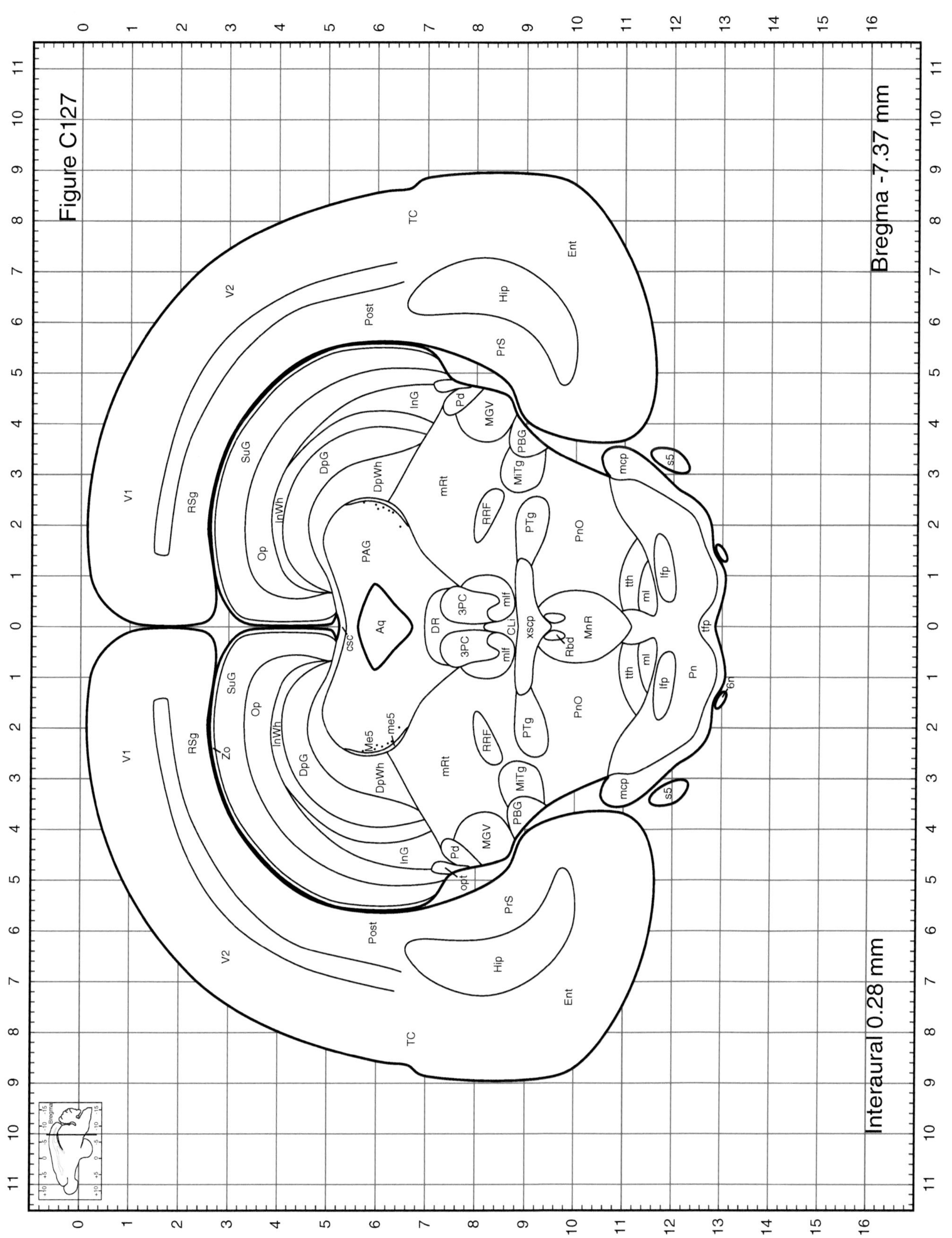
Figure C127
Bregma -7.37 mm
Interaural 0.28 mm
TC
Ent
Hip
V2
Post
PrS
V1
RSg
SuG
Op
InWh
DpG
DpWh
InG
Pd
MGV
PBG
MiTg
mRt
RRF
PTg
PnO
PAG
Aq
csc
DR
3PC
mlf
CLi
xscp
Rbd
MnR
tth
ml
lfp
mcp
s5
tfp
Pn
6n
Me5
me5
Zo
opt

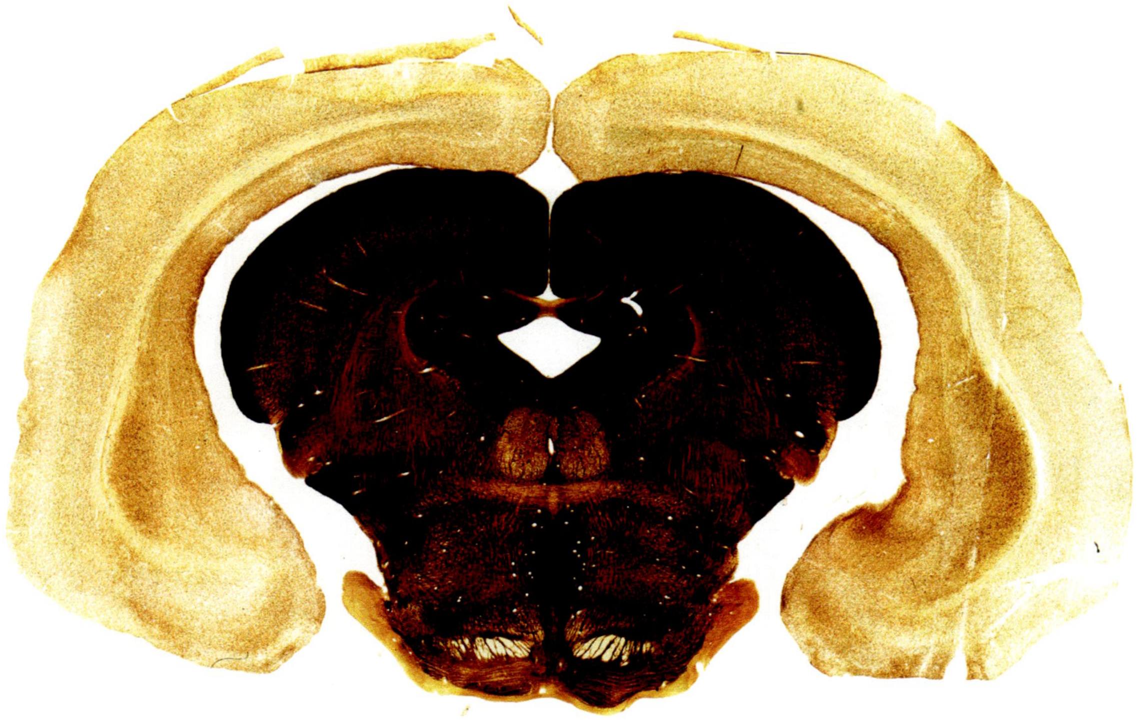

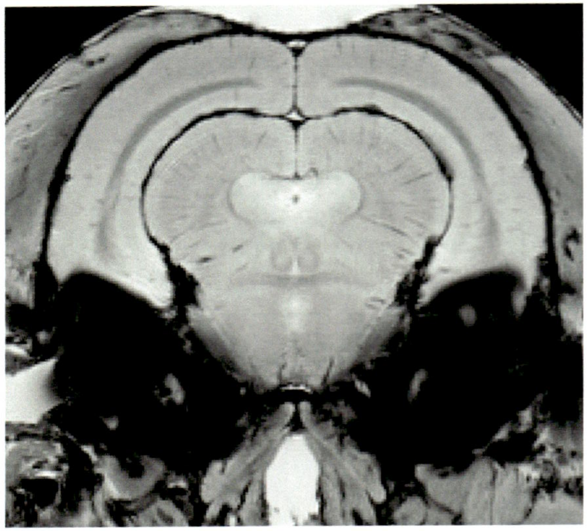

3PC	oculomotor nucleus, parvicellular part
4N	trochlear nucleus
6n	root of abducens nerve
Aq	aqueduct
CLi	caudal linear nucleus of the raphe
csc	commissure of the superior colliculus
DpG	deep gray layer of the superior colliculus
DpWh	deep white layer of the superior colliculus
DR	dorsal raphe nucleus
Ent	entorhinal cortex
Hip	hippocampus
InG	intermediate gray layer of the superior colliculus
InWh	intermediate white layer of the superior colliculus
lfp	longitudinal fasciculus of the pons
mcp	middle cerebellar peduncle
me5	mesencephalic trigeminal tract
Me5	mesencephalic trigeminal nucleus
MGV	medial geniculate nucleus, ventral part
MiTg	microcellular tegmental nucleus
ml	medial lemniscus
mlf	medial longitudinal fasciculus
MnR	median raphe nucleus
mRt	mesencephalic reticular formation
Op	optic nerve layer of the superior colliculus
opt	optic tract
PAG	periaqueductal gray
PBG	parabigeminal nucleus
Pd	dorsal nucleus of the pulvinar
Pn	pontine nuclei
PnO	pontine reticular nucleus, oral part
Post	postsubiculum
PrS	presubiculum
PTg	pedunculopontine tegmental nucleus
Rbd	rhabdoid nucleus
RRF	retrorubral field
RSg	retrosplenial granular cortex
SPTg	subpeduncular tegmental nucleus
SuG	superficial gray layer of the superior colliculus
TC	temporal cortex
tfp	transverse fibers of the pons
tth	trigeminothalamic tract
V1	primary visual cortex
V2	secondary visual cortex
xscp	decussation of the superior cerebellar peduncle
Zo	zonal layer of the superior colliculus aa aa a

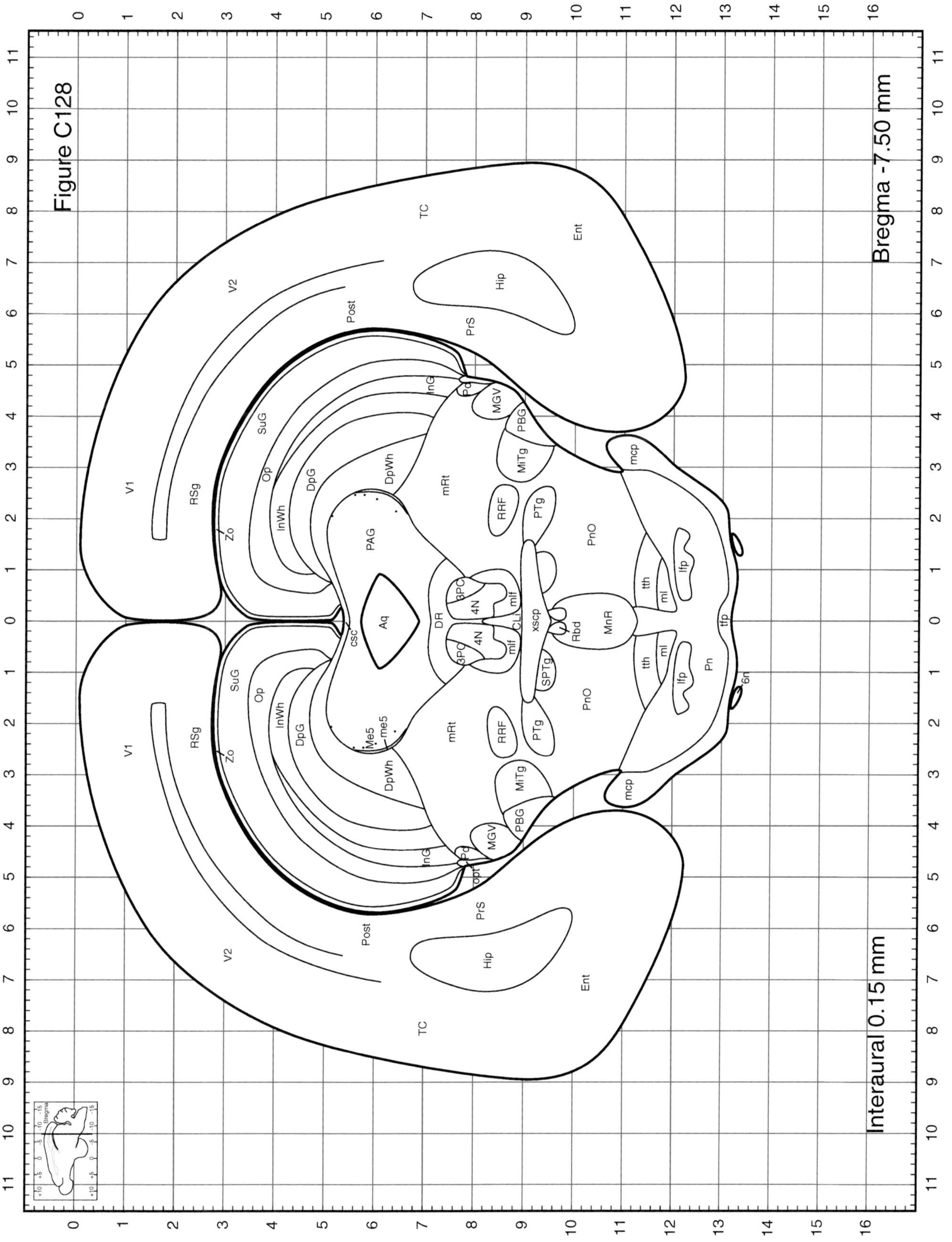

Figure C128
Bregma -7.50 mm
Interaural 0.15 mm
V1
V2
RSg
TC
Ent
Hip
PrS
Post
SuG
Op
InWh
DpG
DpWh
InG
Zo
PAG
Aq
csc
Me5
me5
DR
3PC
4N
mlf
CLi
xscp
mRt
RRF
PTg
SPTg
MiTg
PBG
MGV
Pd
opt
PnO
Rbd
MnR
tth
ml
lfp
tfp
Pn
mcp
6n

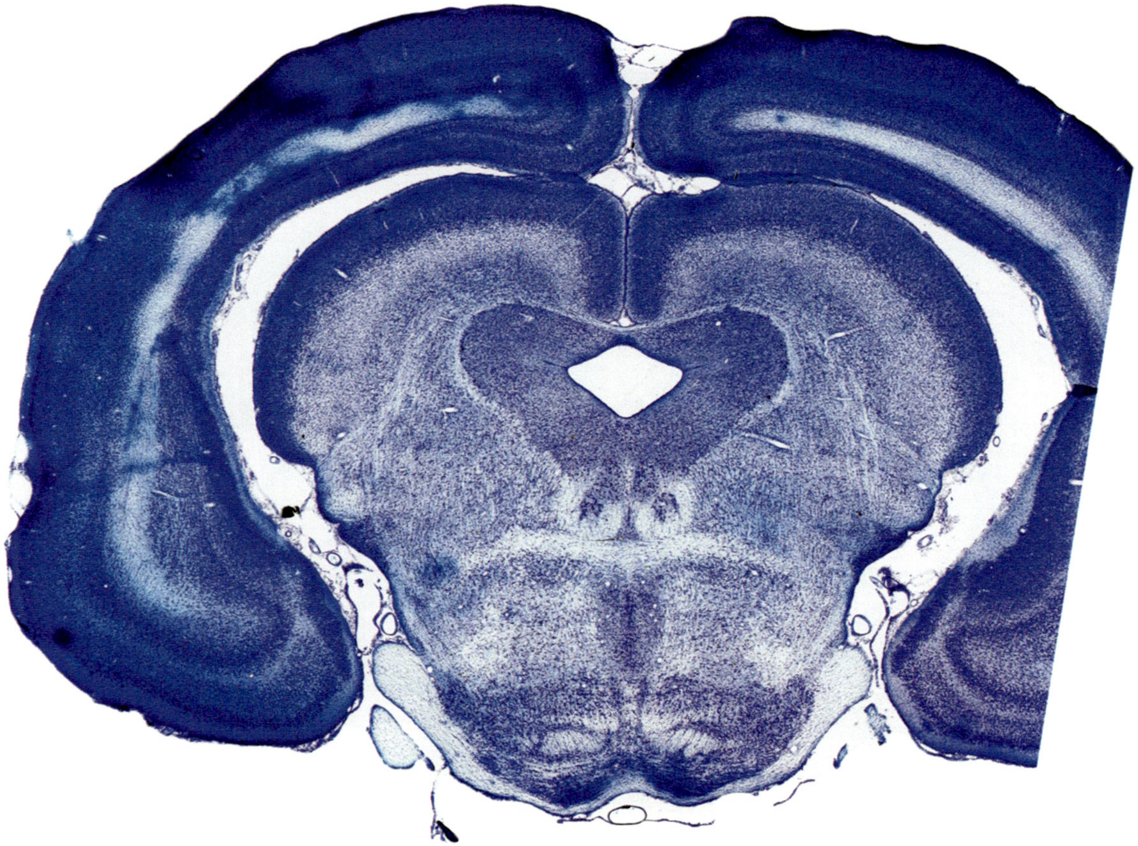

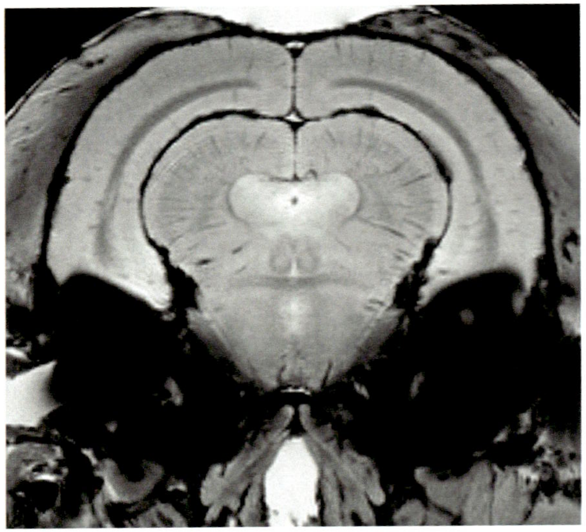

3PC	oculomotor nucleus, parvicellular part
4N	trochlear nucleus
6n	root of abducens nerve
Aq	aqueduct
CLi	caudal linear nucleus of the raphe
csc	commissure of the superior colliculus
DpG	deep gray layer of the superior colliculus
DpWh	deep white layer of the superior colliculus
DR	dorsal raphe nucleus
Ent	entorhinal cortex
Hip	hippocampus
InG	intermediate gray layer of the superior colliculus
InWh	intermediate white layer of the superior colliculus
lfp	longitudinal fasciculus of the pons
mcp	middle cerebellar peduncle
MGV	medial geniculate nucleus, ventral part
me5	mesencephalic trigeminal tract
Me5	mesencephalic trigeminal nucleus
MiTg	microcellular tegmental nucleus
ml	medial lemniscus
mlf	medial longitudinal fasciculus
MnR	median raphe nucleus
mRt	mesencephalic reticular formation
Op	optic nerve layer of the superior colliculus
PAG	periaqueductal gray
PBG	parabigeminal nucleus
Pn	pontine nuclei
PnO	pontine reticular nucleus, oral part
Post	postsubiculum
PrS	presubiculum
PTg	pedunculopontine tegmental nucleus
Rbd	rhabdoid nucleus
RRF	retrorubral field
RSg	retrosplenial granular cortex
s5	sensory root of the trigeminal nerve
SPTg	subpeduncular tegmental nucleus
SuG	superficial gray layer of the superior colliculus
TC	temporal cortex
tfp	transverse fibers of the pons
tth	trigeminothalamic tract
V1	primary visual cortex
V2	secondary visual cortex
xscp	decussation of the superior cerebellar peduncle
Zo	zonal layer of the superior colliculus

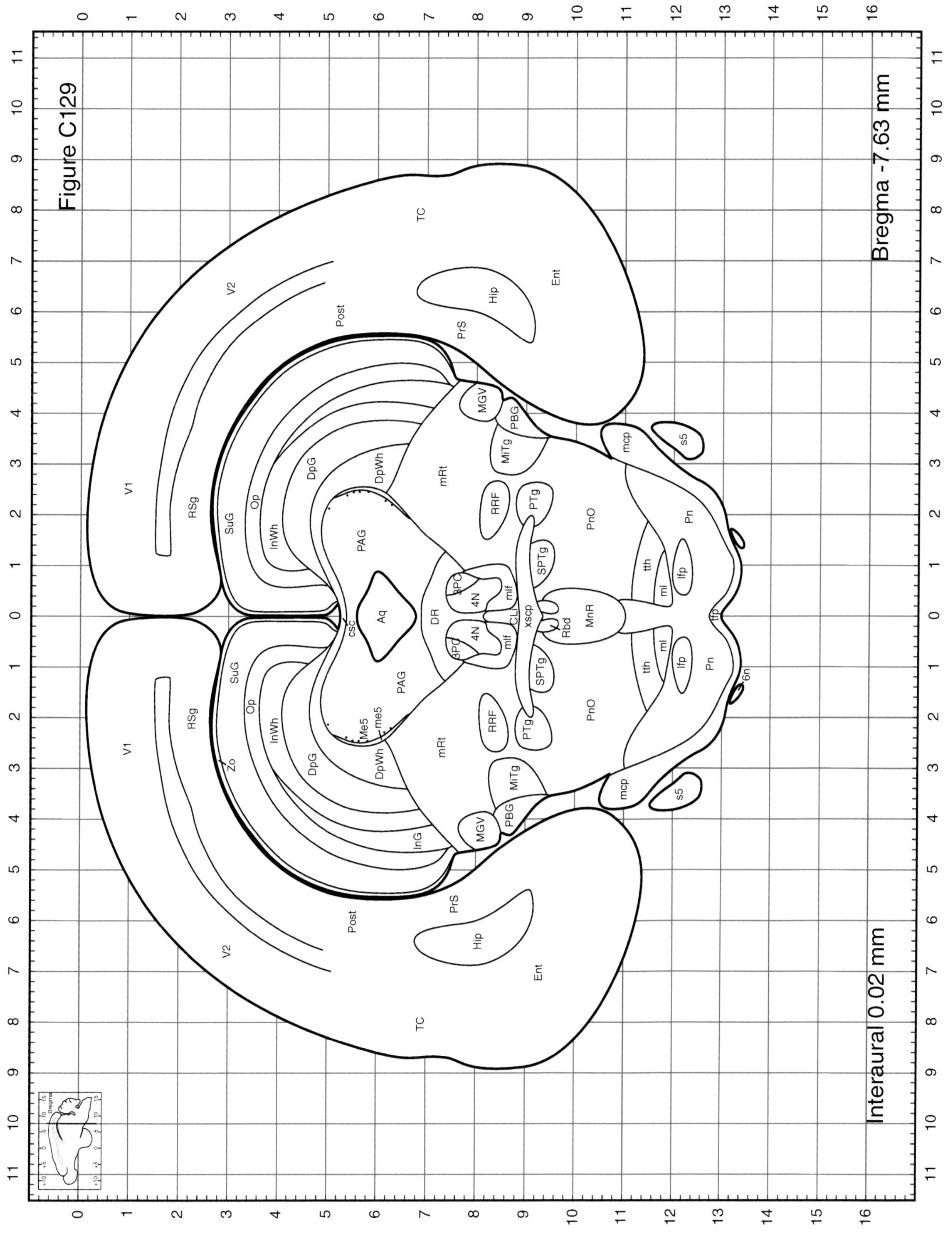
Figure C129
Bregma -7.63 mm
Interaural 0.02 mm
V1
V2
RSg
Post
TC
Ent
Hip
PrS
SuG
Op
InWh
InG
DpG
DpWh
Zo
PAG
Aq
csc
Me5
me5
DR
mRt
MGV
PBG
MiTg
RRF
PTg
SPTg
3PC
4N
mlf
CLi
xscp
Rbd
MnR
PnO
Pn
tth
ml
lfp
tfp
mcp
s5
6n

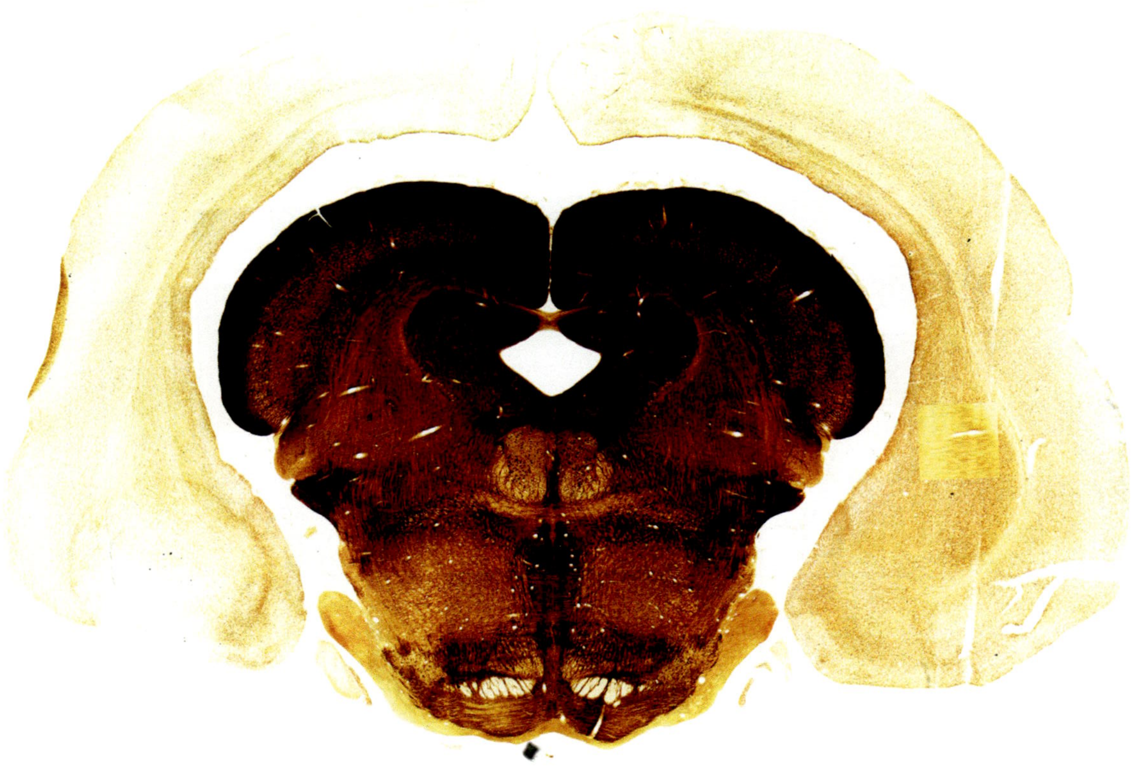

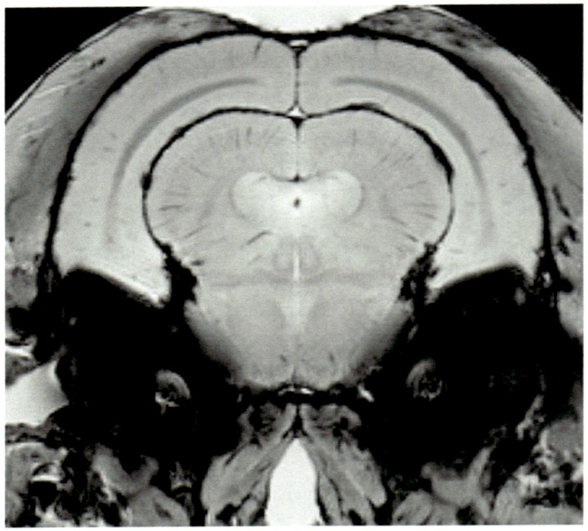

4N	trochlear nucleus
6n	root of abducens nerve
Aq	aqueduct
CLi	caudal linear nucleus of the raphe
csc	commissure of the superior colliculus
DpG	deep gray layer of the superior colliculus
DpWh	deep white layer of the superior colliculus
DR	dorsal raphe nucleus
Ent	entorhinal cortex
Hip	hippocampus
InG	intermediate gray layer of the superior colliculus
InWh	intermediate white layer of the superior colliculus
lfp	longitudinal fasciculus of the pons
LL	nucleus of the lateral lemniscus
mcp	middle cerebellar peduncle
me5	mesencephalic trigeminal tract
Me5	mesencephalic trigeminal nucleus
MGV	medial geniculate nucleus, ventral part
MiTg	microcellular tegmental nucleus
ml	medial lemniscus
mlf	medial longitudinal fasciculus
MnR	median raphe nucleus
mRt	mesencephalic reticular formation
Op	optic nerve layer of the superior colliculus
PAG	periaqueductal gray
PBG	parabigeminal nucleus
Pn	pontine nuclei
PnO	pontine reticular nucleus, oral part
Post	postsubiculum
PrS	presubiculum
PTg	pedunculopontine tegmental nucleus
Rbd	rhabdoid nucleus
RSg	retrosplenial granular cortex
RtTg	reticulotegmental nucleus of the pons
s5	sensory root of the trigeminal nerve
SPTg	subpeduncular tegmental nucleus
SuG	superficial gray layer of the superior colliculus
TC	temporal cortex
tfp	transverse fibers of the pons
tth	trigeminothalamic tract
V1	primary visual cortex
V2	secondary visual cortex
xscp	decussation of the superior cerebellar peduncle
Zo	zonal layer of the superior colliculus

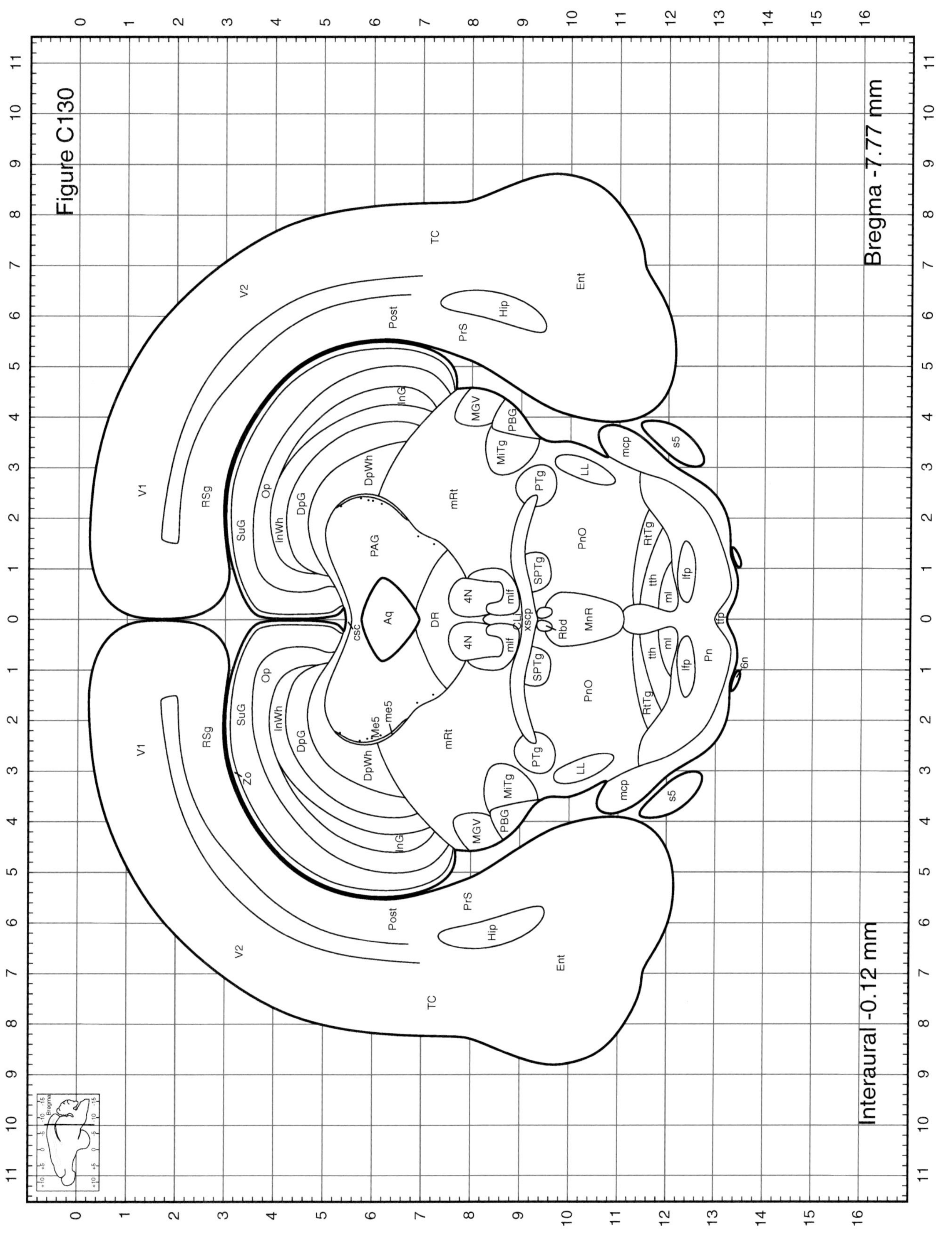
Figure C130
Bregma -7.77 mm
Interaural -0.12 mm
V1
V2
RSg
TC
Post
PrS
Hip
Ent
SuG
Op
InWh
DpG
DpWh
InG
Zo
PAG
Aq
csc
DR
Me5
me5
mRt
MGV
PBG
MiTg
PTg
LL
mcp
s5
4N
mlf
CLi
xscp
SPTg
Rbd
MnR
PnO
RtTg
tth
ml
lfp
Pn
6n

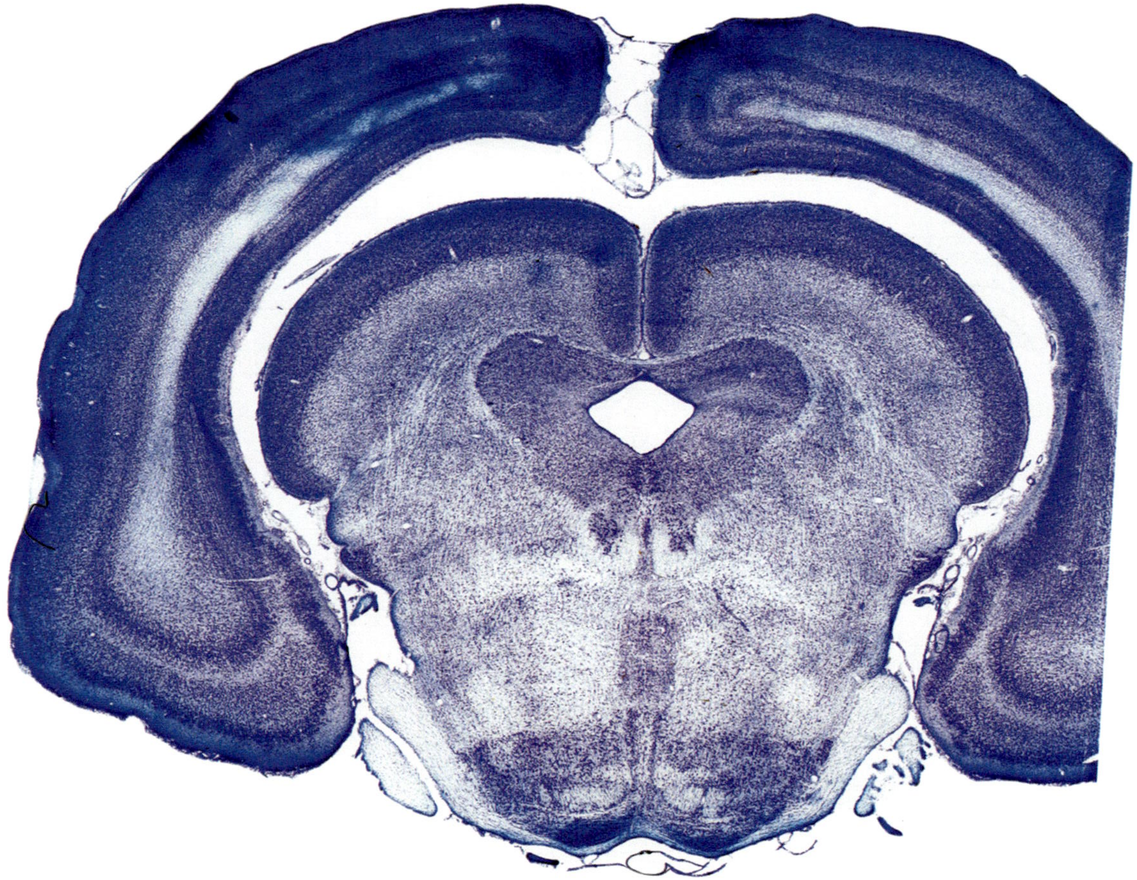

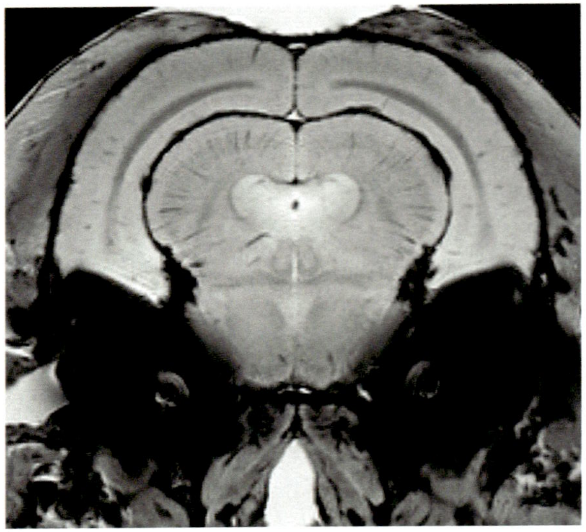

4N	trochlear nucleus
6n	root of abducens nerve
Aq	aqueduct
csc	commissure of the superior colliculus
DpG	deep gray layer of the superior colliculus
DpWh	deep white layer of the superior colliculus
DR	dorsal raphe nucleus
Ent	entorhinal cortex
InG	intermediate gray layer of the superior colliculus
InWh	intermediate white layer of the superior colliculus
lfp	longitudinal fasciculus of the pons
LL	nucleus of the lateral lemniscus
mcp	middle cerebellar peduncle
me5	mesencephalic trigeminal tract
Me5	mesencephalic trigeminal nucleus
MGV	medial geniculate nucleus, ventral part
MiTg	microcellular tegmental nucleus
ml	medial lemniscus
mlf	medial longitudinal fasciculus
MnR	median raphe nucleus
mRt	mesencephalic reticular formation
Op	optic nerve layer of the superior colliculus
PAG	periaqueductal gray
PBG	parabigeminal nucleus
Pn	pontine nuclei
PnO	pontine reticular nucleus, oral part
Post	postsubiculum
PrS	presubiculum
PTg	pedunculopontine tegmental nucleus
Rbd	rhabdoid nucleus
RSg	retrosplenial granular cortex
RtTg	reticulotegmental nucleus of the pons
s5	sensory root of the trigeminal nerve
scp	superior cerebellar peduncle
SPTg	subpeduncular tegmental nucleus
SuG	superficial gray layer of the superior colliculus
TC	temporal cortex
tfp	transverse fibers of the pons
tth	trigeminothalamic tract
V1	primary visual cortex
V2	secondary visual cortex
Zo	zonal layer of the superior colliculus

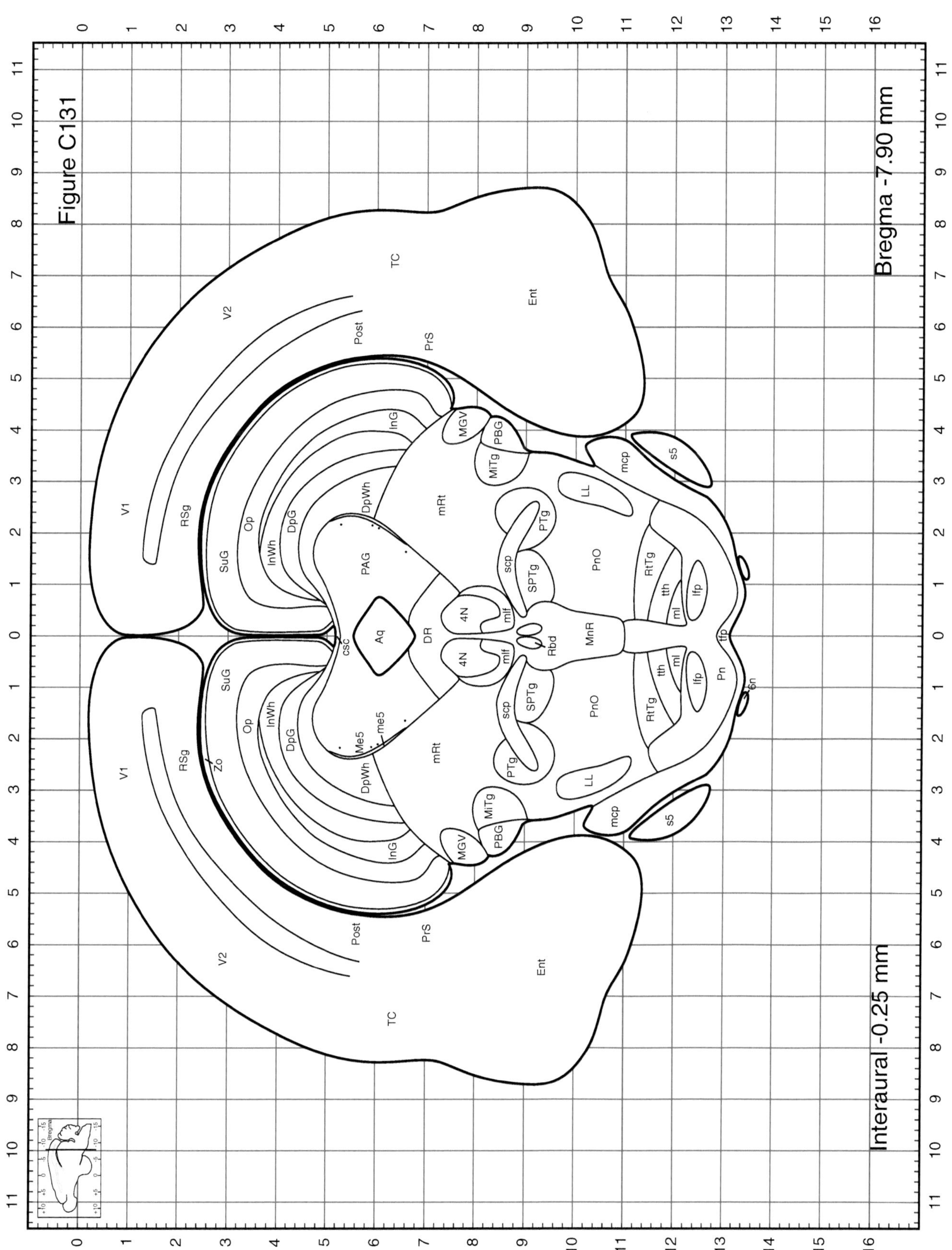
Figure C131
Bregma -7.90 mm
Interaural -0.25 mm
TC
Ent
V2
Post
PrS
V1
RSg
SuG
Op
InWh
DpG
DpWh
InG
PAG
Aq
csc
DR
mRt
Me5
me5
Zo
MGV
PBG
MiTg
4N
mlf
scp
SPTg
PTg
LL
mcp
s5
PnO
MnR
Rbd
RtTg
tth
ml
lfp
tfp
Pn
6n

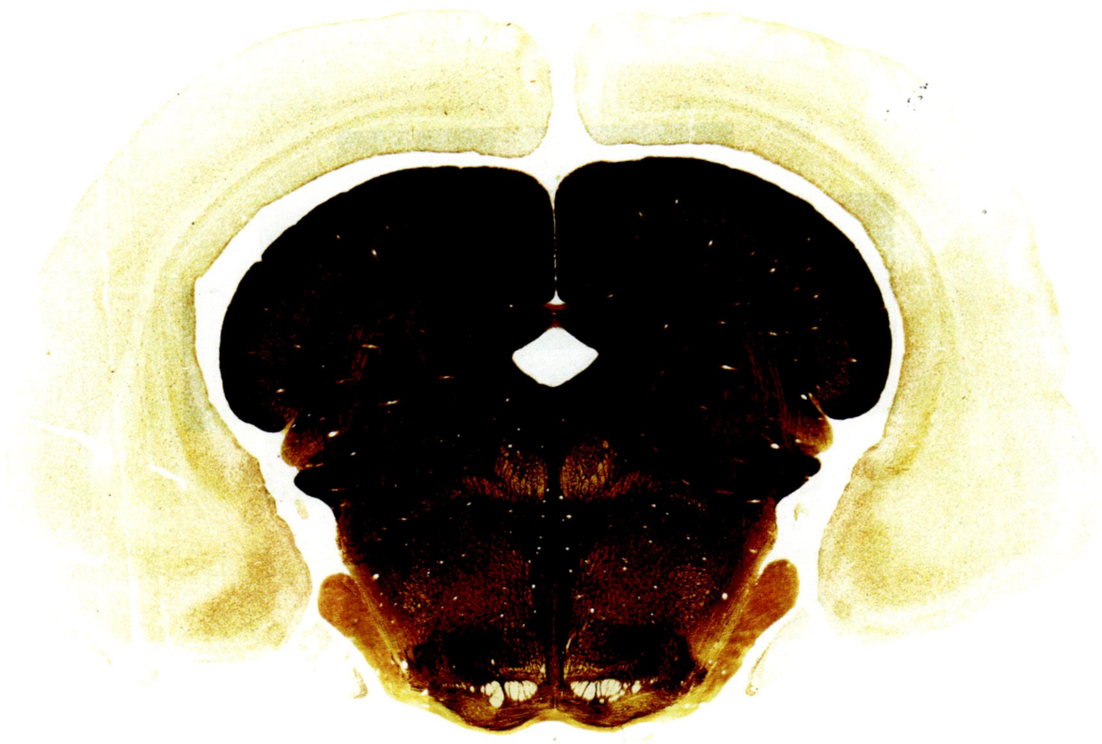

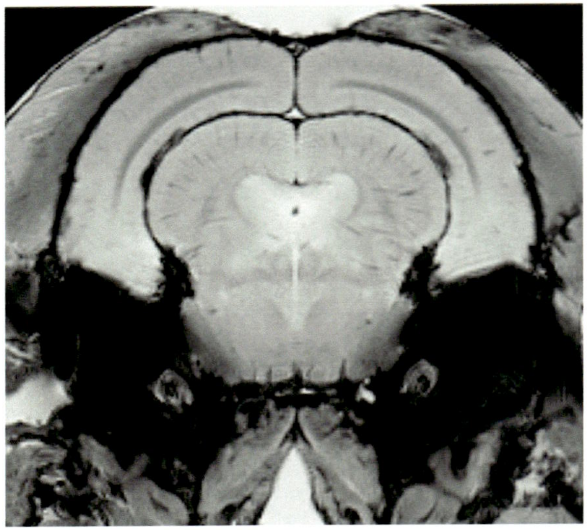

4n	trochlear nerve
4N	trochlear nucleus
6n	root of abducens nerve
Aq	aqueduct
csc	commissure of the superior colliculus
DpG	deep gray layer of the superior colliculus
DpWh	deep white layer of the superior colliculus
DR	dorsal raphe nucleus
Ent	entorhinal cortex
IC	inferior colliculus
InG	intermediate gray layer of the superior colliculus
InWh	intermediate white layer of the superior colliculus
lfp	longitudinal fasciculus of the pons
LL	nucleus of the lateral lemniscus
mcp	middle cerebellar peduncle
me5	mesencephalic trigeminal tract
Me5	mesencephalic trigeminal nucleus
MGV	medial geniculate nucleus, ventral part
MiTg	microcellular tegmental nucleus
ml	medial lemniscus
mlf	medial longitudinal fasciculus
MnR	median raphe nucleus
mRt	mesencephalic reticular formation
Op	optic nerve layer of the superior colliculus
PAG	periaqueductal gray
PBG	parabigeminal nucleus
Pn	pontine nuclei
PnO	pontine reticular nucleus, oral part
Post	postsubiculum
PrS	presubiculum
PTg	pedunculopontine tegmental nucleus
Rbd	rhabdoid nucleus
RSg	retrosplenial granular cortex
RtTg	reticulotegmental nucleus of the pons
s5	sensory root of the trigeminal nerve
scp	superior cerebellar peduncle
SPTg	subpeduncular tegmental nucleus
SuG	superficial gray layer of the superior colliculus
TC	temporal cortex
tfp	transverse fibers of the pons
tth	trigeminothalamic tract
V1	primary visual cortex
V2	secondary visual cortex
VTg	ventral tegmental nucleus
Zo	zonal layer of the superior colliculus

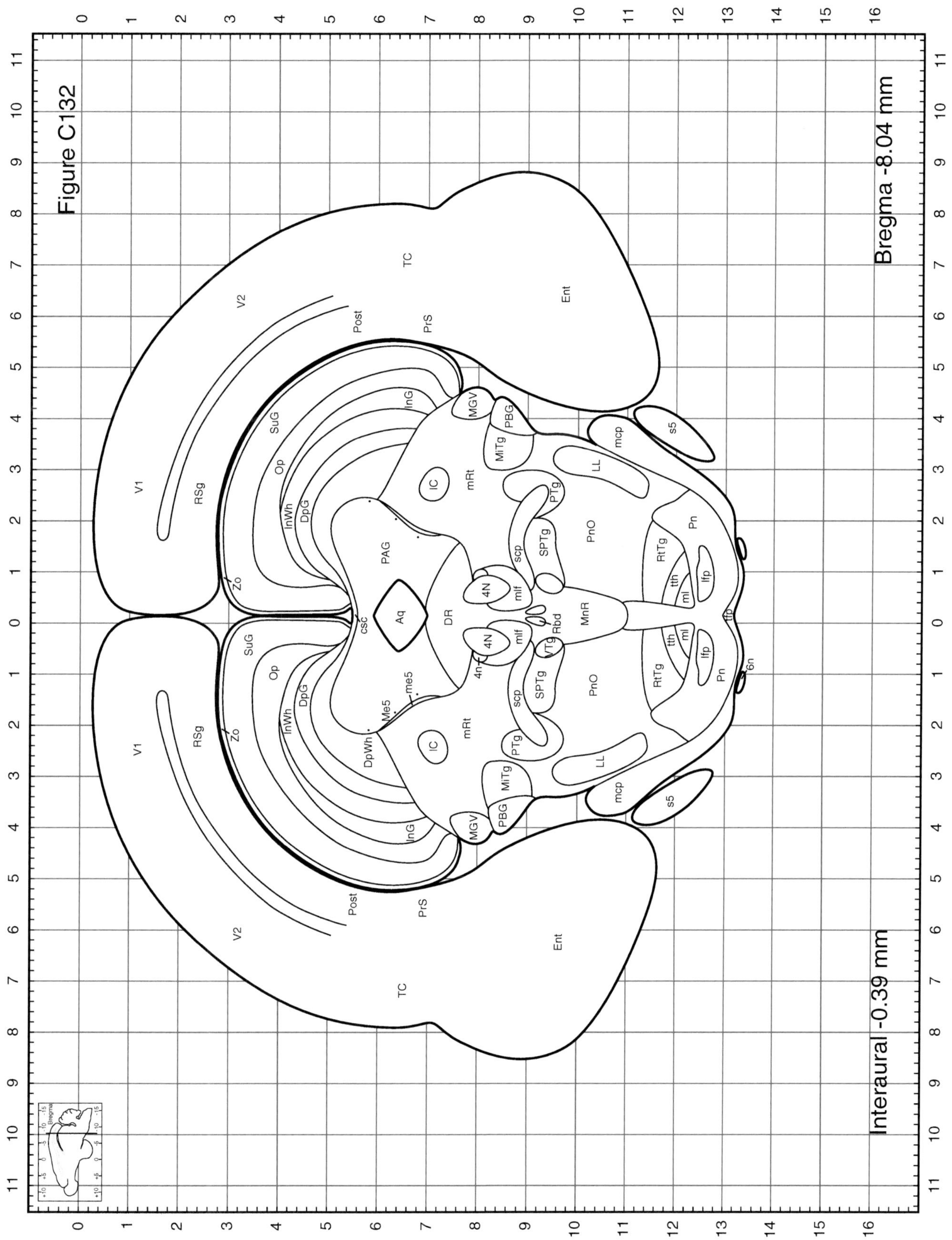

Figure C132
Bregma -8.04 mm
Interaural -0.39 mm
V1
V2
RSg
TC
Post
PrS
Ent
SuG
Op
InWh
DpG
InG
DpWh
Zo
PAG
Aq
csc
DR
Me5
me5
IC
mRt
MGV
PBG
MiTg
PTg
scp
SPTg
4N
4n
mlf
Rbd
VTg
MnR
PnO
LL
mcp
s5
RtTg
Pn
tth
ml
lfp
6n

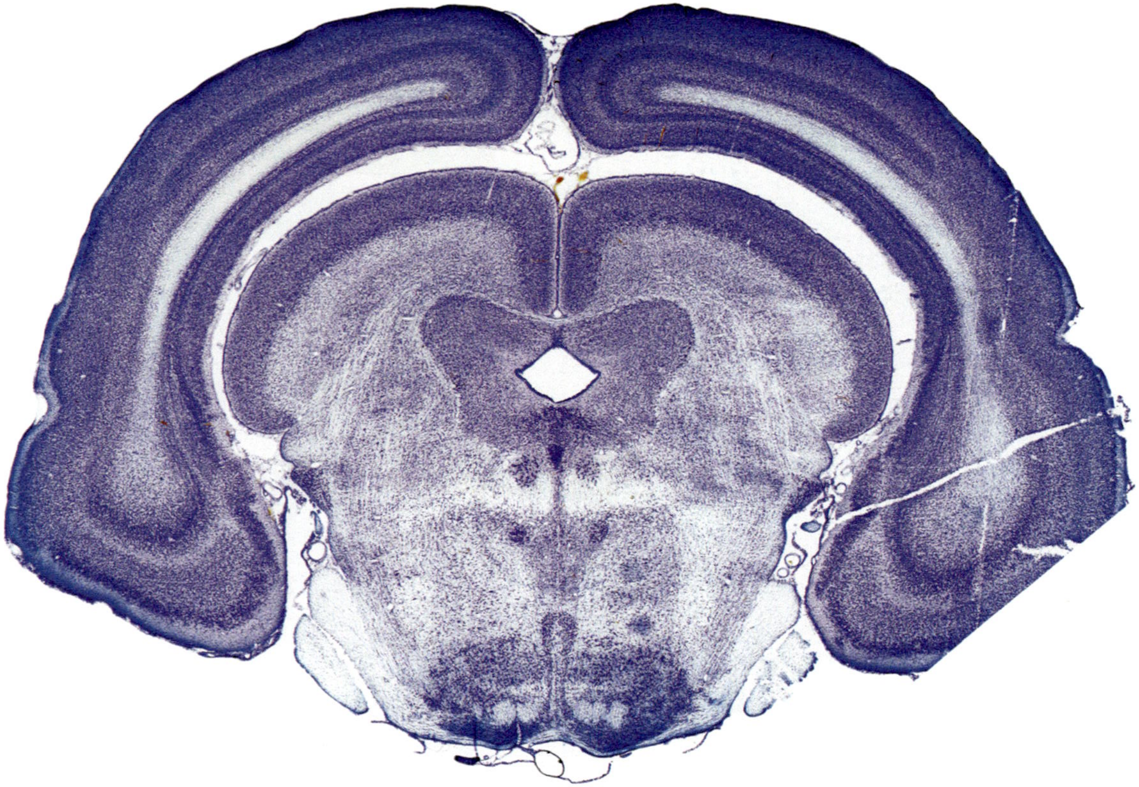

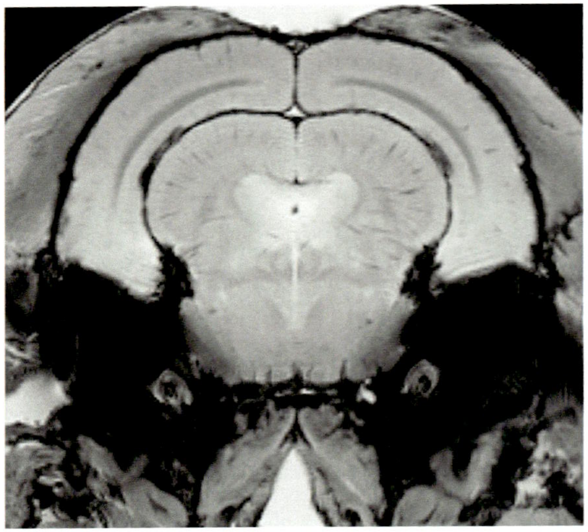

4n	trochlear nerve
4N	trochlear nucleus
6n	root of abducens nerve
Aq	aqueduct
csc	commissure of the superior colliculus
DpG	deep gray layer of the superior colliculus
DpWh	deep white layer of the superior colliculus
DR	dorsal raphe nucleus
Ent	entorhinal cortex
IC	inferior colliculus
InG	intermediate gray layer of the superior colliculus
InWh	intermediate white layer of the superior colliculus
lfp	longitudinal fasciculus of the pons
LL	nucleus of the lateral lemniscus
mcp	middle cerebellar peduncle
me5	mesencephalic trigeminal tract
Me5	mesencephalic trigeminal nucleus
MGV	medial geniculate nucleus, ventral part
MiTg	microcellular tegmental nucleus
ml	medial lemniscus
mlf	medial longitudinal fasciculus
MnR	median raphe nucleus
mRt	mesencephalic reticular formation
Op	optic nerve layer of the superior colliculus
PAG	periaqueductal gray
PBG	parabigeminal nucleus
Pn	pontine nuclei
PnO	pontine reticular nucleus, oral part
PTg	pedunculopontine tegmental nucleus
Rbd	rhabdoid nucleus
RSg	retrosplenial granular cortex
RtTg	reticulotegmental nucleus of the pons
s5	sensory root of the trigeminal nerve
scp	superior cerebellar peduncle
SPTg	subpeduncular tegmental nucleus
SuG	superficial gray layer of the superior colliculus
TC	temporal cortex
tfp	transverse fibers of the pons
tth	trigeminothalamic tract
V1	primary visual cortex
V2	secondary visual cortex
VTg	ventral tegmental nucleus
Zo	zonal layer of the superior colliculus

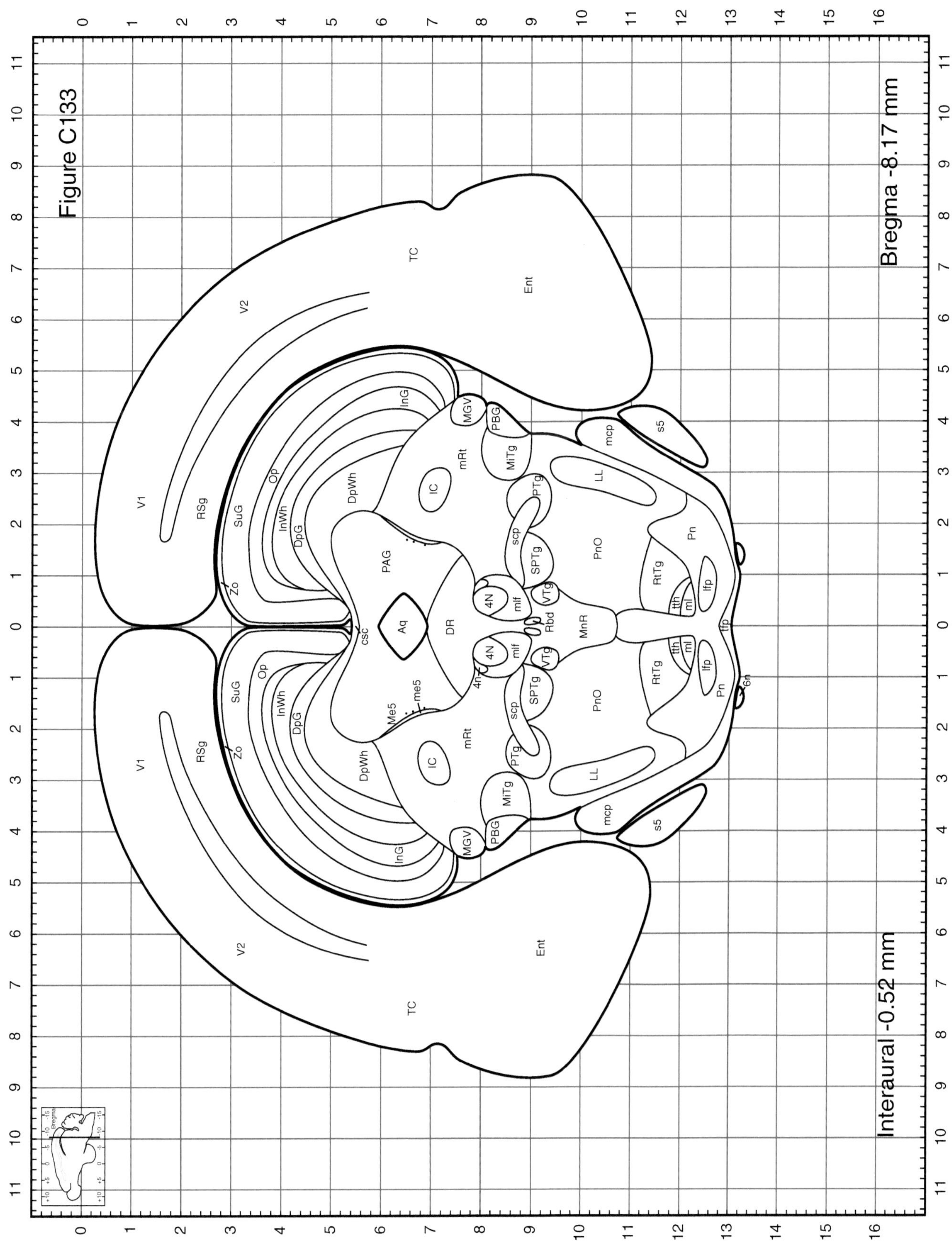

Figure C133
Bregma -8.17 mm
Interaural -0.52 mm
TC
Ent
V2
V1
RSg
SuG
Op
InWh
DpG
Zo
InG
DpWh
MGV
PBG
MiTg
mRt
IC
PAG
Aq
DR
csc
Me5
me5
4N
4n
mlf
VTg
SPTg
scp
PTg
Rbd
MnR
PnO
LL
mcp
s5
RtTg
Pn
tth
ml
lfp
tfp
6n

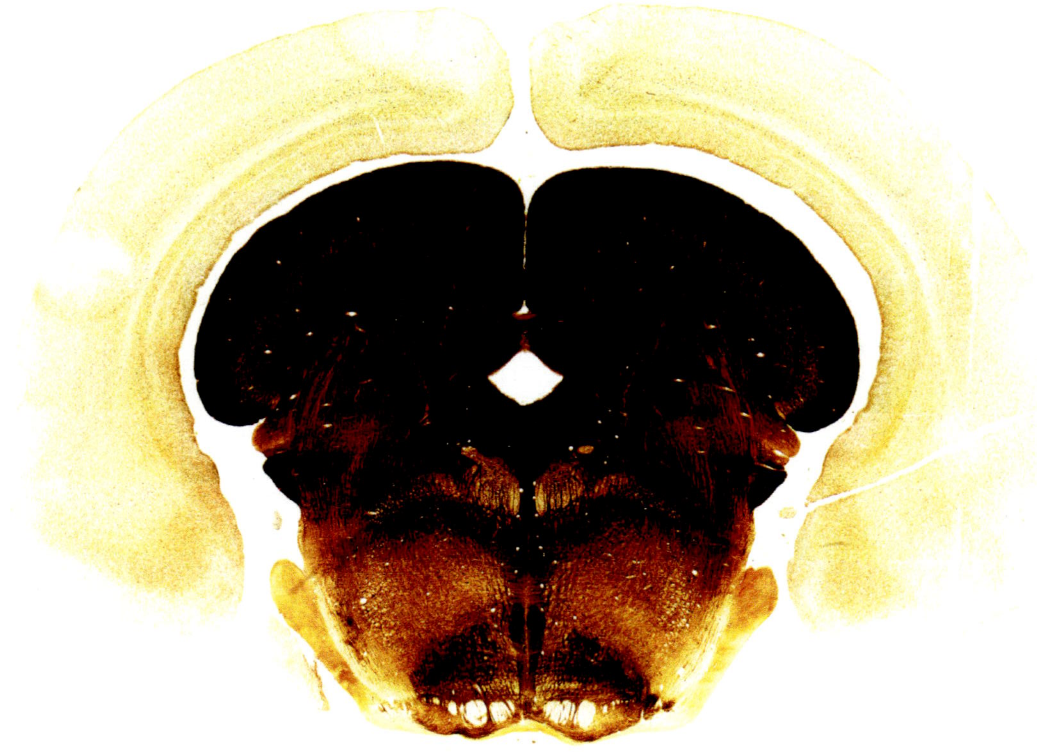

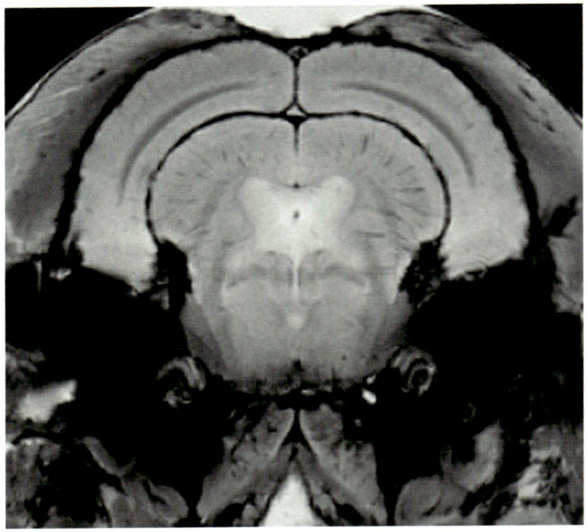

- 4n trochlear nerve
- 4N trochlear nucleus
- 6n root of abducens nerve
- Aq aqueduct
- CnF cuneiform nucleus
- csc commissure of the superior colliculus
- DpG deep gray layer of the superior colliculus
- DpWh deep white layer of the superior colliculus
- DR dorsal raphe nucleus
- Ent entorhinal cortex
- IC inferior colliculus
- InG intermediate gray layer of the superior colliculus
- InWh intermediate white layer of the superior colliculus
- lfp longitudinal fasciculus of the pons
- LL nucleus of the lateral lemniscus
- mcp middle cerebellar peduncle
- me5 mesencephalic trigeminal tract
- Me5 mesencephalic trigeminal nucleus
- MGV medial geniculate nucleus, ventral part
- MiTg microcellular tegmental nucleus
- ml medial lemniscus
- mlf medial longitudinal fasciculus
- MnR median raphe nucleus
- mRt mesencephalic reticular formation
- Op optic nerve layer of the superior colliculus
- PAG periaqueductal gray
- PBG parabigeminal nucleus
- Pn pontine nuclei
- PnO pontine reticular nucleus, oral part
- PTg pedunculopontine tegmental nucleus
- Rbd rhabdoid nucleus
- RSg retrosplenial granular cortex
- RtTg reticulotegmental nucleus of the pons
- s5 sensory root of the trigeminal nerve
- scp superior cerebellar peduncle
- SPTg subpeduncular tegmental nucleus
- SuG superficial gray layer of the superior colliculus
- TC temporal cortex
- tfp transverse fibers of the pons
- tth trigeminothalamic tract
- V1 primary visual cortex
- V2 secondary visual cortex
- VTg ventral tegmental nucleus
- Zo zonal layer of the superior colliculus

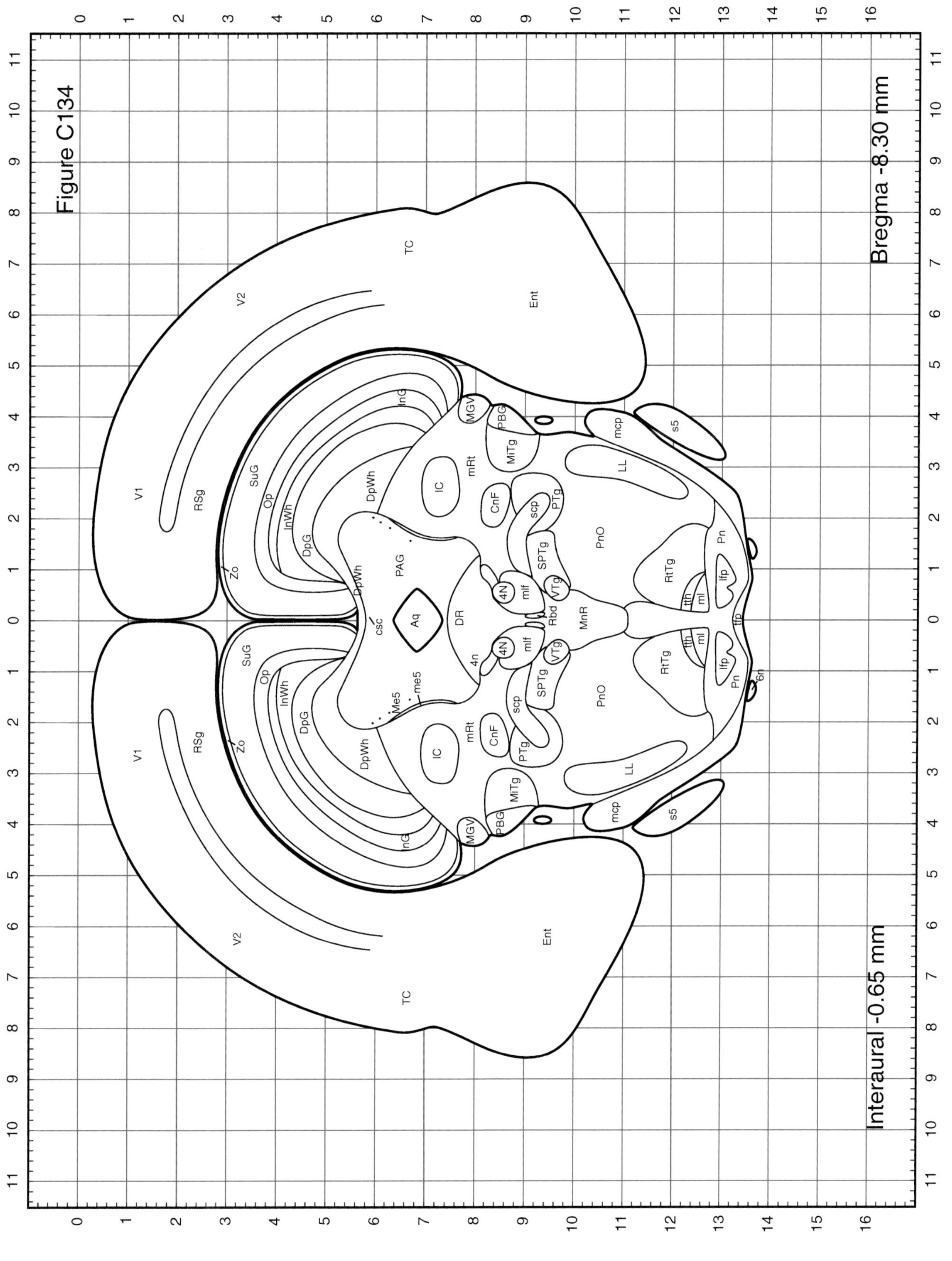
Figure C134
Bregma -8.30 mm
Interaural -0.65 mm
TC
V2
Ent
V1
RSg
SuG
Op
InWh
DpG
Zo
InG
DpWh
PAG
csc
Aq
DR
Me5
me5
4n
MGV
PBG
MiTg
mRt
IC
CnF
scp
PTg
SPTg
4N
mlf
VTg
Rbd
MnR
PnO
LL
mcp
s5
RtTg
tth
ml
lfp
Pn
6n

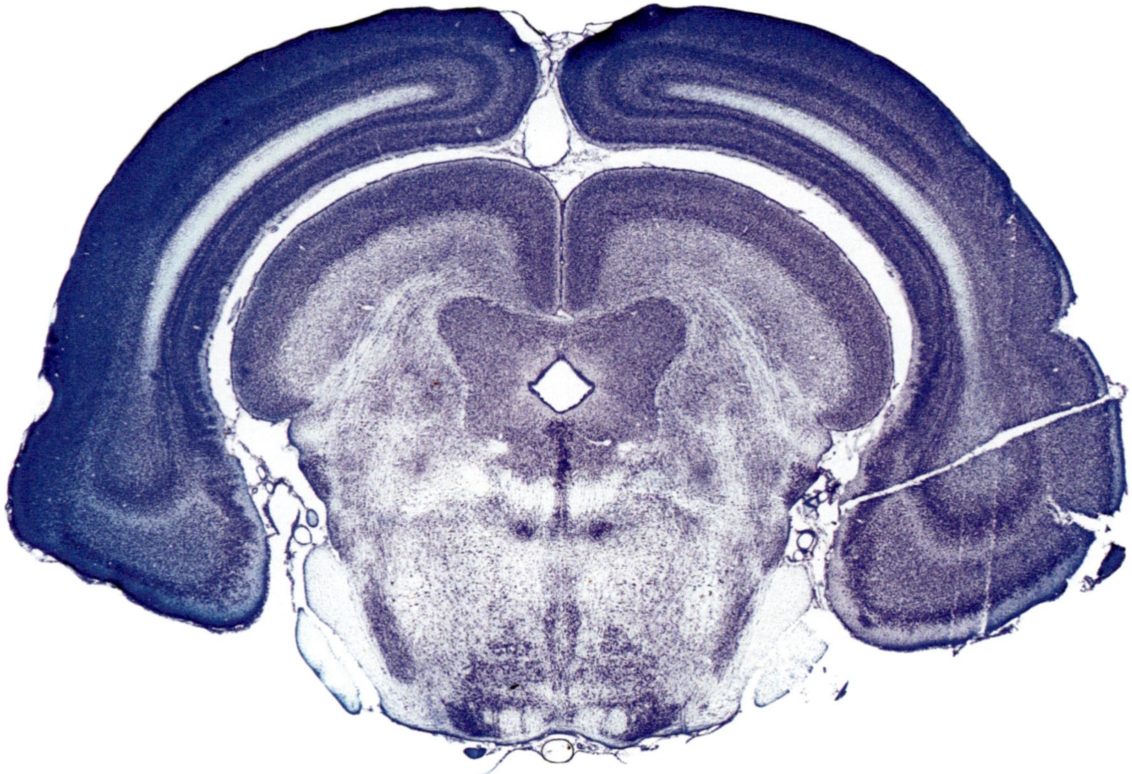

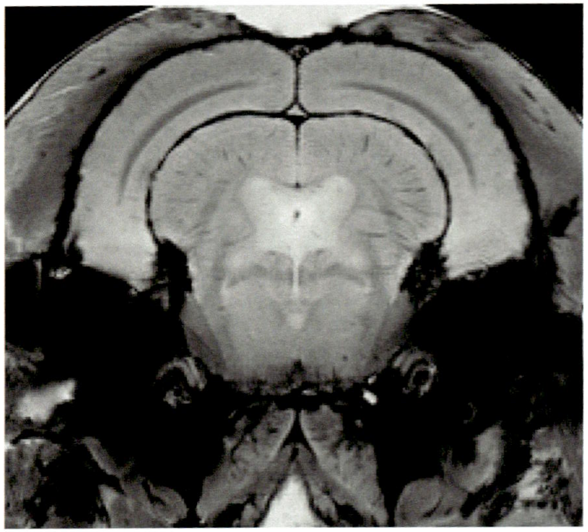

4n	trochlear nerve
6n	root of abducens nerve
Aq	aqueduct
CnF	cuneiform nucleus
csc	commissure of the superior colliculus
DpG	deep gray layer of the superior colliculus
DpWh	deep white layer of the superior colliculus
DR	dorsal raphe nucleus
Ent	entorhinal cortex
IC	inferior colliculus
InG	intermediate gray layer of the superior colliculus
InWh	intermediate white layer of the superior colliculus
lfp	longitudinal fasciculus of the pons
LL	nucleus of the lateral lemniscus
mcp	middle cerebellar peduncle
me5	mesencephalic trigeminal tract
Me5	mesencephalic trigeminal nucleus
MGV	medial geniculate nucleus, ventral part
MiTg	microcellular tegmental nucleus
ml	medial lemniscus
mlf	medial longitudinal fasciculus
MnR	median raphe nucleus
mRt	mesencephalic reticular formation
Op	optic nerve layer of the superior colliculus
PAG	periaqueductal gray
PBG	parabigeminal nucleus
Pn	pontine nuclei
PnO	pontine reticular nucleus, oral part
PTg	pedunculopontine tegmental nucleus
Rbd	rhabdoid nucleus
RSg	retrosplenial granular cortex
RtTg	reticulotegmental nucleus of the pons
s5	sensory root of the trigeminal nerve
scp	superior cerebellar peduncle
SPTg	subpeduncular tegmental nucleus
SuG	superficial gray layer of the superior colliculus
TC	temporal cortex
tfp	transverse fibers of the pons
tth	trigeminothalamic tract
V1	primary visual cortex
V2	secondary visual cortex
VTg	ventral tegmental nucleus
Zo	zonal layer of the superior colliculus

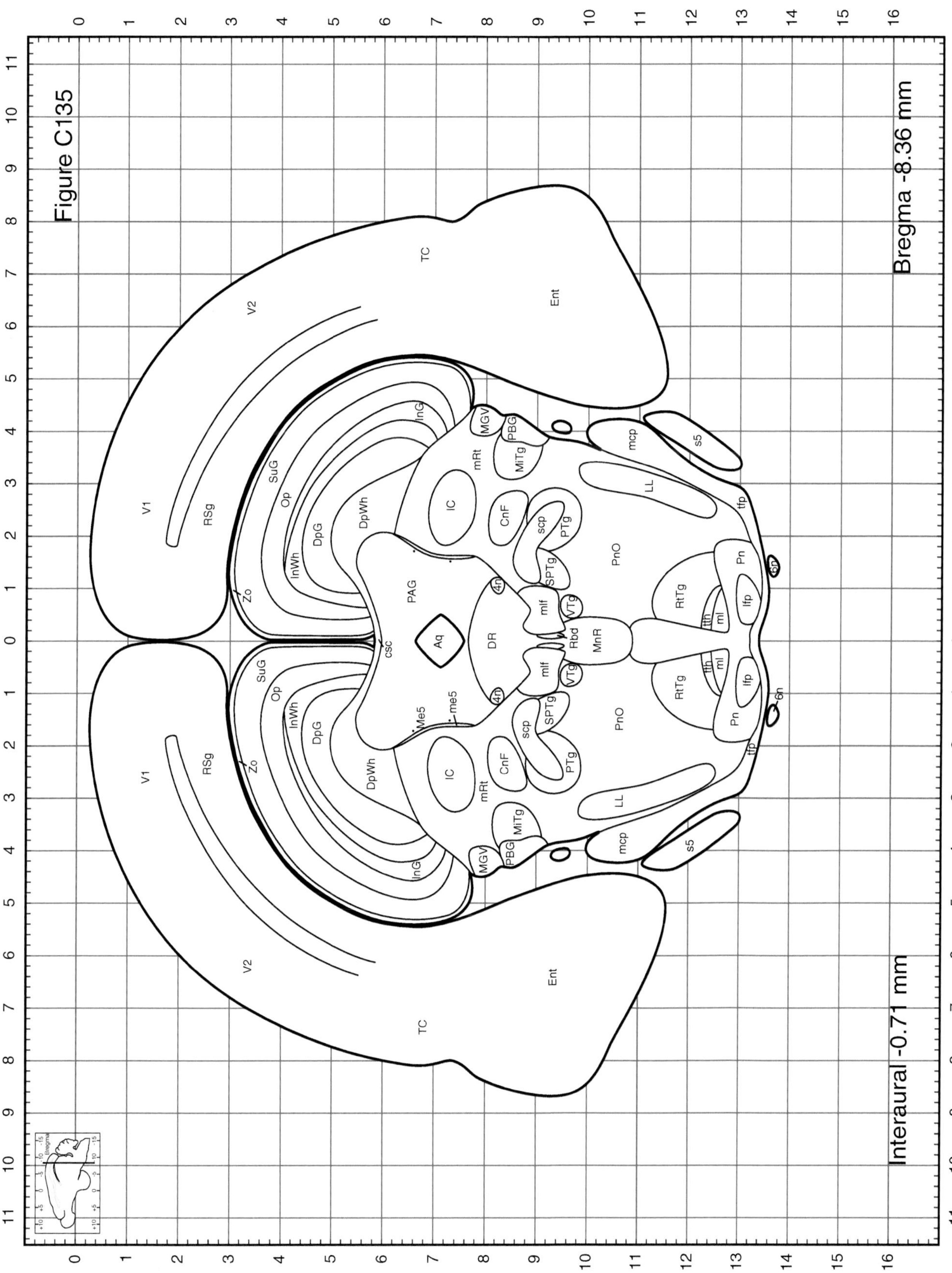
Figure C135
Bregma -8.36 mm
Interaural -0.71 mm
TC
Ent
V2
V1
RSg
SuG
Op
InWh
DpG
DpWh
InG
Zo
PAG
Aq
csc
DR
Me5
me5
4n
mlf
VTg
Rbd
MnR
IC
CnF
scp
PTg
SPTg
mRt
MiTg
MGV
PBG
mcp
s5
LL
tfp
PnO
RtTg
Pn
tth
ml
lfp
6n

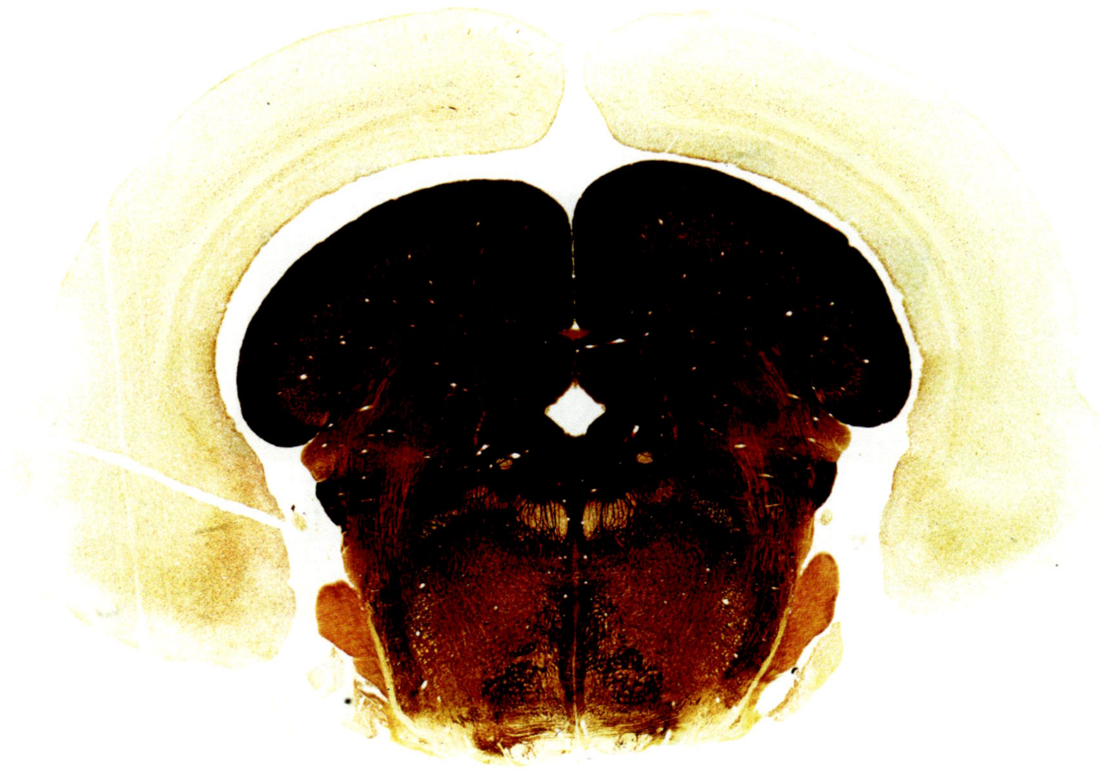

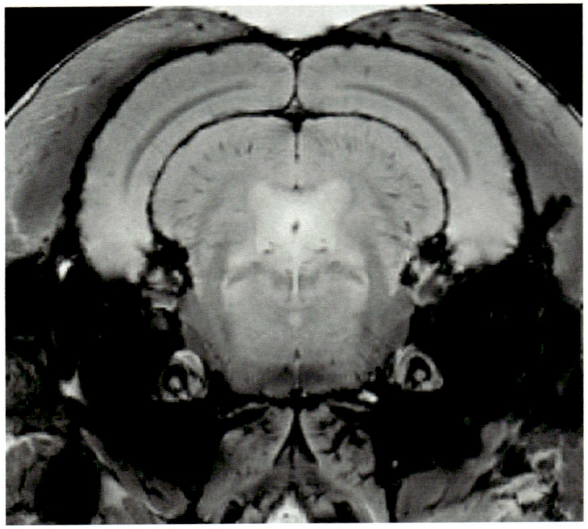

4n	trochlear nerve
6n	root of abducens nerve
Aq	aqueduct
CnF	cuneiform nucleus
csc	commissure of the superior colliculus
DpG	deep gray layer of the superior colliculus
DpWh	deep white layer of the superior colliculus
DR	dorsal raphe nucleus
Ent	entorhinal cortex
IC	inferior colliculus
InG	intermediate gray layer of the superior colliculus
InWh	intermediate white layer of the superior colliculus
LDTg	laterodorsal tegmental nucleus
lfp	longitudinal fasciculus of the pons
LL	nucleus of the lateral lemniscus
mcp	middle cerebellar peduncle
me5	mesencephalic trigeminal tract
Me5	mesencephalic trigeminal nucleus
MGV	medial geniculate nucleus, ventral part
MiTg	microcellular tegmental nucleus
ml	medial lemniscus
mlf	medial longitudinal fasciculus
MnR	median raphe nucleus
mRt	mesencephalic reticular formation
Op	optic nerve layer of the superior colliculus
PAG	periaqueductal gray
PBG	parabigeminal nucleus
Pn	pontine nuclei
PnO	pontine reticular nucleus, oral part
PTg	pedunculopontine tegmental nucleus
Rbd	rhabdoid nucleus
RSg	retrosplenial granular cortex
RtTg	reticulotegmental nucleus of the pons
s5	sensory root of the trigeminal nerve
scp	superior cerebellar peduncle
SPTg	subpeduncular tegmental nucleus
SuG	superficial gray layer of the superior colliculus
TC	temporal cortex
tfp	transverse fibers of the pons
tth	trigeminothalamic tract
V1	primary visual cortex
V2	secondary visual cortex
VTg	ventral tegmental nucleus
Zo	zonal layer of the superior colliculus

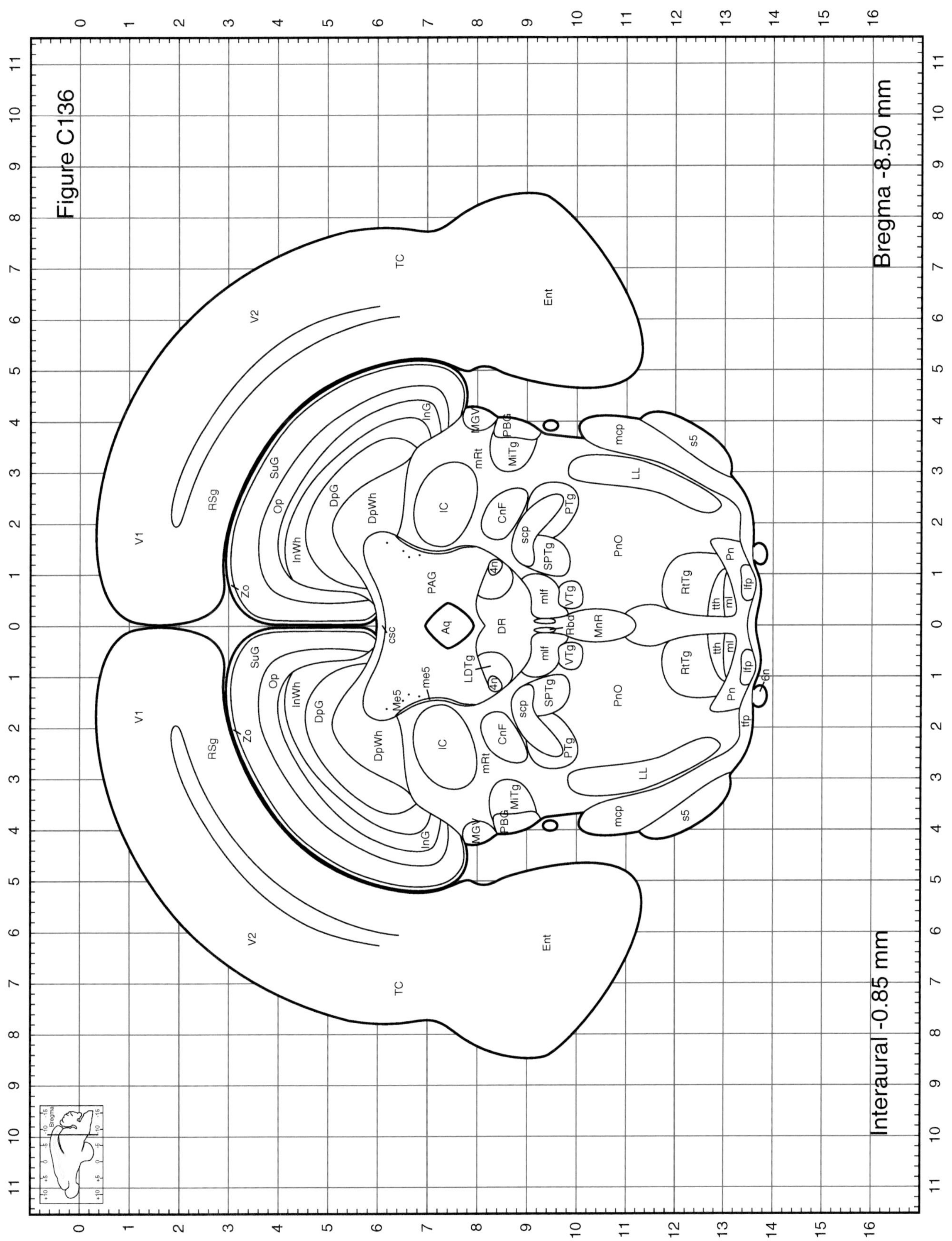
Figure C136
Bregma -8.50 mm
Interaural -0.85 mm
V1
V2
TC
Ent
RSg
Zo
SuG
Op
InWh
DpG
DpWh
InG
IC
MGV
PBG
MiTg
mRt
CnF
scp
PTg
SPTg
PAG
Aq
csc
me5
Me5
4n
LDTg
DR
mlf
VTg
Rbd
MnR
PnO
LL
mcp
s5
RtTg
tth
ml
lfp
Pn
tfp
6n

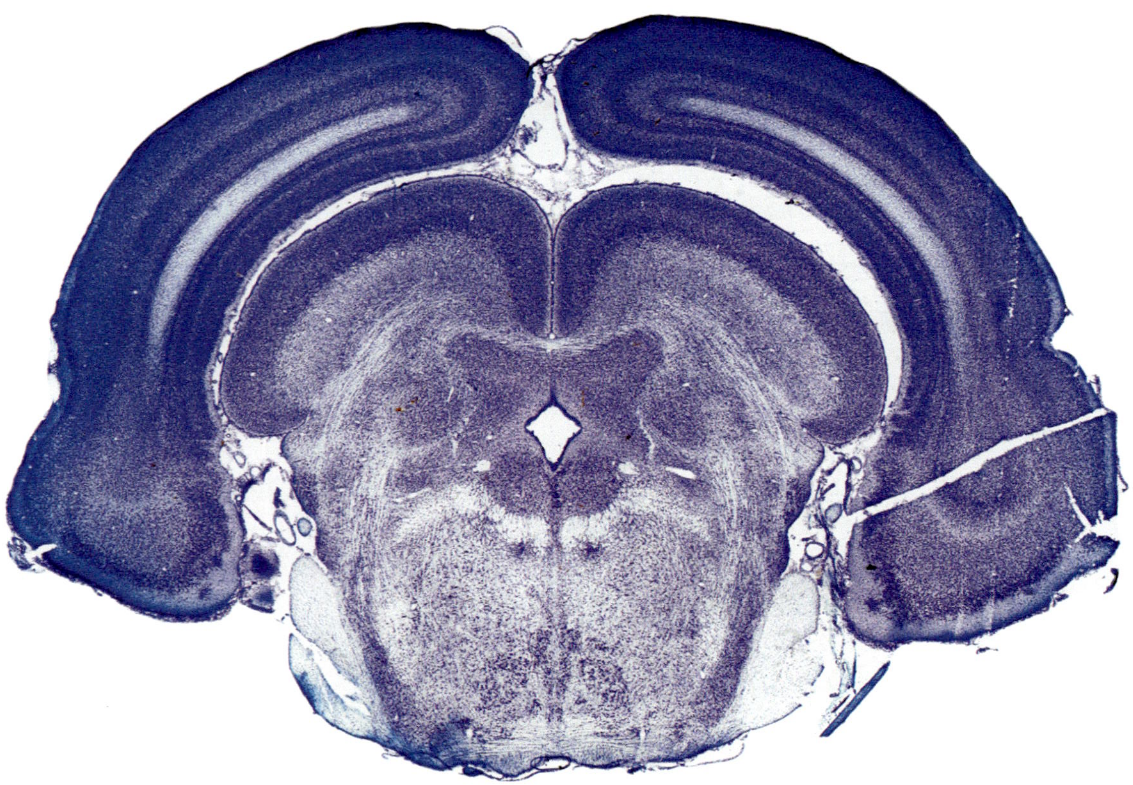

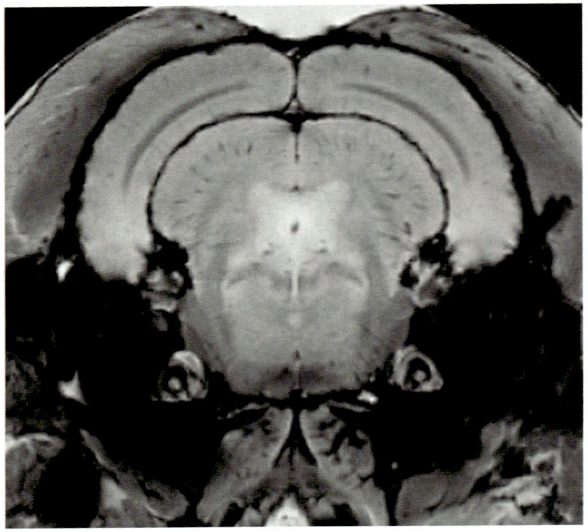

4n	trochlear nerve
6n	root of abducens nerve
Aq	aqueduct
CnF	cuneiform nucleus
csc	commissure of the superior colliculus
DpG	deep gray layer of the superior colliculus
DpWh	deep white layer of the superior colliculus
DR	dorsal raphe nucleus
Ent	entorhinal cortex
IC	inferior colliculus
InG	intermediate gray layer of the superior colliculus
InWh	intermediate white layer of the superior colliculus
LDTg	laterodorsal tegmental nucleus
LL	nucleus of the lateral lemniscus
mcp	middle cerebellar peduncle
me5	mesencephalic trigeminal tract
Me5	mesencephalic trigeminal nucleus
MGV	medial geniculate nucleus, ventral part
MiTg	microcellular tegmental nucleus
ml	medial lemniscus
mlf	medial longitudinal fasciculus
MnR	median raphe nucleus
mRt	mesencephalic reticular formation
Op	optic nerve layer of the superior colliculus
PAG	periaqueductal gray
PBG	parabigeminal nucleus
PnO	pontine reticular nucleus, oral part
PTg	pedunculopontine tegmental nucleus
py	pyramidal tract
Rbd	rhabdoid nucleus
RSg	retrosplenial granular cortex
RtTg	reticulotegmental nucleus of the pons
s5	sensory root of the trigeminal nerve
scp	superior cerebellar peduncle
SPTg	subpeduncular tegmental nucleus
SuG	superficial gray layer of the superior colliculus
TC	temporal cortex
tfp	transverse fibers of the pons
tth	trigeminothalamic tract
tz	trapezoid body
V1	primary visual cortex
V2	secondary visual cortex
VTg	ventral tegmental nucleus
Zo	zonal layer of the superior colliculus

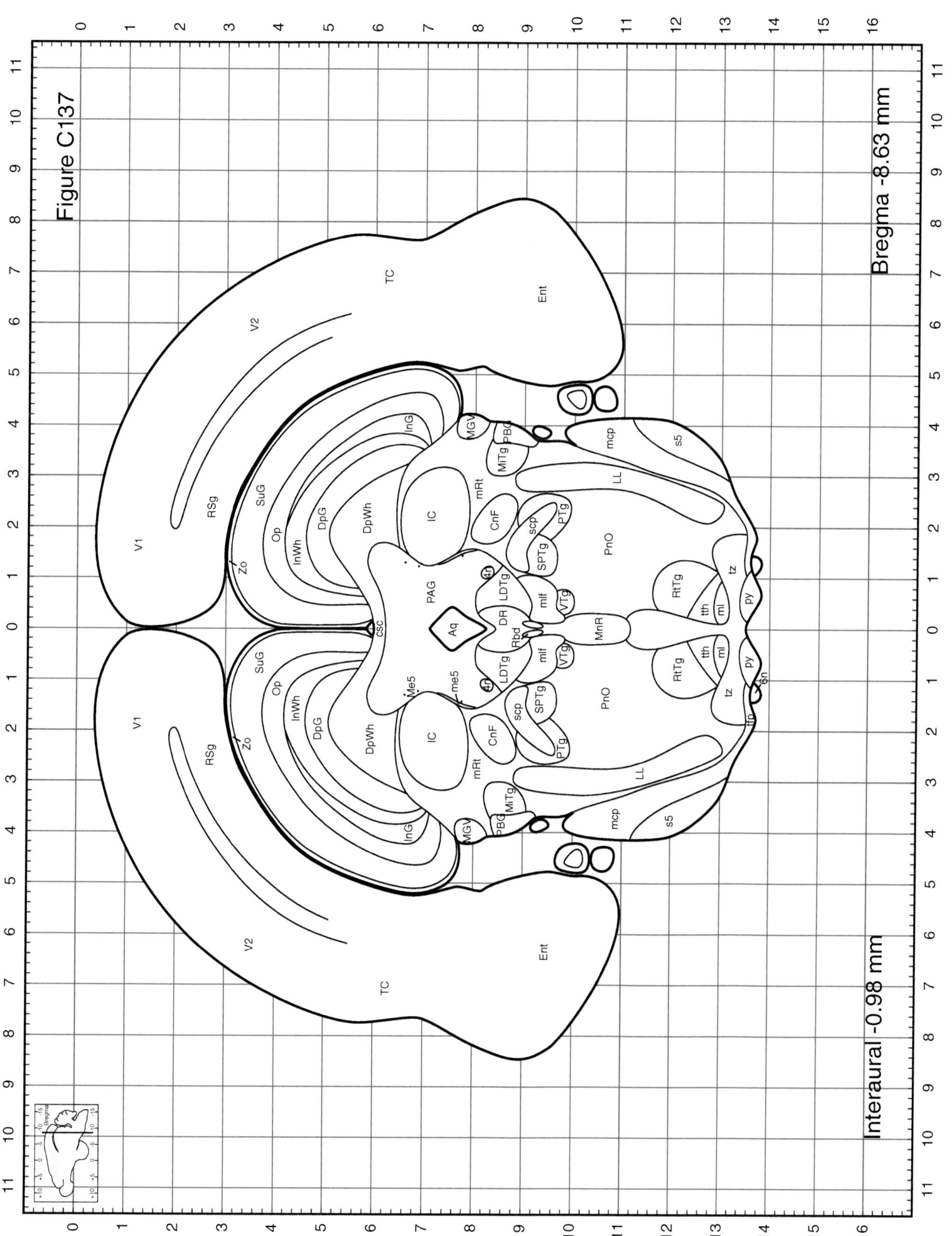
Figure C137
Bregma -8.63 mm
Interaural -0.98 mm
V1
V2
TC
Ent
RSg
SuG
Zo
Op
InWh
DpG
DpWh
InG
IC
PAG
Aq
csc
Me5
me5
4n
mRt
MGV
PBG
MiTg
CnF
scp
PTg
SPTg
LDTg
DR
Rbd
mlf
VTg
MnR
PnO
mcp
LL
s5
RtTg
tth
ml
tz
py
6n
tfp

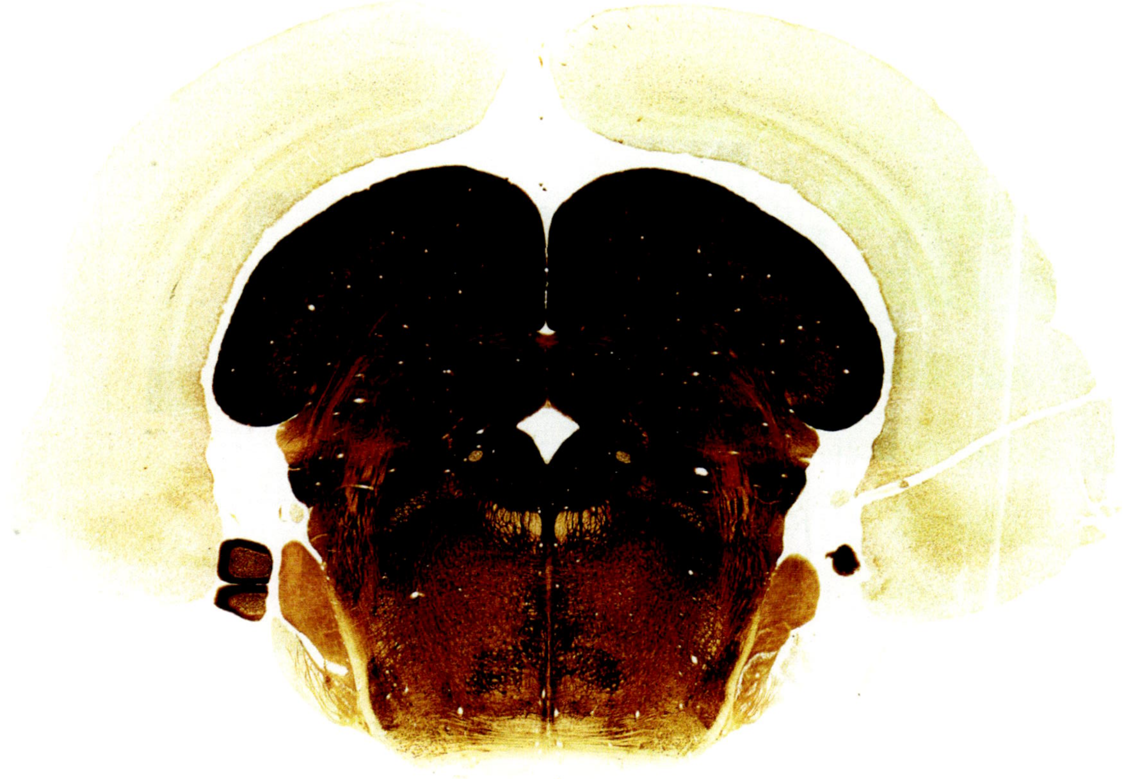

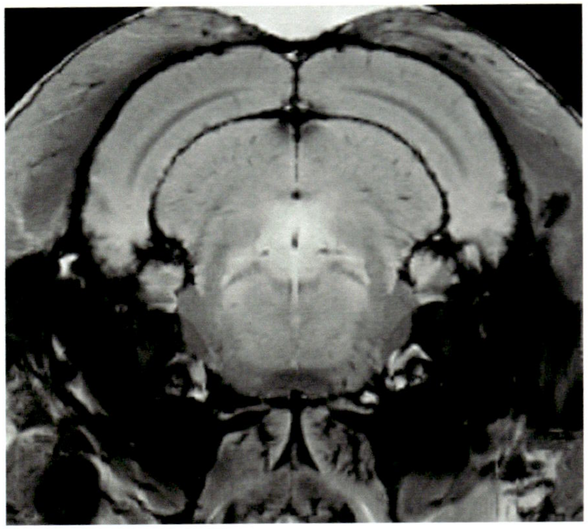

4n	trochlear nerve
6n	root of abducens nerve
Aq	aqueduct
CnF	cuneiform nucleus
csc	commissure of the superior colliculus
DpG	deep gray layer of the superior colliculus
DpWh	deep white layer of the superior colliculus
DR	dorsal raphe nucleus
Ent	entorhinal cortex
IC	inferior colliculus
InG	intermediate gray layer of the superior colliculus
InWh	intermediate white layer of the superior colliculus
LDTg	laterodorsal tegmental nucleus
LL	nucleus of the lateral lemniscus
ll	lateral lemniscus
mcp	middle cerebellar peduncle
me5	mesencephalic trigeminal tract
Me5	mesencephalic trigeminal nucleus
MGV	medial geniculate nucleus, ventral part
MiTg	microcellular tegmental nucleus
ml	medial lemniscus
mlf	medial longitudinal fasciculus
MnR	median raphe nucleus
MPB	medial parabrachial nucleus
Op	optic nerve layer of the superior colliculus
PAG	periaqueductal gray
PBG	parabigeminal nucleus
PnO	pontine reticular nucleus, oral part
PTg	pedunculopontine tegmental nucleus
py	pyramidal tract
Rbd	rhabdoid nucleus
RSg	retrosplenial granular cortex
RtTg	reticulotegmental nucleus of the pons
s5	sensory root of the trigeminal nerve
scp	superior cerebellar peduncle
SPTg	subpeduncular tegmental nucleus
SuG	superficial gray layer of the superior colliculus
TC	temporal cortex
tfp	transverse fibers of the pons
tth	trigeminothalamic tract
tz	trapezoid body
V1	primary visual cortex
V2	secondary visual cortex
VPO	ventral periolivary nucleus
VTg	ventral tegmental nucleus
Zo	zonal layer of the superior colliculus

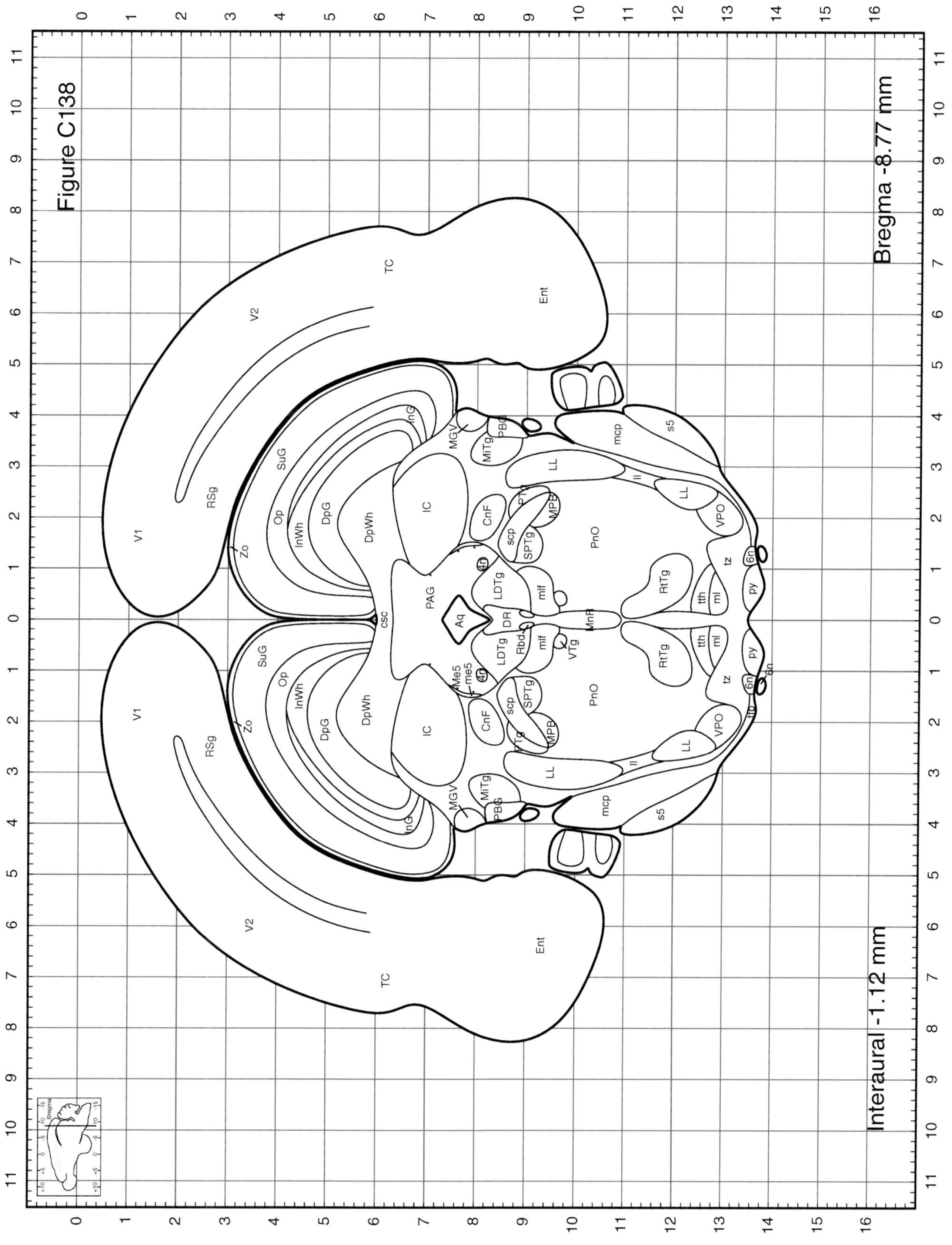
Figure C138
Bregma -8.77 mm
Interaural -1.12 mm
V1
V2
RSg
TC
Ent
SuG
Op
InWh
DpG
DpWh
InG
Zo
IC
csc
PAG
Aq
MGV
MiTg
PBG
CnF
scp
PPTg
SPTg
MPB
LDTg
4n
DR
Rbd
mlf
VTg
MnR
PnO
LL
ll
mcp
s5
RtTg
tth
ml
tz
VPO
py
6n
ttp
Me5
me5

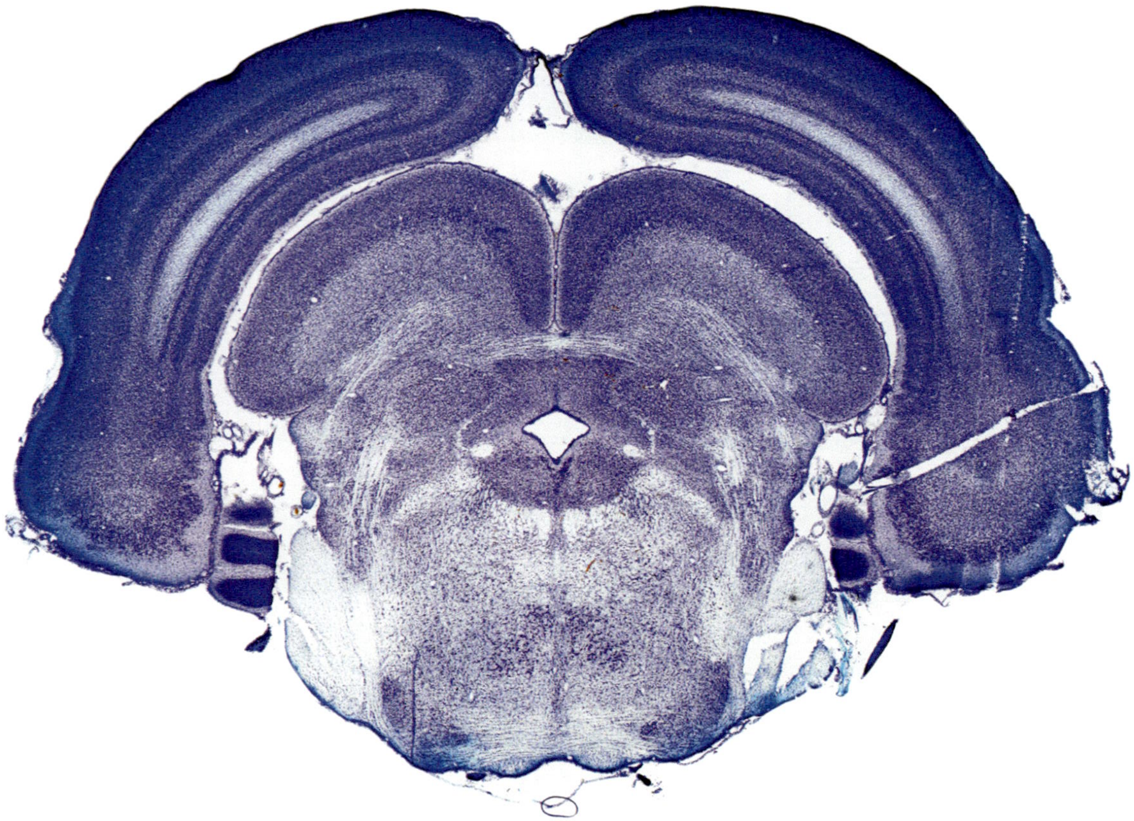

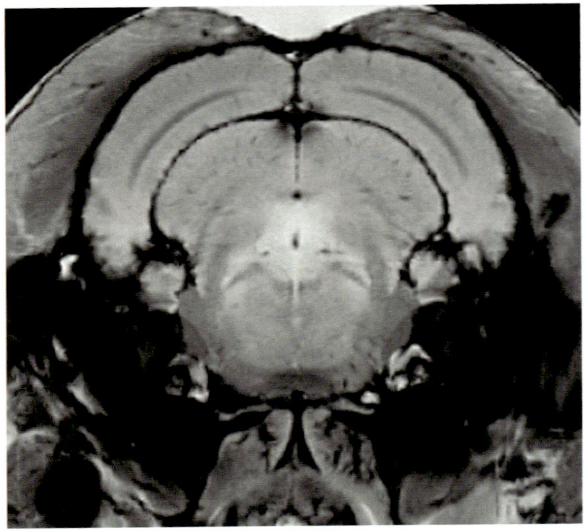

4n	trochlear nerve
6n	root of abducens nerve
Aq	aqueduct
CnF	cuneiform nucleus
csc	commissure of the superior colliculus
DpG	deep gray layer of the superior colliculus
DpWh	deep white layer of the superior colliculus
DR	dorsal raphe nucleus
Ent	entorhinal cortex
IC	inferior colliculus
InG	intermediate gray layer of the superior colliculus
InWh	intermediate white layer of the superior colliculus
LDTg	laterodorsal tegmental nucleus
LL	nucleus of the lateral lemniscus
ll	lateral lemniscus
m5	motor root of the trigeminal nerve
mcp	middle cerebellar peduncle
me5	mesencephalic trigeminal tract
Me5	mesencephalic trigeminal nucleus
MGV	medial geniculate nucleus, ventral part
MiTg	microcellular tegmental nucleus
ml	medial lemniscus
mlf	medial longitudinal fasciculus
MnR	median raphe nucleus
MPB	medial parabrachial nucleus
Op	optic nerve layer of the superior colliculus
P5	peritrigeminal zone
PAG	periaqueductal gray
PBG	parabigeminal nucleus
PnO	pontine reticular nucleus, oral part
PTg	pedunculopontine tegmental nucleus
py	pyramidal tract
RSg	retrosplenial granular cortex
RtTg	reticulotegmental nucleus of the pons
s5	sensory root of the trigeminal nerve
scp	superior cerebellar peduncle
Su5	supratrigeminal nucleus
SuG	superficial gray layer of the superior colliculus
TC	temporal cortex
tfp	transverse fibers of the pons
tth	trigeminothalamic tract
tz	trapezoid body
V1	primary visual cortex
V2	secondary visual cortex
VPO	ventral periolivary nucleus
Zo	zonal layer of the superior colliculus

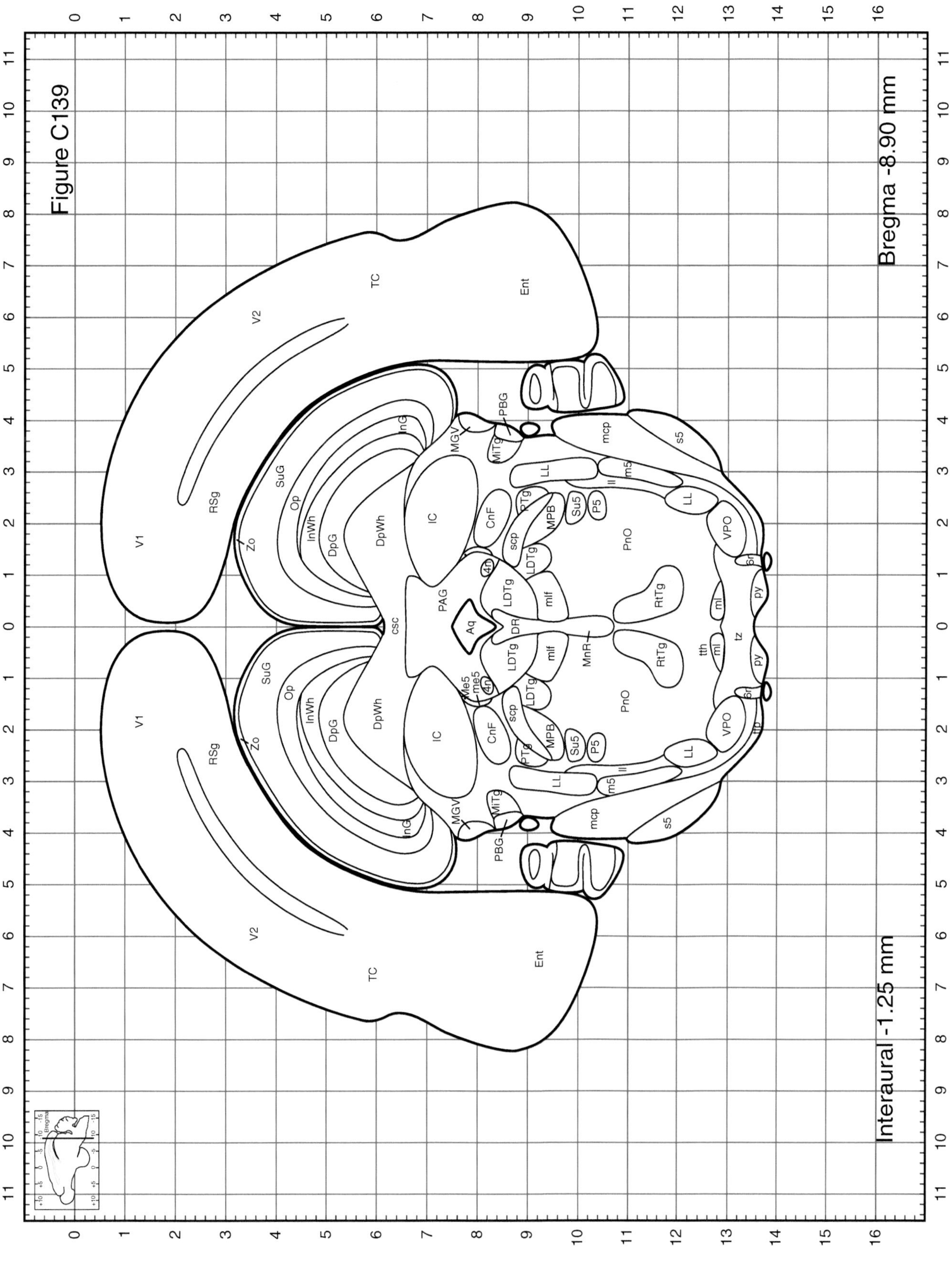
Figure C139
Bregma -8.90 mm
Interaural -1.25 mm
V1
V2
RSg
TC
Ent
Zo
SuG
Op
InWh
DpG
DpWh
InG
IC
csc
PAG
Aq
DR
MGV
PBG
MiTg
CnF
PTg
scp
MPB
LDTg
4n
Me5
me5
mlf
MnR
Su5
P5
LL
ll
m5
mcp
s5
PnO
RtTg
VPO
tth
ml
tz
py
6n
ttp

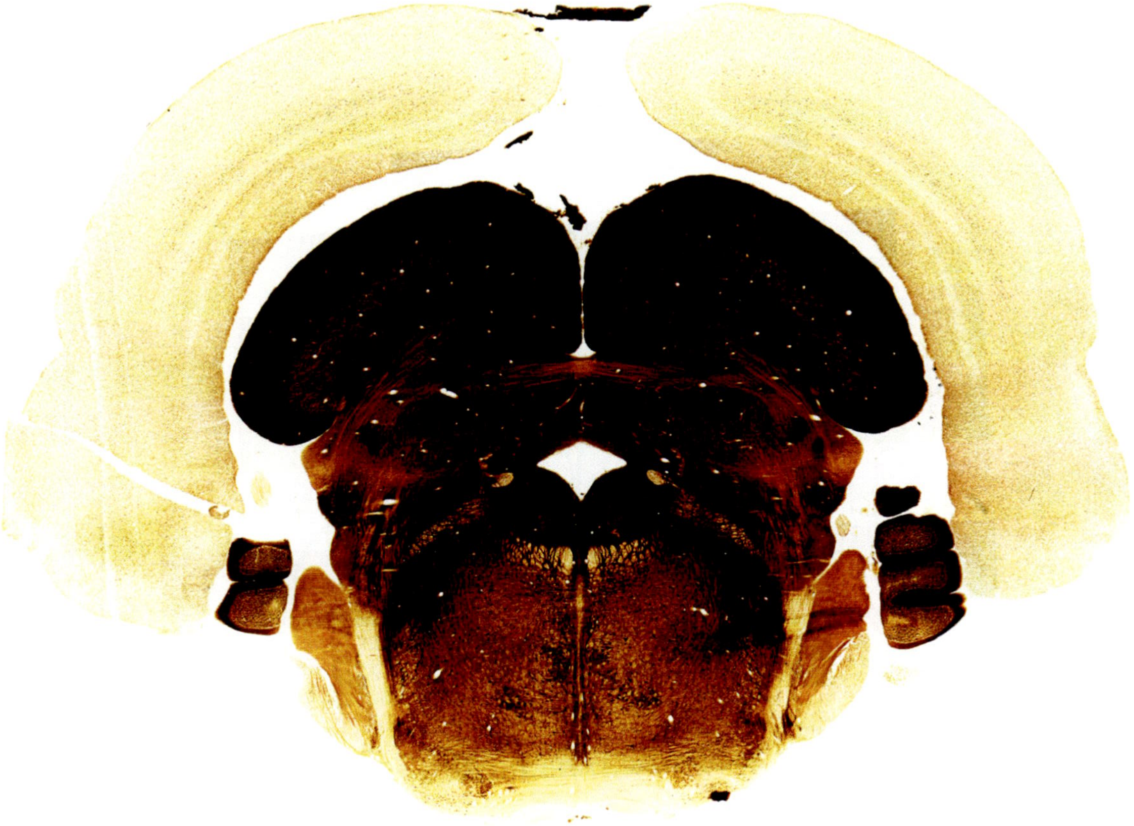

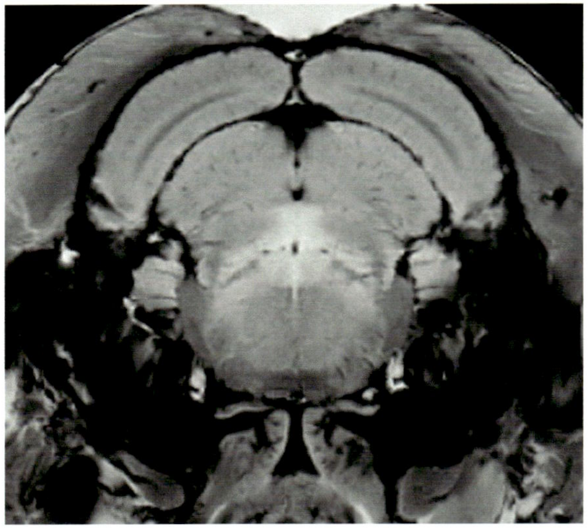

4n trochlear nerve
6n root of abducens nerve
Aq aqueduct
CG central gray
CnF cuneiform nucleus
csc commissure of the superior colliculus
DpG deep gray layer of the superior colliculus
DpWh deep white layer of the superior colliculus
DR dorsal raphe nucleus
DTg dorsal tegmental nucleus
Ent entorhinal cortex
IC inferior colliculus
InG intermediate gray layer of the superior colliculus
InWh intermediate white layer of the superior colliculus
LDTg laterodorsal tegmental nucleus
LL nucleus of the lateral lemniscus
ll lateral lemniscus
LPB lateral parabrachial nucleus
m5 motor root of the trigeminal nerve
mcp middle cerebellar peduncle
me5 mesencephalic trigeminal tract
Me5 mesencephalic trigeminal nucleus
MiTg microcellular tegmental nucleus
ml medial lemniscus
mlf medial longitudinal fasciculus
MnR median raphe nucleus
MPB medial parabrachial nucleus
Op optic nerve layer of the superior colliculus
P5 peritrigeminal zone
PAG periaqueductal gray
PBG parabigeminal nucleus
PnC pontine reticular nucleus, caudal part
PTg pedunculopontine tegmental nucleus
py pyramidal tract
RSg retrosplenial granular cortex
RtTg reticulotegmental nucleus of the pons
s5 sensory root of the trigeminal nerve
scp superior cerebellar peduncle
Su5 supratrigeminal nucleus
SuG superficial gray layer of the superior colliculus
TC temporal cortex
tfp transverse fibers of the pons
tth trigeminothalamic tract
tz trapezoid body
V1 primary visual cortex
V2 secondary visual cortex
VPO ventral periolivary nucleus
Zo zonal layer of the superior colliculus

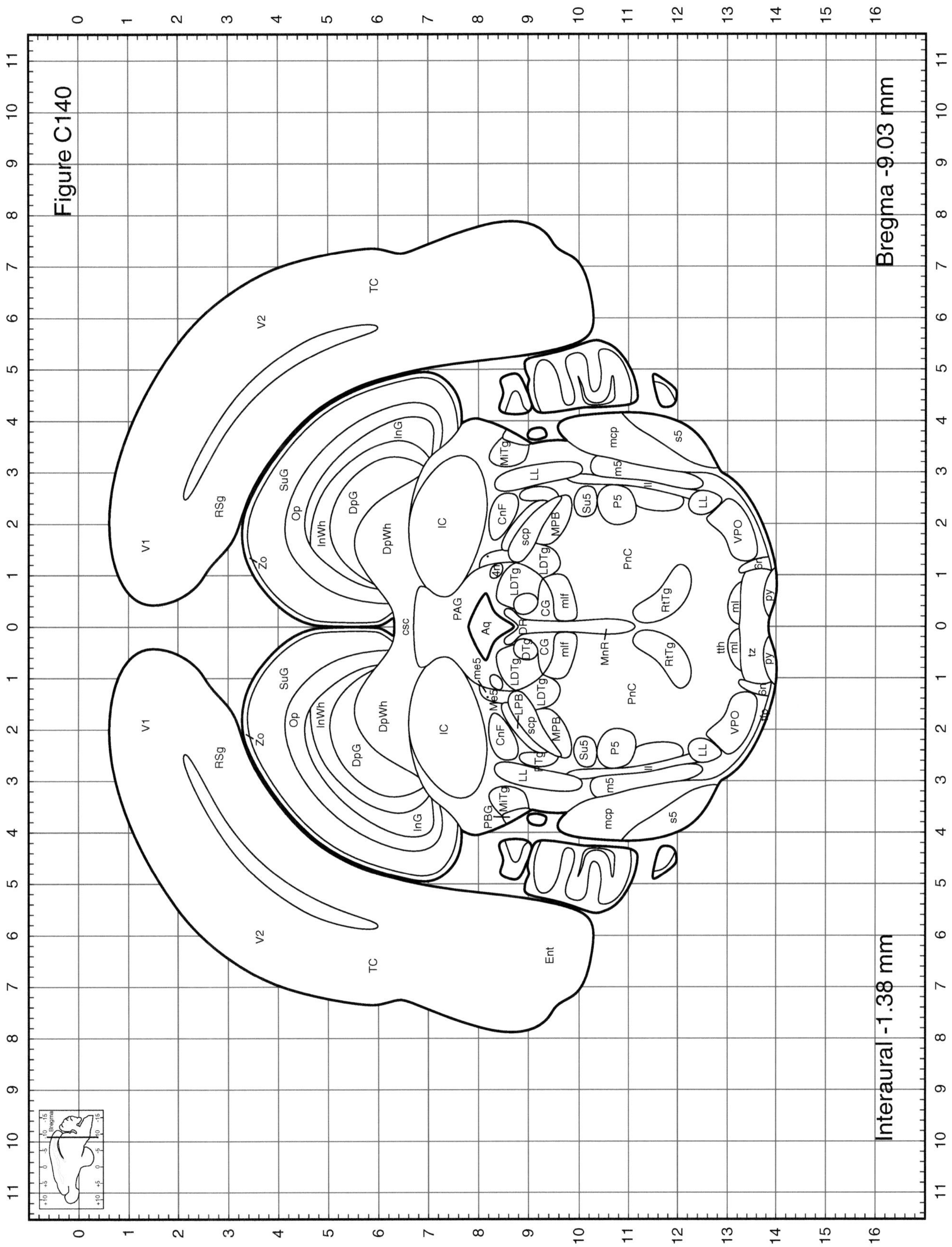
Figure C140
Bregma -9.03 mm
Interaural -1.38 mm
V1
V2
RSg
TC
Ent
Zo
SuG
Op
InWh
InG
DpG
DpWh
IC
csc
PAG
Aq
DR
DTg
LDTg
CnF
scp
MPB
LPB
Me5
me5
4n
CG
mlf
MiTg
LL
PBG
PTg
Su5
P5
m5
mcp
s5
ll
PnC
MnR
RtTg
VPO
ml
tth
tz
py
6n

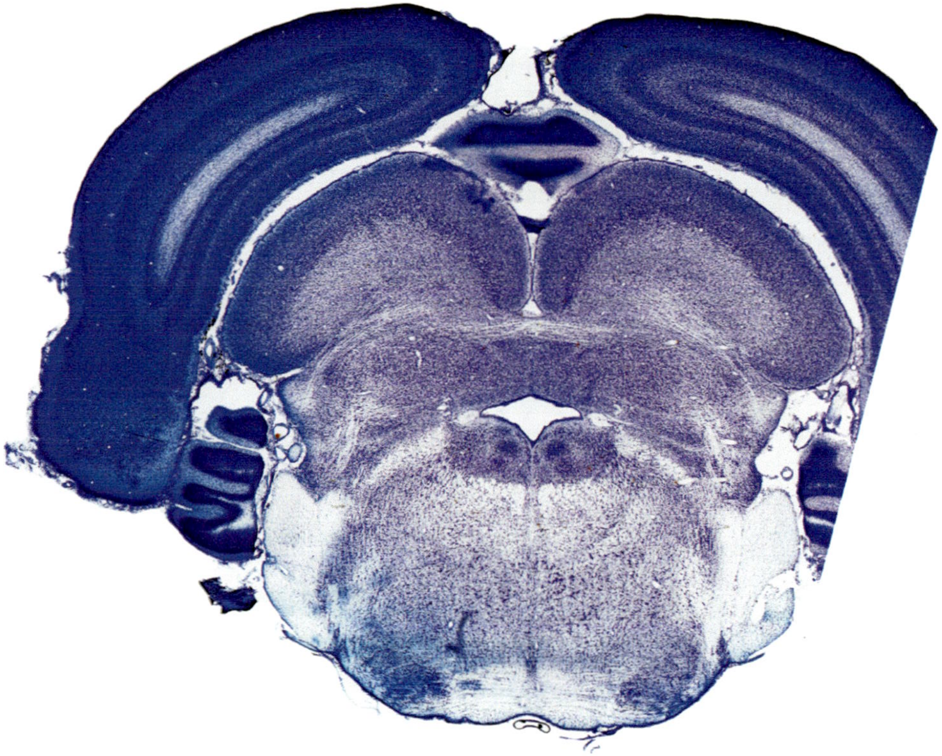

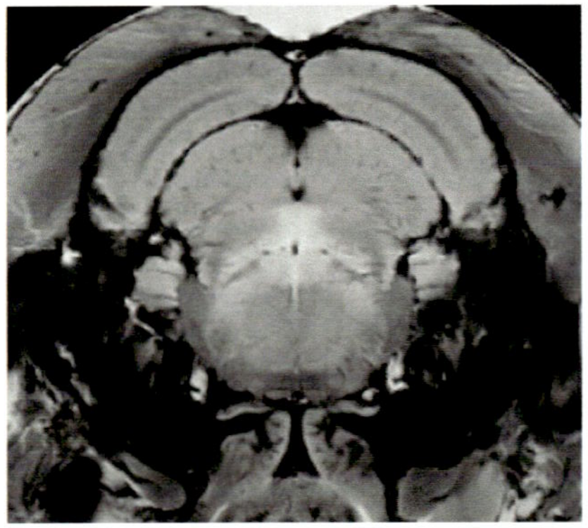

4n	trochlear nerve
6n	root of abducens nerve
7n	facial nerve
Aq	aqueduct
CG	central gray
CnF	cuneiform nucleus
csc	commissure of the superior colliculus
DpG	deep gray layer of the superior colliculus
DpWh	deep white layer of the superior colliculus
DR	dorsal raphe nucleus
DTg	dorsal tegmental nucleus
Ent	entorhinal cortex
IC	inferior colliculus
InG	intermediate gray layer of the superior colliculus
InWh	intermediate white layer of the superior colliculus
LDTg	laterodorsal tegmental nucleus
LL	nucleus of the lateral lemniscus
ll	lateral lemniscus
LPB	lateral parabrachial nucleus
m5	motor root of the trigeminal nerve
mcp	middle cerebellar peduncle
me5	mesencephalic trigeminal tract
Me5	mesencephalic trigeminal nucleus
MiTg	microcellular tegmental nucleus
ml	medial lemniscus
mlf	medial longitudinal fasciculus
MnR	median raphe nucleus
MPB	medial parabrachial nucleus
Op	optic nerve layer of the superior colliculus
P5	peritrigeminal zone
PAG	periaqueductal gray
PnC	pontine reticular nucleus, caudal part
py	pyramidal tract
RSg	retrosplenial granular cortex
RtTg	reticulotegmental nucleus of the pons
s5	sensory root of the trigeminal nerve
scp	superior cerebellar peduncle
Su5	supratrigeminal nucleus
SuG	superficial gray layer of the superior colliculus
TC	temporal cortex
tth	trigeminothalamic tract
tz	trapezoid body
V1	primary visual cortex
V2	secondary visual cortex
VPO	ventral periolivary nucleus
Zo	zonal layer of the superior colliculus

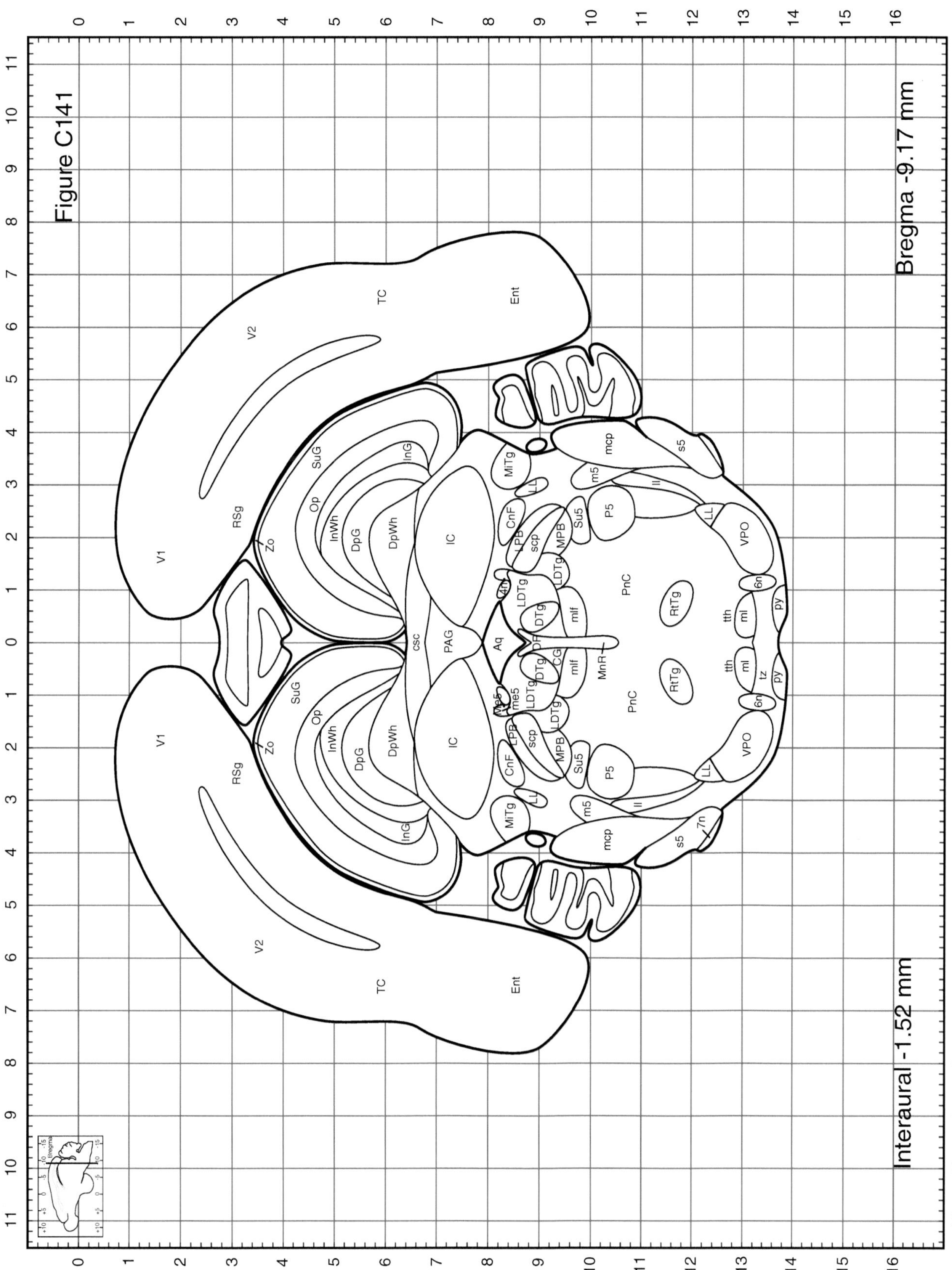
Figure C141
Bregma -9.17 mm
Interaural -1.52 mm
V1
V2
TC
Ent
RSg
Zo
SuG
Op
InWh
DpG
DpWh
InG
IC
csc
PAG
Aq
MiTg
CnF
LPB
scp
MPB
LL
LDTg
DTg
4n
Me5
me5
DP
CG
mlf
MnR
m5
mcp
s5
Su5
P5
ll
PnC
RtTg
tth
ml
tz
py
6n
VPO
7n

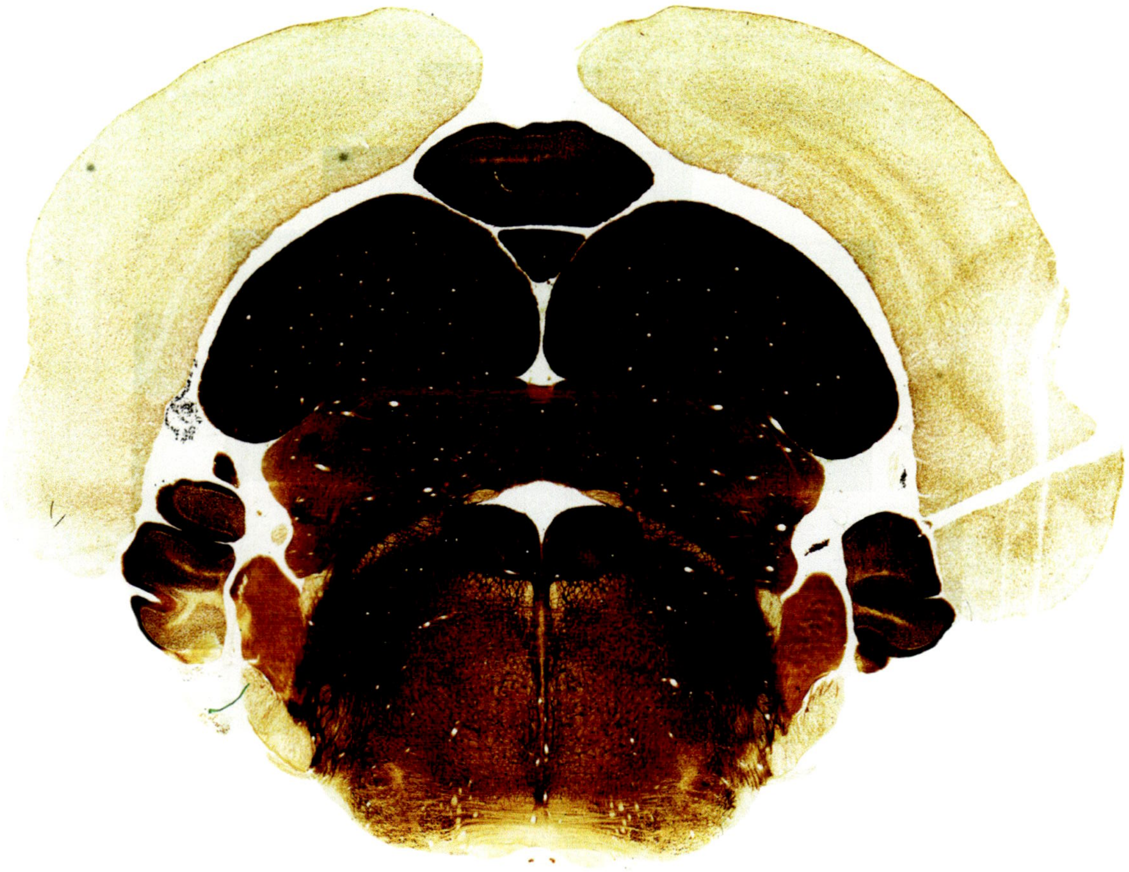

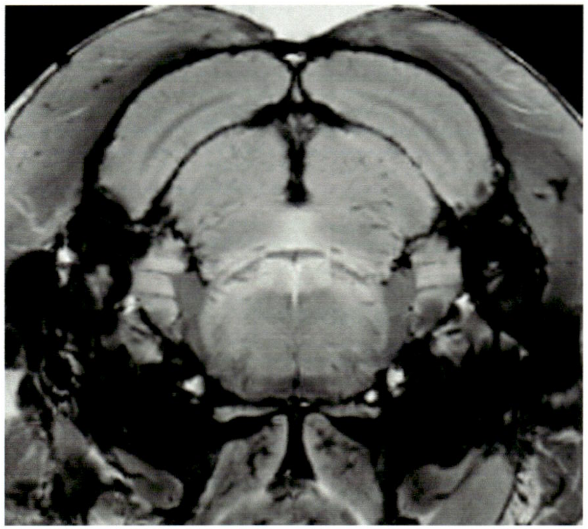

4n	trochlear nerve
6n	root of abducens nerve
7n	facial nerve
A5	A5 noradrenaline cells
Aq	aqueduct
CG	central gray
csc	commissure of the superior colliculus
DpG	deep gray layer of the superior colliculus
DpWh	deep white layer of the superior colliculus
DR	dorsal raphe nucleus
DTg	dorsal tegmental nucleus
Ent	entorhinal cortex
IC	inferior colliculus
InG	intermediate gray layer of the superior colliculus
InWh	intermediate white layer of the superior colliculus
LDTg	laterodorsal tegmental nucleus
LPB	lateral parabrachial nucleus
m5	motor root of the trigeminal nerve
mcp	middle cerebellar peduncle
me5	mesencephalic trigeminal tract
Me5	mesencephalic trigeminal nucleus
MiTg	microcellular tegmental nucleus
ml	medial lemniscus
mlf	medial longitudinal fasciculus
MPB	medial parabrachial nucleus
Op	optic nerve layer of the superior colliculus
P5	peritrigeminal zone
PAG	periaqueductal gray
PnC	pontine reticular nucleus, caudal part
py	pyramidal tract
s5	sensory root of the trigeminal nerve
scp	superior cerebellar peduncle
scpd	superior cerebellar peduncle, descending limb
Su5	supratrigeminal nucleus
SubC	subcoeruleus nucleus
SuG	superficial gray layer of the superior colliculus
TC	temporal cortex
tth	trigeminothalamic tract
Tz	nucleus of the trapezoid body
tz	trapezoid body
V1	primary visual cortex
V2	secondary visual cortex
VPO	ventral periolivary nucleus
Zo	zonal layer of the superior colliculus

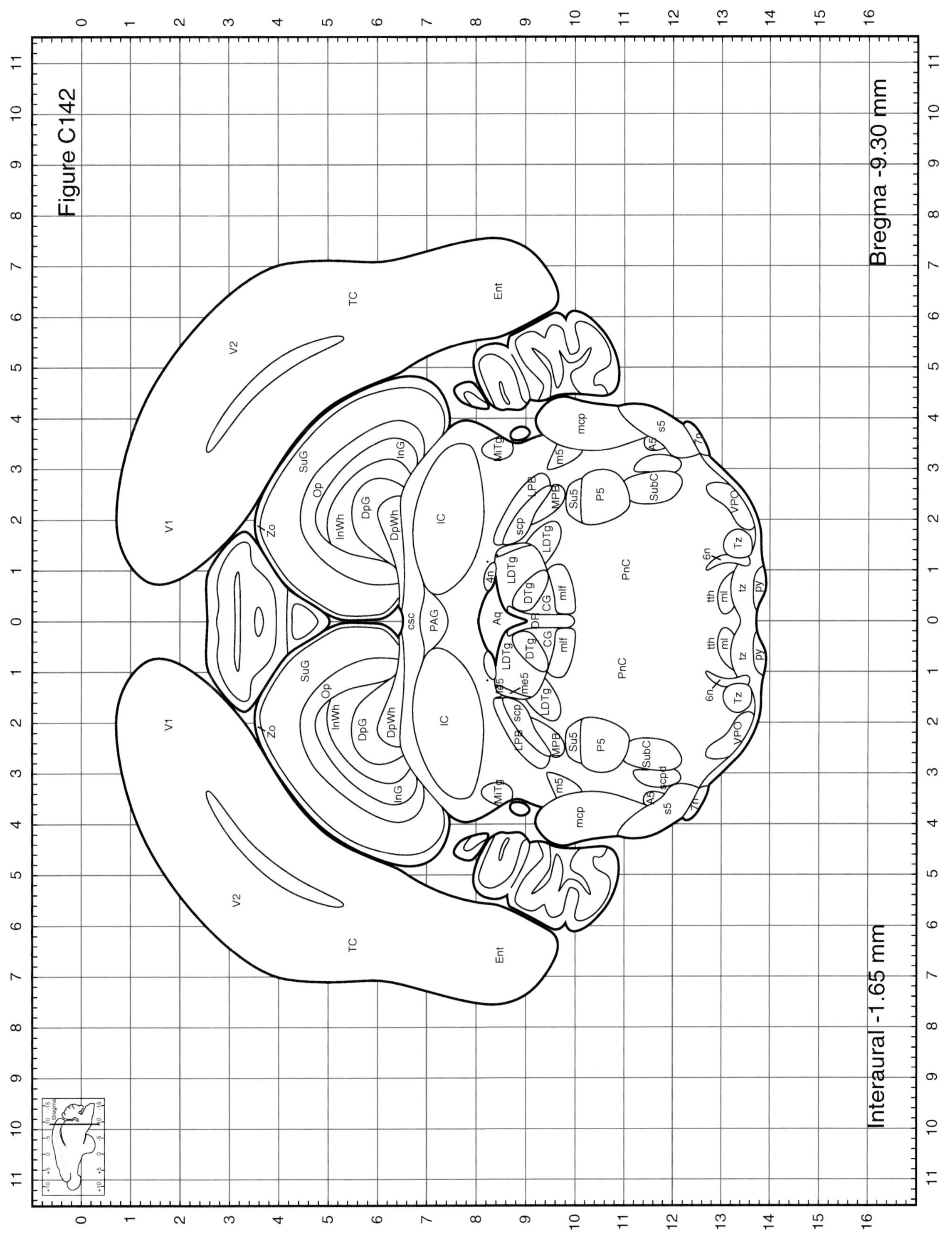
Figure C142
Bregma -9.30 mm
Interaural -1.65 mm
V1
V2
TC
Ent
SuG
Op
InWh
DpG
DpWh
InG
Zo
IC
csc
PAG
Aq
4n
LDTg
DTg
DR
CG
mlf
me5
Me5
scp
LPB
MPB
Su5
P5
MiTg
m5
mcp
s5
A5
SubC
scpd
7n
PnC
6n
Tz
VPO
tth
ml
tz
py

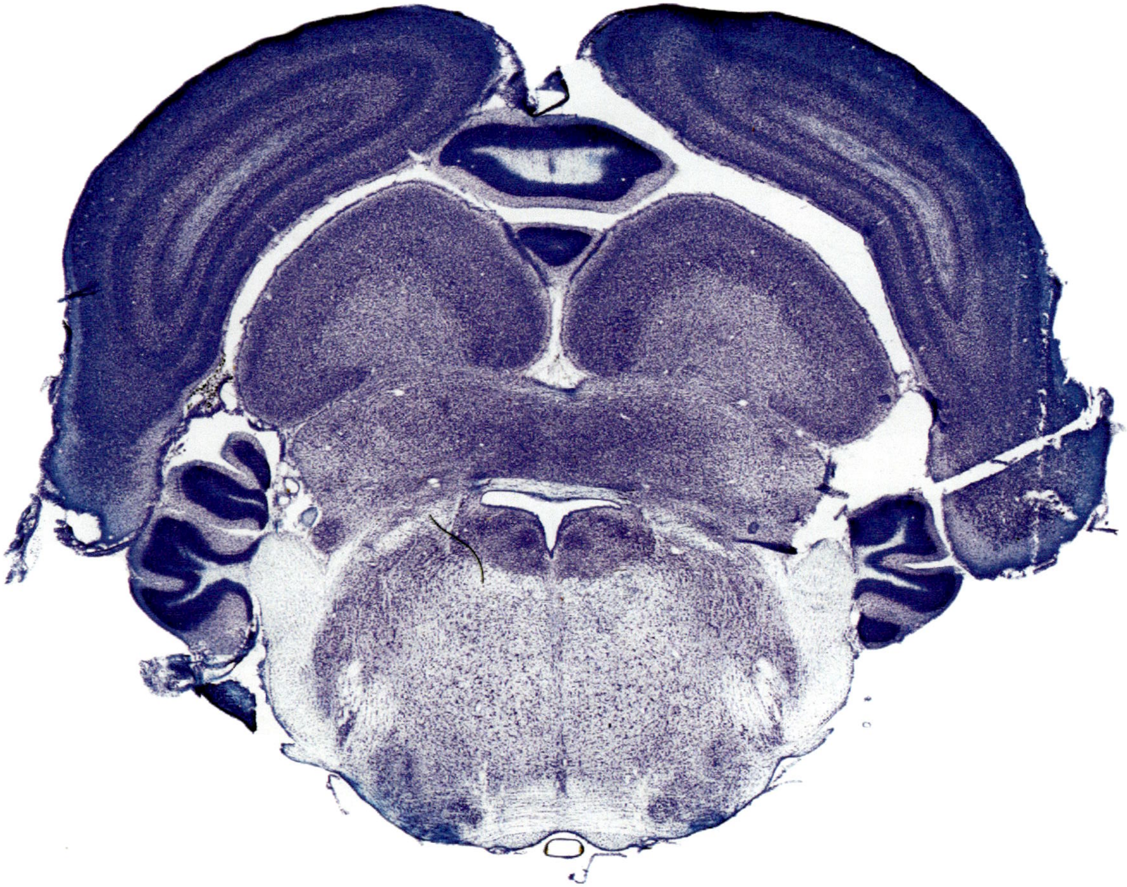

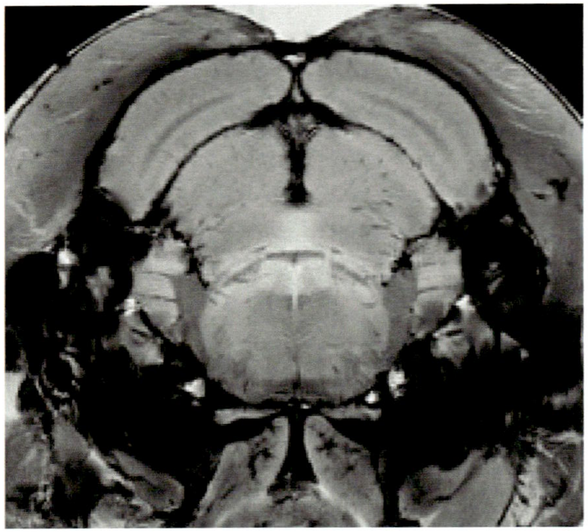

4n	trochlear nerve
4V	4th ventricle
5N	motor trigeminal nucleus
6n	root of abducens nerve
7n	facial nerve
A5	A5 noradrenaline cells
CG	central gray
DpG	deep gray layer of the superior colliculus
DpWh	deep white layer of the superior colliculus
DR	dorsal raphe nucleus
DTg	dorsal tegmental nucleus
IC	inferior colliculus
InG	intermediate gray layer of the superior colliculus
InWh	intermediate white layer of the superior colliculus
LDTg	laterodorsal tegmental nucleus
LPB	lateral parabrachial nucleus
mcp	middle cerebellar peduncle
me5	mesencephalic trigeminal tract
Me5	mesencephalic trigeminal nucleus
ml	medial lemniscus
mlf	medial longitudinal fasciculus
MPB	medial parabrachial nucleus
Op	optic nerve layer of the superior colliculus
P5	peritrigeminal zone
PnC	pontine reticular nucleus, caudal part
Pr5	principal sensory trigeminal nucleus
py	pyramidal tract
s5	sensory root of the trigeminal nerve
scp	superior cerebellar peduncle
scpd	superior cerebellar peduncle, descending limb
SubC	subcoeruleus nucleus
SuG	superficial gray layer of the superior colliculus
TC	temporal cortex
tth	trigeminothalamic tract
Tz	nucleus of the trapezoid body
tz	trapezoid body
V1	primary visual cortex
V2	secondary visual cortex
VPO	ventral periolivary nucleus
vsc	ventral spinocerebellar tract
Zo	zonal layer of the superior colliculus

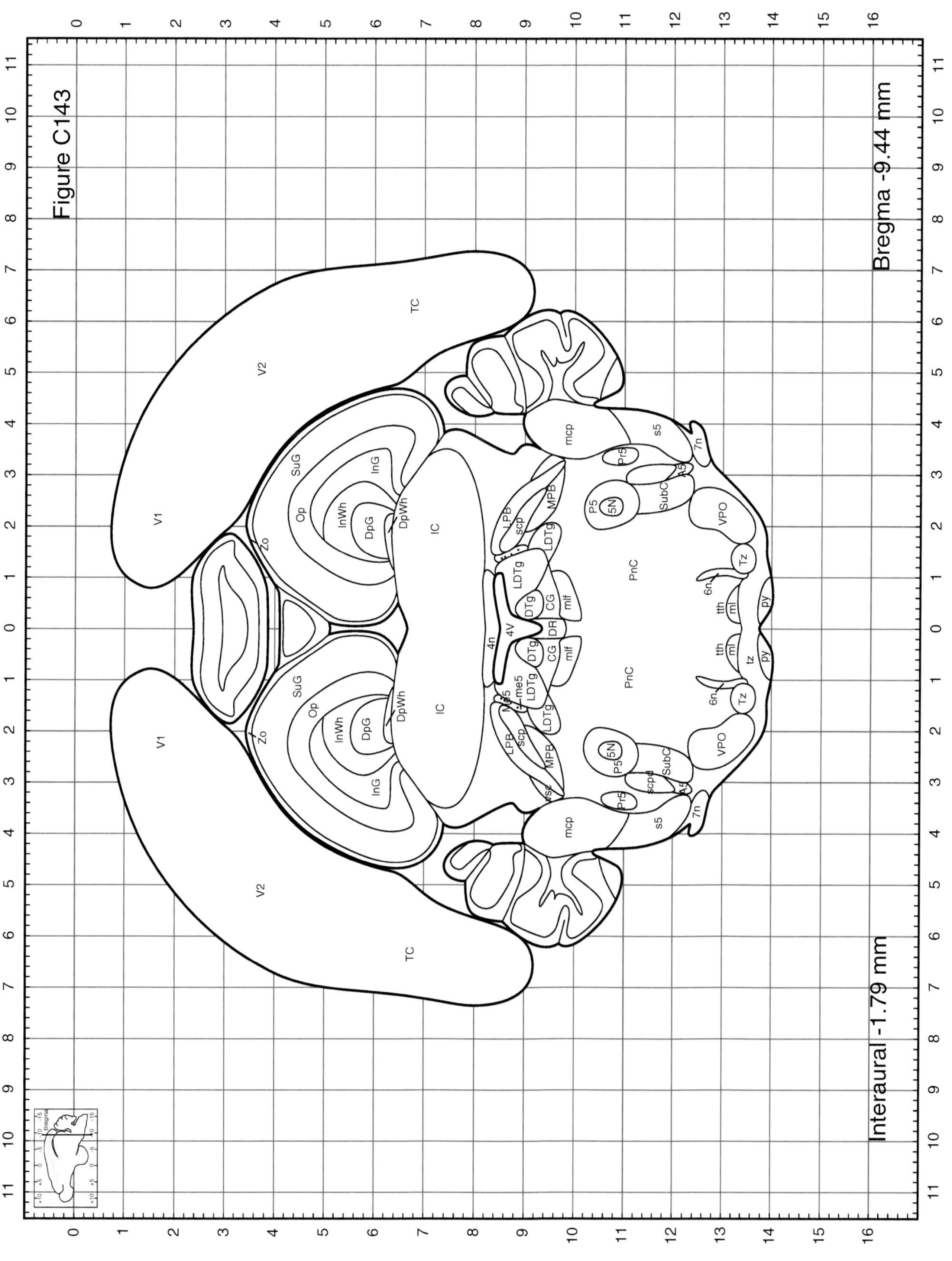
Figure C143
Bregma -9.44 mm
Interaural -1.79 mm
V1
V2
TC
Zo
SuG
Op
InWh
DpG
InG
DpWh
IC
mcp
s5
7n
Pr5
P5
5N
SubC
A5
scp
scpd
LPB
MPB
LDTg
DTg
CG
DR
mlf
4n
4V
Me5
me5
vsc
PnC
VPO
Tz
tz
6n
tth
ml
py

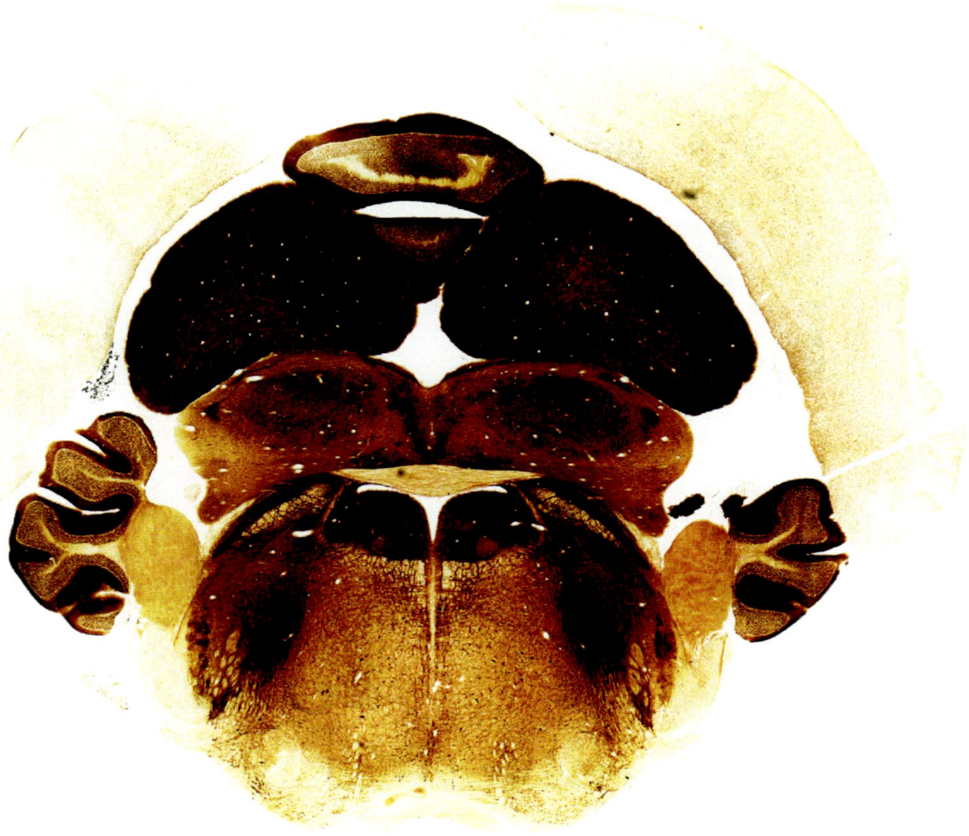

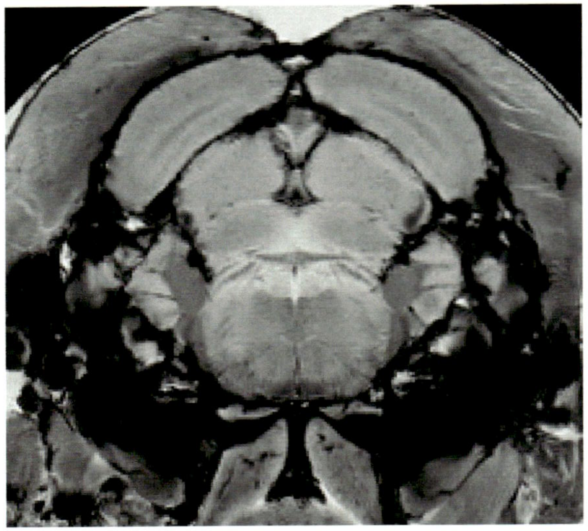

4n	trochlear nerve
4V	4th ventricle
5N	motor trigeminal nucleus
6n	root of abducens nerve
7n	facial nerve
A5	A5 noradrenaline cells
CG	central gray
DTg	dorsal tegmental nucleus
IC	inferior colliculus
InG	intermediate gray layer of the superior colliculus
InWh	intermediate white layer of the superior colliculus
LC	locus coeruleus
LDTg	laterodorsal tegmental nucleus
LPB	lateral parabrachial nucleus
mcp	middle cerebellar peduncle
me5	mesencephalic trigeminal tract
Me5	mesencephalic trigeminal nucleus
ml	medial lemniscus
mlf	medial longitudinal fasciculus
MPB	medial parabrachial nucleus
Op	optic nerve layer of the superior colliculus
P5	peritrigeminal zone
PnC	pontine reticular nucleus, caudal part
Pr5	principal sensory trigeminal nucleus
py	pyramidal tract
rs	rubrospinal tract
s5	sensory root of the trigeminal nerve
scp	superior cerebellar peduncle
scpd	superior cerebellar peduncle, descending limb
SO	superior olive
SuG	superficial gray layer of the superior colliculus
tth	trigeminothalamic tract
Tz	nucleus of the trapezoid body
tz	trapezoid body
V1	primary visual cortex
V2	secondary visual cortex
VPO	ventral periolivary nucleus
vsc	ventral spinocerebellar tract
Zo	zonal layer of the superior colliculus

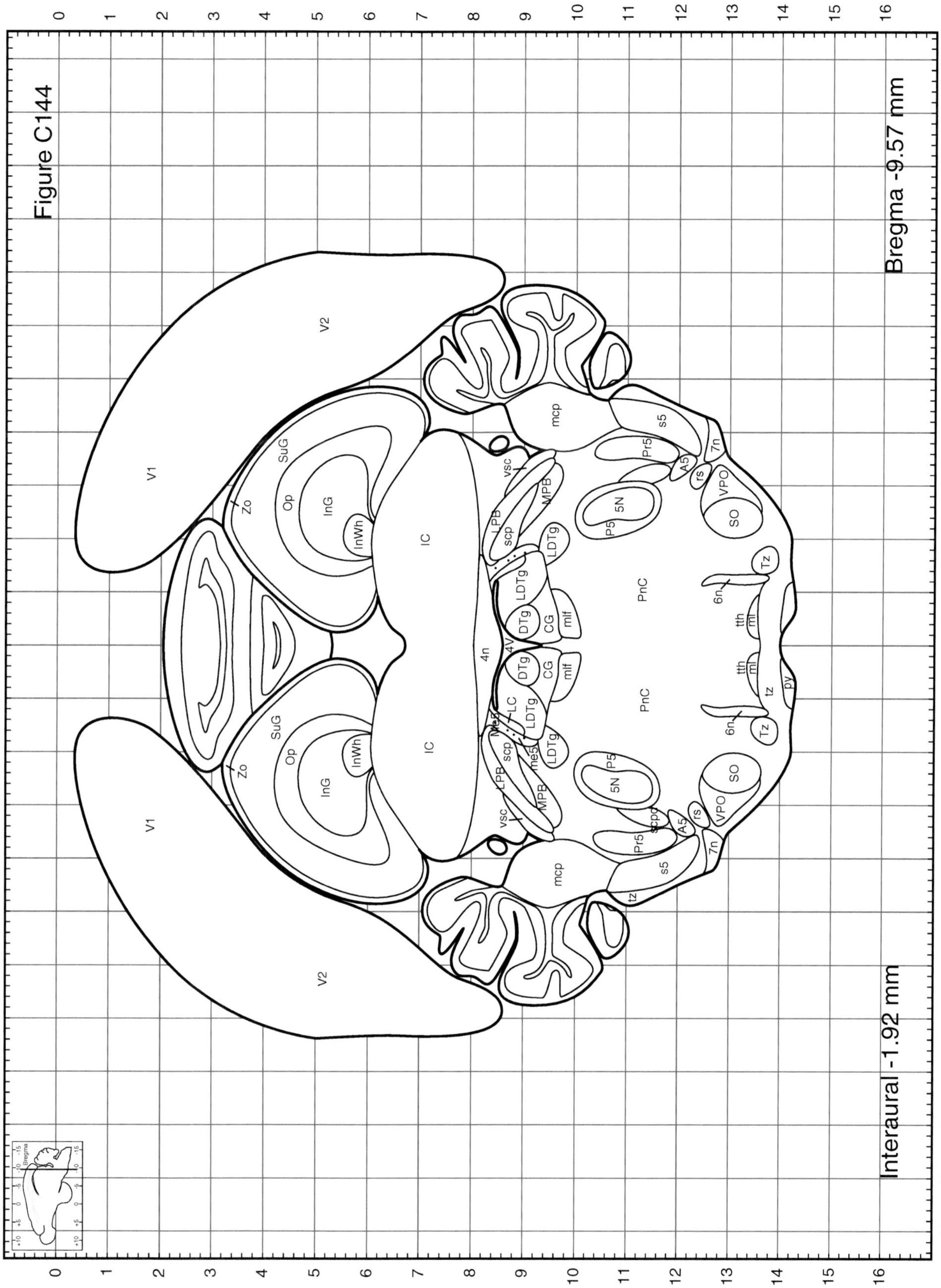
Figure C144
Bregma -9.57 mm
Interaural -1.92 mm
V1
V2
SuG
Op
InG
InWh
Zo
IC
4n
4V
vsc
MPB
LPB
scp
LDTg
DTg
CG
mlf
LC
Me5
me5
mcp
s5
Pr5
P5
5N
PnC
A5
rs
7n
VPO
SO
Tz
tz
6n
tth
ml
py
scpd

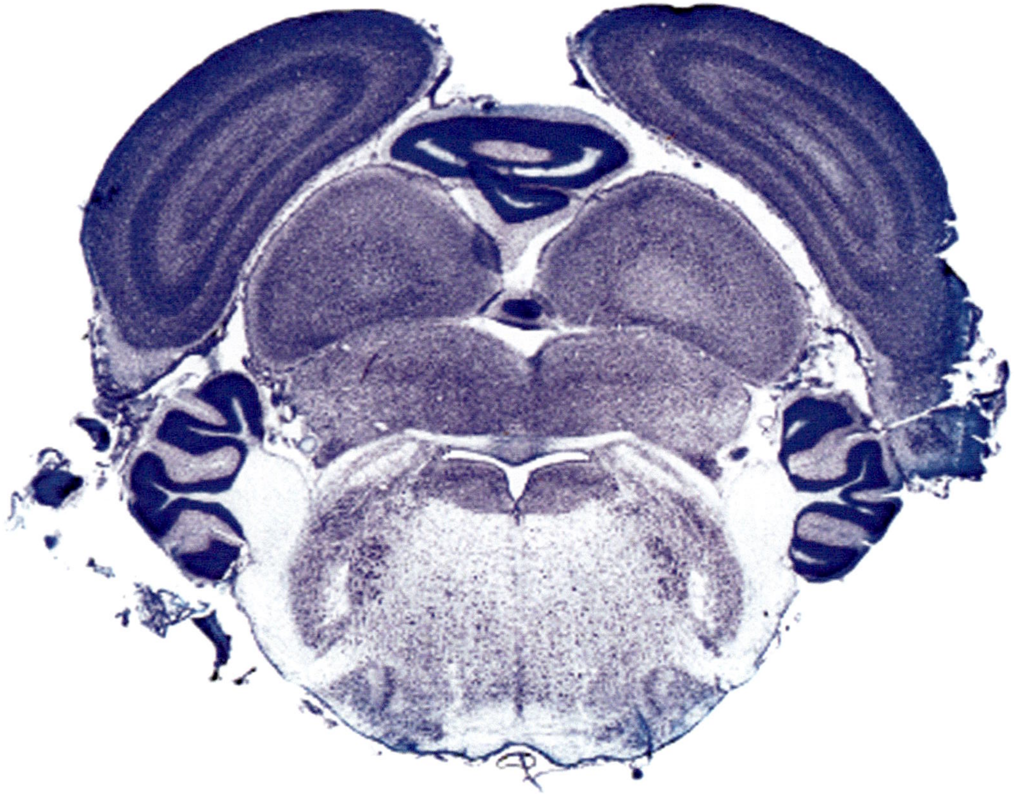

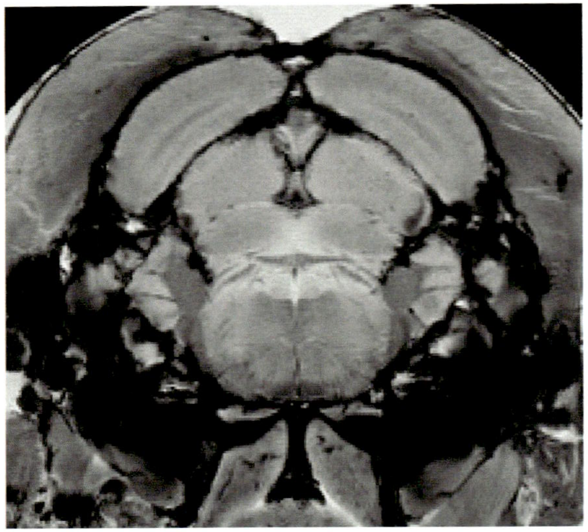

4n	trochlear nerve
4V	4th ventricle
5N	motor trigeminal nucleus
6n	root of abducens nerve
7n	facial nerve
A5	A5 noradrenaline cells
CG	central gray
DTg	dorsal tegmental nucleus
IC	inferior colliculus
InG	intermediate gray layer of the superior colliculus
LC	locus coeruleus
LDTg	laterodorsal tegmental nucleus
LPB	lateral parabrachial nucleus
mcp	middle cerebellar peduncle
me5	mesencephalic trigeminal tract
Me5	mesencephalic trigeminal nucleus
ml	medial lemniscus
mlf	medial longitudinal fasciculus
MPB	medial parabrachial nucleus
Op	optic nerve layer of the superior colliculus
PnC	pontine reticular nucleus, caudal part
Pr5	principal sensory trigeminal nucleus
py	pyramidal tract
rs	rubrospinal tract
s5	sensory root of the trigeminal nerve
scp	superior cerebellar peduncle
scpd	superior cerebellar peduncle, descending limb
SO	superior olive
SubC	subcoeruleus nucleus
SuG	superficial gray layer of the superior colliculus
tth	trigeminothalamic tract
Tz	nucleus of the trapezoid body
tz	trapezoid body
V1	primary visual cortex
V2	secondary visual cortex
VPO	ventral periolivary nucleus
vsc	ventral spinocerebellar tract
Zo	zonal layer of the superior colliculus

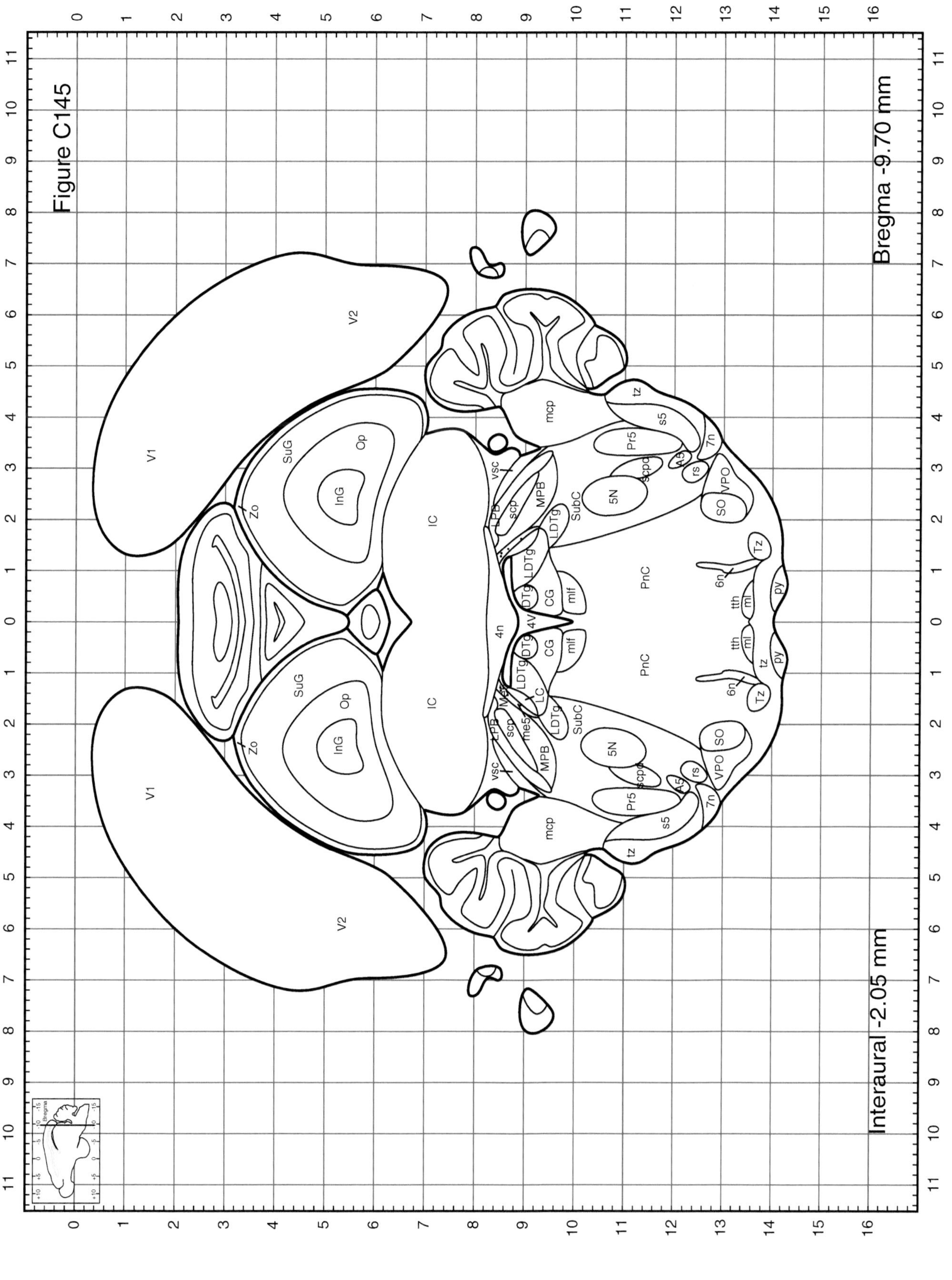
Figure C145
Bregma -9.70 mm
Interaural -2.05 mm
V1
V2
SuG
Op
InG
Zo
IC
4n
4V
vsc
LPB
scp
MPB
me5
Me5
LC
LDTg
DTg
CG
mlf
SubC
mcp
tz
s5
Pr5
scpd
A5
rs
7n
VPO
SO
5N
PnC
6n
tth
ml
py
Tz
0 1 2 3 4 5 6 7 8 9 10 11 12 13 14 15 16
11 10 9 8 7 6 5 4 3 2 1 0 1 2 3 4 5 6 7 8 9 10 11

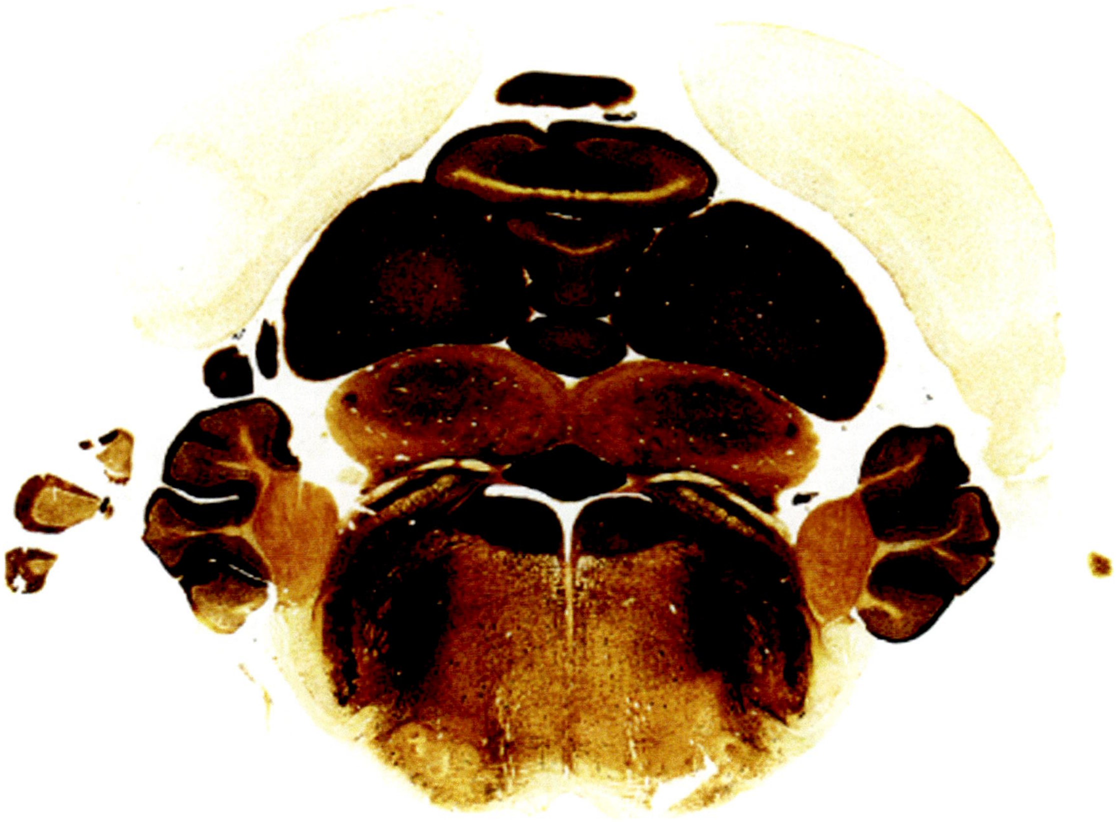

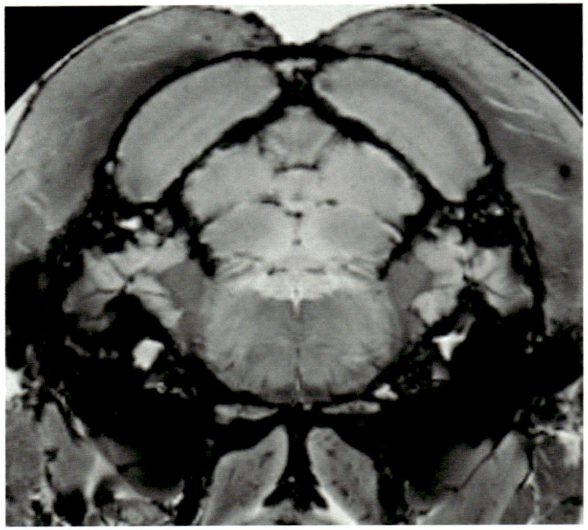

4n trochlear nerve
4V 4th ventricle
5N motor trigeminal nucleus
6n root of abducens nerve
7n facial nerve
CG central gray
DTg dorsal tegmental nucleus
IC inferior colliculus
InG intermediate gray layer of the superior colliculus
LC locus coeruleus
LPB lateral parabrachial nucleus
mcp middle cerebellar peduncle
me5 mesencephalic trigeminal tract
Me5 mesencephalic trigeminal nucleus
ml medial lemniscus
mlf medial longitudinal fasciculus
MPB medial parabrachial nucleus
Op optic nerve layer of the superior colliculus
PnC pontine reticular nucleus, caudal part
Pr5 principal sensory trigeminal nucleus
py pyramidal tract
rs rubrospinal tract
s5 sensory root of the trigeminal nerve
scp superior cerebellar peduncle
scpd superior cerebellar peduncle, descending limb
SO superior olive
SubC subcoeruleus nucleus
SuG superficial gray layer of the superior colliculus
tth trigeminothalamic tract
Tz nucleus of the trapezoid body
tz trapezoid body
V1 primary visual cortex
VPO ventral periolivary nucleus
vsc ventral spinocerebellar tract
Zo zonal layer of the superior colliculus

Figure C146
Bregma -9.84 mm
Interaural -2.19 mm
V1
SuG
Zo
Op
InG
IC
vsc
LPB
scp
MPB
SubC
CG
DTg
4V
LC
Me5
me5
mlf
mcp
tz
s5
Pr5
scpd
5N
PnC
rs
7n
VPO
SO
Tz
6n
tth
ml
py

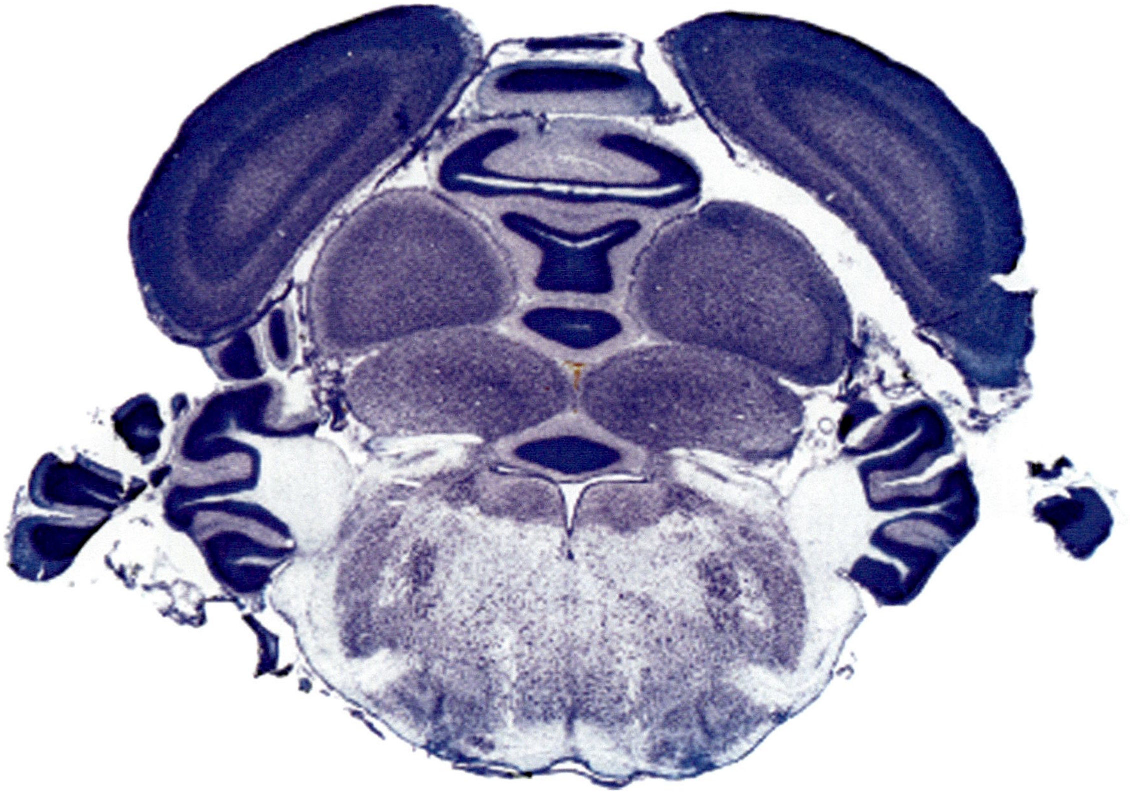

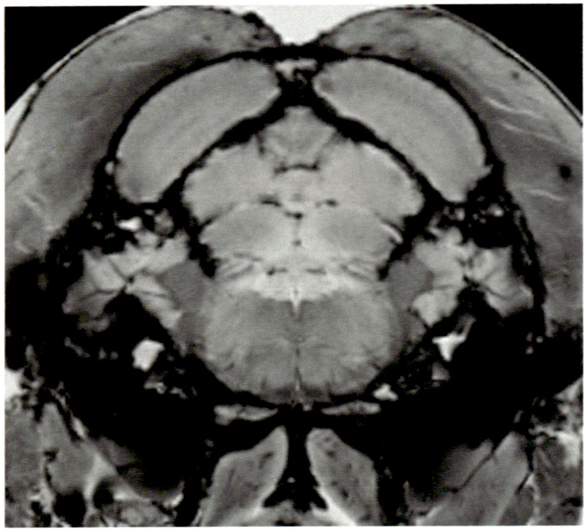

4V	4th ventricle	Me5	mesencephalic trigeminal nucleus	SO	superior olive
5N	motor trigeminal nucleus	ml	medial lemniscus	SubC	subcoeruleus nucleus
6n	root of abducens nerve	mlf	medial longitudinal fasciculus	SuG	superficial gray layer of the superior colliculus
7n	facial nerve	MPB	medial parabrachial nucleus	tth	trigeminothalamic tract
CG	central gray	Op	optic nerve layer of the superior colliculus	Tz	nucleus of the trapezoid body
DTg	dorsal tegmental nucleus	PnC	pontine reticular nucleus, caudal part	tz	trapezoid body
IC	inferior colliculus	Pr5	principal sensory trigeminal nucleus	V1	primary visual cortex
icp	inferior cerebellar peduncle	py	pyramidal tract	Ves	vestibular nucleus
LC	locus coeruleus	rs	rubrospinal tract	VPO	ventral periolivary nucleus
LPB	lateral parabrachial nucleus	s5	sensory root of the trigeminal nerve	vsc	ventral spinocerebellar tract
mcp	middle cerebellar peduncle	scp	superior cerebellar peduncle	Zo	zonal layer of the superior colliculus
me5	mesencephalic trigeminal tract	scpd	superior cerebellar peduncle, descending limb		

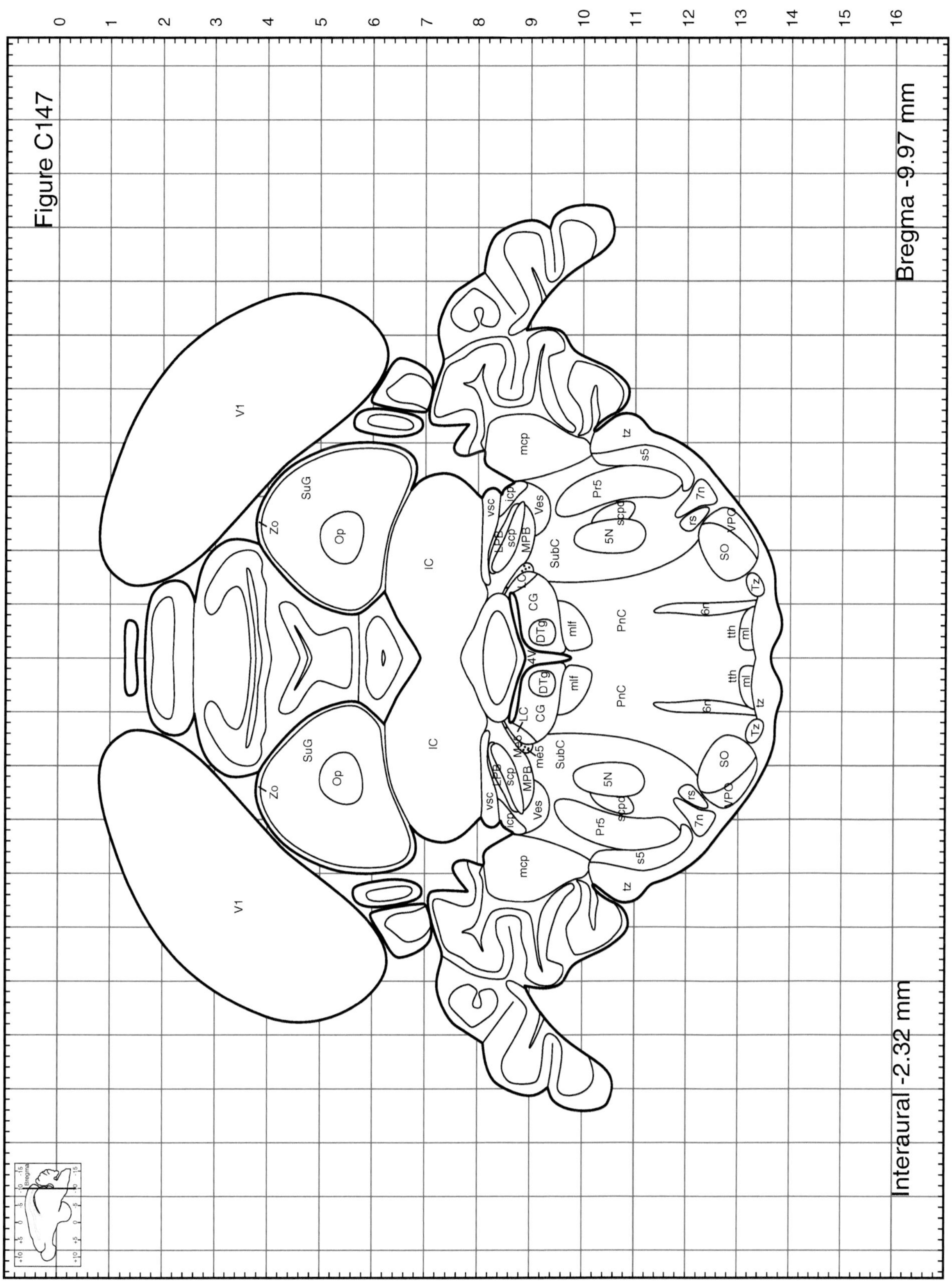

Figure C147
Bregma -9.97 mm
Interaural -2.32 mm
V1
SuG
Zo
Op
IC
vsc
icp
LPB
scp
MPB
Ves
SubC
LC
Me5
me5
CG
DTg
4V
mlf
mcp
tz
s5
Pr5
5N
scpd
PnC
7n
rs
VPO
SO
Tz
6n
tth
ml

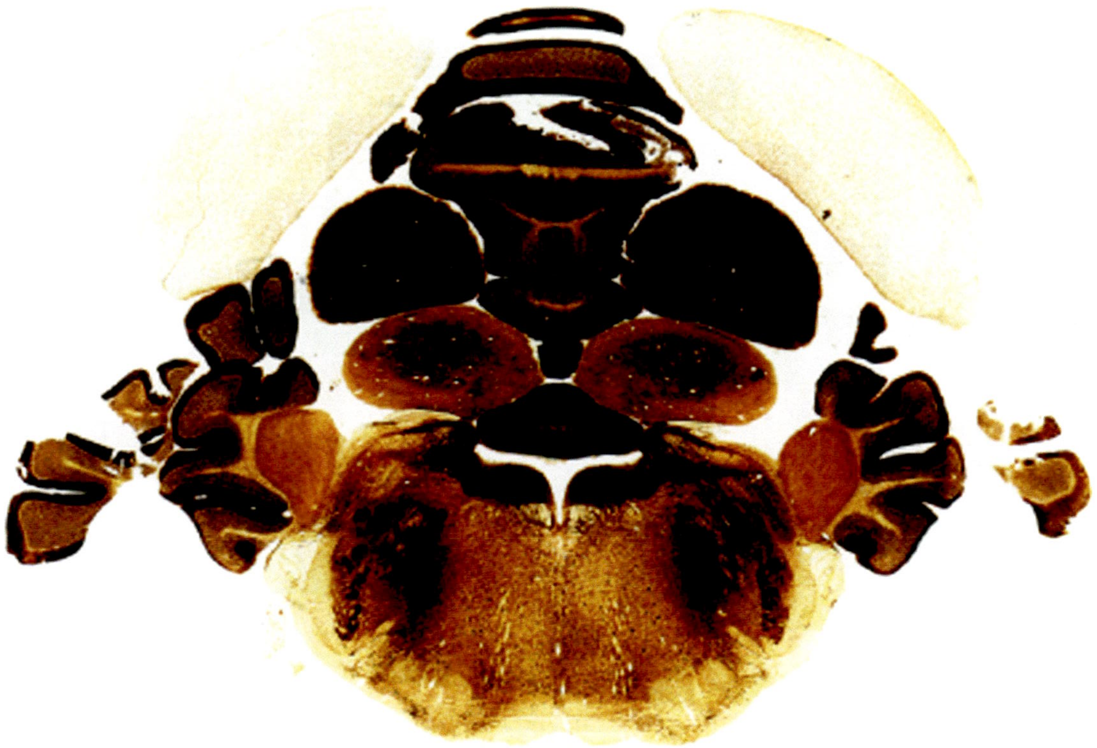

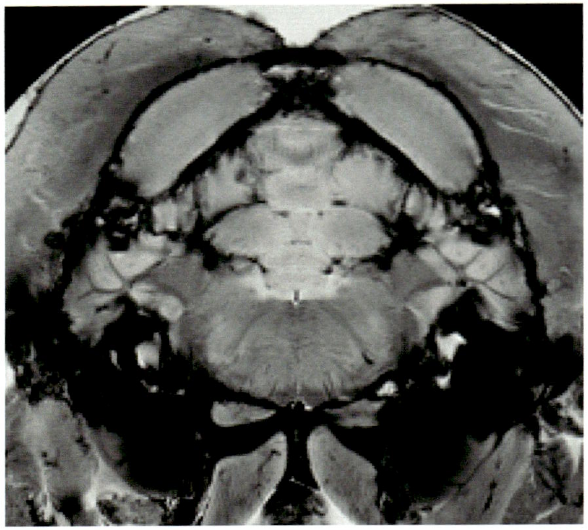

4V	4th ventricle	Me5	mesencephalic trigeminal nucleus	SubC	subcoeruleus nucleus
5N	motor trigeminal nucleus	ml	medial lemniscus	SuG	superficial gray layer of the superior colliculus
6n	root of abducens nerve	mlf	medial longitudinal fasciculus	tth	trigeminothalamic tract
7n	facial nerve	MPB	medial parabrachial nucleus	Tz	nucleus of the trapezoid body
CG	central gray	PnC	pontine reticular nucleus, caudal part	tz	trapezoid body
DTg	dorsal tegmental nucleus	Pr5	principal sensory trigeminal nucleus	V1	primary visual cortex
IC	inferior colliculus	py	pyramidal tract	Ves	vestibular nucleus
icp	inferior cerebellar peduncle	rs	rubrospinal tract	VPO	ventral periolivary nucleus
LC	locus coeruleus	s5	sensory root of the trigeminal nerve	vsc	ventral spinocerebellar tract
LPB	lateral parabrachial nucleus	scp	superior cerebellar peduncle	Zo	zonal layer of the superior colliculus
mcp	middle cerebellar peduncle	scpd	superior cerebellar peduncle, descending limb		
me5	mesencephalic trigeminal tract	SO	superior olive		

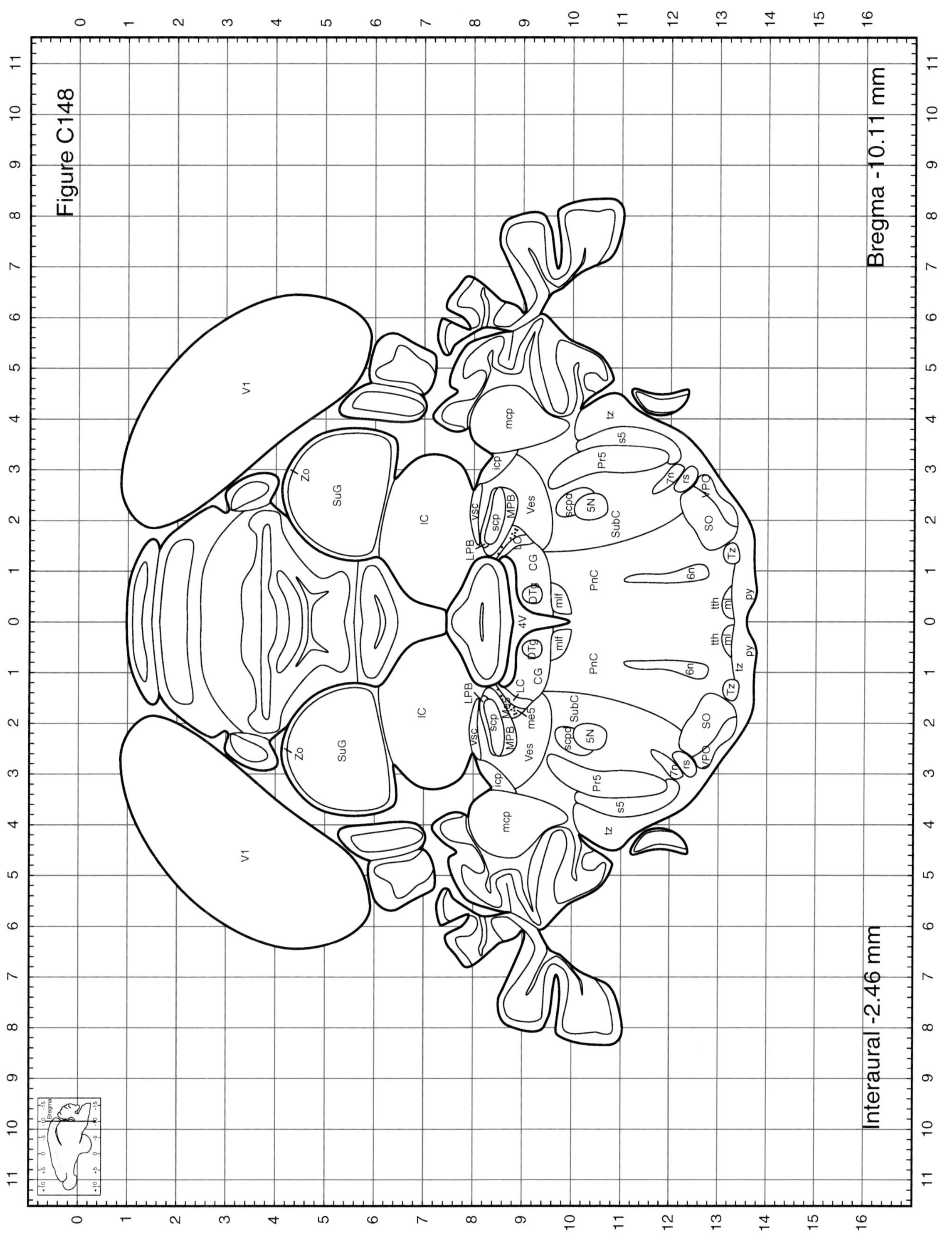
Figure C148
Bregma -10.11 mm
Interaural -2.46 mm
V1
Zo
SuG
IC
LPB
vsc
scp
MPB
LC
Me5
me5
Ves
icp
mcp
CG
DTg
mlf
4V
PnC
scpd
5N
SubC
Pr5
s5
tz
7n
rs
VPO
SO
Tz
6n
tth
ml
py

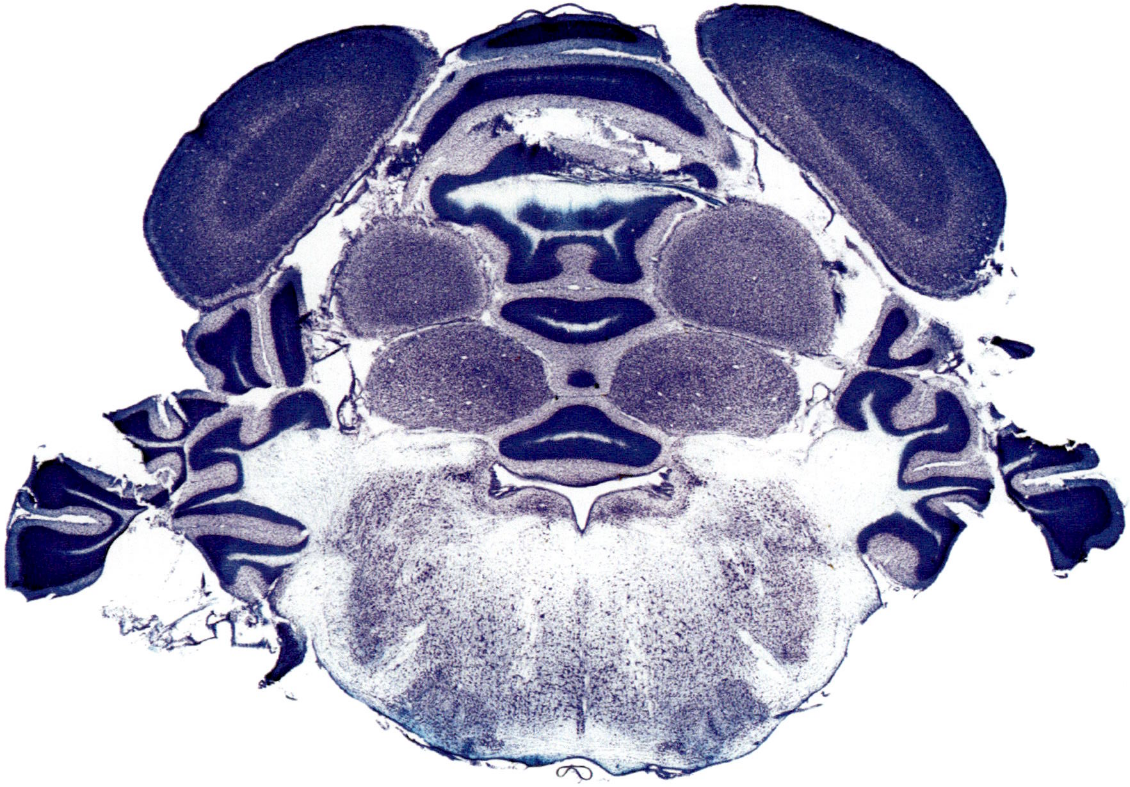

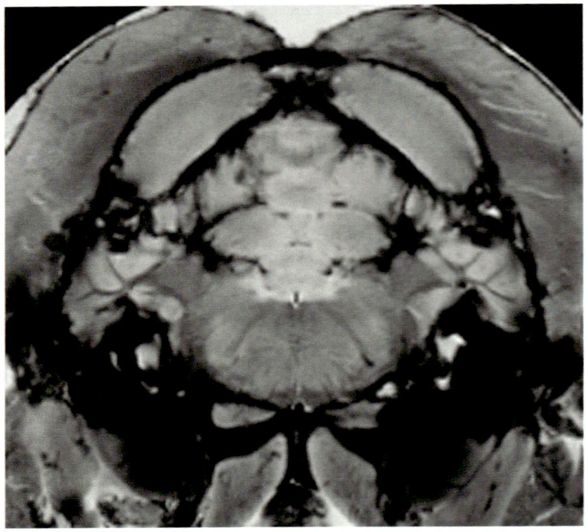

4V	4th ventricle
5N	motor trigeminal nucleus
6n	root of abducens nerve
7n	facial nerve
CG	central gray
IC	inferior colliculus
icp	inferior cerebellar peduncle
LC	locus coeruleus
LPB	lateral parabrachial nucleus
mcp	middle cerebellar peduncle
me5	mesencephalic trigeminal tract
Me5	mesencephalic trigeminal nucleus
ml	medial lemniscus
mlf	medial longitudinal fasciculus
MPB	medial parabrachial nucleus
PnC	pontine reticular nucleus, caudal part
Pr5	principal sensory trigeminal nucleus
py	pyramidal tract
rs	rubrospinal tract
s5	sensory root of the trigeminal nerve
scp	superior cerebellar peduncle
scpd	superior cerebellar peduncle, descending limb
SGe	supragenual nucleus
SO	superior olive
SubC	subcoeruleus nucleus
SuG	superficial gray layer of the superior colliculus
tth	trigeminothalamic tract
Tz	nucleus of the trapezoid body
tz	trapezoid body
V1	primary visual cortex
VCA	ventral cochlear nucleus, anterior part
Ves	vestibular nucleus
VPO	ventral periolivary nucleus
vsc	ventral spinocerebellar tract
Zo	zonal layer of the superior colliculus

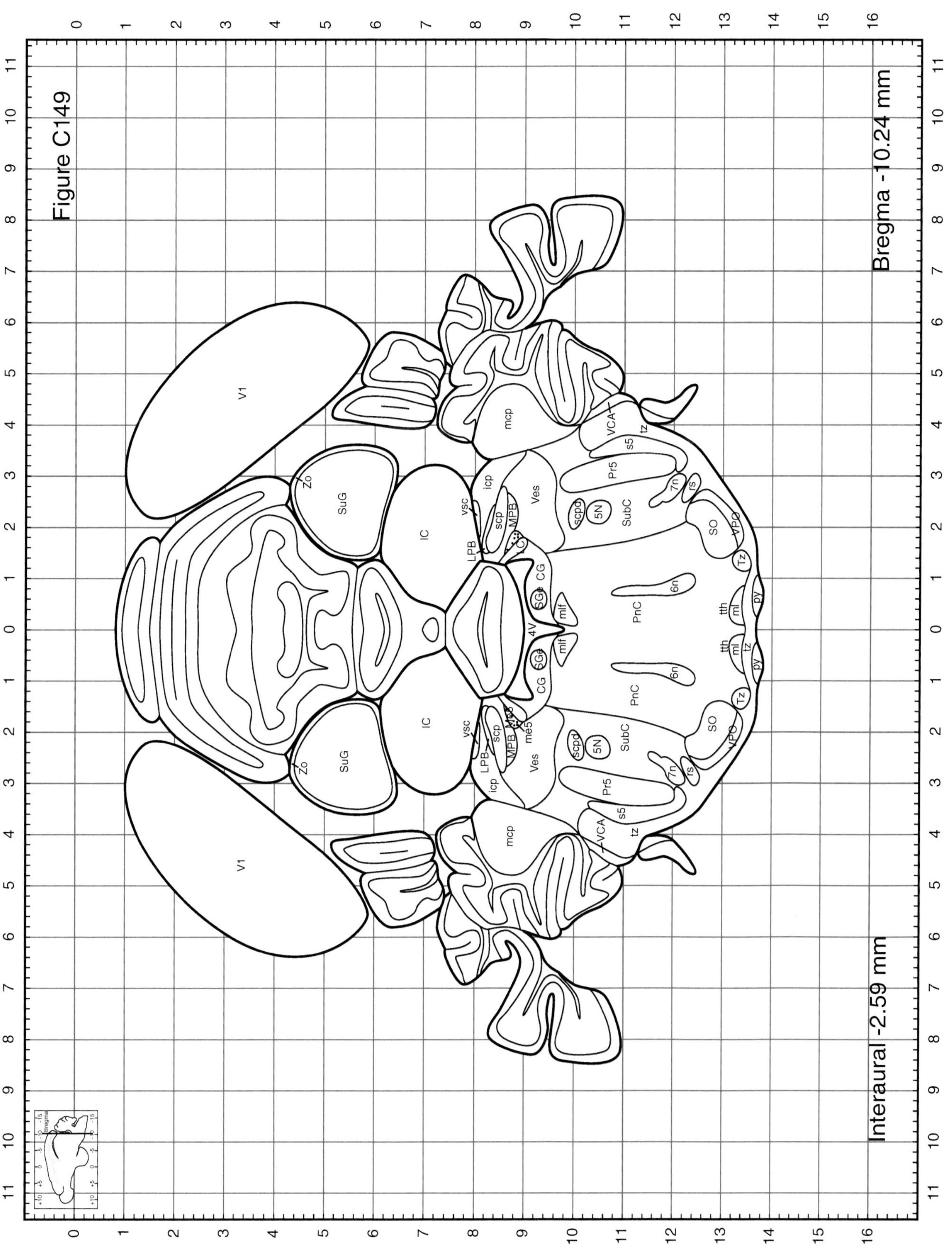
Figure C149
Bregma -10.24 mm
Interaural -2.59 mm
V1
SuG
Zo
IC
vsc
LPB
icp
scp
MPB
Me5
me5
Ves
mcp
VCA
tz
s5
Pr5
scpd
5N
SubC
7n
rs
SO
VPO
Tz
6n
PnC
tth
ml
py
CG
SGe
4V
mlf

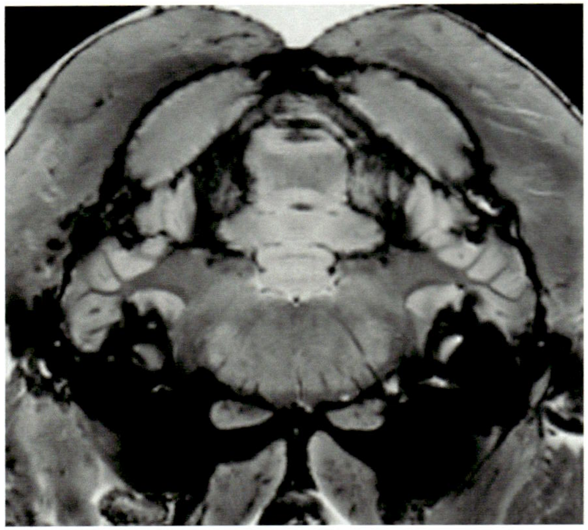

4V	4th ventricle
6n	root of abducens nerve
7n	facial nerve
8cn	cochlear root of the vestibulocochlear nerve
8vn	vestibular root of the vestibulocochlear nerve
CG	central gray
g7	genu of the facial nerve
IC	inferior colliculus
icp	inferior cerebellar peduncle
LC	locus coeruleus
LPB	lateral parabrachial nucleus
mcp	middle cerebellar peduncle
me5	mesencephalic trigeminal tract
Me5	mesencephalic trigeminal nucleus
ml	medial lemniscus
mlf	medial longitudinal fasciculus
MPB	medial parabrachial nucleus
P5	peritrigeminal zone
PCRt	parvicellular reticular nucleus
PnC	pontine reticular nucleus, caudal par
Pr5	principal sensory trigeminal nucleus
py	pyramidal tract
rs	rubrospinal tract
s5	sensory root of the trigeminal nerve
scp	superior cerebellar peduncle
scpd	superior cerebellar peduncle, descending limb
SGe	supragenual nucleus
SO	superior olive
SubC	subcoeruleus nucleus
SuG	superficial gray layer of the superior colliculus
tth	trigeminothalamic tract
Tz	nucleus of the trapezoid body
tz	trapezoid body
V1	primary visual cortex
VCA	ventral cochlear nucleus, anterior part
Ves	vestibular nucleus
VPO	ventral periolivary nucleus
Zo	zonal layer of the superior colliculus

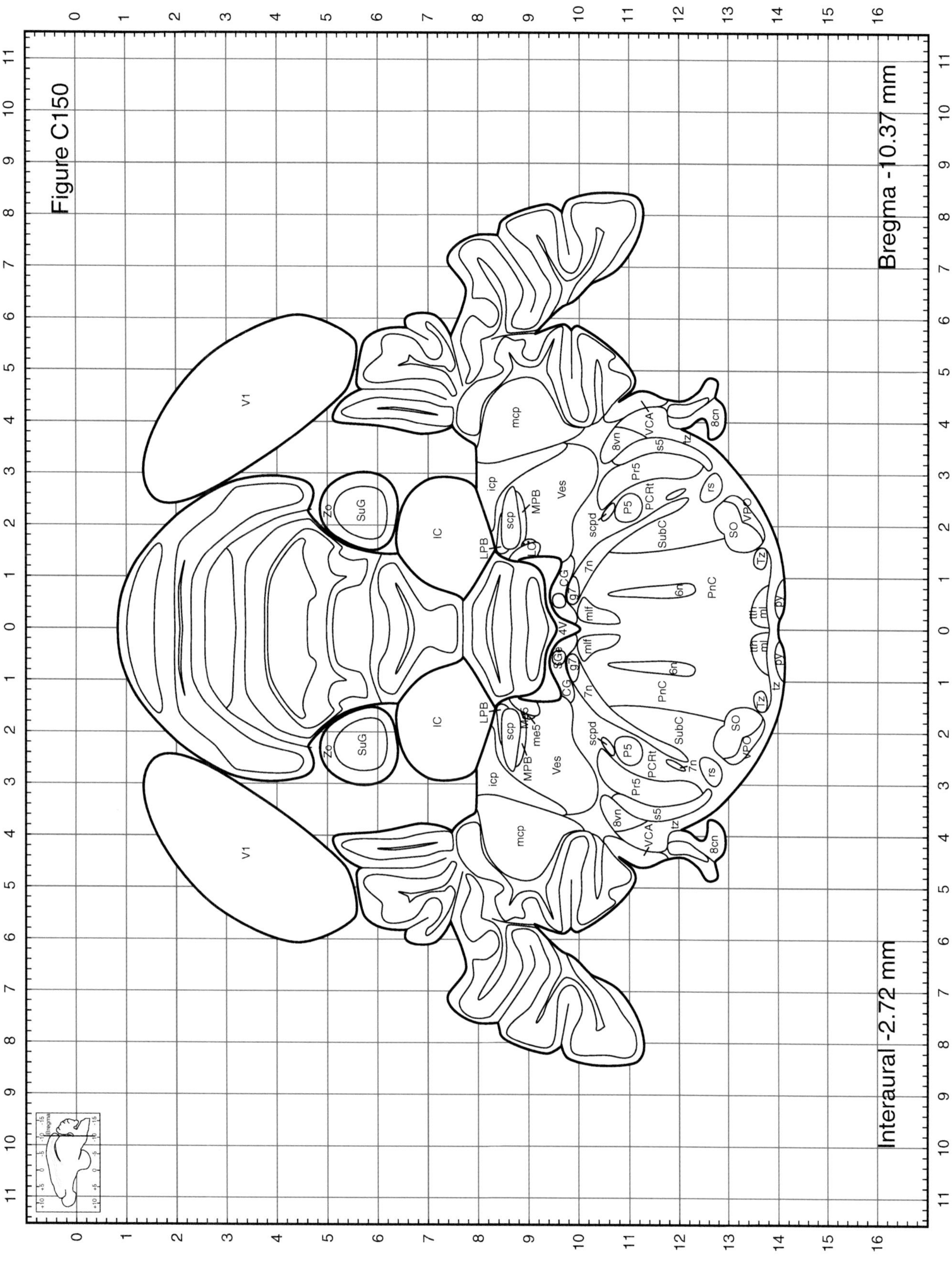
Figure C150
Bregma -10.37 mm
Interaural -2.72 mm
V1
SuG
Zo
IC
LPB
scp
MPB
me5
Me5
icp
mcp
Ves
scpd
P5
Pr5
PCRt
SubC
8vn
VCA
s5
tz
8cn
rs
SO
VPO
Tz
PnC
6n
7n
CG
g7
mlf
4V
SGe
tth
ml
py

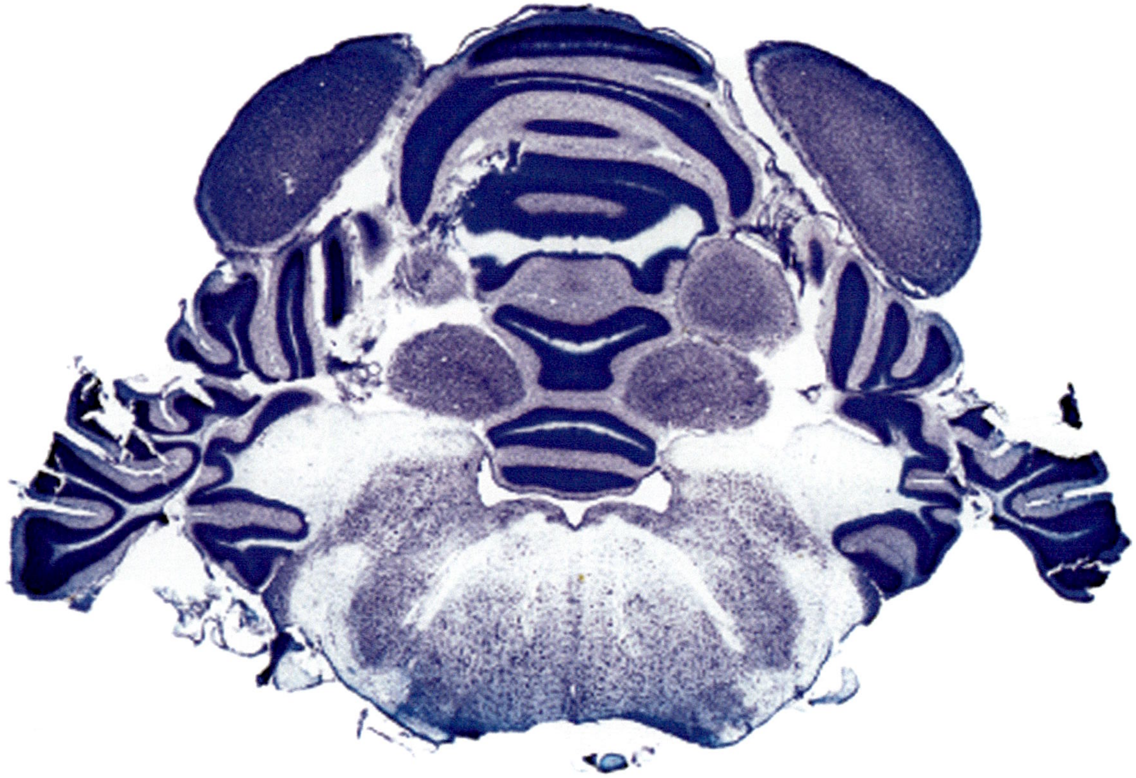

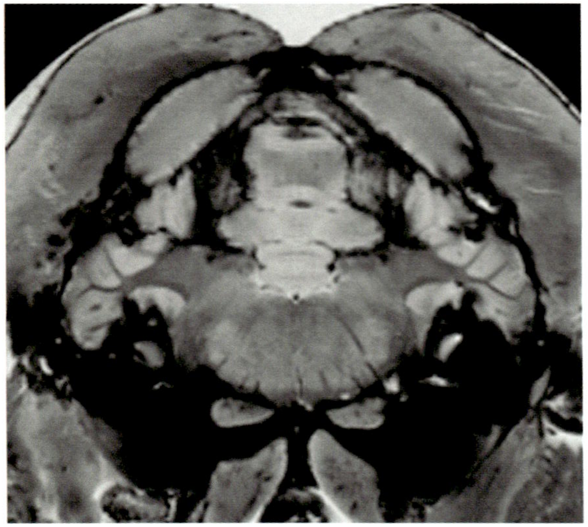

4V	4th ventricle
6n	root of abducens nerve
7N	facial nucleus
7n	facial nerve
8cn	cochlear root of the vestibulocochlear nerve
8vn	vestibular root of the vestibulocochlear nerve
g7	genu of the facial nerve
IC	inferior colliculus
icp	inferior cerebellar peduncle
IRt	intermediate reticular nucleus
LC	locus coeruleus
mcp	middle cerebellar peduncle
ml	medial lemniscus
mlf	medial longitudinal fasciculus
MPB	medial parabrachial nucleus
PCRt	parvicellular reticular nucleus
PnC	pontine reticular nucleus, caudal part
Pr5	principal sensory trigeminal nucleus
py	pyramidal tract
rs	rubrospinal tract
scp	superior cerebellar peduncle
SGe	supragenual nucleus
SO	superior olive
sp	5spinal trigeminal tract
tth	trigeminothalamic tract
Tz	nucleus of the trapezoid body
tz	trapezoid body
V1	primary visual cortex
VCA	ventral cochlear nucleus, anterior part
Ves	vestibular nucleus
VPO	ventral periolivary nucleus
Zo	zonal layer of the superior colliculus

Figure C151
Bregma -10.51 mm
Interaural -2.86 mm
V1
Zo
IC
icp
scp
MPB
LC
Ves
mcp
8vn
sp5
Pr5
PCRt
7n
IRt
7N
SO
VPO
rs
tz
8cn
VCA
PnC
6n
g7
mlf
4V
SGe
tth
ml
py

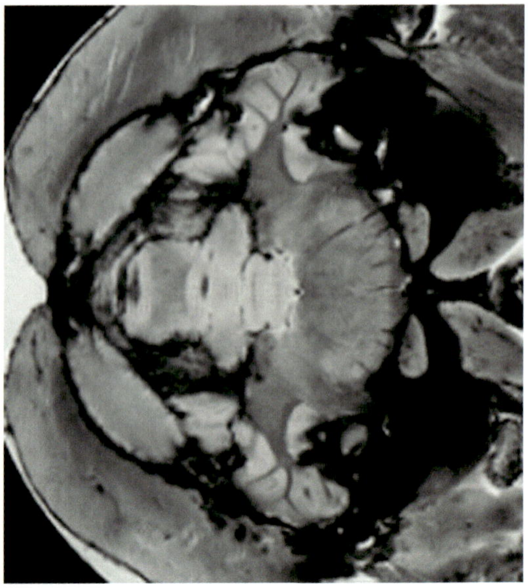

4V 4th ventricle
6n root of abducens nerve
7N facial nucleus
7n facial nerve
8cn cochlear root of the vestibulocochlear nerve
8vn vestibular root of the vestibulocochlear nerve
g7 genu of the facial nerve
IC inferior colliculus
icp inferior cerebellar peduncle
IRt intermediate reticular nucleus
LC locus coeruleus
mcp middle cerebellar peduncle
ml medial lemniscus
mlf medial longitudinal fasciculus
PCRt parvicellular reticular nucleus
PnC pontine reticular nucleus, caudal part
Pr5 principal sensory trigeminal nucleus
py pyramidal tract
rs rubrospinal tract
scp superior cerebellar peduncle
SGe supragenual nucleus
SO superior olive
sp5 spinal trigeminal tract
tth trigeminothalamic tract
tz trapezoid body
V1 primary visual cortex
VCA ventral cochlear nucleus, anterior part
Ves vestibular nucleus
VPO ventral periolivary nucleus

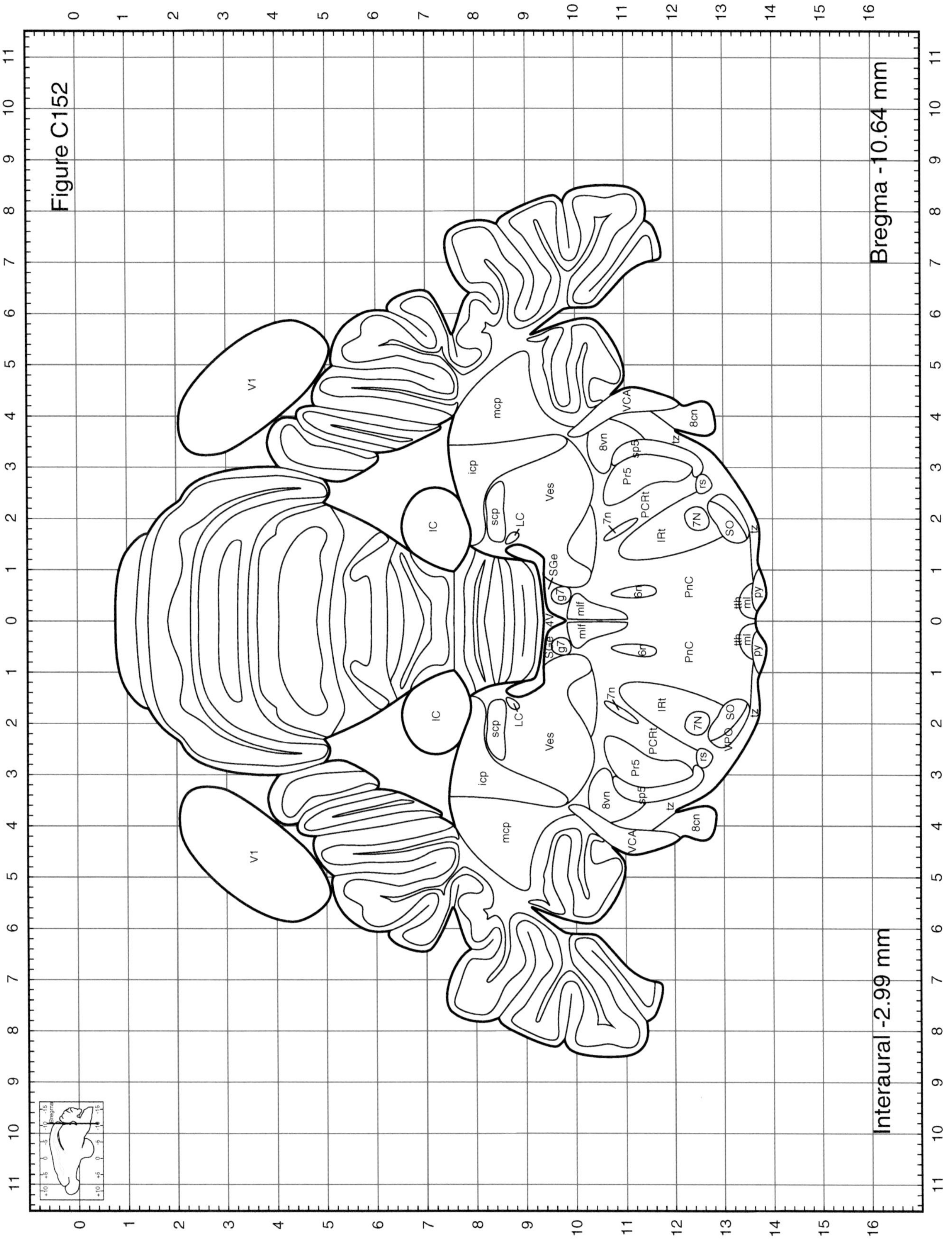
Figure C152
Bregma -10.64 mm
Interaural -2.99 mm
V1
mcp
icp
scp
IC
LC
Ves
SGe
g7
4V
mlf
8vn
sp5
Pr5
PCRt
7n
IRt
6n
PnC
7N
SO
VPO
rs
tz
VCA
8cn
tth
ml
py

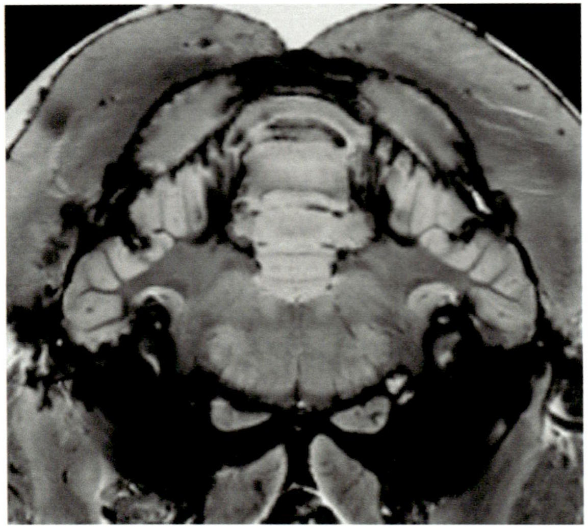

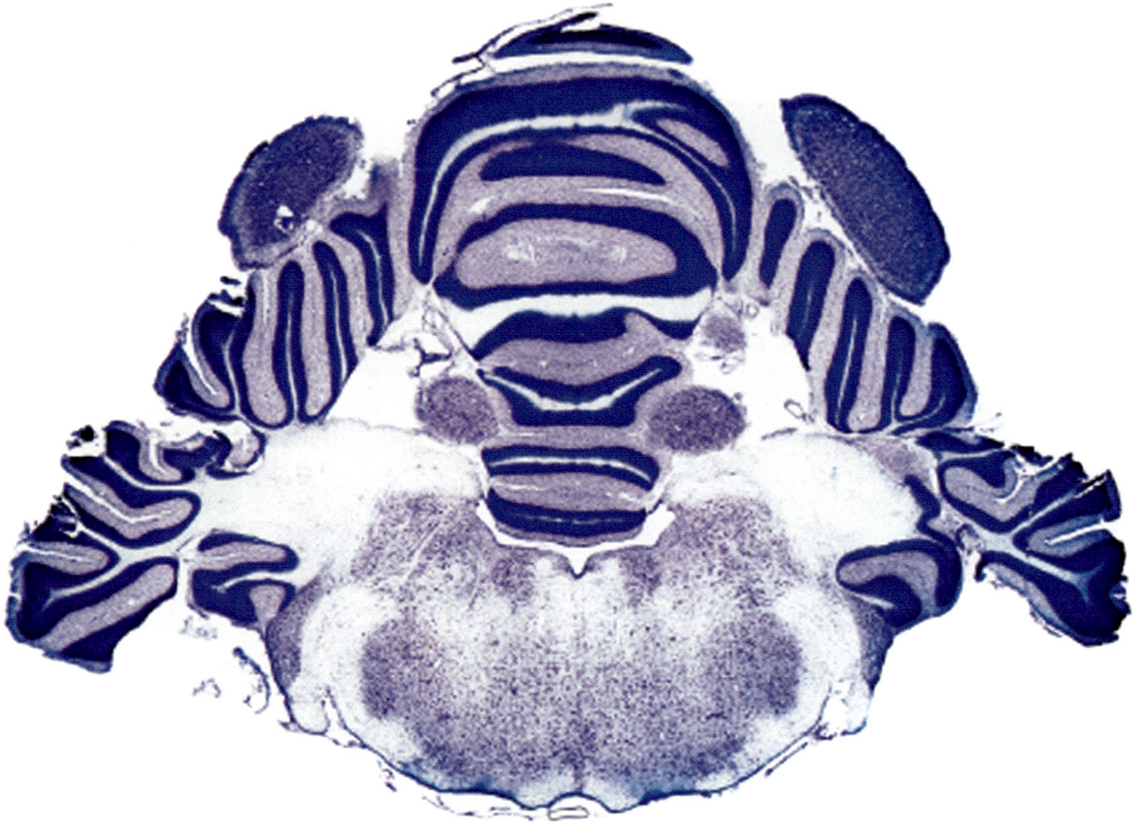

4V	4th ventricle	PnC	pontine reticular nucleus, caudal part
6n	root of abducens nerve	Pr	prepositus nucleus
6N	abducens nucleus	Pr5	principal sensory trigeminal nucleus
7N	facial nucleus	py	pyramidal tract
8cn	cochlear root of the vestibulocochlear nerve	rs	rubrospinal tract
8vn	vestibular root of the vestibulocochlear nerve	scp	superior cerebellar peduncle
g7	genu of the facial nerve	SO	superior olive
IC	inferior colliculus	sp5	spinal trigeminal tract
icp	inferior cerebellar peduncle	tz	trapezoid body
IRt	intermediate reticular nucleus	V1	primary visual cortex
mcp	middle cerebellar peduncle	VCA	ventral cochlear nucleus, anterior part
ml	medial lemniscus	Ves	vestibular nucleus
mlf	medial longitudinal fasciculus	VPO	ventral periolivary nucleus
PCRt	parvicellular reticular nucleus		

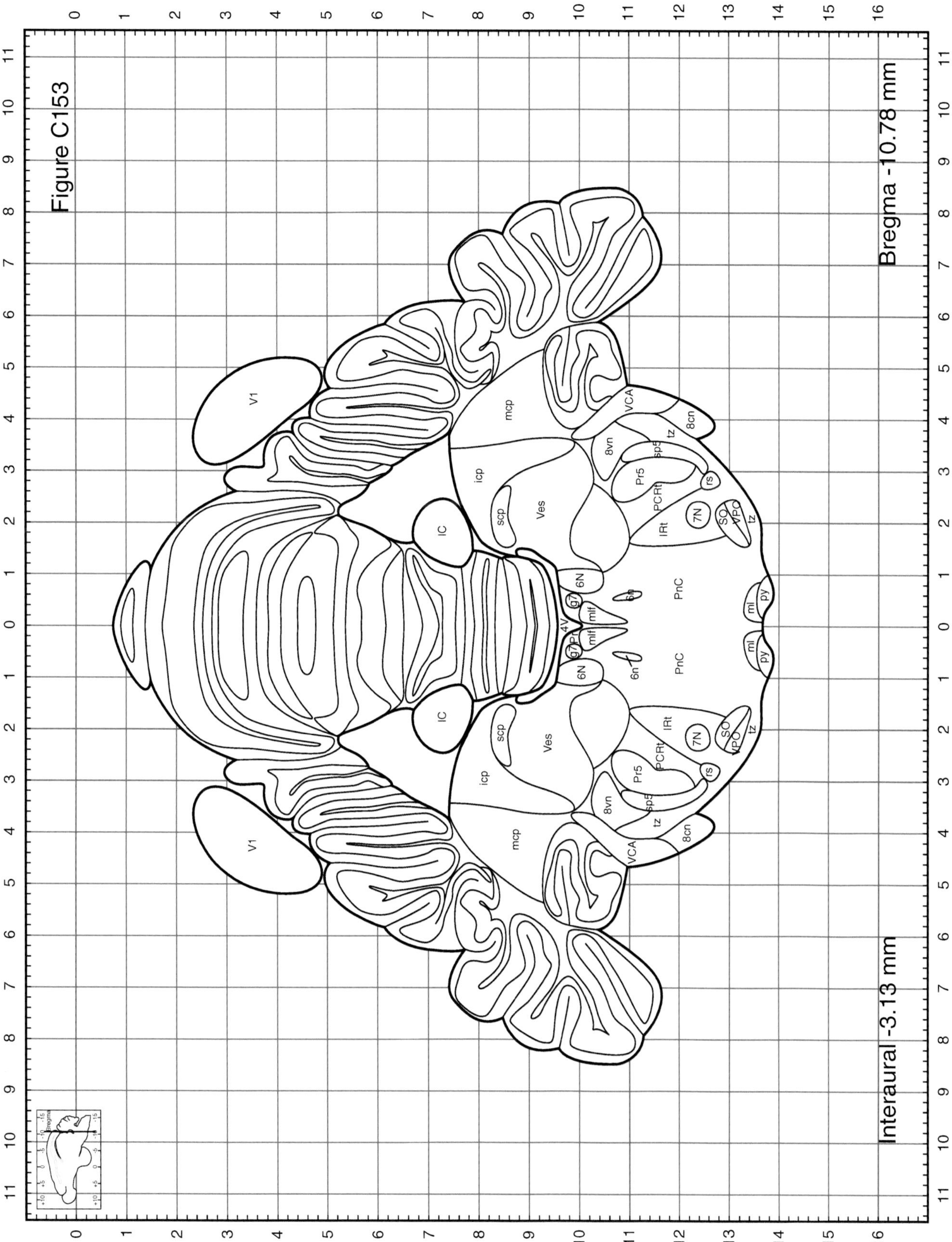
Figure C153
Bregma -10.78 mm
Interaural -3.13 mm
V1
mcp
icp
scp
Ves
IC
VCA
8cn
8vn
sp5
tz
Pr5
PCRt
rs
7N
SO
VPO
IRt
PnC
6N
g7
4V
Pr
mlf
6n
ml
py

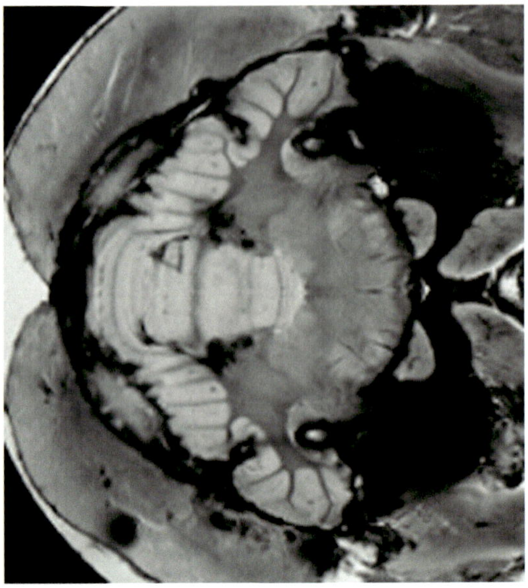

4V	4th ventricle
6N	abducens nucleus
7N	facial nucleus
8cn	cochlear root of the vestibulocochlear nerve
8vn	vestibular root of the vestibulocochlear nerve
g7	genu of the facial nerve
Gi	gigantocellular reticular nucleus
IC	inferior colliculus
icp	inferior cerebellar peduncle
IRt	intermediate reticular nucleus
mcp	middle cerebellar peduncle
ml	medial lemniscus
mlf	medial longitudinal fasciculus
PCRt	parvicellular reticular nucleus
Pr	prepositus nucleus
Pr5	principal sensory trigeminal nucleus
py	pyramidal tract
rs	rubrospinal tract
scp	superior cerebellar peduncle
sp5	spinal trigeminal tract
tth	trigeminothalamic tract
tz	trapezoid body
V1	primary visual cortex
VCA	ventral cochlear nucleus, anterior part
Ves	vestibular nucleus
VPO	ventral periolivary nucleus

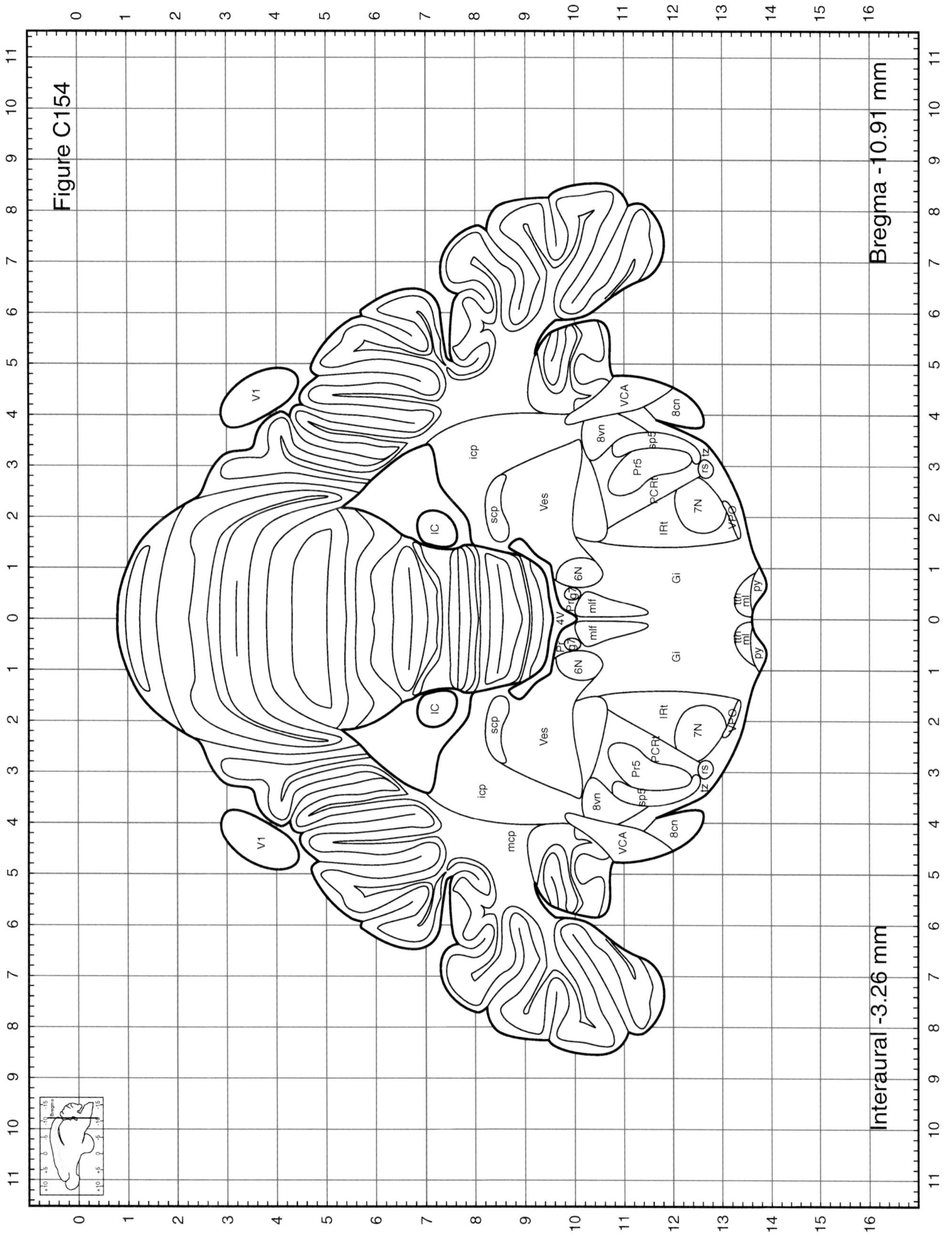
Figure C154
Bregma -10.91 mm
Interaural -3.26 mm
V1
icp
mcp
scp
Ves
IC
4V
6N
g7
Pr
mlf
8vn
8cn
VCA
sp5
Pr5
PCRt
IRt
rs
tz
7N
VPO
Gi
ml
tth
py

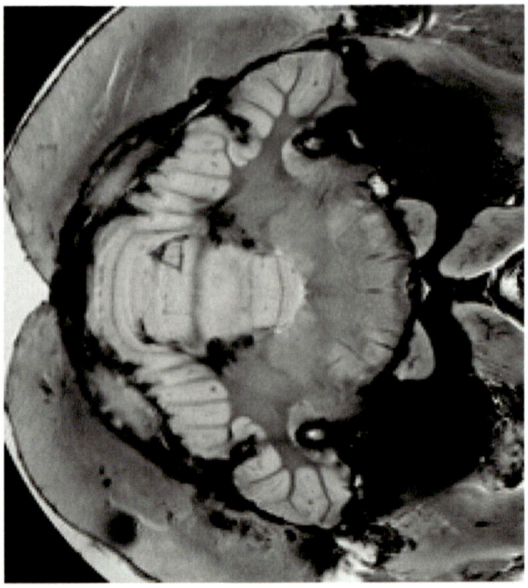

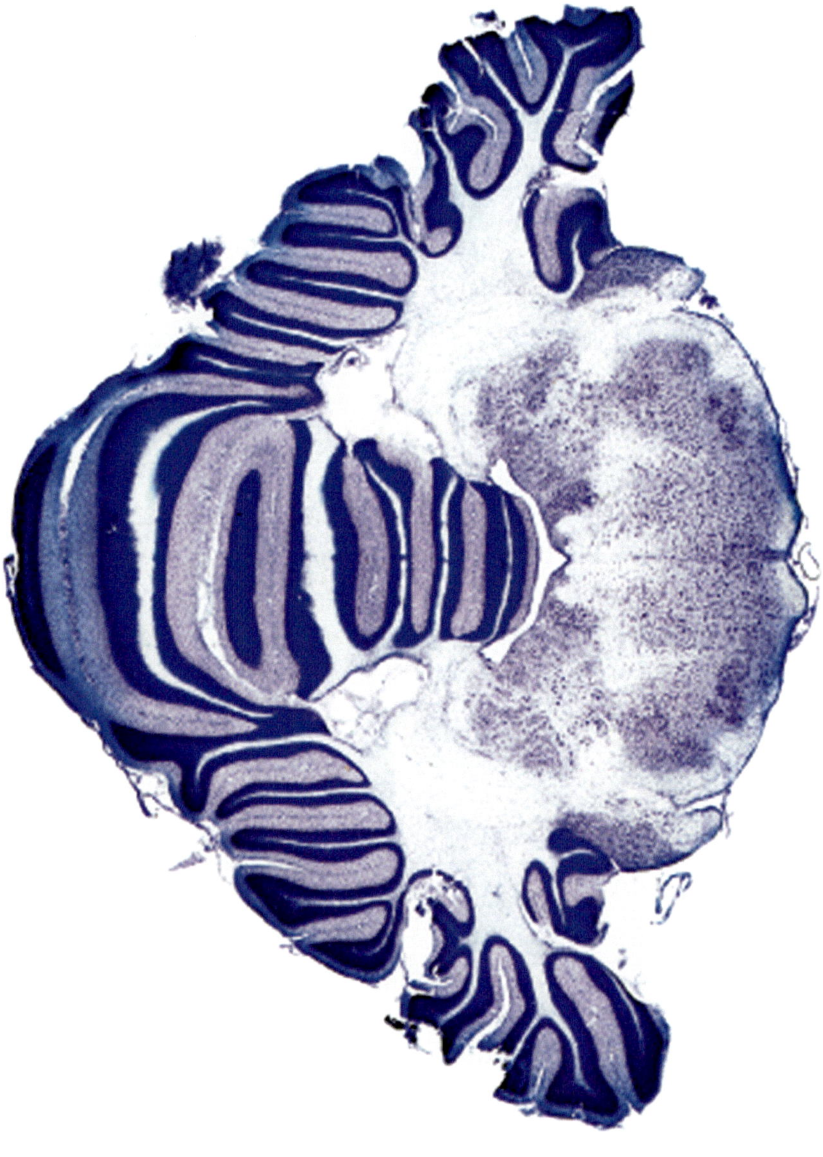

4V 4th ventricle
6N abducens nucleus
7N facial nucleus
8cn cochlear root of the vestibulocochlear nerve
8vn vestibular root of the vestibulocochlear nerve
g7 genu of the facial nerve
Gi gigantocellular reticular nucleus
icp inferior cerebellar peduncle
IRt intermediate reticular nucleus
ml medial lemniscus
mlf medial longitudinal fasciculus

PCRt parvicellular reticular nucleus
Pr prepositus nucleus
py pyramidal tract
rs rubrospinal tract
scp superior cerebellar peduncle
Sp5 spinal trigeminal nucleus
sp5 spinal trigeminal tract
tth trigeminothalamic tract
tz trapezoid body
VCA ventral cochlear nucleus, anterior part
Ves vestibular nucleus

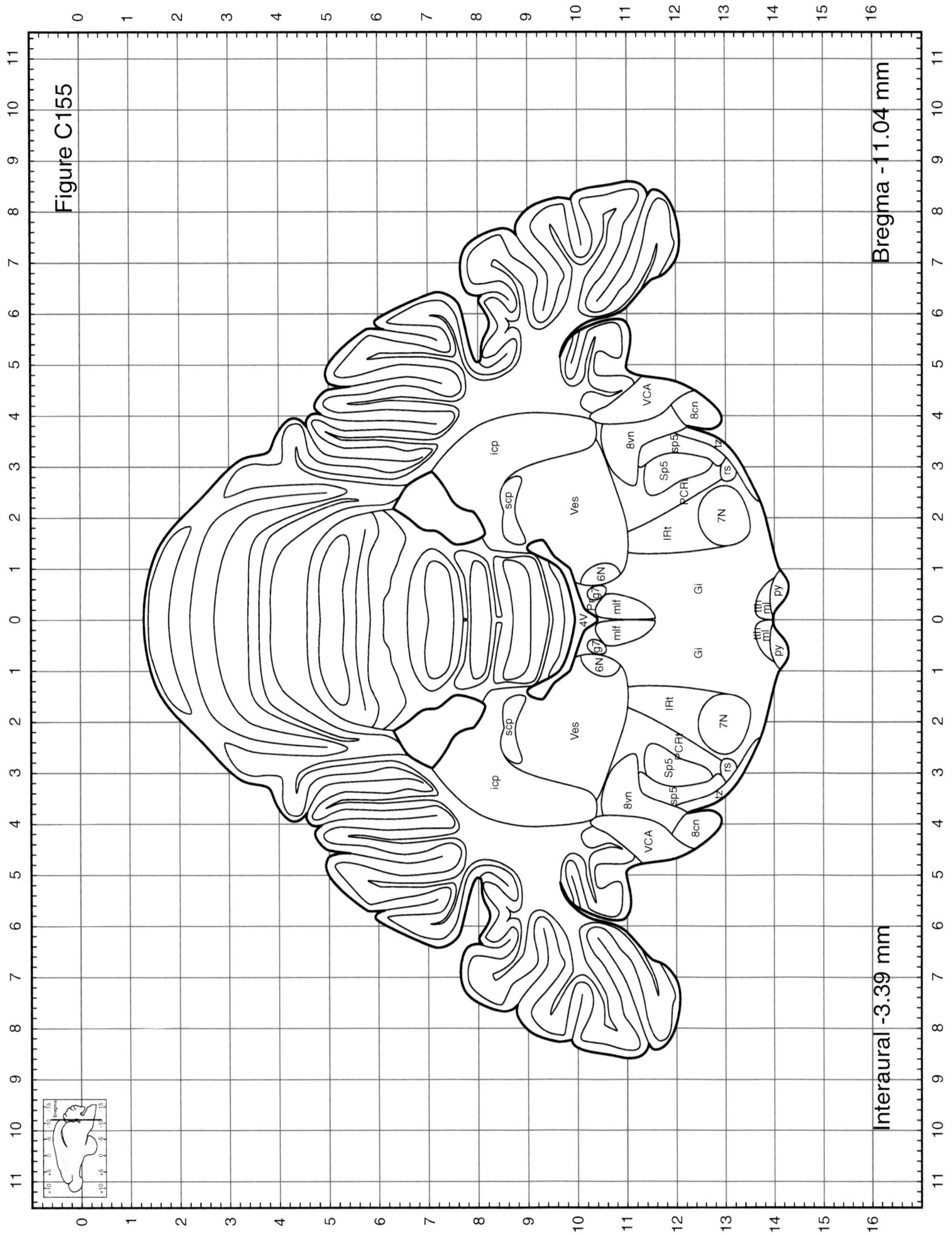
Figure C155
Bregma -11.04 mm
Interaural -3.39 mm
icp
scp
Ves
VCA
8cn
8vn
sp5
Sp5
PCRt
tz
rs
7N
IRt
Gi
6N
g7
Pr
4V
mlf
py
ml
tth

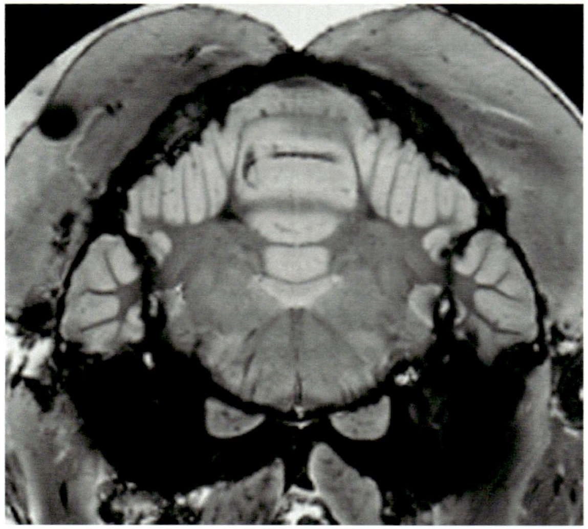

4V	4th ventricle	PCRt	parvicellular reticular nucleus
6N	abducens nucleus	Pr	prepositus nucleus
7N	facial nucleus	py	pyramidal tract
8cn	cochlear root of the vestibulocochlear nerve	rs	rubrospinal tract
8vn	vestibular root of the vestibulocochlear nerve	scp	superior cerebellar peduncle
DCN	dorsal cochlear nucleus	Sp5	spinal trigeminal nucleus
g7	genu of the facial nerve	sp5	spinal trigeminal tract
Gi	gigantocellular reticular nucleus	SpVe	spinal vestibular nucleus
icp	inferior cerebellar peduncle	tth	trigeminothalamic tract
IRt	intermediate reticular nucleus	tz	trapezoid body
ml	medial lemniscus	VCN	ventral cochlear nucleus
mlf	medial longitudinal fasciculus	Ves	vestibular nucleus

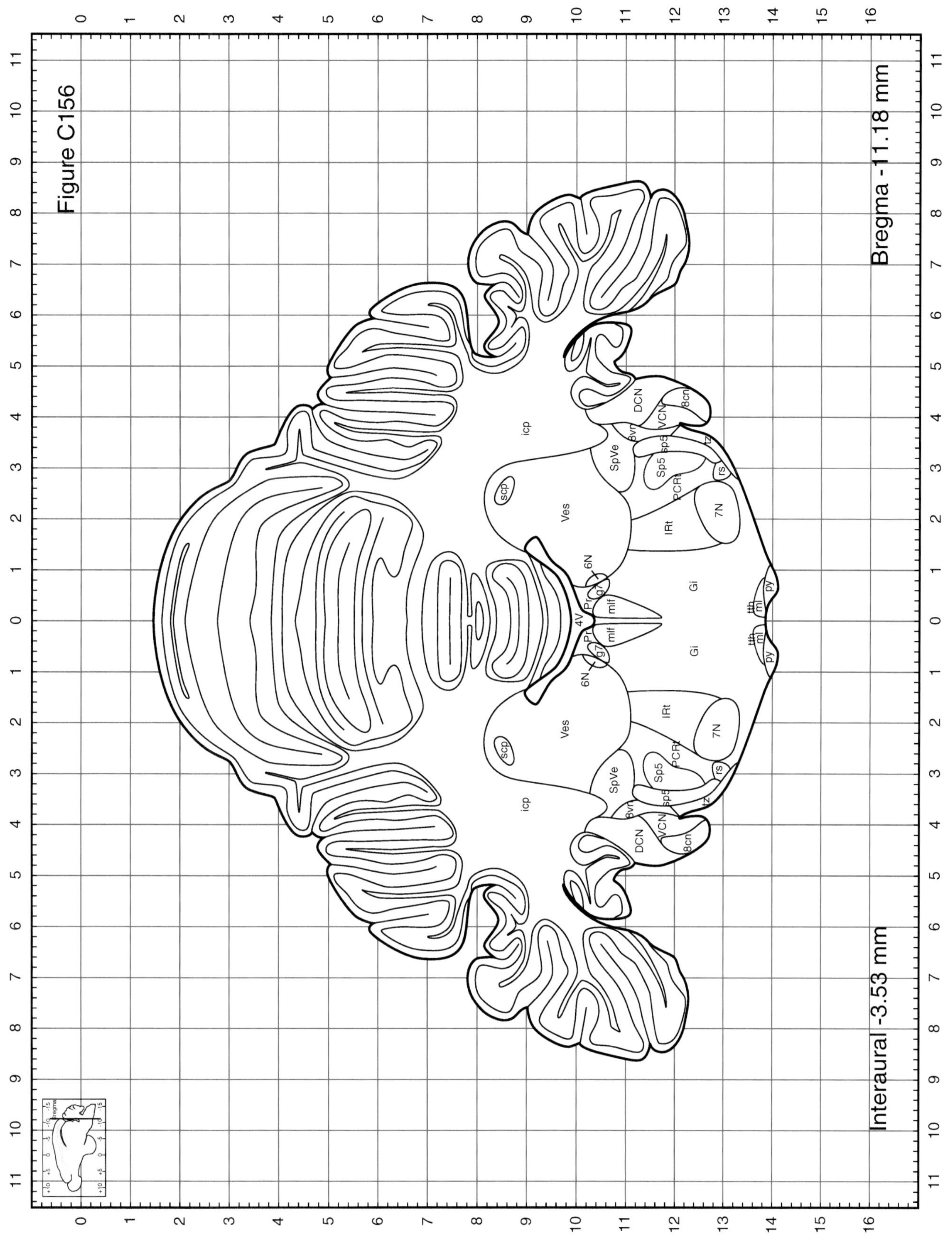
Figure C156
Bregma -11.18 mm
Interaural -3.53 mm
icp
scp
Ves
SpVe
DCN
VCN
8cn
8vn
sp5
Sp5
PCRt
IRt
7N
rs
tz
6N
g7
Pr
4V
mlf
Gi
py
ml
tth

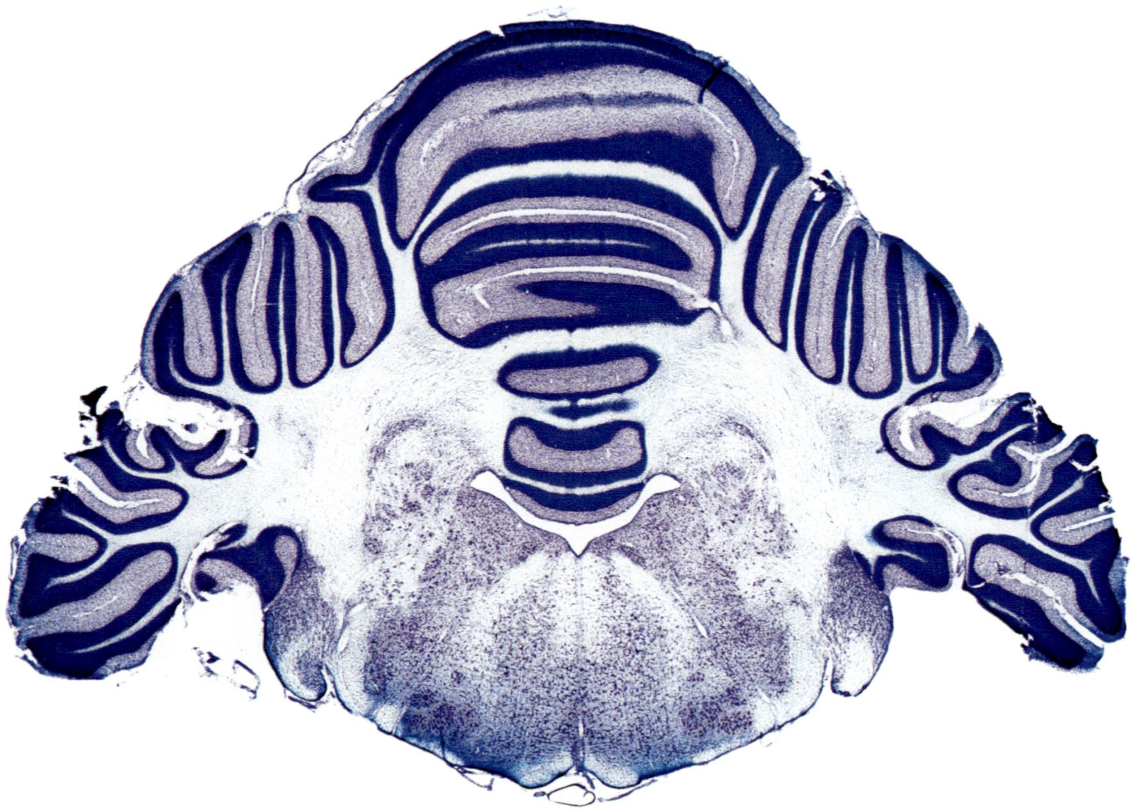

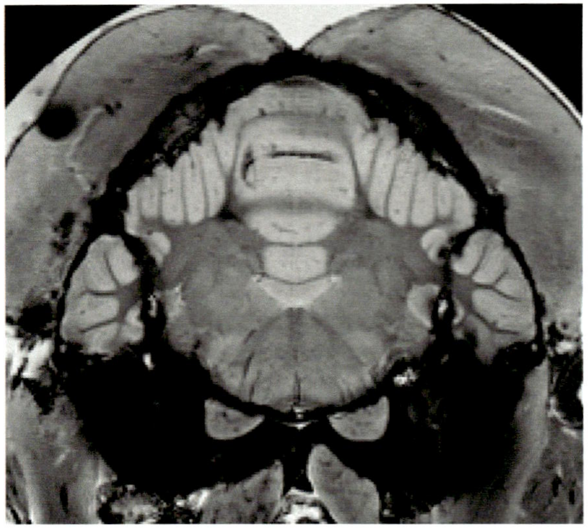

4V	4th ventricle
7N	facial nucleus
8cn	cochlear root of the vestibulocochlear nerve
8vn	vestibular root of the vestibulocochlear nerve
DCN	dorsal cochlear nucleus
g7	genu of the facial nerve
Gi	gigantocellular reticular nucleus
icp	inferior cerebellar peduncle
IRt	intermediate reticular nucleus
ml	medial lemniscus
mlf	medial longitudinal fasciculus
PCRt	parvicellular reticular nucleus
Pr	prepositus nucleus
py	pyramidal tract
rs	rubrospinal tract
scp	superior cerebellar peduncle
Sp5	spinal trigeminal nucleus
sp5	spinal trigeminal tract
SpVe	spinal vestibular nucleus
tth	trigeminothalamic tract
tz	trapezoid body
VCN	ventral cochlear nucleus
Ves	vestibular nucleus

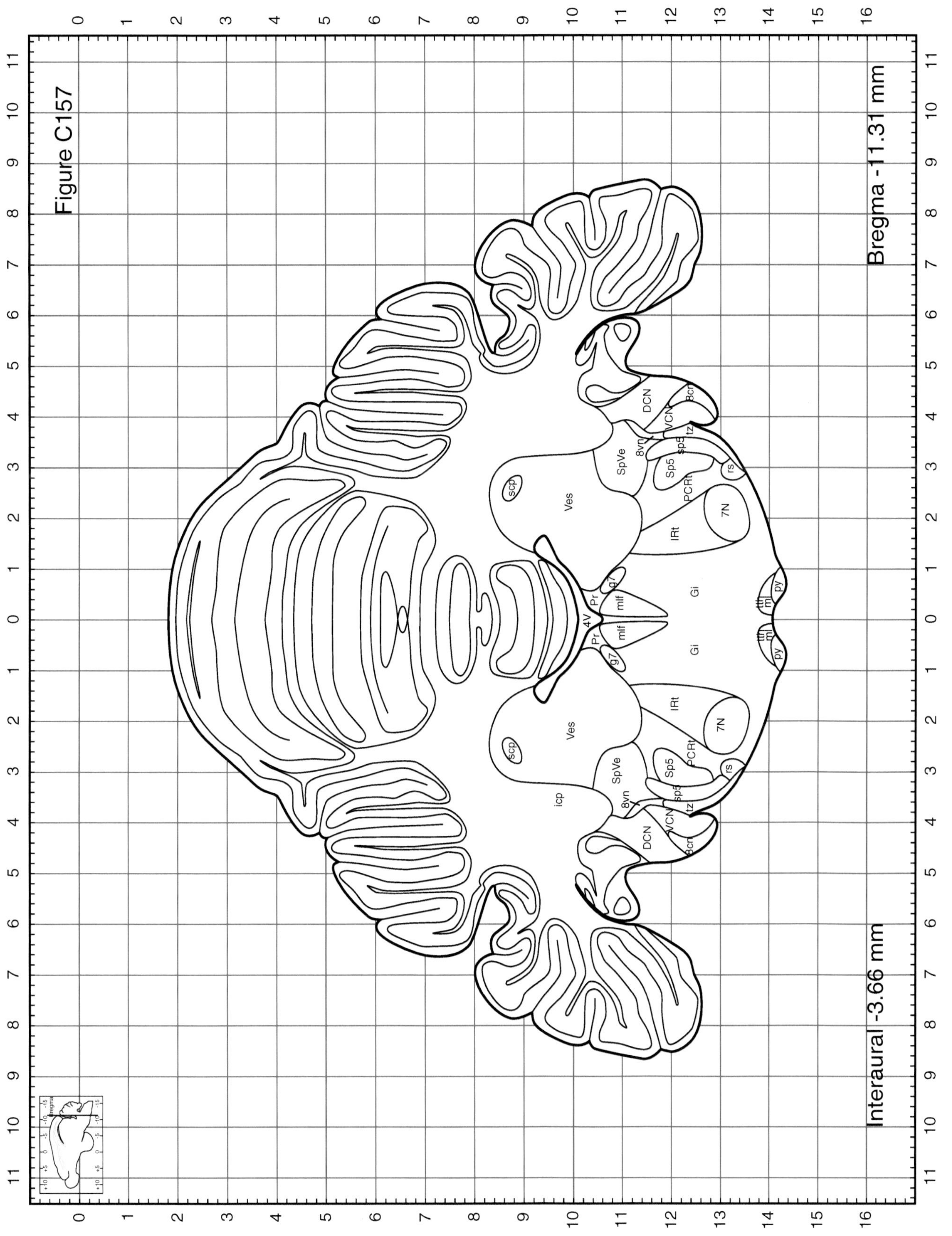
Figure C157
Bregma -11.31 mm
Interaural -3.66 mm
DCN
8cr
VCN
tz
sp5
Sp5
8vn
SpVe
rs
PCRt
7N
scp
Ves
IRt
icp
g7
Pr
4V
mlf
Gi
py
ml
ttt

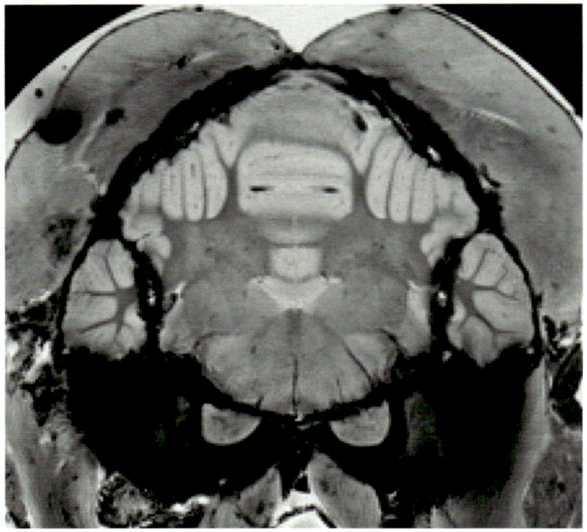

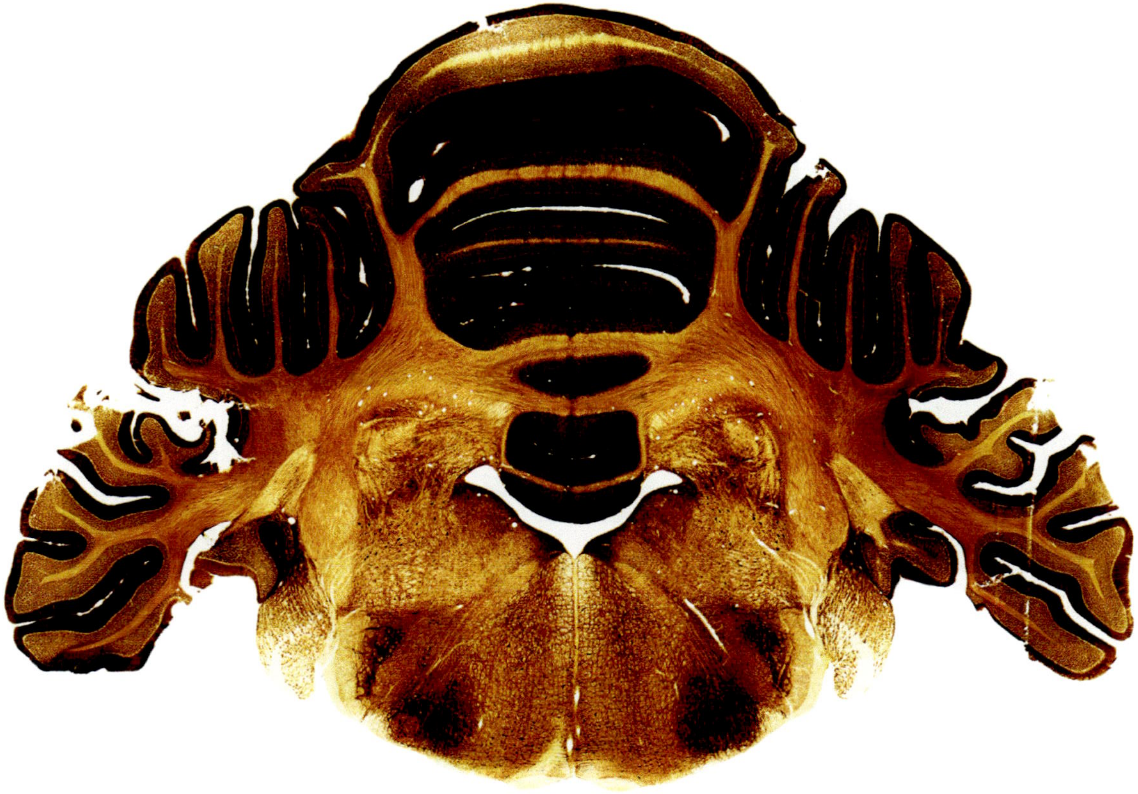

4V	4th ventricle
7N	facial nucleus
8cn	cochlear root of the vestibulocochlear nerve
asc7	ascending fibers of the facial nerve
DCN	dorsal cochlear nucleus
Gi	gigantocellular reticular nucleus
icp	inferior cerebellar peduncle
Int	interposed cerebellar nucleus
IRt	intermediate reticular nucleus
ml	medial lemniscus
mlf	medial longitudinal fasciculus
PCRt	parvicellular reticular nucleus
Pr	prepositus nucleus
py	pyramidal tract
RPa	raphe pallidus nucleus
rs	rubrospinal tract
scp	superior cerebellar peduncle
Sp5	spinal trigeminal nucleus
sp5	spinal trigeminal tract
SpVe	spinal vestibular nucleus
tth	trigeminothalamic tract
tz	trapezoid body
VCN	ventral cochlear nucleus
VeCb	vestibulocerebellar nucleus
Ves	vestibular nucleus

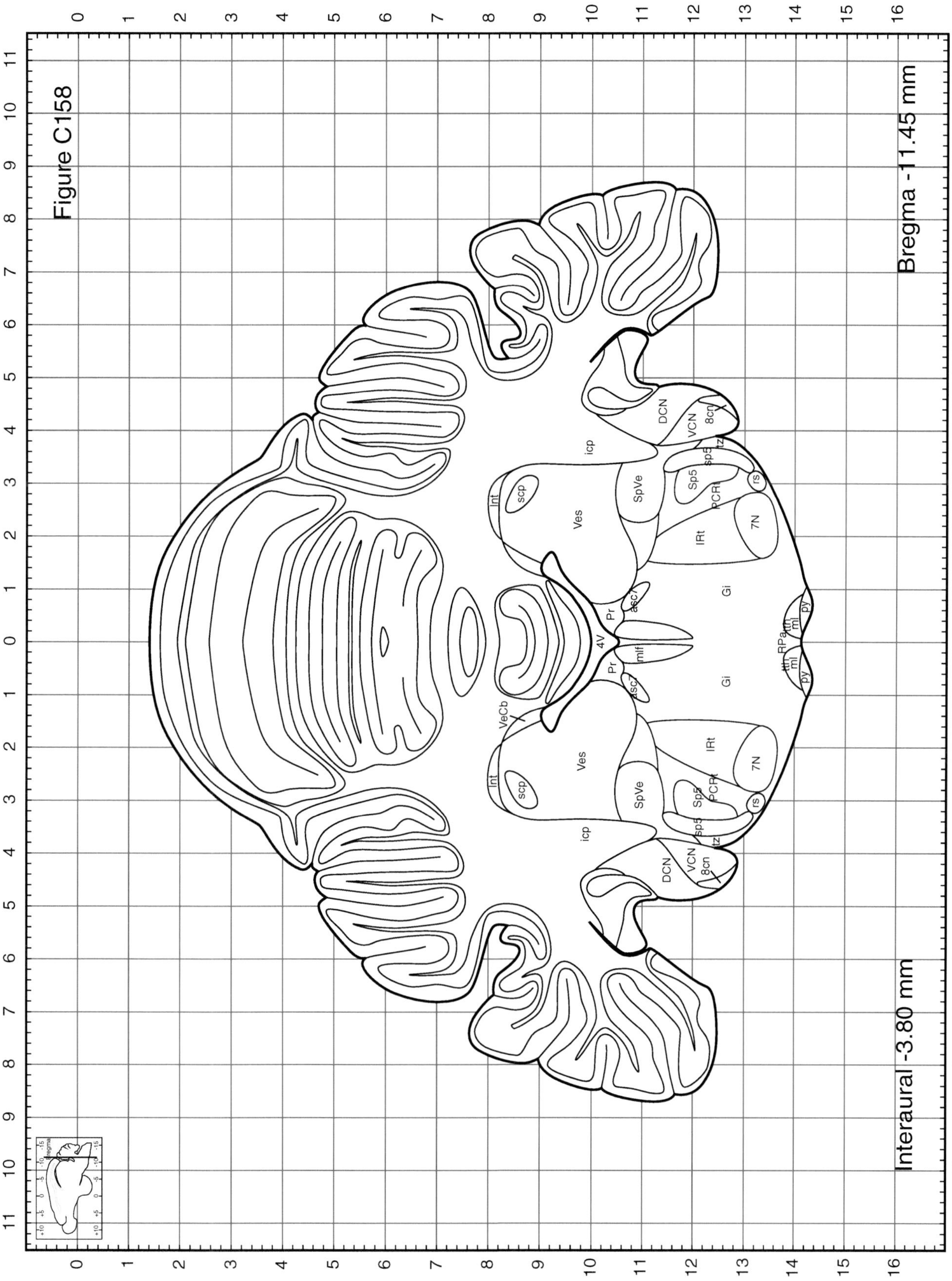
Figure C158
Bregma -11.45 mm
Interaural -3.80 mm
DCN
VCN
8cn
tz
icp
sp5
Sp5
SpVe
scp
Int
Ves
PCRt
rs
7N
IRt
Gi
asc7
Pr
4V
mlf
RPa
tth
ml
py
VeCb

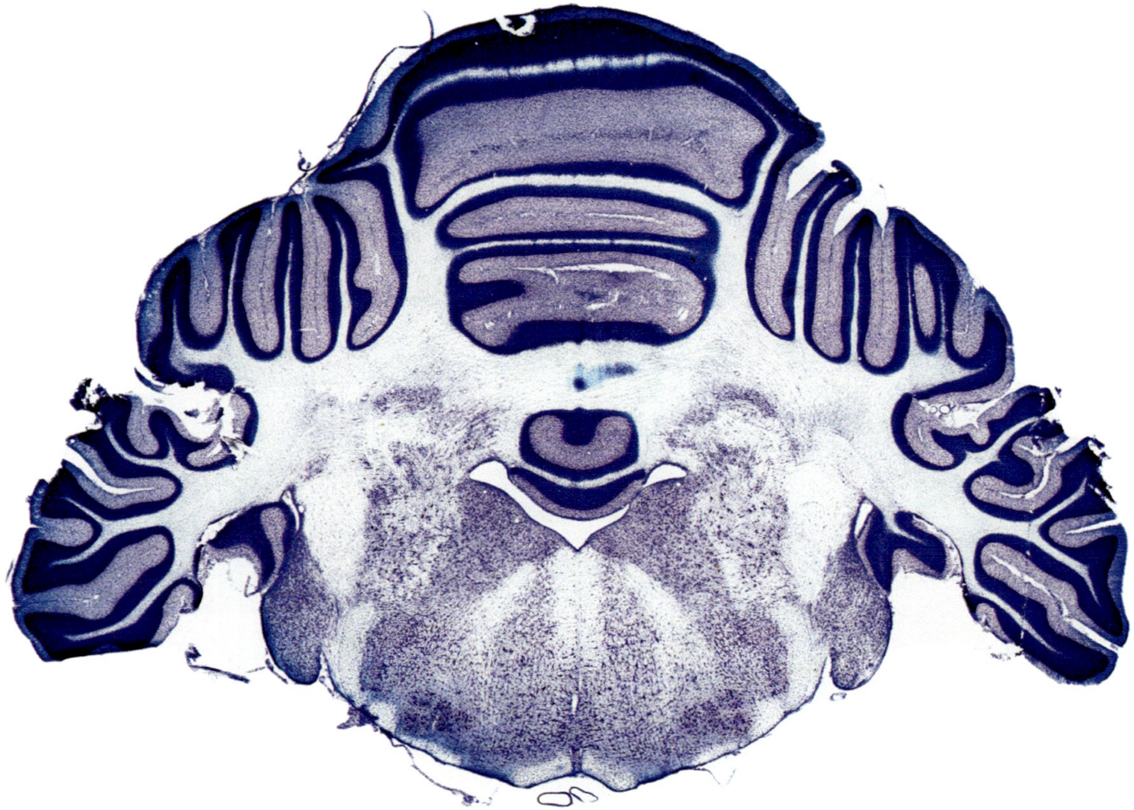

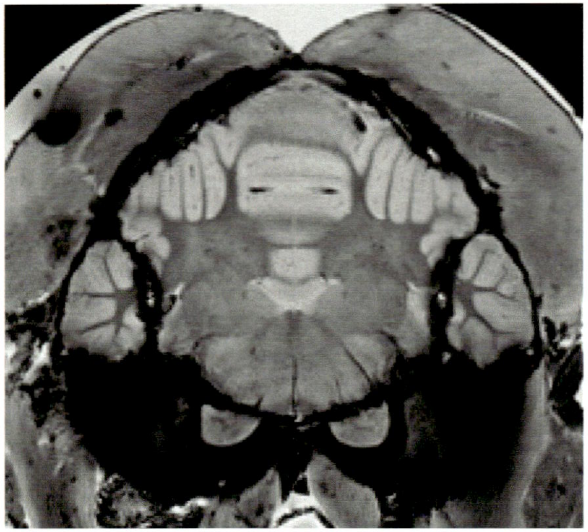

4V	4th ventricle
7N	facial nucleus
asc7	ascending fibers of the facial nerve
DCN	dorsal cochlear nucleus
Gi	gigantocellular reticular nucleus
icp	inferior cerebellar peduncle
Int	interposed cerebellar nucleus
IRt	intermediate reticular nucleus
ml	medial lemniscus
mlf	medial longitudinal fasciculus
PCRt	parvicellular reticular nucleus
Pr	prepositus nucleus
py	pyramidal tract
RPa	raphe pallidus nucleus
rs	rubrospinal tract
scp	superior cerebellar peduncle
Sp5	spinal trigeminal nucleus
sp5	spinal trigeminal tract
SpVe	spinal vestibular nucleus
tth	trigeminothalamic tract
tz	trapezoid body
VCN	ventral cochlear nucleus
VeCb	vestibulocerebellar nucleus
Ves	vestibular nucleus

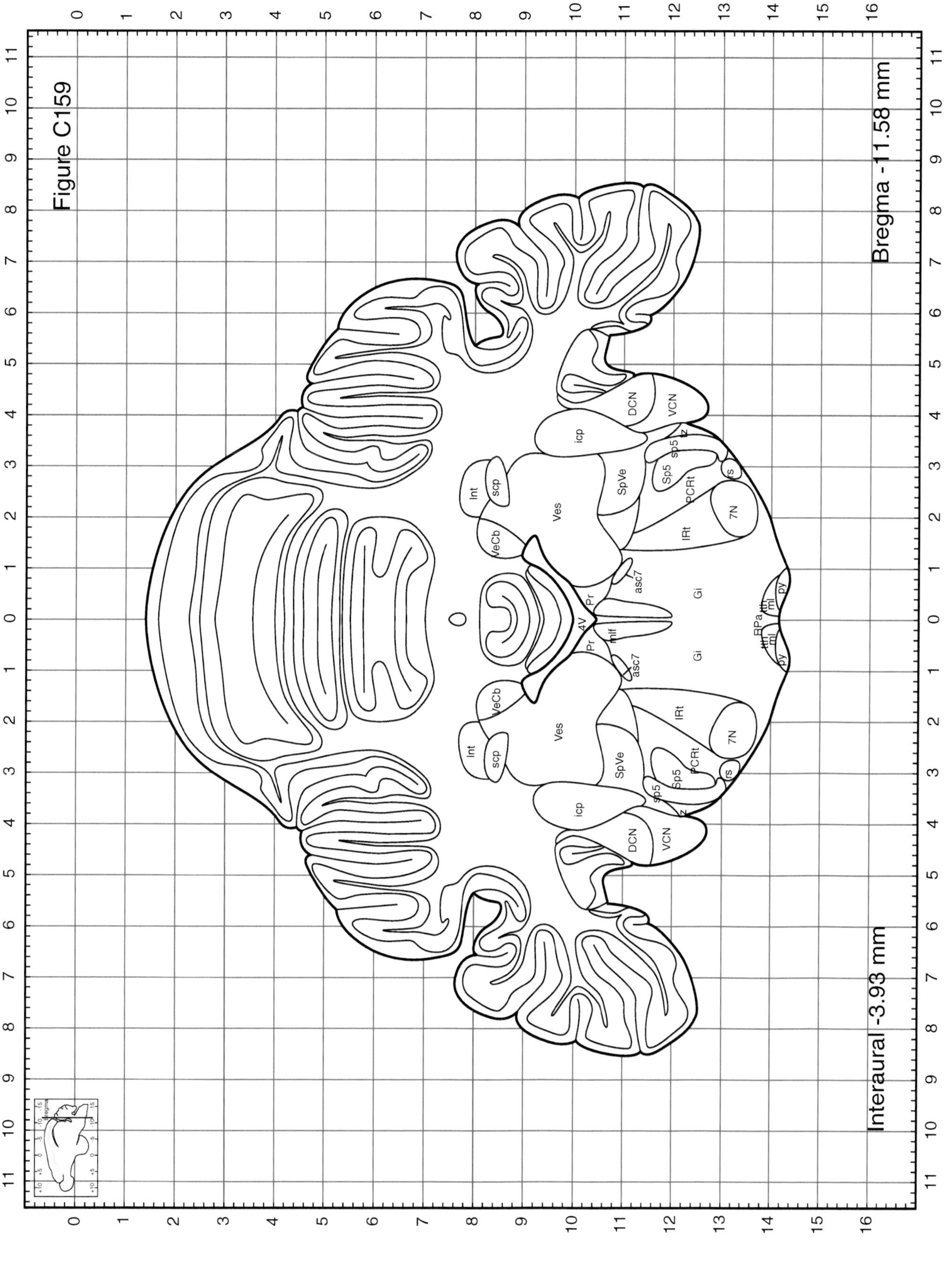
Figure C159
Bregma -11.58 mm
Interaural -3.93 mm
DCN
VCN
icp
sp5
Sp5
SpVe
PCRt
rs
7N
IRt
Ves
Int
scp
VeCb
asc7
Gi
Pr
4V
mlf
RPa
ml
tth
py

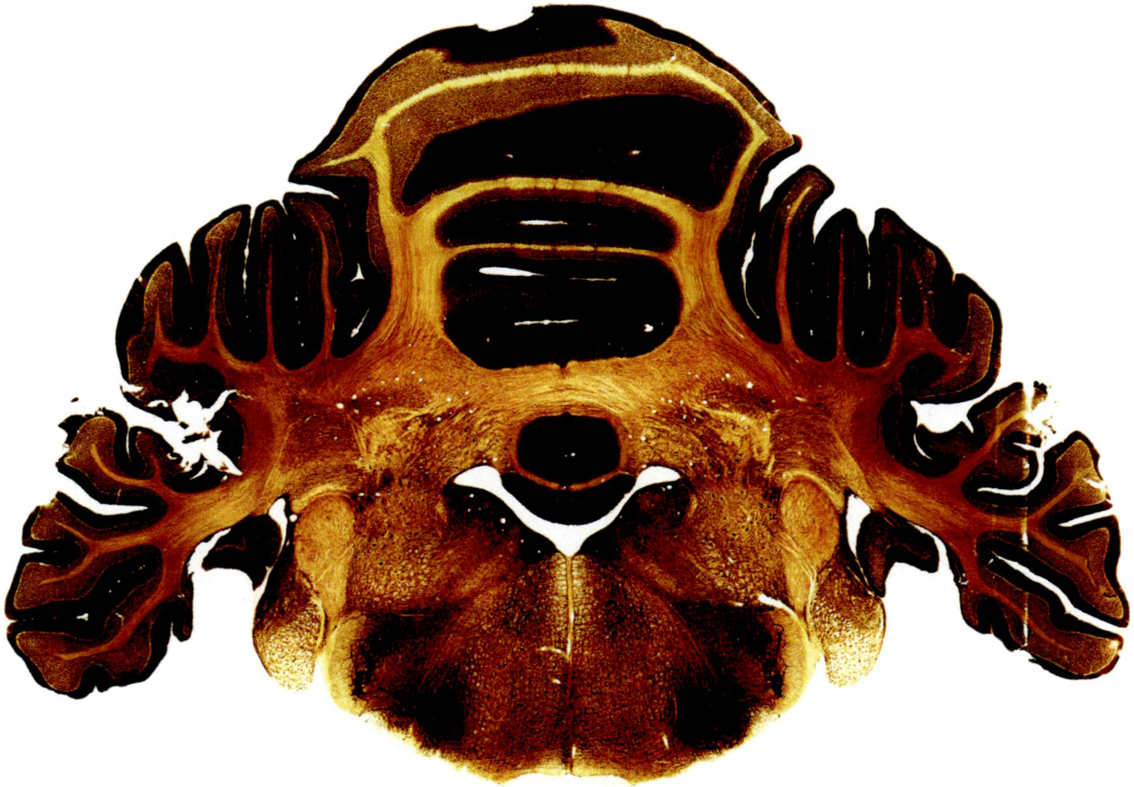

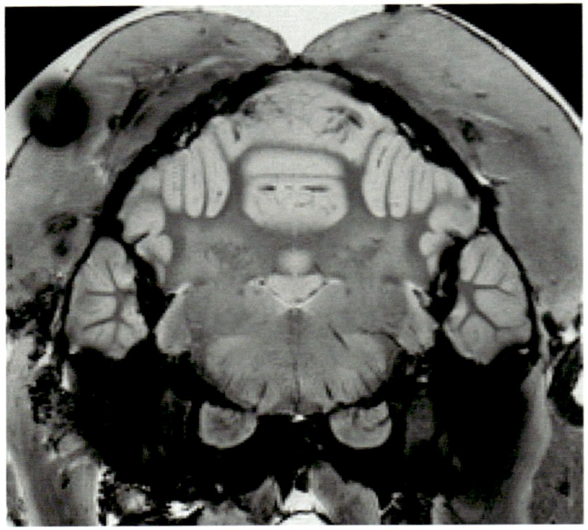

4V	4th ventricle
7N	facial nucleus
DCN	dorsal cochlear nucleus
Gi	gigantocellular reticular nucleus
icp	inferior cerebellar peduncle
Int	interposed cerebellar nucleus
IRt	intermediate reticular nucleus
Lat	lateral cerebellar nucleus
ml	medial lemniscus
mlf	medial longitudinal fasciculus
PCRt	parvicellular reticular nucleus
Pr	prepositus nucleus
py	pyramidal tract
RPa	raphe pallidus nucleus
rs	rubrospinal tract
scp	superior cerebellar peduncle
Sp5	spinal trigeminal nucleus
sp5	spinal trigeminal tract
SpVe	spinal vestibular nucleus
tth	trigeminothalamic tract
tz	trapezoid body
VCN	ventral cochlear nucleus
VeCb	vestibulocerebellar nucleus
Ves	vestibular nucleus

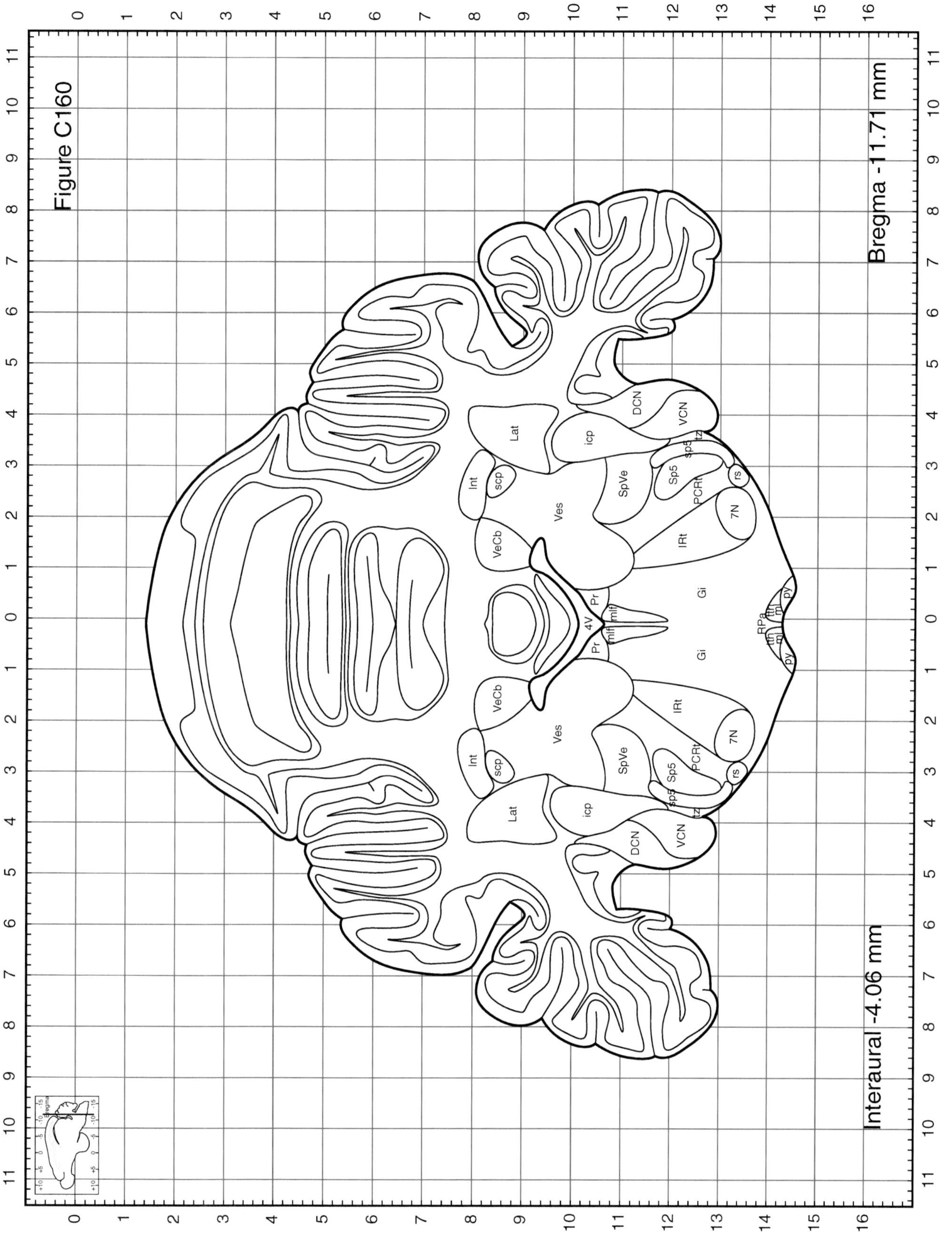
Figure C160
Bregma -11.71 mm
Interaural -4.06 mm
Lat
Int
scp
VeCb
Ves
icp
DCN
VCN
SpVe
Sp5
sp5
tz
PCRt
rs
7N
IRt
Gi
Pr
4V
mlf
RPa
ttl
ml
py

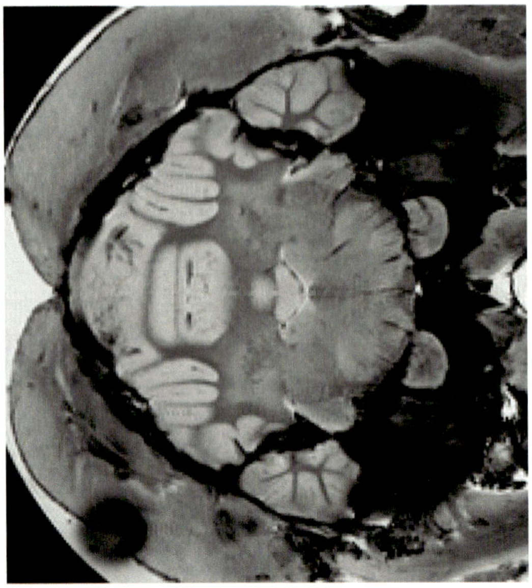

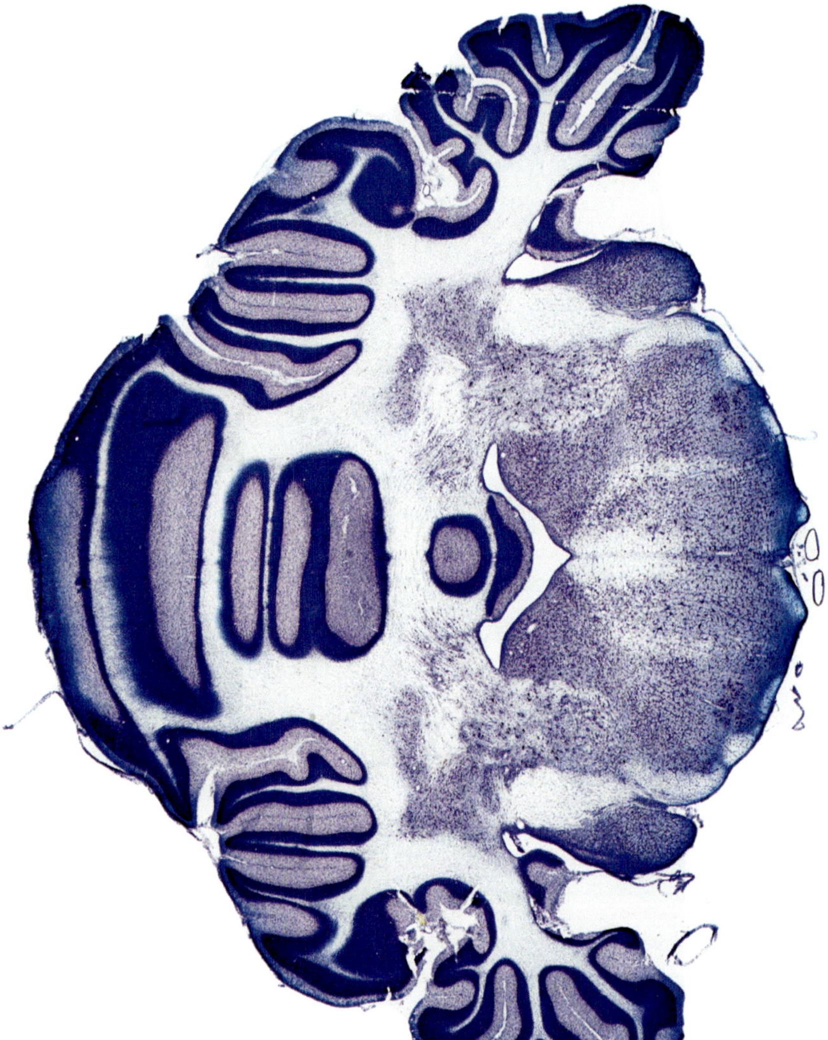

4V 4th ventricle
7N facial nucleus
DCN dorsal cochlear nucleus
Gi gigantocellular reticular nucleus
icp inferior cerebellar peduncle
Int interposed cerebellar nucleus
IRt intermediate reticular nucleus
Lat lateral cerebellar nucleus

ml medial lemniscus
mlf medial longitudinal fasciculus
PCRt parvicellular reticular nucleus
Pr prepositus nucleus
py pyramidal tract
RPa raphe pallidus nucleus
rs rubrospinal tract
scp superior cerebellar peduncle

Sp5 spinal trigeminal nucleus
sp5 spinal trigeminal tract
SpVe spinal vestibular nucleus
tth trigeminothalamic tract
tz trapezoid body
VCN ventral cochlear nucleus
VeCb vestibulocerebellar nucleus
Ves vestibular nucleus

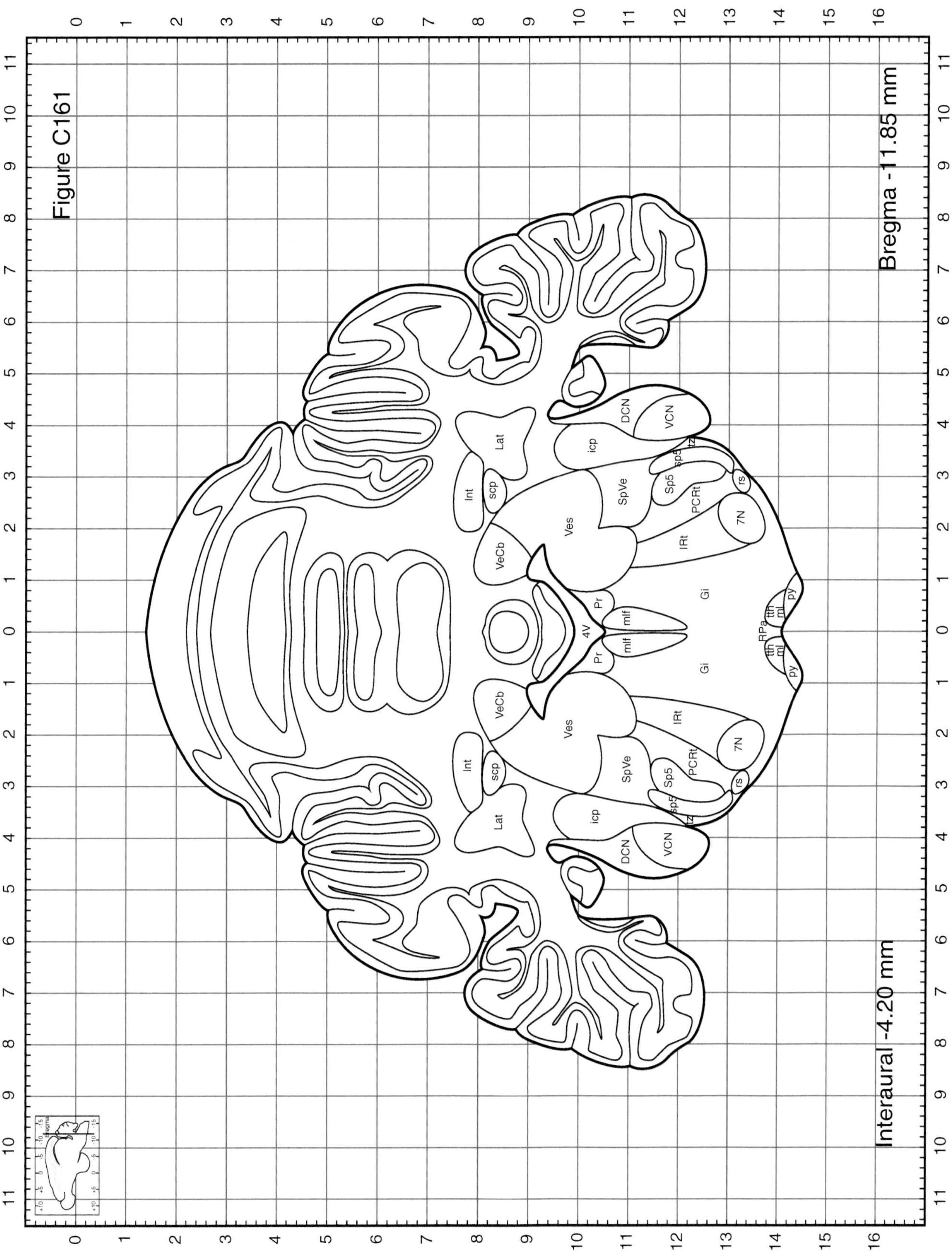

Figure C161
Bregma -11.85 mm
Interaural -4.20 mm
DCN
VCN
icp
Lat
Int
scp
SpVe
Sp5
sp5
tz
PCRt
rs
7N
Ves
VeCb
IRt
Gi
Pr
mlf
4V
RPa
tth
ml
py

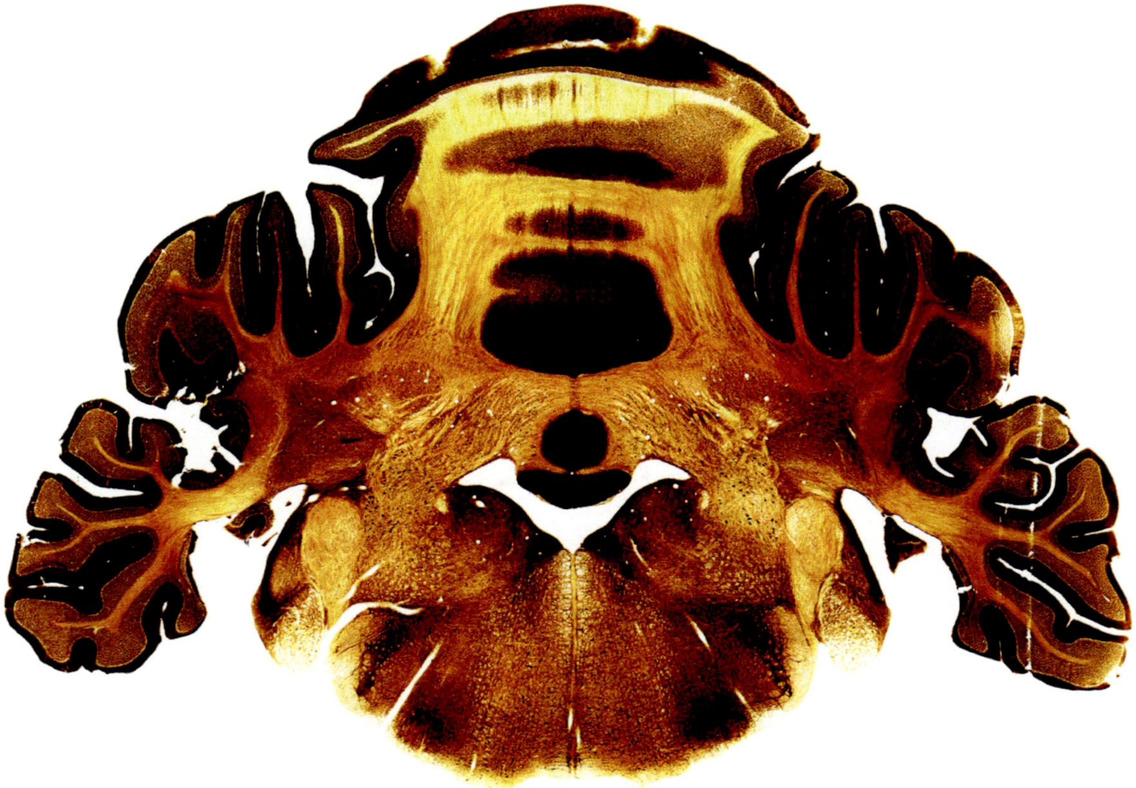

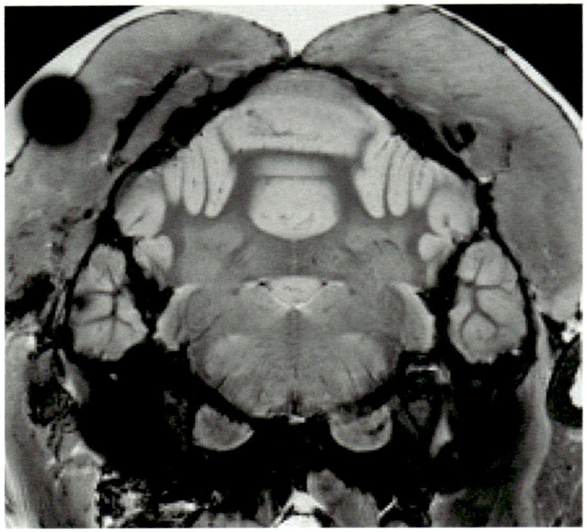

4V	4th ventricle
7N	facial nucleus
das	dorsal acoustic stria
DCN	dorsal cochlear nucleus
Gi	gigantocellular reticular nucleus
icp	inferior cerebellar peduncle
Int	interposed cerebellar nucleus
IRt	intermediate reticular nucleus
Lat	lateral cerebellar nucleus
ml	medial lemniscus
mlf	medial longitudinal fasciculus
PCRt	parvicellular reticular nucleus
Pr	prepositus nucleus
py	pyramidal tract
RPa	raphe pallidus nucleus
rs	rubrospinal tract
scp	superior cerebellar peduncle
Sp5	spinal trigeminal nucleus
sp5	spinal trigeminal tract
SpVe	spinal vestibular nucleus
tth	trigeminothalamic tract
tz	trapezoid body
VCN	ventral cochlear nucleus
VeCb	vestibulocerebellar nucleus
Ves	vestibular nucleus

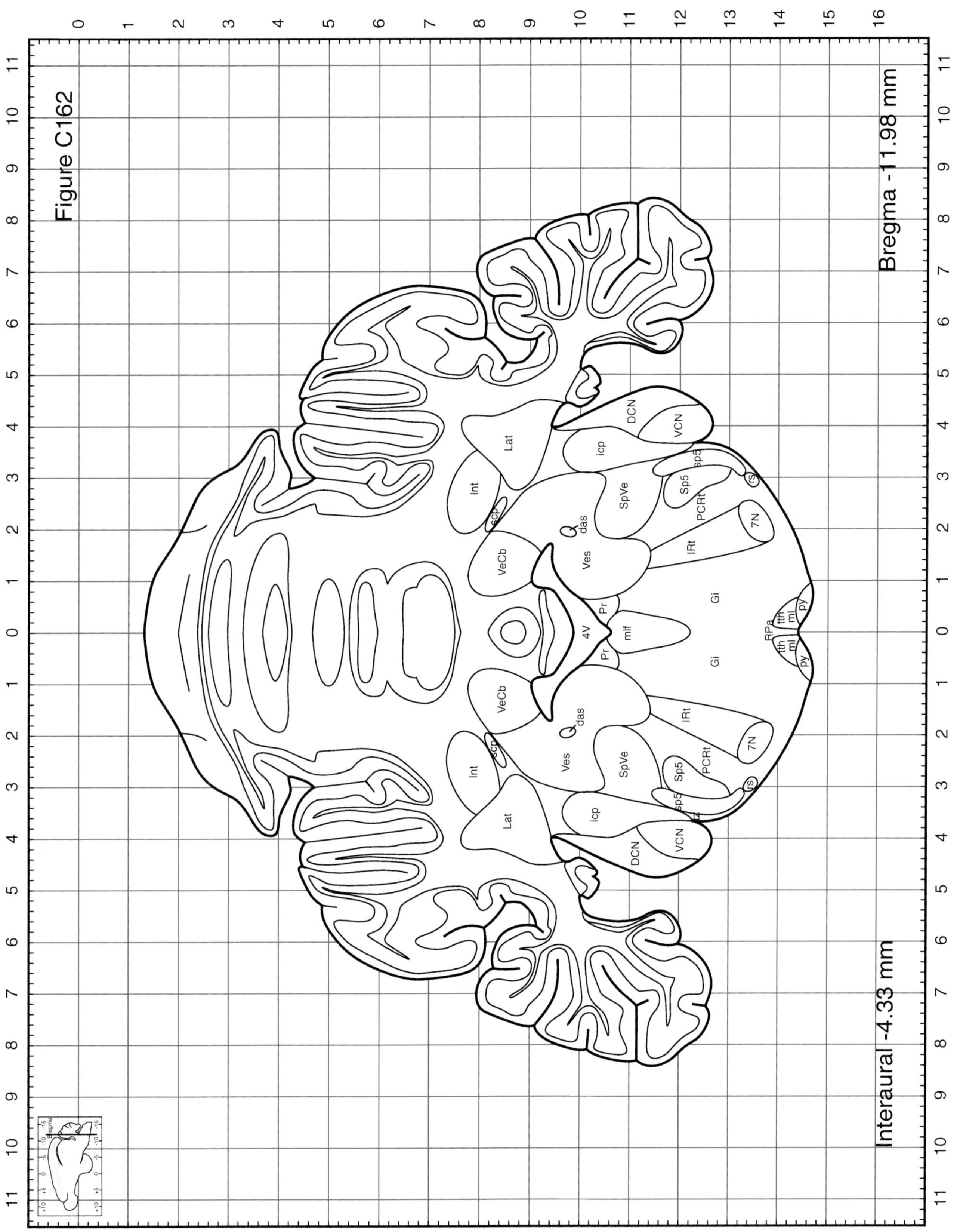
Figure C162
Bregma -11.98 mm
Interaural -4.33 mm
DCN
VCN
icp
sp5
Sp5
rs
Lat
Int
scp
SpVe
PCRt
7N
das
IRt
VeCb
Ves
Gi
Pr
4V
mlf
RPa
tth
ml
py

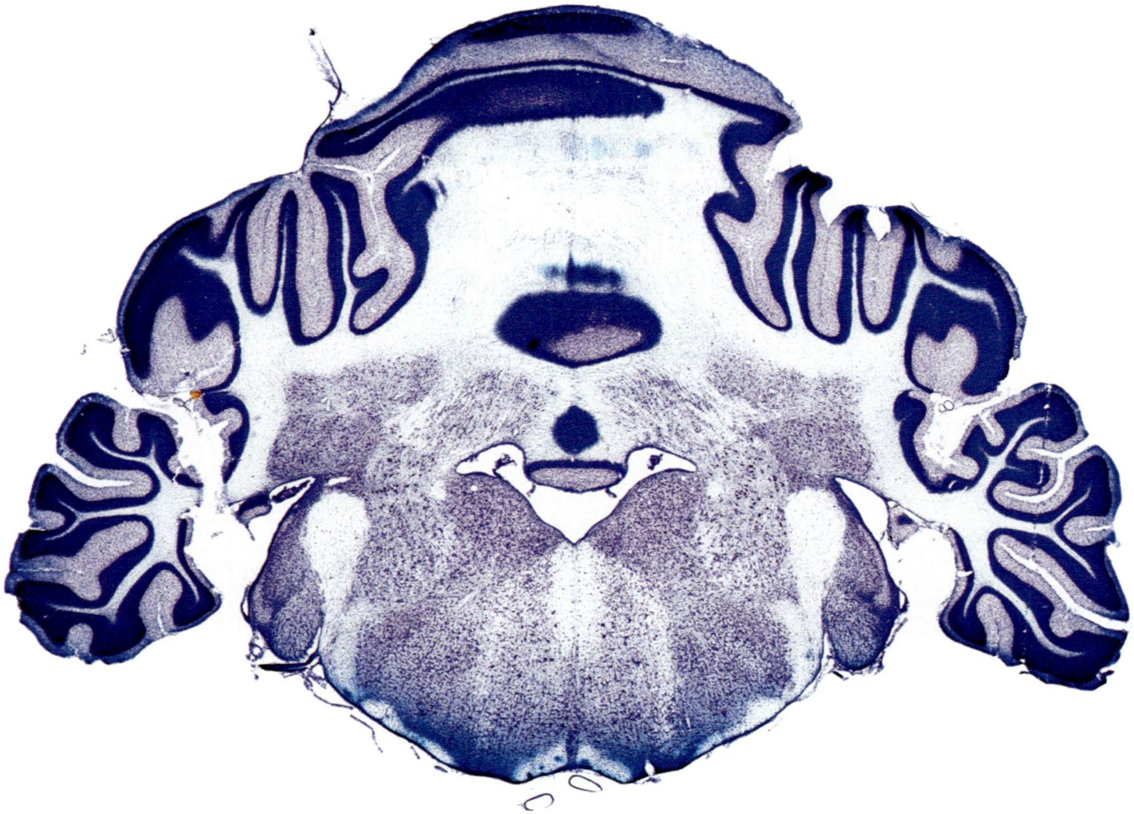

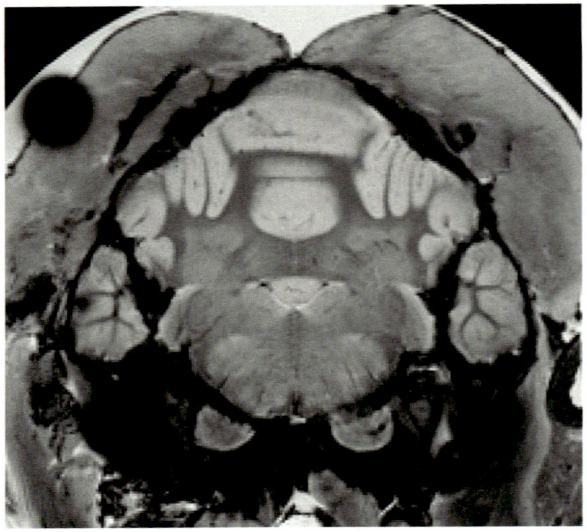

4V	4th ventricle
7N	facial nucleus
das	dorsal acoustic stria
DCN	dorsal cochlear nucleus
Gi	gigantocellular reticular nucleus
icp	inferior cerebellar peduncle
Int	interposed cerebellar nucleus
IRt	intermediate reticular nucleus
Lat	lateral cerebellar nucleus
ml	medial lemniscus
mlf	medial longitudinal fasciculus
PCRt	parvicellular reticular nucleus
Pr	prepositus nucleus
py	pyramidal tract
RPa	raphe pallidus nucleus
rs	rubrospinal tract
scp	superior cerebellar peduncle
Sp5	spinal trigeminal nucleus
sp5	spinal trigeminal tract
SpVe	spinal vestibular nucleus
tth	trigeminothalamic tract
VCN	ventral cochlear nucleus
VeCb	vestibulocerebellar nucleus
Ves	vestibular nucleus

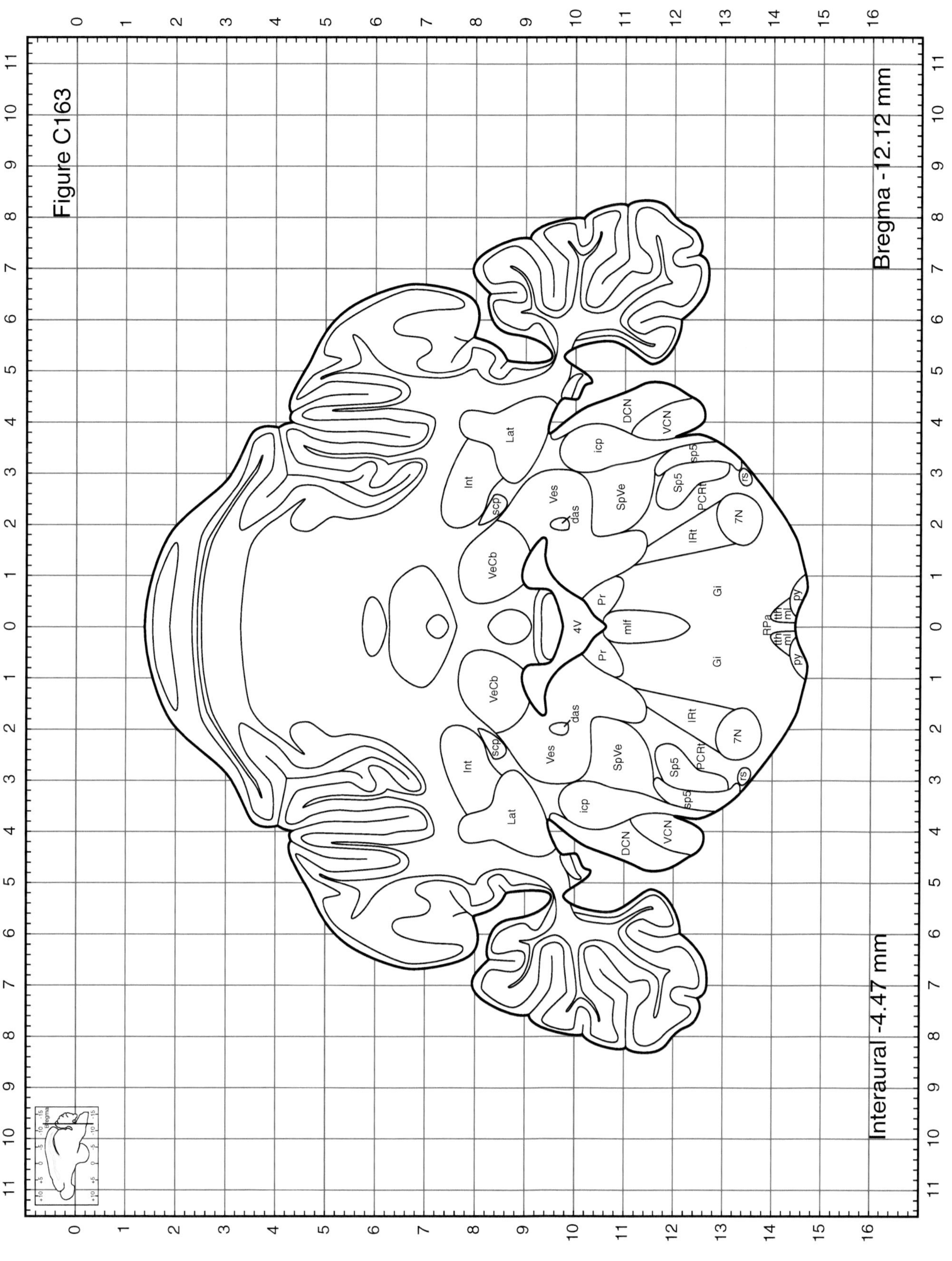
Figure C163
Bregma -12.12 mm
Interaural -4.47 mm
Lat
Int
scp
Ves
das
SpVe
icp
DCN
VCN
sp5
Sp5
PCRt
rs
7N
IRt
Gi
VeCb
Pr
4V
mlf
RPa
tfh
ml
py

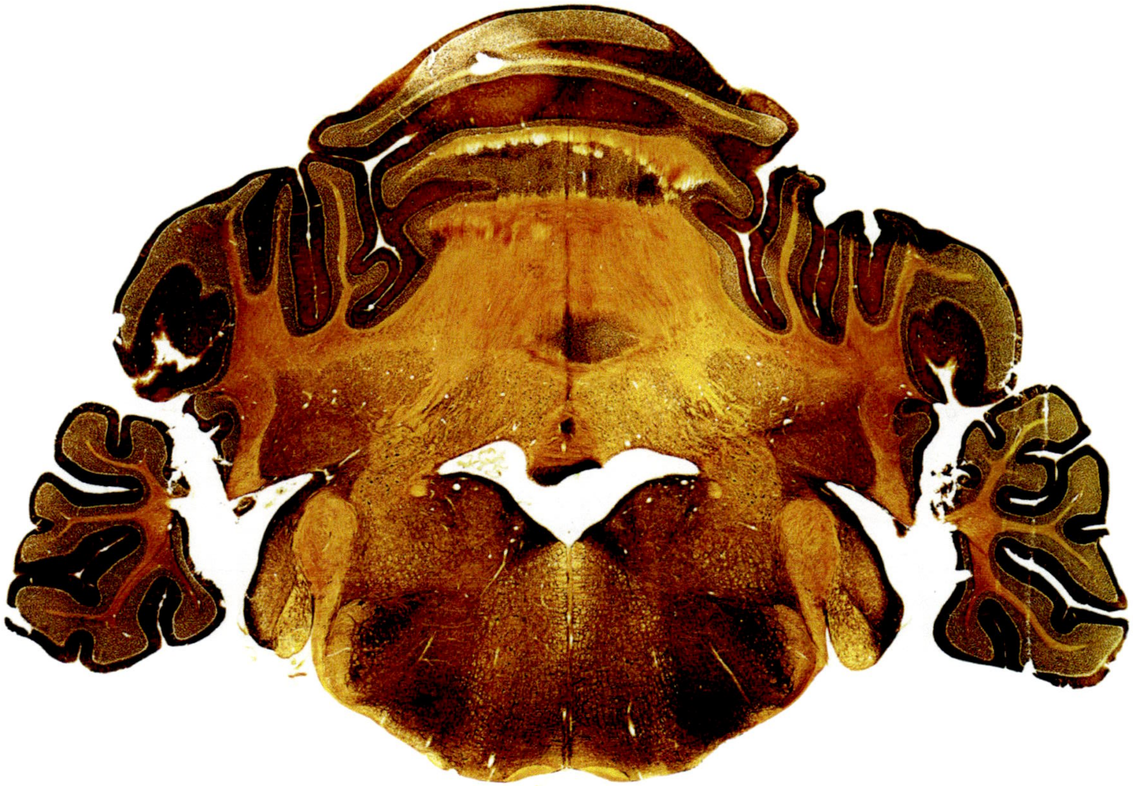

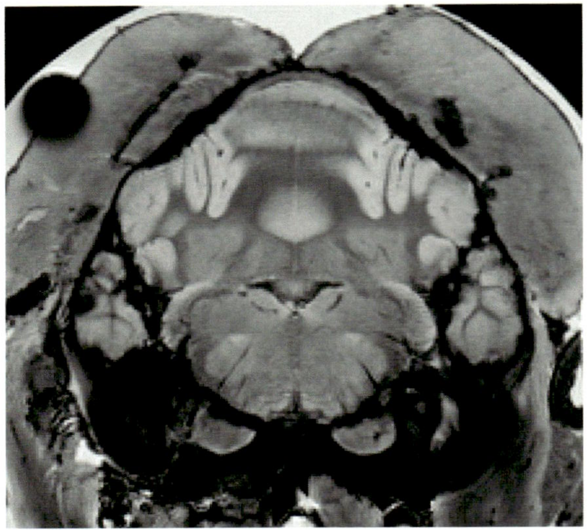

4V	4th ventricle
7N	facial nucleus
das	dorsal acoustic stria
DCN	dorsal cochlear nucleus
Gi	gigantocellular reticular nucleus
icp	inferior cerebellar peduncle
Int	interposed cerebellar nucleus
IRt	intermediate reticular nucleus
Lat	lateral cerebellar nucleus
Med	medial cerebellar nucleus
ml	medial lemniscus
mlf	medial longitudinal fasciculus
PCRt	parvicellular reticular nucleus
Pr	prepositus nucleus
py	pyramidal tract
RPa	raphe pallidus nucleus
rs	rubrospinal tract
scp	superior cerebellar peduncle
Sol	solitary nucleus
Sp5	spinal trigeminal nucleus
sp5	spinal trigeminal tract
SpVe	spinal vestibular nucleus
VCN	ventral cochlear nucleus
VeCb	vestibulocerebellar nucleus
Ves	vestibular nucleus

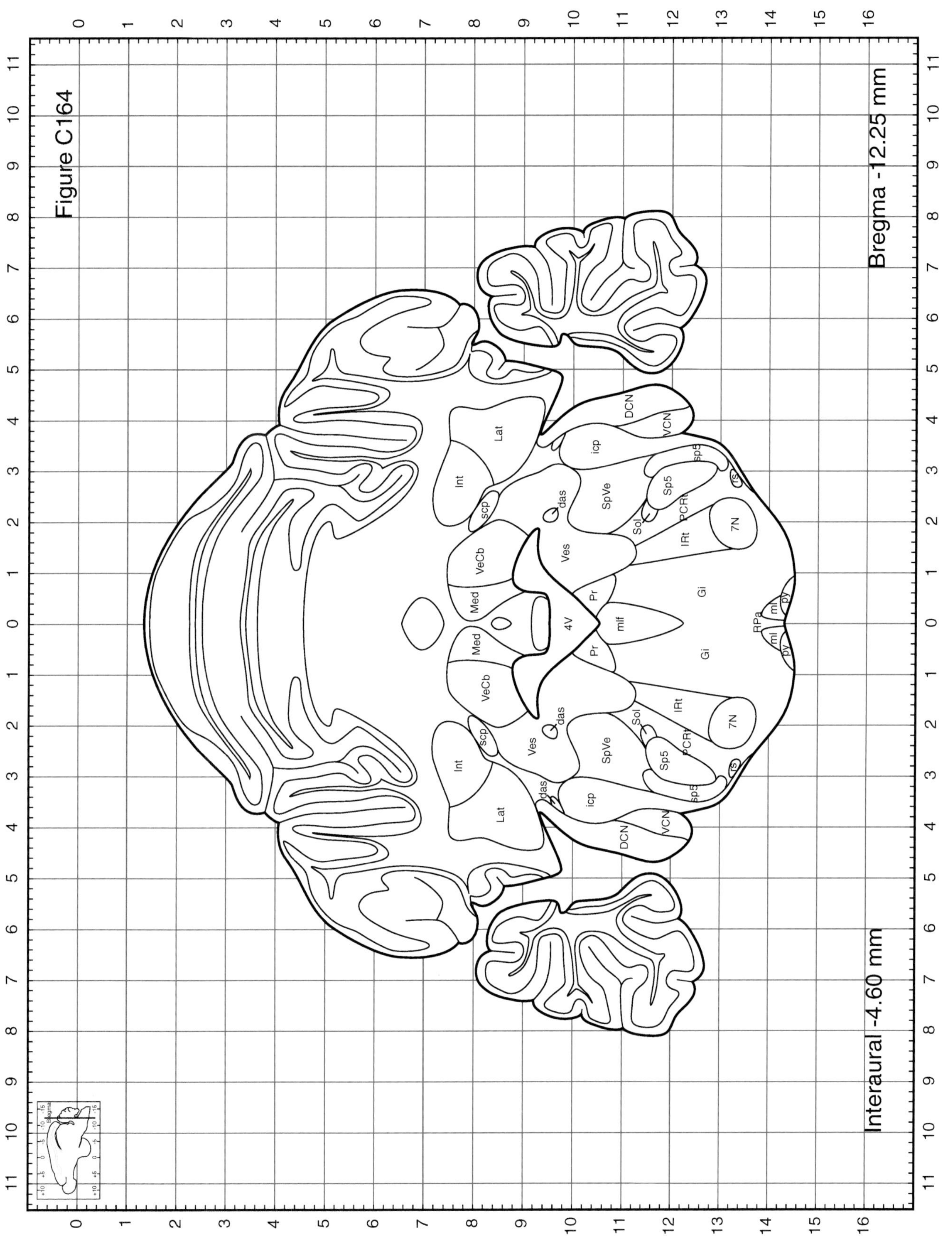
Figure C164
Bregma -12.25 mm
Interaural -4.60 mm
Lat
Int
scp
VeCb
Med
das
Ves
SpVe
icp
DCN
VCN
sp5
Sp5
Sol
PCRt
IRt
7N
Pr
4V
mlf
Gi
RPa
ml
py

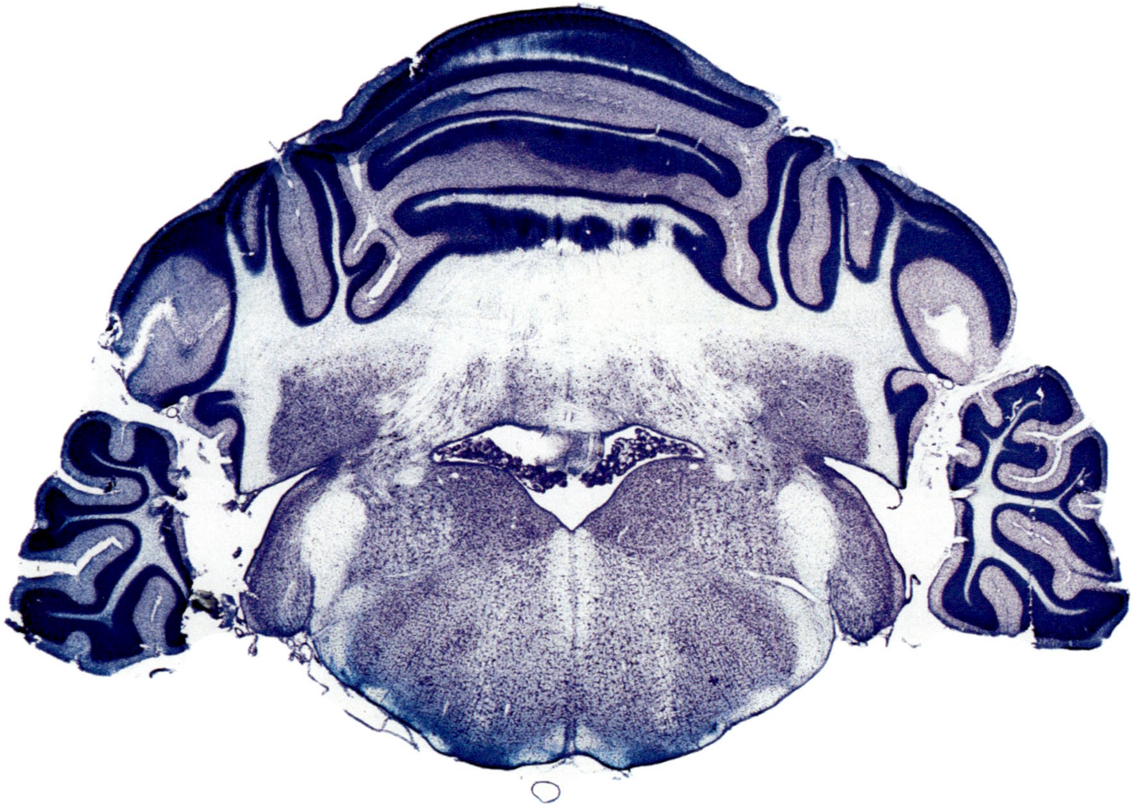

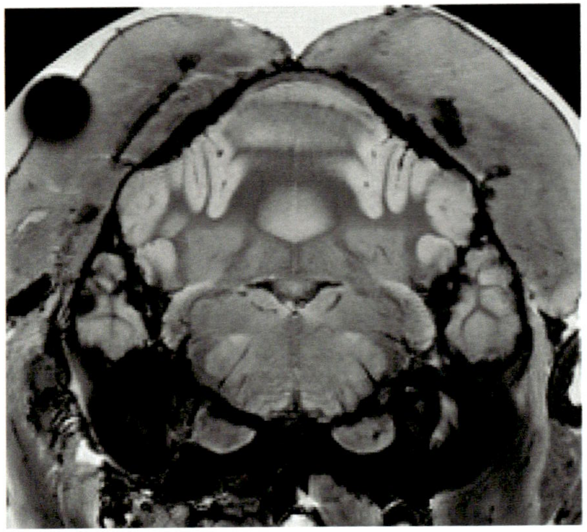

4V	4th ventricle
AmbC	ambiguus nucleus, compact part
das	dorsal acoustic stria
DCN	dorsal cochlear nucleus
Gi	gigantocellular reticular nucleus
icp	inferior cerebellar peduncle
Int	interposed cerebellar nucleus
IRt	intermediate reticular nucleus
Lat	lateral cerebellar nucleus
Med	medial cerebellar nucleus
ml	medial lemniscus
mlf	medial longitudinal fasciculus
oc	olivocerebellar tract
PCRt	parvicellular reticular nucleus
Pr	prepositus nucleus
py	pyramidal tract
RPa	raphe pallidus nucleus
rs	rubrospinal tract
Sol	solitary nucleus
Sp5	spinal trigeminal nucleus
sp5	spinal trigeminal tract
SpVe	spinal vestibular nucleus
VCN	ventral cochlear nucleus
VeCb	vestibulocerebellar nucleus
Ves	vestibular nucleus
VLN	ventrolateral reticular nucleus

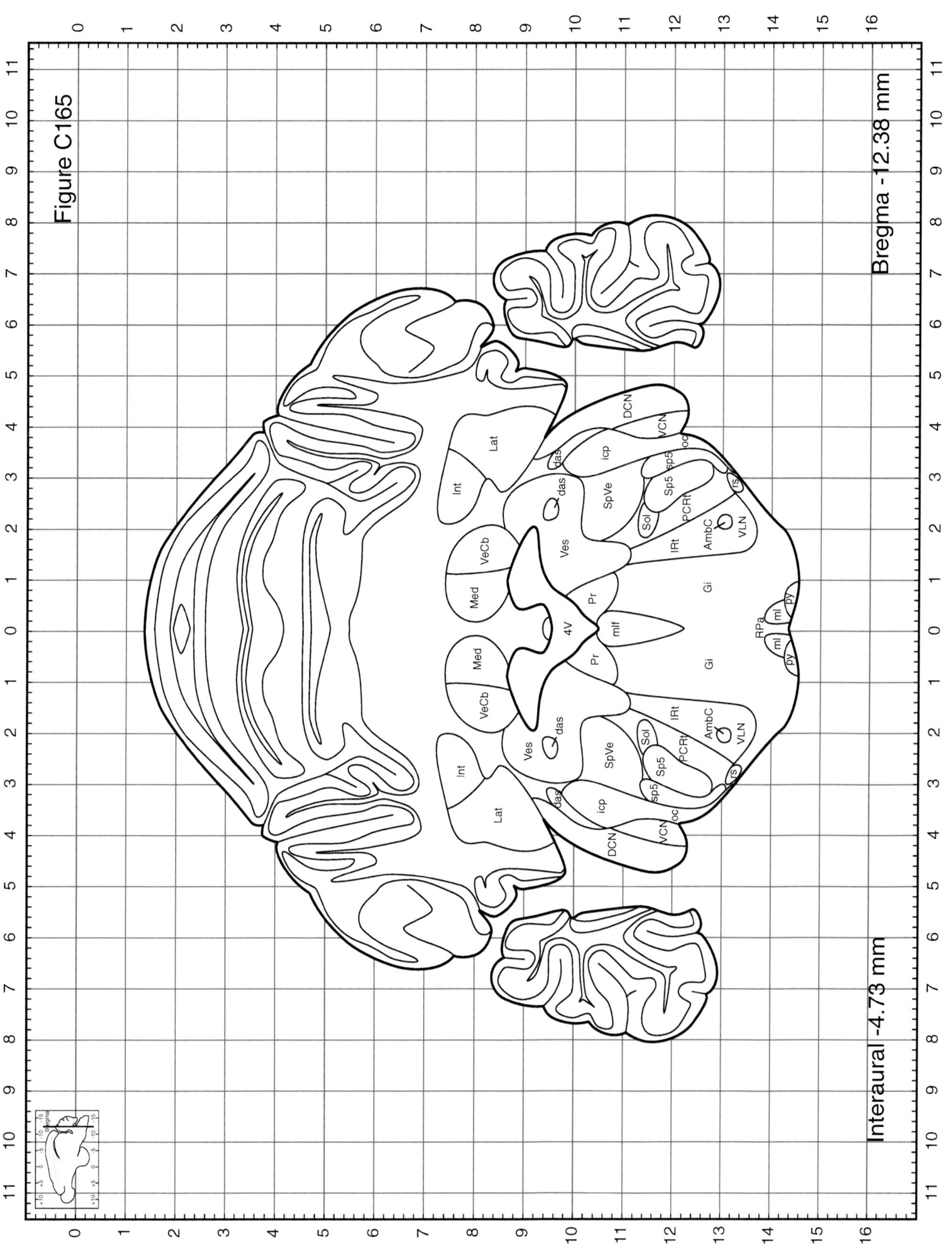
Figure C165
Bregma -12.38 mm
Interaural -4.73 mm
Lat
Int
Med
VeCb
Ves
das
SpVe
icp
DCN
VCN
oc
sp5
Sp5
Sol
PCRt
IRt
AmbC
VLN
rs
Pr
4V
mlf
Gi
RPa
ml
py

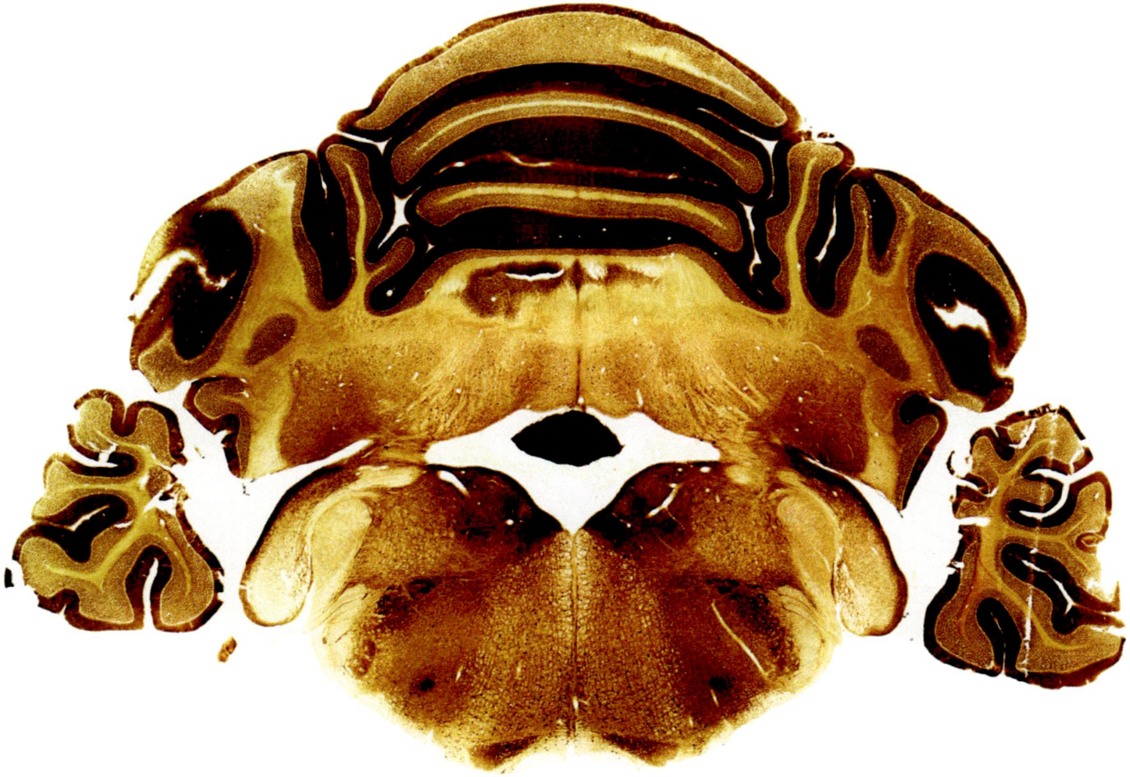

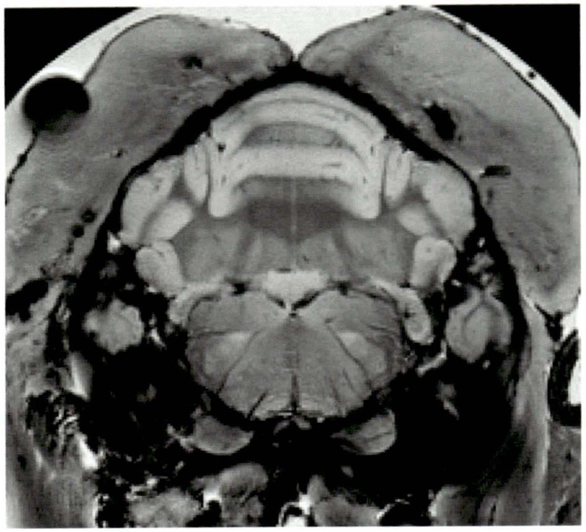

4V	4th ventricle
AmbC	ambiguus nucleus, compact part
das	dorsal acoustic stria
DCN	dorsal cochlear nucleus
Gi	gigantocellular reticular nucleus
icp	inferior cerebellar peduncle
Int	interposed cerebellar nucleus
IO	inferior olive
IRt	intermediate reticular nucleus
Lat	lateral cerebellar nucleus
Med	medial cerebellar nucleus
ml	medial lemniscus
mlf	medial longitudinal fasciculus
oc	olivocerebellar tract
PCRt	parvicellular reticular nucleus
Pr	prepositus nucleus
py	pyramidal tract
RPa	raphe pallidus nucleus
rs	rubrospinal tract
Sol	solitary nucleus
Sp5	spinal trigeminal nucleus
sp5	spinal trigeminal tract
SpVe	spinal vestibular nucleus
VCN	ventral cochlear nucleus
Ves	vestibular nucleus
VLN	ventrolateral reticular nucleus

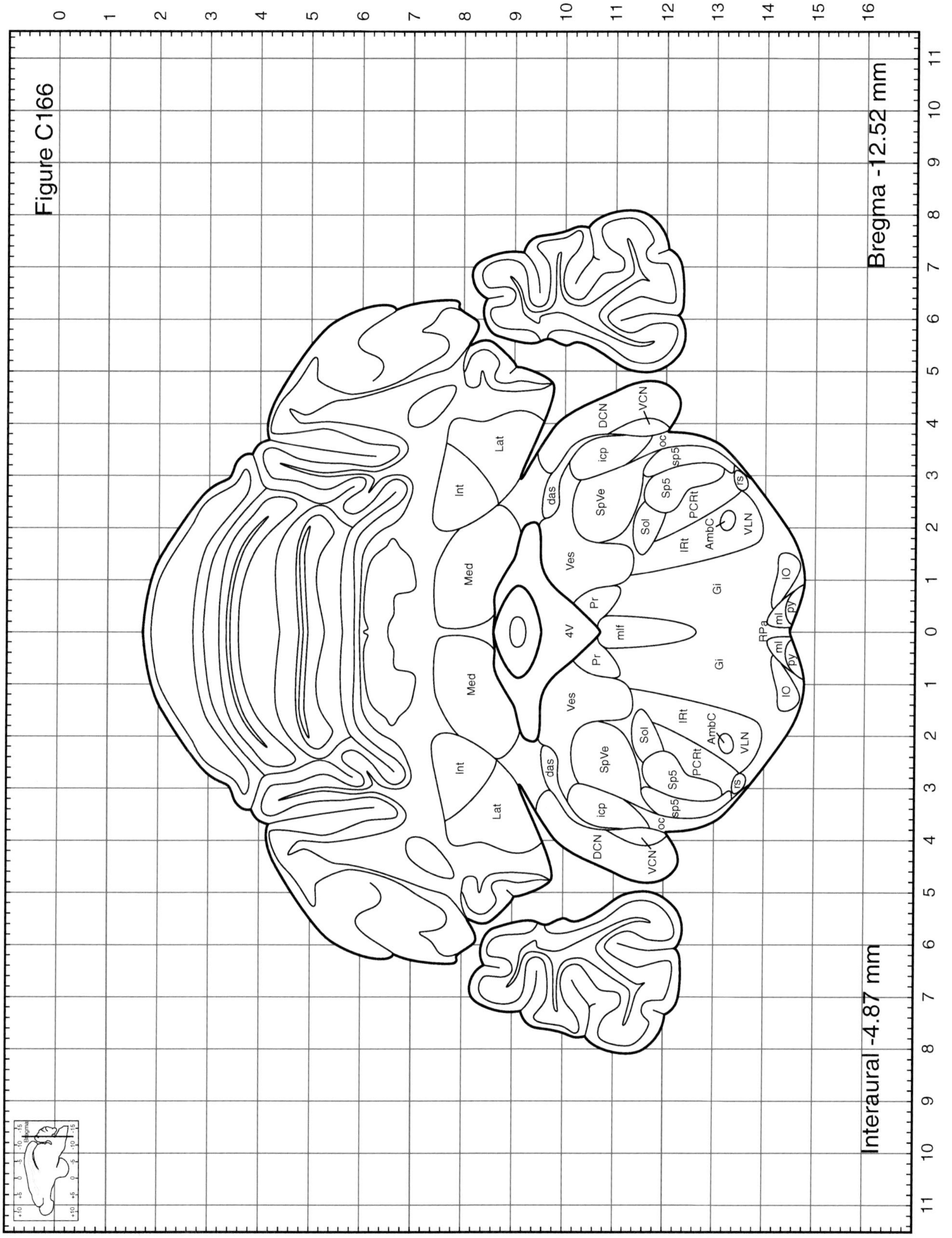
Figure C166
Bregma -12.52 mm
Interaural -4.87 mm
Lat
Int
Med
DCN
VCN
icp
oc
sp5
Sp5
rs
das
SpVe
PCRt
AmbC
VLN
Sol
IRt
Ves
Gi
IO
Pr
py
ml
RPa
4V
mlf

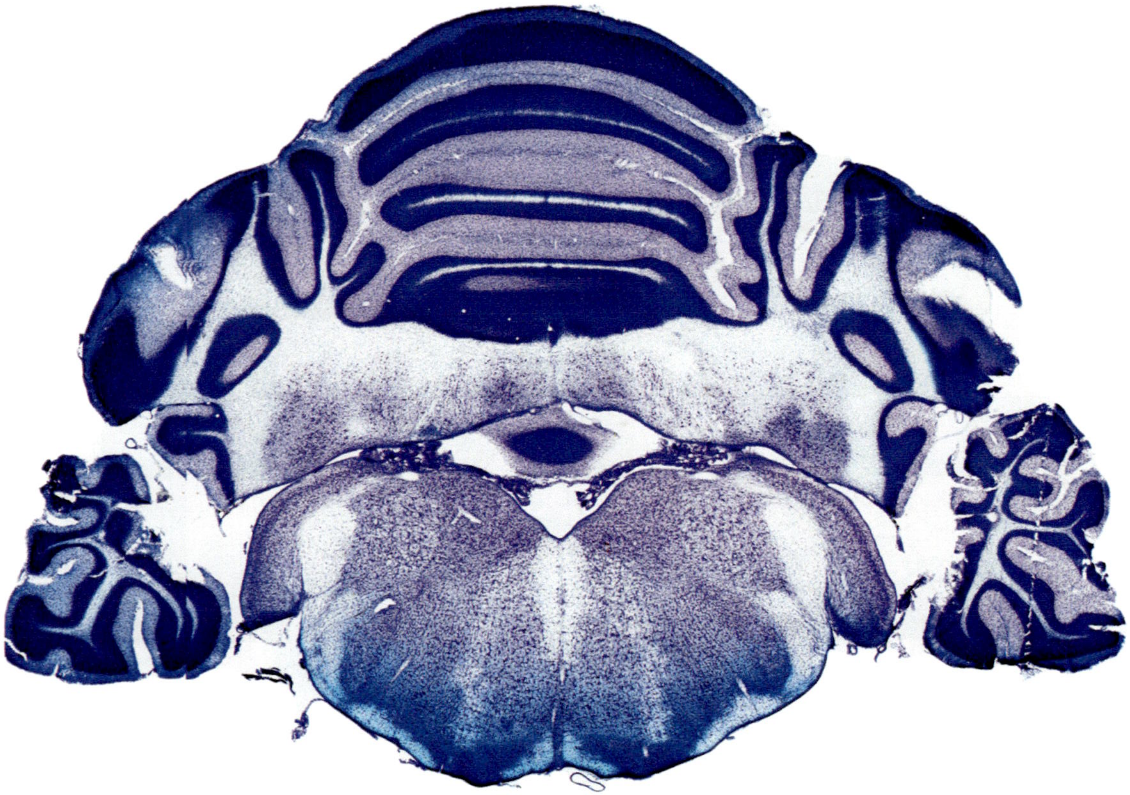

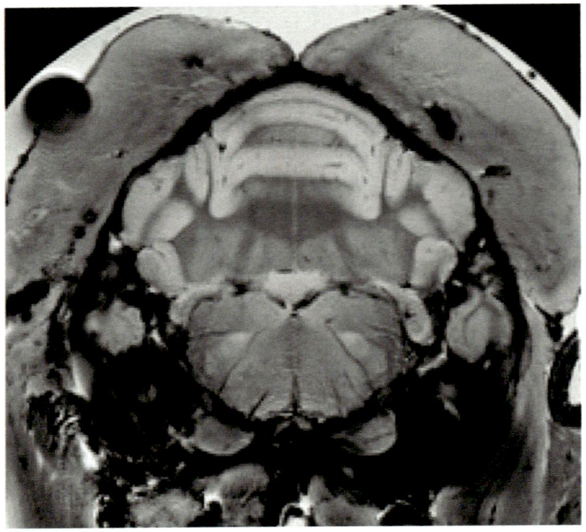

- 4V 4th ventricle
- AmbC ambiguus nucleus, compact part
- das dorsal acoustic stria
- DCN dorsal cochlear nucleus
- Gi gigantocellular reticular nucleus
- icp inferior cerebellar peduncle
- Int interposed cerebellar nucleus
- IO inferior olive
- IRt intermediate reticular nucleus
- Lat lateral cerebellar nucleus
- Med medial cerebellar nucleus
- ml medial lemniscus
- mlf medial longitudinal fasciculus
- oc olivocerebellar tract
- PCRt parvicellular reticular nucleus
- Pr prepositus nucleus
- py pyramidal tract
- RPa raphe pallidus nucleus
- rs rubrospinal tract
- Sol solitary nucleus
- Sp5 spinal trigeminal nucleus
- sp5 spinal trigeminal tract
- SpVe spinal vestibular nucleus
- VCN ventral cochlear nucleus
- Ves vestibular nucleus
- VLN ventrolateral reticular nucleus

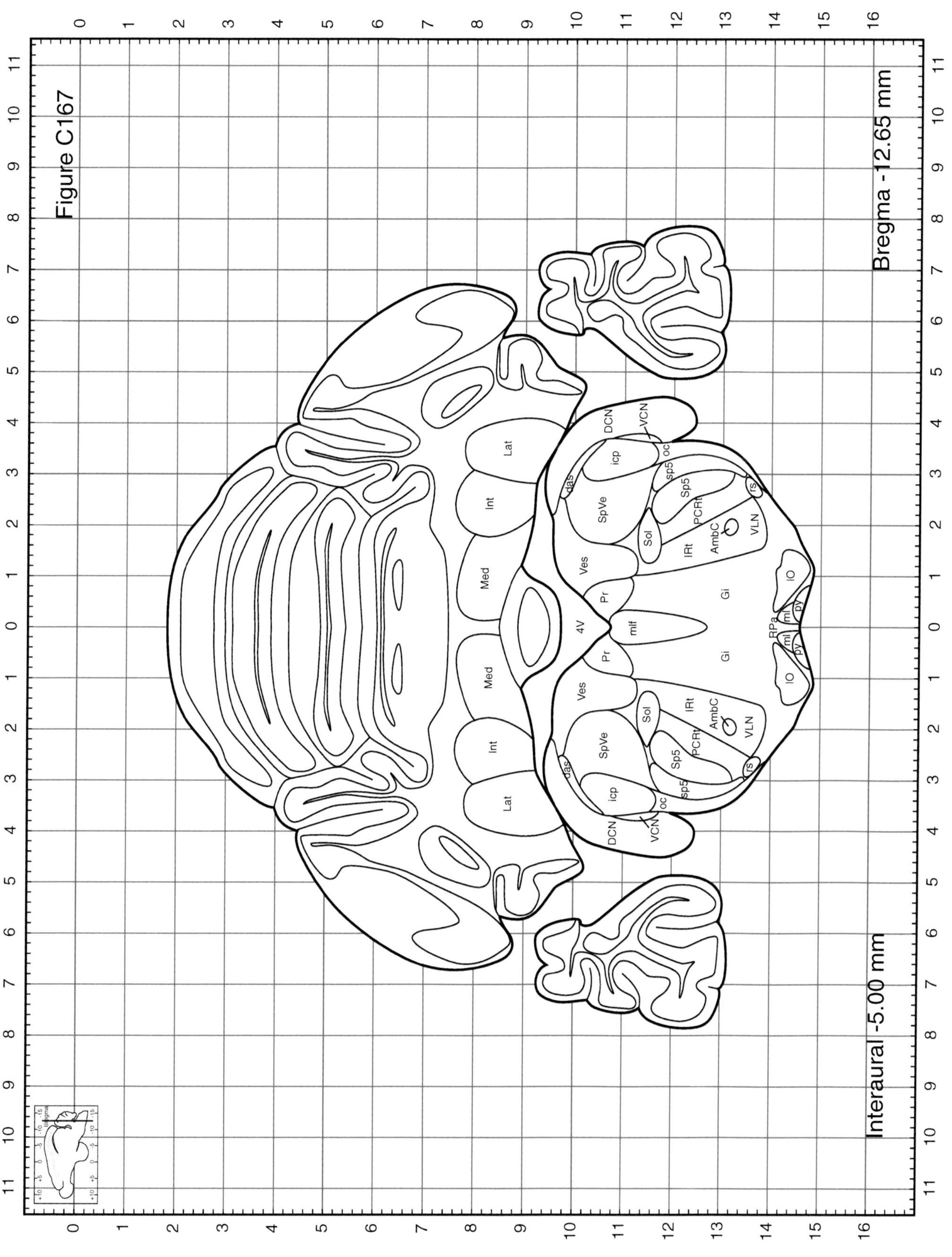
Figure C167
Bregma -12.65 mm
Interaural -5.00 mm
Lat
Int
Med
4V
DCN
VCN
icp
oc
sp5
Sp5
das
SpVe
Sol
PCRt
AmbC
VLN
rs
IRt
Ves
Pr
mlf
Gi
IO
ml
py
RPa

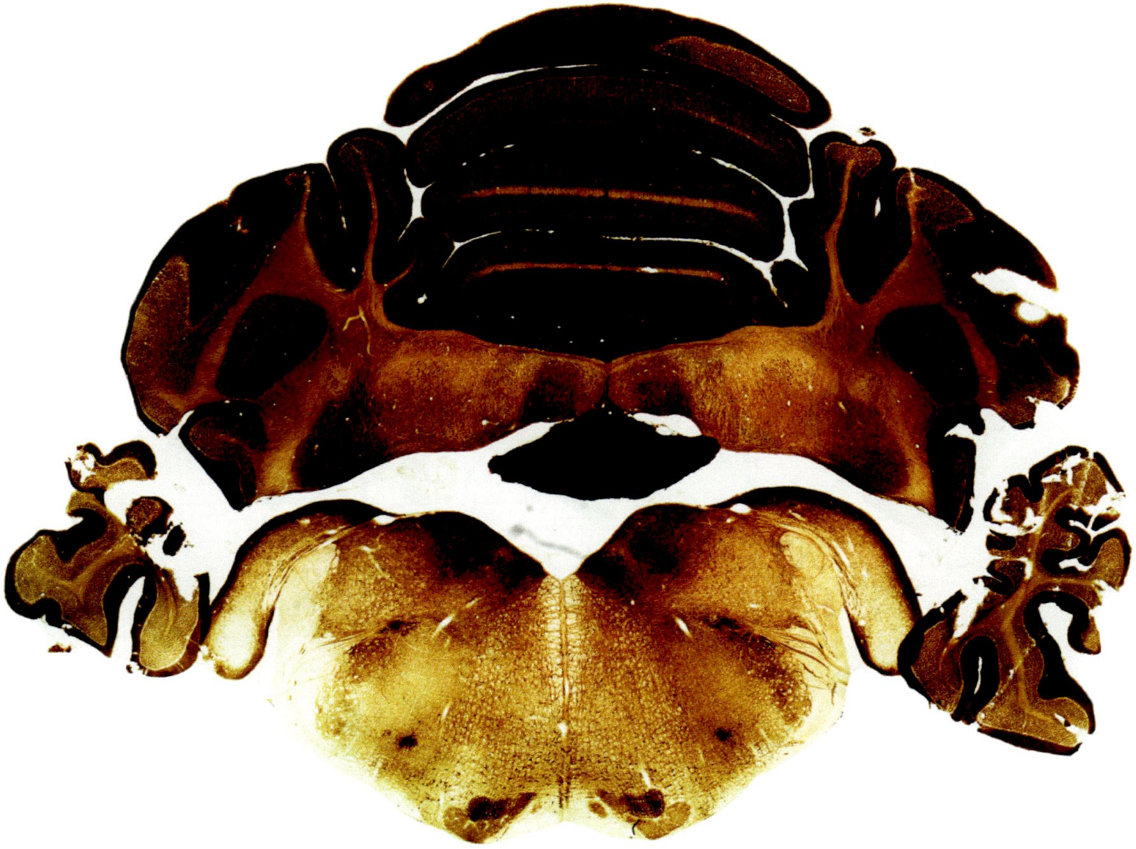

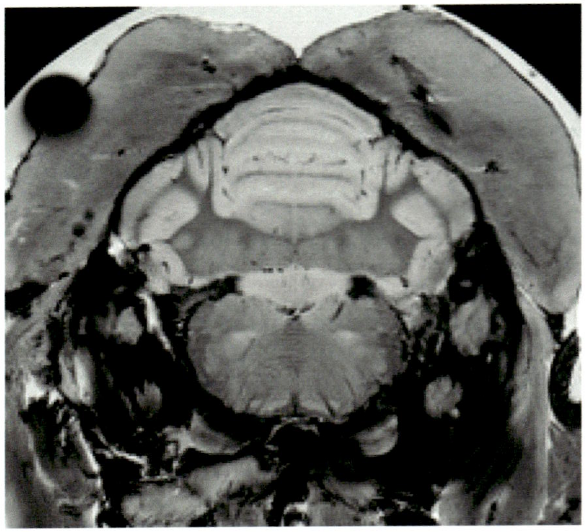

4V	4th ventricle	IRt	intermediate reticular nucleus	RPa	raphe pallidus nucleus
AmbC	ambiguus nucleus, compact part	Lat	lateral cerebellar nucleus	rs	rubrospinal tract
das	dorsal acoustic stria	Med	medial cerebellar nucleus	Sol	solitary nucleus
DCN	dorsal cochlear nucleus	ml	medial lemniscus	Sp5	spinal trigeminal nucleus
dsc	dorsal spinocerebellar tract	mlf	medial longitudinal fasciculus	sp5	spinal trigeminal tract
Gi	gigantocellular reticular nucleus	oc	olivocerebellar tract	SpVe	spinal vestibular nucleus
icp	inferior cerebellar peduncle	PCRt	parvicellular reticular nucleus	Ves	vestibular nucleus
Int	interposed cerebellar nucleus	Pr	prepositus nucleus	VLN	ventrolateral reticular nucleus
IO	inferior olive	py	pyramidal tract		

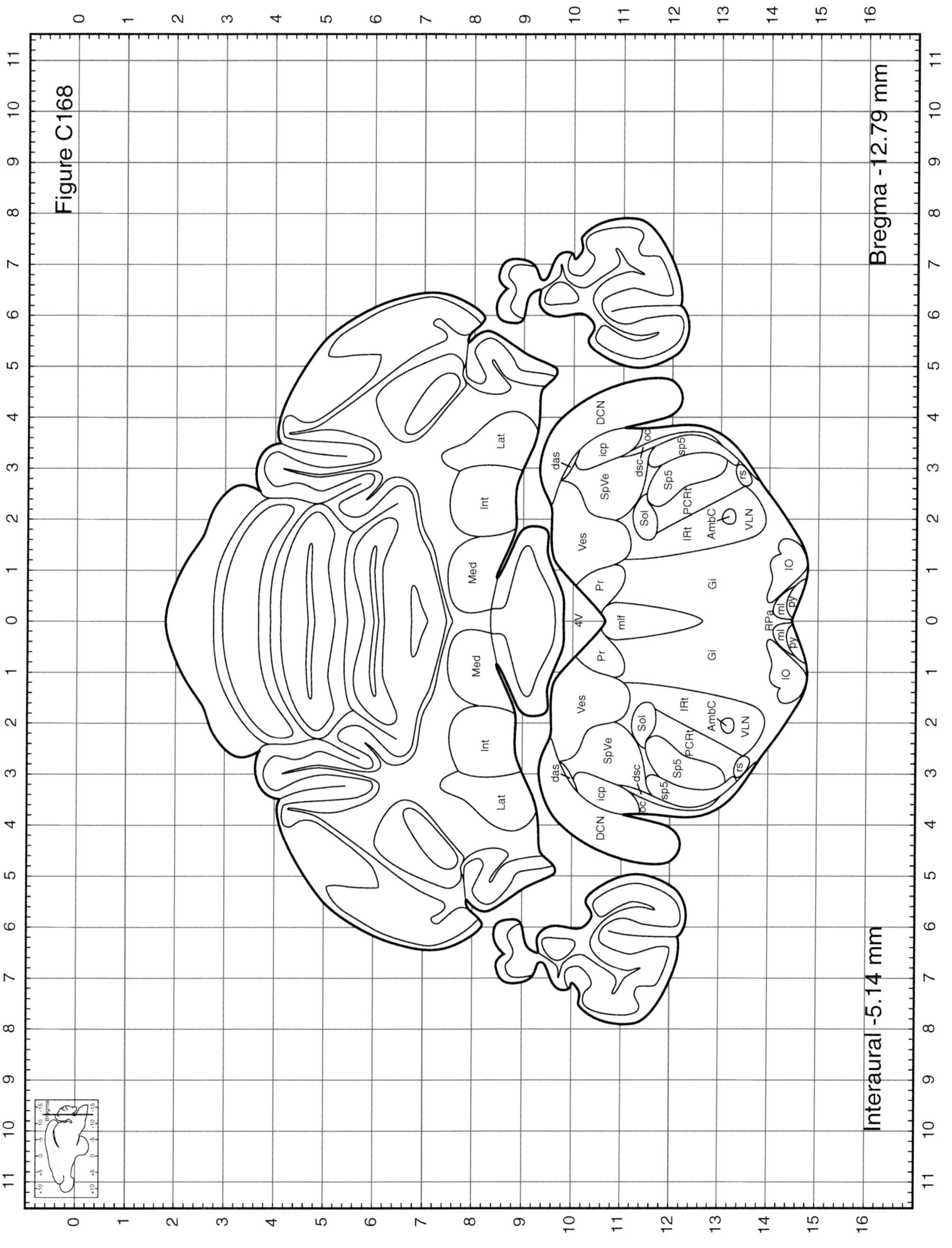
Figure C168
Bregma -12.79 mm
Interaural -5.14 mm
DCN
icp
oc
sp5
das
dsc
Sp5
SpVe
rs
PCRt
Sol
VLN
AmbC
IRt
Ves
IO
Gi
Pr
ml
py
RPa
mlf
4V
Lat
Int
Med

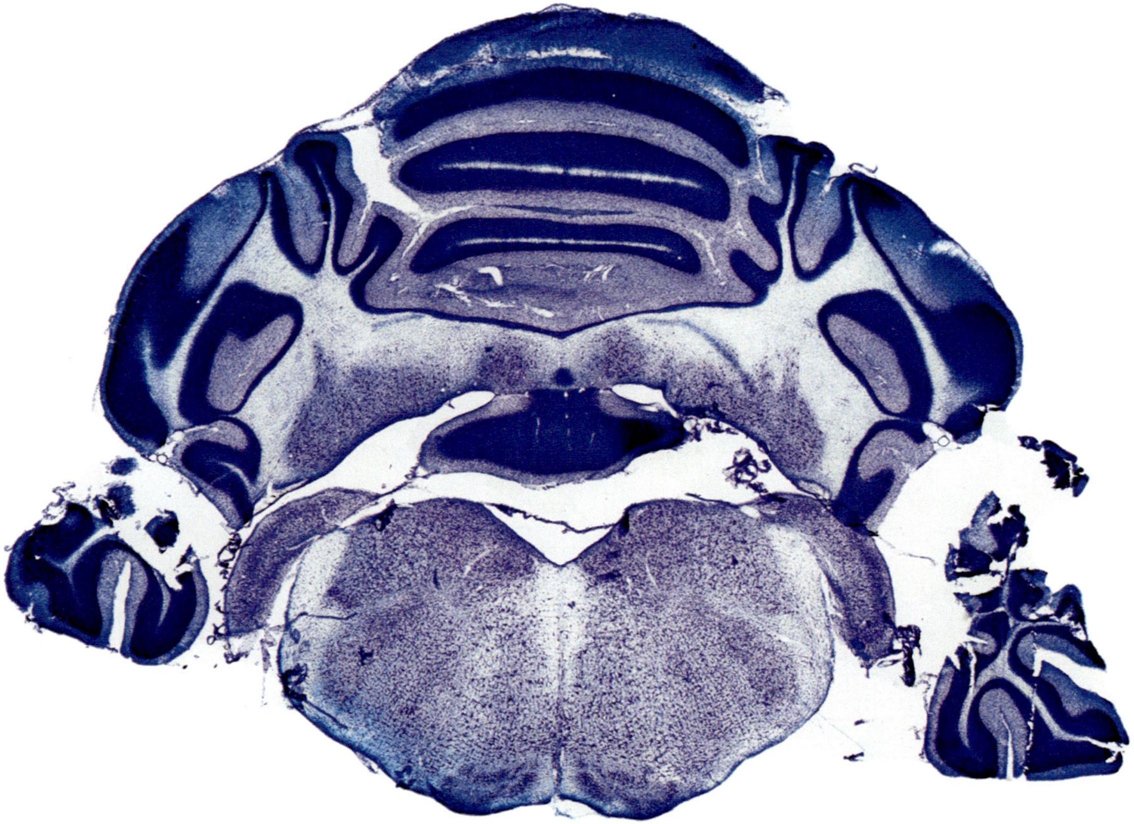

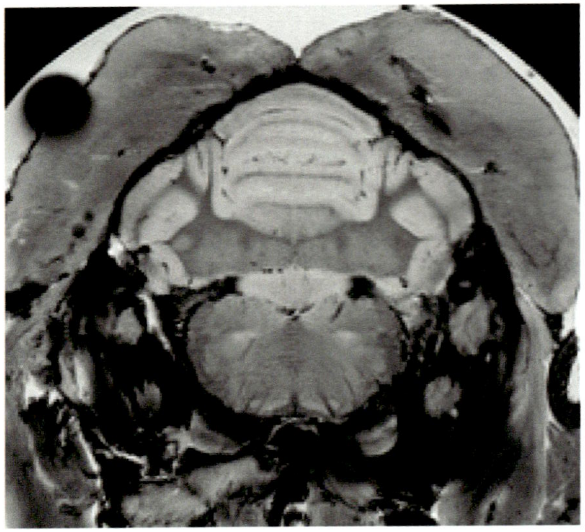

4V	4th ventricle
AmbC	ambiguus nucleus, compact part
DCN	dorsal cochlear nucleus
dsc	dorsal spinocerebellar tract
Gi	gigantocellular reticular nucleus
icp	inferior cerebellar peduncle
Int	interposed cerebellar nucleus
IO	inferior olive
IRt	intermediate reticular nucleus
Lat	lateral cerebellar nucleus
Med	medial cerebellar nucleus
ml	medial lemniscus
mlf	medial longitudinal fasciculus
Mx	matrix region of the medulla
oc	olivocerebellar tract
PCRt	parvicellular reticular nucleus
Pr	prepositus nucleus
py	pyramidal tract
RPa	raphe pallidus nucleus
rs	rubrospinal tract
Sol	solitary nucleus
Sp5	spinal trigeminal nucleus
sp5	spinal trigeminal tract
SpVe	spinal vestibular nucleus
Ves	vestibular nucleus
VLN	ventrolateral reticular nucleus

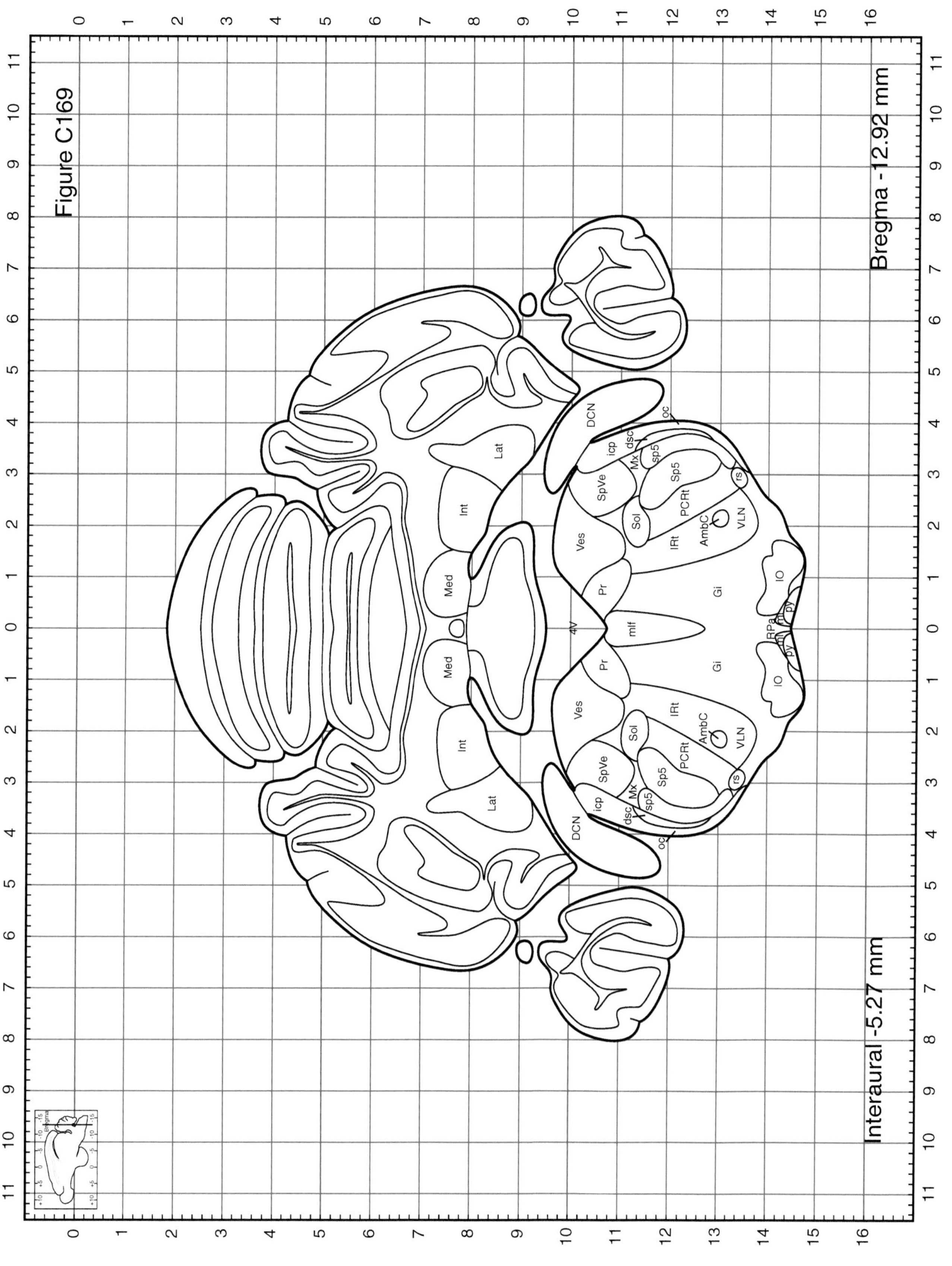

Figure C169
Bregma -12.92 mm
Interaural -5.27 mm
Lat
Int
Med
4V
DCN
oc
dsc
icp
Mx
sp5
Sp5
SpVe
Sol
PCRt
rs
VLN
AmbC
IRt
Ves
Pr
mlf
Gi
IO
RPa
ml
py

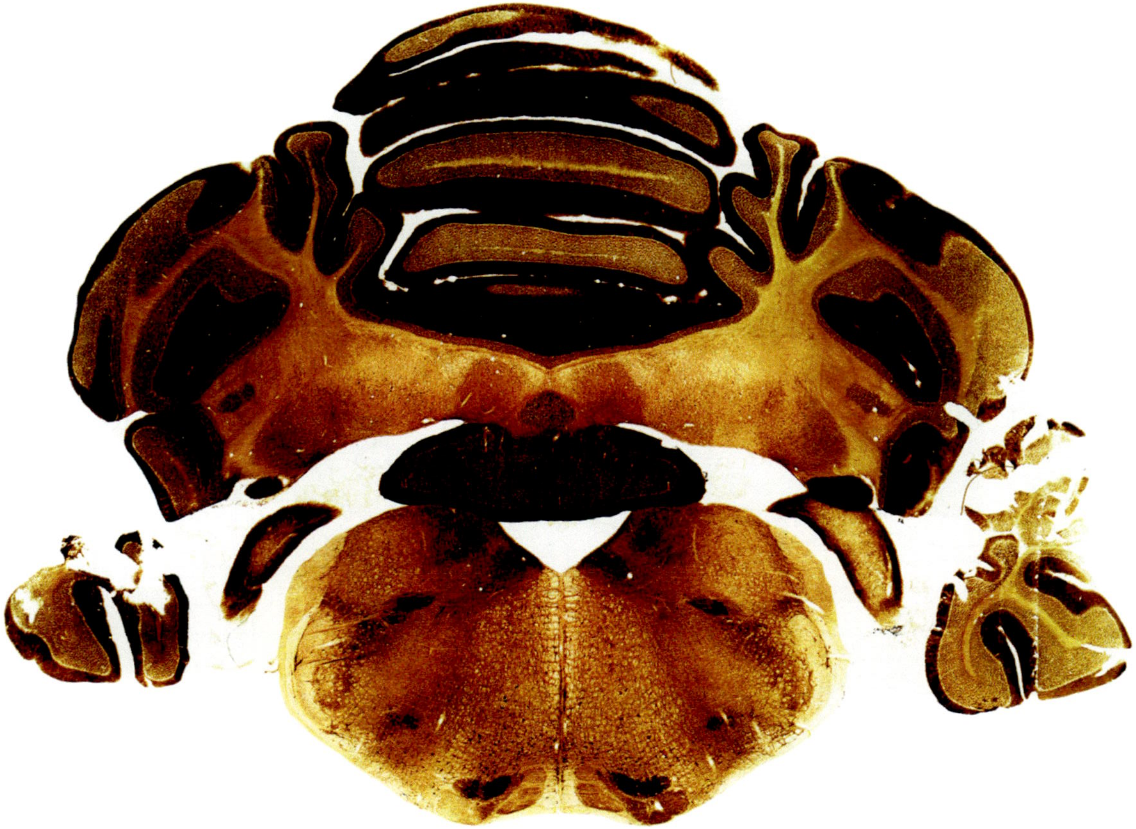

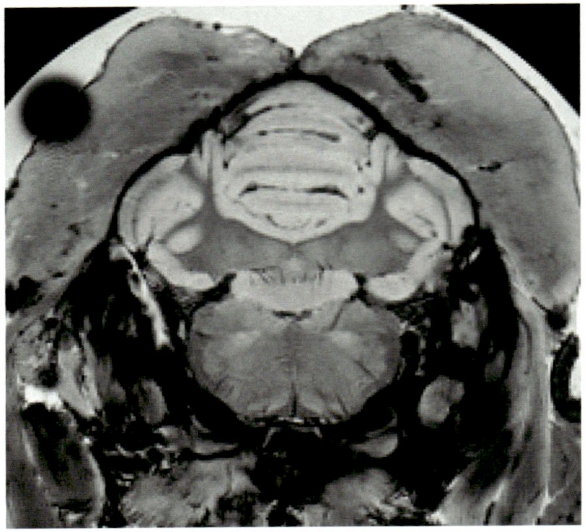

4V	4th ventricle
AmbC	ambiguus nucleus, compact part
DCN	dorsal cochlear nucleus
dsc	dorsal spinocerebellar tract
Gi	gigantocellular reticular nucleus
icp	inferior cerebellar peduncle
Int	interposed cerebellar nucleus
IO	inferior olive
IRt	intermediate reticular nucleus
Lat	lateral cerebellar nucleus
Med	medial cerebellar nucleus
ml	medial lemniscus
mlf	medial longitudinal fasciculus
Mx	matrix region of the medulla
oc	olivocerebellar tract
PCRt	parvicellular reticular nucleus
Pr	prepositus nucleus
py	pyramidal tract
RPa	raphe pallidus nucleus
rs	rubrospinal tract
Sol	solitary nucleus
Sp5	spinal trigeminal nucleus
sp5	spinal trigeminal tract
SpVe	spinal vestibular nucleus
Ves	vestibular nucleus
VLN	ventrolateral reticular nucleus

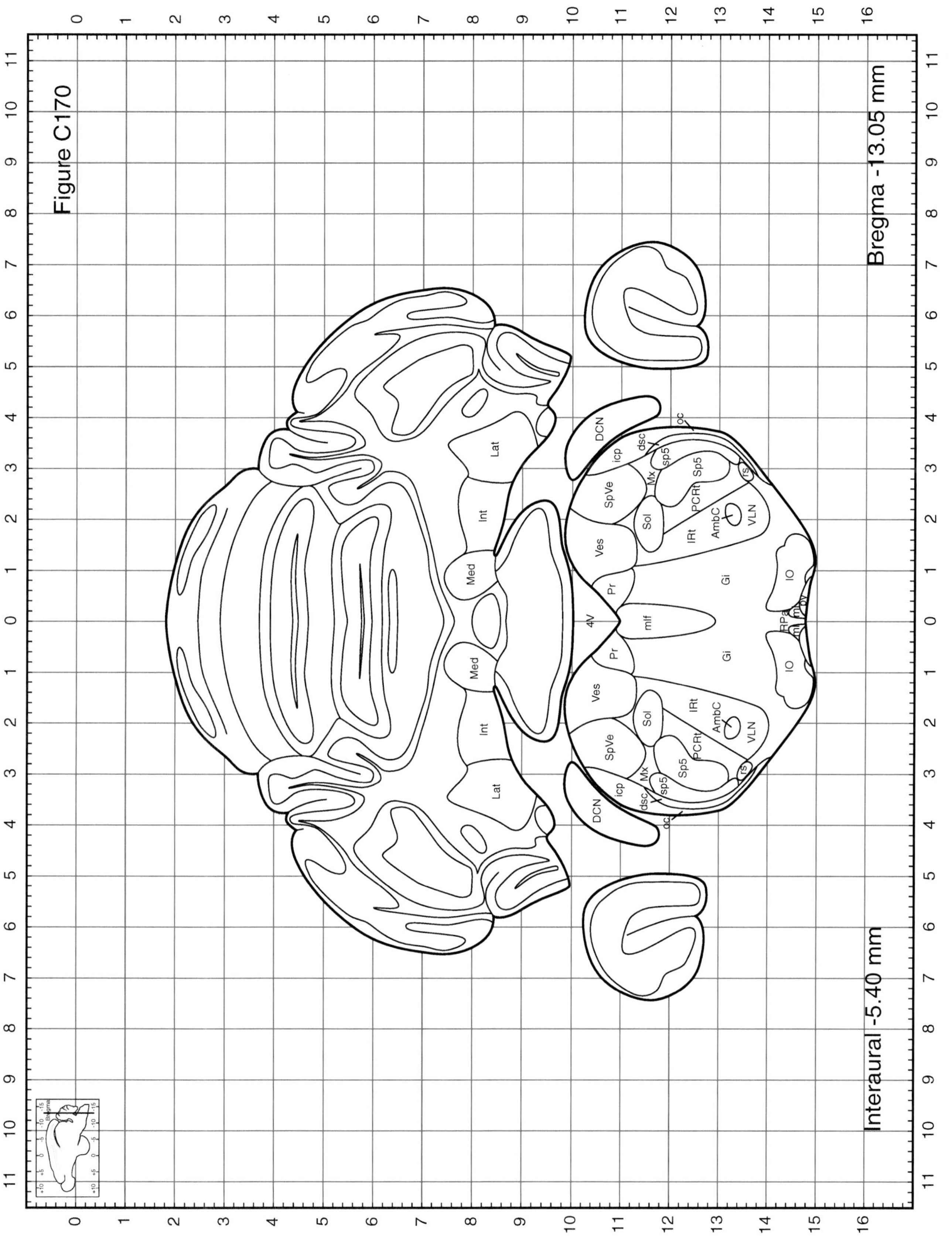
Figure C170
Bregma -13.05 mm
Interaural -5.40 mm
DCN
icp
dsc
oc
sp5
Sp5
Mx
SpVe
Sol
PCRt
AmbC
VLN
IRt
rs
Ves
Pr
4V
mlf
Gi
IO
RPa
py
ml
Lat
Int
Med

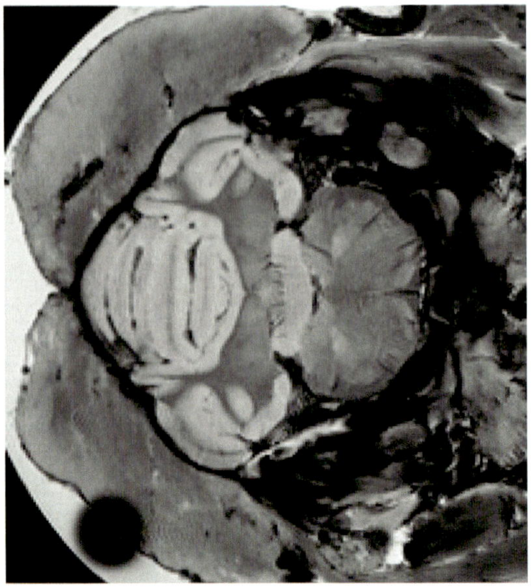

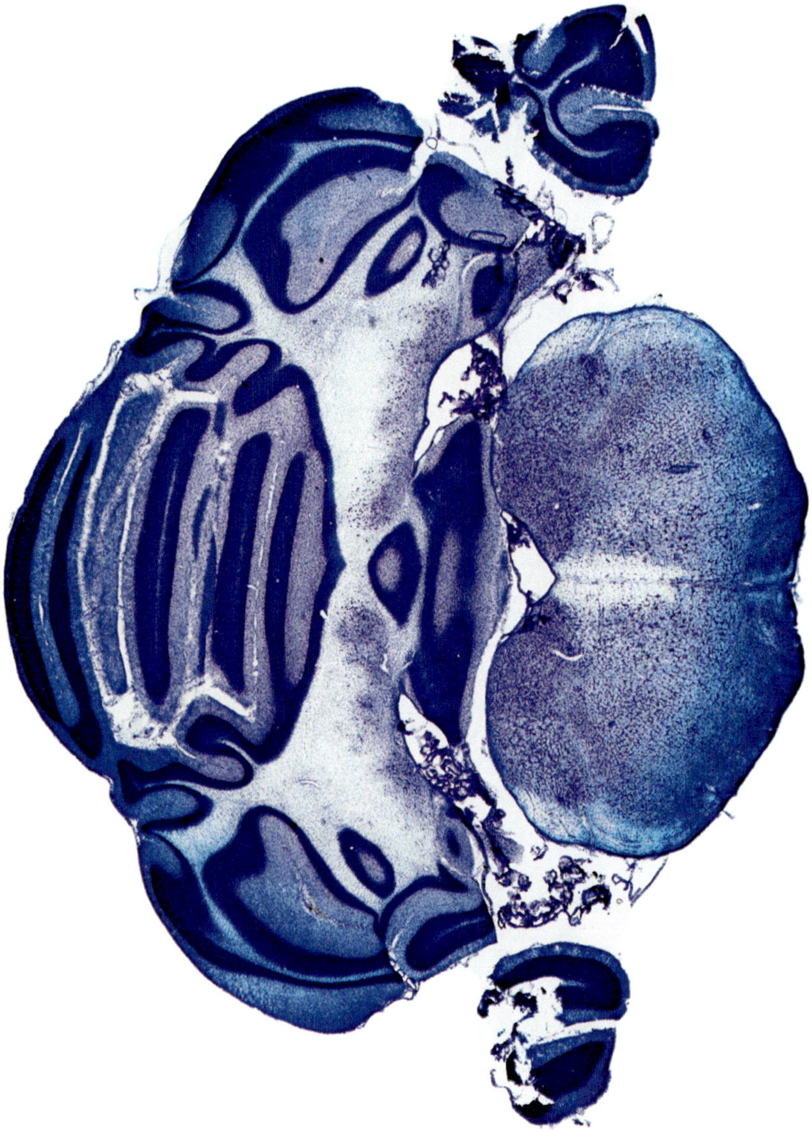

4V 4th ventricle
AmbC ambiguus nucleus, compact part
Gi gigantocellular reticular nucleus
icp inferior cerebellar peduncle
Int interposed cerebellar nucleus
IO inferior olive
IRt intermediate reticular nucleus
Lat lateral cerebellar nucleus
Med medial cerebellar nucleus
ml medial lemniscus
mlf medial longitudinal fasciculus
Mx matrix region of the medulla
oc/dsc olivocerebellar tract and dorsal spinocerebellar tract
PCRt parvicellular reticular nucleus
Pr prepositus nucleus
py pyramidal tract
RPa raphe pallidus nucleus
rs rubrospinal tract
Sol solitary nucleus
Sp5 spinal trigeminal nucleus
sp5 spinal trigeminal tract
SpVe spinal vestibular nucleus
Ves vestibular nucleus
VLN ventrolateral reticular nucleus

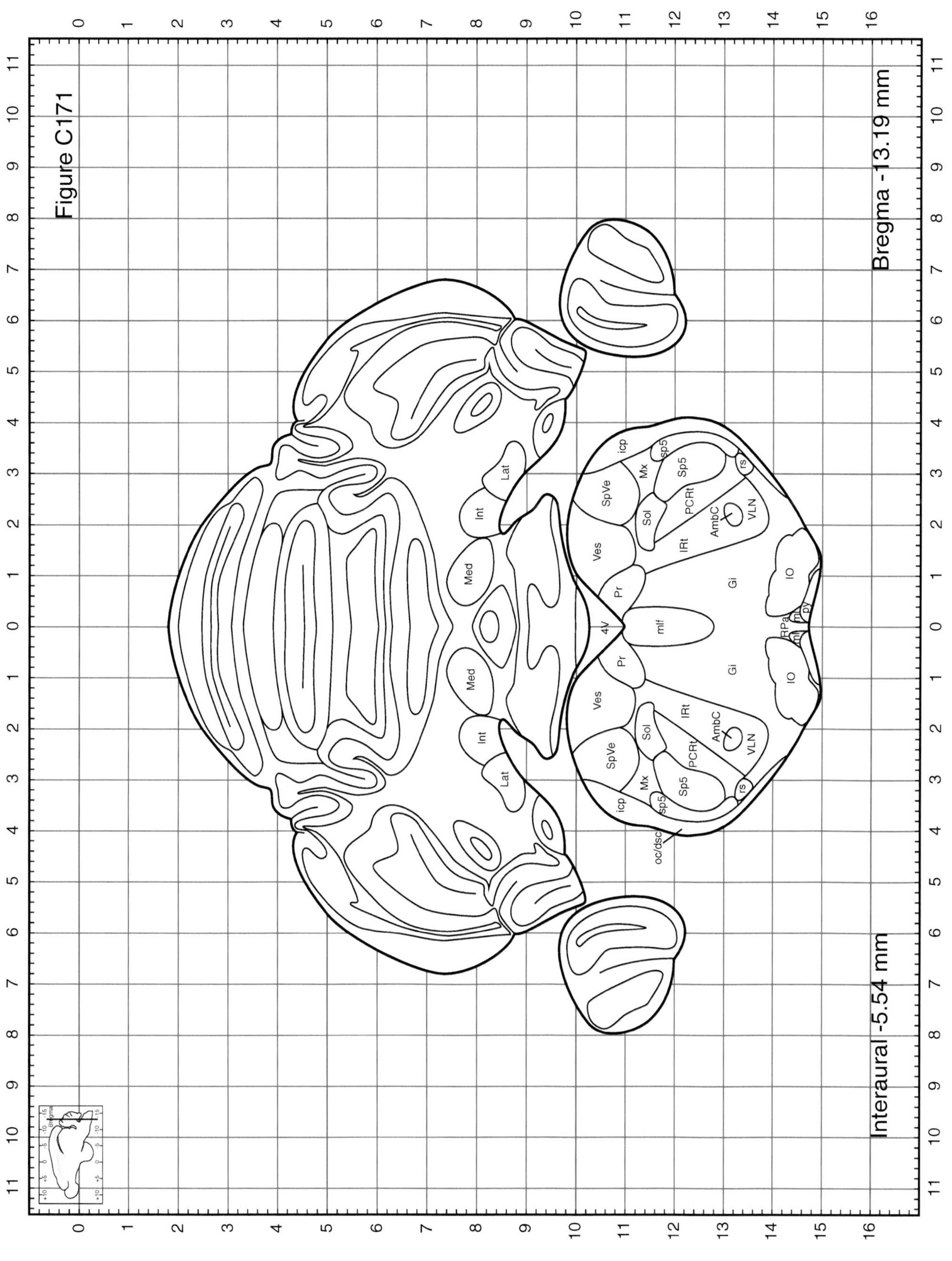
Figure C171
Bregma -13.19 mm
Interaural -5.54 mm
Lat
Int
Med
Lat
Int
Med
icp
sp5
Sp5
Mx
rs
SpVe
PCRt
Sol
AmbC
VLN
Ves
IRt
IO
Gi
Pr
py
4V
mlf
RPa
ml
Pr
Gi
IO
Ves
IRt
Sol
AmbC
VLN
SpVe
PCRt
Mx
Sp5
rs
icp
sp5
oc/dsc

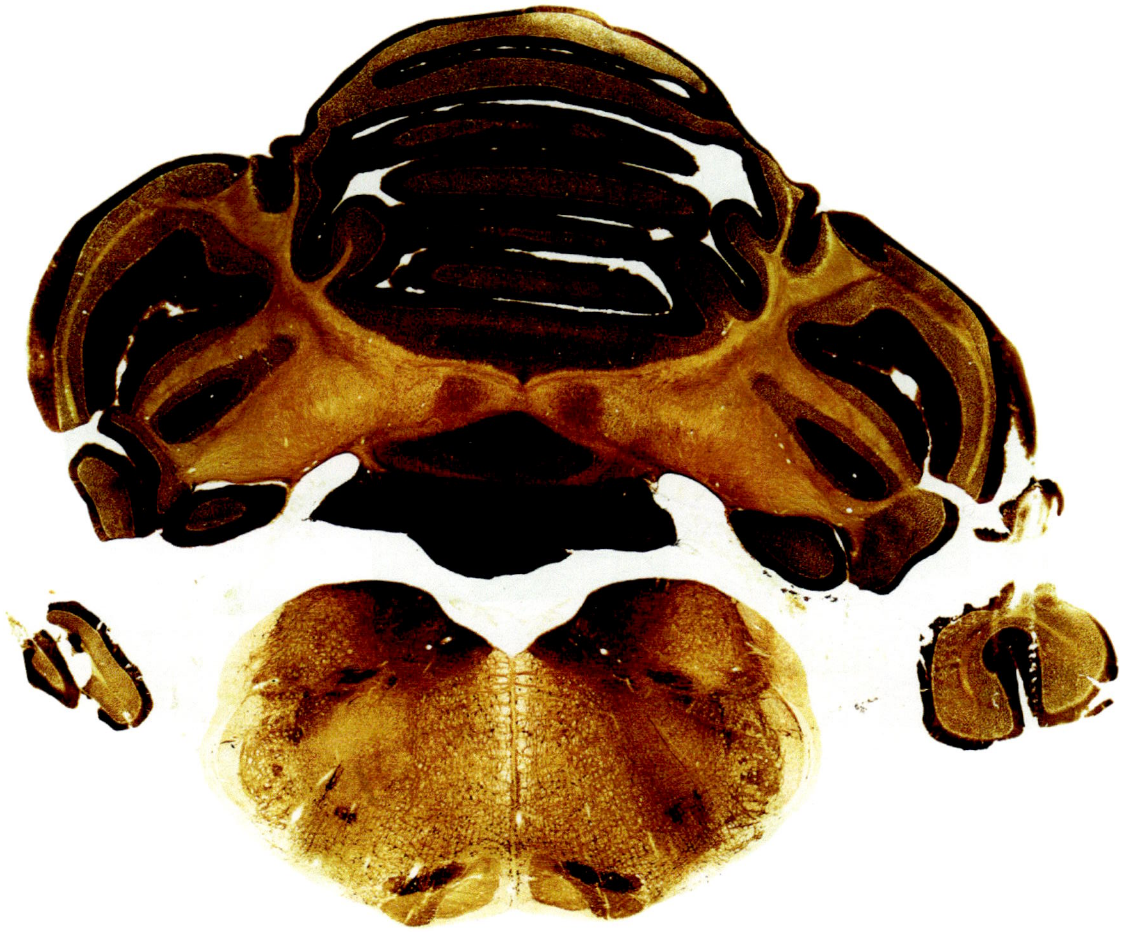

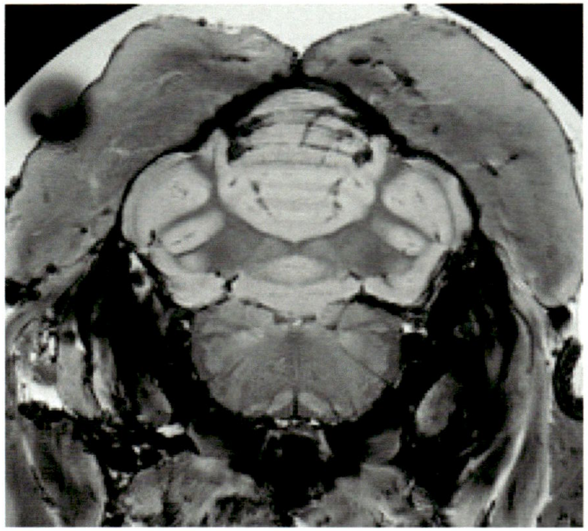

4V	4th ventricle
AmbC	ambiguus nucleus, compact part
Gi	gigantocellular reticular nucleus
icp	inferior cerebellar peduncle
IO	inferior olive
IRt	intermediate reticular nucleus
Med	medial cerebellar nucleus
ml	medial lemniscus
mlf	medial longitudinal fasciculus
Mx	matrix region of the medulla
oc/dsc	olivocerebellar tract and dorsal spinocerebellar tract
PCRt	parvicellular reticular nucleus
Pr	prepositus nucleus
py	pyramidal tract
RPa	raphe pallidus nucleus
rs	rubrospinal tract
Sol	solitary nucleus
Sp5	spinal trigeminal nucleus
sp5	spinal trigeminal tract
SpVe	spinal vestibular nucleus
Ves	vestibular nucleus
VLN	ventrolateral reticular nucleus

Figure C172
Bregma -13.32 mm
Interaural -5.67 mm
oc/dsc
icp
sp5
Mx
Sp5
SpVe
PCRt
rs
Sol
AmbC
VLN
IRt
Ves
IO
Gi
Pr
Med
4V
mlf
RPa
py
ml

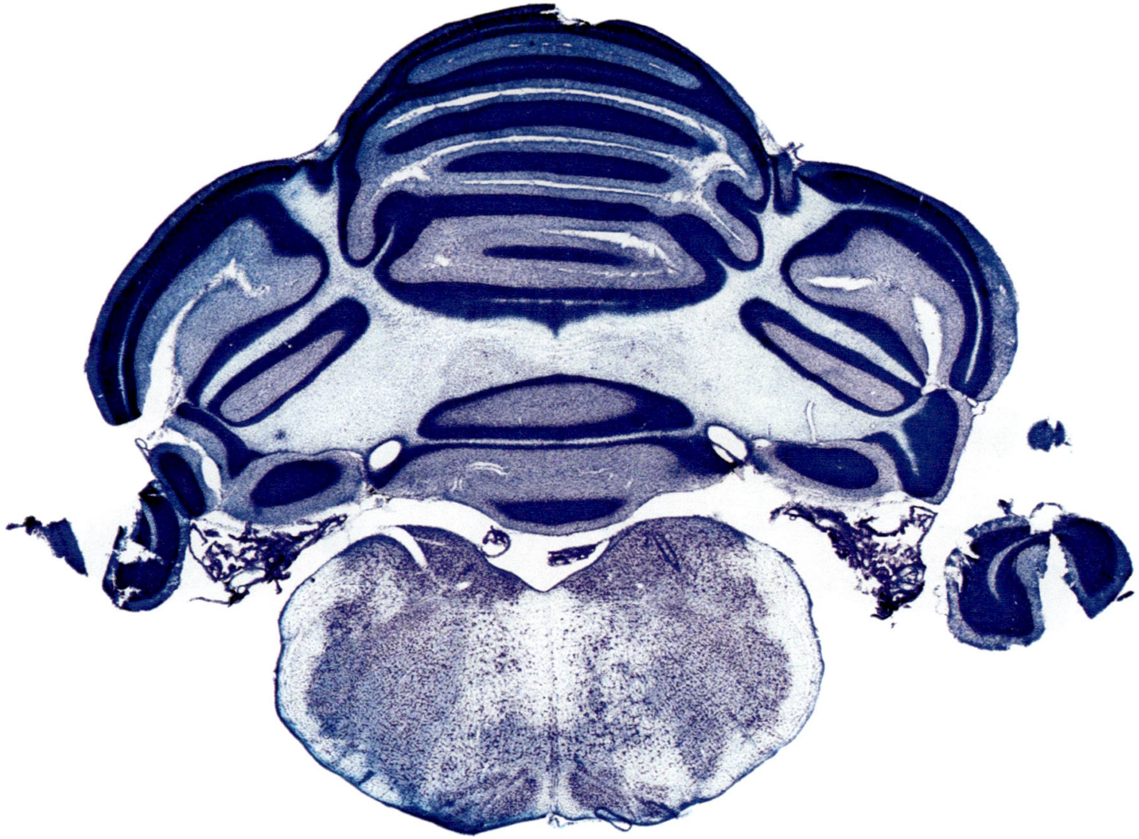

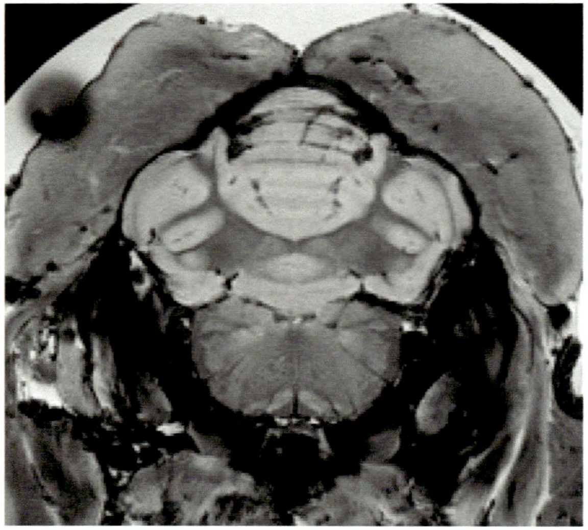

4V	4th ventricle
AmbC	ambiguus nucleus, compact part
Gi	gigantocellular reticular nucleus
icp	inferior cerebellar peduncle
IO	inferior olive
IRt	intermediate reticular nucleus
ml	medial lemniscus
mlf	medial longitudinal fasciculus
Mx	matrix region of the medulla
oc/dsc	olivocerebellar tract and dorsal spinocerebellar tract
PCRt	parvicellular reticular nucleus
Pr	prepositus nucleus
py	pyramidal tract
RPa	raphe pallidus nucleus
rs	rubrospinal tract
Sol	solitary nucleus
Sp5	spinal trigeminal nucleus
sp5	spinal trigeminal tract
SpVe	spinal vestibular nucleus
Ves	vestibular nucleus
VLN	ventrolateral reticular nucleus

Figure C173
Bregma -13.46 mm
Interaural -5.81 mm
oc/dsc
icp
sp5
Sp5
Mx
SpVe
Sol
PCRt
AmbC
VLN
rs
Ves
IRt
Pr
Gi
IO
4V
mlf
RPa
ml
py

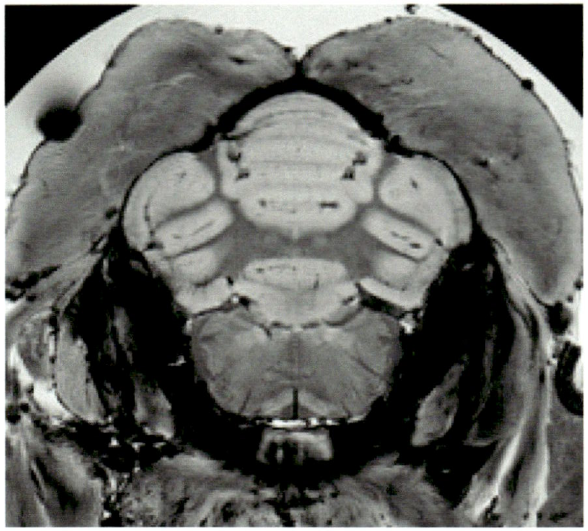

4V	4th ventricle
AmbC	ambiguus nucleus, compact part
ECu	external cuneate nucleus
Gi	gigantocellular reticular nucleus
icp	inferior cerebellar peduncle
IO	inferior olive
IRt	intermediate reticular nucleus
LRt	lateral reticular nucleus
ml	medial lemniscus
mlf	medial longitudinal fasciculus
Mx	matrix region of the medulla
oc/dsc	olivocerebellar tract and dorsal spinocerebellar tract
PCRt	parvicellular reticular nucleus
Pr	prepositus nucleus
py	pyramidal tract
RPa	raphe pallidus nucleus
rs	rubrospinal tract
Sol	solitary nucleus
Sp5	spinal trigeminal nucleus
sp5	spinal trigeminal tract
SpVe	spinal vestibular nucleus
Ves	vestibular nucleus
VLN	ventrolateral reticular nucleus

Bregma -13.59 mm
Interaural -5.94 mm
oc/dsc
icp
ECu
sp5
Mx
Sp5
SpVe
LRt
rs
PCRt
VLN
AmbC
Ves
Sol
IRt
IO
Pr
Gi
4V
mlf
RPa
py
ml

Figure C174

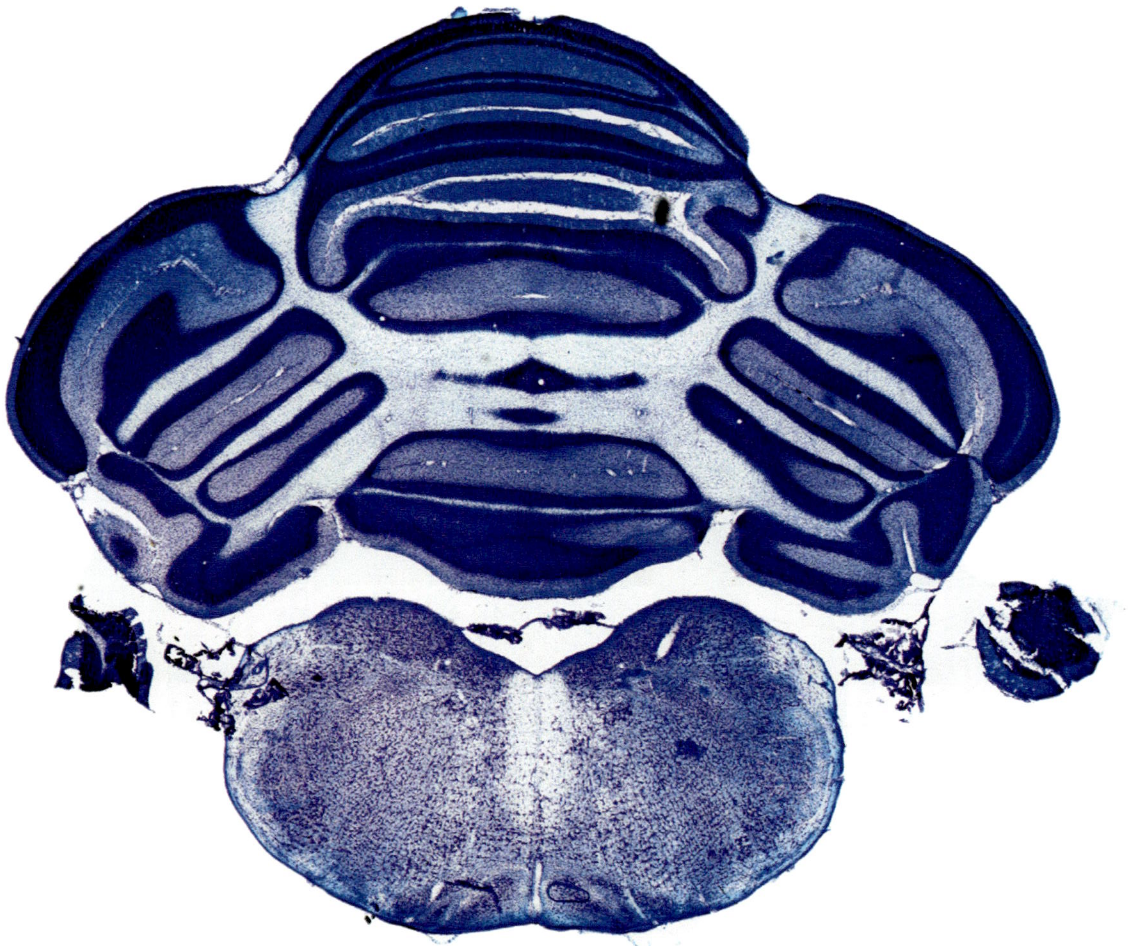

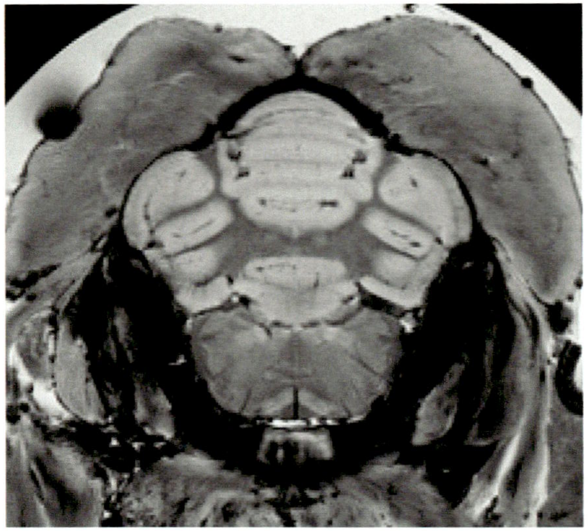

10N	dorsal motor nucleus of vagus		ml	medial lemniscus		Sol	solitary nucleus
4V	4th ventricle		mlf	medial longitudinal fasciculus		Sp5	spinal trigeminal nucleus
AmbC	ambiguus nucleus, compact part		Mx	matrix region of the medulla		sp5	spinal trigeminal tract
ECu	external cuneate nucleus		oc/dsc	olivocerebellar tract and dorsal spinocerebellar tract		SpVe	spinal vestibular nucleus
Gi	gigantocellular reticular nucleus		PCRt	parvicellular reticular nucleus		Ves	vestibular nucleus
icp	inferior cerebellar peduncle		Pr	prepositus nucleus		VLN	ventrolateral reticular nucleus
IO	inferior olive		py	pyramidal tract		vsc	ventral spinocerebellar tract
IRt	intermediate reticular nucleus		RPa	raphe pallidus nucleus			
LRt	lateral reticular nucleus		rs	rubrospinal tract			

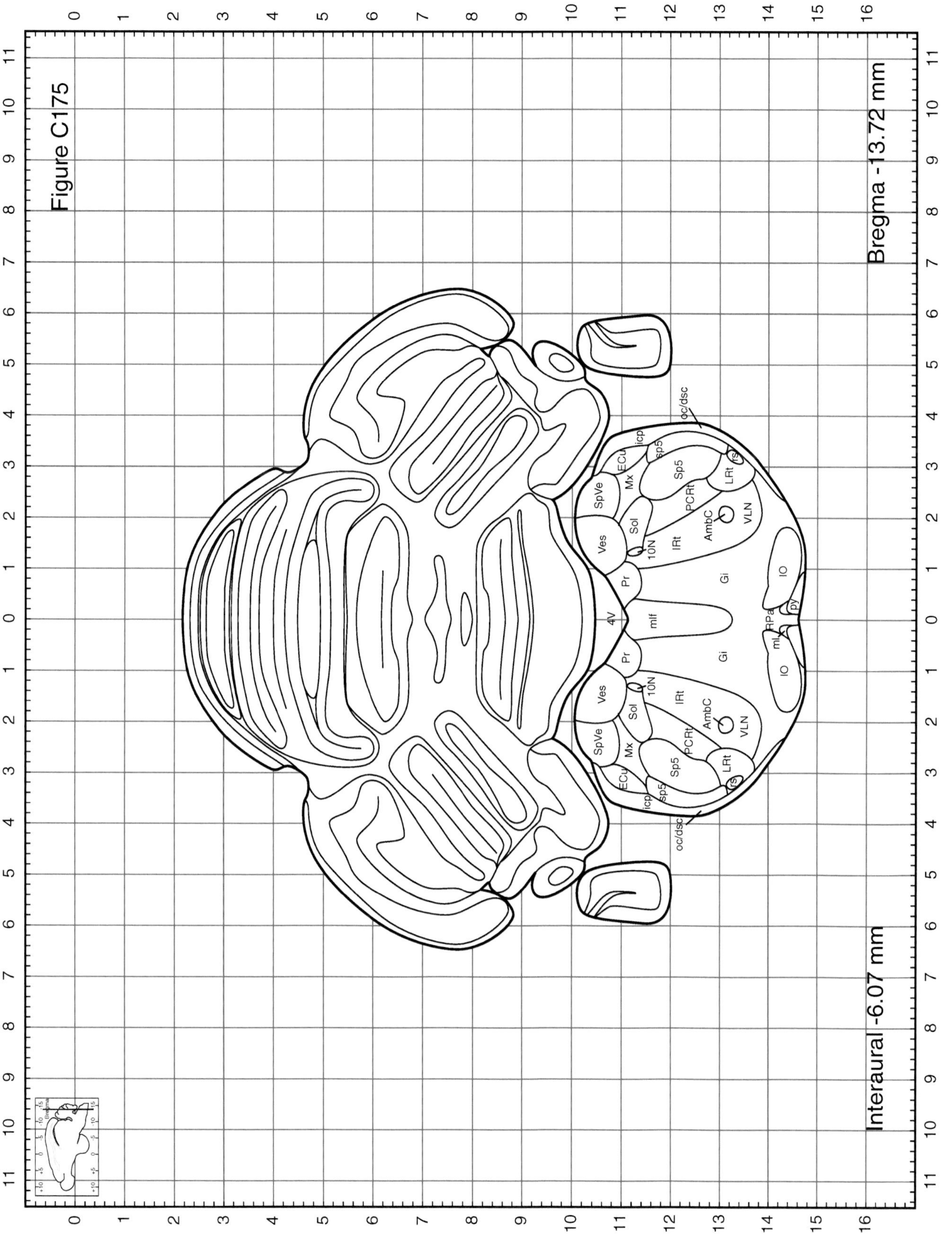
Figure C175
Bregma -13.72 mm
Interaural -6.07 mm
oc/dsc
icp
sp5
ECu
Mx
Sp5
LRt
rs
SpVe
PCRt
Sol
AmbC
VLN
Ves
10N
IRt
Pr
Gi
IO
4V
mlf
py
RPa
ml

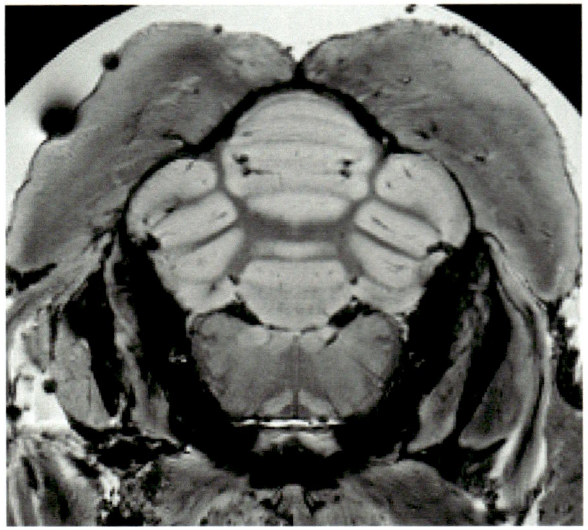

10N	dorsal motor nucleus of vagus	LRt	lateral reticular nucleus	Sol	solitary nucleus
12N	hypoglossal nucleus	ml	medial lemniscus	Sp5	spinal trigeminal nucleus
4V	4th ventricle	mlf	medial longitudinal fasciculus	sp5	spinal trigeminal tract
AmbC	ambiguus nucleus, compact part	Mx	matrix region of the medulla	SpVe	spinal vestibular nucleus
Cu	cuneate nucleus	oc/dsc	olivocerebellar tract and dorsal spinocerebellar tract	Ves	vestibular nucleus
ECu	external cuneate nucleus	PCRt	parvicellular reticular nucleus	VLN	ventrolateral reticular nucleus
Gi	gigantocellular reticular nucleus	Pr	prepositus nucleus	vsc	ventral spinocerebellar tract
icp	inferior cerebellar peduncle	py	pyramidal tract		
IO	inferior olive	RPa	raphe pallidus nucleus		
IRt	intermediate reticular nucleus	rs	rubrospinal tract		

Figure C176
Bregma -13.86 mm
Interaural -6.21 mm
oc/dsc
icp
ECu
Cu
sp5
Sp5
Mx
SpVe
Sol
Ves
PCRt
IRt
AmbC
LRt
rs
VLN
vsc
10N
12N
Pr
4V
mlf
Gi
IO
RPa
py
ml

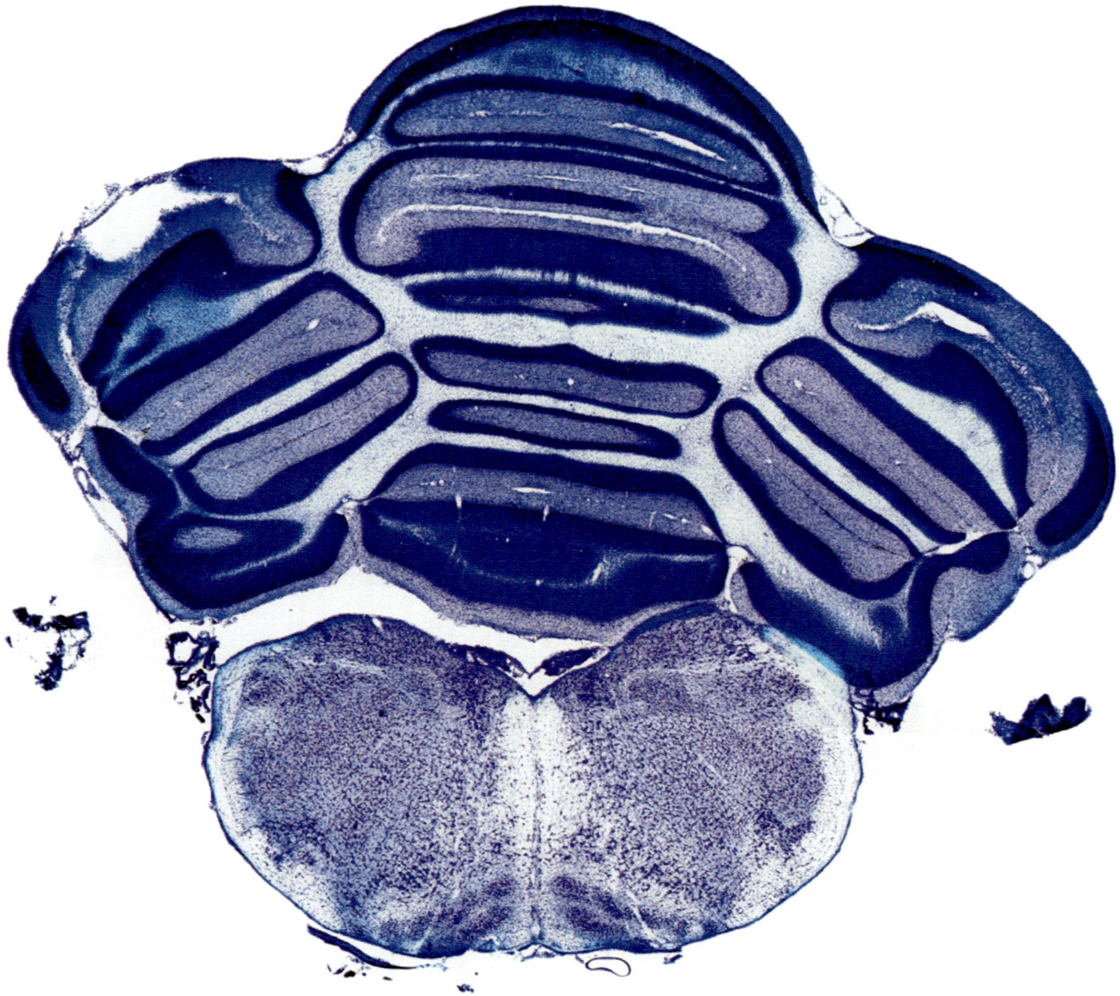

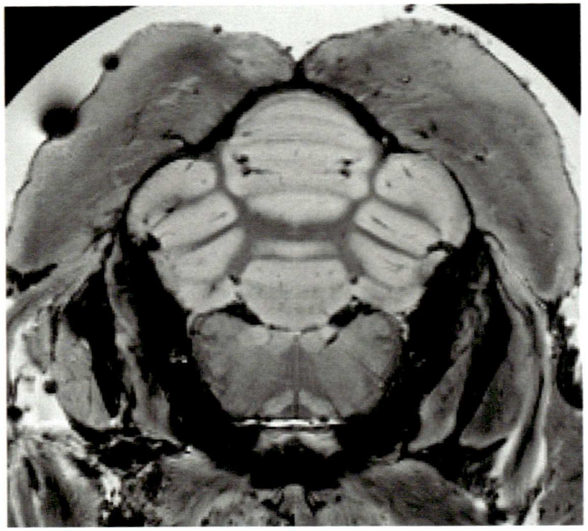

10N	dorsal motor nucleus of vagus	IRt	intermediate reticular nucleus	RPa	raphe pallidus nucleus
12N	hypoglossal nucleus	LRt	lateral reticular nucleus	rs	rubrospinal tract
4V	4th ventricle	ml	medial lemniscus	Sol	solitary nucleus
AmbC	ambiguus nucleus, compact part	mlf	medial longitudinal fasciculus	Sp5	spinal trigeminal nucleus
Cu	cuneate nucleus	Mx	matrix region of the medulla	sp5	spinal trigeminal tract
ECu	external cuneate nucleus	oc/dsc	olivocerebellar tract and dorsal spinocerebellar tract	SpVe	spinal vestibular nucleus
Gi	gigantocellular reticular nucleus	PCRt	parvicellular reticular nucleus	Ves	vestibular nucleus
icp	inferior cerebellar peduncle	Pr	prepositus nucleus	VLN	ventrolateral reticular nucleus
IO	inferior olive	py	pyramidal tract	vsc	ventral spinocerebellar tract

Figure C177
Bregma -13.99 mm
Interaural -6.34 mm
4V
Pr
12N
10N
Sol
Ves
SpVe
Mx
Cu
ECu
icp
sp5
Sp5
PCRt
IRt
AmbC
LRt
rs
VLN
vsc
Gi
mlf
IO
RPa
py
ml
oc/dsc

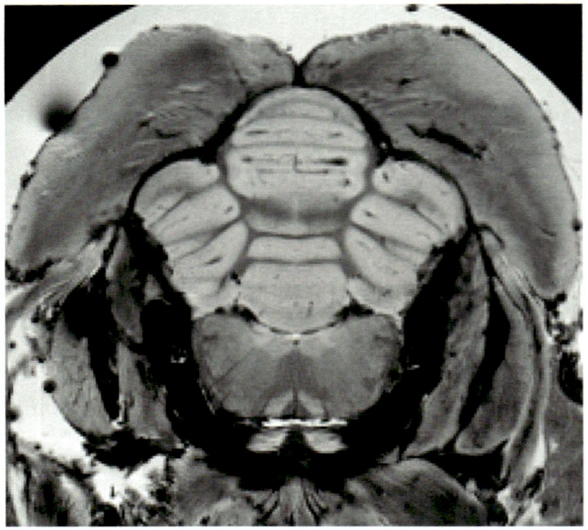

10N	dorsal motor nucleus of vagus
12N	hypoglossal nucleus
4V	4th ventricle
AmbC	ambiguus nucleus, compact part
Cu	cuneate nucleus
ECu	external cuneate nucleus
Gi	gigantocellular reticular nucleus
icp	inferior cerebellar peduncle
IO	inferior olive
IRt	intermediate reticular nucleus
LRt	lateral reticular nucleus
mlf	medial longitudinal fasciculus
Mx	matrix region of the medulla
oc/dsc	olivocerebellar tract and dorsal spinocerebellar tract
PCRt	parvicellular reticular nucleus
Pr	prepositus nucleus
py	pyramidal tract
RPa	raphe pallidus nucleus
rs	rubrospinal tract
Sol	solitary nucleus
Sp5	spinal trigeminal nucleus
sp5	spinal trigeminal tract
SpVe	spinal vestibular nucleus
Ves	vestibular nucleus
VLN	ventrolateral reticular nucleus
vsc	ventral spinocerebellar tract

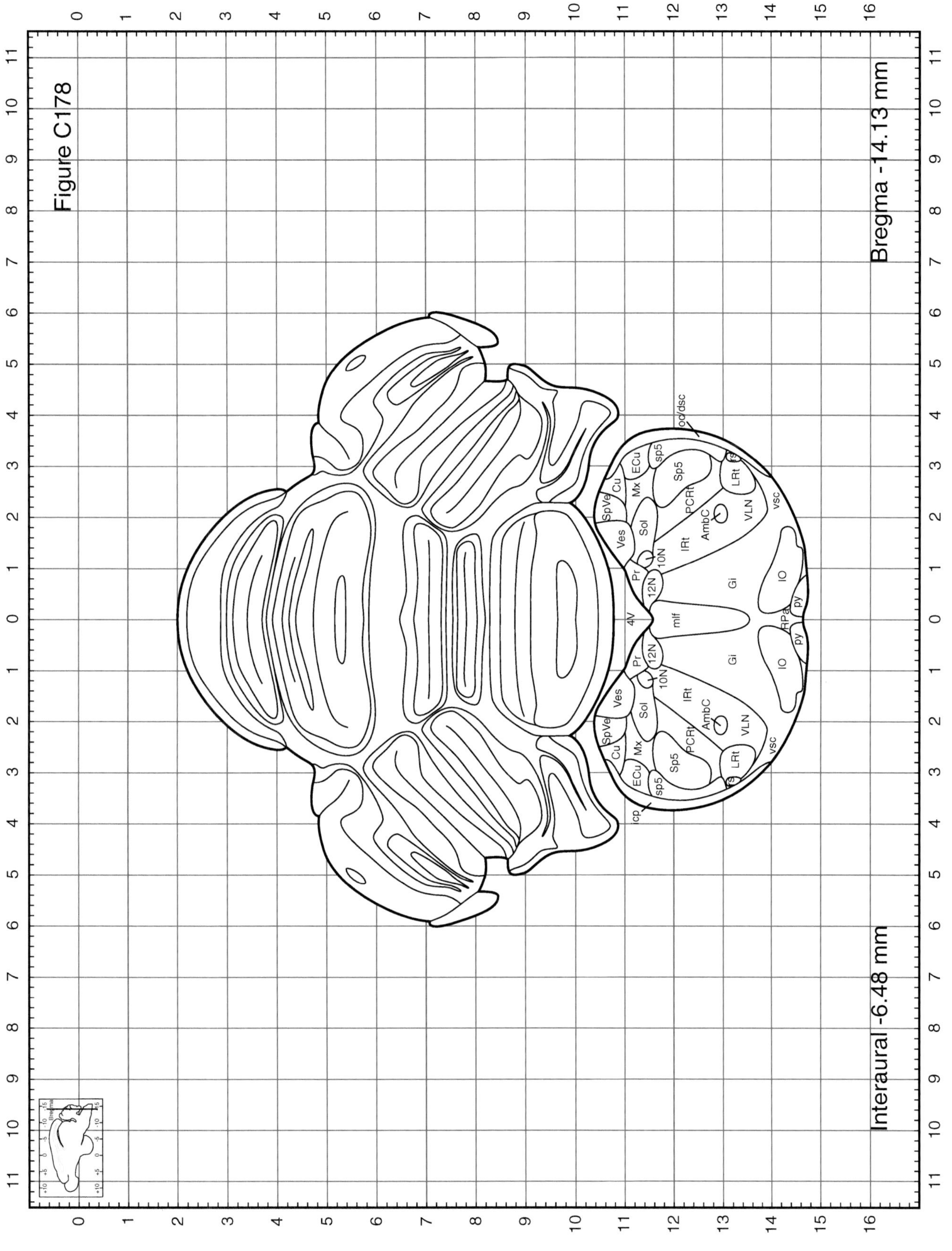
Figure C178
Bregma -14.13 mm
Interaural -6.48 mm
icp
ocd/dsc
SpVe
Cu
ECu
Mx
sp5
Sp5
PCRt
AmbC
LRt
VLN
vsc
Ves
Sol
10N
12N
IRt
Pr
4V
mlf
Gi
IO
py
RPa

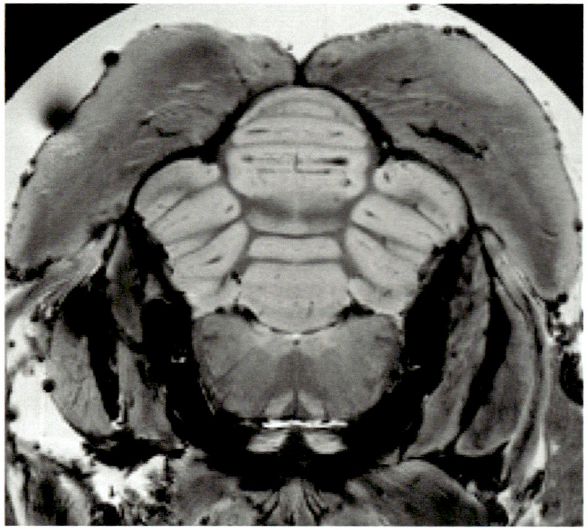

10N	dorsal motor nucleus of vagus		IRt	intermediate reticular nucleus		rs	rubrospinal tract
12N	hypoglossal nucleus		LRt	lateral reticular nucleus		Sol	solitary nucleus
4V	4th ventricle		mlf	medial longitudinal fasciculus		Sp5	spinal trigeminal nucleus
AmbC	ambiguus nucleus, compact part		Mx	matrix region of the medulla		sp5	spinal trigeminal tract
Cu	cuneate nucleus		oc/dsc	olivocerebellar tract and dorsal spinocerebellar tract		SpVe	spinal vestibular nucleus
ECu	external cuneate nucleus		PCRt	parvicellular reticular nucleus		Ves	vestibular nucleus
Gi	gigantocellular reticular nucleus		Pr	prepositus nucleus		VLN	ventrolateral reticular nucleus
icp	inferior cerebellar peduncle		py	pyramidal tract		vsc	ventral spinocerebellar tract
IO	inferior olive		RPa	raphe pallidus nucleus			

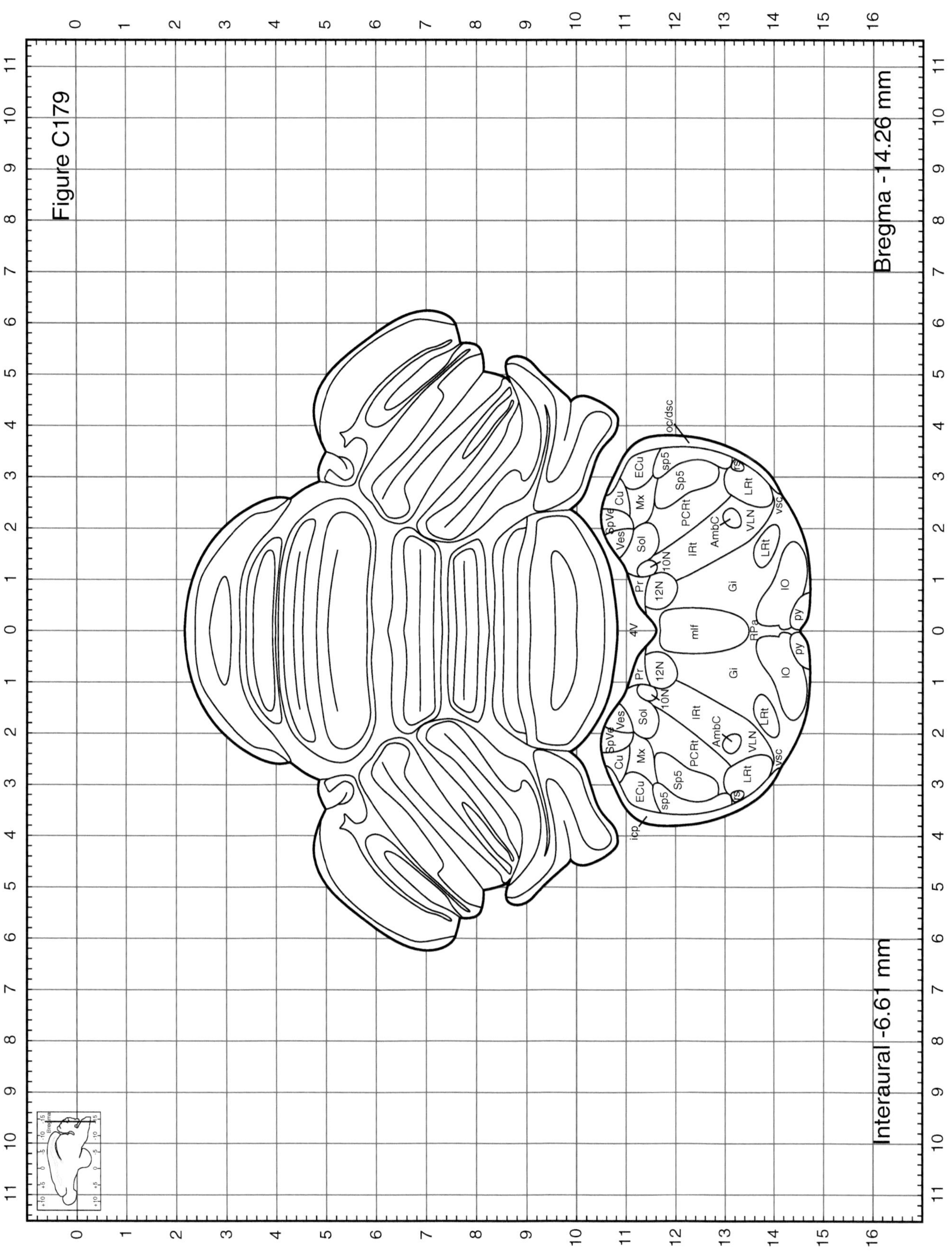
Figure C179
Bregma -14.26 mm
Interaural -6.61 mm
oc/dsc
ECu
sp5
Cu
Mx
Sp5
rs
SpVe
Ves
Sol
PCRt
AmbC
LRt
VLN
vsc
10N
IRt
Pr
12N
Gi
IO
4V
mlf
RPa
py
icp

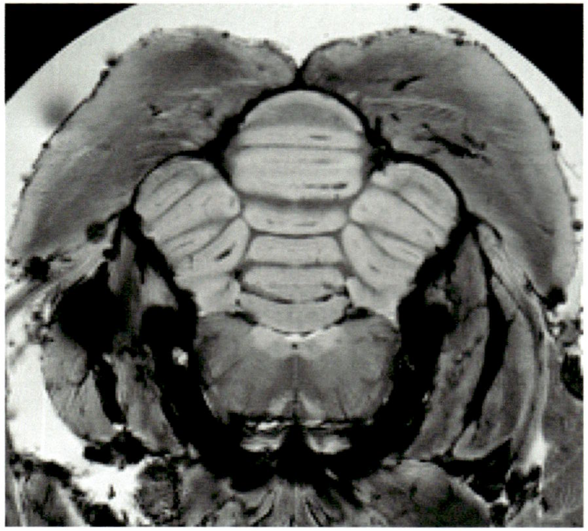

10N	dorsal motor nucleus of vagus
12N	hypoglossal nucleus
4V	4th ventricle
AmbC	ambiguus nucleus, compact part
Cu	cuneate nucleus
ECu	external cuneate nucleus
Gi	gigantocellular reticular nucleus
icp	inferior cerebellar peduncle
IO	inferior olive
IRt	intermediate reticular nucleus
LRt	lateral reticular nucleus
mlf	medial longitudinal fasciculus
Mx	matrix region of the medulla
oc/dsc	olivocerebellar tract and dorsal spinocerebellar tract
PCRt	parvicellular reticular nucleus
Pr	prepositus nucleus
py	pyramidal tract
RPa	raphe pallidus nucleus
rs	rubrospinal tract
Sol	solitary nucleus
Sp5	spinal trigeminal nucleus
sp5	spinal trigeminal tract
SpVe	spinal vestibular nucleus
Ves	vestibular nucleus
VLN	ventrolateral reticular nucleus
vsc	ventral spinocerebellar tract

Figure C180
Bregma -14.39 mm
Interaural -6.74 mm
oc/dsc
ECu
sp5
Cu
Mx
Sp5
Ves
SpVe
Sol
PCRt
AmbC
VLN
LRt
vsc
fs
10N
IRt
Pr
12N
Gi
IO
RPa
py
4V
mlf
icp

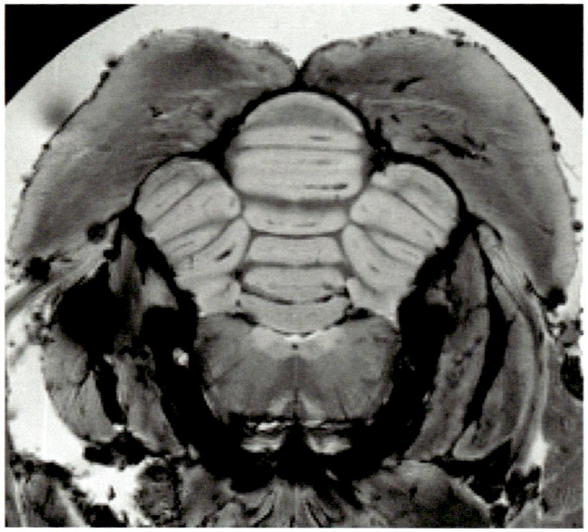

10N	dorsal motor nucleus of vagus	IO	inferior olive	RPa	raphe pallidus nucleus
12N	hypoglossal nucleus	IRt	intermediate reticular nucleus	rs	rubrospinal tract
4V	4th ventricle	LRt	lateral reticular nucleus	Sol	solitary nucleus
AmbC	ambiguus nucleus, compact part	mlf	medial longitudinal fasciculus	Sp5	spinal trigeminal nucleus
Cu	cuneate nucleus	Mx	matrix region of the medulla	sp5	spinal trigeminal tract
ECu	external cuneate nucleus	oc/dsc	olivocerebellar tract and dorsal spinocerebellar tract	VLN	ventrolateral reticular nucleus
Gi	gigantocellular reticular nucleus	PCRt	parvicellular reticular nucleus	vsc	ventral spinocerebellar tract
Gr	gracile nucleus	py	pyramidal tract		

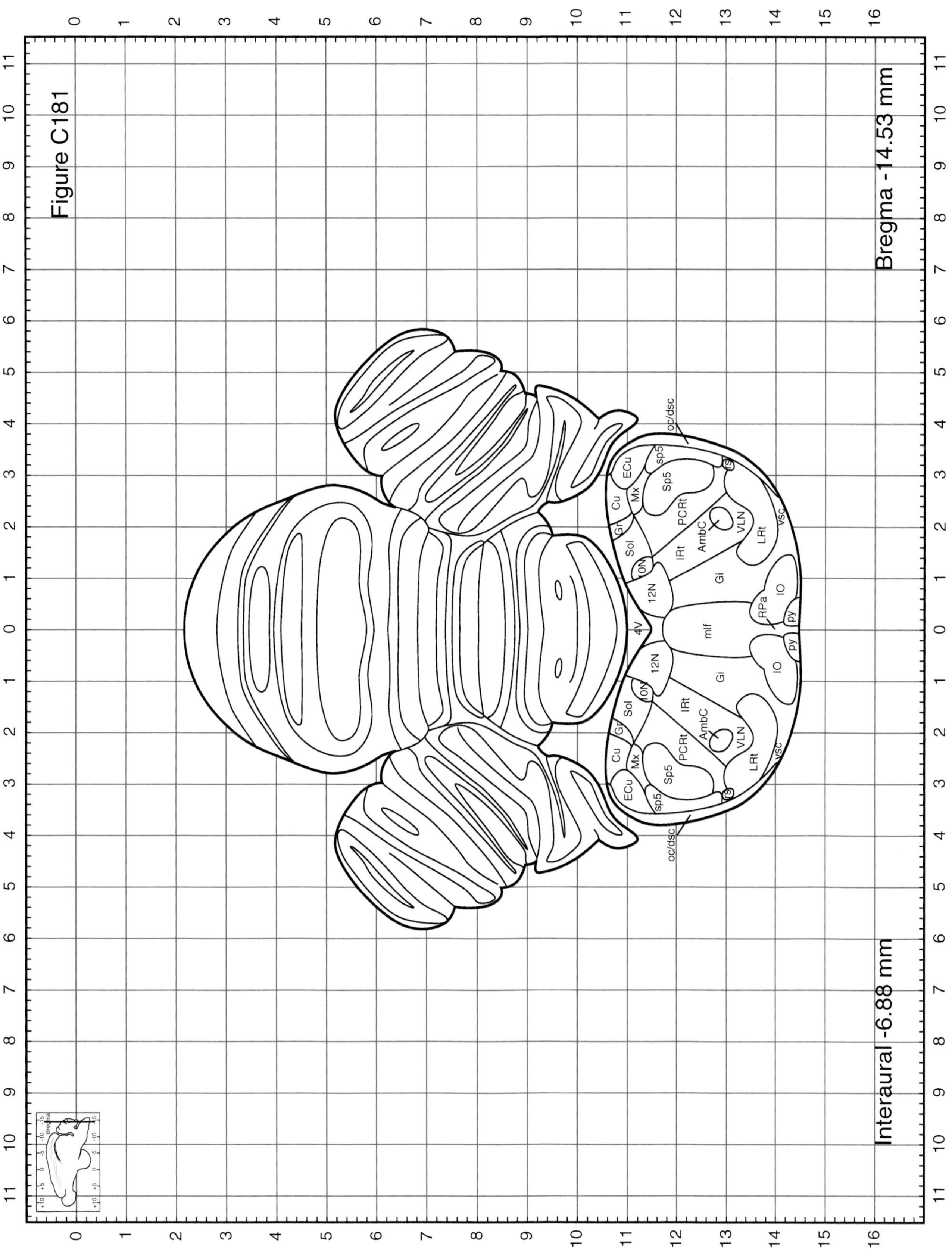

Figure C181
Bregma -14.53 mm
Interaural -6.88 mm
oc/dsc
sp5
Sp5
ECu
Cu
Mx
Gr
Sol
10N
12N
4V
PCRt
IRt
AmbC
VLN
LRt
vsc
rs
Gi
mlf
IO
RPa
py

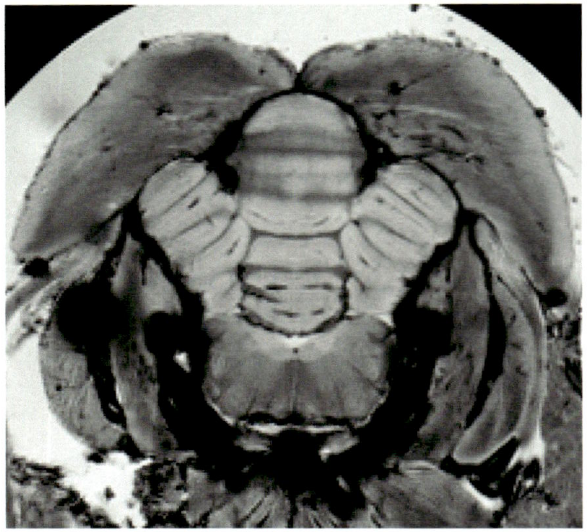

10N	dorsal motor nucleus of vagus
12N	hypoglossal nucleus
4V	4th ventricle
AmbC	ambiguus nucleus, compact part
Cu	cuneate nucleus
ECu	external cuneate nucleus
Gi	gigantocellular reticular nucleus
Gr	gracile nucleus
IO	inferior olive
IRt	intermediate reticular nucleus
LRt	lateral reticular nucleus
mlf	medial longitudinal fasciculus
Mx	matrix region of the medulla
oc/dsc	olivocerebellar tract and dorsal spinocerebellar tract
PCRt	parvicellular reticular nucleus
py	pyramidal tract
RPa	raphe pallidus nucleus
rs	rubrospinal tract
Sol	solitary nucleus
Sp5	spinal trigeminal nucleus
sp5	spinal trigeminal tract
VLN	ventrolateral reticular nucleus
vsc	ventral spinocerebellar tract

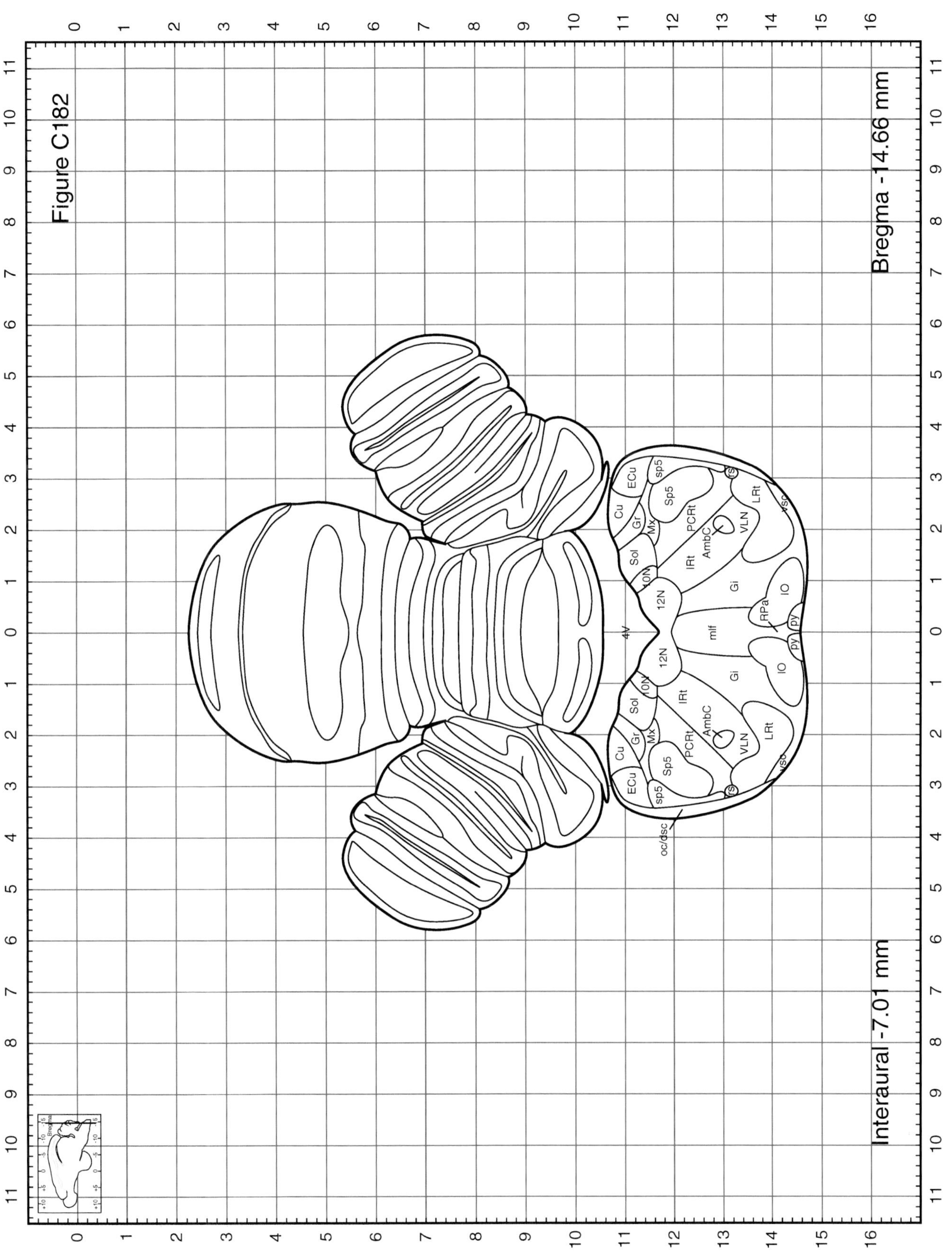
Figure C182
Bregma -14.66 mm
Interaural -7.01 mm
Cu
ECu
sp5
Sp5
Gr
Mx
Sol
10N
12N
4V
PCRt
IRt
AmbC
VLN
LRt
rs
vsc
Gi
IO
RPa
py
mlf
oc/dsc

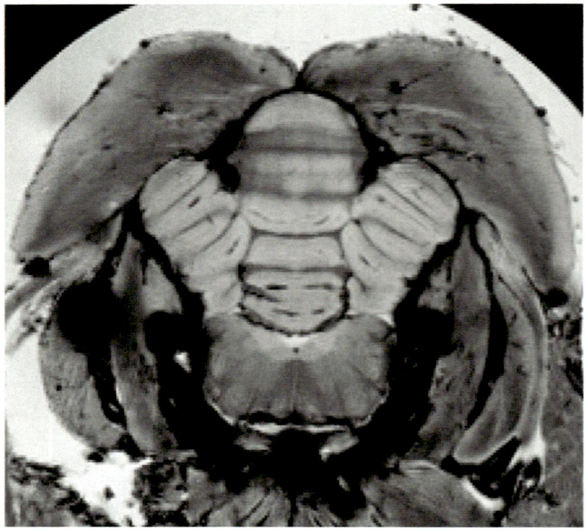

10N dorsal motor nucleus of vagus
12N hypoglossal nucleus
4V 4th ventricle
AmbC ambiguus nucleus, compact part
Cu cuneate nucleus
ECu external cuneate nucleus
Gi gigantocellular reticular nucleus
Gr gracile nucleus
IO inferior olive
IRt intermediate reticular nucleus
LRt lateral reticular nucleus
mlf medial longitudinal fasciculus
Mx matrix region of the medulla
oc/dsc olivocerebellar tract and dorsal spinocerebellar tract
PCRt parvicellular reticular nucleus
py pyramidal tract
RPa raphe pallidus nucleus
rs rubrospinal tract
Sol solitary nucleus
Sp5 spinal trigeminal nucleus
sp5 spinal trigeminal tract
VLN ventrolateral reticular nucleus
vsc ventral spinocerebellar tract

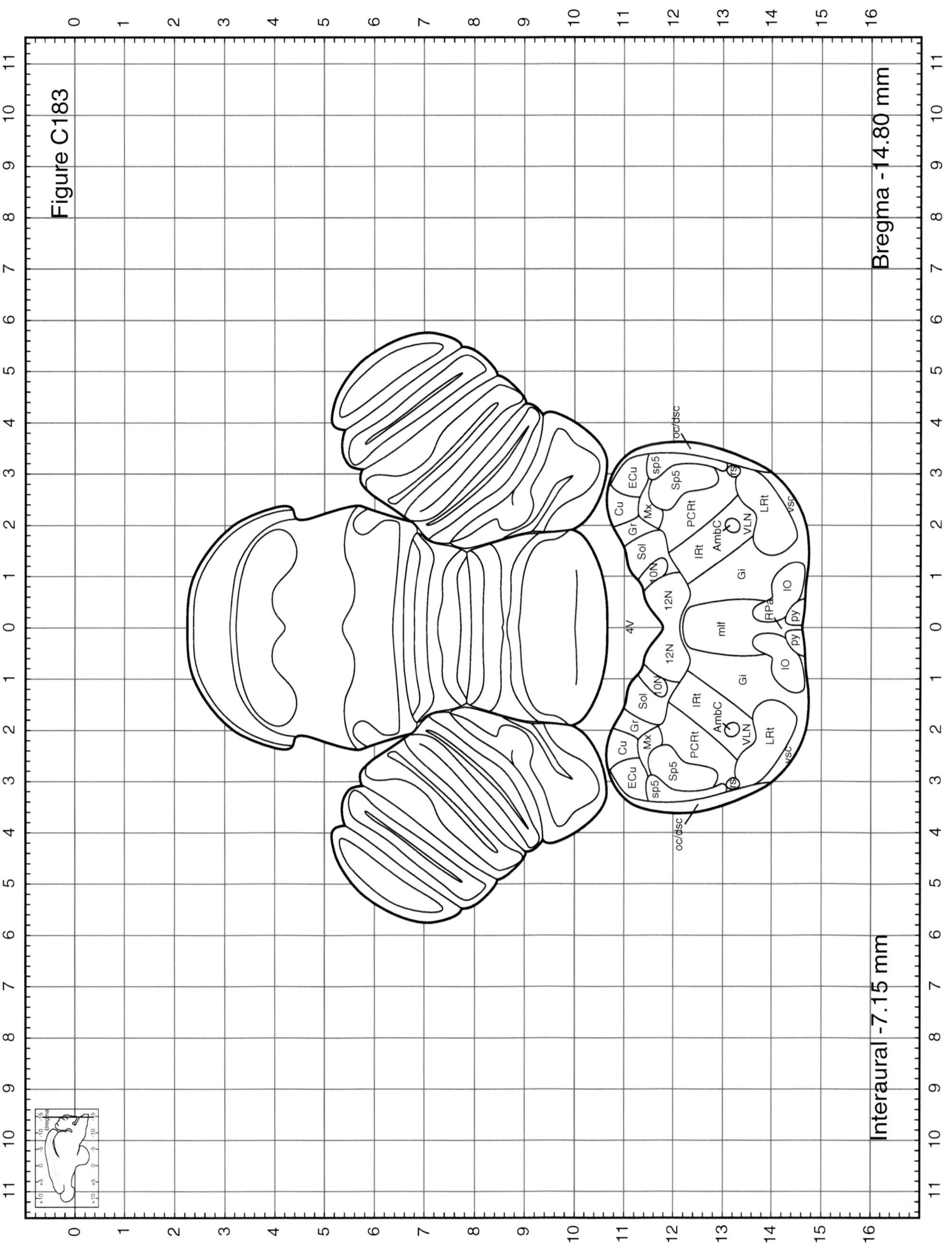

Figure C183
Bregma -14.80 mm
Interaural -7.15 mm
oc/dsc
sp5
ECu
Sp5
rs
Cu
Mx
PCRt
LRt
vsc
Gr
AmbC
VLN
Sol
IRt
10N
Gi
IO
12N
RPa
4V
mlf
py

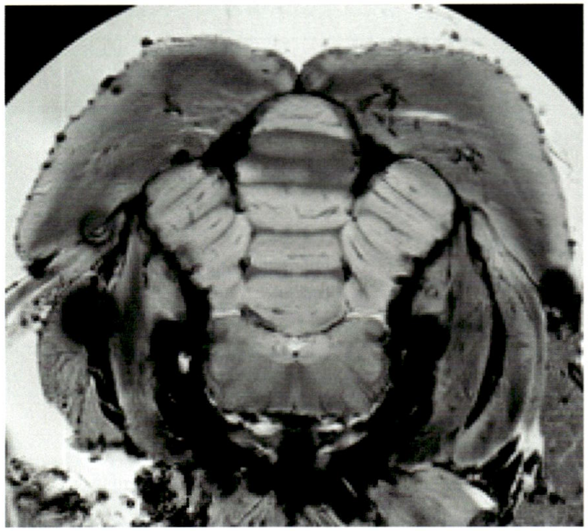

10N	dorsal motor nucleus of vagus
12N	hypoglossal nucleus
4V	4th ventricle
AmbC	ambiguus nucleus, compact part
Cu	cuneate nucleus
ECu	external cuneate nucleus
Gi	gigantocellular reticular nucleus
Gr	gracile nucleus
IO	inferior olive
IRt	intermediate reticular nucleus
LRt	lateral reticular nucleus
mlf	medial longitudinal fasciculus
Mx	matrix region of the medulla
oc/dsc	olivocerebellar tract and dorsal spinocerebellar tract
PCRt	parvicellular reticular nucleus
py	pyramidal tract
Ro	nucleus of Roller
RPa	raphe pallidus nucleus
rs	rubrospinal nucleus
Sol	solitary tract
Sp5	spinal trigeminal nucleus
sp5	spinal trigeminal tract
VLN	ventrolateral reticular nucleus
vsc	ventral spinocerebellar tract

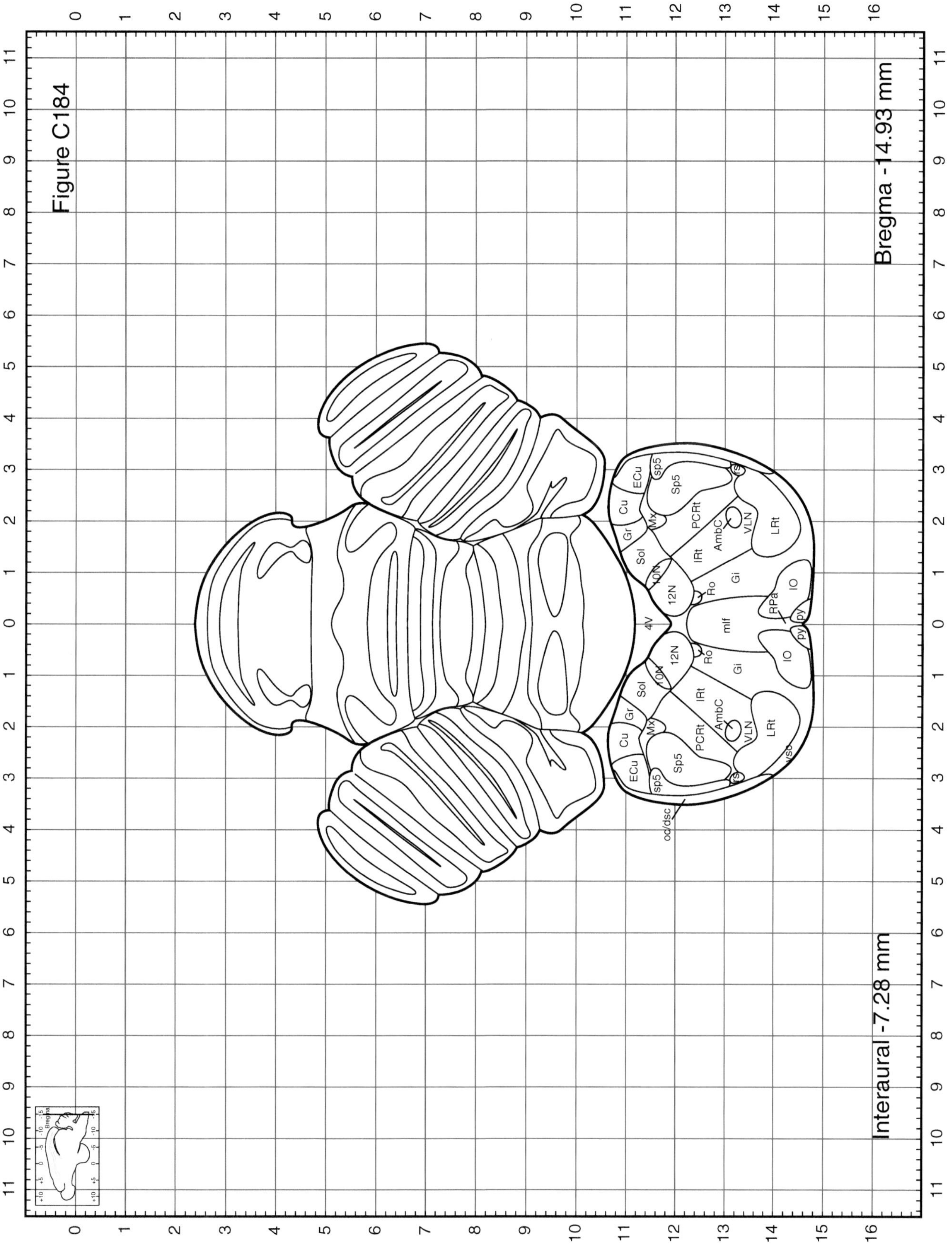
Figure C184
Bregma -14.93 mm
Interaural -7.28 mm
ECu
sp5
Sp5
Cu
Mx
PCRt
AmbC
VLN
LRt
Gr
Sol
IRt
10N
12N
Ro
Gi
IO
RPa
py
4V
mlf
vsc
oc/dsc

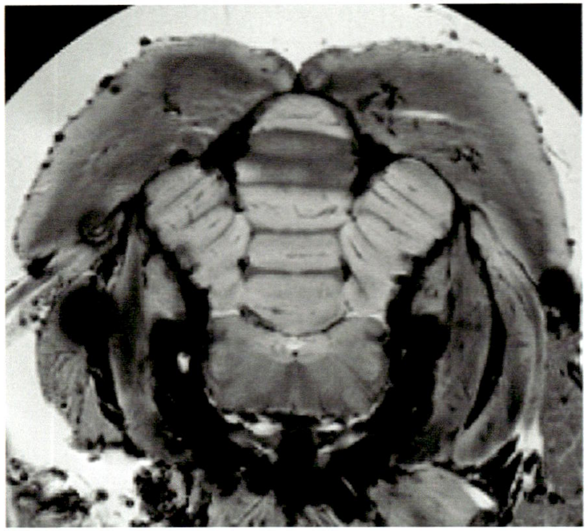

10N	dorsal motor nucleus of vagus
12N	hypoglossal nucleus
4V	4th ventricle
AmbC	ambiguus nucleus, compact part
Cu	cuneate nucleus
ECu	external cuneate nucleus
Gi	gigantocellular reticular nucleus
Gr	gracile nucleus
IO	inferior olive
IRt	intermediate reticular nucleus
LRt	lateral reticular nucleus
mlf	medial longitudinal fasciculus
Mx	matrix region of the medulla
oc/dsc	olivocerebellar tract and dorsal spinocerebellar tract
PCRt	parvicellular reticular nucleus
py	pyramidal tract
Ro	nucleus of Roller
RPa	raphe pallidus nucleus
rs	rubrospinal tract
Sol	solitary nuclues
Sp5	spinal trigeminal nucleus
sp5	spinal trigeminal tract
vsc	ventral spinocerebellar tract

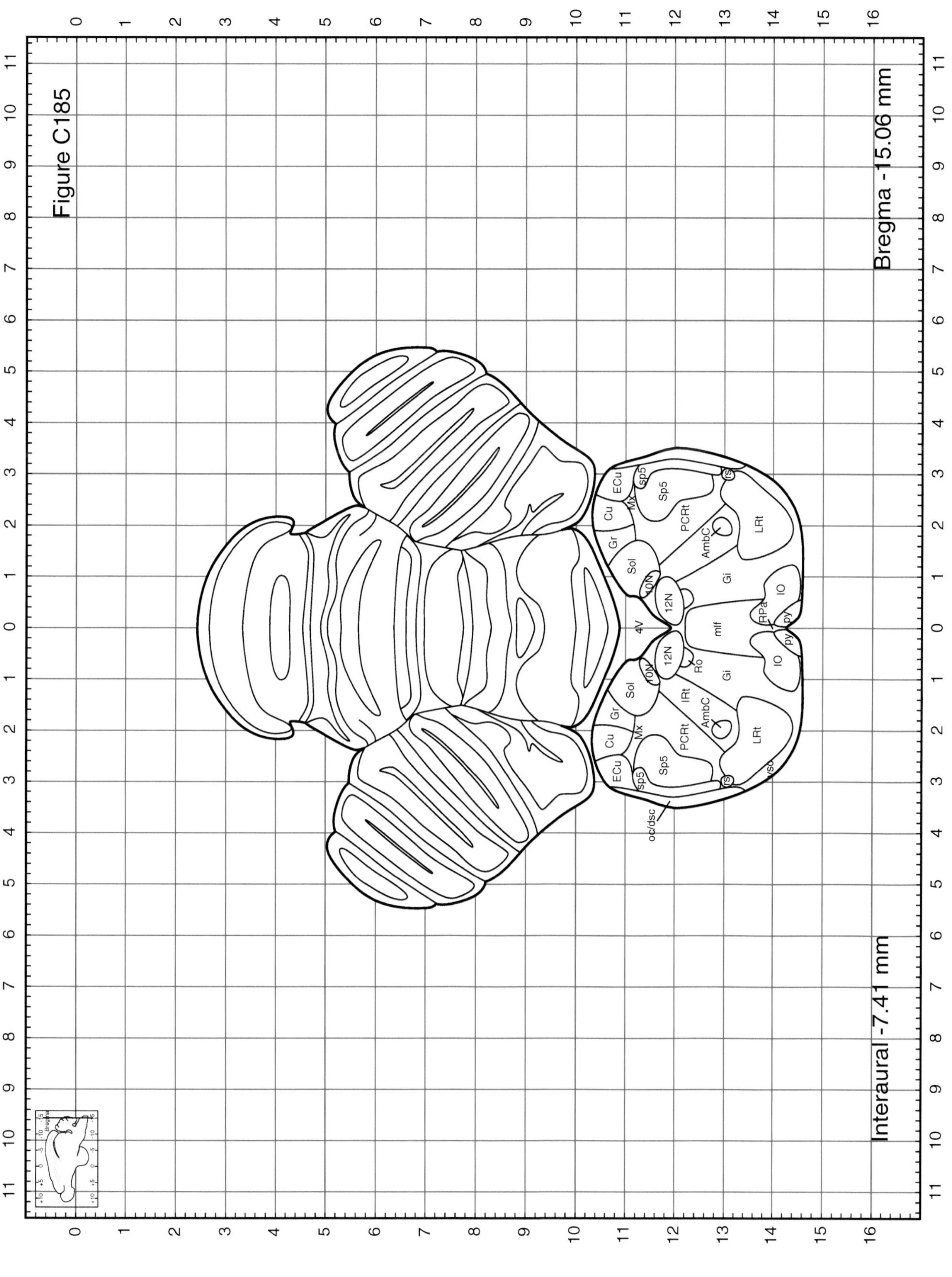
Figure C185
Bregma -15.06 mm
Interaural -7.41 mm
ECu
sp5
Sp5
Cu
Mx
PCRt
AmbC
LRt
Gr
Sol
Gi
10N
12N
IO
4V
mlf
RPa
py
Ro
IRt
rs
vsc
oc/dsc

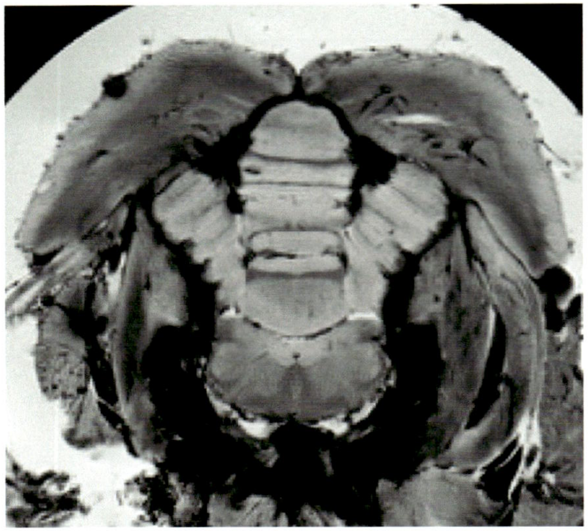

10N	dorsal motor nucleus of vagus
12n	root of hypoglossal nerve
12N	hypoglossal nucleus
4V	4th ventricle
AmbC	ambiguus nucleus, compact part
AP	area postrema
Cu	cuneate nucleus
ECu	external cuneate nucleus
Gi	gigantocellular reticular nucleus
Gr	gracile nucleus
IO	inferior olive
IRt	intermediate reticular nucleus
LRt	lateral reticular nucleus
mlf	medial longitudinal fasciculus
mlx	medial lemniscus decussation
Mx	matrix region of the medulla
oc/dsc	olivocerebellar tract and dorsal spinocerebellar tract
PCRt	parvicellular reticular nucleus
py	pyramidal tract
Ro	nucleus of Roller
rs	rubrospinal tract
Sol	solitary nucleus
Sp5	spinal trigeminal nucleus
sp5	spinal trigeminal tract
vsc	ventral spinocerebellar tract

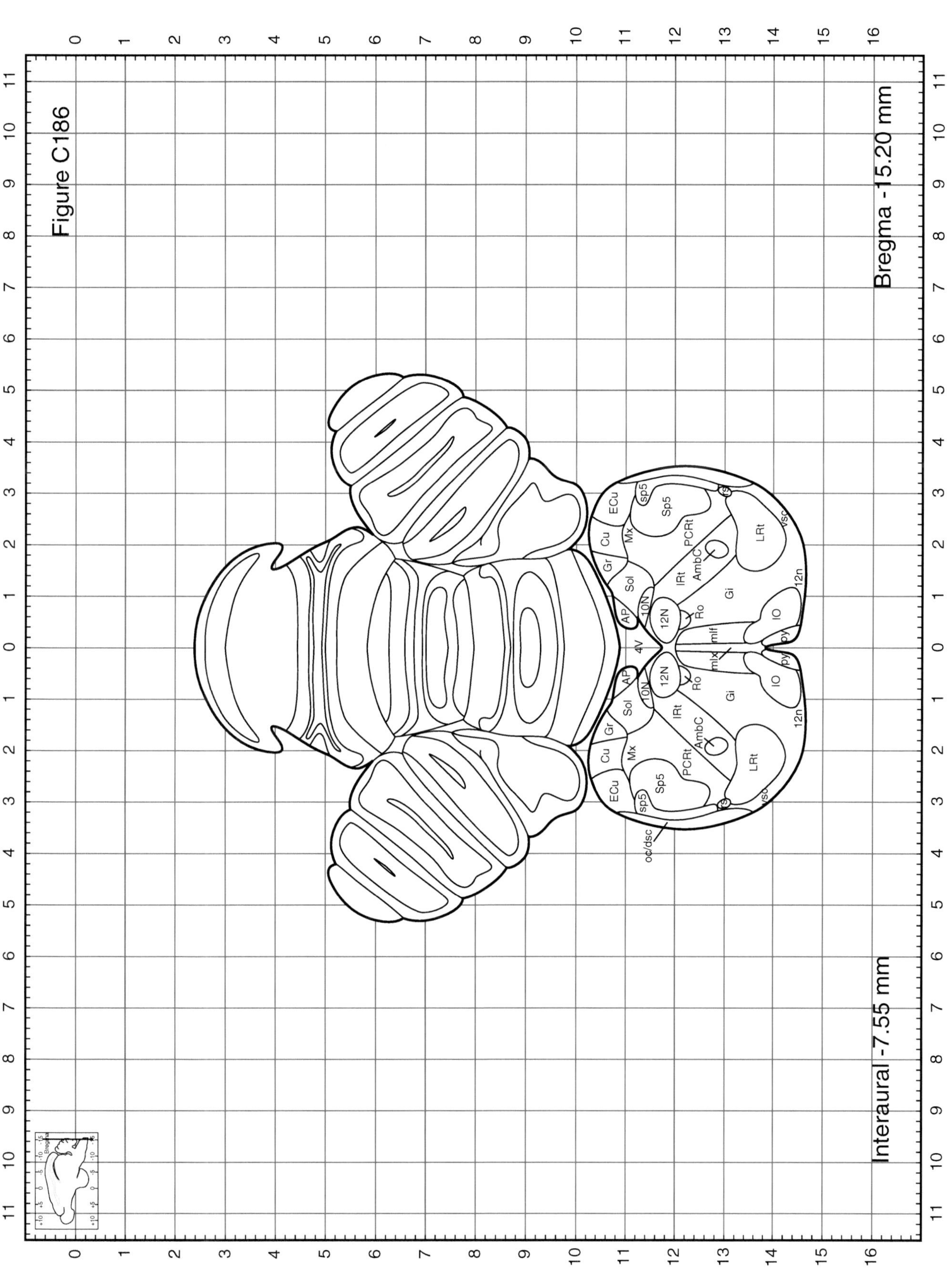
Figure C186
Bregma -15.20 mm
Interaural -7.55 mm
ECu
sp5
Sp5
Cu
Mx
PCRt
Gr
AmbC
IRt
Sol
LRt
vsc
10N
12N
Ro
Gi
AP
12n
IO
4V
mlf
mlx
py
ts
oc/dsc
Bregma

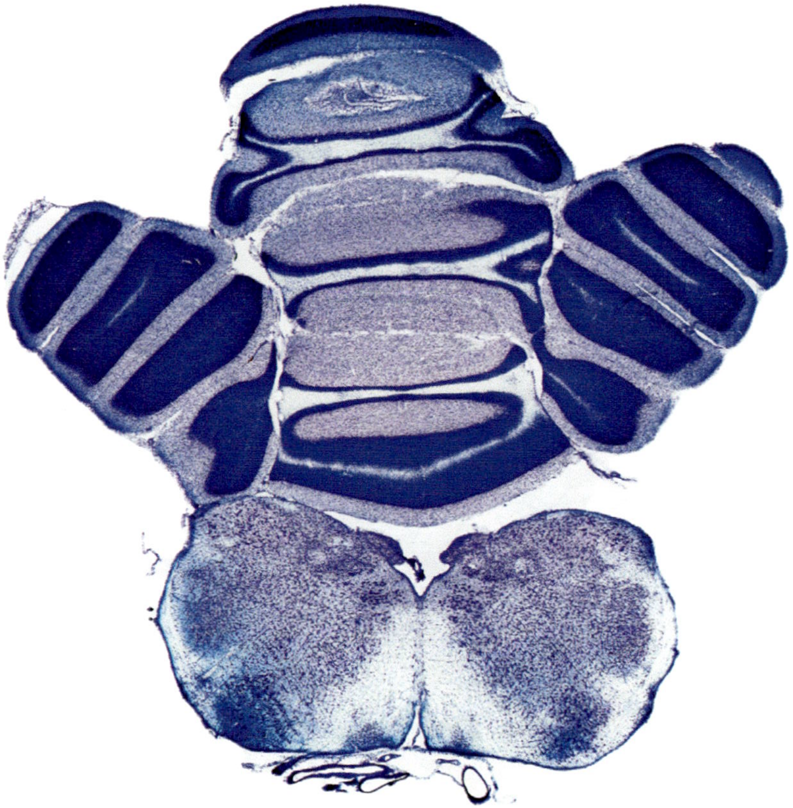

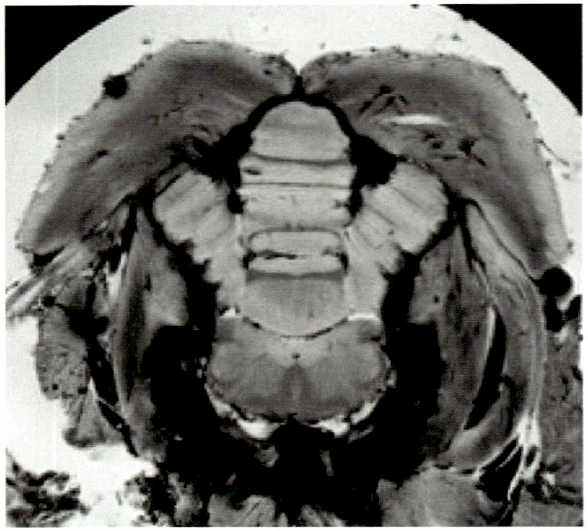

10N	dorsal motor nucleus of vagus
12n	root of hypoglossal nerve
12N	hypoglossal nucleus
4V	4th ventricle
AmbC	ambiguus nucleus, compact part
AP	area postrema
Cu	cuneate nucleus
ECu	external cuneate nucleus
Gi	gigantocellular reticular nucleus
Gr	gracile nucleus
IO	inferior olive
IRt	intermediate reticular nucleus
LRt	lateral reticular nucleus
mlf	medial longitudinal fasciculus
mlx	medial lemniscus decussation
Mx	matrix region of the medulla
oc/dsc	olivocerebellar tract and dorsal spinocerebellar tract
PCRt	parvicellular reticular nucleus
py	pyramidal tract
Ro	nucleus of Roller
rs	rubrospinal tract
Sol	solitary nucleus
Sp5	spinal trigeminal nucleus
sp5	spinal trigeminal tract
vsc	ventral spinocerebellar tract

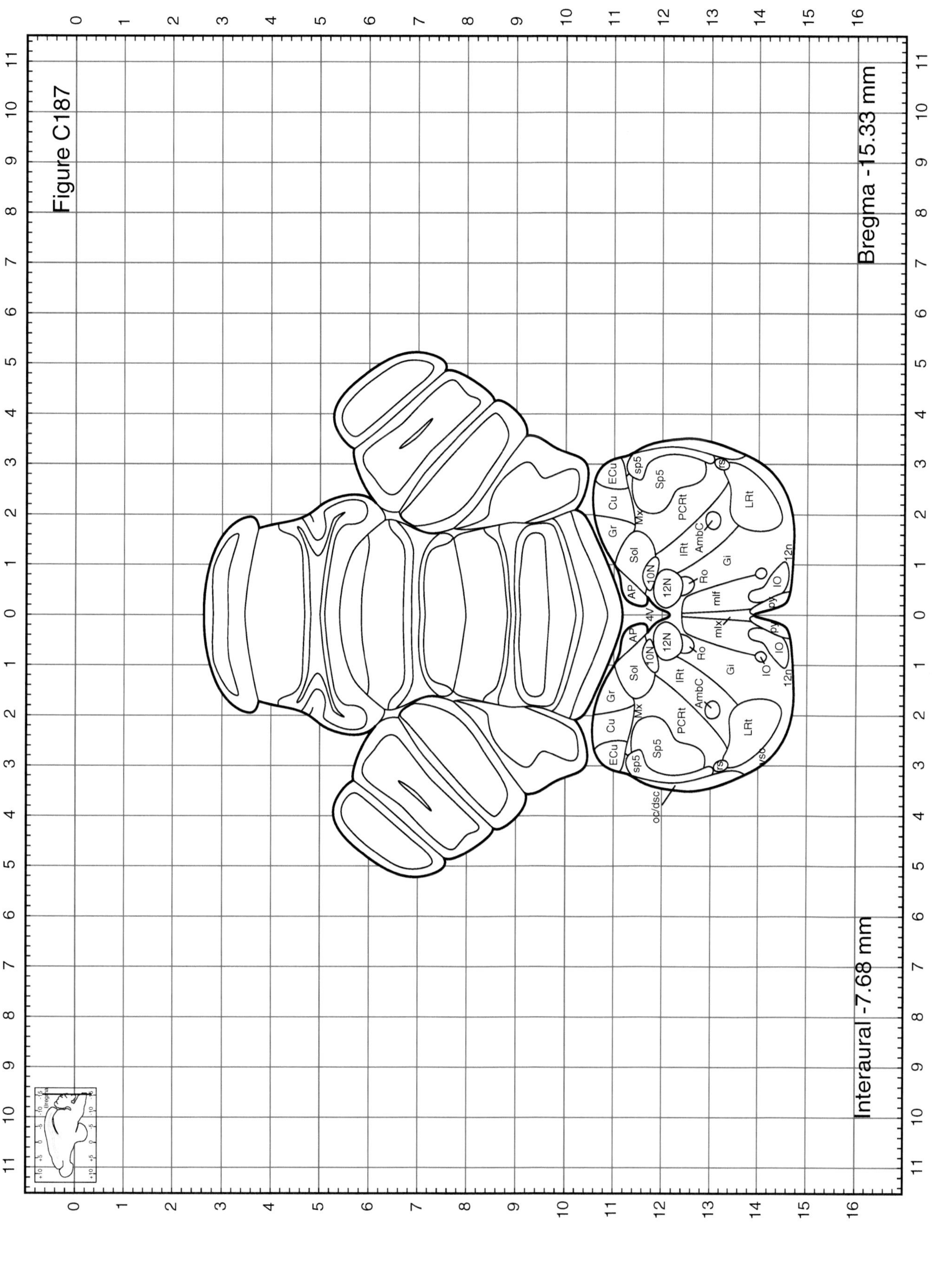
Figure C187
Bregma -15.33 mm
Interaural -7.68 mm
ECu
sp5
Sp5
Cu
Gr
Mx
PCRt
LRt
AmbC
IRt
Sol
10N
12N
AP
4V
Ro
mlf
mlx
Gi
IO
py
12n
rs
vsc
oc/dsc

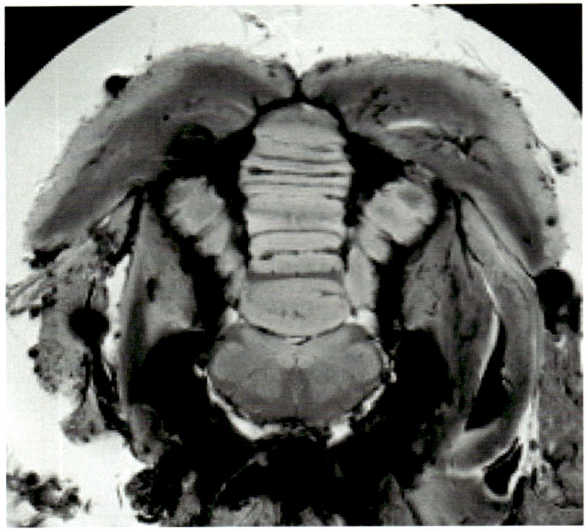

10N dorsal motor nucleus of vagus
12n root of hypoglossal nerve
12N hypoglossal nucleus
Amb ambiguus nucleus
AP area postrema
CC central canal
Cu cuneate nucleus
cu cuneate fasciculus
ECu external cuneate nucleus

Gr gracile nucleus
IO inferior olive
IRt intermediate reticular nucleus
LRt lateral reticular nucleus
MdD medullary reticular nucleus, dorsal part
MdV medullary reticular nucleus, ventral part
mlf medial longitudinal fasciculus
mlx medial lemniscus decussation
Mx matrix region of the medulla

oc/dsc olivocerebellar tract and dorsal spinocerebellar tract
py pyramidal tract
pyx pyramidal decussation
Ro nucleus of Roller
rs rubrospinal tract
Sol solitary nucleus
Sp5 spinal trigeminal nucleus
sp5 spinal trigeminal tract
vsc ventral spinocerebellar tract

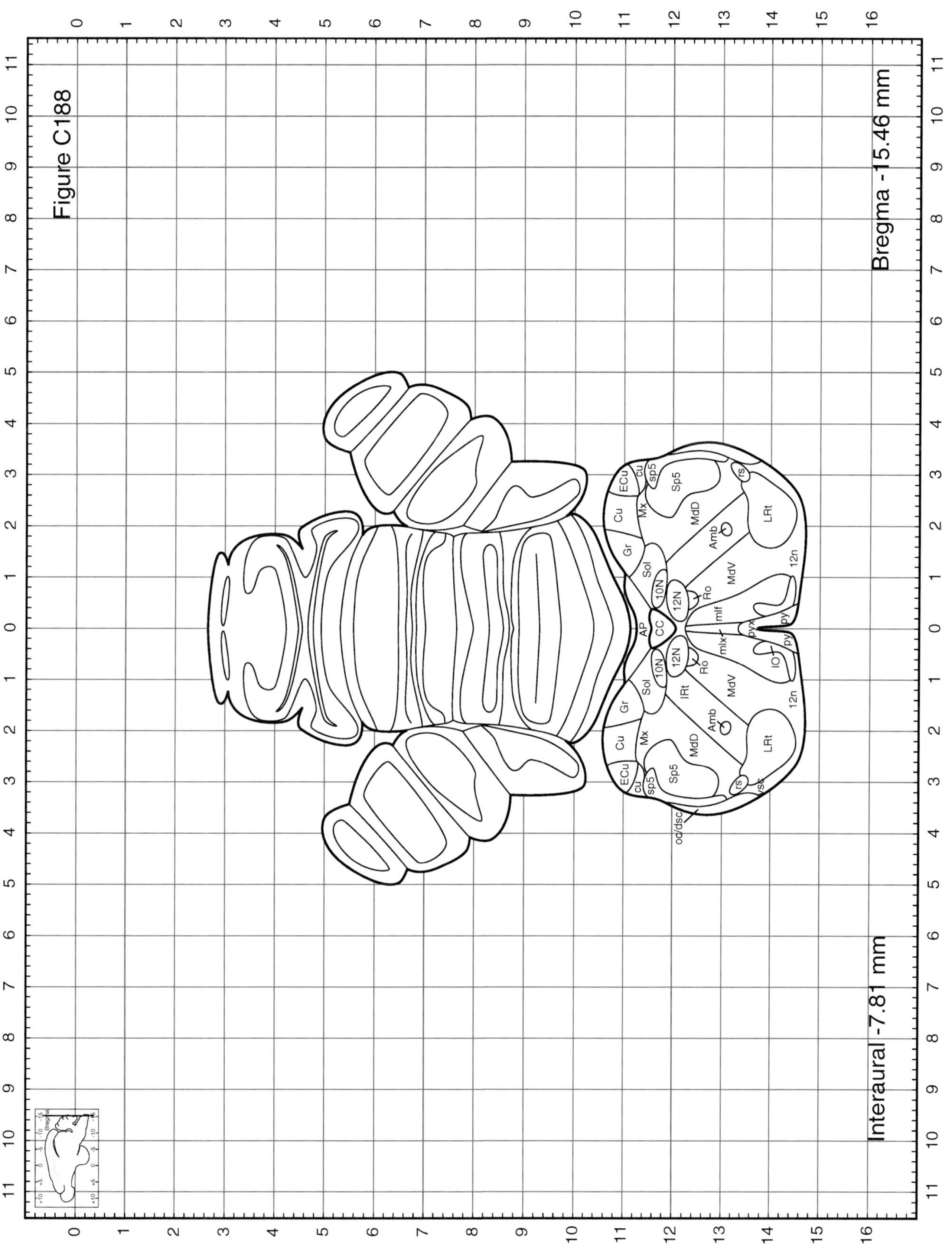
Figure C188
Bregma -15.46 mm
Interaural -7.81 mm
ECu
cu
sp5
Sp5
Cu
Mx
MdD
rs
LRt
Amb
Gr
Sol
MdV
12n
10N
12N
Ro
AP
CC
mlf
pyx
py
mlx
IO
IRt
rss
od/dsc

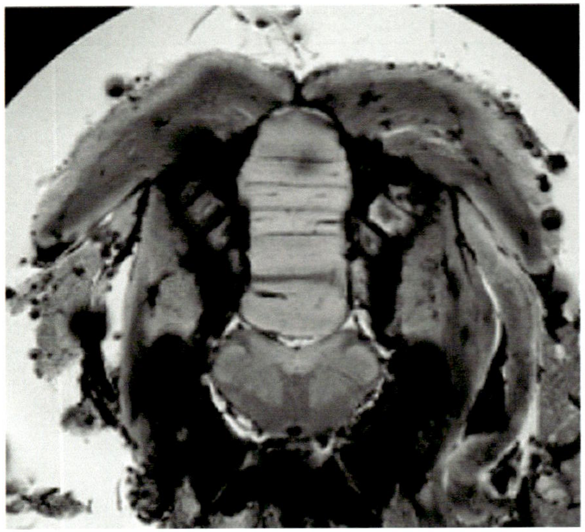

10N	dorsal motor nucleus of vagus	Gr	gracile nucleus	py	pyramidal tract
12n	root of hypoglossal nerve	IRt	intermediate reticular nucleus	pyx	pyramidal decussation
12N	hypoglossal nucleus	LRt	lateral reticular nucleus	Ro	nucleus of Roller
Amb	ambiguus nucleus	MdD	medullary reticular nucleus, dorsal part	rs	rubrospinal tract
AP	area postrema	MdV	medullary reticular nucleus, ventral part	Sol	solitary nucleus
CC	central canal	mlf	medial longitudinal fasciculus	Sp5	spinal trigeminal nucleus
Cu	cuneate nucleus	mlx	medial lemniscus decussation	sp5	spinal trigeminal tract
cu	cuneate fasciculus	Mx	matrix region of the medulla	vsc	ventral spinocerebellar tract
ECu	external cuneate nucleus	oc/dsc	olivocerebellar tract and dorsal spinocerebellar tract		

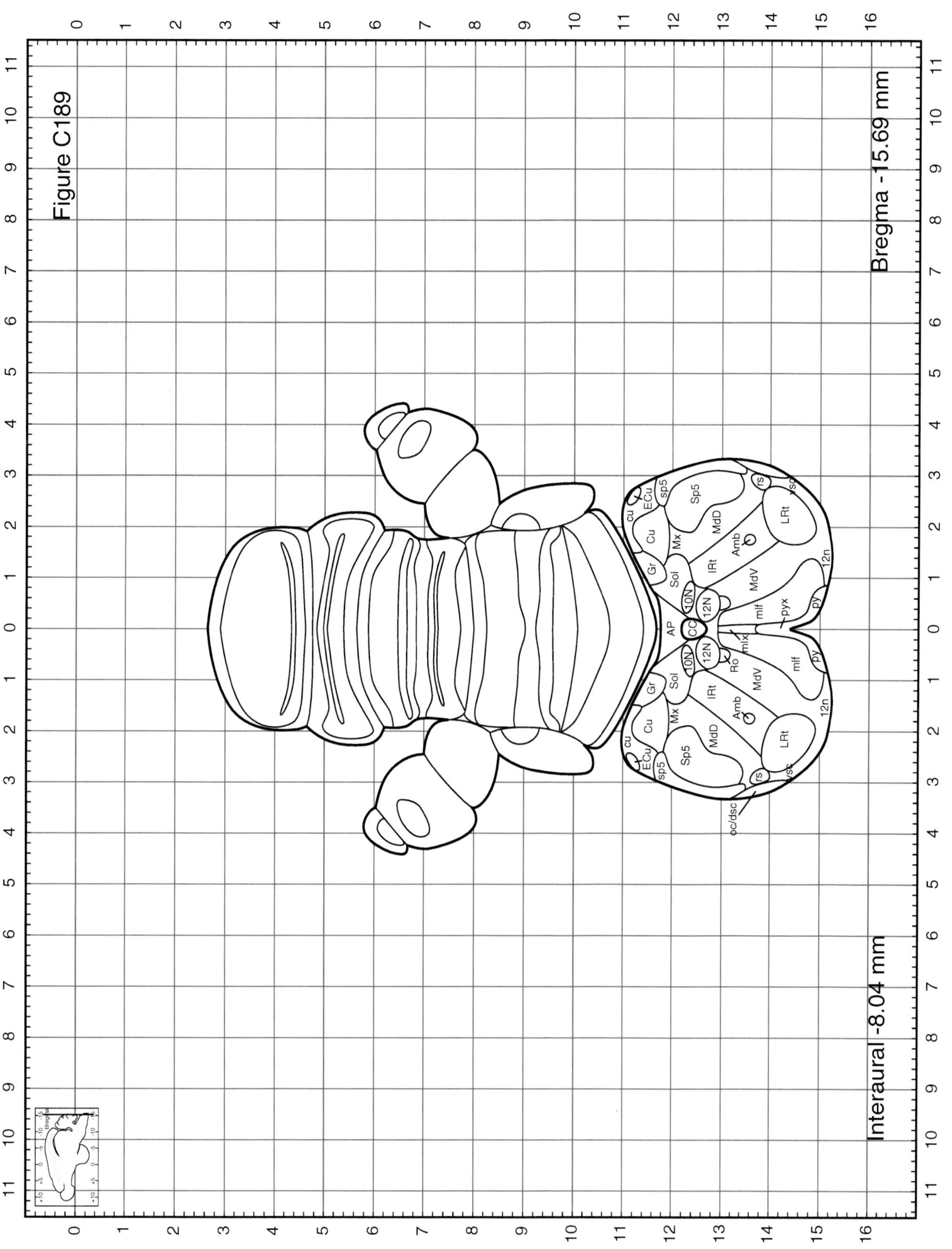

Figure C189
Bregma -15.69 mm
Interaural -8.04 mm
cu
ECu
sp5
Sp5
Cu
Mx
MdD
Gr
Sol
IRt
Amb
MdV
LRt
rs
vsc
10N
12N
AP
CC
mlf
mlx
pyx
py
12n
Ro
oc/dsc

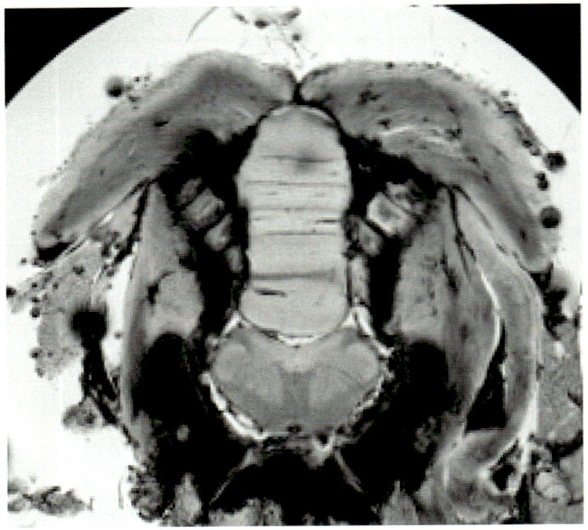

10N	dorsal motor nucleus of vagus
12n	root of hypoglossal nerve
12N	hypoglossal nucleus
AP	area postrema
CC	central canal
Cu	cuneate nucleus
cu	cuneate fasciculus
dsc	dorsal spinocerebellar tract
Gr	gracile nucleus
IRt	intermediate reticular nucleus
LRt	lateral reticular nucleus
MdD	medullary reticular nucleus, dorsal part
MdV	medullary reticular nucleus, ventral part
mlf	medial longitudinal fasciculus
Mx	matrix region of the medulla
pyx	pyramidal decussation
rs	rubrospinal tract
Sol	solitary nucleus
Sp5	spinal trigeminal nucleus
sp5	spinal trigeminal tract
vsc	ventral spinocerebellar tract

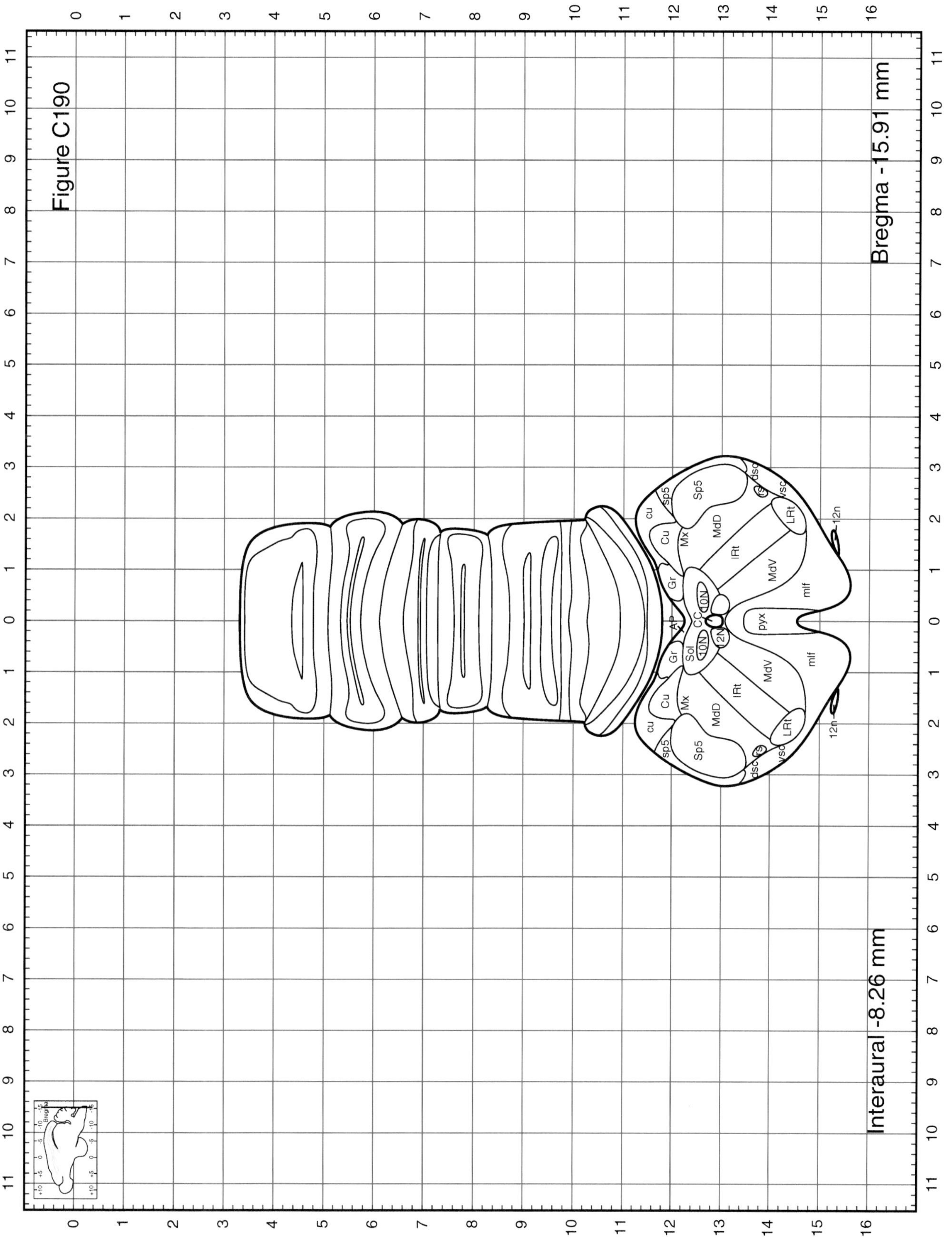
Figure C190
Bregma -15.91 mm
Interaural -8.26 mm
sp5
Sp5
dsc
rs
vsc
cu
Cu
Mx
MdD
IRt
LRt
12n
MdV
mlf
Gr
10N
AP
CC
Sol
12N
pyx

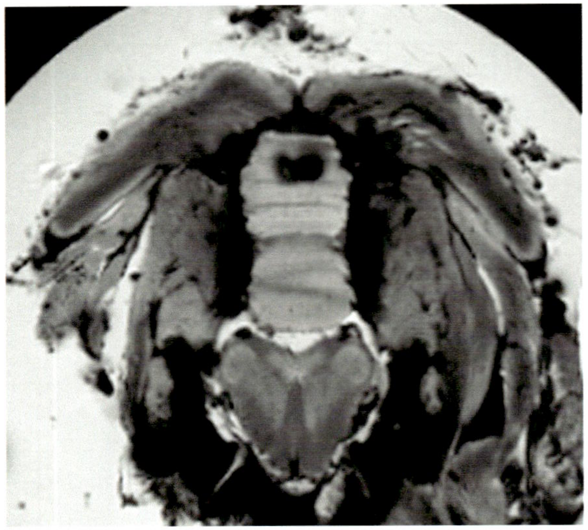

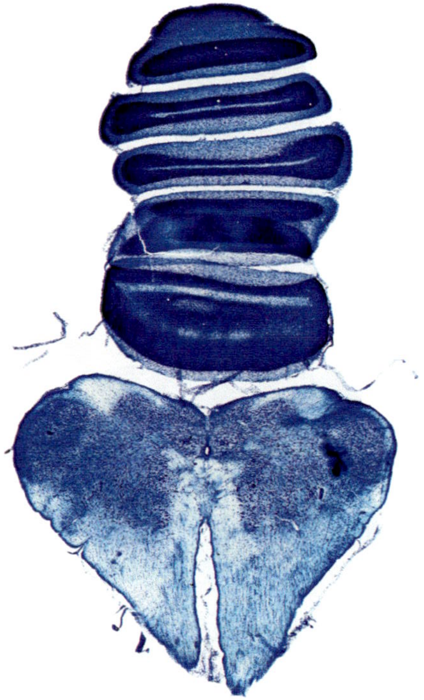

10N	dorsal motor nucleus of vagus	MdD	medullary reticular nucleus, dorsal part
12n	root of hypoglossal nerve	MdV	medullary reticular nucleus, ventral part
12N	hypoglossal nucleus	mlf	medial longitudinal fasciculus
CC	central canal	pyx	pyramidal decussation
Cu	cuneate nucleus	rs	rubrospinal tract
cu	cuneate fasciculus	Sol	solitary nucleus
dsc	dorsal spinocerebellar tract	Sp5	spinal trigeminal nucleus
Gr	gracile nucleus	sp5	spinal trigeminal tract
IRt	intermediate reticular nucleus	vsc	ventral spinocerebellar tract

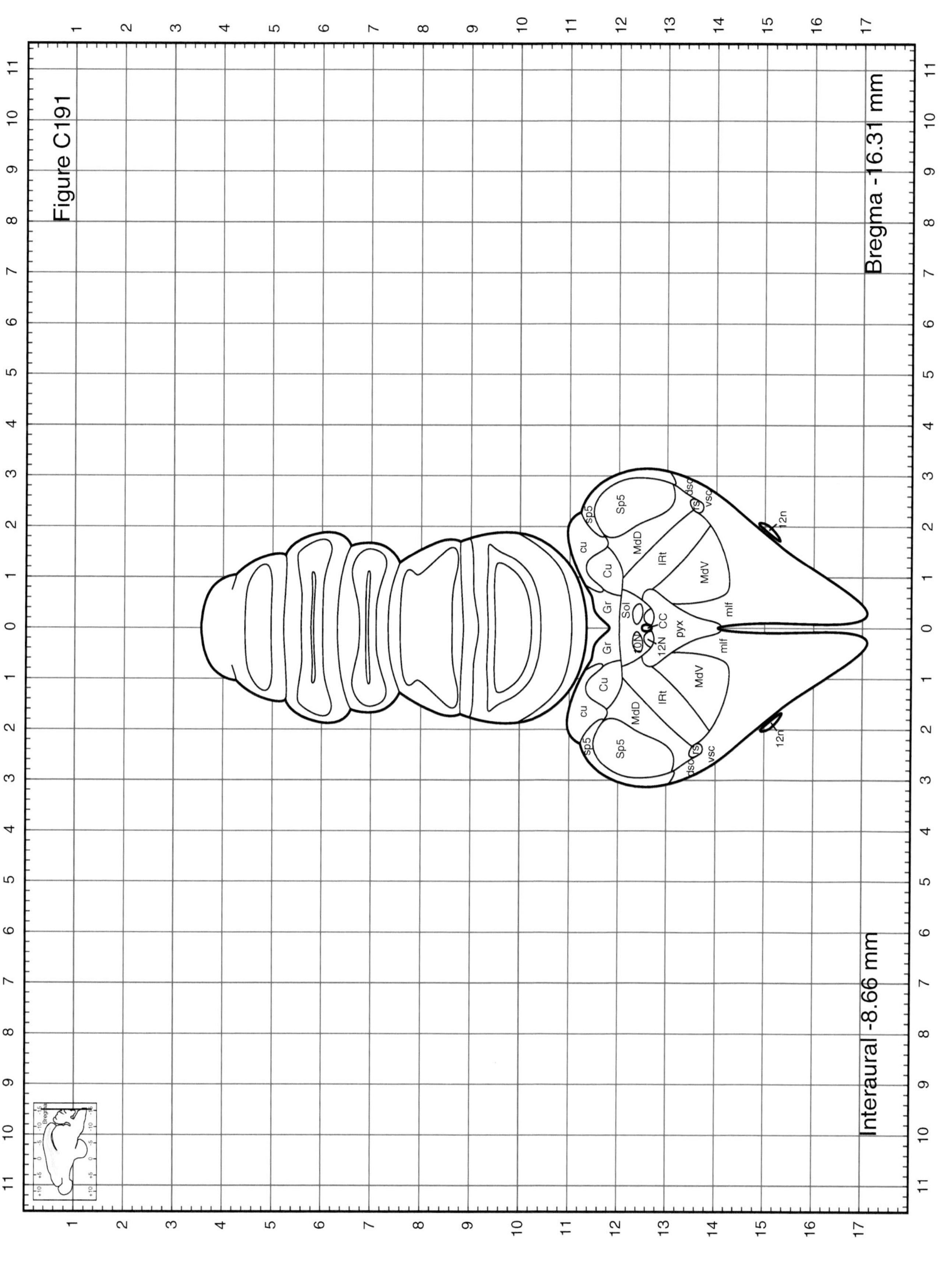
Figure C191
Bregma -16.31 mm
Interaural -8.66 mm
Sp5
sp5
dsc
vsc
rs
12n
cu
Cu
Gr
MdD
IRt
MdV
Sol
10N
12N
CC
pyx
mlf

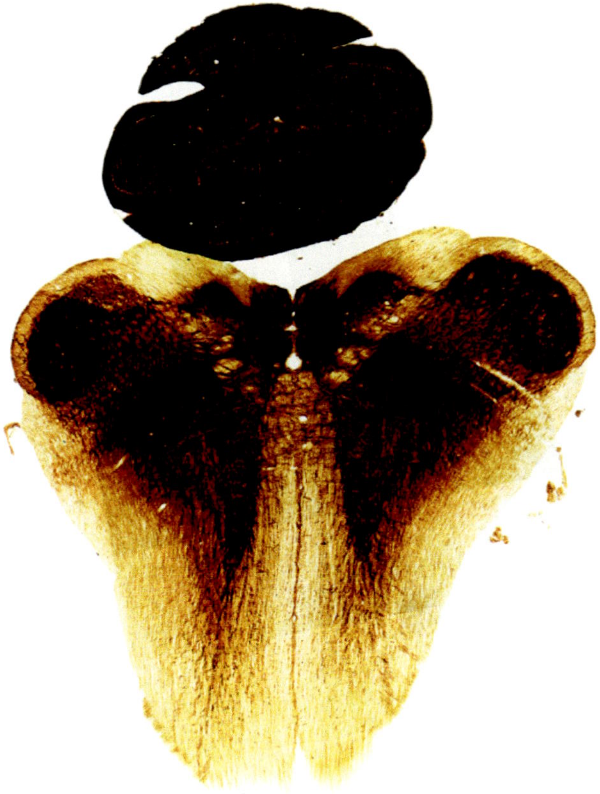

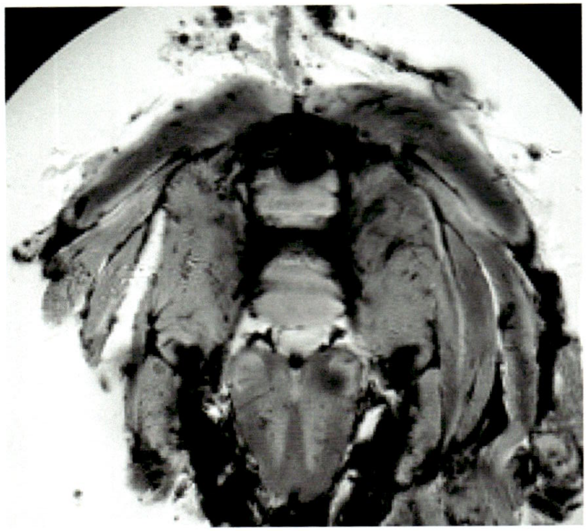

CC	central canal
Cu	cuneate nucleus
cu	cuneate fasciculus
dsc	dorsal spinocerebellar tract
Gr	gracile nucleus
IRt	intermediate reticular nucleus
MdD	medullary reticular nucleus, dorsal part
MdV	medullary reticular nucleus, ventral part
mlf	medial longitudinal fasciculus
pyx	pyramidal decussation
rs	rubrospinal tract
Sol	solitary nucleus
Sp5	spinal trigeminal nucleus
sp5	spinal trigeminal tract
vsc	ventral spinocerebellar tract

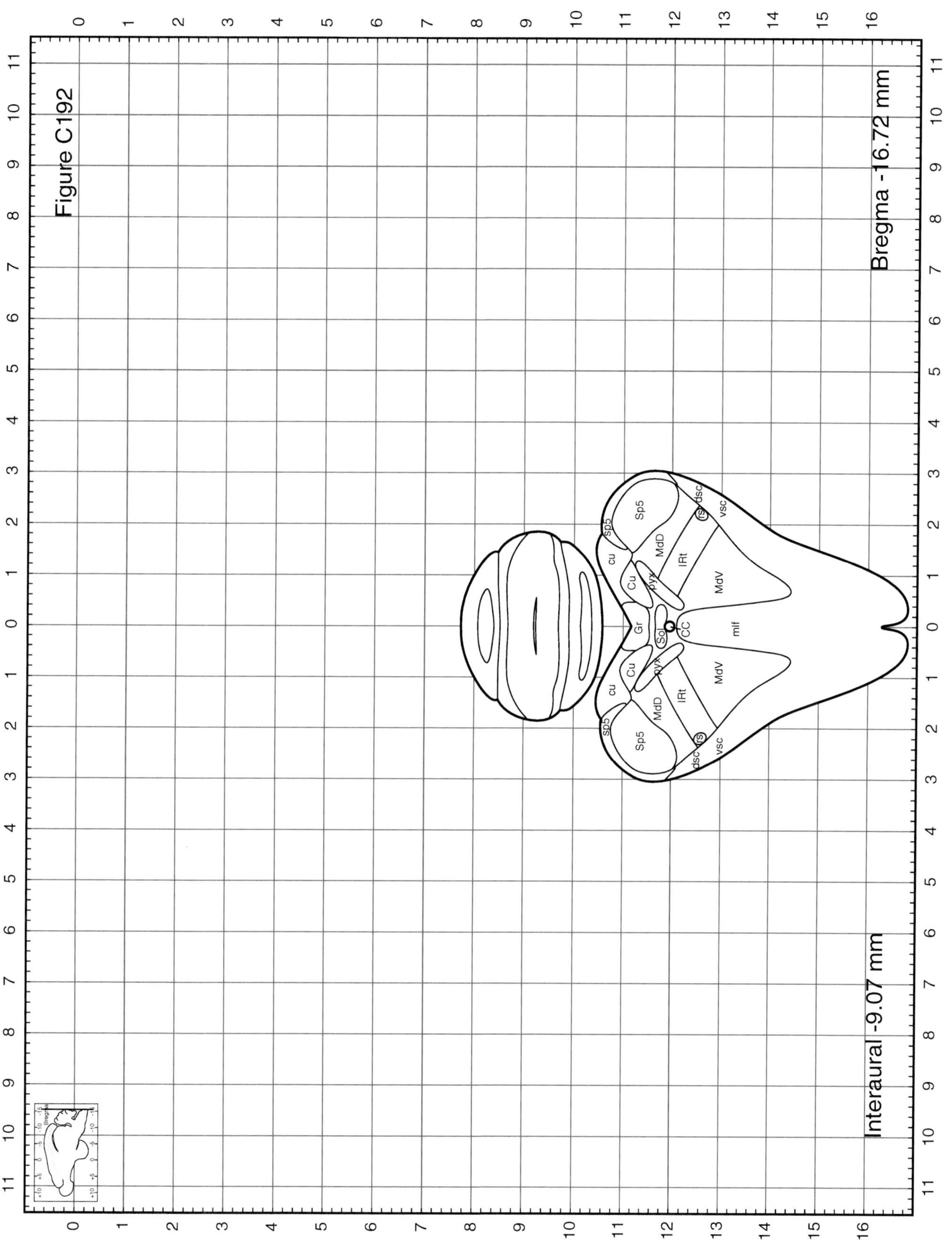
Figure C192
Bregma -16.72 mm
Interaural -9.07 mm
Sp5
sp5
cu
Cu
Gr
MdD
IRt
MdV
pyx
Sol
CC
mlf
rs
dsc
vsc

3 Horizontal Sections of the Tree Shrew Brain

J.-N. Zhou, R.-J. Ni, *The Tree Shrew (Tupaia belangeri chinensis) Brain in Stereotaxic Coordinates*,
DOI 10.1007/978-981-10-0611-1_3

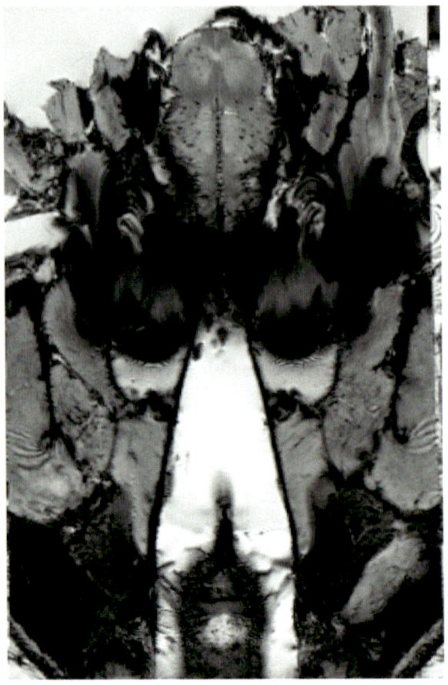

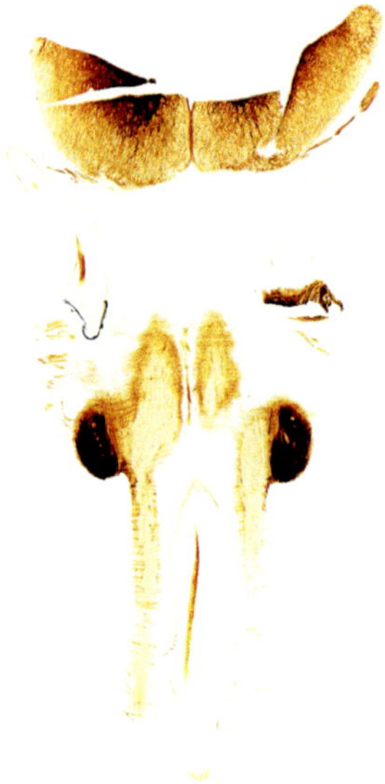

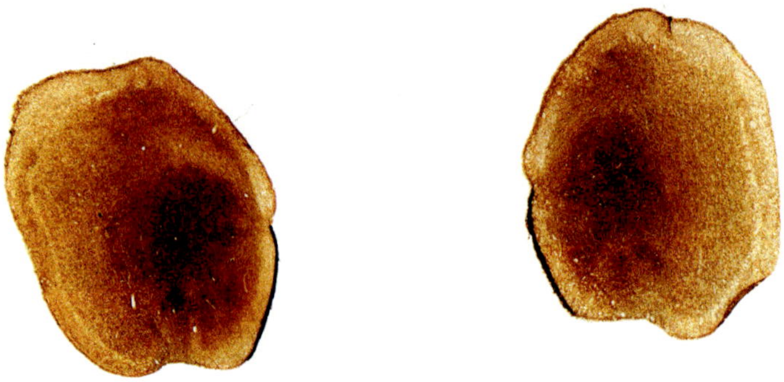

Aco anterior cortical amygdaloid nucleus
IO inferior olive
MdV medullary reticular nucleus, ventral part
mlf medial longitudinal fasciculus
py pyramidal tract

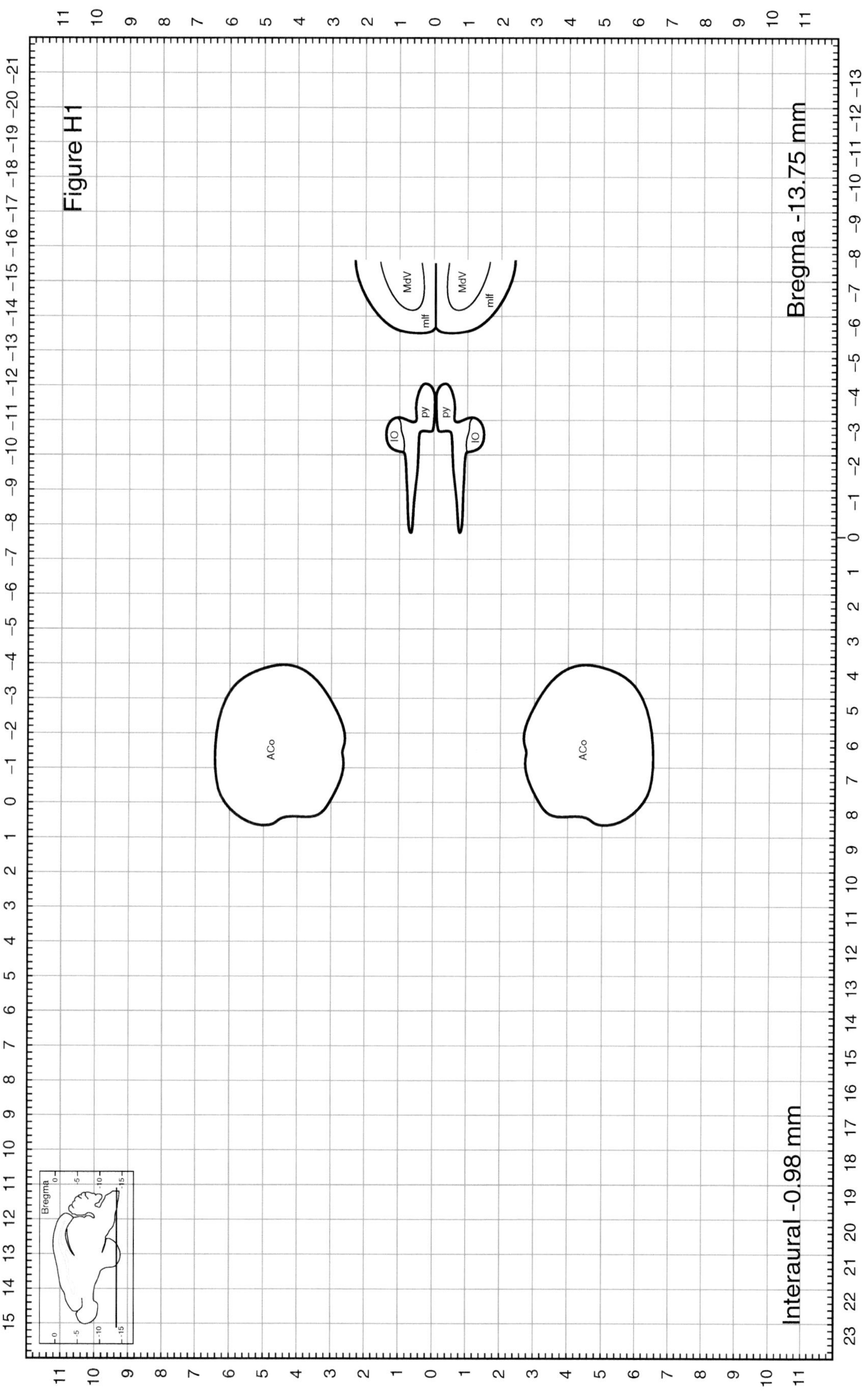
Figure H1
Bregma -13.75 mm
Interaural -0.98 mm
MdV
MdV
mlf
mlf
py
py
IO
IO
ACo
ACo
Bregma

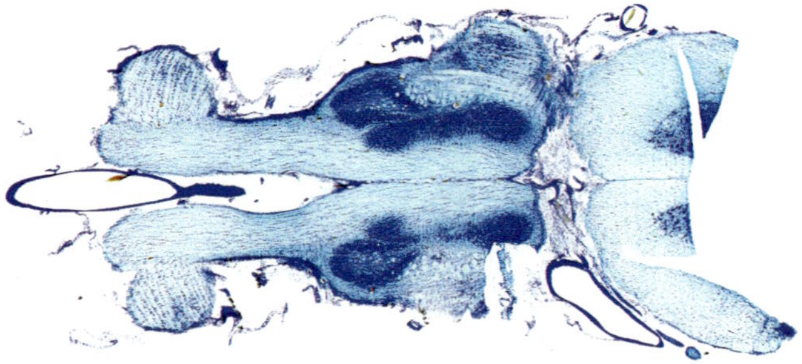

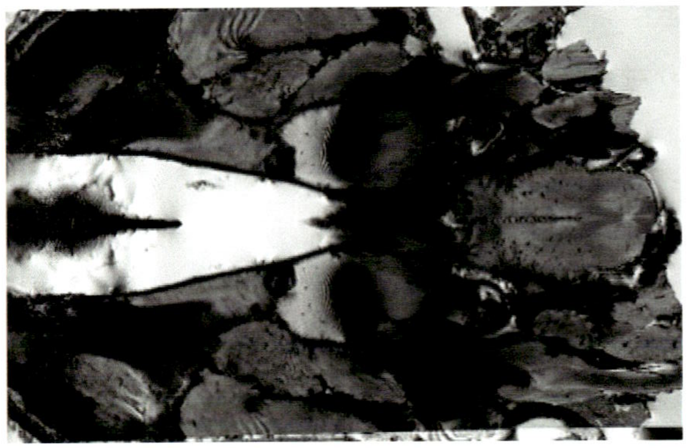

12n	root of hypoglossal nerve	IO	inferior olive	py	pyramidal tract
6n	root of abducens nerve	LRt	lateral reticular nucleus	Sp5	spinal trigeminal nucleus
Aco	anterior cortical amygdaloid nucleus	MdV	medullary reticular nucleus, ventral part	tz	trapezoid body
Gi	gigantocellular reticular nucleus	mlf	medial longitudinal fasciculus		

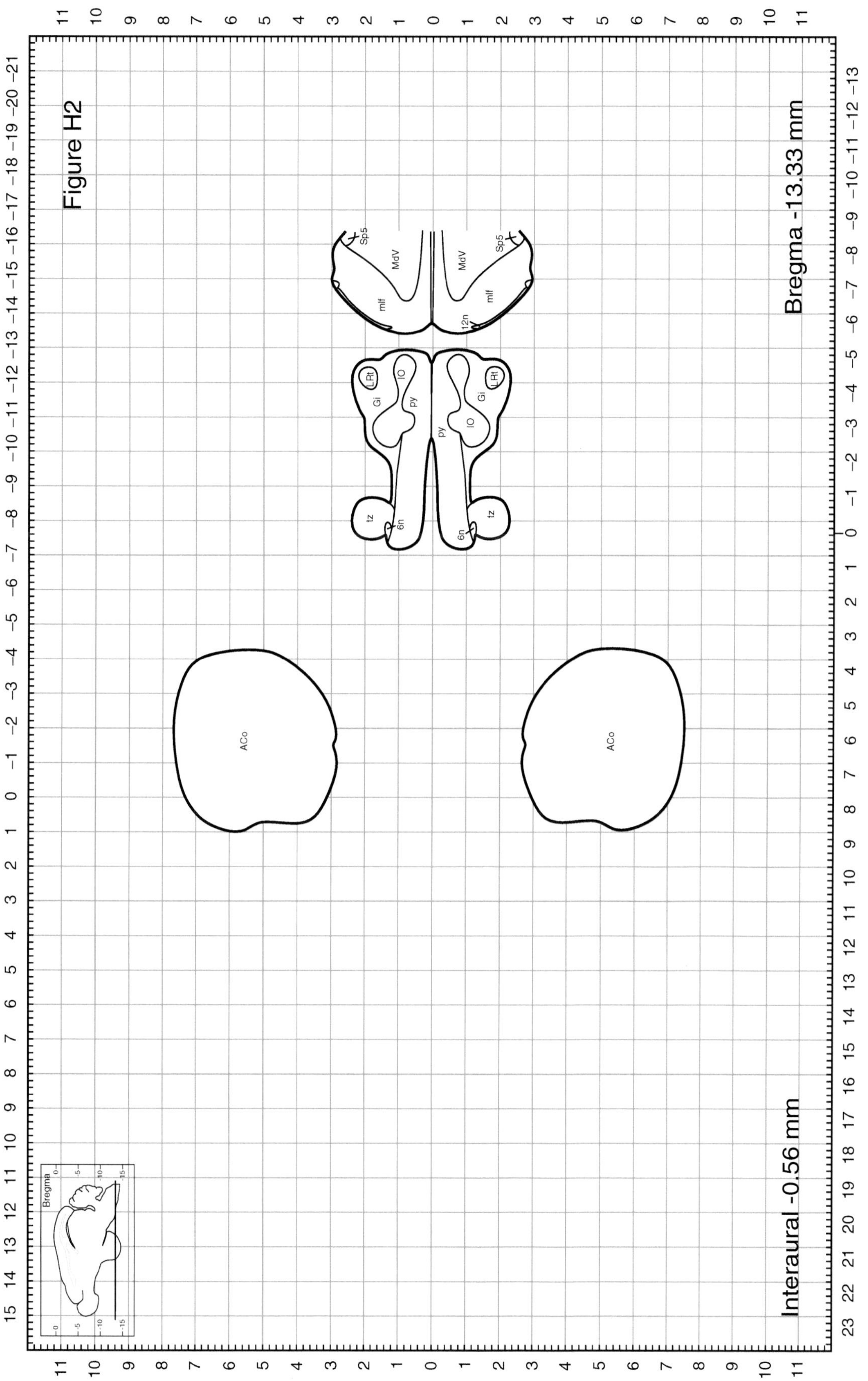
Figure H2
Bregma -13.33 mm
Interaural -0.56 mm
Sp5
MdV
mlf
12n
LRt
IO
Gi
py
tz
6n
ACo
Bregma

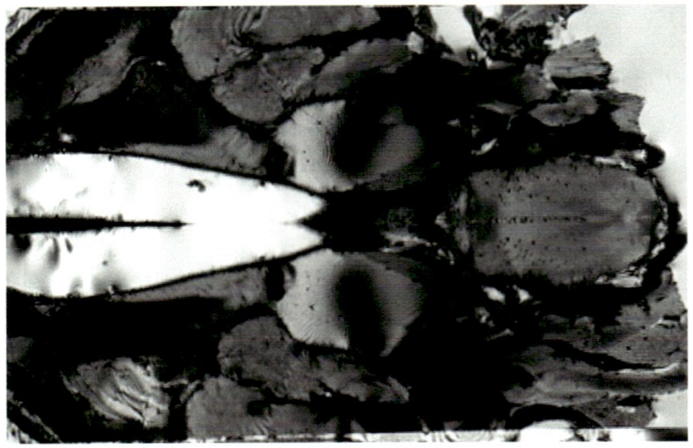

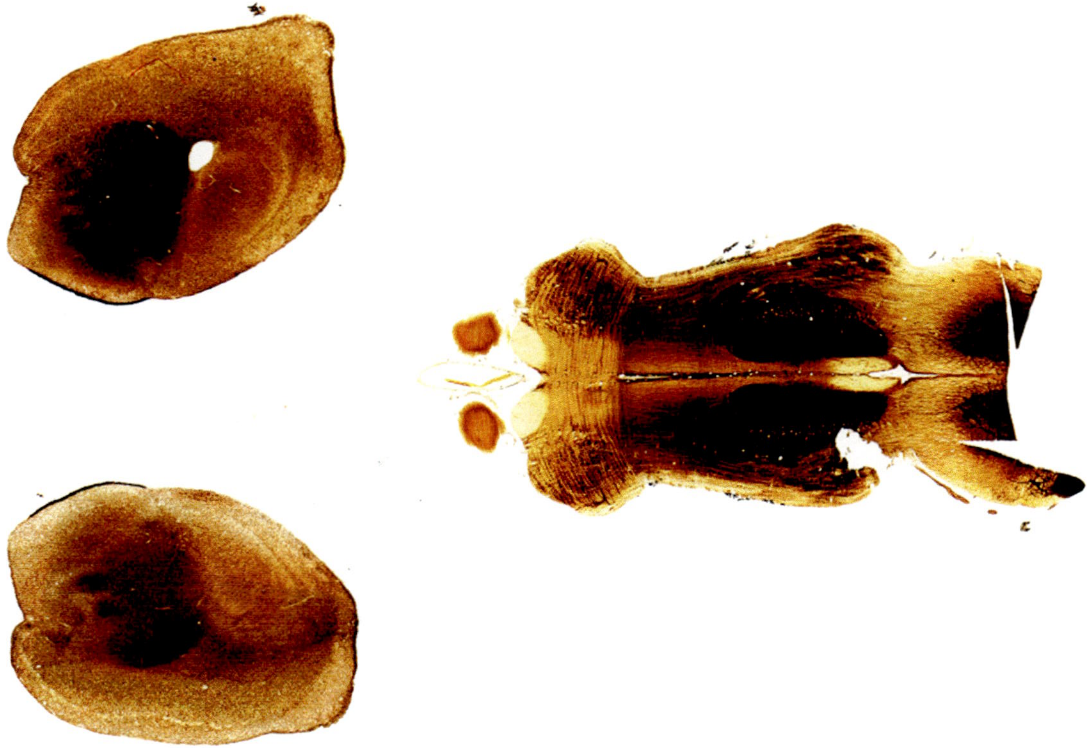

2	layer 2 of cortex	Gi	gigantocellular reticular nucleus	Sp5	spinal trigeminal nucleus
3	layer 3 of cortex	IO	inferior olive	tfp	transverse fibers of the pons
12n	root of hypoglossal nerve	LRt	lateral reticular nucleus	Tz	nucleus of the trapezoid body
6n	root of abducens nerve	MdV	medullary reticular nucleus, ventral part	tz	trapezoid body
AB	accessory basal nucleus	mlf	medial longitudinal fasciculus	VS	ventral subiculum
Aco	anterior cortical amygdaloid nucleus	Pir1	piriform cortex, layer 1		
Ent	entorhinal cortex	py	pyramidal tract		

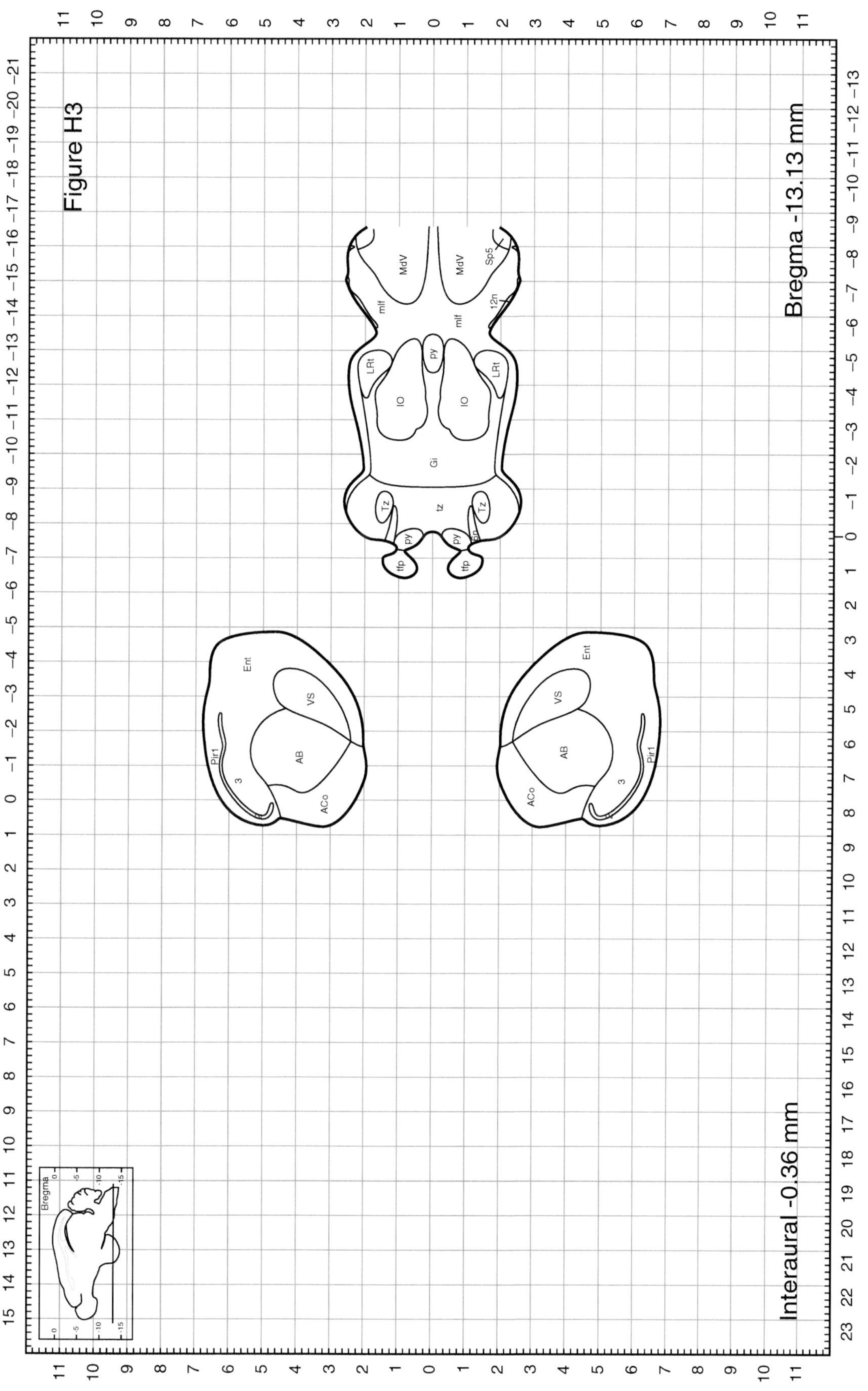

Figure H3
Bregma -13.13 mm
Interaural -0.36 mm
MdV
Sp5
12n
mlf
py
LRt
IO
Gi
Tz
tz
tfp
6n
Ent
VS
AB
Pir1
3
ACo
Bregma

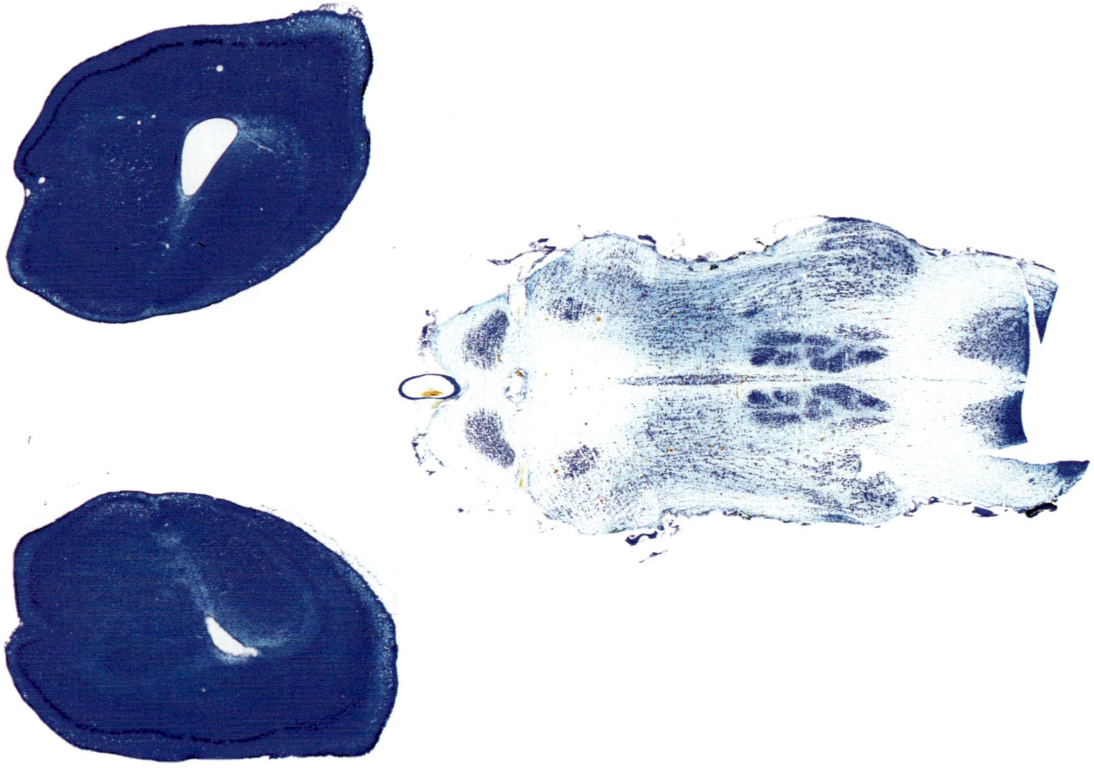

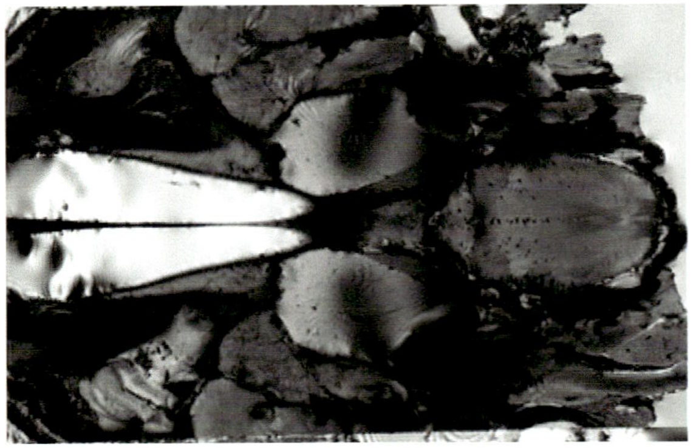

2	layer 2 of cortex	IO	inferior olive	py	pyramidal tract
3	layer 3 of cortex	lfp	longitudinal fasciculus of the pons	RPa	raphe pallidus nucleus
12n	root of hypoglossal nerve	LRt	lateral reticular nucleus	Sp5	spinal trigeminal nucleus
6n	root of abducens nerve	LV	lateral ventricle	tfp	transverse fibers of the pons
AB	accessory basal nucleus	MdV	medullary reticular nucleus, ventral part	Tz	nucleus of the trapezoid body
ACo	anterior cortical amygdaloid nucleus	mlf	medial longitudinal fasciculus	tz	trapezoid body
Ent	entorhinal cortex	Pir1	piriform cortex, layer 1	VS	ventral subiculum
Gi	gigantocellular reticular nucleus	Pn	pontine nuclei		

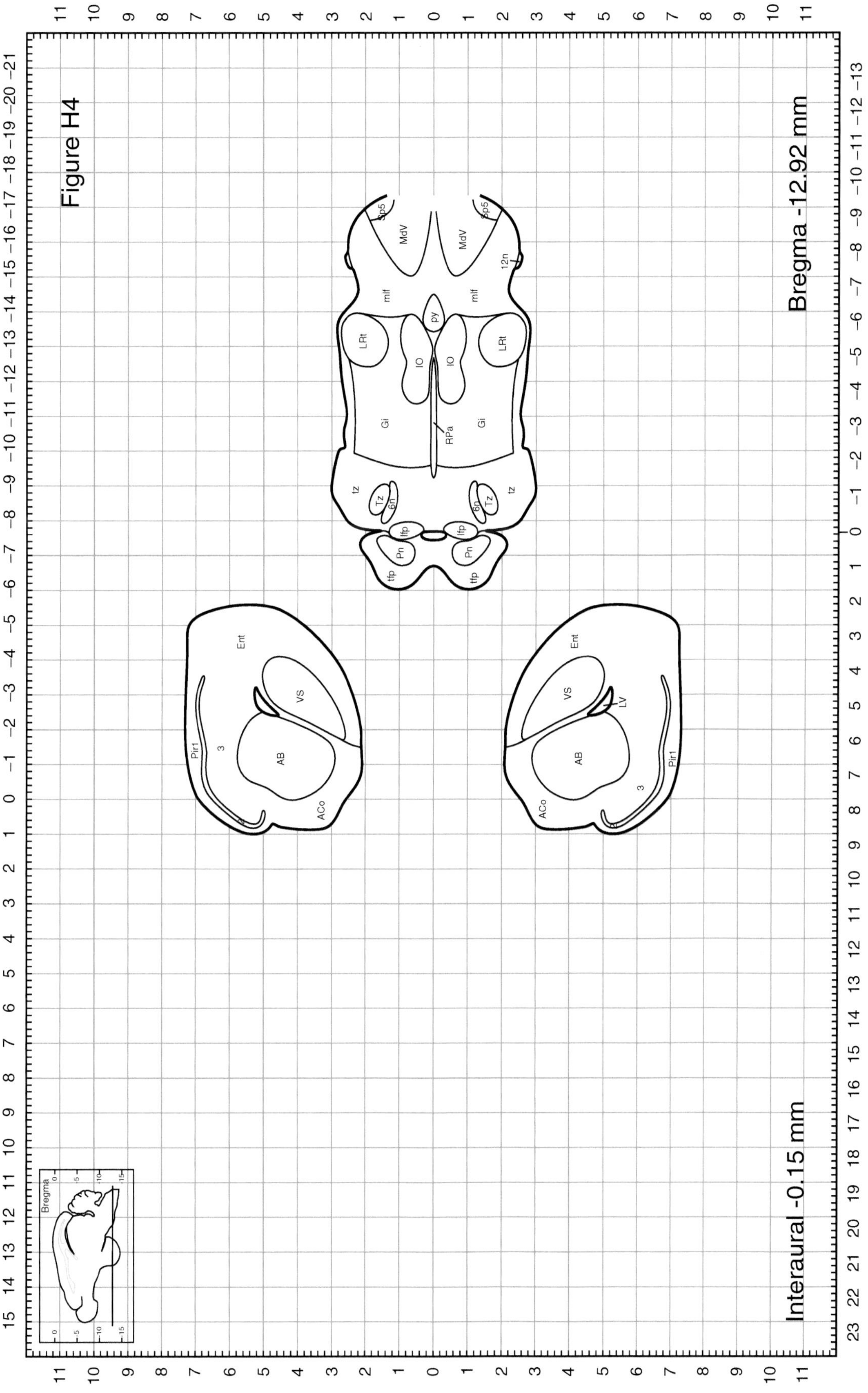
Figure H4
Bregma -12.92 mm
Interaural -0.15 mm
Sp5
MdV
12n
mlf
py
LRt
IO
Gi
RPa
tz
Tz
6n
tfp
Pn
Ent
VS
LV
AB
Pir1
3
ACo
Bregma

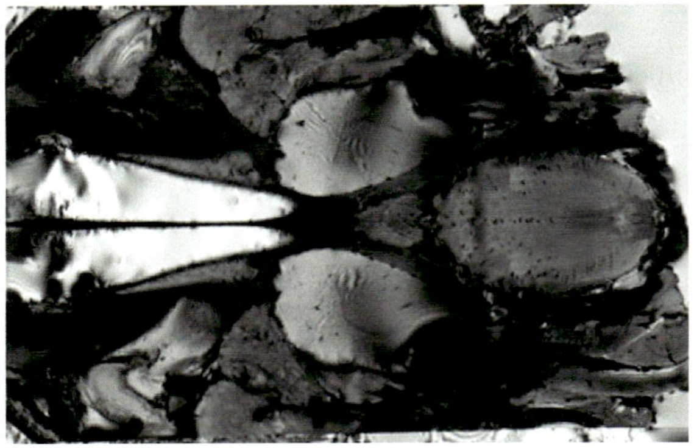

2	layer 2 of cortex	Gi	gigantocellular reticular nucleus	RPa	raphe pallidus nucleus
3	layer 3 of cortex	IO	inferior olive	Sp5	spinal trigeminal nucleus
12n	root of hypoglossal nerve	lfp	longitudinal fasciculus of the pons	tfp	transverse fibers of the pons
6n	root of abducens nerve	LRt	lateral reticular nucleus	Tz	nucleus of the trapezoid body
7N	facial nucleus	LV	lateral ventricle	tz	trapezoid body
AB	accessory basal nucleus	MdV	medullary reticular nucleus, ventral part	VLN	ventrolateral reticular nucleus
ACo	anterior cortical amygdaloid nucleus	mlf	medial longitudinal fasciculus	VPO	ventral periolivary nucleus
AHi	amygdalohippocampal area	Pir1	piriform cortex, layer 1	VS	ventral subiculum
CC	central canal	Pn	pontine nuclei		
Ent	entorhinal cortex	py	pyramidal tract		

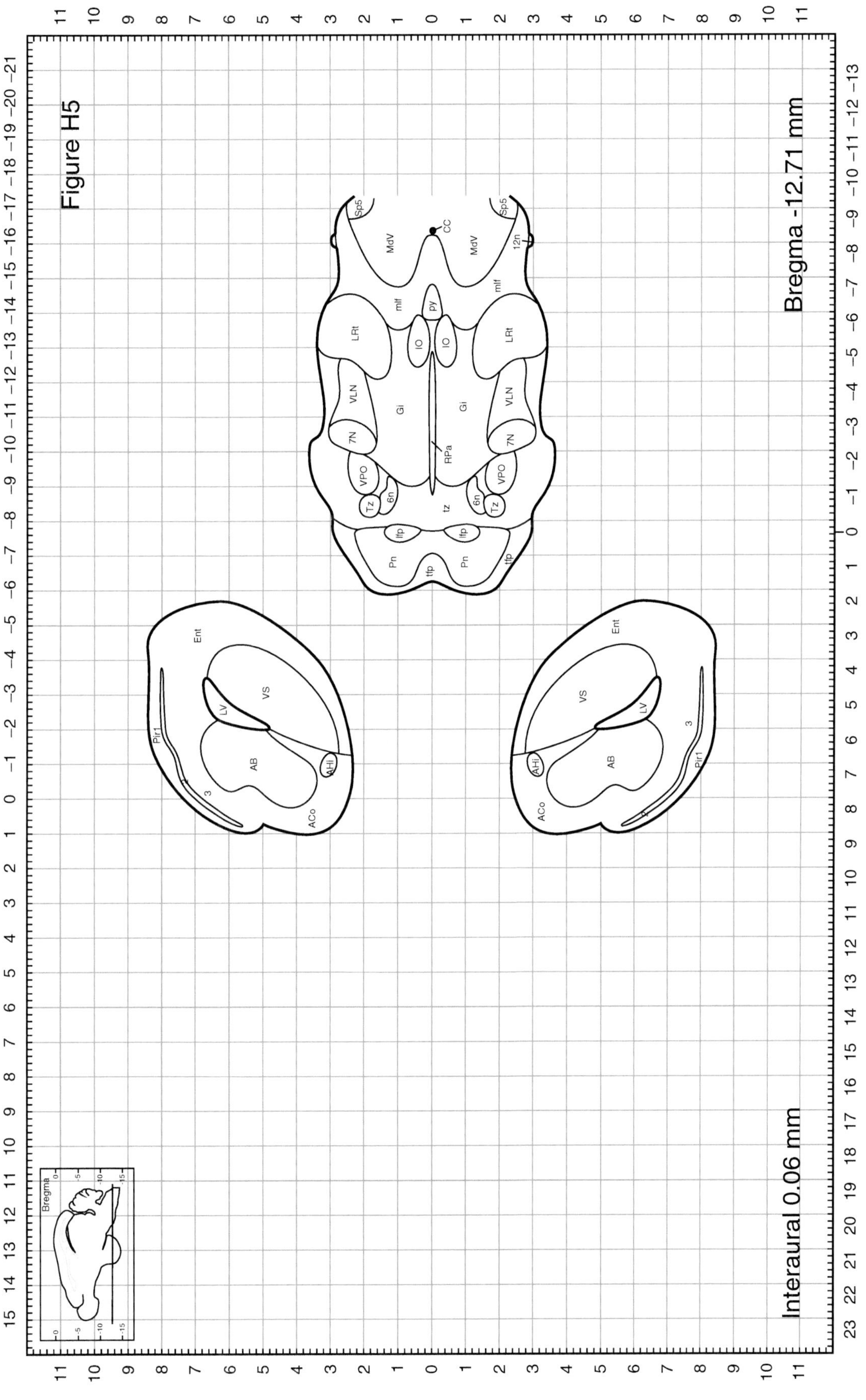
Figure H5
Bregma -12.71 mm
Interaural 0.06 mm
Sp5
CC
MdV
12n
mlf
py
LRt
IO
VLN
Gi
7N
RPa
VPO
6n
Tz
tz
tfp
Pn
Ent
VS
LV
Pir1
AB
AHi
3
ACo
Bregma

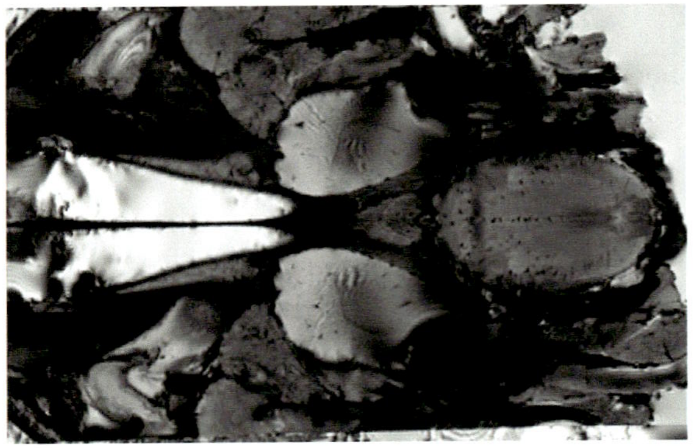

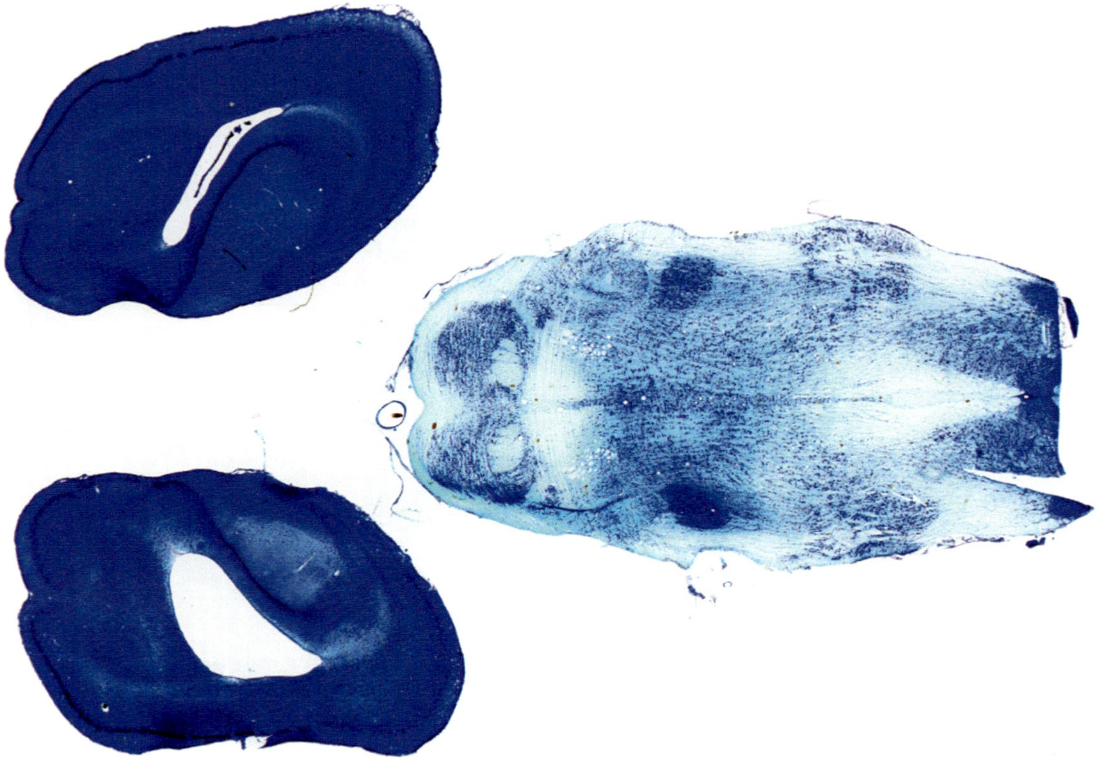

2	layer 2 of cortex	Gi	gigantocellular reticular nucleus	Rad	radiatum layer of the hippocampus
3	layer 3 of cortex	lfp	longitudinal fasciculus of the pons	RPa	raphe pallidus nucleus
12n	root of hypoglossal nerve	LRt	lateral reticular nucleus	Sp5	spinal trigeminal nucleus
6n	root of abducens nerve	LV	lateral ventricle	tfp	transverse fibers of the pons
7N	facial nucleus	MdV	medullary reticular nucleus, ventral part	Tz	nucleus of the trapezoid body
AB	accessory basal nucleus	mlf	medial longitudinal fasciculus	tz	trapezoid body
ACo	anterior cortical amygdaloid nucleus	Or	oriens layer of the hippocampus	VLN	ventrolateral reticular nucleus
AHi	amygdalohippocampal area	Pir1	piriform cortex, layer 1	VPO	ventral periolivary nucleus
CA1	field CA1 of the hippocampus	Pn	pontine nuclei	VS	ventral subiculum
CC	central canal	PnC	pontine reticular nucleus, caudal part	vsc	ventral spinocerebellar tract
cu	cuneate fasciculus	Py	pyramidal cell layer of the hippocampus		
Ent	entorhinal cortex	py	pyramidal tract		

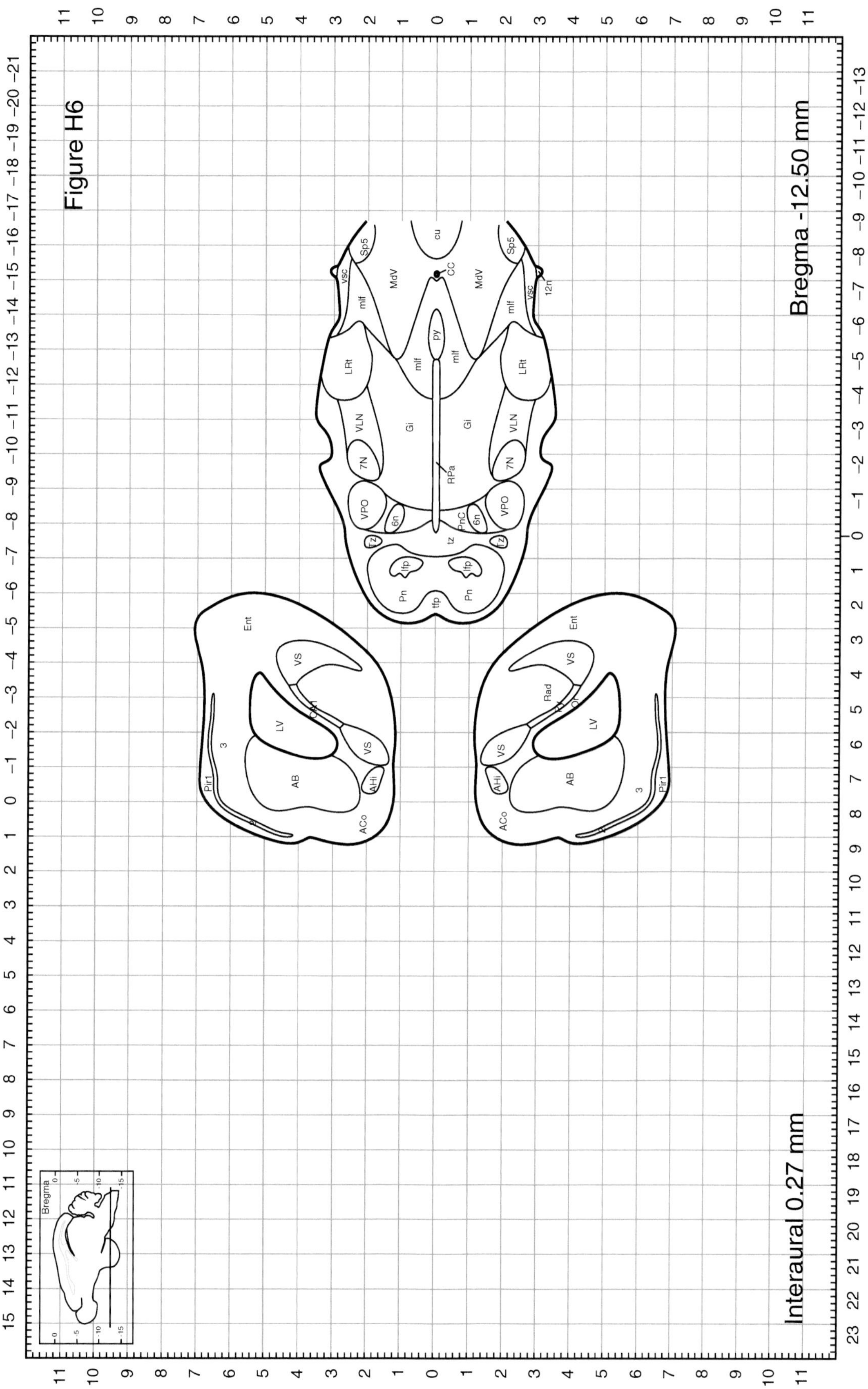
Figure H6
Bregma -12.50 mm
Interaural 0.27 mm
cu
Sp5
vsc
CC
MdV
mlf
12n
py
LRt
Gi
VLN
7N
RPa
VPO
6n
PnC
Tz
tz
tfp
Pn
Ent
VS
LV
Or
Rad
3
Pir1
AB
AHi
ACo
Bregma

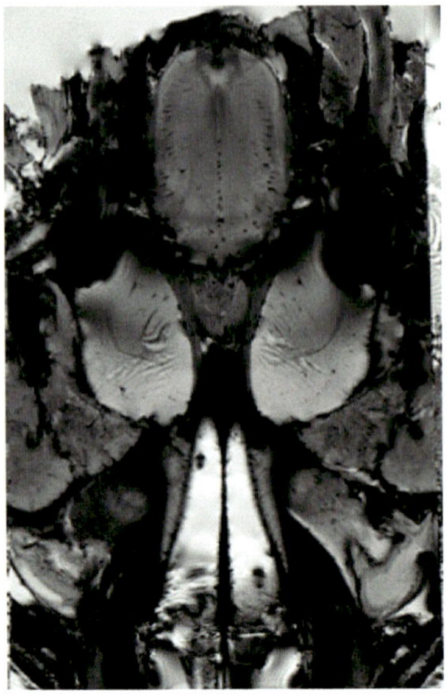

2	layer 2 of cortex
3	layer 3 of cortex
6n	root of abducens nerve
7N	facial nucleus
8cn	cochlear root of the vestibulocochlear nerve
AB	accessory basal nucleus
ACo	anterior cortical amygdaloid nucleus
AHi	amygdalohippocampal area
BMA	basomedial amygdaloid nucleus
CA1	field CA1 of the hippocampus
CC	central canal
cu	cuneate fasciculus
Ent	entorhinal cortex
Gi	gigantocellular reticular nucleus
IRt	intermediate reticular nucleus
lfp	longitudinal fasciculus of the pons
LRt	lateral reticular nucleus
LV	lateral ventricle
MdV	medullary reticular nucleus, ventral part
mlf	medial longitudinal fasciculus
oc	olivocerebellar tract
Or	oriens layer of the hippocampus
Pir1	piriform cortex, layer 1
Pn	pontine nuclei
PnC	pontine reticular nucleus, caudal part
PrS	presubiculum
Py	pyramidal cell layer of the hippocampus
py	pyramidal tract
Rad	radiatum layer of the hippocampus
RPa	raphe pallidus nucleus
rs	rubrospinal tract
RtTg	reticulotegmental nucleus of the pons
Sp5	spinal trigeminal nucleus
tfp	transverse fibers of the pons
Tz	nucleus of the trapezoid body
tz	trapezoid body
VCN	ventral cochlear nucleus
VPO	ventral periolivary nucleus
VS	ventral subiculum
vsc	ventral spinocerebellar tract

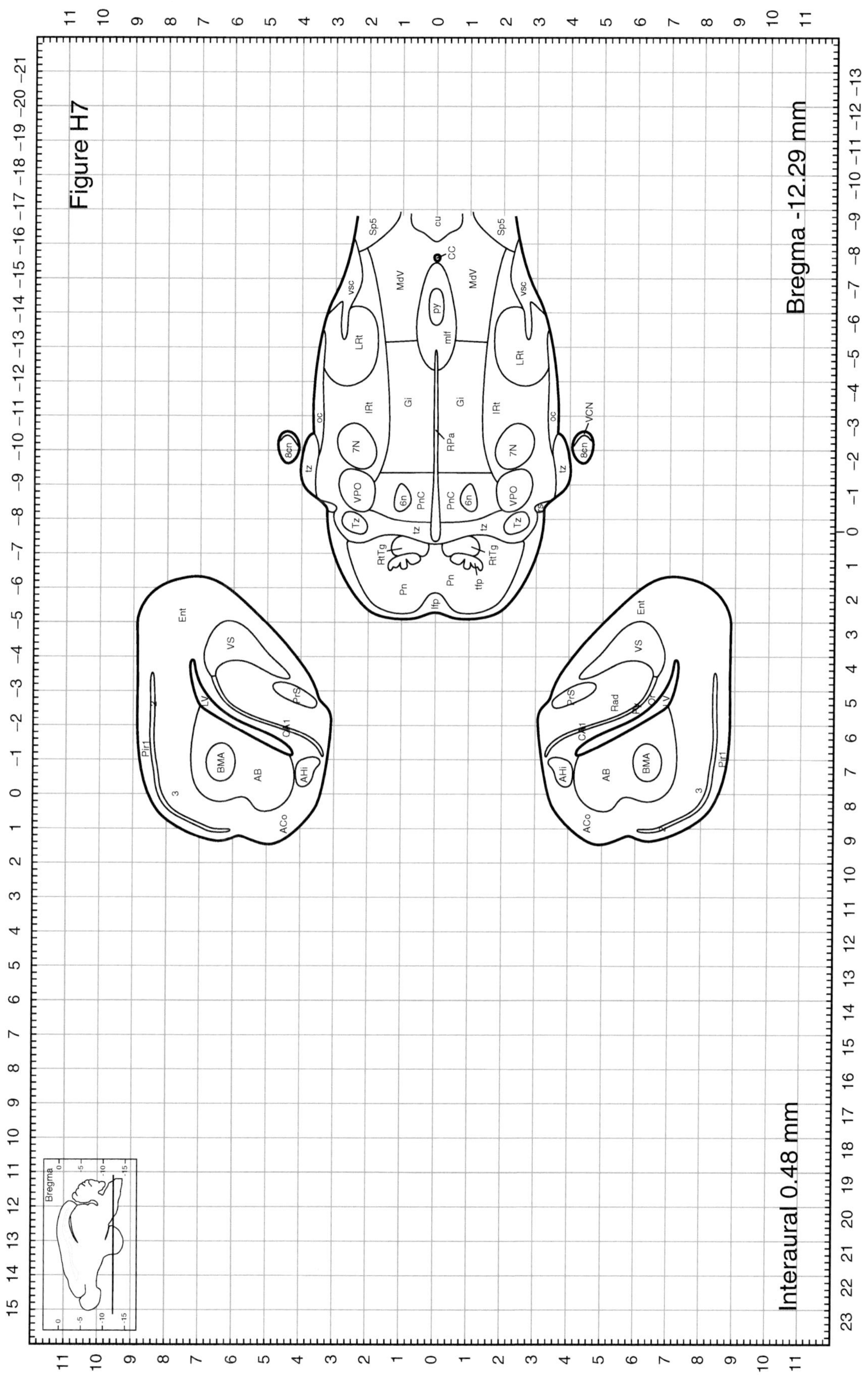
Figure H7
Bregma -12.29 mm
Interaural 0.48 mm
Sp5
cu
CC
vsc
MdV
py
mlf
LRt
Gi
IRt
oc
VCN
8cn
RPa
7N
tz
VPO
6n
PnC
Tz
RtTg
tfp
Pn
tfp
Ent
VS
LV
PrS
Rad
CA1
Pir1
BMA
AB
AHi
ACo
Bregma

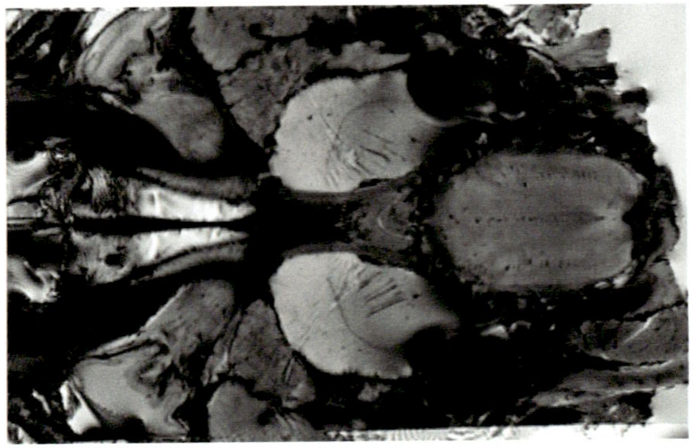

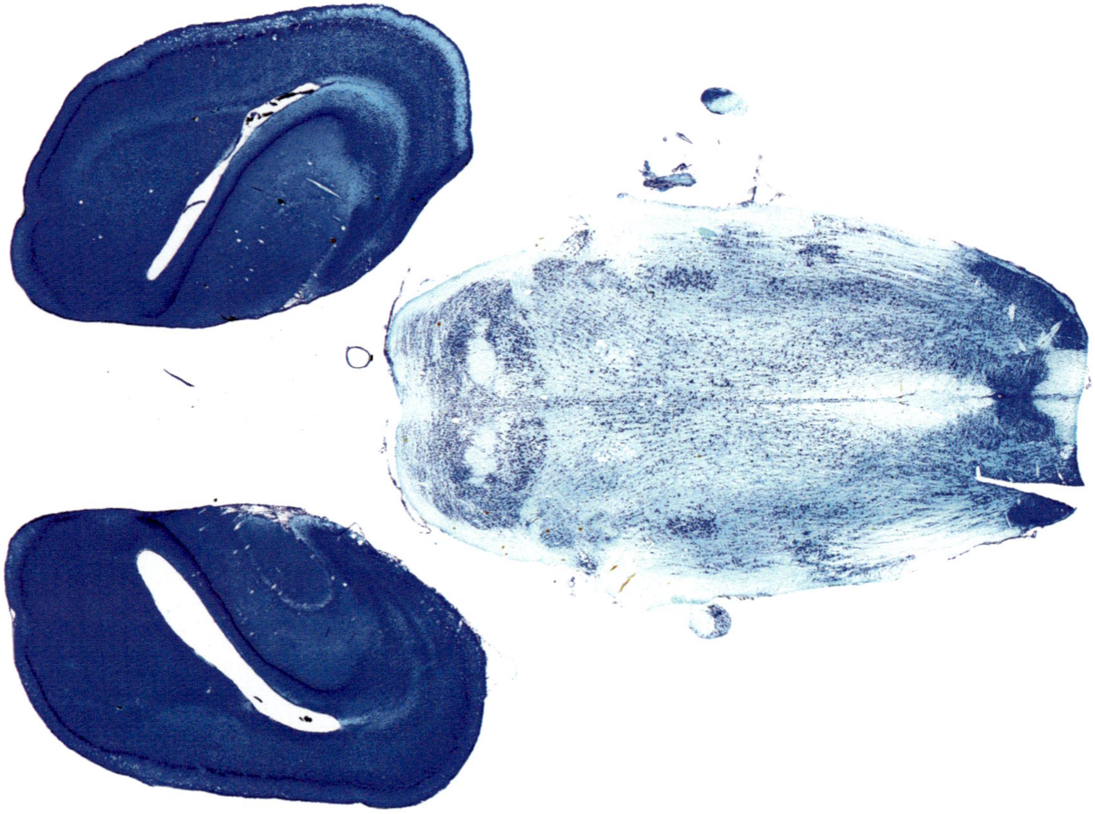

2	layer 2 of cortex
3	layer 3 of cortex
12N	hypoglossal nucleus
6n	root of abducens nerve
7N	facial nucleus
7n	facial nerve
8cn	cochlear root of the vestibulocochlear nerve
AB	accessory basal nucleus
ACo	anterior cortical amygdaloid nucleus
AHi	amygdalohippocampal area
BMA	basomedial amygdaloid nucleus
CA1	field CA1 of the hippocampus
CC	central canal
cu	cuneate fasciculus
Ent	entorhinal cortex
Gi	gigantocellular reticular nucleus
IRt	intermediate reticular nucleus
lfp	longitudinal fasciculus of the pons
LMoL	lacunosum moleculare layer of the hippocampus
LRt	lateral reticular nucleus
LV	lateral ventricle
MdV	medullary reticular nucleus, ventral part
ml	medial lemniscus
mlf	medial longitudinal fasciculus
MoDG	molecular layer of the dentate gyrus
oc/dsc	olivocerebellar tract and dorsal spinocerebellar tract
Or	oriens layer of the hippocampus
PCRt	parvicellular reticular nucleus
Pir1	piriform cortex, layer 1
Pn	pontine nuclei
PnC	pontine reticular nucleus, caudal part
PrS	presubiculum
Py	pyramidal cell layer of the hippocampus
py	pyramidal tract
pyx	pyramidal decussation
Rad	radiatum layer of the hippocampus
RPa	raphe pallidus nucleus
rs	rubrospinal tract
RtTg	reticulotegmental nucleus of the pons
s5	sensory root of the trigeminal nerve
sp5	spinal trigeminal tract
Sp5	spinal trigeminal nucleus
tfp	transverse fibers of the pons
tz	trapezoid body
VCN	ventral cochlear nucleus
VPO	ventral periolivary nucleus
VS	ventral subiculum
vsc	ventral spinocerebellar tract

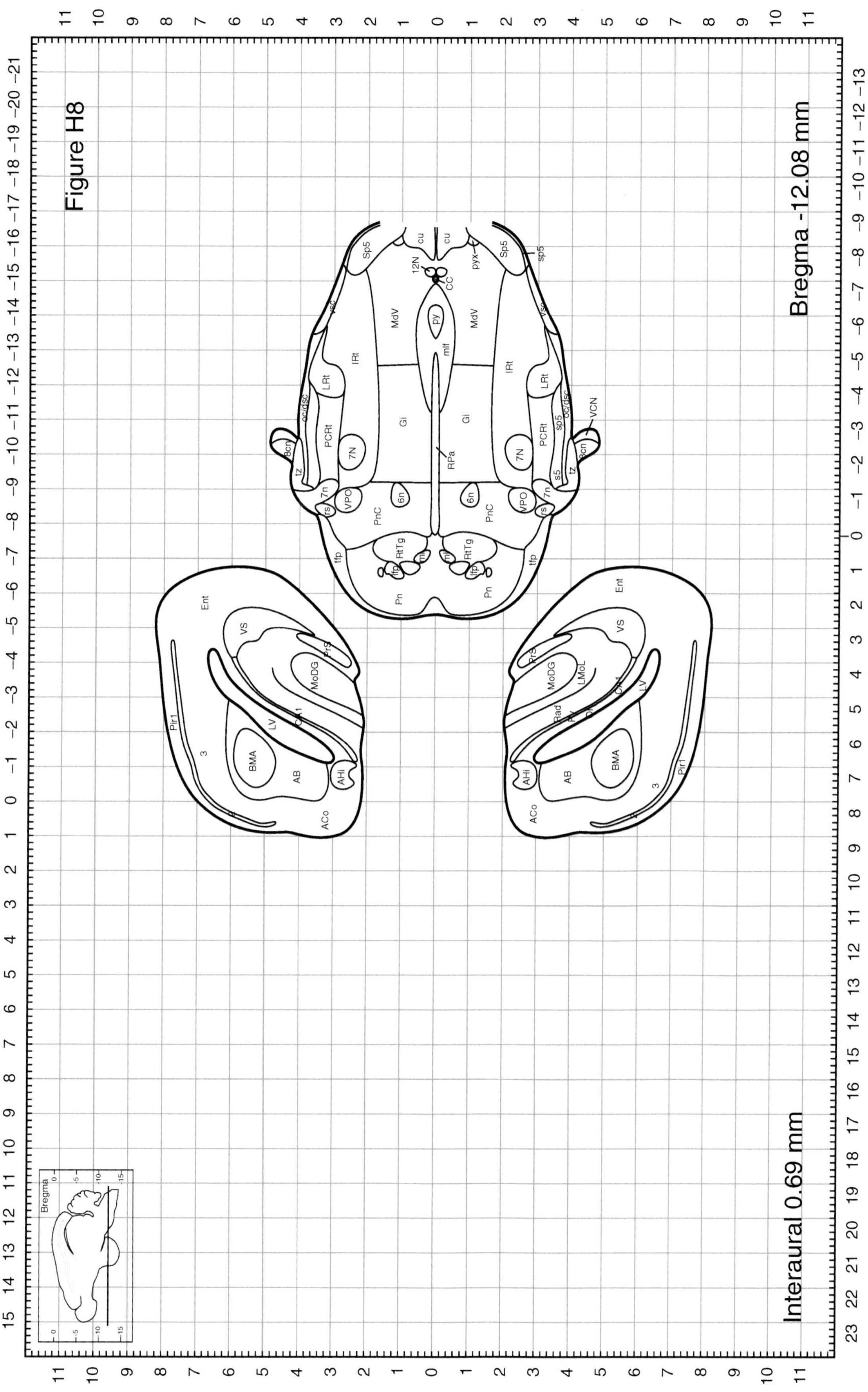
Figure H8
Bregma -12.08 mm
Interaural 0.69 mm
Sp5
cu
12N
pyx
CC
sp5
vsc
MdV
py
mlf
IRt
LRt
Gi
oc/dsc
VCN
PCRt
7N
RPa
8cn
tz
s5
7n
VPO
6n
rs
PnC
RtTg
tfp
ml
Pn
Ent
VS
PrS
MoDG
LMoL
CA1
LV
Rad
Py
Pir1
BMA
AB
AHi
ACo
3
Bregma

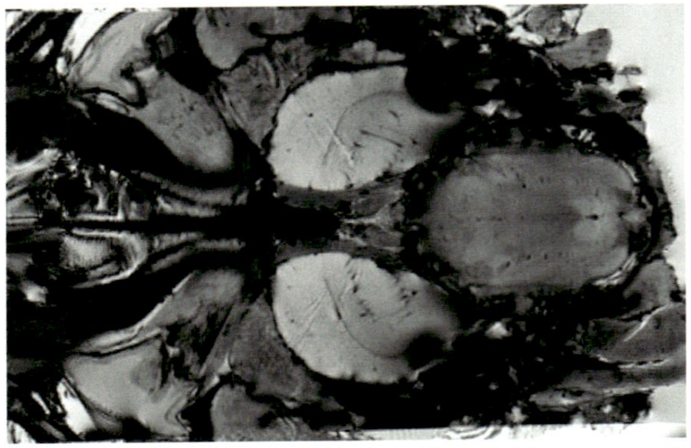

2	layer 2 of cortex		IRt	intermediate reticular nucleus		PnC	pontine reticular nucleus, caudal part
3	layer 3 of cortex		lfp	longitudinal fasciculus of the pons		PoDG	polymorph layer of the dentate gyrus
12N	hypoglossal nucleus		ll	lateral lemniscus		PrS	presubiculum
6n	root of abducens nerve		LL	nucleus of the lateral lemniscus		Py	pyramidal cell layer of the hippocampus
7n	facial nerve		LMoL	lacunosum moleculare layer of the hippocampus		pyx	pyramidal decussation
8cn	cochlear root of the vestibulocochlear nerve		LRt	lateral reticular nucleus		Rad	radiatum layer of the hippocampus
AB	accessory basal nucleus		LV	lateral ventricle		RPa	raphe pallidus nucleus
ACo	anterior cortical amygdaloid nucleus		MdD	medullary reticular nucleus, dorsal part		rs	rubrospinal tract
AHi	amygdalohippocampal area		MdV	medullary reticular nucleus, ventral part		RtTg	reticulotegmental nucleus of the pons
BLA	basolateral amygdaloid nucleus		ml	medial lemniscus		s5	sensory root of the trigeminal nerve
BMA	basomedial amygdaloid nucleus		mlf	medial longitudinal fasciculus		sp5	spinal trigeminal tract
CA1	field CA1 of the hippocampus		MoDG	molecular layer of the dentate gyrus		Sp5	spinal trigeminal nucleus
CC	central canal		oc/dsc	olivocerebellar tract and dorsal spinocerebellar tract		SubC	subcoeruleus nucleus
cu	cuneate fasciculus		Or	oriens layer of the hippocampus		tfp	transverse fibers of the pons
Ent	entorhinal cortex		PCRt	parvicellular reticular nucleus		tz	trapezoid body
Gi	gigantocellular reticular nucleus		Pir1	piriform cortex, layer 1		VCN	ventral cochlear nucleus
GrDG	granular layer of the dentate gyrus		Pn	pontine nuclei		VS	ventral subiculuma

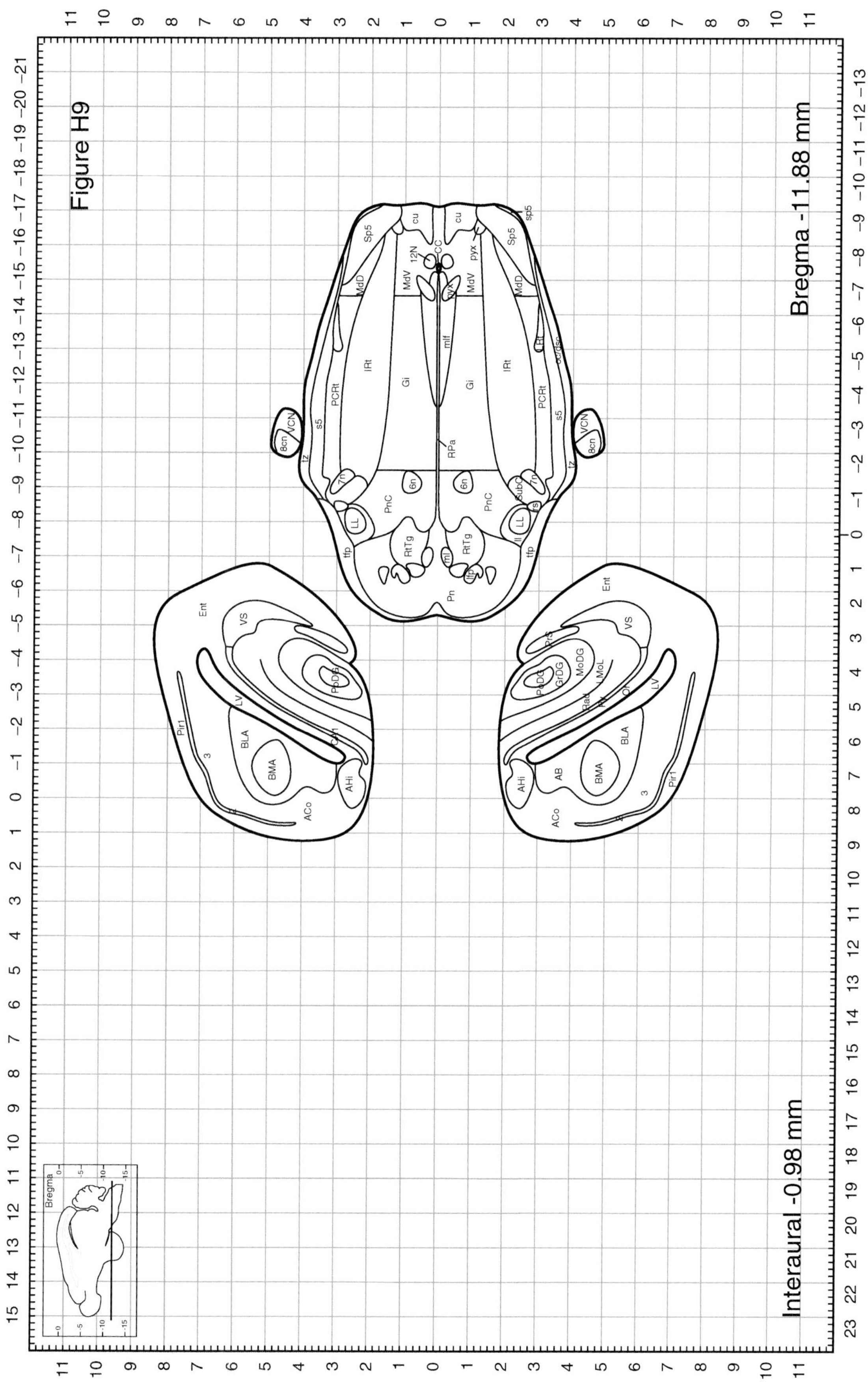
Figure H9
Bregma -11.88 mm
Interaural -0.98 mm
Bregma
cu
Sp5
sp5
12N
CC
pyx
MdD
MdV
mlf
IRt
Gi
PCRt
s5
VCN
8cn
tz
RPa
7n
6n
PnC
SubC
LL
ll
tfp
RtTg
ml
Pn
Ent
VS
PoDG
GrDG
MoDG
MoL
PrS
Rad
Py
Or
LV
Pir1
BLA
BMA
AB
AHi
ACo
CA1
3

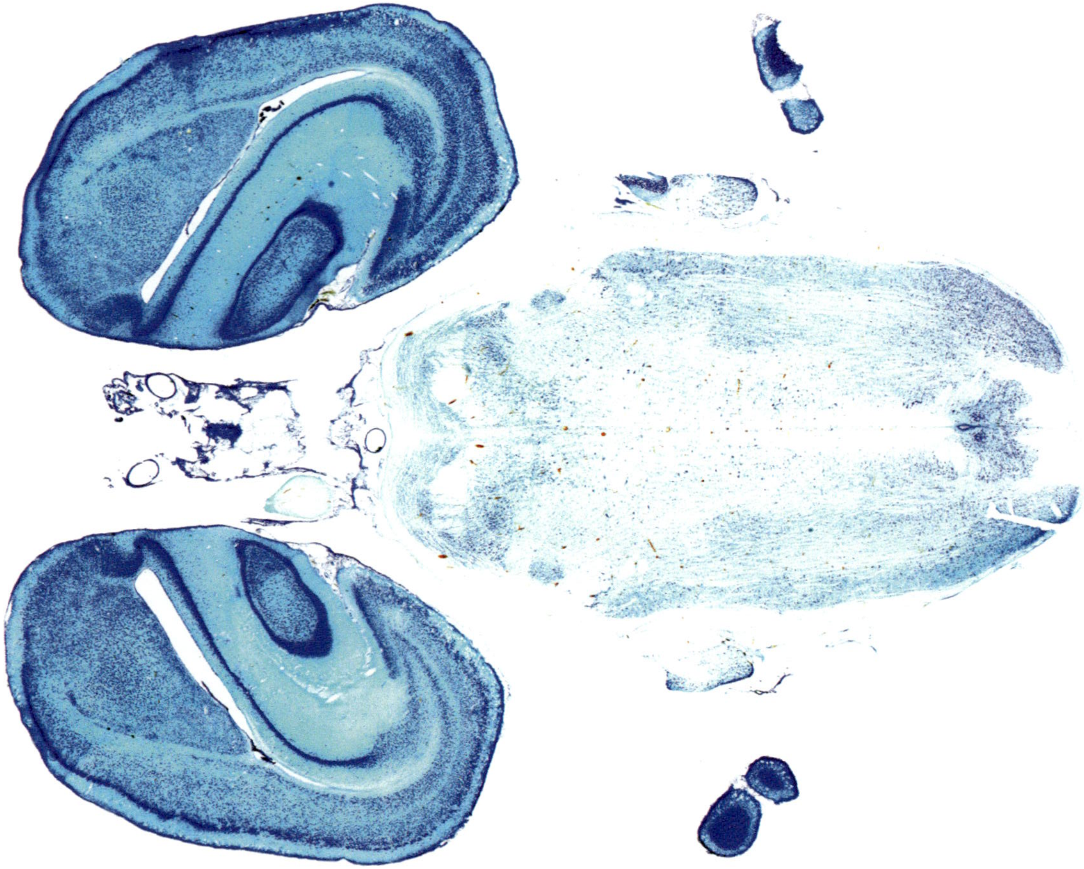

2	layer 2 of cortex
3	layer 3 of cortex
12N	hypoglossal nucleus
6n	root of abducens nerve
7n	facial nerve
8cn	cochlear root of the vestibulocochlear nerve
8vn	vestibular root of the vestibulocochlear nerve
ACo	anterior cortical amygdaloid nucleus
AHi	amygdalohippocampal area
BLA	basolateral amygdaloid nucleus
BMA	basomedial amygdaloid nucleus
CA1	field CA1 of the hippocampus
CC	central canal
cu	cuneate fasciculus
Ent	entorhinal cortex
fr	fasciculus retroflexus
Gi	gigantocellular reticular nucleus
GrDG	granular layer of the dentate gyrus
IRt	intermediate reticular nucleus
lfp	longitudinal fasciculus of the pons
ll	lateral lemniscus
LL	nucleus of the lateral lemniscus
LMoL	lacunosum moleculare layer of the hippocampus
LV	lateral ventricle
MdD	medullary reticular nucleus, dorsal part
MdV	medullary reticular nucleus, ventral part
ML	medial mammillary nucleus, lateral part
ml	medial lemniscus
mlf	medial longitudinal fasciculus
MoDG	molecular layer of the dentate gyrus
oc/dsc	olivocerebellar tract and dorsal spinocerebellar tract
Or	oriens layer of the hippocampus
PCRt	parvicellular reticular nucleus
Pir1	piriform cortex, layer 1
Pn	pontine nuclei
PnC	pontine reticular nucleus, caudal part
PnO	pontine reticular nucleus, oral part
PoDG	polymorph layer of the dentate gyrus
Pr5	principal sensory trigeminal nucleus
PrS	presubiculum
Py	pyramidal cell layer of the hippocampus
pyx	pyramidal decussation
Rad	radiatum layer of the hippocampus
RtTg	reticulotegmental nucleus of the pons
s5	sensory root of the trigeminal nerve
sp5	spinal trigeminal tract
Sp5	spinal trigeminal nucleus
SubC	subcoeruleus nucleus
tfp	transverse fibers of the pons
tz	trapezoid body
VCN	ventral cochlear nucleus
VS	ventral subiculum

Figure H10
Bregma -11.67 mm
Interaural 1.10 mm
cu
pyx
Sp5
MdV
12N
MdD
CC
sp5
cu/dsc
IRt
Gi
mlf
PCRt
tz
VCN
8vn
8cn
s5
Pr5
6n
PnC
SubC
LL
ll
tfp
PnO
RtTg
lfp
ml
Pn
Ent
VS
PrS
LMoL
MoDG
PoDG
CA1
GrDG
Rad
Py
Or
fr
ML
BLA
BMA
LV
AHi
Pir1
ACo
3
Bregma

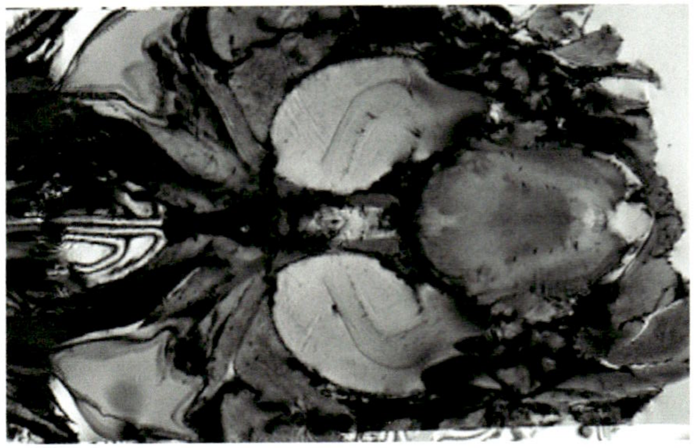

2	layer 2 of cortex
3	layer 3 of cortex
10N	dorsal motor nucleus of vagus
12N	hypoglossal nucleus
3V	3rd ventricle
6n	root of abducens nerve
7n	facial nerve
8cn	cochlear root of the vestibulocochlear nerve
8vn	vestibular root of the vestibulocochlear nerve
ACo	anterior cortical amygdaloid nucleus
AHi	amygdalohippocampal area
Arc	arcuate hypothalamic nucleus
BLA	basolateral amygdaloid nucleus
BMA	basomedial amygdaloid nucleus
CA1	field CA1 of the hippocampus
CA3	field CA3 of the hippocampus
CC	central canal
cu	cuneate fasciculus
DCN	dorsal cochlear nucleus
Ent	entorhinal cortex
fr	fasciculus retroflexus
Gi	gigantocellular reticular nucleus
GrDG	granular layer of the dentate gyrus
IRt	intermediate reticular nucleus
lfp	longitudinal fasciculus of the pons
LHA	lateral hypothalamic area
ll	lateral lemniscus
LL	nucleus of the lateral lemniscus
LMoL	lacunosum moleculare layer of the hippocampus
LV	lateral ventricle
m5	motor root of the trigeminal nerve
mcp	middle cerebellar peduncle
MdD	medullary reticular nucleus, dorsal part
MdV	medullary reticular nucleus, ventral part
MeA	medial amygdaloid nucleus
ML	medial mammillary nucleus, lateral part
ml	medial lemniscus
mlf	medial longitudinal fasciculus
MoDG	molecular layer of the dentate gyrus
MRe	mammillary recess of the 3rd ventricle
oc/dsc	olivocerebellar tract and dorsal spinocerebellar tract
och	optic chiasm
Or	oriens layer of the hippocampus
PCRt	parvicellular reticular nucleus
Pir1	piriform cortex, layer 1
Pn	pontine nuclei
PnC	pontine reticular nucleus, caudal part
PnO	pontine reticular nucleus, oral part
PoDG	polymorph layer of the dentate gyrus
Pr5	principal sensory trigeminal nucleus
PrS	presubiculum
Py	pyramidal cell layer of the hippocampus
pyx	pyramidal decussation
Rad	radiatum layer of the hippocampus
RtTg	reticulotegmental nucleus of the pons
s5	sensory root of the trigeminal nerve
Sol	solitary nucleus
sp5	spinal trigeminal tract
Sp5	spinal trigeminal nucleus
SubC	subcoeruleus nucleus
tfp	transverse fibers of the pons
tz	trapezoid body
VCN	ventral cochlear nucleus
VS	ventral subiculum

Figure H11
Bregma -11.46 mm
Interaural 1.31 mm

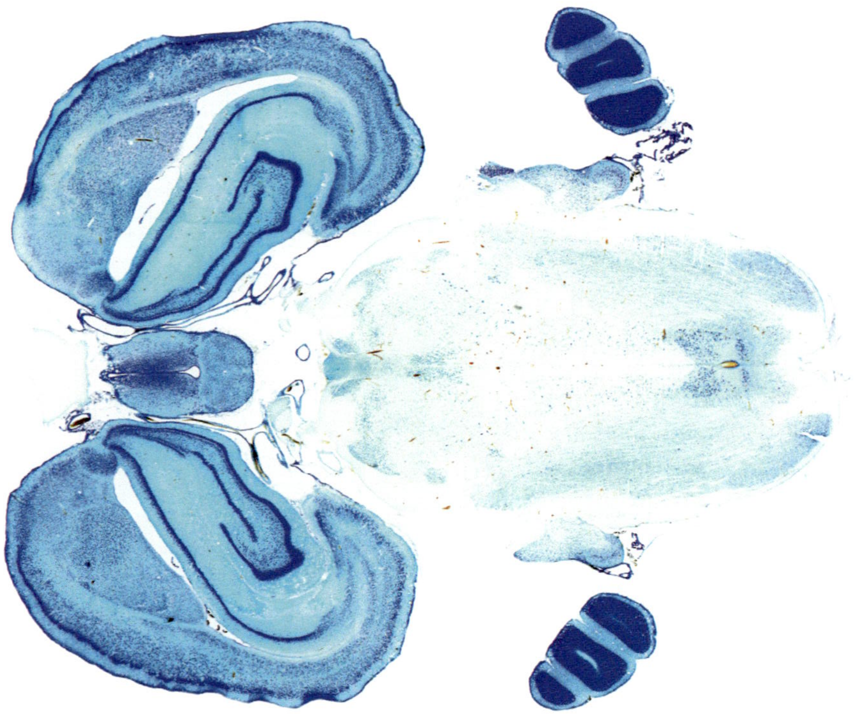

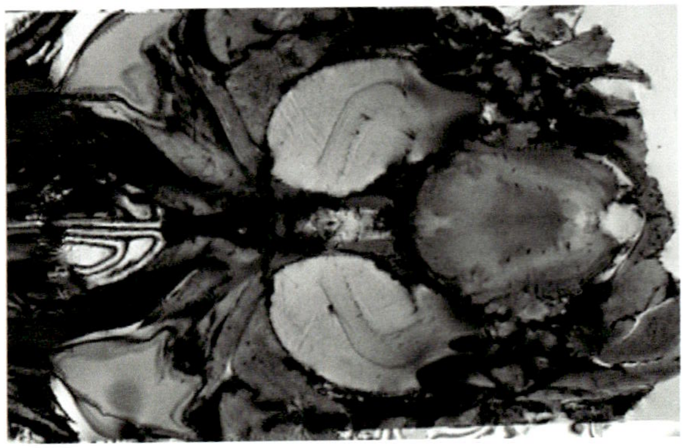

2 layer 2 of cortex
3 layer 3 of cortex
10N dorsal motor nucleus of vagus
12N hypoglossal nucleus
3V 3rd ventricle
5N motor trigeminal nucleus
6n root of abducens nerve
7n facial nerve
8cn cochlear root of the vestibulocochlear nerve
8vn vestibular root of the vestibulocochlear nerve
ACo anterior cortical amygdaloid nucleus
AHi amygdalohippocampal area
Arc arcuate hypothalamic nucleus
BLA basolateral amygdaloid nucleus
BMA basomedial amygdaloid nucleus
CA1 field CA1 of the hippocampus
CA3 field CA3 of the hippocampus
CC central canal
cp cerebral peduncle
cu cuneate fasciculus
Cu cuneate nucleus
DCN dorsal cochlear nucleus
Ent entorhinal cortex
fr fasciculus retroflexus
Gi gigantocellular reticular nucleus
GrDG granular layer of the dentate gyrus
IP interpeduncular nucleus
IRt intermediate reticular nucleus
lfp longitudinal fasciculus of the pons
ll lateral lemniscus
LL nucleus of the lateral lemniscus
LM lateral mammillary nucleus
LMoL lacunosum moleculare layer of the hippocampus
LV lateral ventricle
m5 motor root of the trigeminal nerve
mcp middle cerebellar peduncle
MdD medullary reticular nucleus, dorsal part
MdV medullary reticular nucleus, ventral part
ME median eminence
MeA medial amygdaloid nucleus
ML medial mammillary nucleus, lateral part
ml medial lemniscus
mlf medial longitudinal fasciculus
MnR median raphe nucleus
MoDG molecular layer of the dentate gyrus
MRe mammillary recess of the 3rd ventricle
oc/dsc olivocerebellar tract and dorsal spinocerebellar tract
och optic chiasm
Or oriens layer of the hippocampus
P5 peritrigeminal zone
PCRt parvicellular reticular nucleus
Pir1 piriform cortex, layer 1
PMV premammillary nucleus, ventral part
Pn pontine nuclei
PnC pontine reticular nucleus, caudal part
PnO pontine reticular nucleus, oral part
PoDG polymorph layer of the dentate gyrus
Pr5 principal sensory trigeminal nucleus
PrS presubiculum
Py pyramidal cell layer of the hippocampus
pyx pyramidal decussation
Rad radiatum layer of the hippocampus
RtTg reticulotegmental nucleus of the pons
s5 sensory root of the trigeminal nerve
scpd superior cerebellar peduncle, descending limb
Sol solitary nucleus
sp5 spinal trigeminal tract
Sp5 spinal trigeminal nucleus
SubC subcoeruleus nucleus
tfp transverse fibers of the pons
TuLH tuberal region of lateral hypothalamus
tz trapezoid body
VCN ventral cochlear nucleus
VS ventral subiculum
VTM ventral tuberomammillary nucleus

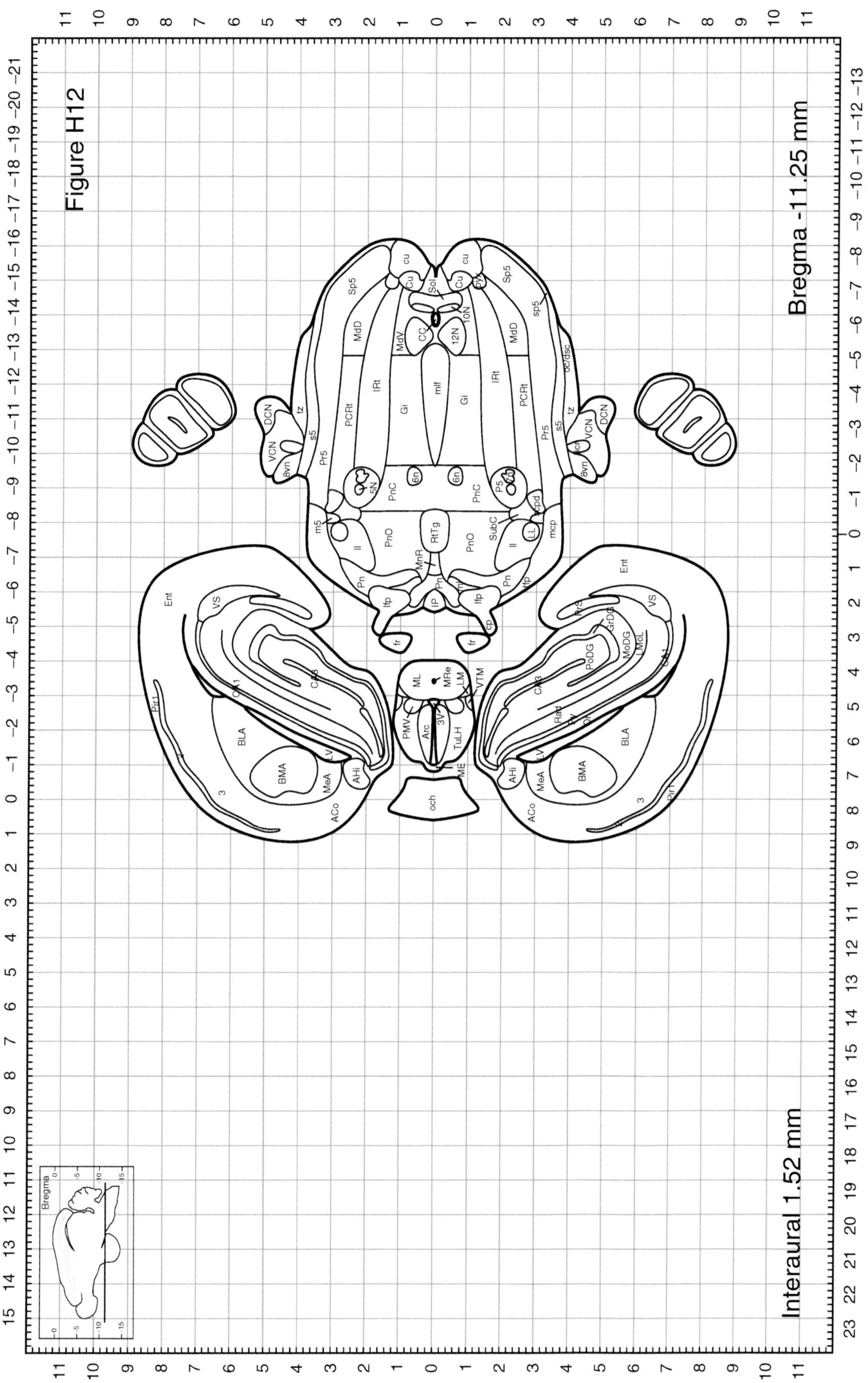
Figure H12
Bregma -11.25 mm
Interaural 1.52 mm
Bregma
cu
Cu
Sol
Sp5
sp5
10N
12N
CC
MdV
MdD
IRt
Gi
mlf
PCRt
Pr5
s5
tz
DCN
VCN
8vn
8cn
ocb/dsc
5N
6n
PnC
P5
m5
scpd
SubC
LL
mcp
ll
PnO
RtTg
MnR
Pn
lfp
IP
ml
cp
fr
Ent
VS
CA1
CA3
Pir1
BLA
BMA
MeA
AHi
ACo
LV
PrS
GrDG
PoDG
MoDG
LMol
Rad
Py
Or
ML
MRe
LM
VTM
PMV
Arc
3V
TuLH
ME
och
3
2

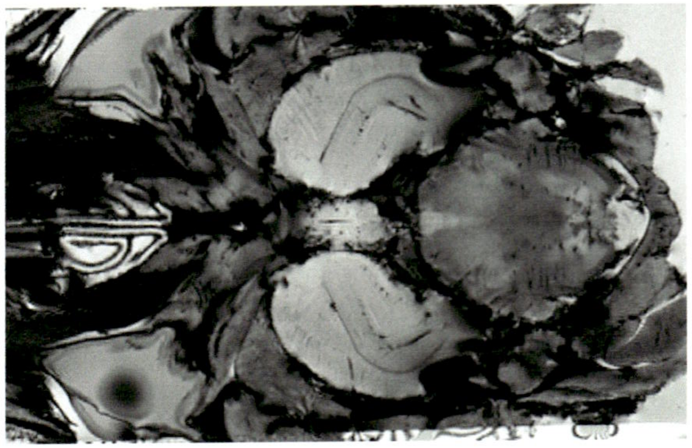

2	layer 2 of cortex
3	layer 3 of cortex
10N	dorsal motor nucleus of vagus
12N	hypoglossal nucleus
3V	3rd ventricle
5N	motor trigeminal nucleus
6n	root of abducens nerve
7n	facial nerve
8vn	vestibular root of the vestibulocochlear nerve
ACo	anterior cortical amygdaloid nucleus
AHi	amygdalohippocampal area
Arc	arcuate hypothalamic nucleus
BLA	basolateral amygdaloid nucleus
BMA	basomedial amygdaloid nucleus
CA1	field CA1 of the hippocampus
CA2	field CA2 of the hippocampus
CA3	field CA3 of the hippocampus
CC	central canal
cp	cerebral peduncle
cu	cuneate fasciculus
Cu	cuneate nucleus
DCN	dorsal cochlear nucleus
Ent	entorhinal cortex
fr	fasciculus retroflexus
Gi	gigantocellular reticular nucleus
GrDG	granular layer of the dentate gyrus
IP	interpeduncular nucleus
IRt	intermediate reticular nucleus
ll	lateral lemniscus
LL	nucleus of the lateral lemniscus
LM	lateral mammillary nucleus
LMoL	lacunosum moleculare layer of the hippocampus
LV	lateral ventricle
m5	motor root of the trigeminal nerve
mcp	middle cerebellar peduncle
MdD	medullary reticular nucleus, dorsal part
MeA	medial amygdaloid nucleus
ML	medial mammillary nucleus, lateral part
mlf	medial longitudinal fasciculus
MnR	median raphe nucleus
MoDG	molecular layer of the dentate gyrus
MRe	mammillary recess of the 3rd ventricle
oc/dsc	olivocerebellar tract and dorsal spinocerebellar tract
och	optic chiasm
opt	optic tract
Or	oriens layer of the hippocampus
P5	peritrigeminal zone
PCRt	parvicellular reticular nucleus
Pir1	piriform cortex, layer 1
PMV	premammillary nucleus, ventral part
Pn	pontine nuclei
PnC	pontine reticular nucleus, caudal part
PnO	pontine reticular nucleus, oral part
PoDG	polymorph layer of the dentate gyrus
Pr5	principal sensory trigeminal nucleus
PrS	presubiculum
Py	pyramidal cell layer of the hippocampus
Rad	radiatum layer of the hippocampus
s5	sensory root of the trigeminal nerve
scpd	superior cerebellar peduncle, descending limb
Sol	solitary nucleus
SON	supraoptic nucleus
sp5	spinal trigeminal tract
Sp5	spinal trigeminal nucleus
SubC	subcoeruleus nucleus
TuLH	tuberal region of lateral hypothalamus
tz	trapezoid body
VCN	ventral cochlear nucleus
VMH	ventromedial hypothalamic nucleus
VS	ventral subiculum
VTM	ventral tuberomammillary nucleus

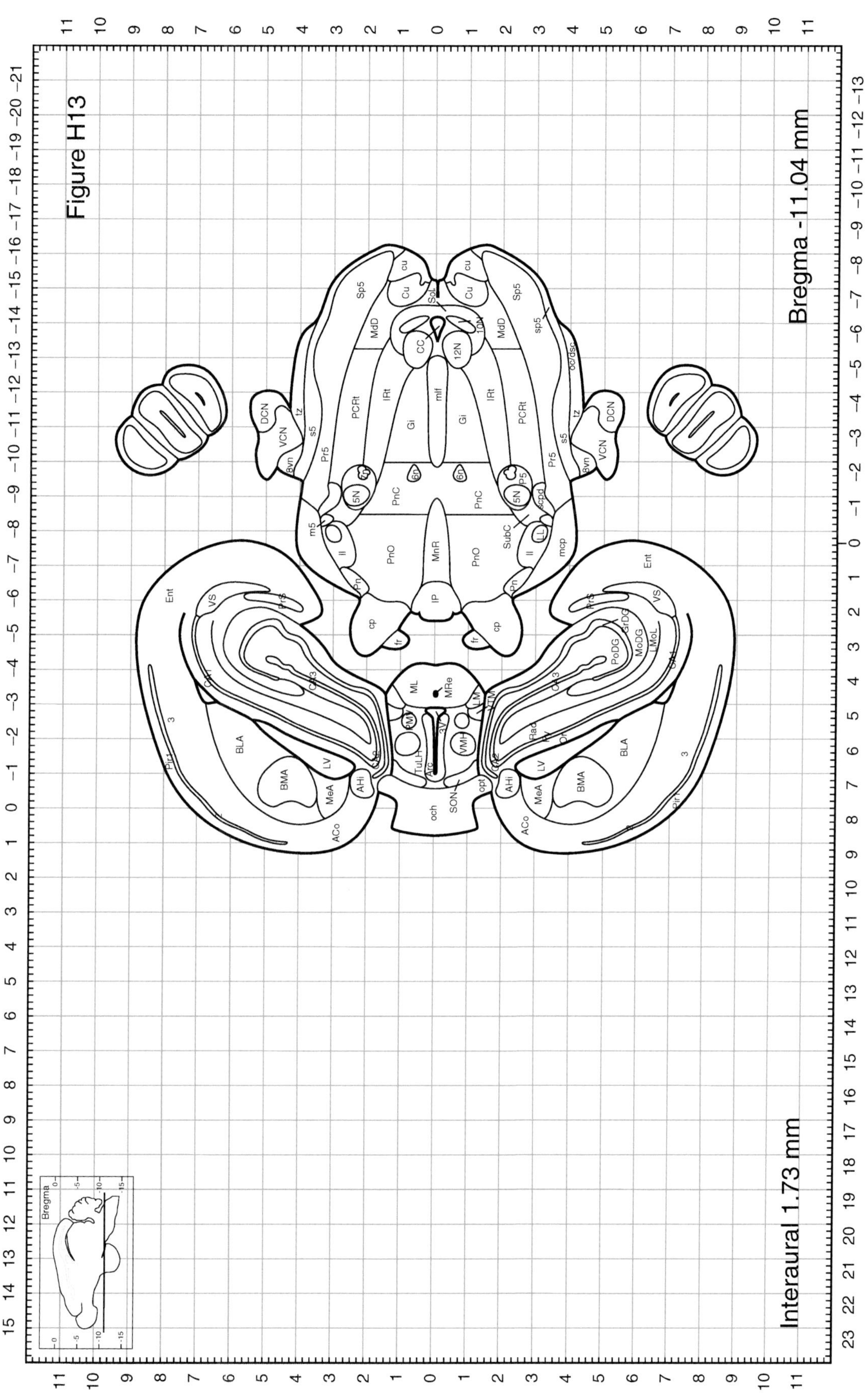
Figure H13
Bregma -11.04 mm
Interaural 1.73 mm
Sp5
cu
Cu
SolL
MdD
10N
12N
CC
sp5
oc/dsc
PCRt
IRt
mlf
Gi
DCN
VCN
tz
s5
Pr5
8vn
6n
5N
P5
PnC
scpd
m5
SubC
ll
LL
mcp
PnO
MnR
Ent
VS
PrS
Pn
IP
cp
fr
GrDG
PoDG
MoDG
LMoL
CA1
CA3
ML
MRe
LM
VTM
PMV
3V
VMH
TuLH
Arc
SON
opt
och
Rad
Py
Or
BLA
LV
BMA
AHi
MeA
ACo
Pir1
3
Bregma

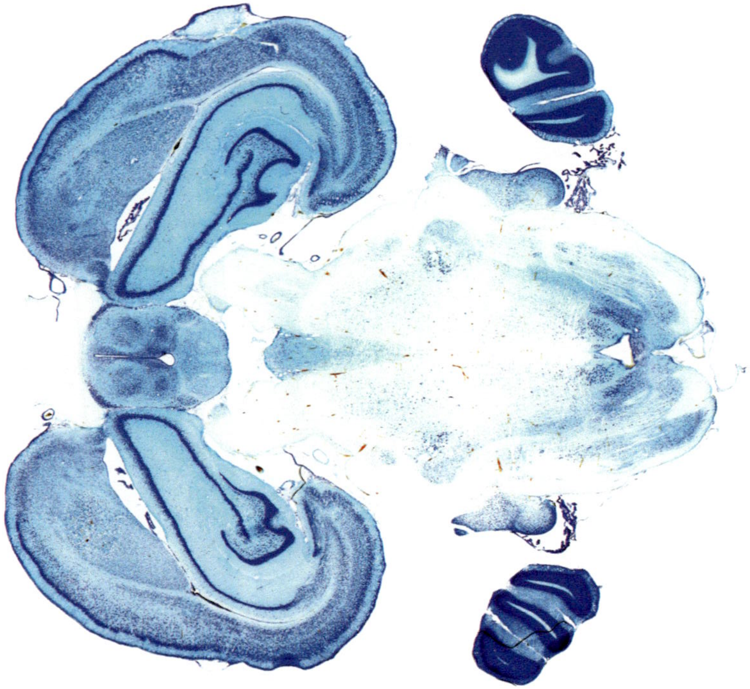

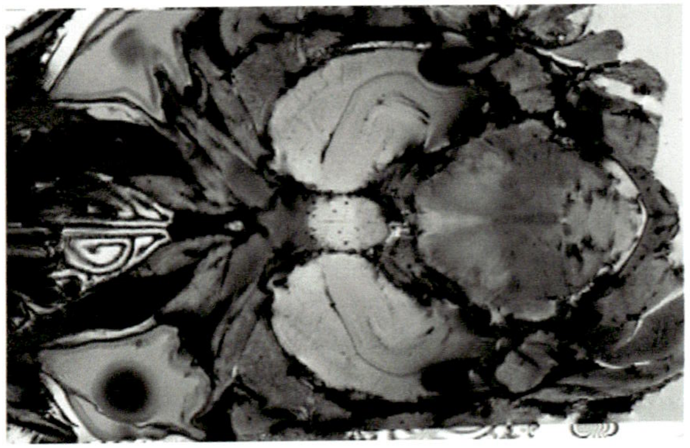

2	layer 2 of cortex
3	layer 3 of cortex
10N	dorsal motor nucleus of vagus
12N	hypoglossal nucleus
3V	3rd ventricle
5N	motor trigeminal nucleus
6n	root of abducens nerve
7n	facial nerve
8vn	vestibular root of the vestibulocochlear nerve
ACo	anterior cortical amygdaloid nucleus
AHi	amygdalohippocampal area
alv	alveus of the hippocampus
AP	area postrema
Arc	arcuate hypothalamic nucleus
BLA	basolateral amygdaloid nucleus
BMA	basomedial amygdaloid nucleus
CA1	field CA1 of the hippocampus
CA2	field CA2 of the hippocampus
CA3	field CA3 of the hippocampus
CC	central canal
cp	cerebral peduncle
cu	cuneate fasciculus
Cu	cuneate nucleus
DCN	dorsal cochlear nucleus
Ent	entorhinal cortex
f	fornix
fr	fasciculus retroflexus
Gi	gigantocellular reticular nucleus
GrDG	granular layer of the dentate gyrus
IP	interpeduncular nucleus
IRt	intermediate reticular nucleus
ll	lateral lemniscus
LL	nucleus of the lateral lemniscus
LM	lateral mammillary nucleus
LMoL	lacunosum moleculare layer of the hippocampus
LV	lateral ventricle
m5	motor root of the trigeminal nerve
mcp	middle cerebellar peduncle
MdD	medullary reticular nucleus, dorsal part
MeA	medial amygdaloid nucleus
ML	medial mammillary nucleus, lateral part
mlf	medial longitudinal fasciculus
MM	medial mammillary nucleus, medial part
MnR	median raphe nucleus
MoDG	molecular layer of the dentate gyrus
mt	mammillothalamic tract
MT	medial terminal nucleus of the accessory optic tract
oc/dsc	olivocerebellar tract and dorsal spinocerebellar tract
och	optic chiasm
opt	optic tract
Or	oriens layer of the hippocampus
P5	peritrigeminal zone
PCRt	parvicellular reticular nucleus
Pir1	piriform cortex, layer 1
PMD	premammillary nucleus, dorsal part
PMV	premammillary nucleus, ventral part
Pn	pontine nuclei
PnC	pontine reticular nucleus, caudal part
PnO	pontine reticular nucleus, oral part
PoDG	polymorph layer of the dentate gyrus
Pr5	principal sensory trigeminal nucleus
PrS	presubiculum
Py	pyramidal cell layer of the hippocampus
Rad	radiatum layer of the hippocampus
s5	sensory root of the trigeminal nerve
scpd	superior cerebellar peduncle, descending limb
SNR	substantia nigra, reticular part
Sol	solitary nucleus
SON	supraoptic nucleus
sp5	spinal trigeminal tract
Sp5	spinal trigeminal nucleus
SubC	subcoeruleus nucleus
TuLH	tuberal region of lateral hypothalamus
tz	trapezoid body
VCN	ventral cochlear nucleus
VMH	ventromedial hypothalamic nucleus
VS	ventral subiculum
VTM	ventral tuberomammillary nucleus

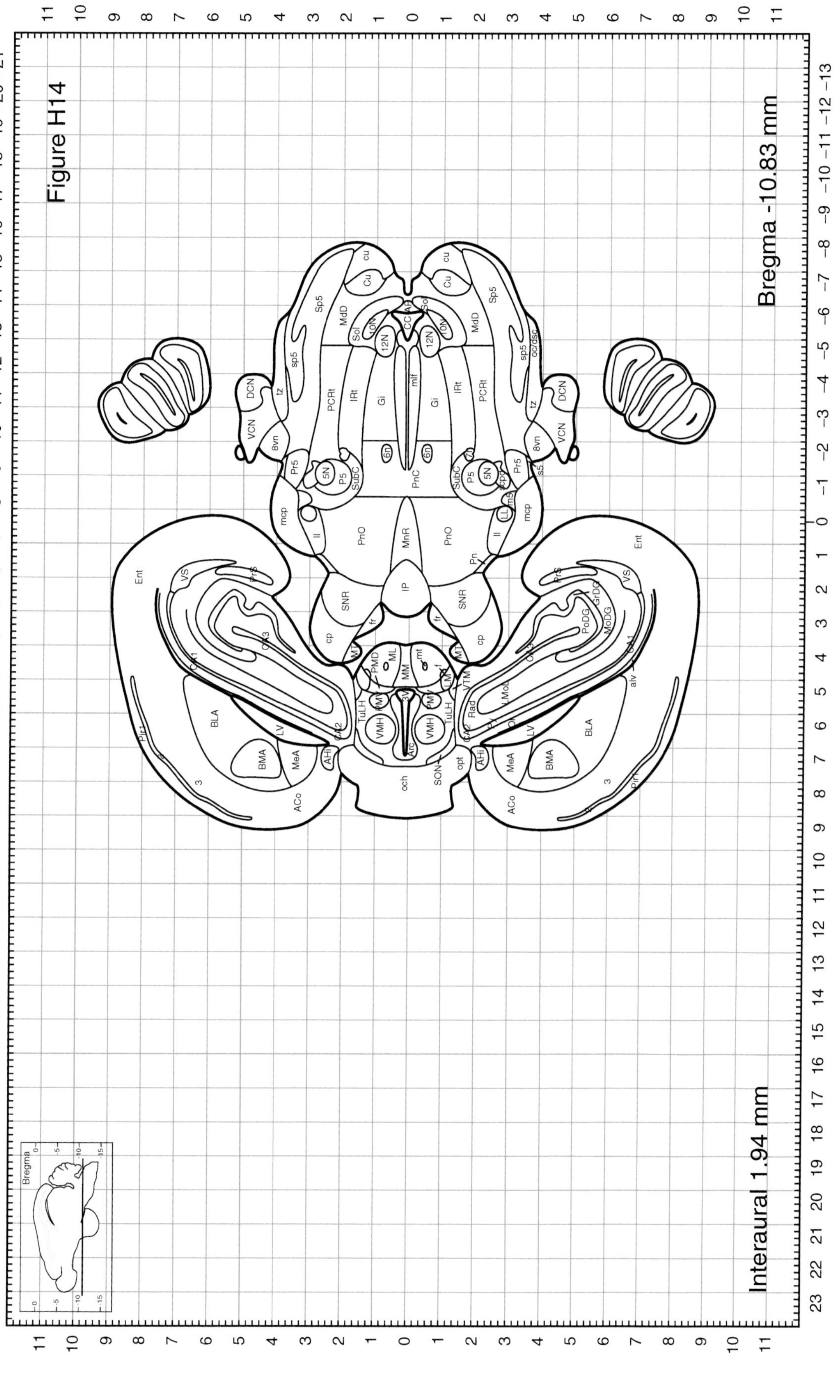
Figure H14
Bregma -10.83 mm
Interaural 1.94 mm
Bregma
cu
Cu
Sp5
MdD
Sol
10N
12N
CC
AP
mlf
sp5
oc/dsc
DCN
VCN
tz
8vn
PCRt
IRt
Gi
6n
Pr5
5N
P5
SubC
PnC
mcp
ll
LL
scp
PnO
MnR
Pn
Ent
VS
PrS
IP
SNR
fr
cp
ML
MM
mt
PMD
PMV
3V
TuLH
VMH
Arc
LM
VTM
CA1
CA2
CA3
PoDG
GrDG
MoDG
alv
LMol
Rad
Or
LV
BLA
BMA
MeA
AHi
ACo
Pir
och
opt
SON

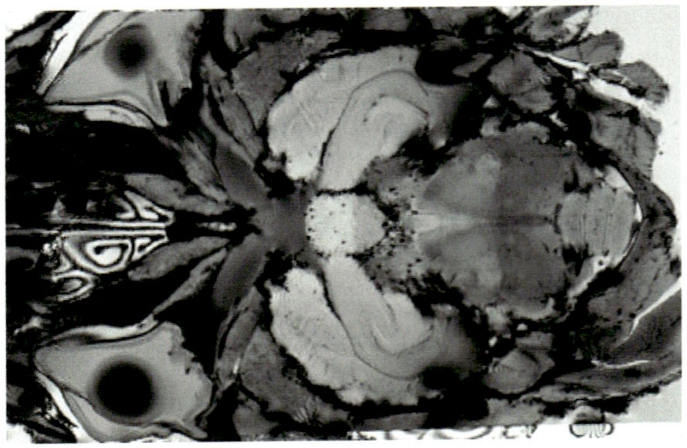

2	layer 2 of cortex
3	layer 3 of cortex
10N	dorsal motor nucleus of vagus
3V	3rd ventricle
4V	4th ventricle
5N	motor trigeminal nucleus
6n	root of abducens nerve
7n	facial nerve
8vn	vestibular root of the vestibulocochlear nerve
ACo	anterior cortical amygdaloid nucleus
alv	alveus of the hippocampus
AP	area postrema
Arc	arcuate hypothalamic nucleus
BLA	basolateral amygdaloid nucleus
CA1	field CA1 of the hippocampus
CA2	field CA2 of the hippocampus
CA3	field CA3 of the hippocampus
CeA	central amygdaloid nucleus
cp	cerebral peduncle
cu	cuneate fasciculus
Cu	cuneate nucleus
DCN	dorsal cochlear nucleus
DMH	dorsomedial hypothalamic nucleus
ECu	external cuneate nucleus
Ent	entorhinal cortex
f	fornix
Gi	gigantocellular reticular nucleus
Gr	gracile nucleus
GrDG	granular layer of the dentate gyrus
icp	inferior cerebellar peduncle
IM	intercalated amygdaloid nucleus, main part
IP	interpeduncular nucleus
IRt	intermediate reticular nucleus
LHA	lateral hypothalamic area
ll	lateral lemniscus
LL	nucleus of the lateral lemniscus
LM	lateral mammillary nucleus
LMoL	lacunosum moleculare layer of the hippocampus
LV	lateral ventricle
m5	motor root of the trigeminal nerve
mcp	middle cerebellar peduncle
MeA	medial amygdaloid nucleus
ML	medial mammillary nucleus, lateral part
mlf	medial longitudinal fasciculus
MM	medial mammillary nucleus, medial part
MnR	median raphe nucleus
MoDG	molecular layer of the dentate gyrus
mt	mammillothalamic tract
MT	medial terminal nucleus of the accessory optic tract
Mx	matrix region of the medulla
och	optic chiasm
opt	optic tract
Or	oriens layer of the hippocampus
P5	peritrigeminal zone
PBP	parabrachial pigmented nucleus of the VTA
PCRt	parvicellular reticular nucleus
Pir1	piriform cortex, layer 1
PMD	premammillary nucleus, dorsal part
PnC	pontine reticular nucleus, caudal part
PnO	pontine reticular nucleus, oral part
PoDG	polymorph layer of the dentate gyrus
Pr	prepositus nucleus
Pr5	principal sensory trigeminal nucleus
PrS	presubiculum
Py	pyramidal cell layer of the hippocampus
Rad	radiatum layer of the hippocampus
s5	sensory root of the trigeminal nerve
scpd	superior cerebellar peduncle, descending limb
SNR	substantia nigra, reticular part
Sol	solitary nucleus
SON	supraoptic nucleus
sp5	spinal trigeminal tract
Sp5	spinal trigeminal nucleus
SpVe	spinal vestibular nucleus
SubC	subcoeruleus nucleus
VCN	ventral cochlear nucleus
VMH	ventromedial hypothalamic nucleus
VS	ventral subiculum
VTM	ventral tuberomammillary nucleus

Figure H15
Bregma -10.63 mm
Interaural 2.15 mm
Bregma
cu
sp5
Cu
Gr
4V
AP
Mx
Sp5
ECu
Sol
Pr
10N
mlf
icp
DCN
SpVe
IRt
Gi
VCN
8vn
PCRt
PnC
S5
5N
Pr5
SubC
m5
mcp
LL
ll
PnO
MnR
Ent
VS
PrS
IP
SNR
PBP
GrDG
PoDG
MoDG
CA1
CA3
cp
MT
ML
mt
LM
MM
alv
LMol
Or
Rad
Pir1
3V
DMH
PMD
VTM
LHA
VMH
CA2
BLA
LV
3
CeA
MeA
opt
SON
IM
ACo
och
Pir

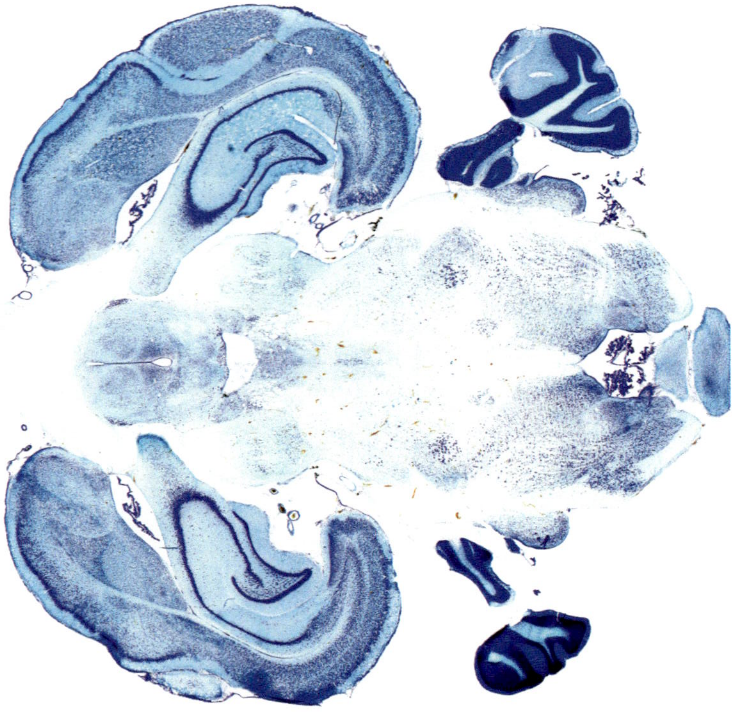

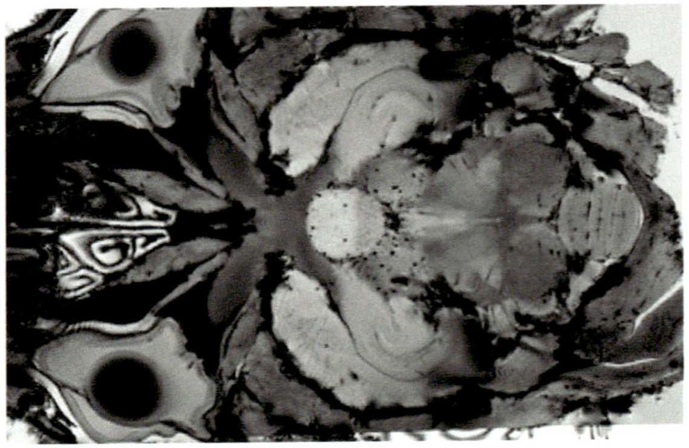

2	layer 2 of cortex
3	layer 3 of cortex
3V	3rd ventricle
4V	4th ventricle
5N	motor trigeminal nucleus
7n	facial nerve
8vn	vestibular root of the vestibulocochlear nerve
AA	anterior amygdaloid area
Ac	auditory core area
alv	alveus of the hippocampus
AP	area postrema
BLA	basolateral amygdaloid nucleus
CA1	field CA1 of the hippocampus
CA2	field CA2 of the hippocampus
CA3	field CA3 of the hippocampus
CeA	central amygdaloid nucleus
cp	cerebral peduncle
cu	cuneate fasciculus
Cu	cuneate nucleus
DCN	dorsal cochlear nucleus
DMH	dorsomedial hypothalamic nucleus
Ect	ectorhinal cortex
ECu	external cuneate nucleus
Ent	entorhinal cortex
f	fornix
g7	genu of the facial nerve
Gi	gigantocellular reticular nucleus
Gr	gracile nucleus
GrDG	granular layer of the dentate gyrus
icp	inferior cerebellar peduncle
IM	intercalated amygdaloid nucleus, main part
IP	interpeduncular nucleus
LHA	lateral hypothalamic area
ll	lateral lemniscus
LL	nucleus of the lateral lemniscus
LMoL	lacunosum moleculare layer of the hippocampus
LV	lateral ventricle
m5	motor root of the trigeminal nerve
mcp	middle cerebellar peduncle
MeA	medial amygdaloid nucleus
ML	medial mammillary nucleus, lateral part
mlf	medial longitudinal fasciculus
MM	medial mammillary nucleus, medial part
MnR	median raphe nucleus
MoDG	molecular layer of the dentate gyrus
mt	mammillothalamic tract
MT	medial terminal nucleus of the accessory optic tract
Mx	matrix region of the medulla
och	optic chiasm
opt	optic tract
Or	oriens layer of the hippocampus
PBP	parabrachial pigmented nucleus of the VTA
Pir1	piriform cortex, layer 1
PLH	peduncular part of lateral hypothalamus
PnC	pontine reticular nucleus, caudal part
PnO	pontine reticular nucleus, oral part
PoDG	polymorph layer of the dentate gyrus
Pr	prepositus nucleus
Pr5	principal sensory trigeminal nucleus
PrS	presubiculum
Py	pyramidal cell layer of the hippocampus
Rad	radiatum layer of the hippocampus
SNR	substantia nigra, reticular part
Sol	solitary nucleus
SON	supraoptic nucleus
SpVe	spinal vestibular nucleus
SubC	subcoeruleus nucleus
SuML	supramammillary nucleus, lateral par
Tu	olfactory tubercle
VCN	ventral cochlear nucleus
Ves	vestibular nucleus
VMH	ventromedial hypothalamic nucleus
VS	ventral subiculum

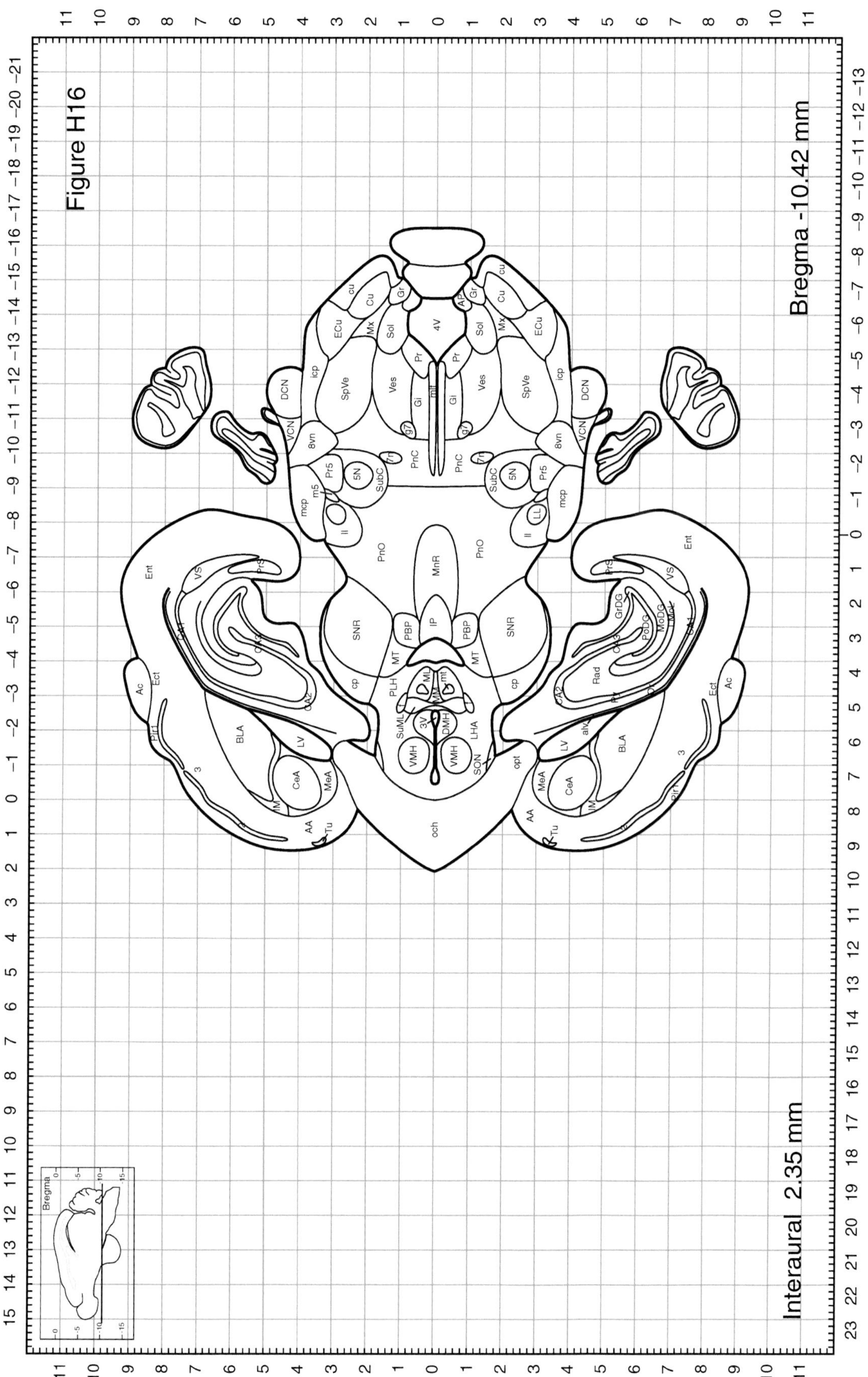

Figure H16
Bregma -10.42 mm
Interaural 2.35 mm
Bregma
cu
Cu
Gr
AP
ECu
Mx
Sol
4V
Pr
icp
DCN
SpVe
Ves
Gi
mlf
VCN
8vn
g7
7n
PnC
Pr5
5N
SubC
m5
mcp
LL
ll
PnO
MnR
Ent
VS
PrS
IP
SNR
PBP
CA1
CA2
CA3
GrDG
PoDG
MoDG
LMol
Rad
Py
Or
MT
cp
PLH
ML
MM
mt
Ac
Ect
SuML
3V
DMH
LHA
alv
LV
BLA
Pir1
VMH
SON
opt
CeA
MeA
IM
AA
Tu
och
3

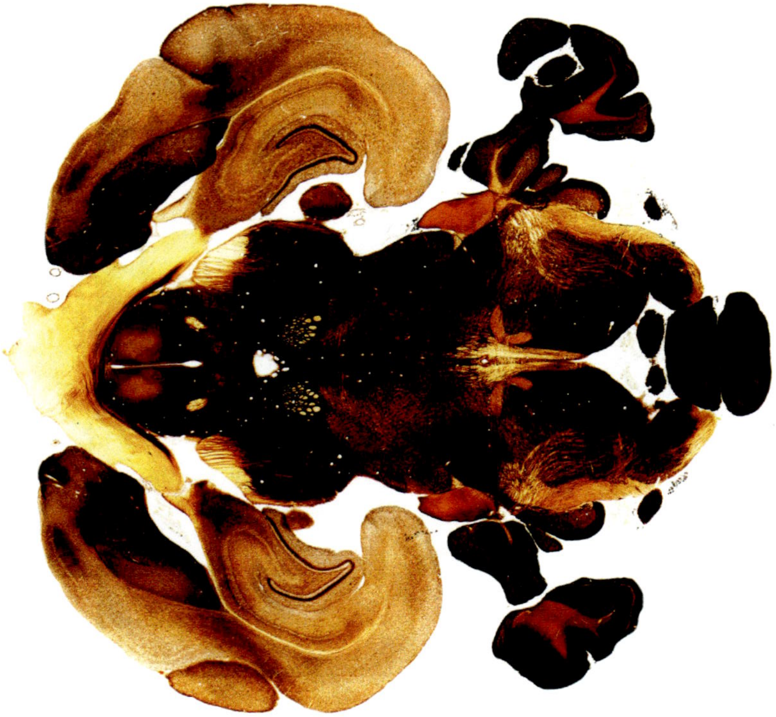

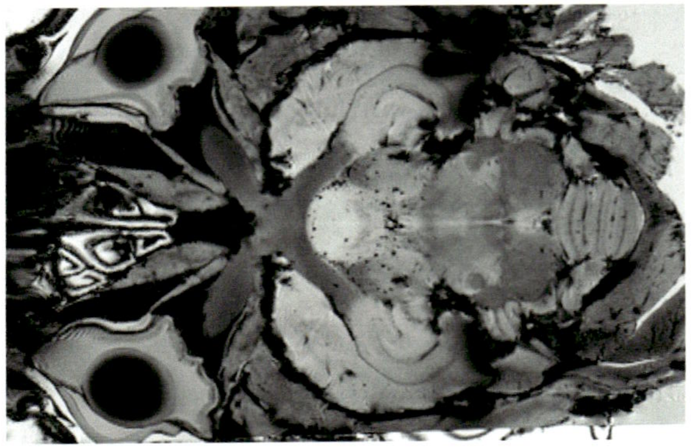

2	layer 2 of cortex
3	layer 3 of cortex
3V	3rd ventricle
4V	4th ventricle
7n	facial nerve
8vn	vestibular root of the vestibulocochlear nerve
AA	anterior amygdaloid area
Ac	auditory core area
alv	alveus of the hippocampus
BLA	basolateral amygdaloid nucleus
CA1	field CA1 of the hippocampus
CA2	field CA2 of the hippocampus
CA3	field CA3 of the hippocampus
CeA	central amygdaloid nucleus
cp	cerebral peduncle
cu	cuneate fasciculus
Cu	cuneate nucleus
DCN	dorsal cochlear nucleus
DMH	dorsomedial hypothalamic nucleus
Ect	ectorhinal cortex
ECu	external cuneate nucleus
Ent	entorhinal cortex
f	fornix
fr	fasciculus retroflexus
g7	genu of the facial nerve
Gr	gracile nucleus
GrDG	granular layer of the dentate gyrus
icp	inferior cerebellar peduncle
IM	intercalated amygdaloid nucleus, main part
IP	interpeduncular nucleus
LHA	lateral hypothalamic area
ll	lateral lemniscus
LL	nucleus of the lateral lemniscus
LMoL	lacunosum moleculare layer of the hippocampus
LOT	nucleus of the lateral olfactory tract
LV	lateral ventricle
m5	motor root of the trigeminal nerve
mcp	middle cerebellar peduncle
MeA	medial amygdaloid nucleus
MGV	medial geniculate nucleus, ventral part
mlf	medial longitudinal fasciculus
MnR	median raphe nucleus
MoDG	molecular layer of the dentate gyrus
mt	mammillothalamic tract
MT	medial terminal nucleus of the accessory optic tract
och	optic chiasm
opt	optic tract
Or	oriens layer of the hippocampus
P5	peritrigeminal zone
PBP	parabrachial pigmented nucleus of the VTA
Pir1	piriform cortex, layer 1
PLH	peduncular part of lateral hypothalamus
PnC	pontine reticular nucleus, caudal part
PnO	pontine reticular nucleus, oral part
PoDG	polymorph layer of the dentate gyrus
Pr	prepositus nucleus
Pr5	principal sensory trigeminal nucleus
PRh	perirhinal cortex
PrS	presubiculum
Py	pyramidal cell layer of the hippocampus
Rad	radiatum layer of the hippocampus
SNR	substantia nigra, reticular part
Sol	solitary nucleus
SON	supraoptic nucleus
SpVe	spinal vestibular nucleus
SubC	subcoeruleus nucleus
SuML	supramammillary nucleus, lateral par
SuMM	supramammillary nucleus, medial part
Tu	olfactory tubercle
VCN	ventral cochlear nucleus
Ves	vestibular nucleus
VMH	ventromedial hypothalamic nucleus
VS	ventral subiculum

Figure H17
Bregma -10.21 mm
Interaural 2.56 mm

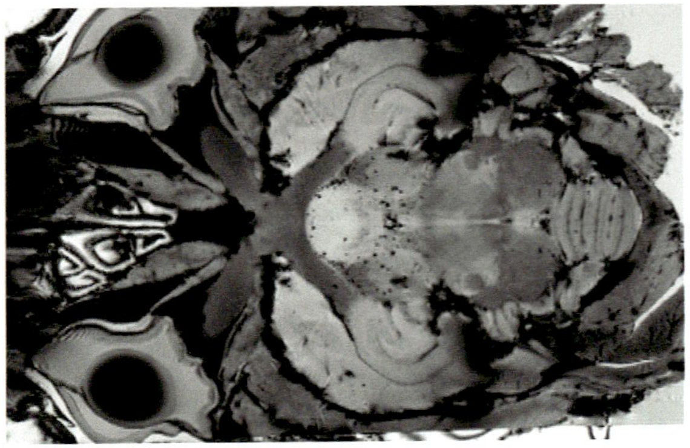

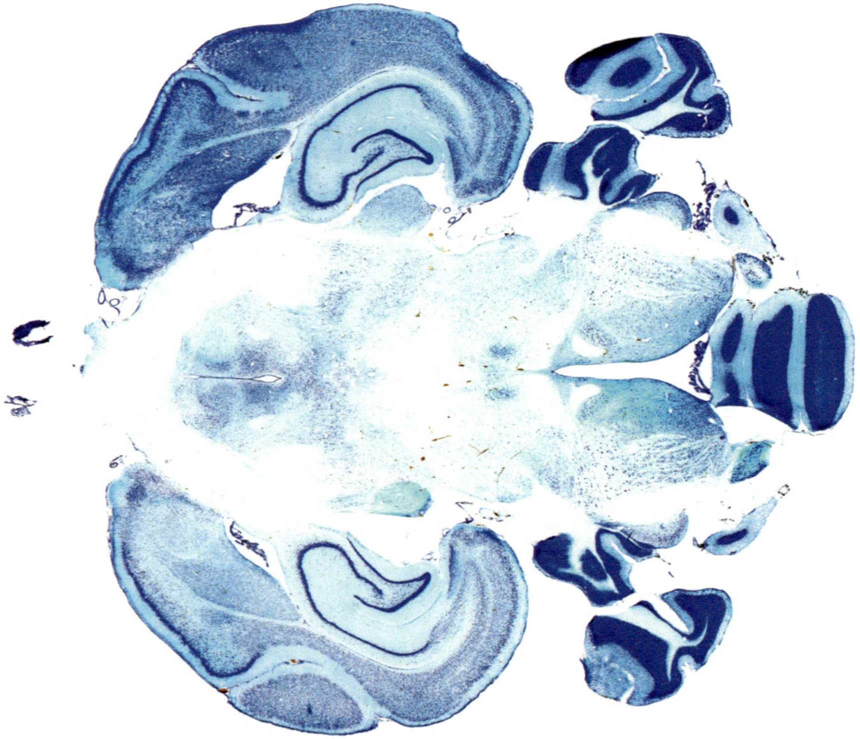

2	layer 2 of cortex
3	layer 3 of cortex
3V	3rd ventricle
4V	4th ventricle
6N	abducens nucleus
AA	anterior amygdaloid area
Ac	auditory core area
alv	alveus of the hippocampus
BLA	basolateral amygdaloid nucleus
CA1	field CA1 of the hippocampus
CA2	field CA2 of the hippocampus
CA3	field CA3 of the hippocampus
CeA	central amygdaloid nucleus
CG	central gray
CLi	caudal linear nucleus of the raphe
cp	cerebral peduncle
cu	cuneate fasciculus
DCN	dorsal cochlear nucleus
DMH	dorsomedial hypothalamic nucleus
Ect	ectorhinal cortex
ECu	external cuneate nucleus
Ent	entorhinal cortex
f	fornix
fi	fimbria of the hippocampus
fr	fasciculus retroflexus
g7	genu of the facial nerve
GrDG	granular layer of the dentate gyrus
icp	inferior cerebellar peduncle
IM	intercalated amygdaloid nucleus, main part
IP	interpeduncular nucleus
LHA	lateral hypothalamic area
ll	lateral lemniscus
LL	nucleus of the lateral lemniscus
LMoL	lacunosum moleculare layer of the hippocampus
LOT	nucleus of the lateral olfactory tract
LV	lateral ventricle
m5	motor root of the trigeminal nerve
mcp	middle cerebellar peduncle
MeA	medial amygdaloid nucleus
MGV	medial geniculate nucleus, ventral part
ml	medial lemniscus
mlf	medial longitudinal fasciculus
MnR	median raphe nucleus
MoDG	molecular layer of the dentate gyrus
mt	mammillothalamic tract
och	optic chiasm
opt	optic tract
Or	oriens layer of the hippocampus
P5	peritrigeminal zone
PBP	parabrachial pigmented nucleus of the VTA
Pir1	piriform cortex, layer 1
PLH	peduncular part of lateral hypothalamus
PnO	pontine reticular nucleus, oral part
PoDG	polymorph layer of the dentate gyrus
Pr	prepositus nucleus
Pr5	principal sensory trigeminal nucleus
PRh	perirhinal cortex
PrS	presubiculum
Py	pyramidal cell layer of the hippocampus
Rad	radiatum layer of the hippocampus
Rbd	rhabdoid nucleus
SNR	substantia nigra, reticular part
SpVe	spinal vestibular nucleus
SuM	supramammillary nucleus
SuVe	superior vestibular nucleus
Tu	olfactory tubercle
Ves	vestibular nucleus
VMH	ventromedial hypothalamic nucleus
VS	ventral subiculum
VTAR	ventral tegmental area, rostral part
vtgx	ventral tegmental decussation
xscp	decussation of the superior cerebellar peduncle

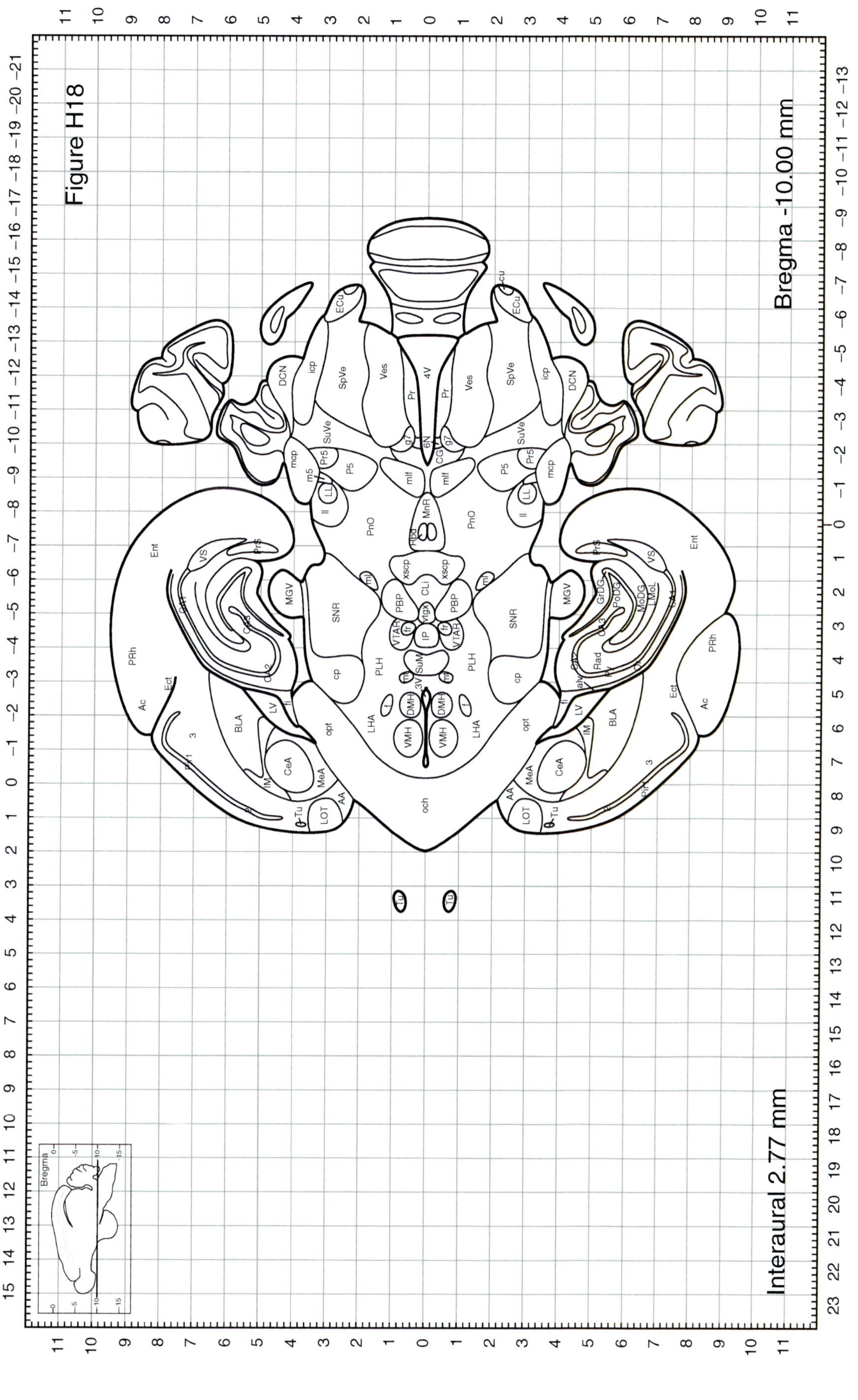

Figure H18
Bregma -10.00 mm
Interaural 2.77 mm
Bregma
DCN
icp
SpVe
Ves
4V
Pr
ECu
SuVe
mcp
m5
Pr5
P5
g7
6N
CG
mlf
LL
ll
MnR
PnO
Rbd
Ent
VS
PrS
xscp
ml
CLi
MGV
SNR
PBP
vtgx
IP
fr
VTAR
PRh
PLH
SuM
3V
cp
Ect
Ac
LV
BLA
opt
DMH
LHA
VMH
f
CeA
MeA
IM
AA
LOT
Tu
och
Rad
GrDG
PoDG
MoDG
LMol
CA1
CA2
CA3
alv
fi
Py
Pir
3

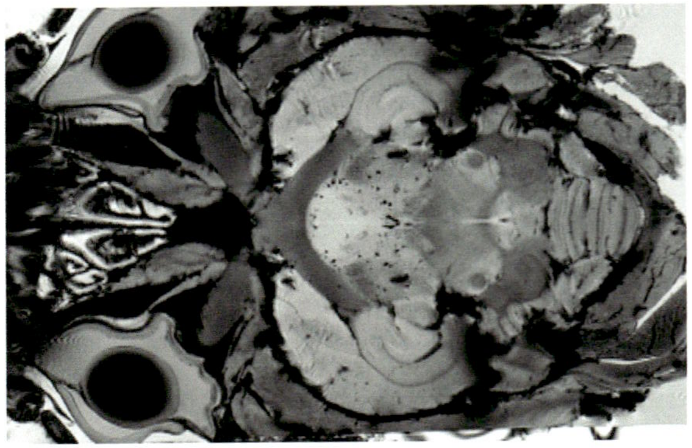

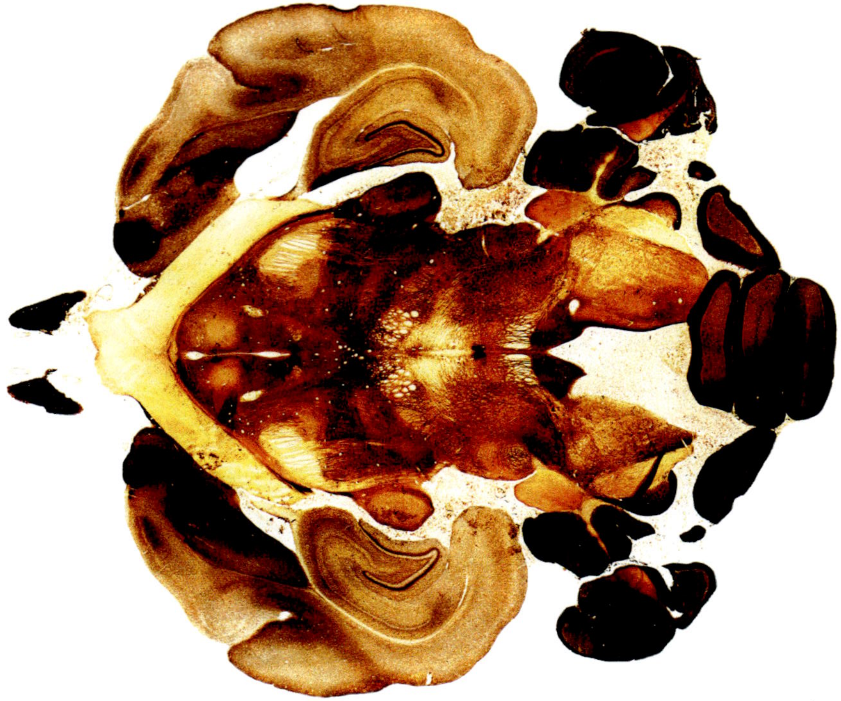

2	layer 2 of cortex
3	layer 3 of cortex
3V	3rd ventricle
4V	4th ventricle
AA	anterior amygdaloid area
Ac	auditory core area
alv	alveus of the hippocampus
BLA	basolateral amygdaloid nucleus
CA1	field CA1 of the hippocampus
CA2	field CA2 of the hippocampus
CA3	field CA3 of the hippocampus
CeA	central amygdaloid nucleus
CG	central gray
CLi	caudal linear nucleus of the raphe
cp	cerebral peduncle
DCN	dorsal cochlear nucleus
DMH	dorsomedial hypothalamic nucleus
DTg	dorsal tegmental nucleus
Ect	ectorhinal cortex
Ent	entorhinal cortex
f	fornix
fi	fimbria of the hippocampus
fr	fasciculus retroflexus
GrDG	granular layer of the dentate gyrus
icp	inferior cerebellar peduncle
IM	intercalated amygdaloid nucleus, main part
LDTg	laterodorsal tegmental nucleus
LHA	lateral hypothalamic area
ll	lateral lemniscus
LL	nucleus of the lateral lemniscus
LMoL	lacunosum moleculare layer of the hippocampus
LOT	nucleus of the lateral olfactory tract
LV	lateral ventricle
m5	motor root of the trigeminal nerve
mcp	middle cerebellar peduncle
MeA	medial amygdaloid nucleus
MGV	medial geniculate nucleus, ventral part
ml	medial lemniscus
mlf	medial longitudinal fasciculus
MnR	median raphe nucleus
MoDG	molecular layer of the dentate gyrus
MPB	medial parabrachial nucleus
mt	mammillothalamic tract
och	optic chiasm
opt	optic tract
Or	oriens layer of the hippocampus
PBP	parabrachial pigmented nucleus of the VTA
PH	posterior hypothalamic nucleus
Pir1	piriform cortex, layer 1
PLH	peduncular part of lateral hypothalamus
PnO	pontine reticular nucleus, oral part
PoDG	polymorph layer of the dentate gyrus
Pr	prepositus nucleus
PRh	perirhinal cortex
PrS	presubiculum
PTg	pedunculopontine tegmental nucleus
Py	pyramidal cell layer of the hippocampus
Rad	radiatum layer of the hippocampus
Rbd	rhabdoid nucleus
RLi	rostral linear nucleus of the raphe
SNC	substantia nigra, compact part
SNR	substantia nigra, reticular part
SpVe	spinal vestibular nucleus
STh	subthalamic nucleus
SuM	supramammillary nucleus
SuVe	superior vestibular nucleus
Tu	olfactory tubercle
Tu1	layer 1 of olfactory tubercle
Ves	vestibular nucleus
VMH	ventromedial hypothalamic nucleus
VS	ventral subiculum
VTAR	ventral tegmental area, rostral part
vtgx	ventral tegmental decussation
xscp	decussation of the superior cerebellar peduncle

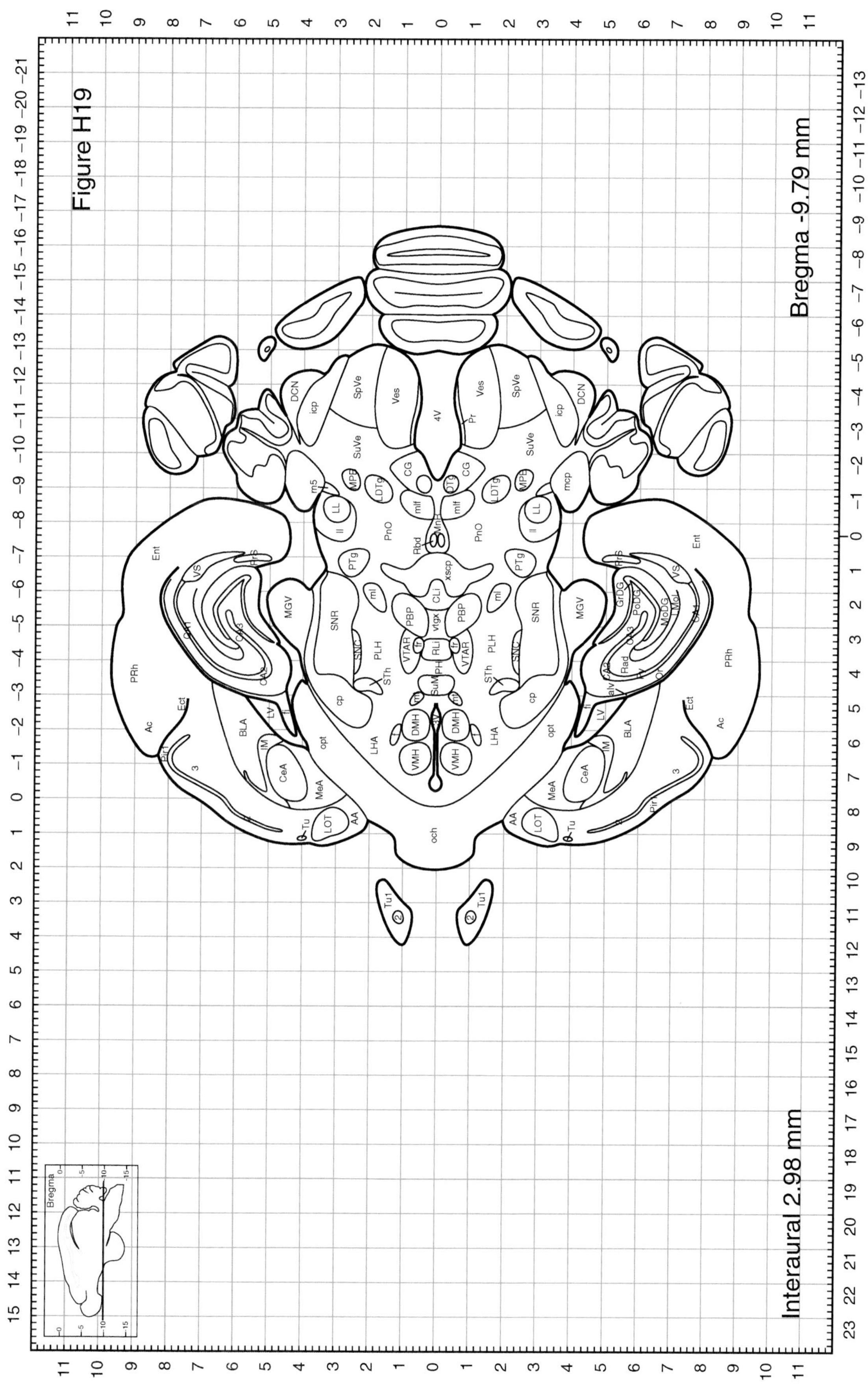
Figure H19
Bregma -9.79 mm
Interaural 2.98 mm
DCN
icp
SpVe
Ves
4V
Pr
SuVe
mcp
m5
MPB
LDTg
CG
DTg
LL
ll
mlf
PnO
MnR
Rbd
PTg
xscp
CLi
ml
MGV
SNR
PBP
vtgx
PLH
SNC
VTAR
fr
RLi
PH
STh
SuM
cp
DMH
VMH
LHA
opt
Ent
PRh
VS
PrS
CA1
CA3
CA2
GrDG
PoDG
MoDG
Mol
Rad
Py
Or
alv
Ect
Ac
LV
fi
BLA
IM
CeA
MeA
Pir1
3
AA
LOT
Tu
och
Tu1
2
Bregma

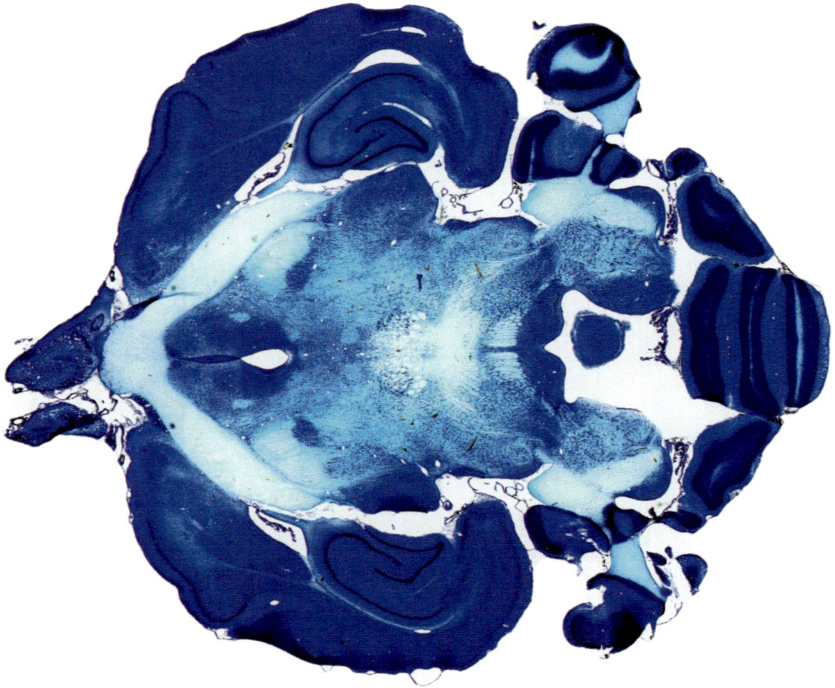

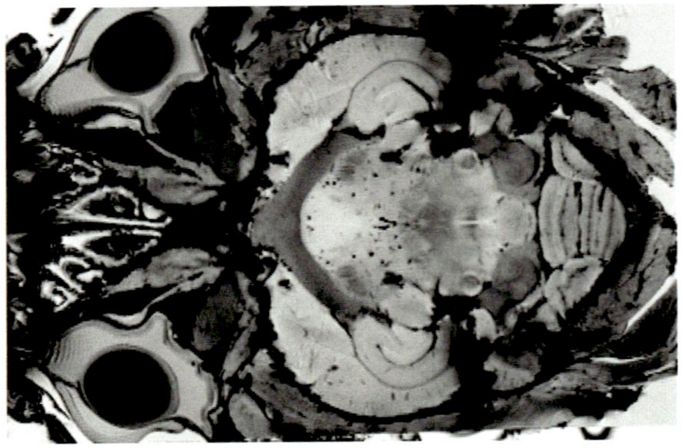

2 layer 2 of cortex
3 layer 3 of cortex
3V 3rd ventricle
4V 4th ventricle
AA anterior amygdaloid area
AHy anterior hypothalamic area
alv alveus of the hippocampus
BLA basolateral amygdaloid nucleus
CA1 field CA1 of the hippocampus
CA2 field CA2 of the hippocampus
CA3 field CA3 of the hippocampus
CeA central amygdaloid nucleus
CG central gray
CLi caudal linear nucleus of the raphe
cp cerebral peduncle
das dorsal acoustic stria
DCN dorsal cochlear nucleus
DMH dorsomedial hypothalamic nucleus
DTg dorsal tegmental nucleus
EA sublenticular extended amygdala
Ect ectorhinal cortex
Ent entorhinal cortex
f fornix
fi fimbria of the hippocampus
fr fasciculus retroflexus
GrDG granular layer of the dentate gyrus
icp inferior cerebellar peduncle
IM intercalated amygdaloid nucleus, main part
LC locus coeruleus
LDTg laterodorsal tegmental nucleus
LHA lateral hypothalamic area
ll lateral lemniscus
LL nucleus of the lateral lemniscus
LMoL lacunosum moleculare layer of the hippocampus
LOT nucleus of the lateral olfactory tract
LPB lateral parabrachial nucleus
LV lateral ventricle
mcp middle cerebellar peduncle
me5 mesencephalic trigeminal tract
Me5 mesencephalic trigeminal nucleus
MeA medial amygdaloid nucleus
MGV medial geniculate nucleus, ventral part
ml medial lemniscus
mlf medial longitudinal fasciculus
MnR median raphe nucleus
MoDG molecular layer of the dentate gyrus
MPB medial parabrachial nucleus
mRt mesencephalic reticular formation
mt mammillothalamic tract
och optic chiasm
opt optic tract
Or oriens layer of the hippocampus
PBG parabigeminal nucleus
PBP parabrachial pigmented nucleus of the VTA
PH posterior hypothalamic nucleus
Pir1 piriform cortex, layer 1
PLH peduncular part of lateral hypothalamus
Po posterior complex of the thalamus
PoDG polymorph layer of the dentate gyrus
PRh perirhinal cortex
PrS presubiculum
pRt p1 reticular formation
PTg pedunculopontine tegmental nucleus
Py pyramidal cell layer of the hippocampus
Rad radiatum layer of the hippocampus
Rbd rhabdoid nucleus
RLi rostral linear nucleus of the raphe
RN red nucleus
SCN suprachiasmatic nucleus
scp superior cerebellar peduncle
SNC substantia nigra, compact part
SNL substantia nigra, lateral part
SNR substantia nigra, reticular part
SON supraoptic nucleus
SPTg subpeduncular tegmental nucleus
SpVe spinal vestibular nucleus
STh subthalamic nucleus
SuVe superior vestibular nucleus
Tu olfactory tubercle
Tu1 layer 1 of olfactory tubercle
Ves vestibular nucleus
VMH ventromedial hypothalamic nucleus
VS ventral subiculum
vsc ventral spinocerebellar tract
VTAR ventral tegmental area, rostral part
vtgx ventral tegmental decussation
xscp decussation of the superior cerebellar peduncle

Figure H20
Bregma -9.58 mm
Interaural 3.19 mm
Bregma
DCN
icp
SpVe
Ves
SuVe
mcp
4V
MPB
LPB
vsc
CG
LDTg
mlf
Rbd
MnR
LL
ll
PTg
SPTg
PBG
xscp
mRt
PrS
VS
Ent
Po
MGV
CLi
PBP
ml
RN
vtgx
SNR
SNC
pRt
VTAR
fr
RLi
SNL
CA1
CA3
CA2
PH
STh
PLH
cp
mt
DMH
3V
VMH
LHA
PRh
LV
fi
opt
BLA
IM
Ect
Ac
MeA
CeA
EA
SCN
AHy
SON
Pir
LOT
AA
Tu
och
Tu1
GrDG
PoDG
MoDG
LMol
Rad
alv
Or
Py

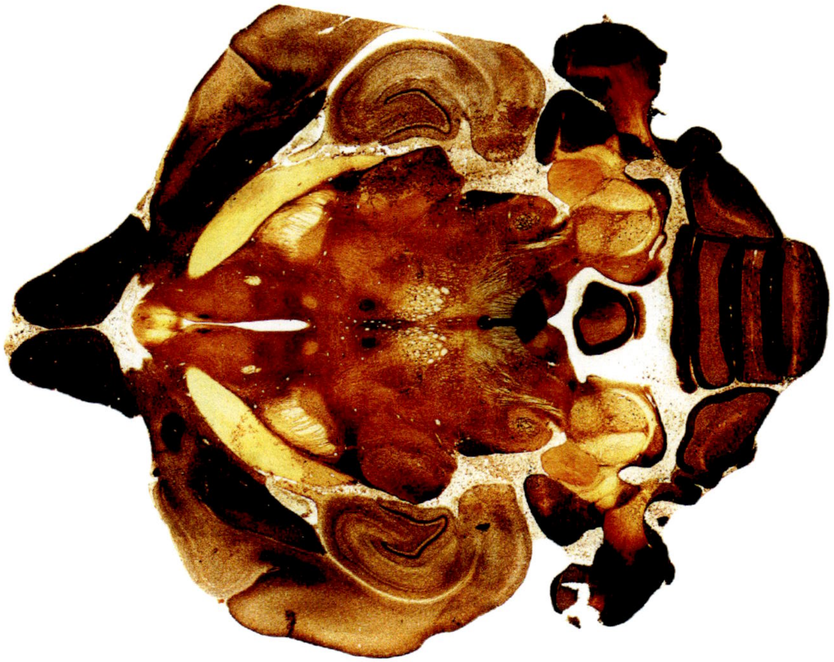

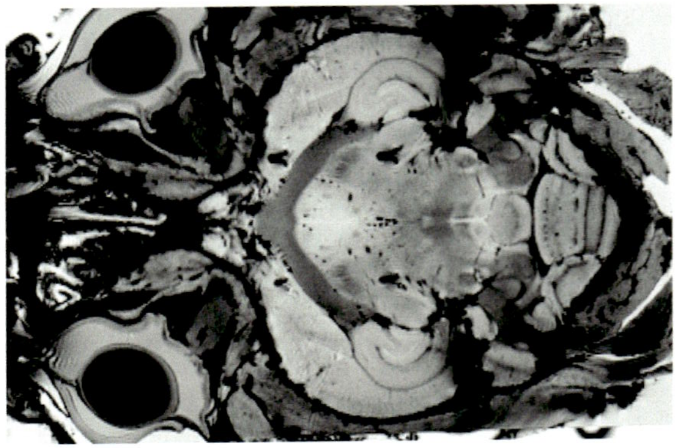

2	layer 2 of cortex
3	layer 3 of cortex
3V	3rd ventricle
4V	4th ventricle
AA	anterior amygdaloid area
Ac	auditory core area
AHy	anterior hypothalamic area
alv	alveus of the hippocampus
BLA	basolateral amygdaloid nucleus
CA1	field CA1 of the hippocampus
CA2	field CA2 of the hippocampus
CA3	field CA3 of the hippocampus
CG	central gray
CLi	caudal linear nucleus of the raphe
cp	cerebral peduncle
das	dorsal acoustic stria
DCN	dorsal cochlear nucleus
DMH	dorsomedial hypothalamic nucleus
DR	dorsal raphe nucleus
DTg	dorsal tegmental nucleus
EA	sublenticular extended amygdala
Ect	ectorhinal cortex
Ent	entorhinal cortex
f	fornix
fi	fimbria of the hippocampus
fr	fasciculus retroflexus
GrDG	granular layer of the dentate gyrus
ICj	islands of Calleja
icp	inferior cerebellar peduncle
IT	inferior temporal cortex
LC	locus coeruleus
LDTg	laterodorsal tegmental nucleus
LHA	lateral hypothalamic area
ll	lateral lemniscus
LL	nucleus of the lateral lemniscus
LMoL	lacunosum moleculare layer of the hippocampus
LOT	nucleus of the lateral olfactory tract
LPB	lateral parabrachial nucleus
LV	lateral ventricle
mcp	middle cerebellar peduncle
me5	mesencephalic trigeminal tract
Me5	mesencephalic trigeminal nucleus
MGV	medial geniculate nucleus, ventral part
MiTg	microcellular tegmental nucleus
ml	medial lemniscus
mlf	medial longitudinal fasciculus
MnR	median raphe nucleus
MoDG	molecular layer of the dentate gyrus
MPA	medial preoptic area
MPB	medial parabrachial nucleus
mRt	mesencephalic reticular formation
mt	mammillothalamic tract
och	optic chiasm
opt	optic tract
Or	oriens layer of the hippocampus
PBG	parabigeminal nucleus
PBP	parabrachial pigmented nucleus of the VTA
PH	posterior hypothalamic nucleus
Pir1	piriform cortex, layer 1
PLH	peduncular part of lateral hypothalamus
Po	posterior complex of the thalamus
PoDG	polymorph layer of the dentate gyrus
PRh	perirhinal cortex
PrS	presubiculum
pRt	p1 reticular formation
PTg	pedunculopontine tegmental nucleus
Pu	putamen
Py	pyramidal cell layer of the hippocampus
Rad	radiatum layer of the hippocampus
Rbd	rhabdoid nucleus
RLi	rostral linear nucleus of the raphe
RN	red nucleus
SCN	suprachiasmatic nucleus
scp	superior cerebellar peduncle
SNL	substantia nigra, lateral part
SON	supraoptic nucleus
SPTg	subpeduncular tegmental nucleus
SpVe	spinal vestibular nucleus
STh	subthalamic nucleus
SuVe	superior vestibular nucleus
Tu	olfactory tubercle
Tu1	layer 1 of olfactory tubercle
Ves	vestibular nucleus
VLG	ventral lateral geniculate nucleus
VS	ventral subiculum
vsc	ventral spinocerebellar tract
vtgx	ventral tegmental decussation
xscp	decussation of the superior cerebellar peduncle

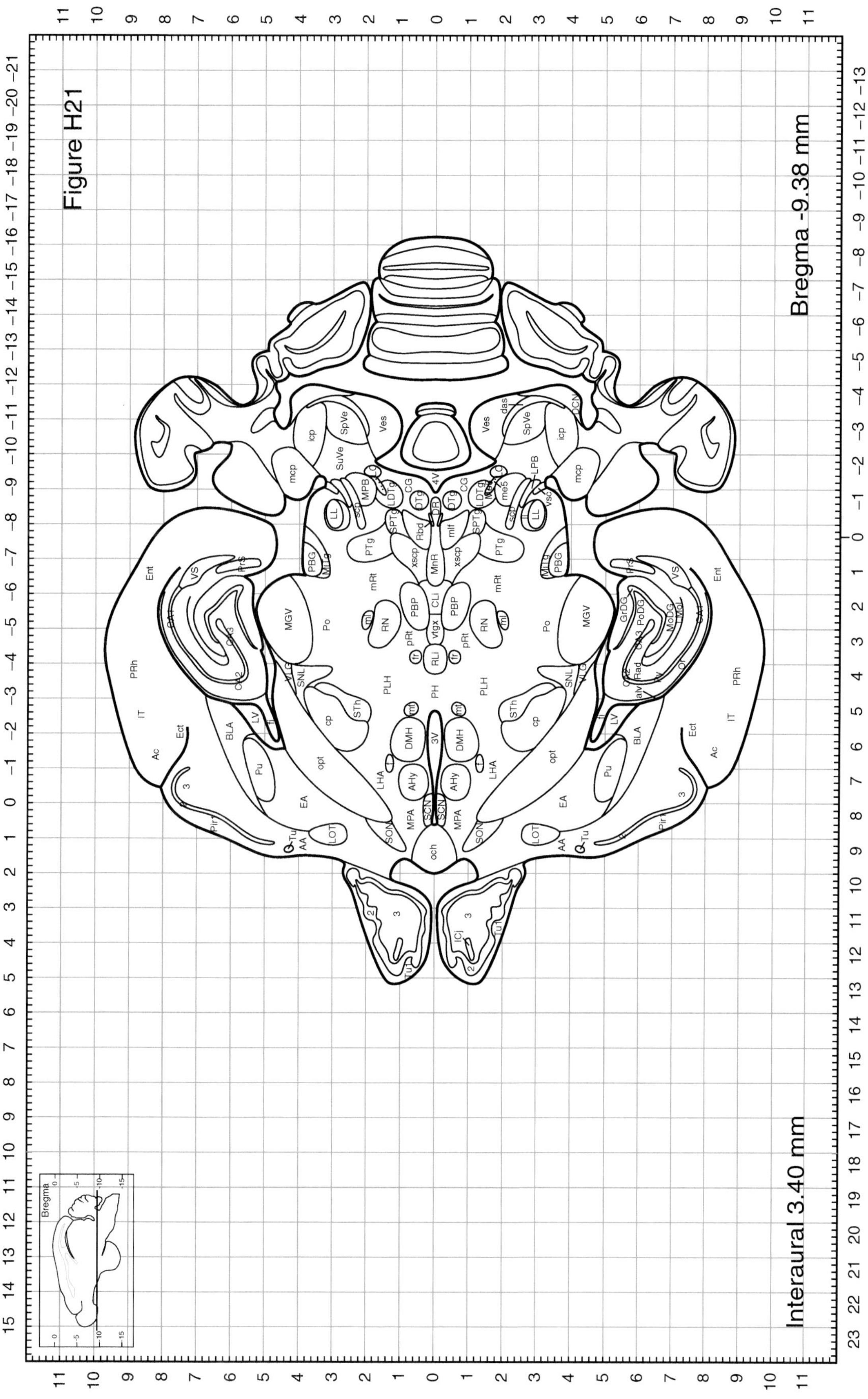
Figure H21
Bregma -9.38 mm
Interaural 3.40 mm
Bregma
Ves
SpVe
icp
SuVe
mcp
das
DCN
LPB
4V
CG
DTg
LDTg
MPB
scp
LL
SPTg
Rbd
DR
mlf
PTg
xscp
MnR
PBG
MiTg
PrS
VS
Ent
mRt
MGV
Po
CLi
PBP
RN
ml
vtgx
pRt
fr
RLi
PLH
PH
GrDG
PoDG
MoDG
CA1
CA2
CA3
LMol
Or
Rad
alv
SNL
VLG
STh
cp
mt
DMH
3V
LHA
AHy
MPA
SCN
SON
och
LOT
EA
AA
Tu
Pir
Pu
BLA
LV
opt
Ect
Ac
IT
PRh
ICj

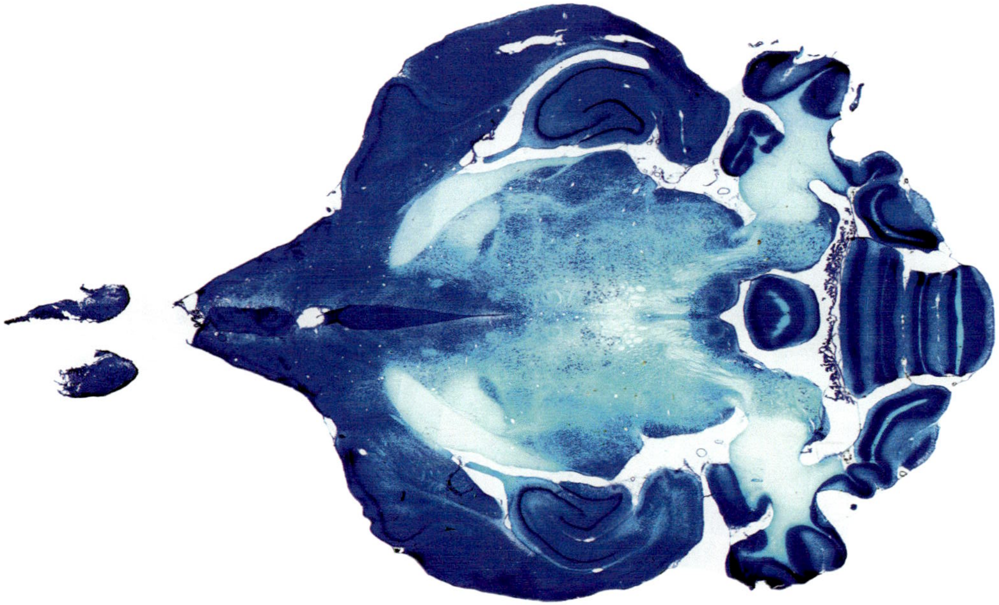

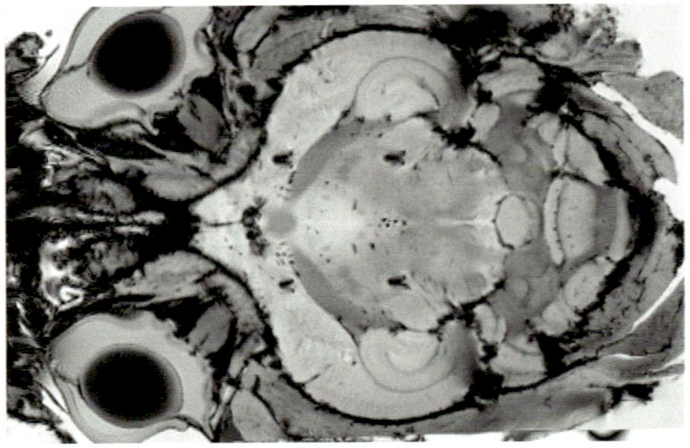

2	layer 2 of cortex
3	layer 3 of cortex
3V	3rd ventricle
4N	trochlear nucleus
4V	4th ventricle
Ab	auditory belt area
Ac	auditory core area
AHy	anterior hypothalamic area
alv	alveus of the hippocampus
BLA	basolateral amygdaloid nucleus
CA1	field CA1 of the hippocampus
CA2	field CA2 of the hippocampus
CA3	field CA3 of the hippocampus
CG	central gray
Cl	claustrum
CLi	caudal linear nucleus of the raphe
cp	cerebral peduncle
das	dorsal acoustic stria
DCN	dorsal cochlear nucleus
DMH	dorsomedial hypothalamic nucleus
DR	dorsal raphe nucleus
DTg	dorsal tegmental nucleus
EA	sublenticular extended amygdala
Ect	ectorhinal cortex
Ent	entorhinal cortex
f	fornix
fi	fimbria of the hippocampus
fr	fasciculus retroflexus
GL	glomerular layer of the olfactory bulb
GrDG	granular layer of the dentate gyrus
ICj	islands of Calleja
icp	inferior cerebellar peduncle
IT	inferior temporal cortex
LC	locus coeruleus
LDTg	laterodorsal tegmental nucleus
LHA	lateral hypothalamic area
ll	lateral lemniscus
LL	nucleus of the lateral lemniscus
LMoL	lacunosum moleculare layer of the hippocampus
lo	lateral olfactory tract
LPB	lateral parabrachial nucleus
LPO	lateral preoptic area
LV	lateral ventricle
mcp	middle cerebellar peduncle
me5	mesencephalic trigeminal tract
Me5	mesencephalic trigeminal nucleus
MGD	medial geniculate nucleus, dorsal part
MGV	medial geniculate nucleus, ventral part
MiTg	microcellular tegmental nucleus
ml	medial lemniscus
mlf	medial longitudinal fasciculus
MnR	median raphe nucleus
MoDG	molecular layer of the dentate gyrus
MPB	medial parabrachial nucleus
MPO	medial preoptic nucleus
mRt	mesencephalic reticular formation
mt	mammillothalamic tract
NDB	nucleus of the diagonal band
ONL	olfactory nerve layer
opt	optic tract
Or	oriens layer of the hippocampus
PBG	parabigeminal nucleus
PBP	parabrachial pigmented nucleus of the VTA
PH	posterior hypothalamic nucleus
Pir1	piriform cortex, layer 1
PLH	peduncular part of lateral hypothalamus
Po	posterior complex of the thalamus
PoDG	polymorph layer of the dentate gyrus
PRh	perirhinal cortex
PrS	presubiculum
pRt	p1 reticular formation
PTg	pedunculopontine tegmental nucleus
Pu	putamen
Py	pyramidal cell layer of the hippocampus
Rad	radiatum layer of the hippocampus
Rbd	rhabdoid nucleus
RLi	rostral linear nucleus of the raphe
RN	red nucleus
scp	superior cerebellar peduncle
SpVe	spinal vestibular nucleus
st	stria terminalis
STh	subthalamic nucleus
SuVe	superior vestibular nucleus
Tu	olfactory tubercle
Tu1	layer 1 of olfactory tubercle
Ves	vestibular nucleus
VLG	ventral lateral geniculate nucleus
VS	ventral subiculum
vsc	ventral spinocerebellar tract
ZI	zona incerta

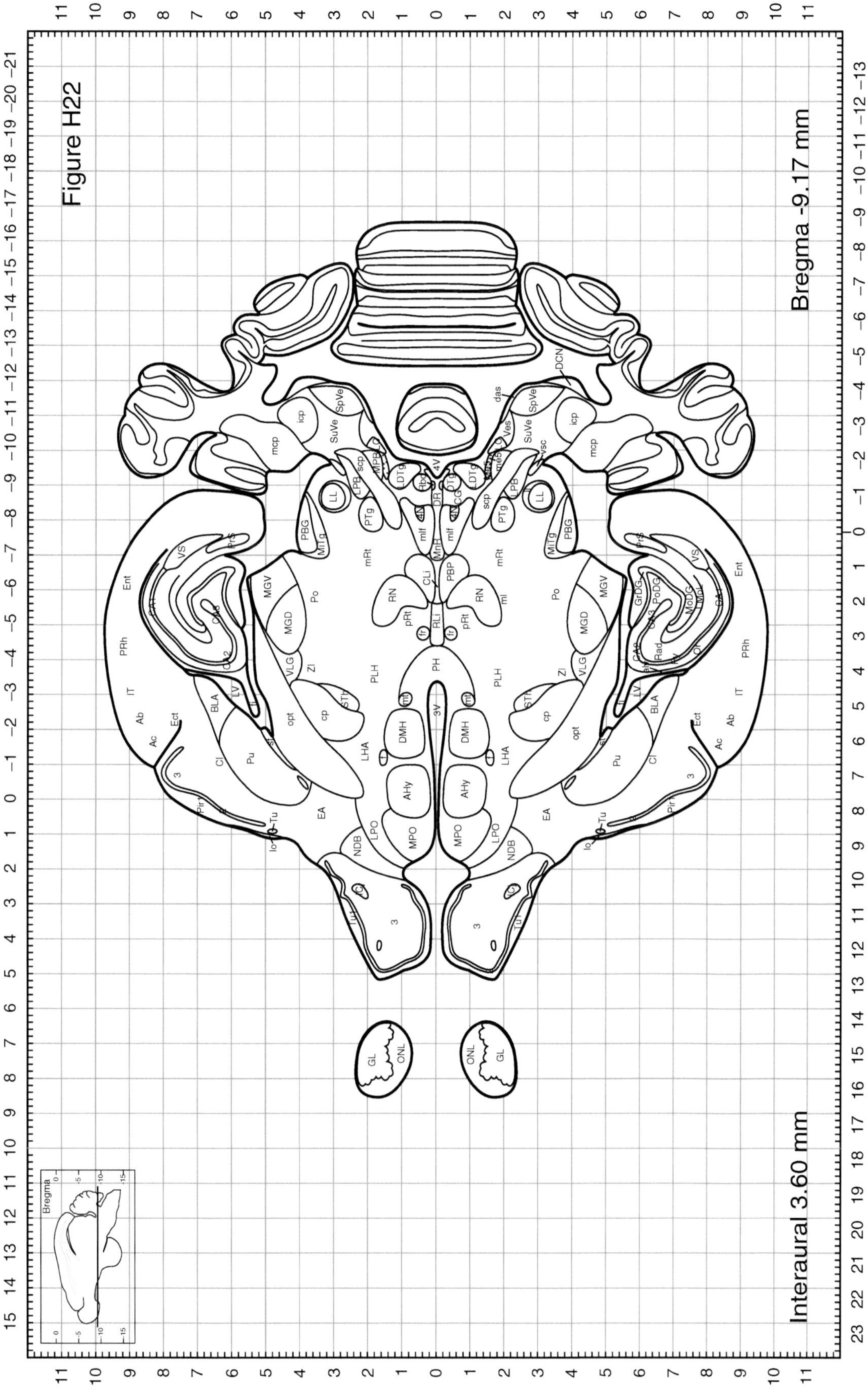

Figure H22
Bregma -9.17 mm
Interaural 3.60 mm

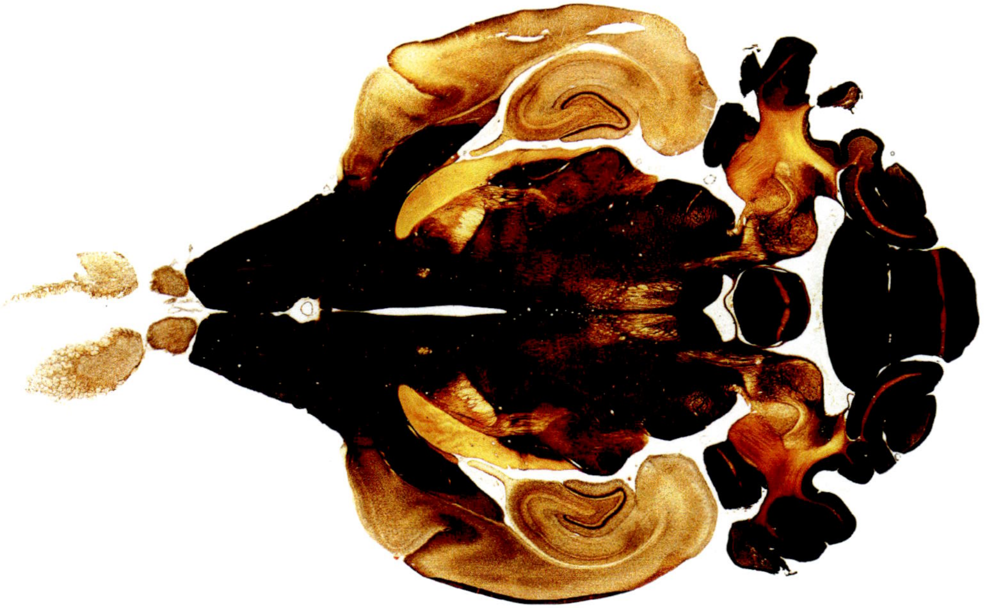

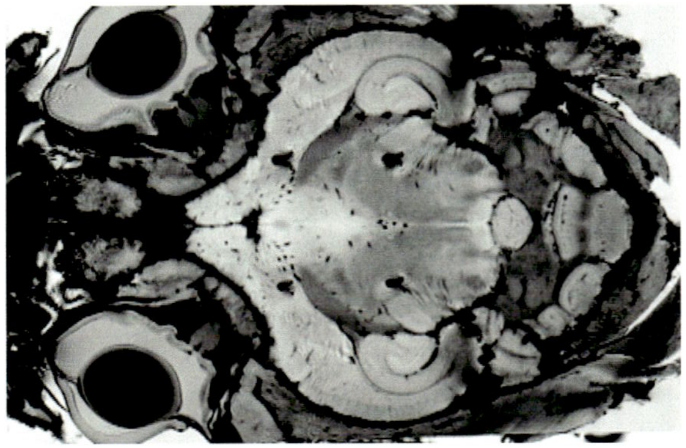

2	layer 2 of cortex
3	layer 3 of cortex
3V	3rd ventricle
4N	trochlear nucleus
4n	trochlear nerve
4V	4th ventricle
Ab	auditory belt area
Ac	auditory core area
Acb	accumbens nucleus
acp	anterior commissure, posterior part
AHy	anterior hypothalamic area
alv	alveus of the hippocampus
AON	anterior olfactory nucleus
BLA	basolateral amygdaloid nucleus
CA1	field CA1 of the hippocampus
CA2	field CA2 of the hippocampus
CA3	field CA3 of the hippocampus
CG	central gray
Cl	claustrum
CLi	caudal linear nucleus of the raphe
cp	cerebral peduncle
das	dorsal acoustic stria
DCN	dorsal cochlear nucleus
DLG	dorsal lateral geniculate nucleus
DMH	dorsomedial hypothalamic nucleus
DR	dorsal raphe nucleus
DTg	dorsal tegmental nucleus
EA	sublenticular extended amygdala
Ect	ectorhinal cortex
Ent	entorhinal cortex
EPl	external plexiform layer of the olfactory bulb
f	fornix
fi	fimbria of the hippocampus
fr	fasciculus retroflexus
GL	glomerular layer of the olfactory bulb
GrDG	granular layer of the dentate gyrus
ICj	islands of Calleja
icp	inferior cerebellar peduncle
IT	inferior temporal cortex
Lat	lateral cerebellar nucleus
LC	locus coeruleus
LDTg	laterodorsal tegmental nucleus
LHA	lateral hypothalamic area
ll	lateral lemniscus
LL	nucleus of the lateral lemniscus
LMoL	lacunosum moleculare layer of the hippocampus
lo	lateral olfactory tract
LPB	lateral parabrachial nucleus
LPO	lateral preoptic area
LV	lateral ventricle
mcp	middle cerebellar peduncle
me5	mesencephalic trigeminal tract
Me5	mesencephalic trigeminal nucleus
MGD	medial geniculate nucleus, dorsal part
MGV	medial geniculate nucleus, ventral part
MiTg	microcellular tegmental nucleus
ml	medial lemniscus
mlf	medial longitudinal fasciculus
MnR	median raphe nucleus
MoDG	molecular layer of the dentate gyrus
MPO	medial preoptic nucleus
mRt	mesencephalic reticular formation
mt	mammillothalamic tract
NDB	nucleus of the diagonal band
ONL	olfactory nerve layer
opt	optic tract
Or	oriens layer of the hippocampus
PaV	paraventricular hypothalamic nucleus, ventral part
PBG	parabigeminal nucleus
PBP	parabrachial pigmented nucleus of the VTA
PH	posterior hypothalamic nucleus
Pir1	piriform cortex, layer 1
PLH	peduncular part of lateral hypothalamus
Po	posterior complex of the thalamus
PoDG	polymorph layer of the dentate gyrus
PrS	presubiculum
pRt	p1 reticular formation
PTg	pedunculopontine tegmental nucleus
Pu	putamen
Pv	ventral nucleus of the pulvinar
Py	pyramidal cell layer of the hippocampus
Rad	radiatum layer of the hippocampus
RLi	rostral linear nucleus of the raphe
RN	red nucleus
Rt	reticular thalamic nucleus
scp	superior cerebellar peduncle
SpVe	spinal vestibular nucleus
st	stria terminalis
str	superior thalamic radiation
SuVe	superior vestibular nucleus
TC	temporal cortex
Tu	olfactory tubercle
Tu1	layer 1 of olfactory tubercle
Ves	vestibular nucleus
VLG	ventral lateral geniculate nucleus
VP	ventral pallidum
VS	ventral subiculum
vsc	ventral spinocerebellar tract
ZI	zona incerta

Figure H23
Bregma -8.96 mm
Interaural 3.81 mm

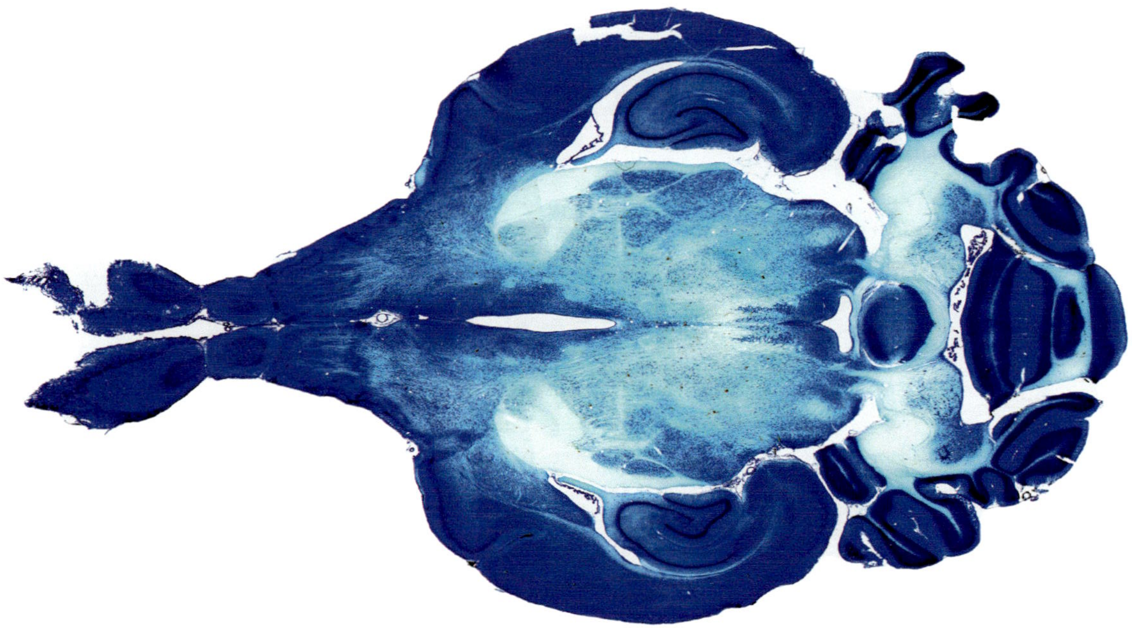

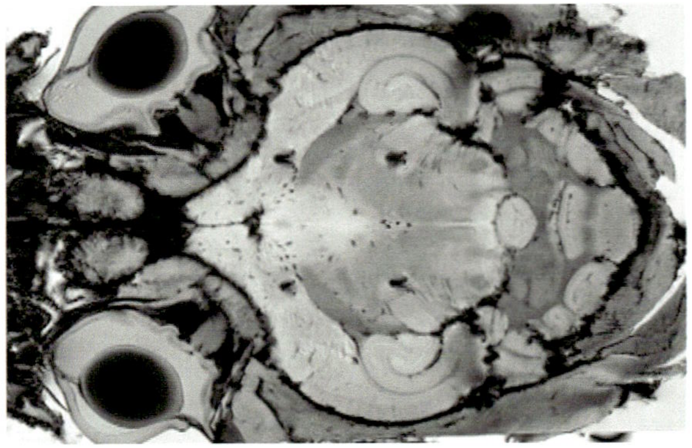

2	layer 2 of cortex
3	layer 3 of cortex
3N	oculomotor nucleus
3V	3rd ventricle
4N	trochlear nucleus
4n	trochlear nerve
4V	4th ventricle
Ab	auditory belt area
Ac	auditory core area
Acb	accumbens nucleus
acp	anterior commissure, posterior part
AHy	anterior hypothalamic area
alv	alveus of the hippocampus
AON	anterior olfactory nucleus
BLA	basolateral amygdaloid nucleus
CA1	field CA1 of the hippocampus
CA2	field CA2 of the hippocampus
CA3	field CA3 of the hippocampus
CG	central gray
Cl	claustrum
CLi	caudal linear nucleus of the raphe
cp	cerebral peduncle
DLG	dorsal lateral geniculate nucleus
DMH	dorsomedial hypothalamic nucleus
DR	dorsal raphe nucleus
EA	sublenticular extended amygdala
Ent	entorhinal cortex
EPl	external plexiform layer of the olfactory bulb
f	fornix
fi	fimbria of the hippocampus
fr	fasciculus retroflexus
GL	glomerular layer of the olfactory bulb
GrDG	granular layer of the dentate gyrus
ic	internal capsule
ICj	islands of Calleja
icp	inferior cerebellar peduncle
IGL	intergeniculate leaf
IGP	internal globus pallidus
Ins	insular cortex
IPAC	interstitial nucleus of the posterior limb of the anterior commissure
IPl	internal plexiform layer of the olfactory bulb
IT	inferior temporal cortex
Lat	lateral cerebellar nucleus
LDTg	laterodorsal tegmental nucleus
LHA	lateral hypothalamic area
ll	lateral lemniscus
LMoL	lacunosum moleculare layer of the hippocampus
lo	lateral olfactory tract
LPB	lateral parabrachial nucleus
LPO	lateral preoptic area
LSS	lateral stripe of the striatum
LV	lateral ventricle
MCL	mitral cell layer of the olfactory bulb
mcp	middle cerebellar peduncle
me5	mesencephalic trigeminal tract
Me5	mesencephalic trigeminal nucleus
MGD	medial geniculate nucleus, dorsal part
MGV	medial geniculate nucleus, ventral part
MiTg	microcellular tegmental nucleus
mlf	medial longitudinal fasciculus
MoDG	molecular layer of the dentate gyrus
MPO	medial preoptic nucleus
mRt	mesencephalic reticular formation
mt	mammillothalamic tract
NDB	nucleus of the diagonal band
ONL	olfactory nerve layer
opt	optic tract
Or	oriens layer of the hippocampus
PaV	paraventricular hypothalamic nucleus, ventral part
PBG	parabigeminal nucleus
Pd	dorsal nucleus of the pulvinar
PH	posterior hypothalamic nucleus
Pir1	piriform cortex, layer 1
PLH	peduncular part of lateral hypothalamus
Po	posterior complex of the thalamus
PoDG	polymorph layer of the dentate gyrus
PrEW	pre-Edinger-Westphal nucleus
PrS	presubiculum
pRt	p1 reticular formation
Pu	putamen
Pv	ventral nucleus of the pulvinar
Py	pyramidal cell layer of the hippocampus
Rad	radiatum layer of the hippocampus
Rt	reticular thalamic nucleus
scp	superior cerebellar peduncle
st	stria terminalis
str	superior thalamic radiation
SuVe	superior vestibular nucleus
TC	temporal cortex
Tu1	layer 1 of olfactory tubercle
Ves	vestibular nucleus
VL	ventrolateral thalamic nucleus
VLG	ventral lateral geniculate nucleus
VP	ventral pallidum
VS	ventral subiculum
vsc	ventral spinocerebellar tract
ZI	zona incerta

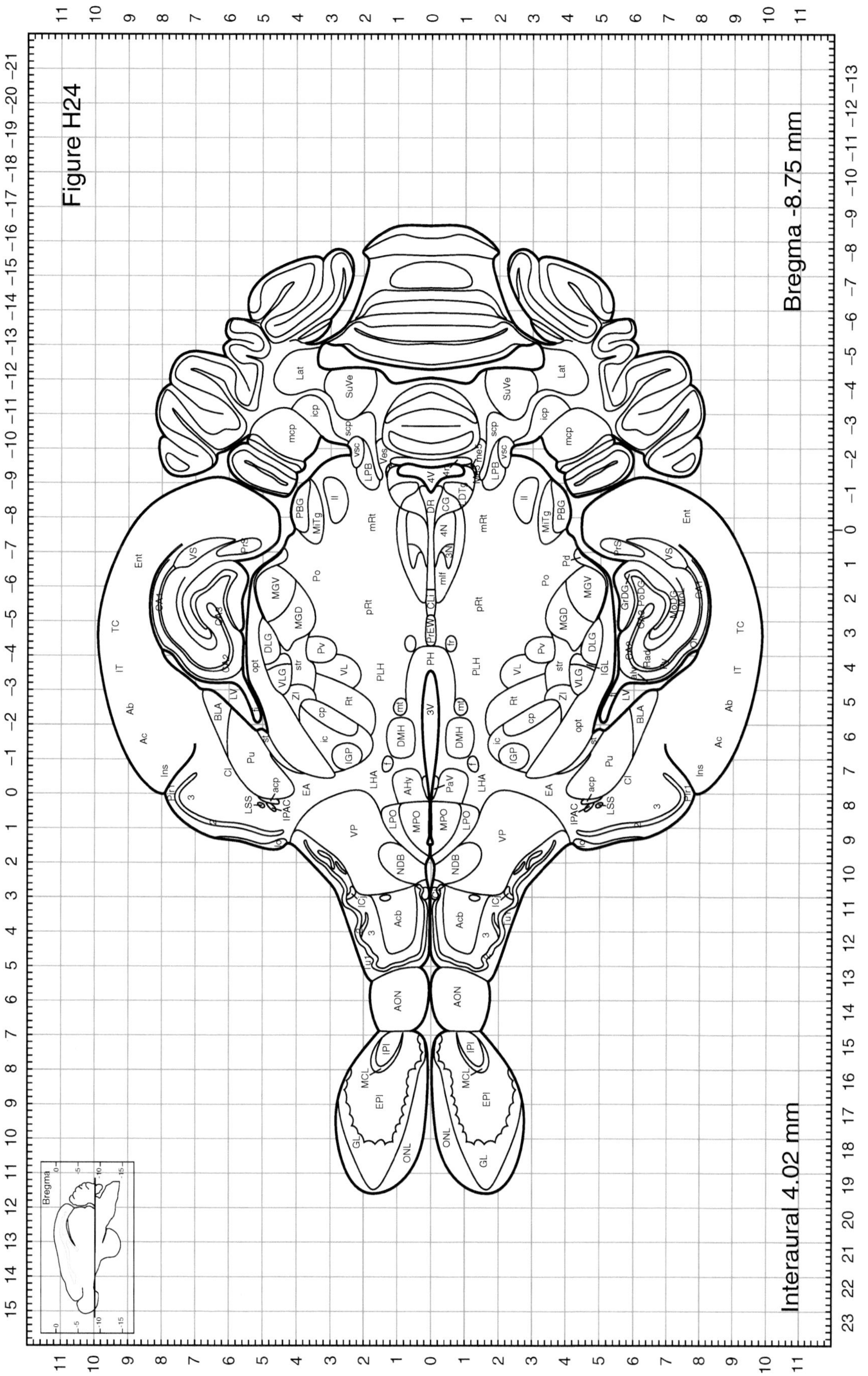
Figure H24
Bregma -8.75 mm
Interaural 4.02 mm
Lat
SuVe
icp
scp
mcp
vsc
LPB
Ves
4V
DR
CG
DTg
PBG
MiTg
ll
mRt
4N
3N
mlf
Ent
VS
PrS
Po
MGV
MGD
pRt
CLi
PrEW
DLG
Pv
PH
fr
opt
str
VLG
VL
PLH
ZI
Rt
mt
3V
cp
DMH
ic
IGP
f
LHA
AHy
PaV
BLA
LV
Pu
Cl
acp
EA
LSS
IPAC
LPO
MPO
VP
NDB
ICj
Acb
AON
IPl
MCL
EPl
GL
ONL
TC
IT
Ab
Ac
Ins
Pir
IGL
Rad
Bregma

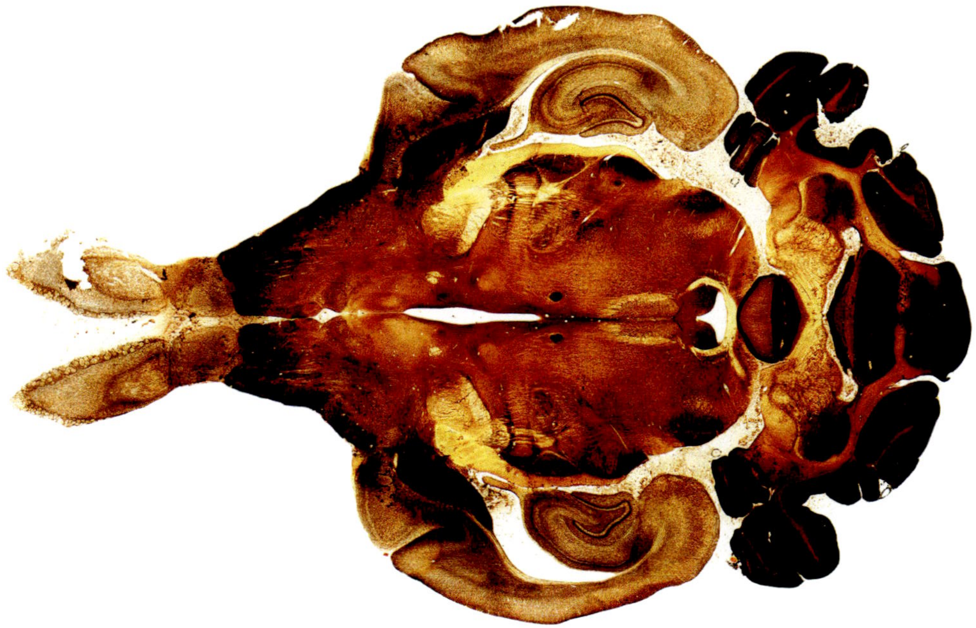

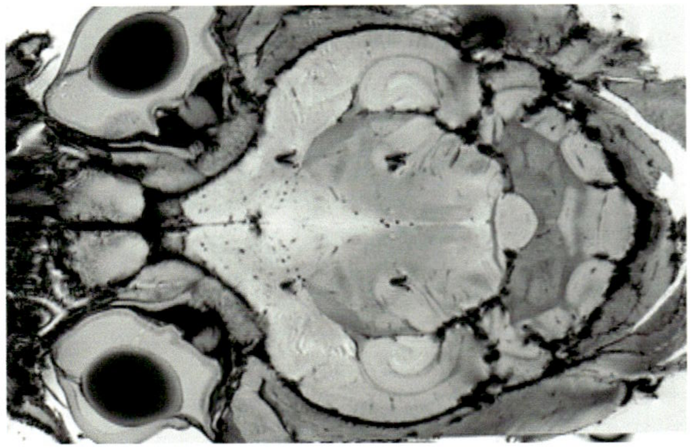

2	layer 2 of cortex
3	layer 3 of cortex
3N	oculomotor nucleus
3PC	oculomotor nucleus, parvicellular part
3V	3rd ventricle
4N	trochlear nucleus
4n	trochlear nerve
4V	4th ventricle
Ab	auditory belt area
Ac	auditory core area
Acb	accumbens nucleus
acp	anterior commissure, posterior part
AHy	anterior hypothalamic area
alv	alveus of the hippocampus
AON	anterior olfactory nucleus
BLA	basolateral amygdaloid nucleus
CA1	field CA1 of the hippocampus
CA2	field CA2 of the hippocampus
CA3	field CA3 of the hippocampus
Cl	claustrum
cp	cerebral peduncle
DLG	dorsal lateral geniculate nucleus
DR	dorsal raphe nucleus
EA	sublenticular extended amygdala
EGP	external globus pallidus
Ent	entorhinal cortex
EPl	external plexiform layer of the olfactory bulb
f	fornix
fi	fimbria of the hippocampus
fr	fasciculus retroflexus
GCL	granule cell layer of the olfactory bulb
GL	glomerular layer of the olfactory bulb
GrDG	granular layer of the dentate gyrus
ic	internal capsule
IC	inferior colliculus
ICj	islands of Calleja
icp	inferior cerebellar peduncle
IGL	intergeniculate leaf
IGP	internal globus pallidus
Ins	insular cortex
IPAC	interstitial nucleus of the posterior limb of the anterior commissure
IPl	internal plexiform layer of the olfactory bulb
IT	inferior temporal cortex
Lat	lateral cerebellar nucleus
LDTg	laterodorsal tegmental nucleus
LHA	lateral hypothalamic area
ll	lateral lemniscus
LMoL	lacunosum moleculare layer of the hippocampus
lo	lateral olfactory tract
LPB	lateral parabrachial nucleus
LPO	lateral preoptic area
LSS	lateral stripe of the striatum
LV	lateral ventricle
MCL	mitral cell layer of the olfactory bulb
mcp	middle cerebellar peduncle
me5	mesencephalic trigeminal tract
Me5	mesencephalic trigeminal nucleus
MGD	medial geniculate nucleus, dorsal part
MGV	medial geniculate nucleus, ventral part
MiTg	microcellular tegmental nucleus
mlf	medial longitudinal fasciculus
MoDG	molecular layer of the dentate gyrus
MPO	medial preoptic nucleus
mRt	mesencephalic reticular formation
mt	mammillothalamic tract
NDB	nucleus of the diagonal band
ONL	olfactory nerve layer
opt	optic tract
Or	oriens layer of the hippocampus
PAG	periaqueductal gray
PBG	parabigeminal nucleus
Pd	dorsal nucleus of the pulvinar
PH	posterior hypothalamic nucleus
Pir1	piriform cortex, layer 1
PLH	peduncular part of lateral hypothalamus
Po	posterior complex of the thalamus
PoDG	polymorph layer of the dentate gyrus
PrEW	pre-Edinger-Westphal nucleus
PrS	presubiculum
pRt	p1 reticular formation
Pu	putamen
Pv	ventral nucleus of the pulvinar
PVN	paraventricular hypothalamic nucleus
Py	pyramidal cell layer of the hippocampus
Rad	radiatum layer of the hippocampus
Rt	reticular thalamic nucleus
scp	superior cerebellar peduncle
st	stria terminalis
str	superior thalamic radiation
SuVe	superior vestibular nucleus
TC	temporal cortex
Tu1	layer 1 of olfactory tubercle
VL	ventrolateral thalamic nucleus
VLG	ventral lateral geniculate nucleus
VP	ventral pallidum
VS	ventral subiculum
vsc	ventral spinocerebellar tract
ZI	zona incerta

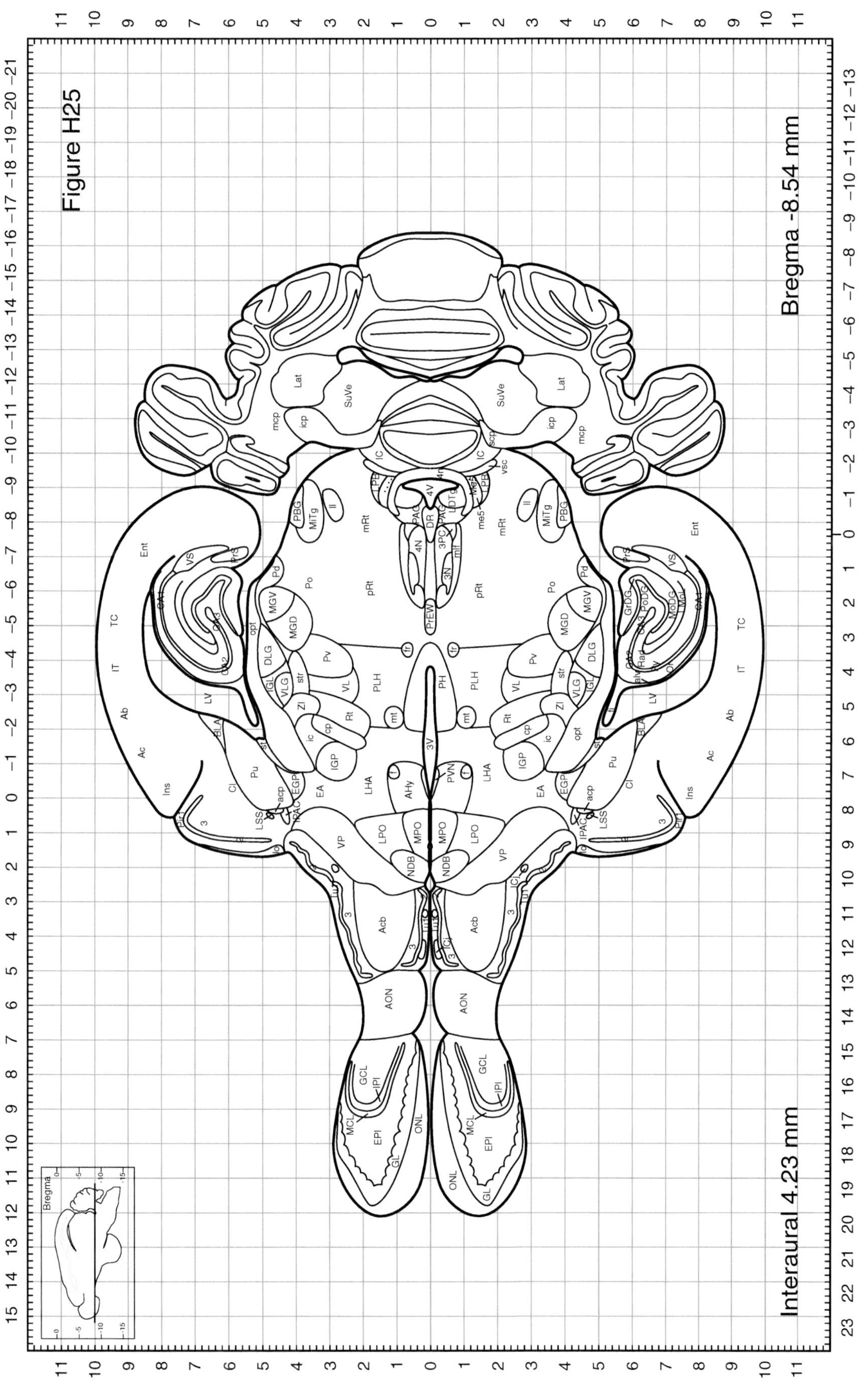
Figure H25
Bregma -8.54 mm
Interaural 4.23 mm
Bregma
Lat
SuVe
mcp
icp
IC
scp
vsc
LPB
4V
4n
Me5
me5
PAG
DR
LDTg
PBG
MiTg
ll
mRt
4N
3PC
mlf
3N
Ent
VS
PrS
Pd
Po
pRt
PrEW
CA1
CA2
CA3
GrDG
PoDG
MoDG
LMol
Rad
alv
Or
TC
MGV
MGD
opt
DLG
Pv
fr
IT
IGL
VLG
str
VL
PLH
PH
LV
ZI
Rt
mt
Ab
BLA
cp
ic
st
3V
Ac
Pu
IGP
EGP
EA
LHA
AHy
PVN
f
Ins
Cl
acp
IPAC
LSS
Pir
3
lo
VP
LPO
MPO
NDB
ICj
Acb
AON
GCL
IPl
MCL
EPl
ONL
GL

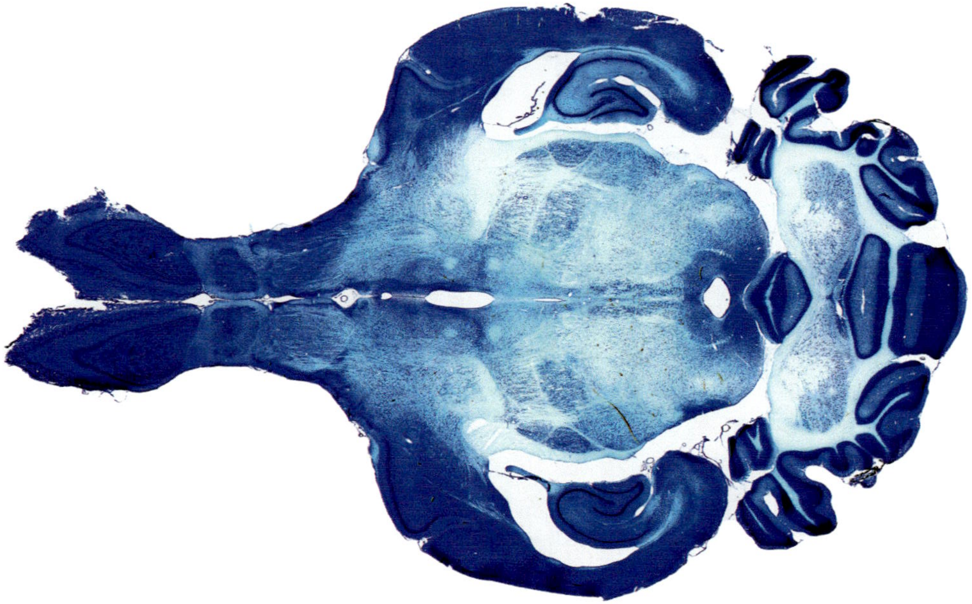

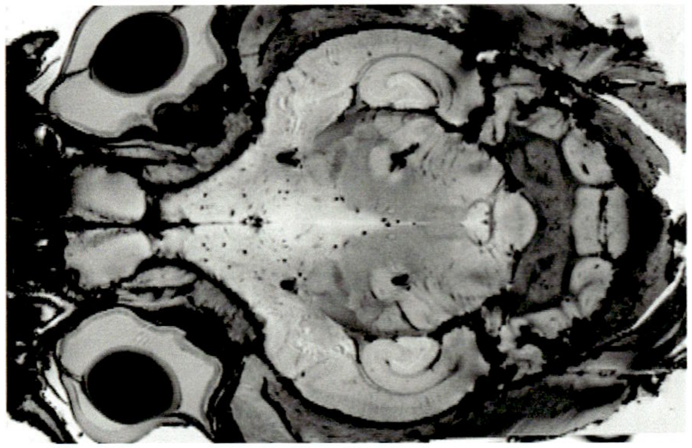

2 layer 2 of cortex
3 layer 3 of cortex
3N oculomotor nucleus
3PC oculomotor nucleus, parvicellular part
3V 3rd ventricle
4N trochlear nucleus
Ab auditory belt area
Ac auditory core area
Acb accumbens nucleus
acp anterior commissure, posterior part
AHy anterior hypothalamic area
alv alveus of the hippocampus
AON anterior olfactory nucleus
APT anterior pretectal nucleus
Aq aqueduct
CA1 field CA1 of the hippocampus
CA2 field CA2 of the hippocampus
CA3 field CA3 of the hippocampus
Cl claustrum
DLG dorsal lateral geniculate nucleus
DR dorsal raphe nucleus
EA sublenticular extended amygdala
EGP external globus pallidus
Ent entorhinal cortex
EPl external plexiform layer of the olfactory bulb
f fornix
fi fimbria of the hippocampus
fr fasciculus retroflexus
GCL granule cell layer of the olfactory bulb
GL glomerular layer of the olfactory bulb
GrDG granular layer of the dentate gyrus
ic internal capsule
IC inferior colliculus
ICj islands of Calleja
icp inferior cerebellar peduncle
IGL intergeniculate leaf
IGP internal globus pallidus
InC interstitial nucleus of Cajal
Ins insular cortex
Int interposed cerebellar nucleus
IPAC interstitial nucleus of the posterior limb of the anterior commissure
IPl internal plexiform layer of the olfactory bulb
IT inferior temporal cortex
JPLH juxtaparaventricular part of lateral hypothalamus
Lat lateral cerebellar nucleus
LDTg laterodorsal tegmental nucleus
LHA lateral hypothalamic area
LMoL lacunosum moleculare layer of the hippocampus
lo lateral olfactory tract
LPO lateral preoptic area
LSS lateral stripe of the striatum
LV lateral ventricle
MCL mitral cell layer of the olfactory bulb
mcp middle cerebellar peduncle
me5 mesencephalic trigeminal tract
Me5 mesencephalic trigeminal nucleus
MGD medial geniculate nucleus, dorsal part
MiTg microcellular tegmental nucleus
mlf medial longitudinal fasciculus
MoDG molecular layer of the dentate gyrus
MPO medial preoptic nucleus
mRt mesencephalic reticular formation
mt mammillothalamic tract
NDB nucleus of the diagonal band
ONL olfactory nerve layer
opt optic tract
Or oriens layer of the hippocampus
PAG periaqueductal gray
PBG parabigeminal nucleus
Pd dorsal nucleus of the pulvinar
PH posterior hypothalamic nucleus
Pir1 piriform cortex, layer 1
PLH peduncular part of lateral hypothalamus
Po posterior complex of the thalamus
PoDG polymorph layer of the dentate gyrus
PrEW pre-Edinger-Westphal nucleus
PrS presubiculum
pRt p1 reticular formation
Pu putamen
Pv ventral nucleus of the pulvinar
PVN paraventricular hypothalamic nucleus
Py pyramidal cell layer of the hippocampus
Rad radiatum layer of the hippocampus
Rt reticular thalamic nucleus
st stria terminalis
str superior thalamic radiation
SuVe superior vestibular nucleus
TC temporal cortex
Tu1 layer 1 of olfactory tubercle
VeCb vestibulocerebellar nucleus
VL ventrolateral thalamic nucleus
VLG ventral lateral geniculate nucleus
VP ventral pallidum
VS ventral subiculum
ZI zona incerta

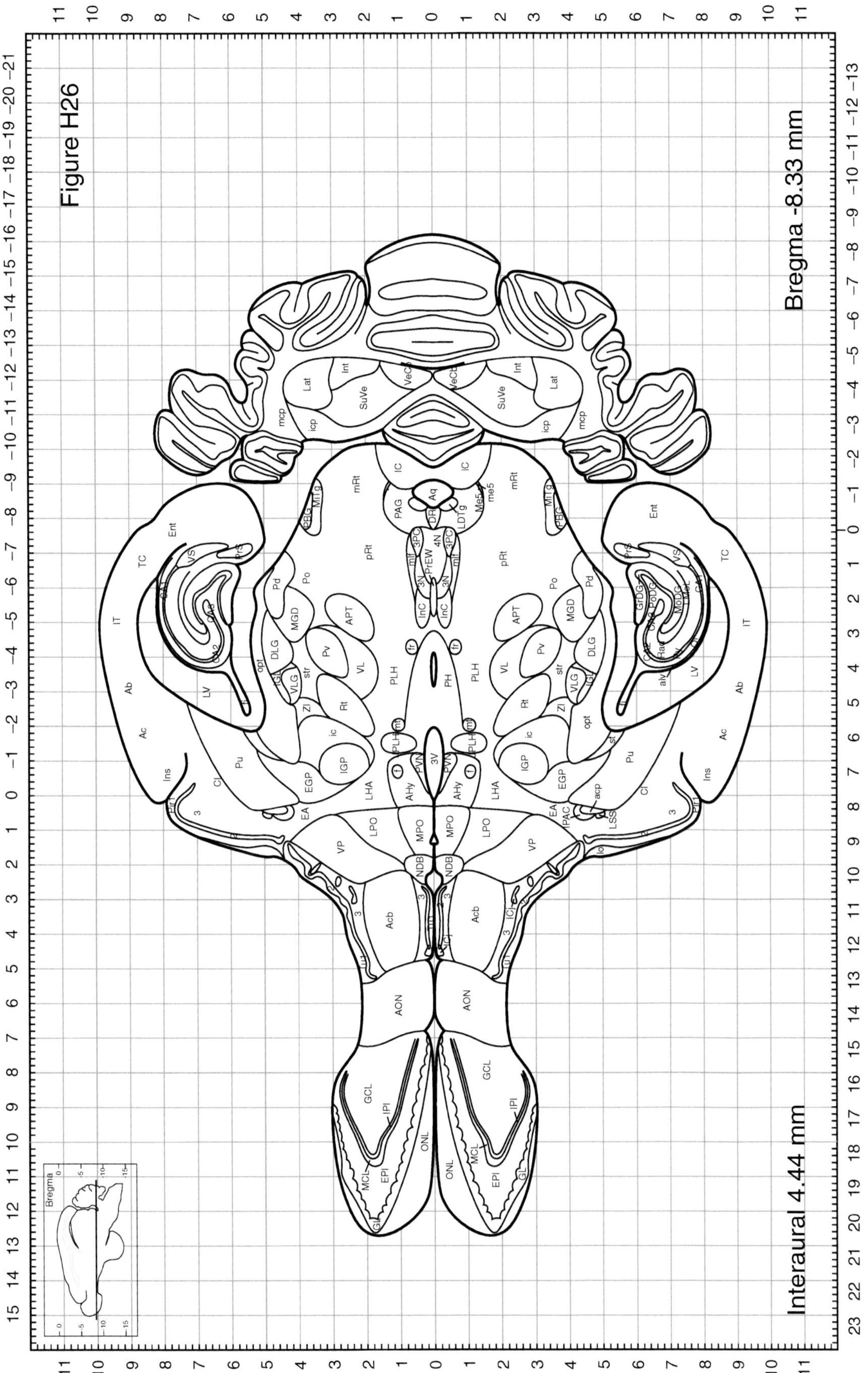
Figure H26
Bregma -8.33 mm
Interaural 4.44 mm

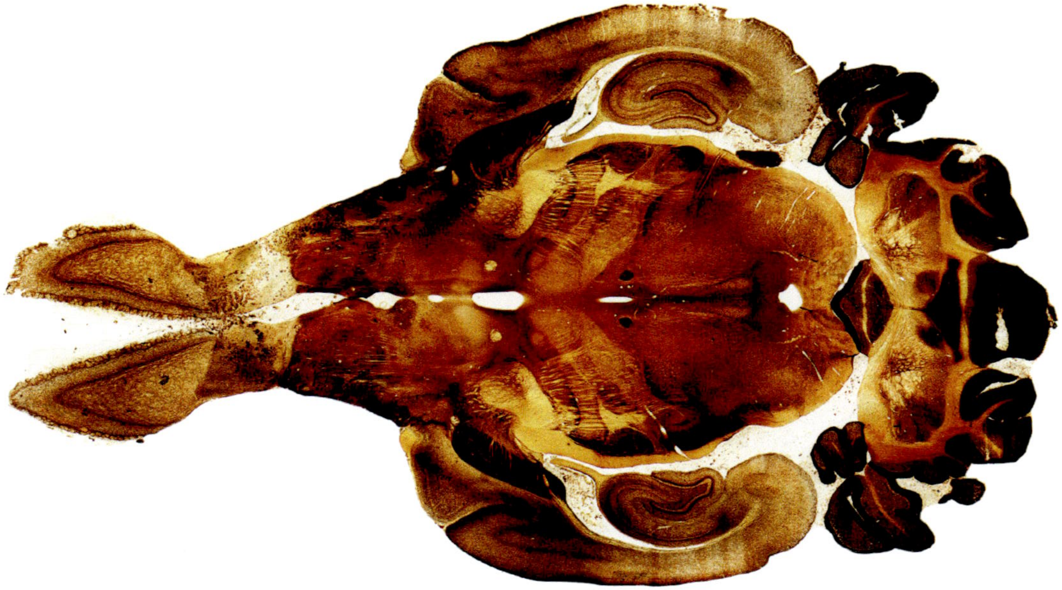

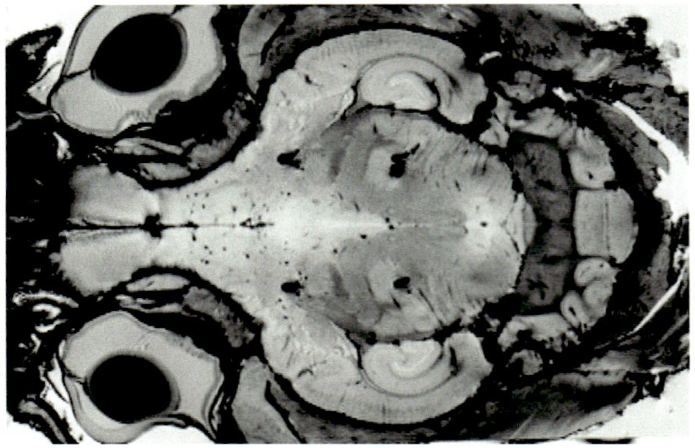

2	layer 2 of cortex
3	layer 3 of cortex
3N	oculomotor nucleus
3PC	oculomotor nucleus, parvicellular part
3V	3rd ventricle
4N	trochlear nucleus
Ab	auditory belt area
Ac	auditory core area
Acb	accumbens nucleus
acp	anterior commissure, posterior part
AHy	anterior hypothalamic area
alv	alveus of the hippocampus
AON	anterior olfactory nucleus
APT	anterior pretectal nucleus
Aq	aqueduct
CA1	field CA1 of the hippocampus
CA2	field CA2 of the hippocampus
CA3	field CA3 of the hippocampus
Cl	claustrum
DLG	dorsal lateral geniculate nucleus
DR	dorsal raphe nucleus
EGP	external globus pallidus
Ent	entorhinal cortex
EPl	external plexiform layer of the olfactory bulb
f	fornix
fi	fimbria of the hippocampus
fr	fasciculus retroflexus
GCL	granule cell layer of the olfactory bulb
GL	glomerular layer of the olfactory bulb
GrDG	granular layer of the dentate gyrus
ic	internal capsule
IC	inferior colliculus
ICj	islands of Calleja
icp	inferior cerebellar peduncle
IGL	intergeniculate leaf
IGP	internal globus pallidus
InC	interstitial nucleus of Cajal
Ins	insular cortex
Int	interposed cerebellar nucleus
IPAC	interstitial nucleus of the posterior limb of the anterior commissure
IPl	internal plexiform layer of the olfactory bulb
IT	inferior temporal cortex
JPLH	juxtaparaventricular part of lateral hypothalamus
Lat	lateral cerebellar nucleus
LHA	lateral hypothalamic area
LMoL	lacunosum moleculare layer of the hippocampus
lo	lateral olfactory tract
LPO	lateral preoptic area
LSS	lateral stripe of the striatum
LV	lateral ventricle
MCL	mitral cell layer of the olfactory bulb
mcp	middle cerebellar peduncle
me5	mesencephalic trigeminal tract
Me5	mesencephalic trigeminal nucleus
Med	medial cerebellar nucleus
MGD	medial geniculate nucleus, dorsal part
MiTg	microcellular tegmental nucleus
mlf	medial longitudinal fasciculus
MoDG	molecular layer of the dentate gyrus
MPO	medial preoptic nucleus
mRt	mesencephalic reticular formation
mt	mammillothalamic tract
NDB	nucleus of the diagonal band
ONL	olfactory nerve layer
opt	optic tract
Or	oriens layer of the hippocampus
PAG	periaqueductal gray
Pc	central nucleus of the pulvinar
PBG	parabigeminal nucleus
Pd	dorsal nucleus of the pulvinar
PH	posterior hypothalamic nucleus
Pir1	piriform cortex, layer 1
PLH	peduncular part of lateral hypothalamus
Po	posterior complex of the thalamus
PoDG	polymorph layer of the dentate gyrus
PrEW	pre-Edinger-Westphal nucleus
PrS	presubiculum
pRt	p1 reticular formation
Pu	putamen
Pv	ventral nucleus of the pulvinar
PVN	paraventricular hypothalamic nucleus
Py	pyramidal cell layer of the hippocampus
Rad	radiatum layer of the hippocampus
Re	reuniens thalamic nucleus
Rt	reticular thalamic nucleus
st	stria terminalis
str	superior thalamic radiation
Sub	submedius thalamic nucleus
SuVe	superior vestibular nucleus
TC	temporal cortex
Tu1	layer 1 of olfactory tubercle
VeCb	vestibulocerebellar nucleus
VL	ventrolateral thalamic nucleus
VLG	ventral lateral geniculate nucleus
VP	ventral pallidum
VPPC	ventral posterior nucleus of the thalamus, parvicellular part
VS	ventral subiculum
ZI	zona incerta

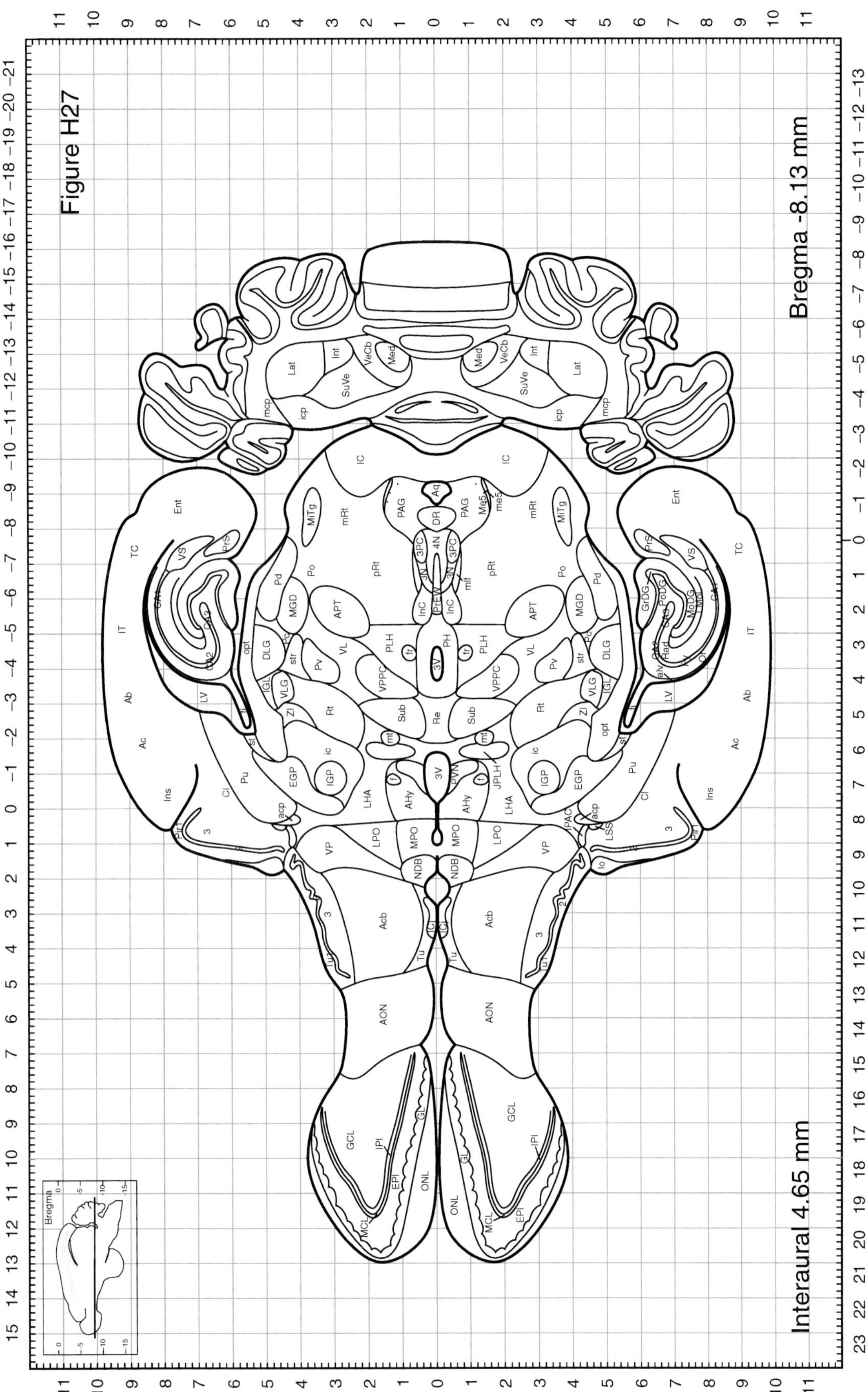
Figure H27
Bregma -8.13 mm
Interaural 4.65 mm

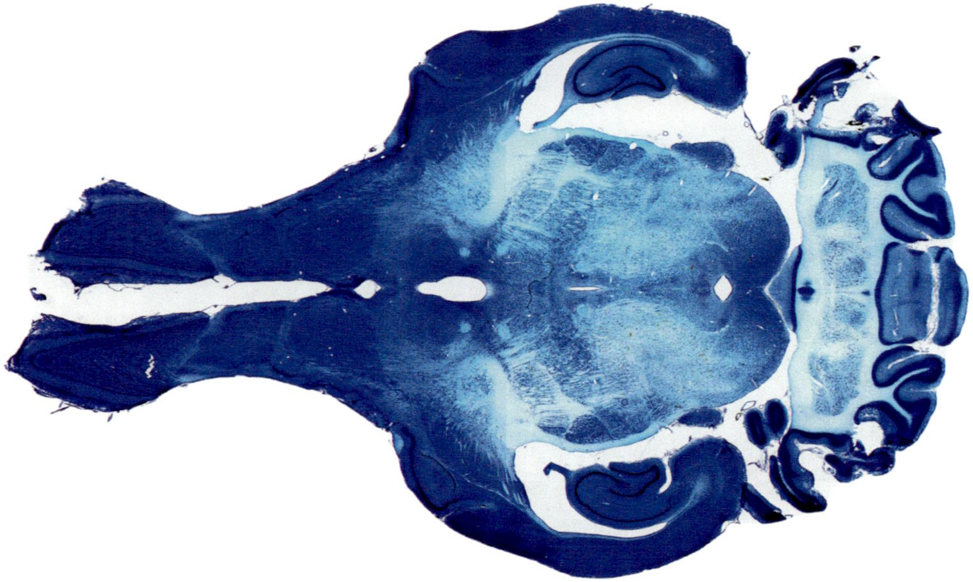

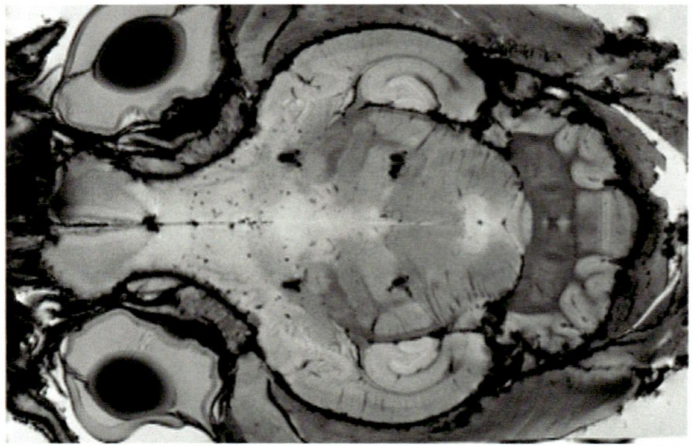

2	layer 2 of cortex
3	layer 3 of cortex
3PC	oculomotor nucleus, parvicellular part
3V	3rd ventricle
4N	trochlear nucleus
Ab	auditory belt are
Acb	accumbens nucleus
acp	anterior commissure, posterior part
AHy	anterior hypothalamic area
alv	alveus of the hippocampus
AON	anterior olfactory nucleus
APT	anterior pretectal nucleus
Aq	aqueduct
CA1	field CA1 of the hippocampus
CA2	field CA2 of the hippocampus
CA3	field CA3 of the hippocampus
Cl	claustrum
CM	central medial thalamic nucleus
DLG	dorsal lateral geniculate nucleus
DR	dorsal raphe nucleus
EGP	external globus pallidus
Ent	entorhinal cortex
EPl	external plexiform layer of the olfactory bulb
EW	Edinger-Westphal nucleus
f	fornix
fi	fimbria of the hippocampus
fr	fasciculus retroflexus
GCL	granule cell layer of the olfactory bulb
GL	glomerular layer of the olfactory bulb
GrDG	granular layer of the dentate gyrus
ic	internal capsule
IC	inferior colliculus
ICj	islands of Calleja
icp	inferior cerebellar peduncle
IGL	intergeniculate leaf
IGP	internal globus pallidus
InC	interstitial nucleus of Cajal
Ins	insular cortex
Int	interposed cerebellar nucleus
IPAC	interstitial nucleus of the posterior limb of the anterior commissure
IPl	internal plexiform layer of the olfactory bulb
IT	inferior temporal cortex
Lat	lateral cerebellar nucleus
LHA	lateral hypothalamic area
LMoL	lacunosum moleculare layer of the hippocampus
lo	lateral olfactory tract
LPO	lateral preoptic area
LV	lateral ventricle
MCL	mitral cell layer of the olfactory bulb
mcp	middle cerebellar peduncle
me5	mesencephalic trigeminal tract
Me5	mesencephalic trigeminal nucleus
Med	medial cerebellar nucleus
MGD	medial geniculate nucleus, dorsal part
MoDG	molecular layer of the dentate gyrus
MPO	medial preoptic nucleus
mRt	mesencephalic reticular formation
mt	mammillothalamic tract
NDB	nucleus of the diagonal band
ONL	olfactory nerve layer
OPT	olivary pretectal nucleus
opt	optic tract
Or	oriens layer of the hippocampus
PAG	periaqueductal gray
Pc	central nucleus of the pulvinar
Pd	dorsal nucleus of the pulvinar
PH	posterior hypothalamic nucleus
Pir1	piriform cortex, layer 1
PLH	peduncular part of lateral hypothalamus
PoDG	polymorph layer of the dentate gyrus
PrS	presubiculum
pRt	p1 reticular formation
Pu	putamen
Pv	ventral nucleus of the pulvinar
PVN	paraventricular hypothalamic nucleus
Py	pyramidal cell layer of the hippocampus
Rad	radiatum layer of the hippocampus
Re	reuniens thalamic nucleus
Rh	rhomboid thalamic nucleus
Rt	reticular thalamic nucleus
st	stria terminalis
str	superior thalamic radiation
Sub	submedius thalamic nucleus
SuG	superficial gray layer of the superior colliculus
TC	temporal cortex
Tu	olfactory tubercle
Tu1	layer 1 of olfactory tubercle
VeCb	vestibulocerebellar nucleus
VL	ventrolateral thalamic nucleus
VLG	ventral lateral geniculate nucleus
VP	ventral pallidum
VPPC	ventral posterior nucleus of the thalamus, parvicellular part
VS	ventral subiculum
ZI	zona incerta
Zo	zonal layer of the superior colliculus

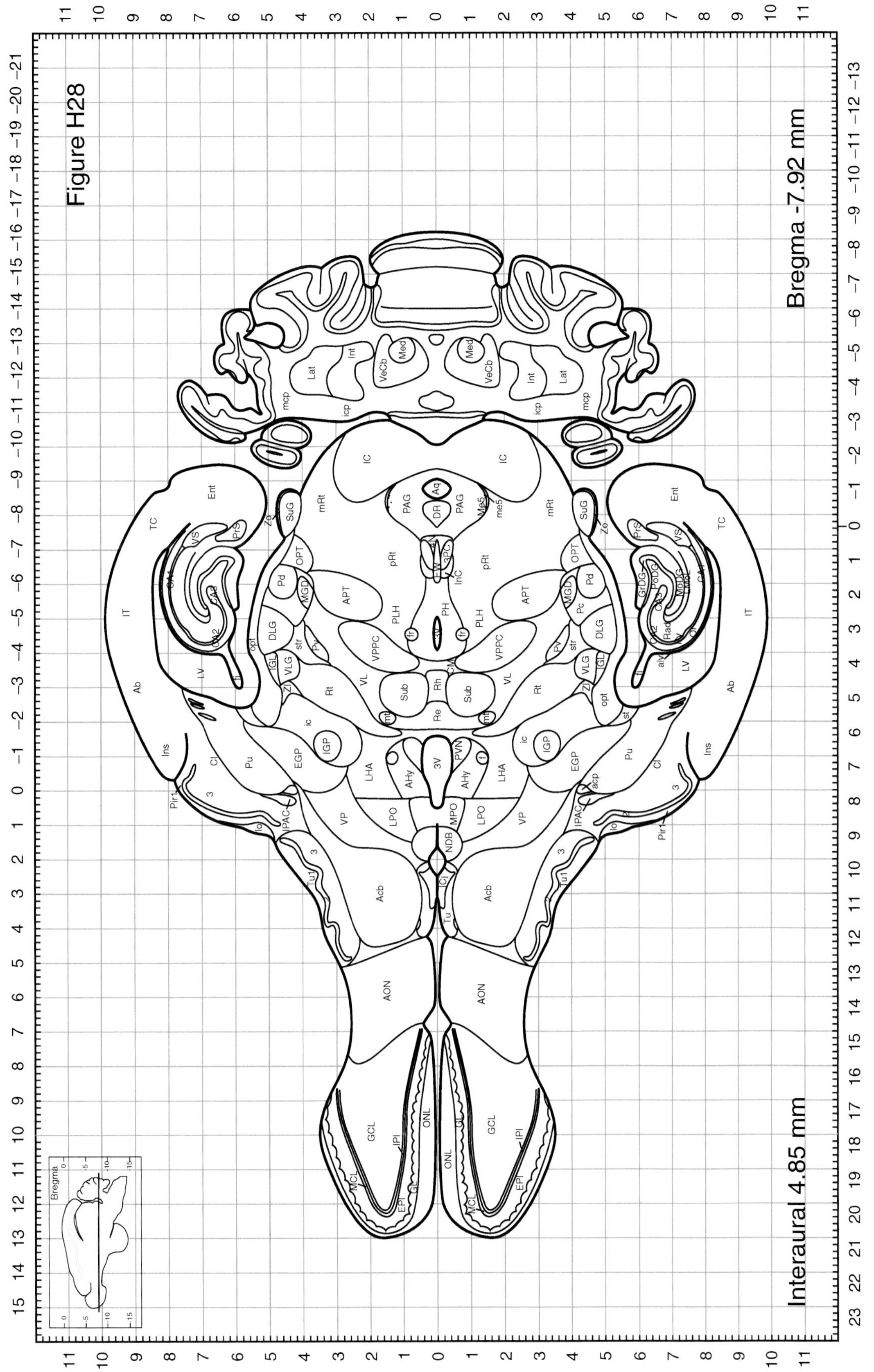
Figure H28
Bregma -7.92 mm
Interaural 4.85 mm
Bregma
Lat
Int
Med
VeCb
mcp
icp
IC
Ent
TC
mRt
SuG
PAG
Aq
DR
Zo
PrS
VS
OPT
pRt
InC
Pd
MGD
APT
IT
PLH
PH
DLG
fr
3V
opt
str
VPPC
LV
IGL
VLG
VL
Rt
ZI
Sub
Rh
Re
Ab
ic
IGP
Ins
Cl
Pu
EGP
LHA
AHy
PVN
Pir1
LPO
MPO
VP
IPAC
acp
NDB
Acb
Tu1
ICj
Tu
AON
GCL
IPl
ONL
GL
MCL
EPl
lo
st
alv
Pc
Rad

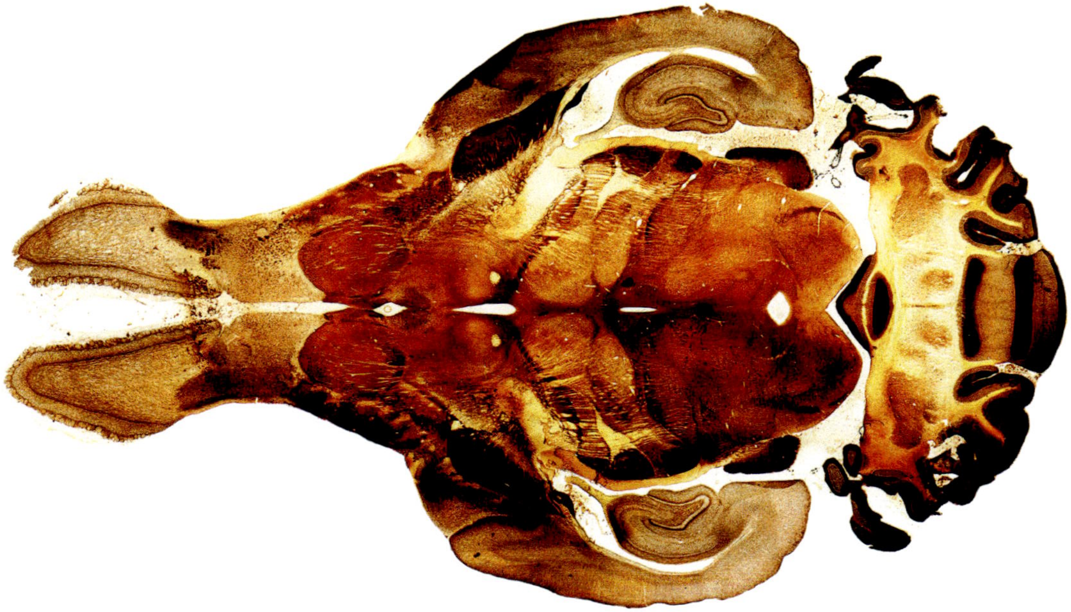

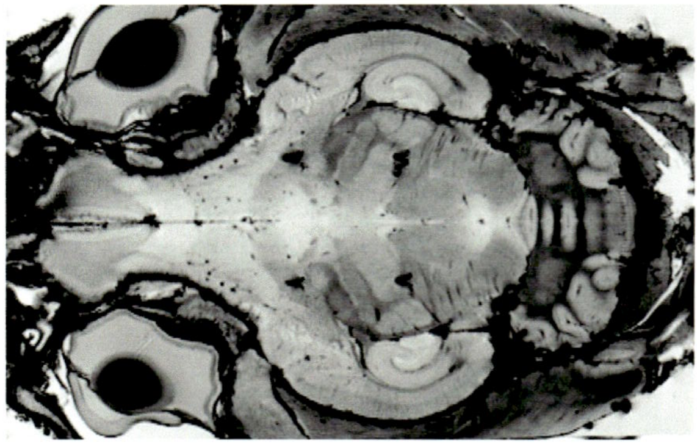

2	layer 2 of cortex
3	layer 3 of cortex
3PC	oculomotor nucleus, parvicellular part
3V	3rd ventricle
Ab	auditory belt area
Acb	accumbens nucleus
acp	anterior commissure, posterior part
AHy	anterior hypothalamic area
alv	alveus of the hippocampus
AON	anterior olfactory nucleus
APT	anterior pretectal nucleus
Aq	aqueduct
B	brachium of the superior colliculus
CA1	field CA1 of the hippocampus
CA2	field CA2 of the hippocampus
CA3	field CA3 of the hippocampus
Cl	claustrum
CM	central medial thalamic nucleus
DLG	dorsal lateral geniculate nucleus
DR	dorsal raphe nucleus
EGP	external globus pallidus
Ent	entorhinal cortex
EPl	external plexiform layer of the olfactory bulb
EW	Edinger-Westphal nucleus
f	fornix
fi	fimbria of the hippocampus
fr	fasciculus retroflexus
GCL	granule cell layer of the olfactory bulb
GL	glomerular layer of the olfactory bulb
GrDG	granular layer of the dentate gyrus
ic	internal capsule
IC	inferior colliculus
ICj	islands of Calleja
icp	inferior cerebellar peduncle
IGL	intergeniculate leaf
IGP	internal globus pallidus
Ins	insular cortex
Int	interposed cerebellar nucleus
IPl	internal plexiform layer of the olfactory bulb
IT	inferior temporal cortex
Lat	lateral cerebellar nucleus
LHA	lateral hypothalamic area
LMoL	lacunosum moleculare layer of the hippocampus
lo	lateral olfactory tract
LPO	lateral preoptic area
LV	lateral ventricle
MCL	mitral cell layer of the olfactory bulb
mcp	middle cerebellar peduncle
me5	mesencephalic trigeminal tract
Me5	mesencephalic trigeminal nucleus
Med	medial cerebellar nucleus
MoDG	molecular layer of the dentate gyrus
MPO	medial preoptic nucleus
mRt	mesencephalic reticular formation
mt	mammillothalamic tract
NDB	nucleus of the diagonal band
ONL	olfactory nerve layer
Op	optic nerve layer of the superior colliculus
OPT	olivary pretectal nucleus
opt	optic tract
Or	oriens layer of the hippocampus
PAG	periaqueductal gray
Pc	central nucleus of the pulvinar
Pd	dorsal nucleus of the pulvinar
PF	parafascicular thalamic nucleus
PH	posterior hypothalamic nucleus
Pir1	piriform cortex, layer 1
PoDG	polymorph layer of the dentate gyrus
pPAG	p1 periaqueductal gray
PrS	presubiculum
pRt	p1 reticular formation
Pu	putamen
Pv	ventral nucleus of the pulvinar
PVN	paraventricular hypothalamic nucleus
Py	pyramidal cell layer of the hippocampus
Rad	radiatum layer of the hippocampus
Re	reuniens thalamic nucleus
Rh	rhomboid thalamic nucleus
Rt	reticular thalamic nucleus
st	stria terminalis
str	superior thalamic radiation
Sub	submedius thalamic nucleus
SuG	superficial gray layer of the superior colliculus
TC	temporal cortex
Tu	olfactory tubercle
Tu1	layer 1 of olfactory tubercle
VeCb	vestibulocerebellar nucleus
VL	ventrolateral thalamic nucleus
VLG	ventral lateral geniculate nucleus
VP	ventral pallidum
VPPC	ventral posterior nucleus of the thalamus, parvicellular part
VS	ventral subiculum
Zo	zonal layer of the superior colliculus

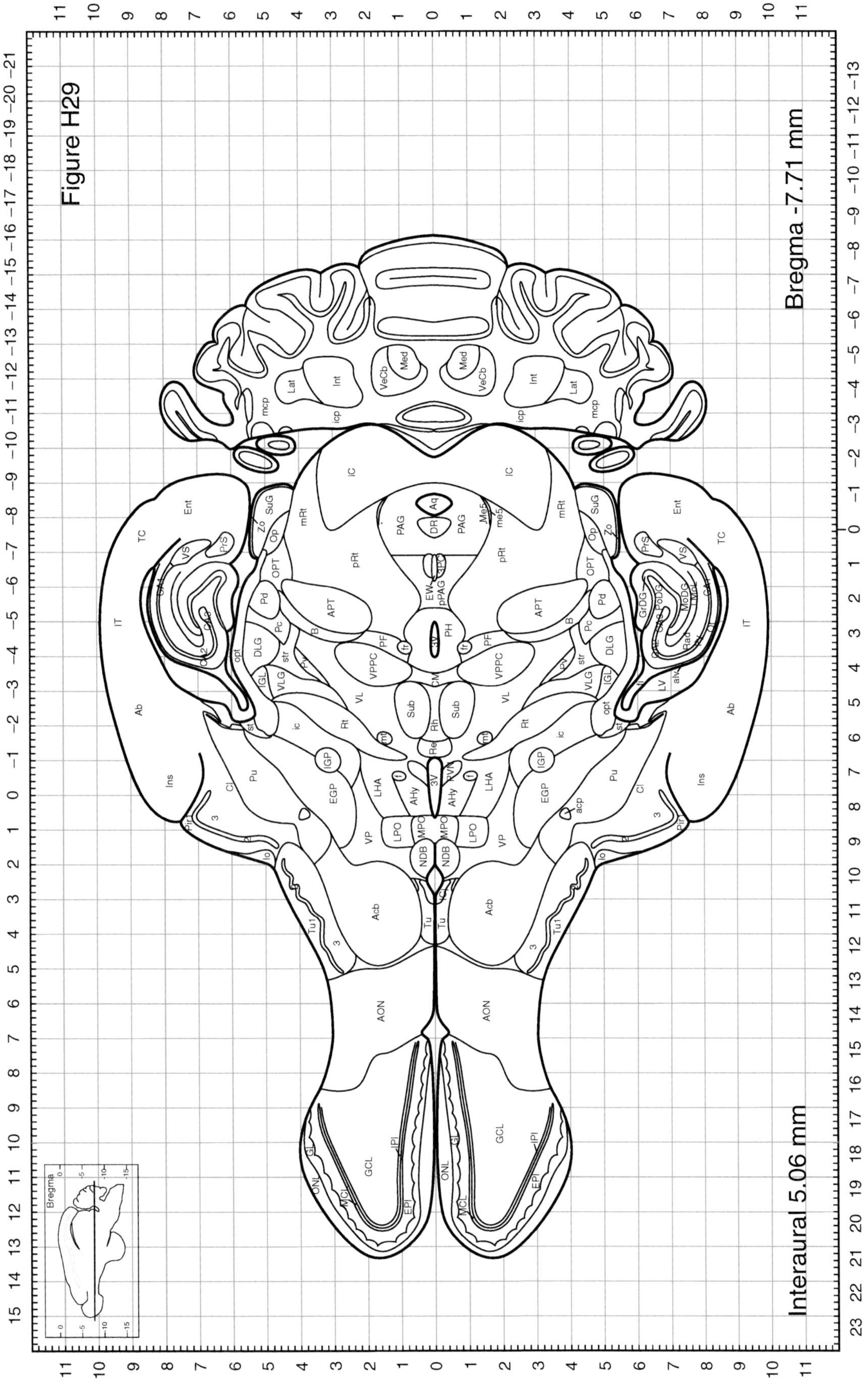
Figure H29
Bregma -7.71 mm
Interaural 5.06 mm

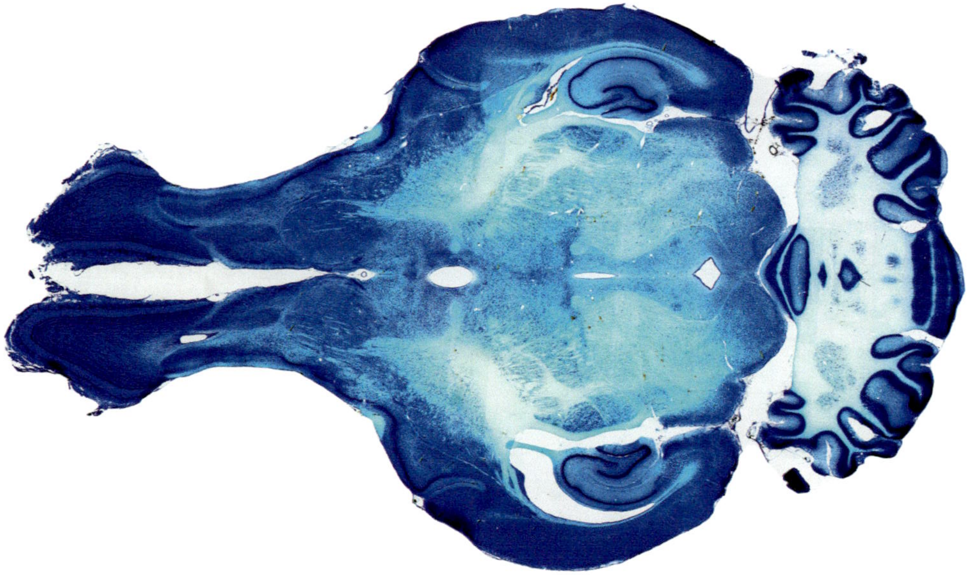

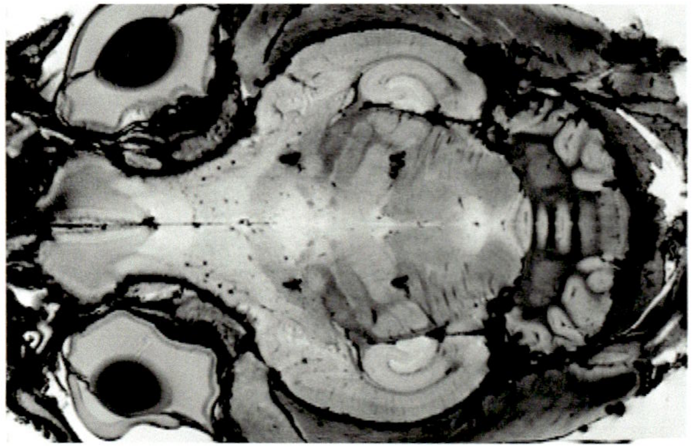

2 ayer 2 of cortex
3 layer 3 of cortex
3V 3rd ventricle
Ab auditory belt area
Ac auditory core area
Acb accumbens nucleus
aci anterior commissure, intrabulbar part
acp anterior commissure, posterior part
AHy anterior hypothalamic area
alv alveus of the hippocampus
AON anterior olfactory nucleus
APT anterior pretectal nucleus
Aq aqueduct
B brachium of the superior colliculus
CA1 field CA1 of the hippocampus
CA2 field CA2 of the hippocampus
CA3 field CA3 of the hippocampus
Cl laustrum
CM central medial thalamic nucleus
DLG dorsal lateral geniculate nucleus
DpG deep gray layer of the superior colliculus
DR dorsal raphe nucleus
E ependyma and subependymal layer
EGP external globus pallidus
Ent entorhinal cortex
EPl external plexiform layer of the olfactory bulb
f fornix
fi fimbria of the hippocampus
fr fasciculus retroflexus
GCL granule cell layer of the olfactory bulb
GL glomerular layer of the olfactory bulb
GrDG granular layer of the dentate gyrus
ic internal capsule
IC inferior colliculus
ICj islands of Calleja
icp inferior cerebellar peduncle
IGL intergeniculate leaf
InG intermediate gray layer of the superior colliculus
Ins insular cortex
Int interposed cerebellar nucleus
InWh intermediate white layer of the superior colliculus
IP internal plexiform layer of the olfactory bulb
IT inferior temporal cortex
Lat lateral cerebellar nucleus
LD laterodorsal thalamic nucleus
LHA lateral hypothalamic area
LMoL lacunosum moleculare layer of the hippocampus
lo lateral olfactory tract
LPO lateral preoptic area
LV lateral ventricle
MCL mitral cell layer of the olfactory bulb
MD mediodorsal thalamic nucleus
me5 mesencephalic trigeminal tract
Me5 mesencephalic trigeminal nucleus
Med medial cerebellar nucleus
MoDG molecular layer of the dentate gyrus
MPO medial preoptic nucleus
mRt mesencephalic reticular formation
mt mammillothalamic tract
NDB nucleus of the diagonal band
ONL olfactory nerve layer
Op optic nerve layer of the superior colliculus
OPT olivary pretectal nucleus
opt optic tract
Or oriens layer of the hippocampus
OV olfactory ventricle (olfactory part of lateral ventricle)
PAG periaqueductal gray
pc posterior commissure
Pc central nucleus of the pulvinar
Pd dorsal nucleus of the pulvinar
PF parafascicular thalamic nucleus
PH posterior hypothalamic nucleus
Pir1 piriform cortex, layer 1
PoDG polymorph layer of the dentate gyrus
pPAG p1 periaqueductal gray
PrS presubiculum
pRt p1 reticular formation
Pu putamen
Pv ventral nucleus of the pulvinar
PVN paraventricular hypothalamic nucleus
Py pyramidal cell layer of the hippocampus
Rad radiatum layer of the hippocampus
Re reuniens thalamic nucleus
Rh rhomboid thalamic nucleus
Rt reticular thalamic nucleus
SHy septohypothalamic nucleus
st stria terminalis
STMA bed nucleus of the stria terminalis, medial division, anterior part
str superior thalamic radiation
Sub submedius thalamic nucleus
SuG superficial gray layer of the superior colliculus
TC temporal cortex
Tu olfactory tubercle
Tu1 layer 1 of olfactory tubercle
VeCb vestibulocerebellar nucleus
VL ventrolateral thalamic nucleus
VLG ventral lateral geniculate nucleus
VPPC ventral posterior nucleus of the thalamus, parvicellular part
VS ventral subiculum
Zo zonal layer of the superior colliculus

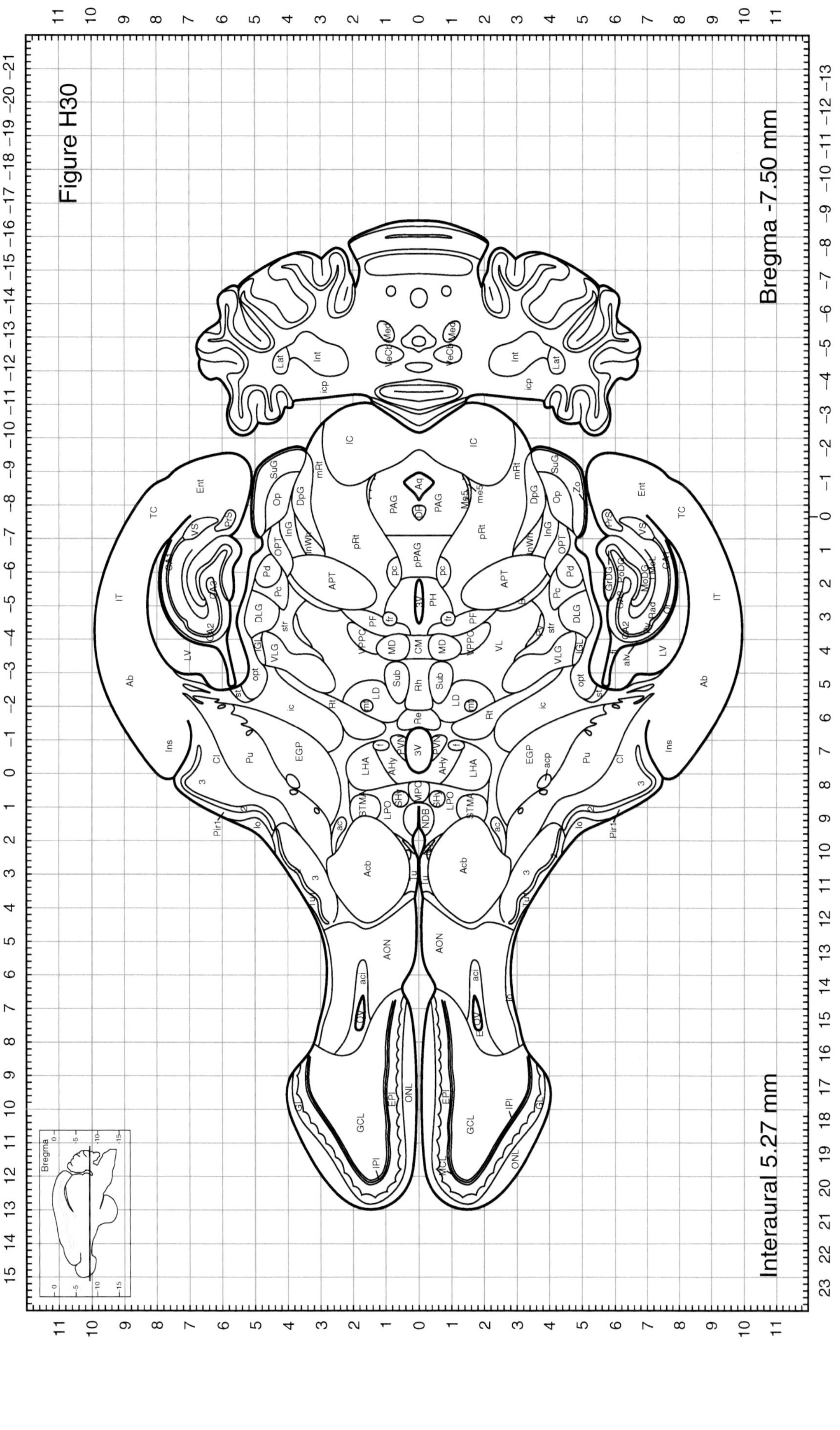
Figure H30
Bregma -7.50 mm
Interaural 5.27 mm
Bregma

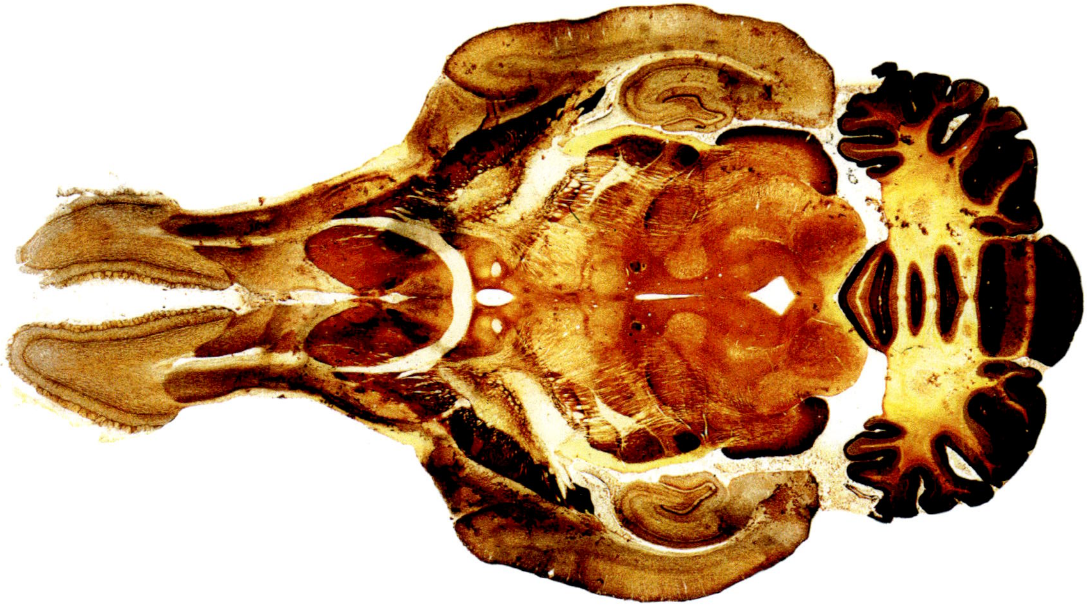

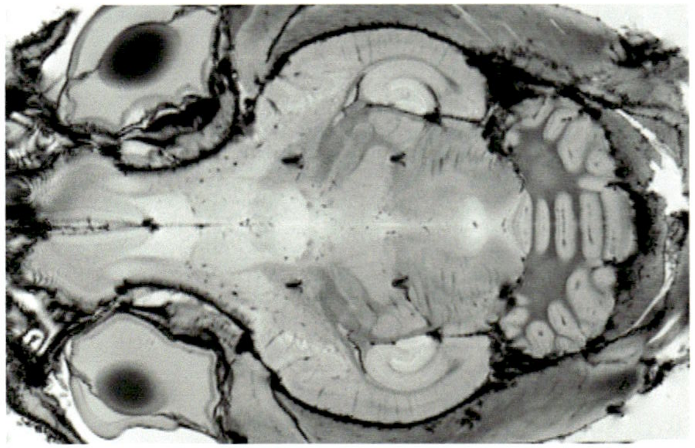

2 layer 2 of cortex
3 layer 3 of cortex
3V 3rd ventricle
Ab auditory belt area
ac anterior commissure
Acb accumbens nucleus
aci anterior commissure, intrabulbar part
acp anterior commissure, posterior part
alv alveus of the hippocampus
AM anteromedial thalamic nucleus
AON anterior olfactory nucleus
APT anterior pretectal nucleus
Aq aqueduct
B brachium of the superior colliculus
CA1 field CA1 of the hippocampus
CA2 field CA2 of the hippocampus
CA3 field CA3 of the hippocampus
Cl claustrum
CM central medial thalamic nucleus
csc commissure of the superior colliculus
DLG dorsal lateral geniculate nucleus
DpG deep gray layer of the superior colliculus
E ependyma and subependymal layer
EGP external globus pallidus
Ent entorhinal cortex
EPl external plexiform layer of the olfactory bulb
f fornix
fi fimbria of the hippocampus
fr fasciculus retroflexus
GCL granule cell layer of the olfactory bulb
GL glomerular layer of the olfactory bulb
GrDG granular layer of the dentate gyrus
IAM interanteromedial thalamic nucleus
ic internal capsule
IC inferior colliculus
ICj islands of Calleja
icp inferior cerebellar peduncle
IGL intergeniculate leaf
InG intermediate gray layer of the superior colliculus
Ins insular cortex
Int interposed cerebellar nucleus
InWh intermediate white layer of the superior colliculus
IPl internal plexiform layer of the olfactory bulb
IT inferior temporal cortex
LD laterodorsal thalamic nucleus
LMoL lacunosum moleculare layer of the hippocampus
lo lateral olfactory tract
LV lateral ventricle
MCL mitral cell layer of the olfactory bulb
MD mediodorsal thalamic nucleus
Me5 mesencephalic trigeminal nucleus
me5 mesencephalic trigeminal tract
MoDG molecular layer of the dentate gyrus
mRt mesencephalic reticular formation
NDB nucleus of the diagonal band
ONL olfactory nerve layer
Op optic nerve layer of the superior colliculus
OPT olivary pretectal nucleus
opt optic tract
Or oriens layer of the hippocampus
OV olfactory ventricle (olfactory part of lateral ventricle)
PAG periaqueductal gray
pc posterior commissure
Pc central nucleus of the pulvinar
Pd dorsal nucleus of the pulvinar
PF parafascicular thalamic nucleus
Pir1 piriform cortex, layer 1
PoDG polymorph layer of the dentate gyrus
pPAG p1 periaqueductal gray
PrS presubiculum
pRt p1 reticular formation
Pu putamen
Pv ventral nucleus of the pulvinar
PVN paraventricular hypothalamic nucleus
Py pyramidal cell layer of the hippocampus
Rt reticular thalamic nucleus
Rad radiatum layer of the hippocampus
Rh rhomboid thalamic nucleus
SHy septohypothalamic nucleus
st stria terminalis
StHy striohypothalamic nucleus
STMA bed nucleus of the stria terminalis, medial division, anterior part
STMPL bed nucleus of the stria terminalis, medial division, anterolateral part
STMPM bed nucleus of the stria terminalis, medial division, anteromedial part
str superior thalamic radiation
Sub submedius thalamic nucleus
SuG superficial gray layer of the superior colliculus
TC temporal cortex
Tu olfactory tubercle
VL ventrolateral thalamic nucleus
VLG ventral lateral geniculate nucleus
VS ventral subiculum
Zo zonal layer of the superior colliculus

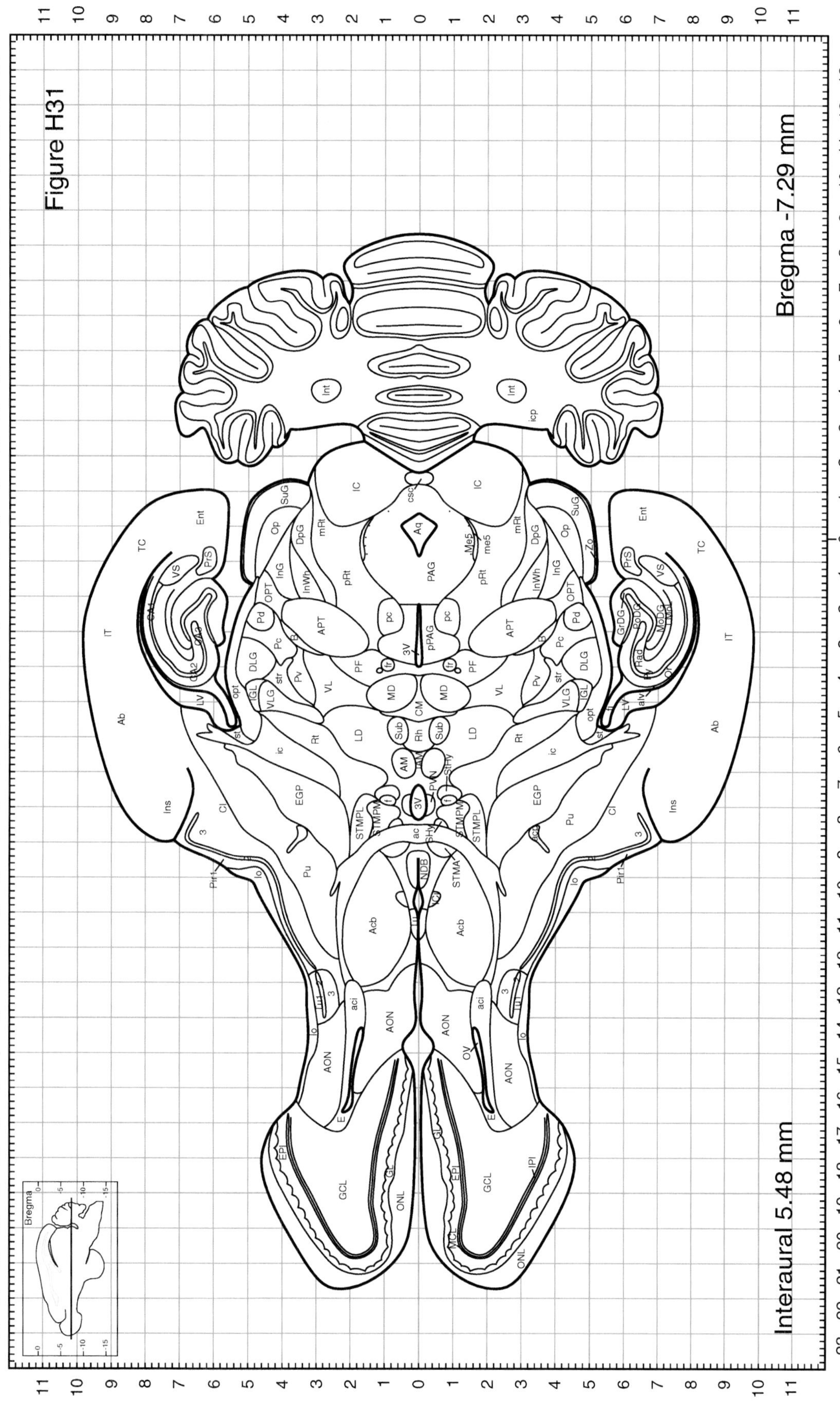
Figure H31
Bregma -7.29 mm
Interaural 5.48 mm
Int
icp
IC
csc
Aq
PAG
Me5
me5
mRt
pRt
DpG
InWh
InG
Op
SuG
OPT
Zo
Ent
TC
PrS
VS
CA1
CA3
CA2
GrDG
PoDG
MoDG
LMol
Rad
Py
Or
alv
fi
LV
IT
Pd
Pc
DLG
IGL
VLG
opt
str
Pv
VL
APT
pc
3V
pPAG
fr
PF
MD
CM
LD
Sub
Rh
AM
PVN
StHy
Rt
ic
st
Ab
EGP
Ins
Cl
STMPL
STMPM
ac
SHy
STMA
NDB
VOLT
Pu
lo
Pir1
Acb
aci
AON
E
OV
EPl
GCL
Gl
ONL
IPl
MCL
Bregma

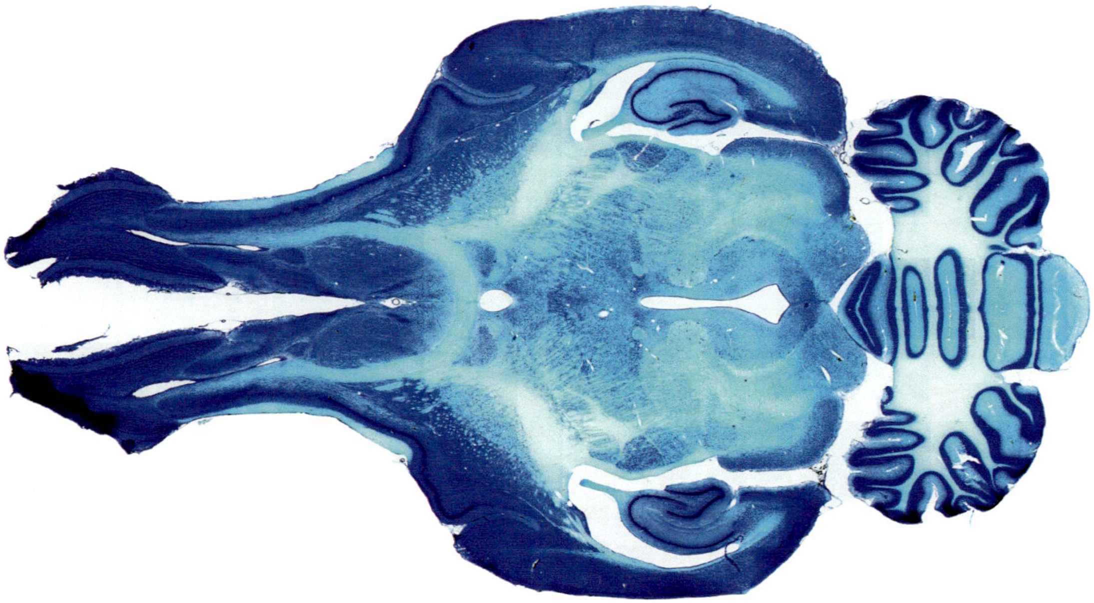

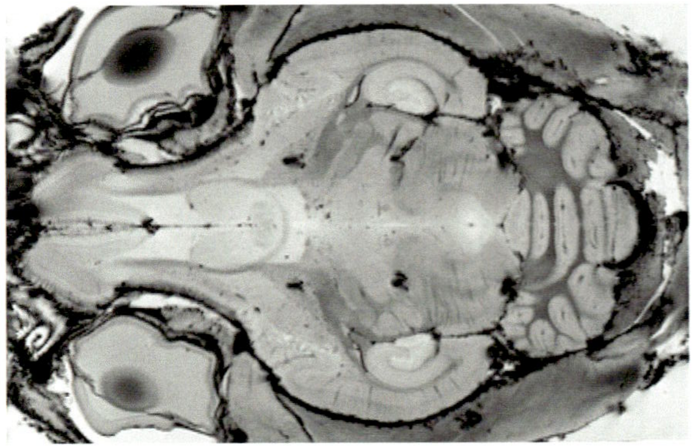

2	layer 2 of cortex
3	layer 3 of cortex
3V	3rd ventricle
Ab	auditory belt area
ac	anterior commissure
Acb	accumbens nucleus
aci	anterior commissure, intrabulbar part
acp	anterior commissure, posterior part
alv	alveus of the hippocampus
AM	anteromedial thalamic nucleus
AON	anterior olfactory nucleus
AOP	anterior olfactory nucleus, posterior part
APT	anterior pretectal nucleus
Aq	aqueduct
AV	anteroventral thalamic nucleus
B	brachium of the superior colliculus
CA1	field CA1 of the hippocampus
CA2	field CA2 of the hippocampus
CA3	field CA3 of the hippocampus
Cl	claustrum
CM	central medial thalamic nucleus
csc	commissure of the superior colliculus
DLG	dorsal lateral geniculate nucleus
DpG	deep gray layer of the superior colliculus
DpWh	deep white layer of the superior colliculus
E	ependyma and subependymal layer
EGP	external globus pallidus
Ent	entorhinal cortex
EPl	external plexiform layer of the olfactory bulb
f	fornix
fi	fimbria of the hippocampus
fr	fasciculus retroflexus
GCL	granule cell layer of the olfactory bulb
GL	glomerular layer of the olfactory bulb
GrDG	granular layer of the dentate gyrus
IAM	interanteromedial thalamic nucleus
ic	internal capsule
IC	inferior colliculus
ICj	islands of Calleja
icp	inferior cerebellar peduncle
IGL	intergeniculate leaf
IMD	intermediodorsal thalamic nucleus
InG	intermediate gray layer of the superior colliculus
Ins	insular cortex
InWh	intermediate white layer of the superior colliculus
IPl	internal plexiform layer of the olfactory bulb
IT	inferior temporal cortex
LD	laterodorsal thalamic nucleus
LMoL	lacunosum moleculare layer of the hippocampus
lo	lateral olfactory tract
LV	lateral ventricle
MCL	mitral cell layer of the olfactory bulb
MD	mediodorsal thalamic nucleus
Me5	mesencephalic trigeminal nucleus
me5	mesencephalic trigeminal tract
MoDG	molecular layer of the dentate gyrus
MS	medial septal nucleus
OFC	orbital frontal cortex
ONL	olfactory nerve layer
Op	optic nerve layer of the superior colliculus
OPT	olivary pretectal nucleus
opt	optic tract
Or	oriens layer of the hippocampus
OV	olfactory ventricle (olfactory part of lateral ventricle)
PAG	periaqueductal gray
PC	paracentral thalamic nucleus
pc	posterior commissure
Pc	central nucleus of the pulvinar
Pd	dorsal nucleus of the pulvinar
PF	parafascicular thalamic nucleus
Pir1	piriform cortex, layer 1
PoDG	polymorph layer of the dentate gyrus
pPAG	p1 periaqueductal gray
PrS	presubiculum
pRt	p1 reticular formation
Pu	putamen
Pv	ventral nucleus of the pulvinar
PVA	paraventricular thalamic nucleus, anterior part
Py	pyramidal cell layer of the hippocampus
Rad	radiatum layer of the hippocampus
Rt	reticular thalamic nucleus
st	stria terminalis
StHy	striohypothalamic nucleus
STMA	bed nucleus of the stria terminalis, medial division, anterior part
STMPL	bed nucleus of the stria terminalis, medial division, anterolateral part
STMPM	bed nucleus of the stria terminalis, medial division, anteromedial part
str	superior thalamic radiation
SuG	superficial gray layer of the superior colliculus
TC	temporal cortex
VL	ventrolateral thalamic nucleus
VLG	ventral lateral geniculate nucleus
VS	ventral subiculum
Zo	zonal layer of the superior colliculus

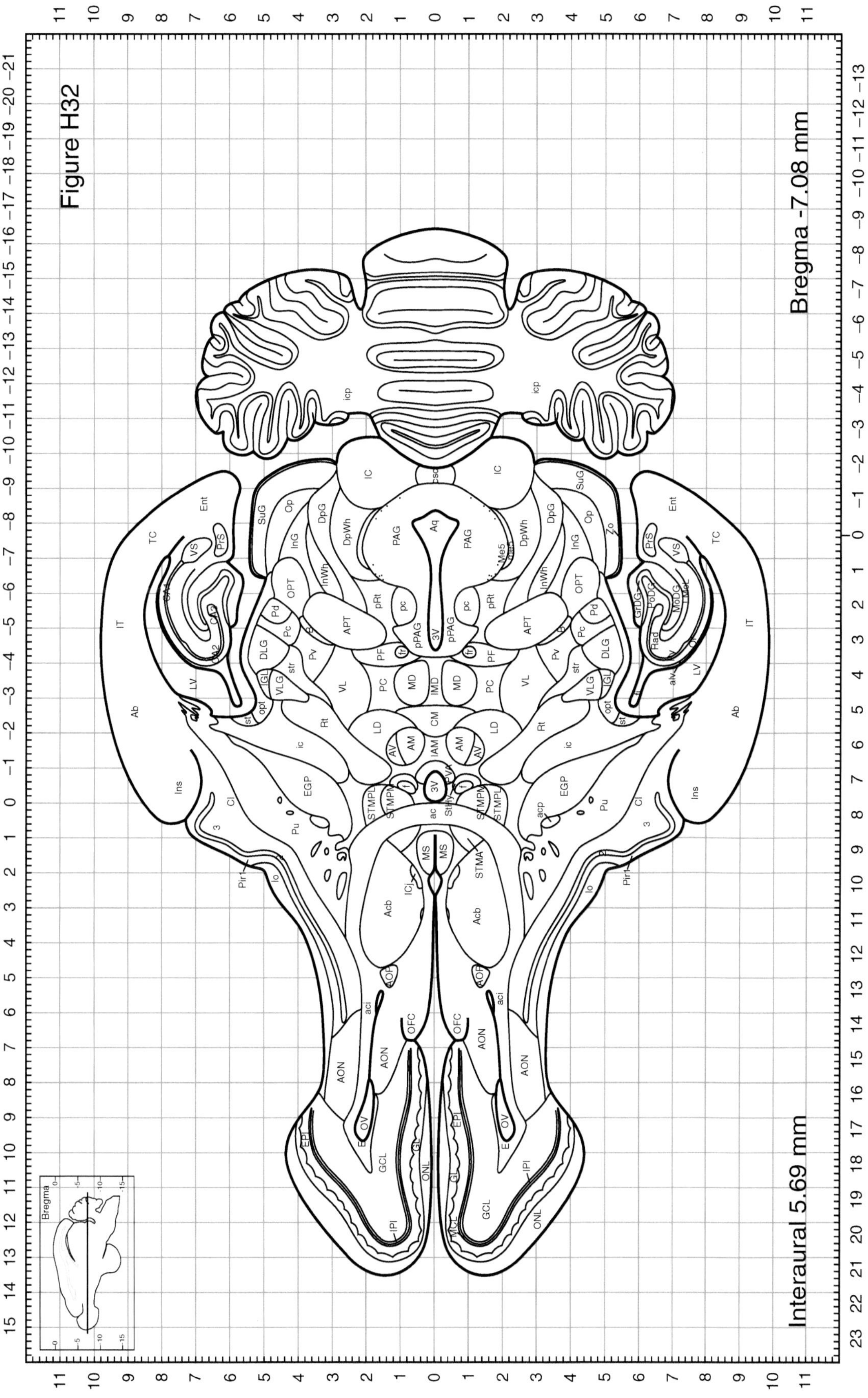
Figure H32
Bregma -7.08 mm
Interaural 5.69 mm

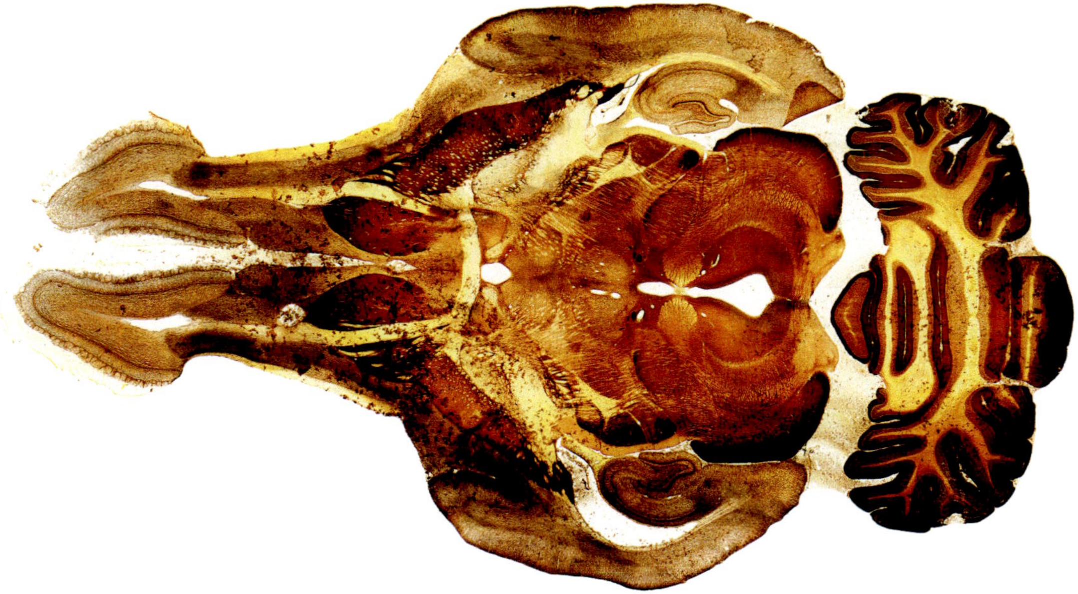

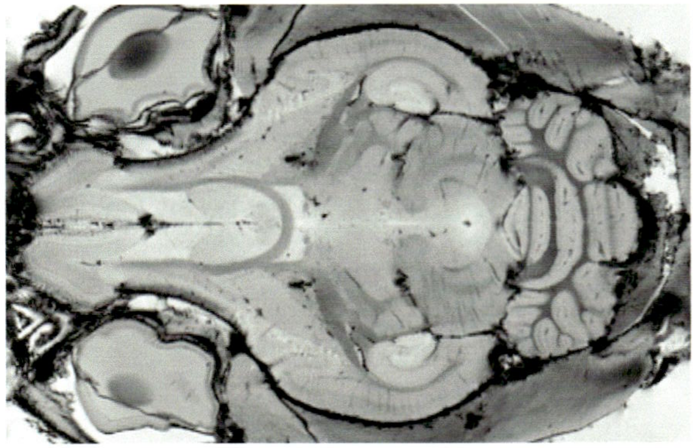

2	layer 2 of cortex
3	layer 3 of cortex
3V	3rd ventricle
Ab	auditory belt area
ac	anterior commissure
Acb	accumbens nucleus
aci	anterior commissure, intrabulbar part
acp	anterior commissure, posterior part
alv	alveus of the hippocampus
AM	anteromedial thalamic nucleus
AON	anterior olfactory nucleus
APT	anterior pretectal nucleus
Aq	aqueduct
AV	anteroventral thalamic nucleus
B	brachium of the superior colliculus
CA1	field CA1 of the hippocampus
CA2	field CA2 of the hippocampus
CA3	field CA3 of the hippocampus
Cl	claustrum
CM	central medial thalamic nucleus
csc	commissure of the superior colliculus
DLG	dorsal lateral geniculate nucleus
DpG	deep gray layer of the superior colliculus
DpWh	deep white layer of the superior colliculus
E	ependyma and subependymal layer
EGP	external globus pallidus
Ent	entorhinal cortex
EPl	external plexiform layer of the olfactory bulb
f	fornix
fi	fimbria of the hippocampus
fr	fasciculus retroflexus
GCL	granule cell layer of the olfactory bulb
GL	glomerular layer of the olfactory bulb
GrDG	granular layer of the dentate gyrus
IAM	interanteromedial thalamic nucleus
ic	internal capsule
IC	inferior colliculus
ICj	islands of Calleja
IGL	intergeniculate leaf
IMD	intermediodorsal thalamic nucleus
InG	intermediate gray layer of the superior colliculus
Ins	insular cortex
InWh	intermediate white layer of the superior colliculus
IPl	internal plexiform layer of the olfactory bulb
IT	inferior temporal cortex
LD	laterodorsal thalamic nucleus
LMoL	lacunosum moleculare layer of the hippocampus
lo	lateral olfactory tract
LV	lateral ventricle
MCL	mitral cell layer of the olfactory bulb
MD	mediodorsal thalamic nucleus
Me5	mesencephalic trigeminal nucleus
me5	mesencephalic trigeminal tract
MoDG	molecular layer of the dentate gyrus
MS	medial septal nucleus
OFC	orbital frontal cortex
ONL	olfactory nerve layer
Op	optic nerve layer of the superior colliculus
OPT	olivary pretectal nucleus
opt	optic tract
Or	oriens layer of the hippocampus
OV	olfactory ventricle (olfactory part of lateral ventricle)
PAG	periaqueductal gray
PC	paracentral thalamic nucleus
pc	posterior commissure
Pc	central nucleus of the pulvinar
Pd	dorsal nucleus of the pulvinar
PF	parafascicular thalamic nucleus
Pir1	piriform cortex, layer 1
PoDG	polymorph layer of the dentate gyrus
pPAG	p1 periaqueductal gray
PrS	presubiculum
pRt	p1 reticular formation
Pu	putamen
Pv	ventral nucleus of the pulvinar
PVA	paraventricular thalamic nucleus, anterior part
Py	pyramidal cell layer of the hippocampus
Rad	radiatum layer of the hippocampus
Rt	reticular thalamic nucleus
sm	stria medullaris of the thalamus
st	stria terminalis
STMA	bed nucleus of the stria terminalis, medial division, anterior part
STMPL	bed nucleus of the stria terminalis, medial division, anterolateral part
STMPM	bed nucleus of the stria terminalis, medial division, anteromedial part
str	superior thalamic radiation
SuG	superficial gray layer of the superior colliculus
TC	temporal cortex
VL	ventrolateral thalamic nucleus
VLG	ventral lateral geniculate nucleus
VS	ventral subiculum
Zo	zonal layer of the superior colliculus

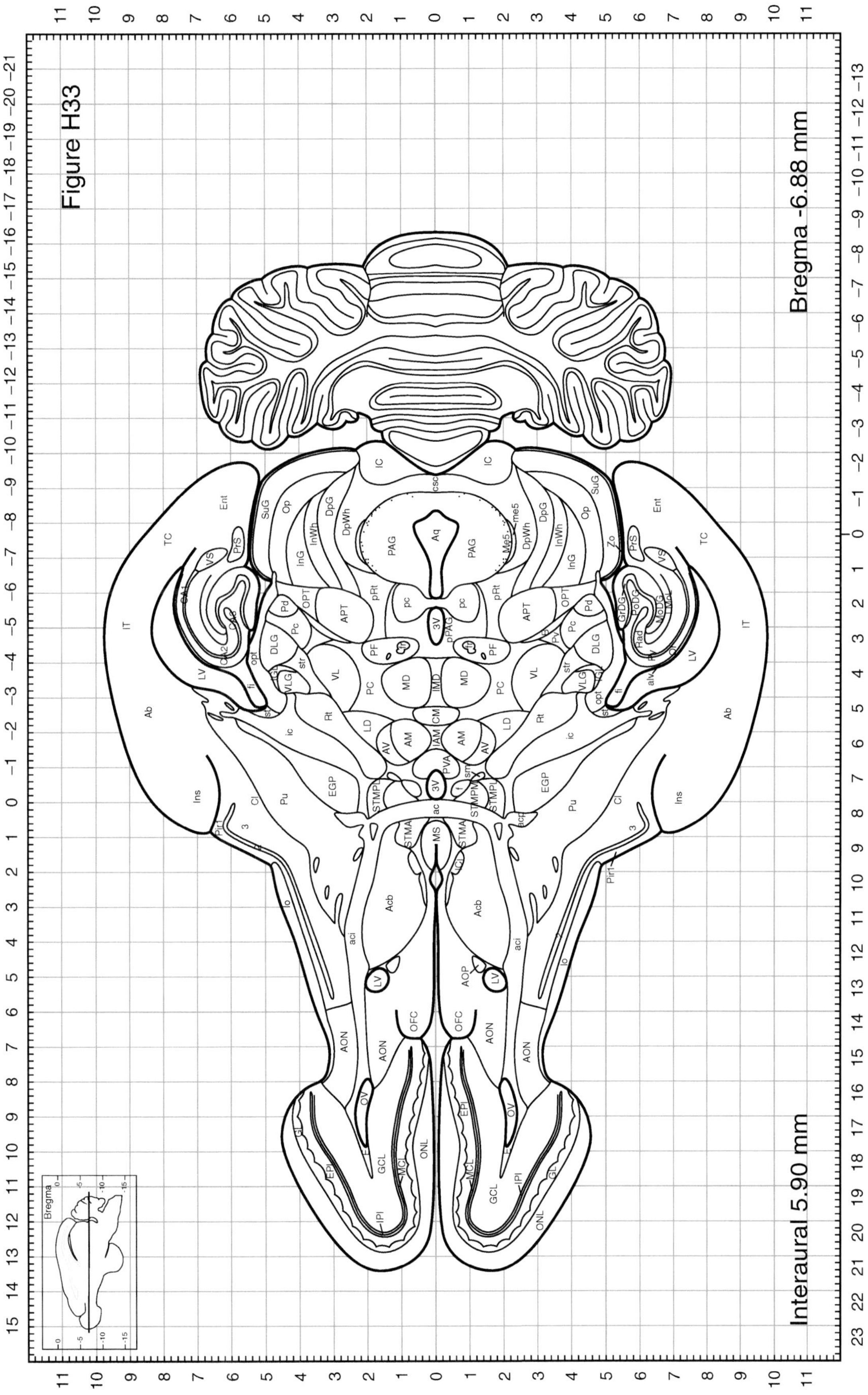
Figure H33
Bregma -6.88 mm
Interaural 5.90 mm
IC
csc
SuG
Op
DpG
DpWh
InWh
InG
PAG
Aq
Me5
me5
Zo
Ent
TC
PrS
VS
CA1
CA3
CA2
DG
IT
LV
Ab
APT
pRt
pc
OPT
Pd
Pc
3V
pPAG
PF
DLG
opt
str
fi
VLG
IGL
VL
PC
MD
IMD
Rt
LD
CM
AM
AV
IAM
PVA
sm
f
ic
EGP
STMPL
STMPM
ac
MS
STMA
acp
ICj
Pu
Cl
Ins
Pir1
lo
Acb
aci
LV
AOP
OFC
AON
OV
E
EPl
GCL
Mi
IPl
Gl
ONL
GrDG
PoDG
MoDG
Rad
Py
alv
Or
Bregma

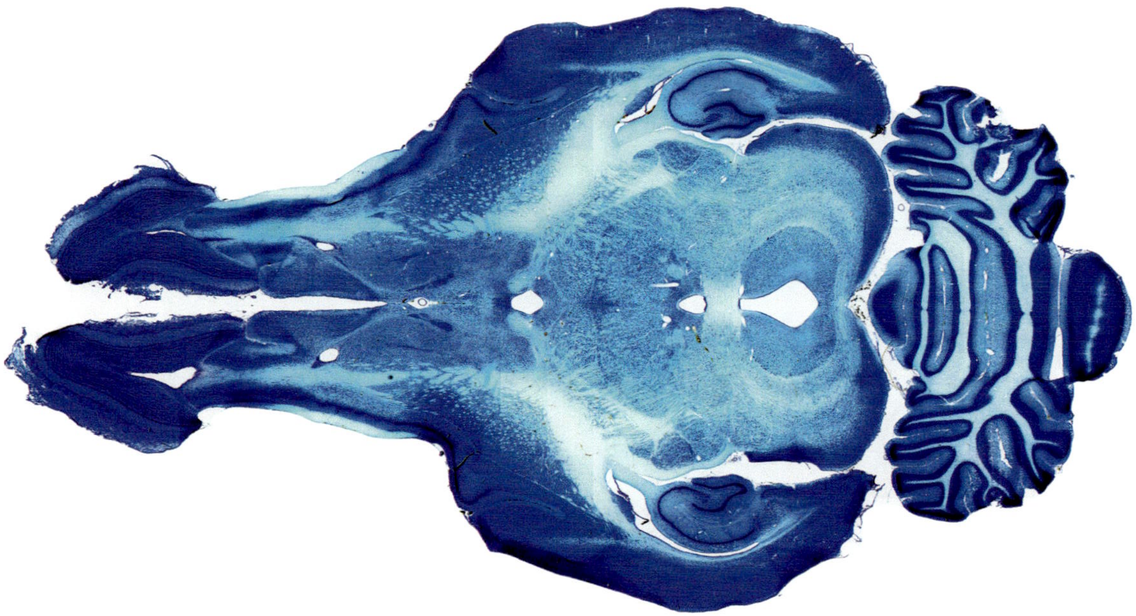

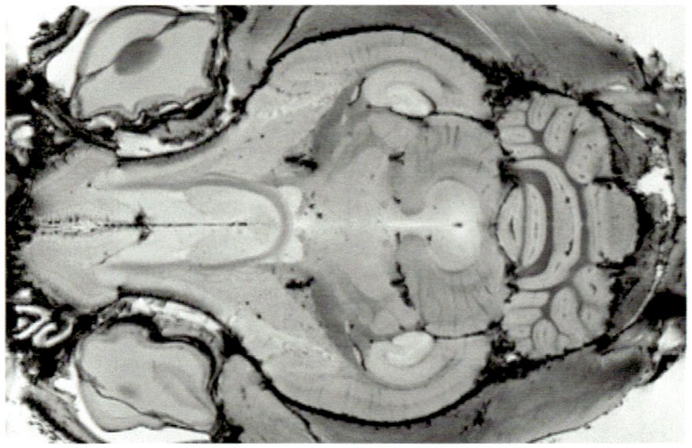

2 layer 2 of cortex
3 layer 3 of cortex
3V 3rd ventricle
ac anterior commissure
Acb accumbens nucleus
aci anterior commissure, intrabulbar part
alv alveus of the hippocampus
AM anteromedial thalamic nucleus
AON anterior olfactory nucleus
APT anterior pretectal nucleus
Aq aqueduct
AV anteroventral thalamic nucleus
B brachium of the superior colliculus
CA1 field CA1 of the hippocampus
CA2 field CA2 of the hippocampus
CA3 field CA3 of the hippocampus
Cd caudate nucleus
CM central medial thalamic nucleus
csc commissure of the superior colliculus
DLG dorsal lateral geniculate nucleus
DpG deep gray layer of the superior colliculus
DpWh deep white layer of the superior colliculus
E ependyma and subependymal layer
EGP external globus pallidus
Ent entorhinal cortex
EPI external plexiform layer of the olfactory bulb
f fornix
fi fimbria of the hippocampus
fr fasciculus retroflexus
GCL granule cell layer of the olfactory bulb
GL glomerular layer of the olfactory bulb
GrDG granular layer of the dentate gyrus
IAM interanteromedial thalamic nucleus
ic internal capsule
Cj islands of Calleja
GL intergeniculate leaf
IL infralimbic cortex
IMD intermediodorsal thalamic nucleus
InG intermediate gray layer of the superior colliculus
Ins insular cortex
InWh intermediate white layer of the superior colliculus
IPl internal plexiform layer of the olfactory bulb
IT inferior temporal cortex
LD laterodorsal thalamic nucleus
LMoL lacunosum moleculare layer of the hippocampus
lo lateral olfactory tract
LSi lateral septal nucleus, intermediate part
LV lateral ventricle
MCL mitral cell layer of the olfactory bulb
MD mediodorsal thalamic nucleus
Me5 mesencephalic trigeminal nucleus
me5 mesencephalic trigeminal tract
MnPo median preoptic nucleus
MoDG molecular layer of the dentate gyrus
MS medial septal nucleus
OFC orbital frontal cortex
ONL olfactory nerve layer
Op optic nerve layer of the superior colliculus
OPT olivary pretectal nucleus
opt optic tract
Or oriens layer of the hippocampus
OV olfactory ventricle (olfactory part of lateral ventricle)
PAG periaqueductal gray
PC paracentral thalamic nucleus
pc posterior commissure
Pc central nucleus of the pulvinar
Pd dorsal nucleus of the pulvinar
PF parafascicular thalamic nucleus
Pir1 piriform cortex, layer 1
PoDG polymorph layer of the dentate gyrus
pPAG p1 periaqueductal gray
PrC precommissural nucleus
PrS presubiculum
Pu putamen
Pv ventral nucleus of the pulvinar
PvA parietal ventral area
PVA paraventricular thalamic nucleus, anterior part
Py pyramidal cell layer of the hippocampus
Rad radiatum layer of the hippocampus
Rt reticular thalamic nucleus
sm stria medullaris of the thalamus
st stria terminalis
STMPL bed nucleus of the stria terminalis, medial division, anterolateral part
STMPM bed nucleus of the stria terminalis, medial division, anteromedial part
str superior thalamic radiation
SuG superficial gray layer of the superior colliculus
TC temporal cortex
VL ventrolateral thalamic nucleus
VLG ventral lateral geniculate nucleus
VS ventral subiculum
Zo zonal layer of the superior colliculus

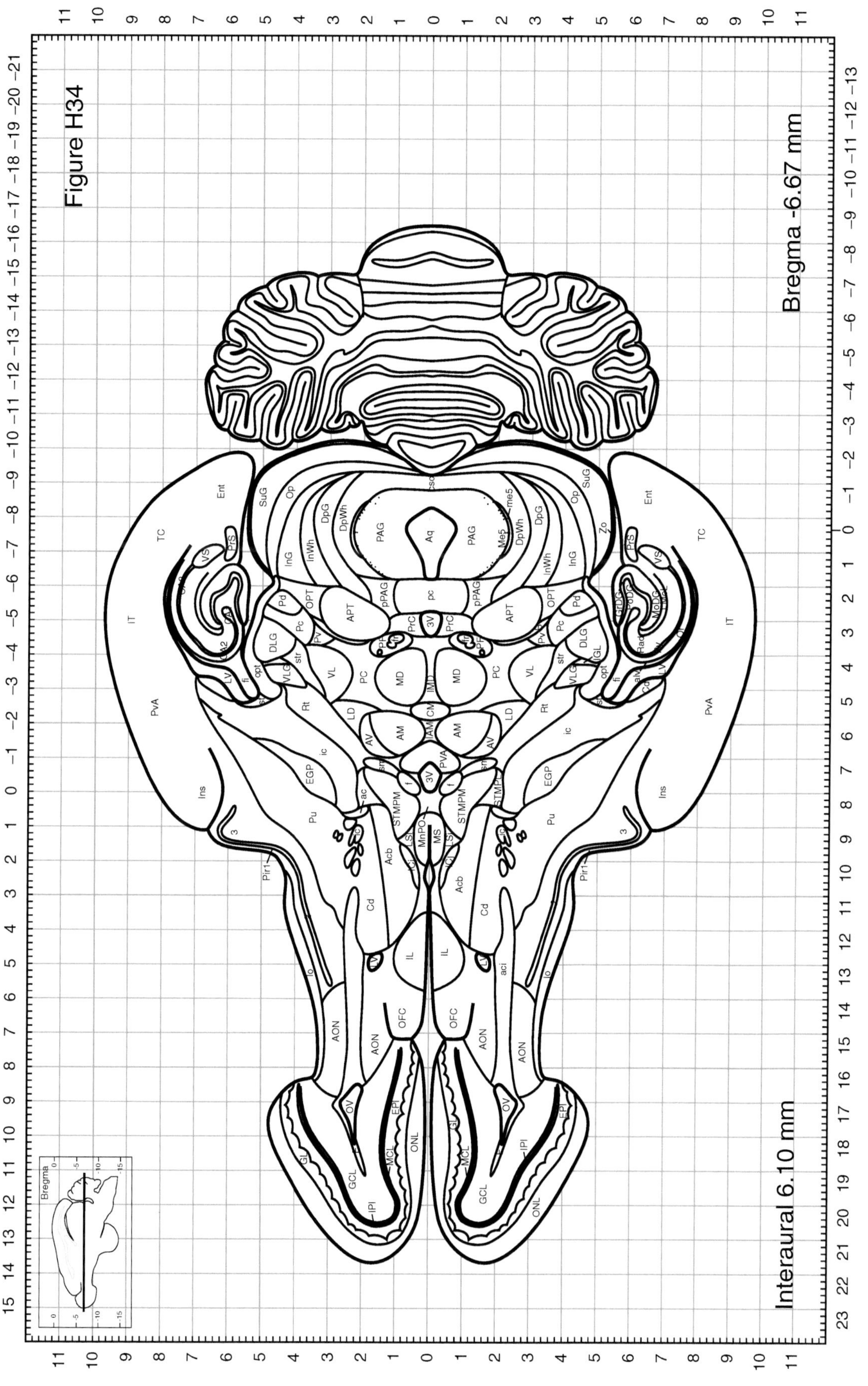
Figure H34
Bregma -6.67 mm
Interaural 6.10 mm

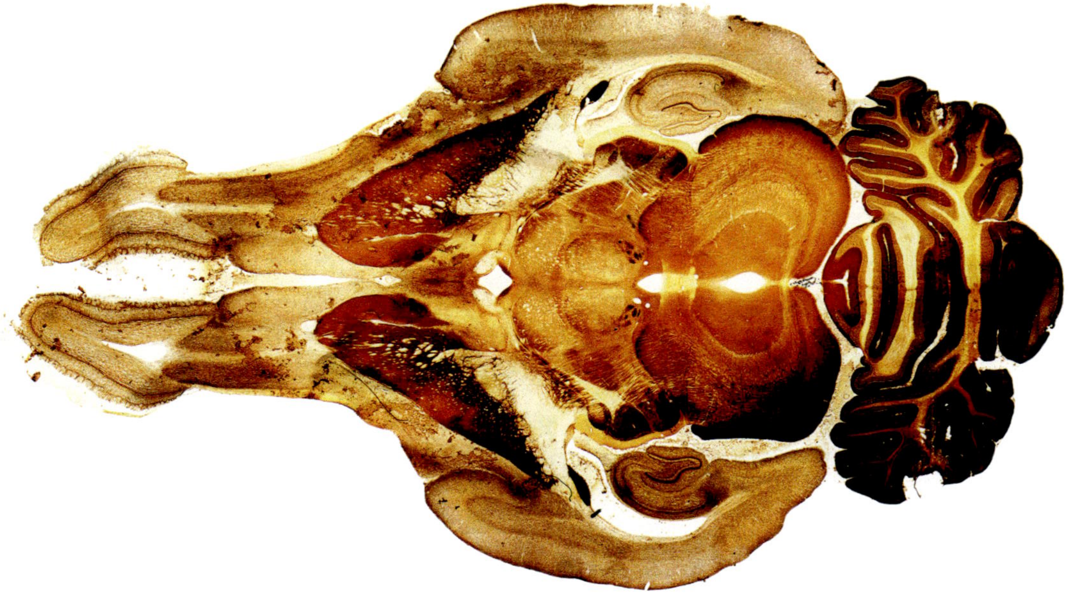

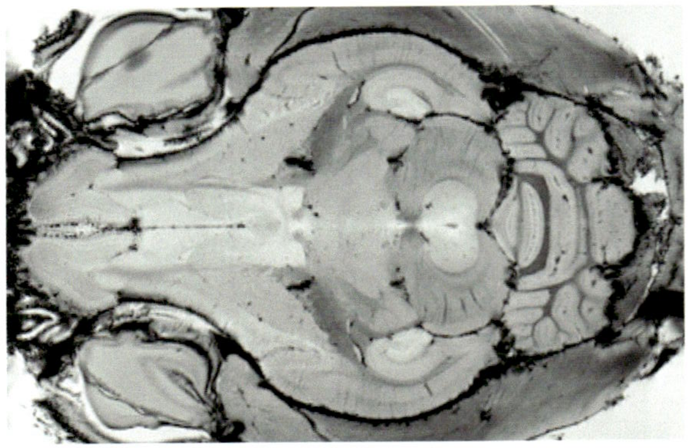

2	layer 2 of cortex
3	layer 3 of cortex
3V	3rd ventricle
Acb	accumbens nucleus
aci	anterior commissure, intrabulbar part
alv	alveus of the hippocampus
AM	anteromedial thalamic nucleus
AON	anterior olfactory nucleus
APT	anterior pretectal nucleus
Aq	aqueduct
AV	anteroventral thalamic nucleus
B	brachium of the superior colliculus
CA1	field CA1 of the hippocampus
CA2	field CA2 of the hippocampus
CA3	field CA3 of the hippocampus
Cd	caudate nucleus
CM	central medial thalamic nucleus
csc	commissure of the superior colliculus
DLG	dorsal lateral geniculate nucleus
DpG	deep gray layer of the superior colliculus
DpWh	deep white layer of the superior colliculus
E	ependyma and subependymal layer
EGP	external globus pallidus
Ent	entorhinal cortex
EPl	external plexiform layer of the olfactory bulb
f	fornix
fi	fimbria of the hippocampus
fr	fasciculus retroflexus
GCL	granule cell layer of the olfactory bulb
GL	glomerular layer of the olfactory bulb
GrDG	granular layer of the dentate gyrus
IAM	interanteromedial thalamic nucleus
ic	internal capsule
ICj	islands of Calleja
IGL	intergeniculate leaf
IL	infralimbic cortex
IMD	intermediodorsal thalamic nucleus
InG	intermediate gray layer of the superior colliculus
Ins	insular cortex
InWh	intermediate white layer of the superior colliculus
IPl	internal plexiform layer of the olfactory bulb
IT	inferior temporal cortex
LD	laterodorsal thalamic nucleus
LMoL	lacunosum moleculare layer of the hippocampus
lo	lateral olfactory tract
LSi	lateral septal nucleus, intermediate part
LV	lateral ventricle
MCL	mitral cell layer of the olfactory bulb
MD	mediodorsal thalamic nucleus
Me5	mesencephalic trigeminal nucleus
me5	mesencephalic trigeminal tract
MnPo	median preoptic nucleus
MoDG	molecular layer of the dentate gyrus
MS	medial septal nucleus
OFC	orbital frontal cortex
ONL	olfactory nerve layer
Op	optic nerve layer of the superior colliculus
OPT	olivary pretectal nucleus
opt	optic tract
Or	oriens layer of the hippocampus
OV	olfactory ventricle (olfactory part of lateral ventricle)
PAG	periaqueductal gray
PC	paracentral thalamic nucleus
pc	posterior commissure
Pc	central nucleus of the pulvinar
Pd	dorsal nucleus of the pulvinar
PF	parafascicular thalamic nucleus
Pir 1	piriform cortex, layer 1
PoDG	polymorph layer of the dentate gyrus
pPAG	p1 periaqueductal gray
PrC	precommissural nucleus
PrS	presubiculum
Pu	putamen
Pv	ventral nucleus of the pulvinar
PvA	parietal ventral area
PVA	paraventricular thalamic nucleus, anterior part
Py	pyramidal cell layer of the hippocampus
Rad	radiatum layer of the hippocampus
Rt	reticular thalamic nucleus
sm	stria medullaris of the thalamus
st	stria terminalis
STMPL	bed nucleus of the stria terminalis, medial division,anterolateral part
STMPM	bed nucleus of the stria terminalis, medial division,anteromedial part
str	superior thalamic radiation
SuG	superficial gray layer of the superior colliculus
TC	temporal cortex
VL	ventrolateral thalamic nucleus
VLG	ventral lateral geniculate nucleus
VS	ventral subiculum
Zo	zonal layer of the superior colliculus

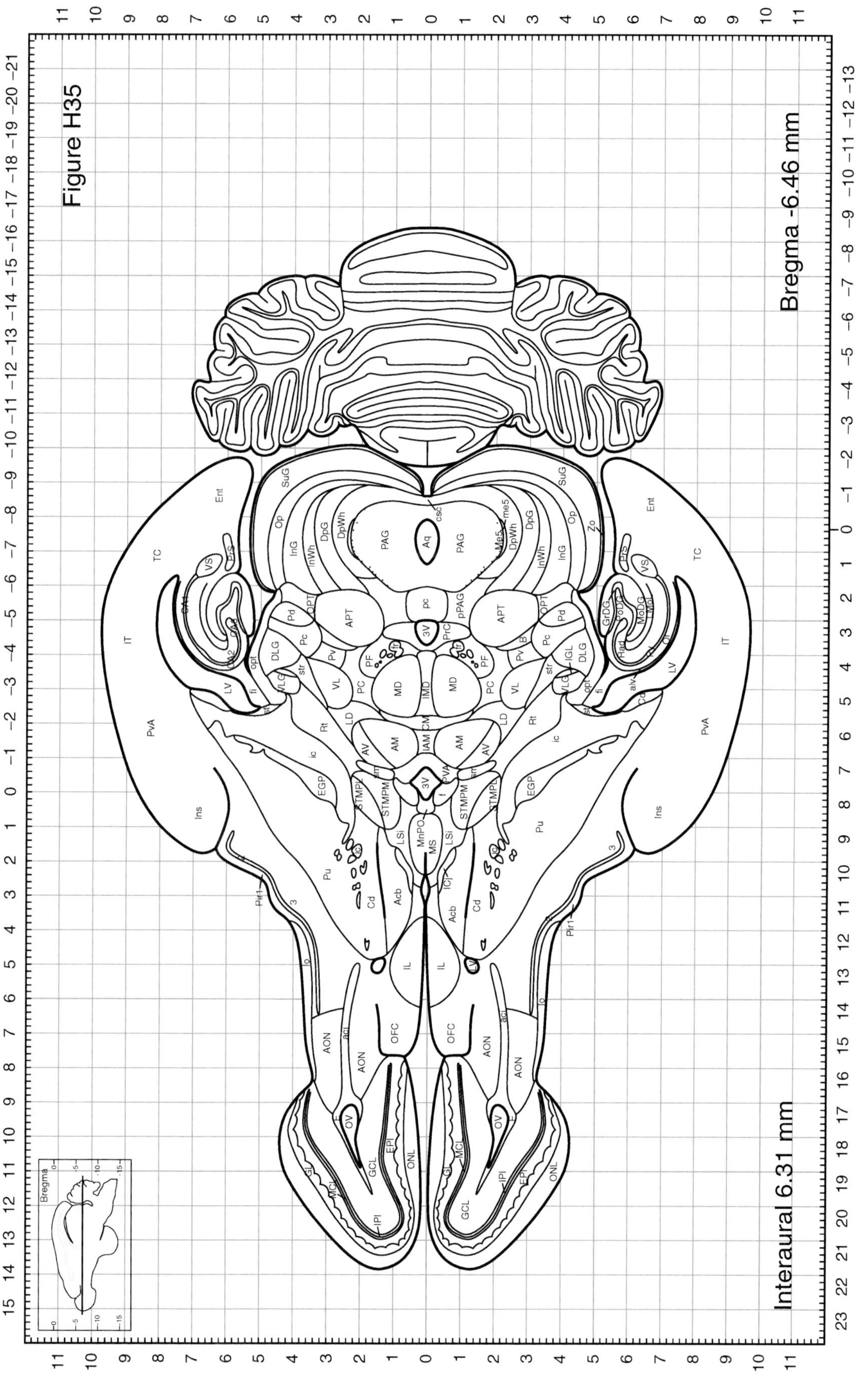

Figure H35
Bregma -6.46 mm
Interaural 6.31 mm
Bregma
SuG
Op
InG
InWh
DpG
DpWh
PAG
Aq
csc
Me5
me5
Zo
Ent
TC
VS
PrS
CA1
CA2
CA3
GrDG
PoDG
MoDG
Mol
Rad
Or
alv
LV
IT
APT
OPT
Pd
Pc
pc
pPAG
3V
PrC
fr
DLG
IGL
VLG
opt
str
Pv
PF
VL
PC
MD
IMD
LD
Rt
CM
IAM
AM
AV
sm
PVA
f
STMPL
STMPM
EGP
ic
fi
st
Cd
PvA
Ins
Pu
LSi
MnPO
MS
ICj
Acb
Pir1
lo
IL
OFC
aci
AON
OV
E
EPl
ONL
GCL
MCL
IPl
Gl

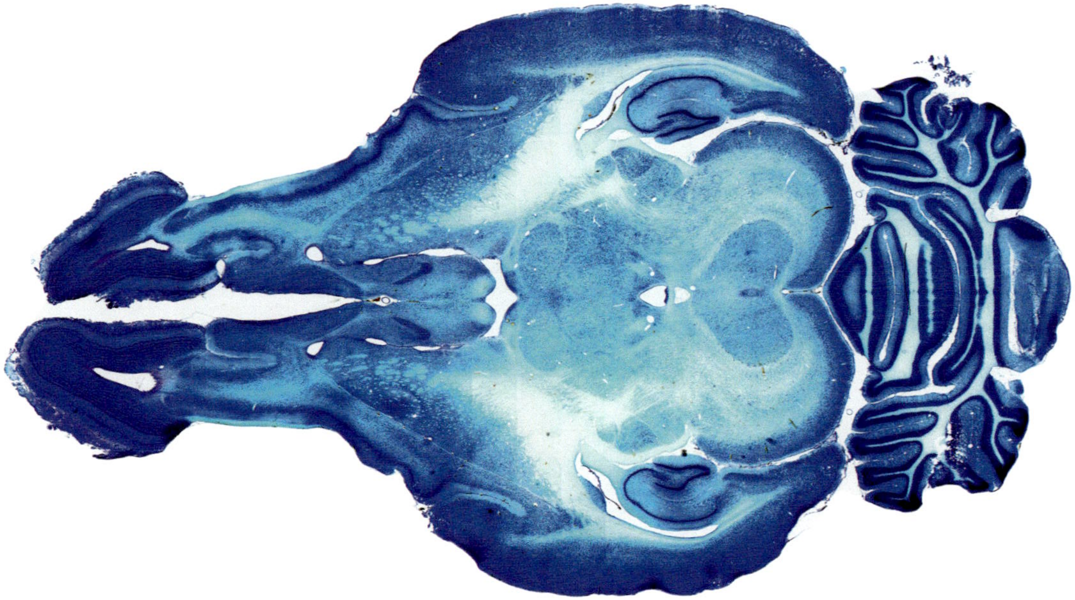

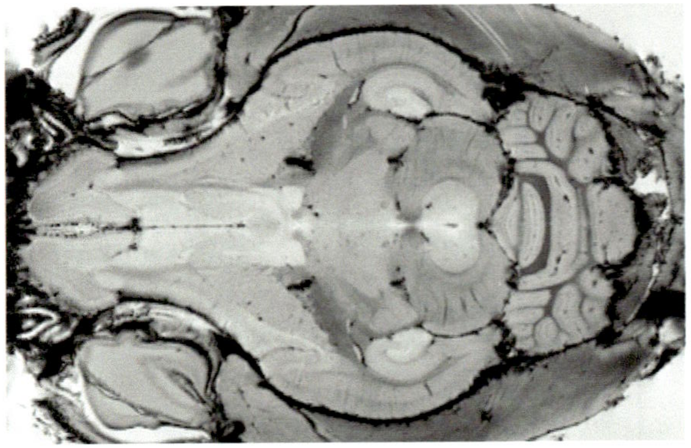

2	layer 2 of cortex
3	layer 3 of cortex
3V	3rd ventricle
Acb	accumbens nucleus
aci	anterior commissure, intrabulbar part
alv	alveus of the hippocampus
AM	anteromedial thalamic nucleus
AON	anterior olfactory nucleus
APT	anterior pretectal nucleus
Aq	aqueduct
AV	anteroventral thalamic nucleus
B	brachium of the superior colliculus
CA1	field CA1 of the hippocampus
CA2	field CA2 of the hippocampus
CA3	field CA3 of the hippocampus
Cd	caudate nucleus
CM	central medial thalamic nucleus
csc	commissure of the superior colliculus
DLG	dorsal lateral geniculate nucleus
DpG	deep gray layer of the superior colliculus
DpWh	deep white layer of the superior colliculus
E	ependyma and subependymal layer
EGP	external globus pallidus
Ent	entorhinal cortex
EPl	external plexiform layer of the olfactory bulb
f	fornix
fi	fimbria of the hippocampus
fr	fasciculus retroflexus
GCL	granule cell layer of the olfactory bulb
GL	glomerular layer of the olfactory bulb
GrA	granule cell layer of the accessory olfactory bulb
GrDG	granular layer of the dentate gyrus
IAM	interanteromedial thalamic nucleus
ic	internal capsule
ICj	islands of Calleja
IGL	intergeniculate leaf
IL	infralimbic cortex
IMD	intermediodorsal thalamic nucleus
InG	intermediate gray layer of the superior colliculus
Ins	insular cortex
InWh	intermediate white layer of the superior colliculus
IPl	internal plexiform layer of the olfactory bulb
IT	inferior temporal cortex
LD	laterodorsal thalamic nucleus
LHb	lateral habenular nucleus
LMoL	lacunosum moleculare layer of the hippocampus
lo	lateral olfactory tract
LSi	lateral septal nucleus, intermediate part
LSv	lateral septal nucleus, ventral part
LV	lateral ventricle
MCL	mitral cell layer of the olfactory bulb
MD	mediodorsal thalamic nucleus
Me5	mesencephalic trigeminal nucleus
me5	mesencephalic trigeminal tract
MnPo	median preoptic nucleus
MoDG	molecular layer of the dentate gyrus
MS	medial septal nucleus
OFC	orbital frontal cortex
ONL	olfactory nerve layer
Op	optic nerve layer of the superior colliculus
OPT	olivary pretectal nucleus
opt	optic tract
Or	oriens layer of the hippocampus
OV	olfactory ventricle (olfactory part of lateral ventricle)
PAG	periaqueductal gray
PC	paracentral thalamic nucleus
pc	posterior commissure
Pc	central nucleus of the pulvinar
Pd	dorsal nucleus of the pulvinar
PF	parafascicular thalamic nucleus
Pir	piriform cortex, layer 1
PoDG	polymorph layer of the dentate gyrus
pPAG	p1 periaqueductal gray
PrC	precommissural nucleus
PrS	presubiculum
Pu	putamen
Pv	ventral nucleus of the pulvinar
PvA	parietal ventral area
PVA	paraventricular thalamic nucleus, anterior part
PVP	paraventricular thalamic nucleus, posterior part
Py	pyramidal cell layer of the hippocampus
Rad	radiatum layer of the hippocampus
Rt	reticular thalamic nucleus
S2	secondary somatosensory cortex
sm	stria medullaris of the thalamus
st	stria terminalis
STMPL	bed nucleus of the stria terminalis, medial division,anterolateral part
STMPM	bed nucleus of the stria terminalis, medial division,anteromedial part
str	superior thalamic radiation
SuG	superficial gray layer of the superior colliculus
TC	temporal cortex
TI	temporal inferior area
VL	ventrolateral thalamic nucleus
VLG	ventral lateral geniculate nucleus
VS	ventral subiculum
Zo	zonal layer of the superior colliculus

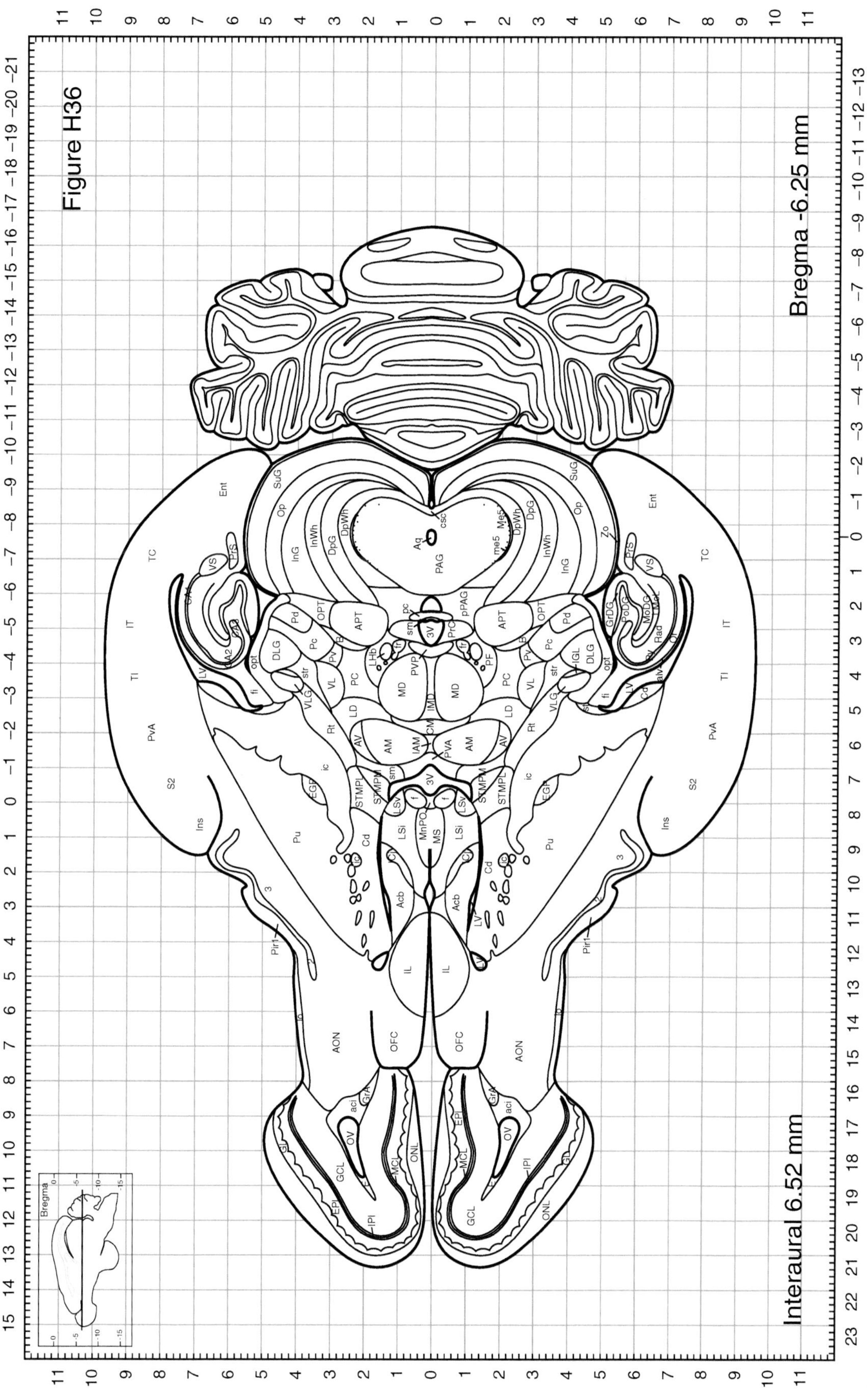
Figure H36
Bregma -6.25 mm
Interaural 6.52 mm
Bregma
SuG
Op
InG
InWh
DpG
DpWh
PAG
Aq
csc
Me5
me5
Ent
TC
IT
TI
PvA
S2
Ins
Pu
Zo
PrS
VS
CA1
CA2
CA3
GrDG
PoDG
MoDG
LMol
Rad
Or
alv
LV
fi
opt
str
DLG
IGL
VLG
Pd
Pc
Pv
OPT
APT
pc
sm
3V
PrC
pPAG
LHb
fr
PF
PVP
PC
VL
MD
IMD
LD
Rt
CM
AV
AM
IAM
PVA
ic
EGP
STMPL
STMPM
LSv
f
MnPO
MS
LSi
Cd
Cl
Acb
IL
Pir1
lo
AON
OFC
aci
GrA
OV
EPl
GCL
MCl
ONL
IPl
Gl
3

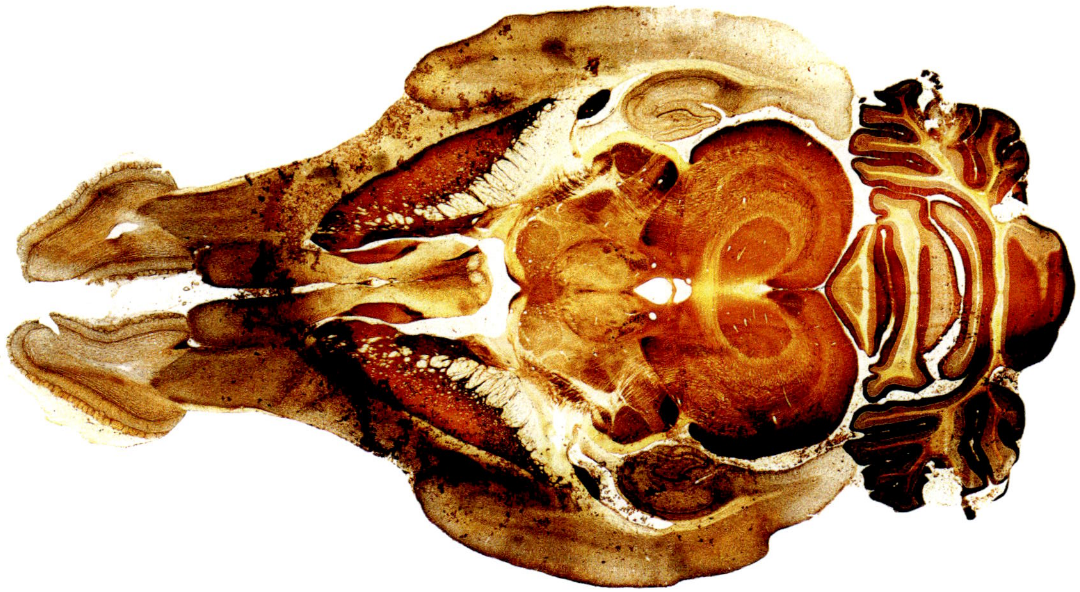

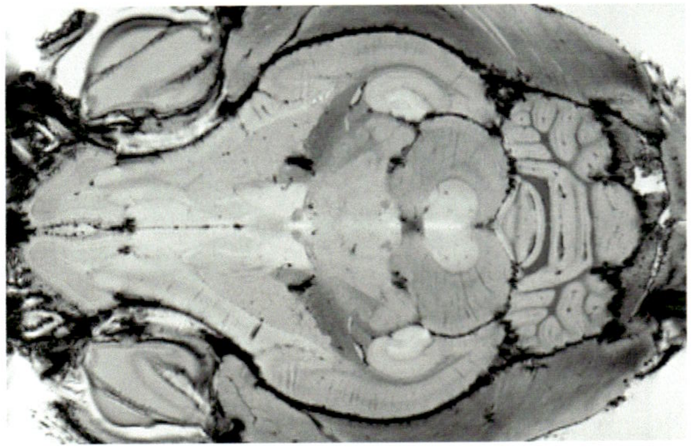

2	layer 2 of cortex
3	layer 3 of cortex
3V	3rd ventricle
Acb	accumbens nucleus
aci	anterior commissure, intrabulbar part
alv	alveus of the hippocampus
AM	anteromedial thalamic nucleus
AON	anterior olfactory nucleus
APT	anterior pretectal nucleus
B	brachium of the superior colliculus
CA1	field CA1 of the hippocampus
CA2	field CA2 of the hippocampus
CA3	field CA3 of the hippocampus
Cd	caudate nucleus
csc	commissure of the superior colliculus
DLG	dorsal lateral geniculate nucleus
DpG	deep gray layer of the superior colliculus
DpWh	deep white layer of the superior colliculus
E	ependyma and subependymal layer
Ent	entorhinal cortex
EPI	external plexiform layer of the olfactory bulb
EPIA	external plexiform layer of the accessory olfactory bulb
f	fornix
fi	fimbria of the hippocampus
fr	fasciculus retroflexus
GCL	granule cell layer of the olfactory bulb
GL	glomerular layer of the olfactory bulb
GrA	granule cell layer of the accessory olfactory bulb
GrDG	granular layer of the dentate gyrus
IAM	interanteromedial thalamic nucleus
ic	internal capsule
ICj	islands of Calleja
IGL	intergeniculate leaf
IL	infralimbic cortex
IMD	intermediodorsal thalamic nucleus
InG	intermediate gray layer of the superior colliculus
Ins	insular cortex
InWh	intermediate white layer of the superior colliculus
IPI	internal plexiform layer of the olfactory bulb
IT	inferior temporal cortex
LD	laterodorsal thalamic nucleus
LHb	lateral habenular nucleus
LMoL	lacunosum moleculare layer of the hippocampus
lo	lateral olfactory tract
LSi	lateral septal nucleus, intermediate part
LSv	lateral septal nucleus, ventral part
LV	lateral ventricle
MCL	mitral cell layer of the olfactory bulb
MD	mediodorsal thalamic nucleus
Me5	mesencephalic trigeminal nucleus
me5	mesencephalic trigeminal tract
MoDG	molecular layer of the dentate gyrus
MS	medial septal nucleus
OFC	orbital frontal cortex
ONL	olfactory nerve layer
Op	optic nerve layer of the superior colliculus
OPT	olivary pretectal nucleus
opt	optic tract
Or	oriens layer of the hippocampus
OV	olfactory ventricle (olfactory part of lateral ventricle)
PAG	periaqueductal gray
PC	paracentral thalamic nucleus
pc	posterior commissure
Pc	central nucleus of the pulvinar
Pd	dorsal nucleus of the pulvinar
PF	parafascicular thalamic nucleus
Pir 1	piriform cortex, layer 1
PoDG	polymorph layer of the dentate gyrus
pPAG	p1 periaqueductal gray
PrC	precommissural nucleus
PrS	presubiculum
pRt	p1 reticular formation
Pu	putamen
Pv	ventral nucleus of the pulvinar
PvA	parietal ventral area
PVA	paraventricular thalamic nucleus, anterior part
PVP	paraventricular thalamic nucleus, posterior part
Py	pyramidal cell layer of the hippocampus
Rad	radiatum layer of the hippocampus
Rt	reticular thalamic nucleus
S2	secondary somatosensory cortex
sm	stria medullaris of the thalamus
st	stria terminalis
STMPL	bed nucleus of the stria terminalis, medial division,anterolateral part
STMPM	bed nucleus of the stria terminalis, medial division,anteromedial part
str	superior thalamic radiation
SuG	superficial gray layer of the superior colliculus
TC	temporal cortex
TI	temporal inferior area
TS	triangular septal nucleus
VL	ventrolateral thalamic nucleus
VLG	ventral lateral geniculate nucleus
VS	ventral subiculum
Zo	zonal layer of the superior colliculus

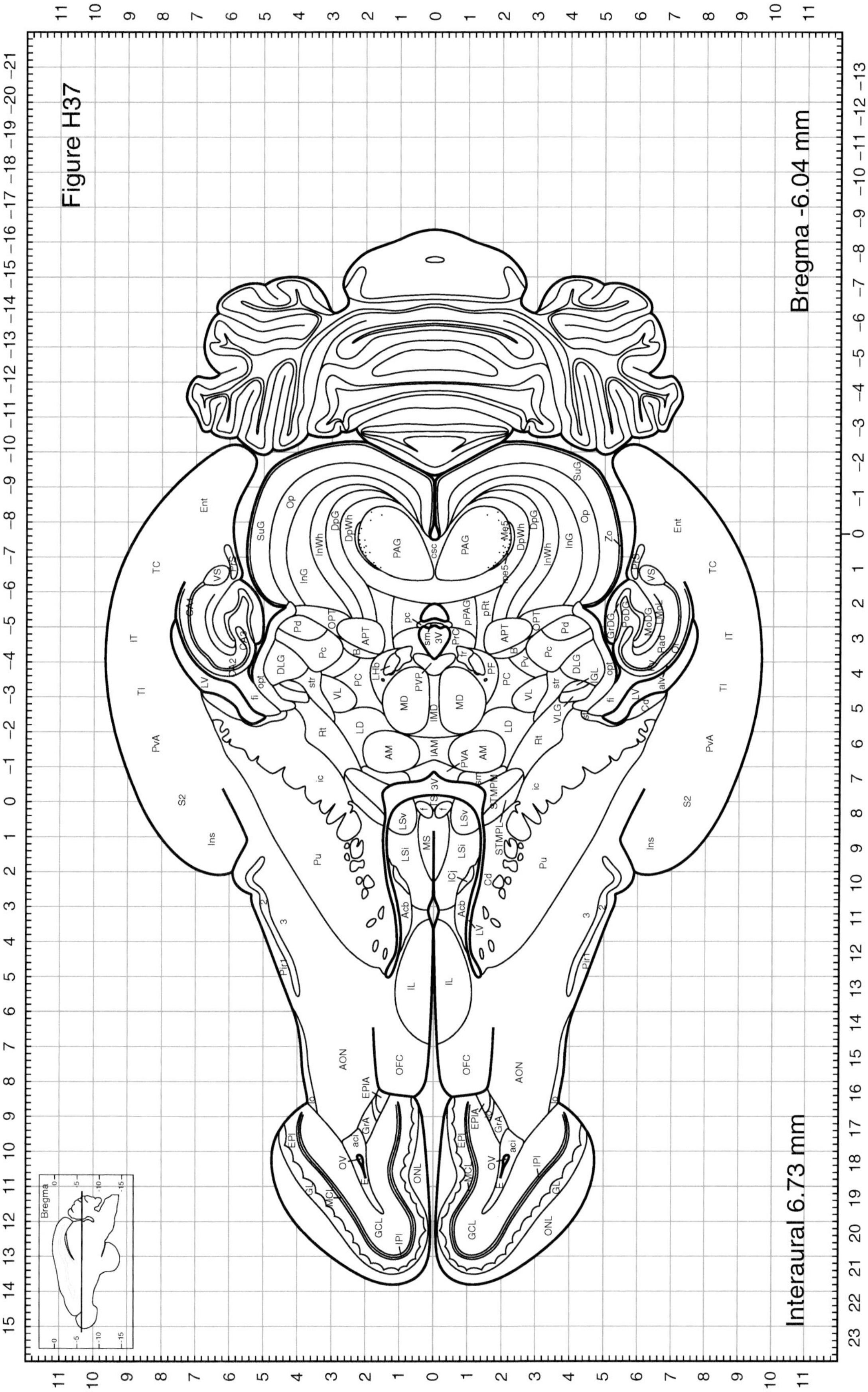

Figure H37
Bregma -6.04 mm
Interaural 6.73 mm
Bregma
Ent
TC
IT
TI
PvA
S2
Ins
Pu
ic
Rt
LD
AM
IAM
PVA
MD
IMD
PVP
VL
PC
Pc
Pd
DLG
VLG
IGL
APT
OPT
LHb
PF
pc
pPAG
pRt
PrC
fr
sm
3V
PAG
csc
DpWh
DpG
InWh
InG
Op
SuG
Zo
Me5
VS
CA1
CA2
PoDG
MoDG
GrDG
LMol
Rad
Or
alv
LV
Cd
fi
opt
str
LSv
LSi
MS
ICj
Acb
STMPM
STMPL
IL
OFC
AON
EPIA
GrA
aci
OV
E
EPl
GL
MCL
GCL
IPl
ONL
Pir1

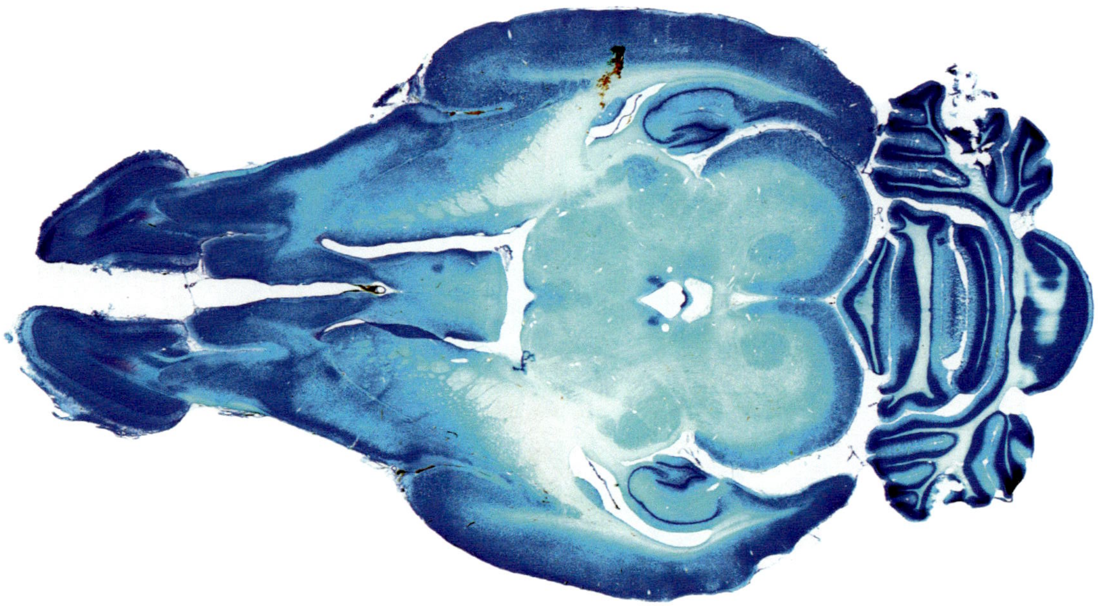

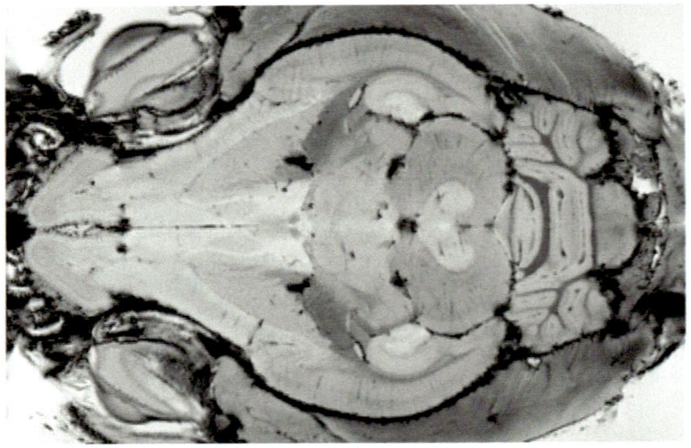

2	layer 2 of cortex
3	layer 3 of cortex
3V	3rd ventricle
Acb	accumbens nucleus
aci	anterior commissure, intrabulbar part
AD	anterodorsal thalamic nucleus
alv	alveus of the hippocampus
AON	anterior olfactory nucleus
APT	anterior pretectal nucleus
B	brachium of the superior colliculus
CA1	field CA1 of the hippocampus
CA2	field CA2 of the hippocampus
CA3	field CA3 of the hippocampus
Cd	caudate nucleus
csc	commissure of the superior colliculus
DLG	dorsal lateral geniculate nucleus
DpG	deep gray layer of the superior colliculus
DpWh	deep white layer of the superior colliculus
E	ependyma and subependymal layer
Ent	entorhinal cortex
EPI	external plexiform layer of the olfactory bulb
EPIA	external plexiform layer of the accessory olfactory bulb
f	fornix
fi	fimbria of the hippocampus
fr	fasciculus retroflexus
GCL	granule cell layer of the olfactory bulb
GL	glomerular layer of the olfactory bulb
GlA	glomerular layer of the accessory olfactory bulb
GrA	granule cell layer of the accessory olfactory bulb
GrDG	granular layer of the dentate gyrus
ic	internal capsule
IGL	intergeniculate leaf
IL	infralimbic cortex
IMD	intermediodorsal thalamic nucleus
InG	intermediate gray layer of the superior colliculus
Ins	insular cortex
InWh	intermediate white layer of the superior colliculus
IPl	internal plexiform layer of the olfactory bulb
IT	inferior temporal cortex
LD	laterodorsal thalamic nucleus
LHb	lateral habenular nucleus
LMoL	lacunosum moleculare layer of the hippocampus
lo	lateral olfactory tract
LSi	lateral septal nucleus, intermediate part
LSv	lateral septal nucleus, ventral part
LV	lateral ventricle
MCL	mitral cell layer of the olfactory bulb
MD	mediodorsal thalamic nucleus
me5	mesencephalic trigeminal tract
Me5	mesencephalic trigeminal nucleus
MHb	medial habenular nucleus
MoDG	molecular layer of the dentate gyrus
MS	medial septal nucleus
OFC	orbital frontal cortex
ONL	olfactory nerve layer
Op	optic nerve layer of the superior colliculus
OPT	olivary pretectal nucleus
opt	optic tract
Or	oriens layer of the hippocampus
PAG	periaqueductal gray
PC	paracentral thalamic nucleus
pc	posterior commissure
Pc	central nucleus of the pulvinar
Pd	dorsal nucleus of the pulvinar
PF	parafascicular thalamic nucleus
Pir 1	piriform cortex, layer 1
PoDG	polymorph layer of the dentate gyrus
pPAG	p1 periaqueductal gray
PrC	precommissural nucleus
PrS	presubiculum
Pu	putamen
PvA	parietal ventral area
PVA	paraventricular thalamic nucleus, anterior part
PVP	paraventricular thalamic nucleus, posterior part
Py	pyramidal cell layer of the hippocampus
Rad	radiatum layer of the hippocampus
Rt	reticular thalamic nucleus
S2	secondary somatosensory cortex
SFi	septofimbrial nucleus
sm	stria medullaris of the thalamus
st	stria terminalis
STMP	bed nucleus of the stria terminalis, medial division, posterior part
str	superior thalamic radiation
SuG	superficial gray layer of the superior colliculus
TC	temporal cortex
TI	temporal inferior area
TS	triangular septal nucleus
VLG	ventral lateral geniculate nucleus
VS	ventral subiculum
Zo	zonal layer of the superior colliculus

Figure H38
Bregma -5.83 mm
Interaural 6.94 mm

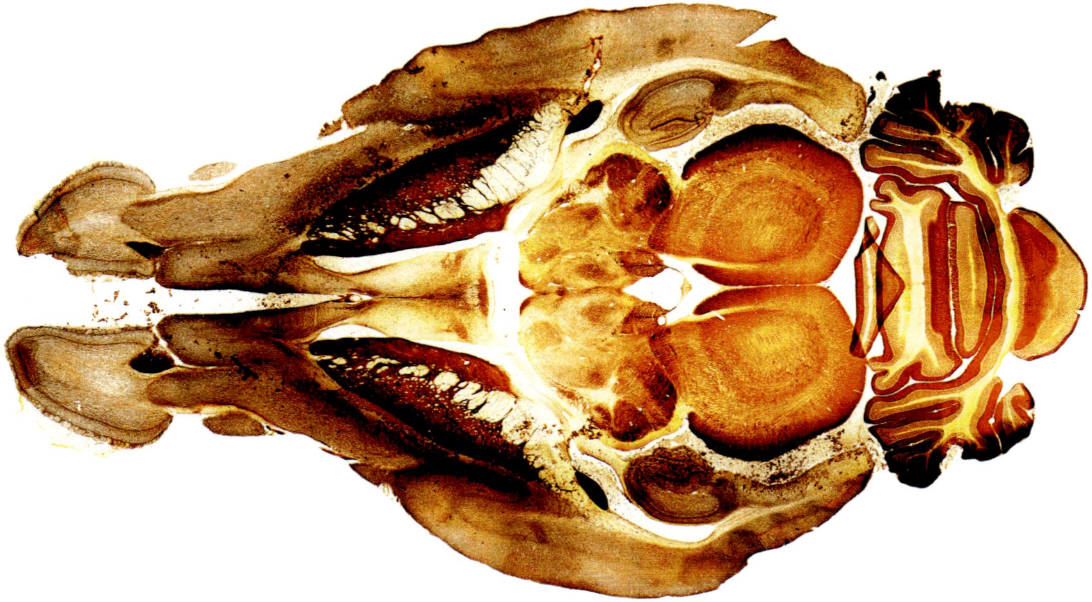

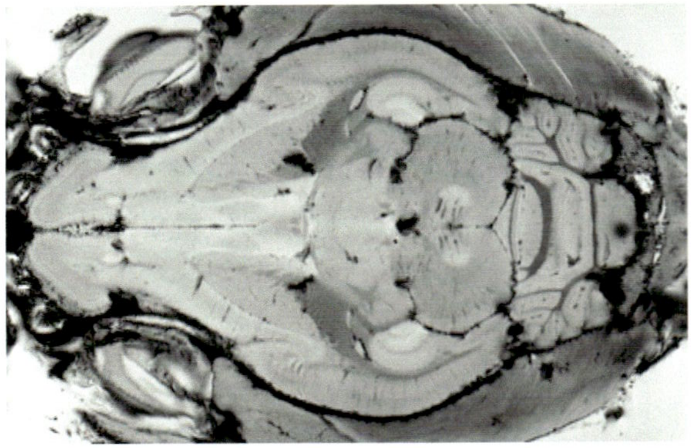

3V	3rd ventricle
aci	anterior commissure, intrabulbar part
AD	anterodorsal thalamic nucleus
alv	alveus of the hippocampus
AON	anterior olfactory nucleus
APT	anterior pretectal nucleus
CA1	field CA1 of the hippocampus
CA2	field CA2 of the hippocampus
CA3	field CA3 of the hippocampus
Cd	caudate nucleus
DLG	dorsal lateral geniculate nucleus
DpG	deep gray layer of the superior colliculus
DpWh	deep white layer of the superior colliculus
Ent	entorhinal cortex
EPl	external plexiform layer of the olfactory bulb
EPlA	external plexiform layer of the accessory olfactory bulb
f	fornix
fi	fimbria of the hippocampus
fr	fasciculus retroflexus
GCL	granule cell layer of the olfactory bulb
GL	glomerular layer of the olfactory bulb
GlA	glomerular layer of the accessory olfactory bulb
GrA	granule cell layer of the accessory olfactory bulb
GrDG	granular layer of the dentate gyrus
ic	internal capsule
IG	indusium griseum
IGL	intergeniculate leaf
IL	infralimbic cortex
IMD	intermediodorsal thalamic nucleus
InG	intermediate gray layer of the superior colliculus
Ins	insular cortex
InWh	intermediate white layer of the superior colliculus
IPl	internal plexiform layer of the olfactory bulb
IT	inferior temporal cortex
LD	laterodorsal thalamic nucleus
LHb	lateral habenular nucleus
LMoL	lacunosum moleculare layer of the hippocampus
lo	lateral olfactory tract
LSi	lateral septal nucleus, intermediate part
LSv	lateral septal nucleus, ventral part
LV	lateral ventricle
MCL	mitral cell layer of the olfactory bulb
MD	mediodorsal thalamic nucleus
me5	mesencephalic trigeminal tract
Me5	mesencephalic trigeminal nucleus
MHb	medial habenular nucleus
MiA	mitral cell layer of the accessory olfactory bulb
MoDG	molecular layer of the dentate gyrus
MPT	medial pretectal nucleus
MS	medial septal nucleus
OFC	orbital frontal cortex
ONL	olfactory nerve layer
Op	optic nerve layer of the superior colliculus
opt	optic tract
Or	oriens layer of the hippocampus
PC	paracentral thalamic nucleus
Pc	central nucleus of the pulvinar
Pd	dorsal nucleus of the pulvinar
PF	parafascicular thalamic nucleus
Pir	piriform cortex
PoDG	polymorph layer of the dentate gyrus
PrS	presubiculum
Pu	putamen
PvA	parietal ventral area
PVA	paraventricular thalamic nucleus, anterior part
PVP	paraventricular thalamic nucleus, posterior part
Py	pyramidal cell layer of the hippocampus
Rad	radiatum layer of the hippocampus
Rt	reticular thalamic nucleus
S2	secondary somatosensory cortex
SFi	septofimbrial nucleus
sm	stria medullaris of the thalamus
st	stria terminalis
str	superior thalamic radiation
SuG	superficial gray layer of the superior colliculus
TC	temporal cortex
TI	temporal inferior area
TS	triangular septal nucleus
V2	secondary visual cortex
VLG	ventral lateral geniculate nucleus
VS	ventral subiculum
Zo	zonal layer of the superior colliculus

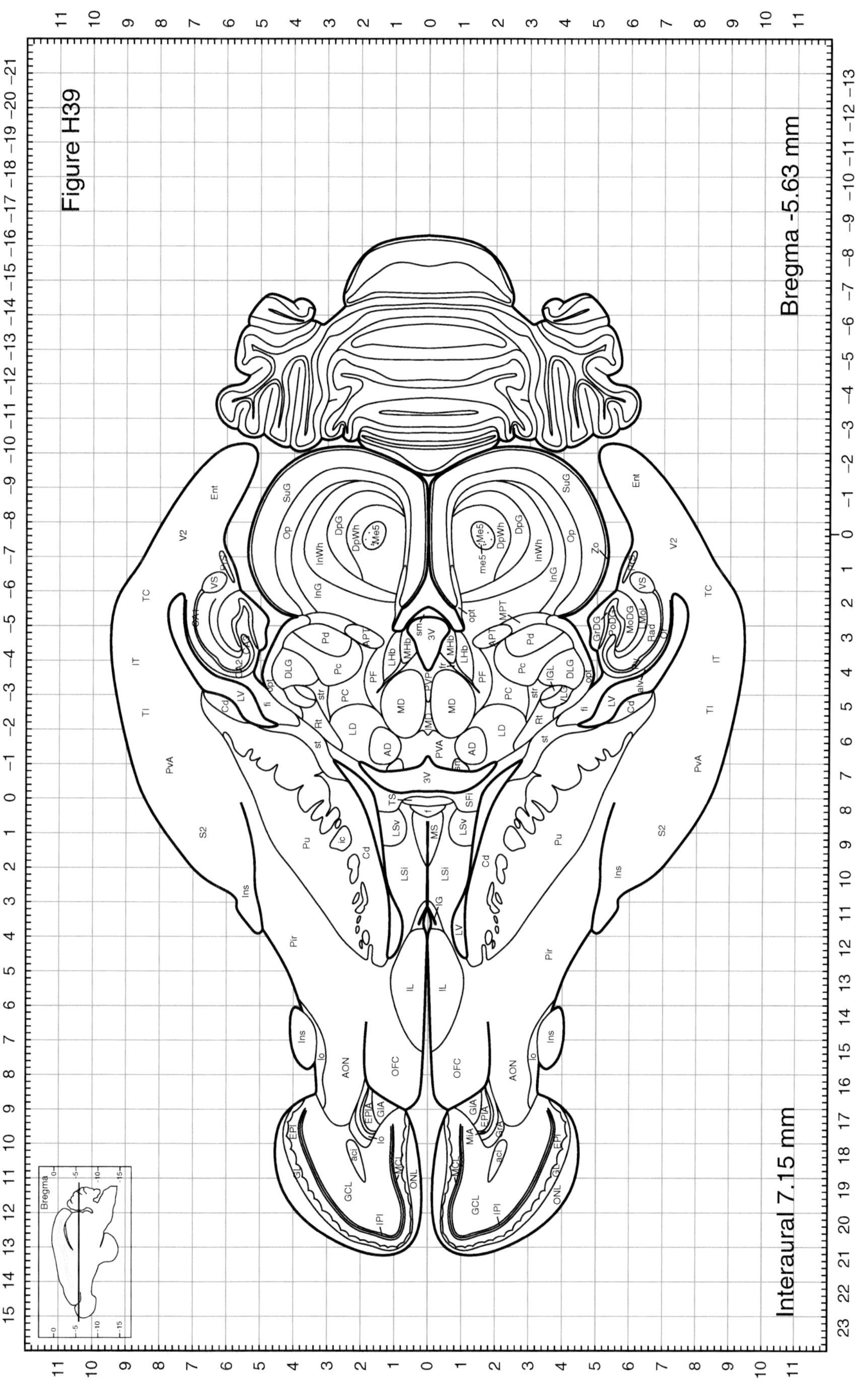
Figure H39
Bregma -5.63 mm
Interaural 7.15 mm
Bregma
Ent
SuG
Op
InWh
DpG
DpWh
Me5
me5
InG
V2
TC
IT
TI
PvA
S2
Ins
Pir
VS
Zo
MPT
APT
Pd
Pc
DLG
LHb
MHb
sm
3V
PF
PVP
fr
MD
PC
IGL
VLG
str
opt
fi
LV
Cd
Rt
st
LD
AD
PVA
TS
SFi
f
LSv
MS
LSi
Pu
ic
IG
IL
OFC
AON
lo
GlA
EPlA
MlA
EPl
Gl
aci
GCL
IPl
MCL
ONL
GrDG
PoDG
MoDG
LMol
Rad
Or
alv
CA1
CA2
CA3

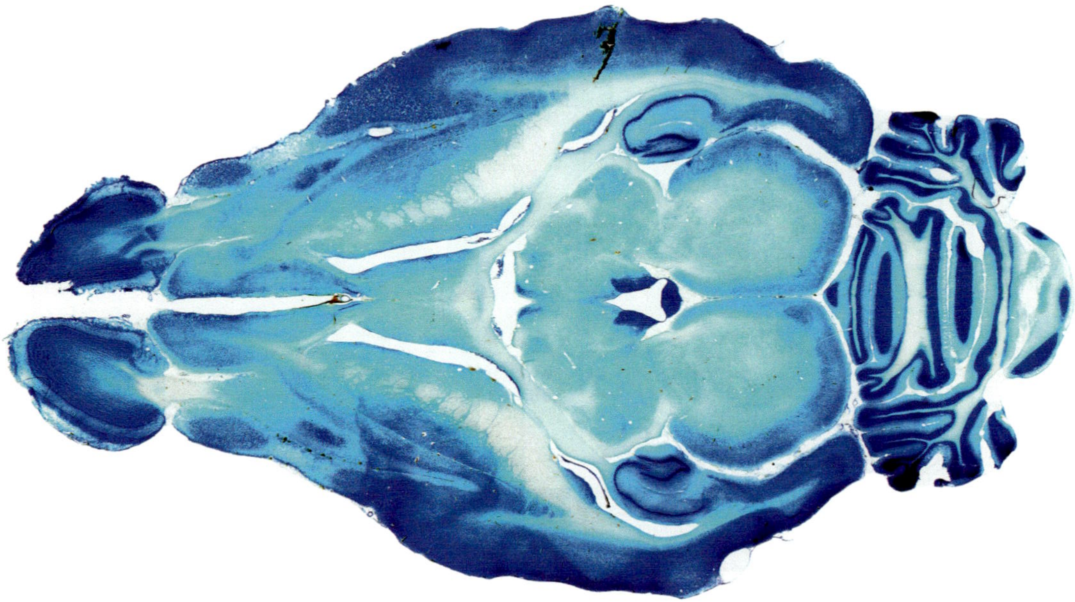

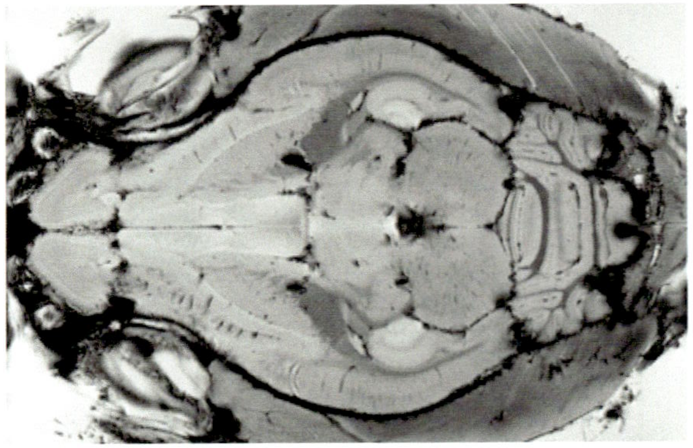

3V	3rd ventricle
AD	anterodorsal thalamic nucleus
alv	alveus of the hippocampus
CA1	field CA1 of the hippocampus
CA2	field CA2 of the hippocampus
CA3	field CA3 of the hippocampus
cc	corpus callosum
Cd	caudate nucleus
DLG	dorsal lateral geniculate nucleus
DpG	deep gray layer of the superior colliculus
Ent	entorhinal cortex
EPl	external plexiform layer of the olfactory bulb
EPlA	external plexiform layer of the accessory olfactory bulb
f	fornix
fi	fimbria of the hippocampus
fr	fasciculus retroflexus
GCL	granule cell layer of the olfactory bulb
GL	glomerular layer of the olfactory bulb
GlA	glomerular layer of the accessory olfactory bulb
GrDG	granular layer of the dentate gyrus
ic	internal capsule
IG	indusium griseum
IMD	intermediodorsal thalamic nucleus
InG	intermediate gray layer of the superior colliculus
Ins	insular cortex
InWh	intermediate white layer of the superior colliculus
IPl	internal plexiform layer of the olfactory bulb
IT	inferior temporal cortex
LD	laterodorsal thalamic nucleus
LHb	lateral habenular nucleus
LMoL	lacunosum moleculare layer of the hippocampus
lo	lateral olfactory tract
LSi	lateral septal nucleus, intermediate part
LSv	lateral septal nucleus, ventral part
LV	lateral ventricle
MCL	mitral cell layer of the olfactory bulb
MD	mediodorsal thalamic nucleus
MHb	medial habenular nucleus
MiA	mitral cell layer of the accessory olfactory bulb
MoDG	molecular layer of the dentate gyrus
MPT	medial pretectal nucleus
MS	medial septal nucleus
OFC	orbital frontal cortex
ONL	olfactory nerve layer
Op	optic nerve layer of the superior colliculus
opt	optic tract
Or	oriens layer of the hippocampus
PC	paracentral thalamic nucleus
Pc	central nucleus of the pulvinar
Pd	dorsal nucleus of the pulvinar
PF	parafascicular thalamic nucleus
PG	pineal gland
Pir	piriform cortex
PoDG	polymorph layer of the dentate gyrus
Post	postsubiculum
PrL	prelimbic cortex
PrS	presubiculum
Pu	putamen
PvA	parietal ventral area
PVA	paraventricular thalamic nucleus, anterior part
PVP	paraventricular thalamic nucleus, posterior part
Py	pyramidal cell layer of the hippocampus
Rad	radiatum layer of the hippocampus
Rt	reticular thalamic nucleus
S2	secondary somatosensory cortex
SFi	septofimbrial nucleus
sm	stria medullaris of the thalamus
st	stria terminalis
str	superior thalamic radiation
SuG	superficial gray layer of the superior colliculus
TC	temporal cortex
TI	temporal inferior area
TS	triangular septal nucleus
V2	secondary visual cortex
VLG	ventral lateral geniculate nucleus
VS	ventral subiculum
Zo	zonal layer of the superior colliculus

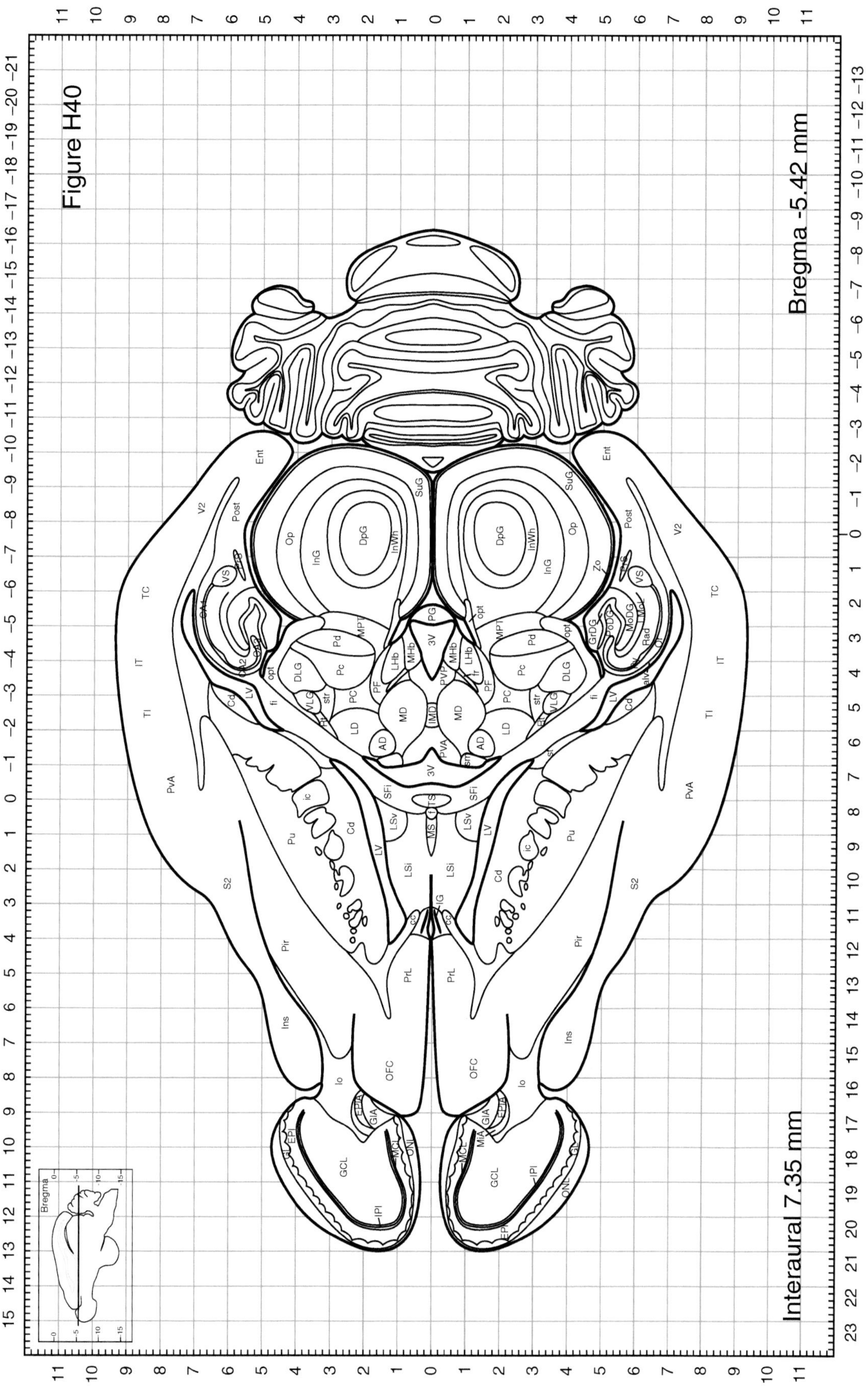

Figure H40
Bregma -5.42 mm
Interaural 7.35 mm
Bregma
Ent
SuG
V2
Post
Op
InG
DpG
InWh
TC
VS
IT
TI
PvA
S2
Pir
Ins
Io
OFC
PrL
Cd
Pu
ic
LV
LSi
LSv
SFi
MS
TS
IG
cc
3V
PG
MPT
Pd
Pc
DLG
VLG
PC
PF
LHb
MHb
PVP
fr
MD
IMD
LD
AD
PVA
sm
st
str
opt
fi
Zo
GrDG
PoDG
MoDG
MoL
Rad
Or
alv
GCL
IPl
MCL
ONL
EPl
GL
GlA
EPlA
MiA

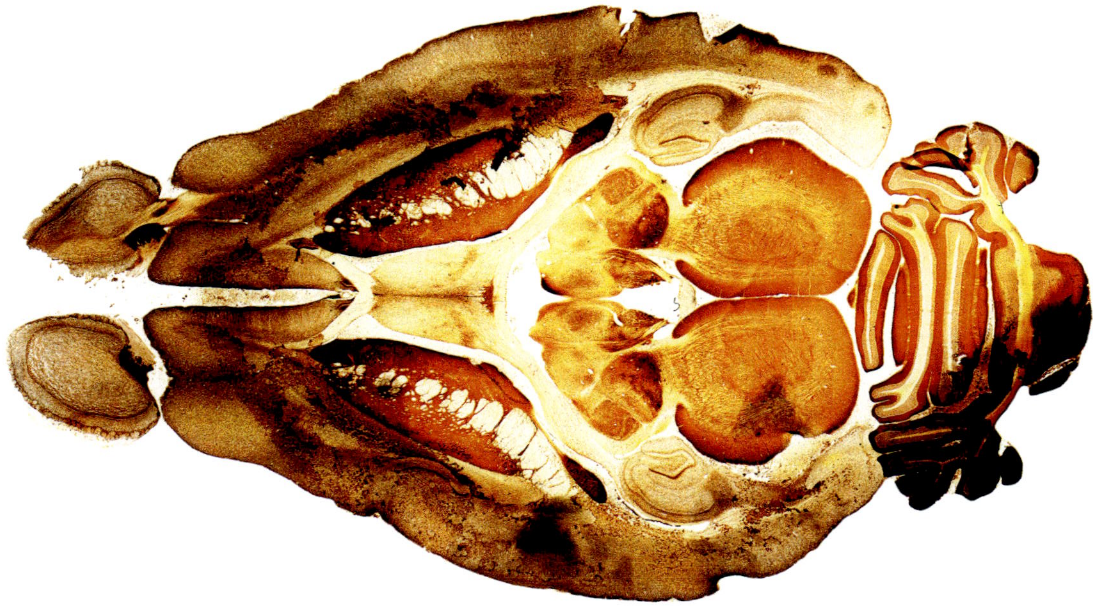

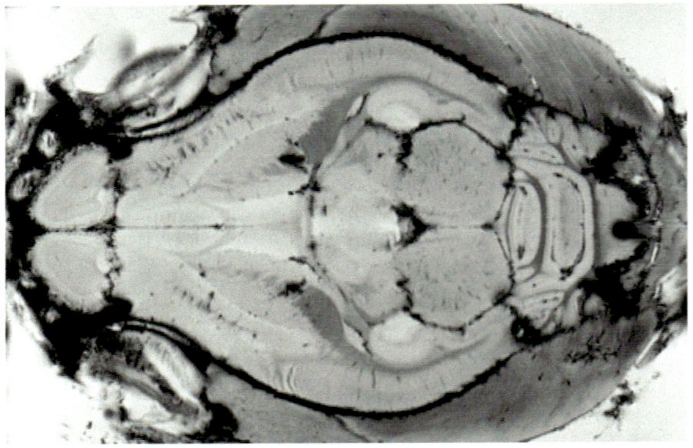

3V	3rd ventricle
AD	anterodorsal thalamic nucleus
alv	alveus of the hippocampus
CA1	field CA1 of the hippocampus
CA2	field CA2 of the hippocampus
CA3	field CA3 of the hippocampus
cc	corpus callosum
Cd	caudate nucleus
DLG	dorsal lateral geniculate nucleus
Ent	entorhinal cortex
EPl	external plexiform layer of the olfactory bulb
EPlA	external plexiform layer of the accessory olfactory bulb
f	fornix
fi	fimbria of the hippocampus
GCL	granule cell layer of the olfactory bulb
GL	glomerular layer of the olfactory bulb
GlA	glomerular layer of the accessory olfactory bulb
GrDG	granular layer of the dentate gyrus
ic	internal capsule
IG	indusium griseum
InG	intermediate gray layer of the superior colliculus
Ins	insular cortex
InWh	intermediate white layer of the superior colliculus
IPl	internal plexiform layer of the olfactory bulb
IT	inferior temporal cortex
LD	laterodorsal thalamic nucleus
LHb	lateral habenular nucleus
LMoL	lacunosum moleculare layer of the hippocampus
LSd	lateral septal nucleus, dorsal part
LSi	lateral septal nucleus, intermediate part
LV	lateral ventricle
MCL	mitral cell layer of the olfactory bulb
MD	mediodorsal thalamic nucleus
MHb	medial habenular nucleus
MoDG	molecular layer of the dentate gyrus
MPT	medial pretectal nucleus
MS	medial septal nucleus
OFC	orbital frontal cortex
ONL	olfactory nerve layer
Op	optic nerve layer of the superior colliculus
opt	optic tract
Or	oriens layer of the hippocampus
PC	paracentral thalamic nucleus
Pc	central nucleus of the pulvinar
Pd	dorsal nucleus of the pulvinar
PG	pineal gland
PoDG	polymorph layer of the dentate gyrus
Post	postsubiculum
PrL	prelimbic cortex
PrS	presubiculum
Pu	putamen
PvA	parietal ventral area
PVA	paraventricular thalamic nucleus, anterior part
PVP	paraventricular thalamic nucleus, posterior part
Py	pyramidal cell layer of the hippocampus
Rad	radiatum layer of the hippocampus
S2	secondary somatosensory cortex
SFi	septofimbrial nucleus
SHi	septohippocampal nucleus
sm	stria medullaris of the thalamus
st	stria terminalis
SuG	superficial gray layer of the superior colliculus
TC	temporal cortex
TI	temporal inferior area
TS	triangular septal nucleus
V2	secondary visual cortex
VLG	ventral lateral geniculate nucleus
VS	ventral subiculum
Zo	zonal layer of the superior colliculus

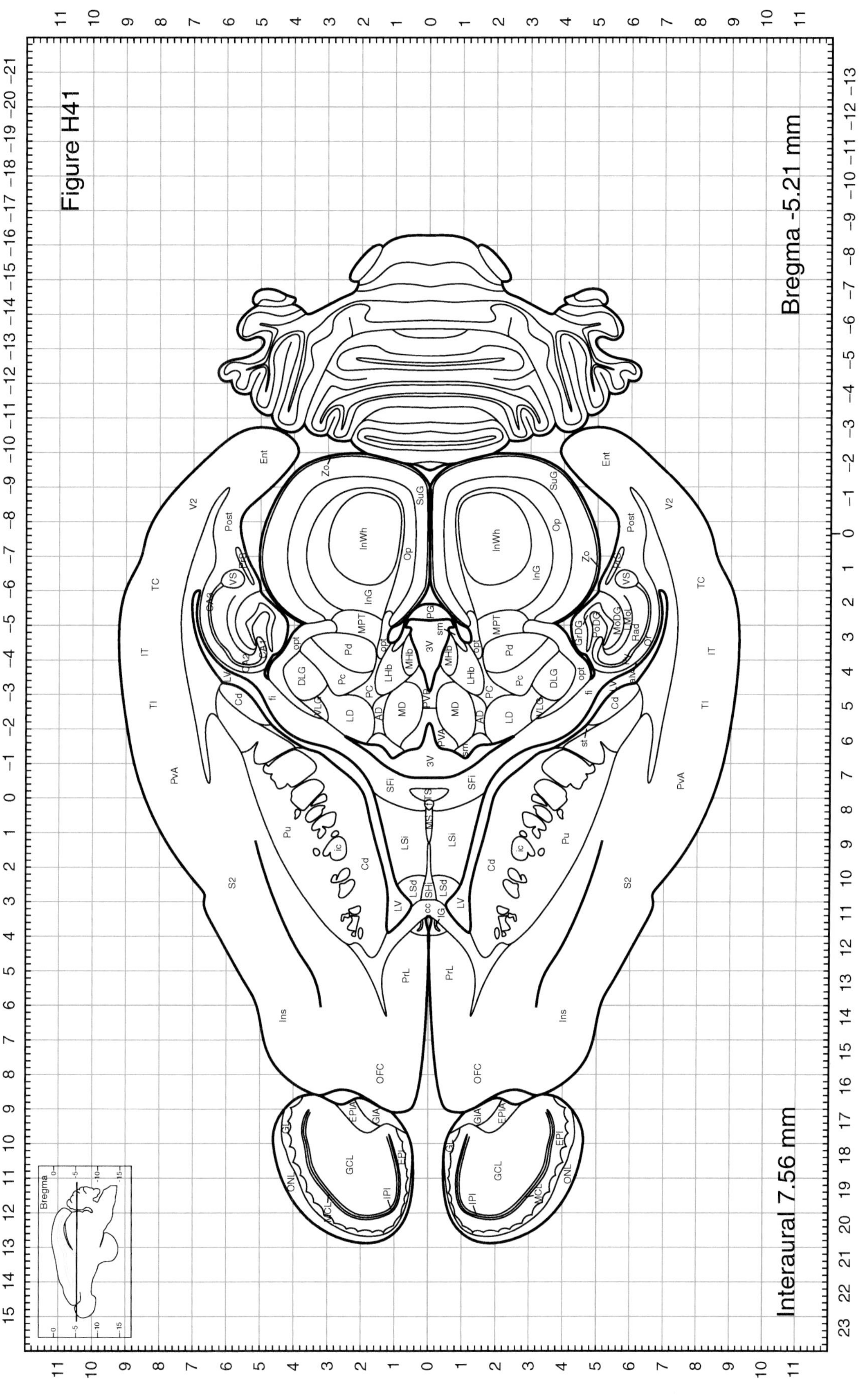
Figure H41
Bregma -5.21 mm
Interaural 7.56 mm
Bregma
Ent
V2
Post
TC
IT
TI
PvA
S2
Ins
OFC
PrL
Pu
Cd
ic
LSi
LSd
LV
SFi
MS
SHi
cc
IG
3V
Zo
SuG
Op
InWh
InG
MPT
Pd
Pc
PC
DLG
VLG
LD
AD
MD
PVA
PVP
LHb
MHb
sm
opt
PG
fi
st
VS
CA2
CA3
GrDG
PoDG
MoDG
LMol
Rad
Or
GCL
IPl
EPl
GlA
EPlA
Gl
ONL
MCL

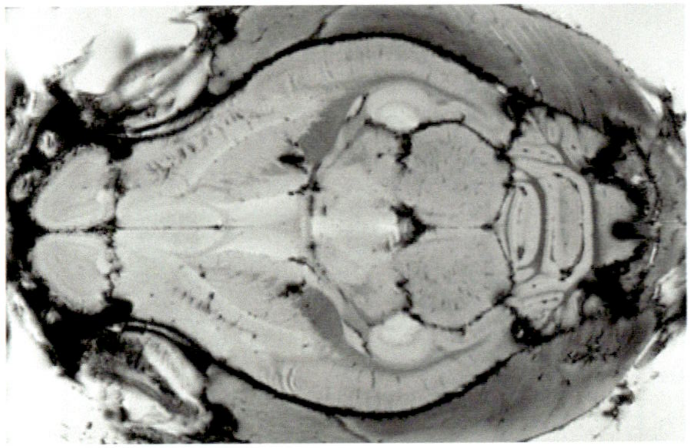

alv	alveus of the hippocampus
CA1	field CA1 of the hippocampus
CA2	field CA2 of the hippocampus
CA3	field CA3 of the hippocampus
cc	corpus callosum
Cd	caudate nucleus
D3V	dorsal 3rd ventricle
DLG	dorsal lateral geniculate nucleus
Ent	entorhinal cortex
EPI	external plexiform layer of the olfactory bulb
EPIA	external plexiform layer of the accessory olfactory bulb
fi	fimbria of the hippocampus
GCL	granule cell layer of the olfactory bulb
GL	glomerular layer of the olfactory bulb
GlA	glomerular layer of the accessory olfactory bulb
GrDG	granular layer of the dentate gyrus
ic	internal capsule
IG	indusium griseum
InG	intermediate gray layer of the superior colliculus
IT	inferior temporal cortex
InWh	intermediate white layer of the superior colliculus
IPl	internal plexiform layer of the olfactory bulb
LD	laterodorsal thalamic nucleus
LHb	lateral habenular nucleus
LMoL	lacunosum moleculare layer of the hippocampus
LSd	lateral septal nucleus, dorsal part
LSi	lateral septal nucleus, intermediate part
LV	lateral ventricle
MCL	mitral cell layer of the olfactory bulb
MD	mediodorsal thalamic nucleus
MHb	medial habenular nucleus
MoDG	molecular layer of the dentate gyrus
MS	medial septal nucleus
OFC	orbital frontal cortex
ONL	olfactory nerve layer
Op	optic nerve layer of the superior colliculus
opt	optic tract
Or	oriens layer of the hippocampus
Pc	central nucleus of the pulvinar
Pd	dorsal nucleus of the pulvinar
PG	pineal gland
PoDG	polymorph layer of the dentate gyrus
Post	postsubiculum
PrL	prelimbic cortex
PrS	presubiculum
Pu	putamen
PVP	paraventricular thalamic nucleus, posterior part
Py	pyramidal cell layer of the hippocampus
Rad	radiatum layer of the hippocampus
S1	primary somatosensory cortex
S2	secondary somatosensory cortex
SFi	septofimbrial nucleus
SHi	septohippocampal nucleus
sm	stria medullaris of the thalamus
SuG	superficial gray layer of the superior colliculus
TC	temporal cortex
TI	temporal inferior area
TS	triangular septal nucleus
V2	secondary visual cortex
VS	ventral subiculum
Zo	zonal layer of the superior colliculus

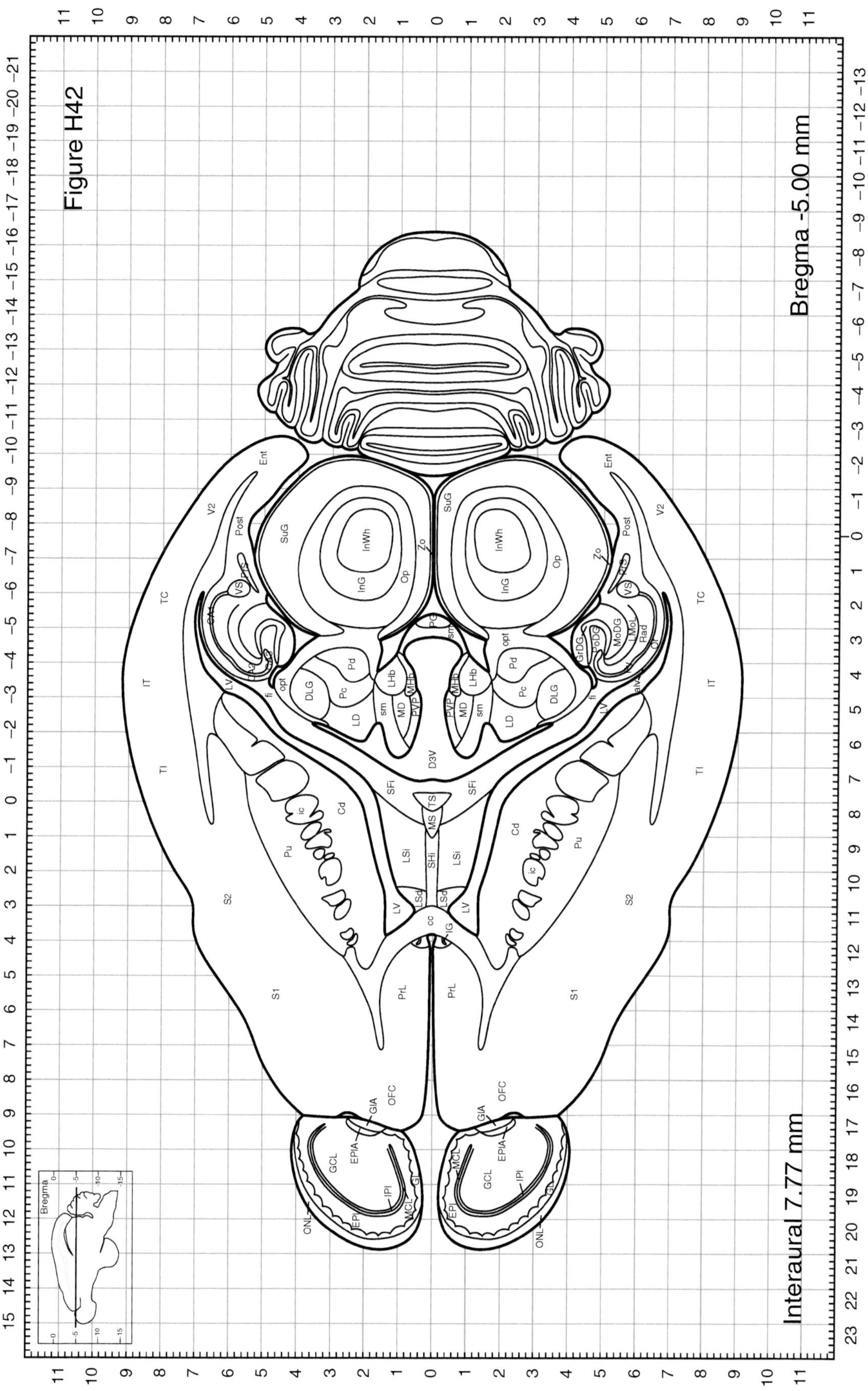
Figure H42
Bregma -5.00 mm
Interaural 7.77 mm
Ent
V2
Post
SuG
InWh
Zo
Op
InG
PrS
VS
TC
CA1
CA2
opt
Pd
DLG
Pc
LD
LHb
MHb
MD
sm
PVP
PG
GrDG
PoDG
MoDG
LMol
Rad
Or
LV
fi
alv
IT
D3V
TI
SFi
TS
MS
Cd
ic
Pu
LSi
SHi
S2
LSd
cc
IG
S1
PrL
OFC
GIA
EPIA
GCL
IPl
MCL
Gl
EPl
ONL
Bregma

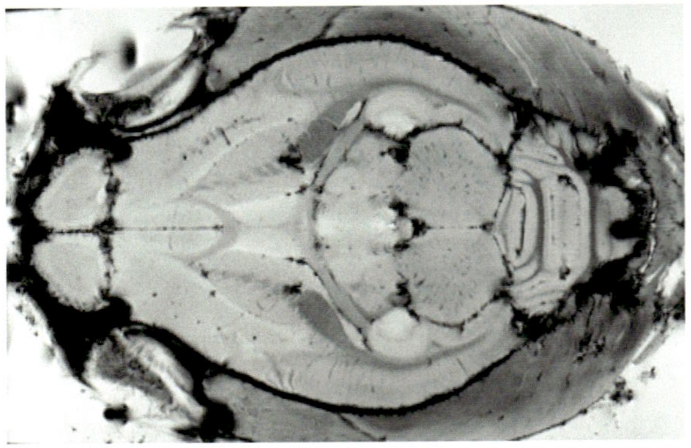

alv	alveus of the hippocampus
CA1	field CA1 of the hippocampus
CA2	field CA2 of the hippocampus
CA3	field CA3 of the hippocampus
cc	corpus callosum
Cd	caudate nucleus
Cg	cingulate cortex
D3V	dorsal 3rd ventricle
DLG	dorsal lateral geniculate nucleus
EPl	external plexiform layer of the olfactory bulb
fi	fimbria of the hippocampus
GCL	granule cell layer of the olfactory bulb
GL	glomerular layer of the olfactory bulb
GrDG	granular layer of the dentate gyrus
ic	internal capsule
IG	indusium griseum
InG	intermediate gray layer of the superior colliculus
IPl	internal plexiform layer of the olfactory bulb
LHb	lateral habenular nucleus
LMoL	lacunosum moleculare layer of the hippocampus
LSd	lateral septal nucleus, dorsal part
LSi	lateral septal nucleus, intermediate part
LV	lateral ventricle
MCL	mitral cell layer of the olfactory bulb
MHb	medial habenular nucleus
MoDG	molecular layer of the dentate gyrus
OFC	orbital frontal cortex
ONL	olfactory nerve layer
Op	optic nerve layer of the superior colliculus
opt	optic tract
Or	oriens layer of the hippocampus
Pc	central nucleus of the pulvinar
Pd	dorsal nucleus of the pulvinar
PoDG	polymorph layer of the dentate gyrus
Post	postsubiculum
PPc	posterior parietal caudal area
PPr	posterior parietal rostral area
Pu	putamen
PVP	paraventricular thalamic nucleus, posterior part
Py	pyramidal cell layer of the hippocampus
Rad	radiatum layer of the hippocampus
S1	primary somatosensory cortex
S2	secondary somatosensory cortex
SFi	septofimbrial nucleus
SFO	subfornical organ
SHi	septohippocampal nucleus
sm	stria medullaris of the thalamus
SuG	superficial gray layer of the superior colliculus
TC	temporal cortex
TS	triangular septal nucleus
V1	primary visual cortex
V2	secondary visual cortex
VS	ventral subiculum
Zo	zonal layer of the superior colliculus

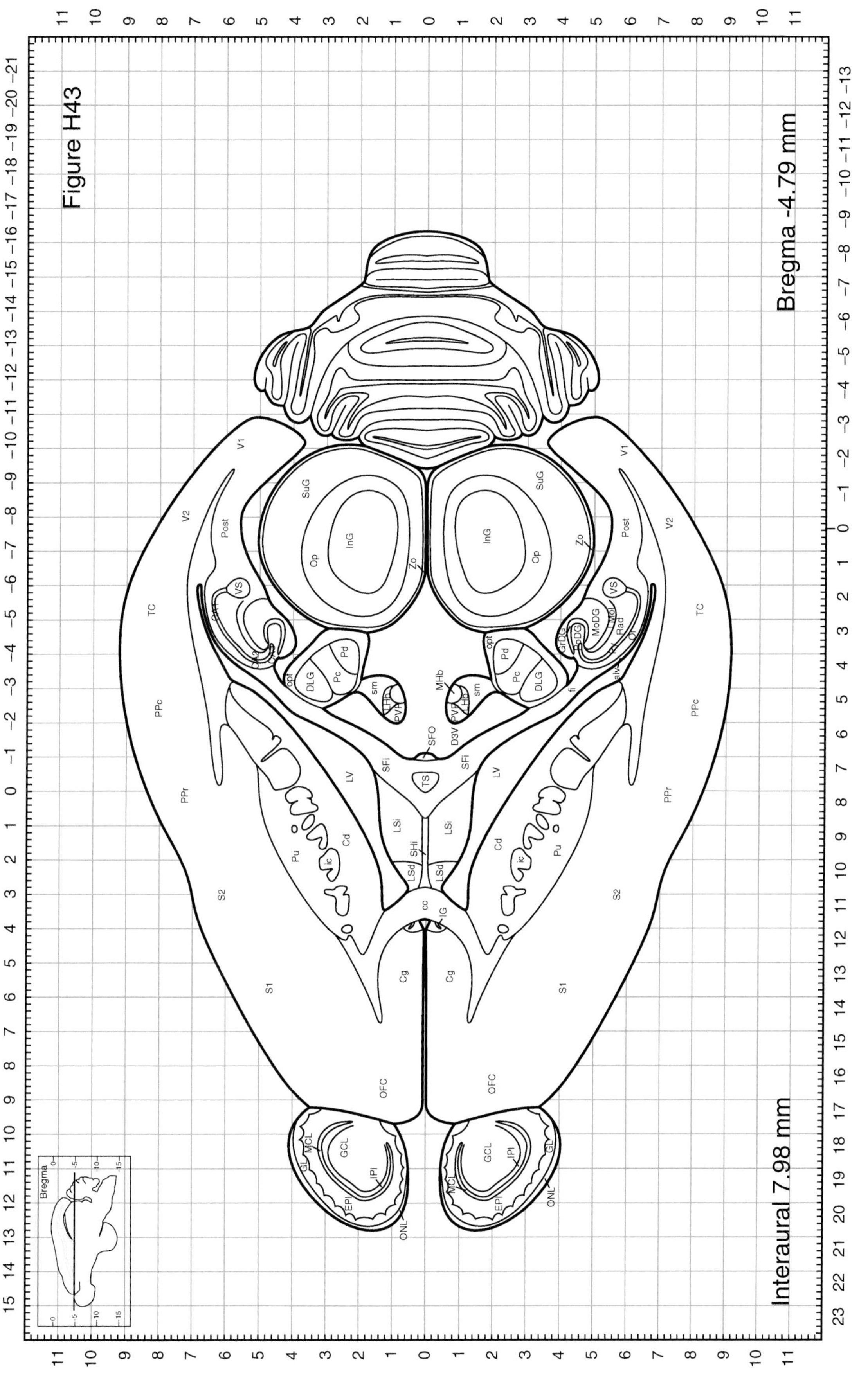

Figure H43
Bregma -4.79 mm
Interaural 7.98 mm
V1
V2
Post
TC
PPc
PPr
S2
S1
SuG
InG
Op
Zo
VS
opt
Pd
Pc
DLG
sm
LHb
PVP
MHb
D3V
SFO
SFi
LV
TS
LSi
SHi
LSd
cc
IG
Cd
Pu
ic
Cg
OFC
GCL
MCL
GL
IPl
EPl
ONL
GrDG
PoDG
MoDG
LMol
Rad
Or
fi
alv
Bregma

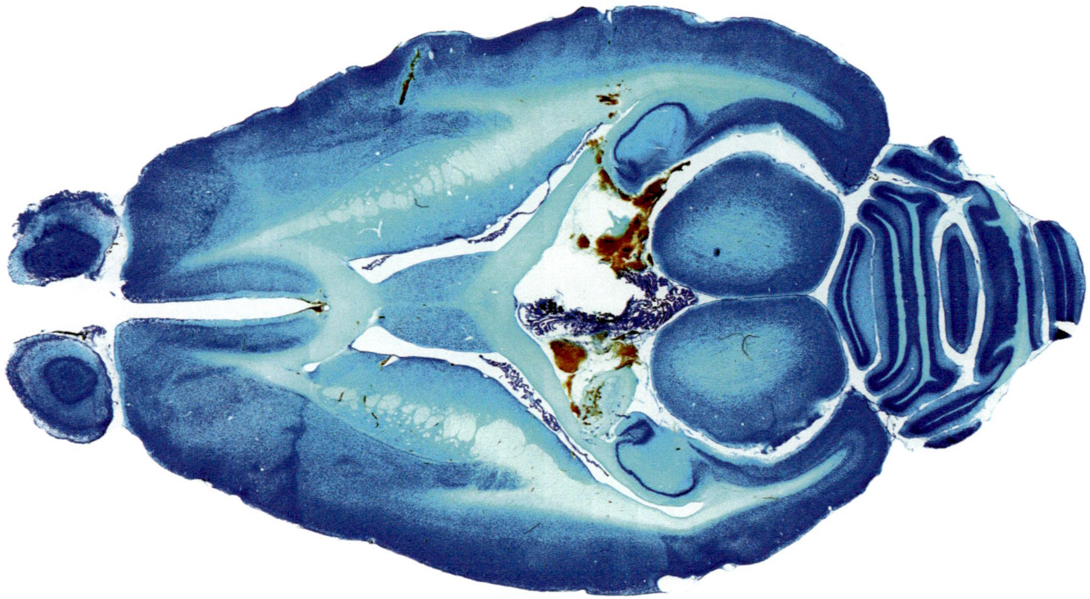

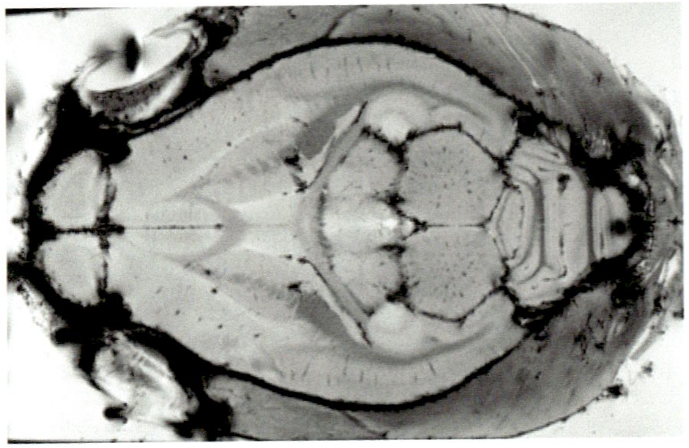

alv	alveus of the hippocampus
CA1	field CA1 of the hippocampus
CA3	field CA3 of the hippocampus
cc	corpus callosum
Cd	caudate nucleus
Cg	cingulate cortex
D3V	dorsal 3rd ventricle
EPl	external plexiform layer of the olfactory bulb
fi	fimbria of the hippocampus
GCL	granule cell layer of the olfactory bulb
GL	glomerular layer of the olfactory bulb
GrDG	granular layer of the dentate gyrus
ic	internal capsule
IG	indusium griseum
InG	intermediate gray layer of the superior colliculus
IPl	internal plexiform layer of the olfactory bulb
LMoL	lacunosum moleculare layer of the hippocampus
LSd	lateral septal nucleus, dorsal part
LSi	lateral septal nucleus, intermediate part
LV	lateral ventricle
MCL	mitral cell layer of the olfactory bulb
MoDG	molecular layer of the dentate gyrus
OFC	orbital frontal cortex
ONL	olfactory nerve layer
Op	optic nerve layer of the superior colliculus
opt	optic tract
Or	oriens layer of the hippocampus
Pc	central nucleus of the pulvinar
Pd	dorsal nucleus of the pulvinar
PoDG	polymorph layer of the dentate gyrus
Post	postsubiculum
PPc	posterior parietal caudal area
PPr	posterior parietal rostral area
Pu	putamen
Py	pyramidal cell layer of the hippocampus
Rad	radiatum layer of the hippocampus
S1	primary somatosensory cortex
S2	secondary somatosensory cortex
SFi	septofimbrial nucleus
SFO	subfornical organ
SHi	septohippocampal nucleus
SuG	superficial gray layer of the superior colliculus
TC	temporal cortex
TS	triangular septal nucleus
V1	primary visual cortex
V2	secondary visual cortex
VS	ventral subiculum
Zo	zonal layer of the superior colliculus

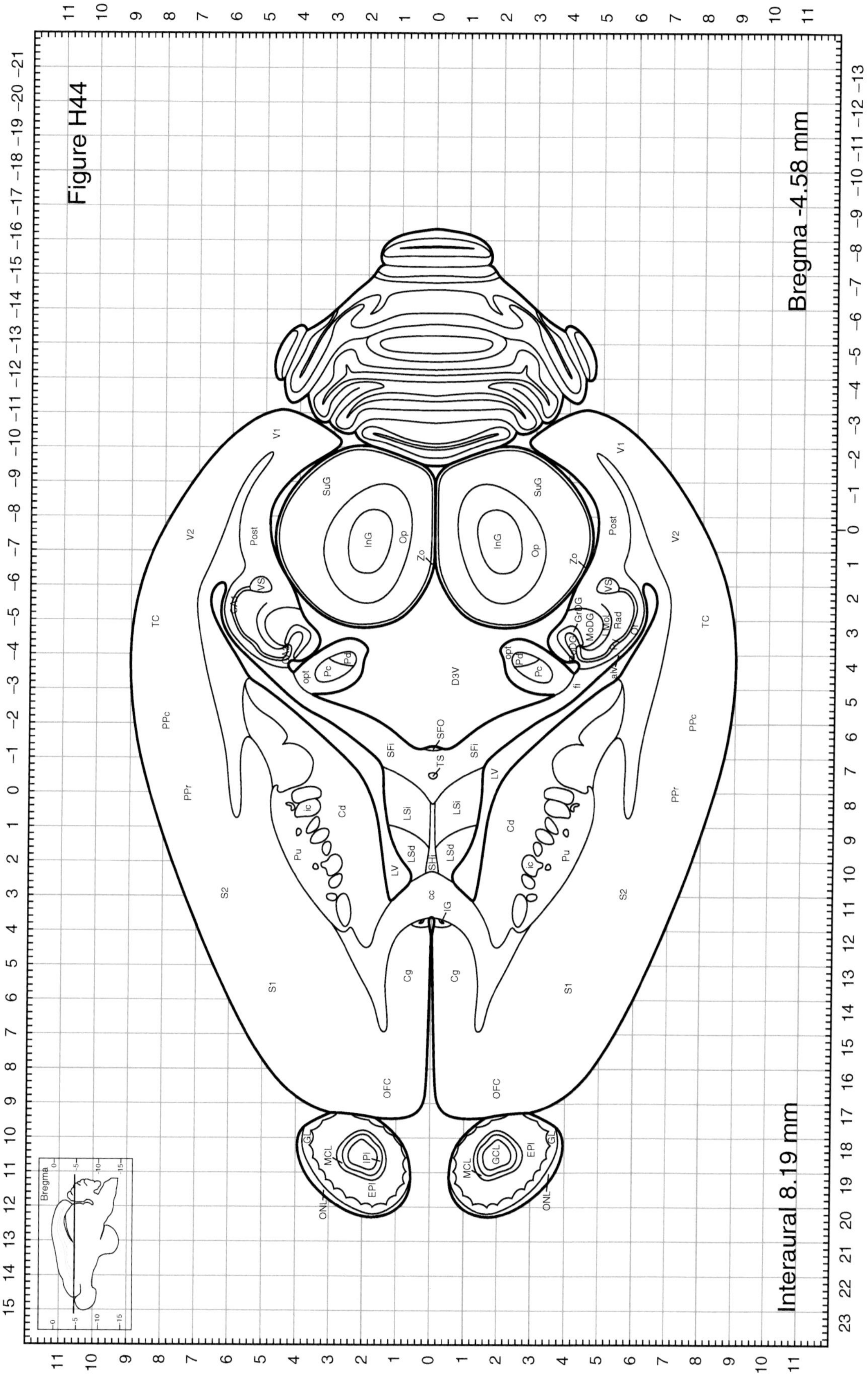

Figure H44
Bregma -4.58 mm
Interaural 8.19 mm
V1
V2
SuG
InG
Op
Zo
Post
TC
VS
GrDG
MoDG
LMol
Rad
Or
opt
Pd
Pc
D3V
fi
alv
PPc
SFO
SFi
TS
LV
PPr
Cd
ic
LSi
LSd
SHi
Pu
S2
cc
IG
Cg
S1
OFC
GL
MCL
IPl
EPl
GCL
ONL
Bregma

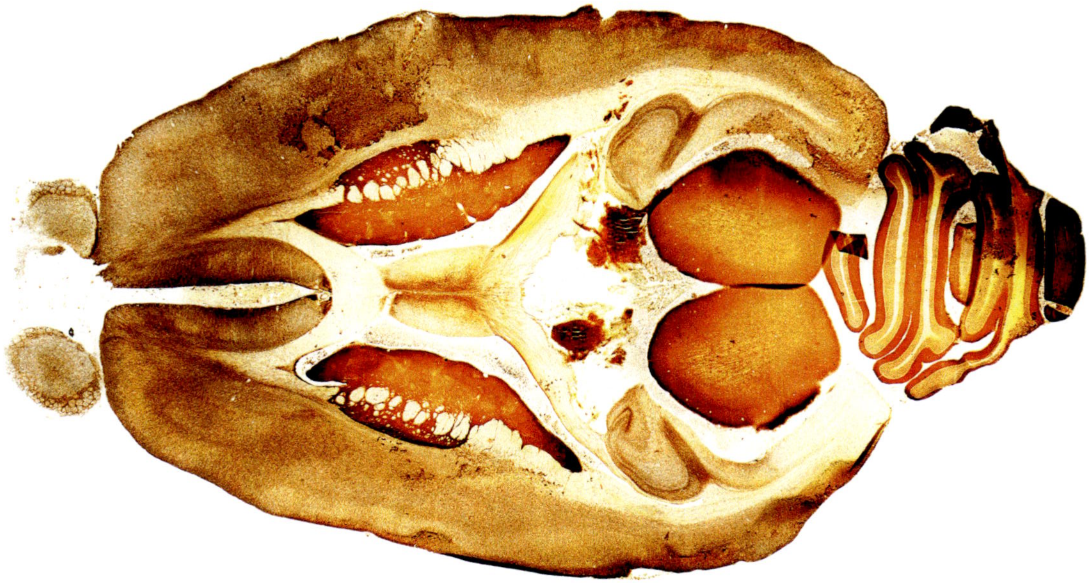

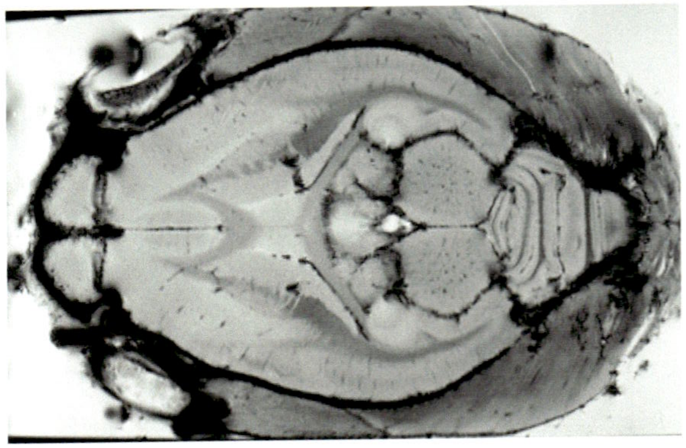

alv	alveus of the hippocampus
CA1	field CA1 of the hippocampus
CA3	field CA3 of the hippocampus
cc	corpus callosum
Cd	caudate nucleus
Cg	cingulate cortex
D3V	dorsal 3rd ventricle
EPl	external plexiform layer of the olfactory bulb
fi	fimbria of the hippocampus
GL	glomerular layer of the olfactory bulb
GrDG	granular layer of the dentate gyrus
ic	internal capsule
IG	indusium griseum
LMoL	lacunosum moleculare layer of the hippocampus
LSd	lateral septal nucleus, dorsal part
LSi	lateral septal nucleus, intermediate part
LV	lateral ventricle
MCL	mitral cell layer of the olfactory bulb
MoDG	molecular layer of the dentate gyrus
OFC	orbital frontal cortex
ONL	olfactory nerve layer
Op	optic nerve layer of the superior colliculus
Or	oriens layer of the hippocampus
PoDG	polymorph layer of the dentate gyrus
Post	postsubiculum
PPc	posterior parietal caudal area
PPr	posterior parietal rostral area
Pu	putamen
Py	pyramidal cell layer of the hippocampus
Rad	radiatum layer of the hippocampus
S1	primary somatosensory cortex
SCC	somatosensory caudal cortex
SFi	septofimbrial nucleus
SHi	septohippocampal nucleus
SuG	superficial gray layer of the superior colliculus
TC	temporal cortex
V1	primary visual cortex
V2	secondary visual cortex
VS	ventral subiculum
Zo	zonal layer of the superior colliculus

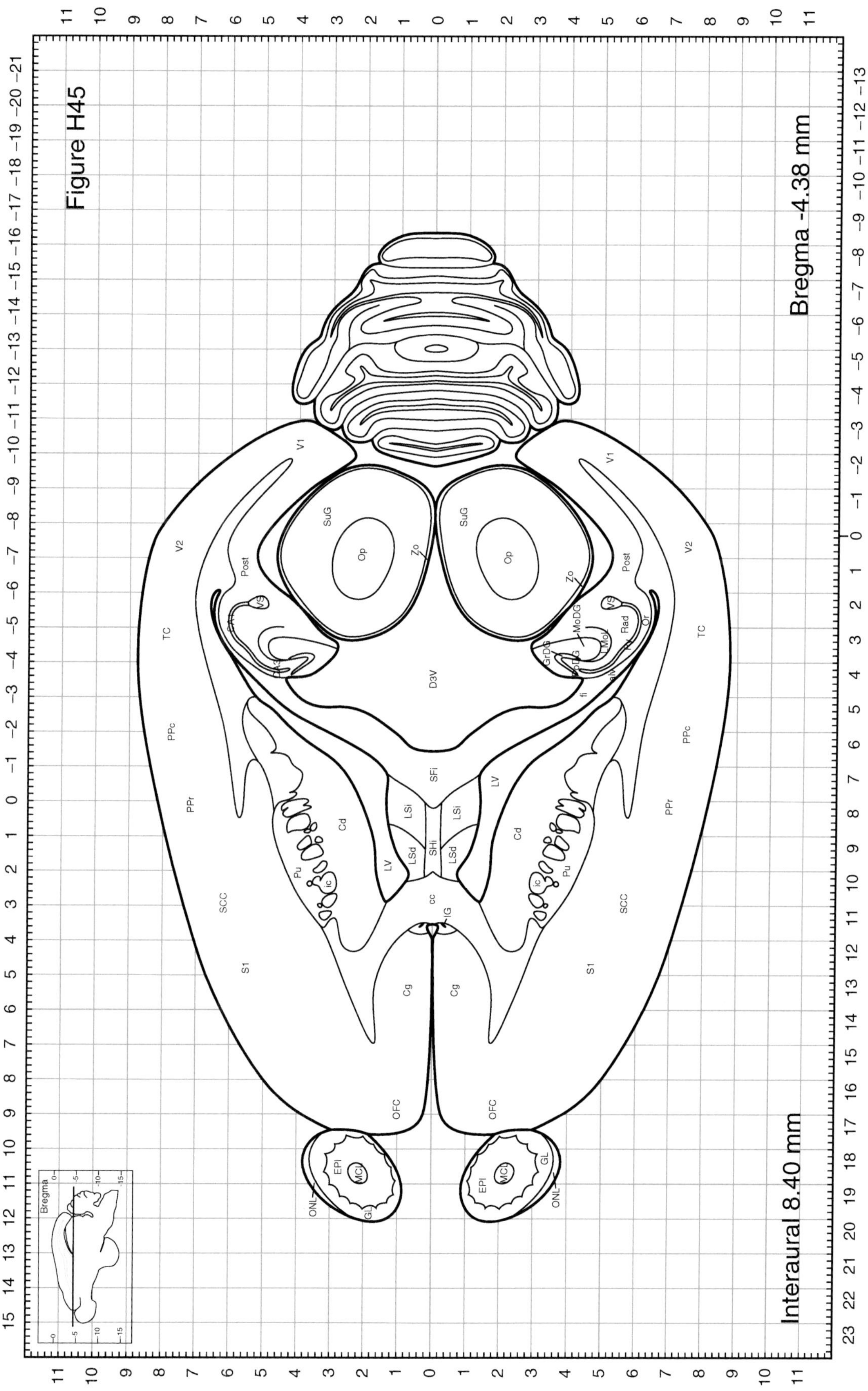

Figure H45
Bregma -4.38 mm
Interaural 8.40 mm
V1
V2
TC
PPc
PPr
SCC
S1
SuG
Op
Zo
Post
D3V
SFi
LSi
LSd
SHi
LV
Cd
Pu
ic
cc
IG
Cg
OFC
EPl
MCl
GL
ONL
MoDG
GrDG
Rad
Or
LMol
fi
Bregma

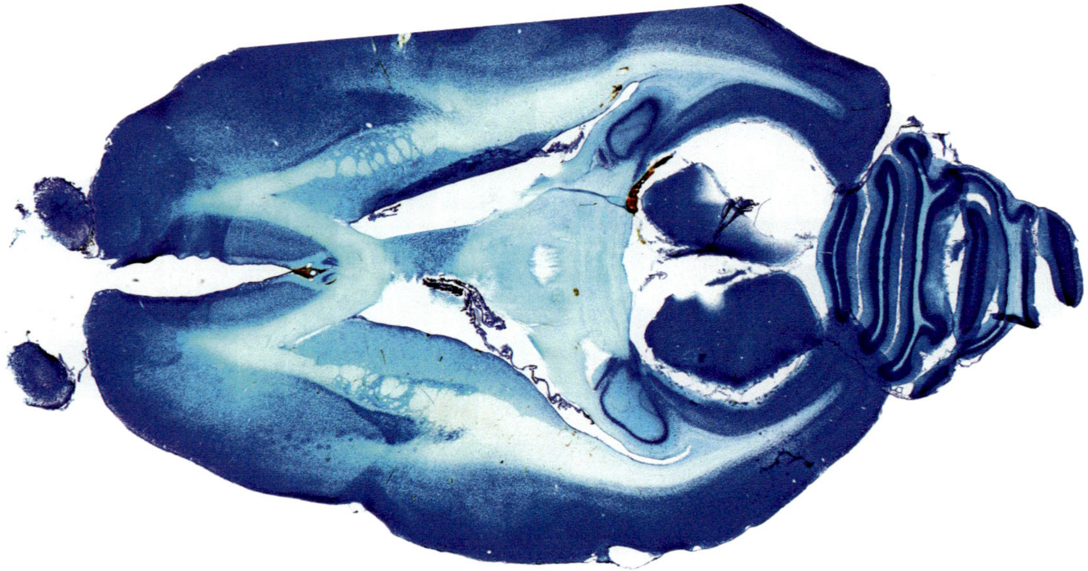

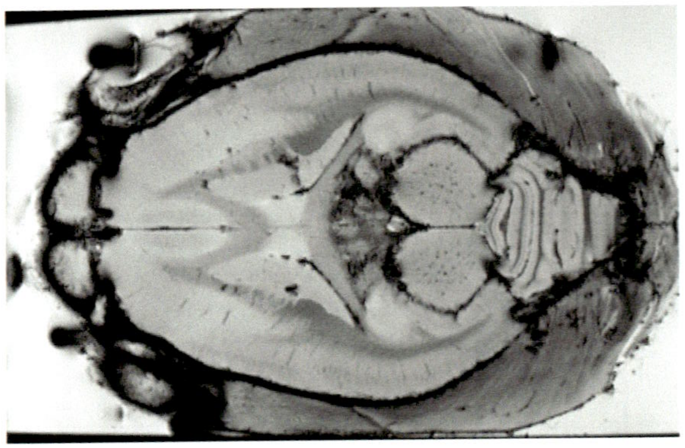

alv alveus of the hippocampus
CA1 field CA1 of the hippocampus
CA3 field CA3 of the hippocampus
cc corpus callosum
Cd caudate nucleus
Cg cingulate cortex
D3V dorsal 3rd ventricle
EPl external plexiform layer of the olfactory bulb
fi fimbria of the hippocampus
GL glomerular layer of the olfactory bulb
GrDG granular layer of the dentate gyrus
ic internal capsule
IG indusium griseum
LSd lateral septal nucleus, dorsal part
LV lateral ventricle
M1 primary motor cortex
MoDG molecular layer of the dentate gyrus
OFC orbital frontal cortex
ONL olfactory nerve layer
Or oriens layer of the hippocampus
PoDG polymorph layer of the dentate gyrus
Post postsubiculum
PPc posterior parietal caudal area
PPr posterior parietal rostral area
Pu putamen
Py pyramidal cell layer of the hippocampus
Rad radiatum layer of the hippocampus
S1 primary somatosensory cortex
SCC somatosensory caudal cortex
SFi septofimbrial nucleus
SHi septohippocampal nucleus
SuG superficial gray layer of the superior colliculus
TC temporal cortex
V1 primary visual cortex
V2 secondary visual cortex
VS ventral subiculum
Zo zonal layer of the superior colliculus

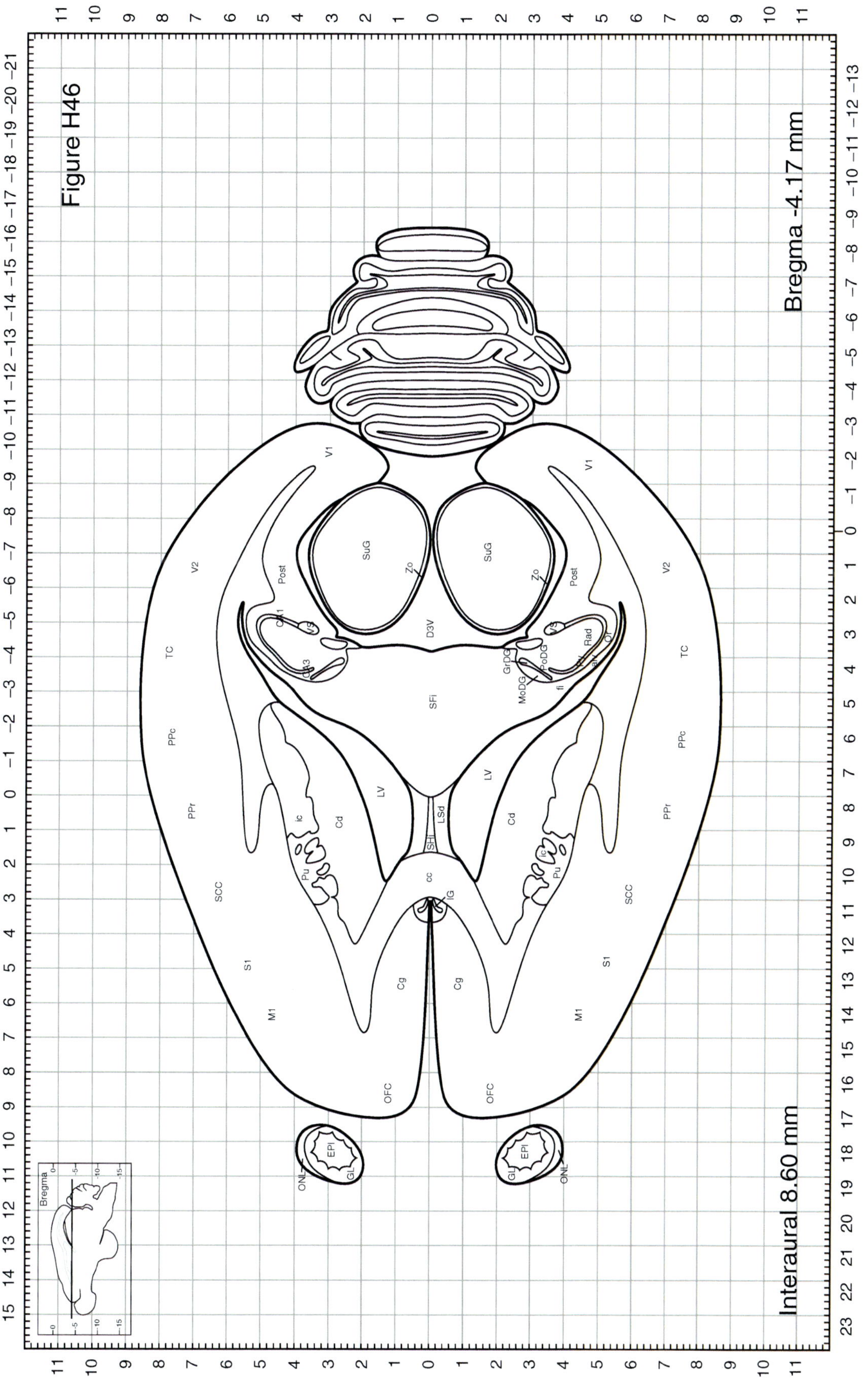

Figure H46
Bregma -4.17 mm
Interaural 8.60 mm
V1
V2
TC
PPc
PPr
SCC
S1
M1
OFC
Cg
SuG
Zo
Post
CA1
CA3
D3V
SFi
LV
Cd
LSd
SHi
cc
IG
Ic
Pu
GrDG
MoDG
PoDG
Rad
Or
fi
EPl
GL
ONL
Bregma

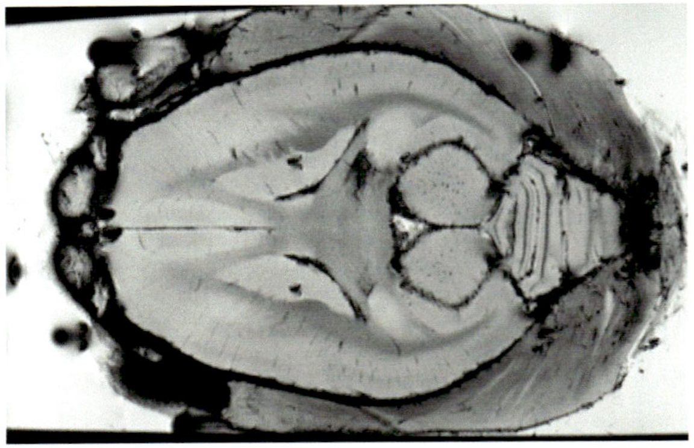

alv alveus of the hippocampus
CA1 field CA1 of the hippocampus
cc corpus callosum
Cd caudate nucleus
Cg cingulate cortex
D3V dorsal 3rd ventricle
fi fimbria of the hippocampus
ic internal capsule
IG indusium griseum
LV lateral ventricle
M1 primary motor cortex
OFC orbital frontal cortex
ONL olfactory nerve layer
Or oriens layer of the hippocampus
Post postsubiculum
PPc posterior parietal caudal area
PPr posterior parietal rostral area
Pu putamen
Py pyramidal cell layer of the hippocampus
Rad radiatum layer of the hippocampus
RSg retrosplenial granular cortex
S1 primary somatosensory cortex
SCC somatosensory caudal cortex
SFi septofimbrial nucleus
SuG superficial gray layer of the superior colliculus
TC temporal cortex
V1 primary visual cortex
V2 secondary visual cortex
Zo zonal layer of the superior colliculus

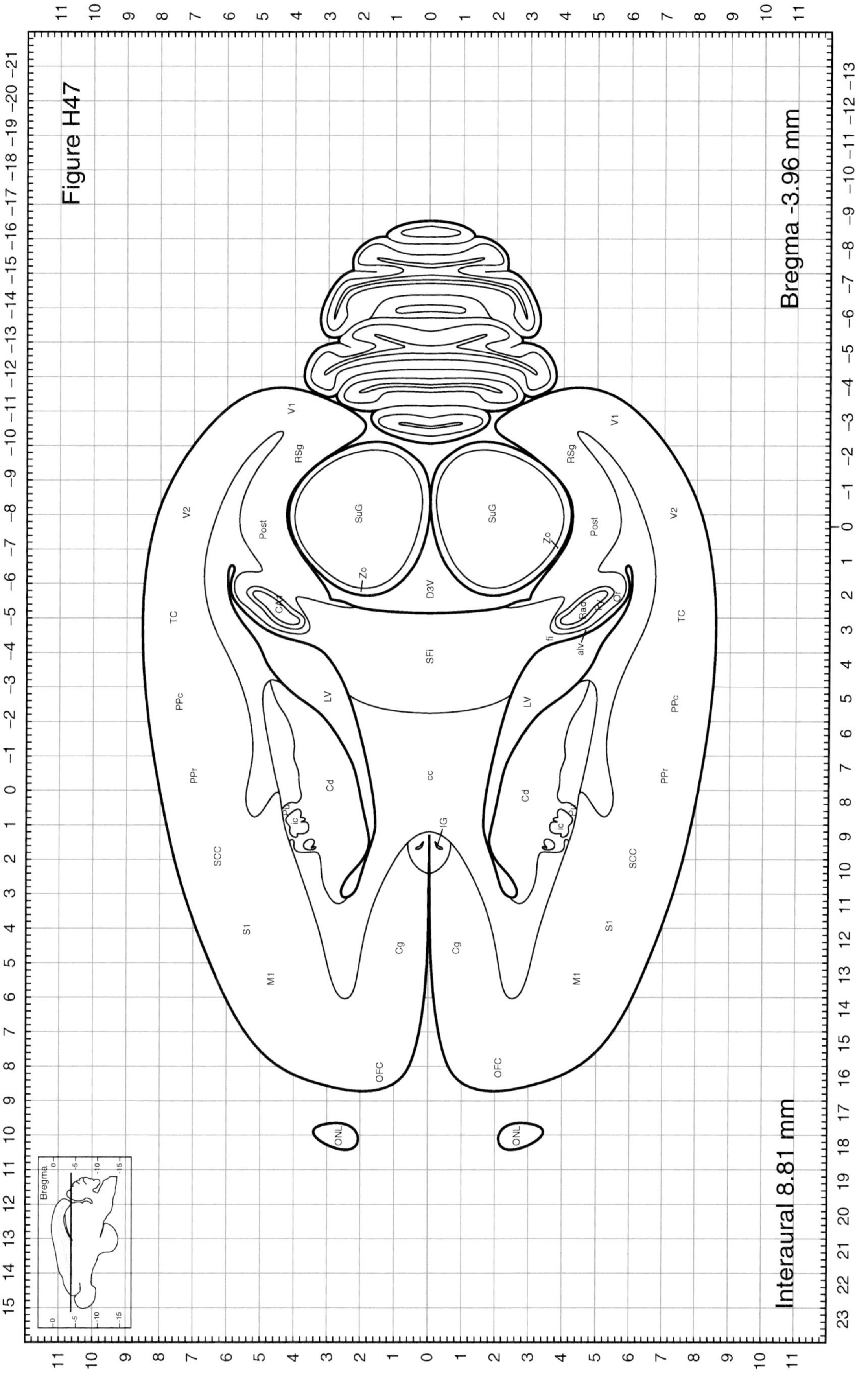
Figure H47
Bregma -3.96 mm
Interaural 8.81 mm
V1
RSg
V2
Post
SuG
Zo
D3V
TC
SFi
fi
alv
Rad
PPc
LV
cc
Cd
PPr
ic
IG
SCC
S1
Cg
M1
OFC
ONL
Bregma

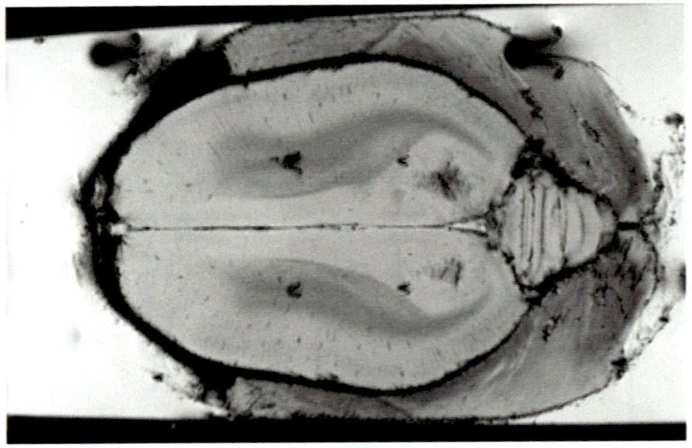

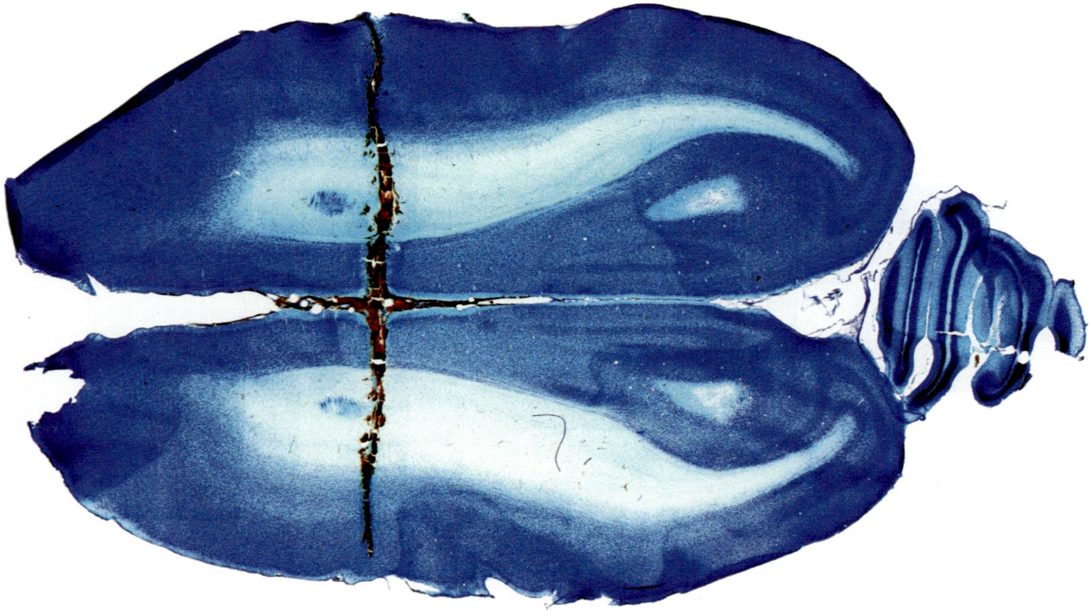

Cd caudate nucleus
Cg cingulate cortex
D3V dorsal 3rd ventricle
IRv infraradiata ventral area
M1 primary motor cortex

M2 secondary motor cortex
OFC orbital frontal cortex
PPc posterior parietal caudal area
PPr posterior parietal rostral area
RSg retrosplenial granular cortex

S1 primary somatosensory cortex
SCC somatosensory caudal cortex
TC temporal cortex
V1 primary visual cortex
V2 secondary visual cortex

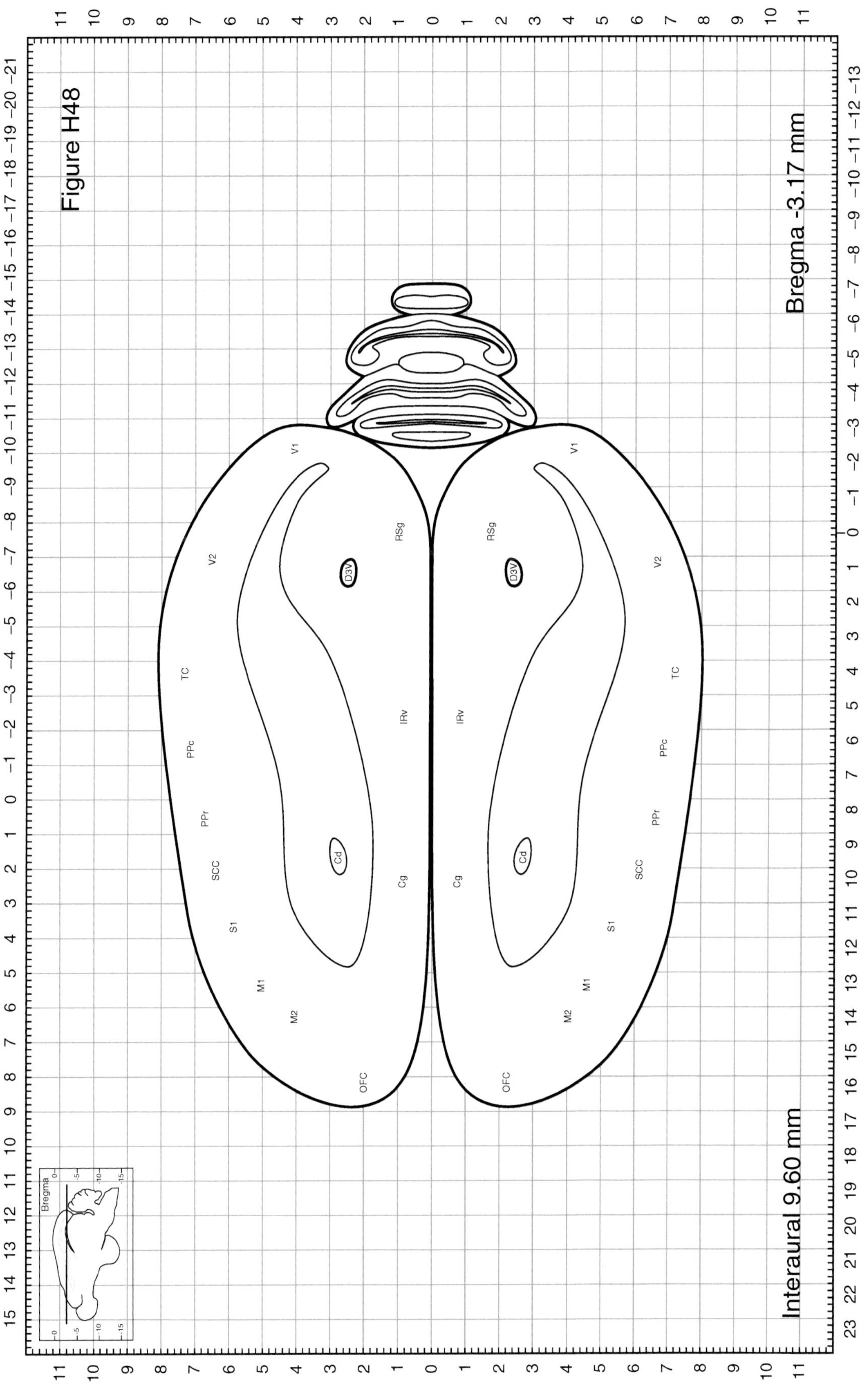
Figure H48
Bregma -3.17 mm
Interaural 9.60 mm
V1
RSg
V2
D3V
TC
IRv
PPc
PPr
Cd
SCC
Cg
S1
M1
M2
OFC

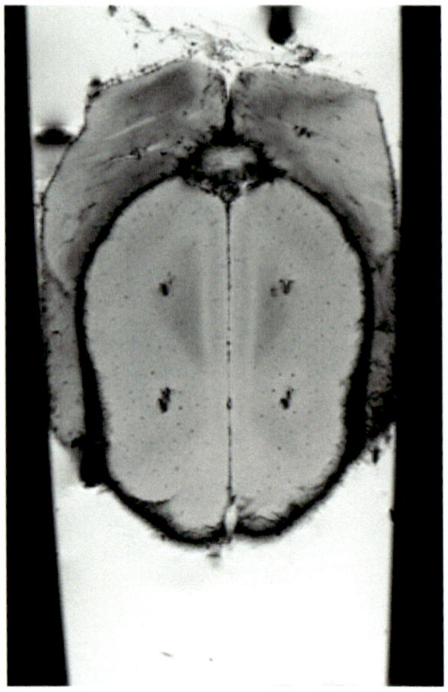

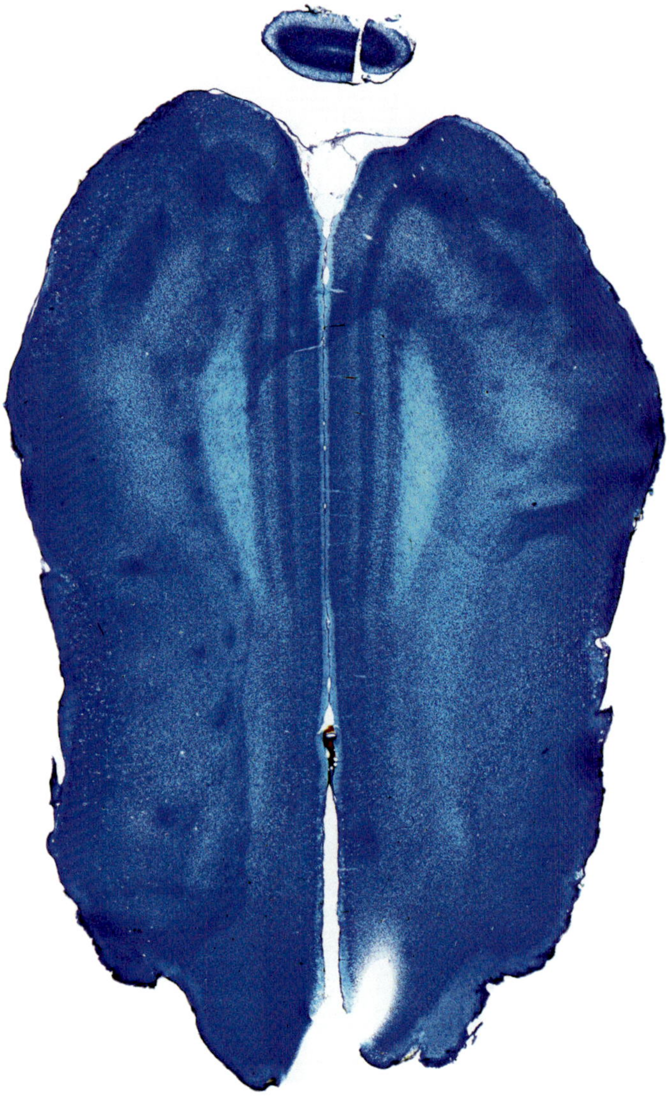

DFC dorsal frontal cortex
IRd infraradiata dorsalis
M1 primary motor cortex
M2 secondary motor cortex
MFC medial frontal cortex
PPd posterior parietal dorsal area
RSg retrosplenial granular cortex
S1 primary somatosensory cortex
SCC somatosensory caudal cortex
TC temporal cortex
V1 primary visual cortex
V2 secondary visual cortex

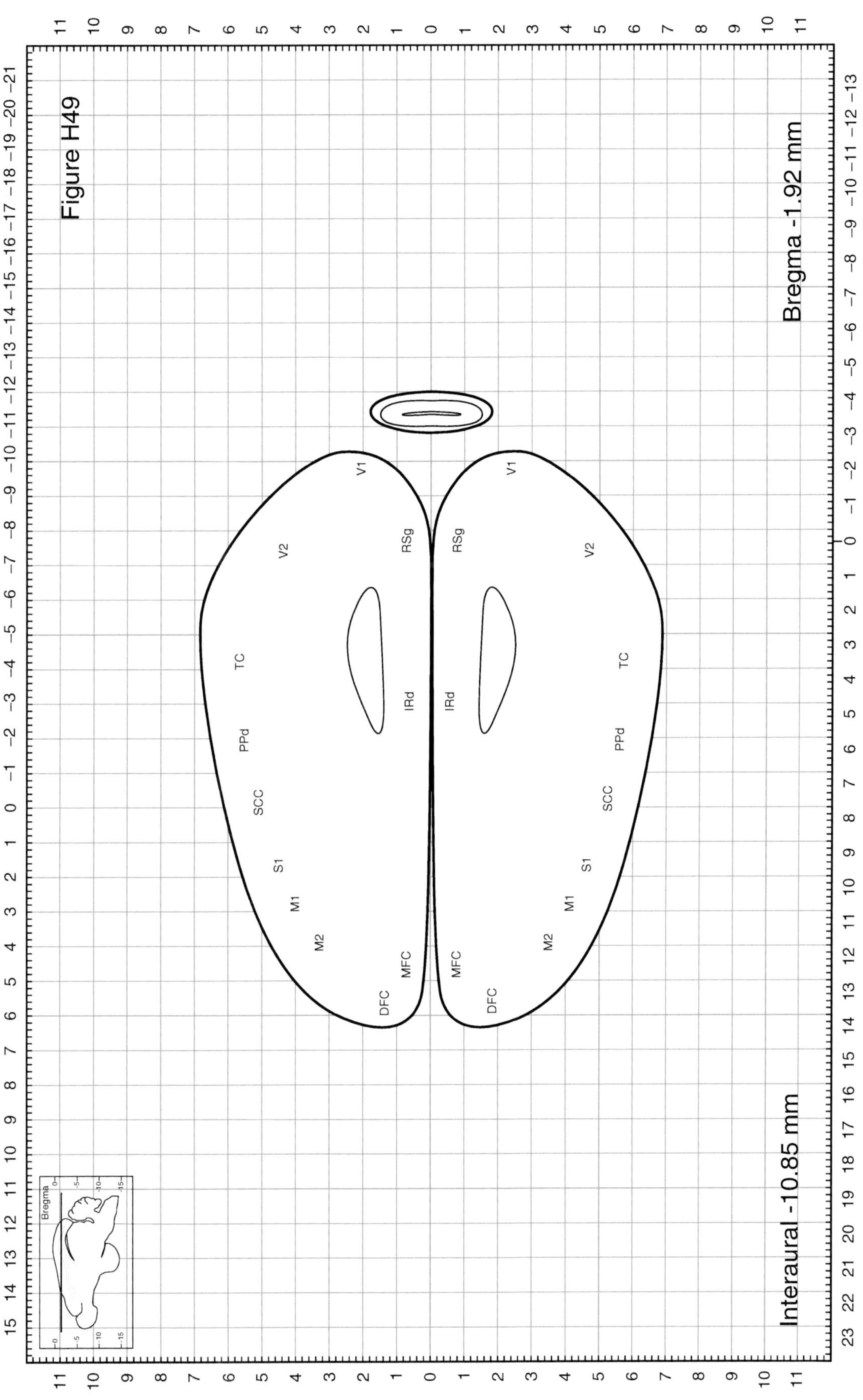
Figure H49
Bregma -1.92 mm
Interaural -10.85 mm
V1
RSg
V2
TC
IRd
PPd
SCC
S1
M1
M2
MFC
DFC
Bregma

4 Sagittal Sections of the Tree Shrew Brain

J.-N. Zhou, R.-J. Ni, *The Tree Shrew (Tupaia belangeri chinensis) Brain in Stereotaxic Coordinates*,
DOI 10.1007/978-981-10-0611-1_4

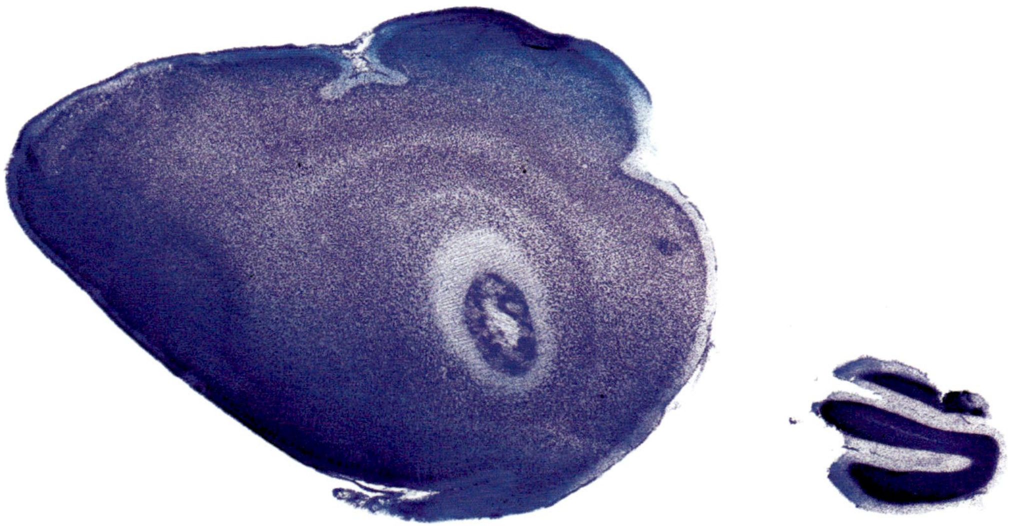

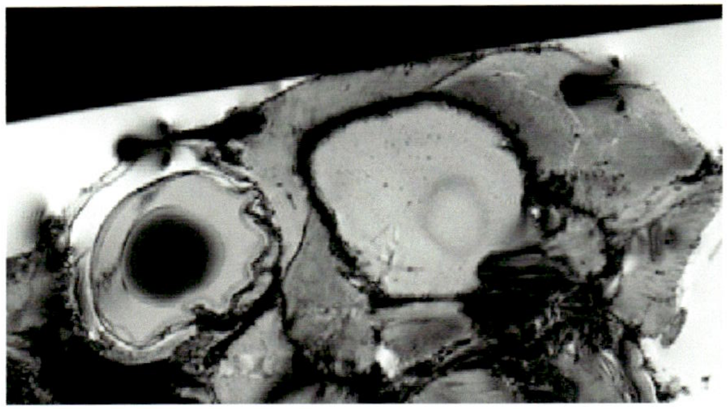

Ab	auditory belt area
Ac	auditory core area
alv	alveus of the hippocampus
Ect	ectorhinal cortex
Ent	entorhinal cortex
IT	inferior temporal cortex
LV	lateral ventricle
PRh	perirhinal cortex
PvA	parietal ventral area
TC	temporal cortex

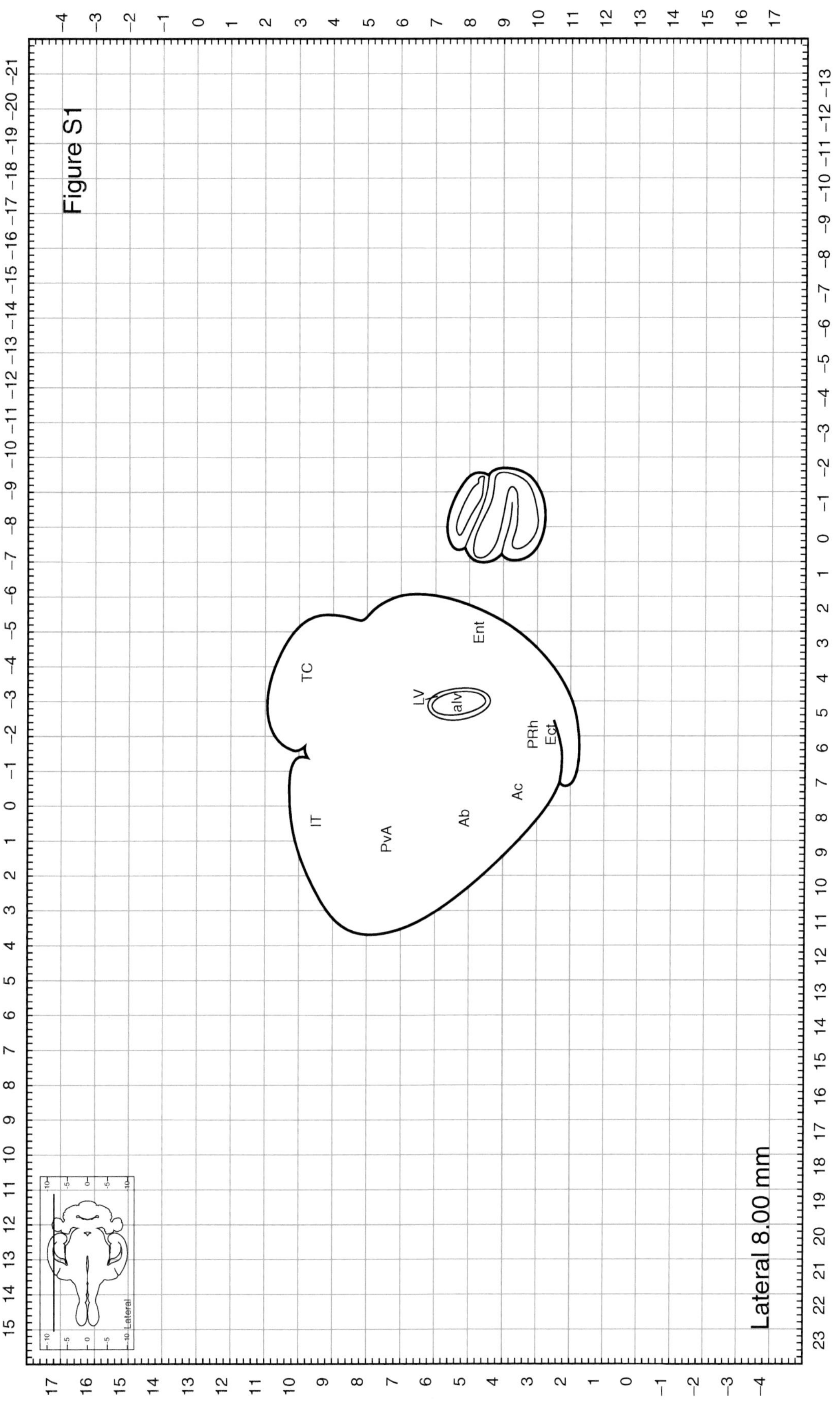
Figure S1
Lateral 8.00 mm
TC
IT
PvA
Ab
Ac
PRh
Ect
Ent
LV
alv

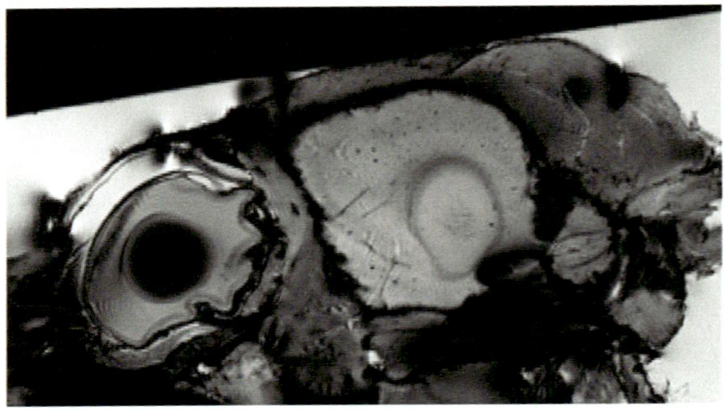

2	layer 2 of cortex
3	layer 3 of cortex
Ab	auditory belt area
Ac	auditory core area
alv	alveus of the hippocampus
Ect	ectorhinal cortex
Ent	entorhinal cortex
IT	inferior temporal cortex
LV	lateral ventricle
Or	oriens layer of the hippocampus
Pir1	piriform cortex, layer 1
PRh	perirhinal cortex
PvA	parietal ventral area
Py	pyramidal cell layer of the hippocampus
Rad	radiatum layer of the hippocampus
TC	temporal cortex

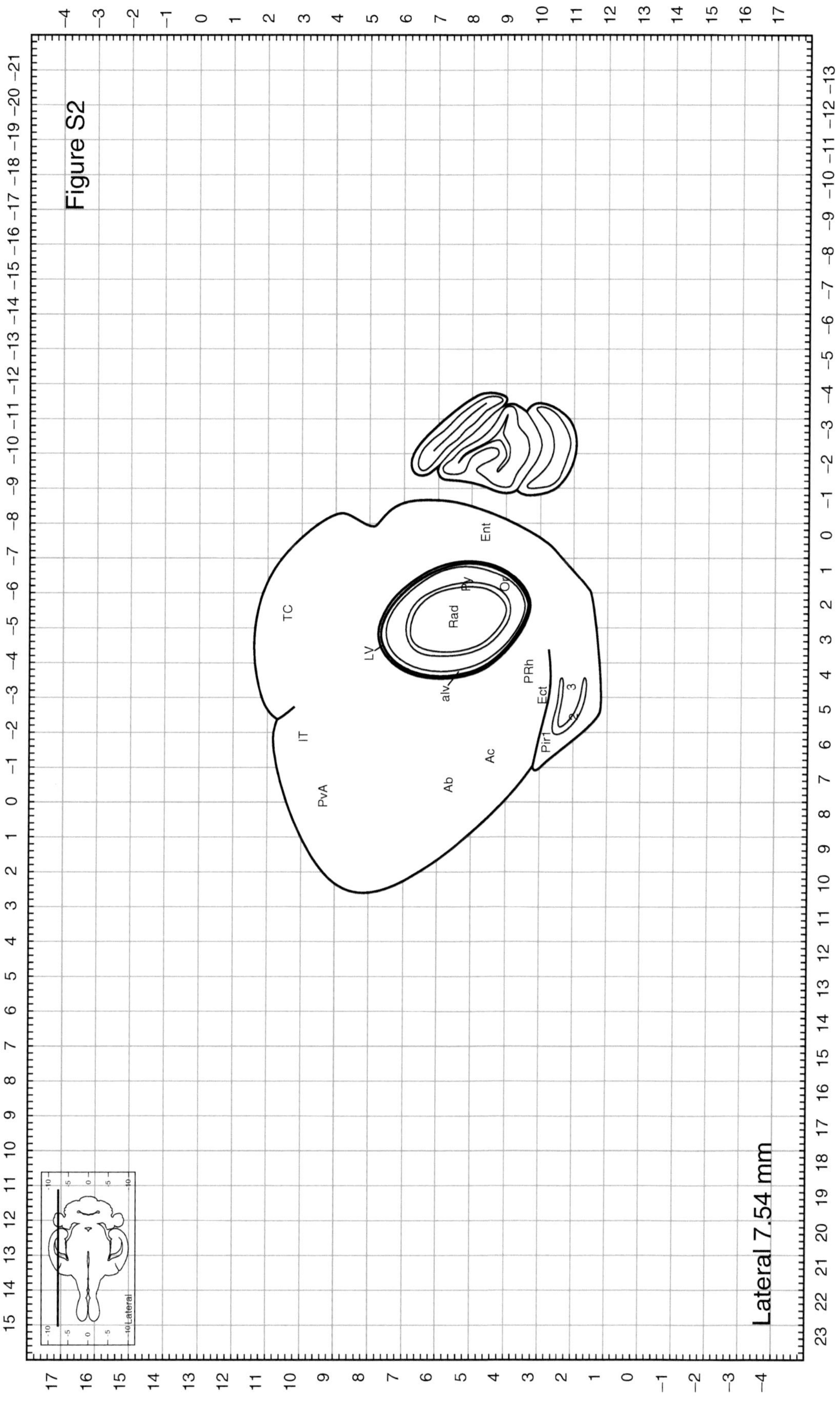
Figure S2
TC
IT
PvA
Ab
Ac
LV
alv
Rad
Py
Or
Ent
PRh
Ect
Pir1
3
2
Lateral 7.54 mm

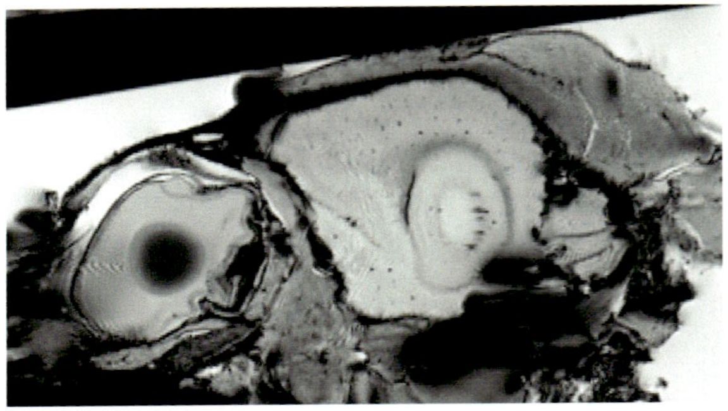

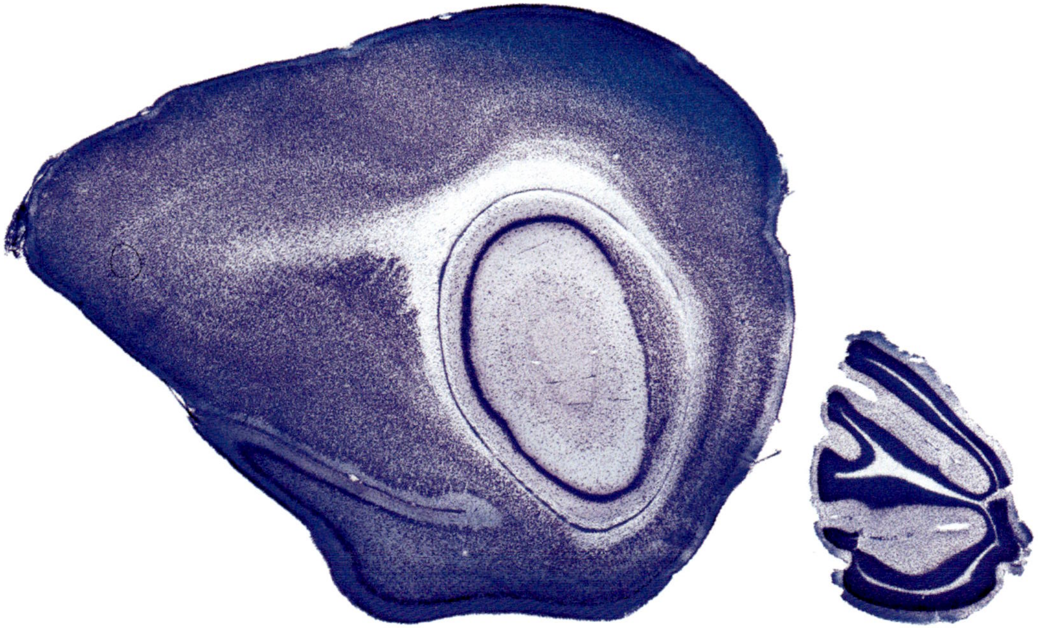

2	layer 2 of cortex
3	layer 3 of cortex
Ab	auditory belt area
Ac	auditory core area
alv	alveus of the hippocampus
DS	dorsal subiculum
Ect	ectorhinal cortex
Ent	entorhinal cortex
IT	inferior temporal cortex
LMoL	lacunosum moleculare layer of the hippocampus
LV	lateral ventricle
MoDG	molecular layer of the dentate gyrus
Or	oriens layer of the hippocampus
Pir1	piriform cortex, layer 1
PRh	perirhinal cortex
PvA	parietal ventral area
Py	pyramidal cell layer of the hippocampus
Rad	radiatum layer of the hippocampus
S2	secondary somatosensory cortex
TC	temporal cortex
TI	temporal inferior area
VS	ventral subiculum

Figure S3
Lateral 7.08 mm
TC
IT
TI
PvA
S2
Ab
Ac
Pir
Ect
PRh
Ent
alv
LV
Or
Py
Rad
LMoL
MoDG
DS
VS
3

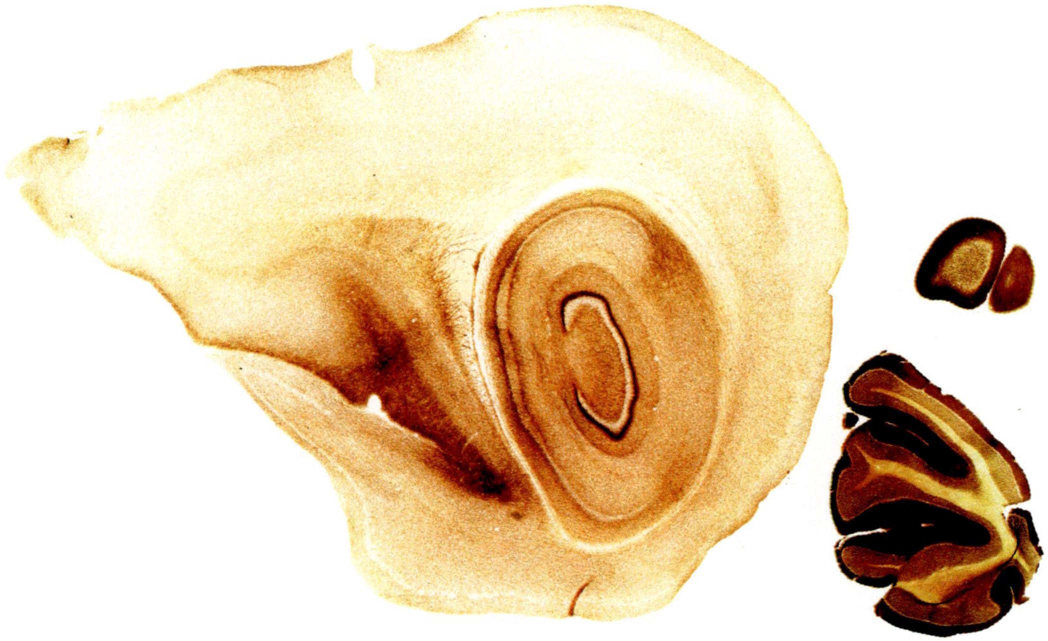

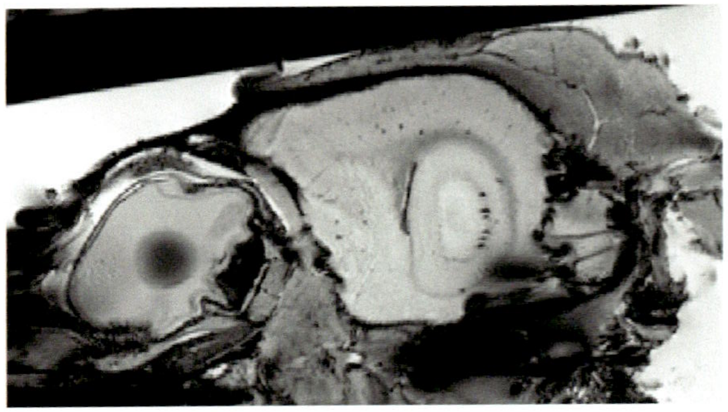

2	layer 2 of cortex
3	layer 3 of cortex
alv	alveus of the hippocampus
CA1	field CA1 of the hippocampus
CA2	field CA2 of the hippocampus
CA3	field CA2 of the hippocampus
DS	dorsal subiculum
Ent	entorhinal cortex
GrDG	granular layer of the dentate gyrus
Ins	insular cortex
IT	inferior temporal cortex
LMoL	lacunosum moleculare layer of the hippocampus
LV	lateral ventricle
MoDG	molecular layer of the dentate gyrus
Or	oriens layer of the hippocampus
Pir1	piriform cortex, layer 1
PoDG	polymorph layer of the dentate gyrus
Py	pyramidal cell layer of the hippocampus
Rad	radiatum layer of the hippocampus
S2	secondary somatosensory cortex
TC	temporal cortex
TI	temporal inferior area
V2	secondary visual cortex
VS	ventral subiculum

Figure S4
Lateral 6.71 mm
V2
TC
IT
TI
S2
Ins
Pir
Ent
VS
DS
CA1
CA2
CA3
LV
Or
Rad
Py
alv
LMol
MoDG
PoDG
GrDG
Lateral

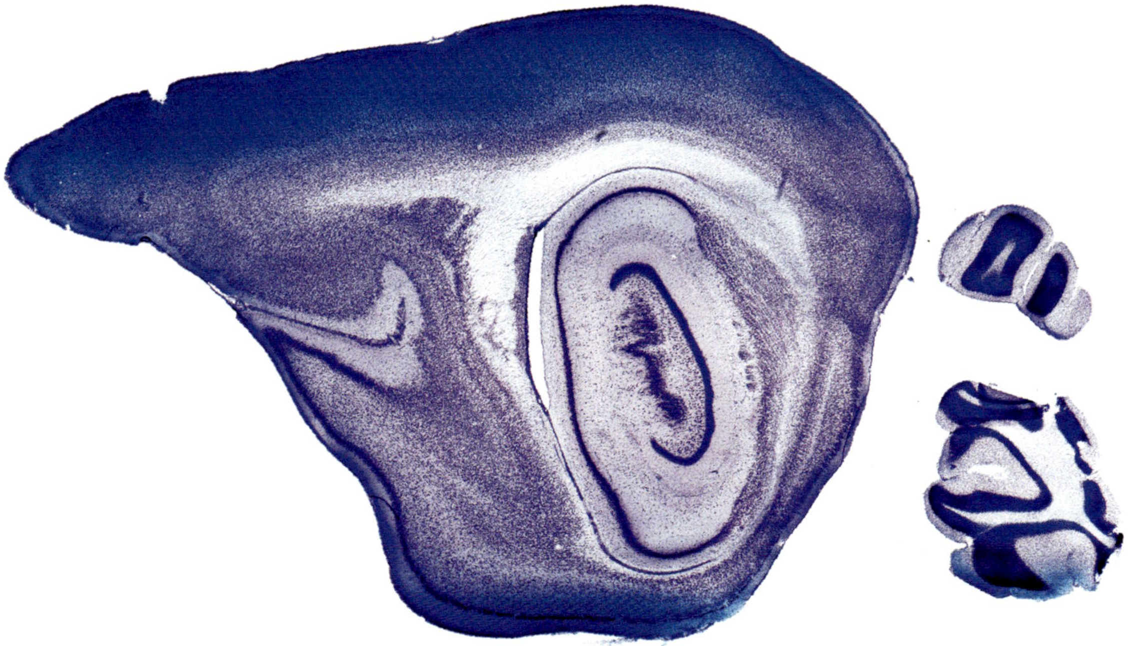

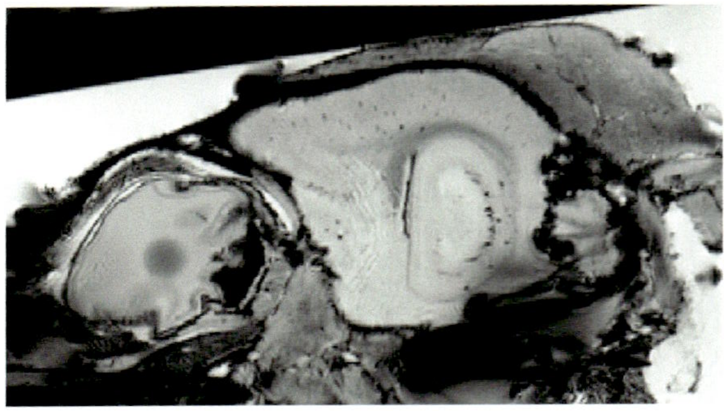

2 layer 2 of cortex
3 layer 3 of cortex
alv alveus of the hippocampus
CA1 field CA1 of the hippocampus
CA2 field CA2 of the hippocampus
CA3 field CA2 of the hippocampus
Cd caudate nucleus
Cl claustrum
DS dorsal subiculum
Ent entorhinal cortex
GrDG granular layer of the dentate gyrus
ic internal capsule
Ins insular cortex
LMoL lacunosum moleculare layer of the hippocampus
LV lateral ventricle
MoDG molecular layer of the dentate gyrus
Or oriens layer of the hippocampus
Pir1 piriform cortex, layer 1
PoDG polymorph layer of the dentate gyrus
PPc posterior parietal caudal area
PPr posterior parietal rostral area
Pu putamen
Py pyramidal cell layer of the hippocampus
Rad radiatum layer of the hippocampus
S1 primary somatosensory cortex
S2 secondary somatosensory cortex
TC temporal cortex
V2 secondary visual cortex
VS ventral subiculum

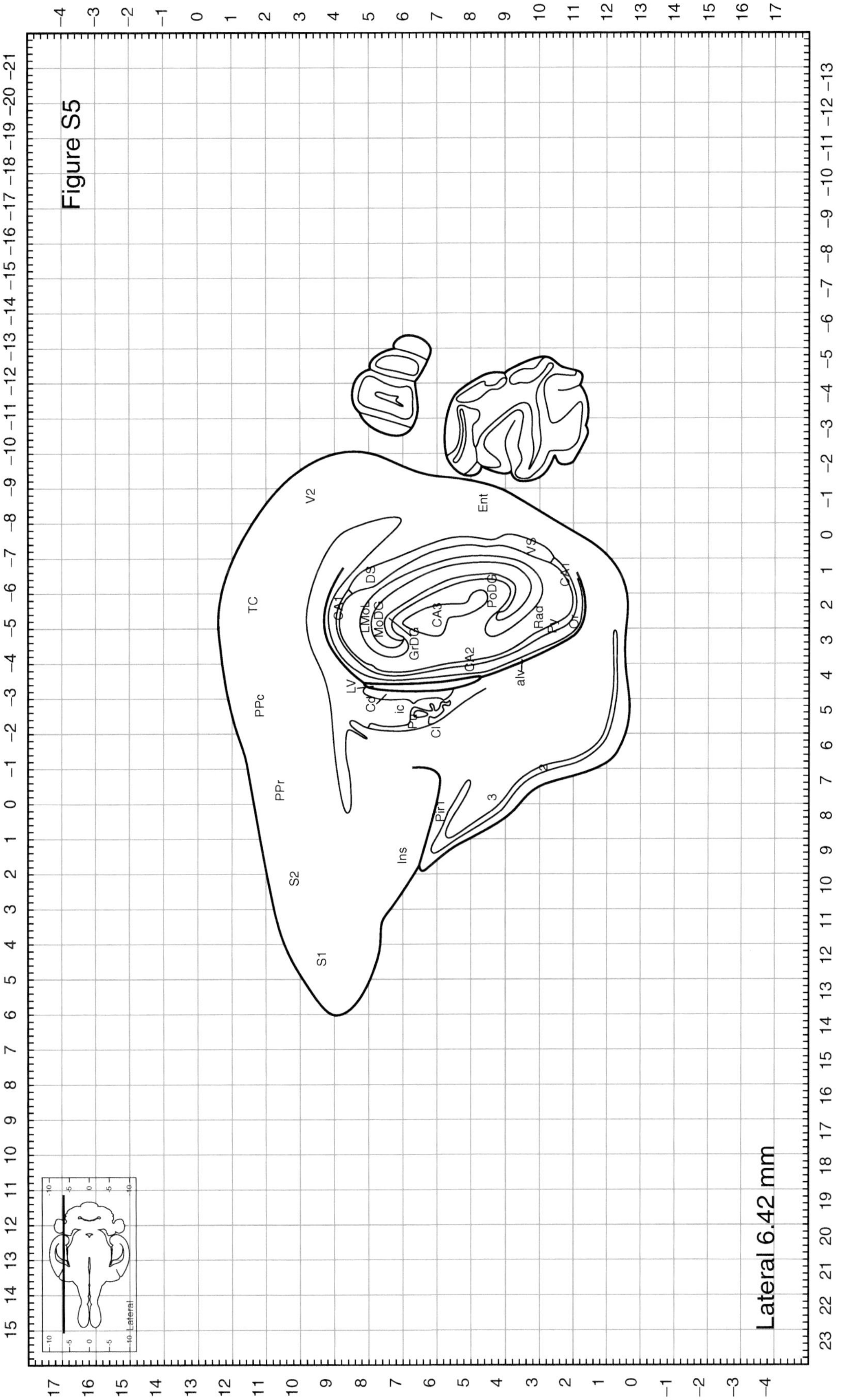
Figure S5
Lateral 6.42 mm
V2
Ent
TC
PPc
PPr
S2
S1
Ins
CA1
DS
LMol
MoDG
GrDG
CA3
PoDG
CA2
VS
Rad
Py
Or
alv
LV
Cd
ic
Cl
3
2

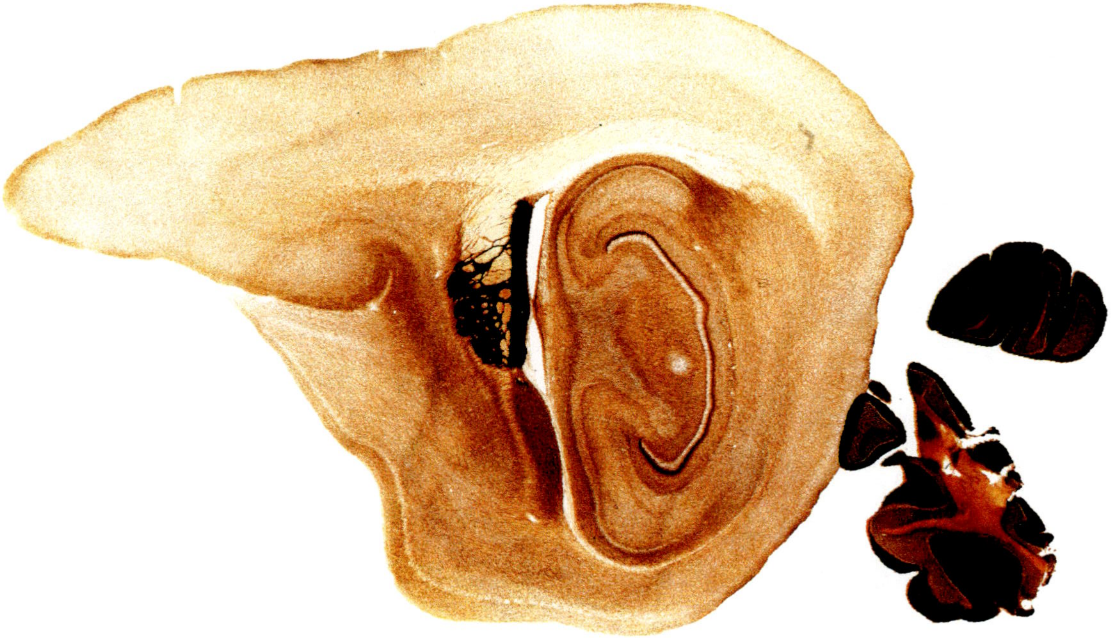

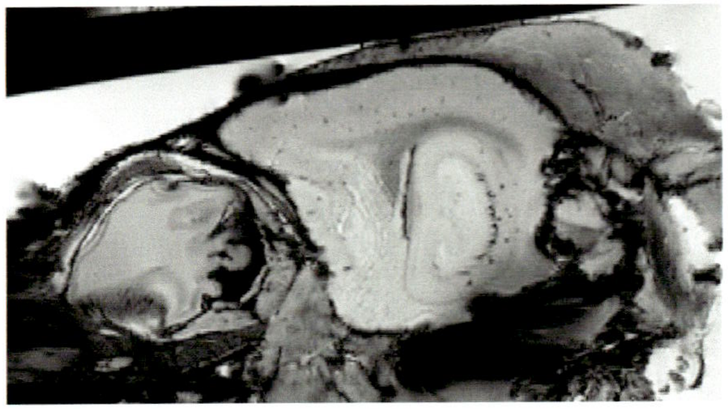

2	layer 2 of cortex
3	layer 3 of cortex
alv	alveus of the hippocampus
BLA	basolateral amygdaloid nucleus
CA1	field CA1 of the hippocampus
CA2	field CA2 of the hippocampus
CA3	field CA2 of the hippocampus
Cd	caudate nucleus
Cl	claustrum
DS	dorsal subiculum
Ent	entorhinal cortex
GrDG	granular layer of the dentate gyrus
ic	internal capsule
Ins	insular cortex
LMoL	lacunosum moleculare layer of the hippocampus
LV	lateral ventricle
MoDG	molecular layer of the dentate gyrus
Or	oriens layer of the hippocampus
Pir1	piriform cortex, layer 1
PoDG	polymorph layer of the dentate gyrus
Post	postsubiculum
PPc	posterior parietal caudal area
PPr	posterior parietal rostral area
PrS	presubiculum
Pu	putamen
Py	pyramidal cell layer of the hippocampus
Rad	radiatum layer of the hippocampus
S1	primary somatosensory cortex
S2	secondary somatosensory cortex
TC	temporal cortex
V2	secondary visual cortex
VS	ventral subiculum

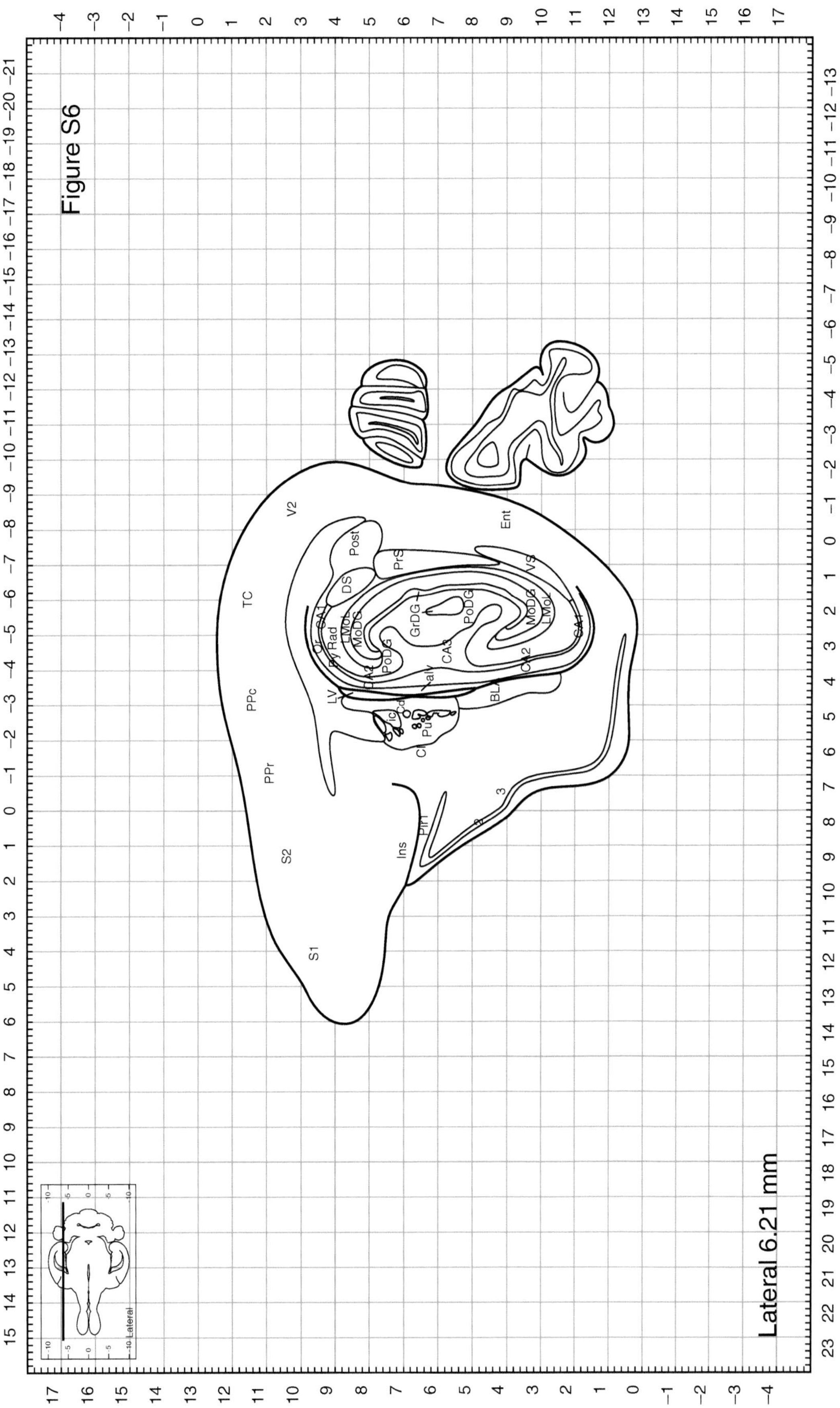
Figure S6
Lateral 6.21 mm
V2
TC
PPc
PPr
S2
S1
Ins
Pir1
Ent
Post
PrS
DS
VS
GrDG
PoDG
MoDG
LMol
CA1
CA2
CA3
Or
Py
Rad
LV
BL
Cd
Pu
Cl
ic
Lateral

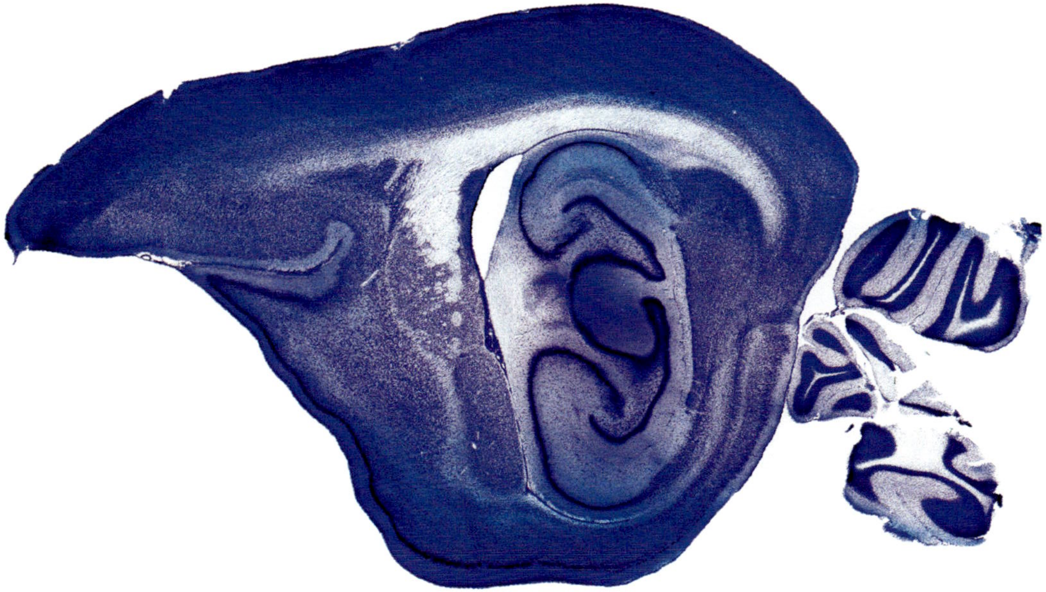

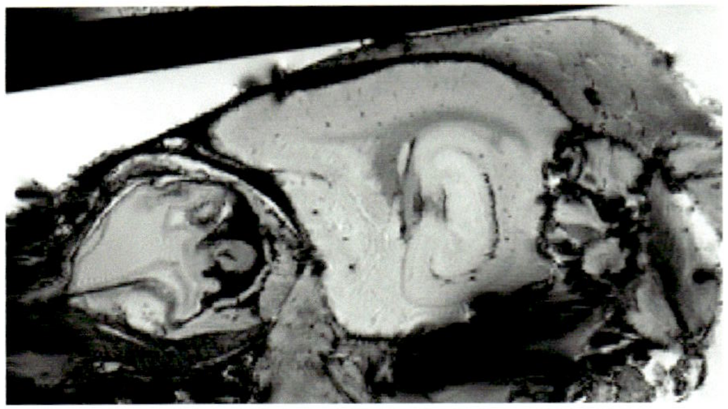

2	layer 2 of cortex
3	layer 3 of cortex
alv	alveus of the hippocampus
BLA	basolateral amygdaloid nucleus
CA1	field CA1 of the hippocampus
CA2	field CA2 of the hippocampus
CA3	field CA2 of the hippocampus
Cd	caudate nucleus
Cl	claustrum
DS	dorsal subiculum
Ent	entorhinal cortex
GrDG	granular layer of the dentate gyrus
ic	internal capsule
Ins	insular cortex
LMoL	lacunosum moleculare layer of the hippocampus
LV	lateral ventricle
M1	primary motor cortex
MoDG	molecular layer of the dentate gyrus
Or	oriens layer of the hippocampus
Pir1	piriform cortex, layer 1
PoDG	polymorph layer of the dentate gyrus
Post	postsubiculum
PPc	posterior parietal caudal area
PPr	posterior parietal rostral area
PrS	presubiculum
Pu	putamen
Py	pyramidal cell layer of the hippocampus
Rad	radiatum layer of the hippocampus
S1	primary somatosensory cortex
SCC	somatosensory caudal cortex
TC	temporal cortex
V2	secondary visual cortex
VS	ventral subiculum

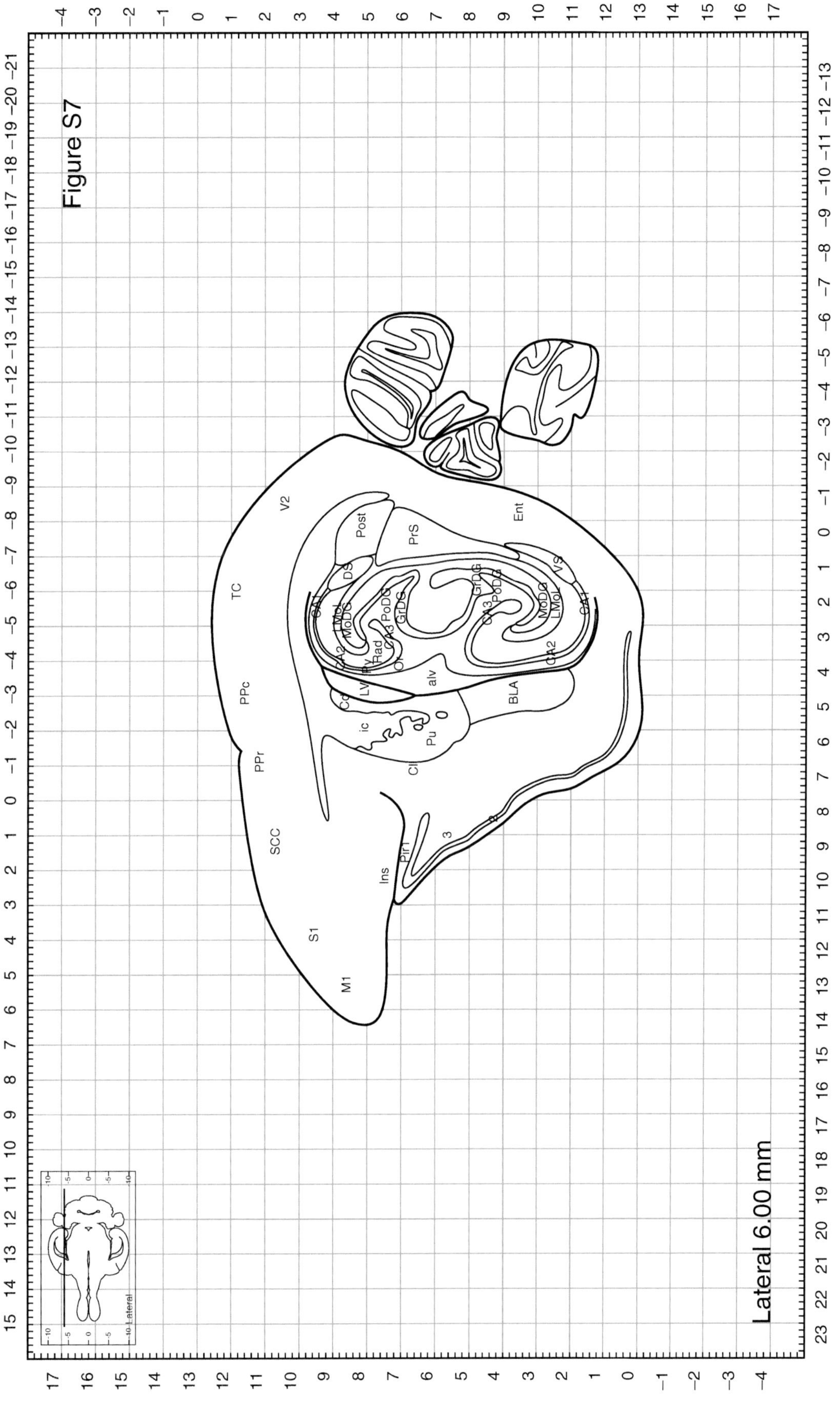
Figure S7
Lateral 6.00 mm
V2
TC
PPc
PPr
SCC
S1
M1
Ins
Pir1
Cl
Pu
ic
LV
alv
BLA
Ent
PrS
Post
DS
VS
CA1
CA2
CA3
GrDG
PoDG
MoDG
LMol
Rad
Or
Py
Lateral

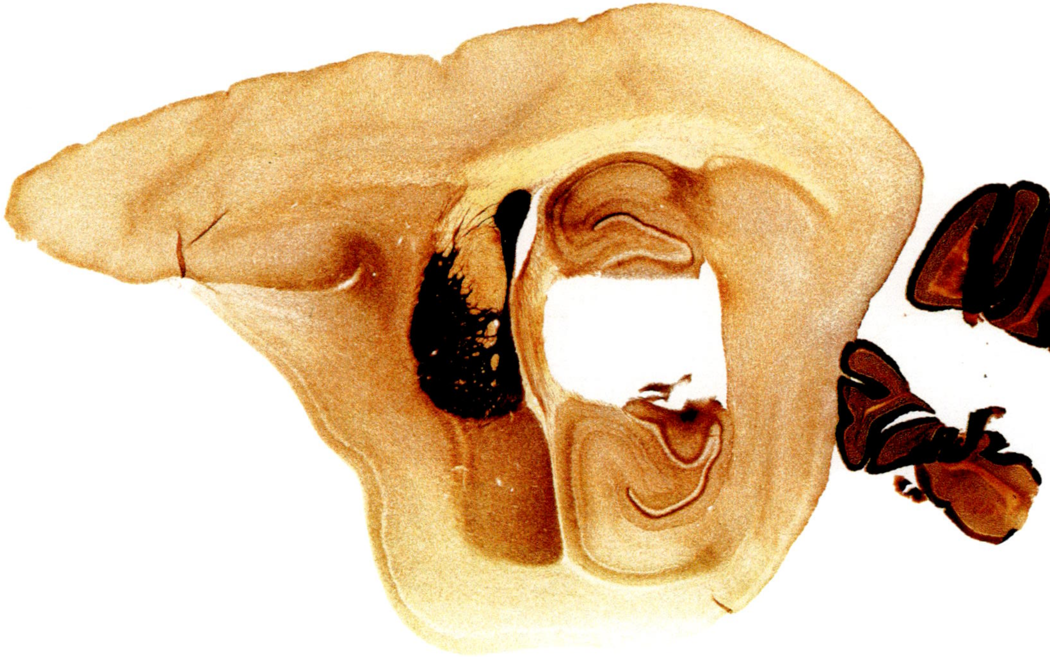

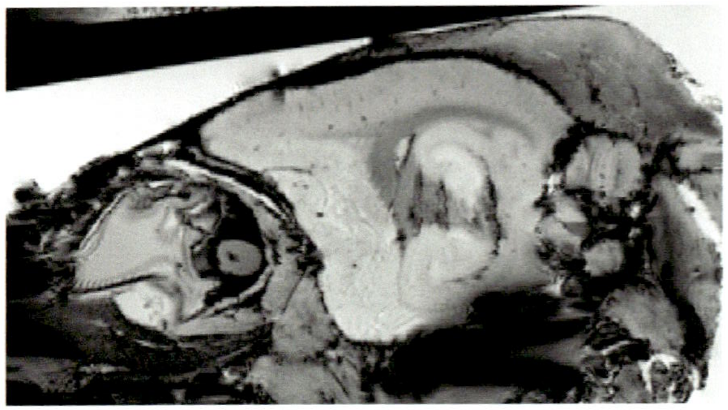

2	layer 2 of cortex	ic	internal capsule	PPr	posterior parietal rostral area
3	layer 3 of cortex	Ins	insular cortex	PrS	presubiculum
alv	alveus of the hippocampus	LMoL	lacunosum moleculare layer of the hippocampus	Pu	putamen
BLA	basolateral amygdaloid nucleus	LV	lateral ventricle	Py	pyramidal cell layer of the hippocampus
CA1	field CA1 of the hippocampus	M1	primary motor cortex	Rad	radiatum layer of the hippocampus
CA2	field CA2 of the hippocampus	MoDG	molecular layer of the dentate gyrus	S1	primary somatosensory cortex
CA3	field CA2 of the hippocampus	opt	optic tract	SCC	somatosensory caudal cortex
Cd	caudate nucleus	Or	oriens layer of the hippocampus	TC	temporal cortex
Cl	claustrum	Pir1	piriform cortex, layer 1	V2	secondary visual cortex
DS	dorsal subiculum	PoDG	polymorph layer of the dentate gyrus	VS	ventral subiculum
Ent	entorhinal cortex	Post	postsubiculum		
GrDG	granular layer of the dentate gyrus	PPc	posterior parietal caudal area		

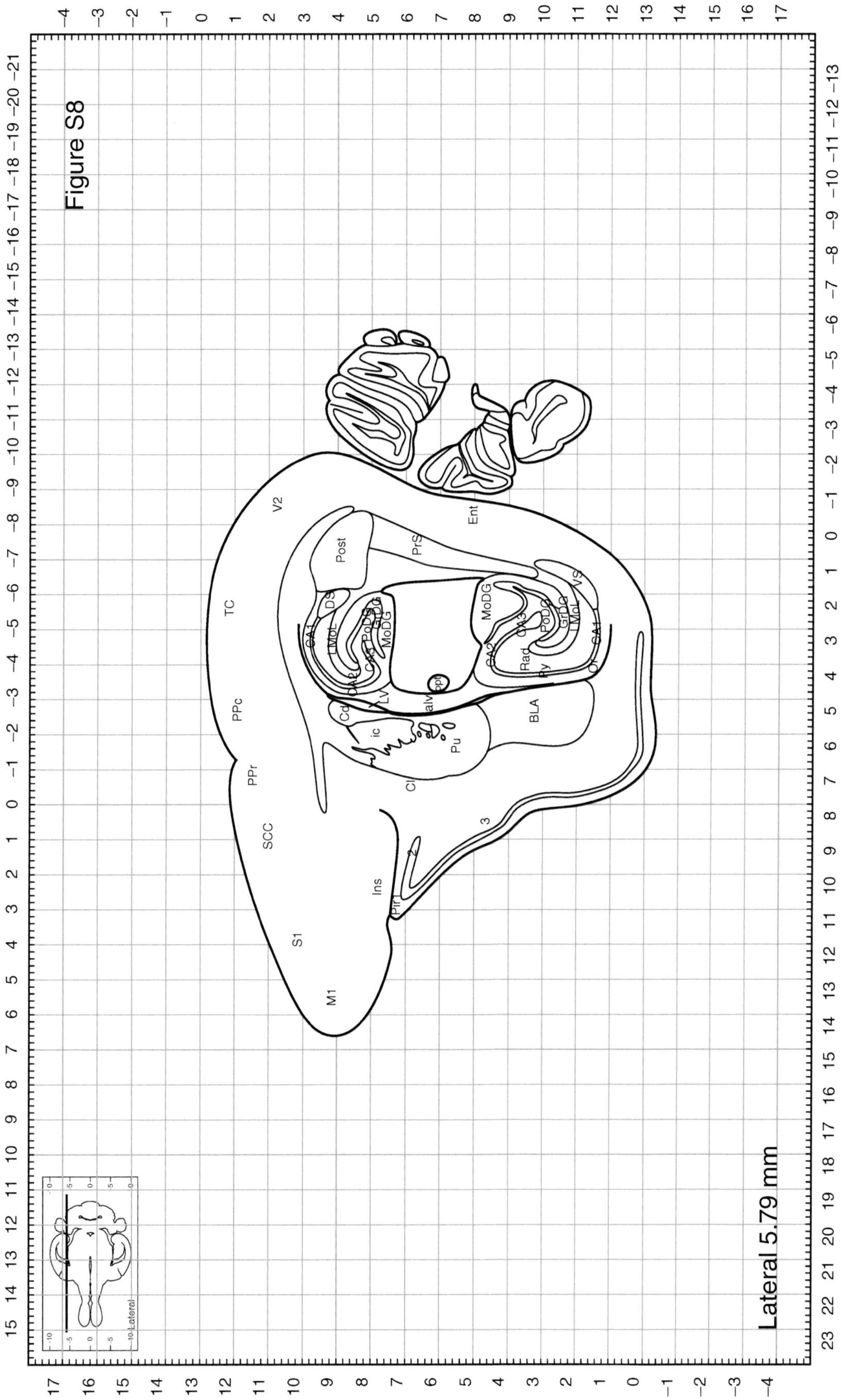
Figure S8
Lateral 5.79 mm
V2
TC
PPc
PPr
SCC
S1
M1
Ins
Pir1
Cl
Ent
PrS
Post
DS
CA1
LMoL
PoDG
GrDG
MoDG
CA3
CA2
LV
Cd
ic
alv
opt
Pu
MoDG
CA3
CA2
Rad
Py
Or
PoDG
GrDG
LMoL
CA1
VS
BLA
2
3

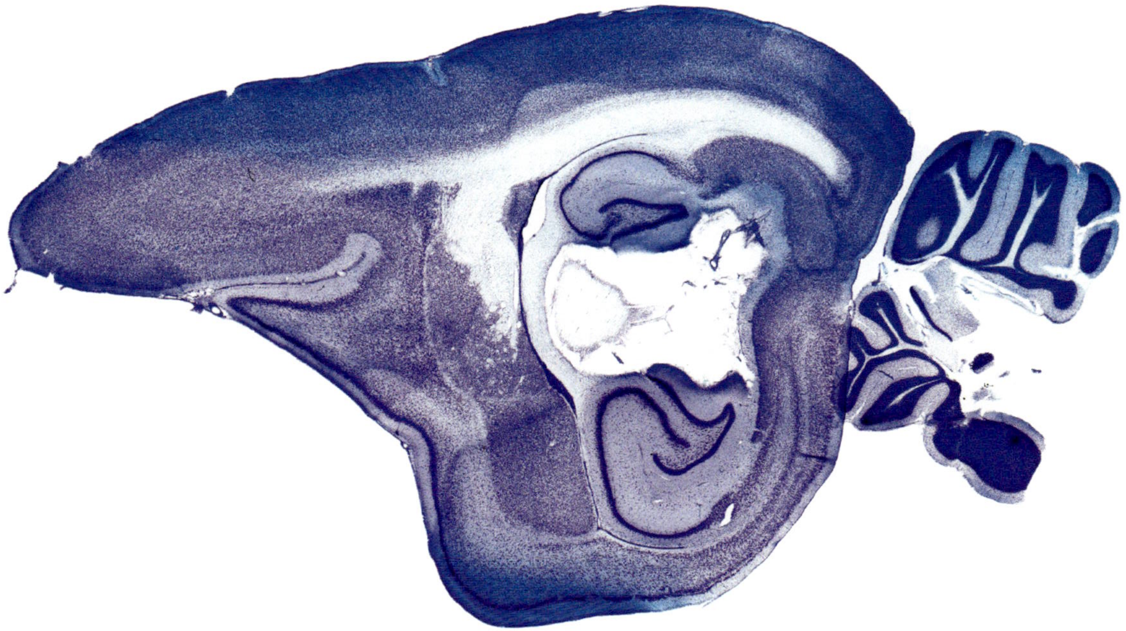

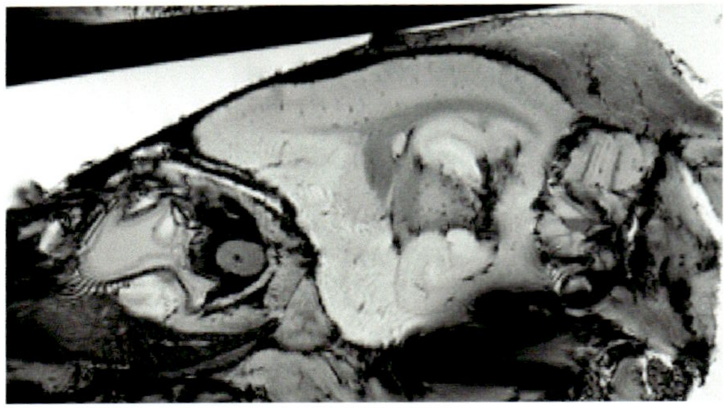

2	layer 2 of cortex	ic	internal capsule	PPc	posterior parietal caudal area
3	layer 3 of cortex	Ins	insular cortex	PPr	posterior parietal rostral area
alv	alveus of the hippocampus	LMoL	lacunosum moleculare layer of the hippocampus	PrS	presubiculum
BLA	basolateral amygdaloid nucleus	lo	lateral olfactory tract	Pu	putamen
CA1	field CA1 of the hippocampus	LV	lateral ventricle	Py	pyramidal cell layer of the hippocampus
CA2	field CA2 of the hippocampus	MoDG	molecular layer of the dentate gyrus	Rad	radiatum layer of the hippocampus
CA3	field CA2 of the hippocampus	M1	primary motor cortex	S1	primary somatosensory cortex
Cd	caudate nucleus	opt	optic tract	SCC	somatosensory caudal cortex
Cl	claustrum	Or	oriens layer of the hippocampus	TC	temporal cortex
DS	dorsal subiculum	Pir1	piriform cortex, layer 1	V2	secondary visual cortex
Ent	entorhinal cortex	PoDG	polymorph layer of the dentate gyrus	VS	ventral subiculum
GrDG	granular layer of the dentate gyrus	Post	postsubiculum	Zo	zonal layer of the superior colliculus

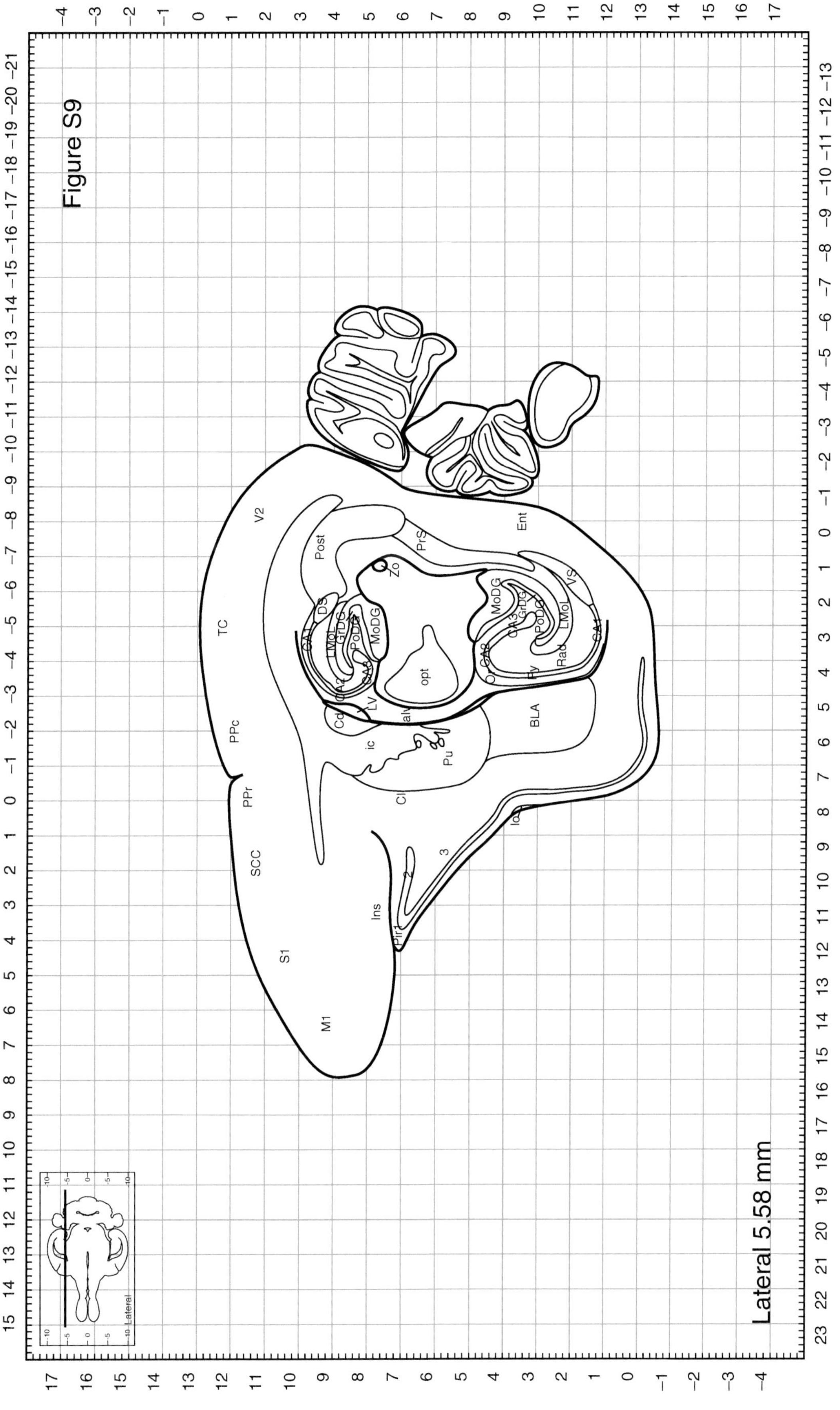
Figure S9
Lateral 5.58 mm
V2
TC
PPc
PPr
SCC
S1
M1
Ins
Pir1
Post
PrS
Ent
Zo
DS
CA1
LMol
GrDG
PoDG
MoDG
CA3
CA2
Cd
LV
alv
ic
Pu
Cl
opt
Or
Rad
BLA
VS
io
2
3

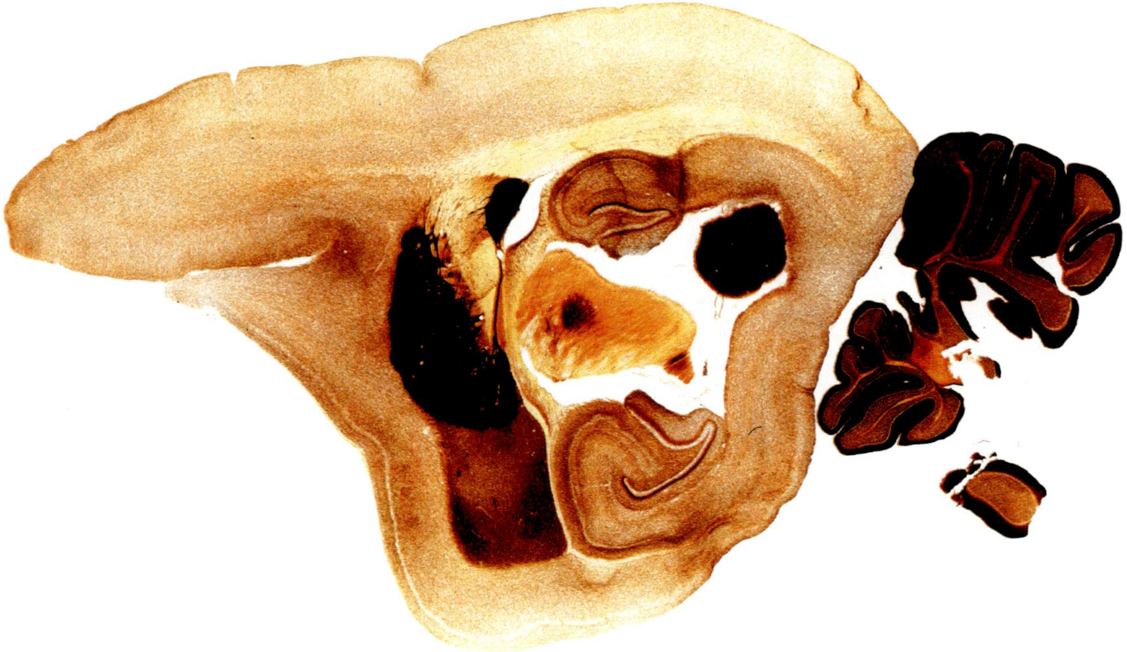

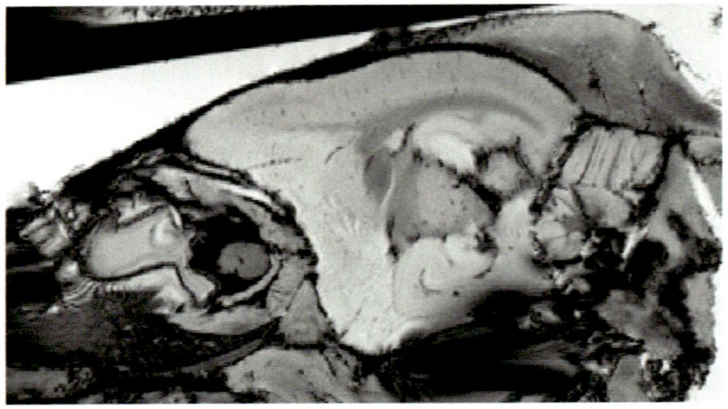

2	layer 2 of cortex
3	layer 3 of cortex
alv	alveus of the hippocampus
BLA	basolateral amygdaloid nucleus
CA1	field CA1 of the hippocampus
CA2	field CA2 of the hippocampus
CA3	field CA2 of the hippocampus
Cd	caudate nucleus
Cl	claustrum
DS	dorsal subiculum
Ent	entorhinal cortex
fi	fimbria of the hippocampus
GrDG	granular layer of the dentate gyrus
ic	internal capsule
Ins	insular cortex
LMoL	lacunosum moleculare layer of the hippocampus
lo	lateral olfactory tract
LV	lateral ventricle
MoDG	molecular layer of the dentate gyrus
M1	primary motor cortex
opt	optic tract
Or	oriens layer of the hippocampus
Pir1	piriform cortex, layer 1
PoDG	polymorph layer of the dentate gyrus
Post	postsubiculum
PPc	posterior parietal caudal area
PPr	posterior parietal rostral area
PrS	presubiculum
Pu	putamen
Py	pyramidal cell layer of the hippocampus
Rad	radiatum layer of the hippocampus
S1	primary somatosensory cortex
SCC	somatosensory caudal cortex
st	stria terminalis
SuG	superficial gray layer of the superior colliculus
TC	temporal cortex
V1	primary visual cortex
V2	secondary visual cortex
VLG	ventral lateral geniculate nucleus
VS	ventral subiculum
Zo	zonal layer of the superior colliculus

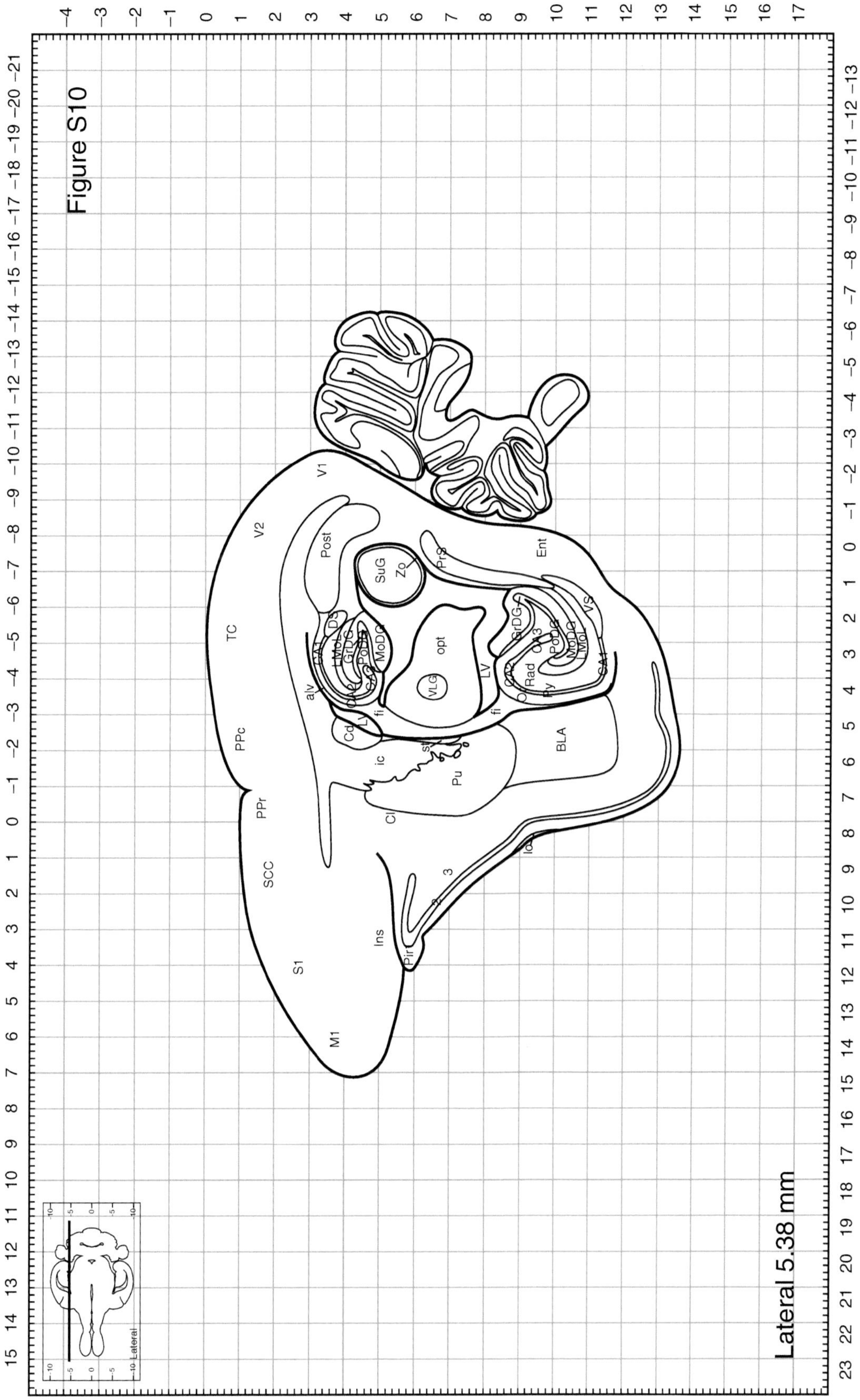

Figure S10
Lateral 5.38 mm
V1
V2
Post
SuG
Zo
PrS
Ent
TC
opt
VLG
LV
alv
fi
Cd
ic
st
Pu
BLA
PPc
PPr
Cl
SCC
S1
Ins
Pir
M1
DS
LMol
GrDG
PoDG
MoDG
CA1
CA2
CA3
Rad
Py
VS
Lateral

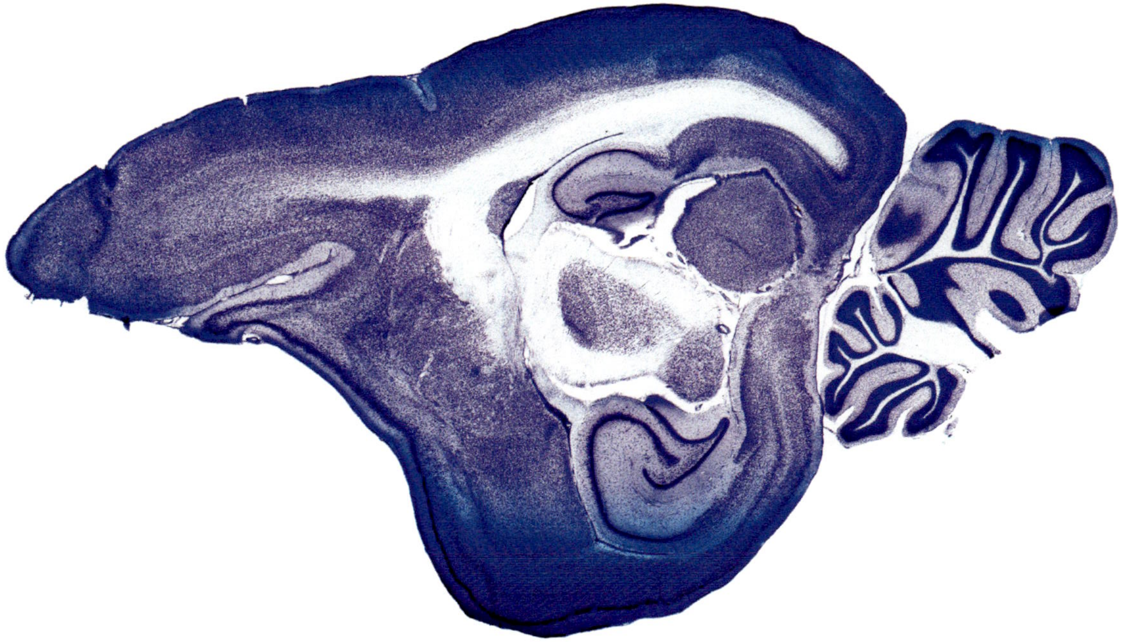

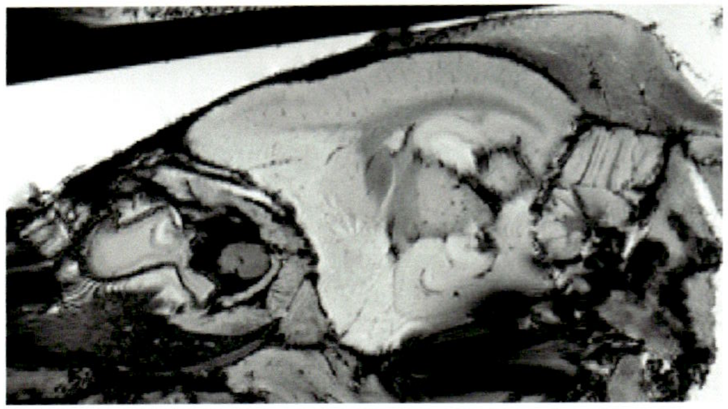

2	layer 2 of cortex
3	layer 3 of cortex
alv	alveus of the hippocampus
BLA	basolateral amygdaloid nucleus
CA1	field CA1 of the hippocampus
CA2	field CA2 of the hippocampus
CA3	field CA2 of the hippocampus
Cd	caudate nucleus
DLG	dorsal lateral geniculate nucleus
DS	dorsal subiculum
Ent	entorhinal cortex
fi	fimbria of the hippocampus
GrDG	granular layer of the dentate gyrus
ic	internal capsule
Ins	insular cortex
LMoL	lacunosum moleculare layer of the hippocampus
lo	lateral olfactory tract
LV	lateral ventricle
M1	primary motor cortex
MGD	medial geniculate nucleus, dorsal part
MGV	medial geniculate nucleus, ventral part
MoDG	molecular layer of the dentate gyrus
opt	optic tract
Or	oriens layer of the hippocampus
Pc	central nucleus of the pulvinar
Pd	dorsal nucleus of the pulvinar
Pir1	piriform cortex, layer 1
PoDG	polymorph layer of the dentate gyrus
Post	postsubiculum
PPc	posterior parietal caudal area
PPr	posterior parietal rostral area
PrS	presubiculum
Pu	putamen
Py	pyramidal cell layer of the hippocampus
Rad	radiatum layer of the hippocampus
S1	primary somatosensory cortex
SCC	somatosensory caudal cortex
st	stria terminalis
SuG	superficial gray layer of the superior colliculus
TC	temporal cortex
V1	primary visual cortex
V2	secondary visual cortex
VLG	ventral lateral geniculate nucleus
VS	ventral subiculum
Zo	zonal layer of the superior colliculus

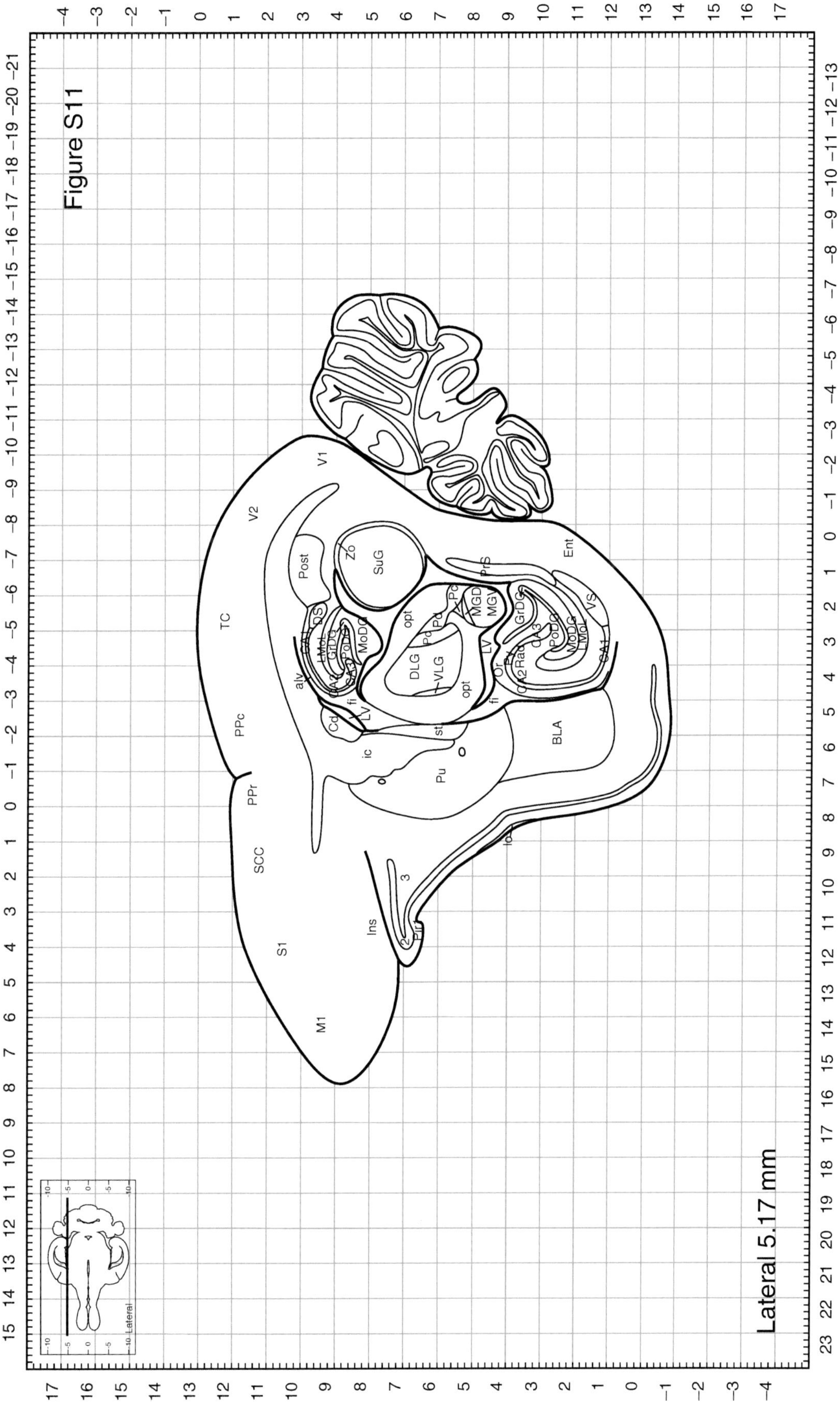

Figure S11
Lateral 5.17 mm
V1
V2
TC
PPc
PPr
SCC
S1
M1
Ins
Post
Zo
SuG
PrS
Ent
VS
DS
CA1
LMol
GrDG
PoDG
MoDG
CA3
CA2
Rad
Or
Py
alv
fi
LV
Cd
ic
st
Pu
DLG
VLG
opt
Pd
Pc
MGD
MGV
BLA
lo
Pir
3
2
Lateral

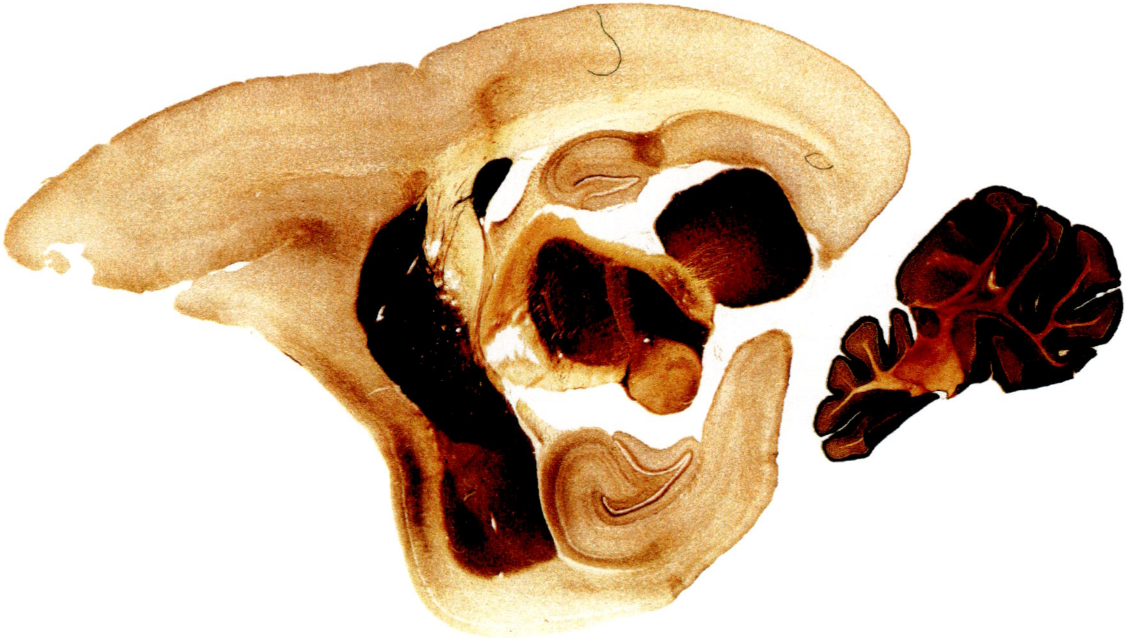

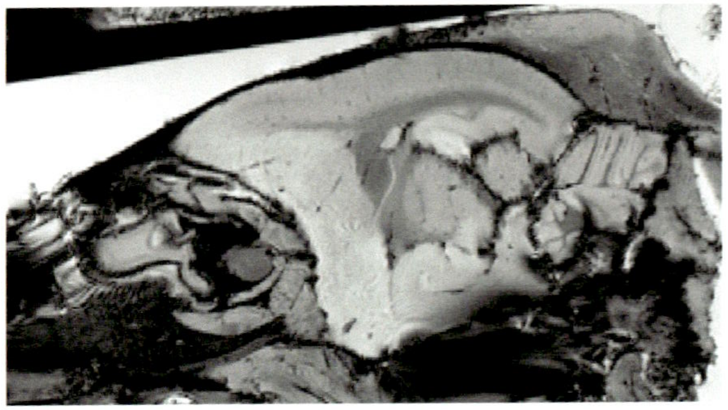

2	layer 2 of cortex
3	layer 3 of cortex
alv	alveus of the hippocampus
BLA	basolateral amygdaloid nucleus
BMA	basomedial amygdaloid nucleus
CA1	field CA1 of the hippocampus
CA2	field CA2 of the hippocampus
CA3	field CA2 of the hippocampus
Cd	caudate nucleus
DLG	dorsal lateral geniculate nucleus
DS	dorsal subiculum
EGP	external globus pallidus
Ent	entorhinal cortex
fi	fimbria of the hippocampus
GrDG	granular layer of the dentate gyrus
ic	internal capsule
IGL	intergeniculate leaf
Ins	insular cortex
LMoL	lacunosum moleculare layer of the hippocampus
lo	lateral olfactory tract
LV	lateral ventricle
M1	primary motor cortex
M2	secondary motor cortex
MGD	medial geniculate nucleus, dorsal part
MGV	medial geniculate nucleus, ventral part
MoDG	molecular layer of the dentate gyrus
Op	optic nerve layer of the superior colliculus
opt	optic tract
Or	oriens layer of the hippocampus
Pc	central nucleus of the pulvinar
Pd	dorsal nucleus of the pulvinar
Pir1	piriform cortex, layer 1
PoDG	polymorph layer of the dentate gyrus
Post	postsubiculum
PPc	posterior parietal caudal area
PPr	posterior parietal rostral area
PrS	presubiculum
Pu	putamen
Py	pyramidal cell layer of the hippocampus
Rad	radiatum layer of the hippocampus
S1	primary somatosensory cortex
SCC	somatosensory caudal cortex
st	stria terminalis
SuG	superficial gray layer of the superior colliculus
TC	temporal cortex
V1	primary visual cortex
V2	secondary visual cortex
VLG	ventral lateral geniculate nucleus
VS	ventral subiculum
Zo	zonal layer of the superior colliculus

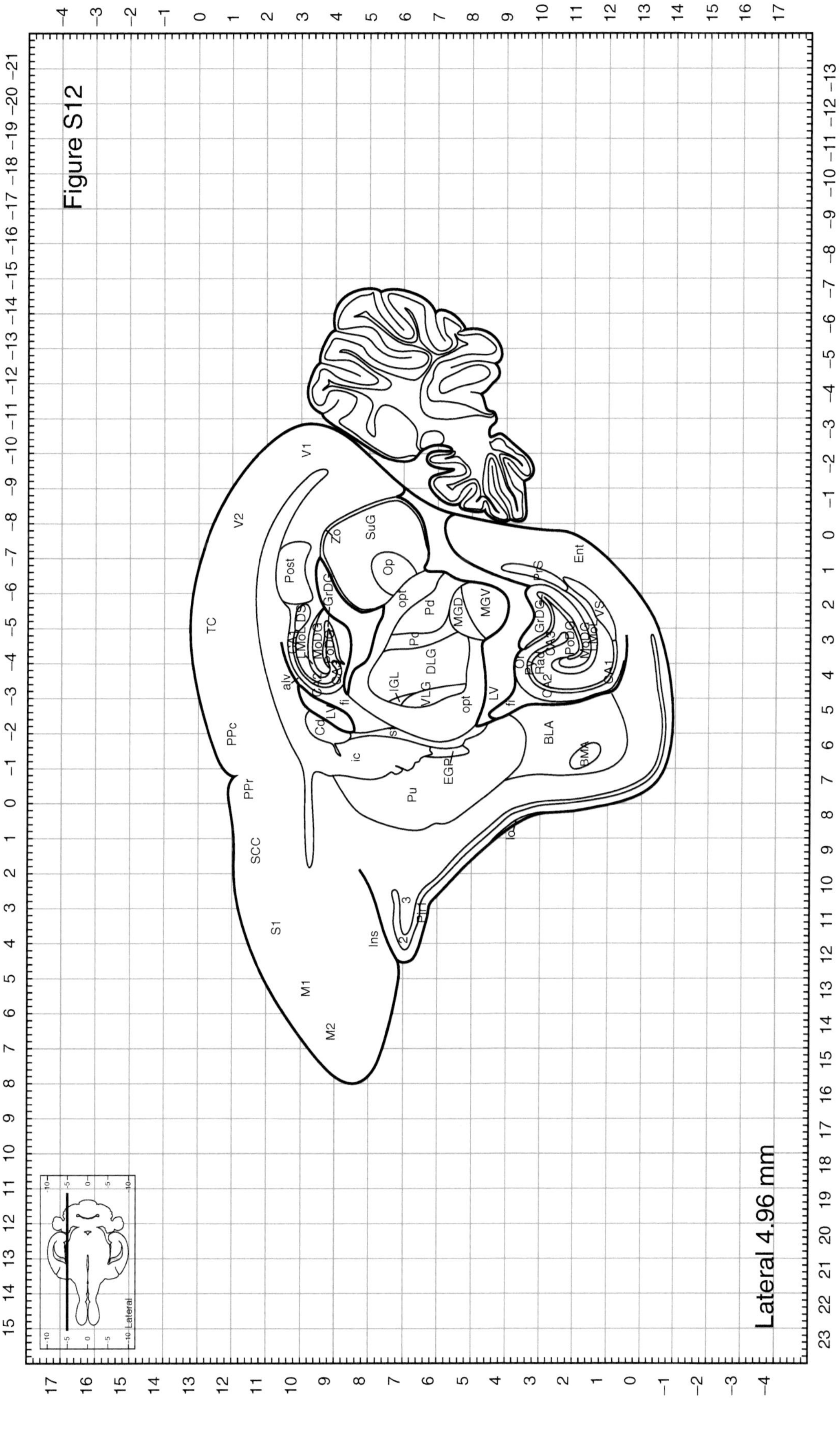
Figure S12
Lateral 4.96 mm
V1
V2
TC
PPc
PPr
SCC
S1
M1
M2
Ins
Post
SuG
Zo
Op
opt
Pd
MGD
MGV
Po
DLG
IGL
VLG
LV
fi
Cd
alv
ic
st
Pu
EGP
Ent
PaS
GrDG
VS
CA1
BLA
BMA
LV
fi
lo
Pir
Lateral

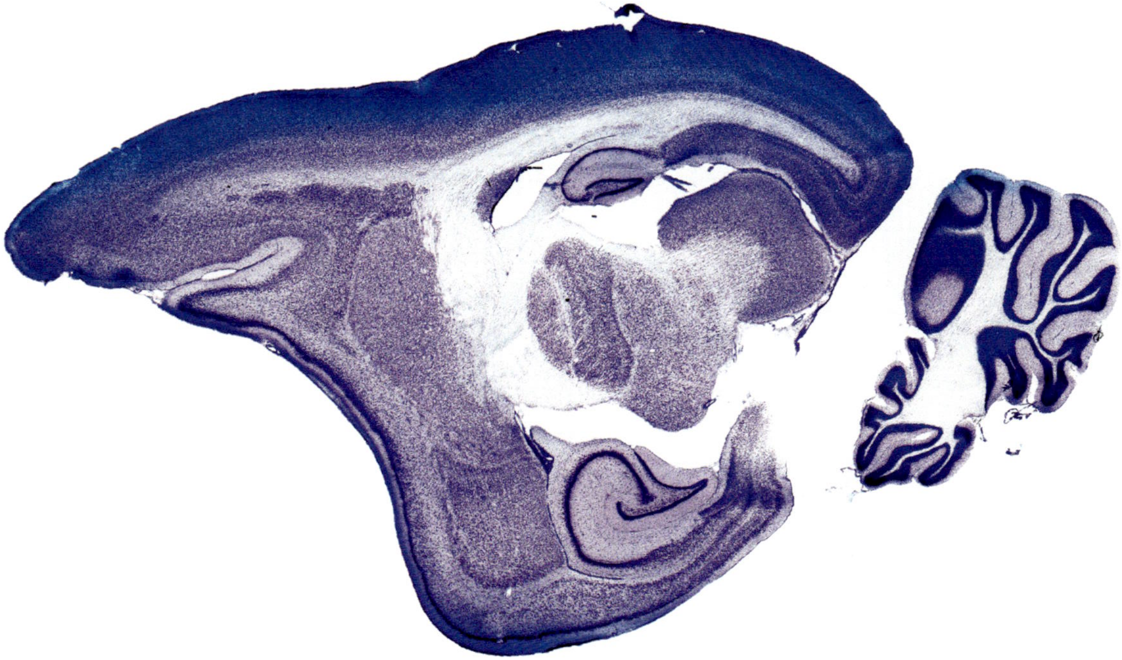

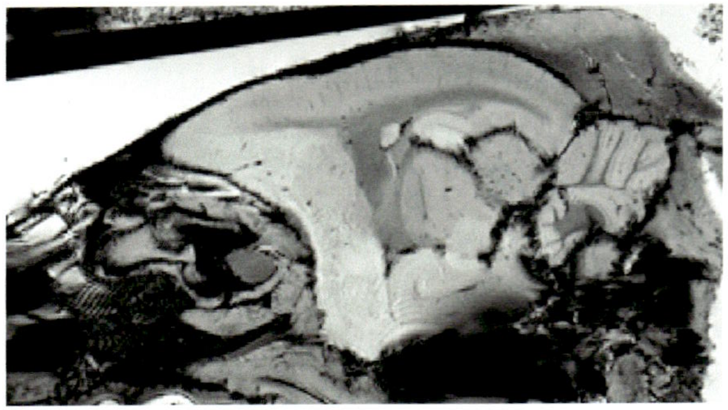

2 layer 2 of cortex
3 layer 3 of cortex
acp anterior commissure, posterior part
alv alveus of the hippocampus
B brachium of the superior colliculus
BLA basolateral amygdaloid nucleus
BMA basomedial amygdaloid nucleus
BN basal nucleus (Meynert)
CA1 field CA1 of the hippocampus
CA2 field CA2 of the hippocampus
CA3 field CA2 of the hippocampus
Cd caudate nucleus
DLG dorsal lateral geniculate nucleus
DS dorsal subiculum
EGP external globus pallidus
Ent entorhinal cortex
fi fimbria of the hippocampus
GrDG granular layer of the dentate gyrus
ic internal capsule
IGL intergeniculate leaf
InG intermediate gray layer of the superior colliculus
Ins insular cortex
IPAC interstitial nucleus of the posterior limb of the anterior commissure
LMoL lacunosum moleculare layer of the hippocampus
lo lateral olfactory tract
LSS lateral stripe of the striatum
LV lateral ventricle
M1 primary motor cortex
M2 secondary motor cortex
MGD medial geniculate nucleus, dorsal part
MGV medial geniculate nucleus, ventral part
MoDG molecular layer of the dentate gyrus
Op optic nerve layer of the superior colliculus
opt optic tract
Or oriens layer of the hippocampus
Pc central nucleus of the pulvinar
Pd dorsal nucleus of the pulvinar
Pir1 piriform cortex, layer 1
PoDG polymorph layer of the dentate gyrus
Post postsubiculum
PPc posterior parietal caudal area
PPr posterior parietal rostral area
PrS presubiculum
Pu putamen
Py pyramidal cell layer of the hippocampus
Rad radiatum layer of the hippocampus
S1 primary somatosensory cortex
SCC somatosensory caudal cortex
st stria terminalis
SuG superficial gray layer of the superior colliculus
TC temporal cortex
V1 primary visual cortex
V2 secondary visual cortex
VLG ventral lateral geniculate nucleus
VS ventral subiculum
Zo zonal layer of the superior colliculus

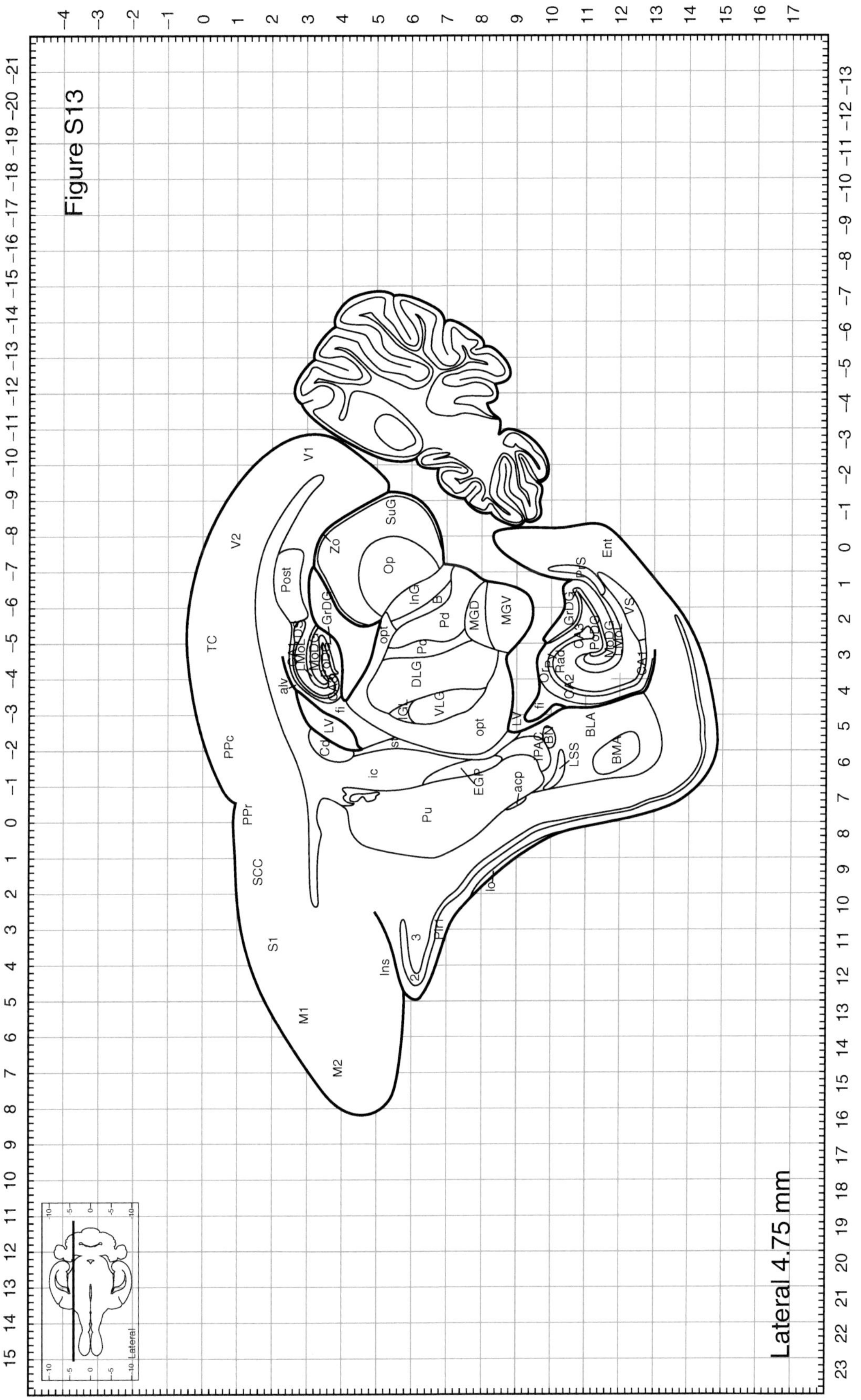
Figure S13
Lateral 4.75 mm
V1
V2
TC
PPc
PPr
SCC
S1
M1
M2
Ins
Post
alv
fi
LV
Cd
st
ic
Pu
EGP
acp
SuG
Zo
Op
InG
opt
Pd
Po
DLG
VLG
IGL
MGD
MGV
Ent
VS
CA1
BLA
BMA
LSS
IPAC
lo
Pir
Lateral

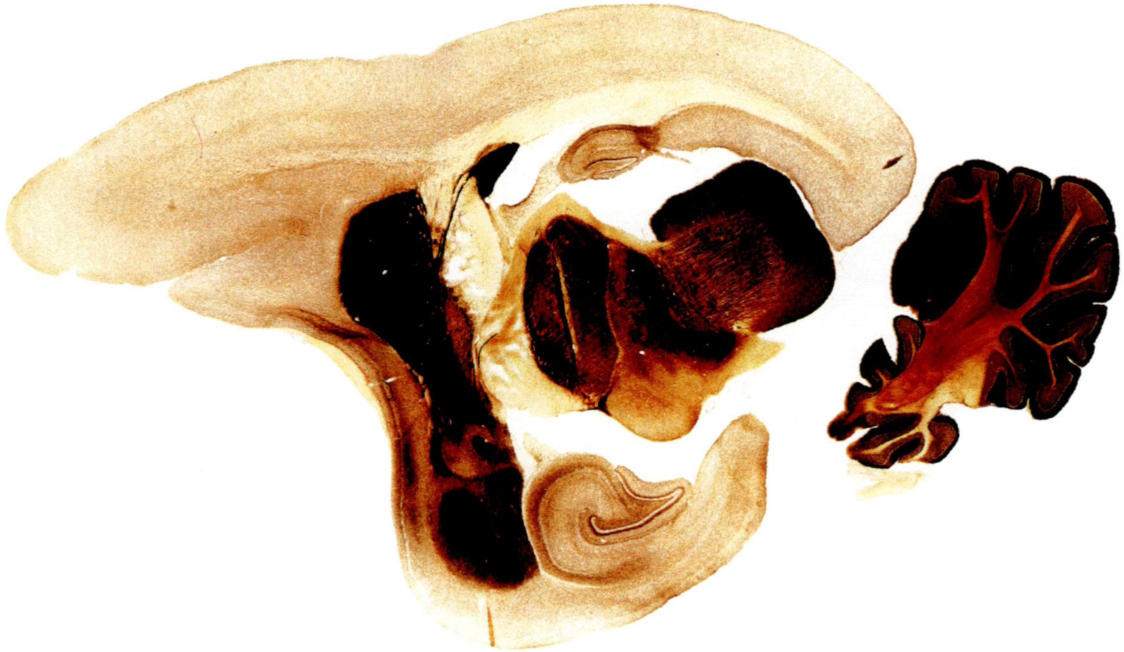

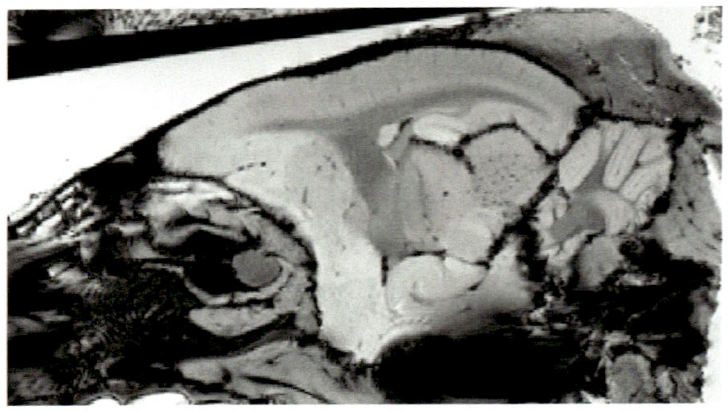

2	layer 2 of cortex
3	layer 3 of cortex
acp	anterior commissure, posterior part
alv	alveus of the hippocampus
B	brachium of the superior colliculus
BLA	basolateral amygdaloid nucleus
BMA	basomedial amygdaloid nucleus
BN	basal nucleus (Meynert)
CA1	field CA1 of the hippocampus
CA2	field CA2 of the hippocampus
CA3	field CA2 of the hippocampus
Cd	caudate nucleus
CeA	central amygdaloid nucleus
DLG	dorsal lateral geniculate nucleus
DS	dorsal subiculum
EGP	external globus pallidus
Ent	entorhinal cortex
fi	fimbria of the hippocampus
GrDG	granular layer of the dentate gyrus
ic	internal capsule
IGL	intergeniculate leaf
IM	intercalated amygdaloid nucleus, main part
InG	intermediate gray layer of the superior colliculus
Ins	insular cortex
IPAC	interstitial nucleus of the posterior limb of the anterior commissure
LMoL	lacunosum moleculare layer of the hippocampus
lo lateral	olfactory tract
LSS	lateral stripe of the striatum
LV	lateral ventricle
M1	primary motor cortex
M2	secondary motor cortex
mcp	middle cerebellar peduncle
MGD	medial geniculate nucleus, dorsal part
MGV	medial geniculate nucleus, ventral part
MoDG	molecular layer of the dentate gyrus
OFC	orbital frontal cortex
ONL	olfactory nerve layer
Op	optic nerve layer of the superior colliculus
opt	optic tract
Or	oriens layer of the hippocampus
Pc	central nucleus of the pulvinar
Pd	dorsal nucleus of the pulvinar
Pir1	piriform cortex, layer 1
PoDG	polymorph layer of the dentate gyrus
Post	postsubiculum
PPc	posterior parietal caudal area
PPr	posterior parietal rostral area
PrS	presubiculum
Pu	putamen
Py	pyramidal cell layer of the hippocampus
Rad	radiatum layer of the hippocampus
S1	primary somatosensory cortex
SCC	somatosensory caudal cortex
st	stria terminalis
str	superior thalamic radiation
SuG	superficial gray layer of the superior colliculus
TC	temporal cortex
V1	primary visual cortex
V2	secondary visual cortex
VLG	ventral lateral geniculate nucleus
VS	ventral subiculum
Zo	zonal layer of the superior colliculus

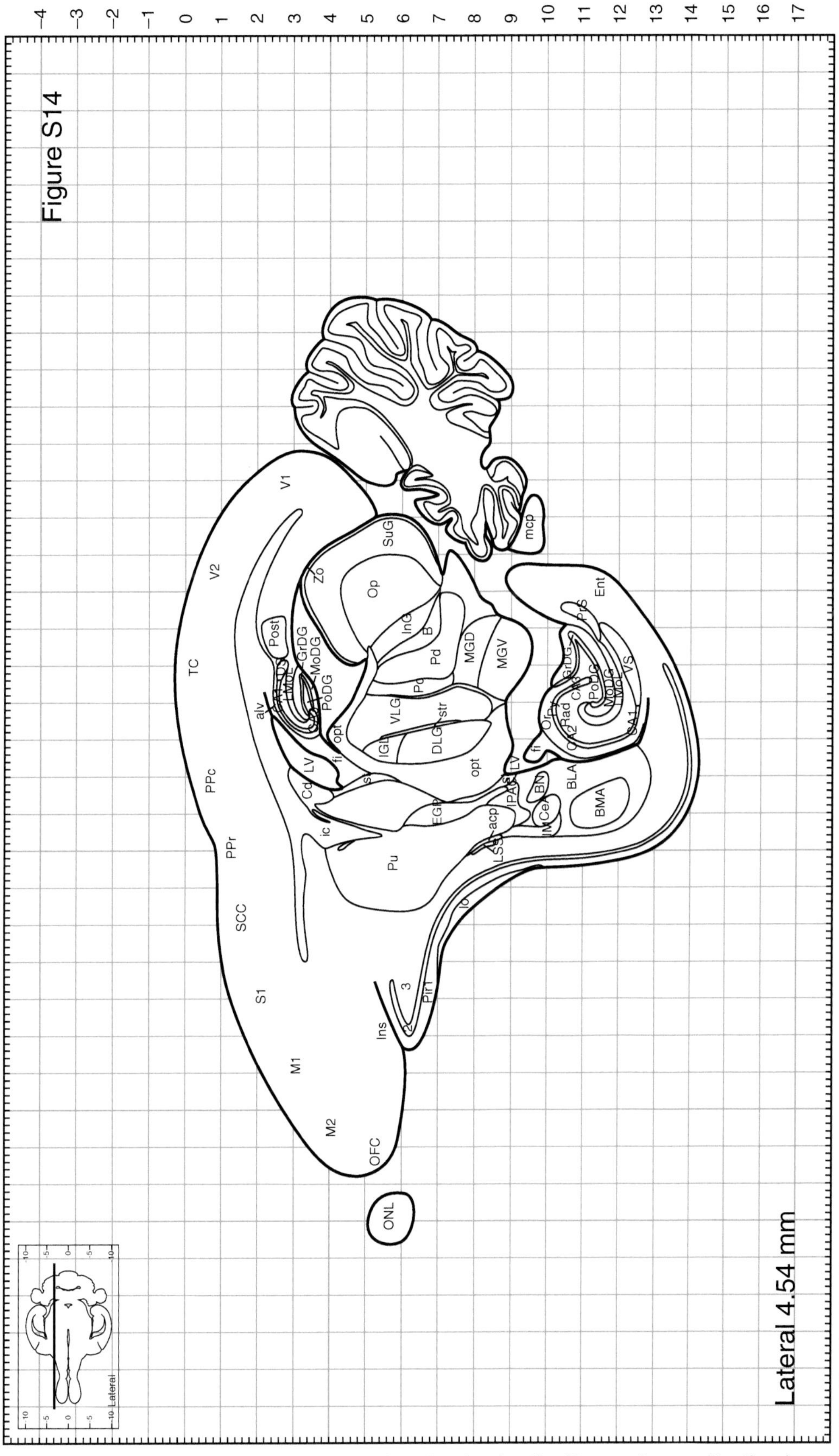
Figure S14
Lateral 4.54 mm
V1
V2
TC
PPc
PPr
SCC
S1
M1
M2
OFC
ONL
Ins
Pu
SuG
Zo
Op
InG
Post
GrDG
MoDG
PoDG
alv
opt
fi
LV
Cd
ic
IGL
VLG
DLG
Pd
MGD
MGV
EGP
acp
LSS
BLA
BMA
BN
IMCe
Ent
PrS
VS
CA1
Or
Rad
mcp
PirT

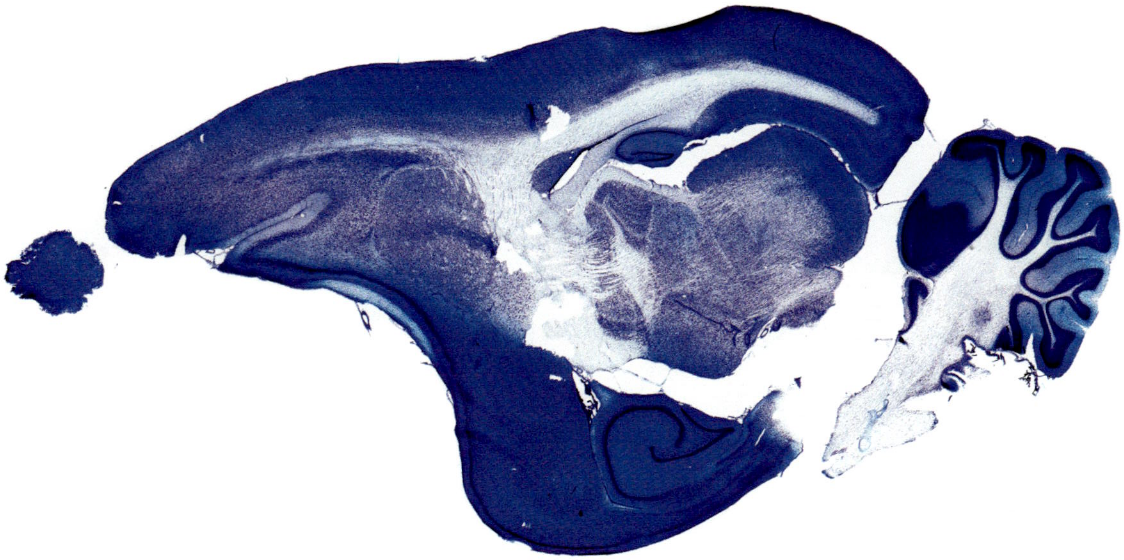

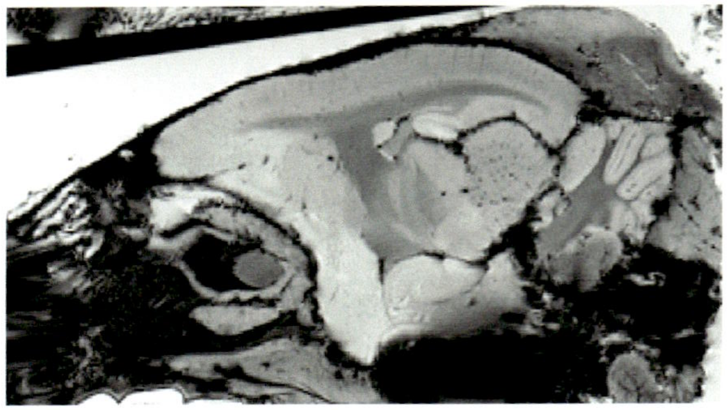

2	layer 2 of cortex
3	layer 3 of cortex
AB	accessory basal nucleus
acp	anterior commissure, posterior part
alv	alveus of the hippocampus
B	brachium of the superior colliculus
BLA	basolateral amygdaloid nucleus
BMA	basomedial amygdaloid nucleus
BN	basal nucleus (Meynert)
CA1	field CA1 of the hippocampus
CA2	field CA2 of the hippocampus
CA3	field CA2 of the hippocampus
Cd	caudate nucleus
CeA	central amygdaloid nucleus
DLG	dorsal lateral geniculate nucleus
DS	dorsal subiculum
EA	sublenticular extended amygdala
EGP	external globus pallidus
Ent	entorhinal cortex
fi	fimbria of the hippocampus
GrDG	granular layer of the dentate gyrus
ic	internal capsule
IGL	intergeniculate leaf
IM	intercalated amygdaloid nucleus, main part
InG	intermediate gray layer of the superior colliculus
Ins	insular cortex
InWh	intermediate white layer of the superior colliculus
IPAC	interstitial nucleus of the posterior limb of the anterior commissure
LMoL	lacunosum moleculare layer of the hippocampus
lo	lateral olfactory tract
LSS	lateral stripe of the striatum
LV	lateral ventricle
M1	primary motor cortex
M2	secondary motor cortex
mcp	middle cerebellar peduncle
MGD	medial geniculate nucleus, dorsal part
MGV	medial geniculate nucleus, ventral part
MoDG	molecular layer of the dentate gyrus
OFC	orbital frontal cortex
ONL	olfactory nerve layer
Op	optic nerve layer of the superior colliculus
opt	optic tract
Or	oriens layer of the hippocampus
PBG	parabigeminal nucleus
Pc	central nucleus of the pulvinar
Pd	dorsal nucleus of the pulvinar
Pir1	piriform cortex, layer 1
PoDG	polymorph layer of the dentate gyrus
Post	postsubiculum
PPc	posterior parietal caudal area
PPr	posterior parietal rostral area
PrS	presubiculum
Pu	putamen
Pv	ventral nucleus of the pulvinar
Py	pyramidal cell layer of the hippocampus
Rad	radiatum layer of the hippocampus
Rt	reticular thalamic nucleus
S1	primary somatosensory cortex
s5	sensory root of the trigeminal nerve
SCC	somatosensory caudal cortex
SNL	substantia nigra, lateral part
st	stria terminalis
str	superior thalamic radiation
SuG	superficial gray layer of the superior colliculus
TC	temporal cortex
V1	primary visual cortex
V2	secondary visual cortex
VLG	ventral lateral geniculate nucleus
VS	ventral subiculum
Zo	zonal layer of the superior colliculus

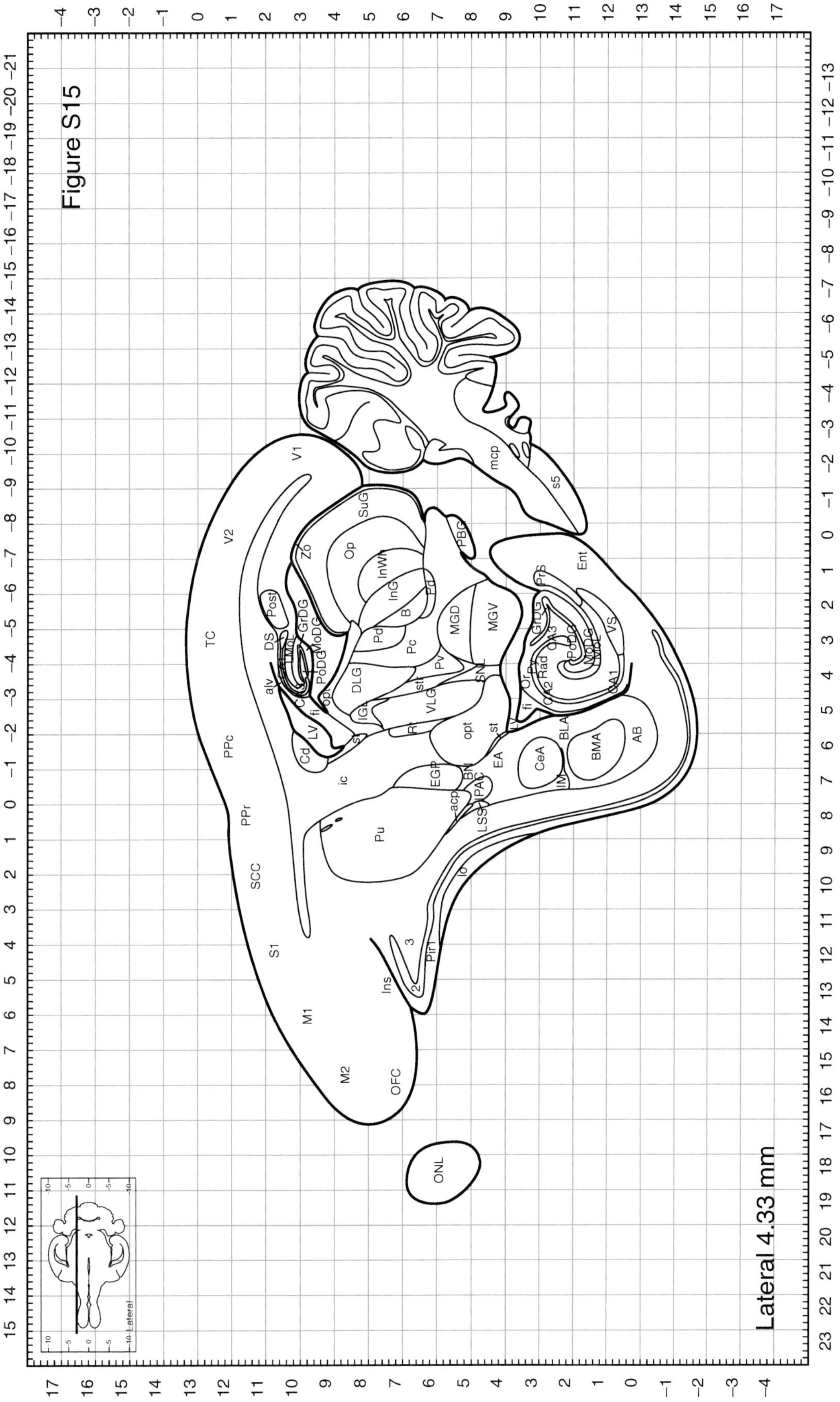
Figure S15
Lateral 4.33 mm
V1
V2
TC
PPc
PPr
SCC
S1
M1
M2
OFC
Ins
Pu
ic
Cd
LV
fi
alv
DS
Post
GrDG
MoDG
PoDG
Zo
SuG
Op
InWh
InG
Pd
B
Pc
DLG
IGL
VLG
Rt
stt
Pv
MGD
MGV
SNL
opt
st
EA
EGP
BN
IPAC
acp
LSS
lo
Pir1
ONL
PBG
mcp
s5
Ent
PrS
CA3
Rad
CA2
Or
Py
CA1
VS
BLA
CeA
IM
BMA
AB

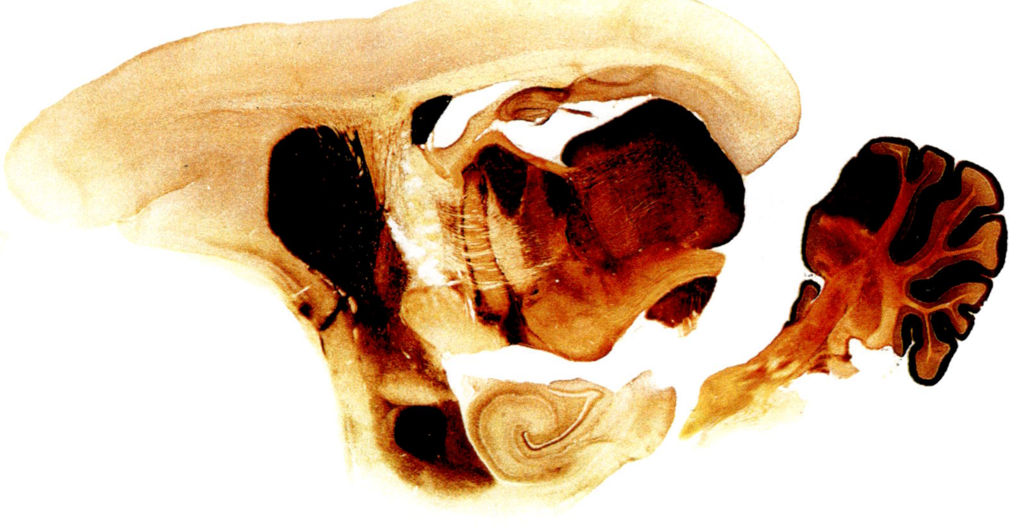

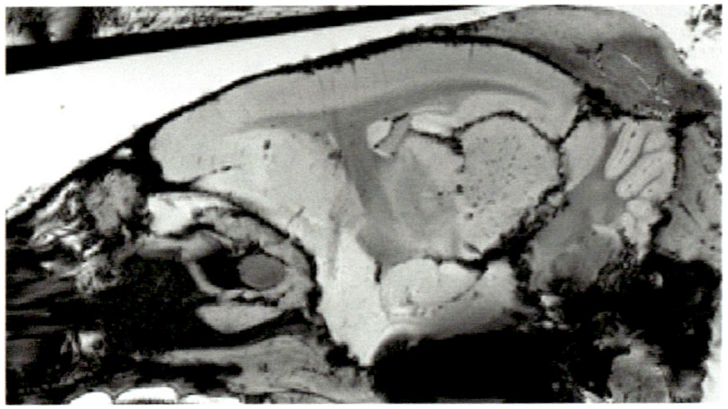

2	layer 2 of cortex
3	layer 3 of cortex
AB	accessory basal nucleus
ACo	anterior cortical amygdaloid nucleus
acp	anterior commissure, posterior part
alv	alveus of the hippocampus
B	brachium of the superior colliculus
BMA	basomedial amygdaloid nucleus
CA1	field CA1 of the hippocampus
CA2	field CA2 of the hippocampus
CA3	field CA2 of the hippocampus
Cd	caudate nucleus
CeA	central amygdaloid nucleus
DFC	dorsal frontal cortex
DLG	dorsal lateral geniculate nucleus
DS	dorsal subiculum
EA	sublenticular extended amygdala
EGP	external globus pallidus
Ent	entorhinal cortex
fi	fimbria of the hippocampus
GL	glomerular layer of the olfactory bulb
GrDG	granular layer of the dentate gyrus
ic	internal capsule
IGL	intergeniculate leaf
IM	intercalated amygdaloid nucleus, main part
InG	intermediate gray layer of the superior colliculus
Ins	insular cortex
InWh	intermediate white layer of the superior colliculus
IPAC	interstitial nucleus of the posterior limb of the anterior commissure
Lat	lateral cerebellar nucleus
LMoL	lacunosum moleculare layer of the hippocampus
lo	lateral olfactory tract
LSS	lateral stripe of the striatum
LV	lateral ventricle
M1	primary motor cortex
M2	secondary motor cortex
mcp	middle cerebellar peduncle
MGD	medial geniculate nucleus, dorsal part
MGV	medial geniculate nucleus, ventral part
MoDG	molecular layer of the dentate gyrus
OFC	orbital frontal cortex
ONL	olfactory nerve layer
Op	optic nerve layer of the superior colliculus
opt	optic tract
Or	oriens layer of the hippocampus
PBG	parabigeminal nucleus
Pc	central nucleus of the pulvinar
Pd	dorsal nucleus of the pulvinar
Pir1	piriform cortex, layer 1
PoDG	polymorph layer of the dentate gyrus
PPd	posterior parietal dorsal area
PrS	presubiculum
Pu	putamen
Pv	ventral nucleus of the pulvinar
Py	pyramidal cell layer of the hippocampus
Rad	radiatum layer of the hippocampus
Rt	reticular thalamic nucleus
S1	primary somatosensory cortex
s5	sensory root of the trigeminal nerve
SCC	somatosensory caudal cortex
SNL	substantia nigra, lateral part
st	stria terminalis
str	superior thalamic radiation
SuG	superficial gray layer of the superior colliculus
TC	temporal cortex
Tu	olfactory tubercle
V1	primary visual cortex
V2	secondary visual cortex
VLG	ventral lateral geniculate nucleus
VP	ventral pallidum
VS	ventral subiculum
Zo	zonal layer of the superior colliculus

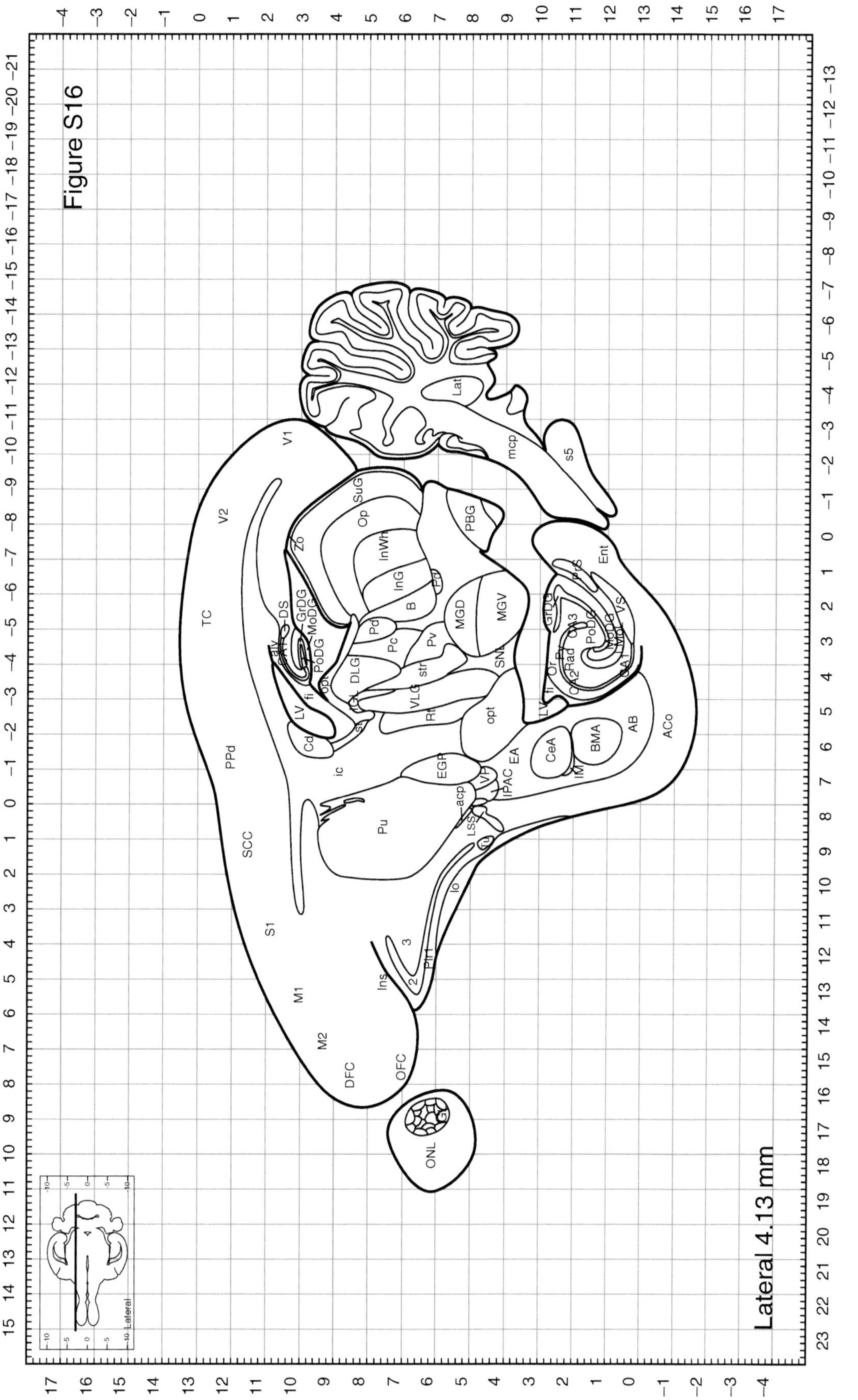
Figure S16
Lateral 4.13 mm
V1
V2
TC
PPd
SCC
S1
M1
M2
DFC
OFC
ONL
Ins
Pir1
lo
Pu
ic
Cd
LV
fi
SuG
Op
InWh
InG
Zo
DS
GrDG
MoDG
PoDG
Pd
Pc
B
DLG
VLG
Rt
str
Pv
MGD
MGV
SNL
opt
EGP
acp
VP
IPAC
EA
LSS
Tu
CeA
BMA
AB
ACo
IM
PBG
Lat
mcp
s5
Ent
PrS
VS
CA1
CA2
CA3
Or
Py
Rad
3
2

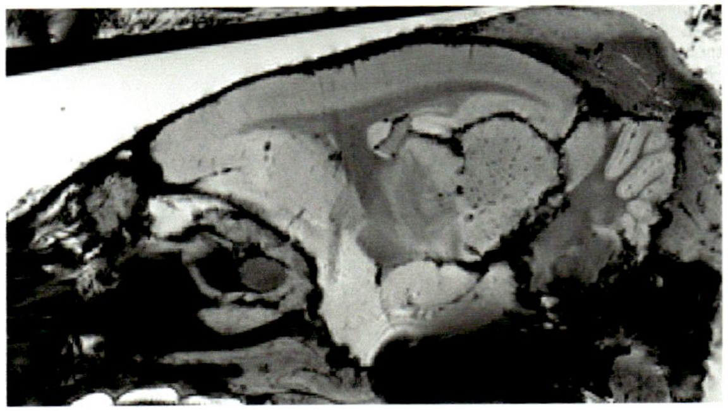

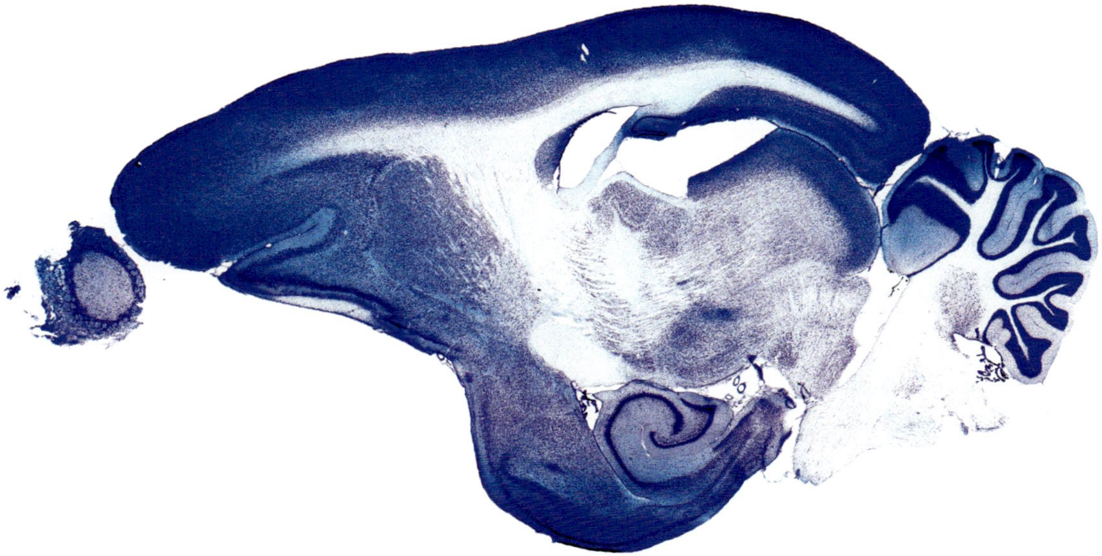

2 layer 2 of cortex
3 layer 3 of cortex
AB accessory basal nucleus
ACo anterior cortical amygdaloid nucleus
acp anterior commissure, posterior part
alv alveus of the hippocampus
B brachium of the superior colliculus
BMA basomedial amygdaloid nucleus
CA1 field CA1 of the hippocampus
CA2 field CA2 of the hippocampus
CA3 field CA2 of the hippocampus
Cd caudate nucleus
CeA central amygdaloid nucleus
DFC dorsal frontal cortex
DLG dorsal lateral geniculate nucleus
DpG deep gray layer of the superior colliculus
DpWh deep white layer of the superior colliculus
DS dorsal subiculum
EA sublenticular extended amygdala
EGP external globus pallidus
Ent entorhinal cortex
EPl external plexiform layer of the olfactory bulb
fi fimbria of the hippocampus
GL glomerular layer of the olfactory bulb
GrDG granular layer of the dentate gyrus
ic internal capsule
IGL intergeniculate leaf
InG intermediate gray layer of the superior colliculus
Ins insular cortex
InWh intermediate white layer of the superior colliculus
Lat lateral cerebellar nucleus
LMoL lacunosum moleculare layer of the hippocampus
lo lateral olfactory tract
LV lateral ventricle
M1 primary motor cortex
M2 secondary motor cortex
mcp middle cerebellar peduncle
MGD medial geniculate nucleus, dorsal part
MGV medial geniculate nucleus, ventral part
MiTg microcellular tegmental nucleus
MoDG molecular layer of the dentate gyrus
mRt mesencephalic reticular formation
OFC orbital frontal cortex
ONL olfactory nerve layer
Op optic nerve layer of the superior colliculus
opt optic tract
Or oriens layer of the hippocampus
PBG parabigeminal nucleus
Pc central nucleus of the pulvinar
Pd dorsal nucleus of the pulvinar
Pir1 piriform cortex, layer 1
Po posterior complex of the thalamus
PoDG polymorph layer of the dentate gyrus
PPd posterior parietal dorsal area
PrS presubiculum
Pu putamen
Pul pulvinar nuclei
Pv ventral nucleus of the pulvinar
Py pyramidal cell layer of the hippocampus
Rad radiatum layer of the hippocampus
Rt reticular thalamic nucleus
S1 primary somatosensory cortex
s5 sensory root of the trigeminal nerve
SCC somatosensory caudal cortex
SNL substantia nigra, lateral part
st stria terminalis
str superior thalamic radiation
SuG superficial gray layer of the superior colliculus
TC temporal cortex
Tu olfactory tubercle
Tu1 layer 1 of olfactory tubercle
V1 primary visual cortex
V2 secondary visual cortex
VLG ventral lateral geniculate nucleus
VP ventral pallidum
VS ventral subiculum
Zo zonal layer of the superior colliculus

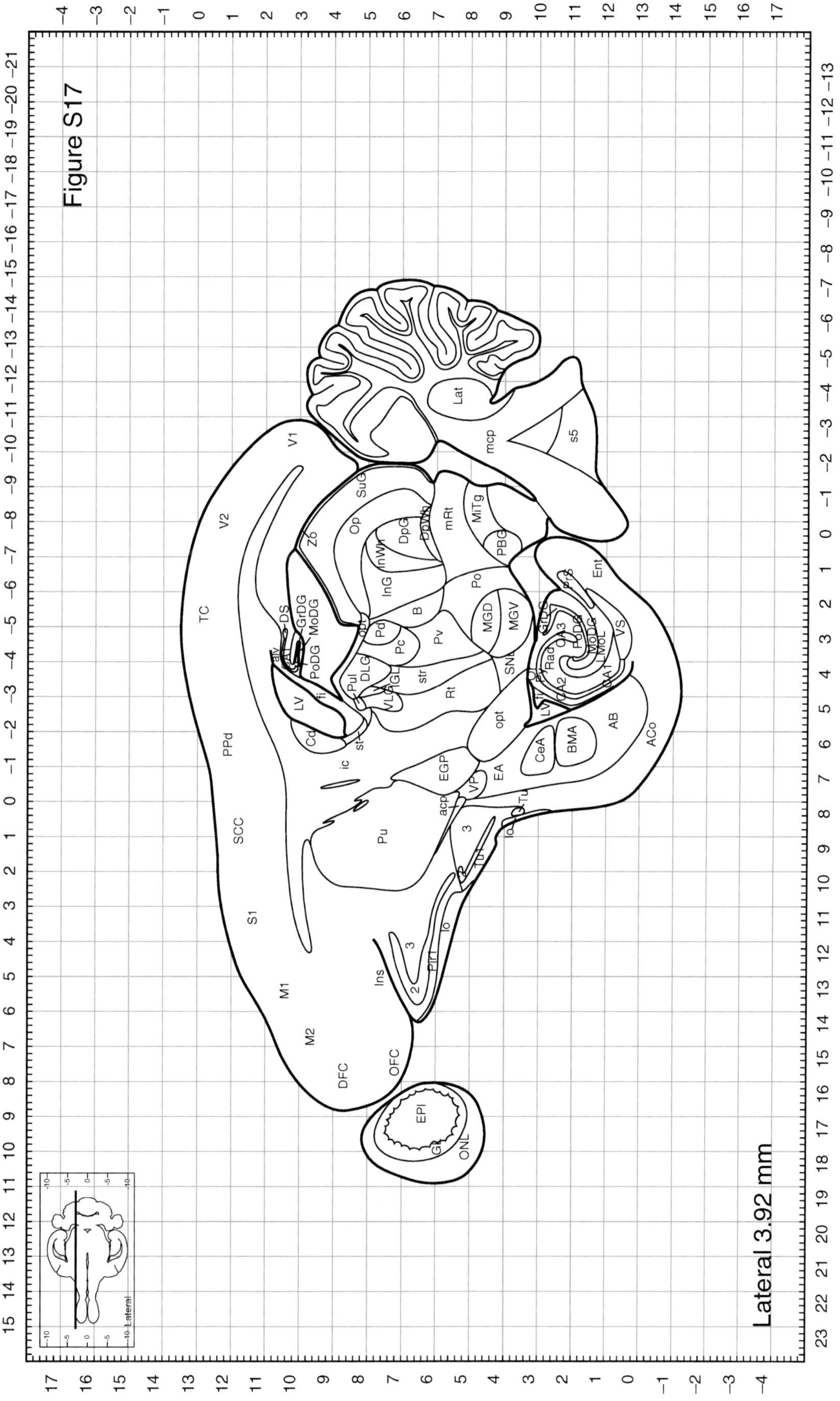
Figure S17
Lateral 3.92 mm
V1
V2
TC
PPd
SCC
S1
M1
M2
DFC
OFC
Ins
EPI
ONL
Gl
Pu
Cd
ic
LV
fi
SuG
Op
Zo
InWh
InG
DpG
DpWh
mRt
MiTg
PBG
Po
MGD
MGV
SNL
Pv
Pc
Pd
B
DLG
VLG
Pul
str
Rt
opt
EGP
VP
EA
acp
lo
Tu
Pir
Lat
mcp
s5
Ent
PrS
VS
CA1
CA2
CA3
PoDG
MoDG
GrDG
DS
LMol
Rad
Or
Py
LV
AB
ACo
BMA
CeA

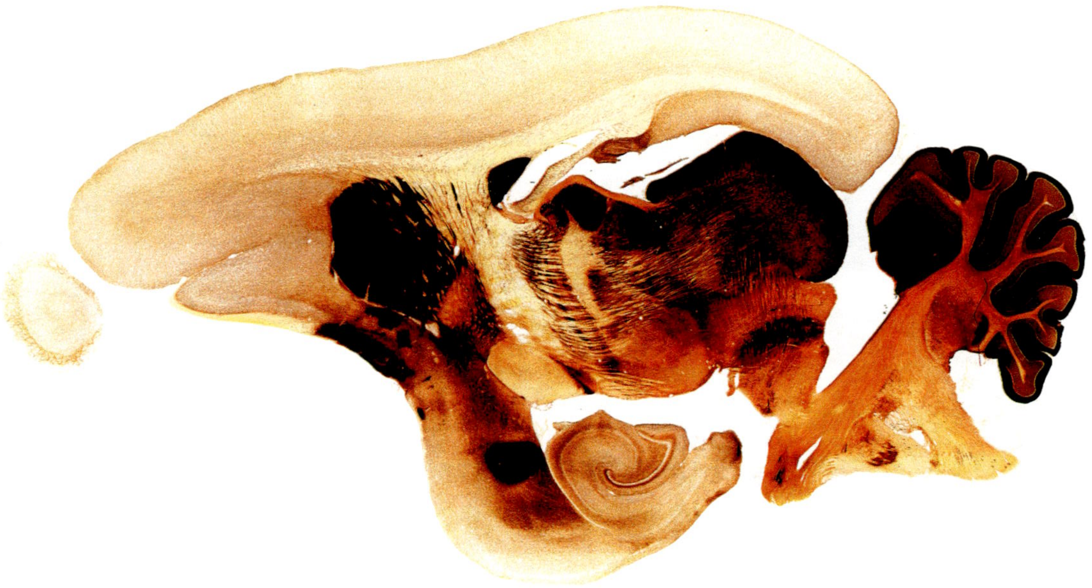

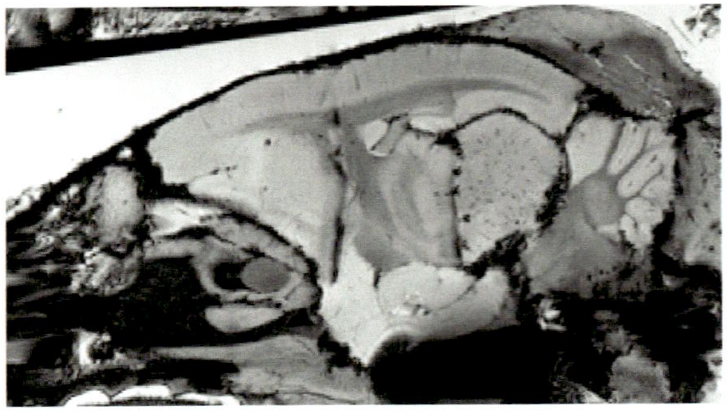

2 layer 2 of cortex
3 layer 3 of cortex
8vn vestibular root of the vestibulocochlear nerve
AB accessory basal nucleus
ACo anterior cortical amygdaloid nucleus
acp anterior commissure, posterior part
APT anterior pretectal nucleus
BMA basomedial amygdaloid nucleus
CA1 field CA1 of the hippocampus
CA2 field CA2 of the hippocampus
CA3 field CA2 of the hippocampus
Cd caudate nucleus
CeA central amygdaloid nucleus
DFC dorsal frontal cortex
DLG dorsal lateral geniculate nucleus
DpG deep gray layer of the superior colliculus
DpWh deep white layer of the superior colliculus
DS dorsal subiculum
EA sublenticular extended amygdala
EGP external globus pallidus
Ent entorhinal cortex
EPl external plexiform layer of the olfactory bulb
fi fimbria of the hippocampus
GL glomerular layer of the olfactory bulb
GrDG granular layer of the dentate gyrus
ic internal capsule
icp inferior cerebellar peduncle
InG intermediate gray layer of the superior colliculus
Int interposed cerebellar nucleus
InWh intermediate white layer of the superior colliculus
Ins insular cortex
IPl internal plexiform layer of the olfactory bulb
Lat lateral cerebellar nucleus
LMoL lacunosum moleculare layer of the hippocampus
lo lateral olfactory tract
LOT nucleus of the lateral olfactory tract
LV lateral ventricle
M1 primary motor cortex
M2 secondary motor cortex
MCL mitral cell layer of the olfactory bulb
mcp middle cerebellar peduncle
MGV medial geniculate nucleus, ventral part
MiTg microcellular tegmental nucleus
MoDG molecular layer of the dentate gyrus
mRt mesencephalic reticular formation
OFC orbital frontal cortex
ONL olfactory nerve layer
Op optic nerve layer of the superior colliculus
opt optic tract
OPT olivary pretectal nucleus
Or oriens layer of the hippocampus
Pc central nucleus of the pulvinar
Pd dorsal nucleus of the pulvinar
Pir1 piriform cortex, layer 1
Pn pontine nuclei
PnO pontine reticular nucleus, oral part
Po posterior complex of the thalamus
PoDG polymorph layer of the dentate gyrus
PPd posterior parietal dorsal area
Pr5 principal sensory trigeminal nucleus
PrS presubiculum
Pu putamen
Pul pulvinar nuclei
Pv ventral nucleus of the pulvinar
Py pyramidal cell layer of the hippocampus
Rad radiatum layer of the hippocampus
Rt reticular thalamic nucleus
S1 primary somatosensory cortex
s5 sensory root of the trigeminal nerve
SCC somatosensory caudal cortex
SNL substantia nigra, lateral part
SNR substantia nigra, reticular part
sp5 spinal trigeminal tract
st stria terminalis
str superior thalamic radiation
SuG superficial gray layer of the superior colliculus
TC temporal cortex
Tu olfactory tubercle
Tu1 layer 1 of olfactory tubercle
V1 primary visual cortex
V2 secondary visual cortex
VL ventrolateral thalamic nucleus
VP ventral pallidum
VS ventral subiculum
ZI zona incerta
Zo zonal layer of the superior colliculus

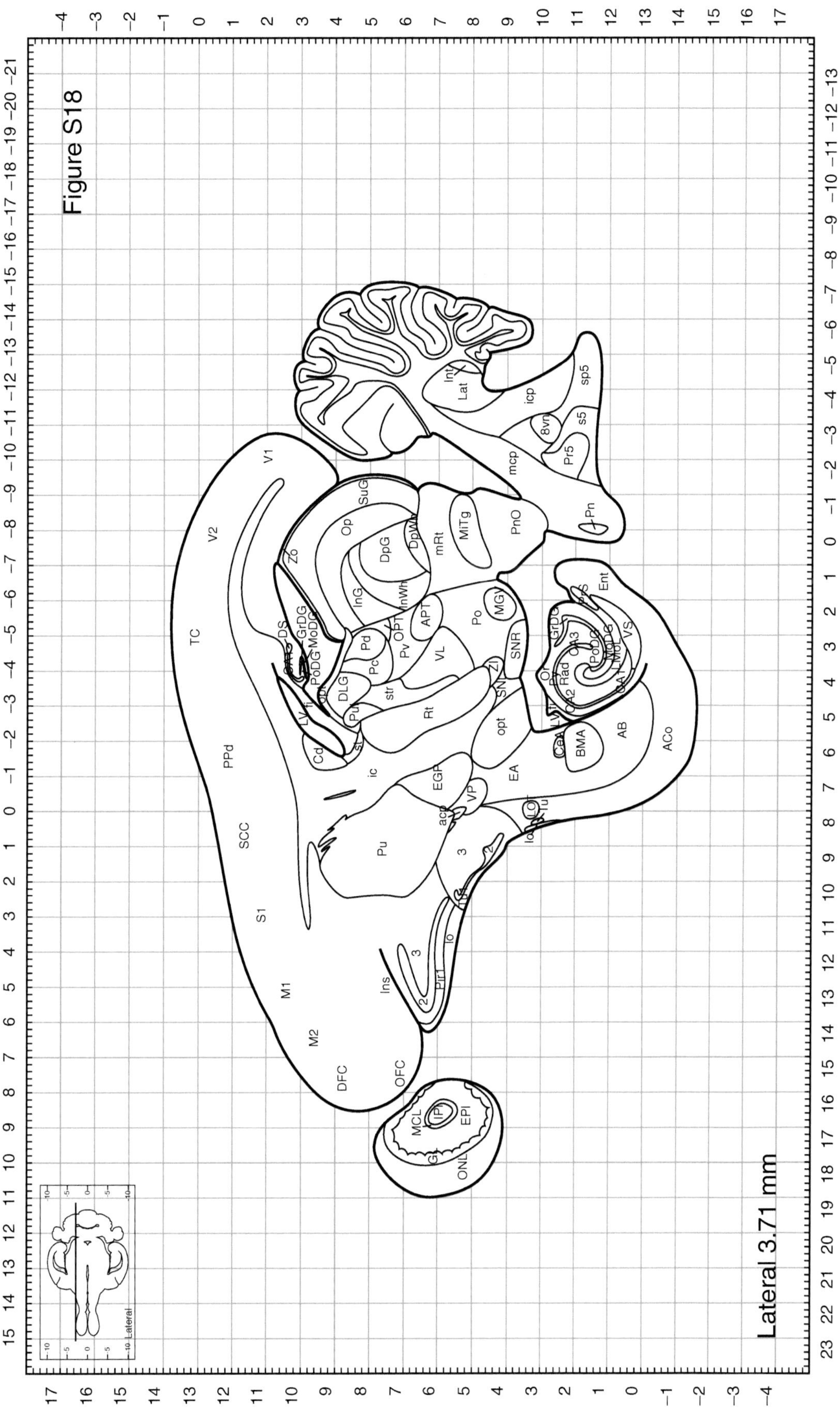
Figure S18
Lateral 3.71 mm
Lateral
V1
V2
TC
PPd
SCC
S1
M1
M2
DFC
OFC
Ins
Pir1
lo
3
2
Pu
ic
EGP
acp
VP
Cd
St
LV
fi
DLG
Pul
Pc
Pd
str
Rt
Pv
OPT
APT
VL
InG
Op
SuG
Zo
DpG
DpWh
InWh
mRt
MiTg
PnO
Po
MG
SNR
ZI
SNL
opt
EA
LOT
Tu
lo
CA1
CA2
CA3
Or
Py
Rad
GrDG
MoDG
PoDG
DS
VS
Ent
PrS
BMA
Ce
AB
ACo
Int
Lat
icp
mcp
8vn
s5
sp5
Pr5
Pn
MCL
IPl
EPl
Gl
ONL

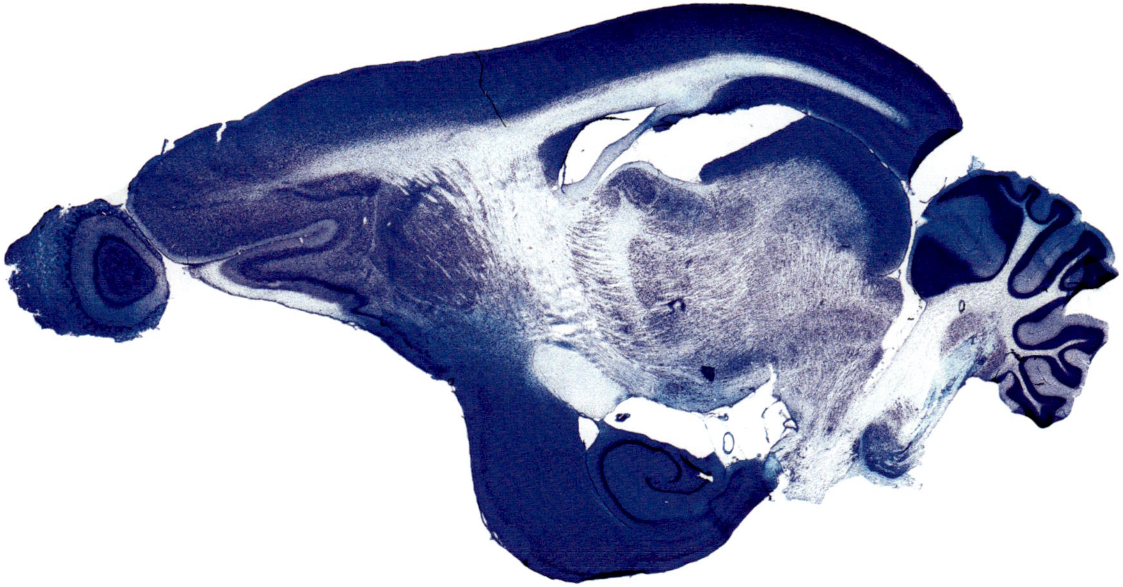

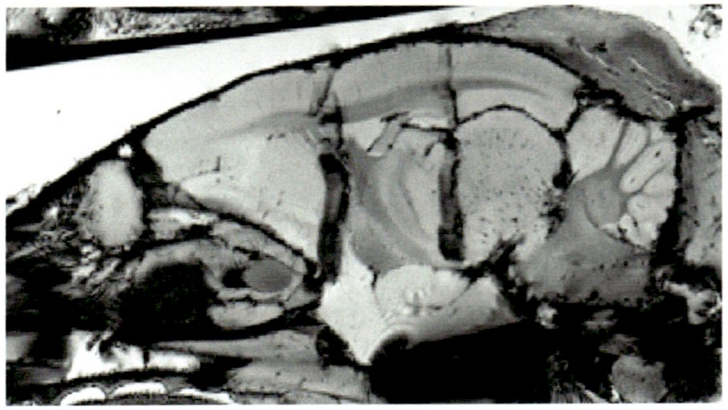

2	layer 2 of cortex
3	layer 3 of cortex
8vn	vestibular root of the vestibulocochlear nerve
AB	accessory basal nucleus
Acb	accumbens nucleus
ACo	anterior cortical amygdaloid nucleus
acp	anterior commissure, posterior part
APT	anterior pretectal nucleus
BMA	basomedial amygdaloid nucleus
CA1	field CA1 of the hippocampus
CA2	field CA2 of the hippocampus
CA3	field CA2 of the hippocampus
Cd	caudate nucleus
cp	cerebral peduncle
DG	dentate gyrus of the hippocampus
DFC	dorsal frontal cortex
DLG	dorsal lateral geniculate nucleus
DpG	deep gray layer of the superior colliculus
DpWh	deep white layer of the superior colliculus
DS	dorsal subiculum
EA	sublenticular extended amygdala
EGP	external globus pallidus
Ent	entorhinal cortex
EPl	external plexiform layer of the olfactory bulb
fi	fimbria of the hippocampus
GCL	granule cell layer of the olfactory bulb
GL	glomerular layer of the olfactory bulb
GrDG	granular layer of the dentate gyrus
ic	internal capsule
IC	inferior colliculus
icp	inferior cerebellar peduncle
IGP	internal globus pallidus
InG	intermediate gray layer of the superior colliculus
Int	interposed cerebellar nucleus
InWh	intermediate white layer of the superior colliculus
IPl	internal plexiform layer of the olfactory bulb
Lat	lateral cerebellar nucleus
LD	laterodorsal thalamic nucleus
LL	nucleus of the lateral lemniscus
LMoL	lacunosum moleculare layer of the hippocampus
lo	lateral olfactory tract
LOT	nucleus of the lateral olfactory tract
LV	lateral ventricle
M1	primary motor cortex
M2	secondary motor cortex
m5	motor root of the trigeminal nerve
MCL	mitral cell layer of the olfactory bulb
mcp	middle cerebellar peduncle
MeA	medial amygdaloid nucleus
MiTg	microcellular tegmental nucleus
MoDG	molecular layer of the dentate gyrus
mRt	mesencephalic reticular formation
OFC	orbital frontal cortex
ONL	olfactory nerve layer
Op	optic nerve layer of the superior colliculus
opt	optic tract
OPT	olivary pretectal nucleus
Or	oriens layer of the hippocampus
Pc	central nucleus of the pulvinar
Pd	dorsal nucleus of the pulvinar
Pir1	piriform cortex, layer 1
Pn	pontine nuclei
PnO	pontine reticular nucleus, oral part
Po	posterior complex of the thalamus
PoDG	polymorph layer of the dentate gyrus
PPd	posterior parietal dorsal area
Pr5	principal sensory trigeminal nucleus
Pu	putamen
Pul	pulvinar nuclei
Pv	ventral nucleus of the pulvinar
Py	pyramidal cell layer of the hippocampus
Rad	radiatum layer of the hippocampus
Rt	reticular thalamic nucleus
S1	primary somatosensory cortex
SCC	somatosensory caudal cortex
scp	superior cerebellar peduncle
scpd	superior cerebellar peduncle, descending limb
SFi	septofimbrial nucleus
SNR	substantia nigra, reticular part
sp5	spinal trigeminal tract
Sp5	spinal trigeminal nucleus
SpVe	spinal vestibular nucleus
st	stria terminalis
str	superior thalamic radiation
SuG	superficial gray layer of the superior colliculus
TC	temporal cortex
tfp	transverse fibers of the pons
Tu	olfactory tubercle
Tu1	layer 1 of olfactory tubercle
tz	trapezoid body
V1	primary visual cortex
V2	secondary visual cortex
Ves	vestibular nucleus
VL	ventrolateral thalamic nucleus
VP	ventral pallidum
VS	ventral subiculum
vsc	ventral spinocerebellar tract
ZI	zona incerta
Zo	zonal layer of the superior colliculus

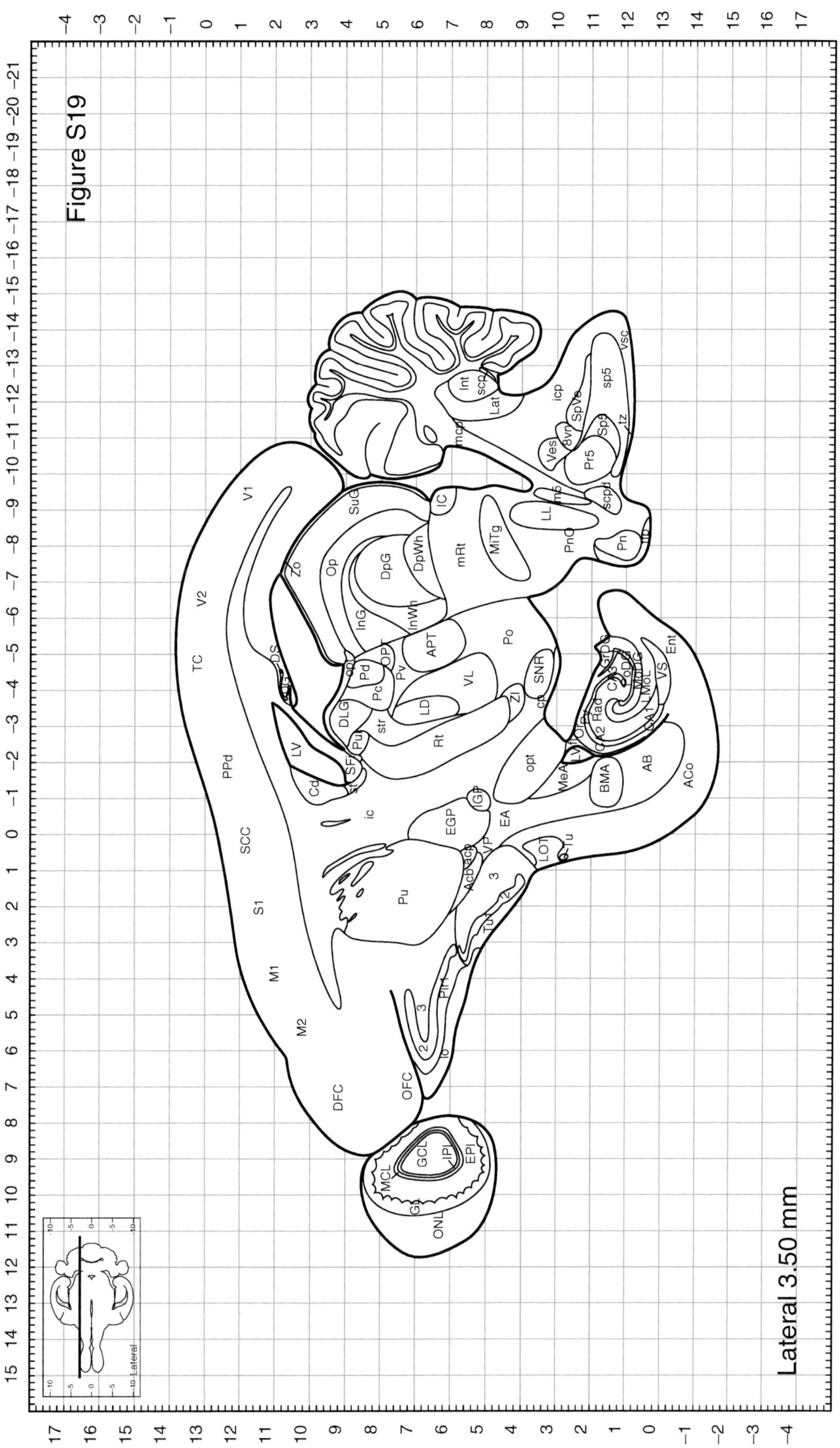
Figure S19
Lateral 3.50 mm
V1
V2
TC
PPd
SCC
S1
M1
M2
DFC
OFC
SuG
Zo
Op
DpG
DpWh
InG
InWh
APT
Po
VL
LD
Rt
ZI
SNR
opt
EA
EGP
IGP
ic
Pu
Cd
LV
DLG
str
Pd
Pc
Pv
OPT
MeA
BMA
AB
ACo
VS
Ent
LOT
Tu
VP
Acb
Pir
lo
GCL
IPl
EPl
MCL
Gl
ONL
mRt
MiTg
IC
LL
PnO
Pn
tfp
scpd
Ves
Pr5
Sp5
SpVe
8vn
icp
sp5
tz
Vsc
Int
scp
Lat
mcp
m5
cp
Rad
CA1
CA2
CA3
GrDG
MoDG
PoDG
MoL
LMol
Or
Lateral

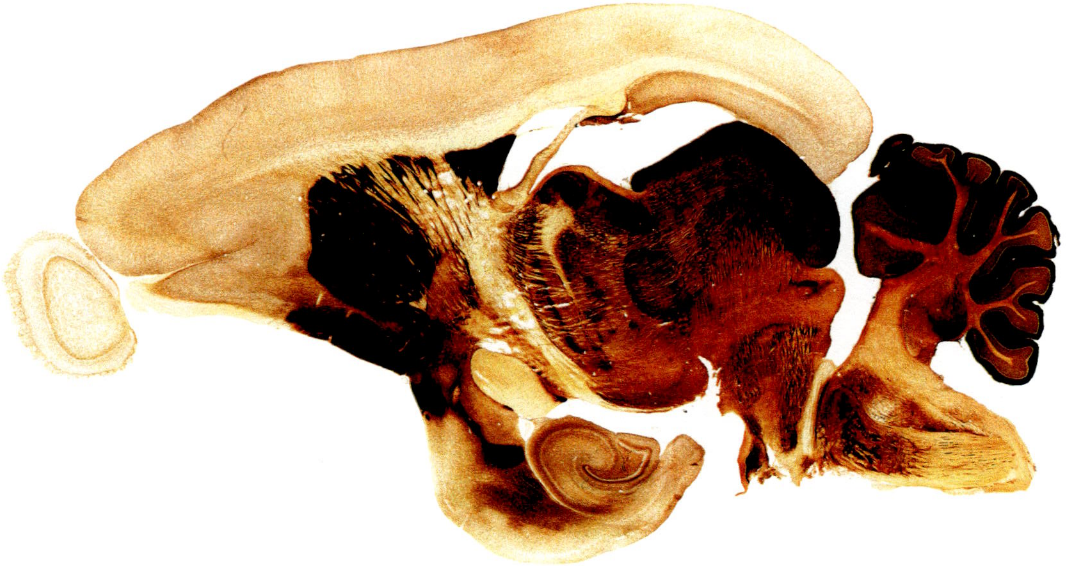

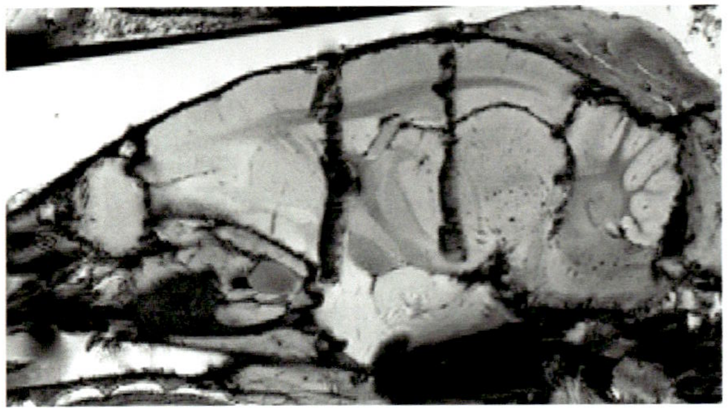

2 layer 2 of cortex
3 layer 3 of cortex
8vn vestibular root of the vestibulocochlear nerve
A5 A5 noradrenaline cells
AB accessory basal nucleus
Acb accumbens nucleus
ACo anterior cortical amygdaloid nucleus
acp anterior commissure, posterior part
APT anterior pretectal nucleus
BMA basomedial amygdaloid nucleus
CA1 field CA1 of the hippocampus
CA2 field CA2 of the hippocampus
CA3 field CA2 of the hippocampus
Cd caudate nucleus
cp cerebral peduncle
DLG dorsal lateral geniculate nucleus
DFC dorsal frontal cortex
DpG deep gray layer of the superior colliculus
DpWh deep white layer of the superior colliculus
EA sublenticular extended amygdala
EGP external globus pallidus
Ent entorhinal cortex
EPI external plexiform layer of the olfactory bulb
fi fimbria of the hippocampus
GCL granule cell layer of the olfactory bulb
GL glomerular layer of the olfactory bulb
GrDG granular layer of the dentate gyrus
Hip hippocampus
ic internal capsule
IC inferior colliculus
icp inferior cerebellar peduncle
IGP internal globus pallidus
IM intercalated amygdaloid nucleus, main part
InG intermediate gray layer of the superior colliculus
Int interposed cerebellar nucleus
InWh intermediate white layer of the superior colliculus
IPl internal plexiform layer of the olfactory bulb
Lat lateral cerebellar nucleus
LD laterodorsal thalamic nucleus
LHA lateral hypothalamic area
LL nucleus of the lateral lemniscus
LMoL lacunosum moleculare layer of the hippocampus
lo lateral olfactory tract
LOT nucleus of the lateral olfactory tract
LV lateral ventricle
M1 primary motor cortex
M2 secondary motor cortex
m5 motor root of the trigeminal nerve
MCL mitral cell layer of the olfactory bulb
mcp middle cerebellar peduncle
MeA medial amygdaloid nucleus
MiTg microcellular tegmental nucleus
MoDG molecular layer of the dentate gyrus
mRt mesencephalic reticular formation
OFC orbital frontal cortex
ONL olfactory nerve layer
Op optic nerve layer of the superior colliculus
opt optic tract
OPT olivary pretectal nucleus
Or oriens layer of the hippocampus
Pc central nucleus of the pulvinar
Pd dorsal nucleus of the pulvinar
Pir1 piriform cortex, layer 1
Pn pontine nuclei
PnO pontine reticular nucleus, oral part
Po posterior complex of the thalamus
PoDG polymorph layer of the dentate gyrus
PPd posterior parietal dorsal area
Pr5 principal sensory trigeminal nucleus
Pu putamen
Pul pulvinar nuclei
Pv ventral nucleus of the pulvinar
Py pyramidal cell layer of the hippocampus
Rad radiatum layer of the hippocampus
Rt reticular thalamic nucleus
S1 primary somatosensory cortex
SCC somatosensory caudal cortex
scpd superior cerebellar peduncle, descending limb
SFi septofimbrial nucleus
SNR substantia nigra, reticular part
sp5 spinal trigeminal tract
Sp5 spinal trigeminal nucleus
SpVe spinal vestibular nucleus
st stria terminalis
str superior thalamic radiation
SuG superficial gray layer of the superior colliculus
TC temporal cortex
tfp transverse fibers of the pons
Tu olfactory tubercle
Tu1 layer 1 of olfactory tubercle
V1 primary visual cortex
V2 secondary visual cortex
Ves vestibular nucleus
VL ventrolateral thalamic nucleus
VS ventral subiculum
ZI zona incerta
Zo zonal layer of the superior colliculus

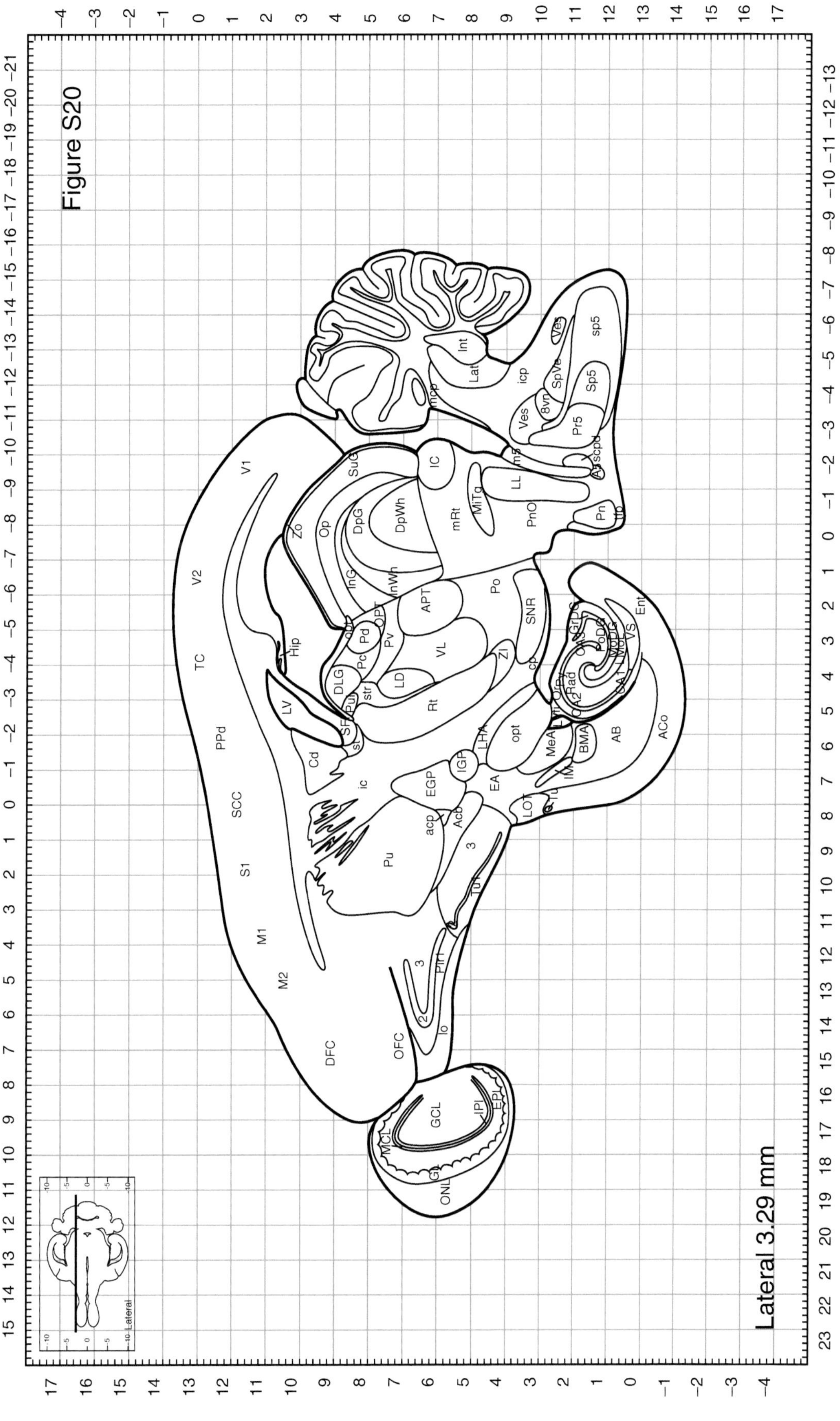
Figure S20
Lateral 3.29 mm
V1
V2
TC
PPd
SCC
S1
M1
M2
DFC
OFC
Cd
LV
Hip
ic
Pu
EGP
IGP
acp
Acb
3
Tu
Pir
lo
GCL
IPl
EPl
MCL
Gl
ONL
SuG
Zo
Op
DpG
DpWh
InG
InWh
APT
OPT
Pv
Pd
Pc
DLG
Pul
LD
str
st
SFi
VL
Rt
Po
ZI
SNR
cp
LHA
opt
EA
MeA
BMA
AB
ACo
Ent
VS
IM
LOT
IC
mRt
MiTg
LL
PnO
Pn
tfp
m5
Pr5
Sp5
sp5
SpVe
Ves
8vn
icp
Int
Lat
mcp
Rad
CA1
PoDG
GrDG
MoDG

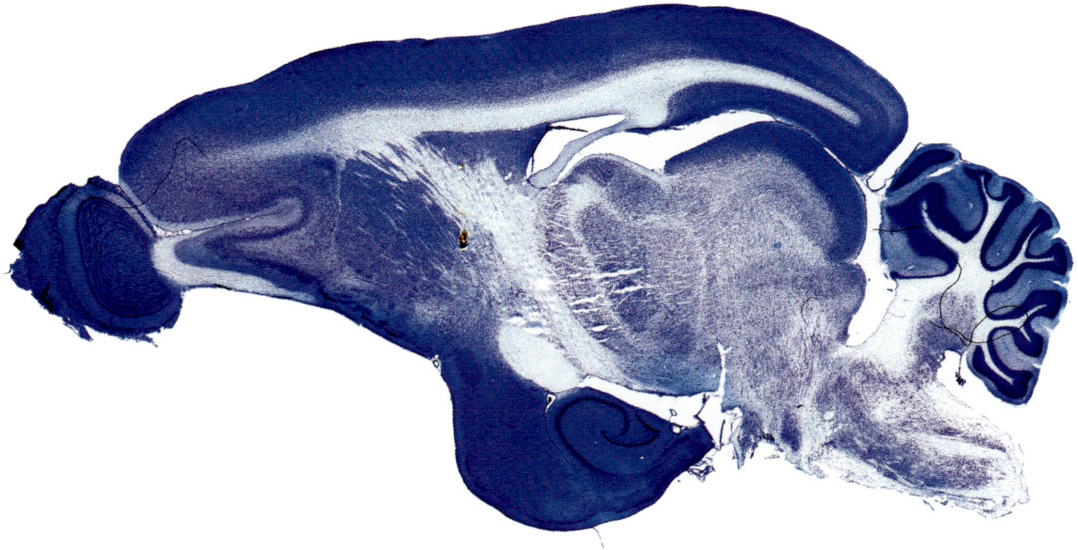

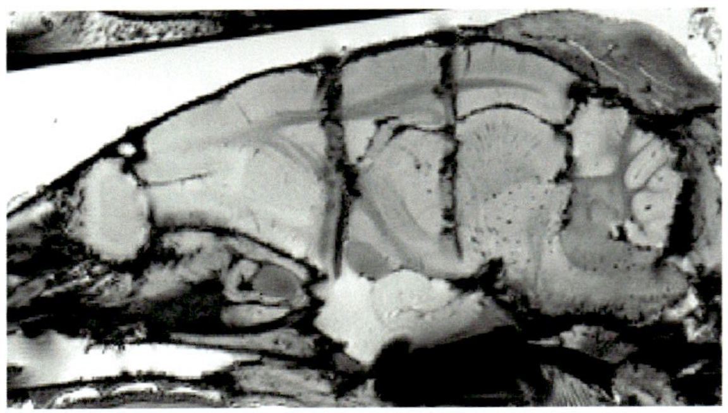

2	layer 2 of cortex
3	layer 3 of cortex
8vn	vestibular root of the vestibulocochlear nerve
A5	A5 noradrenaline cells
AB	accessory basal nucleus
Acb	accumbens nucleus
ACo	anterior cortical amygdaloid nucleus
acp	anterior commissure, posterior part
APT	anterior pretectal nucleus
CA1	field CA1 of the hippocampus
CA2	field CA2 of the hippocampus
CA3	field CA2 of the hippocampus
Cd	caudate nucleus
cp	cerebral peduncle
DFC	dorsal frontal cortex
DLG	dorsal lateral geniculate nucleus
DpG	deep gray layer of the superior colliculus
DpWh	deep white layer of the superior colliculus
EA	sublenticular extended amygdala
ECu	external cuneate nucleus
EGP	external globus pallidus
Ent	entorhinal cortex
EPl	external plexiform layer of the olfactory bulb
fi	fimbria of the hippocampus
GCL	granule cell layer of the olfactory bulb
GL	glomerular layer of the olfactory bulb
GrDG	granular layer of the dentate gyrus
Hip	hippocampus
ic	internal capsule
IC	inferior colliculus
icp	inferior cerebellar peduncle
IGP	internal globus pallidus
InG	intermediate gray layer of the superior colliculus
Int	interposed cerebellar nucleus
InWh	intermediate white layer of the superior colliculus
IPl	internal plexiform layer of the olfactory bulb
Lat	lateral cerebellar nucleus
LD	laterodorsal thalamic nucleus
LHA	lateral hypothalamic area
LL	nucleus of the lateral lemniscus
ll	lateral lemniscus
LMoL	lacunosum moleculare layer of the hippocampus
lo	lateral olfactory tract
LOT	nucleus of the lateral olfactory tract
LPB	lateral parabrachial nucleus
LRt	lateral reticular nucleus
LV	lateral ventricle
M1	primary motor cortex
M2	secondary motor cortex
MCL	mitral cell layer of the olfactory bulb
MeA	medial amygdaloid nucleus
MiTg	microcellular tegmental nucleus
MoDG	molecular layer of the dentate gyrus
mRt	mesencephalic reticular formation
oc/dsc	olivocerebellar tract and dorsal spinocerebellar tract
OFC	orbital frontal cortex
ONL	olfactory nerve layer
Op	optic nerve layer of the superior colliculus
opt	optic tract
OPT	olivary pretectal nucleus
Or	oriens layer of the hippocampus
Pc	central nucleus of the pulvinar
Pd	dorsal nucleus of the pulvinar
Pir1	piriform cortex, layer 1
Pn	pontine nuclei
PnO	pontine reticular nucleus, oral part
Po	posterior complex of the thalamus
PoDG	polymorph layer of the dentate gyrus
PPd	posterior parietal dorsal area
Pr5	principal sensory trigeminal nucleus
Pu	putamen
Pul	pulvinar nuclei
Pv	ventral nucleus of the pulvinar
Py	pyramidal cell layer of the hippocampus
Rad	radiatum layer of the hippocampus
Rt	reticular thalamic nucleus
S1	primary somatosensory cortex
SCC	somatosensory caudal cortex
scpd	superior cerebellar peduncle, descending limb
SFi	septofimbrial nucleus
SNR	substantia nigra, reticular part
sp5	spinal trigeminal tract
Sp5	spinal trigeminal nucleus
SpVe	spinal vestibular nucleus
st	stria terminalis
str	superior thalamic radiation
SuG	superficial gray layer of the superior colliculus
tfp	transverse fibers of the pons
TC	temporal cortex
Tu	olfactory tubercle
Tu1	layer 1 of olfactory tubercle
tz	trapezoid body
V1	primary visual cortex
V2	secondary visual cortex
Ves	vestibular nucleus
VL	ventrolateral thalamic nucleus
VS	ventral subiculum
vsc	ventral spinocerebellar tract
ZI	zona incerta
Zo	zonal layer of the superior colliculus

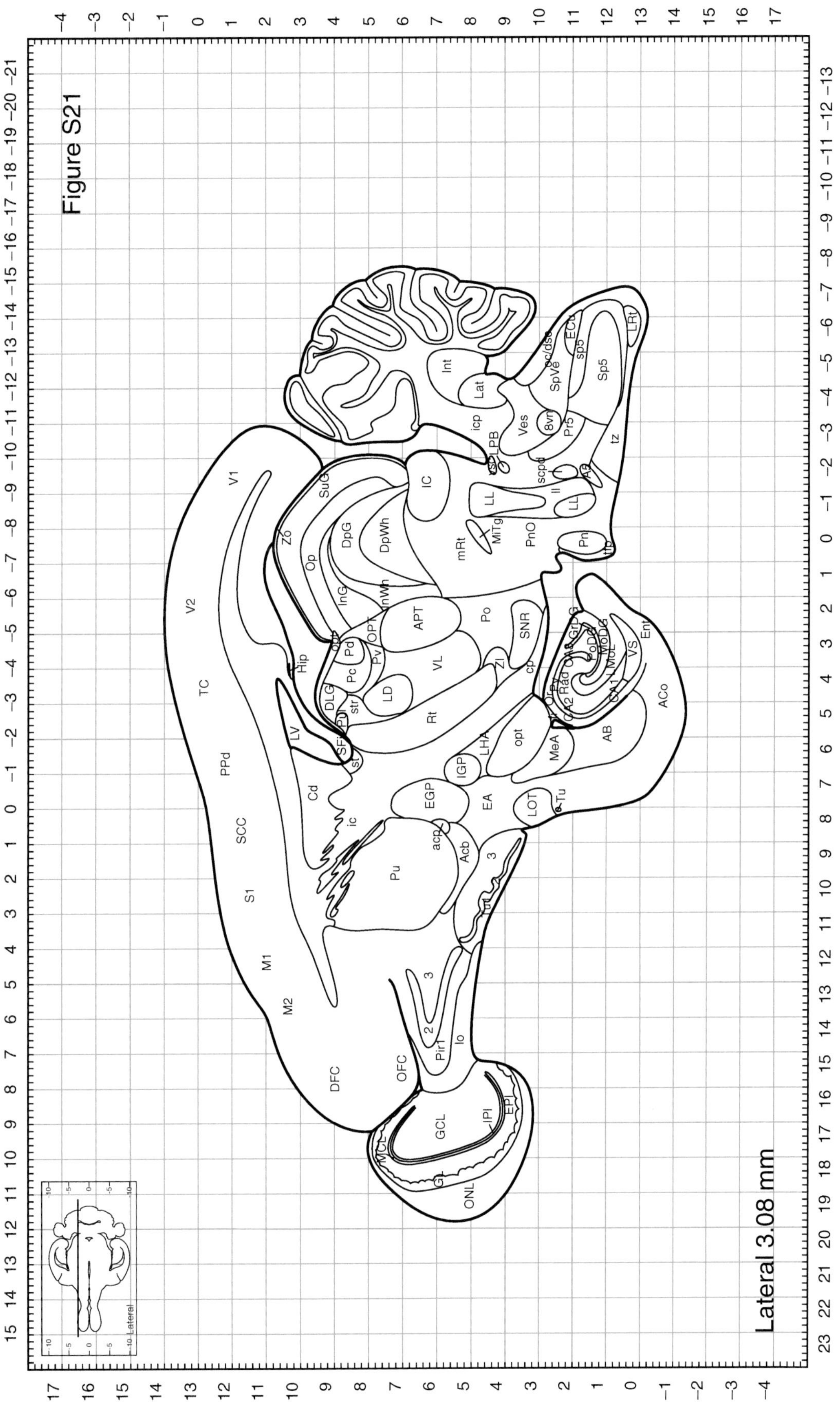
Figure S21
Lateral 3.08 mm
V1
V2
TC
PPd
SCC
S1
M1
M2
DFC
OFC
Pir1
lo
GCL
IPl
EPl
MCL
Gl
ONL
Cd
LV
Hip
SuG
Zo
Op
DpG
DpWh
InG
InWh
APT
Pd
OPT
Pc
Pv
DLG
LD
str
Pul
SFi
VL
Rt
Po
ZI
SNR
cp
LHA
opt
IGP
EGP
EA
LOT
Tu
MeA
AB
ACo
acp
Acb
Pu
ic
IC
mRt
MiTg
LL
icp
Int
Lat
LPB
Ves
8vn
Pr5
SpVe
sp5
Sp5
ECu
cc/dsc
LRt
tz
scpd
ll
PnO
Pn
tfp
Ent
VS
MoDG
MoL
CA1
CA2
GrDG
PoDG
Rad
Or
3
2

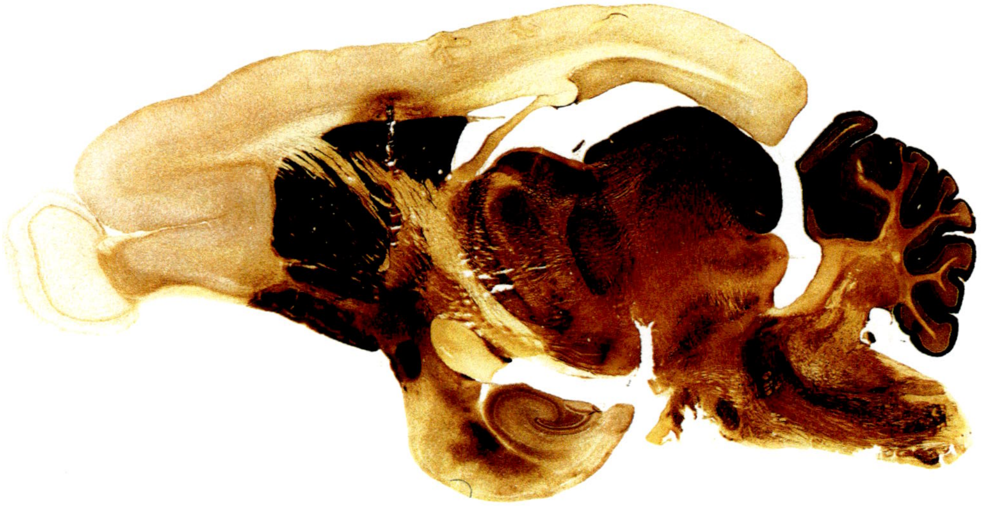

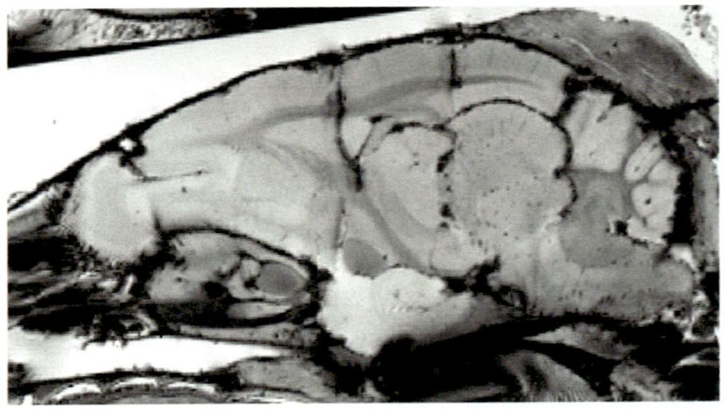

2	layer 2 of cortex
3	layer 3 of cortex
5N	motor trigeminal nucleus
7n	facial nerve
8vn	vestibular root of the vestibulocochlear nerve
AB	accessory basal nucleus
Acb	accumbens nucleus
ACo	anterior cortical amygdaloid nucleus
acp	anterior commissure, posterior part
AON	anterior olfactory nucleus
APT	anterior pretectal nucleus
CA1	field CA1 of the hippocampus
CA2	field CA2 of the hippocampus
CA3	field CA2 of the hippocampus
Cd	caudate nucleus
cp	cerebral peduncle
DCN	dorsal cochlear nucleus
DFC	dorsal frontal cortex
DLG	dorsal lateral geniculate nucleus
DpG	deep gray layer of the superior colliculus
DpWh	deep white layer of the superior colliculus
EA	sublenticular extended amygdala
ECu	external cuneate nucleus
EGP	external globus pallidus
Ent	entorhinal cortex
EPl	external plexiform layer of the olfactory bulb
GCL	granule cell layer of the olfactory bulb
GL	glomerular layer of the olfactory bulb
GrDG	granular layer of the dentate gyrus
ic	internal capsule
IC	inferior colliculus
icp	inferior cerebellar peduncle
IGP	internal globus pallidus
InG	intermediate gray layer of the superior colliculus
Int	interposed cerebellar nucleus
InWh	intermediate white layer of the superior colliculus
IPl	internal plexiform layer of the olfactory bulb
LD	laterodorsal thalamic nucleus
LHA	lateral hypothalamic area
LL	nucleus of the lateral lemniscus
LMoL	lacunosum moleculare layer of the hippocampus
lo	lateral olfactory tract
LOT	nucleus of the lateral olfactory tract
LPB	lateral parabrachial nucleus
LRt	lateral reticular nucleus
LV	lateral ventricle
M1	primary motor cortex
M2	secondary motor cortex
MCL	mitral cell layer of the olfactory bulb
MeA	medial amygdaloid nucleus
MoDG	molecular layer of the dentate gyrus
MPB	medial parabrachial nucleus
mRt	mesencephalic reticular formation
Mx	matrix region of the medulla
oc/dsc	olivocerebellar tract and dorsal spinocerebellar tract
OFC	orbital frontal cortex
ONL	olfactory nerve layer
Op	optic nerve layer of the superior colliculus
opt	optic tract
OPT	olivary pretectal nucleus
Or	oriens layer of the hippocampus
Pc	central nucleus of the pulvinar
PCRt	parvicellular reticular nucleus
Pd	dorsal nucleus of the pulvinar
Pir1	piriform cortex, layer 1
Pn	pontine nuclei
PnO	pontine reticular nucleus, oral part
Po	posterior complex of the thalamus
PoDG	polymorph layer of the dentate gyrus
PPd	posterior parietal dorsal area
Pr5	principal sensory trigeminal nucleus
PTg	pedunculopontine tegmental nucleus
Pu	putamen
Pv	ventral nucleus of the pulvinar
Py	pyramidal cell layer of the hippocampus
Rad	radiatum layer of the hippocampus
Rt	reticular thalamic nucleus
S1	primary somatosensory cortex
SCC	somatosensory caudal cortex
scp	superior cerebellar peduncle
scpd	superior cerebellar peduncle, descending limb
SFi	septofimbrial nucleus
SNC	substantia nigra, compact part
SNR	substantia nigra, reticular part
sp5	spinal trigeminal tract
Sp5	spinal trigeminal nucleus
SpVe	spinal vestibular nucleus
st	stria terminalis
STh	subthalamic nucleus
str	superior thalamic radiation
SuG	superficial gray layer of the superior colliculus
tfp	transverse fibers of the pons
TC	temporal cortex
Tu	olfactory tubercle
Tu1	layer 1 of olfactory tubercle
tz	trapezoid body
V1	primary visual cortex
V2	secondary visual cortex
Ves	vestibular nucleus
VL	ventrolateral thalamic nucleus
VS	ventral subiculum
vsc	ventral spinocerebellar tract
ZI	zona incerta
Zo	zonal layer of the superior colliculus

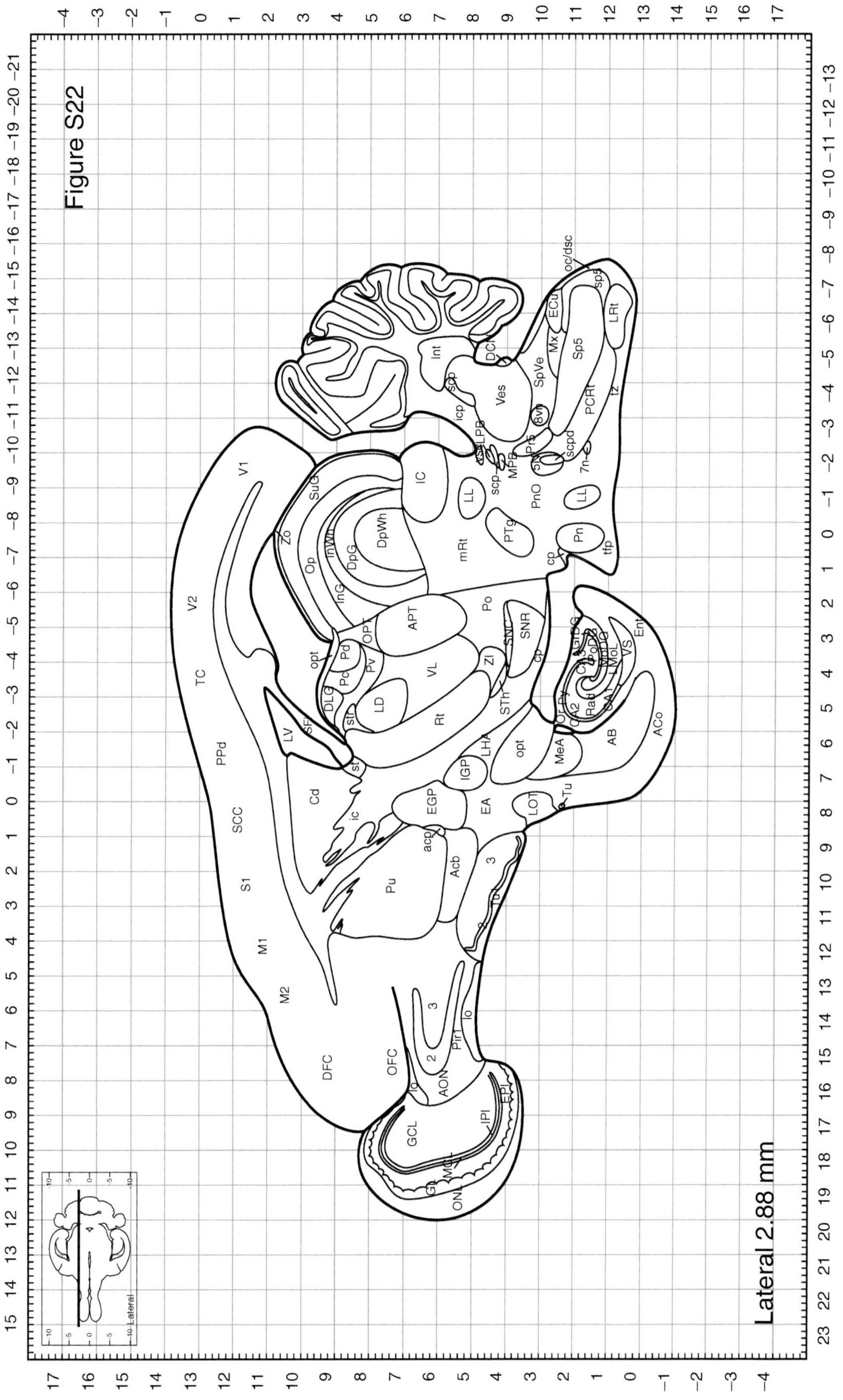
Figure S22
Lateral 2.88 mm
V1
V2
TC
PPd
SCC
S1
M1
M2
DFC
OFC
GCL
Gl
ONL
MOL
IPl
EPl
AON
lo
Pir
Tu
Acb
Pu
acp
EGP
IGP
EA
LOT
Cd
ic
st
LV
SFi
DLG
Pc
Pd
Pv
opt
OPT
APT
LD
VL
Rt
ZI
STh
LHA
MeA
AB
ACo
Ent
VS
LMol
SNR
SNC
cp
Po
mRt
PTg
Pn
tfp
LL
PnO
IC
SuG
Zo
Op
InWh
InG
DpG
DpWh
scp
MPB
LPB
icp
Ves
Int
DCN
SpVe
8vn
Mx
ECu
Sp5
sp5
oc/dsc
LRt
PCRt
tz
7n
scpd
Pr5
5N

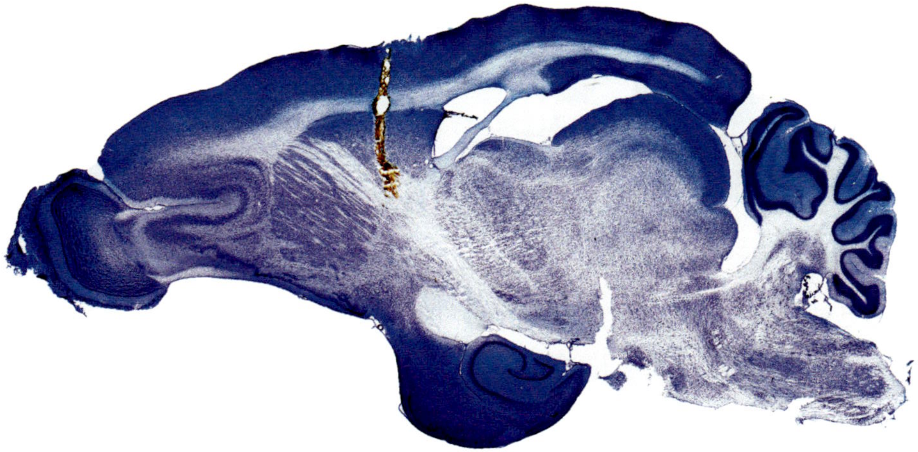

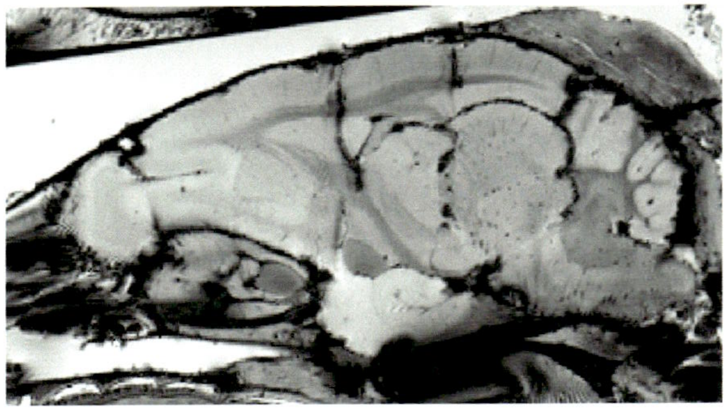

2	layer 2 of cortex
3	layer 3 of cortex
5N	motor trigeminal nucleus
7n	facial nerve
7N	facial nucleus
AB	accessory basal nucleus
Acb	accumbens nucleus
aci	anterior commissure, intrabulbar part
ACo	anterior cortical amygdaloid nucleus
acp	anterior commissure, posterior part
AON	anterior olfactory nucleus
APT	anterior pretectal nucleus
CA1	field CA1 of the hippocampus
CA2	field CA2 of the hippocampus
CA3	field CA2 of the hippocampus
Cd	caudate nucleus
CnF	cuneiform nucleus
cp	cerebral peduncle
DCN	dorsal cochlear nucleus
DFC	dorsal frontal cortex
DLG	dorsal lateral geniculate nucleus
DpG	deep gray layer of the superior colliculus
DpWh	deep white layer of the superior colliculus
EA	sublenticular extended amygdala
ECu	external cuneate nucleus
EGP	external globus pallidus
Ent	entorhinal cortex
EPl	external plexiform layer of the olfactory bulb
GCL	granule cell layer of the olfactory bulb
GL	glomerular layer of the olfactory bulb
GrDG	granular layer of the dentate gyrus
ic	internal capsule
IC	inferior colliculus
icp	inferior cerebellar peduncle
IGP	internal globus pallidus
InG	intermediate gray layer of the superior colliculus
Int	interposed cerebellar nucleus
InWh	intermediate white layer of the superior colliculus
IPl	internal plexiform layer of the olfactory bulb
LD	laterodorsal thalamic nucleus
LHA	lateral hypothalamic area
LL	nucleus of the lateral lemniscus
LMoL	lacunosum moleculare layer of the hippocampus
lo	lateral olfactory tract
LOT	nucleus of the lateral olfactory tract
LPB	lateral parabrachial nucleus
LRt	lateral reticular nucleus
LV	lateral ventricle
M1	primary motor cortex
M2	secondary motor cortex
MCL	mitral cell layer of the olfactory bulb
MeA	medial amygdaloid nucleus
MoDG	molecular layer of the dentate gyrus
MPB	medial parabrachial nucleus
mRt	mesencephalic reticular formation
Mx	matrix region of the medulla
oc/dsc	olivocerebellar tract and dorsal spinocerebellar tract
OFC	orbital frontal cortex
ONL	olfactory nerve layer
Op	optic nerve layer of the superior colliculus
opt	optic tract
OPT	olivary pretectal nucleus
Or	oriens layer of the hippocampus
Pc	central nucleus of the pulvinar
PCRt	parvicellular reticular nucleus
Pd	dorsal nucleus of the pulvinar
Pir1	piriform cortex, layer 1
Pn	pontine nuclei
PnO	pontine reticular nucleus, oral part
Po	posterior complex of the thalamus
PoDG	polymorph layer of the dentate gyrus
PPd	posterior parietal dorsal area
Pr5	principal sensory trigeminal nucleus
PTg	pedunculopontine tegmental nucleus
Pu	putamen
Pv	ventral nucleus of the pulvinar
Py	pyramidal cell layer of the hippocampus
Rad	radiatum layer of the hippocampus
rs	rubrospinal tract
Rt	reticular thalamic nucleus
RtTg	reticulotegmental nucleus of the pons
S1	primary somatosensory cortex
SCC	somatosensory caudal cortex
scp	superior cerebellar peduncle
SFi	septofimbrial nucleus
SNC	substantia nigra, compact part
SNR	substantia nigra, reticular part
sp5	spinal trigeminal tract
Sp5	spinal trigeminal nucleus
SpVe	spinal vestibular nucleus
st	stria terminalis
STh	subthalamic nucleus
str	superior thalamic radiation
SuG	superficial gray layer of the superior colliculus
TC	temporal cortex
tfp	transverse fibers of the pons
Tu	olfactory tubercle
Tu1	layer 1 of olfactory tubercle
tz	trapezoid body
V1	primary visual cortex
V2	secondary visual cortex
Ves	vestibular nucleus
VL	ventrolateral thalamic nucleus
VS	ventral subiculum
vsc	ventral spinocerebellar tract
ZI	zona incerta
Zo	zonal layer of the superior colliculus

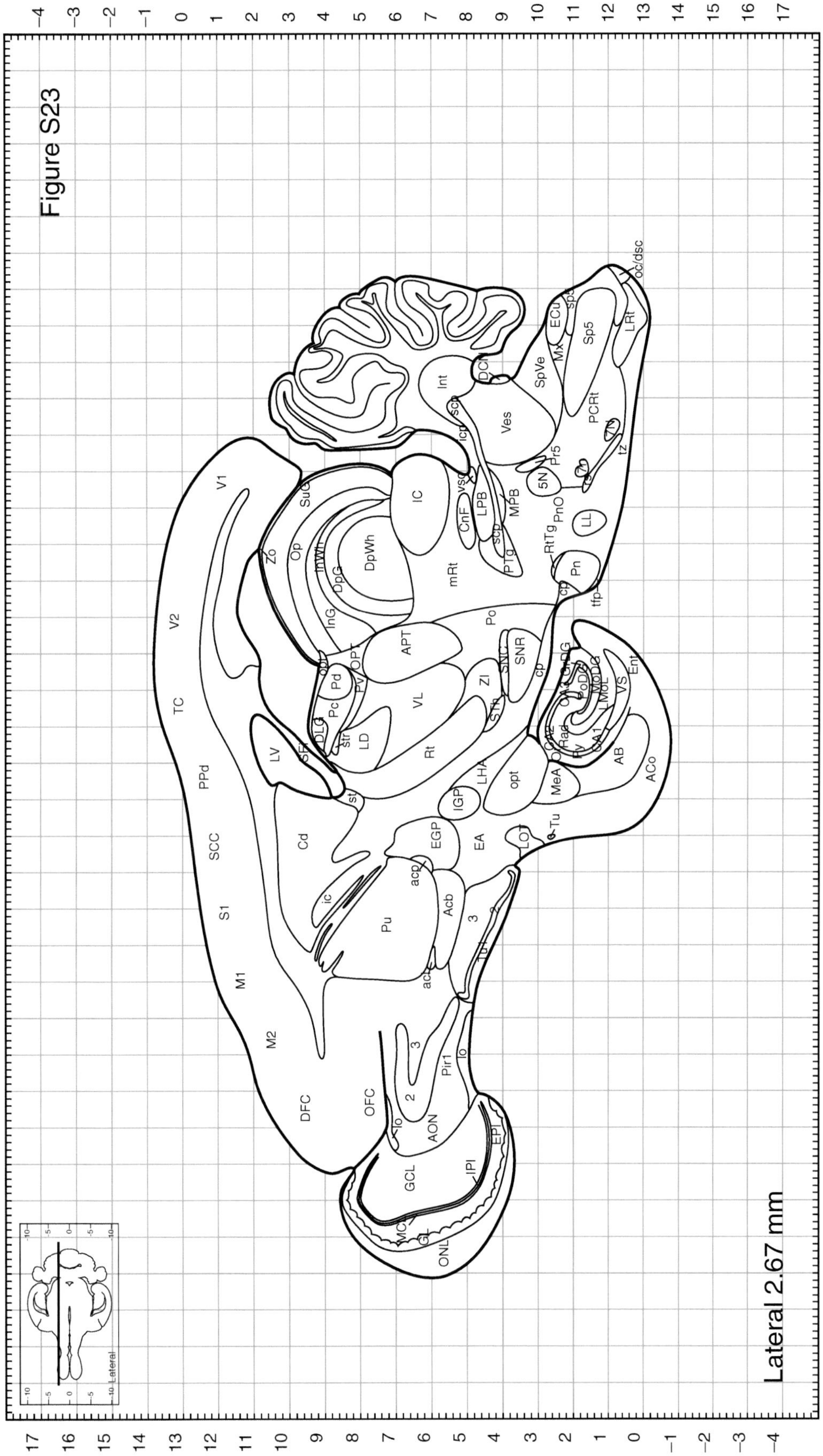
Figure S23
Lateral 2.67 mm
V1
V2
TC
PPd
SCC
S1
M1
M2
DFC
OFC
Cd
LV
SFi
st
ic
Pu
acp
Acb
aca
EGP
IGP
EA
LOT
Tu
3
2
Pir1
lo
AON
GCL
IPl
EPl
Mi
GL
ONL
LHA
opt
MeA
AB
ACo
Rt
VL
LD
Str
Pc
Pd
DLG
APT
PV
OPT
ZI
STh
SNC
SNR
cp
Po
mRt
InG
DpG
InWh
DpWh
Op
SuG
Zo
IC
CnF
LPB
MPB
scp
PTg
RtTg
Pn
tfp
LL
PnO
5N
Pr5
7n
tz
PCRt
Ves
SpVe
Mx
ECu
sp5
Sp5
LRt
oc/dsc
Int
DCN
icp
vsc
CA1
CA2
Py
Rad
Or
LMol
MoDG
PoDG
GrDG
VS
Ent
Lateral

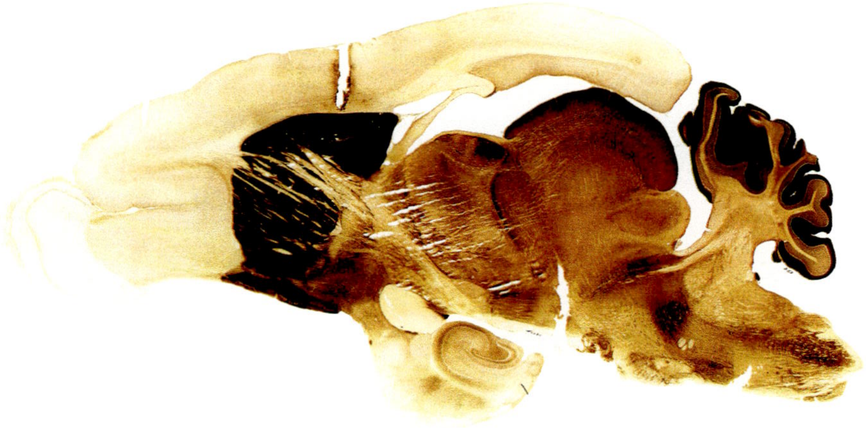

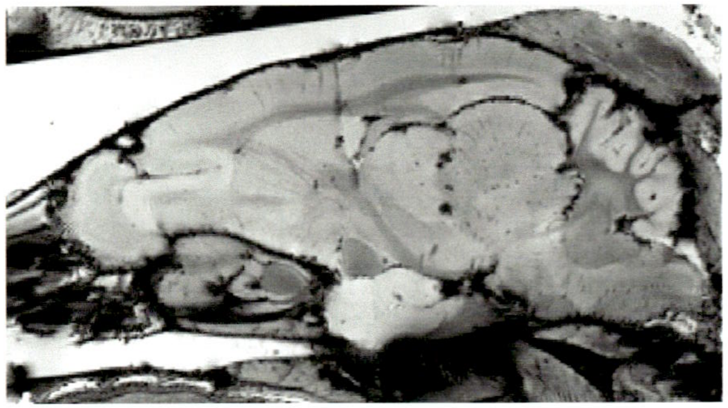

2	layer 2 of cortex
3	layer 3 of cortex
5N	motor trigeminal nucleus
7n	facial nerve
7N	facial nucleus
AB	accessory basal nucleus
Acb	accumbens nucleus
aci	anterior commissure, intrabulbar part
ACo	anterior cortical amygdaloid nucleus
acp	anterior commissure, posterior part
AM	anteromedial thalamic nucleus
AON	anterior olfactory nucleus
APT	anterior pretectal nucleus
CA1	field CA1 of the hippocampus
CA2	field CA2 of the hippocampus
CA3	field CA2 of the hippocampus
cc	corpus callosum
Cd	caudate nucleus
CnF	cuneiform nucleus
cp	cerebral peduncle
DCN	dorsal cochlear nucleus
DFC	dorsal frontal cortex
DLG	dorsal lateral geniculate nucleus
DpG	deep gray layer of the superior colliculus
DpWh	deep white layer of the superior colliculus
EA	sublenticular extended amygdala
ECu	external cuneate nucleus
EGP	external globus pallidus
Ent	entorhinal cortex
EPl	external plexiform layer of the olfactory bulb
GCL	granule cell layer of the olfactory bulb
GL	glomerular layer of the olfactory bulb
GrDG	granular layer of the dentate gyrus
ic	internal capsule
IC	inferior colliculus
icp	inferior cerebellar peduncle
IGP	internal globus pallidus
InG	intermediate gray layer of the superior colliculus
Int	interposed cerebellar nucleus
InWh	intermediate white layer of the superior colliculus
IPl	internal plexiform layer of the olfactory bulb
LD	laterodorsal thalamic nucleus
LHA	lateral hypothalamic area
LL	nucleus of the lateral lemniscus
LMoL	lacunosum moleculare layer of the hippocampus
lo	lateral olfactory tract
LPB	lateral parabrachial nucleus
LRt	lateral reticular nucleus
LV	lateral ventricle
M1	primary motor cortex
M2	secondary motor cortex
MCL	mitral cell layer of the olfactory bulb
me5	mesencephalic trigeminal tract
Me5	mesencephalic trigeminal nucleus
MeA	medial amygdaloid nucleus
ml	medial lemniscus
MoDG	molecular layer of the dentate gyrus
MPB	medial parabrachial nucleus
mRt	mesencephalic reticular formation
Mx	matrix region of the medulla
oc/dsc	olivocerebellar tract and dorsal spinocerebellar tract
OFC	orbital frontal cortex
ONL	olfactory nerve layer
Op	optic nerve layer of the superior colliculus
opt	optic tract
OPT	olivary pretectal nucleus
Or	oriens layer of the hippocampus
P5	peritrigeminal zone
Pc	central nucleus of the pulvinar
PCRt	parvicellular reticular nucleus
Pd	dorsal nucleus of the pulvinar
Pn	pontine nuclei
PnO	pontine reticular nucleus, oral part
Po	posterior complex of the thalamus
PoDG	polymorph layer of the dentate gyrus
PPd	posterior parietal dorsal area
PTg	pedunculopontine tegmental nucleus
Pu	putamen
Pv	ventral nucleus of the pulvinar
Py	pyramidal cell layer of the hippocampus
Rad	radiatum layer of the hippocampus
Rt	reticular thalamic nucleus
RtTg	reticulotegmental nucleus of the pons
S1	primary somatosensory cortex
SCC	somatosensory caudal cortex
scp	superior cerebellar peduncle
SFi	septofimbrial nucleus
SNC	substantia nigra, compact part
SNR	substantia nigra, reticular part
Sol	solitary nucleus
sp5	spinal trigeminal tract
Sp5	spinal trigeminal nucleus
SpVe	spinal vestibular nucleus
st	stria terminalis
STh	subthalamic nucleus
SuG	superficial gray layer of the superior colliculus
TC	temporal cortex
tfp	transverse fibers of the pons
Tu1	layer 1 of olfactory tubercle
V1	primary visual cortex
V2	secondary visual cortex
Ves	vestibular nucleus
VL	ventrolateral thalamic nucleus
VLN	ventrolateral reticular nucleus
VPO	ventral periolivary nucleus
VPPC	ventral posterior nucleus of the thalamus, parvicellular part
VS	ventral subiculum
vsc	ventral spinocerebellar tract
ZI	zona incerta
Zo	zonal layer of the superior colliculus

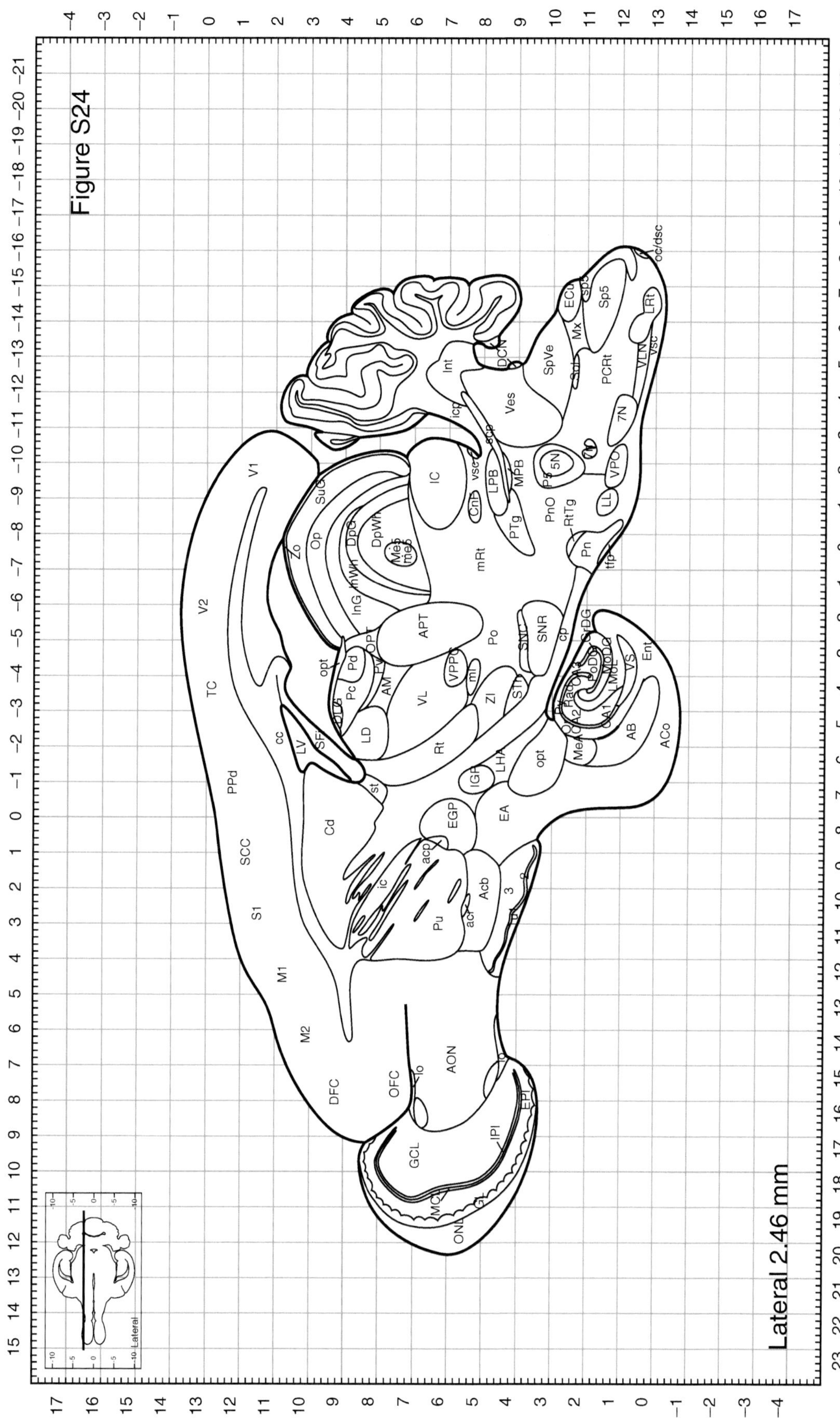
Figure S24
Lateral 2.46 mm
Lateral
V1
V2
TC
PPd
SCC
S1
M1
M2
DFC
OFC
AON
GCL
IPl
EPl
Mi
ONL
cc
LV
SFi
Cd
st
ic
acp
EGP
IGP
Pu
aca
Acb
EA
LHA
Rt
LD
AM
VL
ZI
STh
SNR
SNC
Po
APT
VPPC
ml
opt
Pc
Pd
DLG
OPT
PV
InG
InWh
DpG
DpWh
Op
SuG
Zo
Me5
me5
IC
mRt
CnF
PTg
LPB
MPB
vsc
scp
icp
Int
DCN
Ves
SpVe
Mx
ECu
Sp5
PCRt
7N
VPC
LL
5N
P5
PnO
RtTg
Pn
tfp
VLN
LRt
oc/dsc
cp
MoDG
MoL
CA1
CA2
CA3
Rad
VS
Ent
AB
ACo
MeA

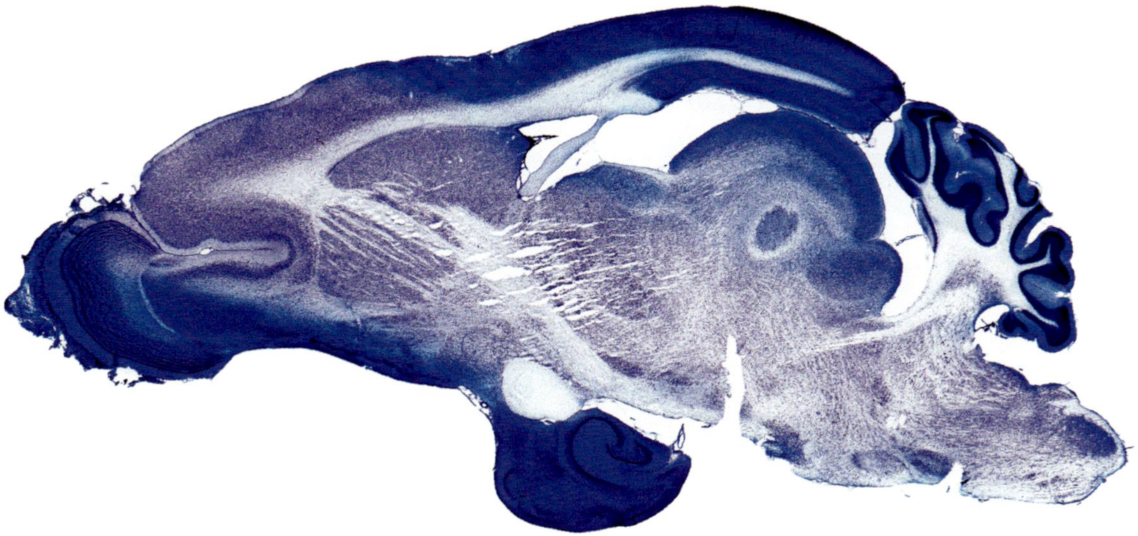

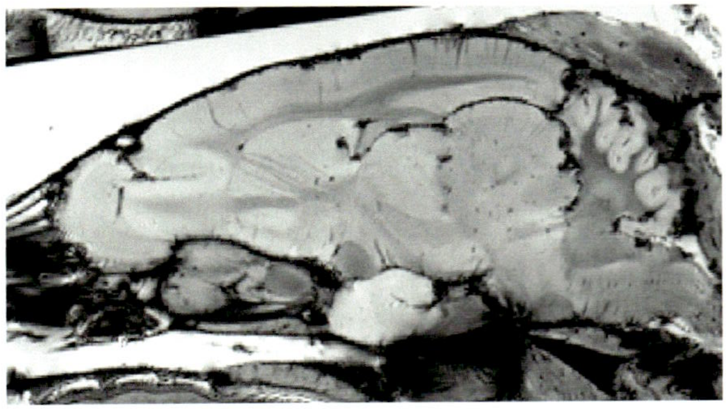

2 layer 2 of cortex
3 layer 3 of cortex
5N motor trigeminal nucleus
7n facial nerve
7N facial nucleus
AB accessory basal nucleus
Acb accumbens nucleus
aci anterior commissure, intrabulbar part
ACo anterior cortical amygdaloid nucleus
acp anterior commissure, posterior part
AM anteromedial thalamic nucleus
AON anterior olfactory nucleus
APT anterior pretectal nucleus
CA1 field CA1 of the hippocampus
CA2 field CA2 of the hippocampus
CA3 field CA2 of the hippocampus
cc corpus callosum
Cd caudate nucleus
CnF cuneiform nucleus
cp cerebral peduncle
cu cuneate fasciculus
Cu cuneate nucleus
das dorsal acoustic stria
DFC dorsal frontal cortex
DLG dorsal lateral geniculate nucleus
DpG deep gray layer of the superior colliculus
DpWh deep white layer of the superior colliculus
EA sublenticular extended amygdala
ECu external cuneate nucleus
EGP external globus pallidus
EPl external plexiform layer of the olfactory bulb
GCL granule cell layer of the olfactory bulb
GL glomerular layer of the olfactory bulb
GrDG granular layer of the dentate gyrus
ic internal capsule
IC inferior colliculusicp inferior cerebellar peduncle
InG intermediate gray layer of the superior colliculus
Int interposed cerebellar nucleus
InWh intermediate white layer of the superior colliculus
IPl internal plexiform layer of the olfactory bulb
IRt intermediate reticular nucleus
LD laterodorsal thalamic nucleus
LHA lateral hypothalamic area
LL nucleus of the lateral lemniscus
LMoL lacunosum moleculare layer of the hippocampus
lo lateral olfactory tract
LPB lateral parabrachial nucleus
LRt lateral reticular nucleus
LV lateral ventricle
M1 primary motor cortex
M2 secondary motor cortex
MCL mitral cell layer of the olfactory bulb
me5 mesencephalic trigeminal tract
Me5 mesencephalic trigeminal nucleus
MeA medial amygdaloid nucleus ml medial lemniscus
MoDG molecular layer of the dentate gyrus
MPB medial parabrachial nucleus
MPT medial pretectal nucleus
mRt mesencephalic reticular formation
Mx matrix region of the medulla
oc/dsc olivocerebellar tract and dorsal spinocerebellar tract
OFC orbital frontal cortex
ONL olfactory nerve layer
Op optic nerve layer of the superior colliculus
opt optic tract
OPT olivary pretectal nucleus
Or oriens layer of the hippocampus
P5 peritrigeminal zone
PAG periaqueductal gray
Pc central nucleus of the pulvinar
PC paracentral thalamic nucleus
PCRt parvicellular reticular nucleus
Pd dorsal nucleus of the pulvinar
PF parafascicular thalamic nucleus
Pn pontine nuclei
PnO pontine reticular nucleus, oral part
Po posterior complex of the thalamus
PoDG polymorph layer of the dentate gyrus
PPd posterior parietal dorsal area
PTg pedunculopontine tegmental nucleus
Pu putamen
Pv ventral nucleus of the pulvinar
Py pyramidal cell layer of the hippocampus
Rad radiatum layer of the hippocampus
RSg retrosplenial granular cortex
Rt reticular thalamic nucleus
RtTg reticulotegmental nucleus of the pons
S1 primary somatosensory cortex
SCC somatosensory caudal cortex
scp superior cerebellar peduncle
SFi septofimbrial nucleus
SNC substantia nigra, compact part
SNR substantia nigra, reticular part
Sol solitary nucleus
sp5 spinal trigeminal tract
Sp5 spinal trigeminal nucleus
SpVe spinal vestibular nucleus
st stria terminalis
STh subthalamic nucleus
STL bed nucleus of the stria terminalis, lateral division
SuG superficial gray layer of the superior colliculus
TC temporal cortex
tfp transverse fibers of the pons
Tu1 layer 1 of olfactory tubercle
V1 primary visual cortex
V2 secondary visual cortex
Ves vestibular nucleus
VL ventrolateral thalamic nucleus
VLN ventrolateral reticular nucleus
VM ventromedial thalamic nucleus
VPO ventral periolivary nucleus
VPPC ventral posterior nucleus of the thalamus, parvicellular part
VS ventral subiculum
vsc ventral spinocerebellar tract
ZI zona incerta
Zo zonal layer of the superior colliculus

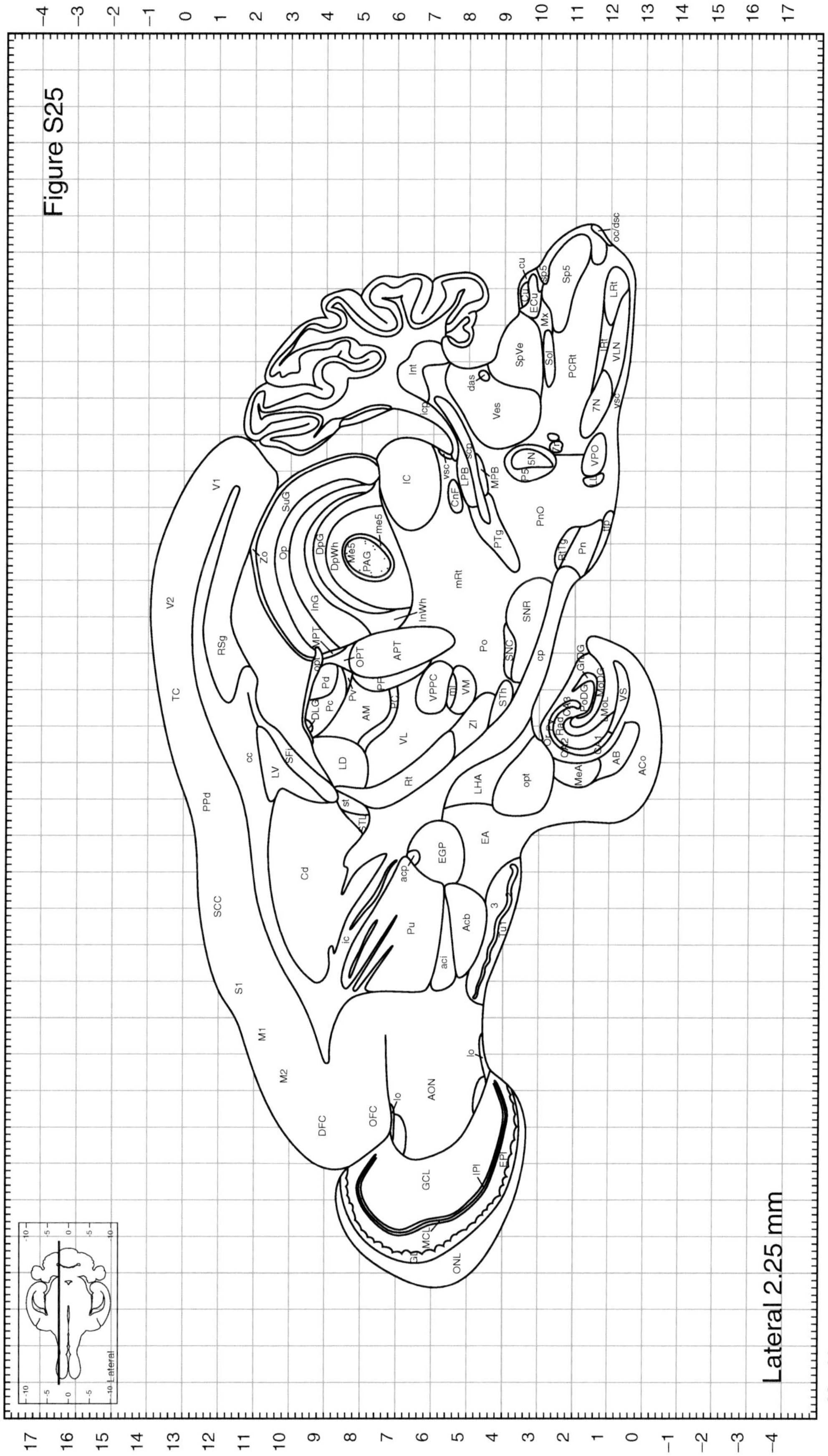

Figure S25
Lateral 2.25 mm
V1
V2
TC
PPd
SCC
S1
M1
M2
DFC
OFC
AON
GCL
IPl
EPl
MCL
Gl
ONL
lo
RSg
cc
LV
SFi
Cd
st
STL
acp
EGP
Pu
ic
Acb
aci
3
Tu1
EA
LHA
opt
Rt
LD
VL
AM
ZI
STh
VPPC
VM
ml
Po
SNC
SNR
cp
APT
OPT
MPT
Pd
Pc
DLG
Pv
InG
InWh
Op
SuG
Zo
DpG
DpWh
Me5
me5
PAG
mRt
IC
CnF
LPB
vsc
scp
MPB
PTg
PnO
Pn
RtTg
tfp
5N
P5
VPO
7N
vsc
Ves
das
Int
icp
SpVe
Sol
PCRt
IRt
VLN
LRt
Sp5
sp5
Mx
ECu
Cu
cu
oc/dsc
GrDG
MoDG
PoDG
CA3
CA2
CA1
Rad
Or
Mol
VS
AB
ACo
MeA

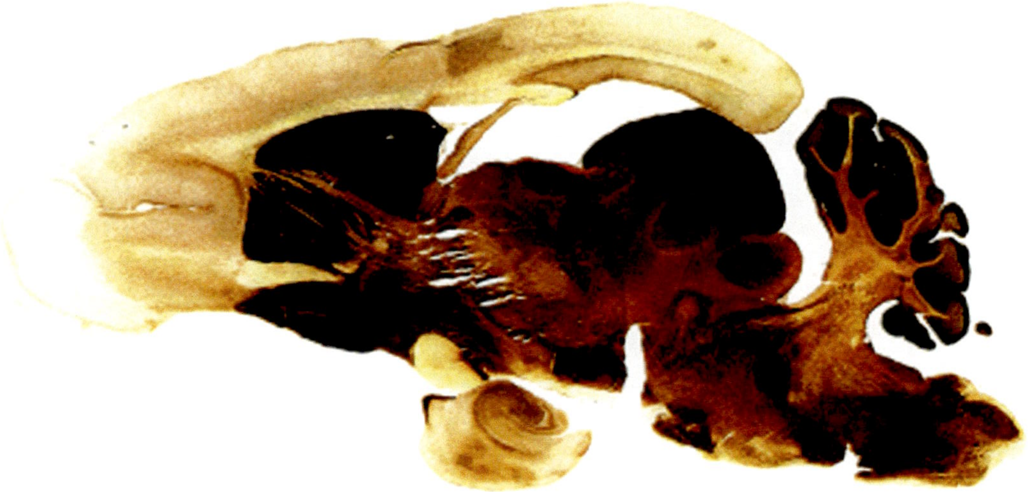

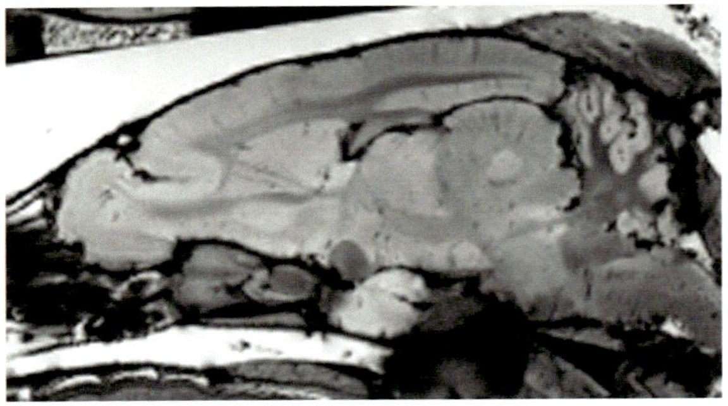

2	layer 2 of cortex
3	layer 3 of cortex
5N	motor trigeminal nucleus
7n	facial nerve
7N	facial nucleus
Acb	accumbens nucleus
aci	anterior commissure, intrabulbar part
ACo	anterior cortical amygdaloid nucleus
acp	anterior commissure, posterior part
AM	anteromedial thalamic nucleus
AON	anterior olfactory nucleus
APT	anterior pretectal nucleus
AV	anteroventral thalamic nucleus
CA1	field CA1 of the hippocampus
CA3	field CA2 of the hippocampus
cc	corpus callosum
Cd	caudate nucleus
CnF	cuneiform nucleus
cp	cerebral peduncle
cu	cuneate fasciculus
Cu	cuneate nucleus
das	dorsal acoustic stria
DFC	dorsal frontal cortex
DpG	deep gray layer of the superior colliculus
DpWh	deep white layer of the superior colliculus
EA	sublenticular extended amygdala
EGP	external globus pallidus
EPl	external plexiform layer of the olfactory bulb
GCL	granule cell layer of the olfactory bulb
Gi	gigantocellular reticular nucleus
GL	glomerular layer of the olfactory bulb
GrDG	granular layer of the dentate gyrus
ic	internal capsule
IC	inferior colliculus
icp	inferior cerebellar peduncle
InG	intermediate gray layer of the superior colliculus
Int	interposed cerebellar nucleus
InWh	intermediate white layer of the superior colliculus
IPl	internal plexiform layer of the olfactory bulb
IRt	intermediate reticular nucleus
LD	laterodorsal thalamic nucleus
LHA	lateral hypothalamic area
LMoL	lacunosum moleculare layer of the hippocampus
lo	lateral olfactory tract
LPB	lateral parabrachial nucleus
LRt	lateral reticular nucleus
LV	lateral ventricle
M1	primary motor cortex
M2	secondary motor cortex
MCL	mitral cell layer of the olfactory bulb
me5	mesencephalic trigeminal tract
Me5	mesencephalic trigeminal nucleus
ml	medial lemniscus
MoDG	molecular layer of the dentate gyrus
MPB	medial parabrachial nucleus
MPT	medial pretectal nucleus
mRt	mesencephalic reticular formation
Mx	matrix region of the medulla
OFC	orbital frontal cortex
ONL	olfactory nerve layer
Op	optic nerve layer of the superior colliculus
opt	optic tract
OPT	olivary pretectal nucleus
Or	oriens layer of the hippocampus
P5	peritrigeminal zone
PAG	periaqueductal gray
Pc	central nucleus of the pulvinar
PC	paracentral thalamic nucleus
PCRt	parvicellular reticular nucleus
Pd	dorsal nucleus of the pulvinar
PF	parafascicular thalamic nucleus
Pn	pontine nuclei
PnO	pontine reticular nucleus, oral part
Po	posterior complex of the thalamus
PoDG	polymorph layer of the dentate gyrus
PPd	posterior parietal dorsal area
PTg	pedunculopontine tegmental nucleus
Pu	putamen
Pv	ventral nucleus of the pulvinar
Py	pyramidal cell layer of the hippocampus
Rad	radiatum layer of the hippocampus
RN	red nucleus
RSg	retrosplenial granular cortex
Rt	reticular thalamic nucleus
RtTg	reticulotegmental nucleus of the pons
S1	primary somatosensory cortex
SCC	somatosensory caudal cortex
scp	superior cerebellar peduncle
SFi	septofimbrial nucleus
SNC	substantia nigra, compact part
SNR	substantia nigra, reticular part
Sol	solitary nucleus
SON	supraoptic nucleus
sp5	spinal trigeminal tract
Sp5	spinal trigeminal nucleus
SpVe	spinal vestibular nucleus
st	stria terminalis
STh	subthalamic nucleus
STL	bed nucleus of the stria terminalis, lateral division
SubC	subcoeruleus nucleus
SuG	superficial gray layer of the superior colliculus
TC	temporal cortex
tfp	transverse fibers of the pons
Tu1	layer 1 of olfactory tubercle
V1	primary visual cortex
V2	secondary visual cortex
VeCb	vestibulocerebellar nucleus
Ves	vestibular nucleus
VL	ventrolateral thalamic nucleus
VLN	ventrolateral reticular nucleus
VM	ventromedial thalamic nucleus
VPO	ventral periolivary nucleus
VPPC	ventral posterior nucleus of the thalamus, parvicellular part
VS	ventral subiculum
vsc	ventral spinocerebellar tract
ZI	zona incerta
Zo	zonal layer of the superior colliculus

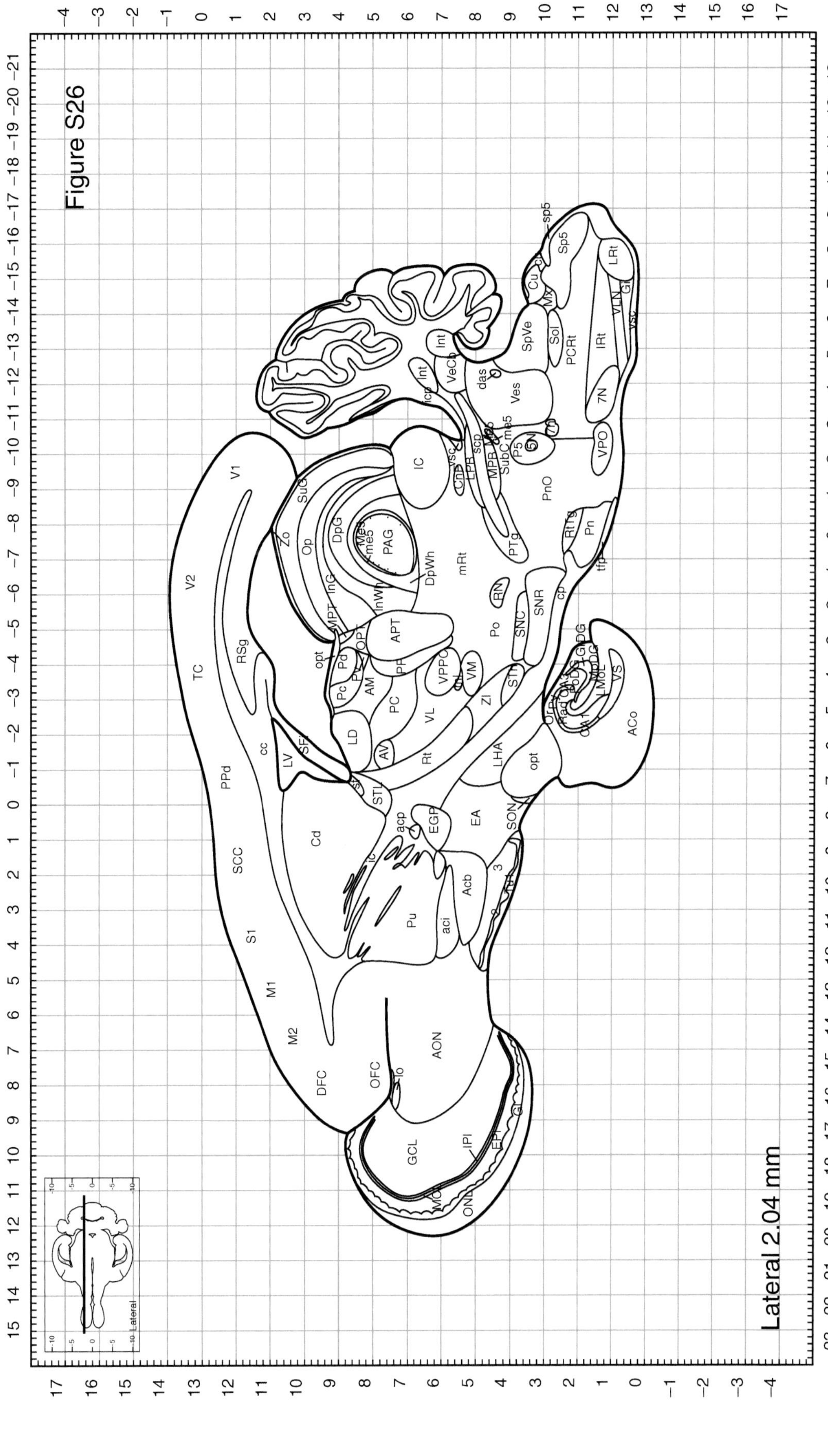
Figure S26
Lateral 2.04 mm
V1
V2
TC
RSg
PPd
cc
LV
SCC
Cd
S1
M1
M2
DFC
OFC
AON
GCL
IPl
EPl
Gl
ONL
SuG
Zo
Op
DpG
InG
InWh
DpWh
PAG
Me5
me5
IC
APT
AM
PC
LD
AV
VL
Rt
VPPC
VM
ZI
LHA
opt
STh
SNC
SNR
RN
Po
mRt
PTg
PnO
RtTg
Pn
tfp
cp
VPO
7N
IRt
PCRt
Sol
SpVe
Ves
VeCb
Int
icp
das
scp
Sp5
sp5
Cu
Mx
LRt
VLN
Gi
vsc
EA
EGP
acp
STL
Acb
aci
Pu
ic
SON
ACo
VS
lo
Lateral

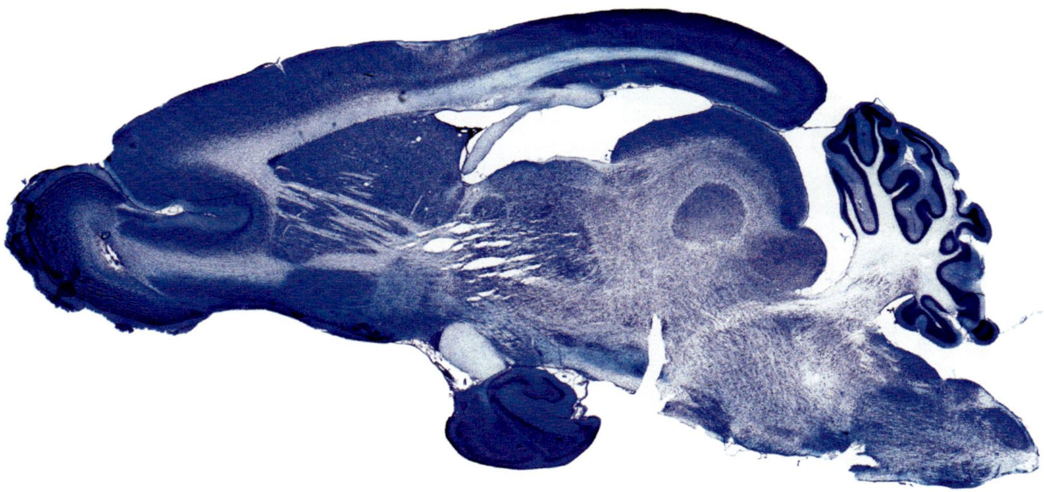

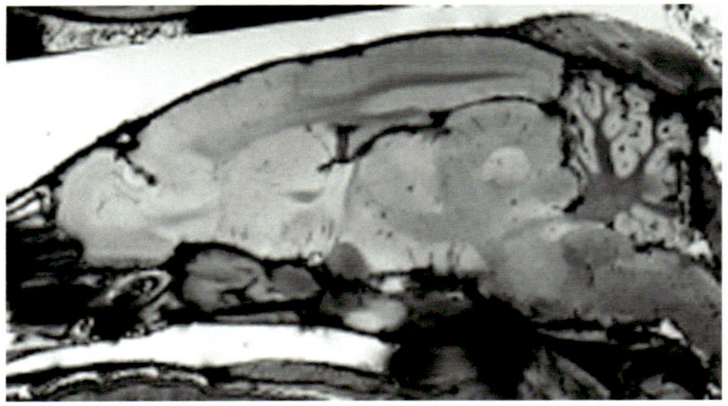

2	layer 2 of cortex
3	layer 3 of cortex
7n	facial nerve
7N	facial nucleus
ac	anterior commissure
Acb	accumbens nucleus
aci	anterior commissure, intrabulbar part
ACo	anterior cortical amygdaloid nucleus
AM	anteromedial thalamic nucleus
AmbC	ambiguus nucleus, compact part
AON	anterior olfactory nucleus
AP	area postrema
AV	anteroventral thalamic nucleus
CA3	field CA2 of the hippocampus
cc	corpus callosum
Cd	caudate nucleus
CnF	cuneiform nucleus
cp	cerebral peduncle
cu	cuneate fasciculus
Cu	cuneate nucleus
das	dorsal acoustic stria
DFC	dorsal frontal cortex
DpG	deep gray layer of the superior colliculus
DpWh	deep white layer of the superior colliculus
EA	sublenticular extended amygdala
EGP	external globus pallidus
EPl	external plexiform layer of the olfactory bulb
GCL	granule cell layer of the olfactory bulb
Gi	gigantocellular reticular nucleus
GL	glomerular layer of the olfactory bulb
GrDG	granular layer of the dentate gyrus
ic	internal capsule
IC	inferior colliculus
icp	inferior cerebellar peduncle
InG	intermediate gray layer of the superior colliculus
Int	interposed cerebellar nucleus
InWh	intermediate white layer of the superior colliculus
IPl	internal plexiform layer of the olfactory bulb
IRt	intermediate reticular nucleus
LC	locus coeruleus
LD	laterodorsal thalamic nucleus
LDTg	laterodorsal tegmental nucleus
LHA	lateral hypothalamic area
lo	lateral olfactory tract
LPB	lateral parabrachial nucleus
LRt	lateral reticular nucleus
LV	lateral ventricle
M1	primary motor cortex
M2	secondary motor cortex
MCL	mitral cell layer of the olfactory bulb
me5	mesencephalic trigeminal tract
Me5	mesencephalic trigeminal nucleus
mlf	medial longitudinal fasciculus
MoDG	molecular layer of the dentate gyrus
MPB	medial parabrachial nucleus
MPT	medial pretectal nucleus
mRt	mesencephalic reticular formation
MT	medial terminal nucleus of the accessory optic tract
Mx	matrix region of the medulla
OFC	orbital frontal cortex
ONL	olfactory nerve layer
Op	optic nerve layer of the superior colliculus
opt	optic tract
OPT	olivary pretectal nucleus
OV	olfactory ventricle (olfactory part of lateral ventricle)
P5	peritrigeminal zone
PAG	periaqueductal gray
Pc	central nucleus of the pulvinar
PC	paracentral thalamic nucleus
PCRt	parvicellular reticular nucleus
Pd	dorsal nucleus of the pulvinar
PF	parafascicular thalamic nucleus
Pn	pontine nuclei
PnO	pontine reticular nucleus, oral part
Po	posterior complex of the thalamus
PoDG	polymorph layer of the dentate gyrus
PPd	posterior parietal dorsal area
PTg	pedunculopontine tegmental nucleus
Pu	putamen
RN	red nucleus
RSg	retrosplenial granular cortex
Rt	reticular thalamic nucleus
RtTg	reticulotegmental nucleus of the pons
S1	primary somatosensory cortex
SCC	somatosensory caudal cortex
scp	superior cerebellar peduncle
SFi	septofimbrial nucleus
sm	stria medullaris of the thalamus
SNC	substantia nigra, compact part
SNR	substantia nigra, reticular part
Sol	solitary nucleus
SON	supraoptic nucleus
sp5	spinal trigeminal tract
Sp5	spinal trigeminal nucleus
SpVe	spinal vestibular nucleus
STh	subthalamic nucleus
STL	bed nucleus of the stria terminalis, lateral division
Sub	submedius thalamic nucleus
SubC	subcoeruleus nucleus
SuG	superficial gray layer of the superior colliculus
TC	temporal cortex
tfp	transverse fibers of the pons
tth	trigeminothalamic tract
Tu1	layer 1 of olfactory tubercle
Tz	nucleus of the trapezoid body
V1	primary visual cortex
V2	secondary visual cortex
VeCb	vestibulocerebellar nucleus
Ves	vestibular nucleus
VL	ventrolateral thalamic nucleus
VLN	ventrolateral reticular nucleus
VM	ventromedial thalamic nucleus
VPO	ventral periolivary nucleus
VPPC	ventral posterior nucleus of the thalamus, parvicellular part
VS	ventral subiculum
vsc	ventral spinocerebellar tract
ZI	zona incerta
Zo	zonal layer of the superior colliculus

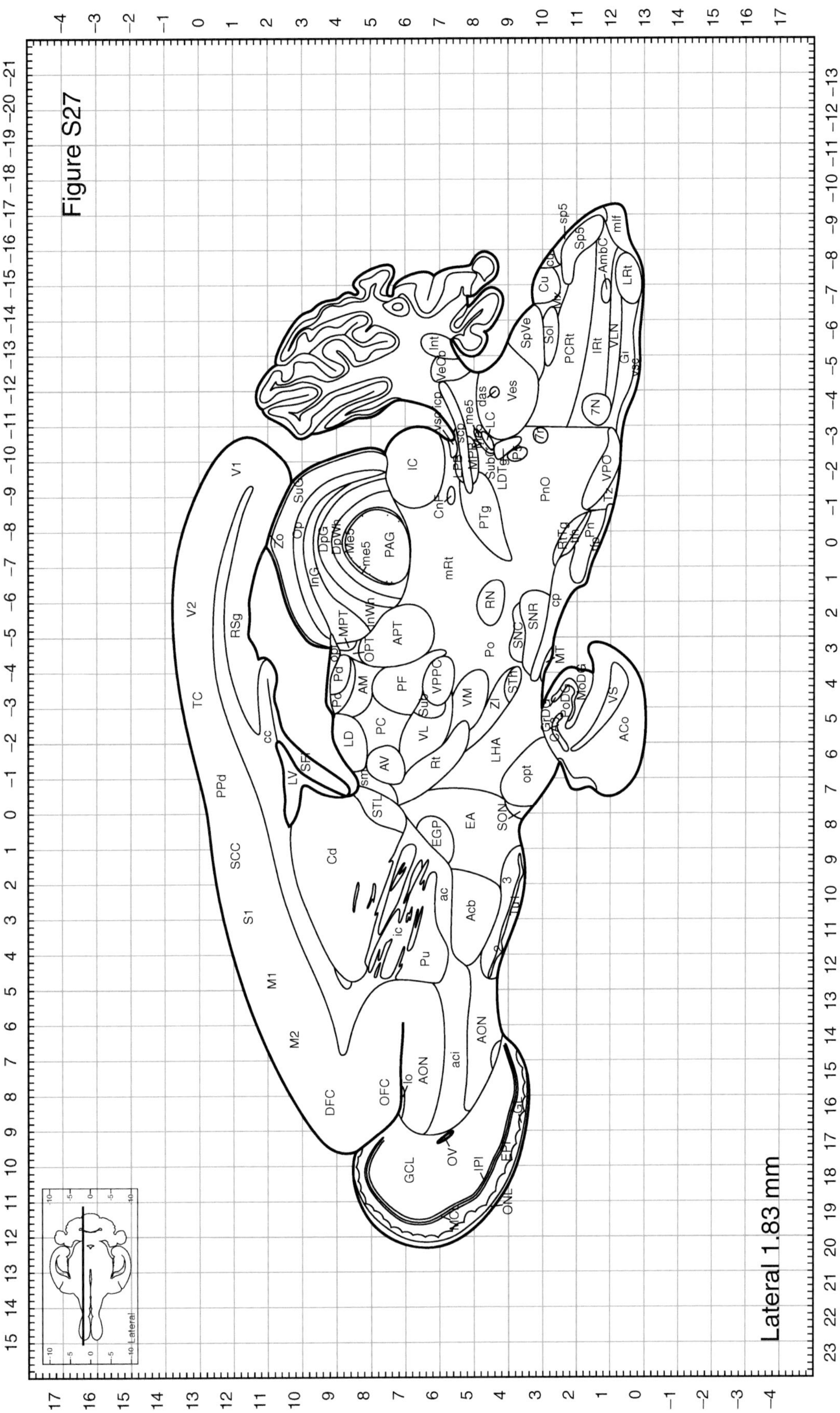
Figure S27
Lateral 1.83 mm
V1
V2
RSg
TC
cc
LV
SFi
PPd
SCC
S1
M1
M2
DFC
OFC
lo
AON
aci
GCL
OV
IPl
EPl
GL
ONL
Mi
Cd
STL
EGP
EA
SON
ac
ic
Pu
Acb
AON
SuG
Op
Zo
InG
DpG
DpWh
Me5
me5
PAG
IC
CnF
mRt
PTg
RN
SNR
SNC
cp
Po
MPT
InWh
OPT
APT
AM
PF
VPPC
VM
ZI
STh
LD
PC
VL
AV
Rt
LHA
opt
sm
MT
ACo
VS
PnO
VPO
Tz
Pn
RtTg
7N
Ves
SpVe
Sol
PCRt
IRt
VLN
Gi
LRt
AmbC
mlf
Sp5
sp5
Cu
Int
VeCb
das
LC
scp
me5
LDTg
SubC

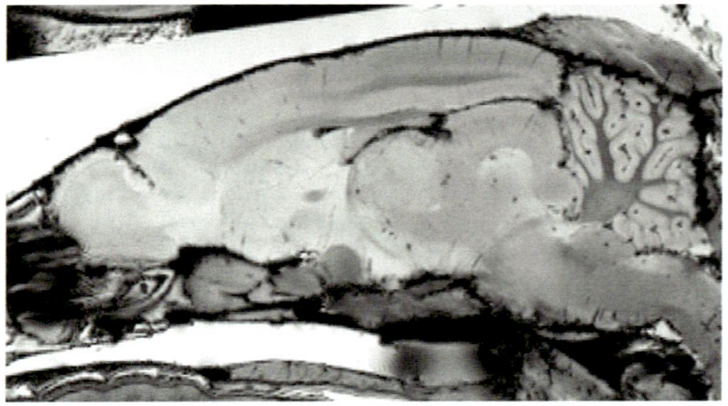

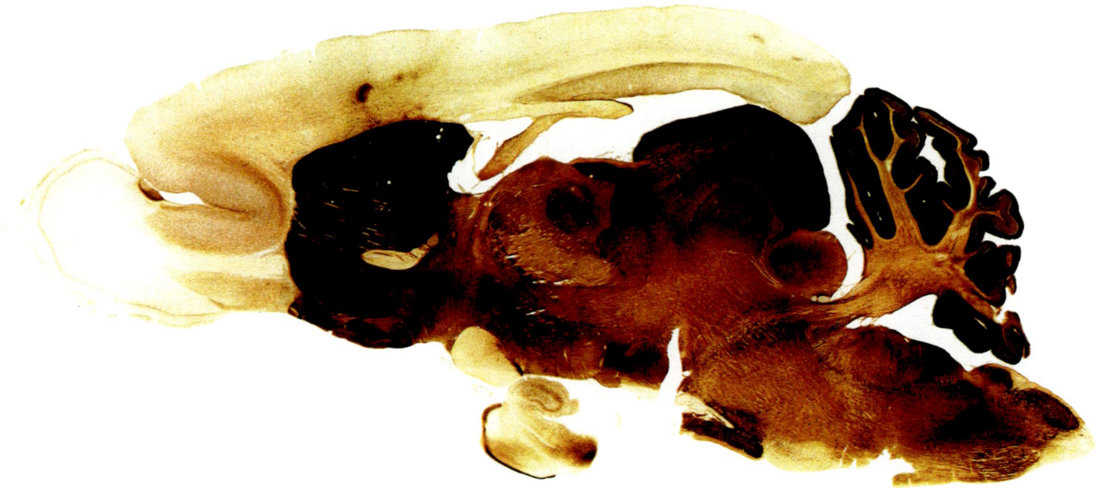

2	layer 2 of cortex
3	layer 3 of cortex
7n	facial nerve
7N	facial nucleus
ac	anterior commissure
Acb	accumbens nucleus
aci	anterior commissure, intrabulbar part
ACo	anterior cortical amygdaloid nucleus
AD	anterodorsal thalamic nucleus
AM	anteromedial thalamic nucleus
AmbC	ambiguus nucleus, compact part
AON	anterior olfactory nucleus
APT	anterior pretectal nucleus
AV	anteroventral thalamic nucleus
cc	corpus callosum
Cd	caudate nucleus
cp	cerebral peduncle
cu	cuneate fasciculus
Cu	cuneate nucleus
DFC	dorsal frontal cortex
DpG	deep gray layer of the superior colliculus
DpWh	deep white layer of the superior colliculus
EA	sublenticular extended amygdala
EGP	external globus pallidus
EPl	external plexiform layer of the olfactory bulb
EPlA	external plexiform layer of the accessory olfactory bulb
fr	fasciculus retroflexus
GCL	granule cell layer of the olfactory bulb
Gi	gigantocellular reticular nucleus
GL	glomerular layer of the olfactory bulb
GlA	glomerular layer of the accessory olfactory bulb
GrA	granule cell layer of the accessory olfactory bulb
GrDG	granular layer of the dentate gyrus
IC	inferior colliculus
icp	inferior cerebellar peduncle
InG	intermediate gray layer of the superior colliculus
InWh	intermediate white layer of the superior colliculus
IPl	internal plexiform layer of the olfactory bulb
IRt	intermediate reticular nucleus
LC	locus coeruleus
LDTg	laterodorsal tegmental nucleus
LHA	lateral hypothalamic area
lo	lateral olfactory tract
LPB	lateral parabrachial nucleus
LPO	lateral preoptic area
LRt	lateral reticular nucleus
LV	lateral ventricle
M1	primary motor cortex
M2	secondary motor cortex
MCL	mitral cell layer of the olfactory bulb
MdD	medullary reticular nucleus, dorsal part
me5	mesencephalic trigeminal tract
Me5	mesencephalic trigeminal nucleus
mlf	medial longitudinal fasciculus
MoDG	molecular layer of the dentate gyrus
MPB	medial parabrachial nucleus
MPT	medial pretectal nucleus
mRt	mesencephalic reticular formation
MT	medial terminal nucleus of the accessory optic tract
Mx	matrix region of the medulla
OFC	orbital frontal cortex
ONL	olfactory nerve layer
Op	optic nerve layer of the superior colliculus
opt	optic tract
OPT	olivary pretectal nucleus
OV	olfactory ventricle (olfactory part of lateral ventricle)
PAG	periaqueductal gray
Pc	central nucleus of the pulvinar
PC	paracentral thalamic nucleus
PCRt	parvicellular reticular nucleus
Pd	dorsal nucleus of the pulvinar
PF	parafascicular thalamic nucleus
PLH	peduncular part of lateral hypothalamus
Pn	pontine nuclei
PnO	pontine reticular nucleus, oral part
PoDG	polymorph layer of the dentate gyrus
PPd	posterior parietal dorsal area
PTg	pedunculopontine tegmental nucleus
RN	red nucleus
RSg	retrosplenial granular cortex
Rt	reticular thalamic nucleus
RtTg	reticulotegmental nucleus of the pons
S1	primary somatosensory cortex
SCC	somatosensory caudal cortex
scp	superior cerebellar peduncle
SFi	septofimbrial nucleus
sm	stria medullaris of the thalamus
SNC	substantia nigra, compact part
SNR	substantia nigra, reticular part
Sol	solitary nucleus
SON	supraoptic nucleus
sp5	spinal trigeminal tract
Sp5	spinal trigeminal nucleus
SpVe	spinal vestibular nucleus
STMA	bed nucleus of the stria terminalis, medial division, anterior part
STMP	bed nucleus of the stria terminalis, medial division, posterior part
Sub	submedius thalamic nucleus
SuG	superficial gray layer of the superior colliculus
TC	temporal cortex
tfp	transverse fibers of the pons
tth	trigeminothalamic tract
Tu1	layer 1 of olfactory tubercle
tz	trapezoid body
Tz	nucleus of the trapezoid body
V1	primary visual cortex
V2	secondary visual cortex
VeCb	vestibulocerebellar nucleus
Ves	vestibular nucleus
VL	ventrolateral thalamic nucleus
VLN	ventrolateral reticular nucleus
VM	ventromedial thalamic nucleus
VPO	ventral periolivary nucleus
VPPC	ventral posterior nucleus of the thalamus, parvicellular part
VS	ventral subiculum
vsc	ventral spinocerebellar tract
VTM	ventral tuberomammillary nucleus
Zo	zonal layer of the superior colliculus

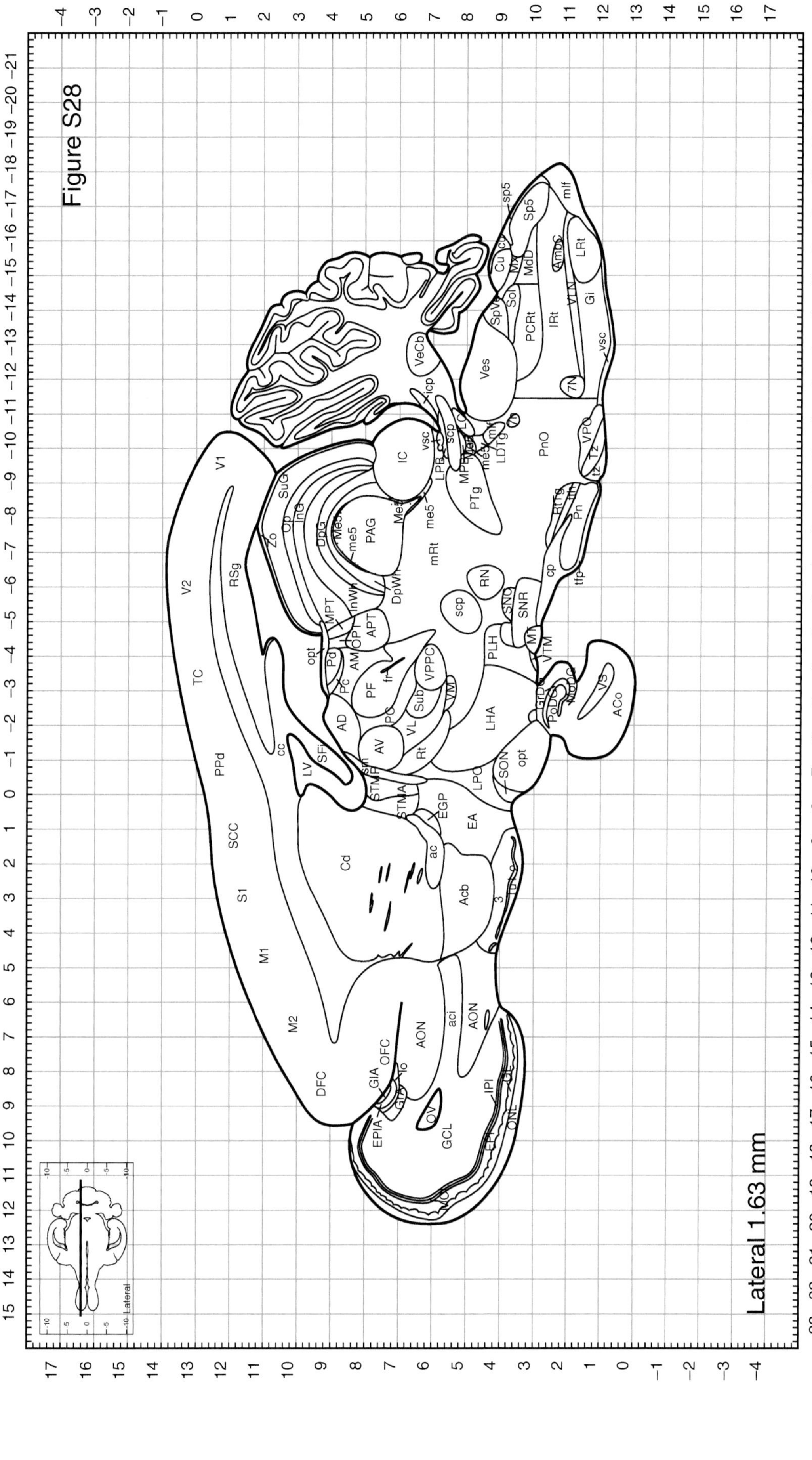
Figure S28
Lateral 1.63 mm
V1
V2
RSg
TC
PPd
SCC
S1
M1
M2
DFC
OFC
GIA
EPIA
GCL
AON
aci
Acb
Cd
EA
LPO
SON
opt
LHA
PLH
SNR
SNC
RN
scp
mRt
PAG
IC
SuG
InG
Op
Zo
MPT
APT
AD
AV
VL
Rt
PF
VPPC
LV
cc
Ves
VeCb
icp
PnO
PCRt
IRt
Gi
LRt
Sp5
sp5
mlf
Cu
Sol
7N
VPO
Tz
Pn
cp
ACo
VTM

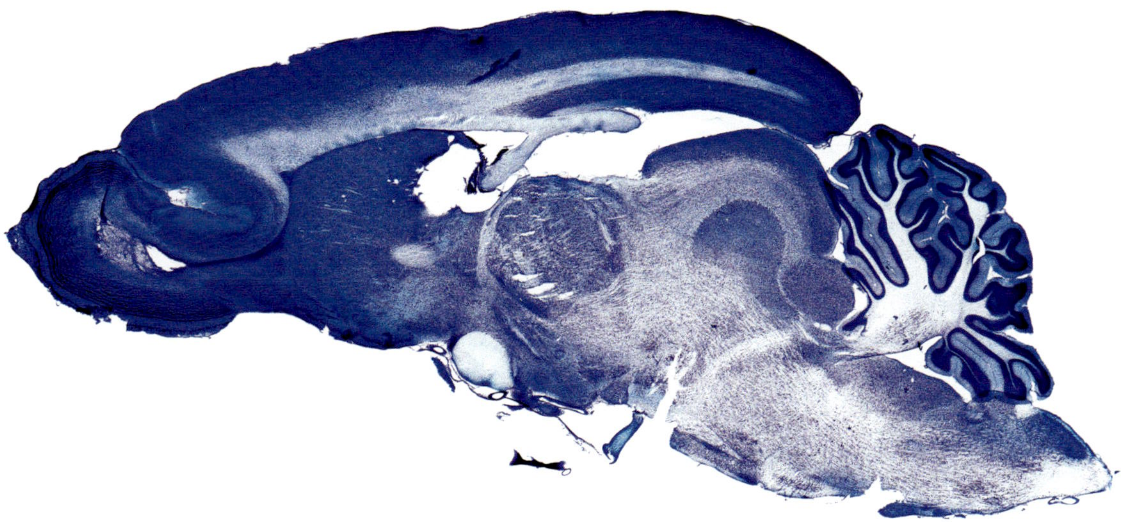

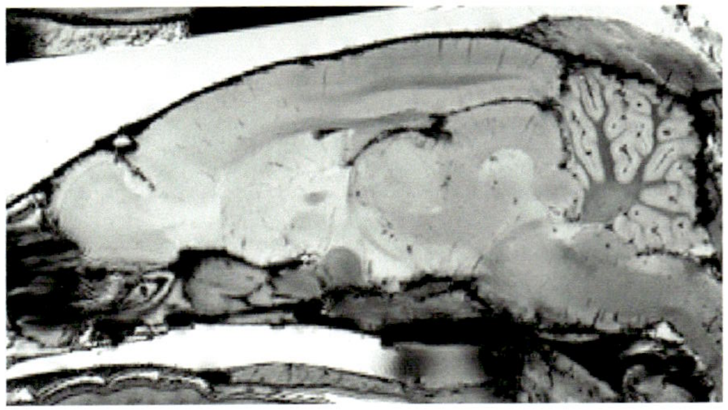

2	layer 2 of cortex
3	layer 3 of cortex
6n	root of abducens nerve
7n	facial nerve
ac	anterior commissure
Acb	accumbens nucleus
AD	anterodorsal thalamic nucleus
AM	anteromedial thalamic nucleus
AON	anterior olfactory nucleus
APT	anterior pretectal nucleus
AV	anteroventral thalamic nucleus
cc	corpus callosum
Cd	caudate nucleus
cp	cerebral peduncle
cu	cuneate fasciculus
Cu	cuneate nucleus
DpG	deep gray layer of the superior colliculus
DpWh	deep white layer of the superior colliculus
EPl	external plexiform layer of the olfactory bulb
EPlA	external plexiform layer of the accessory olfactory bulb
f	fornix
fr	fasciculus retroflexus
GCL	granule cell layer of the olfactory bulb
Gi	gigantocellular reticular nucleus
GL	glomerular layer of the olfactory bulb
GlA	glomerular layer of the accessory olfactory bulb
GrA	granule cell layer of the accessory olfactory bulb
Hip	hippocampus
IC	inferior colliculus
IL	infralimbic cortex
InG	intermediate gray layer of the superior colliculus
InWh	intermediate white layer of the superior colliculus
IO	inferior olive
IPl	internal plexiform layer of the olfactory bulb
IRt	intermediate reticular nucleus
LC	locus coeruleus
LDTg	laterodorsal tegmental nucleus
lfp	longitudinal fasciculus of the pons
LHA	lateral hypothalamic area
lo	lateral olfactory tract
LPB	lateral parabrachial nucleus
LPO	lateral preoptic area
LRt	lateral reticular nucleus
LV	lateral ventricle
M1	primary motor cortex
M2	secondary motor cortex
MCL	mitral cell layer of the olfactory bulb
MdD	medullary reticular nucleus, dorsal part
me5	mesencephalic trigeminal tract
Me5	mesencephalic trigeminal nucleus
MFC	medial frontal cortex
mlf	medial longitudinal fasciculus
MPB	medial parabrachial nucleus
MPT	medial pretectal nucleus
mRt	mesencephalic reticular formation
MT	medial terminal nucleus of the accessory optic tract
OFC	orbital frontal cortex
ONL	olfactory nerve layer
Op	optic nerve layer of the superior colliculus
opt	optic tract
OPT	olivary pretectal nucleus
OV	olfactory ventricle (olfactory part of lateral ventricle)
PAG	periaqueductal gray
Pc	central nucleus of the pulvinar
PC	paracentral thalamic nucleus
PCRt	parvicellular reticular nucleus
PF	parafascicular thalamic nucleus
PLH	peduncular part of lateral hypothalamus
Pn	pontine nuclei
PnC	pontine reticular nucleus, caudal part
PnO	pontine reticular nucleus, oral part
PPd	posterior parietal dorsal area
Pr	prepositus nucleus
PrC	precommissural nucleus
PrL	prelimbic cortex
PTg	pedunculopontine tegmental nucleus
RN	red nucleus
RSg	retrosplenial granular cortex
Rt	reticular thalamic nucleus
RtTg	reticulotegmental nucleus of the pons
S1	primary somatosensory cortex
SCC	somatosensory caudal cortex
scp	superior cerebellar peduncle
SFi	septofimbrial nucleus
sm	stria medullaris of the thalamus
SNC	substantia nigra, compact part
SNR	substantia nigra, reticular part
Sol	solitary nucleus
SON	supraoptic nucleus
sp5	spinal trigeminal tract
Sp5	spinal trigeminal nucleus
STMA	bed nucleus of the stria terminalis, medial division, anterior part
STMP	bed nucleus of the stria terminalis, medial division, posterior part
Sub	submedius thalamic nucleus
SuG	superficial gray layer of the superior colliculus
TC	temporal cortex
tfp	transverse fibers of the pons
tth	trigeminothalamic tract
Tu1	layer 1 of olfactory tubercle
tz	trapezoid body
Tz	nucleus of the trapezoid body
V1	primary visual cortex
V2	secondary visual cortex
VeCb	vestibulocerebellar nucleus
Ves	vestibular nucleus
VL	ventrolateral thalamic nucleus
VP	ventral pallidum
VPPC	ventral posterior nucleus of the thalamus, parvicellular part
vsc	ventral spinocerebellar tract
VTAR	ventral tegmental area, rostral part
VTM	ventral tuberomammillary nucleus
Zo	zonal layer of the superior colliculus

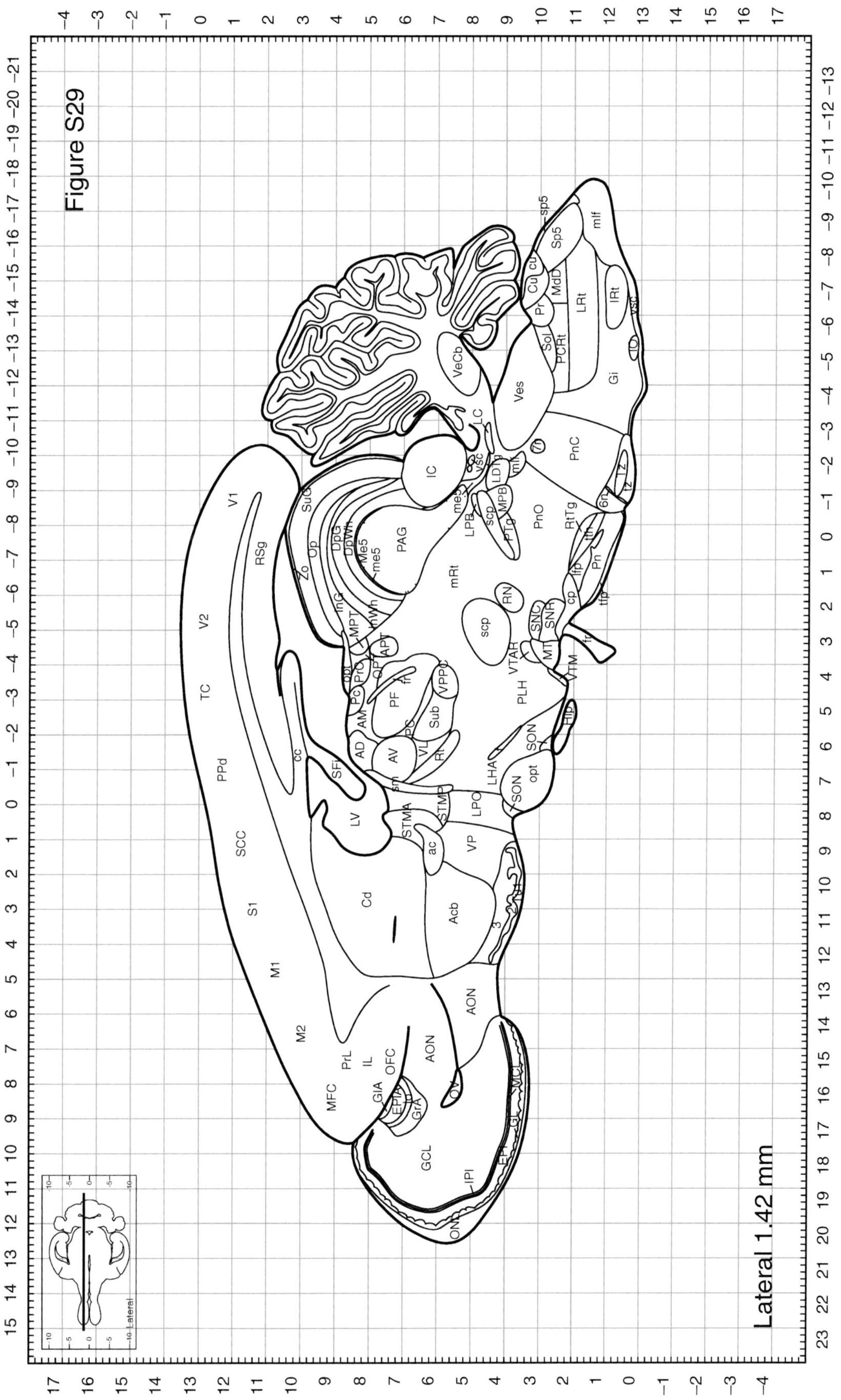
Figure S29
Lateral 1.42 mm
V1
V2
TC
PPd
SCC
S1
M1
M2
MFC
PrL
IL
OFC
GlA
AON
GCL
IPl
EPl
Gl
ON
MCL
RSg
cc
SFi
LV
Cd
Acb
VP
ac
STMA
STMP
LPO
LHA
SON
opt
PLH
VTAR
VTM
SNC
SNR
MT
RN
scp
mRt
PAG
SuG
Op
Zo
InG
InWh
DpG
DpWh
Me5
me5
MPT
APT
OPT
PF
VPPC
Sub
VL
Rt
AV
AD
AM
Pc
IC
LC
LPB
MPB
LDTg
PTg
mlf
PnO
PnC
RtTg
Pn
cp
tfp
Ves
VeCb
Sol
PCRt
Pr
Cu
cu
MdD
LRt
IRt
Gi
IO
vsc
Sp5
sp5
tz
Tz
7n
6n

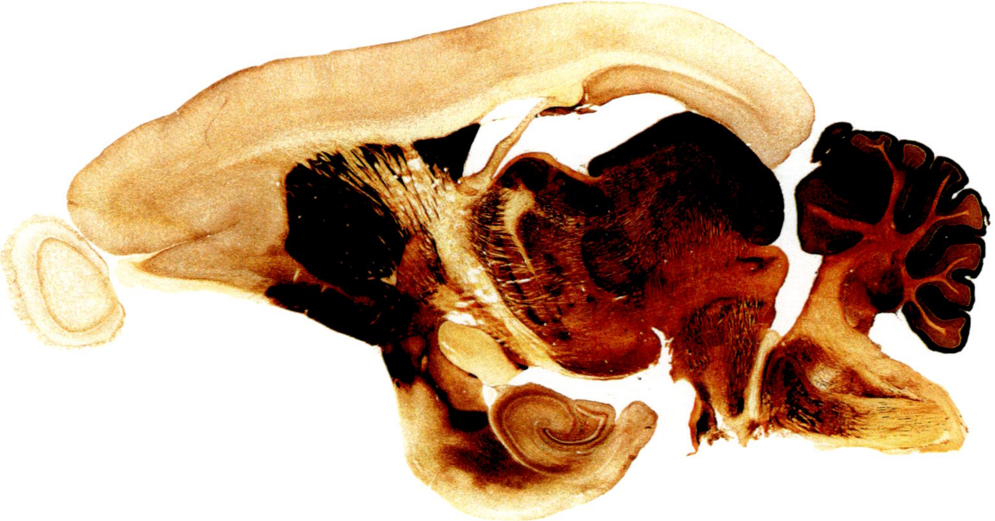

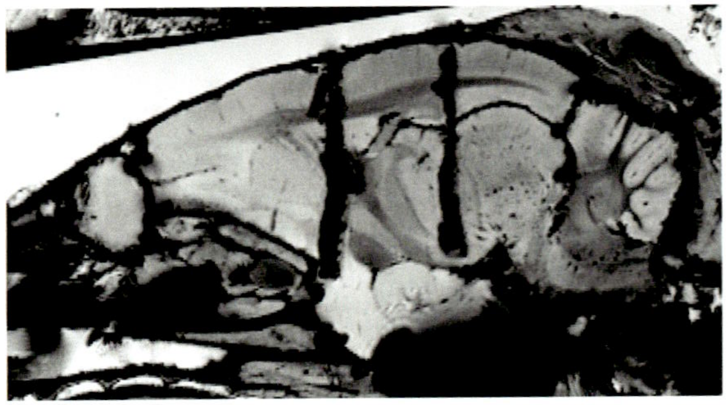

2	layer 2 of cortex
3	layer 3 of cortex
6n	root of abducens nerve
7n	facial nerve
ac	anterior commissure
Acb	accumbens nucleus
AD	anterodorsal thalamic nucleus
AON	anterior olfactory nucleus
AOP	anterior olfactory nucleus, posterior part
APT	anterior pretectal nucleus
AV	anteroventral thalamic nucleus
cc	corpus callosum
Cd	caudate nucleus
Cg	cingulate cortex
cp	cerebral peduncle
cu	cuneate fasciculus
Cu	cuneate nucleus
DpG	deep gray layer of the superior colliculus
DpWh	deep white layer of the superior colliculus
EPl	external plexiform layer of the olfactory bulb
EPlA	external plexiform layer of the accessory olfactory bulb
f	fornix
fr	fasciculus retroflexus
g7	genu of the facial nerve
GCL	granule cell layer of the olfactory bulb
Gi	gigantocellular reticular nucleus
GL	glomerular layer of the olfactory bulb
GlA	glomerular layer of the accessory olfactory bulb
Gr	gracile nucleus
GrA	granule cell layer of the accessory olfactory bulb
IC	inferior colliculus
ICj	islands of Calleja
IL	infralimbic cortex
InG	intermediate gray layer of the superior colliculus
InWh	intermediate white layer of the superior colliculus
IO	inferior olive
IPl	internal plexiform layer of the olfactory bulb
IRt	intermediate reticular nucleus
IRv	infraradiata ventral area
LDTg	laterodorsal tegmental nucleus
lfp	longitudinal fasciculus of the pons
LHA	lateral hypothalamic area
LHb	lateral habenular nucleus
LM	*lateral mammillary nucleus*
lo	lateral olfactory tract
LPO	lateral preoptic area
LSd	lateral septal nucleus, dorsal part
LSi	lateral septal nucleus, intermediate part
LV	lateral ventricle
M1	primary motor cortex
M2	secondary motor cortex
MCL	mitral cell layer of the olfactory bulb
MdD	medullary reticular nucleus, dorsal part
MdV	medullary reticular nucleus, ventral part
me5	mesencephalic trigeminal tract
Me5	mesencephalic trigeminal nucleus
Med	medial cerebellar nucleus
MFC	medial frontal cortex
MiA	mitral cell layer of the accessory olfactory bulb
ml	medial lemniscus
MT	medial terminal nucleus of the accessory optic tract
mlf	medial longitudinal fasciculus
MPT	medial pretectal nucleus
mRt	mesencephalic reticular formation
MT	medial terminal nucleus of the accessory optic
OFC	orbital frontal cortex
ONL	olfactory nerve layer
Op	optic nerve layer of the superior colliculus
opt	optic tract
OV	olfactory ventricle (olfactory part of lateral ventricle)
PAG	periaqueductal gray
PBP	parabrachial pigmented nucleus of the VTA
Pc	central nucleus of the pulvinar
PC	paracentral thalamic nucleus
PCRt	parvicellular reticular nucleus
PeF	perifornical nucleus
PF	parafascicular thalamic nucleus
PLH	peduncular part of lateral hypothalamus
Pn	pontine nuclei
PnC	pontine reticular nucleus, caudal part
PnO	pontine reticular nucleus, oral part
PPd	posterior parietal dorsal area
Pr	prepositus nucleus
PrC	precommissural nucleus
PrL	prelimbic cortex
RSg	retrosplenial granular cortex
Rt	reticular thalamic nucleus
RtTg	reticulotegmental nucleus of the pons
S1	primary somatosensory cortex
SCC	somatosensory caudal cortex
scp	superior cerebellar pedunclea
SFi	septofimbrial nucleus
sm	stria medullaris of the thalamus
SNR	substantia nigra, reticular part
Sol	solitary nucleus
SON	supraoptic nucleus
sp5	spinal trigeminal tract
Sp5	spinal trigeminal nucleus
SPTg	subpeduncular tegmental nucleus
STMA	bed nucleus of the stria terminalis, medial division, anterior part
STMP	bed nucleus of the stria terminalis, medial division, posterior part
Sub	submedius thalamic nucleus
SuG	superficial gray layer of the superior colliculus
SuML	supramammillary nucleus, lateral par
TC	temporal cortex
tfp	transverse fibers of the pons
tth	trigeminothalamic tract
Tu1	layer 1 of olfactory tubercle
tz	trapezoid body
V1	primary visual cortex
V2	secondary visual cortex
Ves	vestibular nucleus
VL	ventrolateral thalamic nucleus
VP	ventral pallidum
VPPC	ventral posterior nucleus of the thalamus, parvicellular part
vsc	ventral spinocerebellar tract
VTAR	ventral tegmental area, rostral part
xscp	decussation of the superior cerebellar peduncle
Zo	zonal layer of the superior colliculus

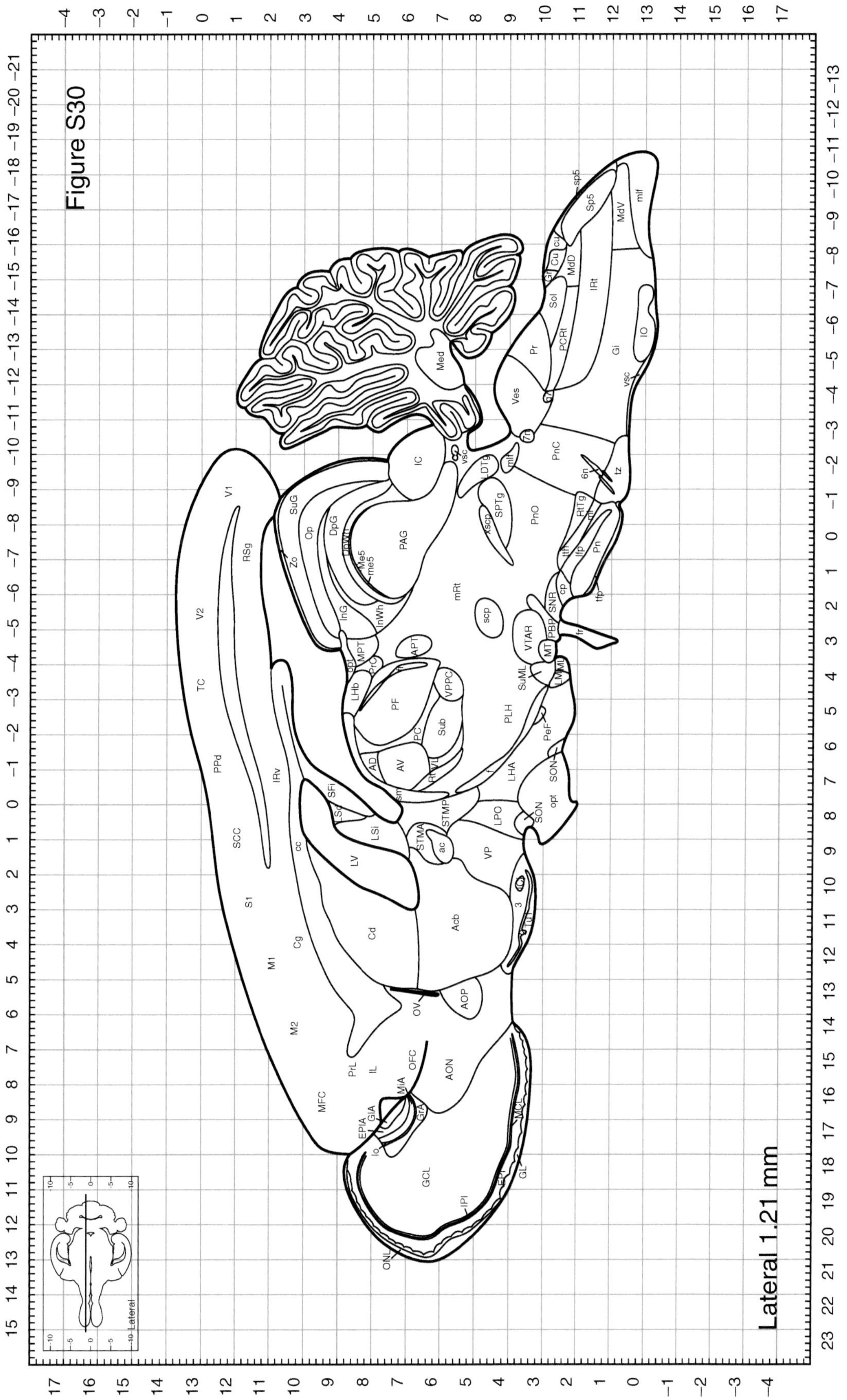
Figure S30
Lateral 1.21 mm

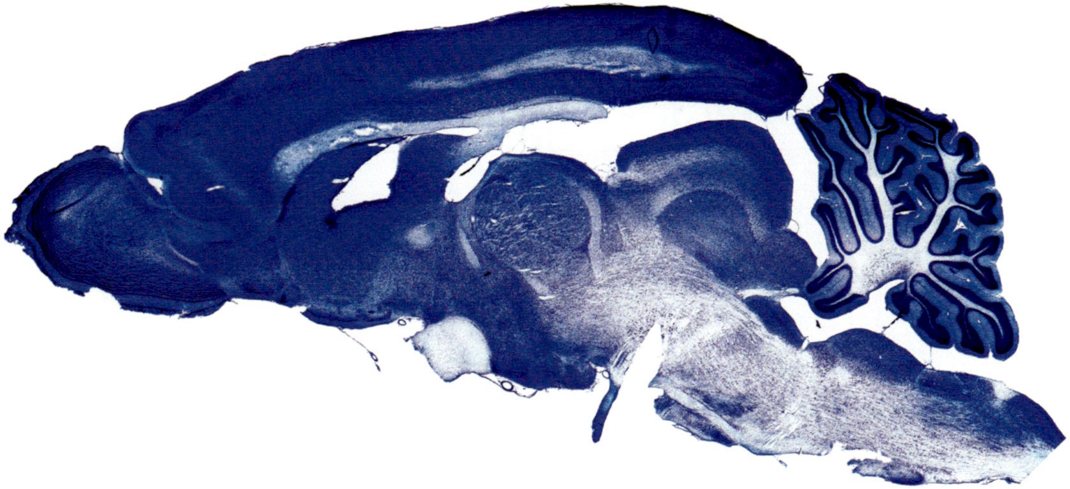

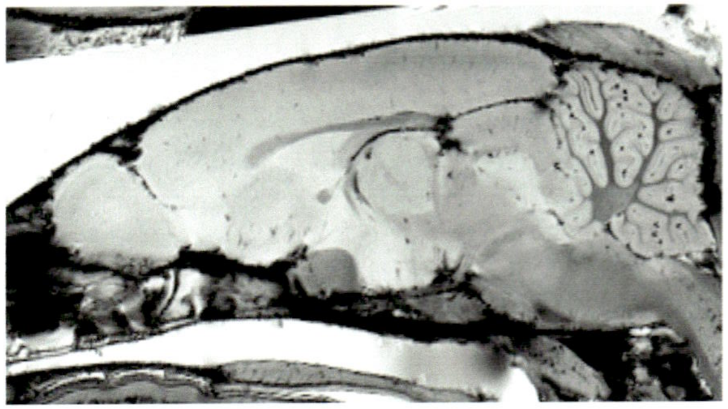

2	layer 2 of cortex
3	layer 3 of cortex
10N	dorsal motor nucleus of vagus
4n	trochlear nerve
4V	4th ventricle
6n	root of abducens nerve
7n	facial nerve
ac	anterior commissure
Acb	accumbens nucleus
AD	anterodorsal thalamic nucleus
AHy	anterior hypothalamic area
AON	anterior olfactory nucleus
AOP	anterior olfactory nucleus, posterior part
AV	anteroventral thalamic nucleus
cc	corpus callosum
Cd	caudate nucleus
Cg	cingulate cortex
cp	cerebral peduncle
cu	cuneate fasciculus
Cu	cuneate nucleus
DpG	deep gray layer of the superior colliculus
DpWh	deep white layer of the superior colliculus
EPl	external plexiform layer of the olfactory bulb
EPlA	external plexiform layer of the accessory olfactory bulb
f	fornix
fr	fasciculus retroflexus
g7	genu of the facial nerve
GCL	granule cell layer of the olfactory bulb
Gi	gigantocellular reticular nucleus
GL	glomerular layer of the olfactory bulb
GlA	glomerular layer of the accessory olfactory bulb
Gr	gracile nucleus
GrA	granule cell layer of the accessory olfactory bulb
IC	inferior colliculus
ICj	islands of Calleja
IG	indusium griseum
IL	infralimbic cortex
InG	intermediate gray layer of the superior colliculus
IO	inferior olive
IPl	internal plexiform layer of the olfactory bulb
IRt	intermediate reticular nucleus
IRv	infraradiata ventral area
LDTg	laterodorsal tegmental nucleus
lfp	longitudinal fasciculus of the pons
LHb	lateral habenular nucleus
LM	lateral mammillary nucleus
LSd	lateral septal nucleus, dorsal part
LSi	lateral septal nucleus, intermediate part
LSv	lateral septal nucleus, ventral part
LV	lateral ventricle
M1	primary motor cortex
M2	secondary motor cortex
MCL	mitral cell layer of the olfactory bulb
MD	mediodorsal thalamic nucleus
MdD	medullary reticular nucleus, dorsal part
MdV	medullary reticular nucleus, ventral part
Med	medial cerebellar nucleus
MFC	medial frontal cortex
ml	medial lemniscus
ML	medial mammillary nucleus, lateral part
mlf	medial longitudinal fasciculus
MPA	medial preoptic area
MPT	medial pretectal nucleus
mRt	mesencephalic reticular formation
MT	medial terminal nucleus of the accessory optic tract
mt	mammillothalamic tract
NDB	nucleus of the diagonal band
OFC	orbital frontal cortexONL olfactory nerve layer
Op	optic nerve layer of the superior colliculus
opt	optic tract
OV	olfactory ventricle (olfactory part of lateral ventricle)
PAG	periaqueductal gray
PBP	parabrachial pigmented nucleus of the
VTAPC	paracentral thalamic nucleus
PeF	perifornical nucleus
PF	parafascicular thalamic nucleus
PLH	peduncular part of lateral hypothalamus
Pn	pontine nuclei
PnC	pontine reticular nucleus, caudal part
PnO	pontine reticular nucleus, oral partp
PAG	p1 periaqueductal gray
PPd	posterior parietal dorsal area
Pr	prepositus nucleus
PrC	precommissural nucleus
PrL	prelimbic cortex
RSg	retrosplenial granular cortex
Rt	reticular thalamic nucleus
RtTg	reticulotegmental nucleus of the pons
S1	primary somatosensory cortex
SCC	somatosensory caudal cortex
scp	superior cerebellar peduncle
SFi	septofimbrial nucleus
sm	stria medullaris of the thalamus
SNR	substantia nigra, reticular part
Sol	solitary nucleus
SON	supraoptic nucleus
sp5	spinal trigeminal tract
Sp5	spinal trigeminal nucleus
SPTg	subpeduncular tegmental nucleus
STMA	bed nucleus of the stria terminalis, medial division, anterior part
STMP	bed nucleus of the stria terminalis, medial division, posterior part
Sub	submedius thalamic nucleus
SuG	superficial gray layer of the superior colliculus
SuML	supramammillary nucleus, lateral par
TC	temporal cortex
tfp	transverse fibers of the pons
tth	trigeminothalamic tract
Tu1	layer 1 of olfactory tubercle
tz	trapezoid body
V1	primary visual cortex
V2	secondary visual cortex
Ves	vestibular nucleus
VP	ventral pallidum
VPPC	ventral posterior nucleus of the thalamus, parvicellular part
vsc	ventral spinocerebellar tract
VTAR	ventral tegmental area, rostral part
xscp	decussation of the superior cerebellar peduncle
Zo	zonal layer of the superior colliculus

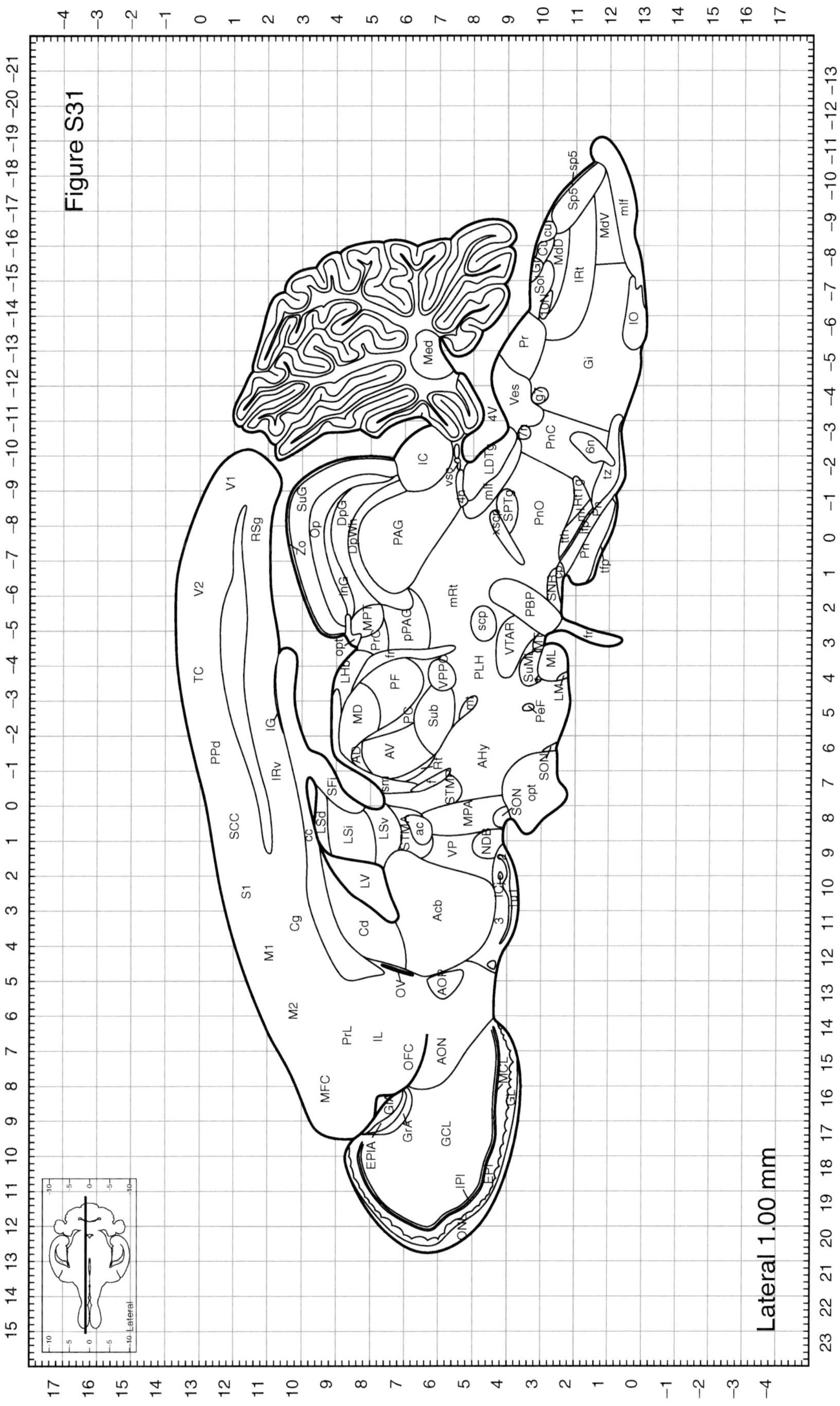
Figure S31
Lateral 1.00 mm
V1
V2
TC
PPd
SCC
S1
M1
M2
MFC
PrL
IL
OFC
AON
GCL
GrA
EPIA
IPI
RSg
IRv
IG
Cg
cc
LSd
LSi
LSv
SFi
LV
Cd
Acb
OV
AOP
VP
ac
STMA
MPA
NDB
SON
opt
SO
AHy
PeF
LM
ML
SuM
VTAR
PBP
SNR
scp
PLH
mt
Sub
PC
AV
AD
MD
PF
VPPC
LHb
PrC
MPT
pPAG
fr
InG
DpWh
DpG
Op
Zo
SuG
PAG
IC
mRt
vsc
LDTg
mlf
SPTg
xscp
PnO
PnC
6n
tz
Pn
ttp
tth
rtTg
4V
Ves
g7
Pr
Gi
IRt
IO
MdD
MdV
Sp5
sp5
Sol
Gr
Cu
cu
Med

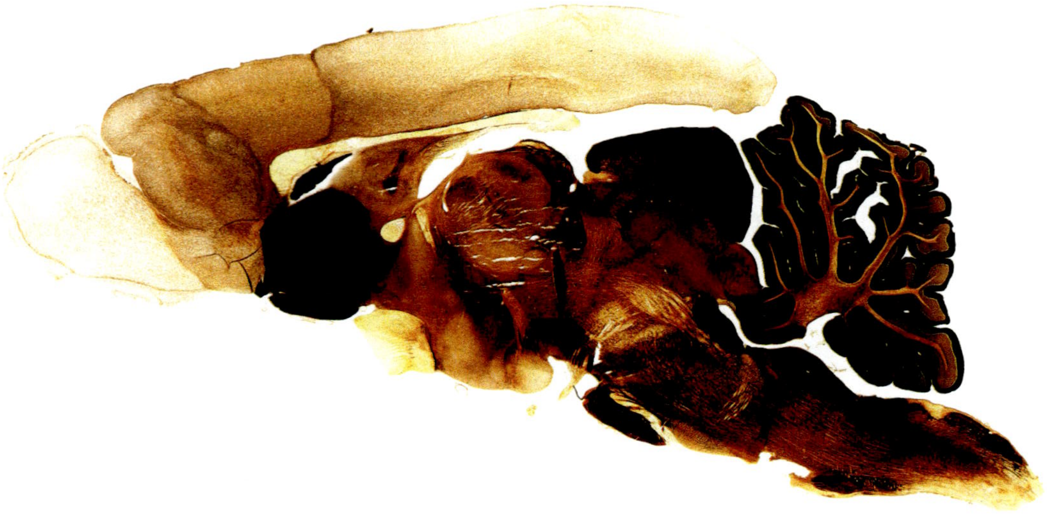

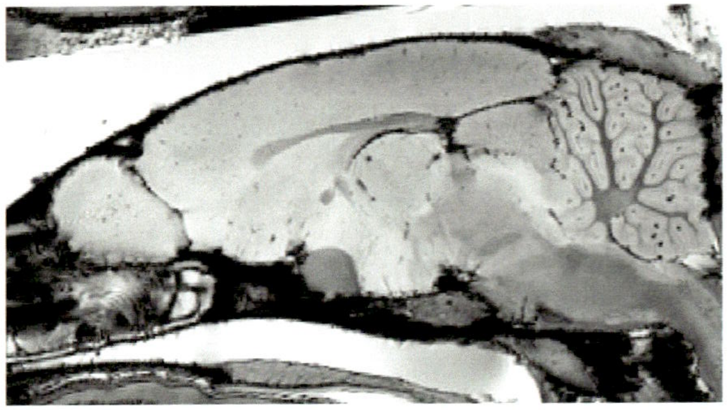

2	layer 2 of cortex
3	ayer 3 of cortex
10N	dorsal motor nucleus of vagus
12N	hypoglossal nucleus
4n	trochlear nerve
4N	trochlear nucleus
4V	4th ventricle
6n	root of abducens nerve
7n	facial nerve
ac	anterior commissure
Acb	accumbens nucleus
AHy	anterior hypothalamic area
AM	anteromedial thalamic nucleus
AON	anterior olfactory nucleus
Arc	arcuate hypothalamic nucleus
cc	corpus callosum
Cd	caudate nucleus
Cg	cingulate cortex
CLi	caudal linear nucleus of the raphe
cu	cuneate fasciculus
Cu	cuneate nucleus
DMH	dorsomedial hypothalamic nucleus
DpG	deep gray layer of the superior colliculus
DpWh	deep white layer of the superior colliculus
DR	dorsal raphe nucleus
EPI	external plexiform layer of the olfactory bulb
f	fornix
fr	fasciculus retroflexus
g7	genu of the facial nerve
GCL	granule cell layer of the olfactory bulb
Gi	gigantocellular reticular nucleus
GL	glomerular layer of the olfactory bulb
GlA	glomerular layer of the accessory olfactory bulb
Gr	gracile nucleus
IC	nferior colliculus
ICj	islands of Calleja
IG	indusium griseum
IL	infralimbic cortex
InC	interstitial nucleus of Cajal
InG	intermediate gray layer of the superior colliculus
IO	inferior olive
IP	interpeduncular nucleus
IPl	internal plexiform layer of the olfactory bulb
IRd	infraradiata dorsalis
IRt	intermediate reticular nucleus
IRv	infraradiata ventral area
LDTg	laterodorsal tegmental nucleus
lfp	longitudinal fasciculus of the pons
LHb	lateral habenular nucleus
LSd	lateral septal nucleus, dorsal part
LSi	lateral septal nucleus, intermediate part
LSv	lateral septal nucleus, ventral part
LV	lateral ventricle
M1	primary motor cortex
M2	secondary motor cortex
MCL	mitral cell layer of the olfactory bulb
MD	mediodorsal thalamic nucleus
MdD	medullary reticular nucleus, dorsal part
MdV	medullary reticular nucleus, ventral part
Med	medial cerebellar nucleus
MFC	medial frontal cortex
ml	medial lemniscus
ML	medial mammillary nucleus, lateral part
mlf	medial longitudinal fasciculus
MnR	median raphe nucleus
MPA	medial preoptic area
mRt	mesencephalic reticular formation
mt	mammillothalamic tract
NDB	nucleus of the diagonal band
OFC	orbital frontal cortex
ONL	olfactory nerve layer
Op	optic nerve layer of the superior colliculus
opt	optic tract
PAG	periaqueductal gray
PBP	parabrachial pigmented nucleus of the VTA
PC	paracentral thalamic nucleus
PF	parafascicular thalamic nucleus
PH	posterior hypothalamic nucleus
PLH	peduncular part of lateral hypothalamus
Pn	pontine nuclei
PnC	pontine reticular nucleus, caudal part
PnO	pontine reticular nucleus, oral part
pPAG	p1 periaqueductal gray
PPd	posterior parietal dorsal area
Pr	prepositus nucleus
PrC	precommissural nucleus
PrL	prelimbic cortex
py	pyramidal tract
Re	reuniens thalamic nucleus
RSg	retrosplenial granular cortex
RtTg	reticulotegmental nucleus of the pons
S1	primary somatosensory cortex
SCC	somatosensory caudal cortex
SFi	septofimbrial nucleus
sm	stria medullaris of the thalamus
Sol	solitary nucleus
SON	supraoptic nucleus
sp5	spinal trigeminal tract
Sp5	spinal trigeminal nucleus
STMA	bed nucleus of the stria terminalis, medial division, anterior part
STMP	bed nucleus of the stria terminalis, medial division, posterior part
Sub	submedius thalamic nucleus
SuG	superficial gray layer of the superior colliculus
SuML	supramammillary nucleus, lateral par
TC	temporal cortex
tfp	transverse fibers of the pons
tth	trigeminothalamic tract
Tu1	layer 1 of olfactory tubercle
tz	trapezoid body
V1	primary visual cortex
V2	secondary visual cortex
Ves	vestibular nucleus
VMH	ventromedial hypothalamic nucleus
VP	ventral pallidum
VPPC	ventral posterior nucleus of the thalamus, parvicellular part
vsc	ventral spinocerebellar tract
VTAR	ventral tegmental area, rostral part
vtgx	ventral tegmental decussation
xscp	decussation of the superior cerebellar peduncle
Zo	zonal layer of the superior colliculus

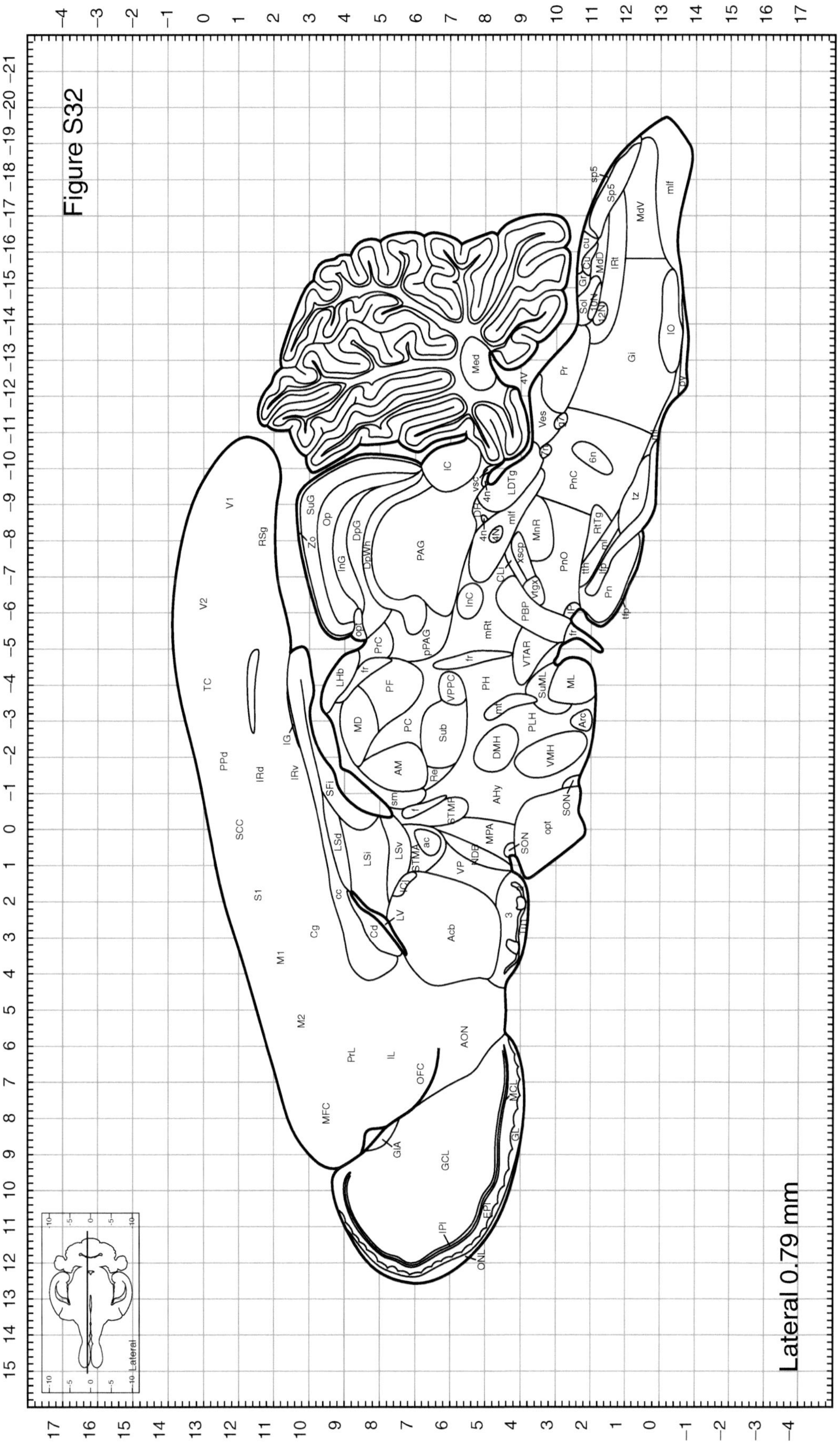
Figure S32
Lateral 0.79 mm

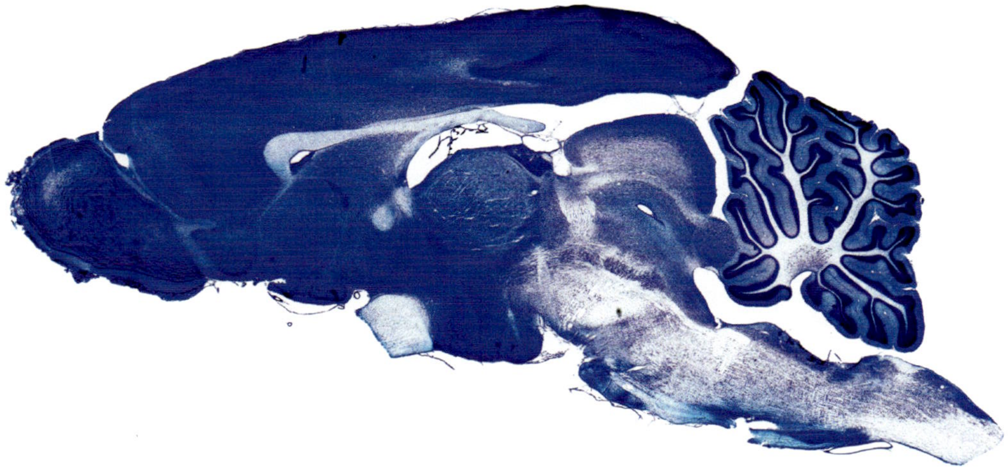

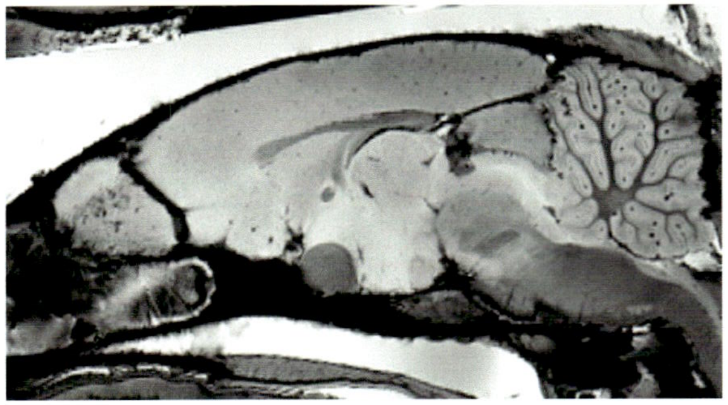

2	layer 2 of cortex
3	layer 3 of cortex
10N	dorsal motor nucleus of vagus
12N	hypoglossal nucleus
3N	oculomotor nucleus
3PC	oculomotor nucleus, parvicellular part
4n	trochlear nerve
4N	trochlear nucleus
4V	4th ventricle
6n	root of abducens nerve
ac	anterior commissure
Acb	accumbens nucleus
AHy	anterior hypothalamic area
AM	anteromedial thalamic nucleus
AON	anterior olfactory nucleus
Aq	aqueduct
Arc	arcuate hypothalamic nucleus
cc	corpus callosum
Cg	cingulate cortex
CG	central gray
CM	central medial thalamic nucleus
cu	cuneate fasciculus
Cu	cuneate nucleus
D3V	dorsal 3rd ventricle
DMH	dorsomedial hypothalamic nucleus
DpG	deep gray layer of the superior colliculus
DpWh	deep white layer of the superior colliculus
DR	dorsal raphe nucleus
DTg	dorsal tegmental nucleus
EPl	external plexiform layer of the olfactory bulb
f	fornix
fr	fasciculus retroflexus
g7	genu of the facial nerve
GCL	granule cell layer of the olfactory bulb
Gi	gigantocellular reticular nucleus
GL	glomerular layer of the olfactory bulb
GlA	glomerular layer of the accessory olfactory bulb
Gr	gracile nucleus
IC	inferior colliculus
ICj	islands of Calleja
IF	interfascicular nucleus
IG	indusium griseum
IL	infralimbic cortex
InC	interstitial nucleus of Cajal
InG	intermediate gray layer of the superior colliculus
IO	inferior olive
IP	interpeduncular nucleus
IPl	internal plexiform layer of the olfactory bulb
IRd	infraradiata dorsalis
IRt	intermediate reticular nucleus
IRv	infraradiata ventral area
LDTg	laterodorsal tegmental nucleus
lfp	longitudinal fasciculus of the pons
LHb	lateral habenular nucleus
LSd	lateral septal nucleus, dorsal part
LSi	lateral septal nucleus, intermediate part
LSv	lateral septal nucleus, ventral part
LV	lateral ventricle
M1	primary motor cortex
M2	secondary motor cortex
MCL	mitral cell layer of the olfactory bulb
MD	mediodorsal thalamic nucleus
MdV	medullary reticular nucleus, ventral part
Med	medial cerebellar nucleus
MFC	medial frontal cortex
MHb	medial habenular nucleus
ml	medial lemniscus
ML	medial mammillary nucleus, lateral part
mlf	medial longitudinal fasciculus
MnR	median raphe nucleus
MPA	medial preoptic area
mt	mammillothalamic tract
NDB	nucleus of the diagonal band
OFC	orbital frontal cortex
ONL	olfactory nerve layer
Op	optic nerve layer of the superior colliculus
opt	optic tract
PAG	periaqueductal gray
PBP	parabrachial pigmented nucleus of the VTA
pc	posterior commissure
PF	parafascicular thalamic nucleus
PH	posterior hypothalamic nucleus
Pn	pontine nuclei
PnC	pontine reticular nucleus, caudal part
pPAG	p1 periaqueductal gray
PPd	posterior parietal dorsal area
Pr	prepositus nucleus
PrC	precommissural nucleus
PnO	pontine reticular nucleus, oral part
PrL	prelimbic cortex
pRt p1	reticular formation
PVA	paraventricular thalamic nucleus, anterior part
PVN	paraventricular hypothalamic nucleus
PVP	paraventricular thalamic nucleus, posterior part
py	pyramidal tract
pyx	pyramidal decussation
Re	reuniens thalamic nucleus
Rh	rhomboid thalamic nucleus
RLi	rostral linear nucleus of the raphe
Ro	nucleus of Roller
RSg	retrosplenial granular cortex
RtTg	reticulotegmental nucleus of the pons
S1	primary somatosensory cortex
SCC	somatosensory caudal cortex
SFi	septofimbrial nucleus
sm	stria medullaris of the thalamus
Sol	solitary nucleus
SON	supraoptic nucleus
sp5	spinal trigeminal tract
Sp5	spinal trigeminal nucleus
STMA	bed nucleus of the stria terminalis, medial division, anterior part
STMP	bed nucleus of the stria terminalis, medial division, posterior part
Sub	submedius thalamic nucleus
SuG	superficial gray layer of the superior colliculus
SuML	supramammillary nucleus, lateral par
TC	temporal cortex
tfp	transverse fibers of the pons
TS	triangular septal nucleus
tth	trigeminothalamic tract
Tu1	layer 1 of olfactory tubercle
tz	trapezoid body
V1	primary visual cortex
V2	secondary visual cortex
Ves	vestibular nucleus
VMH	ventromedial hypothalamic nucleus
VP	ventral pallidum
VTAR	ventral tegmental area, rostral part
vtgx	ventral tegmental decussation
Zo	zonal layer of the superior colliculus

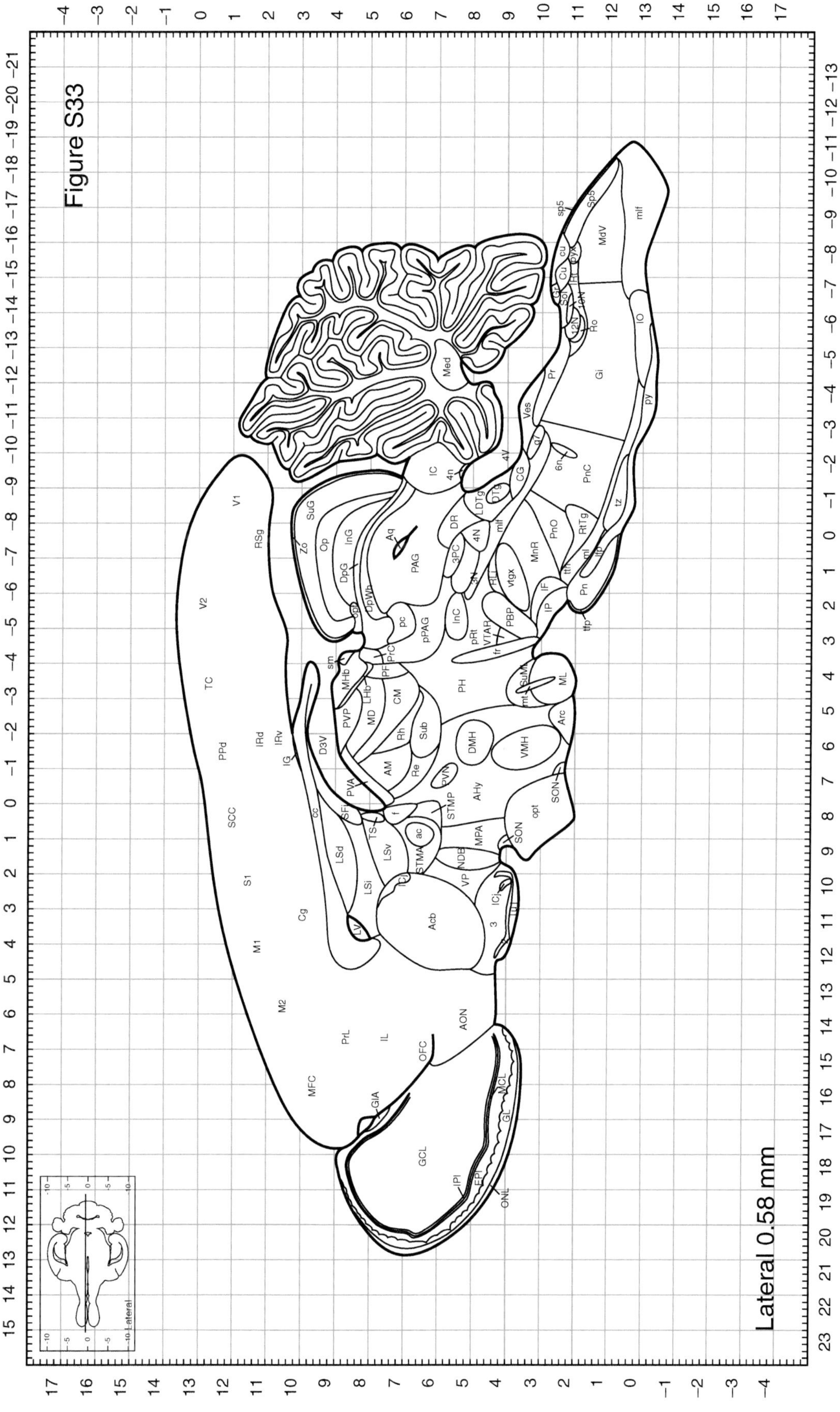
Figure S33
Lateral 0.58 mm
V1
V2
RSg
TC
PPd
SCC
S1
M1
M2
MFC
PrL
IL
OFC
AON
GlA
GCL
IPl
EPl
GL
MCL
ONL
Cg
cc
IG
lRd
lRv
D3V
LSd
LSv
LSi
LV
SFi
TS
Acb
ICj
VP
STMA
ac
STMP
MPA
NDB
AHy
SON
opt
PVA
AM
Re
PVN
DMH
VMH
Arc
ML
SuM
PH
Sub
Rh
CM
MD
PVP
MHb
LHb
PF
PrC
sm
pc
pPAG
PAG
Aq
InC
pRt
VTAR
PBP
fr
vtgx
IF
IP
Pn
tth
ml
MnR
PnO
RtTg
PnC
tz
DR
3PC
4N
3N
mlf
LDTg
DTg
CG
4V
4n
IC
SuG
Zo
Op
InG
DpG
DpWh
Med
Ves
g7
6n
Pr
Gi
Ro
12N
10N
Sol
Cu
cu
Dyx
SpT
MdV
sp5
Sp5
IO
py
mlf

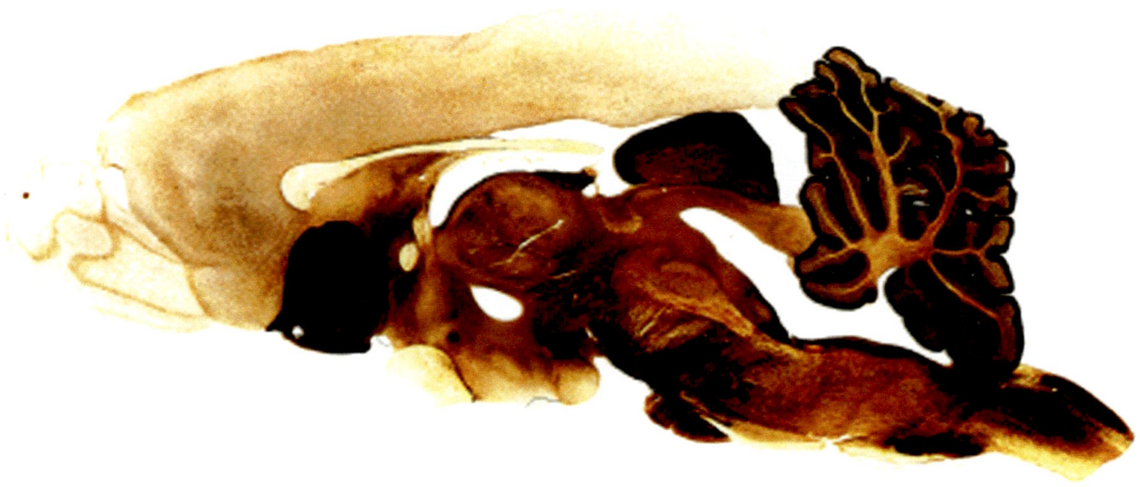

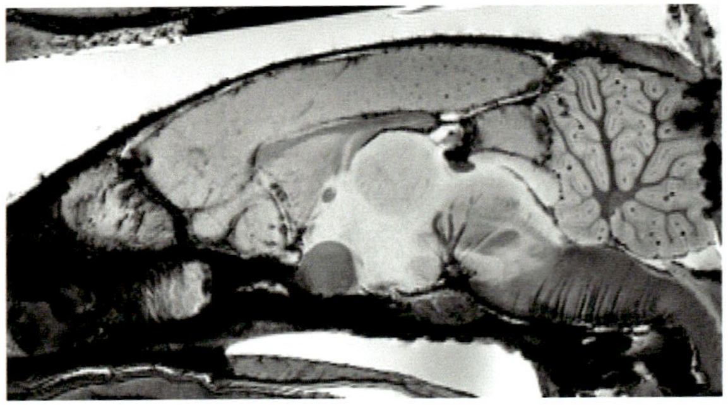

2	layer 2 of cortex
3	layer 3 of cortex
10	Ndorsal motor nucleus of vagus
12	Nhypoglossal nucleus
3N	oculomotor nucleus
3PC	oculomotor nucleus, parvicellular part
3V	3rd ventricle
4n	trochlear nerve
4N	trochlear nucleus
4V	4th ventricle
ac	anterior commissure
Acb	accumbens nucleus
AM	anteromedial thalamic nucleus
AON	anterior olfactory nucleus
AP	area postrema
Aq	aqueduct
Arc	arcuate hypothalamic nucleus
cc	corpus callosum
Cg	cingulate cortex
CG	central gray
CM	central medial thalamic nucleus
cu	cuneate fasciculus
Cu	cuneate nucleus
D3V	dorsal 3rd ventricle
DMH	dorsomedial hypothalamic nucleus
DpG	deep gray layer of the superior colliculus
DpWh	deep white layer of the superior colliculus
DR	dorsal raphe nucleus
DTg	dorsal tegmental nucleus
EPl	external plexiform layer of the olfactory bulb
f	fornix
g7	genu of the facial nerve
GCL	granule cell layer of the olfactory bulb
Gi	gigantocellular reticular nucleus
GL	glomerular layer of the olfactory bulb
Gr	gracile nucleus
IAM	interanteromedial thalamic nucleus
IC	inferior colliculus
ICj	islands of Calleja
IF	interfascicular nucleus
IG	indusium griseum
IL	infralimbic cortex
IMD	intermediodorsal thalamic nucleus
InC	interstitial nucleus of Cajal
IO	inferior olive
IPC	interpeduncular nucleus, caudal subnucleus
IPl	internal plexiform layer of the olfactory bulb
IPR	interpeduncular nucleus, rostral subnucleus
IRd	infraradiata dorsalis
IRv	infraradiata ventral area
LDTg	laterodorsal tegmental nucleus
lfp	longitudinal fasciculus of the pons
LSd	lateral septal nucleus, dorsal part
LSi	lateral septal nucleus, intermediate part
LSv	lateral septal nucleus, ventral part
LV	lateral ventricle
M1	primary motor cortex
M2	secondary motor cortex
MCL	mitral cell layer of the olfactory bulb
MdV	medullary reticular nucleus, ventral part
Med	medial cerebellar nucleus
MFC	medial frontal cortex
MHb	medial habenular nucleus
ml	medial lemniscus
ML	medial mammillary nucleus, lateral part
mlf	medial longitudinal fasciculus
MnR	median raphe nucleus
MPA	medial preoptic area
MS	medial septal nucleus
NDB	nucleus of the diagonal band
och	optic chiasm
OFC	orbital frontal cortex
ONL	olfactory nerve layer
Op	optic nerve layer of the superior colliculus
PAG	periaqueductal gray
PBP	parabrachial pigmented nucleus of the VTA
pc	posterior commissure
PH	posterior hypothalamic nucleus
Pn	pontine nuclei
PnC	pontine reticular nucleus, caudal part
pPAG	p1 periaqueductal gray
PPd	posterior parietal dorsal area
Pr	prepositus nucleus
PrC	precommissural nucleus
PrL	prelimbic cortex
PVA	paraventricular thalamic nucleus, anterior part
PVN	paraventricular hypothalamic nucleus
PVP	paraventricular thalamic nucleus, posterior part
py	pyramidal tract
pyx	pyramidal decussation
Re	reuniens thalamic nucleus
Rh	rhomboid thalamic nucleus
RLi	rostral linear nucleus of the raphe
RSg	retrosplenial granular cortex
RtTg	reticulotegmental nucleus of the pons
S1	primary somatosensory cortex
SCC	somatosensory caudal cortex
SCN	suprachiasmatic nucleus
SFi	septofimbrial nucleus
SHy	septohypothalamic nucleus
sm	stria medullaris of the thalamus
Sol	solitary nucleus
SON	supraoptic nucleus
sp5	spinal trigeminal tract
Sp5	spinal trigeminal nucleus
SPa	subparaventricular zone of the hypothalamus
STMA	bed nucleus of the stria terminalis, medial division, anterior part
Sub	submedius thalamic nucleus
SuG	superficial gray layer of the superior colliculus
SuMM	supramammillary nucleus, medial part
sumx	supramammillary decussation
TC	temporal cortex
tfp	transverse fibers of the pons
TS	triangular septal nucleus
tth	trigeminothalamic tract
Tu1	layer 1 of olfactory tubercle
tz	trapezoid body
V1	primary visual cortex
V2	secondary visual cortex
Ves	vestibular nucleus
VMH	ventromedial hypothalamic nucleus
VP	ventral pallidum
vtgx	ventral tegmental decussation
Zo	zonal layer of the superior colliculus

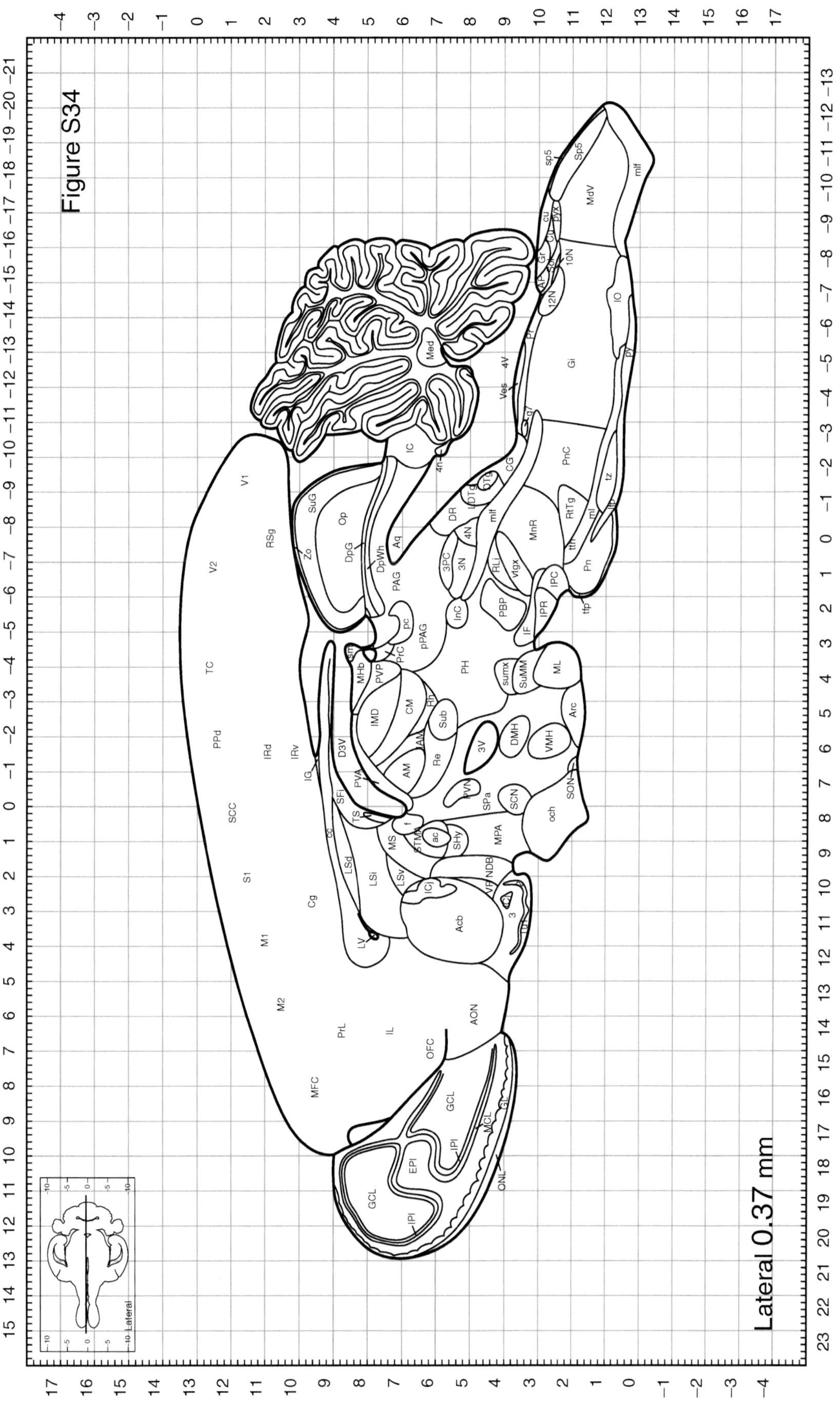
Figure S34
Lateral 0.37 mm
V1
V2
RSg
TC
PPd
IRd
IRv
SCC
S1
Cg
M1
M2
PrL
IL
MFC
OFC
AON
GCL
EPI
IPl
MCL
GL
ONL
Acb
LV
LSd
LSi
LSv
MS
cc
IG
SFi
TS
D3V
PVA
IMD
CM
AM
Re
Sub
MHb
PVP
PrC
pc
pPAG
PAG
PH
DpG
DpWh
Op
SuG
Zo
Aq
IC
Med
3PC
3N
4N
DR
LDTg
DTg
CG
mlf
RLi
vtgx
IPC
IPR
PBP
InC
IF
sumx
SuMM
ML
Arc
VMH
DMH
3V
PVN
SPa
SCN
och
SON
MPA
SHy
ac
STMA
VP
NDB
ICj
MnR
RtTg
PnC
Pn
ml
tz
tfp
Gi
IO
py
4V
Ves
PrH
12N
10N
AP
Gr
Cu
cu
DVX
MdV
Sp5
sp5
mlf

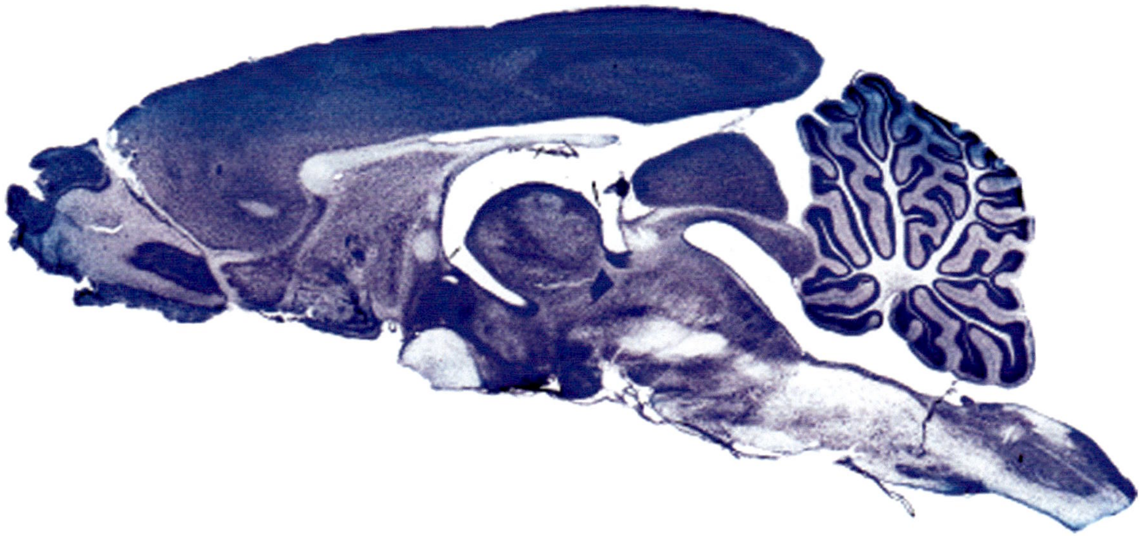

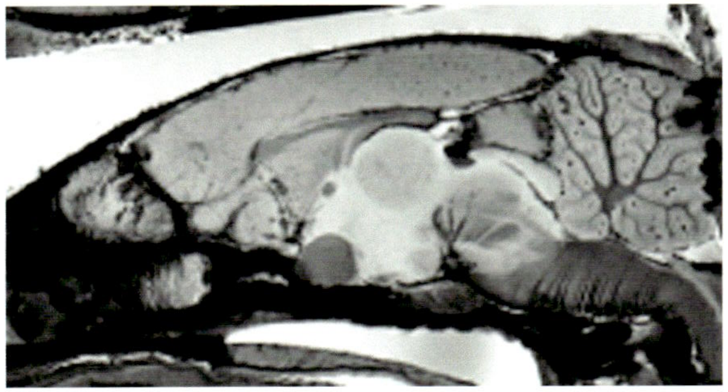

2	layer 2 of cortex
3	layer 3 of cortex
10	Ndorsal motor nucleus of vagus
12	Nhypoglossal nucleus
3N	oculomotor nucleus
3PC	oculomotor nucleus, parvicellular part
3V	3rd ventricle
4n	trochlear nerve
4N	trochlear nucleus
4V	4th ventricle
ac	anterior commissure
Acb	accumbens nucleus
AM	anteromedial thalamic nucleus
AON	anterior olfactory nucleus
AP	area postrema
Aq	aqueduct
Arc	arcuate hypothalamic nucleus
cc	corpus callosum
Cg	cingulate cortex
CG	central gray
CLi	caudal linear nucleus of the raphe
CM	central medial thalamic nucleus
cu	cuneate fasciculus
Cu	cuneate nucleus
D3V	dorsal 3rd ventricle
DMH	dorsomedial hypothalamic nucleus
DpWh	deep white layer of the superior colliculus
DR	dorsal raphe nucleus
dtgx	dorsal tegmental decussation
EPl	external plexiform layer of the olfactory bulb
EW	Edinger-Westphal nucleus
f	fornix
GCL	granule cell layer of the olfactory bulb
Gi	gigantocellular reticular nucleus
GL	glomerular layer of the olfactory bulb
Gr	gracile nucleus
IAM	interanteromedial thalamic nucleus
IC	inferior colliculus
ICj	islands of Calleja
IF	interfascicular nucleus
IL	infralimbic cortex
IMD	intermediodorsal thalamic nucleus
IO	inferior olive
IPC	interpeduncular nucleus, caudal subnucleus
IPl	internal plexiform layer of the olfactory bulb
IPR	interpeduncular nucleus, rostral subnucleus
IRd	infraradiata dorsalis
IRv	infraradiata ventral area
LSd	lateral septal nucleus, dorsal part
LSi	lateral septal nucleus, intermediate part
LV	lateral ventricle
M1	primary motor cortex
M2	secondary motor cortex
MCL	mitral cell layer of the olfactory bulb
MdV	medullary reticular nucleus, ventral part
ME	median eminence
Med	medial cerebellar nucleus
MFC	medial frontal cortex
ml	medial lemniscus
ML	medial mammillary nucleus, lateral part
mlf	medial longitudinal fasciculus
MnR	median raphe nucleus
MPA	medial preoptic area
MRe	mammillary recess of the 3rd ventricle
MS	medial septal nucleus
NDB	nucleus of the diagonal band
och	optic chiasm
OFC	orbital frontal cortex
ONL	olfactory nerve layer
PAG	periaqueductal gray
pc	posterior commissure
PG	pineal gland
PH	posterior hypothalamic nucleus
Pn	pontine nuclei
PnC	pontine reticular nucleus, caudal part
pPAG	p1 periaqueductal gray
PPd	posterior parietal dorsal area
Pr	prepositus nucleus
PrL	prelimbic cortex
PVA	paraventricular thalamic nucleus, anterior part
PVN	paraventricular hypothalamic nucleus
PVP	paraventricular thalamic nucleus, posterior part
py	pyramidal tract
pyx	pyramidal decussation
Rbd	rhabdoid nucleus
Re	reuniens thalamic nucleus
Rh	rhomboid thalamic nucleus
RLi	rostral linear nucleus of the raphe
RSg	retrosplenial granular cortex
RtTg	reticulotegmental nucleus of the pons
S1	primary somatosensory cortex
SCC	somatosensory caudal cortex
SCN	suprachiasmatic nucleus
SFi	septofimbrial nucleus
SHi	septohippocampal nucleus
SHy	septohypothalamic nucleus
sm	stria medullaris of the thalamus
Sol	solitary nucleus
SON	supraoptic nucleus
sp5	spinal trigeminal tract
Sp5	spinal trigeminal nucleus
SPa	subparaventricular zone of the hypothalamus
SuG	superficial gray layer of the superior colliculus
SuMM	supramammillary nucleus, medial part
sumx	supramammillary decussation
TC	temporal cortex
tfp	transverse fibers of the pons
TS	triangular septal nucleus
tth	trigeminothalamic tract
Tu1	layer 1 of olfactory tubercle
tz	trapezoid body
V1	primary visual cortex
V2	secondary visual cortex
VMH	ventromedial hypothalamic nucleus
vtgx	ventral tegmental decussation
Zo	zonal layer of the superior colliculus

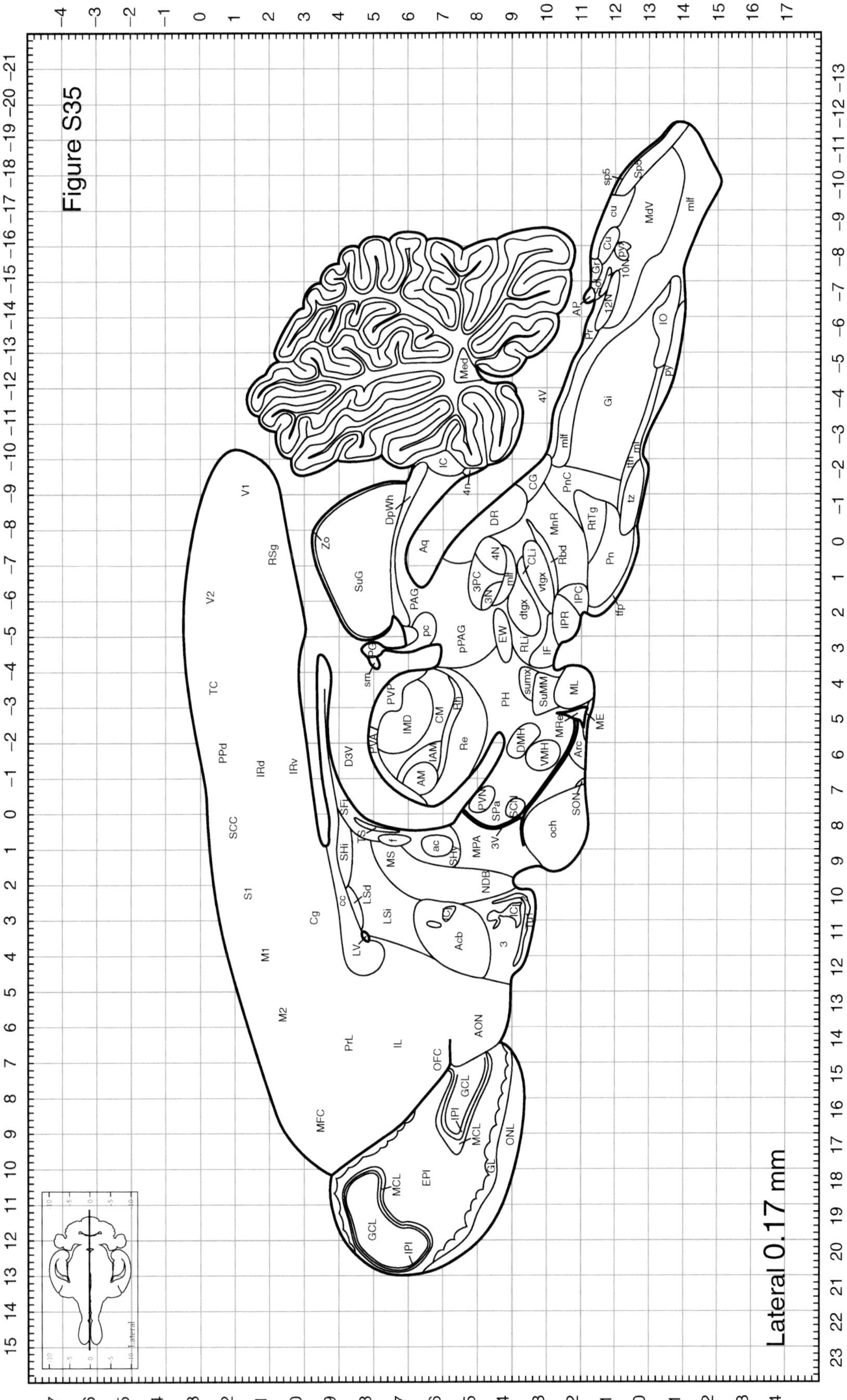
Figure S35
V1
RSg
V2
TC
PPd
IRd
IRv
SCC
S1
M1
M2
Cg
PrL
IL
MFC
OFC
AON
EPI
MCL
GCL
IPl
ONL
GL
Acb
3
LV
cc
LSd
LSi
SHi
SFi
TS
MS
f
ac
SHy
MPA
3V
NDB
D3V
PVA
PVP
IMD
CM
Rh
Re
IAM
AM
sm
PVN
SPa
SCh
och
SON
Arc
ME
MRe
VMH
DMH
PH
SuMM
sumx
ML
IF
RLi
EW
pPAG
pc
PAG
SuG
Zo
DpWh
Aq
IC
4n
Med
4V
CG
DR
PnC
MnR
RtTg
tz
ml
tth
mlf
3PC
3N
4N
CLi
Rbd
vtgx
dtgx
IPR
IPC
Pn
tfp
Gi
py
IO
Pr
AP
Sol
Gr
12N
10N
Cu
cu
sp5
Sp5
MdV
Lateral 0.17 mm

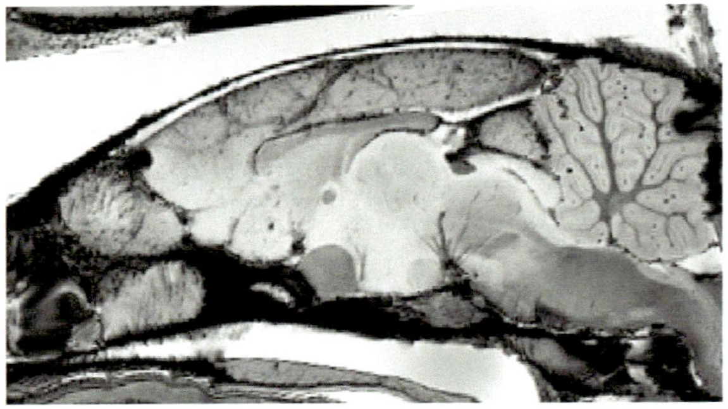

10N	dorsal motor nucleus of vagus
12N	hypoglossal nucleus
3PC	oculomotor nucleus, parvicellular part
3V	3rd ventricle
4n	trochlear nerve
4N	trochlear nucleus
4V	4th ventricle
ac	anterior commissure
AM	anteromedial thalamic nucleus
AON	anterior olfactory nucleus
AP	area postrema
Arc	arcuate hypothalamic nucleus
cc	corpus callosum
Cg	cingulate cortex
CG	central gray
CLi	caudal linear nucleus of the raphe
CM	central medial thalamic nucleus
csc	commissure of the superior colliculus
cu	cuneate fasciculus
D3V	dorsal 3rd ventricle
DMH	dorsomedial hypothalamic nucleus
DpWh	deep white layer of the superior colliculus
DR	dorsal raphe nucleus
dtgx	dorsal tegmental decussation
EPl	external plexiform layer of the olfactory bulb
EW	Edinger-Westphal nucleus
GCL	granule cell layer of the olfactory bulb
Gi	gigantocellular reticular nucleus
GL	glomerular layer of the olfactory bulb
Gr	gracile nucleus
IAM	interanteromedial thalamic nucleus
IC	inferior colliculus
ICj	islands of Calleja
IF	interfascicular nucleus
IG	indusium griseum
IL	infralimbic cortex
IMD	intermediodorsal thalamic nucleus
IO	inferior olive
IPC	interpeduncular nucleus, caudal subnucleus
IPl	internal plexiform layer of the olfactory bulb
IPR	interpeduncular nucleus, rostral subnucleus
IRd	infraradiata dorsalis
IRv	infraradiata ventral area
LSi	lateral septal nucleus, intermediate part
M1	primary motor cortex
M2	secondary motor cortex
MCL	mitral cell layer of the olfactory bulb
MdV	medullary reticular nucleus, ventral part
ME	median eminence
Med	medial cerebellar nucleus
MFC	medial frontal cortex
ml	medial lemniscus
ML	medial mammillary nucleus, lateral part
mlf	medial longitudinal fasciculus
MnR	median raphe nucleus
MPA	medial preoptic area
MS	medial septal nucleus
NDB	nucleus of the diagonal band
och	optic chiasm
OFC	orbital frontal cortex
ONL	olfactory nerve layer
PAG	periaqueductal gray
pc	posterior commissure
PG	pineal gland
PH	posterior hypothalamic nucleus
Pn	pontine nuclei
PnC	pontine reticular nucleus, caudal part
pPAG	p1 periaqueductal gray
PPd	posterior parietal dorsal area
Pr	prepositus nucleus
PrL	prelimbic cortex
PVA	paraventricular thalamic nucleus, anterior part
PVN	paraventricular hypothalamic nucleus
PVP	paraventricular thalamic nucleus, posterior part
py	pyramidal tract
pyx	pyramidal decussation
Re	reuniens thalamic nucleus
Rh	rhomboid thalamic nucleus
RLi	rostral linear nucleus of the raphe
RSg	retrosplenial granular cortex
RtTg	reticulotegmental nucleus of the pons
S1	primary somatosensory cortex
SCC	somatosensory caudal cortex
SCN	suprachiasmatic nucleus
SFi	septofimbrial nucleus
SHi	septohippocampal nucleus
SHy	septohypothalamic nucleus
Sol	solitary nucleus
sp5	spinal trigeminal tract
Sp5	spinal trigeminal nucleus
SPa	subparaventricular zone of the hypothalamus
SuG	superficial gray layer of the superior colliculus
SuMM	supramammillary nucleus, medial partsumx supramammillary decussation
TC	temporal corte
xtfp	transverse fibers of the pons
TS	triangular septal nucleus
Tu	olfactory tubercle
tz	trapezoid body
V1	primary visual cortex
V2	secondary visual cortex
VMH	ventromedial hypothalamic nucleus
vtgx	ventral tegmental decussation
Zo	zonal layer of the superior colliculus

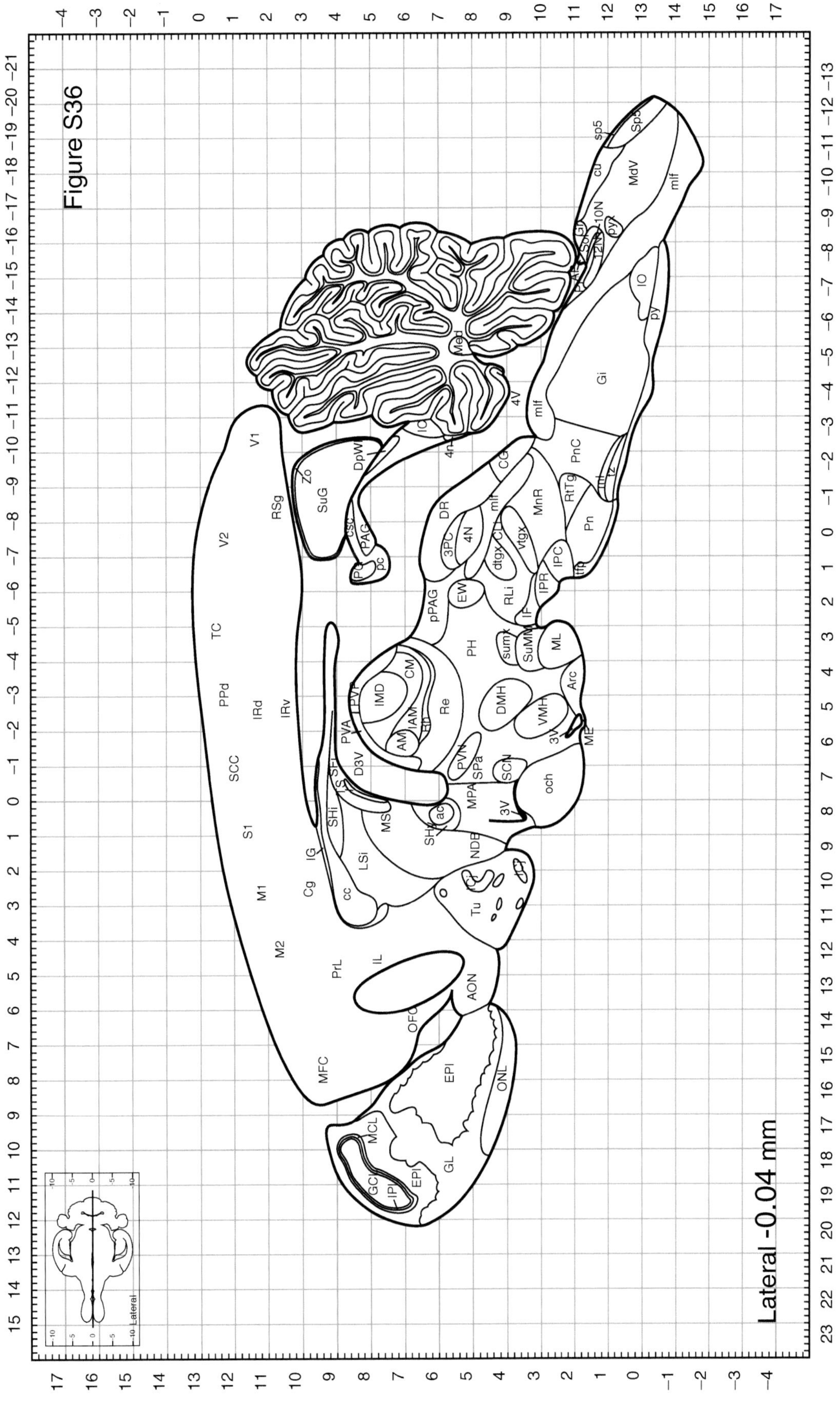
Figure S36
Lateral -0.04 mm

List of Brain Regions

Rhinencephalon			
	Main olfactory bulb	Ependyma and subependymal layer	E
		External plexiform layer of the olfactory bulb	EPl
		Glomerular layer of the olfactory bulb	GL
		Granule cell layer of the olfactory bulb	GCL
		Internal plexiform layer of the olfactory bulb	IPl
		Mitral cell layer of the olfactory bulb	MCL
		Olfactory nerve layer	ONL
	Accessory olfactory bulb	External plexiform layer of the accessory olfactory bulb	EPlA
		Glomerular layer of the accessory olfactory bulb	GlA
		Granule cell layer of the accessory olfactory bulb	GrA
		Mitral cell layer of the accessory olfactory bulb	MiA
	Anterior olfactory nucleus	Anterior olfactory nucleus, dorsal part	AOD
		Anterior olfactory nucleus, external part	AOE
		Anterior olfactory nucleus, lateral part	AOL
		Anterior olfactory nucleus, medial part	AOM
		Anterior olfactory nucleus, posterior part	AOP
		Anterior olfactory nucleus, ventral part	AOV
	Layer 1 of olfactory tubercle		Tu1
	Layer 2 of olfactory tubercle		2
	Layer 3 of olfactory tubercle		3
	Nucleus of the lateral olfactory tract		LOT
	Olfactory tubercle		Tu

Telencephalon			
	Neocortex	Alveus of the hippocampus	alv
		Auditory belt area	Ab
		Auditory core area	Ac
		Cingulate cortex	Cg
		Dentate gyrus of the hippocampus	DG
		Dorsal frontal cortex	DFC
		Dorsal peduncular cortex	DP
		Dorsal subiculum	DS
		Ectorhinal cortex	Ect
		Entorhinal cortex	Ent
		Field CA1 of the hippocampus	CA1
		Field CA2 of the hippocampus	CA2
		Field CA2 of the hippocampus	CA3

J.-N. Zhou, R.-J. Ni, *The Tree Shrew (Tupaia belangeri chinensis) Brain in Stereotaxic Coordinates*,
DOI 10.1007/978-981-10-0611-1

		Granular layer of the dentate gyrus	GrDG
		Hippocampus	Hip
		Indusium griseum	IG
		Inferior temporal cortex	IT
		Infralimbic cortex	IL
		Infraradiata dorsalis	IRd
		Infraradiata ventral area	IRv
		Insular cortex	Ins
		Lacunosum moleculare layer of the hippocampus	LMoL
		Layer 2 of cortex	2
		Layer 3 of cortex	3
		Medial frontal cortex	MFC
		Molecular layer of the dentate gyrus	MoDG
		Orbital frontal cortex	OFC
		Oriens layer of the hippocampus	Or
		Parietal ventral area	PvA
		Perirhinal cortex	PRh
		Piriform cortex, layer 1	Pir1
		Polymorph layer of the dentate gyrus	PoDG
		Posterior parietal caudal area	PPc
		Posterior parietal dorsal area	PPd
		Posterior parietal rostral area	PPr
		Postsubiculum	Post
		Prelimbic cortex	PrL
		Presubiculum	PrS
		Primary motor cortex	M1
		Primary somatosensory cortex	S1
		Primary visual cortex	V1
		Pyramidal cell layer of the hippocampus	Py
		Radiatum layer of the hippocampus	Rad
		Retrosplenial granular cortex	RSg
		Secondary motor cortex	M2
		Secondary somatosensory cortex	S2
		Secondary visual cortex	V2
		Somatosensory caudal cortex	SCC
		Subiculum, transition area	STr
		Temporal cortex	TC
		Temporal inferior area	TI
		Ventral subiculum	VS
	Basal ganglia	Accumbens nucleus	Acb
		Accumbens nucleus, core	AcbC
		Accumbens nucleus, shell	AcbSh
		Basal nucleus (Meynert)	BN
		Caudate nucleus	Cd
		Claustrum	Cl
		External globus pallidus	EGP
		Globus pallidus	GP
		Internal globus pallidus	IGP
		Interstitial nucleus of the posterior limb of the anterior commissure	IPAC
		Islands of Calleja	ICj
		Lateral stripe of the striatum	LSS
		Putamen	Pu
		Ventral pallidum	VP

	Amygdala	Accessory basal nucleus	AB
		Amygdalohippocampal area	AHi
		Amygdalostriatal transition area	ASt
		Anterior amygdaloid area	AA
		Anterior cortical amygdaloid nucleus	ACo
		Basolateral amygdaloid nucleus	BLA
		Basomedial amygdaloid nucleus	BMA
		Central amygdaloid nucleus	CeA
		Intercalated amygdaloid nucleus, main part	IM
		Medial amygdaloid nucleus	MeA
		Sublenticular extended amygdala	EA
		Sublenticular extended amygdala, central part	EAC
	Septal and preoptic regions	Bed nucleus of the stria terminalis, lateral division	STL
		Bed nucleus of the stria terminalis, medial division, anterolateral part	STMAL
		Bed nucleus of the stria terminalis, medial division, anteromedial part	STMAM
		Bed nucleus of the stria terminalis, medial division, posterior part	STMP
		Bed nucleus of the stria terminalis, medial division, posterolateral part	STMPL
		Bed nucleus of the stria terminalis, medial division, posteromedial part	STMPM
		Bed nucleus of the stria terminalis, medial division, ventral part	STMV
		Lateral preoptic area	LPO
		Lateral septal nucleus, dorsal part	LSd
		Lateral septal nucleus, intermediate part	LSi
		Lateral septal nucleus, ventral part	LSv
		Medial preoptic area	MPA
		Medial preoptic nucleus	MPO
		Medial septal nucleus	MS
		Median preoptic nucleus	MnPO
		Nucleus of the diagonal band	NDB
		Nucleus of the horizontal limb of the diagonal band	HDB
		Nucleus of the vertical limb of the diagonal band	VDB
		Septofimbrial nucleus	SFi
		Septohippocampal nucleus	SHi
		Septohypothalamic nucleus	SHy
		Subfornical organ	SFO
		Triangular septal nucleus	TS
Diencephalon			
	Hypothalamus	Anterior hypothalamic area	AHy
		Anteroventral periventricular nucleus	AVPe
		Arcuate hypothalamic nucleus	Arc
		Arcuate hypothalamic nucleus, posterior part	ArcP
		Dorsomedial hypothalamic nucleus	DMH
		Gemini hypothalamic nucleus	Gem
		Juxtaparaventricular part of lateral hypothalamus	JPLH
		Lateral hypothalamic area	LHA
		Lateral mammillary nucleus	LM
		Medial mammillary nucleus, lateral part	ML
		Medial mammillary nucleus, medial part	MM
		Median eminence	ME
		Paraventricular hypothalamic nucleus	PVN
		Paraventricular hypothalamic nucleus, anterior part	PaA
		Paraventricular hypothalamic nucleus, posterior part	PaP
		Paraventricular hypothalamic nucleus, ventral part	PaV

		Peduncular part of lateral hypothalamus	PLH
		Perifornical nucleus	PeF
		Periventricular hypothalamic nucleus	Pe
		Posterior hypothalamic area	PHA
		Posterior hypothalamic nucleus	PH
		Premammillary nucleus, dorsal part	PMD
		Premammillary nucleus, ventral part	PMV
		Striohypothalamic nucleus	StHy
		Subparaventricular zone of the hypothalamus	SPa
		Suprachiasmatic nucleus	SCN
		Supramammillary nucleus	SuM
		Supramammillary nucleus, lateral par	SuML
		Supramammillary nucleus, medial part	SuMM
		Supraoptic nucleus	SON
		Tuberal region of lateral hypothalamus	TuLH
		Ventral tuberomammillary nucleus	VTM
		Ventromedial hypothalamic nucleus	VMH
	Thalamus	Anterior pretectal nucleus	APT
		Anterior pretectal nucleus, dorsal part	APTD
		Anterior pretectal nucleus, ventral part	APTV
		Anterodorsal thalamic nucleus	AD
		Anteromedial thalamic nucleus	AM
		Anteroventral thalamic nucleus	AV
		Brachium of the superior colliculus	B
		Central medial thalamic nucleus	CM
		Central nucleus of the pulvinar	Pc
		Dorsal lateral geniculate nucleus	DLG
		Dorsal nucleus of the pulvinar	Pd
		Interanteromedial thalamic nucleus	IAM
		Intergeniculate leaf	IGL
		Intermediodorsal thalamic nucleus	IMD
		Lateral habenular nucleus	LHb
		Lateral habenular nucleus, lateral part	LHbL
		Lateral habenular nucleus, medial part	LHbM
		Laterodorsal thalamic nucleus	LD
		Medial geniculate nucleus	MG
		Medial geniculate nucleus, dorsal part	MGD
		Medial geniculate nucleus, ventral part	MGV
		Medial habenular nucleus	MHb
		Medial pretectal nucleus	MPT
		Mediodorsal thalamic nucleus	MD
		Olivary pretectal nucleus	OPT
		Paracentral thalamic nucleus	PC
		Parafascicular thalamic nucleus	PF
		Paratenial thalamic nucleus	PT
		Paraventricular thalamic nucleus, anterior part	PVA
		Paraventricular thalamic nucleus, posterior part	PVP
		Posterior complex of the thalamus	Po
		Posteromedian thalamic nucleus	PoMn
		Precommissural nucleus	PrC
		Pulvinar nuclei	Pul
		Reticular thalamic nucleus	Rt
		Reuniens thalamic nucleus	Re

		Rhomboid thalamic nucleus	Rh
		Stria medullaris of the thalamus	sm
		Submedius thalamic nucleus	Sub
		Superior thalamic radiation	str
		Ventral lateral geniculate nucleus	VLG
		Ventral nucleus of the pulvinar	Pv
		Ventral posterior nucleus of the thalamus, parvicellular part	VPPC
		Ventral posterior thalamic nucleus	VPT
		Ventral posteromedial thalamic nucleus	VPM
		Ventrolateral thalamic nucleus	VL
		Ventromedial thalamic nucleus	VM
	Pineal gland		PG
	Subthalamic nucleus		STh
	Zona incerta		ZI
Mesencephalon		Aqueduct	Aq
		Caudal linear nucleus of the raphe	CLi
		Central gray	CG
		Deep gray layer of the superior colliculus	DpG
		Deep white layer of the superior colliculus	DpWh
		Dorsal raphe nucleus	DR
		Edinger-Westphal nucleus	EW
		Hypoglossal nucleus	12N
		Inferior colliculus	IC
		Interfascicular nucleus	IF
		Intermediate gray layer of the superior colliculus	InG
		Intermediate white layer of the superior colliculus	InWh
		Interpeduncular nucleus	IP
		Interpeduncular nucleus, apical subnucleus	IPA
		Interpeduncular nucleus, caudal subnucleus	IPC
		Interpeduncular nucleus, lateral subnucleus	IPL
		Interpeduncular nucleus, rostral subnucleus	IPR
		Interstitial nucleus of Cajal	InC
		Medial terminal nucleus of the accessory optic tract	MT
		Median raphe nucleus	MnR
		Mesencephalic reticular formation	mRt
		Mesencephalic trigeminal nucleus	Me5
		Microcellular tegmental nucleus	MiTg
		Nucleus of the lateral lemniscus	LL
		Nucleus of the trapezoid body	Tz
		Oculomotor nucleus, parvicellular part	3PC
		Optic nerve layer of the superior colliculus	Op
		p1 periaqueductal gray	pPAG
		p1 reticular formation	pRt
		Parabigeminal nucleus	PBG
		Parabrachial pigmented nucleus of the VTA	PBP
		Paranigral nucleus of the VTA	PN
		Pedunculopontine tegmental nucleus	PTg
		Periaqueductal gray	PAG
		Pre-Edinger-Westphal nucleus	PrEW
		Red nucleus	RN
		Red nucleus, magnocellular part	RMC
		Red nucleus, parvicellular part	RPC
		Retrorubral field	RRF
		Rhabdoid nucleus	Rbd

		Rostral linear nucleus of the raphe	RLi
		Subpeduncular tegmental nucleus	SPTg
		Substantia nigra, compact part	SNC
		Substantia nigra, lateral part	SNL
		Substantia nigra, reticular part	SNR
		Superficial gray layer of the superior colliculus	SuG
		Superior colliculus	SC
		Trochlear nucleus	4N
		Ventral tegmental area	VTA
		Ventral tegmental area, rostral part	VTAR
		Ventral tegmental nucleus	VTg
		Zonal layer of the superior colliculus	Zo
Metencephalon/Myelencephalon		A5 noradrenaline cells	A5
		Abducens nucleus	6N
		Ambiguus nucleus	Amb
		Ambiguus nucleus, compact part	AmbC
		Area postrema	AP
		Central canal	CC
		Cuneate nucleus	Cu
		Cuneiform nucleus	CnF
		Dorsal cochlear nucleus	DCN
		Dorsal motor nucleus of vagus	10N
		Dorsal tegmental nucleus	DTg
		External cuneate nucleus	ECu
		Facial nucleus	7N
		Gigantocellular reticular nucleus	Gi
		Gracile nucleus	Gr
		Inferior olive	IO
		Intermediate reticular nucleus	IRt
		Interposed cerebellar nucleus	Int
		Lateral cerebellar nucleus	Lat
		Lateral parabrachial nucleus	LPB
		Lateral reticular nucleus	LRt
		Laterodorsal tegmental nucleus	LDTg
		Locus coeruleus	LC
		Matrix region of the medulla	Mx
		Medial cerebellar nucleus	Med
		Medial parabrachial nucleus	MPB
		Medullary reticular nucleus, dorsal part	MdD
		Medullary reticular nucleus, ventral part	MdV
		Motor trigeminal nucleus	5N
		Nucleus of Roller	Ro
		Oculomotor nucleus	3N
		Parvicellular reticular nucleus	PCRt
		Peritrigeminal zone	P5
		Pontine nuclei	Pn
		Pontine reticular nucleus, caudal part	PnC
		Pontine reticular nucleus, oral part	PnO
		Prepositus nucleus	Pr
		Principal sensory trigeminal nucleus	Pr5
		Raphe pallidus nucleus	RPa
		Reticulotegmental nucleus of the pons	RtTg
		Solitary nucleus	Sol

		Spinal trigeminal nucleus	Sp5
		Spinal vestibular nucleus	SpVe
		Subcoeruleus nucleus	SubC
		Superior olive	SO
		Superior vestibular nucleus	SuVe
		Supragenual nucleus	SGe
		Supratrigeminal nucleus	Su5
		Ventral cochlear nucleus	VCN
		Ventral cochlear nucleus, anterior part	VCA
		Ventral periolivary nucleus	VPO
		Ventrolateral reticular nucleus	VLN
		Vestibular nucleus	Ves
		Vestibulocerebellar nucleus	VeCb

List of Structures

3rd ventricle	3V
4th ventricle	4V
A	
A5 noradrenaline cells	A5
Abducens nucleus	6N
Accessory basal nucleus	AB
Accumbens nucleus	Acb
Accumbens nucleus, core	AcbC
Accumbens nucleus, shell	AcbSh
Alveus of the hippocampus	alv
Ambiguus nucleus	Amb
Ambiguus nucleus, compact part	AmbC
Amygdalohippocampal area	AHi
Amygdalostriatal transition area	ASt
Anterior amygdaloid area	AA
Anterior commissure	ac
Anterior commissure, intrabulbar part	aci
Anterior commissure, posterior part	acp
Anterior cortical amygdaloid nucleus	ACo
Anterior hypothalamic area	AHy
Anterior olfactory nucleus	AON
Anterior olfactory nucleus, dorsal part	AOD
Anterior olfactory nucleus, external part	AOE
Anterior olfactory nucleus, lateral part	AOL
Anterior olfactory nucleus, medial part	AOM
Anterior olfactory nucleus, posterior part	AOP
Anterior olfactory nucleus, ventral part	AOV
Anterior pretectal nucleus	APT
Anterior pretectal nucleus, dorsal part	APTD
Anterior pretectal nucleus, ventral part	APTV
Anterodorsal thalamic nucleus	AD
Anteromedial thalamic nucleus	AM
Anteroventral periventricular nucleus	AVPe
Anteroventral thalamic nucleus	AV
Aqueduct	Aq
Arcuate hypothalamic nucleus	Arc
Arcuate hypothalamic nucleus, posterior part	ArcP
Area postrema	AP
Ascending fibers of the facial nerve	asc7
Auditory belt area	Ab
Auditory core area	Ac

B	
Basal nucleus (Meynert)	BN
Basolateral amygdaloid nucleus	BLA
Basomedial amygdaloid nucleus	BMA
Bed nucleus of the stria terminalis, lateral division	STL
Bed nucleus of the stria terminalis, medial division, anterior part	STMA
Bed nucleus of the stria terminalis, medial division, anterolateral part	STMAL
Bed nucleus of the stria terminalis, medial division, anteromedial part	STMAM
Bed nucleus of the stria terminalis, medial division, posterior part	STMP
Bed nucleus of the stria terminalis, medial division, posterolateral part	STMPL
Bed nucleus of the stria terminalis, medial division, posteromedial part	STMPM
Bed nucleus of the stria terminalis, medial division, ventral part	STMV
Brachium of the superior colliculus	B
C	
Caudal linear nucleus of the raphe	CLi
Caudate nucleus	Cd
Central amygdaloid nucleus	CeA
Central canal	CC
Central gray	CG
Central medial thalamic nucleus	CM
Central nucleus of the pulvinar	Pc
Cerebral peduncle	cp
Cingulate cortex	Cg
Claustrum	Cl
Cochlear root of the vestibulocochlear nerve	8cn
Commissure of the superior colliculus	csc
Corpus callosum	cc
Cuneate fasciculus	cu
Cuneate nucleus	Cu
Cuneiform nucleus	CnF
D	
Decussation of the superior cerebellar peduncle	xscp
Deep gray layer of the superior colliculus	DpG
Deep white layer of the superior colliculus	DpWh
Dentate gyrus of the hippocampus	DG
Dorsal 3rd ventricle	D3V

Dorsal acoustic stria	das
Dorsal cochlear nucleus	DCN
Dorsal frontal cortex	DFC
Dorsal lateral geniculate nucleus	DLG
Dorsal motor nucleus of vagus	10N
Dorsal nucleus of the pulvinar	Pd
Dorsal peduncular cortex	DP
Dorsal raphe nucleus	DR
Dorsal spinocerebellar tract	dsc
Dorsal subiculum	DS
Dorsal tegmental decussation	dtgx
Dorsal tegmental nucleus	DTg
Dorsomedial hypothalamic nucleus	DMH
E	
Ectorhinal cortex	Ect
Edinger-Westphal nucleus	EW
Entorhinal cortex	Ent
Ependyma and subependymal layer	E
External cuneate nucleus	ECu
External globus pallidus	EGP
External plexiform layer of the accessory olfactory bulb	EPlA
External plexiform layer of the olfactory bulb	EPl
F	
Facial nerve	7n
Facial nucleus	7N
Fasciculus retroflexus	fr
Field CA1 of the hippocampus	CA1
Field CA2 of the hippocampus	CA2
Field CA2 of the hippocampus	CA3
Fimbria of the hippocampus	fi
Forceps minor of the corpus callosum	fmi
Fornix	f
G	
Gemini hypothalamic nucleus	Gem
Genu of the facial nerve	g7
Gigantocellular reticular nucleus	Gi
Globus pallidus	GP
Glomerular layer of the accessory olfactory bulb	GlA
Glomerular layer of the olfactory bulb	GL
Gracile nucleus	Gr
Granular layer of the dentate gyrus	GrDG
Granule cell layer of the accessory olfactory bulb	GrA
Granule cell layer of the olfactory bulb	GCL
H	
Hippocampus	Hip
Hypoglossal nucleus	12N
I	
Indusium griseum	IG
Inferior cerebellar peduncle	icp
Inferior colliculus	IC
Inferior olive	IO
Inferior temporal cortex	IT
Infralimbic cortex	IL
Infraradiata dorsalis	IRd

Infraradiata ventral area	IRv
Insular cortex	Ins
Interanteromedial thalamic nucleus	IAM
Intercalated amygdaloid nucleus, main part	IM
Interfascicular nucleus	IF
Intergeniculate leaf	IGL
Intermediate gray layer of the superior colliculus	InG
Intermediate reticular nucleus	IRt
Intermediate white layer of the superior colliculus	InWh
Intermediodorsal thalamic nucleus	IMD
Internal capsule	ic
Internal globus pallidus	IGP
Internal plexiform layer of the olfactory bulb	IPl
Interpeduncular nucleus	IP
Interpeduncular nucleus, apical subnucleus	IPA
Interpeduncular nucleus, caudal subnucleus	IPC
Interpeduncular nucleus, lateral subnucleus	IPL
Interpeduncular nucleus, rostral subnucleus	IPR
Interposed cerebellar nucleus	Int
Interstitial nucleus of Cajal	InC
Interstitial nucleus of the posterior limb of the anterior commissure	IPAC
Islands of Calleja	ICj
J	
Juxtaparaventricular part of lateral hypothalamus	JPLH
L	
Lacunosum moleculare layer of the hippocampus	LMoL
Lateral cerebellar nucleus	Lat
Lateral habenular nucleus	LHb
Lateral habenular nucleus, lateral part	LHbL
Lateral habenular nucleus, medial part	LHbM
Lateral hypothalamic area	LHA
Lateral lemniscus	ll
Lateral mammillary nucleus	LM
Lateral olfactory tract	lo
Lateral parabrachial nucleus	LPB
Lateral preoptic area	LPO
Lateral reticular nucleus	LRt
Lateral septal nucleus, dorsal part	LSd
Lateral septal nucleus, intermediate part	LSi
Lateral septal nucleus, ventral part	LSv
Lateral stripe of the striatum	LSS
Lateral ventricle	LV
Laterodorsal tegmental nucleus	LDTg
Laterodorsal thalamic nucleus	LD
Layer 1 of olfactory tubercle	Tu1
Layer 2 of cortex	2
Layer 3 of cortex	3
Locus coeruleus	LC
Longitudinal fasciculus of the pons	lfp
M	
Mammillary recess of the 3rd ventricle	MRe
Mammillothalamic tract	mt
Matrix region of the medulla	Mx

Medial amygdaloid nucleus	MeA
Medial cerebellar nucleus	Med
Medial frontal cortex	MFC
Medial geniculate nucleus	MG
Medial geniculate nucleus, dorsal part	MGD
Medial geniculate nucleus, ventral part	MGV
Medial habenular nucleus	MHb
Medial lemniscus	ml
Medial lemniscus decussation	mlx
Medial longitudinal fasciculus	mlf
Medial mammillary nucleus, lateral part	ML
Medial mammillary nucleus, medial part	MM
Medial parabrachial nucleus	MPB
Medial preoptic area	MPA
Medial preoptic nucleus	MPO
Medial pretectal nucleus	MPT
Medial septal nucleus	MS
Medial terminal nucleus of the accessory optic tract	MT
Median eminence	ME
Median preoptic nucleus	MnPO
Median raphe nucleus	MnR
Mediodorsal thalamic nucleus	MD
Medullary reticular nucleus, dorsal part	MdD
Medullary reticular nucleus, ventral part	MdV
Mesencephalic reticular formation	mRt
Mesencephalic trigeminal nucleus	Me5
Mesencephalic trigeminal tract	me5
Microcellular tegmental nucleus	MiTg
Middle cerebellar peduncle	mcp
Mitral cell layer of the accessory olfactory bulb	MiA
Mitral cell layer of the olfactory bulb	MCL
Molecular layer of the dentate gyrus	MoDG
Motor root of the trigeminal nerve	m5
Motor trigeminal nucleus	5N
N	
Nucleus of Roller	Ro
Nucleus of the diagonal band	NDB
Nucleus of the horizontal limb of the diagonal band	HDB
Nucleus of the lateral lemniscus	LL
Nucleus of the lateral olfactory tract	LOT
Nucleus of the trapezoid body	Tz
Nucleus of the vertical limb of the diagonal band	VDB
O	
Oculomotor nucleus	3N
Oculomotor nucleus, parvicellular part	3PC
Olfactory nerve layer	ONL
Olfactory tubercle	Tu
Olfactory ventricle (olfactory part of lateral ventricle)	OV
Olivary pretectal nucleus	OPT
Olivocerebellar tract	oc
Olivocerebellar tract and dorsal spinocerebellar tract	oc/dsc
Optic chiasm	och
Optic nerve layer of the superior colliculus	Op
Optic tract	opt
Orbital frontal cortex	OFC
Oriens layer of the hippocampus	Or
P	
p1 periaqueductal gray	pPAG
p1 reticular formation	pRt
Parabigeminal nucleus	PBG
Parabrachial pigmented nucleus of the VTA	PBP
Paracentral thalamic nucleus	PC
Parafascicular thalamic nucleus	PF
Paranigral nucleus of the VTA	PN
Paratenial thalamic nucleus	PT
Paraventricular hypothalamic nucleus	PVN
Paraventricular hypothalamic nucleus, anterior part	PaA
Paraventricular hypothalamic nucleus, posterior part	PaP
Paraventricular hypothalamic nucleus, ventral part	PaV
Paraventricular thalamic nucleus, anterior part	PVA
Paraventricular thalamic nucleus, posterior part	PVP
Parietal ventral area	PvA
Parvicellular reticular nucleus	PCRt
Peduncular part of lateral hypothalamus	PLH
Pedunculopontine tegmental nucleus	PTg
Periaqueductal gray	PAG
Perifornical nucleus	PeF
Perirhinal cortex	PRh
Peritrigeminal zone	P5
Periventricular hypothalamic nucleus	Pe
Pineal gland	PG
Piriform cortex, layer 1	Pir1
Polymorph layer of the dentate gyrus	PoDG
Pontine nuclei	Pn
Pontine reticular nucleus, caudal part	PnC
Pontine reticular nucleus, oral part	PnO
Posterior commissure	pc
Posterior complex of the thalamus	Po
Posterior hypothalamic area	PHA
Posterior hypothalamic nucleus	PH
Posterior parietal caudal area	PPc
Posterior parietal dorsal area	PPd
Posterior parietal rostral area	PPr
Posteromedian thalamic nucleus	PoMn
Postsubiculum	Post
Precommissural nucleus	PrC
Pre-Edinger-Westphal nucleus	PrEW
Prelimbic cortex	PrL
Premammillary nucleus, dorsal part	PMD
Premammillary nucleus, ventral part	PMV
Prepositus nucleus	Pr
Presubiculum	PrS
Primary motor cortex	M1
Primary somatosensory cortex	S1
Primary visual cortex	V1
Principal sensory trigeminal nucleus	Pr5
Pulvinar nuclei	Pul
Putamen	Pu

Pyramidal cell layer of the hippocampus	Py
Pyramidal decussation	pyx
Pyramidal tract	py
R	
Radiatum layer of the hippocampus	Rad
Raphe pallidus nucleus	RPa
Red nucleus	RN
Red nucleus, magnocellular part	RMC
Red nucleus, parvicellular part	RPC
Reticular thalamic nucleus	Rt
Reticulotegmental nucleus of the pons	RtTg
Retrorubral field	RRF
Retrosplenial granular cortex	RSg
Reuniens thalamic nucleus	Re
Rhabdoid nucleus	Rbd
Rhomboid thalamic nucleus	Rh
Root of abducens nerve	6n
Root of hypoglossal nerve	12n
Rostral linear nucleus of the raphe	RLi
Rubrospinal tract	rs
S	
Secondary motor cortex	M2
Secondary somatosensory cortex	S2
Secondary visual cortex	V2
Sensory root of the trigeminal nerve	s5
Septofimbrial nucleus	SFi
Septohippocampal nucleus	SHi
Septohypothalamic nucleus	SHy
Solitary nucleus	Sol
Somatosensory caudal cortex	SCC
Spinal trigeminal nucleus	Sp5
Spinal trigeminal tract	sp5
Spinal vestibular nucleus	SpVe
Stria medullaris of the thalamus	sm
Stria terminalis	st
Striohypothalamic nucleus	StHy
Subcoeruleus nucleus	SubC
Subfornical organ	SFO
Subiculum, transition area	STr
Sublenticular extended amygdala	EA
Sublenticular extended amygdala, central part	EAC
Submedius thalamic nucleus	Sub
Subparaventricular zone of the hypothalamus	SPa
Subpeduncular tegmental nucleus	SPTg
Substantia nigra, compact part	SNC
Substantia nigra, lateral part	SNL
Substantia nigra, reticular part	SNR
Subthalamic nucleus	STh
Superficial gray layer of the superior colliculus	SuG
Superior cerebellar peduncle	scp
Superior cerebellar peduncle, descending limb	scpd
Superior colliculus	SC
Superior olive	SO
Superior thalamic radiation	str
Superior vestibular nucleus	SuVe
Suprachiasmatic nucleus	SCN
Supragenual nucleus	SGe
Supramammillary decussation	sumx
Supramammillary nucleus	SuM
Supramammillary nucleus, lateral par	SuML
Supramammillary nucleus, medial part	SuMM
Supraoptic nucleus	SON
Supratrigeminal nucleus	Su5
T	
Temporal cortex	TC
Temporal inferior area	TI
Transverse fibers of the pons	tfp
Trapezoid body	tz
Triangular septal nucleus	TS
Trigeminothalamic tract	tth
Trochlear nerve	4n
Trochlear nucleus	4N
Tuberal region of lateral hypothalamus	TuLH
V	
Ventral cochlear nucleus	VCN
Ventral cochlear nucleus, anterior part	VCA
Ventral lateral geniculate nucleus	VLG
Ventral nucleus of the pulvinar	Pv
Ventral pallidum	VP
Ventral periolivary nucleus	VPO
Ventral posterior nucleus of the thalamus, parvicellular part	VPPC
Ventral posterior thalamic nucleus	VPT
Ventral posteromedial thalamic nucleus	VPM
Ventral spinocerebellar tract	vsc
Ventral subiculum	VS
Ventral tegmental area	VTA
Ventral tegmental area, rostral part	VTAR
Ventral tegmental decussation	vtgx
Ventral tegmental nucleus	VTg
Ventral tuberomammillary nucleus	VTM
Ventrolateral reticular nucleus	VLN
Ventrolateral thalamic nucleus	VL
Ventromedial hypothalamic nucleus	VMH
Ventromedial thalamic nucleus	VM
Vestibular nucleus	Ves
Vestibular root of the vestibulocochlear nerve	8vn
Vestibulocerebellar nucleus	VeCb
Z	
Zona incerta	ZI
Zonal layer of the superior colliculus	Zo

List of Abbreviations

2	Layer 2 of cortex
3	Layer 3 of cortex
10N	Dorsal motor nucleus of vagus
12n	Root of hypoglossal nerve
12N	Hypoglossal nucleus
3N	Oculomotor nucleus
3PC	Oculomotor nucleus, parvicellular part
3V	3rd ventricle
4n	Trochlear nerve
4N	Trochlear nucleus
4V	4th ventricle
5N	Motor trigeminal nucleus
6n	Root of abducens nerve
6N	Abducens nucleus
7N	Facial nucleus
7n	Facial nerve
8cn	Cochlear root of the vestibulocochlear nerve
8vn	Vestibular root of the vestibulocochlear nerve
A	
A5	A5 noradrenaline cells
AA	Anterior amygdaloid area
Ab	Auditory belt area
AB	Accessory basal nucleus
Ac	Auditory core area
ac	Anterior commissure
Acb	Accumbens nucleus
AcbC	Accumbens nucleus, core
AcbSh	Accumbens nucleus, shell
aci	Anterior commissure, intrabulbar part
ACo	Anterior cortical amygdaloid nucleus
acp	Anterior commissure, posterior part
AD	Anterodorsal thalamic nucleus
AHi	Amygdalohippocampal area
AHy	Anterior hypothalamic area
alv	Alveus of the hippocampus
AM	Anteromedial thalamic nucleus
Amb	Ambiguus nucleus
AmbC	Ambiguus nucleus, compact part
AOD	Anterior olfactory nucleus, dorsal part
AOE	Anterior olfactory nucleus, external part
AOL	Anterior olfactory nucleus, lateral part
AOM	Anterior olfactory nucleus, medial part
AON	Anterior olfactory nucleus
AOP	Anterior olfactory nucleus, posterior part
AOV	Anterior olfactory nucleus, ventral part
AP	Area postrema
APT	Anterior pretectal nucleus
APTD	Anterior pretectal nucleus, dorsal part
APTV	Anterior pretectal nucleus, ventral part
Aq	Aqueduct
Arc	Arcuate hypothalamic nucleus
ArcP	Arcuate hypothalamic nucleus, posterior part
asc7	Ascending fibers of the facial nerve
ASt	Amygdalostriatal transition area
AV	Anteroventral thalamic nucleus
AVPe	Anteroventral periventricular nucleus
B	
B	Brachium of the superior colliculus
BLA	Basolateral amygdaloid nucleus
BMA	Basomedial amygdaloid nucleus
BN	Basal nucleus (Meynert)
C	
CA1	Field CA1 of the hippocampus
CA2	Field CA2 of the hippocampus
CA3	Field CA2 of the hippocampus
cc	Corpus callosum
CC	Central canal
Cd	Caudate nucleus
CeA	Central amygdaloid nucleus
Cg	Cingulate cortex
CG	Central gray
Cl	Claustrum
CLi	Caudal linear nucleus of the raphe
CM	Central medial thalamic nucleus
CnF	Cuneiform nucleus
cp	Cerebral peduncle
csc	Commissure of the superior colliculus
Cu	Cuneate nucleus
cu	Cuneate fasciculus
D	
D3V	Dorsal 3rd ventricle
das	Dorsal acoustic stria
DCN	Dorsal cochlear nucleus
DFC	Dorsal frontal cortex

DG	Dentate gyrus of the hippocampus
DLG	Dorsal lateral geniculate nucleus
DMH	Dorsomedial hypothalamic nucleus
DP	Dorsal peduncular cortex
DpG	Deep gray layer of the superior colliculus
DpWh	Deep white layer of the superior colliculus
DR	Dorsal raphe nucleus
DS	Dorsal subiculum
dsc	Dorsal spinocerebellar tract
DTg	Dorsal tegmental nucleus
dtgx	Dorsal tegmental decussation
E	
E	Ependyma and subependymal layer
EA	Sublenticular extended amygdala
EAC	Sublenticular extended amygdala, central part
Ect	Ectorhinal cortex
ECu	External cuneate nucleus
EGP	External globus pallidus
Ent	Entorhinal cortex
EPl	External plexiform layer of the olfactory bulb
EPlA	External plexiform layer of the accessory olfactory bulb
EW	Edinger-Westphal nucleus
F	
f	Fornix
fi	Fimbria of the hippocampus
fmi	Forceps minor of the corpus callosum
fr	Fasciculus retroflexus
G	
g7	Genu of the facial nerve
GCL	Granule cell layer of the olfactory bulb
Gem	Gemini hypothalamic nucleus
Gi	Gigantocellular reticular nucleus
GL	Glomerular layer of the olfactory bulb
GlA	Glomerular layer of the accessory olfactory bulb
GP	Globus pallidus
Gr	Gracile nucleus
GrA	Granule cell layer of the accessory olfactory bulb
GrDG	Granular layer of the dentate gyrus
H	
HDB	Nucleus of the horizontal limb of the diagonal band
Hip	Hippocampus
I	
IAM	Interanteromedial thalamic nucleus
ic	Internal capsule
IC	Inferior colliculus
ICj	Islands of Calleja
icp	Inferior cerebellar peduncle
IF	Interfascicular nucleus
IG	Indusium griseum
IGL	Intergeniculate leaf
IGP	Internal globus pallidus
IL	Infralimbic cortex
IM	Intercalated amygdaloid nucleus, main part
IMD	Intermediodorsal thalamic nucleus

InC	Interstitial nucleus of Cajal
InG	Intermediate gray layer of the superior colliculus
Ins	Insular cortex
Int	Interposed cerebellar nucleus
InWh	Intermediate white layer of the superior colliculus
IO	Inferior olive
IP	Interpeduncular nucleus
IPA	Interpeduncular nucleus, apical subnucleus
IPAC	Interstitial nucleus of the posterior limb of the anterior commissure
IPC	Interpeduncular nucleus, caudal subnucleus
IPl	Internal plexiform layer of the olfactory bulb
IPL	Interpeduncular nucleus, lateral subnucleus
IPR	Interpeduncular nucleus, rostral subnucleus
IRd	Infraradiata dorsalis
IRt	Intermediate reticular nucleus
IRv	Infraradiata ventral area
IT	Inferior temporal cortex
J	
JPLH	Juxtaparaventricular part of lateral hypothalamus
L	
Lat	Lateral cerebellar nucleus
LC	Locus coeruleus
LD	Laterodorsal thalamic nucleus
LDTg	Laterodorsal tegmental nucleus
lfp	Longitudinal fasciculus of the pons
LHA	Lateral hypothalamic area
LHb	Lateral habenular nucleus
LHbL	Lateral habenular nucleus, lateral part
LHbM	Lateral habenular nucleus, medial part
LL	Nucleus of the lateral lemniscus
ll	Lateral lemniscus
LM	Lateral mammillary nucleus
LMoL	Lacunosum moleculare layer of the hippocampus
lo	Lateral olfactory tract
LOT	Nucleus of the lateral olfactory tract
LPB	Lateral parabrachial nucleus
LPO	Lateral preoptic area
LRt	Lateral reticular nucleus
LSd	Lateral septal nucleus, dorsal part
LSi	Lateral septal nucleus, intermediate part
LSS	Lateral stripe of the striatum
LSv	Lateral septal nucleus, ventral part
LV	Lateral ventricle
M	
M1	Primary motor cortex
M2	Secondary motor cortex
m5	Motor root of the trigeminal nerve
MCL	Mitral cell layer of the olfactory bulb
mcp	Middle cerebellar peduncle
MD	Mediodorsal thalamic nucleus
MdD	Medullary reticular nucleus, dorsal part
MdV	Medullary reticular nucleus, ventral part
ME	Median eminence
me5	Mesencephalic trigeminal tract

Me5	Mesencephalic trigeminal nucleus
MeA	Medial amygdaloid nucleus
Med	Medial cerebellar nucleus
MFC	Medial frontal cortex
MG	Medial geniculate nucleus
MGD	Medial geniculate nucleus, dorsal part
MGV	Medial geniculate nucleus, ventral part
MHb	Medial habenular nucleus
MiA	Mitral cell layer of the accessory olfactory bulb
MiTg	Microcellular tegmental nucleus
ml	Medial lemniscus
ML	Medial mammillary nucleus, lateral part
mlf	Medial longitudinal fasciculus
mlx	Medial lemniscus decussation
MM	Medial mammillary nucleus, medial part
MnPO	Median preoptic nucleus
MnR	Median raphe nucleus
MoDG	Molecular layer of the dentate gyrus
MPA	Medial preoptic area
MPB	Medial parabrachial nucleus
MPO	Medial preoptic nucleus
MPT	Medial pretectal nucleus
MRe	Mammillary recess of the 3rd ventricle
mRt	Mesencephalic reticular formation
MS	Medial septal nucleus
mt	Mammillothalamic tract
MT	Medial terminal nucleus of the accessory optic tract
Mx	Matrix region of the medulla
N	
NDB	Nucleus of the diagonal band
O	
oc	Olivocerebellar tract
oc/dsc	Olivocerebellar tract and dorsal spinocerebellar tract
och	Optic chiasm
OFC	Orbital frontal cortex
ONL	Olfactory nerve layer
Op	Optic nerve layer of the superior colliculus
opt	Optic tract
OPT	Olivary pretectal nucleus
Or	Oriens layer of the hippocampus
OV	Olfactory ventricle (olfactory part of lateral ventricle)
P	
P5	Peritrigeminal zone
PaA	Paraventricular hypothalamic nucleus, anterior part
PAG	Periaqueductal gray
PaP	Paraventricular hypothalamic nucleus, posterior part
PaV	Paraventricular hypothalamic nucleus, ventral part
PBG	Parabigeminal nucleus
PBP	Parabrachial pigmented nucleus of the VTA
PC	Paracentral thalamic nucleus
Pc	Central nucleus of the pulvinar
pc	Posterior commissure
PCRt	Parvicellular reticular nucleus

Pd	Dorsal nucleus of the pulvinar
Pe	Periventricular hypothalamic nucleus
PeF	Perifornical nucleus
PF	Parafascicular thalamic nucleus
PG	Pineal gland
PH	Posterior hypothalamic nucleus
PHA	Posterior hypothalamic area
Pir1	Piriform cortex, layer 1
PLH	Peduncular part of lateral hypothalamus
PMD	Premammillary nucleus, dorsal part
PMV	Premammillary nucleus, ventral part
Pn	Pontine nuclei
PN	Paranigral nucleus of the VTA
PnC	Pontine reticular nucleus, caudal part
PnO	Pontine reticular nucleus, oral part
Po	Posterior complex of the thalamus
PoDG	Polymorph layer of the dentate gyrus
PoMn	Posteromedian thalamic nucleus
Post	Postsubiculum
pPAG	p1 periaqueductal gray
PPc	Posterior parietal caudal area
PPd	Posterior parietal dorsal area
PPr	Posterior parietal rostral area
Pr	Prepositus nucleus
Pr5	Principal sensory trigeminal nucleus
PrC	Precommissural nucleus
PrEW	Pre-Edinger-Westphal nucleus
PRh	Perirhinal cortex
PrL	Prelimbic cortex
PrS	Presubiculum
pRt	p1 reticular formation
PT	Paratenial thalamic nucleus
PTg	Pedunculopontine tegmental nucleus
Pu	Putamen
Pul	Pulvinar nuclei
Pv	Ventral nucleus of the pulvinar
PvA	Parietal ventral area
PVA	Paraventricular thalamic nucleus, anterior part
PVN	Paraventricular hypothalamic nucleus
PVP	Paraventricular thalamic nucleus, posterior part
py	Pyramidal tract
Py	Pyramidal cell layer of the hippocampus
pyx	Pyramidal decussation
R	
Rad	Radiatum layer of the hippocampus
Rbd	Rhabdoid nucleus
Re	Reuniens thalamic nucleus
Rh	Rhomboid thalamic nucleus
RLi	Rostral linear nucleus of the raphe
RMC	Red nucleus, magnocellular part
RN	Red nucleus
Ro	Nucleus of Roller
RPa	Raphe pallidus nucleus

RPC	Red nucleus, parvicellular part
RRF	Retrorubral field
rs	Rubrospinal tract
RSg	Retrosplenial granular cortex
Rt	Reticular thalamic nucleus
RtTg	Reticulotegmental nucleus of the pons
S	
S1	Primary somatosensory cortex
S2	Secondary somatosensory cortex
s5	Sensory root of the trigeminal nerve
SC	Superior colliculus
SCC	Somatosensory caudal cortex
SCN	Suprachiasmatic nucleus
scp	Superior cerebellar peduncle
scpd	Superior cerebellar peduncle, descending limb
SFi	Septofimbrial nucleus
SFO	Subfornical organ
SGe	Supragenual nucleus
SHi	Septohippocampal nucleus
SHy	Septohypothalamic nucleus
sm	Stria medullaris of the thalamus
SNC	Substantia nigra, compact part
SNL	Substantia nigra, lateral part
SNR	Substantia nigra, reticular part
SO	Superior olive
Sol	Solitary nucleus
SON	Supraoptic nucleus
Sp5	Spinal trigeminal nucleus
sp5	Spinal trigeminal tract
SPa	Subparaventricular zone of the hypothalamus
SPTg	Subpeduncular tegmental nucleus
SpVe	Spinal vestibular nucleus
st	Stria terminalis
STh	Subthalamic nucleus
StHy	Striohypothalamic nucleus
STL	Bed nucleus of the stria terminalis, lateral division
STMA	Bed nucleus of the stria terminalis, medial division, anterior part
STMAL	Bed nucleus of the stria terminalis, medial division, anterolateral part
STMAM	Bed nucleus of the stria terminalis, medial division, anteromedial part
STMP	Bed nucleus of the stria terminalis, medial division, posterior part
STMPL	Bed nucleus of the stria terminalis, medial division, posterolateral part
STMPM	Bed nucleus of the stria terminalis, medial division, posteromedial part
STMV	Bed nucleus of the stria terminalis, medial division, ventral part
STr	Subiculum, transition area
str	Superior thalamic radiation

Su5	Supratrigeminal nucleus
Sub	Submedius thalamic nucleus
SubC	Subcoeruleus nucleus
SuG	Superficial gray layer of the superior colliculus
SuM	Supramammillary nucleus
SuML	Supramammillary nucleus, lateral par
SuMM	Supramammillary nucleus, medial part
sumx	Supramammillary decussation
SuVe	Superior vestibular nucleus
T	
TC	Temporal cortex
tfp	Transverse fibers of the pons
TI	Temporal inferior area
TS	Triangular septal nucleus
tth	Trigeminothalamic tract
Tu	Olfactory tubercle
Tu1	Layer 1 of olfactory tubercle
TuLH	Tuberal region of lateral hypothalamus
tz	Trapezoid body
Tz	Nucleus of the trapezoid body
V	
V1	Primary visual cortex
V2	Secondary visual cortex
VCA	Ventral cochlear nucleus, anterior part
VCN	Ventral cochlear nucleus
VDB	Nucleus of the vertical limb of the diagonal band
VeCb	Vestibulocerebellar nucleus
Ves	Vestibular nucleus
VL	Ventrolateral thalamic nucleus
VLG	Ventral lateral geniculate nucleus
VLN	Ventrolateral reticular nucleus
VM	Ventromedial thalamic nucleus
VMH	Ventromedial hypothalamic nucleus
VP	Ventral pallidum
VPM	Ventral posteromedial thalamic nucleus
VPO	Ventral periolivary nucleus
VPPC	Ventral posterior nucleus of the thalamus, parvicellular part
VPT	Ventral posterior thalamic nucleus
VS	Ventral subiculum
vsc	Ventral spinocerebellar tract
VTA	Ventral tegmental area
VTAR	Ventral tegmental area, rostral part
VTg	Ventral tegmental nucleus
vtgx	Ventral tegmental decussation
VTM	Ventral tuberomammillary nucleus
X	
xscp	Decussation of the superior cerebellar peduncle
Z	
ZI	Zona incerta
Zo	Zonal layer of the superior colliculus

GPSR Compliance

The European Union's (EU) General Product Safety Regulation (GPSR) is a set of rules that requires consumer products to be safe and our obligations to ensure this.

If you have any concerns about our products, you can contact us on ProductSafety@springernature.com

In case Publisher is established outside the EU, the EU authorized representative is:

Springer Nature Customer Service Center GmbH
Europaplatz 3
69115 Heidelberg, Germany

Batch number: 10756632

Printed by Printforce, the Netherlands